ENCOMPASSING NATURE

A Sourcebook

Edited by

Robert M. Torrance

COUNTERPOINT
WASHINGTON, D.C.

First paperback edition 1999

Acknowledgments of copyright and permission to reprint are listed
on pages 1187–91. Every reasonable effort has been made to clear
the use of materials in this volume with the copyright owners.

Library of Congress Cataloging-in-Publication Data
Encompassing nature : a sourcebook / edited by Robert M. Torrance.
Includes bibliographical references and index.
1. Nature—Literary collections. 2. Nature—Folklore.
3. Nature—Mythology. 4. Tales.
I. Torrance, Robert M. (Robert Mitchell), 1939– .
PN6071.N3E53 1998
508—DC21 97-44481
ISBN 1-58243-009-8 (alk. paper)

Printed in the United States of America on acid-free paper that meets
the American National Standards Institute Z39-48 Standard

Design by David Bullen
Typesetting by Wilsted & Taylor Publishing Services

COUNTERPOINT
P.O. Box 65793
Washington, D.C. 20035-5793

1 3 5 7 9 8 6 4 2

To

Scott McLean
David Robertson
Gary Snyder
Lenora Timm
Mark Wheelis

and the faculty and students
of the Nature and Culture Program
of the University of California, Davis

Preface

This treasury of writings on experiences and concepts of nature from around the world and throughout the ages originated some fifteen years ago when I put together, at the University of California, Davis, a reader of selections from Western writers on nature to supplement the books I was assigning for a new course, "Man and the Natural World." This reader continued to be used when the course was taught by Dr. Scott McLean; but only after I finished *The Spiritual Quest* in 1994 could I give undivided attention to compiling a wide-ranging sourcebook of materials both Western and non-Western from the earliest times to the late eighteenth century. The intense labor of researching and arranging these varied readings (and typing a manuscript of more than two thousand pages), translating poems and prose of every description, and writing introductions to works ranging from children's stories, tribal myths, and sacred scriptures to philosophical and scientific treatises, has entailed both exhilarating reacquaintance, in unexpected new contexts, with many books I had read as long as twenty or forty years ago, and the exciting discovery of others I had barely heard of before.

I am deeply grateful to the Program in Nature and Culture for providing me with the computer and printer with whose help the bulk of this very long manuscript was composed;

to the graduate students in my seminar Comparative Literature 210 in the spring of 1996—Julian Nelson, Rod Romesburg, Eric Smith, and Patrick Vincent—who worked their way through a preliminary version of the manuscript with me;

to three other graduate students, Jianguo Chen, Mariko Enomura, and Mary Frances Fahey, who read selected chapters and gave me helpful comments;

and above all, as always, to Donna, for supporting me through intense labors with unfailing radiance and warmth, in sickness and in health.

Contents

Complete listings of chapter contents appear at the beginning of each of the eight parts.

Introduction

How is it, if "nature" (comprising not only birds and beasts, trees and flowers, meadows, gardens, and farmlands, but also rocks, streams, and oceans, sun, moon, and innumerable galaxies and stars) is coeval in origin with the cosmos, that "nature writing" is said to be little more than two centuries old, as if our species, in three thousand years or more of its literate history, had been blind to the nonhuman world that lay in profligate if rapidly diminishing abundance around us? Surely we have defined the term too narrowly or stand in need of another.

Nature writing as a distinct literary genre, whether in the form of poetic reverie or of reflective prose, whose great prototypes in English are respectively Wordsworth and Thoreau, may indeed date back no farther in the West than the mid to late eighteenth century, when the present anthology comes to an end. Of the writers included here, only James Thomson, Jean-Jacques Rousseau, Gilbert White, and William Bartram might qualify as nature writers in this conventional sense. But whether or not, as Edward O. Wilson hypothesizes in *Biophilia* (1984), "the urge to affiliate with other forms of life is to some degree innate" in human beings, the world of other living things—far from having been suddenly discovered two hundred years ago by Romantic and Transcendentalist writers of England, Germany, and America—has been of supreme importance throughout human history. Indeed, less urbanized and industrial civilizations of the past were far more urgently in contact than ours with that world and tended to include among living things, besides plants and animals, rivers and mountains, clouds, rainbows, and thunderbolts, sun, moon, and stars.

The English and German poets loosely grouped together as Romantic who gave birth to modern nature writing around the turn of the nineteenth century—the crucial time at which our anthology ends—were intensely aware not of given intimacy with the natural world but of troubled separation from it, which they strove through memory, rapture, or yearning to overcome. William Blake, for whom "every thing that lives is holy," lived amid the chartered streets of a city blighted by chimney-sweeper's cry and youthful harlot's curse. William Wordsworth, poet of meadows, groves, and streams, observed, as "a transient visitant" in that same "monstrous ant-hill," London, that men could live, he wrote in *The Prelude*, as "strangers, not knowing each the other's name." So too Friedrich Schiller, in *Letters on the Aesthetic Education of Man* (1795): "Eternally chained to only one single little fragment of the whole, Man himself grew to be only a fragment; . . . he never develops the harmony of his being, and instead of imprinting humanity upon his nature he becomes merely the imprint of his occupation, of his science." And Friedrich Hölderlin, who was "reared by the euphony / Of the rustling grove / And learned to love / Among the flowers," lamented, through the young poet of *Hyperion* (1799), that in Germany "divine Nature and her artists are so insulted" that "men grow ever more sterile, ever more empty," until "the blessing of each year becomes a curse, and all gods flee."

The Romantic Age was thus an age of aggravated division of the human from the natural world, and of intense longing for reparation of their shattered communion: an age, Wordsworth wrote, when "little we see in nature that is ours." Not from greater accord with nature, but from the *unnaturalness* of their perceived condition, Schiller contended in his seminal essay of 1795–96, *On Naïve and Sentimental Poetry*, did writers of his time seek in the external world what they failed to find within themselves. The ancient Greeks "felt naturally; we feel the natural. . . . Our feeling for nature is like the feeling of an invalid for health."

Schiller's dichotomy expresses a critically important insight. For the ancients in general, as H. and H. A. Frankfort write in *The Intellectual Adventure of Western Man* (1946) —as also for tribal cultures throughout the world—"nature and man did not stand in opposition." Because it was truly encompassing, the world of animals and plants, rocks, rivers, and stars had no existence in isolation from the human and the divine. Only when refined urban civilizations begin to undergo breakdown or turbulent change—as in Hellenistic Greece, medieval Japan, and Europe at the convergence of industrial and political revolution—have writers searched in the natural world for a wholeness the human world appears to have lost, and evolved new forms of writing to overcome division from it.

For most of human history, however, writing about nature has taken forms very different from the self-conscious nature writing of Romantic and modern Europe and America, which would have been puzzling if not inconceivable to peoples more fully integrated with the world around them. Prominent among these forms in most early cultures are hymns, myths, and songs. Invocations of divinities are often celebrations of the divine forces manifested in nature—as in the *molimo* ceremony of the BaMbuti of Congo, Sumerian hymns to the stormy love goddess Inanna, Vedic hymns to the Primordial Man or the Mighty Earth, the shamanistic "Nine Songs" of China, certain Hebrew psalms, the Homeric hymns to Demeter or Pan, Caedmon's hymn to the Creator, and St. Francis of Assisi's Praise of Created Things. Tales of the origin or creation of the universe or humankind—in Maidu myth, the Hebrew Genesis, the Indian *Rig Veda*, the Chinese *Huai-nan Tzu*, Hesiod's *Theogony*, Plato's *Timaeus*, and Ovid's *Metamorphoses*—are among the most revealing accounts of any culture's conception of the origin and composition (that is, the nature) of the world. And stories of adventures and travels of heroes and saints from the Maidu To'lowim-Woman and the Blackfoot Scarface to Homer's Odysseus, Virgil's Aeneas, the Irish Bran and St. Brendan, Sir Gawain, Christopher Columbus, and Robinson Crusoe give legendary expression to the discoveries repeatedly made by human beings in their perpetually renewed encounters with that inexhaustible world. Epic and drama frequently portray a human or partly human hero who comes into conflict with divine forces ruling in or through nature, as Enkidu and Gilgamesh do in the Mesopotamian *Epic of Gilgamesh*, Job in the Hebrew Bible, Arjuna in the *Bhagavad Gītā*, or Pentheus in the *Bacchae* of Euripides; from that conflict may follow resignation or acceptance, accommodation or destruction. Often songs and lyrics of love or death, friendship or festivity, seasonal change or the fertility of the land—whether in the *Vedas* or the Song of Solomon, the Chinese *Book of Poetry* or the Japanese *Man'yōshū*, the poems of Li Po or Tu Fu, Saigyō or Bashō, Sappho or Ibycus, Catullus or Horace, Shakespeare or Ronsard—bring the intertwined human and natural realms into poignant conjunction in a world where all flesh is grass and the generations of men pass away like the generations of leaves.

There can be, as a rule, no distinct genre of nature writing in such cultures for the very reason that nature is always nearby, whatever the subject may be. Homer's *Iliad*, a stringent saga of unrelenting war, is interspersed with similes evoking every realm of the nonhuman world from snowstorms to mosquitoes. Howling winter gales and freezing rains frame Alcaeus's invitation to drink by a blazing fire. And in the midst of Oedipus's anguish at Colonus, the chorus sings of the god's untrodden vale where wine-dark ivy climbs the bough and "the sweet, sojourning nightingale / Murmurs all night long." No image is more emblematic of encompassing nature than Homer's depiction of the shield of Achilles, on which the divine smith Hephaestus has placed earth, sea, and sky, sun, moon, and stars at the center, then encircled two cities of men, one at war, one at peace, with plowland, vineyard, and pastures, all surrounded by the mighty river ocean: a golden world in which nature and man, like nature and art, are forever united. And in the *Works and Days*, a manual for fellow farmers, the grouchy Boeotian poet Hesiod created a highly personal, locally rooted, graphically realistic account not of aristocratic or monkish seclusion from urban turmoil or bureaucratic drudgery, as in T'ao Ch'ien or Chōmei, Horace or Fray Luis de León, but of the harsh peasant life of subsistence farming—a life in inescapable daily contact with the natural world—such as the great majority of our species has lived for much of the last several millennia. His myths of Prometheus and Pandora, and of the degenerative races of man, explain the hard condition natural to humankind in this ruthless age of iron. Later writers on agriculture, from Xenophon to Cato, Varro, and Columella, and above all Virgil in the *Georgics*, praise the virtues inculcated by farming, but none brings its sweat and toil to life with the stark immediacy of Hesiod. Here is nature stripped of all idealization: not the virgin wilderness or the forest primeval, but the crippling cold of winter, the searing heat of the Dog Star, and the incessantly wearying work of every season.

The other major work attributed to Hesiod, the *Theogony*, is more conventional in stringing together tales about the divinities of earth, sea, and sky; but its account of their spontaneous genesis out of primeval chaos through the union of Ouranos and Gaia contributed, as Aristotle recognized, to Greek philosophical speculation about the origin of the world. It thus suggests a tendency, apparent in other civilizations as well, for cosmogonic myths to be accompanied by religious or metaphysical reflection on the world and man's place within it. Thus the Chinese *T'ien Wen*, or *Heavenly Questions*, poses riddles concerning the origin and structure of the cosmos to which a long philosophical tradition—extending from the *Tao Te Ching* and *Chuang-tzu*, through centuries of bold speculation on nature and nonexistence, to the Neo-Confucian "Great Harmony" in the Sung—gives expression, much as the classics of Indian thought from the *Upanishads* to the sūtras of Mahāyāna Buddhism and the *Bhagavad Gītā* respond to questions posed by the late Rig-Vedic "Hymn of the Origins" or "The Unknown God."

In Chinese thought, several terms (including *tzu-jan*, the "self-so"; *t'ien*, "heaven"; and *tao*, "the way") overlap with our concept of "nature," yet that concept remains distinctively Western, and a significant part of the present anthology is devoted to its evolving meanings. For if the trees and flowers, birds and beasts, hills, valleys, and heavenly bodies that we loosely call "nature"—designating everything apart from our unnatural selves and what we have made—have existed throughout our sojourn on earth, the Greek concept of *physis* (from which, through Latin *natura*, "nature" derives) was something new under the sun when the pre-Socratic thinkers of the seventh to sixth centuries B.C. began their momentous speculations on coming-to-be and passing-away in the divinely living cosmos to which we belong.

The meanings given to this crucial term by the *physikoi*, or "natural philosophers," from Thales to Democritus—and by their successors in later Greek and Roman antiquity, the Middle Ages, and the Renaissance and Enlightenment—have been many. To Aristotle's definitions in the *Physics* and *Metaphysics* and Cicero's in *On the Nature of the Gods*, others were added by Boëthius in late antiquity, by Hugh of St. Victor in the High Middle Ages, by Robert Boyle in the seventeenth century, and by Samuel Johnson's *Dictionary* and Diderot's *Encyclopedia* in the eighteenth; in our own century, A. O. Lovejoy and George Boas, in *Primitivism and Related Ideas in Antiquity* (1935), list no fewer than sixty-six. In Greek natural philosophy from its Milesian origins in Thales to its culmination in Aristotle nearly two centuries later, *physis*, which apparently derived from a verb meaning "to grow," signified not only the primal substance of things, and not only the sum of things composed of that substance (the self-regulated *kosmos*), but still more importantly the spontaneous *process* by which all things in our perpetually changing world repeatedly come to be and again pass away. This is the "source or cause of being moved and of being at rest" by which Aristotle defines nature in the *Physics*, not as a material substance but as an internal principle of both change and stasis, thus combining the incessant flux and the stabilizing *logos* (ratio or law)—"living each other's death, dying each other's life" in "harmonious tension, as of the bow and the lyre"—of his cryptic predecessor Heraclitus. The process of nature further involves for Aristotle the continual actualization through form of what is potential in matter, and although this process is certainly teleological (the acorn is actualized in the full-grown oak), it implies no consciously purposeful volition, but rather the working through of an innate "hypothetical" or conditional necessity: that is, of a program.

This fundamentally biological conception of nature—more of Aristotle's writings, including the longest, concern animals than any other subject—was one central development of pre-Socratic thought. Another, the mathematical conception associated with the Pythagoreans and with Plato and his school, reached fulfillment in geometers such as Euclid, Archimedes, and Apollonius of Pergê, and astronomers from Philolaus in the fifth century B.C. to Hipparchus three hundred and Ptolemy six hundred years later. In various combinations, the complementary Milesian-Aristotelian and Pythagorean-Platonic conceptions of nature, alternating or intermingling at times with the Epicurean materialism expounded in Lucretius's *On the Nature of Things*, have for millennia been central to the Western—and now to the worldwide—concept of an autonomous nature on which natural philosophy, or science, has been founded, and without which it could never have come into being and could never continue.

Judeo-Christian thinkers since Philo of Alexandria have found it necessary to bring this Greek conception into accord with biblical authority and with the doctrine of an omnipotent God. Here Plato's story, in the *Timaeus*, of a divine (though not omnipotent) demiurge shaping the *materials* of nature into a rationally ordered cosmos provided a bridge between seemingly incompatible views. The medieval conception of Nature as the Lord's vicar, acting in accord with her own laws but subject to God's overriding authority and occasional miraculous intervention, permitted believers to reconcile faith and reason, as Boyle and Newton could do even at the height of the scientific revolution. In this way, the seminal concept of autonomous nature was renewed by the discoveries of that "century of genius," whose accomplishments would have been inconceivable without the recovery of Aristotle and Ptolemy, Galen and Plato, in the "renaissances" of the twelfth to sixteenth centuries. Far from signifying the "death of nature," sacrificed to a lifelessly mechanized conception of the universe, then, the scientific revolution signaled its robust continuity into the centuries that followed. Charles Darwin, for whom even Linnaeus and Cuvier, his "gods" among biologists, were "mere schoolboys to old Aristotle," saw nature, in *The Origin of Species*, as exercising a limitless power, both of stability and of change, "in slowly and beautifully adapting each form to the most complex relations of life": a concept that Anaximander or Aristotle might have recognized as the heir of their own. So too might Plato have applauded a science that not only, like Kepler's or Newton's, discerned the underlying mathematical structure of the universe but also, like Heisenberg's, affirmed that ultimate knowledge of the physical world can only be a probable story, not final certainty. And so might Epicurus have seen in quantum mechanics a confirmation of his much-derided theory of the unpredictable swerve of the atoms. The more the Greek conception of nature changes the more it remains, by so changing, the same.

So protean a term cannot be pinned down to a fixed definition, since its *logos* is that of flux itself; one way to apprehend some of its shifting dimensions, however, is to consider terms to which it has (or has not) been opposed. In the classic Greek period nature was not opposed to the divine or "supernatural" (as it frequently was by later Christian writers), since nature and the cosmos were themselves considered divine. Nor was it opposed (as in recent times) to the human, since human beings no less than plants and animals were included in nature. Plato and the Platonists appear to be exceptions, since in the *Timaeus* divine reason is distinguished from mere natural matter, which it shapes, and in the *Laws* the human soul is said to be prior to the four natural elements of fire, air, earth, and water, and to be "the chief author of their changes and transpositions." Yet for Plato these distinctions are only apparent, for once the priority of soul is understood, it may truly be seen to exist by nature, which *is* divine reason. Similarly, after asserting the priority of art over nature, Plato undercuts the distinction: "Let us suppose," the Athenian Stranger says in the *Sophist* (265E), "that things which are said to be made by nature are the work of divine art, and that things which are made by man out of these are the work of human art." Aristotle, too, in the *Physics*, rejects the dichotomy of nature and art by seeing both as directed toward an end, so that human art is not the antithesis but the fulfillment of nature: "generally art partly completes what nature cannot bring to a finish, and partly imitates her." Cicero, in *On the Nature of the Gods*, speaks of "nature's art"—a Platonic concept widely adopted by the Stoics, who likewise identified nature with divine reason. And in the subtlest expression of the interplay of natural and human creativity, Polixenes in Shakespeare's *Winter's Tale* rebuts Perdita's preference for unadulterated nature to the hybrid products of human art by affirming that "the art itself is nature."

The Greek opposition between *physis* (nature) and *nomos* (law, custom, or convention), which anticipates our later dichotomy of nature and culture, is a variation on this theme, since convention or culture, like art, is a human product. For the Sophist Callicles in Plato's *Gorgias*, "convention and nature are generally at variance with one another," and natural justice consists "in the superior ruling over and having more than the inferior." Yet Plato, in repudiating this view, by no means simply upholds convention over nature, for the very opposition between them is a sophistry. Rightly understood, human laws or conventions, when they accord with reason, are one with nature, and the lawgiver, as Cleinias declares in the *Laws*, "ought to support the law and also art, and acknowledge that both alike exist by nature . . . if they are the creations of mind in accordance with right reason." Here again the Stoics adopted the Platonic doctrine that nature and convention are in fundamental accord.

Both Stoics and Epicureans concurred that a fully human life was "in accord with nature." Yet so elastic a term suggested very different norms to different schools. To some, nature was identified with instinct or desire, to others, with reason, or suppression of passion. The ambiguities of life in accord with nature persist throughout Western history. For the Christian of late antiquity, as for Plato and Plotinus, ultimate truth lay in another realm than that of this earth—the realm of form as opposed to matter, the intelligible as opposed to the sensible, spirit as opposed to flesh. Yet for both Christian and Neoplatonist, in contrast to dualistic Manichaeans and Gnostics, nature and the physical cosmos, though requiring completion or transcendence by divine grace or human contemplation of the divine, were essentially good—as God's creation or as an image of intelligible beauty. The Christian life could not be counter to nature (not natural but "unnatural" acts were sinful), but it could accord with nature only in part, since the home of the soul lay finally in another world.

With the revival of ancient Greek and Roman philosophies, including Platonism and Stoicism, in the Renaissance and Enlightenment, life in accord with nature identified with reason again became a widespread ideal. Yet against this abstractly general idea of nature later writers such as Blake and D. H. Lawrence vehemently reacted by extolling nature as instinct or desire. For Plato, *physis* and *nomos* could be reconciled; since the Romantics (and since Nietzsche and Freud), nature and culture have often been opposed, the first associated with vital animal instinct and the unconscious, the second with repressive reason and empty artifice. All too often, the injunction to live in accord with nature has become, Basil Willey observes in "Naturam Sequere" ("Follow Nature," a quotation from Seneca) in *The English Moralists* (1964), "a mere attempt to secure illegitimate sanction for moral principles of which we approve on other grounds."

Despite this danger, the richly varied concept of nature bequeathed by the pre-Socratics has enriched not only scientific discovery but also Western responsiveness to other living things. The sharp division that took extreme form in Gnostic condemnations of the material universe as the creation of an evil God was also implicit in Platonic, Hermetic, and Christian views of the world, finding expression in the *contemptus mundi*, or "contempt for the world," of monastic ascetics. In such views the human being is a composite of body and soul who shares in both the phenomenal and the noumenal, the physical and the spiritual, the natural and the supernatural realms—and these are often in conflict. Yet for the Christian, *both* dimensions, the animal and the divine, characterize humanity during life on earth. Man, in Alexander Pope's phrase, remains "in doubt to deem himself a god, or beast" because he contains elements of both in uncertain and fluctuating combination; he can no more be separated from the natural world, before death, than from the divine.

For the ancient Greeks, what we call nature was everywhere, in the forests and on the peaks where Artemis and Pan roam with their nymphs and satyrs and in the tilled furrows and vineyards sacred to Demeter and Dionysus. So, too, the poets of China praise both the simple life of Wang Wei's farmhouse and the jagged scarps and misty ravines of Han-shan's isolated Cold Mountain. The "pathless woods" and sparsely inhabited wilderness celebrated by the Romantics and their successors, seeking in wildness, with Thoreau, the preservation of an urbanized world, are no more "nature" than the farmer's fields or shepherd's pastures; as Theophrastus affirms, both spontaneously growing plants and those that are cultivated are natural, since cultivation, like other arts, completes what nature produces. And if the Greek concept of nature embraces stars and planets (thought, as in tribal myths, to be living things) no less than flora and fauna, the latter, which are born, grow, and die, are nonetheless its special domain. Aristotle wrongly thought that the realm beyond the moon was exempt by its aethereal purity from the mutability of earthly things; yet our "greater nearness and affinity" to the sublunary world in which we live most absorbed his attention. "Every realm of nature is marvellous," he writes in the *Parts of Animals*: "and as Heraclitus, when the strangers who came to visit him found him warming himself at the furnace in the kitchen and hesitated to go in, is reported to have bidden them not to be afraid to enter, as even in that kitchen divinities were present, so we should venture on the study of every kind of animal without distaste; for each and all will reveal to us something natural and something beautiful."

The natural world is very much the present world all around us. But just as tribal ritual and myth frequently celebrate an earlier time when human beings and animals (and sometimes plants) were more fully and interchangeably one; just as the Mesopotamian *Epic of Gilgamesh* tells of Enkidu's rejection by the animals, and the Hebrew Bible of Adam and Eve's expulsion from the garden; and just as Chinese poetry is haunted by T'ao Ch'ien's story of the carefree Peach Blossom Spring lost scarcely sooner than found, so has Western literature since Hesiod (or indeed since Homer's description of the orchard of Alcinous) repeatedly told of a past golden age or a now inaccessible earthly paradise where man and nature once coexisted in closer union than has since been possible anywhere on earth. Under the Titan Kronos (Roman Saturn), Hesiod writes in the *Works and Days*, a golden race of human beings feasted and danced free of toil and trouble, fed by the bounty of uncultivated fields. The legend descended through Virgil's *Georgics* and Ovid's *Metamorphoses* to late antiquity, the Middle Ages, and the Renaissance. In many of these stories, the abundance of nature is a defining characteristic of the Golden Age, which early Christian poets identified with the earthly paradise of Eden—an identification reaffirmed, centuries later, in Dante's *Purgatorio* and Milton's *Paradise Lost*. For these and other poets of an "eternal spring" not subject to change or death, nature as divine work of art is both the perfection and the antithesis of earthly nature: a contradiction transcended in the Garden of Adonis of Spenser's *Faerie Queene*, where life is not exempted from death but perpetually renewed by it, and the only eternity is that of mutability itself. Here, for once, nature as process and growth, such as we know it on earth, is no stranger in paradise.

The "soft primitivism" (in Lovejoy and Boas's phrase) exemplified by fabled stories of unfailing natural abundance found a counterpart in the pastoral idylls of Theocritus, which exerted enormous influence for millennia to come. In Theocritus's fictional Sicily and Cos (in contrast to the never-land of *Daphnis and Chloe* centuries later), herdsmen's daily lives are still exposed to the occasional hardships of a recognizably physical world. Later pastoral poets, emulating Virgil's *Eclogues*, placed their shepherds in an "Arcadian" landscape where artifice thoroughly eclipses a natural setting relegated to a scenic backdrop. For this reason, few conventional pastoral poems are included in this anthology.

Not every ancient account of the earliest stages of human existence portrayed them in Hesiod's purely glowing colors. In his division of previous history into three successive stages, the natural, the pastoral, and the agricultural, Dicaearchus (a late-fourth-century-B.C. Greek philosopher whose views were summarized by the Roman Varro centuries later) apparently gave no special preference to the natural stage, but considered the pastoral the most illustrious. Cicero, in *Pro Sestio* (xlii.91–92), as translated in Lovejoy and Boas, describes the state of nature as one "such that men, in the days before either natural or civil law had been drawn up, wandered dispersed and scattered about the fields, and that each possessed no more than he could seize or keep by his own strength, through killing or wounding others." And Lucretius, in Book V of *On the Nature of Things*, depicts the hard race of early human beings as living on acorns and wild strawberries and drinking from mountain streams, free of the soft vices and destructive wars of later times—but also as the prey of savage beasts and of one another, since "they could not contemplate the common good / or rule themselves by custom or by law."

A putative "state of nature," then (whether thought to have existed somewhere else or sometime before, or advanced as a theoretical construct against which to measure the deficiencies of the here and now), was a very equivocal one, ranging in later accounts from Thomas Hobbes's brutal warfare of every man against every man to nostalgic resuscitations of a Golden Age. The very ambiguities of a state of nature said to be characterized by oneness with other living things, but lacking the security provided by contracts and laws, were a major source of its fascination for writers such as Rousseau and Vico, who, like Lucretius, knew that earlier conditions of humanity, whatever they may have been, were conditions to which there was no return.

In contrast, the "hard primitivists" of Greek and Roman antiquity, including many Cynics and Stoics, professed to find in contemporary (and safely remote) "savages" such as the Scythians of northeastern Europe and western Asia—and sometimes in wild animals—a way of life more in accord than their own with nature as they conceived it. (Similarly, in China, Confucius, disillusioned with his degenerate age, expressed a wish to settle among the Nine Wild Tribes of the East.) Tacitus in the *Germania* depicts the strict morals of Germanic tribes who "live scattered and apart, just as a spring, a meadow, or a wood has attracted them," in laudable contrast to those of his own decadent urban world. Fifteen hundred years later, Montaigne provocatively suggested that the cannibals of a recently discovered Brazil seem "barbarous" only because "the laws of nature still rule them, very little corrupted by ours; and they are in such a state of purity that . . . what we actually see in these nations surpasses not only all the pictures in which poets have idealized the golden age . . . but also the conceptions and the very desire of philosophy."

The ironical Montaigne, who could never have doffed his breeches to run naked in the jungles of Brazil, intended, of course, like Confucius and Tacitus before and Rousseau (still more ambivalently) after him, by extolling primitive virtues to challenge his contemporaries to question their complacent assumptions of "civilized" superiority. His claim that these virtues are "natural" for all human beings gives his challenge its cutting edge. A similar claim, in his "Apology for Raymond Sebond," for the equality if not superiority of animals to human beings, not merely in "natural and obligatory instinct" but in reason and method, likewise echoes an ancient tradition—expressed most boldly in Plutarch's *Gryllus*, where one of Odysseus's shipmates, transformed into a swine by Circe, rejects Odysseus's offer to restore his human form, arguing that animals are superior to human beings in every respect, physical, moral, and intellectual. These two works alone are enough to remind us how much in the occidental tradition challenges facile assumptions of the intrinsic superiority of our species to others with which we share the planet.

There can be no single literary genre, then, that encompasses writing about nature throughout the ages, since such a genre could no more be isolated from the myths, hymns, and songs, epics and dramas, religious scriptures and philosophical or scientific treatises that make up the classics of literature than experience of the extrahuman world can be isolated from our humanity. The Western tradition, like others represented in this anthology, is immensely diverse, and the attitudes toward nature that it embodies are irreducible to any formula or agenda. To view the Hebrew Bible, for example, as sanctioning ecological destruction because God bids Adam and Eve "subdue" the earth and "have dominion" over living things is to ignore the sense of wonder in God's workmanship ("and God saw that it was good") celebrated in the Psalms ("from thy lofty abode thou waterest the mountains; the earth is satisfied with the fruit of thy work"); in the account of creation, "when the morning stars sang together, and all the sons of God shouted for joy," pronounced from the whirlwind in Job; and in the rhapsodically erotic imagery of nature in the Song of Solomon.

To portray Christianity as an insistently antinatural religion indifferent if not antagonistic to nature is to be equally blind to repeated affirmations of the goodness of the natural world from St. Augustine and St. Gregory of Nyssa, Pseudo-Dionysius the Areopagite and Boëthius in the patristic period through John the Scot in the ninth century, Bernard Silvestris in the twelfth, and St. Francis of Assisi in the thirteenth, to Sir Thomas Browne in the seventeenth and parson Gilbert White in the eighteenth. "Let it not suffice us to be book-learned, to read what others have written, and to take upon trust more falsehood than truth," John Ray wrote in *The Wisdom of God Manifested in the Works of Creation* (1691), "but let us ourselves examine things as we have opportunity, and converse with nature as well as books. . . . The treasures of nature are inexhaustible."

Like the Judeo-Christian tradition, the scientific revolution and Enlightenment have frequently been portrayed as contributing by their mechanization of the cosmos to the "death of nature" as a living power, or even to violation of the earth goddess Gaia: a picture far too one-sided to withstand scrutiny. The great scientific advances of the sixteenth and seventeenth centuries were virtually contemporary with the European literary and artistic Renaissance and the intensified revival of ancient Greek and Latin learning that fueled natural philosophy no less than letters. Not only the *Almagest* of Ptolemy and the medical writings of Galen—available in translations from Greek and Arabic since the twelfth century—but also the Greek *Corpus Hermeticum* and the writings of Plato and Plotinus, translated by the Florentine humanist Marsilio Ficino, played a key role in the ferment associated with the "new philosophy." Natural philosophy and natural magic were not clearly distinguished. The "chemical" physician Paracelsus combined Hermetic mysticism with empirical observation; the Hermetist Giordano Bruno defended Copernicus and upheld an infinite universe; and the English "magus" Robert Fludd defended both the mysteries of the Rosicrucians and William Harvey's discovery of the circulation of the blood. Nicolaus Copernicus cited Hermes Trismegistus along with Ptolemy; Johannes Kepler endeavored to correlate planetary orbits with the geometric solids of Plato and Euclid; and Isaac Newton pursued alchemical studies for much of his life. Far from seeming merely mechanical (for mathematics was still often associated with Pythagorean mysteries and astrology), Newton's theory of universal gravitation was rejected by many because of its seemingly occult affirmation of attraction at a distance. Amid all these influences, nature was very much alive—nowhere more than in the notebooks of Leonardo da Vinci. Even Francis Bacon, often portrayed as the arch-dominator if not the assassin of nature, insisted that "Nature to be commanded must be obeyed," and wrote that the scientist should be like the bee that "gathers its material from the flowers of the garden and of the field," not like the mechanically experimental ant or the dogmatically theoretical spider.

The eighteenth century, too, though often travestied as a pedestrian "Age of Reason," embraced attitudes toward the natural world ranging from Locke's cool empiricism to Shaftesbury's rhapsodic apostrophes. It was an age that could define the rules of poetry, with Pope, as "nature methodized," or imagine its primeval origins, with Vico, as responses to the "great animated body" of a "Sympathetic Nature" incomprehensible to civilized minds detached by abstraction from the senses. The men and women of this age discovered, with Addison, the imagination and the sublime in nature, explored with Diderot the illuminating nightmares of reason, and shared the solitary reveries of Rousseau revisiting the Island of Saint-Pierre in the transports of memory. The rigorous classification of plants in Linnaeus's *System of Nature*, far from laying the deadening hand of an imperialistic taxonomy on a defenseless world, was an inspiration to Rousseau on his botanical outings, and to the meticulous observations of Gilbert White and William Bartram (as it later was to Thoreau, who ranked Linnaeus with Homer and Chaucer among those who had expressed "the purest and deepest love of nature"). If nature was dead, the news had reached neither these seminal eighteenth-century writers nor the Romantics who both reacted against and drew upon them.

This is not to suggest that the Western concept of nature has not been subject to serious abuses or given birth to unfortunate consequences. Aristotle considered men superior by nature to women, free men to slaves, and Greeks to other free men. Both Galen and Paracelsus found supposed medical reasons to rationalize the prejudice of woman's inferiority as a "mutilated" gender existing only for the sake of the womb. Christian thinkers since St. Augustine denied a soul to animals, and Descartes considered them, in the absence of mind, mere machines that could not even suffer pain. In the hierarchical ladder of creation, or great chain of being, descending from Plato and Aristotle through the Christian Middle Ages to the Renaissance and eighteenth century, plants were beneath animals and animals beneath humans (who in turn were beneath the angels and God), providing a convenient excuse for mistreatment of "lower" beings. In the marquis de Sade, the ancient Sophists' claim that nature justifies the stronger in tyrannizing over the weaker achieves its crudest (or most refined) expression before the fascisms of our own century. And to the extent that a mechanized conception of nature did arise as a result, or distortion, of Newtonian science, it has served (as "social Darwinism" also has) to rationalize the increasingly widespread ecological destruction made possible by expanding industrialization and technology.

Yet to emphasize only these aspects of Western attitudes to nature—and indiscriminately condemn some parody of the Greco-Roman/Judeo-Christian/scientific-rational view of the world as predatory and destructive, "patriarchal," "anthropocentric," and so on, usually in contrast with some other putative attitude (tribal, Taoist or Buddhist, or prelapsarian/utopian/fantastic) considered to be pure of these corrupting impedimenta of history and human imperfection—is to distort beyond recognition the rich diversity of the Western tradition as represented in the present anthology. Ever since human beings began to hoe and plow the earth, or perhaps since they began to sharpen stone tools, they have been doing violence to nature (as beavers, tree blights, boll weevils, and locusts have been known to do); and the depredations and deforestations wrought by acquisitive and aggressive civilizations from China to Peru have been incalculable over the millennia. Only within the last two centuries, however, have the explosions of industrialization, technology, and population become a threat to the ecological balance and biodiversity of the planet, even without the ultimate human catastrophe of nuclear war.

No lover of nature, or of humanity—indeed, no sane person—will minimize the gravity of the threat, or the urgent need to combat it by conservation, reduction of pollution, population control, and wiser use of resources. Insofar as the conditions that brought about these dangers first arose in the West and were made possible by Western science, the West is rightly held accountable for them. But it is in large part through this same science, now global—the science that has brought humankind so many benefits along with the devastations resulting from its abuse—that the dangers have been recognized and can most effectively be addressed. Western attitudes toward nature, as expressed in our myths and poems, our philosophy and our science since the time of the ancient Greeks, far from being merely destructive, have within them—especially when enriched by openness to the great traditions of other peoples in a multicultural world—the potential for reviving our awareness of the natural world as a living organism of which we are inseparably part. We find such a world in Hesiod's *gaia*, "wide-bosomed earth, foundation of all things," born out of chaos; a world of continual flux, of coming-to-be and passing-away, of harmonious tension, ruled by nature as an autonomous "source of being moved and being at rest"; the first seminary, as Spenser called his Garden of Adonis, of all things that are born to live and die.

Those who would claim that all of Western "alphabetic civilization" since the ancient Hebrews and Greeks, if not all civilization in any form, has sundered us from the unity with nature enjoyed by Paleolithic cave dwellers, tribal hunters and gatherers, and the animals with whom we no longer recognize our kinship are retelling a myth that goes back at least to the *Epic of Gilgamesh* and the Book of Genesis—but telling it with a portentous solemnity sorely lacking the complex ironic awareness of Plutarch, Montaigne, or even Rousseau, and therefore perpetrating a distorted half-truth. Such reductive simplifications, far from advancing environmental awareness, may instead undermine it by self-evident falsity to the wealth of our cultural heritage—by "an antipathy toward Western culture," Wendell Berry remarks in *Standing by Words* (1983), "that is widespread, ill-founded, and destructive." As Simon Schama writes in *Landscape and Memory* (1995), spoofing this widespread ecological fable, "Once the archaic cosmology in which the whole earth was held to be sacred, and man but a single link in the long chain of creation, was broken, it was all over, give or take a few millennia. Ancient Mesopotamia, all unknowing, begat global warming."

Yet surely this half-truth of a ruinous civilization fanatically devoted to dominating and pillaging the natural world is among the "prevailing myths concerning our predatory actions toward each other and the environment"—as Edward O. Wilson writes—that have become "obsolete, unreliable, and destructive." Wilson urges us each to regain "the old excitement of the untrammeled world" by learning to feel like a naturalist the splendor that "awaits in minute proportions" among the mysterious organisms always within walking distance of us; and nature's own book must be our primary text for discovering its ways. But in addition, the primordial nature myths of the West, as Schama writes, have "never gone away. . . . The cults which we are told to seek in other native cultures—of the primitive forest, of the river of life, of the sacred mountain—are in fact alive and well and all about us if only we know where to look for them." *Encompassing Nature* provides a few places to begin to look for some of these myths—no less of plowland and vineyard than of forest, river, and mountain—and for the ever-changing concept of nature that evolved from them: on the shield of Achilles, for example, and in the orchard of Alcinous, on Hesiod's rocky Boeotian farm, in virgin Artemis's shadowy forests, on goat-footed Pan's towering crags, and by the invitingly warm furnace in Heraclitus's humble but divinely inhabited kitchen.

*

This large book contains a wide diversity of readings from mythical, religious, poetic, philosophical, and scientific writings of different cultures from ancient times until the threshold of Romanticism in the late eighteenth century; a few selections, including the children's stories of the Prelude and the accounts of tribal cultures in Part One, date from the nineteenth or twentieth. Parts Two and Three contain representative readings from the ancient civilizations of Mesopotamia and Israel, and from India, China, and Japan. The remainder of the anthology consists of readings in the Western tradition from ancient Greece to eighteenth-century Europe and America. Comprehensive as it is, it necessarily excludes, through limitations of space and knowledge, many of the world's great cultural traditions. The readings have been chosen for their variety, literary quality, and pertinence to central themes of changing human experiences and conceptions of nature. The book tells a story and can be read at a leisurely pace from start to finish or dipped into from time to time in different places as the reader's interest inclines; cross-references and index entries suggest some possible connections to follow.

Excerpts from religious, philosophical, scientific, and medical writings have been included to the extent that they raise major questions pertaining to broader concerns; technical discussions have been omitted. Descriptions and classifications of flora and fauna, such as those that make up much of Aristotle's writings on animals and Theophrastus's on plants, and continue, from Pliny's *Naturalis Historia* down to such works as Francis Willughby's *Ornithology* of 1708 and Linnaeus's epochal taxonomies, important though they are for the history of botany and zoology, are represented sparsely if at all; thus John Gerard's *Of the Historie of Plants* of 1597 exemplifies a long tradition of herbals dating back to Dioscorides in the first century A.D. A few passages, such as Pliny's and Aelian's gullible accounts of supposedly natural marvels, or the naïve allegories and fantastic descriptions of medieval bestiaries and encyclopedias, will be mainly historical curiosities, but most can stand as penetrating observations or provocative comments in their own right, and many are among the masterpieces of world literature and thought. Given the inexhaustible wealth and diversity of writings concerning human experiences and perceptions of nature, it is obvious that vastly more has been omitted than could have been included, and that no two persons' selections would have been the same.

The introductions to each part, chapter, and selection are intended to provide historical background helpful to understanding the readings, and in some cases to enrich these with brief reference to important writings not otherwise included. Along with the selections themselves, these introductions constitute a kind of informal (and of course very partial) history of responses to nature in different ages and different parts of the world. That they will contain occasional mistakes and inaccuracies, despite scrupulous efforts, is regrettably a foregone conclusion in a work of this scope. Selections from earlier English writers have generally been modernized in spelling and capitalization (leaving "Nature" capitalized mainly when personified), and to a lesser extent in punctuation (without breaking up long sentences); exceptions, with proper disregard for the hobgoblin of consistency, include titles such as Newton's *Opticks* and Walton's *Compleat Angler*, and the deliberately archaic language of Spenser. Poetry from other languages than English makes up a substantial portion of the selections; among translators are some of the foremost practitioners of this demanding art from earlier centuries and our own, including John Brough, Witter Bynner, Abraham Cowley, John Dryden, Robert Fitzgerald, Donald Keene, Henry Wadsworth Longfellow, Kenneth Rexroth, George Santayana, Gary Snyder, Josuah Sylvester, Helen Waddell, Arthur Waley, Burton Watson, Oscar Wilde, and William Wordsworth. In my own extensive verse translations, mainly from Greek and Latin, but also from Old English, French, Italian, and Spanish, with snippets from Provençal and Middle High German, I have attempted to suggest the metrical form or rhyme scheme of the original by an exacting line-for-line English equivalent while making as close an approximation to literal meaning as the demands of poetry will allow. This has been a challenging but immensely rewarding task, as indeed everything has been in the labor of love involved in compiling a treasury that has drawn not only on readings over a lifetime (since my father read me the haunting story of Pan from *The Wind in the Willows*) but also on unexpected and exciting new discoveries of the last few years.

In addition to the hundreds of writers included in the anthology, I have profited from many scholarly works, cited in my introductions in the hope that readers may wish to pursue some of them further; those mentioned can of course be only a few of those I have read or consulted, to say nothing of those I have not. Books of general interest (some of which have been cited or referred to in this introduction) include Arthur O. Lovejoy and George Boas, *Primitivism and Related Ideas in Antiquity* (1935); Arthur O. Lovejoy, *The Great Chain of Being: A Study in the History of an Idea* (1936); R. G. Collingwood, *The Idea of Nature* (1945); Joseph Wood Krutch, *The Great Chain of Life* (1956); Basil Willey, *The English Moralists* (1964); Clarence J. Glacken, *Traces on the Rhodian Shore* (1967); Raymond Williams, *The Country and the City* (1973); William Leiss, *The Domination of Nature* (1974); Donald Worster, *Nature's Economy: A History of Ecological Ideas* (1977; 2nd ed. 1994); Carolyn Merchant, *The Death of Nature* (1980); Wendell Berry, *Standing by Words* (1983); Keith Thomas, *Man and the Natural World: A History of the Modern Sensibility* (1983); Edward O. Wilson, *Biophilia* (1984); Daniel Halpern, ed., *On Nature: Nature, Landscape, and Natural History* (1987); Gary Snyder, *The Practice of the Wild* (1990) and *A Place in Space: Ethics, Aesthetics, and Watersheds* (1995); Robert Pogue Harrison, *Forests: The Shadow of Civilization* (1992); Peter Marshall, *Nature's Web: An Exploration of Ecological Thinking* (1992); John Torrance, ed., *The Concept of Nature* (1992); Matt Cartmill, *A View to a Death in the Morning: Hunting and Nature through History* (1993); Roger French, *Ancient Natural History* (1994); T. C. McLuhan, *The Way of the Earth: Encounters with Nature in Ancient and Contemporary Thought* (1994); Michael E. Soulé and Gary Lease, eds., *Reinventing Nature? Responses to Postmodern Deconstruction* (1995); and David Abram, *The Spell of the Sensuous: Perception and Language in a More-Than-Human World* (1996). Other anthologies of writings on nature, all mainly or entirely from the late eighteenth century and after (and mostly from English and American, or only American, writers), include Joseph Wood Krutch, ed., *Great American Nature Writing* (1950); Hal Borland, ed., *Our Natural World* (1965); Frank Bergon, ed., *A Wilderness Reader* (1980);

Thomas J. Lyon, eds., *This Incomperable Lande: A Book of American Nature Writing* (1989); Robert Finch and John Elder, ed., *The Norton Book of Nature Writing* (1990); and Richard Mabey, ed., *The Oxford Book of Nature Writing* (1995). Of these, only the last contains excerpts from non-English-language writers and writers before the eighteenth century, as does Joseph Wood Krutch's *The World of Animals: A Treasury of Lore, Legend, and Literature by Great Writers and Naturalists from the 5th Century* B.C. *to the Present* (1961).

PRELUDE

Dumbbells and Fellow Creatures: Children's Stories

The deep affinity in our culture between children and animals—some children, at least, and some animals—is attested not only by a profusion of pets and teddy bears but also by the perennial popularity of stories, films, and comic strips about more or less humanoid animals. These animals are portrayed either in isolation from an alien human world or in communion with children or childlike adults who speak their language and understand their ways. Since the time of the legendary Aesop, whose animal fables—like those of the Indian *Panchatantra* and others of different cultures throughout the world—have continued to be told and retold through the ages, a teeming menagerie of fictional creatures has delighted children and others in every generation. Though the names and species are continually changing, most English-speaking people at the turn of the twenty-first century can hardly have been unacquainted at some time with the likes of Brer Rabbit and Brer Fox; Mr. Toad and Mole; Stuart Little the mouse and Charlotte the spider; Babar the elephant; Freddy the pig and Mrs. Wiggins the cow; Curious George; Pooh and Piglet, Kanga and Eeyore (*stuffed* animals, to be sure); Lassie and Rin-Tin-Tin; Black Beauty and Misty; Krazy Kat; Pogo, Albert, and Miz Hepzibah; Snoopy and Garfield; Mickey, Minnie, and Donald; Bugs, Petunia, and Porky; Sylvester and Tweety; Bambi, the Lion King—and others past numbering. Without them and their kind our childhoods, if not our very culture, would be difficult to imagine.

Many of these beasts, to be sure, whether of household, barnyard, or forest, may have served, from the time of father Aesop to that of Peter Rabbit (or the Berenstain Bears), as little more than allegorical stand-ins to point a moral concerning another species: our own. And some may be seen merely as cuddly surrogates for less adorable, quite possibly savage members of the human or animal kingdom (but how many, come to think of it? most are canny and feisty). Even so, it tells us a great deal if children learn lessons and form relationships most easily by identifying with animals they often know, outside these fictions, only in zoos, dreams, or the untamed forests of the imagination. For what is really at issue is relationships, not primarily of animal to animal but—even when no humans appear on the scene—of human to animal and ultimately, through the enlargement this primal relation can bring, of every human and animal being to every other in a world of which all are citizens alike.

For not all animal stories for children are even ostensibly about animals only. In some, human beings appear not as superfluous extras or menacing outsiders (like the ominous specter of Man to Bambi and his mother) but as pupils or participants in this world that humans and animals share. What John Dolittle, M.D., of Puddleby-on-the-Marsh learns from Polynesia Parrot and her fellow creatures in Hugh Lofting's story (though "'M.D.' means that he was a proper doctor and knew a whole lot") is not only the Birds' ABC nor even the more comprehensive language animals speak with ears, feet, and tails no less than with mouths: he learns, above all, in switching from people to animal doctor, to be more humane. For in this as in other stories it is our relationship to the world of other living things that fulfills and completes us, thus making us fully human. Without Hobbes (a *stuffed* tiger only to unseeing grown-ups) and the wild imaginings that Hobbes, skeptic though he can be, helps provoke, Calvin would be at most half of himself, half of the Calvin and Hobbes he and we, for better and worse, all potentially are.

All of us, not children alone; for it goes without saying that adults, who are grown (or overgrown) children, have also been entranced by animal stories throughout the ages. If Kipling's *Jungle Book*, written mainly for children, told of a boy reared by wolves, the Roman historian Livy two thousand years before had recounted a similar tale, already ancient, of the suckling twins Romulus and Remus, revered as the founders of a mighty empire by the Senate and People of Rome. From the pseudo-Homeric *Battle of the Frogs and Mice* through the wasps, birds, and frogs of Aristophanes' comic choruses to the *Golden Ass* of Apuleius in the late Roman Empire, animal tales found a place in classical Western literature. The immensely popular beast epic of the outlaw Reynard the Fox's endless struggle with the vassals of King Noble the Lion rivaled the stories of King Arthur's Round Table in the various languages of medieval Europe. Few would contend that our comic strips are only for children when Pogo has run for President, Pooh had his Tao, and Snoopy his gospel. Aesop's fables have been retold, to name only a few of his avatars, by Phaedrus in Roman times, Marie de France in the Middle Ages, and La Fontaine in the seventeenth century, and revisited by Marianne Moore, among others, in the twentieth. Most telling of all, perhaps, is Plato's revelation, in the *Phaedo*, that Socrates, who had never written a line of poetry before, occupied himself during the last weeks of his life in versifying the fables of Aesop, thus fulfilling a dream that instructed him to make music by composing "a few verses before I departed." Nothing could be more childish than to think such stories only for children.

Yet it is often children in these stories—and often children slighted by the adult world—who are most in touch, like Dickon in *The Secret Garden*, with animals and other natural beings. Such stories give voice to a tenacious myth of lost innocence (or better, of lost experience fuller than later life will ever give) that is both Romantic and Platonic: what is lost in growing up is an inborn remembrance of oneness with the surrounding world which we gradually, almost inexorably relinquish—all but the childlike few who are madmen, lovers, or poets—as we relentlessly leave childhood behind. The myth appears, for example, in P. L. Travers's *Mary Poppins Comes Back*, where the newborn Annabel is destined soon to forget the words she has wordlessly spoken in infancy, but which only the birds and her shamanistic nanny understand: "I come from the sea and its tides. I come from the sky and its stars. I come from the sun and its brightness . . ." It hauntingly appears in Ratty and Mole's mysterious encounter with Pan (whom we shall be meeting again) in *The Wind in the Willows*, a song-dream no sooner glimpsed or overheard, with painful remembrance, than lost, as the gift of forgetfulness permits lightheartedness to return. It is of course a myth: there is death in Arcadia, and pirates and crocodiles in the Neverland; paradises are true only after, and only *because*, they are lost. Yet this myth, like others, conveys a deep truth so long as it is not fatally mistaken for fact.

The children's books from which our excerpts, after the two Grimm fairy tales, come are embedded, as all books are, in their own place and time. In these three Anglo-American stories of the late nineteenth and early twentieth centuries it is easy to recognize the muscular Social Darwinism of Rudyard Kipling's "Law of the Jungle"; the longing for status and security of Frances Hodgson Burnett, the impoverished Englishwoman who wrote *Little Lord Fauntleroy* and *A Little Princess* after being transplanted to Tennessee; and the sentimental Edwardian insouciance that pervades much of Kenneth Grahame's world, with its boating excursions, motor car tours, and picnics on the grass. In no simple sense, then (if such a sense could exist), can these stories be called "universal": children and adults of different cultures might well find much that is puzzling or even parochial in them.

Yet allowing for this, since no writer's outlook can be independent of the cultural matrix from which it arose, there remains in all three a mythic or "archetypal" dimension deeper than mere personal nostalgia for a fantasized "lost innocence" of childhood—of which both Kipling's Mowgli and Burnett's unhappy Mary Lennox know little. For the longing for oneness (perpetually threatened if never wholly lost) between human beings and other inhabitants of our planet to which these stories give expression is one that runs, often beneath the surface, throughout human history. Perhaps a few isolated bands of hunting-and-gathering peoples, like the BaMbuti pygmies whom we shall meet in Part One, still at times precariously experience this communion with their world, at least in the enchanted visitor's wistful perception. But among tribal peoples, too, it is a oneness widely thought to have existed more fully in a remote mythical past—*in illo tempore*, in Mircea Eliade's phrase, "in that time" of tribal origins when human beings and the animals who in so many ways transcend us could transform themselves into one another at will, as only rare and powerful shamans now can, and only in spirit, no longer in body. What Mowgli in the Indian jungles, Mary Lennox and her crippled cousin in their secret garden, and Rat and Mole on the island where they encounter the god whose name is "All" partly—though only partly—achieve is restoration of this imperiled oneness: even if Mowgli must separate himself from the wolves who reared him, Mary must leave behind the peasant boy who charms rabbits and pheasants, and Rat and Mole will remember only a trace of what they have hauntingly seen and heard. This mythic restitution of a revitalizing communion with the natural world is surely a principal source of the fascination these distant descendants of tribal folktales exert on children of all ages and perhaps—in our time, when women are exhorted to howl with wolves and men to be sensitive (like Ferdinand the bull in another children's story) to flowers—of both sexes as well.

The parallel with tribal folktales is most evident in the fairy tales of the Brothers Grimm, with which our encompassing sourcebook begins. For Jacob and Wilhelm Grimm set out, in the early years of the nineteenth century, to record tales *(Märchen* in German; "fairies" are mainly conspicuous by their absence) told by the peasant *Volk* of the villages and farms around Kassel in the central German state of Hesse. These people, in their view, lived lives and told stories much as their Germanic forebears had done for countless centuries before. Scholars have questioned some of the Grimms' Romantic assumptions—the tales are no doubt not so purely Germanic, nor so purely popular, as they thought—yet their scrupulous effort to retell the stories in simple words approximating those of their sources helped preserve a rich trove of oral folklore in a rapidly vanishing peasant tradition. In these and other European folktales we seldom find the interchangeable oneness of the human and animal worlds celebrated in many tribal rituals and myths; humans now take animal form only when bewitched, and must be restored, like the Frog King of the Grimms' first tale or the Beast of the French "Beauty and the Beast" (or Apuleius's ass long before), to human shape in order to fulfill their suspended potential. Yet the fact that they realize their full humanity only after taking animal form is perhaps a residual tribute to the often despised animal kingdom; and throughout these folktales animals frequently appear as helpers and guides (like the duck that ferries Hänsel and Gretel homeward across the water) to those who are lost or in danger. In the two stories that follow, "The Three Languages" and "The Queen Bee," those whom the animals help in their need are explicitly those who have learned their languages and felt a compassionate oneness with them: those whom their human kin, to begin with, not coincidentally disdain as dumbbells and fools. Communion with nature in an unnatural world is a priceless treasure that is not without its price.

Natives of the Kingdom of Heaven

Two Tales from the Brothers Grimm, translated by Robert M. Torrance

Jacob Grimm (1785–1863) and Wilhelm Grimm (1786–1859) were the leading German philologists and folklorists of their age. Their Kinder- und Hausmärchen *first appeared in two volumes in 1812 and 1815; subsequent revised and expanded editions appeared as late as 1857. The following two tales are number 33 ("Die drei Sprachen") and number 62 ("Die Bienenkönigin").*

The Three Languages

In Switzerland there once lived an old Count, who had an only son, who was stupid and could learn nothing. So the father said: "Listen to me, my son, I can get nothing into your head, no matter how I begin. You must go away from here; I will hand you over to a celebrated teacher who will give you a try."

The boy was sent to a foreign city and remained a whole year with the teacher. When this time had gone by he came home again, and the father asked: "Well, my son, what have you learned?" "Father, I have learned what the dogs bark," he answered. "God have mercy on us!" the father cried out, "is that all you have learned? I will send you to another city to another teacher."

The boy was taken there and stayed with this teacher for another year. When he came back the father asked again, "My son, what have you learned?" "Father, I have learned what the birds speak." The father flew into a rage and said, "Why, you worthless fellow! Have you wasted this precious time and learned nothing, and are not even ashamed to come into my sight? I will send you to a third teacher, but if you learn nothing again this time, I will no longer be your father."

The son stayed with the third teacher a whole year also, and when he came home again, and the father asked, "My son, what have you learned?" he answered, "Dear father, this year I have learned what the frogs croak." Then the father flew into his utmost rage, jumped up, called his people to him, and said: "This man is no longer my son, I turn him out and command you to lead him into the forest and take his life." They led him out, but when they were about to kill him, they could not do it for pity, and let him go. They cut out the eyes and tongue of a deer to take back as tokens to the old man.

The young man wandered on and after a while came to a castle, where he begged for a night's lodging. "Yes," said the lord of the castle, "if you are willing to spend the night down there in the old tower, go ahead, but I warn you it will be at the risk of your life, for it is full of savage dogs that bark and howl incessantly and at certain hours must have a man thrown to them, whom they devour at once." The whole district was in mourning and anguish because of this, yet no one could help. But the young man was not afraid and said, "Just let me go down to these barking dogs and give me something to throw to them; they will do nothing to me."

Because he himself wished nothing else, they gave him some food for the savage beasts and took him down to the tower. When he entered, the dogs did not bark at him but wagged their tails in a most friendly way around him, ate what he set out for them and did not muss a hair on his head. Next morning, to everyone's astonishment, he appeared again safe and unharmed, and said to the lord of the castle: "The dogs have revealed to me in their language why they dwell there and bring trouble to the land. They are bewitched and must guard a great treasure that lies below in the tower, and will not rest until it is removed, and how this must happen I have understood from their talk."

Then all who heard this rejoiced, and the lord of the castle said he would adopt him as his son if he accomplished this successfully. He went down again, and because he knew what he had to do, he accomplished it and brought up a chest filled with gold. The howling of the savage dogs was henceforth heard no more; they had vanished, and the land was freed of their scourge.

After a time it entered his head that he would like to travel to Rome. On the way he passed by a swamp in which frogs were sitting and croaking. He listened, and when he heard what they were saying he became very pensive and sad. At last he reached Rome, where the Pope had just died and there was great doubt among the Cardinals whom they should pick as his successor. They at last agreed that the one to whom a divine miracle should be revealed ought to be chosen as Pope. And just as this was decided, at that very moment the young Count entered the church and suddenly two snow-white doves flew down onto his shoulders and remained sitting there.

The churchmen recognized in this the sign of God and asked him on the spot if he would be Pope. He was undecided, and did not know whether he was worthy of this, but the doves told him that he might do it, and at last he said yes. Then he was anointed and consecrated, and thus was fulfilled what he had heard from the frogs on the way, which had disturbed him so much—that he was to be the holy Pope. Then he had to sing a mass, of which he knew not a word, but the two doves continued to sit on his shoulders and told him everything in his ear.

The Queen Bee

Two princes once went in search of adventure and fell into a wild and unruly way of life, so that they never came home again. The youngest, who was called Dumbbell, set out to look for his brothers. But when he found them at last, they poked fun at him for wanting, in his simplicity, to make his way in the world when the two of them could not manage it, though they were much cleverer.

All three of them started out together and came to an anthill. The two eldest wanted to dig it up and watch the little ants crawl around in anguish and carry their eggs away, but Dumbbell said, "Leave the creatures in peace; I will not suffer you to destroy them." Then they went on and came to a lake on which many, many ducks were swimming. The two brothers wanted to catch a couple and roast them, but Dumbbell would not allow it and said, "Leave the creatures in peace; I will not suffer you to kill them." At last they came to a beehive in which there was so much honey that it ran down the tree trunk. The two wanted to set a fire under the tree and suffocate the bees so that they could take the honey away. But Dumbbell stopped them again and said, "Leave the creatures in peace; I will not suffer you to burn them."

At last the three brothers came to a castle where horses all of stone were standing in the stables, and no human being was in sight; and they went through all the corridors until they came to a door at the very end on which were three locks, but in the middle of the door was a little shutter through which they could see into the room. There they saw a little grey man sitting at a table. They called to him once, twice, but he did not hear them; at last they called to him a third time and he stood up, opened the locks, and came out. He spoke not a word, but led them to a richly laid table; and when they had eaten and drunk took each of them to his own bedroom.

Next morning the little grey man came to the eldest, beckoned, and led him to a stone tablet on which were written three tasks by which the castle could be delivered. The first was: in the forest under the moss lay the princess's pearls, a thousand in number, which must be sought out, and if by sunset a single one was still missing, he who had looked for them would be turned to stone. The eldest went there and looked for them all day, but when the day ended, he had found only a hundred; it came to pass, as it said on the tablet, that he was turned to stone. Next day, the second brother undertook the adventure, but with him it went little better than with the eldest; he found no more than two hundred pearls, and was turned to stone.

At last came Dumbbell's turn to seek in the moss, but it was so hard to find the pearls, and went so slow, that he sat down on a stone and cried. And as he sat there the king of the ants, whose life he had once preserved, came with five thousand ants, and before long the little creatures had brought all the pearls together and laid them in a heap. But the second task was to fetch out of the lake the key to the princess's bedchamber. When Dumbbell came to the lake, the ducks which he had once saved swam to him, dived down, and fetched the key from the bottom.

But the third task was the hardest, to pick out the youngest and dearest of the king's three sleeping daughters. But they looked exactly alike, and were distinguished by nothing except that they had eaten different sweets before falling asleep: the eldest a lump of sugar, the second a little syrup, the youngest a spoonful of honey. Then the queen bee of the bees, whom Dumbbell had protected from the fire, came and tasted the mouths of all three; at last she remained sitting on the mouth which had eaten honey, and so the prince recognized the right one. Then the enchantment was over; everything was delivered from sleep, and all who were made of stone regained their human form. And Dumbbell married the youngest and dearest and became king after his father's death; but his two brothers took the other two sisters.

Animal Brother and Outcast

The Jungle Book, by Rudyard Kipling

Rudyard Kipling (1865–1936) was born in Bombay, India, under the British Raj; after being educated in England, he returned to India from 1882 to 1889. One of the leading writers of his age, of both poems and fiction (including many short stories set in both India and England), he was awarded the Nobel Prize for Literature in 1907. Among his books written for children, in addition to The Jungle Book *(1894) and* Second Jungle Book *(1895), were* Kim *(1901),* Just So Stories *(1902), and* Puck of Pook's Hill *(1906). Chapter 1 of* The Jungle Book *begins by telling how Mother and Father Wolf adopted the naked man's cub Mowgli (whose parents the tiger Shere Khan had frightened away), with the approval of Akela, "the great grey lone Wolf, who led all the pack by strength and cunning," and the support of the Bear Baloo and the Black Panther Bagheera, who bought him at the price of a slaughtered bull. The story continues as follows.*

FROM *Chapter 1: Mowgli's Brothers*

Now you must be content to skip ten or eleven whole years, and only guess at all the wonderful life that Mowgli led among the wolves, because if it were written out it would fill ever so many books. He grew up with the cubs, though they, of course, were grown wolves almost before he was a child, and Father Wolf taught him his business, and the meaning of things in the jungle, till every rustle in the grass, every breath of the warm night air, every note of the owls above his head, every scratch of a bat's claws as it roosted for a while in a tree, and every splash of every little fish jumping in a pool, meant just as much to him as the work of his office means to a business man. When he was not learning he sat out in the sun and slept, and ate and went to sleep again; when he felt dirty or hot he swam in the forest pools; and when he wanted honey (Baloo told him that honey and nuts were just as pleasant to eat as raw meat) he climbed up for it, and that Bagheera showed him how to do. Bagheera would lie out on a branch and call, "Come along, Little Brother," and at first Mowgli would cling like the sloth, but afterward he would fling himself through the branches almost as boldly as the gray ape. He took his place at the Council Rock, too, when the Pack met, and there he discovered that if he stared hard at any wolf, the wolf would be forced to drop his eyes, and so he used to stare for fun. At other times he would pick the long thorns out of the pads of his friends, for wolves suffer terribly from thorns and burs in their coats. He would go down the hillside into the cultivated lands by night, and look very curiously at the villagers in their huts, but he had a mistrust of men because Bagheera showed him a square box with a drop-gate so cunningly hidden in the jungle that he nearly walked into it, and told him that it was a trap. He loved better than anything else to go with Bagheera into the dark warm heart of the forest, to sleep all through the drowsy day, and at night see how Bagheera did his killing. Bagheera killed right and left as he felt hungry, and so did Mowgli—with one exception. As soon as he was old enough to understand things, Bagheera told him that he must never touch cattle because he had been bought into the Pack at the price of a bull's life. "All the jungle is thine," said Bagheera, "and thou canst kill everything that thou art strong enough to kill; but for the sake of the bull that bought thee thou must never kill or eat any cattle young or old. That is the Law of the Jungle." Mowgli obeyed faithfully.

And he grew and grew strong as a boy must grow who does not know that he is learning any lessons, and who has nothing in the world to think of except things to eat.

Mother Wolf told him once or twice that Shere Khan was not a creature to be trusted, and that some day he must kill Shere Khan; but though a young wolf would have remembered that advice every hour, Mowgli forgot it because he was only a boy—though he would have called himself a wolf if he had been able to speak in any human tongue.

Shere Khan was always crossing his path in the jungle, for as Akela grew older and feebler the lame tiger had come to be great friends with the younger wolves of the Pack, who followed him for scraps, a thing Akela would never have allowed if he had dared to push his authority to the proper bounds. Then Shere Khan would flatter them and wonder that such fine young hunters were content to be led by a dying wolf and a man's cub. "They tell me," Shere Khan would say, "that at Council ye dare not look him between the eyes"; and the young wolves would growl and bristle.

Bagheera, who had eyes and ears everywhere, knew something of this, and once or twice he told Mowgli in so many words that Shere Khan would kill him some day; and

Mowgli would laugh and answer: "I have the Pack and I have thee; and Baloo, though he is so lazy, might strike a blow or two for my sake. Why should I be afraid?"

It was one very warm day that a new notion came to Bagheera—born of something that he had heard. Perhaps Sahi the Porcupine had told him; but he said to Mowgli when they were deep in the jungle, as the boy lay with his head on Bagheera's beautiful black skin: "Little Brother, how often have I told thee that Shere Khan is thy enemy?"

"As many times as there are nuts on that palm," said Mowgli, who, naturally, could not count. "What of it? I am sleepy, Bagheera, and Shere Khan is all long tail and loud talk—like Mor the Peacock."

"But this is no time for sleeping. Baloo knows it; I know it; the Pack know it; and even the foolish, foolish deer know. Tabaqui[1] has told thee, too."

"Ho! ho!" said Mowgli. "Tabaqui came to me not long ago with some rude talk that I was a naked man's cub and not fit to dig pig-nuts; but I caught Tabaqui by the tail and swung him twice against a palm-tree to teach him better manners."

"That was foolishness; for though Tabaqui is a mischief-maker, he would have told thee of something that concerned thee closely. Open those eyes, Little Brother. Shere Khan dare not kill thee in the jungle; but remember, Akela is very old, and soon the day comes when he cannot kill his buck, and then he will be leader no more. Many of the wolves that looked thee over when thou wast brought to the Council first are old too, and the young wolves believe, as Shere Khan has taught them, that a man-cub has no place with the Pack. In a little time thou wilt be a man."

"And what is a man that he should not run with his brothers?" said Mowgli. "I was born in the jungle. I have obeyed the Law of the Jungle, and there is no wolf of ours from whose paws I have not pulled a thorn. Surely they are my brothers!"

Bagheera stretched himself at full length and half shut his eyes. "Little Brother," he said, "feel under my jaw."

Mowgli put up his strong brown hand, and just under Bagheera's silky chin, where the giant rolling muscles were all hid by the glossy hair, he came upon a little bald spot.

"There is no one in the jungle that knows that I, Bagheera, carry that mark—the mark of the collar; and yet, Little Brother, I was born among men, and it was among men that my mother died—in the cages of the King's Palace at Oodeypore. It was because of this that I paid the price for thee at the Council when thou wast a little naked cub. Yes, I too was born among men. I had never seen the jungle. They fed me behind bars from an iron pan till one night I felt that I was Bagheera—the Panther—and no man's plaything, and I broke the silly lock with one blow of my paw and came away; and because I had learned the ways of men, I became more terrible in the jungle than Shere Khan. Is it not so?"

"Yes," said Mowgli; "all the jungle fear Bagheera—all except Mowgli."

"Oh, *thou* art a man's cub," said the Black Panther, very tenderly; "and even as I returned to my jungle, so thou must go back to men at last,—to the men who are thy brothers,—if thou art not killed in the Council."

"But why—but why should any wish to kill me?" said Mowgli.

"Look at me," said Bagheera; and Mowgli looked at him steadily between the eyes. The big panther turned his head away in half a minute.

"*That* is why," he said, shifting his paw on the leaves. "Not even I can look thee be-

[1]The jackal.

tween the eyes, and I was born among men, and I love thee, Little Brother. The others they hate thee because their eyes cannot meet thine; because thou art wise; because thou hast pulled out thorns from their feet—because thou art a man."

"I did not know these things," said Mowgli, sullenly; and he frowned under his heavy black eyebrows.

"What is the Law of the Jungle? Strike first and then give tongue. By thy very carelessness they know that thou art a man. But be wise. It is in my heart that when Akela misses his next kill,—and at each hunt it costs him more to pin the buck,—the Pack will turn against him and against thee. They will hold a jungle Council at the Rock, and then—and then—I have it!" said Bagheera, leaping up. "Go thou down quickly to the men's huts in the valley, and take some of the Red Flower which they grow there, so that when the time comes thou mayest have even a stronger friend than I or Baloo or those of the Pack that love thee. Get the Red Flower."

By Red Flower Bagheera meant fire, only no creature in the jungle will call fire by its proper name. Every beast lives in deadly fear of it, and invents a hundred ways of describing it.

"The Red Flower?" said Mowgli. "That grows outside their huts in the twilight. I will get some."

"There speaks the man's cub," said Bagheera, proudly. "Remember that it grows in little pots. Get one swiftly, and keep it by thee for time of need."

"Good!" said Mowgli. "I go. But art thou sure, O my Bagheera"—he slipped his arm round the splendid neck, and looked deep into the big eyes—"art thou sure that all this is Shere Khan's doing?"

"By the Broken Lock that freed me, I am sure, Little Brother."

"Then, by the Bull that bought me, I will pay Shere Khan full tale for this, and it may be a little over," said Mowgli; and he bounded away.

"That is a man. That is all a man," said Bagheera to himself, lying down again. "Oh, Shere Khan, never was a blacker hunting than that froghunt of thine ten years ago!"

Mowgli was far and far through the forest, running hard, and his heart was hot in him. He came to the cave as the evening mist rose, and drew breath, and looked down the valley. The cubs were out, but Mother Wolf, at the back of the cave, knew by his breathing that something was troubling her frog.

"What is it, Son?" she asked.

"Some bat's chatter of Shere Khan," he called back. "I hunt among the plowed fields tonight"; and he plunged downward through bushes, to the stream at the bottom of the valley. There he checked, for he heard the yell of the Pack hunting, heard the bellow of a hunted Sambhur, and the snort as the buck turned at bay. Then there were wicked, bitter howls from the young wolves: "Akela! Akela! Let the Lone Wolf show his strength. Room for the leader of the Pack! Spring, Akela!"

The Lone Wolf must have sprung and missed his hold, for Mowgli heard the snap of his teeth and then a yelp as the Sambhur knocked him over with his fore foot.

He did not wait for anything more, but dashed on; and the yells grew fainter behind him as he ran into the crop-lands where the villagers lived.

"Bagheera spoke truth," he panted, as he nestled down in some cattle-fodder by the window of a hut. "Tomorrow is one day both for Akela and for me."

Then he pressed his face close to the window and watched the fire on the hearth. He saw the husbandman's wife get up and feed it in the night with black lumps; and when the morning came and the mists were all white and cold, he saw the man's child pick up a wicker pot plastered inside with earth, fill it with lumps of red-hot charcoal, put it under his blanket, and go out to tend the cows in the byre.

"Is that all?" said Mowgli. "If a cub can do it, there is nothing to fear"; so he strode round the corner and met the boy, took the pot from his hand, and disappeared into the mist while the boy howled with fear.

"They are very like me," said Mowgli, blowing into the pot, as he had seen the woman do. "This thing will die if I do not give it things to eat"; and he dropped twigs and dried bark on the red stuff. Half-way up the hill he met Bagheera with the morning dew shining like moonstones on his coat.

"Akela has missed," said the Panther. "They would have killed him last night, but they needed thee also. They were looking for thee on the hill."

"I was among the plowed lands. I am ready. See!" Mowgli held up the fire-pot.

"Good! Now, I have seen men thrust a dry branch into that stuff, and presently the Red Flower blossomed at the end of it. Art thou not afraid?"

"No. Why should I fear? I remember now—if it is not a dream—how, before I was a Wolf, I lay beside the Red Flower, and it was warm and pleasant."

All that day Mowgli sat in the cave tending his fire-pot and dipping dry branches into it to see how they looked. He found a branch that satisfied him, and in the evening when Tabaqui came to the cave and told him rudely enough that he was wanted at the Council Rock, he laughed till Tabaqui ran away. Then Mowgli went to the Council, still laughing.

Akela the Lone Wolf lay by the side of his rock as a sign that the leadership of the Pack was open, and Shere Khan with his following of scrap-fed wolves walked to and fro openly being flattered. Bagheera lay close to Mowgli, and the fire-pot was between Mowgli's knees. When they were all gathered together, Shere Khan began to speak—a thing he would never have dared to do when Akela was in his prime.

"He has no right," whispered Bagheera. "Say so. He is a dog's son. He will be frightened."

Mowgli sprang to his feet. "Free People," he cried, "does Shere Khan lead the Pack? What has a tiger to do with our leadership?"

"Seeing that the leadership is yet open, and being asked to speak—" Shere Khan began.

"By whom?" said Mowgli. "Are we *all* jackals, to fawn on this cattle-butcher? The leadership of the Pack is with the Pack alone."

There were yells of "Silence, thou man's cub!" "Let him speak. He has kept our Law"; and at last the seniors of the Pack thundered: "Let the Dead Wolf speak." When a leader of the Pack has missed his kill, he is called the Dead Wolf as long as he lives, which is not long.

Akela raised his old head wearily:—

"Free People, and ye too, jackals of Shere Khan, for twelve seasons I have led ye to and from the kill, and in all that time not one has been trapped or maimed. Now I have missed my kill. Ye know how that plot was made. Ye know how ye brought me up to an untried buck to make my weakness known. It was cleverly done. Your right is to kill me here on the Council Rock, now. Therefore, I ask, who comes to make an end of the Lone Wolf? For it is my right, by the Law of the Jungle, that ye come one by one."

There was a long hush, for no single wolf cared to fight Akela to the death. Then Shere Khan roared: "Bah! what have we to do with this toothless fool? He is doomed to die! It is the man-cub who has lived too long. Free people, he was my meat from the first. Give him to me. I am weary of this man-wolf folly. He has troubled the jungle for ten seasons. Give me the man-cub, or I will hunt here always, and not give you one bone. He is a man, a man's child, and from the marrow of my bones I hate him!"

Then more than half the Pack yelled: "A man! a man! What has a man to do with us? Let him go to his own place."

"And turn all the people of the villages against us?" clamored Shere Khan. "No; give him to me. He is a man, and none of us can look him between the eyes."

Akela lifted his head again, and said: "He has eaten our food. He has slept with us. He has driven game for us. He has broken no word of the Law of the Jungle."

"Also, I paid for him with a bull when he was accepted. The worth of a bull is little, but Bagheera's honor is something that he will perhaps fight for," said Bagheera, in his gentlest voice.

"A bull paid ten years ago!" the Pack snarled. "What do we care for bones ten years old?"

"Or for a pledge?" said Bagheera, his white teeth bared under his lip. "Well are ye called the Free People!"

"No man's cub can run with the people of the jungle," howled Shere Khan. "Give him to me!"

"He is our brother in all but blood," Akela went on; "and ye would kill him here! In truth, I have lived too long. Some of ye are eaters of cattle, and of others I have heard that, under Shere Khan's teaching, ye go by dark night and snatch children from the villager's door-step. Therefore I know ye to be cowards, and it is to cowards I speak. It is certain that I must die, and my life is of no worth, or I would offer that in the man-cub's place. But for the sake of the Honor of the Pack,—a little matter that by being without a leader ye have forgotten,—I promise that if ye let the man-cub go to his own place, I will not, when my time comes to die, bare one tooth against ye. I will die without fighting. That will at least save the Pack three lives. More I cannot do; but if ye will, I can save ye the shame that comes of killing a brother against whom there is no fault,— a brother spoken for and bought into the Pack according to the Law of the Jungle."

"He is a man—a man—a man!" snarled the Pack; and most of the wolves began to gather round Shere Khan, whose tail was beginning to switch.

"Now the business is in thy hands," said Bagheera to Mowgli. "*We* can do no more except fight."

Mowgli stood upright—the firepot in his hands. Then he stretched out his arms, and yawned in the face of the Council; but he was furious with rage and sorrow, for, wolf-like, the wolves had never told him how they hated him. "Listen you!" he cried. "There is no need for this dog's jabber. Ye have told me so often tonight that I am a man (and indeed I would have been a wolf with you to my life's end), that I feel your words are true. So I do not call ye my brothers any more, but *sag* [dogs], as a man should. What ye will do, and what ye will not do, is not yours to say. That matter is with *me*; and that we may see the matter more plainly, I, the man, have brought here a little of the Red Flower which ye, dogs, fear."

He flung the fire-pot on the ground, and some of the red coals lit a tuft of dried moss that flared up, as all the Council drew back in terror before the leaping flames.

Mowgli thrust his dead branch into the fire till the twigs lit and crackled, and whirled it above his head among the cowering wolves.

"Thou art the master," said Bagheera, in an undertone. "Save Akela from the death. He was ever thy friend."

Akela, the grim old wolf who had never asked for mercy in his life, gave one piteous look at Mowgli as the boy stood all naked, his long black hair tossing over his shoulders in the light of the blazing branch that made the shadows jump and quiver.

"Good!" said Mowgli, staring round slowly. "I see that ye are dogs. I go from you to my own people—if they be my own people. The Jungle is shut to me, and I must forget your talk and your companionship; but I will be more merciful than ye are. Because I was all but your brother in blood, I promise that when I am a man among men I will not betray ye to men as ye have betrayed me." He kicked the fire with his foot, and the sparks flew up. "There shall be no war between any of us in the Pack. But here is a debt to pay before I go." He strode forward to where Shere Khan sat blinking stupidly at the flames, and caught him by the tuft on his chin. Bagheera followed in case of accidents. "Up, dog!" Mowgli cried. "Up, when a man speaks, or I will set that coat ablaze!"

Shere Khan's ears lay flat back on his head, and he shut his eyes, for the blazing branch was very near.

"This cattle-killer said he would kill me in the Council because he had not killed me when I was a cub. Thus and thus, then, do we beat dogs when we are men. Stir a whisker, Lungri, and I ram the Red Flower down thy gullet!" He beat Shere Khan over the head with the branch, and the tiger whimpered and whined in an agony of fear.

"Pah! Singed jungle-cat—go now! But remember when next I come to the Council Rock, as a man should come, it will be with Shere Khan's hide on my head. For the rest, Akela goes free to live as he pleases. Ye will *not* kill him, because that is not my will. Nor do I think that ye will sit here any longer, lolling out your tongues as though ye were somebodies, instead of dogs whom I drive out—thus! Go!" The fire was burning furiously at the end of the branch, and Mowgli struck right and left round the circle, and the wolves ran howling with the sparks burning their fur. At last there were only Akela, Bagheera, and perhaps ten wolves that had taken Mowgli's part. Then something began to hurt Mowgli inside him, as he had never been hurt in his life before, and he caught his breath and sobbed, and the tears ran down his face.

"What is it? What is it?" he said. "I do not wish to leave the jungle, and I do not know what this is. Am I dying, Bagheera?"

"No, Little Brother. That is only tears such as men use," said Bagheera. "Now I know thou art a man, and a man's cub no longer. The Jungle is shut indeed to thee henceforward. Let them fall, Mowgli. They are only tears." So Mowgli sat and cried as though his heart would break; and he had never cried in all his life before.

"Now," he said, "I will go to men. But first I must say farewell to my mother"; and he went to the cave where she lived with Father Wolf, and he cried on her coat, while the four cubs howled miserably.

"Ye will not forget me?" said Mowgli.

"Never while we can follow a trail," said the cubs. "Come to the foot of the hill when thou art a man, and we will talk to thee; and we will come into the crop-lands to play with thee by night."

"Come soon!" said Father Wolf. "Oh, wise little frog, come again soon; for we be old, thy mother and I."

"Come soon," said Mother Wolf, "little naked son of mine; for, listen, child of man, I loved thee more than ever I loved my cubs."

"I will surely come," said Mowgli; "and when I come it will be to lay out Shere Khan's hide upon the Council Rock. Do not forget me! Tell them in the jungle never to forget me!"

The dawn was beginning to break when Mowgli went down the hillside alone, to meet those mysterious things that are called men.

In the Secret Garden

The Secret Garden, by Frances Hodgson Burnett

Frances Hodgson Burnett (1849–1924) was born in Manchester, England, where her father died when she was five, forcing her family to live in a tenement. In 1865, the family moved to Knoxville, Tennessee, where in 1873 she married a doctor, Swan Burnett. The first of her well-known children's novels, Little Lord Fauntleroy *(1886), tells of an American boy made wealthy after his rich English grandfather discovers him. In* A Little Princess *(1905)—expanded from* Sara Crewe *(1887)—Sara becomes a "beggar" after her father loses his fortune and dies; in the shabby attic where she lives before she is magically restored to fairy-tale fortune, she finds compensation in imagination, remaining "a princess inside," and makes friends with a rat, Melchisedec. "How it is that animals understand things," the author writes, "I do not know, but it is certain that they do understand. Perhaps there is a language which is not made of words and everything in the world understands it." In* The Secret Garden *(1911), the disagreeable Mary Lennox, orphaned when her parents die of cholera in India, is sent to her morosely secluded uncle's somber Misselthwaite Manor on the Yorkshire moors; there, with the aid of an obliging robin pointed out by the old gardener Ben Weatherstaff, she finds the key to a walled garden locked up since the death of her uncle's wife there ten years before. In the chapters from which our selections are taken, she meets the peasant boy Dickon, brother of the servant girl Martha, and introduces him to her secret; with his help the unweeded garden comes to flower, and in it, through the revitalizing "magic" of nature, her bedridden cousin Colin will regain his health and learn to walk. In the end, Dickon will quietly fade from the picture when Colin's father is happily reunited with his son in the garden where the boy's mother had died.*

FROM *Chapter 10: Dickon*

. . . There was a laurel-hedged walk which curved round the secret garden and ended at a gate which opened into a wood in the park. She thought she would skip round this walk and look into the wood and see if there were any rabbits hopping about. She enjoyed the skipping very much, and when she reached the little gate she opened it and went through because she heard a low, peculiar whistling sound and wanted to find out what it was.

It was a very strange thing indeed. She quite caught her breath as she stopped to look at it. A boy was sitting under a tree, with his back against it, playing on a rough wooden pipe. He was a funny-looking boy about twelve. He looked very clean and his nose turned up and his cheeks were as red as poppies, and never had Mistress Mary seen such round and such blue eyes in any boy's face. And on the trunk of the tree he leaned against, a brown squirrel was clinging and watching him, and from behind a bush near by a cock pheasant was delicately stretching his neck to peep out, and quite near him were two rabbits sitting up and sniffing with tremulous noses—and actually it appeared as if they were all drawing near to watch him and listen to the strange, low, little call his pipe seemed to make.

When he saw Mary he held up his hand and spoke to her in a voice almost as low as and rather like his piping.

"Don't tha' move," he said. "It'd flight 'em."

Mary remained motionless. He stopped playing his pipe and began to rise from the ground. He moved so slowly that it scarcely seemed as though he were moving at all, but at last he stood on his feet and then the squirrel scampered back up into the branches of his tree, the pheasant withdrew his head, and the rabbits dropped on all fours and began to hop away, though not at all as if they were frightened.

"I'm Dickon," the boy said. "I know tha'rt Miss Mary."

Then Mary realized that somehow she had known at first that he was Dickon. Who else could have been charming rabbits and pheasants as the natives charm snakes in India? He had a wide, red, curving mouth and his smile spread all over his face.

"I got up slow," he explained, "because if tha' makes a quick move it startles 'em. A body 'as to move gentle an' speak low when wild things is about."

He did not speak to her as if they had never seen each other before, but as if he knew her quite well. Mary knew nothing about boys, and she spoke to him a little stiffly because she felt rather shy.

"Did you get Martha's letter?" she asked.

He nodded his curly, rust-colored head.

"That's why I come."

He stooped to pick up something which had been lying on the ground beside him when he piped.

"I've got th' garden tools. There's a little spade an' rake an' a fork an' hoe. Eh! they are good 'uns. There's a trowel, too. An' th' woman in th' shop threw in a packet o' white poppy an' one o' blue larkspur when I bought th' other seeds."

"Will you show the seeds to me?" Mary said.

She wished she could talk as he did. His speech was so quick and easy. It sounded as if he liked her and was not the least afraid she would not like him, though he was only a common moor boy, in patched clothes and with a funny face and a rough, rusty-red head. As she came closer to him she noticed that there was a clean fresh scent of heather and grass and leaves about him, almost as if he were made of them. She liked it very much, and when she looked into his funny face with the red cheeks and round blue eyes she forgot that she had felt shy.

"Let us sit down on this log and look at them," she said.

They sat down and he took a clumsy little brown-paper package out of his coat pocket. He untied the string and inside there were ever so many neater and small packages, with a picture of a flower on each one.

"There's a lot o' mignonette an' poppies," he said. "Mignonette's th' sweetest smellin' thing as grows an' it'll grow wherever you cast it, same as poppies will. Them as'll come up an' bloom if you just whistle to 'em, them's th' nicest of all."

He stopped and turned his head quickly, his poppy-cheeked face lighting up.

"Where's that robin as is callin' us?" he said.

The chirp came from a thick holly bush, bright with scarlet berries, and Mary thought she knew whose it was.

"Is it really calling us?" she asked.

"Aye," said Dickon, as if it was the most natural thing in the world, "he's callin' someone he's friends with. That's the same as sayin' 'Here I am. Look at me. I want a bit of a chat.' There he is in the bush. Whose is he?"

"He's Ben Weatherstaff's, but he knows me a little," answered Mary.

"Aye, he knows thee," said Dickon in his low voice again. "An' he likes thee. He's took thee on. He'll tell me all about thee in a minute."

He moved quite close to the bush with the slow movement Mary had noticed before and then he made a sound almost like the robin's own twitter. The robin listened a few seconds, intently, and then answered quite as if he were replying to a question.

"Aye, he's a friend o' yours," chuckled Dickon.

"Do you think he is?" cried Mary eagerly. She did so want to know. "Do you think he really likes me?"

"He wouldn't come near thee if he didn't," answered Dickon. "Birds is rare choosers an' a robin can flout a body worse than a man. See, he's making up to thee now. 'Cannot tha' see a chap?' he's sayin'."

And it really seemed as if it must be true. He so sidled and twittered and tilted as he hopped on his bush.

"Do you understand everything birds say?" said Mary.

Dickon's grin spread until he seemed all wide, red, curving mouth, and he rubbed his rough head.

"I think I do, and they think I do," he said. "I've lived on th' moor with 'em so long. I've watched 'em break shell and come out an' fledge an' learn to fly an' begin to sing, till I think I'm one of 'em. Sometimes I think p'raps I'm a bird, or a fox, or a rabbit, or a squirrel, or even a beetle, an' I don't know it."

He laughed and came back to the log and began to talk about the flower seeds again. He told her what they looked like when they were flowers; he told her how to plant them, and watch them, and feed and water them.

"See here," he said suddenly, turning round to look at her. "I'll plant them for thee myself. Where is tha' garden?"

Mary's thin hands clutched each other as they lay on her lap. She did not know what to say, so for a whole minute she said nothing. She had never thought of this. She felt miserable. And she felt as if she went red and then pale.

"Tha's got a bit o' garden, hasn't tha?" Dickon said.

It was true that she had turned red and then pale. Dickon saw her do it, and as she still said nothing, he began to be puzzled.

"Wouldn't they give thee a bit?" he asked. "Hasn't tha' got any yet?"

She held her hands even tighter and turned her eyes towards him.

"I don't know anything about boys," she said slowly. "Could you keep a secret, if I told you one? It's a great secret, I don't know what I should do if anyone found it out. I believe I should die!" She said the last sentence quite fiercely.

Dickon looked more puzzled than ever and even rubbed his hand over his rough head again, but he answered good-humoredly.

"I'm keepin' secrets all th' time," he said. "If I couldn't keep secrets from th' other lads, secrets about foxes' cubs, an' birds' nests, an' wild things' holes, there'd be naught safe on th' moor. Aye, I can keep secrets."

Mistress Mary did not mean to put out her hand and clutch his sleeve, but she did it.

"I've stolen a garden," she said very fast. "It isn't mine. It isn't anybody's. Nobody wants it, nobody cares for it, nobody ever goes into it. Perhaps everything is dead in it already; I don't know."

She began to feel hot and as contrary as she had ever felt in her life.

"I don't care, I don't care! Nobody has any right to take it from me when I care about it and they don't. They're letting it die, all shut in by itself," she ended passionately, and she threw her arms over her face and burst out crying—poor little Mistress Mary.

Dickon's curious blue eyes grew rounder and rounder.

"Eh-h-h!" he said, drawing his exclamation out slowly, and the way he did it meant both wonder and sympathy.

"I've nothing to do," said Mary. "Nothing belongs to me. I found it myself and I got into it myself. I was only just like the robin, and they wouldn't take it from the robin."

"Where is it?" asked Dickon in a dropped voice.

Mistress Mary got up from the log at once. She knew she felt contrary again, and obstinate, and she did not care at all. She was imperious and Indian, and at the same time hot and sorrowful.

"Come with me and I'll show you," she said.

She led him round the laurel path and to the walk where the ivy grew so thickly. Dickon followed her with a queer, almost pitying, look on his face. He felt as if he were being led to look at some strange bird's nest and must move softly. When she stepped to the wall and lifted the hanging ivy he started. There was a door and Mary pushed it slowly open and they passed in together, and then Mary stood and waved her hand round defiantly.

"It's this," she said. "It's a secret garden, and I'm the only one in the world who wants it to be alive."

Dickon looked round and round about it, and round and round again.

"Eh!" he almost whispered, "it is a queer, pretty place. It's like as if a body was in a dream."

FROM *Chapter 23: Magic*

. . . They always called it Magic, and indeed it seemed like it in the months that followed—the wonderful months—the radiant months—the amazing ones. Oh! the things which happened in that garden! If you have never had a garden, you cannot understand, and if you have had a garden, you will know that it would take a whole book to describe all that came to pass there. At first it seemed that green things would never cease pushing their way through the earth, in the grass, in the beds, even in the crevices of the walls. Then the green things began to show buds, and the buds began to unfurl and show color, every shade of blue, every shade of purple, every tint and hue of crimson. In its happy days flowers had been tucked away into every inch and hole and corner. Ben Weatherstaff had seen it done and had himself scraped out mortar from between the bricks of the wall and made pockets of earth for lovely clinging things to grow on. Irises and white lilies rose out of the grass in sheaves, and the green alcoves filled themselves with amazing armies of the blue and white flower lances of tall delphiniums or columbines or campanulas.

"She was main fond o' them—she was," Ben Weatherstaff said. "She liked them things as was allus pointin' up to th' blue sky, she used to tell. Not as she was one o' them as looked down on th' earth—not her. She just loved it, but she said as th' blue sky allus looked so joyful."

The seeds Dickon and Mary had planted grew as if fairies had tended them. Satiny poppies of all tints danced in the breeze by the score, gaily defying flowers which had lived in the garden for years, and which it might be confessed seemed rather to wonder how such new people had got there. And the roses—the roses! Rising out of the grass, tangled round the sun-dial, wreathing the tree-trunks, and hanging from their branches, climbing up the walls and spreading over them with long garlands falling in cascades—they came alive day by day, hour by hour. Fair, fresh leaves, and buds—and buds—tiny at first, but swelling and working Magic until they burst and uncurled into cups of scent delicately spilling themselves over their brims and filling the garden air.

Colin saw it all, watching each change as it took place. Every morning he was brought out and every hour of each day, when it didn't rain, he spent in the garden. Even grey days pleased him. He would lie on the grass, "watching things growing," he said. If you watched long enough, he declared, you could see buds unsheathe themselves. Also you could make the acquaintance of strange, busy insect things running about on various unknown but evidently serious errands, sometimes carrying tiny scraps of straw or feather or food, or climbing blades of grass as if they were trees from whose tops one could look out to explore the country. A mole throwing up its mound at the end of its burrow and making its way out at last with the long-nailed paws, which looked so like elfish hands, had absorbed him one whole morning. Ants' ways, beetles' ways, bees' ways, frogs' ways, birds' ways, plants' ways, gave him a new world to explore, and when Dickon revealed them all and added foxes' ways, otters' ways, ferrets' ways, squirrels' ways, and trouts' and water-rats' and badgers' ways, there was no end to the things to talk about and think over. . . .

The Piper at the Gates of Dawn

The Wind in the Willows, by Kenneth Grahame

Kenneth Grahame (1859–1932) was born in Edinburgh, Scotland, but went to live with his grandmother in England after the death of his mother when he was five. He published two books of sketches of childhood, The Golden Age *(1895) and* Dream Days *(1898), but is best known for* The Wind in the Willows *(1908), which began as bedtime stories and letters for his only son. In the opening chapter, "The River Bank," the Mole tires of spring-cleaning his subterranean home and wondrously discovers the blossoming world outside, above all the world of the "sleek, sinuous, full-bodied animal" the river, where he strikes up a friendship with the Water Rat. The theme of expansive discovery and freedom continues throughout much of the book, both in Mole's frightening exploration of the Wild Wood where the Badger lives and in the comical Mr. Toad's bungling adventures on the Open Road. In Chapter 7, here given in its entirety, Mole and Rat have their most mysterious encounter with the timeless divinity of nature that transcends and embraces them all.*

Chapter 7: The Piper at the Gates of Dawn

The Willow-Wren was piping his thin little song, hidden himself in the dark selvedge of the river bank. Though it was past ten o'clock at night, the sky still clung to and retained some lingering skirts of light from the departed day; and the sullen heats of the torrid afternoon broke up and rolled away at the dispersing touch of the cool fingers of the short midsummer night. Mole lay stretched on the bank, still panting from the stress of the fierce day that had been cloudless from dawn to late sunset, and waited for his friend to return. He had been on the river with some companions, leaving the Water Rat free to keep an engagement of long standing with Otter; and he had come back to find the house dark and deserted, and no sign of Rat, who was doubtless keeping it up late with his old comrade. It was still too hot to think of staying indoors, so he lay on some cool dock leaves, and thought over the past day and its doings, and how very good they all had been.

The Rat's light footfall was presently heard approaching over the parched grass. "O, the blessed coolness!" he said, and sat down, gazing thoughtfully into the river, silent and preoccupied.

"You stayed to supper, of course?" said the Mole presently.

"Simply had to," said the Rat. "They wouldn't hear of my going before. You know how kind they always are. And they made things as jolly for me as ever they could, right up to the moment I left. But I felt a brute all the time, as it was clear to me they were very unhappy, though they tried to hide it. Mole, I'm afraid they're in trouble. Little Portly is missing again; and you know what a lot his father thinks of him, though he never says much about it."

"What, that child?" said the Mole lightly. "Well, suppose he is; why worry about it? He's always straying off and getting lost, and turning up again; he's so adventurous. But no harm ever happens to him. Everybody hereabouts knows him and likes him, just as they do old Otter, and you may be sure some animal or other will come across him and bring him back again all right. Why, we've found him ourselves, miles from home and quite self-possessed and cheerful!"

"Yes; but this time it's more serious," said the Rat gravely. "He's been missing for some days now, and the Otters have hunted everywhere, high and low, without finding the slightest trace. And they've asked every animal, too, for miles around, and no one knows anything about him. Otter's evidently more anxious than he'll admit. I got out of him that young Portly hasn't learned to swim very well yet, and I can see he's thinking of the weir. There's a lot of water coming down still, considering the time of year, and the place always had a fascination for the child. And then there are—well, traps and things—*you* know. Otter's not the fellow to be nervous about any son of his before it's time. And now he *is* nervous. When I left, he came out with me—said he wanted some air, and talked about stretching his legs. But I could see it wasn't that, so I drew him out and pumped him, and got it all from him at last. He was going to spend the night by the ford. You know the place where the old ford used to be, in bygone days before they built the bridge?"

"I know it well," said the Mole. "But why should Otter choose to watch there?"

"Well, it seems that it was there he gave Portly his first swimming lesson," continued the Rat. "From that shallow, gravelly spit near the bank. And it was there he used to teach him fishing, and there young Portly caught his first fish, of which he was so very proud. The child loved the spot, and Otter thinks that if he came wandering back from wherever he is—if he *is* anywhere by this time, poor little chap—he might make for the ford he was so fond of; or if he came across it he'd remember it well, and stop there and play, perhaps. So Otter goes there every night and watches—on the chance, you know, just on the chance!"

They were silent for a time, both thinking of the same thing—the lonely, heart-sore animal, crouched by the ford, watching and waiting, the long night through, on the chance.

"Well, well," said the Rat presently, "I suppose we ought to be thinking about turning in." But he never offered to move.

"Rat," said the Mole, "I simply can't go and turn in, and go to sleep, and *do* nothing, even though there doesn't seem to be anything to be done. We'll get the boat out, and paddle upstream. The moon will be up in an hour or so, and then we will search as well as we can—anyhow, it will be better than going to bed and doing *nothing*."

"Just what I was thinking myself," said the Rat. "It's not the sort of night for bed anyhow; and daybreak is not so very far off, and then we may pick up some news of him from early risers as we go along."

They got the boat out, and the Rat took the sculls, paddling with caution. Out in midstream there was a clear, narrow track that faintly reflected the sky; but wherever shadows fell on the water from bank, bush, or tree, they were as solid to all appearance as the banks themselves, and the Mole had to steer with judgment accordingly. Dark and deserted as it was, the night was full of small noises, song and chatter and rustling, telling of the busy little population who were up and about, plying their trades and vocations through the night till sunshine should fall on them at last and send them off to their well-earned repose. The water's own noises, too, were more apparent than by day, its gurglings and "cloops" more unexpected and near at hand; and constantly they started at what seemed a sudden clear call from an actual articulate voice.

The line of the horizon was clear and hard against the sky, and in one particular quarter it showed black against a silvery climbing phosphorescence that grew and grew. At last, over the rim of the waiting earth the moon lifted with slow majesty till it swung clear of the horizon and rode off, free of moorings; and once more they began to see surfaces—meadows wide spread, and quiet gardens, and the river itself from bank to bank, all softly disclosed, all washed clean of mystery and terror, all radiant again as by day, but with a difference that was tremendous. Their old haunts greeted them again in other raiment, as if they had slipped away and put on this pure new apparel and come quietly back, smiling as they shyly waited to see if they would be recognized again under it.

Fastening their boat to a willow, the friends landed in this silent, silver kingdom, and patiently explored the hedges, the hollow trees, the runnels and their little culverts, the ditches and dry waterways. Embarking again and crossing over, they worked their way up the stream in this manner, while the moon, serene and detached in a cloudless sky, did what she could, though so far off, to help them in their quest; till her hour came and she sank earthwards reluctantly, and left them, and mystery once more held field and river.

Then a change began slowly to declare itself. The horizon became clearer, field and tree came more into sight, and somehow with a different look; the mystery began to drop away from them. A bird piped suddenly, and was still; and a light breeze sprang up and set the reeds and bulrushes rustling. Rat, who was in the stern of the boat, while Mole sculled, sat up suddenly and listened with a passionate intentness. Mole, who with gentle strokes was just keeping the boat moving while he scanned the banks with care, looked at him with curiosity.

"It's gone!" sighed the Rat, sinking back in his seat again. "So beautiful and strange and new! Since it was to end so soon, I almost wish I had never heard it. For it has roused a longing in me that is pain, and nothing seems worth while but just to hear that sound once more and go on listening to it for ever. No! There it is again!" he cried, alert once more. Entranced, he was silent for a long space, spellbound.

"Now it passes on and I begin to lose it," he said presently. "O, Mole! the beauty of it! The merry bubble and joy, the thin, clear, happy call of the distant piping! Such music I never dreamed of, and the call in it is stronger even than the music is sweet! Row on, Mole, row! For the music and the call must be for us."

The Mole, greatly wondering, obeyed. "I hear nothing myself," he said, "but the wind playing in the reeds and rushes and osiers."

The Rat never answered, if indeed he heard. Rapt, transported, trembling, he was possessed in all his senses by this new divine thing that caught up his helpless soul and swung and dandled it, a powerless but happy infant, in a strong sustaining grasp.

In silence Mole rowed steadily, and soon they came to a point where the river divided, a long backwater branching off to one side. With a slight movement of his head Rat, who had long dropped the rudder lines, directed the rower to take the backwater. The creeping tide of light gained and gained, and now they could see the color of the flowers that gemmed the water's edge.

"Clearer and nearer still," cried the Rat joyously. "Now you must surely hear it! Ah—at last—I see you do!"

Breathless and transfixed the Mole stopped rowing as the liquid run of that glad piping broke on him like a wave, caught him up, and possessed him utterly. He saw the tears on his comrade's cheeks, and bowed his head and understood. For a space they hung there, brushed by the purple loosestrife that fringed the bank; then the clear imperious summons that marched hand-in-hand with the intoxicating melody imposed its will on Mole, and mechanically he bent to his oars again. And the light grew steadily stronger, but no birds sang as they were wont to do at the approach of dawn; and but for the heavenly music all was marvellously still.

On either side of them, as they glided onwards, the rich meadow grass seemed that morning of a freshness and a greenness unsurpassable. Never had they noticed the roses so vivid, the willow-herb so riotous, the meadow-sweet so odorous and pervading. Then the murmur of the approaching weir began to hold the air, and they felt a consciousness that they were nearing the end, whatever it might be, that surely awaited their expedition.

A wide half-circle of foam and glinting lights and shining shoulders of green water, the great weir closed the backwater from bank to bank, troubled all the quiet surface with twirling eddies and floating foam streaks, and deadened all other sounds with its solemn and soothing rumble. In midmost of the stream, embraced in the weir's shimmering armspread, a small island lay anchored, fringed close with willow and silver birch and alder. Reserved, shy, but full of significance, it hid whatever it might hold behind a veil, keeping it till the hour should come, and, with the hour, those who were called and chosen.

Slowly, but with no doubt or hesitation whatever, and in something of a solemn expectancy, the two animals passed through the broken, tumultuous water and moored their boat at the flowery margin of the island. In silence they landed, and pushed through the blossom and scented herbage and undergrowth that led up to the level ground, till they stood on a little lawn of a marvelous green, set round with Nature's own orchard trees—crab apple, wild cherry, and sloe.

"This is the place of my song-dream, the place the music played to me," whispered the Rat, as if in a trance. "Here, in this holy place, here if anywhere, surely we shall find Him!"

Then suddenly the Mole felt a great Awe fall upon him, an awe that turned his muscles to water, bowed his head, and rooted his feet to the ground. It was no panic terror—indeed he felt wonderfully at peace and happy—but it was an awe that smote and held him and, without seeing, he knew it could only mean that some august Presence was very, very near. With difficulty he turned to look for his friend and saw him at his side cowed, stricken, and trembling violently. And still there was utter silence in the populous bird-haunted branches around them; and still the light grew and grew.

Perhaps he would never have dared to raise his eyes, but that, though the piping was now hushed, the call and the summons seemed still dominant and imperious. He might not refuse, were Death himself waiting to strike him instantly, once he had looked with mortal eye on things rightly kept hidden. Trembling he obeyed, and raised his humble head; and then, in that utter clearness of the imminent dawn, while Nature, flushed with fullness of incredible color, seemed to hold her breath for the event, he looked in the very eyes of the Friend and Helper; saw the backward sweep of the curved horns, gleaming in the growing daylight; saw the stern, hooked nose between the kindly eyes that were looking down on them humorously, while the bearded mouth broke into a half-smile at the corners; saw the rippling muscles on the arm that lay across the broad chest, the long supple hand still holding the panpipes only just fallen away from the parted lips; saw the splendid curves of the shaggy limbs disposed in majestic ease on the sward; saw, last of all, nestling between his very hooves, sleeping soundly in entire peace and contentment, the little, round, podgy, childish form of the baby otter. All this he saw, for one moment breathless and intense, vivid on the morning sky; and still, as he looked, he lived; and still, as he lived, he wondered.

"Rat!" he found breath to whisper, shaking. "Are you afraid?"

"Afraid?" murmured the Rat, his eyes shining with unutterable love. "Afraid! Of *Him?* O, never, never! And yet—and yet—O, Mole, I am afraid!"

Then the two animals, crouching to the earth, bowed their heads and did worship.

Sudden and magnificent, the sun's broad golden disc showed itself over the horizon facing them; and the first rays, shooting across the level water meadows, took the animals full in the eyes and dazzled them. When they were able to look once more, the Vision had vanished, and the air was full of the carol of birds that hailed the dawn.

As they stared blankly, in dumb misery deepening as they slowly realized all they had seen and all they had lost, a capricious little breeze, dancing up from the surface of the water, tossed the aspens, shook the dewy roses, and blew lightly and caressingly in their faces, and with its soft touch came instant oblivion. For this is the last best gift that the kindly demigod is careful to bestow on those to whom he has revealed himself in their helping: the gift of forgetfulness. Lest the awful remembrance should remain and grow, and overshadow mirth and pleasure, and the great haunting memory should spoil all the after-lives of little animals helped out of difficulties, in order that they should be happy and light-hearted as before.

Mole rubbed his eyes and stared at Rat, who was looking about him in a puzzled sort of way. "I beg your pardon; what did you say, Rat?" he asked.

"I think I was only remarking," said Rat slowly, "that this was the right sort of place, and that here, if anywhere, we should find him. And look! Why, there he is, the little fellow!" And with a cry of delight he ran towards the slumbering Portly.

But Mole stood still a moment, held in thought. As one wakened suddenly from a beautiful dream, who struggles to recall it, and can recapture nothing but a dim sense of the beauty of it, the beauty! Till that, too, fades away in its turn, and the dreamer bitterly accepts the hard, cold waking and all its penalties; so Mole, after struggling with his memory for a brief space, shook his head sadly and followed the Rat.

Portly woke up with a joyous squeak, and wriggled with pleasure at the sight of his father's friends, who had played with him so often in past days. In a moment, however, his face grew blank, and he fell to hunting round in a circle with pleading whine. As a child that has fallen happily asleep in its nurse's arms, and wakes to find itself alone and laid in a strange place, and searches corners and cupboards, and runs from room to room, despair growing silently in its heart, even so Portly searched the island and searched, dogged and unwearying, till at last the black moment came for giving it up, and sitting down and crying bitterly.

The Mole ran quickly to comfort the little animal; but Rat, lingering, looked long and doubtfully at certain hoof marks deep in the sward.

"Some—great—animal—has been here," he murmured slowly and thoughtfully; and stood musing, musing; his mind strangely stirred.

"Come along, Rat!" called the Mole. "Think of poor Otter, waiting up there by the ford!"

Portly had soon been comforted by the promise of a treat—a jaunt on the river in Mr. Rat's real boat; and the two animals conducted him to the water's side, placed him securely between them in the bottom of the boat, and paddled off down the backwater. The sun was fully up by now, and hot on them, birds sang lustily and without restraint, and flowers smiled and nodded from either bank, but somehow—so thought the animals—with less of richness and blaze of color than they seemed to remember seeing quite recently somewhere—they wondered where.

The main river reached again, they turned the boat's head upstream, toward the point where they knew their friend was keeping his lonely vigil. As they drew near the familiar ford, the Mole took the boat in to the bank, and they lifted Portly out and set him on his legs on the towpath, gave him his marching orders and a friendly farewell pat on the back, and shoved out into mid-stream. They watched the little animal as he waddled along the path contentedly and with importance; watched him till they saw his muzzle suddenly lift and his waddle break into a clumsy amble as he quickened his pace with shrill whines and wriggles of recognition. Looking up the river, they could see Otter start up, tense and rigid, from out of the shallows where he crouched in dumb patience, and could hear his amazed and joyous bark as he bounded up through the osiers on to the path. Then the Mole, with a strong pull on one oar, swung the boat round and let the full stream bear them down again whither it would, their quest now happily ended.

"I feel strangely tired, Rat," said the Mole, leaning wearily over his oars as the boat drifted. "It's being up all night, you'll say, perhaps; but that's nothing. We do as much half the nights of the week, at this time of the year. No; I feel as if I had been through something very exciting and rather terrible, and it was just over; and yet nothing particular has happened."

"Or something very surprising and splendid and beautiful," murmured the Rat, leaning back and closing his eyes. "I feel just as you do, Mole; simply dead tired, though not body-tired. It's lucky we've got the stream with us, to take us home. Isn't it jolly to feel the sun again, soaking into one's bones! And hark to the wind playing in the reeds!"

"It's like music—far-away music," said the Mole, nodding drowsily.

"So I was thinking," murmured the Rat, dreamful and languid. "Dance music—the lilting sort that runs on without a stop—but with words in it, too—it passes into words and out of them again—I catch them at intervals—then it is dance music once more, and then nothing but the reeds' soft thin whispering."

"You hear better than I," said the Mole sadly. "I cannot catch the words."

"Let me try and give you them," said the Rat softly, his eyes still closed. "Now it is turning into words again—faint but clear—*Lest the awe should dwell—And turn your frolic to fret—You shall look on my power at the helping hour—But then you shall forget!* Now the reeds take it up—*forget, forget,* they sigh, and it dies away in a rustle and a whisper. Then the voice returns—

"*Lest limbs be reddened and rent—I spring the trap that is set—As I loose the snare you may glimpse me there—For surely you shall forget!* Row nearer, Mole, nearer to the reeds! It is hard to catch, and grows each minute fainter.

"*Helper and healer, I cheer—Small waifs in the woodland wet—Strays I find in it, wounds I bind in it—Bidding them all forget!* Nearer, Mole, nearer! No, it is no good; the song has died away into reed-talk."

"But what do the words mean?" asked the wondering Mole.

"That I do not know," said the Rat simply. "I passed them on to you as they reached me. Ah! now they return again, and this time full and clear! This time, at last, it is the real, the unmistakable thing, simple—passionate—perfect—"

"Well, let's have it, then," said the Mole, after he had waited patiently for a few minutes, half dozing in the hot sun.

But no answer came. He looked, and understood the silence. With a smile of much happiness on his face, and something of a listening look still lingering there, the weary Rat was fast asleep.

ONE

All Our Relatives: Tribal Ritual and Myth

"Civilized" peoples, before all else, have been those who live, as only small minorities did until recent centuries, in a city—our very word for which derives from the Latin *civitas.* But the city as a permanent and populous settlement is a relatively new thing in human history, going back scarcely more than five thousand years, and in only a very few places. Not even the walls that enclosed and strove to protect it could long shut out the view, or the memory, of the unbounded, immeasurably vaster extramural world. From the beginning, "urbane" citizens, however much they may have looked *down* on the "rustic" countryman or "savage" forest-dweller (words whose origin clearly betrays an urbanite prejudice), also looked continually *out* both on the rapidly expanding farmlands that supported the island city's existence and on the untamed, all-environing woodlands or wastelands further and further beyond. They also no doubt looked frequently *back* on ways of life, associated with forest and farm, that had been until recently the only ones the upstart human species, for all its newfound civility, had known.

Even many tribal cultures, as we have already remarked, projected into a distant (yet also a timeless) past the oneness with other forms of life that myth celebrates and ritual repeatedly reenacts. But the growth of specialized hierarchical societies that attended the agricultural revolution, especially as cities arose and expanded from ceremonial to economic and military centers, seems to have brought with it—along with immense increases of wealth, at least for those at the top of the social pyramid, and of knowledge of both good and evil—a sense of *loss* of that mythical oneness. That oneness could now be recaptured, if at all, only by more and more violent rituals, above all by blood sacrifices of humans or animals to appease the capriciously arbitrary divinities of nature on whom life itself continued to depend. The intense, if ambivalent and intermittent, yearning for earlier and seemingly simpler ways of life did not by any means begin with Rousseau and his Romantic and ecological descendants in our industrialized and technological age; we will find it in T'ao Ch'ien's Peach Blossom Spring, in Hesiod's Golden Age, in Virgil's fortunate farmers, and throughout much of our literary and philosophical tradition. Civilization does indeed appear to have always been inseparable from radical discontents, but perhaps a simpler explanation for this than Freud's lies in the severance often experienced by urban man and woman between the "artificial" world of the city and the retrospectively "natural," truly encompassing, even "oceanic" world of other living things beyond the recurrently claustrophobic enclosure of the walls that brought civilization into being.

Only in these new urban civilizations did writing arise (originally, it appears, to facilitate inventories of expanding wealth and trade) as a scribal instrument of knowledge and of the power knowledge entails: a superpersonal extension of the individual memory so highly developed in oral societies. In consequence, although cities have existed for barely a tenth of the fifty thousand years or more since our species began its hegemonic ascent, we have little or no direct testimony, such as writing alone could have preserved, of the ways that our distant and not-so-distant ancestors lived for most of human history, both as migratory hunters and gatherers during the unnumbered millennia since self-styled *homo sapiens* diverged from (and eliminated) less adaptable hominid rivals and as village agriculturalists in pre-urban societies during much of the last ten or twelve thousand years.

In addition to archeological traces and a few splendid monuments such as the cave-paintings of paleolithic Europe, some early portrayals of what would later be called "natural peoples" *(Naturvölker)* have been handed down by literate cultures. Among these are myths glorifying "civilized" man's victories over bestial creatures like the Babylonian Humbaba and the Cyclopes, Centaurs, and Amazons of Greek myth. Bestial creatures, no doubt: yet the felling of Humbaba by Gilgamesh brought desolation; the brute Cyclops Polyphemus, blinded by Odysseus, became a paragon of pastoral simplicity; and the wise Centaur Chiron was so skilled in medicinal lore that he instructed both the physician Asclepius (later worshipped as the god of healing) and the greatest of Greek heroes, Achilles, who in turn fell in love with, and slew, Penthesilea, queen of the Amazons—suggesting how far from simply despised the monstrous beings and domineering women of primitive legend could be. Similar tall tales would of course continue through the ages—as in the stories, with which Othello won Desdemona's heart, of

Cannibals that each other eat,
The Anthropophagi, and men whose heads
Do grow beneath their shoulders—

but more soberly factual accounts of preliterate tribal peoples by ancient historians and travelers, from China to Greece, also occasionally survive (notably including Herodotus's of the Scythians, Caesar's of the Gauls, and Tacitus's of the Britons and Germans [Chapter 12 below]), and have proliferated since the European explorations of the Renaissance revealed unexpected "new worlds."

Yet only since the late nineteenth century, when anthropology began to establish itself as a social science, have trained ethnographers undertaken to dwell for months or years among tribes whose languages they learned and whose lives they shared, and to study with attentive respect the material culture, kinship systems, myths, rituals, and beliefs of rapidly vanishing hunters and gatherers in Australia, the Andaman Islands, the American Arctic, or Tierra del Fuego, and of tribal agriculturalists from Oceania and New Guinea to Southwest North America and the Amazon. The sad paradox, on which many have commented, is that sustained efforts by "civilized" peoples to understand and memorialize tribal cultures began in earnest only when these were threatened with imminent extinction by the depredations those very peoples had unleashed upon them. It has recently become fashionable to expose the unwitting "ethnocentric" prejudices and denigrate or even ridicule the imperfectly achieved objectivity of the great nineteenth- and early twentieth-century ethnographers—as if any human observer could ever find an extracultural Archimedean point from which to contemplate with divine apathy any culture, including one's own. But without the heroic efforts of these devoted men and women—for an astonishing number of leading American ethnographers in particular, from Matilda Coxe Stevenson, Alice Fletcher, and Elsie Clews Parsons to Ruth Bunzel, Ruth Benedict, and Ruth Underhill, to name but a few, have been women—to learn about the lives of seemingly remote peoples long dismissed as "primitive" or peripheral, we would surely know far less of intense importance and interest today not only about others but also about ourselves as fellow human beings.

A large part of this importance and interest lies in the unconformable particularity, the unmistakable individuality of each of these varied cultures, the ingenuity each has shown in carving out an ecological niche and adopting lifeways and traditions uniquely appropriate to it, thereby making a local habitation and a name for itself on our richly multicultural planet. To attempt to restrict this diversity to only a few "patterns of culture" would be to distort and falsify that particularity beyond recognition. Of course, no tribal society can be equated with any other; each has always been embedded in history and is no less subject to change (although normally to less frenetic and disruptive change) than our own; none can be reduced, as "evolutionary" anthropologists once fondly believed, to an early stage toward attainment of the climactic civilization that is ours. Yet it would surely be equally false to recognize no significant commonalities either among tribal civilizations or between them and ourselves. The remaining few who may still live in the Congo, the Amazon forests, the Australian outback, or the Arctic tundra, much as their ancestors lived thousands of years before them, have a lot to teach us as valued contemporaries irreducibly distinct from ourselves, and a lot to teach us about ways of life in communion with the world around us that we of the urbanized world risk losing but have not entirely forgotten: about that closeness to nature, cultivated or wild, we continue to need in our often troubled and alienated condition. To say this is not to romanticize the "noble savage" of condescending myth but to acknowledge that from fellow human beings no doubt as imperfect as we, but far more expert than we in the ways of plant and animal beings—of "all our relatives," in the Lakota phrase—we have much of enormous value to learn (though perhaps we once knew it), if we can but listen, or remember.

I

In the Words of Others

In the absence of writing, and for the most part of spoken languages understood outside their immediate regions, the richly complex traditions of tribal cultures remained largely unknown to others, except as bizarre curiosities of "savages," until the last two centuries, when both dedicated amateurs and trained ethnographers have labored to understand and make them known. (Among exceptions to the general dearth of detailed studies were the meticulous efforts of certain Roman Catholic priests of New Spain and New France, especially in the "Jesuit Relations" of seventeenth-century French Canada, to record the practices of peoples whom they were attempting to convert.) Until very recently, then, when tribal peoples began to write their own languages in scripts learned from outsiders or devised, like that of the Cherokee chief Sequoiah, by themselves, they have depended on the words of others to communicate (on the rare occasions when they wished to do so) their myths and rituals to the world beyond that of their own people.

An immense ethnographic literature exists describing tribal peoples of every inhabited region of the globe. Because of the exceptional wealth of the American material, however, and in order to focus on the traditions of a single continent, both this chapter (which consists of accounts by outside observers) and the next (in which observers record the tribespeople's own words) are devoted, after the first selection, to lifeways and religious practices of indigenous North America. In tribal cultures, still more than in others, these lifeways are interfused with myths and rituals concerned with the vital relation between the people and the cosmos to which they and "all their relatives" belong. There can ultimately be no division in such cultures, varied though their beliefs and practices otherwise are, between the secular and the religious, the human and the nonhuman, the animate and the inanimate, for all are equally part of one living universe. So myth affirms, celebrating a primal unity that may once have been more complete but remains attainable by all who authentically seek it and through participation in ritual repeatedly bring it back into being.

Nowhere is this oneness more vividly reenacted than in the molimo *ceremony of the BaMbuti pygmies of Congo, as described by Colin Turnbull: a celebration so apparently free of fixed prescriptions and rules as to be "devoid of ritual" as we commonly understand the word. The BaMbuti spend much of their year as servants in the settled agricultural villages of neighboring Bantu peoples, but periodically return to their preferred existence as hunters and gatherers in the Ituri forest where they are truly at home. From its forest hiding place they fetch, from time to time, the molimo, a long trumpet formerly made from wood but now, to the "conservative" ethnographer's initial dismay, from pieces of metal drainpipe (thus turning the implements of civilization to the uses of nature!). On this hybrid instrument the males—but is it their monopoly, after all?—sing to the forest that has bestowed its blessings upon them. The climate of the northern forest inhabited by the Alaskan Koyukon, as Richard K. Nelson portrays them, could hardly be further from that of the Congo; yet here too we find the sense of a vibrantly "watchful world" to which all living creatures belong, the fusion of "natural" and "supernatural," and the hunter's reverence for the animal he slays, that characterize many tribal peoples throughout the world. Finally, among our selections in this chapter, Edmund Wilson describes the performance, in a kitchen in upstate New York, of the "Little Water Ceremony" of the Seneca, one of the six Iroquois Nations of northeastern North America. In this incongruous setting, too, is reenacted a tribal quest for unity with the world of nature that remains essential to life itself.*

The portrayals of these three cultures are of course deeply colored—how could it be otherwise?—by the preconceptions and temperaments of those who portray them. Their words are inevitably "the words of others," and sophisticated readers will be aware of Turnbull's openly professed nostalgia for a world—as he writes in dedicating his book to Kenge—"that is still kind and good . . . And without evil"; of Nelson's reproachful contrast between the ecological sensitivity of the Koyukon and the predatory wastefulness of our Western ways; and of the guilty conscience expressed in Wilson's very title, Apologies to the Iroquois. *Yet when all the differences, the presuppositions, the subjectivities and the partialities have been remarked and discounted, what surely remains from these accounts, and many like them, is a powerful tribute to ways of life and ancient traditions that embody, or seek to re-embody, a communion with encompassing nature that we of the "civilized" world can forgo only at extreme peril to both body and soul.*

The Goodness of the Forest

The Forest People: A Study of the Pygmies of the Congo,
by Colin Turnbull

Colin Turnbull (1924–94) was born in London and studied philosophy and politics at Oxford. After serving in the Second World War, he spent two years at Banaras Hindu University in India, where "I was taught the doctrine of truth, goodness, and beauty" before returning to Ox-

ford to study anthropology. His fieldwork in Africa led to his two best-known books, The Forest People *(1962), written after three years among the BaMbuti pygmies of the then Belgian Congo, and his darkly contrasting study of the dysfunctional society of the Ik of northern Uganda,* The Mountain People *(1973). After coming to the United States in 1959, he was associate curator of African ethnology of the American Museum of Natural History in New York until 1969 and later professor at Virginia Commonwealth University and at George Washington University. Among his other books are* Man in Africa *(1976) and* The Human Cycle *(1983). His account of the molimo ceremony follows.*

FROM *Chapter 4: The Song of the Forest*

. . . The molimo of the Pygmies is not concerned with ritual or magic. In fact, it is so devoid of ritual, expressed either in action or words, that it is difficult to see what it *is* concerned with. Every day, around midday, a couple of youths would go around the camp, just as they had done this day, collecting offerings of food and firewood from hut to hut, for the molimo concerns everyone, and everyone must contribute. And each evening the women and children shut themselves up in their huts after the evening meal, for the molimo is mainly the concern of the men. And when the women have retired the men sit around the kumamolimo—the hearth of the molimo—and gaze into the molimo fire. Nearby a basket hangs, full of the offerings of food that will be eaten later. But first the men must sing, for this is the real work of the molimo, as they say; to eat and to sing, to eat and to sing. Yet behind these apparently simple outer trappings of the festival there was an atmosphere of almost overwhelming expectancy.

That first night, after I had eaten my meal, I went over to the molimo fire, halfway between Manyalibo's hut and that of his daughter Kondabate and her husband Ausu, where the men were already gathering. I noticed that Amabosu, the singer, was not there. I knew why when, a few minutes after the singing had begun, I heard the voice of the molimo answering, way off by itself in the forest. It no longer worried me that the trumpet was a metal drainpipe instead of a piece of bamboo or wood, because now that I could not see it I realized that Ausu was right. It was the sound that mattered. As the men sang in the camp, the voice of the molimo echoed their song, moving about continually so that it seemed to be everywhere at once. During a lull in the singing it started giving animal growls, and the men looked around to make sure that all the women were safely in their huts. Almost immediately Amabosu and Madyadya came running into the camp from the far end, past Cephu's camp, picking their feet up high in the air so that more than ever they seemed to be floating above the ground they trod. They came right up to the fire and squatted down either side of it, the front of the trumpet still held on Madyadya's shoulder as Amabosu warbled softly into the other end.

I noticed that the trumpet was still wet; it must have been given a "drink" just before it was carried into the camp. Now both youths stood up, and with Amabosu singing all the time, in reply to the song of the men, Madyadya took the end of the trumpet in his hands and waved it up and down, passing it through the flames of the molimo fire and over the heads of all of us sitting there. Some of the men took up hot ashes and rubbed them over the trumpet, and some even put live coals in at the far end. As Amabosu sang harder and louder, sparks flew out over Madyadya's shoulder, disappearing into the night with the song. . . .

The molimo was often referred to as "the animal of the forest," and the women were supposed to believe that it really was an animal, and that to see it would bring death. That of course is why they were all bundled off to bed with the children before the trumpet was ever brought into camp. And even when it *was* brought in it was often shielded by a number of youths so that if any woman should happen to look, she would see nothing. The animal sounds it produced were certainly realistic, but I wondered what the women thought when it sang. What kind of animal was it that one moment could make such threatening sounds, and the next instant sing more beautifully than anything else in the whole forest?

I remembered again Ausu's saying that the only important thing about the trumpet was the fact that it had a good voice and could sing well. I was reminded of the Pygmy legend of the Bird with the Most Beautiful Song. This bird was found by a young boy who heard such a Beautiful Song that he had to go and see who was singing. When he found the Bird he brought it back to the camp to feed it. His father was annoyed at having to give food to the Bird, but the son pleaded and the Bird was fed. The next day the Bird sang again; it sang the Most Beautiful Song in the Forest, and again the boy went to it and brought it back to feed it. This time the father was even more angered, but once again he gave in and fed the Bird. The third day (most Pygmy stories repeat themselves at least three times) the same thing happened. But this time the father took the Bird from his son and told his son to go away. When his son had left, the father killed the Bird, the Bird with the Most Beautiful Song in the Forest, and with the Bird he killed the Song, and with the Song he killed himself and he dropped dead, completely dead, dead for ever.

There are other legends about song, all telling how important it is, and I wondered just why. I looked at the others sitting with me around the kumamolimo. The flames from the fire flickered, lighting up the serious faces, shining brightly in their large, honest eyes. Some gazed straight into the fire, others leaned back and looked up into the trees as they sang. A few lay down on their hunting nets, but they were awake and they sang with the rest. Then I saw that Makubasi had his infant son on his lap, holding the boy tightly against him, wrapped in his powerful arms, rocking backward and forward. He was singing quietly, with his mouth up against his son's ear. The words of the song, like the words of most molimo songs, were few. They simply said, "The Forest is Good." . . .

For a month I sat every evening at the kumamolimo; listening, watching, and feeling—above all, feeling. If I still had little idea of what was going on, at least I felt that air of importance and expectancy. Every evening, when the women shut themselves up, pretending that they were afraid to see "the animal of the forest"; every evening, when the men gathered around the fire, pretending they thought that the women thought the drainpipes were animals; every evening, when the trumpet drainpipes imitated leopards and elephants and buffalos—every evening, when all this make-believe was going on, I felt that something very real and very great was going on beneath it, something that everyone else took for granted, and about which only I was ignorant.

It was as though the songs which lured the "animal" to the fireside also invoked some other kind of presence. As the evenings wore on toward morning, the songs got more and more serious, and the atmosphere not exactly tense but charged with an emotion powerful enough to send the dancers swirling through the molimo fire as though the flames and red-hot coals held no heat, as though the glowing embers were cold ashes. Yet there was nothing fanatic or frenzied about their action in dancing through the fire. Then there was always that point when the "animal" left the camp and returned to the forest, taking the presence with it. I could feel it departing as the mellow, wistful voice of the molimo got farther and farther away. . . .

That same night, just before the singing started, Moke was sitting outside his hut, whittling away at a new bow he was making, and talking to himself because there was nobody else for him to talk to. Kenge had spoken to him when we got back to the camp at dusk, but I did not expect anything to come of it, and Kenge had said nothing to me. Moke looked up and stared in my direction, then he looked back at his bow and continued whittling away with a rough stone. But he spoke a little louder, so that I would hear.

"Ebamunyama," he said, using the name I had been given, "*pika'i to*." ("Come here.")

He spoke as if he was addressing his bow, and when I got to him he still did not look up but just told me to sit down. Talking to the bow in his soft old voice, he said that he had been told I was asking about the molimo, and that this was a good thing and he would tell me whatever he could.

He went on to tell that the Pygmies call out their molimo whenever things seem to be going wrong. "It may be that the hunting is bad," he said, "or that someone is ill, or, as now, that someone has died. These are not good things, and we like things to be good. So we call out the molimo and it makes them good, as they should be."

Kind, quiet old Moke, all alone and without a wife to look after him, still working away at his bow, occasionally looking along the shank to make sure he was keeping the line true—he told me many things that evening. But, most important, he told me, or rather showed me, how the Pygmies believe in the goodness of the forest.

"The forest is a father and mother to us," he said, "and like a father or mother it gives us everything we need—food, clothing, shelter, warmth . . . and affection. Normally everything goes well, because the forest is good to its children, but when things go wrong there must be a reason."

I wondered what he would say now, because I knew that the village people, in times of crisis, believe that they have been cursed either by some evil spirit or by a witch or sorcerer. But not the Pygmies; their logic is simpler and their faith stronger, because their world is kinder.

Moke showed me this when he said, "Normally everything goes well in our world. But at night when we are sleeping, sometimes things go wrong, because we are not awake to stop them from going wrong. Army ants invade the camp; leopards may come in and steal a hunting dog or even a child. If we were awake these things would not happen. So when something big goes wrong, like illness or bad hunting or death, it must be because the forest is sleeping and not looking after its children. So what do we do? We wake it up. We wake it up by singing to it, and we do this because we want it to awaken happy. Then everything will be well and good again. So when our world is going well then also we sing to the forest because we want it to share our happiness."

All this I had heard before, but I had not realized quite so clearly that this was what the molimo was all about. It was as though the nightly chorus were an intimate communion between a people and their god, the forest. Moke even talked about this, but when he did so he stopped working on his bow and turned his wrinkled old face to stare at me with his deep, brown, smiling eyes. He told me how all Pygmies have different names for their god, but how they all know that it is really the same one. Just what it is, of course, they don't know, and that is why the name really does not matter very much. "How can we know?" he asked. "We can't see him; perhaps only when we die will we know and then we can't tell anyone. So how can we say what he is like or what his name is? But he must be good to give us so many things. He must be of the forest. So when we sing, we sing to the forest."

The complete faith of the Pygmies in the goodness of their forest world is perhaps best of all expressed in one of their great molimo songs, one of the songs that is sung fully only when someone has died. At no time do their songs ask for this or that to be done, for the hunt to be made better or for someone's illness to be cured; it is not necessary. All that is needful is to awaken the forest, and everything will come right. But suppose it does not, suppose that someone dies, then what? Then the men sit around their evening fire, as I had been doing with them for the past month, and they sing songs of devotion, songs of praise, to wake up the forest and rejoice it, to make it happy again. Of the disaster that has befallen them they sing, in this one great song, "There is darkness all around us; but if darkness *is*, and the darkness is of the forest, then the darkness must be good."

As Moke looked at me and spoke I wondered if this was how he had once explained things to his own son. Now I began to understand what happened at night when fifteen feet of drainpipe was carried into the camp and someone blew or sang into it; and I began to understand what the "presence" was—it was surely the presence of the Forest itself, in all its beauty and goodness.

Moke turned back to his bow and began whittling again.

"You will soon see things of which you have never heard, and which you have never seen. Then you will understand things that I can never tell you. But you must stay awake—you may see them only once."

He started humming to himself, and I left.

FROM *Chapter 8:* Molimo: *The Dance of Death*

. . . At first I had thought that the molimo was virtually the sole concern of the men. In the evenings, when the singing first started, there were sometimes a few girls who stood nearby and danced among themselves as we sang, but they were soon sent to join the rest of the women and children in their huts—certainly long before the molimo made its appearance. But one day I felt that something different was going on.

I had noticed that Kondabate, Manyalibo's daughter, who was married to Ausu, was building an extension at the back of her hut. She and Ausu had no children, and when I asked her why she was enlarging her home she said there was going to be an *elima*, and that all the girls would stay in her hut, under her care. An elima is a celebration that takes place when a girl reaches maturity, as marked by the first appearance of menstrual blood. At such a time she enters a special hut with her girl friends and stays there for a month. But I had heard none of the usual discussion about the event, and I felt this could not be the full answer.

Then, that afternoon, an old, old woman arrived with her husband. They stopped first in Cephu's camp, where they had relatives, but then came on into the main camp, where they were received with the greatest respect. The husband sat down at the kumamolimo and was offered food; the old woman disappeared into Kondabate's house. She stayed there until evening, but all I could hear from inside were occasional snatches of song. It was plain that there were several girls in there as well, and it sounded as though they were either being taught new songs or else rehearsing songs they already knew. They never sang more than a few bars at a time, and I was continually being startled, thinking that I heard something resembling one of the molimo songs. . . .

After we had eaten the evening meal Manyalibo went around the camp himself and took a piece of wood from the fire outside each hut, placing them all on the molimo fire. Then, before it was completely dark, the men gathered to sing. Even Cephu was there, and the singing was light and cheerful. The girls that had been in Kondabate's hut came out as soon as we started. They had painted their bodies with black kangay juice, and in their hair they had twined circlets of vine, one strand of which stuck way out in front, over their foreheads, with a small bunch of feathers dangling from the end. They kept together in a tight little group, their bodies swaying in unison to the songs of the men. Then they began to dance, still so close together that I could not see between them. They danced all around the group, forcing us closer and closer to the fire. In a snaking, graceful line they seemed to take charge of the music, increasing the tempo until the men were as exhausted as they were, and we all stopped for a rest.

By then it was dark, and I expected the girls and the rest of the women and children to retire to their huts. But they had built a fire of their own at the edge of the molimo group, and there they all sat, chatting inconsequentially as though they did this every night. It was then that I first noticed the old woman. Under a great headdress of vine and feathers her hatchet face looked more angular than ever, thin and straight, with eyes still wide and unblinking and, it seemed, unseeing. She sat in the middle of the group, hugging her knees. Her skin sagged in withered folds, her bones stuck out in ugly lines and bumps. She made a strange contrast with the background of young feminine beauty and vitality. The only time she so much as moved was when Kondabate joined the group.

Kondabate, the belle with filed teeth and great diamond-shaped scarifications on her stomach; Kondabate the beauty, who after two years of married life still preserved her fine upstanding breasts and trim figure—but also Kondabate the childless. She had on a headdress like the old woman's, and her eyes, too, seemed unnaturally wide and staring. She was smiling and happy—until she sat down opposite the old lady. That was when the old woman moved, looking up from the fire into Kondabate's eyes. In an instant Kondabate seemed to lose her smile and her happiness. She became as empty as the withered old shell of a woman in front of her, and she sat as quietly, staring into the flickering flames.

After a while the men started singing again, but gently, and then with a shock I realized that the women were singing as well, the sacred songs of the molimo. And they were not just joining in, they were leading the singing. Songs that I had thought only the men knew and were allowed to sing—all of a sudden the women were showing that they not only knew them but could sing them with just as much intensity. Kondabate suddenly sprang up and seized the banja sticks from Kelemoke, and began beating the urgent molimo rhythm: *one* two, *one* two, *one* two three. She beat the banja faster and faster, and I saw the old woman begin to twitch in every muscle. Kondabate was already half dancing, and it was as though she was passing the life and force from her body into the dried-up skinny old body in front of her. The old crone's legs started stamping up and down, and she sat back on her buttocks so that she could lift her feet higher. Then her arms began moving, pointing in this direction and that; even her fingers each seemed to have a dance of its own. But her head was still, and her eyes were fixed on those of Kondabate.

At this point Kondabate handed the banja to another of the women standing nearby and led the young girls right into the center of the men's group, dancing in a close circle around the molimo fire. As they moved around, the men stood up and made room for them, and now the singing came almost entirely from the women. The men just stood and stamped and clapped and growled a bass accompaniment. The girls danced more and more wildly, but their heads and eyes were always turned toward the molimo fire, no matter how violent the body movement. After nearly half an hour the old woman, still crouching, ran into the center after the girls, and turning to face them from the far side of the fire she drove them slowly backward, back until they were lost among the others at the edge of the group. The singing was quieter, and in a few minutes it faded away altogether. Not a word was spoken.

After a short rest the singing started again, and this time there was no mistaking it—the women were leading. Manyalibo's wife picked up the banja and beat them, sharply and decisively, and the girls began the song. The men followed in obedient chorus. Once again, as the song grew faster, the old woman began to twitch until her whole body was alive with movement. Then she sprang up and came into the center of the kumamolimo alone. For an instant she stood there, and her head moved sharply from side to side. It was like the movement of a bird, keeping a wary eye open for possible foes. As she began to dance she seemed to sink slowly into the ground, her knees folding up like a concertina, her body bending at the waist; but always her head stuck out in front, jerking from side to side, eyes staring and expressionless.

She circled the fire three times, looking out and upward into the night, but unseeing; then as if satisfied that she was alone and unobserved she flung her head and her whole body around and faced the fire directly. She was about seven or eight feet away from it. She advanced slowly and cautiously, two steps forward, one step back, her attention riveted on the flames. The singing was louder and faster, but she refused to be hurried. At her own pace she approached closer and closer until it seemed impossible that she could stand the heat. She hesitated on the brink of the fire, her frail body gleaming with sweat, her face glowing and trembling. She made as if she was going to pounce on the fire, but whirled around at the last moment and danced quickly back to where the women were singing, watching her, eagerly and anxiously. She sat down on her haunches and accepted a steaming bowl of liko.

The singing did not stop this time but went on quietly at a slower tempo. The men made no attempt to move back closer to the fire. After about ten minutes the old lady, apparently refreshed and as full of vigor as ever, danced back into the men's circle, with Kondabate at her side. Their movements were perfectly co-ordinated, even to the twitching of their heads; as they sank down close to the ground they might have been one person. But as they neared the fire, again having circled it several times, the old woman sprang to her feet and ran swiftly around to the far side and crouched there, staring at Kondabate through the flames. They started circling once more, each keeping her eyes on the other, moving from side to side so that they were always opposite each other. But now they were closer to the fire than they had ever been and on their hands and knees as they passed through the hot ashes, as though trying to reach each other through the flames.

Finally Kondabate retreated and squatted down at the edge of the circle, her body twitching just as the old woman's had done, her eyes staring and empty as she gazed at the molimo fire. The men heaped more wood on the flames, which were beginning to die down. Now they were singing as loudly as the women, stamping on the ground with their feet and dancing with their arms and bodies, but never moving from where they stood. The old woman grew taller and more upright. Her bent old frame straightened out and she stood proudly, arms at her side, bent at the elbows so that her hands were out in front. Those hands had a life of their own, pointing and gesticulating and dancing with infinite grace at the end of motionless arms. The old woman no longer looked from side to side. She knew she had nothing to fear now, that she was the victor. And when she stooped once more to approach the fire she did not stoop so low, and each step was a step forward. As the singing quickened so did her steps, until with a burst of frenzied shouting she was driven right into the flames.

She seemed to hover there an instant, that skinny old crone who should have been burned to a cinder in a flash. Then she whirled around and kicked out with her feet, scattering the sacred molimo fire in all directions. Blazing logs and glowing embers alike she scattered, right among the circle of men surrounding her. And then she danced away, erect and proud. The men, without even faltering in their song, quickly gathered all the scattered embers and threw them back onto the remaining coals, and then for the first time they moved in a dance of their own. They danced in a wild circle, and as they danced their bodies swayed backward and forward, facing the fire, as though by imitation of the act of generation they were giving the fire new life. And as they danced the flames slowly began to rise, and as they rose the men danced all the more violently, until they had brought the fire of the molimo back to life.

Twice more this happened. Each time the old woman made a more determined effort to stamp the fire out of existence. And each time the strangely beautiful and exciting erotic dance of the men gave it new life. Finally the old woman conceded defeat and retreated among the others. Shortly afterward all the women disappeared, and the men rearranged themselves at the kumamolimo without making any mention of what had happened.

I looked at Moke, and found him staring at me curiously. He understood my question and nodded. Yes, this was what he had told me I was yet to see.

There is an old legend that once it was the women who "owned" the molimo, but the men stole it from them and ever since the women have been forbidden to see it. Perhaps this was a way of reminding the men of the origin of their molimo. There is another old legend which tells that it was a woman who stole fire from the chimpanzees or, in yet another version, from the great forest spirit. Perhaps the dance had been in imitation of this. I did not understand it by any means, but somehow it seemed to make sense.

The woman is not discriminated against in BaMbuti society as she is in some African societies. She has a full and important role to play. There is relatively little specialization according to sex. Even the hunt is a joint effort. A man is not ashamed to pick mushrooms and nuts if he finds them, or to wash and clean a baby. A woman is free to take part in the discussions of men, if she has something relevant to say. In fact, it was the apparent absolute dominance of the male in the molimo that had seemed to be the exception.

But now even in this the woman had come into her own. She had asserted her prior claim to the fire of life, and her ability to destroy it, to extinguish life. Or had she been destroying it? Perhaps when she was kicking the fire out in all directions among the menfolk she was giving it to them, for them to gather and rebuild and revitalize with the dance of life.

I was trying to puzzle it all out when the singing started again, and I saw that the old woman had returned, alone. All the others were in their huts, and the men seemed to be singing as though by their song they could drive the last remaining female away. They sang louder and harder, but she kept circling the kumamolimo, keeping in the shadows, until with surprisingly swift and agile strides she was once more in our midst. In her hands she held a long roll of nkusa twine. The men continued singing, and as they sang, the old woman went around knotting a loop around each man's neck, so that in the end we were all tied together. The men made no attempt to resist; rather, they ignored what was going on. When they were all tied in this way, however, they stopped singing.

Moke spoke—I am not sure whether it was for my benefit or because it was his place to say what he did. He said, "This woman has tied us up. She has bound the men, bound the hunt, and bound the molimo. We can do nothing." Then Manyalibo said that we had to admit we had been bound, and that we should give the woman something as a token of our defeat; then she would let us go. A certain quantity of food and cigarettes was agreed upon, and the old woman solemnly went among us again, untying each man. Nobody attempted to loose himself, but as each man was untied he began to sing once more—the molimo was free. The old woman received her gifts and went back to Cephu's camp, where she and her husband were staying. The couple stayed for another week or two, but she danced only once again. Before she left us she went to every man, giving him her hand to touch as though it were some kind of blessing. . . .

FROM *Chapter 15: The Dream World*

. . . It was, I knew, to be my last trek into the forest with the people I had come to think of as such close friends, and even though I still had another two months to spend with them I was sad. It was partly because of this, and partly because, like Kenge, I was so profoundly content to be back in our own part of the forest, that I found the forest more beautiful than it had ever been. Kenge said it was just because it was the honey season, but as we passed the plantation and entered once more into the shade of the great friendly old trees, he added, "This is the *real* world . . . this is a good world, our forest." . . .

One day Kenge had gone off on some unknown mission and I decided to make a final pilgrimage to Apa Lelo, where the cycle had begun. This day of all days I wanted to be by myself, but there is something about the forest, not exactly threatening, but challenging, that dares you to travel alone. I had to refuse several Pygmies who wanted to come with me, and once I was well away I was glad. I knew what the challenge was, for to be alone was as though you were daring to look on the face of the great God of the Forest himself, so overpowering was the goodness and beauty of the world all around. Every trembling leaf, every weathered stone, every cry of an animal or chirp of a cricket tells you that the forest is alive with some presence.

As I splashed my way through the last stream and climbed to the top of the rise above Apa Lelo, the sounds of the forest grew louder. This was a friendly sign, but somehow it made me feel even more lonely. Then I came down into the old camp. The huts were rotted and collapsed, the clearing overgrown. As I roamed about, tracing where huts had been and remembering who had lived there, kicking idly with my feet in the hopes of finding some familiar object, it all began to live again. I wandered down to the Lelo, as calm and placid and beautiful as ever, and I thought of the molimo. As I looked into the waters where the trumpet had once been hidden, it was as if I heard its voice in the distance, gentle but insistent.

When I had bade my farewell to that very special river and returned to the camp, I saw it as it really was in the timelessness of the forest world—the huts strong and green, the fires burning brightly and the air heavy with the scent of cooking. Lizabeti was running about with the other children, swinging happily on her crutches. The men helped one another to examine and repair their nets, while the women cooked and gossiped. Masisi was shouting at someone, and Manyalibo was trying to pacify him in his usual gruff, good-natured way. Old Moke sat back in his chair and puffed at his pipe, gazing up at the tree tops so very high above. I heard him murmur once more, "You will see things you have never seen before. . . . You will understand why we are called People of the Forest. . . . When the Forest dies, we shall die." And for the last time I heard the chorus of that great song of praise: "If Darkness *is*, Darkness is Good."

The camp and the people disappeared, leaving only a shimmering haze over Apa Lelo, but the song remained, for the song is the soul of the people and the soul of the forest. The molimo trumpet, now infinitely wistful and far away, took up the song, and the forest echoed it on with its myriad magic sounds.

It echoes on and on, and it will still be there when our short lives are silenced . . . until, perhaps like us, it comes to rest in the deepest distance of some other world beyond . . . the dream world that is so real to the People of the Forest.

The Watchful World

Make Prayers to the Raven: A Koyukon View of the Northern Forest, by Richard K. Nelson

Richard K. Nelson, a cultural anthropologist, consultant, and writer, lives in Sitka, Alaska. Make Prayers to the Raven *(1983) was based on sixteen months of living among the Koyukon, a Northern Athapaskan people, in the villages of Huslia and Hughes on the Koyukuk River not far below the Arctic Circle in central Alaska, mainly from September 1976 to July 1977—a time when "I learned a different perception not only of the raven, but of every living and nonliving thing in the northern forest." Among his other books are* Hunters of the Northern Forest *(1973; 2nd ed. 1986) and* The Island Within *(1989).*

FROM *Chapter 2: The Watchful World*

There's always things in the air that watch us.

A WAY OF SEEING This chapter describes the nature of nature, as it is understood by the Koyukon people. . . .

Traditional Koyukon people live in a world that watches, in a forest of eyes. A person moving through nature—however wild, remote, even desolate the place may be—is never truly alone. The surroundings are aware, sensate, personified. They feel. They can be offended. And they must, at every moment, be treated with proper respect. All things in nature have a special kind of life, something unknown to contemporary Euro-Americans, something powerful. . . .

Koyukon perceptions of nature are aligned on two interconnected levels. The first of these is empirical knowledge. The practical challenges of survival by hunting, fishing, and gathering require a deep objective understanding of the environment and the methods for utilizing its resources. In short, the Koyukon people are sophisticated natural historians, especially well versed in animal behavior and ecology.

But their perception of the natural environment extends beyond what Westerners define as the empirical level, into the realm of the spiritual. The Koyukon inherit an elaborate system of supernatural concepts for explaining and manipulating the environment. From this perspective the natural and supernatural worlds are inseparable, and environmental events are often caused or influenced by spiritual forces. Detailed explanations are provided for the origin of natural entities and for the causation of natural events (which seldom, if ever, take place purely by chance). Furthermore, behavior toward nature is governed by an array of supernaturally based rules that ensure the well-being of both humans and the environment.

It is important to understand that Koyukon beliefs about nature are as logical and consistent as they are powerful, but that they differ substantially from those prevailing in modern Western societies. Our own tradition envisions the universe as a system whose functioning can be explained through rationalistic and scientific means. The natural and supernatural worlds are clearly separated. Environmental events are caused by ongoing evolutionary and ecological processes, or else they happen purely by chance. Finally, modern Western cultures regulate human behavior toward nature and its resources primarily on the basis of practical rather than religious considerations.

For the traditional Koyukon Athapaskans, ideology is a fundamental element of subsistence, as important as the more tangible practicalities of harvesting and utilizing natural resources. Most interactions with natural entities are governed in some way by a moral code that maintains a proper spiritual balance between the human and nonhuman worlds. This is not an esoteric abstraction, but a matter of direct, daily concern to the Koyukon people. Failure to behave according to the dictates of this code can have an immediate impact on the violator's health or success. And so, when Koyukon people carry out their subsistence activities they make many decisions on the basis of supernatural concerns. The world is ever aware.

FROM THE DISTANT TIME As the Koyukon reckon it, all things human and natural go back to a time called *Kk'adonts'idnee*, which is so remote that no one can explain or understand how long ago it really was. But however ancient this time may be, its events are recounted accurately and in great detail through a prodigious number of stories. *Kk'adonts'idnee* (literally, "in Distant Time it is said") is the Koyukon word for these stories, but following from its conversational use I will translate it simply as Distant Time.

The stories constitute an oral history of the Koyukon people and their environment, beginning in an age before the present order of existence was established. During this age "the animals were human"—that is, they had human form, they lived in a human society, and they spoke human (Koyukon) language. At some point in the Distant Time certain humans died and were transformed into animal or plant beings, the species that inhabit Koyukon country today. These dreamlike metamorphoses left a residue of human qualities and personality traits in the north-woods creatures. . . .

A central figure in this ancient world was the Raven (it is unclear, perhaps irrelevant, whether there was one Raven or many), who was its creator and who engineered many of its metamorphoses. Raven, the contradiction—omnipotent clown, benevolent mischief-maker, buffoon, and deity. It was he, transformed into a spruce needle, who was swallowed by a woman so she would give birth to him as a boy. When the boy was old enough to play, he took from beneath a blanket in her house the missing sun and rolled it to the door. Once outside, he became Raven again and flew up to return the sun to the sky, making the earth light again.

And it was he who manipulated the natural design to suit his whim or fancy. When he first created the earth, for example, the rivers ran both ways, upstream on one side and downstream on the other. But this made life too easy for humans, he decided, because their boats could drift along in either direction without paddling. So Raven altered his creation and made the rivers flow only one way, which is how they remain today.

There are hundreds of stories explaining the behavior and appearance of living things. Most of these are about animals and a few are about plants. No species is too insignificant to be mentioned, but importance in the Koyukon economy does not assure a prominent place in the stories. Many of the stories about animal origins are like this one:

> When the burbot *[ling cod]* was human, he decided to leave the land and become a water animal. So he started down the bank, taking a piece of bear fat with him. But the other animal people wanted him to stay and tried to hold him back, stretching him all out of shape in the process. This is why the burbot has such a long, stretched-out body, and why its liver is rich and oily like the bear fat its ancestor carried to the water long ago.

At the end of the Distant Time there was a great catastrophe. The entire earth was covered by a flood, and under the Raven's supervision a pair of each species went aboard a raft. These plants and animals survived, but when the flood ended they could no longer behave like people. All the Distant Time humans had been killed, and so Raven recreated people in their present form. My Koyukon teachers were well aware of the biblical parallel in this story, and they took it as added evidence of the story's accuracy. None suggested that it might be a reinterpretation of Christian teaching.

Distant Time stories were usually told by older people who had memorized the lengthy epics and could best interpret them. But children were also taught stories, simpler ones that they were encouraged to tell, especially as they began to catch game. Doing this after setting out their traps or snares would please the animals and make them willing to be caught. . . .

Distant Time stories also provide the Koyukon with a foundation for understanding the natural world and humanity's proper relationship to it. When people discuss the plants, animals, or physical environment they often refer to the stories. Here they find explanations for the full range of natural phenomena, down to the smallest details. In one story a snowshoe hare was attacked by the hawk owl, which was so small that it only managed to make a little wound in its victim's shoulder. Koyukon people point out a tiny notch in the hare's scapula as evidence that the Distant Time events really took place.

The narratives also provide an extensive code of proper behavior toward the environment and its resources. They contain many episodes showing that certain kinds of actions toward nature can have bad consequences, and these are taken as guidelines to follow today. Stories therefore serve as a medium for instructing young people in the traditional code and as an infallible standard of conduct for everyone.

> Nobody made it up, these things we're supposed to do. It came from the stories; it's just like our Bible. My grandfather said he told the stories because they would bring the people good luck, keep them healthy, and make a good life. When he came to songs in the stories, he sang them like they were hymns.

The most important parts of the code are taboos (*hutlaanee*), prohibitions against acting certain ways toward nature. For example, in one story a salmon-woman was scraping skins at night with her upper jaw, and while doing this she was killed. This is why it is taboo for women to scrape hides during the night. Hundreds of such taboos exist, and a person who violates them (or someone in the immediate family) may suffer bad luck in subsistence activities, clumsiness, illness, accident, or early death. In Koyukuk River villages it is a rare day when someone is not heard saying, "*Hutlaanee!*" ("It's taboo!") . . .

The Place of Humans in a Natural Order

> When Raven created humans, he first used rock for the raw materials, and people never died. But this was too easy so he recreated them, using dust instead. In this way humans became mortal, as they remain today.

How does humanity fit into the world of nature and the scheme of living things? For the Koyukon, humans and animals are clearly and qualitatively separated. Only the human possesses a soul (*nukk'ubidza*, "eye flutterer"), which people say is different from the animals' spirits. I never understood the differences, except that the human soul seems less vengeful and it alone enjoys immortality in a special place after death. The distinction between animals and people is less sharply drawn than in Western thought—the human organism, after all, was created by an animal's power.

The Koyukon seem to conceptualize humans and animals as very similar beings. This derives not so much from the animal nature of humans as from the human nature of animals. I noted earlier, for example, that today's animals once belonged to an essentially human society, and that transmutations between human and animal form were common. One of my Koyukon teachers said, however, that after the Distant Time people and animals became completely separated and unrelated.

Animals still possess qualities that Westerners consider exclusively human, though—they have a range of emotions, they have distinct personalities, they communicate among themselves, and they understand human behavior and language. They are constantly aware of what people say and do, and their presiding spirits are easily offended by disrespectful behavior. The interaction here is very intense, and the two orders of being coexist far more closely than in our own tradition. But animals do not use human language among themselves. They communicate with sounds which are considered their own form of language.

The closeness of animals to humans is reinforced by the fact that some animals are given funeral rituals following the basic form of those held for people, only on a smaller scale. Wolverines have a fairly elaborate rite, and bears are given a potlatchlike feast. In these cases, at least, animal spirits are placated much as human souls are after death.

Most interesting of all is animal behavior interpreted to be religious. "Even animals have their taboos," a woman once told me. From her grandfather, she learned that gestating female beavers will not eat bark from the fork of a branch, because it is apparently tabooed for them. The late Chief Henry had told her of seeing a brown bear kill a ground squirrel, then tear out its heart, lungs, and windpipe and leave them on a rock. Again, the organs must have been taboo (*hutlaanee*). . . .

Nature Spirits and Their Treatment From the Distant Time stories, Koyukon people learn rules for proper conduct toward nature. But punishment for offenses against these rules is given by powerful spirits that are part of the living, present-day world. All animals, some plants, and some inanimate things have spirits, vaguely conceptualized essences that protect the welfare of their material counterparts. They are especially watchful for irreverent, insulting, or wasteful behavior toward living things. The spirits are not offended when people kill animals and use them, but they insist that these beings (or their remains) be treated with the deference owed to the sources of human life. . . .

Proper treatment of natural spirits involves hundreds of rules or taboos (*hutlaanee*), some applying to just one species and others having much more general effects. The rules fall into three main categories—first, treatment of living organisms; second, treatment of organisms (or parts of organisms) that are no longer alive; and third, treatment of nonliving entities or objects. . . .

Koyukon people believe that animals must be treated humanely. The spirits are not offended because humans live by hunting, but people must try to kill without causing suffering and to avoid losing wounded animals. . . . The rules for showing respect to killed animals and harvested plants are myriad. . . . The remains of animals and plants are treated with the deference owed to something sacred. . . . There are also rules for proper butchering of game—for example, certain cuts that should be made or avoided for a particular species. . . . Finally, there are regulations to ensure that unusable parts of animals are respectfully disposed of. . . .

Elements of the earth and sky are imbued with spirits and consciousness, much in the way of living things, and there are codes of proper behavior toward them. Certain landforms have special powers that must be placated or shown deference, for example. Even the weather is aware: if a man brags that storms or cold cannot stop him from doing something, "the weather will take care of him good!" It will humble him with its power, "because it knows."

> In falltime you'll hear the lakes make loud cracking noises after they freeze. It means they're asking for snow to cover them up, to protect them from the cold. When my father told me this, he said *everything* has life in it. He always used to tell us that.

The earth itself is the source of a preeminent spiritual power called *sinh taala'* in Koyukon. This is the foundation of medicine power once used by shamans, and because of it the earth must be shown utmost respect. One person who was cured by medicine power years ago, for example, still abides by the shaman's instructions to avoid digging in the earth. Berry plants have special power because they are nurtured directly from the earth. "People are careful about things that grow close to the ground," I was told, "because the earth is so great."

THE MANIFESTATIONS OF LUCK Luck is the powerful force that binds humanity to the nature spirits and their moral imperatives. For the Koyukon people, luck is a nearly tangible essence, an aura or condition that is "with" someone in certain circumstances or for particular purposes. Luck can be held permanently or it can be fleeting and elusive. It is an essential qualification for success—regardless of a person's skill, in the absence of luck there is no destiny except failure.

The source of luck is not clearly explained, but most people are apparently born with a certain measure of it. The difficulty is not so much in getting it as in keeping it. Luck is sustained by strictly following the rules of conduct toward natural things. People who lose their luck have clearly been punished by an offended spirit; people who possess luck are the beneficiaries of some force that creates it. Koyukon people express luck in the hunt by saying *bik'uhnaatlonh*—literally, "he has been taken care of." . . .

GIFTS FROM THE SPIRITS OF NATURE The Koyukon people live in a world full of signs, directed toward them by the omniscient spirits. The extraordinary power of nature spirits allows them to reveal or determine future events that will affect humans. This understanding is sometimes divulged to watchful human eyes through the behavior of animals or other natural entities.

Rare or unusual events in nature are generally interpreted as signs, often foretelling bad fortune. People say that events like this are taboo (*hutlaanee*), and they encounter them with fear. For example, a Huslia woman said that it is ominous to hear a raven calling in the night. She had heard it only once in her life, and two weeks later her brother-in-law suddenly died. It is also a bad sign to find an animal that has died in strange or bizarre circumstances. A woman found an owl dead in the entangling meshes of her fishnet, and later that year her daughter died. Another woman discovered a ptarmigan hanging dead by a single toe from a willow branch. Her grandfather, a shaman, warned that it was powerfully ominous; and death came to her newborn child the following spring. . . .

Although Koyukon people are helpless to change signs given them through natural spirits, they can sometimes use these powers to influence the course of events. Spirits can be propitiated, asked to benefit people or to contravene an evil sign. One way of doing this is to make "prayers" to certain animals, entreating them for good health or good luck. Such prayers are given especially to ravens, because their powerful spirits often show benevolence toward people. Appeals may be specific or general. For example, when people see certain birds migrating southward in the fall, they may speak to them: "I hope you will return again and that we will be here to see you." It is a request that birds and people may survive the uncertainties of winter. . . .

By understanding the manifestations of spirit powers in nature, Koyukon people are able to foresee and sometimes change the course of events. They can help to create good fortune, they can avoid hardship or shortage, they can prepare themselves for preordained happenings that lie ahead, and they can sometimes directly influence the environment to their own benefit.

Harnessing the Powers of Nature   Modern Koyukon views of nature are strongly influenced by a cultural tradition that is probably not practiced today or that exists only as a remnant at best. This is the tradition of shamanism, the use of medicine power to control nature spirits directly. Although shamanism apparently is seldom, if ever, practiced today, most adult Koyukon have seen and experienced it many times in their lives. Medicine power has been used to cure many people of illnesses that they believe would otherwise have killed them. Today the old medicine people seem to have vanished without passing their skills along, but the concepts and beliefs surrounding them remain intact.

Koyukon shamans (*diyhinyoo;* singular *diyininh*) did not have power themselves, but they knew how to use the spirit forces that surrounded them in nature. With this they could do good or evil, according to their personal inclinations. Each shaman—who might be either a man or a woman—had special associations with a number of familiar spirits. For example, one man "called for" Wood Frog, Birch Woman, Raven, Northern Lights, and others when he made medicine. Some spirit associations were begun in dreams . . . , but this man inherited his animal helpers from an uncle. Often he used the raven spirit to "scare away the sickness in someone," mimicking a raven's melodious cawing, spreading his arms like wings, and bouncing on both feet as a raven would. One of my Koyukon instructors who was sickly in his youth said that this man had cured him many times.

Aside from curing (or causing) sickness, shamans used spirit power to manipulate the environment for their own or someone else's benefit. Before caribou hunting, for example, they made medicine to bring animals to the hunters, to foretell their chances of success, or to show them where to find game. Spirit helpers assisted them by communicating with a protective spirit of the caribou. . . . As I mentioned earlier, shamans could also attract animal spirits to a given area of land, creating abundance there.

When shamans were active, they sometimes sent their spirit animals to the villages or dwellings of persons they wanted to harm. Unusual appearances of wild animals in settlements were presumed to be shaman's work. These animals were never killed because it would cause grave danger to the one who did it. It was always a bad sign to see a creature of the wildlands in a settlement, because it indicated danger from medicine power. . . .

THE KOYUKON VIEW OF NATURE For traditional Koyukon people, the environment is both a natural and a supernatural realm. All that exists in nature is imbued with awareness and power; all events in nature are potentially manifestations of this power; all actions toward nature are mediated by consideration of its consciousness and sensitivity. The interchange between humans and environment is based on an elaborate code of respect and morality, without which survival would be jeopardized. The Koyukon, while they are bound by the strictures of this system, can also manipulate its powers for their own benefit. Nature is a second society in which people live, a watchful and possessive one whose bounty is wrested as much by placation as by cleverness and craft.

Moving across the sprawl of wildland, through the forest and open muskeg, Koyukon people are ever conscious that they are among spirits. Each animal is far more than what can be seen; it is a personage and a personality, known from its legacy in stories of the Distant Time. It is a figure in the community of beings, once at least partially human, and even now possessed of attributes beyond outsiders' perception.

Not only the animals, but also the plants, the earth and landforms, the air, weather, and sky are spiritually invested. For each, the hunter knows an array of respectful gestures and deferential taboos that demand obedience. Violations against them will offend and alienate their spirits, bringing bad luck or illness, or worse if a powerful and vindictive being is treated irreverently.

Aware of these invisible forces and their manifestations, the Koyukon can protect and enhance their good fortune, can understand signs or warnings given them through natural events, and can sometimes influence the complexion of the environment to suit their desires. Everything in the Koyukon world lies partly in the realm beyond the senses, in the realm we would call the supernatural.

The Gift of the Animals

Apologies to the Iroquois, by Edmund Wilson

Edmund Wilson (1895–1972), who was born in Red Bank, New Jersey, educated at Princeton, and served in the American army in World War I, was a leading American man of letters throughout the mid twentieth century. His literary criticism ranges from an early study of Sym-

bolism and Modernism, Axel's Castle *(1931), to* Patriotic Gore: Studies in the Literature of the American Civil War *(1962) and numerous essays on the literature of his own age and others. Among his many other books are* To the Finland Station *(1940), a study of revolutionary Marxism, and* The Scrolls from the Dead Sea *(1955; expanded in 1969). He also wrote poems, plays, short stories, and novels, and was a prolific book reviewer, cultural and social critic, memoirist, and letter writer.* Apologies to the Iroquois *(1960) recounts his belated discovery in 1957, at a time of rising Iroquois nationalism, of the Indians of the Six Nations (Mohawk, Oneida, Onondaga, Cayuga, Seneca, and Tuscarora) living not far from his home in upstate New York and in neighboring Ontario. "I have come to believe," he writes, "that there are many white Americans who now have something important in common with these recalcitrant Indians," since "by defending their rights as Indians, they remind us of our rights as citizens." The final chapter of the book narrates a performance of the Little Water Ceremony—described decades earlier in the ethnographic writings of Arthur C. Parker—which Wilson attended, along with the anthropologist William Fenton, in the Tonawanda Seneca reservation on June 6, 1959. "The Little Water Company is the most important of the Iroquois medicine societies," Wilson writes in introducing his account: "the most sacred and the most secret and the one that has been most rigorously cultivated," for its members "are the guardians of a miraculous medicine which has the power to revive the dying, and they must sing to it three or four times a year in order to keep up its strength . . ." The following selections describe both the legend of the Little Water, in which the animals and birds combine to bring back to life the Good Hunter killed by his human enemies, and key parts of the all-night ceremony based upon it; this begins after the Seneca Chief Corbett Sundown solemnly invokes the Divinity and his deputies, the Sun, Moon, Stars, and Wind. Those who perform the ceremony, Wilson reflects as it comes to an end—just as "the first soft and misty daylight" is topping the forest outside the house in whose darkened kitchen it has taken place—"have mastered the principle of life" and surpassed themselves in their "affirmation of the will of the Iroquois people, of their vitality, their force to persist."*

FROM *Chapter 9: The Little Water Ceremony*

The theme of the Little Water legend is the familiar one of death and resurrection. There was once a young hunter—a chief—who was greatly respected by the animals on account of the kindness he showed them. He never killed a deer that was swimming or a doe that had a fawn, or an animal that he took unawares or that was tired from long pursuit. For the predatory birds and animals he would always first kill a deer and skin it and cut it open and cry, "Hear, all you meat-eaters. This is for you." He would always leave some honey for the bears and some corn in the fields for the crows. The tripes of the game he killed he threw into the lakes and streams for the fish and the water animals. One day, when cut off from his party, he was captured and scalped by the Cherokees. When they had gone, the Wolf smelled blood, and he came up and licked the bloody head. But when he recognized the Good Hunter, he howled for the rest of the animals, who all came and gathered around and mourned for their lost friend. The Bear felt the body with his paws and on the chest found a spot that was warm. The birds had come, too, and they all held a council, while the Bear kept him warm in his arms, as to how they could bring the young hunter to life. The only dissenting voice was that of the Turkey Buzzard, who said, "Let's wait till he gets ripe and eat him." And they compounded a powerful medicine, to which each one contributed a "life-spark"—in some versions a bit of the brain, in others a bit of the heart. But the mixture of all these ingredients made an essence so concentrated that it could all be contained in an acorn shell, and it is called for this reason the Little Water.

Now the Owl said, "A live man must have a scalp," and who was to go for the scalp? In certain of the tellings of the story, a long discussion takes place. One animal after another is decided unfit. The quadrupeds will never do: what is needed is a clever bird. The Dew Eagle is sometimes chosen. He is the ranking bird of the Iroquois, who carries a pool of dew on his back and, when rain fails, can spill it as mist. One variant makes it the Hummingbird, who moves quickly and is almost invisible. But it is more likely to be the Giant Crow (Gáh-ga-go-wa), the messenger for all the birds, who plays a prominent rôle in the ceremony. He flies to the Cherokee village, and there he sees the good hunter's scalp, which has been stretched on a hoop and hung up to dry in the smoke that comes out of the smoke-hole. He swoops down and snatches it. The Cherokees see him and shoot their arrows at him, but the Crow flies so high that they cannot reach him.

When he has brought the scalp back to the council, they find it is too much dried out to be worn by a living man. The Great Crow has to vomit on it and the Dew Eagle sprinkle on it some drops of his dew before it is supple and moist enough to be made to grow back on the hunter's head. He has already been given the medicine, and he feels himself coming to life. As he lies with his eyes still closed, he realizes that he is now able to understand the language of the animals and birds. They are singing a wonderful song, and he listens to it and finds later on that he is able to remember every word. They tell him to form a company and to sing it when their help is needed. He asks how they make the medicine, and they say that they cannot tell him because he is not a virgin. But someone will be given the secret, and they will notify this person by singing their song. The Bear helps the youth to his feet, and when he opens his eyes, there is nobody there: only a circle of tracks. It is dawn. He goes back to his people. . . .

Now the door to the next room was closed, the switch in the wall was turned off, and the ceremony proper began. The room with its Corn Flakes had vanished: you were at once in a different world. The single beat of a rattle is heard in the sudden blackness like the striking of a gigantic match, and it is answered by other such flashes that make rippings of sound as startling as a large-scale electric spark. Then the first of the two chief singers cries "*Wee yoh!*" and the second "*Yoh wee!*," and the first of them now sets the rhythm for the rattles, which is picked up by the rest of the company. In this section the tempo is uniform, and it reminds one of the rapid jogging that is heard by the passenger on an express train. The rhythm is kept up, without raggedness or flagging, for a little less than an hour. Arthur Parker says that the pace is 150 a minute; but I do not know whether this applies to the section I am now describing or to the even faster passages in the later ones. The songs themselves are in slower time. Their structure has something in common with that of old English and Scottish ballads, as well as their nonsense refrains—like the English, "Benorio, benory"—and the wistfulness of their accent, as of human beings alone with nature, singing in unpeopled spaces. The first singer gives the couplet, and this is repeated by the second; then all take it up in a powerful chorus. A man and a woman are searching for the magical Little Water that has brought the Good Hunter to life but the receipt for compounding which, because he was not a virgin, the animals could not teach him. Now their song has been heard, and the man and the woman are obeying its summons to search. The first line of every couplet always applies to the woman—since the woman, among the Iroquois, is always given first place—and the second line to the man:

"She went to the village.
He went to the village."

—to the fields, to the little spring, to the edge of the wood, etc. In the climactic final couplet, they go to the top of a hill, and there they are left "standing under the clouds."

In the second section, the rhythms of the rattles are different. While the first and second singers are introducing the couplets, the rattles are going so fast that they seem to weave a kind of veil or screen—a scratching almost visible on the darkness—that hangs before the lyric voices; but when the chorus takes up the theme, this changes to a slow heavy beat that has something of the pound of a march. The shift is extremely effective. When it occurs, this accompaniment of the rattles—contrary to our convention—does not quite coincide with the song and the chorus but always overlaps a little. The big shimmer of the solo begins before the pound of the chorus has ended: in a moment, you are given notice, a fresh song will be springing up; and in the same way the lusty thump will commence before the song has quite finished. In this section, the animals assemble: "The She-Owl," and then "the He-Owl," "came bringing tobacco and joined the song." And presently these creatures begin to speak, as they are mentioned pair by pair in the couplets. They are mimicked by one or more singers—who have had their parts assigned them—as the arrival of each is announced. The effect of this is startling and eerie. The Gáhgagowa caws—the first half of his name is evidently onomatopoeic; the Bear and the Panther roar; the Owl has a soft four-note hoot: "Wu-wu, Wu-wu." There are many archaic words in the ritual that nobody now understands. Besides the Giant Crow, the Great Eagle, the Dew Eagle, the Great Phoebe, the Great Woodpecker and a bird which, after a moment's thought, had been identified for us by Sundown as the Kingbird, there are birds which can no longer be identified, and when one of these birds was mentioned, the cackle of a hen who has just laid an egg—followed by quiet laughter—was heard from the prehistoric darkness. A climax of animation occurs toward the end of this section. The She- and He-Wolves arrive. They are heralded as running along the rim of the hills, and afterwards through the meadows. They do not howl; since they are running, they bark like dogs. A queer kind of excitement is created here. The animals are supposed to be present, and I learn from Arthur Parker that the adepts are believed to see them. He hints at further mysteries which he is not allowed to reveal; and from reading Fenton's unpublished notes on the Little Water Ceremony, I get the impression that the singers are supposed to *become* the animals. They are forbidden to keep time with their feet—as they do in all the other ceremonies—since the quadrupeds go on pads and the tread of the birds is soundless. The men who serve the syrup are called Eagles, and the man who presides is the Giant Crow. The Wolf has also some special rôle. I did not see any animals nor did the medicine become visibly luminous, as it was formerly supposed to do when its power was renewed by the singing. I could not see that the boxes were even opened. Nor was I conscious of another phenomenon that Chief Sundown had told me to listen for. "You'll hear a woman's voice," he had said, "and there's no woman in the room." I have learned that the best first singers—who always sing the feminine lines—have a practice of outdoing the chorus by pitching each song in a sequence a little higher than the one before. If this singer starts the sequence off so high that it is sure at a later stage to get out of the range of the others, he is stopped, and they have to begin again—the worst *gaffe* that can be committed except for a singer's forgetting the words. (I noticed, in an intermission, that the first singer, who was younger than the others, was studying his lines in a notebook.)

It was delightful to go out in these intermissions. One was here in the cool June night of the green Tonawanda woodland. There was a fragrance of something blooming. The back lawn, where the grass had been cut, was revealed by a bulb with a reflector behind it that directed the light from the top of the house. The peelings of asbestos shingling, in imitation of brick, that were hanging away from this wall contrasted with the beauty of the landscape and the music that had been liberated inside the house. The Indians talked softly and smoked a little. The hoot of a real owl was heard; a mudhen croaked from nearby Salt Creek; a nighthawk flew away through the lighted air to the forest that made the background.

The third section is the climax of the symphony. Now the medicine at last is found. The questers have discovered that the marvellous song emanates from the top of a mountain—the mountain, I suppose, that the man and the woman have ascended at the end of the first sequence. But it ought to be explained that this couple do not appear in all versions of the ceremony, which differs somewhat in the different reservations. There is a version in which it is simply a man; and the account I shall give here of the search is taken from a telling of the *legend*—not always consistent with the songs—in which the searchers are two young men. These questers set out in the dark, and they are to endure a whole set of ordeals, to surmount a whole series of obstacles. At the start they are trapped in a "windfall," a place where the trees have been blown down, and they must break their way out of this. They feel that there are presences about them, that the animals are guiding them to the source of the song. Next they must plod through a swamp. Then they have to cross a ravine; then they hear the roaring of a cataract and are confronted with another ravine through which a swift river runs. One of the young men is "almost afraid"; but they go down and swim across it, though the waters are terribly cold. A mountain now rises before them, so steep that they think they can never climb it. The less hardy of the searchers proposes a rest, but the other says they have to go on, and when he has spoken, they hear a voice that sings "Follow me, follow me!," and a light comes flying over them. This is the Whippoorwill, whose song is the "flourish of the flute" mentioned to me by Nicodemus Bailey. Each of the sequences is opened and closed by this cry, but it is only in this third part, I think, that the flute intervenes in the narrative. It plays a rôle like the bird in [Wagner's] *Siegfried* that will lead the hero to his lost mother. With effort, the young men scale the mountain, and at the top they find a great stalk of corn growing out of the barren rock, and from this stalk comes the song that has drawn them. They are told by the winged light to burn a tobacco offering and then to cut the root of the corn. When they do so, it bleeds human blood, but immediately the wound heals. They are now told by the voices of the animals how to collect the ingredients of the medicine, how to mix it and how to apply it.

This section is more complicated than the others, and it is said to be difficult to sing. In each of the first two sequences, the songs all follow a pattern; but in the third, they begin on unexpected notes and follow unfamiliar courses. This is magic, a force beyond nature is tearing itself free. There is a passage of reiteration that entirely departs from the ballad stanzas and that sounds like some sort of litany. A great structure is raised by the rattles that is neither the big shimmer, the express train nor the grand triumphal march. And a paean is let loose: it fills the room with its volume. One finds oneself surrounded, almost stupefied, as if the space between the four walls had become one of the tubes of a pipe organ and as if one were sitting inside it, almost drowned in a sustained diapason. How strange when the lights are turned on—strange, apparently, for the singers, too, who sit blinking and dazed at first, having to bring themselves up short in this kitchen, in this new electric-lighted world—to find oneself there in the room with these ten men who work in the daytime in the gypsum mines and plants of the neighboring town of Akron, dressed in their unceremonious clothes: an assortment of physical types, some handsome, some not so handsome, some young and some old, some fat and some lean, some sallow, one almost black, some with spectacles, some with their teeth gone, who have just given body in the darkness to a creation which absorbed and transcended them all. A car had driven up to the house while the singing was going on; its headlights had glared toward the window. I saw the profile against it of a man turning round to look; but the singing was not interrupted, and the driver, when he heard it, withdrew. . . .

2

In Their Own Words

The myths and ceremonies through which native peoples of North America gave expression to their reverential oneness with the natural world exerted a strong fascination —initially intermixed with pious horror at their supposed heathenism and savagery—on the early white settlers who were all too often, by deliberation or accident, their destroyers. Before long only remnants of Indian populations survived east of the Appalachians, or indeed of the Mississippi, but for much of the nineteenth century large expanses of Canada and Alaska, parts of California and the Southwest, and most of the Great Plains remained lands where the peoples who had inhabited the continent for thousands of years could still hope, for a while, to find refuge from the encroachments of white civilization and to live their own lives and practice their own ways. The Plains Indians in particular—Sioux, Cheyenne, Blackfoot, Crow, Comanche, and others—appealed to the romantic imagination as embodiments of the wildness and freedom that civilization, constrained to repress them, both feared and admired. The buffalo hunter of the Plains has continued to be the paradigmatic Indian for many, even though he hunted the buffalo on the white man's horse and with the white man's rifle—until the invading whites drove both Indians and buffalo to the verge of extinction.

Ever since Lewis and Clark's exploration of the Louisiana Territory for President Jefferson, white travelers and explorers have left vivid accounts of the religious beliefs and practices of these Western Indians; among the most important are George Catlin's, Prince Maximilian of Wied-Neuwied's, and Francis Parkman's (in The Oregon Trail). *But none of these visitors, valuable though their observations are, could enter into the lives of peoples whose languages they did not know and whose ways remained, at best, exotic to them. (Catlin, who described the Mandan of North Dakota not long before smallpox and cholera nearly annihilated them, was convinced they were descendants of the Welsh, as earlier whites had thought all Indians must be descended from the lost tribes of Israel.) But toward the end of the nineteenth century—just as the profligately slaughtered buffalo were vanishing from the Plains and the last Sioux uprising was being bloodily suppressed at Wounded Knee—American ethnography, under the leadership of the German-born Franz Boas, who focused on the Northwest Coast and the Eskimo, and of Alfred Kroeber, who focused on California but also wrote on the Plains Arapaho, entered its great age of disciplined, detailed, yet sympathetic study of the languages, rites, and customs of the original human inhabitants of our continent. In a remarkable series of monographs and papers published during the half-century and more after 1880 in the* Bulletin of the American Museum of Natural History, *the* University of California Publications in American Archaeology and Ethnology, *the* American Anthropologist, *the* Journal of American Folk-Lore, *and above all the Bulletins and Annual Reports of the Smithsonian Institution Bureau of American Ethnology, these dedicated scholars and their colleagues and students, men and women alike, recorded and thereby helped to preserve ways of life and thought that seemed destined to eternal oblivion.*

Perhaps most significantly of all, because they learned the languages and respected the "mentalities" (as a later generation would say) of the peoples they studied, these ethnographers frequently undertook to transcribe directly and translate faithfully the words in which these peoples told their myths and legends, and to describe in punctilious detail the rituals and ceremonies they performed and the explanations they gave of them. In some cases, at least, they established so deep a relation of trust that chiefs and holy men, normally intent on keeping secret the religious practices of their people, confided to them rites they decided, after much thought and even anguish, to communicate to others. Thus when the Pawnee priest Tahirussawichi explained the Hako Ceremony to Alice C. Fletcher in the closing years of the nineteenth century, "It was the first time," her entry in the Dictionary of American Biography *proclaims, "that any white observer had been permitted to step behind the veil of the Red Man's esoteric mysteries, and set down for scientific study religious beliefs and observances hitherto impossible to witness." And when the Oglala Holy Man Black Elk, over thirty years later, told his most treasured visions to John G. Neihardt, he confessed that "I have lain awake at night worrying and wondering if I was doing right; for I know I have given away my vision, and maybe I cannot live very long now. But I think I have done right to save the vision in this way, even though I may die sooner because I did it; for I know the meaning of the vision is wise and beautiful and good . . ."*

In such intimate accounts we are given, "in their own words"—more or less, to be sure, for no translation can be entirely literal, and Neihardt was not alone in taking the licenses of a poet—the Indian people's own myths and their own understanding of the rites most profoundly important to lives lived in accord, so far as possible, with fellow creatures of the world around them. That communion, whether with Mother Corn or the animals of land, air, or sea, the sun or the stars, was not an easy or automatic one. Its maintenance might involve an arduous quest, as in the collective pilgrimage of the Pawnee Hako or the stories of the Maidu To'lowim-Woman's pursuit of the Butterfly Man and the Blackfoot Scarface's search for the healing Sun; and in a world disrupted by the cruel afflictions of the Eskimo goddess Sedna, or the depredations inflicted by the Wasichus of Black Elk's dog vision—a world of sickness and death, as any world misshaped by the irrepressibly bungling Coyote must be—success was by no means guaranteed. But the supreme importance of maintaining oneness with the life-giving forces of nature, even in so dislocated a world, was never in doubt, and the effort to achieve or renew it never abandoned, so long as the sacred myths and rituals of immemorial tribal inheritances retained their ancient authority and conviction.

Mother Breathing Forth Life

The Hako: A Pawnee Ceremony, by Alice C. Fletcher

Alice C. Fletcher (1838–1923), born in Cuba of American parents, was one of the first of a distinguished line of American woman ethnographers, a pioneer in the study of Indian music, and an important friend and advocate of Native Americans. Her many publications concerning the Sioux, Winnebago, Omaha, Pawnee, and others include Indian Song and Story from North America *(1900) and a two-volume study,* The Omaha Tribe *(1911), co-authored with her adoptive son Francis La Flesche, son of the half-French Omaha chief Joseph La Flesche who was a leading defender of Indian rights, especially those of the displaced Ponca, in disputes with the U.S. government. Her major monograph,* The Hako, *published in 1904 as Part 2 of the Twenty-second Annual Report of the Bureau of American Ethnology, gives a detailed account, with musical scores, Pawnee transcriptions, and English translations of its songs, of this elaborate ceremony as the Pawnee Ku'rahus (or priest) Tahirussawichi, after overcoming his early scruples, explained it to her and her Pawnee assistant, James R. Murie, between 1899 and 1901. (On a journey to Washington, D.C., in 1898, to argue for preservation of this ceremony, the old man admired but declined to enter the Washington Monument, explaining: "The white man likes to pile up stones, and he may go to the top of them; I will not. I have ascended the mountains made by Tira'wa," the Creator.) The ceremony was performed, the Ku'rahus explained, at no prescribed time: "We take up the Hako in the spring when the birds are mating, or in the summer when the birds are nesting and caring for their young, or in the fall when the birds are flocking, but not in the winter when all things are asleep. With the Hako we are praying for the gift of life, of strength, of plenty, and of peace, so we must pray when life is stirring everywhere."*

The following excerpts give only a fragmentary and disconnected account, with many long omissions, of the twenty rituals of this ceremony (which has been compared to the Eleusinian Mysteries of ancient Greece and similar fertility rites of other peoples)—a ceremony in which, as Fletcher remarks, "every article is symbolic and every movement has a meaning." See the discussion of this and of other myths and rituals of the quest (including those of the Little Water Ceremony, above, and of Scarface and Black Elk, below) in Torrance, The Spiritual Quest *(1994).*

The ceremony of the Hako is a prayer for children, in order that the tribe may increase and be strong; and also that the people may have long life, enjoy plenty, and be happy and at peace. . . .

Shakuru, the Sun, is the first of the visible powers to be mentioned. It is very potent; it gives man health, vitality, and strength. Because of its power to make things grow, Shakuru is sometimes spoken of as atius, father. The Sun comes direct from the mighty power above; that gives it its great potency. . . .

Blue is the color of the sky, the dwelling place of Tira'wahut, that great circle of the powers which watch over man. As the man paints the stick blue we sing. We ask as we sing that life be given to this symbol of the dwelling place of Tira'wa. . . .

The ear of corn represents the supernatural power that dwells in H'Uraru, the earth which brings forth the food that sustains life; so we speak of the ear of corn as h'Atira, mother breathing forth life.

The power in the earth which enables it to bring forth comes from above; for that reason we paint the ear of corn with blue. Blue is the color of the sky, the dwelling place of Tira'wahut. . . .

All things live on the earth, Mother Corn knows and can reach all things, can reach all men, so her spirit is to lead our spirits in this search over the earth. When Mother Corn went up to Tira'wahut at the time she was painted, power was given her to lead the spirits of all things in the air and to command the birds and the animals connected with the Hako. Endowed with power from Tira'wahut above and from h'Uraru (Mother Earth) below, Mother Corn leads and we must follow her, our spirits must follow her spirit. We must fix our minds upon Mother Corn and upon the Son, who is the object of our search. It is a very difficult thing to do. All our spirits must become united as one spirit, and as one spirit we must approach the spirit of Mother Corn. This is a very hard thing to do. . . .

We fix our minds upon Mother Corn and upon the Son; if we are in earnest he will respond to her touch. He will not waken, he will not see her, but he will see in a dream that which her touch will bring to him, one of the birds that attend the Hako, for all the spirits of those birds are with Mother Corn and they do her bidding, and he may hear the bird call to him. Then, when he awakens, he will remember his dream, and as he thinks upon it, he will know that he has been chosen to be a Son, and that all the good things that come with the ceremony which will make him a Son are now promised to him.

By touching the Son Mother Corn opened his mind, and prepared the way for our messengers to him, so that he would be willing to receive them, and later to receive us.

Mother Corn has now found the Son; she has made straight and safe the path from our country to his land, and she has made his mind ready to receive us and to carry out his part of this ceremony of the Hako. . . .

Mother Corn, who led our spirits over the path we are now to travel, leads us again as we walk, in our bodies, over the land.

When we were selecting the Son we had to fix our minds on Mother Corn and make our spirits as one spirit with her. We must do so now, as we are about to start on this journey; we must be as one mind, one person, with Mother Corn (h'Atira shira); we, as one person, must walk with her over the devious, winding path (tiware) which leads to the land of the Son.

We speak of this path as devious, not merely because we must go over hills and through valleys and wind around gulches to reach the land of the Son, but because we are thinking of the way by which, through the Hako, we can make a man who is not of our blood a Son; a way which has come down to us from our far-away ancestors like a winding path. . . .

Before us lies a wide pathless stretch of country. We are standing alone and unarmed, facing a land of strangers, and we call upon Mother Corn and we ask her: "Is there a path through this long stretch of country before us where we can see nothing? Does your path, the one which you opened for us, wherein is safety, lie here?" . . . She answers our appeal; she says that here, right before us, stretches out the path she has made straight. Then our eyes are opened and we see the way we are to go.

But although we see our way we are not to take the path by ourselves; we must follow Mother Corn; she must lead us, must direct and guide our steps. . . .

The journey we are taking is for a sacred purpose, and as we are led by the supernatural power in Mother Corn we must address with song every object we meet, because Tira'wa is in all things. Everything we come to as we travel can give us help, and send help by us to the Children.

Trees are among the lesser powers, and they are represented on the Hako which we carry, so when we see trees we must sing to them.

Trees grow along the banks of the streams; we can see them at a distance, like a long line, and we can see the river glistening in the sunlight in its length. We sing to the river, and when we come nearer and see the water and hear it rippling, then we sing to the water, the water that ripples as it runs. . . .

Hills were made by Tira'wa. We ascend hills when we go away alone to pray. From the top of a hill we can look over the country to see if there are enemies in sight or if any danger is near us; we can see if we are to meet friends. The hills can help man, so we sing to them. . . .

All the powers that are in the heavens and all those that are upon the earth are derived from the mighty power, Tira'wa atius. He[1] is the father of all things visible and invisible. He is the father of all the powers represented by the Hako. He is the father of all the lesser powers, those which can approach man. He is the father of all the people, and perpetuates the life of the tribe, through the gift of children. So we sing, your father, meaning the father of all people everywhere, the father of all things that we see and hear and feel. . . .

We sing about the visions which the birds on the feathered stems are to bring to the Children.

Visions come from above, they are sent by Tira'wa atius. The lesser powers come to us in visions. We receive help through the visions. All the promises which attend the Hako will be made good to us in this way.

Visions can come most readily at night; spirits travel better at that time. Now when we are met together we, the Fathers, call upon the visions to come to the Children. . . .

We call to Mother Earth, who is represented by the ear of corn. She has been asleep and resting during the night. We ask her to awake, to move, to arise, for the signs of the dawn are seen in the east and the breath of the new life is here.

H'Atira means Mother breathing forth life; this life is received from Tira'wa atius with the breath of the new-born Dawn.

Mother Earth is the first to be called to awake, that she may receive the breath of the new day. . . .

Mother Earth hears the call; she moves, she awakes, she arises, she feels the breath of the new-born Dawn. The leaves and the grass stir; all things move with the breath of the new day; everywhere life is renewed.

This is very mysterious; we are speaking of something very sacred, although it happens every day. . . .

Kawas, the brown eagle, the messenger of the powers above, now stands within the lodge and speaks. The Ku'rahus hears her voice as she tells him what the signs in the east mean.

She tells him that Tira'wa atius there moves upon Darkness, the Night, and causes her to bring forth the Dawn. It is the breath of the new-born Dawn, the child of Night and Tira'wa atius, which is felt by all the powers and all things above and below and which gives them new life for the new day. . . .

[1]The Pawnee pronoun here translated "he" does not in the original indicate sex, nor is it equivalent to "it" as the word relates to a person. (Fletcher)

The Morning Star is one of the lesser powers. Life and strength and fruitfulness are with the Morning Star. We are reverent toward it. Our fathers performed sacred ceremonies in its honor.[2]

The Morning Star is like a man; he is painted red all over; that is the color of life. He is clad in leggings and a robe is wrapped about him. On his head is a soft downy eagle's feather, painted red. This feather represents the soft, light cloud that is high in the heavens, and the red is the touch of a ray of the coming sun. The soft, downy feather is the symbol of breath and life.

The star comes from a great distance, too far away for us to see the place where it starts. At first we can hardly see it; we lose sight of it, it is so far off; then we see it again, for it is coming steadily toward us all the time. We watch it approach; it comes nearer and nearer; its light grows brighter and brighter. . . .

Still we sing and shout, "Day is here! Daylight has come!" We tell the Children that all the animals are awake. They come forth from the places where they have been sleeping. The deer leads them. She comes from her cover, bringing her young into the light of day. Our hearts are glad as we sing, "Daylight has come! The light of day is here!" . . .

This song to the Pleiades is to remind the people that Tira'wa has appointed the stars to guide their steps. It is very old and belongs to the time when this ceremony was being made. This is the story to explain its meaning which has been handed down to us from our fathers:

A man set out upon a journey; he traveled far; then he thought he would return to his own country, so he turned about. He traveled long, yet at night he was always in the same place. He lay down and slept and a vision came. A man spoke to him; he was the leader of the seven stars. He said: "Tira'wa made these seven stars to remain together, and he fixed a path from east to west for them to travel over. He named the seven stars Chaka. If the people will look at these stars they will be guided aright."

When the man awoke he saw the Pleiades rising; he was glad, and he watched the stars travel. Then he turned to the north and reached his own country.

The stars have many things to teach us, and the Pleiades can guide us and teach us how to keep together. . . .

One day a man whose mind was open to the teaching of the powers wandered on the prairie. As he walked, his eyes upon the ground, he spied a bird's nest hidden in the grass, and arrested his feet just in time to prevent stepping on it. He paused to look at the little nest tucked away so snug and warm, and noted that it held six eggs and that a peeping sound came from some of them. While he watched, one moved and soon a tiny bill pushed through the shell, uttering a shrill cry. At once the parent birds answered and he looked up to see where they were. They were not far off; they were flying about in search of food, chirping the while to each other and now and then calling to the little one in the nest.

The homely scene stirred the heart and the thoughts of the man as he stood there under the clear sky, glancing upward toward the old birds and then down to the helpless young in the nest at his feet. As he looked he thought of his people, who were so often careless and thoughtless of their children's needs, and his mind brooded over the matter. After many days he desired to see the nest again. So he went to the place where he had found it, and there it was as safe as when he left it. But a change had taken place. It was

[2]As recently as 1838 the Pawnee had ritually sacrificed a maiden to the Morning Star to promote the fertility of the corn.

now full to overflowing with little birds, who were stretching their wings, balancing on their little legs and making ready to fly, while the parents with encouraging calls were coaxing the fledglings to venture forth.

"Ah!" said the man, "if my people would only learn of the birds, and, like them, care for their young and provide for their future, homes would be full and happy, and our tribe be strong and prosperous."

When this man became a priest, he told the story of the bird's nest and sang its song; and so it has come down to us from the days of our fathers. . . .

In the early spring the birds lay their eggs in their nests, in the summer they rear their young, in the fall all the young ones are grown, the nests are deserted and the birds fly in flocks over the country. One can hear the fluttering of a startled flock, the birds suddenly rise and their wings make a noise like distant thunder. Everywhere the flocks are flying. In the fall it seems as though new life were put into the people as well as into the birds; there is much activity in coming and going. . . .

Water is for sustenance and the maintenance of health; it is one of the great gifts of Tira'wa atius.

The white man speaks of a heavenly Father; we say Tira'wa atius, the Father above, but we do not think of Tira'wa as a person. We think of Tira'wa as in everything, as the power which has arranged and thrown down from above everything that man needs. What the power above, Tira'wa atius, is like, no one knows; no one has been there. . . .

The circle represents a nest, and is drawn by the toe because the eagle builds its nest with its claws. Although we are imitating the bird making its nest, there is another meaning to the action; we are thinking of Tira'wa making the world for the people to live in. If you go on a high hill and look around, you will see the sky touching the earth on every side, and within this circular inclosure the people live. So the circles we have made are not only nests, but they also represent the circle Tira'wa atius has made for the dwelling place of all the people. The circles also stand for the kinship group, the clan, and the tribe. . . .

The child represents the young generation, the continuation of life, and when it is put in the circle it typifies the bird laying its eggs. The child is covered up, for no one knows when a bird lays its eggs or when a new birth takes place; only Tira'wa can know when life is given. The putting of the child's feet upon the oriole's nest means promised security to the new life, the fat is a promise of plenty of food, and the tobacco is an offering in recognition that all things come from Tira'wa. The entire act means that the clan or tribe of the Son shall increase, that there shall be peace and security, and that the land shall be covered with fatness. This is the promise of Tira'wa through the Hako. . . .

When I sing this song I pray to Tira'wa to come down and touch with his breath the symbol of his face and all the other symbols on the little child. I pray with all my spirit that Tira'wa atius will let the child grow up and become strong and find favor in its life.

This is a very solemn act, because we believe that Tira'wa atius, although not seen by us, sends down his breath as we pray, calling on him to come.

As I sing this song here with you I can not help shedding tears. I have never sung it before except as I stood looking upon the little child and praying for it in my heart. There is no little child here, but you are here writing all these things down that they may not be lost and that our children may know what their fathers believed and practiced in this ceremony. So, as I sing, I am calling to Tira'wa atius to send down his breath upon you, to give you strength and long life. I am praying for you with all my spirit. . . .

I have done what has never been done before, I have given you all the songs of this ceremony and explained them to you. I never thought that I, of all my people, should be the one to give this ancient ceremony to be preserved, and I wonder over it as I sit here.

I think over my long life with its many experiences; of the great number of Pawnees who have been with me in war, nearly all of whom have been killed in battle. I have been severely wounded many times—see this scar over my eye. I was with those who went to the Rocky Mountains to the Cheyennes, when so many soldiers were slain that their dead bodies lying there looked like a great blue blanket spread over the ground. When I think of all the people of my own tribe who have died during my lifetime and then of those in other tribes that have fallen by our hands, they are so many they make a vast cover over Mother Earth. I once walked with these prostrate forms. I did not fall but I passed on, wounded sometimes but not to death, until I am here today doing this thing, singing these sacred songs into that great pipe (the graphophone) and telling you of these ancient rites of my people. It must be that I have been preserved for this purpose, otherwise I should be lying back there among the dead.

World of Butterflies and Coyotes

Maidu Myths, by Roland B. Dixon

Roland Burrage Dixon (1875–1934) was born in Worcester, Massachusetts, and taught at Harvard University (where he also served as librarian of the Peabody Museum) from 1901 until his death, becoming a professor of anthropology in 1915. Maidu Myths *was published in 1902 as Part II of Volume XVII of the* Bulletin of the American Museum of Natural History. *In 1905 Dixon published a general monograph on the Maidu of the Sacramento Valley, and in 1907 another on the Shasta of northern California; among his other books were* Oceanic Mythology *(1916) and* The Building of Cultures *(1928). His sometime collaborator Alfred L. Kroeber, foremost of California ethnologists, wrote in the* Dictionary of American Biography *that "Dixon acquired a fund of ethnological information as great probably as that ever possessed by any one man." The stories that follow exemplify the ethnographer's straightforward retelling of tribal myths more or less as he heard them told, without extensive revision for readers unaccustomed to oral storytelling conventions. (For other examples, see Stith Thompson's excellent annotated selection,* Tales of the North American Indians *[1929].)*

The Maidu Creation Myth is the first of several creation stories that we will encounter; readers may wish to compare it with those of the Hebrew Bible (Chapter 3 below), the Indian Vedas (Chapter 4), the Chinese Huai-nan Tzu *(Chapter 5), Hesiod's* Theogony *(Chapter 8), Plato's* Timaeus *(Chapter 9), Ovid's* Metamorphoses *(Chapter 11), and Bernard Silvestris's* Cosmographia *(Chapter 15). These tales are far from "nature writing" as we normally understand it, yet nothing could be more important to a culture's conception of nature than its account of the origin of the cosmos and its inhabitants. What most strikes us, perhaps, in the Maidu version, as in many tribal myths, is its seeming disconnectedness and lack of any evident rationale or order; things of momentous significance seem to happen helter-skelter and almost by accident: "That does not look well: can't you make it some other way?" The different creators here include Turtle the "earth-diver" (after whom the earth was sometimes called Turtle Island), who appears in myths of North American peoples as far away as the Iroquois of the northeast; Earth-Initiate and Father-of-the-Secret-Society, figures presumably connected with the initiatory Ku'ksū cult of north central California named after the first man of this myth; and the irre-*

pressibly extemporaneous trickster Coyote, who is always getting into and out of messes of his own making. Together they seem like too many cooks spoiling—or is it spicing up?—the stew of creation.

Yet these qualities, far from being defects of "primitive" storytelling or thought, vividly suggest a cosmos without rational plan or purposeful design, a chaotic ever-changing world where "nature" is a kaleidoscopic collage continually being reimprovised according to need and whim, above all by the arch-improviser and troublemaker, the crazy, vital, incessantly self-transformative Coyote, who invents death but "can never be killed off." After our initial bewilderment at such apparently aimless stories, which even so veteran a folklorist as Stith Thompson could find "rather pointless and tedious"—they are seldom the same, in any case, from one telling to the next!—we may, on second thought, find them strangely familiar, as if we had encountered this world somewhere before, perhaps in our dreams . . . or is it our daily lives? It is a makeshift, senseless world, to be sure, but also a world of endless fascination and beauty in which there are countless butterflies to pursue, even if we may never seize one for longer than a moment or make it more than fleetingly ours. For who but Coyote, shifting shapes as rapidly as the moments themselves, could hope to hold on to the butterflies of such a world?

The Maidu Creation Myth

In the beginning there was no sun, no moon, no stars. All was dark, and everywhere there was only water. A raft came floating on the water. It came from the north, and in it were two persons,—Turtle and Father-of-the-Secret-Society. The stream flowed very rapidly. Then from the sky a rope of feathers was let down, and down it came Earth-Initiate. When he reached the end of the rope, he tied it to the bow of the raft, and stepped in. His face was covered and was never seen, but his body shone like the sun. He sat down, and for a long time said nothing.

At last Turtle said, "Where do you come from?" and Earth-Initiate answered, "I come from above." Then Turtle said, "Brother, can you not make for me some good dry land, so that I may sometimes come up out of the water?" Then he asked another time, "Are there going to be any people in the world?" Earth-Initiate thought awhile, then said, "Yes." Turtle asked, "How long before you are going to make people?" Earth-Initiate replied, "I don't know. You want to have some dry land: well, how am I going to get any earth to make it of?"

Turtle answered, "If you will tie a rock about my left arm, I'll dive for some." Earth-Initiate did as Turtle asked, and then, reaching around, took the end of a rope from somewhere, and tied it to Turtle. When Earth-Initiate came to the raft, there was no rope there: he just reached out and found one. Turtle said, "If the rope is not long enough, I'll jerk it once, and you must haul me up; if it is long enough, I'll give two jerks, and then you must pull me up quickly, as I shall have all the earth that I can carry." Just as Turtle went over the side of the boat, Father-of-the-Secret-Society began to shout loudly.

Turtle was gone a long time. He was gone six years; and when he came up, he was covered with green slime, he had been down so long. When he reached the top of the water, the only earth he had was a very little under his nails: the rest had all washed away. Earth-Initiate took with his right hand a stone knife from under his left armpit, and carefully scraped the earth out from under Turtle's nails. He put the earth in the palm of his hand, and rolled it about till it was round; it was as large as a small pebble. He laid it on the stern of the raft. By and by he went to look at it: it had not grown at all. The third time that he went to look at it, it had grown so that it could be spanned by the arms.

The fourth time he looked, it was as big as the world, the raft was aground, and all around were mountains as far as he could see. The raft came ashore at Ta'doikö, and the place can be seen today.

When the raft had come to land, Turtle said, "I can't stay in the dark all the time. Can't you make a light, so that I can see?" Earth-Initiate replied, "Let us get out of the raft, and then we will see what we can do." So all three got out. Then Earth-Initiate said, "Look that way, to the east! I am going to tell my sister to come up." Then it began to grow light, and day began to break; then Father-of-the-Secret-Society began to shout loudly, and the sun came up. Turtle said, "Which way is the sun going to travel?" Earth-Initiate answered, "I'll tell her to go this way, and go down there." After the sun went down, Father-of-the-Secret-Society began to cry and shout again, and it grew very dark. Earth-Initiate said, "I'll tell my brother to come up." Then the moon rose. Then Earth-Initiate asked Turtle and Father-of-the-Secret-Society, "How do you like it?" and they both answered, "It is very good." Then Turtle asked, "Is that all you are going to do for us?" and Earth-Initiate answered, "No, I am going to do more yet." Then he called the stars each by its name, and they came out. When this was done, Turtle asked, "Now what shall we do?" Earth-Initiate replied, "Wait, and I'll show you." Then he made a tree grow at Ta'doikö,—the tree called Hu'kīmtsa; and Earth-Initiate and Turtle and Father-of-the-Secret-Society sat in its shade for two days. The tree was very large, and had twelve different kinds of acorns growing on it.

After they had sat for two days under the tree, they all went off to see the world that Earth-Initiate had made. They started at sunrise, and were back by sunset. Earth-Initiate travelled so fast that all they could see was a ball of fire flashing about under the ground and the water. While they were gone, Coyote and his dog Rattlesnake came up out of the ground. It is said that Coyote could see Earth-Initiate's face. When Earth-Initiate and the others came back, they found Coyote at Ta'doikö. All five of them then built huts for themselves, and lived there at Ta'doikö, but no one could go inside of Earth-Initiate's house. Soon after the travellers came back, Earth-Initiate called the birds from the air, and made the trees and then the animals. He took some mud, and of this made first a deer; after that, he made all the other animals. Sometimes Turtle would say, "That does not look well: can't you make it some other way?"

Some time after this, Earth-Initiate and Coyote were at Marysville Buttes. Earth-Initiate said, "I am going to make people." In the middle of the afternoon he began, for he had returned to Ta'doikö. He took dark red earth, mixed it with water, and made two figures,—one a man, and one a woman. He laid the man on his right side, and the woman on his left, inside his house. Then he lay down himself, flat on his back, with his arms stretched out. He lay thus and sweated all the afternoon and night. Early in the morning the woman began to tickle him in the side. He kept very still, did not laugh. By and by he got up, thrust a piece of pitch-wood into the ground, and fire burst out. The two people were very white. No one today is as white as they were. Their eyes were pink, their hair was black, their teeth shone brightly, and they were very handsome. It is said that Earth-Initiate did not finish the hands of the people, as he did not know how it would be best to do it. Coyote saw the people, and suggested that they ought to have hands like his. Earth-Initiate said, "No, their hands shall be like mine." Then he finished them. When Coyote asked why their hands were to be like that, Earth-Initiate answered, "So that, if they are chased by bears, they can climb trees." This first man was called Ku'ksū; and the woman, Morning-Star Woman.

When Coyote had seen the two people, he asked Earth-Initiate how he had made them. When he was told, he thought, "That is not difficult. I'll do it myself." He did just as Earth-Initiate had told him, but could not help laughing, when, early in the morning, the woman poked him in the ribs. As a result of his failing to keep still, the people were glass-eyed. Earth-Initiate said, "I told you not to laugh," but Coyote declared he had not. This was the first lie.

By and by there came to be a good many people. Earth-Initiate had wanted to have everything comfortable and easy for people, so that none of them should have to work. All fruits were easy to obtain, no one was ever to get sick and die. As the people grew numerous, Earth-Initiate did not come as often as formerly, he only came to see Ku'ksū in the night. One night he said to him, "Tomorrow morning you must go to the little lake near here. Take all the people with you. I'll make you a very old man before you get to the lake." So in the morning Ku'ksū collected all the people, and went to the lake. By the time he had reached it, he was a very old man. He fell into the lake, and sank down out of sight. Pretty soon the ground began to shake, the waves overflowed the shore, and there was a great roaring under the water, like thunder. By and by Ku'ksū came up out of the water, but young again, just like a young man. Then Earth-Initiate came and spoke to the people, and said, "If you do as I tell you, everything will be well. When any of you grow old, so old that you cannot walk, come to this lake, or get someone to bring you here. You must then go down into the water as you have seen Ku'ksū do, and you will come out young again." When he had said this, he went away. He left in the night, and went up above.

All this time food had been easy to get, as Earth-Initiate had wished. The women set out baskets at night, and in the morning they found them full of food, all ready to eat, and lukewarm. One day Coyote came along. He asked the people how they lived, and they told him that all they had to do was to eat and sleep. Coyote replied, "That is no way to do: I can show you something better." Then he told them how he and Earth-Initiate had had a discussion before men had been made; how Earth-Initiate wanted everything easy, and that there should be no sickness or death, but how he had thought it would be better to have people work, get sick, and die. He said, "We'll have a burning." The people did not know what he meant; but Coyote said, "I'll show you. It is better to have a burning, for then the widows can be free." So he took all the baskets and things that the people had, hung them up on poles, made everything all ready. When all was prepared, Coyote said, "At this time you must always have games." So he fixed the moon during which these games were to be played.

Coyote told them to start the games with a foot-race, and everyone got ready to run. Ku'ksū did not come, however. He sat in his hut alone, and was sad, for he knew what was going to occur. Just at this moment Rattlesnake came to Ku'ksū , and said, "What shall we do now? Everything is spoiled!" Ku'ksū did not answer, so Rattlesnake said, "Well, I'll do what I think is best." Then he went out and along the course that the racers were to go over, and hid himself, leaving his head just sticking out of a hole. By this time all the racers had started, and among them Coyote's son. He was Coyote's only child, and was very quick. He soon began to outstrip all the runners, and was in the lead. As he passed the spot where Rattlesnake had hidden himself, however, Rattlesnake raised his head and bit the boy in the ankle. In a minute the boy was dead.

Coyote was dancing about the home-stake. He was very happy, and was shouting at his son and praising him. When Rattlesnake bit the boy, and he fell dead, everyone laughed at Coyote, and said, "Your son has fallen down, and is so ashamed that he does not dare to get up." Coyote said, "No, that is not it. He is dead." This was the first death.

The people, however, did not understand, and picked the boy up, and brought him to Coyote. Then Coyote began to cry, and everyone did the same. These were the first tears. Then Coyote took his son's body and carried it to the lake of which Earth-Initiate had told them, and threw the body in. But there was no noise, and nothing happened, and the body drifted about for four days on the surface, like a log. On the fifth day Coyote took four sacks of beads and brought them to Ku'ksū, begging him to restore his son to life. Ku'ksū did not answer. For five days Coyote begged, then Ku'ksū came out of his house bringing all his beads and bear-skins, and calling to all the people to come and watch him. He laid the body on a bear-skin, dressed it, and wrapped it up carefully. Then he dug a grave, put the body into it, and covered it up. Then he told the people, "From now on, this is what you must do. This is the way you must do till the world shall be made over."

About a year after this, in the spring, all was changed. Up to this time everybody spoke the same language. The people were having a burning, everything was ready for the next day, when in the night everybody suddenly began to speak a different language. Each man and his wife, however, spoke the same. Earth-Initiate had come in the night to Ku'ksū, and had told him about it all, and given him instructions for the next day. So, when morning came, Ku'ksū called all the people together, for he was able to speak all the languages. He told them each the names of the different animals, etc., in their languages, taught them how to cook and to hunt, gave them all their laws, and set the time for all their dances and festivals. Then he called each tribe by name, and sent them off in different directions, telling them where they were to live. He sent the warriors to the north, the singers to the west, the flute-players to the east, and the dancers to the south. So all the people went away, and left Ku'ksū and his wife alone at Ta'doikö. By and by his wife went away, leaving in the night, and going first to Marysville Buttes. Ku'ksū stayed a little while longer, and then he also left. He too went to the Buttes, went into the spirit house, and sat down on the south side. He found Coyote's son there, sitting on the north side. The door was on the west.

Coyote had been trying to find out where Ku'ksū had gone, and where his own son had gone, and at last found the tracks, and followed them to the spirit house. Here he saw Ku'ksū and his son, the latter eating spirit food. Coyote wanted to go in, but Ku'ksū said, "No, wait there. You have just what you wanted, it is your own fault. Every man will now have all kinds of troubles and accidents, will have to work to get his food, and will die and be buried. This must go on till the time is out, and Earth-Initiate comes again, and everything will be made over. You must go home, and tell all the people that you have seen your son, that he is not dead." Coyote said he would go, but that he was hungry, and wanted some of the food. Ku'ksū replied, "You cannot eat that. Only ghosts may eat that food." Then Coyote went away and told all the people, "I saw my son and Ku'ksū, and he told me to kill myself." So he climbed up to the top of a tall tree, jumped off, and was killed. Then he went to the spirit house, thinking he could now have some of the food; but there was no one there, nothing at all, and so he went out and walked away to the west, and was never seen again. Ku'ksū and Coyote's son, however, had gone up above.

Two Coyote Stories

1) People were angry with Coyote. They all agreed that everyone should come in from north and east, from south and west, and crowd all the Coyotes into the center of the country, and then they would kill every Coyote. They did this, but overlooked one.

He was an Initiate, and the chief of all. They hunted everywhere for him, and at last they found him. Then they looked everywhere for the largest tree they could find, and finally found it to the west. It was a great yellow-pine, and, having split it open, they put Coyote inside, and let the tree close together over him. They thought they had killed him this way. The chief called all the people together and said, "As you spread out to go home, see if you can hear any noise like a Coyote." The pine-tree was hollow, and so Coyote was not crushed to death, as the people thought. He was merely imprisoned. By and by Red-headed Woodpecker came, and began tapping on the log, as it sounded hollow. He worked away for two days, and all this time Coyote lay still and listened. At the end of the next two days he could see a faint spot of light. Next day Woodpecker came again, and enlarged the hole he had made, so that Coyote could see quite a little light. By and by Coyote said, "Cousin, make the hole bigger, please"; but Woodpecker was frightened, and flew away. Then Coyote got angry, and said, "The reason people call me crazy is that I don't know enough to keep quiet."

The bird did not return, and Coyote wondered how he was to get out. At last he defecated, and inquired how he could get out. He was answered that he would never get out, but that he would die in the tree. Angry at this prophecy, he defecated a second time, and, on questioning his faeces, was told to transform himself into a fog, and thus pass through the small hole that Woodpecker had made. He did so; and as soon as he came out, he again became a Coyote. He said, "I'm a Coyote, and can never die. People may kill me, but there will always be Coyotes left."

When the people had put Coyote into the tree, the chief had said, "If we hear nothing of him for six days, we may be sure that he is dead." Coyote, however, got out on the fifth day, and started back toward this country. On the sixth morning, just at daylight, he began to howl, just to let people know that there were some Coyotes left. The people heard him, and said, "We hear Coyotes crying. They are still alive." They hunted for him again, caught him, and took him to the west, to a great lake in the middle of which was a rock, from which he could not swim away. They put him on the rock; and the chief said, "If another six days go by, and we hear nothing of him, he will be dead." Coyote thought a long time as to how he should get away, but could not think of any way. So he asked the advice of his excrement, as before. The first time he asked, it said, "How do you think you can get away from here? You will have to stay till you die." The second time he was, as before, more successful, and was told, "You will live. In the morning, if you watch, the fog will rise. When it does, get off on it, and travel to the east, back to the land." Coyote followed this advice, and on the sixth day he reached land again, and, coming back to this country, began to howl. He said, "People can say the Coyote will never die. The Coyote can never be killed off. Wherever I urinate, even if I am killed, there will be another Coyote again." The people heard him howling, and said, "He has got the best of us. He has beaten us. Let us give him up." Then Coyote went off, saying to himself, "I'm going to travel through the middle of this world; and in every valley I come to, I'll catch mice for my living. I'll be a Coyote."

2) Coyote once wanted a woman, but could get no one who would have anything to do with him. So he resolved on a trick. He built himself a sweat-house, cut off his membrum virile [penis], and made a baby of it. He made himself look like a woman, and invited a lot of women to his house for a big feast. When they came, all danced for a while, then ate, and then lay down and went to sleep. As soon as all the women were asleep,

Coyote turned himself back into a man, cohabited with the women, then went away. In the morning the women woke up. They found that some had children, others were in the pains of childbirth. There was no dance-house, everything was gone.

The To'lowim-Woman and the Butterfly-Man

A To'lowim-Woman went out to gather food. She had her child with her; and while she gathered the food, she stuck the point of the cradle-board in the ground, and left the child thus alone. As she was busy, a large butterfly flew past. The woman said to the child, "You stay here while I go and catch the butterfly." She ran after it, and chased it for a long time. She would almost catch it, and then just miss it. She wore a deer-skin robe. She thought, "Perhaps the reason why I cannot catch the butterfly is because I have this on." So she threw it away. Still she could not catch the butterfly, and finally threw away her apron, and hurried on. She had forgotten all about her child, and kept on chasing the butterfly till night came. Then she lay down under a tree and went to sleep. When she awoke in the morning, she found a man lying beside her. He said, "You have followed me thus far, perhaps you would like to follow me always. If you would, you must pass through a lot of my people." All this time the child was where the woman had left it, and she had not thought of it at all. She got up, and followed the butterfly-man. By and by they came to a large valley, the southern side of which was full of butterflies. When the two travellers reached the edge of the valley, the man said, "No one has ever got through this valley. People die before they get through. Don't lose sight of me. Follow me closely." They started, and travelled for a long time. The butterfly-man said, "Keep tight hold of me, don't let go." When they had got halfway through, other butterflies came flying about in great numbers. They flew every way, about their heads, and in their faces. They were fine fellows, and wanted to get the To'lowim-Woman for themselves. She saw them, watched them for a long time, and finally let go of her husband, and tried to seize one of these others. She missed him, and ran after him. There were thousands of others floating about; and she tried to seize, now one, now the other, but always failed, and so was lost in the valley. She said, "When people speak of the olden times by and by, people will say that this woman lost her husband, and tried to get others, but lost them, and went crazy and died." She went on then, and died before she got out of the valley. The butterfly-man she had lost went on, got through the valley, and came to his home.

The Angered Sea Mistress

The Central Eskimo, by Franz Boas

Franz Boas (1858–1942), one of the founders of American ethnology, was born in Minden, Germany, and took his doctorate at Kiel, specializing in physics, mathematics, and geography. After expeditions to the Arctic and British Columbia, he emigrated to the United States in 1886, teaching at Clark University from 1889 and at Columbia from 1896; in 1899 he became Columbia's first professor of anthropology, a position he held for 37 years. His influential publications in ethnology, physical anthropology, and linguistics include many studies of the Northwest Coast Indians of British Columbia and Alaska, especially the Kwakiutl (see Kwakiutl Ethnog-

raphy, ed. Helen Codere [1966]); The Mind of Primitive Man *(1911; rev. ed. 1938);* Primitive Art *(1927); and the papers collected in* Race, Language and Culture *(1940).* The Central Eskimo, *published in 1888 in the Sixth Annual Report of the Bureau of American Ethnology, was one of the first studies of the Eskimo, or Inuit, ever made. Here he recounted the story of Sedna and the fulmar (a gull-like Arctic bird), told with many variations throughout the vast lands of the Inuit. Sedna, as she is known on Baffin Island, was widely worshipped under different names as mistress of the sea animals on which Inuit life depends; in times of shortage, when Sedna, in resentment at her hard fate, withheld the seals and whales that had sprung from her severed fingers, shamans would descend to her undersea realm, comb her hair, and beseech her to release them again. This powerful myth is a reminder of how unsparing both animal and human beings can sometimes be, and how precariously difficult oneness between them may be to sustain in a world where survival hangs continuously in the balance.*

Sedna and the Fulmar

Once upon a time there lived on a solitary shore an Inung[3] with his daughter Sedna. His wife had been dead for some time and the two led a quiet life. Sedna grew up to be a handsome girl and the youths came from all around to sue for her hand, but none of them could touch her proud heart. Finally, at the breaking up of the ice in the spring a fulmar flew from over the ice and wooed Sedna with enticing song. "Come to me," it said; "come into the land of the birds, where there is never hunger, where my tent is made of the most beautiful skins. You shall rest on soft bearskins. My fellows, the fulmars, shall bring you all your heart may desire; their feathers shall clothe you; your lamp shall always be filled with oil, your pot with meat." Sedna could not long resist such wooing and they went together over the vast sea. When at last they reached the country of the fulmar, after a long and hard journey, Sedna discovered that her spouse had shamefully deceived her. Her new home was not built of beautiful pelts, but was covered with wretched fishskins, full of holes, that gave free entrance to wind and snow. Instead of soft reindeer skins her bed was made of hard walrus hides and she had to live on miserable fish, which the birds brought her. Too soon she discovered that she had thrown away her opportunities when in her foolish pride she had rejected the Inuit youth. In her woe she sang: "Aja. O father, if you knew how wretched I am you would come to me and we would hurry away in your boat over the waters. The birds look unkindly upon me the stranger; cold winds roar about my bed; they give me but miserable food. O come and take me back home. Aja."

When a year had passed and the sea was again stirred by warmer winds, the father left his country to visit Sedna. His daughter greeted him joyfully and besought him to take her back home. The father hearing of the outrages wrought upon his daughter determined upon revenge. He killed the fulmar, took Sedna into his boat, and they quickly left the country which had brought so much sorrow to Sedna. When the other fulmars came home and found their companion dead and his wife gone, they all flew away in search of the fugitives. They were very sad over the death of their poor murdered comrade and continue to mourn and cry until this day.

Having flown a short distance they discerned the boat and stirred up a heavy storm. The sea rose in immense waves that threatened the pair with destruction. In this mortal peril the father determined to offer Sedna to the birds and flung her overboard. She clung to the edge of the boat with a death grip. The cruel father then took a knife and

[3]A man, Eskimo.

cut off the first joints of her fingers. Falling into the sea they were transformed into whales, the nails turning into whalebone. Sedna holding on to the boat more tightly, the second finger joints fell under the sharp knife and swam away as seals; when the father cut off the stumps of the fingers they became ground seals. Meantime the storm subsided, for the fulmars thought Sedna was drowned. The father then allowed her to come into the boat again. But from that time she cherished a deadly hatred against him and swore bitter revenge. After they got ashore, she called her dogs and let them gnaw off the feet and hands of her father while he was asleep. Upon this he cursed himself, his daughter, and the dogs which had maimed him; whereupon the earth opened and swallowed the hut, the father, the daughter, and the dogs. They have since lived in the land of Adlivun,[4] of which Sedna is the mistress.

The Healing Sun

Blackfoot Lodge Tales, by George Bird Grinnell

George Bird Grinnell (1849–1938) was born in Brooklyn and studied at Yale, where he took his Ph.D. in 1880. He accompanied Custer's expedition to the Black Hills as a naturalist in 1874 and William Ludlow's to Yellowstone in 1875, and in 1899 he took part in Edward Hamilton's expedition to Alaska. Between 1876 and 1911 he edited Forest and Stream *magazine, and throughout his life was an energetic conservationist, founding the first Audubon Society and helping organize the New York Zoological Society. A leading advocate of national parks, he discovered the glacier in Montana named after him and promoted establishment of Glacier National Park in 1910. He served both President Cleveland and Theodore Roosevelt as a negotiator with Indian tribes, and his intense interest in Plains Indian culture resulted in many books, including* Pawnee Hero Stories and Folk Tales *(1889),* When Buffalo Ran *(1920), and* The Cheyenne Indians *(1923). The story of Scarface, from* Blackfoot Lodge Tales *(1892), is a celestial quest myth of a kind found among peoples as widely separated as the Iroquois and the Navajo: to be healed, and united with his love, the impoverished Scarface ventures forth, with the aid of animal helpers, on a shamanistic journey to the Sun. (Grinnell's version was a principal source for Jamake Highwater's modern retelling,* Anpao: An American Indian Odyssey *[1977].) Through communion with the celestial powers he renews not only himself but also his people: this story tells the origin of the Medicine Lodge, the Blackfoot variation of the Plains Indian ceremony more widely known as the Sun Dance, a ritual of collective renewal closely connected with the individual vision quest for attainment of knowledge and power from the transcendent forces of encompassing nature.*

FROM *Scarface: Origin of the Medicine Lodge*

I ... There was a poor young man, very poor. His father, mother, and all his relations, had gone to the Sand Hills.[5] He had no lodge, no wife to tan his robes or sew his moccasins. He stopped in one lodge today, and tomorrow he ate and slept in another; thus he lived. He was a good-looking young man, except that on his cheek he had a scar, and his clothes were always old and poor.

[4]The world below.
[5]Sand Hills: the shadow land; place of ghosts; the Blackfoot future world. (Grinnell)

After those dances some of the young men met this poor Scarface, and they laughed at him, and said: "Why don't you ask that girl to marry you? You are so rich and handsome!" Scarface did not laugh; he replied: "Ah! I will do as you say. I will go and ask her." All the young men thought this was funny. They laughed a great deal. But Scarface went down by the river. He waited by the river, where the women came to get water, and by and by the girl came along. "Girl," he said, "wait. I want to speak with you. Not as a designing person do I ask you, but openly, where the Sun looks down, and all may see."

"Speak then," said the girl.

"I have seen the days," continued the young man. "You have refused those who are young, and rich, and brave. Now, today, they laughed and said to me, 'Why do you not ask her?' I am poor, very poor. I have no lodge, no food, no clothes, no robes and warm furs. I have no relations; all have gone to the Sand Hills; yet, now, today, I ask you, take pity, be my wife."

The girl hid her face in her robe and brushed the ground with the point of her moccasin, back and forth, back and forth; for she was thinking. After a time she said: "True, I have refused all those rich young men, yet now the poor one asks me, and I am glad. I will be your wife, and my people will be happy. You are poor, but it does not matter. My father will give you dogs. My mother will make us a lodge. My people will give us robes and furs. You will be poor no longer."

Then the young man was happy, and he started to kiss her, but she held him back, and said: "Wait! The Sun has spoken to me. He says I may not marry; that I belong to him. He says if I listen to him, I shall live to great age. But now I say: Go to the Sun. Tell him, 'She whom you spoke with heeds your words. She has never done wrong, but now she wants to marry. I want her for my wife.' Ask him to take that scar from your face. That will be his sign. I will know he is pleased. But if he refuses, or if you fail to find his lodge, then do not return to me."

"Oh!" cried the young man, "at first your words were good. I was glad. But now it is dark. My heart is dead. Where is that far-off lodge? where the trail, which no one yet has traveled?"

"Take courage, take courage!" said the girl; and she went to her lodge.

II Scarface was very sad. He sat down and covered his head with his robe and tried to think what to do. After a while he got up, and went to an old woman who had been kind to him. "Pity me," he said. "I am very poor. I am going away now on a long journey. Make me some moccasins."

"Where are you going?" asked the old woman. "There is no war; we are very peaceful here."

"I do not know where I shall go," replied Scarface. "I am in trouble, but I cannot tell you now what it is."

So the old woman made him some moccasins, seven pairs, with parfleche soles, and also she gave him a sack of food,—pemmican of berries, pounded meat, and dried back fat; for this old woman had a good heart. She liked the young man.

All alone, and with a sad heart, he climbed the bluffs and stopped to take a last look at the camp. He wondered if he would ever see his sweetheart and the people again. "*Hai'-yu!* Pity me, O Sun," he prayed, and turning, he started to find the trail.

For many days he traveled on, over great prairies, along timbered rivers and among the mountains, and every day his sack of food grew lighter; but he saved it as much as he could, and ate berries, and roots, and sometimes he killed an animal of some kind. One night he stopped by the home of a wolf. "*Hai-yah!*" said that one; "what is my brother doing so far from home?"

"Ah!" replied Scarface, "I seek the place where the Sun lives; I am sent to speak with him."

"I have traveled far," said the wolf. "I know all the prairies, the valleys, and the mountains, but I have never seen the Sun's home. Wait; I know one who is very wise. Ask the bear. He may tell you."

The next day the man traveled on again, stopping now and then to pick a few berries, and when night came he arrived at the bear's lodge.

"Where is your home?" asked the bear. "Why are you traveling alone, my brother?"

"Help me! Pity me!" replied the young man; "because of her words[6] I seek the Sun. I go to ask him for her."

"I know not where he stops," replied the bear. "I have traveled by many rivers, and I know the mountains, yet I have never seen his lodge. There is someone beyond, that striped-face, who is very smart. Go and ask him."

The badger was in his hole. Stooping over, the young man shouted: "Oh, cunning striped-face! Oh, generous animal! I wish to speak with you."

"What do you want?" said the badger, poking his head out of the hole.

"I want to find the Sun's home," replied Scarface. "I want to speak with him."

"I do not know where he lives," replied the badger. "I never travel very far. Over there in the timber is a wolverine. He is always traveling around, and is of much knowledge. Maybe he can tell you."

Then Scarface went to the woods and looked all around for the wolverine, but could not find him. So he sat down to rest. "*Hai'-yu! Hai'-yu!*" he cried. "Wolverine, take pity on me. My food is gone, my moccasins worn out. Now I must die."

"What is it, my brother?" he heard, and looking around, he saw the animal sitting near.

"She whom I would marry," said Scarface, "belongs to the Sun; I am trying to find where he lives, to ask him for her."

"Ah!" said the wolverine. "I know where he lives. Wait; it is nearly night. Tomorrow I will show you the trail to the big water. He lives on the other side of it."

Early in the morning, the wolverine showed him the trail, and Scarface followed it until he came to the water's edge. He looked out over it, and his heart almost stopped. Never before had anyone seen such a big water. The other side could not be seen, and there was no end to it. Scarface sat down on the shore. His food was all gone, his moccasins worn out. His heart was sick. "I cannot cross this big water," he said. "I cannot return to the people. Here, by this water, I shall die."

Not so. His Helpers were there. Two swans came swimming up to the shore. "Why have you come here?" they asked him. "What are you doing? It is very far to the place where your people live."

"I am here," replied Scarface, "to die. Far away, in my country, is a beautiful girl. I want to marry her, but she belongs to the Sun. So I started to find him and ask for her. I have traveled many days. My food is gone. I cannot go back. I cannot cross this big water, so I am going to die."

"No," said the swans; "it shall not be so. Across the water is the home of that Above Person. Get on our backs, and we will take you there."

[6]A Blackfoot often talks of what this or that person said, without mentioning names. (Grinnell)

Scarface quickly arose. He felt strong again. He waded out into the water and lay down on the swans' backs, and they started off. Very deep and black is that fearful water. Strange people live there, mighty animals which often seize and drown a person. The swans carried him safely, and took him to the other side. Here was a broad hard trail leading back from the water's edge.

"*Kyi*," said the swans. "You are now close to the Sun's lodge. Follow that trail, and you will soon see it."

III Scarface started up the trail, and pretty soon he came to some beautiful things, lying in it. There was a war shirt, a shield, and a bow and arrows. He had never seen such pretty weapons; but he did not touch them. He walked carefully around them, and traveled on. A little way further on, he met a young man, the handsomest person he had ever seen. His hair was very long, and he wore clothing made of strange skins. His moccasins were sewn with bright colored feathers. The young man said to him, "Did you see some weapons lying on the trail?"

"Yes," replied Scarface; "I saw them."

"But did you not touch them?" asked the young man.

"No; I thought someone had left them there, so I did not take them."

"You are not a thief," said the young man. "What is your name?"

"Scarface."

"Where are you going?"

"To the Sun."

"My name," said the young man, "is A-pi-su'-ahts.[7] The Sun is my father; come, I will take you to our lodge. My father is not now at home, but he will come in at night."

Soon they came to the lodge. It was very large and handsome; strange medicine animals were painted on it. Behind, on a tripod, were strange weapons and beautiful clothes—the Sun's. Scarface was ashamed to go in, but Morning Star said, "Do not be afraid, my friend; we are glad you have come."

They entered. One person was sitting there, Ko-ko-mik'-e-is,[8] the Sun's wife, Morning Star's mother. She spoke to Scarface kindly, and gave him something to eat. "Why have you come so far from your people?" she asked.

Then Scarface told her about the beautiful girl he wanted to marry. "She belongs to the Sun," he said. "I have come to ask him for her."

When it was time for the Sun to come home, the Moon hid Scarface under a pile of robes. As soon as the Sun got to the doorway, he stopped, and said, "I smell a person."

"Yes, father," said Morning Star; "a good young man has come to see you. I know he is good, for he found some of my things on the trail and did not touch them."

Then Scarface came out from under the robes, and the Sun entered and sat down. "I am glad you have come to our lodge," he said. "Stay with us as long as you think best. My son is lonesome sometimes; be his friend."

The next day the Moon called Scarface out of the lodge, and said to him: "Go with Morning Star where you please, but never hunt near that big water; do not let him go there. It is the home of great birds which have long sharp bills; they kill people. I have had many sons, but these birds have killed them all. Morning Star is the only one left."

So Scarface stayed there a long time and hunted with Morning Star. One day they came near the water, and saw the big birds.

[7]Early Riser, *i.e.* The Morning Star. (Grinnell) [8]Night red light, the Moon. (Grinnell)

"Come," said Morning Star; "let us go and kill those birds."

"No, no!" replied Scarface; "we must not go there. Those are very terrible birds; they will kill us."

Morning Star would not listen. He ran towards the water, and Scarface followed. He knew that he must kill the birds and save the boy. If not, the Sun would be angry and might kill him. He ran ahead and met the birds, which were coming towards him to fight, and killed every one of them with his spear: not one was left. Then the young men cut off their heads, and carried them home. Morning Star's mother was glad when they told her what they had done, and showed her the birds' heads. She cried, and called Scarface "my son." When the Sun came home at night, she told him about it, and he too was glad. "My son," he said to Scarface, "I will not forget what you have this day done for me. Tell me now, what can I do for you?"

"*Hai'-yu*," replied Scarface. "*Hai'-yu*, pity me. I am here to ask you for that girl. I want to marry her. I asked her, and she was glad; but she says you own her, that you told her not to marry."

"What you say is true," said the Sun. "I have watched the days, so I know it. Now, then, I give her to you; she is yours. I am glad she has been wise. I know she has never done wrong. The Sun pities good women. They shall live a long time. So shall their husbands and children. Now you will soon go home. Let me tell you something. Be wise and listen: I am the only chief. Everything is mine. I made the earth, the mountains, prairies, rivers, and forests. I made the people and all the animals. This is why I say I alone am the chief. I can never die. True, the winter makes me old and weak, but every summer I grow young again."

Then said the Sun: "What one of all animals is smartest? The raven is, for he always finds food. He is never hungry. Which one of all the animals is most *Nat-o'-ye?*[9] The buffalo is. Of all animals, I like him best. He is for the people. He is your food and your shelter. What part of his body is sacred? The tongue is. That is mine. What else is sacred? Berries are. They are mine too. Come with me and see the world." He took Scarface to the edge of the sky, and they looked down and saw it. It is round and flat, and all around the edge is the jumping-off place [or walls straight down]. Then said the Sun: "When any man is sick or in danger, his wife may promise to build me a lodge, if he recovers. If the woman is pure and true, then I will be pleased and help the man. But if she is bad, if she lies, then I will be angry. You shall build the lodge like the world, round, with walls, but first you must build a sweat house of a hundred sticks. It shall be like the sky [a hemisphere], and half of it shall be painted red. That is me. The other half you will paint black. That is the night."

Further said the Sun: "Which is the best, the heart or the brain? The brain is. The heart often lies, the brain never." Then he told Scarface everything about making the Medicine Lodge, and when he had finished, he rubbed a powerful medicine on his face, and the scar disappeared. Then he gave him two raven feathers, saying: "These are the sign for the girl, that I give her to you. They must always be worn by the husband of the woman who builds a Medicine Lodge."

The young man was now ready to return home. Morning Star and the Sun gave him many beautiful presents. The Moon cried and kissed him, and called him "my son." Then the Sun showed him the short trail. It was the Wolf Road (Milky Way). He followed it, and soon reached the ground.

[9]This word may be translated as "of the Sun," "having Sun power," or more properly, something sacred. (Grinnell)

IV It was a very hot day. All the lodge skins were raised, and the people sat in the shade. There was a chief, a very generous man, and all day long people kept coming to his lodge to feast and smoke with him. Early in the morning this chief saw a person sitting out on a butte near by, close wrapped in his robe. The chief's friends came and went, the sun reached the middle, and passed on, down towards the mountains. Still this person did not move. When it was almost night, the chief said: "Why does that person sit there so long? The heat has been strong, but he has never eaten nor drunk. He may be a stranger; go and ask him in."

So some young men went up to him, and said: "Why do you sit here in the great heat all day? Come to the shade of the lodges. The chief asks you to feast with him."

Then the person arose and threw off his robe, and they were surprised. He wore beautiful clothes. His bow, shield, and other weapons were of strange make. But they knew his face, although the scar was gone, and they ran ahead, shouting, "The scarface poor young man has come. He is poor no longer. The scar on his face is gone."

All the people rushed out to see him. "Where have you been?" they asked. "Where did you get all these pretty things?" He did not answer. There in the crowd stood that young woman; and taking the two raven feathers from his head, he gave them to her, and said: "The trail was very long, and I nearly died, but by those Helpers, I found his lodge. He is glad. He sends these feathers to you. They are the sign."

Great was her gladness then. They were married, and made the first Medicine Lodge, as the Sun had said. The Sun was glad. He gave them great age. They were never sick. When they were very old, one morning, their children said: "Awake! Rise and eat." They did not move. In the night, in sleep, without pain, their shadows had departed for the Sand Hills.

Thunder Beings and Dogs' Heads

Black Elk Speaks, by John G. Neihardt

John G. Neihardt (1881–1973) was born in Sharpsburg, Illinois, taught at the University of Nebraska and later at the University of Missouri, and was literary editor of the St. Louis Post-Dispatch. *He was a poet whose works include* The Divine Enchantment *(1900), the five-part* Cycle of the West *beginning with* The Song of Three Friends *(1919) and ending with* The Song of the Messiah *(1935), and the novel* When the Tree Flowered *(1951). In August 1930, while working on* The Song of the Messiah—*a poem concerned with the messianic "Ghost Dance" movement that culminated in the massacre of the Sioux at Wounded Knee on December 29, 1890—he visited the Pine Ridge Reservation of the Oglala branch of the Teton Sioux in South Dakota, where he met the Oglala holy man Nicholas Black Elk (1863–1950). In 1931, at Black Elk's invitation (he was then sixty-seven), Neihardt returned to record the old man's life and to perpetuate the "great vision" he had been given during sickness at the age of nine.* Black Elk Speaks: Being the Life of a Holy Man of the Oglala Sioux, *published in 1932, was at first largely ignored but eventually became known as a classic account of a Native American shaman's visionary experience; the original transcripts of the interviews from which Neihardt elaborated his story were published in 1984 in Raymond J. DeMallie's* The Sixth Grandfather. *"It is the story," in Black Elk's words, "of all life that is holy and is good to tell, and of us two-leggeds sharing in it with the four-leggeds and the wings of the air and all green things; for these are children of one mother and their father is one Spirit."*

*In the following selection Black Elk tells of his first vision quest, in the spring of 1882, when he was eighteen, a time of terrible troubles for his people following the capture and killing in 1877 of their great leader Crazy Horse, who had led the Sioux and Cheyenne in their annihilation of Custer's army at the Little Bighorn in 1876—a victory that only momentarily slowed the inexorable advance of the Wasichus (whites) who were destroying their bison, slaying their warriors, and expropriating their land. The vision quest, which the Sioux called "lamenting" for a vision—a "crying and praying for understanding," in Black Elk's words—was a widespread practice among many Native American peoples, especially of the Northeast, the Great Plains, and the Northwest. Among the Sioux and some other Plains peoples it was undertaken, once or repeatedly, by either sex and at almost any age, but was especially important for young men coming to manhood; the quester, after a purificatory sweat bath and instruction by a holy man, would normally be left alone (sometimes for four days and nights) on an isolated hilltop where he would cry to the Great Spirit (*Wakan-tanka *in the Lakota language of the Teton Sioux) and the Grandfathers (*Tunkashila*) for a vision, communing in solitude with the sun and stars, animals and birds, butterflies and thunderclouds. The shape in which the vision appeared (if any did appear) would normally become the quester's guardian spirit, although in Black Elk's case the thunder beings had long before guided him to the spontaneous "great vision" of his childhood. Black Elk would later (DeMallie informs us) be simultaneously an Oglala holy man and a Roman Catholic catechist, building bridges to the whites by telling his story to Neihardt and confiding the rites of his people to Joseph Epes Brown, author of *The Sacred Pipe *(1953). But in this vision the hated Wasichus take the form of dogs' heads, destructive enemies who (the original transcripts make clear) "should be killed without pity like dogs." Not everything in nature, or in man, after all, is harmony, peace, and love.*

FROM *Chapter 15: The Dog Vision*

. . . When the grasses began to show their faces again, I was happy, for I could hear the thunder beings coming in the earth and I could hear them saying: "It is time to do the work of your Grandfathers."

After the long winter of waiting, it was my first duty to go out lamenting. So after the first rain storm I began to get ready.

When going out to lament it is necessary to choose a wise old medicine man, who is quiet and generous, to help. He must fill and offer the pipe to the Six Powers and to the four-leggeds and the wings of the air, and he must go along to watch. There was a good and wise old medicine man by the name of Few Tails, who was glad to help me. First he told me to fast four days, and I could have only water during that time. Then, after he had offered the pipe, I had to purify myself in a sweat lodge, which we made with willow boughs set in the ground and bent down to make a round top. Over this we tied a bison robe. In the middle we put hot stones, and when I was in there, Few Tails poured water on the stones. I sang to the spirits while I was in there being purified. Then the old man rubbed me all over with the sacred sage. He then braided my hair, and I was naked except that I had a bison robe to wrap around me while lamenting in the night, for although the days were warm, the nights were cold yet. All I carried was the sacred pipe.

It is necessary to go far away from people to lament, so Few Tails and I started from Pine Ridge toward where we are now.

We came to a high hill close to Grass Creek, which is just a little way west from here. There was nobody there but the old man and myself and the sky and the earth. But the place was full of people; for the spirits were there.

The sun was almost setting when we came to the hill, and the old man helped me make the place where I was to stand. We went to the highest point of the hill and made the ground there sacred by spreading sage upon it. Then Few Tails set a flowering stick in the middle of the place, and on the west, the north, the east, and the south sides of it he placed offerings of red willow bark tied into little bundles with scarlet cloth.

Few Tails now told me what I was to do so that the spirits would hear me and make clear my next duty. I was to stand in the middle, crying and praying for understanding. Then I was to advance from the center to the quarter of the west and mourn there awhile. Then I was to back up to the center, and from there approach the quarter of the north, wailing and praying there, and so on all around the circle. This I had to do all night long.

It was time for me to begin lamenting, so Few Tails went away somewhere and left me there all alone on the hill with the spirits and the dying light.

Standing in the center of the sacred place and facing the sunset, I began to cry, and while crying I had to say: "O Great Spirit, accept my offerings! O make me understand!"

As I was crying and saying this, there soared a spotted eagle from the west and whistled shrill and sat upon a pine tree east of me.

I walked backwards to the center, and from there approached the north, crying and saying: "O Great Spirit, accept my offerings and make me understand!" Then a chicken hawk came hovering and stopped upon a bush towards the south.

I walked backwards to the center once again and from there approached the east, crying and asking the Great Spirit to help me understand, and there came a black swallow flying all around me, singing, and stopped upon a bush not far away.

Walking backwards to the center, I advanced upon the south. Until now I had only been trying to weep, but now I really wept, and the tears ran down my face; for as I looked yonder towards the place whence come the life of things, the nation's hoop and the flowering tree, I thought of the days when my relatives, now dead, were living and young, and of Crazy Horse who was our strength and would never come back to help us any more.

I cried very hard, and I thought it might be better if my crying would kill me; then I could be in the outer world where nothing is ever in despair.

And while I was crying, something was coming from the south. It looked like dust far off, but when it came closer, I saw it was a cloud of beautiful butterflies of all colors. They swarmed around me so thick that I could see nothing else.

I walked backwards to the flowering stick again, and the spotted eagle on the pine tree spoke and said: "Behold these! They are your people. They are in great difficulty and you shall help them." Then I could hear all the butterflies that were swarming over me, and they were all making a pitiful, whimpering noise as though they too were weeping.

Then they all arose and flew back into the south.

Now the chicken hawk spoke from its bush and said: "Behold! Your Grandfathers shall come forth and you shall hear them!"

Hearing this, I lifted up my eyes, and there was a big storm coming from the west. It was the thunder being nation, and I could hear the neighing of horses and the sending of great voices.

It was very dark now, and all the roaring west was streaked fearfully with swift fire.

And as I stood there looking, a vision broke out of the shouting blackness torn with fire, and I saw the two men who had come to me first in my great vision. They came head first like arrows slanting earthward from a long flight; and when they neared the ground, I could see a dust rising there and out of the dust the heads of dogs were peeping. Then suddenly I saw that the dust was the swarm of many-colored butterflies hovering all around and over the dogs.

By now the two men were riding sorrel horses, streaked with black lightning, and they charged with bows and arrows down upon the dogs, while the thunder beings cheered for them with roaring voices.

Then suddenly the butterflies changed, and were storm-driven swallows, swooping and whirling in a great cloud behind the charging riders.

The first of these now plunged upon a dog's head and arose with it hanging bloody on his arrow point, while the whole west roared with cheering. The second did the same; and the black west flashed and cheered again. Then as the two arose together, I saw that the dogs' heads had changed to the heads of Wasichus; and as I saw, the vision went out and the storm was close upon me, terrible to see, and roaring.

I cried harder than ever now, for I was much afraid. The night was black about me and terrible with swift fire and the sending of great voices and the roaring of the hail. And as I cried, I begged the Grandfathers to pity me and spare me and told them that I knew now what they wanted me to do on earth, and I would do it if I could.

All at once I was not afraid any more, and I thought that if I was killed, probably I might be better off in the other world. So I lay down there in the center of the sacred place and offered the pipe again. Then I drew the bison robe over me and waited. All around me growled and roared the voices, and the hail was like the drums of many giants beating while the giants sang: "Hey-a-hey!"

No hail fell there in the sacred circle where I lay, nor any rain. And when the storm was passed, I raised my robe and listened; and in the stillness I could hear the rain-flood singing in the gulches all around me in the darkness, and far away to eastward there were dying voices calling: "Hey-a-hey!"

The night was old by now, and soon I fell asleep. And as I slept I saw my people sitting sad and troubled all around a sacred tepee, and there were many who were sick. And as I looked on them and wept, a strange light leaped upward from the ground close by—a light of many colors, sparkling, with rays that touched the heavens. Then it was gone, and in the place from whence it sprang a herb was growing and I saw the leaves it had. And as I was looking at the herb so that I might not forget it, there was a voice that 'woke me, and it said: "Make haste! Your people need you!"

I looked and saw the east was just beginning to turn white. Standing up, I faced the young light and began to mourn again and pray. Then the daybreak star came slowly, very beautiful and still; and all around it there were clouds of baby faces smiling at me, the faces of the people not yet born. The stars about them now were beautiful with many colors, and beneath these there were heads of men and women moving around, and birds were singing somewhere yonder and there were horses nickering and blowing as they do when they are happy, and somewhere deer were whistling and there were bison mooing too. What I could not see of this, I heard.

I think I fell asleep again, for afterwhile I was startled by a voice that said: "Get up, I have come after you!" I looked to see a spirit, but it was the good old man, Few Tails, standing over me. And now the sun was rising. . . .

TWO

Praise and Renunciation: Mesopotamia, Israel, India

The continuum of animate and inanimate, natural and supernatural beings, characteristic of tribal societies did not of course cease with the onset of urban civilizations and written traditions. On the contrary, the "animism" that the pioneering nineteenth-century British anthropologist Edward Tylor found central to "primitive" religion—the belief that every natural being is endowed with a living soul, so that nature itself, from heavenly bodies to animals, trees, and crops, shares in the divine—pervades the rites and myths of many early literate cultures as well, and has by no means entirely vanished even now. The fusion of human, animal, and divine is graphically evident in ancient Egypt, where a god might be pictured with a man's body and a falcon's head, and a stone scarab beetle be put in place of the human heart as a talisman of immortal life. Even the loftiest anthropomorphic sky god of classical Greece could manifest himself not only in thunder and lightning but also as a bull or a swan, while forest and countryside teemed with countless dryads, naiads, and satyrs, imbuing earth, sea, and sky with divinity. For most early civilized peoples the forces of nature—whether experienced through cyclical changes of the rising and setting sun, the waxing and waning moon, and the seasonal death and rebirth of the crops, or through the unpredictable incursions of storm and flood, famine and disease—were either divinities themselves or products of their words or actions. The "nature writing" of these peoples therefore inevitably takes the form of myths and scriptures that recount or explicate the regular or extraordinary, reassuring or terrifying manifestations of the divine in the world of which human beings are part.

Even so embracing a sourcebook as this one can represent only a few of the varied civilizations that have arisen and flourished, declined and vanished, in different parts of the world, leaving more or less fragmentary written records behind, over the last five millennia of recorded history. (Many others—Egyptian, Persian, Arabic, Armenian, Turkic, South Indian, Southeast Asian, Mayan, to name but a few—might have greatly enriched our understanding but too greatly lengthened our book.) Part Two contains selected writings from the closely interconnected civilizations of ancient Mesopotamia, the Sumerian—perhaps the earliest on earth—and the Akkadian, or Babylonian/Assyrian; from the Hebrew Bible; and from the Sanskritic cultures of northern India. Views of the "natural" world vary immensely in these writings, which express such extreme contrasts as that between the annual death and resurrection of the Sumerian Dumuzi and the omnipotence of the eternal God of Israel, or between the robust creative fervor of Vedic celebration and the ascetic world-denial of early Buddhism. But in each of these cultures, the natural (like the human) world is all but inconceivable apart from the divine forces—immanent, transcendent, or both at once—that embody or govern it: at least until it becomes, in Buddhist and Hindu *samsāra*, a wearisome cycle of repetitiously fated reincarnations from which only release or extinction, *moksha* or *nirvāna*, offers hope for those trapped within it.

At the opposite pole from such world-weary devaluation of the phenomenal world was the culture of ancient Egypt. For some 2,500 years between its union under the legendary first Pharaoh Menes and its conquest by Cambyses of Persia in 525 B.C., it maintained, to all appearances, a serene accord (relative to more turbulent cultures) between the human and the natural world. Each year the Nile would flood its rich alluvial valley, and its waters, through times of greater or lesser plenty, would spread through the network of canals by which men assisted benevolent nature in fertilizing the crops and feeding a grateful people. "In climate Egypt is a veritable paradise," James Henry Breasted affirms in his classic *History of Egypt* (1905), and those who farmed this happy land could live, through centuries of seeming stability, in remarkable harmony with plants and animals, with the endless river and the unfailing sun.

The ancient Egyptian conception of the human and natural world has often been called a static one, but this was a "stasis" of regular and recurrent change,[1] the rhythm of a world ruled by the sun god—Re, Atum, or Amon—in heaven, and by his regent the Pharaoh, divine embodiment of the god Horus, on earth. Through worship of a multitude of gods immanent in nature, Henri Frankfort writes in *Ancient Egyptian Religion* (1948), the Egyptians "could combine a profound awareness of the complexity of the phenomenal world with that of a mystic bond uniting man and nature." The very multiplicity of natural forces worshipped as gods precluded any single canon of scripture or coherent account of creation: in one version the world is formed by the sun god Re emerging from the primeval ocean Nun; in another by the masturbation of Atum; in a third, the so-called "Memphite theology," by pronouncement of the god Ptah. But the continuity of Pharaoh—who at death became Horus's father Osiris, divine ruler of the dead who bestowed immortality on both human beings and the grain—and the preeminence of one or another of the manifold gods, notably Amon-Re of Thebes, tended at times toward unity as well. Never was this more striking than in the endeavor of Pharaoh Amenhotep IV (who took the name Akhenaten or Ikhnaton) to establish, after 1375 B.C., worship of the sun disk Aten as the sole religion of his land. This short-lived experiment in monotheism, some have speculated, may have deeply impressed Moses and his fellow Hebrews during their harsh captivity in the pleasant land of Egypt, which for them was the land of death.

Even so brief a summary of the unusually harmonious, almost placid bond between the human and natural world envisaged by one of the earliest and most enduring of ancient cultures should highlight by contrast the more erratic, tumultuously intoxicating or menacingly violent experience of awesomely superhuman natural forces memorialized in different ways by the peoples of Mesopotamia, Israel, and India. (It is perhaps the relative absence of these qualities that makes much of the literature of Egypt, despite the splendor of some of its hymns, generally less compelling to modern readers.) Certainly the contrasts among—and indeed within—these three cultures are enormous, but in each, fervid celebration of the gods that embody or govern nature is liable to alternate with disquiet or revulsion (as it does of course at moments in Egypt's long history, as in the despondent "Song of the Harper") that may culminate, or threaten to culminate, in partial or total repudiation of the phenomenal world and the forces that rule it: repudiation, in the extreme, of life itself. Seldom has such ardent praise been lavished on nature and nature's gods as in the Sumerian hymns to Inanna, the Hebrew Psalms, and the Indian Vedas: yet Gilgamesh will scorn Ishtar's love, Job will long for death and the Preacher find weariness in all things, and the Buddha will teach that wisdom lies in cessation of misery through extinction of desire for this world. Attachment and withdrawal,

[1]In *Myth and Symbol in Ancient Egypt* (1959), R. T. Rundle Clark suggests that the fundamental Egyptian concept of Mayet (or Maat), usually translated "order," "truth," or "justice," "is probably the earliest approach to the concept of 'Nature' as understood in Western thought."

praise and renunciation, may prove to be opposite poles of a single impassioned perception of a natural world to which we willy-nilly belong, even when it seems suddenly emptied of godhead and meaning.

3

Mesopotamia and Israel: Gods of This World and Beyond

The contrast between the transcendent God of Israel who makes a binding covenant with His people and the immanent Mesopotamian gods who feud among themselves and attempt (in the Gilgamesh *epic) to destroy humankind for disturbing their sleep suggests the immensity of the difference between them, and the momentous significance for human history of the development of Jewish monotheism. Yet the contrast is by no means so categorical as it first appears, for Mesopotamian civilization was in large part the matrix out of which Hebrew civilization grew, both by drawing on it and by reacting against it.*

The biblical garden of Eden (from which the rivers Tigris and Euphrates flow) may itself have been modeled on the Sumerian paradise of Dilmun; survival of the flood by boat was told of Sumerian Ziusudra and Akkadian Utnapishtim thousands of years before biblical Noah; the great ziggurat *or temple of Babylon was the original tower of Babel; and Abram, father of the Jewish people, came to the land of Canaan from "Ur of the Chaldees," the ancient Sumerian city inhabited also by Semitic peoples (the later Babylonians or Chaldeans) to whom the Hebrews were closely related by blood and language. No doubt the Hebrews vehemently repudiated the religious and moral values of their neighbors, as they understood them, yet these too left their mark; the greatest Mesopotamian gods embodied a high conception of order and justice and exerted authority through the creative power of the divine word, and even Jewish law owed much to the great Babylonian code of Hammurabi. Monotheism itself was slow to be born (the very word for the God who created the heavens and earth,* elōhim, *is grammatically plural in Hebrew). For centuries, from Moses through the Babylonian exile, when Ezekiel (8:14) witnessed the abomination of women weeping for Tammuz in the house of the* LORD, *the national God of Israel was worshipped, despite repeated prophetic denunciations, alongside those of Canaanites and Phoenicians, above all Baal and Asherah, or Astarte. Radically though these beliefs and practices were rejected, many traces of them survive in the Hebrew Bible.*

Although walled towns of mud brick had been built long since—Jericho, north of the Dead Sea, may date back before 7000 B.C.*—the first truly urban civilization arose in the mid to late fourth millennium after the Sumerians arrived in the "fertile crescent" of the Tigris-Euphrates valley in southern Mesopotamia, now in Iraq, and began to develop the cuneiform ("wedge-shaped") writing system, originally pictographic but increasingly phonetic, which they incised on clay tablets. For perhaps a thousand years the independent city-states of Sumer—Kish, Nippur, Ur, Uruk, Lagash, Eridu, and others—coexisted, in peace or at war, among themselves and with neighbors (including local Semites) who shared in their civilization. No one city governor* (ensi) *or king* (lugal, *"big man") domi-*

nated the region for long until the Semitic conqueror Sargon of Agade (or Akkad), in about 2360 B.C., united Mesopotamia for nearly two centuries in the empire of Sumer-Akkad. After this the Semitic Akkadian language, adapted to cuneiform writing, slowly displaced Sumerian as the dominant tongue. Sumerian culture flourished again, however, under Gudea of Lagash and the third dynasty of Ur (ca. 2050–1950), when the ziggurat of Ur was built, the law code of Ur-nammu promulgated, and much Sumerian literature composed. It was perhaps in the troubled time after this dynasty fell to the Elamites and west Semitic Amorites that Abram departed from Ur for Canaan.

Around 1750 B.C. Hammurabi of Babylon again united the region, dominated thereafter—with countless conflicts both among themselves and with Kassites, Hurrians, Hittites, and others—by successive dynasties of Akkadian-speaking Babylonians and Assyrians until the fall of Nineveh to Medes and Babylonians in 612 and of Babylon itself to Cyrus of Persia in 539 brought an end to the ancient empires of Mesopotamia. The very existence of Sumerian culture was wholly forgotten until its language began to be deciphered and its cities excavated within the last hundred years. Meanwhile, in the buffer zone of Palestine between the great powers of Egypt and Mesopotamia, the nomadic Hebrews, after exodus from Egypt in perhaps the early thirteenth century, had carved out a place for themselves through continual warfare with Semitic Canaanites, Philistine sea peoples, and others during the chaotic rule of their charismatic "Judges." After being united for a century (ca. 1020–922) under the monarchy of Saul, David, and Solomon, Israel then split into two contentious realms until (after dire warnings by the prophets) Samaria, capital of the northern kingdom of Israel (or Ephraim), fell to Tiglath-pileser III of Assyria in 721 and Jerusalem, capital of the southern kingdom of Judah, fell to Nebuchadnezzar II of Babylon in 597 and again, after a futile revolt, ten years later. The temple of Solomon was destroyed and survivors of the siege were carried into exile in Babylon. Cyrus of Persia, who permitted the Jews to return from "Babylonian captivity" and rebuild their temple, was hailed in the Book of Isaiah (45:1) as the LORD*'s anointed, or messiah.*

These tumultuous events harshly contrast with the relatively tranquil history of Egypt, where for centuries on end times of trouble seemed the exception rather than the norm. And if history exacerbated insecurity, nature too played its part: neither the arid soil of Palestine nor the laboriously irrigated fields of the Tigris-Euphrates valley produced the all but unfailing bounty of the Nile in regions where fierce storms, sudden floods, and lingering drought were perpetual threats. In Mesopotamia, as in Egypt, such superhuman forces were embodied in a multitude of divinities who ruled man and nature alike. "Our concepts of the 'natural' and of 'natural law' did not exist," Henri Frankfort writes in Kingship and the Gods *(1948); "the life of nature was the life of the gods and hence full of significance; the movements of the constellations and the planets, changes in the weather, the behavior of animals—in short, all normal and recurring phenomena—involved divine activities and betokened divine intentions no less than extraordinary occurrences like eclipses, earthquakes, or plagues." But whereas in Egypt "it was unthinkable that nature and society should follow different courses, for both alike were ruled by* maat*— 'right, truth, justice, cosmic order,' " the Mesopotamians "did not presume that the gods themselves were bound by any order which man could comprehend," for this was a world of unpredictable flux continually in need of being sustained and renewed.*

Unlike the Egyptian gods, those of Mesopotamia took consistently human shapes. These "were imposed relatively late," Thorkild Jacobsen writes in Toward the Image of Tammuz *(1970), "and supplanted the older forms only slowly and with difficulty"; from time to time the familiar human image suddenly dis-*

closed the "terrifying nonhuman form" behind it. Among the countless deities in the Sumerians' pantheon—many of whom the Akkadians adopted, changing some of their names or syncretically identifying them with gods of their own—the principal four were the sky god An (Akkadian Anu); the air and storm god Enlil, destroyer and creator, who by earliest recorded times had displaced An as "father of the gods" and "king of heaven and earth"; Enki (Akkadian Ea), god of the abyss, of the sweet waters fertilizing the earth, and of wisdom; and the mother goddess Ninhursag or Ninmah—once probably identified with Ki, the Earth—wife of An, mistress of animals, creatress of vegetation, and, as Nintu, the birth-giving power of the womb. Other major divinities included the moon god Nanna (Akkadian Sin) and his children, the sun god Utu (Akkadian Shamash), and the "Queen of Heaven" Inanna (Akkadian Ishtar), goddess of love and harlots, of rain and fertility, of the morning and evening star, and of war. Through sexual union, in the new year's festival of spring or autumn, with this fearfully potent goddess, who took the form of a temple priestess in a bed chamber atop the city ziggurat, a king of Sumer, in the role of her husband, the annually dying shepherd god Dumuzi (Akkadian Tammuz), might each year renew the imperiled fertility of his land.

Such eroticism, which they beheld with fascinated revulsion as their own people went "whoring after the Baalim" of Canaan, was among the manifestations of Semitic "nature worship" most reviled by the Hebrews, along with its plural gods and its graven images of things in heaven, earth, or the waters below. The L*ORD* *God of Israel was not only one but transcendent, a power not within but above and beyond the forces of nature which he had created, which he could alter at will, and over which he ruled supreme and unrivaled. In comparison with so almighty a God it is no doubt true, as Frankfort affirms in* Kingship and the Gods, *that "man and nature were devaluated" by the Hebrew prophets. But it is not therefore necessarily true that "every attempt to establish a harmony with nature was a futile dissipation of effort," much less that in Hebrew religion alone "the ancient bond between man and nature was destroyed." For this* L*ORD* *too is a shepherd, whose glory the heavens declare and to whom earth belongs, and the fulness thereof: in the psalms that praise him with gladness of soul (as birds sing by the springs he sends running among the hills), or in the nuptial song celebrating union with the beloved as flowers appear on the earth and the voice of the turtledove is heard in the land, we perhaps find a bond with nature more abiding than any that Sumer or Akkad had known. The security of Hebrew belief, in contrast to the uncertainty of Canaanite and Mesopotamian myth, John L. Mackenzie remarks in* A Theology of the Old Testament *(1974), stems from assurance that God's "creative power was constantly active to prevent nature from relapsing into chaos."*

Goddess of the Heights and the Depths

Inanna, Queen of Heaven and Earth,
translated by Diane Wolkstein and Samuel Noah Kramer

The great divinities of polytheistic religions embody not one quality alone—the "god of war," the "goddess of wisdom"—but a mixture of diverse and even contradictory attributes. Nowhere is this more apparent than in Sumerian Inanna, the tender yet savage goddess of both love and war whom Jacobsen, in Treasures of Darkness *(1976), equates (in Shakespeare's phrase for*

Cleopatra) with "infinite variety." As daughter of the moon god Nanna she is above all a celestial deity, appearing—like love goddesses who would follow her, Akkadian Ishtar, Greek Aphrodite, and Roman Venus—as the morning and evening star; but from sky she descends to earth in rain, sometimes with thundering storms, and through marriage with Dumuzi her insatiable sexuality becomes inseparably linked with fertility of the crops on which human life depends. Dumuzi too is a complex (if less infinitely varied) deity combining, Jacobsen suggests in Toward the Image of Tammuz, *four main powers: that of the sap in trees and plants; that of the date palm and its fruits; that of grain and beer; and, as Dumuzi the shepherd, that of milk.*

The tales and hymns that follow survive on broken tablets probably dating back to the early second millennium B.C., *and variants of them had no doubt been told and sung for centuries or even millennia before. Diane Wolkstein's contemporary English versions, published in 1983, were made—with careful adherence to Sumerian texts recombined and conjecturally filled in on occasion, for the sake of intelligibility, and abbreviated when repetition threatened to become excessive—in close consultation with the leading American Sumerologist Samuel Noah Kramer. Kramer's books, including* Sumerian Mythology *(1944; 3rd ed. 1972),* The Sumerians: Their History, Culture, and Character *(1963), and* From the Poetry of Sumer *(1979), contain authoritative translations and discussions of many religious and mythological narratives. Our first selection is taken from Wolkstein's version of "The Courtship of Inanna and Dumuzi," recounting the erotic union of these life-giving divinities of fertility. The second is from "The Descent of Inanna," in which the immortal goddess of the heights longs for the depths, is killed by her sister Ereshkigal, queen of the underworld, but allowed to return on condition that she send a substitute to the underworld in her place. In rage at her insouciant husband, who seems to have neglected to notice her absence, she condemns him to die, but after the demons have hauled him away, she joins his mother and his sister Geshtinanna in lamenting him and suggests that sister and brother alternate, every half-year, in dwelling below. It was Dumuzi's annual death, or captivity—the apparent death each searing Mesopotamian summer of all that grows—that was mourned throughout the centuries by wailing women in Mesopotamia and beyond, and his rebirth and reunion with Inanna that was joyously celebrated in the new year's rites. Both selections (and also the hymn "Loud Thundering Storm") contain major omissions; in Wolkstein's version the "Descent" is divided, for example, into three parts ("From the Great Above to the Great Below"; "The Dream of Dumuzi"; and "The Return"), here abridged and conflated. For other versions, see Jacobsen,* The Harps That Once . . . : Sumerian Poetry in Translation *(1987).*

FROM *The Courtship of Inanna and Dumuzi*

The shepherd went to the royal house with cream.
Dumuzi went to the royal house with milk.
Before the door he called out:
    My daughter, the young man will be your mother.

Inanna ran to Ningal, the mother who bore her.
Ningal counseled her daughter, saying:
    "My child, the young man will be your father.
    My daughter, the young man will be your mother.
    He will treat you like a father.
    He will care for you like a mother.
    Open the house, My Lady, open the house!"

*

Inanna, at her mother's command,
Bathed and anointed herself with scented oil.
She covered her body with the royal white robe.
She readied her dowry.
She arranged her precious lapis beads around her neck.
She took her seal in her hand.

Dumuzi waited expectantly.
Inanna opened the door for him.
Inside the house she shone before him
Like the light of the moon.

Dumuzi looked at her joyously.
He pressed his neck close against hers.
He kissed her.

*

Inanna spoke:
"What I tell you
Let the singer weave into song.
What I tell you,
Let it flow from ear to mouth,
Let it pass from old to young:

My vulva, the horn,
The Boat of Heaven,
Is full of eagerness like the young moon.
My untilled land lies fallow.

As for me, Inanna,
Who will plow my vulva?
Who will plow my high field?
Who will plow my wet ground?

As for me, the young woman,
Who will plow my vulva?
Who will station the ox there?
Who will plow my vulva?

Dumuzi replied:
"Great Lady, the king will plow your vulva.
I, Dumuzi the King, will plow your vulva."

Inanna:
"Then plow my vulva, man of my heart!
Plow my vulva!"

At the king's lap stood the rising cedar.
Plants grew high by their side.
Grains grew high by their side.
Gardens flourished luxuriantly.

*

Inanna sang:

"He has sprouted; he has burgeoned;
He is lettuce planted by the water.
He is the one my womb loves best.

My well-stocked garden of the plain,
My barley growing high in its furrow,
My apple tree which bears fruit up to its crown,
He is lettuce planted by the water.

My honey-man, my honey-man sweetens me always.
My lord, the honey-man of the gods,
He is the one my womb loves best.
His hand is honey, his foot is honey,
He sweetens me always.

My eager impetuous caresser of the navel,
My caresser of the soft thighs,
He is the one my womb loves best,
He is lettuce planted by the water."

*

Dumuzi sang:

"O Lady, your breast is your field.
Inanna, your breast is your field.
Your broad field pours out plants.
Your broad field pours out grain.
Water flows from on high for your servant.
Bread flows from on high for your servant.
Pour it out for me, Inanna.
I will drink all you offer."

*

Inanna sang:

"Make your milk sweet and thick, my bridegroom.
My shepherd, I will drink your fresh milk.
Wild bull, Dumuzi, make your milk sweet and thick.
I will drink your fresh milk.
Let the milk of the goat flow in my sheepfold.
Fill my holy churn with honey cheese.
Lord Dumuzi, I will drink your fresh milk.

My husband, I will guard my sheepfold for you.
I will watch over your house of life, the storehouse,
The shining quivering place which delights Sumer—
The house which decides the fates of the land,
The house which gives the breath of life to the people,
I, the queen of the palace, will watch over your house."

*

Dumuzi spoke:

"My sister, I would go with you to my garden.
Inanna, I would go with you to my garden.

I would go with you to my orchard.
I would go with you to my apple tree.
There I would plant the sweet, honey-covered seed."

Inanna spoke:
"He brought me into his garden.
My brother, Dumuzi, brought me into his garden.
I strolled with him among the standing trees,
I stood with him among the fallen trees,
By an apple tree I knelt as is proper.
Before my brother coming in song,
Who rose out of the poplar leaves,
Who came to me in the midday heat,
Before my lord Dumuzi,
I poured out plants from my womb.
I placed plants before him,
I poured out plants before him.
I placed grain before him,
I poured out grain before him.
I poured out grain from my womb."

Inanna sang:
"Last night as I, the queen, was shining bright,
Last night as I, the Queen of Heaven, was shining bright,
As I was shining bright and dancing,
Singing praises at the coming of the night—

He met me—he met me!
My lord Dumuzi met me.
He put his hand into my hand.
He pressed his neck close against mine.

My high priest is ready for the holy loins.
My lord Dumuzi is ready for the holy loins.
The plants and herbs in his field are ripe.
O Dumuzi! Your fullness is my delight!" . . .

FROM *The Descent of Inanna*

From the Great Above she opened her ear to the Great Below.
From the Great Above the goddess opened her ear to the Great Below.
From the Great Above Inanna opened her ear to the Great Below.

My Lady abandoned heaven and earth to descend to the underworld.
Inanna abandoned heaven and earth to descend to the underworld.
She abandoned her office of holy priestess to descend to the underworld.

*

Inanna set out for the underworld.
Ninshubur, her faithful servant, went with her.
Inanna spoke to her, saying:

"Ninshubur, my constant support,
My *sukkal* who gives me wise advice,
My warrior who fights by my side,
I am descending to the *kur*, to the underworld.
If I do not return,
Set up a lament for me by the ruins.
Beat the drum for me in the assembly places.
Circle the houses of the gods.
Tear at your eyes, at your mouth, at your thighs.
Dress yourself in a single garment like a beggar.
Go to Nippur, to the temple of Enlil.
When you enter his holy shrine, cry out:
'O Father Enlil, do not let your daughter
Be put to death in the underworld.
Do not let your bright silver
Be covered with the dust of the underworld.
Do not let your precious lapis
Be broken into stone for the stoneworker.
Do not let your fragrant boxwood
Be cut into wood for the woodworker.
Do not let the holy priestess of heaven
Be put to death in the underworld.'
If Enlil will not help you,
Go to Ur, to the temple of Nanna.
Weep before Father Nanna.
If Nanna will not help you,
Go to Eridu, to the temple of Enki.
Weep before Father Enki.
Father Enki, the God of Wisdom, knows the food of life,
He knows the water of life;
He knows the secrets.
Surely he will not let me die."

Inanna continued on her way to the underworld.

*

Naked and bowed low, Inanna entered the throne room.
Ereshkigal rose from her throne.
Inanna started toward the throne.
The Annuna, the judges of the underworld, surrounded her.
They passed judgment against her.

Then Ereshkigal fastened on Inanna the eye of death.
She spoke against her the word of wrath.
She uttered against her the cry of guilt.

She struck her.

Inanna was turned into a corpse.
A piece of rotting meat,
And was hung from a hook on the wall.

When, after three days and three nights, Inanna had not returned,
Ninshubur set up a lament for her by the ruins.
She beat the drum for her in the assembly places.
She circled the houses of the gods.
She tore at her eyes; she tore at her mouth; she tore at her thighs.
She dressed herself in a single garment like a beggar.
Alone, she set out for Nippur and the temple of Enlil.

*

Father Enlil would not help.

Ninshubur went to Ur and the temple of Nanna.

*

Father Nanna would not help.

Ninshubur went to Eridu and the temple of Enki.
When she entered the holy shrine,
She cried out:
"O Father Enki, do not let your daughter
Be put to death in the underworld.
Do not let your bright silver
Be covered with the dust of the underworld." . . .

Father Enki said:
"What has happened?
What has my daughter done?
Inanna! Queen of All the Lands! Holy Priestess of Heaven!
What has happened?
I am troubled. I am grieved."

From under his fingernail Father Enki brought forth dirt.
He fashioned the dirt into a *kurgarra*, a creature neither male nor female.
From under the fingernail of his other hand he brought forth dirt.
He fashioned the dirt into a *galatur*, a creature neither male nor female.
He gave the food of life to the *kurgarra*.
He gave the water of life to the *galatur*.
Enki spoke to the *kurgarra* and *galatur*, saying:
"Go to the underworld,
Enter the door like flies.
Ereshkigal, the Queen of the Underworld, is moaning
With the cries of a woman about to give birth. . . .
The queen will be pleased.
She will offer you a gift.
Ask her only for the corpse that hangs from the hook on the wall.
One of you will sprinkle the food of life on it.
The other will sprinkle the water of life.
Inanna will arise."

*

The corpse was given to them.

The *kurgarra* sprinkled the food of life on the corpse.
The *galatur* sprinkled the water of life on the corpse.
Inanna arose. . . .

Inanna was about to ascend from the underworld
When the Anunna, the judges of the underworld, seized her.
They said:
"No one ascends from the underworld unmarked.
If Inanna wishes to return from the underworld,
She must provide someone in her place."

As Inanna ascended from the underworld,
The *galla*, the demons of the underworld, clung to her side.
The *galla* were demons who know no food, who know no drink,
Who eat no offerings, who drink no libations,
Who accept no gifts.
They enjoy no lovemaking.
They have no sweet children to kiss.
They tear the wife from the husband's arms,
They tear the child from the father's knees,
They steal the bride from her marriage home.

*

The *galla* said:
"Walk on, Inanna,
We will take Ninshubur in your place."

Inanna cried:
"No! Ninshubur is my constant support. . . .
Because of her, my life was saved.
I will never give Ninshubur to you."

*

The *galla* said:
"Walk on to your city, Inanna,
We will take Shara in your place."

Inanna cried:
"No! Not Shara!
He is my son who sings hymns to me.
He is my son who cuts my nails and smooths my hair.
I will never give Shara to you."

*

The *galla* said:
"Walk on to your city, Inanna,
We will take Lulal in your place."

Inanna cried:

"Not Lulal! He is my son.
He is a leader among men.
He is my right arm. He is my left arm.
I will never give Lulal to you."

The *galla* said:

"Walk on to your city, Inanna.
We will go with you to the big apple tree in Uruk."

In Uruk, by the big apple tree,
Dumuzi, the husband of Inanna, was dressed in his shining *me*-garments.
He sat on his magnificent throne; (he did not move).

The *galla* seized him by his thighs.
They poured milk out of his seven churns.
They broke the reed pipe which the shepherd was playing.

Inanna fastened on Dumuzi with the eye of death.
She spoke against him the word of wrath.
She uttered against him the cry of guilt:

"*Take him! Take Dumuzi away!*"

The *galla*, who know no food, who know no drink,
Who eat no offerings, who drink no libations,
Who accept no gifts, seized Dumuzi.
They made him stand up; they made him sit down.
They beat the husband of Inanna.
They gashed him with axes.

Dumuzi let out a wail.
He raised his hands to heaven to Utu, the God of Justice, and beseeched him:

". . . Utu, you who are a just god, a merciful god,
Change my hands into the hands of a snake.
Change my feet into the feet of a snake.
Let me escape from my demons;
Do not let them hold me."

The merciful Utu accepted Dumuzi's tears. . . .
Dumuzi escaped from his demons.
They could not hold him. . . .

*

The *galla* clapped their hands gleefully.
They went searching for Dumuzi.
They came to the home of Geshtinanna. They cried out:

"Show us where your brother is!"

Geshtinanna would not speak.

They offered her the water-gift.
She refused it.
They offered her the grain-gift.
She refused it.

Heaven was brought close.
Earth was brought close.
Geshtinanna would not speak.

They tore her clothes.
They poured pitch into her vulva.
Geshtinanna would not speak.

The small *galla* said to the large *galla:*
"Who since the beginning of time
Has ever known a sister to reveal a brother's hiding place?
Come, let us look for Dumuzi in the home of his friend."

The *galla* went to Dumuzi's friend.
They offered him the water-gift.
He accepted it. . . .
They offered him the grain-gift.
He accepted it. . . .

*

The *galla* seized Dumuzi.
They surrounded him.
They bound his hands. They bound his neck.

The churn was silent. No milk was poured.
The cup was shattered. Dumuzi was no more.
The sheepfold was given to the winds.

A lament was raised in the city:
"My Lady weeps bitterly for her young husband.
Inanna weeps bitterly for her young husband.
Woe for her husband! Woe for her young love!
Woe for her house! Woe for her city! . . .

Great is the grief of those who mourn for Dumuzi."

Inanna wept for Dumuzi:
"Gone is my husband, my sweet husband.
Gone is my love, my sweet love.
My beloved has been taken from the city. . . .

The wild bull lives no more.
The shepherd, the wild bull lives no more.
Dumuzi, the wild bull, lives no more.

I ask the hills and valleys:
'Where is my husband?'
I say to them:
'I can no longer bring him food.
I can no longer serve him drink.'

The jackal lies down in his bed.
The raven dwells in his sheepfold.
You ask me about his reed pipe?

The wind must play it for him.
You ask me about his sweet songs?
The wind must sing them for him."

Sirtur, the mother of Dumuzi, wept for her son:
"My heart plays the reed pipe of mourning.
Once my boy wandered so freely on the steppe,
Now he is captive.
Once Dumuzi wandered so freely on the steppe,
Now he is bound. . . ."

The mother walked to the desolate place.
Sirtur walked to where Dumuzi lay.
She looked at the slain wild bull.
She looked into his face. She said:
"My child, the face is yours.
The spirit has fled."

There is mourning in the house.
There is grief in the inner chambers.

The sister wandered about the city, weeping for her brother.
Geshtinanna wandered about the city, weeping for Dumuzi:
"O my brother! Who is your sister?
I am your sister.
O Dumuzi! Who is your mother?
I am your mother.
The day that dawns for you will also dawn for me.
The day that you will see I also will see.

I would find my brother! I would comfort him!
I would share his fate!"

*

Inanna and Geshtinanna went to the edges of the steppe.
They found Dumuzi weeping.
Inanna took Dumuzi by the hand and said:
"You will go to the underworld
Half the year.
Your sister, since she has asked,
Will go the other half.
On the day you are called,
That day you will be taken.
On the day Geshtinanna is called,
That day you will be set free."
Inanna placed Dumuzi in the hands of the eternal.

Holy Ereshkigal! Great is your renown!
Holy Ereshkigal! I sing your praises!

TWO HYMNS TO INANNA

FROM *Loud Thundering Storm*

Proud Queen of the Earth Gods, Supreme Among the Heaven Gods,
Loud Thundering Storm, you pour your rain over all the lands and all the people.
You make the heavens tremble and the earth quake.
Great Priestess, who can soothe your troubled heart?

You flash like lightning over the highlands; you throw your firebrands across the earth.
Your deafening command, whistling like the South Wind, splits apart great mountains.
You trample the disobedient like a wild bull; heaven and earth tremble.
Holy Priestess, who can soothe your troubled heart?

Your frightful cry descending from the heavens devours its victims.
Your quivering hand causes the midday heat to hover over the sea.
Your nighttime stalking of the heavens chills the land with its dark breeze.
Holy Inanna, the riverbanks overflow with the flood-waves of your heart. . . .

The Lady of the Evening

At the end of the day, the Radiant Star, the Great Light, that fills the sky,
The Lady of the Evening appears in the heavens.
The people in all the lands lift their eyes to her.
The men purify themselves; the women cleanse themselves.
The ox in his yoke lows to her.
The sheep stir up the dust in their fold.
All the living creatures of the steppe,
The four-footed creatures of the high steppe,
The lush gardens and orchards, the green reeds and trees,
The fish of the deep and the birds in the heavens—
My Lady makes them all hurry to their sleeping places.

The living creatures and the numerous people of Sumer kneel before her.
Those chosen by the old women prepare great platters of food and drink for her.
The Lady refreshes herself in the land.
There is great joy in Sumer.
The young man makes love with his beloved.

My Lady looks in sweet wonder from heaven.
The people of Sumer parade before the holy Inanna.
Inanna, the Lady of the Evening, is radiant.
I sing your praises, holy Inanna.
The Lady of the Evening is radiant on the horizon.

Nature and Culture in Conflict

Gilgamesh,
translated by John Gardner and John Maier

The great Mesopotamian hero Gilgamesh (originally Bilgamesh) may well have been a historical person. In the inscriptions of the semi-legendary Sumerian "king list" he is named as having ruled Uruk (biblical Erech) as its fifth king for 126 years, probably sometime after 2700 B.C. He was famed as a great builder, and stories no doubt soon grew up around his name; the five surviving fragmentary Sumerian poems recounting his exploits (three of which Kramer translates in The Sumerians) *date from around a millennium later. Several of these—along with tales of two of Gilgamesh's predecessors, Enmerkar, the second king of Uruk, and Lugulbanda, the third; and along with an independent Sumerian poem about the flood—contributed, in some cases nearly word for word, to the Akkadian epic of Gilgamesh that grew up over the next thousand years or so. The earliest Akkadian fragments from the Old Babylonian period (ca. 2000–1600) are roughly contemporary with the extant Sumerian poems; by the Middle Babylonian period (ca. 1600–1000) versions survive not only from Mesopotamia but also from as far away as Megiddo in Canaan and the Hittite capital of Hattusha in Asia Minor, where translations into Hittite and Hurrian were also unearthed.*

From the first millennium a number of more or less identical fragmentary copies of the "standard version," probably composed by one Sîn-leqi-unninnī in the late second millennium, have been found. The most complete, from the great library of the Assyrian king Ashurbanipal in Nineveh, destroyed in 612, was the first to be discovered and deciphered, 2,500 years later, in the mid nineteenth century A.D. (For a detailed study, see Jeffrey H. Tigay, The Evolution of the Gilgamesh Epic *[1982].) The standard version consists of twelve tablets, some of which, despite the existence of multiple copies, remain extremely fragmentary, even when collated with earlier versions; the twelfth tablet gives the appearance of being a separate and wholly unintegrated story, translating the Sumerian "Gilgamesh, Enkidu, and the Nether World." Numerous English versions have been made—some literal and scholarly, as by Maureen Gallery Kovacs (1989); some retold in readable prose, as by N. K. Sandars (1960; rev. ed. 1972); and some poetic, as by David Ferry (1992).*

The version from which the following selections come was completed by the novelist John Gardner (in collaboration with John Maier) shortly before his death in 1982 and published in 1984; it is a fairly close rendition of the Sîn-leqi-unninnī text. The first selection, from Tablet I, tells how the sky god Anu, alarmed at Gilgamesh's initial abuse of power as king, has the goddess Aruru (one manifestation of the Sumerian mother goddess Ninhursag) create Enkidu as a rival to Gilgamesh. Enkidu lives as a wild man, ignorant of civilization and one with the animals of forest and field, until a terrified stalker or trapper tames him by bringing a prostitute to him: she drains his strength (as Delilah does Samson's) but also (like Eve's apple) brings him knowledge or wisdom. In subsequent episodes, Enkidu, cast out from primal union with the animal world, goes to the city where Gilgamesh defeats him in single combat, after which they are bound together in deep love and friendship. Gilgamesh now longs to establish himself as a great king, and sets out with Enkidu on a long journey to the forbidding cedar forest, where he vanquishes and (at Enkidu's urging) slays its guardian Humbaba, bringing his curse down upon them; unfortunately, this episode is among the most fragmentary of the poem.

In our second selection, from Tablet VI, Gilgamesh scornfully rejects the advances of Ishtar, the very goddess of love and fertility celebrated by the Sumerians as Inanna. In doing so he deliberately affirms (as Enkidu had involuntarily done) that as a human *being he can no longer be one with nature or with gods who amorally embody nature's forces, for this would be his destruction; by slaying the Bull of Heaven sent by the vengeful Ishtar, he and Enkidu defiantly declare independence from submission to such gods. In consequence, in the remainder of the poem, Enkidu must die, and the grieving Gilgamesh must learn, through his futile journey in search of immortality, that (except for Utnapishtim and his wife, sole survivors of the flood) death is the inescapable lot of all. Only with this knowledge will he fulfill his destiny as a great and wise king, and thus as a human being who is the product of history and culture as much as of nature.*

FROM *Tablet I: The Taming of Enkidu*

Gilgamesh does not allow the son to go with his father;
day and night he oppresses the weak—
Gilgamesh, who is shepherd of Uruk of the Sheepfold.
Is this our shepherd, strong, shining, full of thought?
Gilgamesh does not let the young woman go to her mother,
the girl to the warrior, the bride to the young groom.

The gods heard their lamentation;
the gods of the above [addressed] the keeper of Uruk:[1]

"Did you not make this mighty wild bull?
The raising of his weapon has no equal;
with the drum his citizens are raised.
He, Gilgamesh, keeps the son from his father day and night.
Is this the shepherd of Uruk of the Sheepfold . . .
strong, shining, full of thought? . . .

When [Anu the sky god] heard their lamentation
he called to Aruru the Mother, Great Lady: "You, Aruru, who created humanity,
create now a second image of Gilgamesh: may the image be equal to the time of his heart.
Let them square off one against the other, that Uruk may have peace."

When Aruru heard this, she formed an image of Anu in her heart.
Aruru washed her hands, pinched off clay and threw it into the wilderness:
In the wilderness she made Enkidu the fighter; she gave birth in darkness and silence to one like the war god Ninurta.
His whole body was covered thickly with hair, his head covered with hair like a woman's;
the locks of his hair grew abundantly, like those of the grain god Nisaba.
He knew neither people nor homeland; he was clothed in the clothing of Sumuqan the cattle god.
He fed with the gazelles on grass;
with the wild animals he drank at waterholes;
with hurrying animals his heart grew light in the waters.

[1]Anu, father of the gods and chief god of Uruk.

The Stalker, man-and-hunter,[2]
met him at the watering place
one day—a second, a third—at the watering place.
Seeing him, the Stalker's face went still.
He, Enkidu, and his beasts had intruded on the Stalker's place.
Worried, troubled, quiet,
the Stalker's heart rushed; his face grew dark.
Woe entered his heart.
His face was like that of one who travels a long road.

The Stalker shaped his mouth and spoke, saying to his father

"Father, there is a man who has come from the hills.
In all the land he is the most powerful; power belongs to him.
Like a shooting star of the god Anu, he has awesome strength.
He ranges endlessly over the hills,
endlessly feeds on grass with the animals,
endlessly sets his feet in the direction of the watering place.
For terror I cannot go near him.
He fills up the pits I dig;
he tears out the traps I set;
he allows the beasts to slip through my hands, the hurrying creatures of the abandon;
in the wilderness he does not let me work."
His father shaped his mouth and spoke, saying to the Stalker

"My son, in Uruk lives a man, Gilgamesh:
no one has greater strength than his.
In all the land he is the most powerful; power belongs to him. Like a shooting star of
Anu, he has awesome strength.
Go, set your face toward Uruk.
Let him, the knowing one, hear of it.
He will say, 'Go, Stalker, and take with you a love-priestess, a temple courtesan,
[let her conquer him with] power [equal to his own].
When he waters the animals at the watering place,
have her take off her clothes, let her show him her strong beauty.
When he sees her, he will come near her.
His animals, who grew up in the wilderness, will turn from him.'"

He listened to the counsel of his father.
The Stalker went to Gilgamesh. . . .

Gilgamesh said to him, to the Stalker,

"Go, Stalker, and take with you a love-priestess, a temple courtesan.
When he waters the animals at the watering place,
have her take off her clothes, have her show him her strong beauty.
When he sees her, he will come near her.

[2]The Stalker . . . is not given a proper name. He stands as a mediating figure between the wilderness and the city. (Gardner and Maier) Others translate this word as "trapper."

His animals, who grew up in his wilderness, will turn from him."
The Stalker went, taking with him a love-priestess, a temple courtesan.
They mounted the road, went on their journey.
On the third day, in the wilderness, the set time arrived.
The Stalker and the woman sat in their places and waited.
One day, a second day, they sat at the watering place.
Then the wild animals came to the watering place to drink.

The animals came; their hearts grew light in the waters.
And as for him, Enkidu, child of the mountain,
he who fed with gazelles on grass,
he drank with the wild beasts at the watering place
and with the hurrying animals his heart grew light in the waters.

The woman saw him, the man-as-he-was-in-the-beginning,
the man-and-killer from the deep wilderness.

"Here he is, courtesan; get ready to embrace him.
Open your legs, show him your beauty.
Do not hold back, take his wind away.
Seeing you, he will come near.
Strip off your clothes so he can mount you.
Make him know, this man-as-he-was, what a woman is.
His beasts who grew up in his wilderness will turn from him.
He will press his body over your wildness."

The courtesan untied her wide belt and spread her legs, and he struck her wildness like
a storm.
She was not shy; she took his wind away.
Her clothing she spread out, and he lay upon her.
She made him know, the man-as-he-was, what a woman is.
His body lay on her;
six days and seven nights Enkidu attacked, fucking the priestess.

After Enkidu was glutted on her richness
he set his face toward his animals.
Seeing him, Enkidu, the gazelles scattered, wheeling:
the beasts of the wildernesss fled from his body.
Enkidu tried to rise up, but his body pulled back.
His knees froze. His animals had turned from him.
Enkidu grew weak; he could not gallop as before.
Yet he had knowledge, wider mind.

Turned around, Enkidu knelt at the knees of the prostitute.
He looked up at her face,
and as the woman spoke, his ears heard.

The woman said to him, to Enkidu:
"You have become wise, like a god, Enkidu.
Why did you range the wilderness with animals?
Come, let me lead you to the heart of Uruk of the Sheepfold,

to the stainless house, holy place of Anu and Ishtar,
where Gilgamesh lives, completely powerful,
and like a wild bull stands supreme, mounted above his people."

She speaks to him, and they look at one another.
With his heart's knowledge, he longs for a deeply loving friend. . . .

FROM *Tablet VI: The Rejection of Ishtar*

To Gilgamesh's beauty great Ishtar lifted her eyes.
"Come, Gilgamesh, be my lover!
Give me the taste of your body.
Would that you were my husband, and I your wife!
I'd order harnessed for you a chariot of lapis lazuli and gold,
its wheels of gold and its horns of precious amber.
You will drive storm demons—powerful mules!
Enter our house, into the sweet scent of cedarwood.
As you enter our house
the purification priests will kiss your feet the way they do in Aratta.
Kings, rulers, princes will bend down before you.
Mountains and lands will bring their yield to you.
Your goats will drop triplets, your ewes twins.
Even loaded down, your donkey will overtake the mule.
Your horses will win fame for their running.
Your ox under its yoke will have no rival."

Gilgamesh shaped his mouth to speak,
saying to great Ishtar:
"What could I give you if I should take you as a wife?
Would I give you oil for the body, and fine wrappings?
Would I give you bread and victuals?—
you who eat food of the gods,
you who drink wine fit for royalty?
[For you] they pour out [libations];
[you are clothed with the Great] Garment.
[Ah,] the gap [between us], if I take you in marriage!

You're a cooking fire that goes out in the cold,
a back door that keeps out neither wind nor storm,
a palace that crushes the brave ones defending it,
a well whose lid collapses,
pitch that defiles the one carrying it,
a waterskin that soaks the one who lifts it,
limestone that crumbles in the stone wall,
a battering ram that shatters in the land of the enemy,
a shoe that bites the owner's foot!

Which of your lovers have you loved forever?
Which of your little shepherds has continued to please you?
Come, let me name your lovers for you.

. . . for Tammuz, the lover of your youth.
Year after year you set up a wailing for him.
You loved the mauve-colored shepherd bird:
you seized him and broke his wing.
In the forest he stands crying, 'Kappi! My wing!'
You loved the lion, full of spry power;
you dug for him seven pits and seven pits.
You loved the stallion glorious in battle:
you ordained for him the whip, the goad, the halter. . . .
You loved a shepherd, a herdsman,
who endlessly put up cakes for you
and every day slaughtered kids for you.
You struck him, turned him into a wolf.
His own boys drove him away,
and his dogs tore his hide to bits.

You loved also Ishullanu, your father's gardener,
who endlessly brought you baskets of dates
and every day made the table jubilant.
You lifted your eyes to him and went to him:
'My Ishullanu, let us take pleasure in your strength.
Reach out your hand and touch my vulva!'
Ishullanu said to you,
'What do you want from me? . . .
Should I eat the bread of bad faith, the food of curses? . . .'
You heard his answer.
You struck him, turned him into a frog.
You set him to dwell in the middle of the garden,
where he can move neither upward nor downward.

So you'd love me in my turn and, as with them, set my fate."

When Ishtar heard this
Ishtar was furious and flew up to the heavens
and went before Anu the father.
Before Antum, her mother, she wept.
"Father, Gilgamesh has insulted me. . . .
Father, make the Bull of Heaven. Let him kill Gilgamesh in the very place where he lives;
let the bull glut himself on Gilgamesh.
If you do not give me the bull,
I will smash in the gates of the netherworld;
I will set up the [ruler] of the great below,
and I will make the dead rise, and they will devour the living,
and the dead will increase beyond the number of the living." . . .

Enkidu leaped up, seized the Bull of Heaven, took hold of his horns. . . .

And Gilgamesh, like a matador . . .
struck [with his sword] in the neck [behind the horns].

After they had killed the Bull they tore out his heart.
They set it up before Shamash [the sun god].
They withdrew and worshipped Shamash.
They sat down, blood-brothers, the two of them.

Ishtar went up on the walls of Uruk of the Sheepfold.
Disguised as a mourner she let loose a curse:
"Curse Gilgamesh, who has besmeared me, killing the Bull of Heaven!"

When Enkidu heard this, the words of Ishtar,
he tore out the thigh of the bull and threw it in her face.
"If I could reach you, as I can him,
it would have been done to you:
I'd hang his guts around your arm!"
Ishtar called together the hair-curled high priestesses,
the love-priestesses and temple whores,
and over the thigh of the Bull of Heaven she set up a wailing.

Gilgamesh called together [the city's] experts, the craftsmen,
all of them. . . .
He brought [the horns] in and hung them in the shrine of his ancestors.

God and Man in a World of Paradise Lost

The Hebrew Bible, Revised Standard Version

The canon of the Hebrew Bible—essentially identical, except for important differences in the arrangement of its books, with the Christian Old Testament—took its present form in the decades following the first Jewish war against Rome and the destruction of the Third Temple (the Temple of Herod) by Titus in A.D. *70, but the books that compose it had been written from at least three hundred to more than a thousand years before. Among the oldest, and the most sacred for Jews, are the first five, called the Pentateuch in Greek or the Torah (the "Law," broadly understood) in Hebrew, traditionally ascribed to Moses himself.*

Whatever its ultimate oral or written origins, most scholars now believe that the Pentateuch (and some other parts of the Hebrew Bible) combines several sources written at different places and times. As long ago as 1753, Jean Astruc, physician to King Louis XV of France, noted that parts of the Pentateuch differ from others both in style and in the word by which they refer to God. In some passages he is called elōhim, *"God," from a common Semitic root; in others, by the personal name Yahweh (or Jahweh)—the name he reveals to Moses in Exodus 3:15 and 6:2–3—which the King James and Revised Standard versions usually translate as "the* LORD.*" (Through misunderstanding of the ineffable Name, written, like all classical Hebrew words, only with consonants—the four sacred letters, or "Tetragrammaton," YHWH or JHWH—the hybrid English form "Jehovah" arose.) Astruc's insight was developed, by Julius Wellhausen and others, into the "documentary hypothesis" of nineteenth- and twentieth-century "higher criticism," which distinguishes four principal sources of the Pentateuch.*

The oldest of these, the Yahwist (labeled J*), was perhaps written down, drawing on long oral tradition, in Judah as long ago as the tenth century* B.C.*, before or after breakup of the unified* kingdom; it is characterized by its dramatic narrative and "earth-centered" portrayal of the interactions between a humanely accessible God and his sharply individualized people. A second source, the Elohist (E), which presents God as more remote and indirect in his interventions in the world, is thought to have been written in the northern kingdom around the ninth to eighth century and to have been combined with the predominant J source in the southern kingdom after the fall of Samaria in 721. A third source, D, consists mainly of the book of Deuteronomy ("the second law"), probably written in the seventh century and generally identified with the "book of the law" (torah) found in 622 by the high priest Hilkiah in the house of the LORD and presented to the reformist King Josiah (2 Kings 22–23). Finally, redactors of the composite, sometimes drily legalistic "Priestly" source, P, appear to have collated these three sources with other materials, some of them very old, in the period during or after the exile, producing the Pentateuch more or less as we know it by about 400 B.C.

This hypothesis, disputable though details (including the date and cohesion of P) remain, is accepted by most Jewish and Christian biblical scholars, but risks losing sight, through excessive attention to parts, of the grandeur of the whole. For the Bible is of course not merely a compilation of sources but a Book whose richly complex unity, achieved in large part by extraordinary openness to incorporation of unresolved contradictions (see chapter one of Erich Auerbach's *Mimesis* [1946]), is essential to it. In our first selection, the famous opening chapters of Genesis, the account of P (through verse three and a half of chapter 2) undoubtedly differs from that of J in the remainder both in significant details such as the order of creation and in its primary emphasis on "the heavens and the earth" as opposed to J's on "the earth and the heavens." But what readers have surely found most impressive throughout the centuries is not the trace of seams but the magnificence of their fusion as the stately process of day-by-day creation from God's lofty vantage point passes over into the humanly centered story that follows.

Very general parallels with the biblical account of creation have been found in scattered traces of a Sumerian cosmogony, discussed by Kramer in *History Begins at Sumer* (1956), in which Enlil divides his parents, Heaven and Earth (An and Ki), to begin the formation of the universe. More distantly, Shu, in an Egyptian myth paralleled in many parts of the world, separates his daughter Nut, the sky, from her sexual union with Geb, the earth. Much more striking are the specific correspondences, especially in the order of events (as Alexander Heidel demonstrates in *The Babylonian Genesis* [1942; 2nd ed. 1951]), between Genesis 1 and the Babylonian creation epic of perhaps a thousand years before, known from its opening words as *enuma elish* ("when on high"), which was recited each year at the new year's festival of ancient Babylon. But anyone reading this epic—in which, after the wise god Ea has slain Apsu, the primordial sweet-water ocean, Ea's son Marduk, city god of Babylon, attains supremacy among the gods by joining in mighty combat with Apsu's ferocious consort Tiamat, the salt-water ocean, whom he kills with an arrow through her mouth and heart, creating the universe by dividing her colossal body into sky and earth—will be far more impressed by the immense contrasts between this violent saga and the majestic account of Genesis 1. In it, an omnipotent God, acting with purposeful deliberation and determining at every stage that his work is good, creates from nothing or from the tohu-bohu of primeval chaos the orderly natural world of heaven and earth, land and sea, plants and trees, sun and moon, fish and birds, animals of all kinds, and man and woman to have dominion over them.

Nature in this universe has a plan, laid down by an all-seeing God; "dominion" by human beings made in His image, far from justifying destructive exploitation of the earth (as some contend), must clearly be exercised, by anyone who knows good from evil, in accord with God's plan. Yet in Eden, that paradise of presumably blissful oneness with all God's creatures, Adam and Eve exercise little judgment of any kind until tasting of the tree of the knowledge of good and *evil which paradoxically, as the serpent foretold and the* LORD *confirms, makes them—except that now they must die—* more *like God. Like Enkidu in the Gilgamesh epic, they become godlike and wise at the price of death, or perhaps of the knowledge of death. This is the paradox of the "fortunate fall,"* or felix culpa: *only when they make distinctions, between good and evil and between their clothed (therefore cultural) human condition and the naked state of the animals from which (again like Enkidu) they must now be severed, can they become fully human—that is, simultaneously "natural" and "cultural," animal and divine. To decide their own destiny in a world to which they no longer fully belong will hereafter be the curse, and the blessing, that paradise lost has bestowed upon them.*

The Creation of the World and the Loss of Paradise: Genesis

1 In the beginning God created the heavens and the earth. 2 The earth was without form and void, and darkness was upon the face of the deep; and the Spirit of God was moving over the face of the waters.

3 And God said, "Let there be light"; and there was light. 4 And God saw that the light was good; and God separated the light from the darkness. 5 God called the light Day, and the darkness he called Night. And there was evening and there was morning, one day.

6 And God said, "Let there be a firmament in the midst of the waters, and let it separate the waters from the waters." 7 And God made the firmament and separated the waters which were under the firmament from the waters which were above the firmament. And it was so. 8 And God called the firmament Heaven. And there was evening and there was morning, a second day.

9 And God said, "Let the waters under the heavens be gathered together into one place, and let the dry land appear." And it was so. 10 God called the dry land Earth, and the waters that were gathered together he called Seas. And God saw that it was good. 11 And God said, "Let the earth put forth vegetation, plants yielding seed, and fruit trees bearing fruit in which is their seed, each according to its kind, upon the earth." And it was so. 12 The earth brought forth vegetation, plants yielding seed according to their own kinds, and trees bearing fruit in which is their seed, each according to its kind. And God saw that it was good. 13 And there was evening and there was morning, a third day.

14 And God said, "Let there be lights in the firmament of the heavens to separate the day from the night; and let them be for signs and for seasons and for days and years, 15 and let them be lights in the firmament of the heavens to give light upon the earth." And it was so. 16 And God made the two great lights, the greater light to rule the day, and the lesser light to rule the night; he made the stars also. 17 And God set them in the firmament of the heavens to give light upon the earth, 18 to rule over the day and over the night, and to separate the light from the darkness. And God saw that it was good. 19 And there was evening and there was morning, a fourth day.

20 And God said, "Let the waters bring forth swarms of living creatures, and let birds fly above the earth across the firmament of the heavens." 21 So God created the great sea monsters and every living creature that moves, with which the waters swarm, according to their kinds, and every winged bird according to its kind. And God saw that it was good. 22 And God blessed them, saying, "Be fruitful and multiply and fill the waters in the seas, and let birds multiply on the earth." 23 And there was evening and there was morning, a fifth day.

24 And God said, "Let the earth bring forth living creatures according to their kinds: cattle and creeping things and beasts of the earth according to their kinds." And it was so. 25 And God made the beasts of the earth according to their kinds and the cattle according to their kinds, and everything that creeps upon the ground according to its kind. And God saw that it was good.

26 Then God said, "Let us make man in our image, after our likeness; and let them have dominion over the fish of the sea, and over the birds of the air, and over the cattle, and over all the earth, and over every creeping thing that creeps upon the earth." 27 So God created man in his own image, in the image of God he created him; male and female he created them. 28 And God blessed them, and God said to them, "Be fruitful and multiply, and fill the earth and subdue it; and have dominion over the fish of the sea and over the birds of the air and over every living thing that moves upon the earth." 29 And God said, "Behold, I have given you every plant yielding seed which is upon the face of all the earth, and every tree with seed in its fruit; you shall have them for food. 30 And to every beast of the earth, and to every bird of the air, and to everything that creeps on the earth, everything that has the breath of life, I have given every green plant for food." And it was so. 31 And God saw everything that he had made, and behold, it was very good. And there was evening and there was morning, a sixth day.

2 Thus the heavens and the earth were finished, and all the host of them. 2 And on the seventh day God finished his work which he had done, and he rested on the seventh day from all his work which he had done. 3 So God blessed the seventh day and hallowed it, because on it God rested from all his work which he had done in creation.

4 These are the generations of the heavens and the earth when they were created.

In the day that the LORD God made the earth and the heavens, 5 when no plant of the field was yet in the earth and no herb of the field had yet sprung up—for the LORD God had not caused it to rain upon the earth, and there was no man to till the ground; 6 but a mist went up from the earth and watered the whole face of the ground— 7 then the LORD God formed man of dust from the ground, and breathed into his nostrils the breath of life; and man became a living being. 8 And the LORD God planted a garden in Eden, in the east; and there he put the man whom he had formed. 9 And out of the ground the LORD God made to grow every tree that is pleasant to the sight and good for food, the tree of life also in the midst of the garden, and the tree of the knowledge of good and evil.

10 A river flowed out of Eden to water the garden, and there it divided and became four rivers. 11 The name of the first is Pishon; it is the one which flows around the whole land of Havilah, where there is gold; 12 and the gold of that land is good; bdellium and onyx stone are there. 13 The name of the second river is Gihon; it is the one which flows around the whole land of Cush. 14 And the name of the third river is Tigris, which flows east of Assyria. And the fourth river is the Euphrates.

15 The LORD God took the man and put him in the garden of Eden to till it and keep it. 16 And the LORD God commanded the man, saying, "You may freely eat of every tree of the garden; 17 but of the tree of the knowledge of good and evil you shall not eat, for in the day that you eat of it you shall die."

18 Then the LORD God said, "It is not good that the man should be alone; I will make him a helper fit for him." 19 So out of the ground the LORD God formed every beast of the field and every bird of the air, and brought them to the man to see what he

would call them; and whatever the man called every living creature, that was its name. 20 The man gave names to all cattle, and to the birds of the air, and to every beast of the field; but for the man there was not found a helper fit for him. 21 So the LORD God caused a deep sleep to fall upon the man, and while he slept took one of his ribs and closed up its place with flesh; 22 and the rib which the LORD God had taken from the man he made into a woman and brought her to the man. 23 Then the man said,

"This at last is bone of my bones
  and flesh of my flesh;
she shall be called Woman,
  because she was taken out of Man."[3]

24 Therefore a man leaves his father and his mother and cleaves to his wife, and they become one flesh. 25 And the man and his wife were both naked, and were not ashamed.

3 Now the serpent was more subtle than any other wild creature that the LORD God had made. He said to the woman, "Did God say, 'You shall not eat of any tree of the garden'?" 2 And the woman said to the serpent, "We may eat of the fruit of the trees of the garden; 3 but God said, 'You shall not eat of the fruit of the tree which is in the midst of the garden, neither shall you touch it, lest you die.'" 4 But the serpent said to the woman, "You will not die. 5 For God knows that when you eat of it your eyes will be opened, and you will be like God, knowing good and evil." 6 So when the woman saw that the tree was good for food, and that it was a delight to the eyes, and that the tree was to be desired to make one wise, she took of its fruit and ate; and she also gave some to her husband, and he ate. 7 Then the eyes of both were opened, and they knew that they were naked; and they sewed fig leaves together and made themselves aprons.

8 And they heard the sound of the LORD God walking in the garden in the cool of the day, and the man and his wife hid themselves from the presence of the LORD God among the trees of the garden. 9 But the LORD God called to the man, and said to him, "Where are you?" 10 And he said, "I heard the sound of thee in the garden, and I was afraid, because I was naked; and I hid myself." 11 He said, "Who told you that you were naked? Have you eaten of the tree of which I commanded you not to eat?" 12 The man said, "The woman whom thou gavest to be with me, she gave me fruit of the tree, and I ate." 13 Then the LORD God said to the woman, "What is this that you have done?" The woman said, "The serpent beguiled me, and I ate." 14 The LORD God said to the serpent,

"Because you have done this,
  cursed are you above all cattle,
  and above all wild animals;
upon your belly you shall go,
  and dust you shall eat
  all the days of your life.
15 I will put enmity between you and the woman,
  and between your seed and her seed;
he shall bruise your head,
  and you shall bruise his heel."

[3]"Woman" is Hebrew *ishshah*, "Man" Hebrew *ish*.

16 To the woman he said,

> "I will greatly multiply your pain in childbearing;
> in pain you shall bring forth children,
> yet your desire shall be for your husband,
> and he shall rule over you."

17 And to Adam he said,

> "Because you have listened to the voice of your wife,
> and have eaten of the tree
> of which I commanded you,
> 'You shall not eat of it,'
> cursed is the ground because of you;
> in toil you shall eat of it all the days of your life;
> 18 thorns and thistles it shall bring forth to you;
> and you shall eat the plants of the field.
> 19 In the sweat of your face
> you shall eat bread
> till you return to the ground,
> for out of it you were taken;
> you are dust,
> and to dust you shall return."

20 The man called his wife's name Eve,[4] because she was the mother of all living. 21 And the LORD God made for Adam and for his wife garments of skins, and clothed them.

22 Then the LORD God said, "Behold, the man has become like one of us, knowing good and evil; and now, lest he put forth his hand and take also of the tree of life, and eat, and live for ever"— 23 therefore the LORD God sent him forth from the garden of Eden, to till the ground from which he was taken. 24 He drove out the man; and at the east of the garden of Eden he placed the cherubim, and a flaming sword which turned every way, to guard the way to the tree of life.

PSALMS IN PRAISE OF THE LORD

The Hebrew Psalms, some of which are explicitly ascribed to King David, vary greatly in form and style, ranging from exultant celebration to mournful supplication, and appear to have been composed over many centuries, from the earliest to the latest biblical age. Some at least, as Hermann Gunkel was among the first to demonstrate, were evidently cultic hymns sung to instrumental accompaniment during religious observances at the House of the LORD *in Jerusalem and other sanctuaries. Like many other poetic passages in the Hebrew Bible, the Psalms are generally characterized, as Bishop Robert Lowth discerned as early as 1753, by "parallelism," the restatement of one line by a second (and sometimes a third) expressing a similar thought in different words and images, as in Psalm 24:2: "for he has founded it upon the seas, / and established it upon the rivers." (Texts from the ancient Canaanite city of Ugarit excavated, beginning in 1929, at Ras-Shamra in Syria show that this poetic device was not unique to the Hebrews.) The predominant theme of the Psalms is praise of the power and glory, the justice and mercy, of the almighty*

[4]The name in Hebrew resembles the word for *living*. (RSV)

LORD God of Israel, sole refuge of his suffering people, so that here if anywhere we might expect to find the "devaluation of nature" often thought to accompany such a faith. Yet in a small number of Psalms represented in our selection the opposite is the case: far from being disdained or downgraded, the natural world—from the stars above to the grass below—is rapturously celebrated as embodying and bearing witness to the glory of a transcendent God who has not therefore ceased to be immanent in the world he created and for which he continues to care. Scholars have pointed out suggestive parallels between, for example, Psalm 19 and a Babylonian sun-hymn; Psalm 29 and a Canaanite hymn from Ras-Shamra; and Psalm 104 and an Egyptian hymn to the heretical Pharaoh Akhenaten's god, the sun-disk Aten. But the inclusion of these and other fervent celebrations of the splendor of nature among the sacred Psalms proves that worship of the LORD did not always or inevitably imply depreciation of the surrounding world.

8 O LORD, our Lord,
how majestic is thy name in all the earth!
Thou whose glory above the heavens is chanted
2 by the mouth of babes and infants,
thou hast founded a bulwark because of thy foes,
to still the enemy and the avenger.

3 When I look at thy heavens, the work of thy fingers,
the moon and the stars which thou hast established;
4 what is man that thou art mindful of him,
and the son of man that thou dost care for him?

5 Yet thou hast made him little less than God,
and dost crown him with glory and honor.
6 Thou hast given him dominion over the works of thy hands;
thou hast put all things under his feet,
7 all sheep and oxen
and also the beasts of the field,
8 the birds of the air, and the fish of the sea,
whatever passes along the paths of the sea.

9 O LORD, our Lord,
how majestic is thy name in all the earth!

19 The heavens are telling the glory of God;
and the firmament proclaims his handiwork.
2 Day to day pours forth speech,
and night to night declares knowledge.
3 There is no speech, nor are there words;
their voice is not heard;
4 yet their voice goes out through all the earth,
and their words to the end of the world.

In them he has set a tent for the sun,
5 which comes forth like a bridegroom leaving his chamber,
and like a strong man runs its course with joy.

6 Its rising is from the end of the heavens,
and its circuit to the end of them;
and there is nothing hid from its heat. . . .

14 Let the words of my mouth and the meditation of my heart
be acceptable in thy sight,
O LORD, my rock and my redeemer.

24 The earth is the LORD's and the fulness thereof,
the world and those who dwell therein;
2 for he has founded it upon the seas,
and established it upon the rivers. . . .

65 Praise is due to thee,
O God, in Zion;
and to thee shall vows be performed. . . .

9 Thou visitest the earth and waterest it,
thou greatly enrichest it;
the river of God is full of water;
thou providest their grain,
for so thou hast prepared it.
10 Thou waterest its furrows abundantly,
settling its ridges,
softening it with showers,
and blessing its growth.
11 Thou crownest the year with thy bounty;
the tracks of thy chariot drip with fatness.
12 The pastures of the wilderness drip,
the hills gird themselves with joy,
13 the meadows clothe themselves with flocks,
the valleys deck themselves with grain,
they shout and sing together for joy.

104 Bless the LORD, O my soul!
O LORD my God, thou art very great!
Thou art clothed with honor and majesty,
2 who coverest thyself with light as with a garment,
who hast stretched out the heavens like a tent,
3 who hast laid the beams of thy chambers on the waters,
who makest the clouds thy chariot,
who ridest on the wings of the wind,
4 who makest the winds thy messengers,
fire and flame thy ministers.

5 Thou didst set the earth on its foundations,
so that it should never be shaken.

6 Thou didst cover it with the deep as with a garment;
the waters stood above the mountains.
7 At thy rebuke they fled;
at the sound of thy thunder they took to flight.
8 The mountains rose, the valleys sank down
to the place which thou didst appoint for them.
9 Thou didst set a bound which they should not pass,
so that they might not again cover the earth.

10 Thou makest springs gush forth in the valleys;
they flow between the hills,
11 they give drink to every beast of the field;
the wild asses quench their thirst.
12 By them the birds of the air have their habitation;
they sing among the branches.
13 From thy lofty abode thou waterest the mountains;
the earth is satisfied with the fruit of thy work.

14 Thou dost cause the grass to grow for the cattle,
and plants for man to cultivate,
that he may bring forth food from the earth,
15 and wine to gladden the heart of man,
oil to make his face shine,
and bread to strengthen man's heart.
16 The trees of the Lord are watered abundantly,
the cedars of Lebanon which he planted.
17 In them the birds build their nests;
the stork has her home in the fir trees.
18 The high mountains are for the wild goats;
the rocks are a refuge for the badgers.
19 Thou hast made the moon to mark the seasons;
the sun knows its time for setting.
20 Thou makest darkness, and it is night,
when all the beasts of the forest creep forth.
21 The young lions roar for their prey,
seeking their food from God.
22 When the sun rises, they get them away
and lie down in their dens.
23 Man goes forth to his work
and to his labor until the evening.
24 O Lord, how manifold are thy works!
In wisdom hast thou made them all;
the earth is full of thy creatures.
25 Yonder is the sea, great and wide,
which teems with things innumerable,
living things both small and great.
26 There go the ships,
and Leviathan which thou didst form to sport in it.

27 These all look to thee,
 to give them their food in due season.
28 When thou givest to them, they gather it up;
 when thou openest thy hand, they are filled with good things.
29 When thou hidest thy face, they are dismayed;
 when thou takest away their breath, they die
 and return to their dust.
30 When thou sendest forth thy Spirit,[5] they are created;
 and thou renewest the face of the ground.

31 May the glory of the LORD endure for ever,
 may the LORD rejoice in his works,
32 who looks on the earth and it trembles,
 who touches the mountains and they smoke!
33 I will sing to the LORD as long as I live;
 I will sing praise to my God while I have being.
34 May my meditation be pleasing to him,
 for I rejoice in the LORD.
35 Let sinners be consumed from the earth,
 and let the wicked be no more!
Bless the LORD, O my soul!
Praise the LORD!

148 Praise the LORD!
 Praise the LORD from the heavens,
 praise him in the heights!
2 Praise him, all his angels,
 praise him, all his host!

3 Praise him, sun and moon,
 praise him, all you shining stars!
4 Praise him, you highest heavens,
 and you waters above the heavens!

5 Let them praise the name of the LORD!
 For he commanded and they were created.
6 And he established them for ever and ever;
 he fixed their bounds which cannot be passed.[6]

7 Praise the LORD from the earth,
 you sea monsters and all deeps,
8 fire and hail, snow and frost,
 stormy wind fulfilling his command!

9 Mountains and all hills,
 fruit trees and all cedars!
10 Beasts and all cattle,
 creeping things and flying birds!

[5]Or *breath*. (RSV) [6]Or *he set a law which cannot pass away.* (RSV)

11 Kings of the earth and all peoples,
princes and all rulers of the earth!
12 Young men and maidens together,
old men and children!

13 Let them praise the name of the LORD,
for his name alone is exalted;
his glory is above earth and heaven.
14 He has raised up a horn for his people,
praise for all his saints,
for the people of Israel who are near to him.
Praise the LORD!

The Voice from the Whirlwind: Job

The Hebrew text of the Book of Job is extremely difficult and in places nearly unintelligible, and its date (perhaps between the sixth and fourth century B.C.) highly uncertain; large sections, moreover, are considered by many to be later interpolations. Although some scholars defend the integrity of the book as it stands, others believe that the prose "prologue" and "epilogue," narrating the patient suffering and final reward of a pious man sorely tested by God at the instigation of one of his counselors, Satan ("the adversary"), was an ancient folktale that served as a framework for the central series of poetic dialogues between a bitterly angry Job and his three complacently self-assured friends or "comforters," Eliphaz, Bildad, and Zophar, into which a number of extraneous passages, including the sententious speech of a fourth "comforter," Elihu, and parts of the voice from the whirlwind, have been inserted. Here again we find important parallels in Egyptian, Sumerian, and Akkadian, all raising the troubled question of unjust suffering; in the Akkadian Ludlul bel nemeqi *("I will praise the lord of wisdom"), for example, a righteous man laments that he is wrongly afflicted and learns, before he is restored to health and happiness, that (as Jacobsen writes in* The Intellectual Adventure of Western Man *[1946]) "Man is too small, too limited in outlook, to pass judgement on things that are divine. He has no right to set up his human values against the values which the gods hold." In the end Job's anguished and humanly unanswerable questions are answered not by the conventional pieties of his so-called friends, who essentially tell him that since God is just, Job must deserve what he has got, but by the divine revelation, or theophany, of God's voice from the whirlwind; and God's "answer" is itself a series of questions. These searingly ironical questions, repeatedly asking whether any man can rival or even comprehend the immeasurably greater powers of the LORD, serve of course to remind Job of his infinitesimal smallness in comparison with this inscrutably omnipotent God, and thus to place in perspective his suffering in a universe wholly beyond human understanding. Yet here too, perhaps, as in some of the Psalms, the very assertion of God's absolute transcendence over the world he created, "when the morning stars sang together, and all the sons of God shouted for joy," and forever infallibly rules becomes a splendid hymn to the glory of that creation itself. Job's response will indeed be to repent in sackcloth and ashes—but also to express his utter awe at "things too wonderful for me, which I did not know."*

38 Then the LORD answered Job out of the whirlwind:
2 "Who is this that darkens counsel by words without knowledge?
3 Gird up your loins like a man,
I will question you, and you shall declare to me.

4 "Where were you when I laid the foundation of the earth?
Tell me, if you have understanding.
5 Who determined its measurements—surely you know!
Or who stretched the line upon it?
6 On what were its bases sunk,
or who laid its cornerstone,
7 when the morning stars sang together,
and all the sons of God shouted for joy?

8 "Or who shut in the sea with doors,
when it burst forth from the womb;
9 when I made clouds its garment,
and thick darkness its swaddling band,
10 and prescribed bounds for it,
and set bars and doors,
11 and said, 'Thus far shall you come, and no farther,
and here shall your proud waves be stayed'?

12 "Have you commanded the morning since your days began,
and caused the dawn to know its place,
13 that it might take hold of the skirts of the earth,
and the wicked be shaken out of it?
14 It is changed like clay under the seal,
and it is dyed like a garment.
15 From the wicked their light is withheld,
and their uplifted arm is broken.

16 "Have you entered into the springs of the sea,
or walked in the recesses of the deep?
17 Have the gates of death been revealed to you,
or have you seen the gates of deep darkness?
18 Have you comprehended the expanse of the earth?
Declare, if you know all this.

19 "Where is the way to the dwelling of light,
and where is the place of darkness,
20 that you may take it to its territory
and that you may discern the paths to its home?
21 You know, for you were born then,
and the number of your days is great!

22 "Have you entered the storehouses of the snow,
or have you seen the storehouses of the hail,
23 which I have reserved for the time of trouble,
for the day of battle and war?
24 What is the way to the place where the light is distributed,
or where the east wind is scattered upon the earth?

25 "Who has cleft a channel for the torrents of rain,
and a way for the thunderbolt,
26 to bring rain on a land where no man is,

on the desert in which there is no man;
27 to satisfy the waste and desolate land,
and to make the ground put forth grass?

28 "Has the rain a father,
or who has begotten the drops of dew?
29 From whose womb did the ice come forth,
and who has given birth to the hoarfrost of heaven?
30 The waters become hard like stone,
and the face of the deep is frozen.

31 "Can you bind the chains of the Pleiades,
or loose the cords of Orion?
32 Can you lead forth the Mazzaroth[7] in their season,
or can you guide the Bear with its children?
33 Do you know the ordinances of the heavens?
Can you establish their rule on the earth?

34 "Can you lift up your voice to the clouds,
that a flood of waters may cover you?
35 Can you send forth lightnings, that they may go
and say to you, 'Here we are'?
36 Who has put wisdom in the clouds,
or given understanding to the mists?[8]
37 Who can number the clouds by wisdom?
Or who can tilt the waterskins of the heavens,
38 when the dust runs into a mass
and the clods cleave fast together?

39 "Can you hunt the prey for the lion,
or satisfy the appetite of the young lions,
40 when they crouch in their dens,
or lie in wait in their covert?
41 Who provides for the raven its prey,
when its young ones cry to God,
and wander about for lack of food? . . .

39 26 "Is it by your wisdom that the hawk soars,
and spreads his wings toward the south?
27 Is it at your command that the eagle mounts up
and makes his nest on high?
28 On the rock he dwells and makes his home
in the fastness of the rocky crag.
29 Thence he spies out the prey;

[7]Probably the twelve signs of the zodiac.
[8]The meaning of the Hebrew words translated "clouds" and "mists" is uncertain. The Authorized (King James) Version reads: "Who hath put wisdom in the inward parts? or who hath given understanding to the heart?"

his eyes behold it afar off.
30 His young ones suck up blood;
and where the slain are, there is he. . . .

41 "Can you draw out Leviathan[9] with a fishhook,
or press down his tongue with a cord?
2 Can you put a rope in his nose,
or pierce his jaw with a hook?
3 Will he make many supplications to you?
Will he speak to you soft words?
4 Will he make a covenant with you
to take him for your servant for ever?
5 Will you play with him as with a bird,
or will you put him on leash for your maidens?
6 Will traders bargain over him?
Will they divide him up among the merchants?
7 Can you fill his skin with harpoons,
or his head with fishing spears?
8 Lay hands on him;
think of the battle; you will not do it again!
9 Behold, the hope of a man is disappointed;
he is laid low even at the sight of him.
10 No one is so fierce that he dares to stir him up.
Who then is he that can stand before me?
11 Who has given to me,[10] that I should repay him?
Whatever is under the whole heaven is mine. . . ."

All Things Are Full of Weariness: Ecclesiastes

Ecclesiastes and the Song of Solomon, neither of which is explicitly a religious work—the Song of Solomon is the only book of the Hebrew Bible, along with Esther, in which the word God or LORD *(Yahweh) never appears—were among the last books accepted into the canon; both are attributed to King Solomon, but the difference in tone between them could hardly be greater. The Hebrew word Qohelet, translated in Greek as Ecclesiastes—the Greek titles of this and other books of our Bible date back to the Greek translation, the Septuagint, undertaken by Jews of Alexandria, Egypt, in the third century* B.C.*—and in the Authorized and Revised Standard versions as "the Preacher," occurs only in this book (usually dated between the sixth and third centuries* B.C.*), and may mean more nearly "speaker," "arguer," or "debater." In any case, the skeptical, world-weary author expresses an attitude strikingly at odds with both the fervor of prophetic faith and the strict adhesion to law of priestly or rabbinical Judaism, as he pursues the paths of pleasure and wealth, of folly and wisdom, and finds them all "vanity and a striving after wind." For him, "There is nothing better for a man than that he should eat and drink, and find enjoyment in his toil." In the opening chapter, which follows, the eternal cycles of the natural world—the rising*

[9] Or *the crocodile*. (RSV)
[10] The meaning of the Hebrew is uncertain. (RSV) The Authorized Version reads: "Who hath prevented me . . ."

and setting of the sun, the blowing of the wind, the endless flowing of streams to the sea—are seen as wearisomely repetitive processes seemingly devoid of all human meaning in a world in which God, though he surely exists, can never be known.

1 The words of the Preacher, the son of David, king in Jerusalem.
2 Vanity of vanities, says the Preacher,
vanity of vanities! All is vanity.
3 What does man gain by all the toil
at which he toils under the sun?
4 A generation goes, and a generation comes,
but the earth remains for ever.
5 The sun rises and the sun goes down,
and hastens to the place where it rises.
6 The wind blows to the south,
and goes round to the north;
round and round goes the wind,
and on its circuits the wind returns.
7 All streams run to the sea,
but the sea is not full;
to the place where the streams flow,
there they flow again.
8 All things are full of weariness;
a man cannot utter it;
the eye is not satisfied with seeing,
nor the ear filled with hearing.
9 What has been is what will be,
and what has been done is what will be done,
and there is nothing new under the sun.
10 Is there a thing of which it is said,
"See, this is new"?
It has been already
In the ages before us.
11 There is no remembrance of former things,
nor will there be any remembrance
of later things yet to happen
among those who come after.

12 I the Preacher have been king over Israel in Jerusalem. 13 And I applied my mind to seek and to search out by wisdom all that is done under heaven; it is an unhappy business that God has given to the sons of men to be busy with. 14 I have seen everything that is done under the sun; and behold, all is vanity and a striving after wind.

15 What is crooked cannot be made straight,
and what is lacking cannot be numbered.

16 I said to myself, "I have acquired great wisdom, surpassing all who were over Jerusalem before me; and my mind has had great experience of wisdom and knowledge." 17 And I applied my mind to know wisdom and to know madness and folly. I perceived that this also is but a striving after wind.

18 For in much wisdom is much vexation,
and he who increases knowledge increases sorrow.

How Sweet Is Your Love!: Song of Solomon

The Song of Solomon, or Song of Songs (as it is called in its opening verse), usually dated from the sixth to the third century B.C., or even later, has been more variously interpreted than any other book of the Hebrew Bible: as a purely secular (even "pagan") celebration, lyrical or dramatic, of love or marriage; as a Hebrew adaptation of Babylonian or Canaanite fertility rituals; as an allegory of God's love for His people, or of man's love of divine Wisdom, or (by Christians) of Christ's love for his Church—and so on. Though the question can never be finally decided, it is surely a mistake to fasten exclusively on a single interpretation, for these dimensions need not be contradictory. Allegorical interpretations, which dominated much of biblical exegesis for two millennia and frequently became arbitrary and far-fetched, are now widely out of favor. Misguided as the "disguised fertility ritual" notion may be, however, the new year's marriage of Dumuzi or Tammuz, personified in a king of Sumer or Babylon, with the goddess Inanna or Ishtar, or of Canaanite El or Baal with Asherah, was a powerful ancient Semitic paradigm of erotic union with the divine; and many post-biblical poets, both Jewish and Christian (notably St. John of the Cross in sixteenth-century Spain), have found the sensuous imagery of the Song of Songs the most suitable vehicle in which to express their ardent love of God. More immediately, the words placed in the LORD'*s mouth in Isaiah 62:5, probably written in the sixth to fifth century B.C., demonstrate that even a Hebrew prophet could envisage religious and national union in nuptial terms:*

For as a young man marries a virgin,
so shall your sons marry you,
and as the bridegroom rejoices over the bride,
so shall your God rejoice over you.

Yet if dimensions symbolic of the divine are entirely possible, erotically human dimensions in this lyrical exchange between woman and man are certain: no poem has ever more passionately rendered the delight and expectation of sensual love. The lavish profusion of natural images in which this love—human, divine, or both—finds expression makes it clear that worshippers of the God of Sinai were by no means always fanatical puritans blind to the apples and raisins, the gazelles and stags, the goats and ewes, the pomegranates and lilies, of the sweet and beautiful world all around them.

2 I am a rose of Sharon,
a lily of the valleys.

2 As a lily among brambles,
so is my love among maidens.

3 As an apple tree among the trees of the wood,
so is my beloved among young men.
With great delight I sat in his shadow,
and his fruit was sweet to my taste.
4 He brought me to the banqueting house,
and his banner over me was love.
5 Sustain me with raisins,
refresh me with apples;
for I am sick with love.
6 O that his left hand were under my head,
and that his right hand embraced me!
7 I adjure you, O daughters of Jerusalem,
by the gazelles or the hinds of the field,
that you stir not up nor awaken love until it please.

8 The voice of my beloved!
Behold, he comes,
leaping upon the mountains,
bounding over the hills.
9 My beloved is like a gazelle,
or a young stag.
Behold, there he stands
behind our wall,
gazing in at the windows,
looking through the lattice.
10 My beloved speaks and says to me:
"Arise, my love, my fair one,
and come away;
11 for lo, the winter is past,
the rain is over and gone.
12 The flowers appear on the earth,
the time of singing has come,
and the voice of the turtledove
is heard in our land.
13 The fig tree puts forth its figs,
and the vines are in blossom;
they give forth fragrance.
Arise, my love, my fair one,
and come away.
14 O my dove, in the clefts of the rock,
in the covert of the cliff,
let me see your face,
let me hear your voice,
for your voice is sweet,
and your face is comely.
15 Catch us the foxes,
the little foxes,

that spoil the vineyards,
    for our vineyards are in blossom."
16 My beloved is mine and I am his,
    he pastures his flock among the lilies.
17 Until the day breathes
    and the shadows flee,
turn, my beloved, be like a gazelle,
    or a young stag upon rugged mountains.

4 Behold, you are beautiful, my love,
    behold, you are beautiful!
Your eyes are doves
    behind your veil.
Your hair is like a flock of goats,
    moving down the slope of Gilead.
2 Your teeth are like a flock of shorn ewes
    that have come up from the washing,
all of which bear twins,
    and not one among them is bereaved.
3 Your lips are like a scarlet thread,
    and your mouth is lovely.
Your cheeks are like halves of a pomegranate
    behind your veil.
4 Your neck is like the tower of David,
    built for an arsenal,
whereon hang a thousand bucklers,
    all of them shields of warriors.
5 Your two breasts are like two fawns,
    twins of a gazelle,
    that feed among the lilies.
6 Until the day breathes
    and the shadows flee,
I will hie me to the mountain of myrrh
    and the hill of frankincense.
7 You are all fair, my love;
    there is no flaw in you.
8 Come with me from Lebanon, my bride;
    come with me from Lebanon.
Depart from the peak of Amana,
    from the peak of Senir and Hermon,
from the dens of lions,
    from the mountains of leopards.

9 You have ravished my heart, my sister, my bride,
    you have ravished my heart with a glance of your eyes,
    with one jewel of your necklace.

10 How sweet is your love, my sister, my bride!
how much better is your love than wine,
and the fragrance of your oils than any spice!
11 Your lips distil nectar, my bride;
honey and milk are under your tongue;
the scent of your garments is like the scent of Lebanon.
12 A garden locked is my sister, my bride,
a garden locked, a fountain sealed.
13 Your shoots are an orchard of pomegranates
with all choicest fruits,
henna with nard,
14 nard and saffron, calamus and cinnamon,
with all trees of frankincense,
myrrh and aloes,
with all chief spices—
15 a garden fountain, a well of living water,
and flowing streams from Lebanon.

16 Awake, O north wind,
and come, O south wind!
Blow upon my garden,
let its fragrance be wafted abroad.
Let my beloved come to his garden,
and eat its choicest fruits.

4

India: Creative Fervor, Extinction, and Love

The vast subcontinent of India (including modern Pakistan and Bangladesh), though immensely varied and seldom united politically during its long history, has always seemed, to observers from both East and West, a world to itself—a world of extreme contrasts held in precarious balance. From the towering Himālayas, to the great rivers descending from them, to the tropical jungles pullulating with exotic floras and faunas, this land of searing heat and drenching monsoon, of tiger and elephant, cobra and mongoose, "was blessed by a bounteous Nature, who demanded little of man in return for sustenance," A. L. Basham writes in The Wonder That Was India *(1954), "but in her terrible anger could not be appeased by any human effort." Such a land gave rise to fabulous legends of gods or demigods combining human and animal, like the monkey Hanūmat of the epic* Rāmāyana *or the elephant-headed Ganesha and half-human Nāga serpent spirits still worshipped by Hindus. The extremes of this world might indeed inspire terrors. In the northern mountains, the Chinese traveler Fa-Hsien writes in* A Record of Buddhistic Kingdoms *of his journey to India in A.D. 399–414, are "venomous drag-*

ons, which, when provoked, spit forth poisonous winds, and cause showers of snow and storms of sand and gravel," so that not one of ten thousand who encounter them survives; and the "terrible phantasms of the dream-forest" of the swampy Sundarbans in the Ganges delta, in Salman Rushdie's Midnight's Children (1980), draw those entangled in "the miasmic state of mind" they induce deeper and deeper into the thraldom of their "livid green world." But to a people for whom, since ancient times, "the whole of nature was in some sense divine," as Basham writes, not only its fearfully awesome or extraordinarily grand manifestations were sacred—Mount Meru, axle of the world, or Gangā, streaming down to earth from the Milky Way—but also the most common and humble: the cows, the snakes, the trees, and the grasses familiar to all from the everyday life of even the smallest village.

Indian religious practices, and their expressions in literature and art, display contrasts no less extreme than those of nature: this land of meditation and yoga, of ascetic self-denial in the quest for extinction of all earthly desire, is also the land of tantric sexual ritual, of the Kāma Sūtra, and of the most lushly erotic sculpture ever to have decorated temple walls. Basham is surely right to emphasize that the peoples of ancient (as indeed of modern) India, far from being sunk in lethargic gloom, "enjoyed life, passionately delighting both in the things of the senses and the things of the spirit." Yet when senses and spirit came into conflict, as they often have through the long centuries of Indian history, the outcome—by no means unique, of course, to India, but perhaps especially dramatic in this land of dramatic extremes—might well turn out to be uncompromising repudiation, by some, of the world so passionately celebrated by others.

India is unusual among the great civilizations of the world in that its first great literary and religious text—the 1,080 hymns of the Rig Veda, composed roughly between 1500 and 900 B.C., which celebrate the heroic deeds of the rain god Indra and others, and exultantly affirm the splendor of the physical world—was the product not of its first urban culture but of the illiterate cattle-herders and horsemen who were probably its destroyers. The very existence of the ancient Indus Valley civilization had been only dimly suspected until excavations beginning in the 1920s laid bare the two great cities of Harappā and Mohenjo Daro, whose "unimaginative but comfortable civilization," in Basham's phrase, had continued with little evident change for perhaps a thousand years after these and other cities were founded sometime in the mid third millennium B.C. By about 1700 they were in evident decline, and the marauding warriors who swept into northwest India over the centuries after about 1500—probably from the north Iranian plateau through the Hindu Kush, and who called themselves Aryans—must have conquered whatever remained of a civilization subsequently forgotten even more entirely (the few remains of its written language have not yet been deciphered) than the Sumerian would be.

The Aryans spoke but could not yet write an Indo-European language, the ancestor of classical Sanskrit, similar in structure to ancient Iranian, Hittite, Greek, and Latin (the word "Aryan" is cognate with both "Iran" and "Ireland"); their orally composed hymns were memorized and recited, as they continue to be even now, by priests who officiated over their sacrificial rites. In the centuries that followed, the Aryan conquerors subdued virtually all of India (though more thoroughly in the north, whose major modern languages, including Hindi, Urdu, and Bengali, descend from Sanskrit, than in the south, where Tamil, Telegu, and other languages stem from non-Indo-European tongues of the indigenous Dravidian peoples); settled in cities; developed writing; and imposed the rigid system of hereditary classes and occupational castes—never as rigid in practice as in theory—that has continued into the *twentieth century. The highest "class" (literally "color") in this system were the priestly brahmins;*[1] *second, the warriors* (kshatriyas); *third, and far lower in standing, the merchants* (vaishyas); *fourth, the menials or "serfs"* (shūdras), *probably remnants of the conquered native peoples, who were not initiated ("twice-born") to full Aryan status; below these were the outcastes and untouchables on the outer fringes of society, if not indeed of humanity.*

Toward the end of the late Vedic period (roughly 900–500 B.C.), the ritualistic religion of the Aryan conquerors turned increasingly toward meditation and ascetic withdrawal, as brahmins became not only sacrificial officiants but also forest-dwelling hermits whose philosophical speculations on the relation of the human soul to the universal spirit culminated in the early Upanishads; *significantly, the word* tapas, *which in the* Rig Veda *can designate the "cosmic heat" of sacrificial rites and the "creative fervor" of the gods they honor, comes increasingly to denote the ascetic's power of austere self-denial. Then, in the sixth century—a time roughly contemporary with that of the Hebrew prophets and Greek pre-Socratic philosophers and of Zoroaster in Iran and Confucius in China—a series of Indian reformers, most notably Vardhamāna Mahāvīra, the founder of Jainism, and Gautama Buddha, repudiated not only the sacred* Vedas *and the class religion on which Brahmanism was founded but also, in large measure, the natural world of drearily determined cyclical repetitions, from which they sought release or even extinction* (nirvāna) *through renunciation of earthly desires and attachments. The doctrine, or* dharma, *of Buddhism was destined to spread to Ceylon (Sri Lanka) and much of Southeast Asia and, in different and evolving forms, to Tibet, China, and Japan.*

In India itself, over the next fifteen hundred years, Buddhism eventually died out as the old Brahmanical religion, in the centuries before and after Christ—the period of the great epics and of the Bhagavad Gītā—*gradually transformed itself into the richly multiple complex of popular Hinduism, centered on cults of Vishnu, Shiva, and Devī, the Mother Goddess, embodiment of the female power of* shakti *(whose many forms and names included those of Pārvatī and Kālī). The conquerors' religion had at length evolved into that of the people, above all through new emphasis on love—both the fervid religious devotion of* bhakti *and the erotic sexuality of* kāma, *to which the gods are no more immune than humans. After an era of widespread renunciation, India had found a faith no less capable than that of the* Vedas *of passionate delight in the sensual beauty and pulsating energy of the phenomenal world.*

The selections in this chapter, though drawn from a wide range of writings over a period of nearly two thousand years, represent only works written in the Sanskritic languages of India before the Muslim conquests, beginning around A.D. 1200; they therefore necessarily exclude both writings in classical or modern Tamil and other Dravidian languages of the south and writings of Muslims, Sikhs, Hindus, and others in the modern prākrits, or Sanskritic vernaculars, of the north.

---

[1]I adopt the derivative English word "brahmin" instead of "brahman" for the Sanskrit brāhmana to avoid confusion with the impersonal magical force or creative power called brahman, later identified with the universal world spirit. To compound the confusion, the sacrificial ritual texts appended as commentaries to the Vedic hymns are called Brāhmanas, and the creator god of post-Vedic times is Brahmā. All these words are of course related.

In Praise of the Splendors of Heaven and Earth

Hymns of the *Vedas*, translated by Raimundo Panikkar

The word Veda, *meaning literally "knowledge," is cognate with other Indo-European words of seeing or knowing such as Latin* video, *Greek* idea, *and English "wit" and "wisdom." Most broadly, it refers to the entire body of Brahmanical or Vedic scripture, from the* Rig Veda *through the* Upanishads, *which remains the foundation of "orthodox" Hindu religions (as opposed to Jainism and Buddhism, which rejected the Veda); more narrowly, to the four collections of Vedic hymns that include its earliest and most sacred parts. Earliest of all is the* Rig Veda *itself, composed, as we have seen, between about 1500 and 900* B.C.*; the others, composed in the centuries that followed, are the* Sāma Veda, *a rearrangement of selected Rig-Vedic verses for liturgical purposes; the* Yajur Veda, *a compilation of sacrificial formulas; and the most recent, the* Atharva Veda, *a collection of magical spells and incantations.*

Many hymns of the Rig Veda *commemorate exploits of the gods of the conquering Aryans, above all of the king of their gods, the rain-maker and wielder of the thunderbolt, Indra. One famous hymn, for example (I.32), recounting his combat with the cloud dragon Vritra, "was probably a variant," Basham suggests, "of the creation myth of Mesopotamia, in which the god Marduk slays the demon of chaos, Tiamat, and creates the universe." Creation is indeed a frequent theme in myths not only of the* Rig Veda *(especially in its two latest books, the first and the tenth) but also of the prose* Brāhmanas *appended to the Vedic hymns, of the* Upanishads, *and of the post-Brahmanical epics and* Purānas; *these myths, like those of ancient Egypt, give no single canonical account but a dizzying range of contradictory versions from cosmic battles to divine incest and sacrificial dismemberment.*

One of the most extraordinary hymns, the first in our selection, shows that even in the late Rig-Vedic period the philosophical speculation that would characterize much of Indian religion was already well advanced, for here creation results not from slaying monsters but from cosmic heat or "ardor," tapas *(as in our second selection also), from which the One arises through the power of love or desire. Yet this explanation of cosmic evolution is no more than a possibility, for perhaps not even the god of the highest heaven knows whence creation arose! The next hymn, called "To the Unknown God" by the nineteenth-century Indologist Max Müller, continues this vein of speculative celebration, seeking in the "Golden Germ" or "Golden Embryo" the unknown and unknowable One God from whom all else arose, some single divine source (like the* Brahman *of the* Upanishads*) underlying the multiplicity of the divine and human world. A seemingly more archaic view is embodied in the hymn "The Primordial Man," which tells of dismemberment of the cosmic giant or "Person"* (Purusha), *who gives birth to the universe, in a Vedic act of sacrifice (as Manu, the first human being, in other myths gives birth to the four classes of men), through his death. The world so created is one where all that we call nature—moon and sun, fire and wind, air, sky, and earth—is inherently divine, and the last four hymns in our selection, from the* Rig *and* Atharva Vedas, *are fervent celebrations of the glories of that creation: above all, of the uncounted splendors copiously shed on her human children by the multifarious, invincible, golden-breasted Earth.*

The translations are from Raimundo Panikkar's The Vedic Experience: Mantramañjarī *(1977). See also the annotated prose renditions of 108 hymns of the* Rig Veda *(1981) translated by Wendy Doniger (O'Flaherty).*

The Hymn of the Origins
Rig Veda *X.129*

1 At first was neither Being nor Nonbeing.
There was not air nor yet sky beyond.
What was its wrapping? Where? In whose protection?
Was Water there, unfathomable and deep?

2 There was no death then, nor yet deathlessness;
of night or day there was not any sign.
The One breathed without breath, by its own impulse.
Other than that was nothing else at all.

3 Darkness was there, all wrapped around by darkness,
and all was Water indiscriminate. Then
that which was hidden by the Void, that One, emerging,
stirring, through power of Ardor, came to be.

4 In the beginning Love arose,
which was the primal germ cell of the mind.
The Seers, searching in their hearts with wisdom,
discovered the connection of Being in Nonbeing.

5 A crosswise line cut Being from Nonbeing.
What was described above it, what below?
Bearers of seed there were and mighty forces,
thrust from below and forward move above.

6 Who really knows? Who can presume to tell it?
Whence was it born? Whence issued this creation?
Even the Gods came after its emergence.
Then who can tell from whence it came to be?

7 That out of which creation has arisen,
whether it held it firm or it did not,
He who surveys it in the highest heaven,
He surely knows—or maybe He does not!

Creative Fervor
Rig Veda *X.190*

1 From blazing Ardor Cosmic Order came
and Truth; from thence was born the obscure night;
from thence the Ocean with its billowing waves.

2 From Ocean with its waves was born the year
which marshals the succession of nights and days,
controlling everything that blinks the eye.

3 Then, as before, did the creator fashion
the Sun and Moon, the Heaven and the Earth,
the atmosphere and the domain of light.

The Unknown God

Rig Veda *X.121*

1 In the beginning arose the Golden Germ:
he was, as soon as born, the Lord of Being,
sustainer of the Earth and of this Heaven.
What God shall we adore with our oblation?

2 He who bestows life-force and hardy vigor,
whose ordinances even the Gods obey,
whose shadow is immortal life—and death—
What God shall we adore with our oblation?

3 Who by his grandeur has emerged sole sovereign
of every living thing that breathes and slumbers,
he who is Lord of man and four-legged creatures—
What God shall we adore with our oblation?

4 To him of right belong, by his own power,
the snow-clad mountains, the world-stream, and the sea.
His arms are the four quarters of the sky.
What God shall we adore with our oblation?

5 Who held secure the mighty Heavens and Earth,
who established light and sky's vast vault above,
who measured out the ether in mid-spheres—
What God shall we adore with our oblation?

6 Toward him, trembling, the embattled forces,
riveted by his glory, direct their gaze.
Through him the risen sun sheds forth its light.
What God shall we adore with our oblation?

7 When came the mighty Waters, bringing with them
the universal Germ, whence sprang the Fire,
thence leapt the God's One Spirit into being.
What God shall we adore with our oblation?

8 This One who in his might surveyed the Waters
pregnant with vital forces, producing sacrifice,
he is the God of Gods and none beside him.
What God shall we adore with our oblation?

9 O Father of the Earth, by fixed laws ruling,
O Father of the Heavens, pray protect us,
O Father of the great and shining Waters!
What God shall we adore with our oblation?

10 O Lord of creatures, Father of all beings,
you alone pervade all that has come to birth.
Grant us our heart's desire for which we pray.
May we become the lords of many treasures!

FROM *The Primordial Man*
Rig Veda *X.90*

6 Using the Man as their oblation,
the Gods performed the sacrifice.
Spring served them for the clarified butter,
Summer for the fuel, and Autumn for the offering. . . .

10 From this were horses born, all creatures
such as have teeth in either jaw;
from this were born the breeds of cattle;
from this were born sheep and goats.

11 When they divided up the Man,
into how many parts did they divide him?
What did his mouth become? What his arms?
What are his legs called? What his feet?

12 His mouth became the brahmin; his arms
became the warrior-prince; his legs
the common man who plies his trade.
The lowly serf was born from his feet.

13 The Moon was born from his mind; the Sun
came into being from his eye;
from his mouth came Indra and Agni,
while from his breath the Wind was born.

14 From his navel issued the Air;
from his head unfurled the Sky,
the Earth from his feet, from his ear the four directions.
Thus have the worlds been organized. . . .

The Mighty Earth
Rig Veda *V.84*

1 The mighty burden of the mountains' bulk
rests, Earth, upon your shoulders; rich in torrents,
you germinate the seed with quickening power.

2 Our hymns of praise resounding now invoke you,
O far-flung Earth, the bright one.
Like a neighing steed you drive abroad your storm clouds.

3 You in your sturdy strength hold fast the forests,
clamping the trees all firmly to the ground,
when rains and lightning issue from your clouds.

FROM *Hymn to the Earth*
Atharva Veda *XII.1*

1 High Truth, unyielding Order, Consecration,
Ardor and Prayer and Holy Ritual
uphold the Earth; may she, the ruling Mistress
of what has been and what will come to be,
for us spread wide a limitless domain.

2 Untrammeled in the midst of men, the Earth,
adorned with heights and gentle slopes and plains,
bears plants and herbs of various healing powers.
May she spread wide for us, afford us joy!

3 On whom are ocean, river, and all waters,
on whom have sprung up food and ploughman's crops,
on whom moves all that breathes and stirs abroad—
Earth, may she grant to us the long first draught!

4 To Earth belong the four directions of space.
On her grows food; on her the ploughman toils.
She carries likewise all that breathes and stirs.
Earth, may she grant us cattle and food in plenty! . . .

7 Limitless Earth, whom the Gods, never sleeping,
protect forever with unflagging care,
may she exude for us the well-loved honey,
shed upon us her splendor copiously! . . .

9 On whom the flowing Waters, ever the same,
course without cease or failure night and day,
may she yield milk, this Earth of many streams,
and shed on us her splendor copiously! . . .

11 Your hills, O Earth, your snow-clad mountain peaks,
your forests, may they show us kindliness!
Brown, black, red, multifarious in hue
and solid is this vast Earth, guarded by Indra.
Invincible, unconquered, and unharmed,
I have on her established my abode.

12 Impart to us those vitalizing forces
that come, O Earth, from deep within your body,
your central point, your navel; purify us wholly.
The Earth is mother; I am son of Earth.
The Rain-giver is my father; may he shower on us blessings! . . .

15 All creatures, born from you, move round upon you.
You carry all that has two legs, three, or four.
To you, O Earth, belong the five human races,
those mortals upon whom the rising sun
sheds the immortal splendor of his rays.

16 May the creatures of earth, united together,
let flow for me the honey of speech!
Grant to me this boon, O Earth!

17 Mother of plants and begetter of all things,
firm far-flung Earth, sustained by Heavenly Law,
kindly and pleasant is she. May we ever
dwell on her bosom, passing to and fro! . . .

23 Instill in me abundantly that fragrance,
O Mother Earth, which emanates from you
and from your plants and waters, that sweet perfume
that all celestial beings are wont to emit,
and let no enemy ever wish us ill! . . .

26 Earth is composed of rock, of stone, of dust;
Earth is compactly held, consolidated.
I venerate this mighty Earth, the golden-breasted!

27 Her upon whom the trees, lords of the forest,
stand firm, unshakable in every place,
this long-enduring Earth we now invoke,
the giver of all manner of delights. . . .

59 Peaceful and fragrant, gracious to the touch,
may Earth, swollen with milk, her breasts overflowing,
grant me her blessing together with her milk! . . .

Lord of the Field

Rig Veda *IV.57*

1 We, with the Lord of the Field as our friend
and helper, obtain for our cattle and horses
food in plenty, that they may be sleek and well-fed.
May he graciously grant us his favor!

2 O Lord of the Field, like a cow yielding milk,
pour forth for us copious rivers of sweetness,
dripping honey like nectar and pure as pure ghee.[2]
May the Lords of the Law grant us mercy!

3 Sweet be the plants for us, sweet be the heavens,
sweet be the waters and the air of the sky!
May the Lord of the Field show us honeylike sweetness,
May we follow his furrow unharmed!

4 In contentment may men and oxen both plough,
in contentment the plough cleave the furrow,
in contentment the yoke be securely attached
and the ploughman urge on his oxen!

[2]Clarified butter, used as an oblation in Vedic sacrifices.

5 Ploughshare and Plough, to our chant be propitious!
Take of the milk you have made in heaven
and let it fall here on this earth!

6 Auspicious Furrow, we venerate you.
We pray you, come near us to prosper and bless
and bring us abundant harvests.

7 May Indra draw the Furrow, may Pūsan[3]
guide well its course! May she yield us milk
in each succeeding year!

8 In contentment may the ploughshare turn up the sod,
in contentment the ploughman follow the oxen,
celestial Rain pour down honey and water.
Ploughshare and Plough, grant us joy!

As a Spring Gushes Forth in a Thousand Streams

Atharva Veda *III.24*

1 Brimful of sweetness is the grain,
brimful of sweetness are my words;
when everything is a thousand times sweet,
how can I not prosper?

2 I know one who is brimful of sweetness,
the one who has given abundant corn,
the God whose name is Reaper-God;
him we invoke with our song.
He dwells in the home of even the lowly
who are debarred from sacrifice.
The God whose name is Reaper-God,
him we invoke with our song.

3 Let the five directions and races of men
bring to our doors prosperity,
as after the rains (in a swollen flood)
a river carries down driftwood.

4 As a spring gushes forth in a hundred, a thousand
streams, and yet stays inexhaustible,
so in a thousand streams may our corn
flow inexhaustibly!

5 Reap, you workers, one hundred hands,
garner, you workers, one thousand hands!
Gather in the bounteous corn that is cut
or still waits on the stalk.

[3]Charioteer of the Sun, herdsman, guide, and nourisher.

6 Three measures I apportion to the Spirits,
four measures to the mistress of the house,
while you I touch with the amplest measure
(of all that the field has yielded).

7 Reaper and Garnerer are your two
distributors, O Lord of creation.
May they convey hither an ample store
of riches never decreasing!

The Mustard Seed Enfolding the Universe

Upanishads,
translated by Juan Mascaró

Continuing the commentaries appended to the four collections of Vedic hymns as Brāhmanas *and* Aranyakas *("forest books"), the* Upanishads *were regarded as the "end of the Veda"* (Vedānta)*; and to these densely speculative metaphysical treatises the much later Hindu philosophical school of Vedānta—formulated in the* Brahma Sūtras *of Bādarāyana early in the Christian era, and expounded by the brilliant South Indian brahmin Shankara (among others) in the early ninth century—looked back for its principal inspiration. After the Sanskrit classics were belatedly translated into European languages beginning in the late eighteenth century, the* Upanishads *exerted considerable influence in the West as well, notably on the nineteenth-century German philosopher Arthur Schopenhauer.*

The word upanishad *means literally "sitting" or "session," probably because these treatises, some of which are partly in dialogue form, convey doctrines originally imparted by brahmin teachers (possibly forest hermits) to pupils sitting at their feet; the word has also been interpreted to mean "secret," suggesting their esoteric nature. They total the sacred number of 108 (one tenth the number of hymns in the* Rig Veda*), but the large majority are late and of minor importance. The dozen or so principal* Upanishads, *including the two generally considered among the oldest and greatest, the* Brihadāranyaka *and* Chāndogya, *were perhaps written down, after long oral transmission, by the sixth century* B.C. *(the dates are highly conjectural), but in their present form, some scholars believe, contain later material showing posssible Buddhist influence. Their spirit of ascetic withdrawal and inner contemplation could hardly be more different from the rhapsodic affirmation of the world characteristic of the* Rig Veda. *In some passages we find, apparently for the first time, the doctrine of* samsāra *—repeated transmigration of the soul through different life forms as a result of* karma *("action"), the accumulated weight of deeds in previous lives—which would be central to Jain, Buddhist, and Hindu thought, finding a counterpart in the metempsychosis of Pythagoras in Greece. Release from this seemingly inescapable cycle of unredeemed nature can come through knowledge, derived from instruction and meditation, of the mystical oneness of the individual soul or self* (ātman, *"breath") with* Brahman, *transformed from its Rig-Vedic meaning—magic power inherent in the ritual spell—into the divine creative principle or "world-soul" pervading the universe.*

In our first selection, from the Katha Upanishad, *Yama, the god of death, instructs Nachiketas, who descends to his realm for the sake of his father, that ātman is in the heart of all beings, "smaller than the smallest atom, greater than the vast spaces"; in the two passages from the* Chāndogya Upanishad *that follow, the central upanishadic doctrine of the oneness of ātman and Brahman, individual and universal soul, becomes explicit, especially when Svetaketu Aruneya's father in the final passage demonstrates to his son that the invisible essence permeating the world is that of his own soul:* tat tvam asi, *he succinctly explains, "*THAT *you are." Thus the* Upanishads, *too, end in affirmation—Nachiketas in the* Katha *finally attains Brahman and freedom from passion and death, as all may who know the ātman—but this is emphatically not affirmation, like the* Rig Veda*'s, of the natural world. On the contrary, "as sharers in the continuity of nature," Paul Deussen writes in* The Philosophy of the Upanishads *(1899), "we are, like it, subject to necessity; but we are free from it as soon as, by virtue of the knowledge of our identity with the ātman, we are set free from this continuity of nature," within which is nothing but human bondage.*

Juan Mascaró has aimed to make "clear and simple" translations of these often obscure and difficult texts for the modern reader. For more scholarly and literal translations, see Robert Ernest Hume's The Thirteen Principal Upanishads *(1921; 2nd ed. 1931).*

The Eternal in Man

Katha Upanishad *II.6, 18–20*

What lies beyond life shines not to those who are childish, or careless, or deluded by wealth. "This is the only world: there is no other," they say; and thus they go from death to death. . . .

Atman, the Spirit of vision, is never born and never dies. Before him there was nothing, and he is ONE for evermore. Never-born and eternal, beyond times gone or to come, he does not die when the body dies.

If the slayer thinks that he kills, and if the slain thinks that he dies, neither knows the ways of truth. The Eternal in man cannot kill: the Eternal in man cannot die.

Concealed in the heart of all beings is the Atman, the Spirit, the Self; smaller than the smallest atom, greater than the vast spaces. The man who surrenders his human will leaves sorrows behind, and beholds the glory of the Atman by the grace of the Creator.

The Invisible Essence

Chāndogya Upanishad *III.14; VI.12–13*

All this universe is in truth Brahman. He is the beginning and end and life of all. As such, in silence, give unto him adoration.

Man in truth is made of faith. As his faith is in this life, so he becomes in the beyond: with faith and vision let him work.

There is a Spirit that is mind and life, light and truth and vast spaces. He contains all works and desires and all perfumes and all tastes. He enfolds the whole universe, and in silence is loving to all.

This is the Spirit that is in my heart, smaller than a grain of rice, or a grain of barley, or a grain of mustard-seed, or a grain of canary-seed, or the kernel of a grain of canary-seed. This is the Spirit that is in my heart, greater than the earth, greater than the sky, greater than heaven itself, greater than all these worlds.

He contains all works and desires and all perfumes and all tastes. He enfolds the whole universe and in silence is loving to all. This is the Spirit that is in my heart, this is Brahman.

*

"Bring me a fruit from this banyan tree."

"Here it is, father."

"Break it."

"It is broken, Sir."

"What do you see in it?"

"Very small seeds, Sir."

"Break one of them, my son."

"It is broken, Sir."

"What do you see in it?"

"Nothing at all, Sir."

Then his father spoke to him: "My son, from the very essence in the seed which you cannot see comes in truth this vast banyan tree.

Believe me, my son, an invisible and subtle essence is the Spirit of the whole universe. That is Reality. That is Atman. THOU ART THAT."

"Explain more to me, father," said Svetaketu.

"So be it, my son.

Place this salt in water and come to me tomorrow morning."

Svetaketu did as he was commanded, and in the morning his father said to him: "Bring me the salt you put into the water last night."

Svetaketu looked into the water, but could not find it, for it had dissolved.

His father then said: "Taste the water from this side. How is it?"

"It is salt."

"Taste it from the middle. How is it?"

"It is salt."

"Taste it from that side. How is it?"

"It is salt."

"Look for the salt again and come again to me."

The son did so, saying: "I cannot see the salt. I only see water."

His father then said: "In the same way, O my son, you cannot see the Spirit. But in truth he is here.

An invisible and subtle essence is the Spirit of the whole universe. That is Reality. That is Truth. THOU ART THAT."

A World of Creatures Groaning in Torment

Teachings of Jainism

Of the great renunciatory religions originating in the sixth century B.C., *Jainism, the most uncompromising, has been the most enduring in India itself, where it retains several million adherents nearly a thousand years after Buddhism all but vanished. Its founder was a near contemporary of the Buddha, Vardhamāna, called Mahāvīra ("Great Hero"), probably born in northern India around 540; according to Jain tradition he was the twenty-fourth and last* Tīrthankara *("Fordmaker"), whose immediate predecessor, Pārshva, lived and taught two and a half centuries before him. At age thirty Mahāvīra he began his austere life as a wandering ascetic, soon casting aside his one garment and living in total nudity; after twelve years he attained* nirvāna *(literally the "blowing out" of earthly desires), and finally died of self-starvation at the age of seventy-two, probably in 468. His followers were not restricted by class, as those of the Vedic religion had been, but so demanding was their discipline of monastic self-denial in a world trapped in relentless cycles of rebirth and weighed down by massive accumulations of karma that few could hope for even eventual release. The great division among later Jains was between those who wore a white garment and those who wore none; the stricter sect no doubt gave rise to the stories of "naked philosophers," or gymnosophists, that reached classical Greece in the wake of Alexander the Great's incursion in the late fourth century.*

But if no Indian religion has been more extreme in repudiating all attachment to the irremediably natural world of samsāra, with its anguished cycles of endlessly repeated transmigration — Jainism, like Theravāda Buddhism (and like the philosophy of Epicurus), denies any significant role to whatever gods may be — none has placed greater emphasis on the kinship of all living things in a universe of life-monads continually interchanging one form for another; and none has more rigorously adhered to the widespread Indian ideal of nonviolence (ahimsā) *to which that kinship and eternal interchange logically give rise. The following brief selections from Jain scriptures compiled (in both Sanskrit and Prākrit) many centuries after Mahāvīra are taken from* Sources of Indian Tradition *(1958), ed. Wm. Theodore de Bary.*

The Anguish of Eternal Rebirth

Book of Latter Instructions (Uttarādhyayana Sūtra) *19.61–67, 71, 74*

From clubs and knives, stakes and maces, breaking my limbs,
An infinite number of times I have suffered without hope.
By keen-edged razors, by knives and shears,
Many times I have been drawn and quartered, torn apart and skinned,
Helpless in snares and traps, a deer,
I have been caught and bound and fastened, and often I have been killed.
A helpless fish, I have been caught with hooks and nets;
An infinite number of times I have been killed and scraped, split and gutted.
A bird, I have been caught by hawks or trapped in nets,
Or held fast by birdlime, and I have been killed an infinite number of times.
A tree, with axes and adzes by the carpenters
An infinite number of times I have been felled, stripped of my bark, cut up, and sawn into planks.
As iron, with hammer and tongs by blacksmiths
An infinite number of times I have been struck and beaten, split and filed. . . .
Ever afraid, trembling, in pain and suffering,
I have felt the utmost sorrow and agony. . . .
In every kind of existence I have suffered
Pains which have scarcely known reprieve for a moment.

The Kinship of Living Things

Book of Sermons (Sūtrakritānga) *I.1–9*

Earth and water, fire and wind,
    Grass, trees, and plants, and all creatures that move,
Born of the egg, born of the womb,
    Born of dung, born of liquids—

These are the classes of living beings.
Know that they all seek happiness.
In hurting them men hurt themselves,
And will be born again among them. . . .

Some men leave mother and father for the life of a monk,
But still make use of fire;
But He[4] has said, "their principles are base
Who hurt for their own pleasure."

The man who lights a fire kills living things,
While he who puts it out kills the fire;
Thus a wise man who understands the Law
Should never light a fire.

There are lives in earth and lives in water,
Hopping insects leap into the fire,
And worms dwell in rotten wood.
All are burned when a fire is lighted.

Even plants are beings, capable of growth,
Their bodies need food, they are individuals.
The reckless cut them for their own pleasure
And slay many living things in doing so.

He who carelessly destroys plants, whether sprouted or full grown,
Provides a rod for his own back.
He has said, "Their principles are ignoble
Who harm plants for their own pleasure."

A World of Misery and Compassion

Teachings of Theravāda Buddhism, translated by Henry Clarke Warren

The anguish of life in a world of endlessly repetitive and unredeemable suffering is no less central to the teaching of Buddha than to that of his near contemporary, Mahāvīra, but in Buddhism the rigorous severity of Jain repudiation of both nature and the divine was increasingly tempered by the compassionate oneness for fellow creatures already implicit in Jainism, and by realization of the illusory nature of the world itself and hence of suffering within it. To begin with, indeed, this latter tendency—which Mahāyāna sūtras (those of the "Greater Vessel," which subsequently spread to Tibet, China, and Japan) would later elaborate in metaphysical speculations of immensely subtle complexity—was at most latent: in teachings of the historical Buddha, as recorded in the Hīnayāna ("Lesser Vessel") or Theravāda ("Way of the Elders") tradition, which spread to Ceylon, Burma, Thailand, and Southeast Asia, it is far more often the intensely experienced reality of the world and of suffering that is stressed.

[4]Mahāvīra.

In scriptures of the "Pāli canon" (written down, according to tradition, in Ceylon in the Pāli vernacular, after oral transmission through many generations, early in the first century B.C.*), Siddhārtha Gautama Shākyamuni, most recent of many Buddhas ("Enlightened Ones") throughout the millennia, took up his spiritual vocation only when this pampered but discontented North Indian prince successively encountered, outside the imperfectly protective walls of his father's palace, old age, sickness, death, and a wandering monk who suggested the possibility of release from the inescapable reality of these sorrows of life. Therewith he began his years of wandering, fasting, and penance as an ascetic, nearly dying of hunger until, at the age of thirty-five, he vowed to sit meditating under a tree until he learned the secret of human suffering. At dawn of the forty-ninth day, after countless temptations and afflictions, he attained* bodhi, *or enlightenment, thus becoming a Buddha. In the long years that remained before his death and final attainment of nirvāna (Pāli* nibbāna*), at the age of eighty, in about 480* B.C.*, he taught to the growing band of disciples that constituted his earliest community, or* sangha*—prototype of the Buddhist monastic order, membership in which was unrestricted by class—the law, or doctrine, or truth* (dharma; *Pāli* dhamma) *of the Middle Way. The Middle Way rejected extreme penitential asceticism but affirmed the need to turn unrelentingly away from the natural world of samsāra, a perpetually decaying (and thus a transitory, perhaps in the end, after all, an illusory) world forever aflame with desire, and seek release from it through right conduct, as defined by the noble eightfold path, and through disciplined meditation culminating in extinction of desire itself, and thereby in conquest of the world.*

"Be not a friend of the world," the Buddha exhorts in an early collection of sayings, the Dhammapada *("Way of Truth"), and in the first of our selections (all three from Henry Clarke Warren's* Buddhism in Translations *[1896]), he unequivocally rejects consideration of all questions concerning the nature of this world—of all questions whatsoever not directly elucidating misery and its cessation—as distractions from the attainment of nirvāna. But the Buddha's enlightenment brought not only repudiation of the world but also infinite compassion (expressed in our remaining selections) for all beings trapped within it. This deep compassion for other living things will remain, in contrast with Jainism's more sternly aloof assertion of kinship in torment, a fundamental hallmark of Buddhism throughout its long and continuing history.*

Questions Not Tending to Edification

Majjhima-Nikāya, *Sutta 63*

"Accordingly, Māluñkyāputta, bear always in mind what it is that I have not elucidated, and what it is that I have elucidated. And what, Māluñkyāputta, have I not elucidated? I have not elucidated, Māluñkyāputta, that the world is eternal; I have not elucidated that the world is not eternal; I have not elucidated that the world is finite; I have not elucidated that the world is infinite; I have not elucidated that the soul and the body are identical; I have not elucidated that the soul is one thing and the body another; I have not elucidated that the saint exists after death; I have not elucidated that the saint does not exist after death; I have not elucidated that the saint both exists and does not exist after death; I have not elucidated that the saint neither exists nor does not exist after death. And why, Māluñkyāputta, have I not elucidated this? Because, Māluñkyāputta, this profits not nor has to do with the fundamentals of religion, nor tends to aversion, absence of passion, cessation, quiescence, the supernatural faculties, supreme wisdom, and Nirvana; therefore have I not elucidated it.

"And what, Māluñkyāputta, have I elucidated? Misery, Māluñkyāputta, have I elucidated; the origin of misery have I elucidated; the cessation of misery have I elucidated;

and the path leading to the cessation of misery have I elucidated. And why, Māluñkyaputta, have I elucidated this? Because, Māluñkyāputta, this does profit, has to do with the fundamentals of religion, and tends to aversion, absence of passion, cessation, quiescence, knowledge, supreme wisdom, and Nirvana; therefore have I elucidated it. Accordingly, Māluñkyāputta, bear always in mind what it is that I have not elucidated, and what it is that I have elucidated."

Thus spake The Blessed One; and, delighted, the venerable Māluñkyāputta applauded the speech of the Blessed One.

The Conversion of Animals

Visuddhi-Magga, *Chapter 7*

The Blessed One, moreover, was The Teacher, because he gave instruction also to animals. These, by listening to the Doctrine of the Blessed One, became destined to conversion, and in the second or third existence would enter the Paths. The frog who became a god is an illustration.

As tradition relates, The Blessed One was teaching the Doctrine to the inhabitants of the town of Campā, on the banks of Lake Gaggarā; and a certain frog, at the sound of The Blessed One's voice, obtained the mental reflex. And a certain cowherd, as he stood leaning on his staff, pinned him down fast by the head. The frog straightway died, and like a person awaking from sleep, he was reborn in the Heaven of the Thirty-three, in a golden palace twelve leagues in length. And when he beheld himself surrounded by throngs of houris, he began to consider: "To think that I should be born here! I wonder what ever I did to bring me here." And he could perceive nothing else than that he had obtained the mental reflex at the sound of the voice of The Blessed One. And straightway he came with his palace, and worshiped at the feet of The Blessed One. And The Blessed One asked him:—

"Who is it worships at my feet,
And flames with glorious, magic power,
And in such sweet and winning guise,
Lights up the quarters all around?"

"A frog was I in former times,
And wandered in the waters free,
And while I listened to thy Law,
A cowherd crushed me, and I died."

Then The Blessed One taught him the Doctrine, and the conversion of eighty-four thousand living beings took place. And the frog, who had become a god, became established in the fruit of conversion, and with a pleased smile on his face departed.

Love for Animals

Culla-Vagga, *V.6*

Now at that time a certain priest had been killed by the bite of a snake, and when they announced the matter to The Blessed One, he said:

"Surely now, O priests, that priest never suffused the four royal families of the snakes with his friendliness. For if, O priests, that priest had suffused the four royal families of the snakes with his friendliness, that priest, O priests, would not have been killed by the bite of a snake. . . . I enjoin, O priests, that ye suffuse these four royal families of the snakes with your friendliness; and that ye sing a song of defense for your protection and safeguard. After this manner, O priests, shall ye sing:

"'. . . Creatures without feet have my love,
And likewise those that have two feet,
And those that have four feet I love,
And those, too, that have many feet.

"'May those without feet harm me not,
And those with two feet cause no hurt;
May those with four feet harm me not,
Nor those who many feet possess.

"'Let creatures all, all things that live,
All beings of whatever kind,
See nothing that will bode them ill!
May naught of evil come to them!

"'Infinite is The Buddha, infinite the Doctrine, infinite the Order! Finite are creeping things: snakes, scorpions, centipedes, spiders, lizards, and mice! I have now made my protection, and sung my song of defense. Let all living beings retreat! I revere The Blessed One, and the seven Supreme Buddhas!'"

Foam on the Ocean and Raindrops from Heaven

Teachings of Mahāyāna Buddhism

In the centuries after the Buddha attained nirvāna, his dharma of suffering and salvation steadily drew adherents to his order, aided no doubt by the openness to all classes that it shared with Jainism and other heterodox sects, by its high ethical standards of conduct, and by its compassion for a sorrowing world. Especially after Candragupta Maurya united almost all of India, for one of the few times in its history, in the years after 320 B.C., repulsing the successor to Alexander the Great's eastern conquests, and his son Ashoka—whom Basham calls "one of the greatest and noblest rulers India has known, and indeed one of the great kings of the world"—ruled the Mauryan Empire in accord with Buddhist principles from about 269 to 232, Buddhism became preeminent in India and "began its career as a world religion."

Like any large-scale religious movement, Buddhism in these centuries had undergone divisions in doctrine and practice, but not until perhaps the late first or early second century A.D., at the fourth Buddhist council in Kashmīr, were the tendencies grouped together as Mahāyāna ("Greater Vessel") recognized, at a time when Buddhism was making inroads in China and many great Sanskrit sūtras were being written, as distinct from Hīnayāna ("Lesser Vessel") or Theravāda. Mahāyāna Buddhism itself soon divided into many schools of immense complexity; among major developments that distinguished it from Theravāda was the doctrine of the bodhisattva, *who from compassion for others voluntarily defers nirvāna until, through transfer of spiritual merit, all living things can be brought to that blissful state. Thus Mahāyāna Buddhism, though fully acknowledging the sorrow of the world, "is fundamentally optimistic," Basham writes, since most of its schools "maintain implicitly or explicitly that ultimately all beings will attain Nirvāna and become Buddha." This "optimism" does not, however, imply affirmation of the physical world of samsāra, which generally remains a condition to be escaped. The Buddha himself is no longer primarily conceived as an enlightened human being, the historically embodied Siddhārtha Gautama, but as the earthly expression of a spiritual being possessing three "bodies": the Body of Essence* (Dharmakāya), *permeating the universe like the Brahman of the Upanishads and identified in some sūtras with the plenitude of ultimate Emptiness; the Body of Bliss* (Samboghakāya), *ruling in the form of Amitābha ("Immeasurable Glory") over the Buddhist heaven; and the Created Body* (Nirmānakāya), *the visible emanation of the Body of Bliss on earth.*

Thus in our first selection, from the Shūrangama Sūtra, *or* Buddha's Great Crown Sutra—*said to have been written in Sanskrit (though some have thought it an indigenous Chinese composition) in the first century* A.D. *and translated around 705 by Pāramiti into Chinese, from which the English translation here reprinted from Dwight Goddard's* A Buddhist Bible *(1938) was made—understanding comes only when the mind is liberated from thinking and desire and recognizes its oneness (as Nachiketas and Svetaketu in the* Upanishads *recognized the oneness of ātman with Brahman) with the great unity of the Mind of Enlightenment, the Body of Essence here conceived as the "Womb of Tathāgata," a title of the Buddha meaning "He Who Has Fully Arrived" or "The Perfect One." In contrast to this supreme reality, the physical world of deaths and rebirths, where bodies float like specks of foam on the trackless ocean, is "a medley of unreal and transitory diversities that contaminate the mind" through multiplication of categories in the "discrimination of Ignorance."*

Despite this unequivocal affirmation of the illusoriness of the world, however, questions concerning its nature are by no means summarily dismissed in this sūtra as not tending to edification, for the Buddha offers an account of the changing phenomena of this unreal universe through continual interaction of the four elements of earth, water, fire, and wind, which are emanations of the indestructible Mind of Enlightenment. Thus the phenomenal world, ultimately illusory though it be, is not unworthy of attention. In our second selection—a verse parable from the most influential of all Mahāyāna sutras, the Lotus of the Wonderful Law, *or "Lotus Sūtra" (composed in Sanskrit at an unknown date and translated several times into Chinese between the mid third and early seventh centuries* A.D.*)—the ubiquitous grace of the Buddha's dharma is conveyed through the extended simile of a rain-cloud fertilizing the earth: an image clearly bestowing dignity and value on the things of this impermanent but beautiful world. This rehabilitation of the natural world would be taken to an extreme conclusion when Tantric Buddhism of the "Thunderbolt Vessel" equated it with ultimate reality itself, so that drinking of alcohol, eating of meat, and violation of other taboos culminating in ritualized sexual intercourse and even incest, whose practitioners were supposed to become Buddha and his spouse, the savior goddess Tārās, might be employed as means of salvation. Here, as Heinrich Zimmer writes in* Philosophies of India *(1951), man (and of course woman) "must rise through and by means of nature, not by rejection of nature": for the wheel of affirmation and rejection of the world has now at last, many centuries after Gautama preached the ascetic gospel of the Four Noble Truths of sorrow and its cessation, come full circle again.*

Since W. E. Soothill's abbreviated version of the Lotus Sūtra was published in 1930, more complete and scholarly translations have been made by Leon Hurvitz (1976) and Burton Watson (1993).

The Illusoriness of the Phenomenal World,
translated by Bhikshu Wai-tao and Dwight Goddard
Shūrangama Sūtra, *or* Buddha's Great Crown Sūtra

Ananda! If perception of sight, hearing, understanding, etc., has a nature that is unlimited and perfect, permeating everywhere throughout the universes, and by nature unchangeable, then you should know that the natures of the multitudinous and different perceptions, infinite, unmovable space, together with the movable elements of earth, water, fire, and wind, are to be regarded as the Six Great Elements. Their essential natures are perfect and in one unity, and they all belong to the Womb of Tathagata, and are devoid of deaths and rebirths. As your mind-essence has fallen out of attention, you have failed to realize that all your perceptions of sight, hearing, understanding and feeling, belong by reason of their nature to Tathagata's Womb. You should meditate upon this and note whether your perceptions of sight, hearing, understanding and feeling, belong to deaths and rebirths, or belong to one great unity. . . .

Thereupon Ananda and all the assembly, having received this wonderful and profound instruction from the Lord Tathagata and having attained to a state of perfect accommodation of mind and perfect emancipation of mind from all remembrances, thinking and desires, became perfectly free in both body and mind. Each one of them understood clearly that the Mind can reach to all the ten quarters of the universes and that their perception of sight can reach to all the ten quarters also. It was just as clear to them as though it was a blade of grass held in their hand. They saw that all the worldly phenomena were nothing but their own wonderful, intelligent, original Mind of Enlightenment embracing all the ten quarters of the universes. In contrast to this wonderful, all-embracing Mind of Enlightenment, their physical bodies begotten from their parents seemed like specks of dust blowing about in the open space of the ten quarters of the universes. Who would notice their existence or their non-existence? Their physical bodies were like a speck of foam floating about on a vast and trackless ocean, with nothing distinctive about them to indicate from whence they came, and if they disappeared, whither they went. They realized very clearly, that they, at last, had acquired their own wonderful Mind, a Mind that was Permanent and Indestructible. . . .

Thereupon, Purna Metaluniputra rose from his seat, arranged his robe, knelt upon his right knee, with the palms of his hands together, addressed the Lord Buddha, saying:—

. . . Blessed Lord! The thing that troubles us, that we do not fully understand, is this: If all of the sense organs, objects of sense, ingredients of sensation, location of perception between objects and consciousness, and spheres of mentation about objects, are all to be considered as manifestations of the Womb of Tathagata which by its essential nature ever abides in freshness and purity, how have all the conditional phenomena of rivers, mountains, earth, etc., which from beginningless time have been going on in successively changing processes, how have they ever come into manifestation?

And one more question. The Lord Tathagata has said that all of the four great Elements of Earth, Water, Fire, and Wind, are perfect and accommodating in their nature and are permeating everywhere throughout the phenomenal universes, how can they be permeating everywhere and, at the same time, be in perfect tranquillity?

Blessed Lord! If the nature of Earth is universal, how can it, at the same time and in the same space, co-exist with water? Or, if the nature of water is universal, how can fire exist at the same time; that is, if water and fire are present universally in the same space and at the same time, how is it that they do not destroy each other? Or, as the nature of earth is dense and impassable and the nature of space is empty and passable, how can the two different and opposing natures be mutually universal at the same time? These things are too difficult for us to comprehend, I pray the Lord Tathagata to have compassion upon us and solve these puzzling questions. . . .

The Lord Buddha said:—Purna! . . . In my previous instruction, I made clear to you that the Essential Intuitive Mind possessed its own mysterious Enlightening Nature, and that the attainment to this Essential Intuitive Mind unveils this mysterious Enlightening Nature. . . . As soon as it is supposed to be enlightened by something else, there rise false conceptions as to this something else and then following there rise fantastic conceptions of functions and processes. Because of this from the perfect unity of Intuitive Mind innumerable varieties have been manifested and as there are distinctions among them so classifications among these varieties are developed. From this arise conceptions of likes and unlikes, and then conceptions of non-likes and non-unlikes, and the mind is thrown into a medley of bewildering puzzles which in time become attached to the mind and contaminate it. At the end, these attachments and contaminations within your mind develop the consciousness of differences between self and the not-self of objects and thus the pure mind becomes entangled in the snarls of attachments and contaminations. Because of their defilement, there rises the disturbing manifestation of an external world, but when they are stilled, there remains only empty space, abiding in perfect unity. The world is a medley of unreal and transitory diversities that contaminate the mind and it is out of these arbitrary conceptions of phenomena that the very conception of unity and diversity arises. But Essential Mind is wholly devoid of conceptions and therefore recognizes neither unity nor diversity.

Moreover, Purna, these two opposites—the Pure Reality of Intuitive Mind by its very self-nature ceaselessly drawing everything into its perfect Unity and Tranquillity, and the unreal and transitory medley of diverse and conflicting differences forever tending to variety and multiplicities—these two opposing conceptions arising from the discrimination of Ignorance bring into existence a vibratory motion that by reason of desire and grasping and the perpetuating influence of habit energy, accounts for all the basic conceptions of the primary Elements, the solidity of Earth, the fluidity of Water, the heat of Fire and the motivity of Wind. Amid them it is the nature of Fire to move upward and the nature of Water to move downward, and from these two Elements being in reciprocal development there are the manifestations of rivers, and volcanoes and land. As Water takes precedence, oceans appear and when Fire takes precedence, continents and islands. The great ocean is also in reciprocal development with the illusive conception of fire within the mind, and reveals the fact that the blazing Fire is arising continuously. The continents and islands are also in a reciprocal development with the false conception of water within the mind, revealing the fact that rivers and streams are ever flowing. Or if the false conception of water is running very slowly within the mind, and the flame of fire is in a high state of activity, then there rise the high mountains and volcanoes which after all are only combinations of the false conceptions of water and fire within the mind. So if we strike flint, sparks of fire shoot out, and if we melt rocks, they will become liquid. If the false conception of water within the mind predominates over the false conception of earth, then the phenomena of grass and trees rise. As grass and trees are also false conceptions of the mind, as soon as they are compressed they become water again. Thus all these false conceptions of phenomena have their reciprocal developments and successive manifestations within the mind, and by means of causes and conditions there rises the false conception of the reciprocal continuance of the world's existence. . . . Thus all these conditional phenomena of rivers, mountains, earth, etc., and their successive and endless changes arise, and all are based upon the illusions of the thinking mind without any other interpretation. . . .

From the beginning you must recollect that all conditional phenomena are transitory and passing. Have you any doubt of this? Can you think of any exception? Take, for example, pure space. Have any of you ever heard of space coming to corruption and destruction? No, because, pure space is free from conditions and, therefore, is indestructible.

Compare this with your body, Ananda. Within your body there is an element of hardness, of Earth; there is an element of fluidity, of Water; there is an element of warmth, of Fire; and an element of breathing and motion, the element of Wind. The body is in bondage to these Four Great Elements, and these four bonds divide your tranquil, mysterious, intuitive, enlightening Mind into such divisions as the sensations and perceptions of seeing, hearing, tasting, smelling and touching, and of the following conceptions and discriminations of thought, that cause your enlightening Mind to fall into the corresponding five defilements of this evil world from its beginning and will continue to do so to its end. . . . When the mind becomes tranquillized and concentrated into perfect unity, then all things will be seen, not in their separateness, but in their unity wherein there is no place for the evil passions to enter, and which is in full conformity with the mysterious and indescribable purity of Nirvana. . . .

Ananda! When you have cut off all dependence upon the sense organs, your inner awareness will become as clear as crystal, manifesting its authentic brightness. Then all vagrant thoughts and transitory objects and the ever varying phenomena of this terrestrial world will melt away like ice when boiling water is poured upon it. . . . When objects of sense experience are all ignored, then the transcendental brightness of Intuition will shine forth mysteriously, and you will have found the true source of cognition and tranquillity. . . .

The Buddha's Law as a Rain Cloud,
translated by W. E. Soothill
Lotus Sūtra

Know, Kasyapa!
It is like unto a great cloud
Rising above the world,
Covering all things everywhere—
A gracious cloud full of moisture;
Lightning-flames flash and dazzle,
Voice of thunder vibrates afar,
Bringing joy and ease to all.
The sun's rays are veiled,
And the earth is cooled;
The cloud lowers and spreads
As if it might be caught and gathered;
Its rain everywhere equally
Descends on all sides,

Streaming and pouring unstinted,
Permeating the land.
On mountains, by rivers, in valleys,
In hidden recesses, there grow
The plants, trees, and herbs;
Trees, both great and small,
The shoots of the ripening grain,
Grape vine and sugar-cane.
Fertilized are these by the rain
And abundantly enriched;
The dry ground is soaked,
Herbs and trees flourish together.
From the one water which
Issued from that cloud,
Plants, trees, thickets, forests,
According to need receive moisture.
All the various trees,
Lofty, medium, low,
Each according to its size,
Grows and develops
Roots, stalks, branches, leaves,
Blossoms and fruits in their brilliant colors;
Wherever the one rain reaches,
All become fresh and glossy.
According as their bodies, forms
And natures are great or small,
So the enriching (rain),
Though it is one and the same,
Yet makes each of them flourish.
In like manner also the Buddha
Appears here in the World,
Like unto a great cloud
Universally covering all things;
And having appeared in the world,
He, for the sake of the living,
Discriminates and proclaims
The truth in regard to all laws.
The Great Holy World-honored One,
Among the gods and men
And among the other beings,
Proclaims abroad this word:
"I am the Tathagata,
The Most Honored among men;
I appear in the world
Like unto this great cloud,
To pour enrichment on all
Parched living beings,
To free men from their misery

To attain the joy of peace,
Joy of the present world,
And joy of Nirvana.
Gods, men, and every one!
Hearken well with your mind,
Come you here to me,
Behold the Peerless Honored One!
I am the World-honored,
Who cannot be equalled.
To give rest to every creature,
I appear in the world,
And, to the hosts of the living,
Preach the pure Law, sweet as dew;
The one and only Law
Of Deliverance and Nirvana. . . .
Going, coming, sitting, standing,
Never am I weary of
Pouring it copious on the world,
Like the all-enriching rain. . . .

Fragments of Immeasurable Splendor

Bhagavad Gītā,
translated by Ann Stanford

The foremost gods of the Rig Veda *generally embodied forces visible in the world around them, as Indra embodies the power of the storm and Agni the force of sacrificial fire; theirs was a dynamic world, to be sure, yet they themselves seemed—while they lasted—essentially unchanging. The Brahman of the* Upanishads, *on the other hand, though pervading the universe as salt pervades water in which it is invisibly dissolved, was a reality perceptible only to disciplined contemplation (like the Mahāyāna Body of Essence). In an illusory phenomenal world of continual transmigration and change, this seeming Emptiness, beyond all diversity and division, was the one unchanging and eternal—therefore the one true—reality. Increasing renunciation of the physical world, and of action within it, was a logical consequence of this conviction that reality lay elsewhere. In this context the composite religion of modern Hinduism arose in the centuries before and after Christ, both drawing on and reacting against the transcendental rigors of Brahmanical and Buddhist world-denial.*

The gods who hold center stage in the immensely long and intricate popular epics, the Mahābhārata *and* Rāmāyana *(probably composed over a period of centuries between about 300 B.C. and A.D. 300), and in the still later prose* Purānas, *though some are descended from Vedic ancestors, are the opposite of single or unchanging; and though all partake of a world inextricably entwined in samsāra and in the web of divine illusion called* māyā, *far from austerely renouncing this world they characteristically celebrate its infinite multiplicity and mutability even while affirming the mystical unity that eternally underlies it. In so kaleidoscopic a universe the less adaptable of the great Vedic gods, Indra, Agni, and Varuna (the guardian of sacred law) among them,* slip quietly into the background. Among the principal incarnations of this newly conceived divinity is Krishna ("the Black One"), apparently a god of tribal origin, who stands at the center of the great religious poem the Bhagavad Gītā ("Lord's Song"), probably composed between about 100 B.C. and A.D. 100—though the date is speculative as usual, and some have placed it centuries earlier or later—and inserted into the Mahābhārata. At a critical point in this great epic, which tells of the conflict between the Kauravas (or Kurus) and their cousins the Pāndavas that will culminate in the great battle of Kurukshetra (the "Kuru Field," which is also the Field of Dharma), the great warrior Arjuna, leader of the Pāndavas, surveys the armies of his kinsmen and friends arrayed on the field just before the decisive battle begins, and in despair and compassion drops his bow and arrows, telling his charioteer, Krishna, that he cannot take part in the fight. The rest of the Gītā (which is being narrated by the charioteer Sanjaya to the blind king Dhritarāshtra, father of the Kauravas and uncle of the Pāndavas) consists primarily of dialogue between Arjuna and his divine charioteer, who will finally convince him to take up arms and participate in the battle.

In part Krishna's appeal is a "conservative" one, for Arjuna belongs to the class of warriors, or kshatriyas, and it is his divinely ordained duty (another meaning of "dharma") to fight. Beyond this, Krishna decisively affirms, against the passivity of purely contemplative or ascetic faiths, the inherent value of action itself—that is, of the very karma whose accumulated weight, both Jains and Buddhists believed, inexorably binds human beings to the wearisome round of this world of samsāra—and of the "discipline of action" (karma yoga), which entails renunciation not of action itself but of involvement in its outcome. But the Gītā is no mere appeal to duty, nor even to action; it is also a divine revelation of the nature of the universe in which our actions take place. For Krishna is the principal avatar of the great god Vishnu, or Hari, and this fact is immensely significant. In this universe, gods no less than mortals are perpetually being reborn in ever-changing forms, but far from being enslaved by inescapable reincarnation, the myriad variousness of their forms suffuses a world no less glorious for its shimmering dome of many-colored soap bubbles than for the white radiance of eternity they refract.

Krishna remembers his many births, he tells Arjuna, as Arjuna does not. Through awareness of the god's splendor, revealed not only in the dazzling brightness of a thousand suns blazing at once in the sky—the verse recalled by J. Robert Oppenheimer as the first atomic bomb exploded at Alamogordo—but also in every fragment of the phenomenal world that mirrors that splendor, Arjuna can progress from action to understanding and knowledge, including the knowledge that he too, no less than sun and moon, horse and elephant, thunderbolt and Ganges, is one of Krishna's forms and therefore, through his very participation in the human and natural world to which he belongs, is immortal. The slayer does not slay, Krishna asserts, echoing a verse of the Katha Upanishad, nor does the slain die, for in this divinely creative universe all that are have been forever and can never cease to be. Such a realization, growing out of Krishna's overwhelming epiphany, leads Arjuna beyond both dutiful action and meditative knowledge to a third discipline, the yoga of devotion, or bhakti: for this god who both permeates the phenomenal world and is the silence of secret things and the knowledge of them that know is not only, like the Rig-Vedic gods, a force of nature nor, like the Brahman of the Upanishads, a spiritual principle, but a deeply personal, involved, and compassionate god whose devotion is returned by those devoted to him. The Gītā played a central part in renewing the ancient Vedic religion in an age when Jains and Buddhists had renounced it. Still, it is not by the Vedas, or sacrifice, or study, or the "grim rigor" of ascetic denial that this god can be known, but by selfless devotion and joyous wonder.

Since first translated by Charles Wilkins in 1784 the Bhagavad Gītā has been the subject of countless translations and studies. The passages that follow are from Ann Stanford's verse translation of 1970. For a scholarly translation with commentary and transliterated Sanskrit text, see the edition of R. C. Zaehner (1969).

The Lord Said:

2.12 Never was a time when I was not
Nor you, nor any of these kings.
Never will come a time hereafter
When all of us shall cease to be. . . .

16 What is not can never come to be.
What is can never disappear.
Those who see truth know the border
Between the two of these.

17 But the whole world is pervaded
By that which none can destroy—
The Imperishable
No one can bring it to nothing.

18 It is said these are temporal bodies
Of an embodied one—eternal,
Indestructible, beyond measure.
Therefore fight, son of Bharata.

19 Who thinks of him as slayer
And who thinks that he is slain
Do not rightly understand.
He slays not, nor is slain.

20 He is never born, nor does he ever die
Nor having been, will he ever cease to be.
Not born, constant, eternal, this ancient one
Is not slain, though the body is slain. . . .

3.27 Actions are being performed
Through nature's forces everywhere.
The self, confused by self-consciousness
Imagines *I am the doer.*

28 But he who knows the principle
Of the division of forces and works—
And how the forces move one another—
Is not caught in them, mighty-armed. . . .

4.5 I have passed through many births
And so have you, Arjuna,
But I remember all of these.
Scourge of the Foe, you do not remember.

6 Though I am unborn and my self
Unchanging, though I am lord of beings,
Ruling my own realm of nature
I am born through my own creative power. . . .

9 He who knows my divine birth
And my works as they truly are
Does not go to rebirth, Arjuna,
Leaving the body, he comes to me. . . .

10.19 Listen, best of Kurus,
My forms are of heaven.
I will tell you the first of them
Though I extend on without end.

20 I am the self, Thick-haired,
That lies in the heart of all creatures,
I am the beginning
The enduring, and the end of all.

21 Of the sons of space, I am Vishnu
Of heavenly lights, I am the sun
I am chief among gods of storm
I am the moon shining among stars.

22 Of Vedas, I am the book of chants
Of gods, I am Indra, lord of the air
I am the mind among the senses
I am awareness in living things . . .

27 Know me Indra's steed among horses
Sprung from the nectar of the sea,
His prince of elephants, Airavata.
Among men, I am the king.

28 Of weapons, I am the thunderbolt
Of cows, I am the Cow-of-wishes
Begetting, I am god of love
I am Vasuki, king of serpents. . . .

31 Of purifiers, I am the wind,
Rama among bearers of weapons
I am the sea-monster among fishes
Ganges among the rivers.

32 I am the beginning and the end
And the middle of all creations
I am the knowledge of the soul.
I am the discourse of those who speak. . . .

37 Among the Vrishnis I am Krishna
I am Arjuna of the Pandavas
I am Vyasa among hermits
The poet Ushanas among the wise.

38 I am the rod of the master
The statecraft of those who seek conquest
I am the silence of secret things
The knowledge of them that know.

39 And whatever is the seed of all beings
That I am, Arjuna.
No creature that moves or does not move
Could exist without me.

40 My divine appearances
Are without end, Scorcher of the Foe,
I have declared but a few
Of the multitude of my glories.

41 Whatever creature comes forth in glory
In vigor and in beauty
Know that that being has sprung
From but a fragment of my splendor.

42 Or rather—for what to you
Is such a multitude of knowing, Arjuna?—
I stand and hold up this universe
With a single portion of myself. . . .

Arjuna Said:

11.3 What you have called yourself, you are,
Supreme Lord. Greatest spirit,
I wish to behold your outward form
In majesty as the mighty god.

4 Master, if you believe
That I may behold you thus
Then, god of mysteries,
Show me yourself everlasting.

The Lord Said:

5 Son of Pritha, behold
The variousness of my forms,
Hundreds, more, thousands of them, marvels,
Of different shapes and colors

6 Son of Bharata, behold the gods
Those who live in celestial light, or in
Nature, storm gods, harbingers of dawn.
Look at wonders never seen before.

7 Today behold the whole world
All things that move or do not move
And whatever else you wish to see.
They stand as one within my body.

8 But you cannot see me
With your own eye alone.
I give you the eye of a god.
Behold my mystery as the lord.

Sanjaya Said:

9 Then, O King, when he had spoken thus
Hari, the great lord of mystery
Revealed to the son of Pritha
His supreme form as a god.

10 With a multitude of mouths and eyes
Displaying wonders
Adorned with the very heavens
With divine weapons raised as for battle.

11 Crowned with the skies, in ethereal raiment
Scented with divine perfumes and balms
Made of all wonders was this limitless god,
And his face looked in all directions.

12 If a thousand suns
Should at once blaze up in the sky
The light of that mighty soul
Would equal all their brightness.

13 So the son of Pandu beheld
The world with its myriad divisions
Standing there together as one
In the body of the god of gods.

14 Then amazement filled the Winner of Wealth
The hair of his body stood on end
He bowed his head before the god
And clasped his hands and spoke.

Arjuna Said:

15 Lord, I see in your body all gods
And the varied hosts of beings
Lord Brahma seated on the lotus throne
And all the sages and celestial serpents

16 Endless in form! I see you everywhere
Myriad of arms, eyes, bellies, mouths,
Without end, without center, without
Beginning. I behold you, All-form, All-god,

17 Bearing the crown, the mace, and the disk,
A mass of brightness, ablaze, everywhere.
I behold you, hard to look upon,
In a glory of flame and sun, immeasurable.

18 You are the imperishable, the farthest to be known
Solid, the last resting-place of all,
Unchanging, the eternal keeper of right.
I deem you the ancient of spirits

19 Without beginning, middle, or end, endless in strength
With innumerable arms, your eyes the moon and sun.
I behold your face a blazing fire
Whose radiance warms the whole universe. . . .

22 The gods of storm, light, nature, prayer,
The All-gods, gods of dawn and wind, household gods,
Hosts of heavenly singers, sprites, demons, perfected ones,
They look at you and are all amazed. . . .

24 You reach the sky, blazing with many colors,
With gaping mouths and broad eyes flaming.
Seeing you my soul shakes within me
And I can find no steadiness or peace. . . .

30 You lick up whole worlds in your fiery jaws
And devour them completely.
Your dreadful rays fill the whole universe,
Vishnu, and scorch it with their brightness.

31 Tell me, you of awful form, who are you?
Glory be to you, highest of gods, have mercy.
I wish to know you, who have been from the beginning,
For I do not know what you have set out to do.

The Lord Said:

32 I am time, destroyer of worlds, grown old
Setting out to gather in the worlds.
These warriors drawn up, facing the arrows,
Even without you, they shall all cease to be.

33 Therefore, stand up. Seize honor.
Conquer your foes. Enjoy the rich kingdom.
They were already killed by me long ago.
Be but the means, left-handed archer. . . .

53 Not by the Vedas or sacrifice
Or austerity or alms-giving
Can I be seen in such form
As you have beheld me.

54 But by devotion to me and no other
May that form of mine be known
And seen in its essence
And entered into, Arjuna.

55 Who does my work and sets me above all
And worships me without desire
And has no hate for any being
He comes to me, son of Pandu.

Intoxications of Springtime and Love

Sanskrit Court Poetry, translated by John Brough

Both Vishnu the preserver, in his incarnation as Krishna, and Shiva the destroyer seem to have assimilated aspects of tribal fertility gods in their complex personae, very probably from cults of indigenous Dravidian peoples. In the guise of Govinda, the god of cowherds, Krishna was famed for erotic exploits, though the notes of his flute, "calling the women to leave their husbands' beds and dance with him in the moonlight," Basham states, were interpreted, much as Jews and Christians have interpreted the Song of Solomon, as representing "the voice of God, calling man to leave earthly things and turn to the joys of divine love." And Shiva (who is also preserver no less than destroyer) was both the patron god of ascetics, maintaining the world's existence by meditating in austere solitude high in the Himālayas, and the husband and lover, worshipped in the form of the phallic linga, of the Mother Goddess under the name of Pārvatī. To these divinities, poets both male and female would for centuries write verses of longing in both Sanskrit and the vernaculars of north and south, for spiritual bhakti *could readily merge, for god and human alike, into sexual* kāma: *the distinction, like that between sacred and secular, was arbitrary at best.*

Ecstatic devotional sects of Hinduism were especially prominent in the Tamil-speaking region of southern India at least from the seventh century A.D. on, and after the great Tamil theologian Rāmānuja (said to have lived from 1017 to 1137!) spread his impassioned gospel of bhakti yoga, *deeply influenced by the* Bhagavad Gītā, *and preached salvation through unrestrained self-abandonment to a god of grace and love, these sects gained new prominence and fervor throughout India. Much earlier, however, at the time of the Gupta Empire after about A.D. 320, and especially under Candra Gupta II (376–415), when India, in Basham's words, "was perhaps the happiest and most civilized region of the world," court poets of northern India writing in a stylized classical Sanskrit that had not been spoken (except by a highly educated few) for centuries were composing poems extraordinary for their celebration of the ravishing beauties of a world tinged, at its most sensual, with the eroticism of the divine.*

The greatest was Kālidāsa, author, probably in the years around A.D. 400 (though he too has been dated centuries earlier), of the famed romantic drama Shakuntalā, *two other plays, an epic, and three long lyrical poems,* Ritusamhāra ("The Gathering of the Seasons"), Meghadūta ("The Cloud Messenger"), *and* Kumāra-sambhava ("The Birth of Kumara"). *The beauties of nature, which manifest the procreative female power of shakti and thus merge with those of woman and of the Goddess who is the consort of Shiva, pervade these three poems. Their rich particularity of closely observed and keenly felt detail—Kālidāsa's verses teem with the names of countless varieties of flora and fauna—is new in Indian poetry and rare in any poetry of the world. In Kālidāsa, no less than in the hymns of the* Rig Veda, *the fertile earth throbs with a vibrant energy manifesting the glory of the divine, since to him, as Arthur W. Ryder remarks, rivers, mountains, and trees "have a conscious individuality as truly and certainly as animals or men or gods."*

The early poem "The Gathering of the Seasons" lushly evokes qualities characteristic of each season throughout the year; thus in autumn (in Chandra Rajan's translation),

> The breathtaking beauty of rippling lakes
> breathed on by a passing wind at daybreak,
> where lotus and lily glow brilliantly
> and pairs of love-drunk wild geese float entrancing,
> suddenly grips the heart with longing.

But it is of course spring that is most associated with love; and in spring—rhapsodically described in our first selection, from the Kumāra-sambhava, *masterfully translated by John Brough—the god of love himself, Kāma, sets out to enamor Shiva. In the myth that provides the background for this stunning evocation of spring and desire, in Brough's summary, "The gods have realized that the only hope against [the demon] Tāraka is a general who shall be born as the son of Shiva and Umā (Pārvatī). But Shiva has no thoughts of marriage, and is engaged in profound meditation in the remote mountain forest of the Himālaya. The third canto opens with Indra, who has come from the council of the gods, seeking the help of Kāma, the god of love, whose task it now is to cause Shiva to fall in love with Umā. Taking with him his wife Rati, and Spring as his companion, Kāma makes his way to the mountain hermitage of Shiva. Then follows a description of the coming of spring to the forest." This first attempt will fail when Shiva catches sight of Kāma aiming his arrow, and reduces him to ashes with a glance of fire from his third eye; but subsequent attempts will succeed, and Kumāra (or Skanda), the war-god born of the union of Shiva and Pārvatī, will slay the demonic Tāraka in single combat.*

The brief selection of Sanskrit lyrics that follows—mostly anonymous, but one by Kālidāsa, one by Dharmakīrti (ca A.D. *700?), and one by Bhartrihari (probably seventh century* A.D.*, but dated as early as 400)—is also taken from Brough's* Poems from the Sanskrit *(1968). The most important collection of classical Sanskrit poetry, Vidyākara's* Subhāshitaratnakosha ("Treasury of Well Turned Verse"), *which gathered together 1,738 poems shortly before* A.D. *1100, has been translated by Daniel H. H. Ingalls as* An Anthology of Sanskrit Court Poetry *(1965). Other translations of Sanskrit lyric poetry are included in Kālidāsa's* Shakuntala and Other Writings *(1912), translated by Arthur W. Ryder, and* The Loom of Time *(1989), translated by Chandra Rajan; in* Bhartrihari: Poems *(1967), translated by Barbara Stoler Miller; and in* The Love of Krishna: The Krishnakarnāmrita of Lālāshuka Bilvamanngala *(1975), ed. Frances Wilson. Collections of vernacular poetry, all deeply influenced by bhakti devotionalism, include* Speaking of Shiva *(1973), translated from the South Indian Kannada language by A. K. Ramanujan;* Songs of the Saints of India *(1988), translated from Hindi by John Stratton Hawley and Mark Juergensmeyer; and* In Praise of Krishna: Songs from the Bengali *(1967), translated by Edward C. Dimock, Jr., and Denise Levertov.*

The Temptation of Shiva, by Kālidāsa

Kumāra-sambhava

> And then within these mountain-forest reaches,
> Skilled to distract saints' thoughts from heaven above
> The young awakening Spring now yawns and stretches,
> Belov'd companion of the god of love.

While the hot sun, untimely, came to waken
The North to be his love, the gentle South
Exhaled a sigh, thus to have been forsaken,
A breath warm-scented from her fragrant mouth. 5

The Ashóka then, its trunk and branches laden,
Full-flowered, with foil of many a green leaf-shoot, 10
Impatient, quite forgot to expect a maiden
To wake its flowers with anklet-tinkling foot.

On every blossom-arrow he created,
Feathered with leaves and tipped with mango-flame,
The fletcher Spring the owner designated, 15
Writing with bees the god of love's own name.

The Karnikāra's blossom, brightly glowing,
While by its scentlessness it grieved the mind,
Showed how God's will is set against bestowing
All excellences in one place combined. 20

Curved like the crescent moon, deep crimson traces
Glowed as Palāsha-buds began to swell,
As if the nail-marks of the Spring's embraces
Flushed on the forest-lands he loved so well.

Spring's Loveliness, with woman's wiles acquainted, 25
With flowers of Tilaka[5] adorned her head,
Made beauty-spots of clinging bees, and painted
Fresh mango-blossom lips with morning-red.

Piyāla-blossom clusters shed their pollen
In smoke-clouds; and the deer, bewildered, blind, 30
Through forest-glades where rustling leaves had fallen,
Made rash by springtime, coursed against the wind.

The cuckoo's song, hoarsened to gentle cooing,
When food of mango-sprouts tightened his throat,
Became the voice of Love, to work the undoing 35
Of maids cold-hearted, by its magic note.

The fairy-women, with their winter faces
Devoid of lipstick, saw their color fade,
While with the Spring the rising sweat left traces,
Smearing the beauty-marks so carefully made. 40

[5] *Tilaka*, besides being the name of a flowering shrub, denotes an ornamental (or sectarian) mark painted on the forehead. (Brough)

When, in the forest of their meditation,
The holy hermits saw the untimely spring,
Their minds were hard-pressed to resist temptation,
To keep their thoughts from Love's imagining.

When Love came there, his flower-bow ready stringing, 45
With fair Desire, his consort, at his side,
The forest creatures showed the passion springing
In every bridegroom's heart towards his bride.

From the same flower-cup which his love had savored
The black bee sipped the nectar as a kiss; 50
While the black doe, by her own consort favored,
Scratched by his antlers, closed her eyes in bliss.

The elephant with water lotus-scented
Sprayed her own lord, giving of love a token;
The wheel-drake, honoring his wife, presented 55
A half-chewed lotus-stalk which he had broken.

When nectar-wine had set her eyes a-dancing,
And sweat had smudged the fairy's painted face,
Her fairy-lover found her more entrancing,
And checked his song to seek a fresh embrace. 60

When trembling petal-lips made laughing faces,
And blossom-breasts the slender stems were bending,
Even the forest-trees received the embraces
Of creeper-wives, from their bough-arms depending.

Yet Shiva still remained in meditation, 65
Absorbed, although he heard the singing elves:
Can anything have power of perturbation
Of souls completely masters of themselves?

Then at the doorway of his forest dwelling
With rod of gold his servant Nandi stood, 70
Frowning, with finger to his lips, thus quelling
The unseemly conduct in the springtime wood.

Throughout the forest, at his simple stricture,
Dumb were the birds, and silent were the bees:
As in a scene fixed in a painted picture, 75
Stilled were the deer and motionless the trees.

As on a journey under baleful omen,
Trying in fear from Shiva's eyes to hide,
To the Lord's sacred ground came Love the bowman,
Entering through creeper-thickets from the side. 80

THE DANCER AT THE FEAST: SELECTED LYRIC POEMS

You are pale, friend moon, and do not sleep at night,
    And day by day you waste away.
    Can it be that you also
    Think only of her, as I do?

—Anonymous

When the East
Gave birth to the Moon,
Love was the dancer at the feast;
The heavens smiled for joy;
And the Wind strewed the perfumed dust
Of lotus-pollen in the courtyard of the sky.

—Dharmakīrti

Who was artificer at her creation?
Was it the moon, bestowing its own charm?
Was it the graceful month of spring, itself
Compact with love, a garden full of flowers?
That ancient saint there, sitting in his trance,
Bemused by prayers and dull theology,
Cares naught for beauty: how could *he* create
Such loveliness, the old religious fool?

—Kālidāsa

In this vain fleeting universe, a man
Of wisdom has two courses: first, he can
Direct his time to pray, to save his soul,
And wallow in religion's nectar-bowl;
But, if he cannot, it is surely best
To touch and hold a lovely woman's breast,
And to caress her warm round hips, and thighs,
And to possess that which between them lies.

—Bhartrihari

I've gazed upon the river's waves, and seen
A bunch of water-weed, and shining pearls
All set in gems. But now the moon is cleansed
Of all her stains; and still the lotus blooms
Unsleeping; and a loving pair of birds
Are nestling close. And who has seen the like?

—Anonymous

With tail-fans spread, and undulating wings
With whose vibrating pulse the air now sings,
Their voices lifted and their beaks stretched wide,
Treading the rhythmic dance from side to side,
Eying the raincloud's dark, majestic hue,
Richer in color than their own throats' blue,
With necks upraised, to which their tails advance,
Now in the rains the screaming peacocks dance.

—Anonymous

The heron seeking supper in the lake
Darts his sharp eye around; his careful feet
Move gently, gently, every step they take.
Now with one leg withdrawn into the air
He twists his neck awry to spy his prey,
Alert for every moving lotus-leaf,
In case it should turn out to be a fish.

—Anonymous

With tumbled hair of swarms of bees
And flower-robes dancing in the breeze,
With sweet, unsteady lotus-glances,
Intoxicated, Spring advances.

—Anonymous

THREE

Transformations of the Way: China, Japan

China and Japan, the two great civilizations, along with Korea, of the Far East, have been linked for centuries by history, religion, writing, and architecture, and by painting and poetry exceptional for rich evocations of nature. Similarities between them are immediately evident. Boldly drawn characters raised to a fine art by the calligraphy of both cultures may, like pagodas and statues of Buddha, be difficult for outsiders to differentiate as Chinese or Japanese; landscape paintings, however individual, may seem alike, by contrast with Western conventions, in their sparely suggestive use of line on largely empty backgrounds of silk; and lyric poems wholly dissimilar in form and style may, by the deeply felt communion with the natural world to which they give expression, seem equally characteristic of either country.

These important affinities are grounded in a long history of Chinese influence and Japanese adaptation. No less important, however, are the very great differences between the two cultures. To begin with, China's written tradition is at least fifteen hundred years older than Japan's. Confucius, in the sixth century B.C., already looked back to a Chinese civilization he thought of as ancient (and superior to his own); the ancient civilization to which Nara Japan looked back, lacking one of its own, twelve hundred years later was that of Confucian (and by then Buddhist) China, whose seemingly autonomous culture serenely regarded itself as manifestly superior to that of the upstart barbarians of the eastern islands—or anywhere else. China was the Middle Kingdom, to which all others paid homage if not tribute; with the momentous exception of Buddhist India, it barely acknowledged the existence, much less conceded the value, of cultural traditions that were not Chinese. Japan, on the contrary, was from the beginning of its history a culture self-consciously adopting and adapting that of others: its written language, its literature, its religion, its art, however completely it made them its own, had origins, as everyone knew, somewhere else. Deeply (though ambivalently) as they venerated the imperial foreign civilization whose most radical changes typically took place in the name of the ancients, as those of Confucius had done, and brashly assertive as they might be of their own tribal and courtly traditions, the Japanese—notwithstanding long periods of exclusion, the most recent of which, in the Tokugawa (1615–1867), lasted over two hundred years—repeatedly embraced, as the Chinese seldom did, innovations from beyond their borders. Nowhere, G. B. Sansom writes in *Japan: A Short Cultural History* (1931; rev. ed. 1943), "have men more eagerly, nay recklessly, leaped to welcome new things and new notions."

One basic difference, which the introduction of Chinese writing to Japan in the early seventh century A.D. has sometimes obscured, is linguistic. Japanese, an agglutinative language of polysyllabic words formed from a small number of consonant-vowel combinations, is wholly unrelated to Chinese, a largely uninflected language whose predominantly monosyllabic and tonally distinct words, each syllable represented by a single character, are composed of complex vowel and consonant sounds. No writing system could thus have been less suited than the Chinese to the Japanese language, and so unwieldy did it prove that for centuries, until a practical phonetic syllabary *(kana)* gradually evolved to supplement the Chinese ideograms, writing appears to have remained a scribal monopoly, used mainly for accounts and registers. The linguistic difference resulted in extremely divergent poetic forms. Those of Chinese vary greatly in number of characters per line, rhyme scheme, tonal pattern, and length, whereas Japanese poets, writing in a language lacking such variation, soon entirely abandoned the loosely constructed longer form, the *chōka* (used extensively only in the earliest poetry collection, the *Man'yōshū*, compiled in the mid eighth century), in favor of the thirty-one syllable *tanka* or *waka*, and its much later seventeen-syllable offshoot, the *haiku*—almost the only forms in which Japanese poetry was written for more than a thousand years. One consequence of this major distinction was that Japanese poems tended to focus intensively on a single evocative image, typically of seasonal changes, flowers, dewdrops, birds in flight, or the moon, whereas those in Chinese could develop a far wider range of feelings and reflections concerning relationships between human beings and the world around them.

Not, of course, that language and prosody alone could account for these differences: from its beginnings, the Chinese tradition has been philosophic and intellectual in contrast to the less contemplative and more pragmatic bias of the Japanese. For centuries and millennia Confucians, Taoists, Legalists, Buddhists, Neo-Taoists, Neo-Confucians, and others disputed basic questions concerning society and the cosmos; for the most part, the Japanese left such speculations to others, adopting (and continually altering) whatever outlook—now Buddhist, now Neo-Confucian—suited their needs. They "have not as a rule," Sansom writes, "been disturbed by a fervid interest in transcendental problems. Their genius on the whole has been empiric and practical, their sentiment romantic rather than passionate." Early Japanese culture, moreover, was overwhelmingly centered on the aristocratic court of the capital city: first Nara, then Kyōto in the highly refined Heian period (794–1185). The first Chinese collection of poetry, the *Shih Ching [Shi Jing]*[1] or *Book of Poetry*, traditionally compiled by Confucius (himself a scholar of modest means wandering from court to court of feudal China's petty kingdoms), consists of anonymous folk songs; and later poets came from all ranks of society, living sometimes in cities, sometimes in the countryside, sometimes as hermits on isolated mountains. In Japan, by contrast, the poems of the *Man'yōshū*, though some are ascribed to soldiers and peasants, were mainly written by courtiers, even by emperors and princesses; after the first official collection, the *Kokinshū*, was published in 905, poetry long remained an aristocratic preserve.

In both China and Japan—as of course in many other societies, from Virgil's and Horace's Rome to Blake's and Wordsworth's London, and beyond—appreciation of the beauties and restorative powers of nature was heightened by contrast with the artifice and the bustle of urban life. But in Japan especially, where "almost all authors and the majority of readers lived (and live) in cities" (Shuichi Katō remarks in *A History of Japanese Literature: The First Thousand Years* [English trans. 1979]), it was not the robust nature worship of the popular indigenous Shintō religion—later transformed into an imperial cult—that found direct expression, however profound its indirect influence may have been, in courtly poetry, but a refined nostalgia bordering at times on escapism. "In Japan at all times," Katō writes, "it was one of the ideals of the writer to escape from the cities into the tranquillity of the mountains," and the less the ideal was realized the more appealing it seemed: hence "love of nature" in Japanese poetry derived from "refinement of the sensitivities of city people." The distinctive aestheticism of a country where (in Sansom's phrase) "religion became an art and art a religion" spread, Donald Keene

[1]Transliterations of Chinese names and words are in the Wade-Giles system widely used by English-language translators for most of the twentieth century; transliterations in brackets are in the *pinyin* system now adopted by the People's Republic of China.

observes in *Appreciations of Japanese Culture* (1971), "from the court to the provinces, and from the upper classes to the commoners."

Distinctly Japanese though it was, this poetry was inspired by the lyric poets of China, above all of the T'ang [Tang] dynasty (618–907) and the Six Dynasties before it, and for many centuries Japanese poets wrote in the Chinese language as well as in their own. But in China, more than a thousand years of indigenous philosophical and religious speculation lay behind and continually enriched a poetry that would have been inconceivable without it. Densely allusive as Chinese poetry is to earlier literary tradition, it is also permeated by currents springing from the "Hundred Schools" of philosophy in the late Chou [Zhou] dynasty, after about 600 B.C., and by the many schools of Mahāyāna Buddhism that made their way into China in the early centuries after Christ. Especially since its official adoption by the Han dynasty around 135 B.C., Confucianism, with its emphasis on education of the "gentleman" in service of the state, was generally dominant among these schools. No less important were the many currents, both philosophical and popular, that flowed into Taoism, where the claims of nature made themselves heard against the insistent demands of society: it was to Taoism that both poetry and an incipient science were most deeply indebted. These currents and others merged in the T'ang with Buddhism at the height of its influence in China. From this rich mix of potentially conflicting yet precariously harmonized views of life, out of which the Neo-Confucianism of the Sung [Song] dynasty (960–1279) would later develop, the poetry of the T'ang—poetry not of nature alone but of the complex and perpetually shifting interdependencies of the human and natural worlds—drew much of its incomparable beauty and power. Through T'ang poetry, as much as through Zen meditation or Confucian ethics, the manifold and incessantly evolving Way of the ancient Chinese sages was transmitted to the receptively transformative aesthetic culture of Japan.

5

The One and the Many: The Ways of Chinese Thought

From the proverbial Hundred Schools of the disintegrating Chou dynasty down to the Neo-Confucian synthesis of the Sung fifteen hundred years later, the transformations of philosophical speculation in China were many, yet the Way was always thought of as one. Chinese thought, Joseph Needham writes in his monumental Science and Civilisation in China *(1956), "refused to separate Man from Nature, or individual man from social man," so that every school, however different its conclusions might be from another's, presupposed a "Great Harmony" in the universe of which man was inseparably part.*

This conviction of cosmic "multëity in unity" (in Coleridge's phrase) may well have originated in religious beliefs long antedating Confucius, the first known Chou philosopher. Like those of Sumer, Egypt, the Indus valley, and other ancient agrarian civilizations, the earliest divinities of China appear to have been fertility gods, and their cult (C. P. Fitzgerald remarks in China: A Short Cultural History *[1935; 2nd ed. 1950]) "was there-*

fore directed to maintaining, by magical forces, the harmonious balance of nature, which alone made possible man's life upon the earth." The ruler by "mandate of Heaven" was responsible for that balance, and himself officiated at the seasonal rites and sacrifices (including, in earliest times, human sacrifices) on which fertility of the land depended: "charged with the maintenance of Natural Order and calendric Regularity," Marcel Granet writes in The Religion of the Chinese People *(1922), "the kings of the Chou family, feudal sovereigns of an agricultural people, were specifically appointed to be in hereditary control of the harvest."*

But although there was certainly no dearth of Chinese divinities now or later, as proliferating legions of Taoist and Buddhist deities attest, this early religion strikingly differed from that of Egypt, Mesopotamia, or India (and more closely resembled that of ancient Rome before it fell captive to the Greek Olympians) in lacking both a pantheon of officially venerated gods and canonically sanctioned myths of cosmic and national origin to validate their authority. As in the legendary Rome of Numa Pompilius and the early kings, worship appears to have been mainly directed toward local divinities of household and field, notably including the ancestral dead who shared responsibility for prosperity of the soil, on which the family's and people's depended. This ancestor worship, like that of tribal societies from Melanesia to Africa, reinforced the "sense that the natural world and human society are closely bonded," which, Granet affirms, "has been the basic element of all Chinese beliefs" in every age. Overarching these local deities, however, was the supreme heavenly Ancestor of the first historically attested dynasty, the Shang or Yin (dating from perhaps the late sixteenth to the late eleventh century B.C.), personified as Shang Ti [Shang Di]; under the Chou, who defeated the Shang, worship of Shang Ti seems to have been displaced by veneration of T'ien [Tian], *or Heaven. In other ancient civilizations, when an ancestral sky deity receded, like Sumerian An, Indo-Aryan Dyaus, or Greek Uranus, to the status of* deus otiosus *or do-nothing god, he was soon pushed aside by more vigorous progeny. But in China, where such divine succession was lacking and the Son of Heaven was a human king—albeit, in the case of the legendary Yellow Emperor and the sage kings Yao and Shun, hedged with divinity—*T'ien *was gradually transformed into an impersonal power identified, for all practical purposes, with Nature, the ultimate source of unity in the cosmos.*

Confucius (the latinized name given by seventeenth-century Jesuits to the Chinese K'ung Fu-tzu [Kung Fuzi], or "Master K'ung") was born in the small state of Lu in 551 B.C. and died in 479. At that time the petty feudal kingdoms of the Yellow River basin of northern China, loosely confederated in the Chou dynasty that had succeeded the Shang half a millennium before, were entering the "period of warring states," which would end with the victory in 221 of the half-barbarian state of Ch'in [Qin] and the unification, under its First Emperor, of the country that would thereafter bear its name. Throughout his long life as a wandering scholar, Confucius failed (as Plato would fail in the following century in Greece) to win the influential government position that he hoped might have allowed him to prevent the disintegration he saw taking place around him. In his veneration of the tao [dao], *or Way, of the ancients, and of the books of rites, poetry, music, divination, and history thereafter called the Confucian classics, of which legend made him the compiler or even author, he was a conservative thinker lamenting the lapse of sacrificial ritual, hierarchical authority, and filial piety in a degenerate age. Yet by insistently affirming that the superiority of the "gentleman"* (chün-tzu [junzi]) *lay not in aristocratic birth but in the civilizing moral force* (te [de]) *of goodness or benevolence* (jen [ren]), *which diligent study could foster in those who sincerely pursued it, he was, as Herbert Fingarette writes in* Confucius: The

Secular As Sacred *(1972), "a great cultural innovator" and "the creator of a new ideal, not an apologist for an old one."*

In Confucius himself, however, as we know him from the Lun-yü [Lunyu] *(called in English, since the nineteenth-century translation by James Legge, the* Analects, *or "Collected Readings"), the ideal remained almost entirely a social and cultural one: neither he nor his most influential follower, Mencius (Meng Tzu [Meng Zi]), in the late fourth century* B.C., *paid more than passing notice to the surrounding nonhuman world. Among the Hundred Schools of this intellectually fertile if troubled period—reduced to six in the later classification of the Han dynasty historian Ssu-ma T'an [Sima Tan] in the late second century* B.C.*—it was above all the Taoists who repaired the imbalance, stressing the need to follow, by the "actionless action"* (wu wei) *of intuitive communion rather than learning, an unnamable* tao *that was not the Way of the ancients alone, but of the universe, the inexhaustible mother of all things. As Taoism in the late Chou and Han assimilated to its philosophical base in the* Lao-tzu [Laozi] *and* Chuang-tzu [Zhuangzi] *an influx of currents from other schools (like that of Yin-Yang and the Five Elements) and from popular superstition, alchemy, hygiene, and the quest for physical immortality, these six (or rather the predominant outlooks to which they gave voice) in effect became two: just as everyone in the West has been called a Platonist or an Aristotelian, so everyone in China could be seen as a Confucian or a Taoist.*

Yet if Confucian Yang was opposed to Taoist Yin as light to shadow, all Taoists know that opposites periodically interact and merge before parting and merging again. In the Analects, *Confucius, the apostle of ceremonious ritual* (li) *and of cultured discourse and learning, sometimes expresses a discontent with civilization, even "a certain idealization of the 'noble savage'" (Arthur Waley suggests), such as we often find in Taoist writing. "The Master said, The Way makes no progress. I shall get upon a raft and float out to sea" (V.6). Or again: "The Master wanted to settle among the Nine Wild Tribes of the East. Someone said, I am afraid you would find it hard to put up with their lack of refinement. The Master said, Were a true gentleman to settle among them there would soon be no trouble about lack of refinement" (IX.13). In seeking the Middle Way, he sought far more than prudential moderation, knowing that the "true gentleman" can exist neither when rustic substance prevails over scholarly ornamentation—"when nature prevails over culture," in Waley's paraphrase—nor vice versa, but only when the two blend (VI.16). And at one point (VI.21) he suggests a polarity of complementary opposites through whose interaction Confucian and Taoist outlooks might merge into one: "The wise man delights in water, the Good man delights in mountains. For the wise move; but the Good stay still. The wise are happy; but the Good, secure."*[1]

For the most part, however, Confucius might logically have seemed the proponent of everything the Taoists opposed, and in the Chuang-tzu *(where he appears more often than Lao Tzu or Chuang Tzu himself!), he is sometimes portrayed as a slightly comical and dull-witted philistine, unable to grasp the Way of Lao Tzu and his followers, which he vainly seeks in rules and regulations. "Why these flags of benevolence and righteousness so bravely upraised, as though you were beating a drum and searching for a lost child?" Lao Tzu asks him (Chapter 13): "Ah, you will bring confusion to the nature of man!" Small wonder that this befuddled Confucius falls silent after encountering such a "dragon":*

[1]In Waley's view, "This saying, in the form in which it now occurs, is completely Taoistic, save that the word Good (Goodness) [jên] has been substituted for 'Tao.' ... The dictum easily passes into the vocabulary of full-blown, systematic quietism." Elsewhere he suggests that "jên is a mystic entity not merely analogous to but in certain sayings practically identical with the Tao of the Quietists. Like Tao, it is contrasted with 'knowledge.'"

"My mouth fell open and I couldn't close it; my tongue flew up and I couldn't even stammer" (14). Yet with perseverance, the pupil finally grasps that he must "take my own place as a man along with the process of change"—"Good, Ch'iu—now you've got it!" Lao Tzu exclaims (14)—and even retires, like a Taoist hermit, to the great swamp, "wearing furs and coarse cloth and living on acorns and chestnuts" (20). Thus we should not be surprised, since Taoist logic is preeminently one of contradictions, to find him appearing elsewhere in the Chuang-tzu *as the very paragon of a Taoist sage, advising* his *pupil, for example, to listen not with mind but with spirit, since "spirit is empty and waits on all things. The Way gathers in emptiness alone" (4). Nor should we be too surprised that Neo-Taoists of later centuries, in a triumphant display of Taoist logic, "considered Confucius to be even greater than Lao Tzu or Chuang Tzu. Confucius, they maintained, did not speak about forgetfulness," Fung Yu-lan remarks in* A Short History of Chinese Philosophy *(1948), "because he had already forgotten that he had learned to forget. Nor did he speak about absence of desire, because he had already reached the stage of lacking any desire to be without desire." Taoism, the philosophy of communion with the natural world, could after all blend in Ch'an [Chan], or Zen, with Mahāyāna Buddhism, which repudiated that world; and from its merger with Confucian ethics, finally, could arise the Neo-Confucian cosmology in which Chinese thought achieved its ultimate affirmation of the unity of nature and man. If the Hundred Schools had been reduced to six, and the six to two, it was surely logical that the two, all along, had been one.*

The Bottomless Vessel

Tao Te Ching [Dao De Jing],
translated by Arthur Waley

Lao Tzu [Lao Zi] was said to have been an older contemporary of Confucius; the book of eighty-one chapters that bears his name, Lao-tzu [Laozi], *was later also called* Tao Te Ching [Dao De Jing], *or* The Book of the Way and Its Power. *"At the age of 160," as Holmes Welch tells the legend in* Taoism: The Parting of the Way *(1957; 2nd ed. 1965), "Lao Tzu grew disgusted with the decay of the Chou dynasty and resolved to pursue virtue in a more congenial atmosphere. Riding in a chariot drawn by a black ox, he left the Middle Kingdom through the Han-ku Pass which leads westward from Loyang. The Keeper of the Pass, Yin Hsi, who, from the state of the weather, had expected a sage, addressed him as follows: 'You are about to withdraw yourself from sight. I pray you to compose a book for me.' Lao Tzu thereupon wrote the 5,000 characters which we call the* Tao Te Ching. *After completing the book, he departed for the west." (Some later Taoists were to claim that in India the Buddha became one of Lao Tzu's disciples!)*

Whether a sage named Lao Tan [Lao Dan] or Lao Tzu (literally, "Old Master") actually lived at this time or later, leaving the memory of sayings that became the nucleus of the book, is at best doubtful; in any case, most scholars agree that the Tao Te Ching *as we have it, whether composed by several hands or by one (with perhaps some later additions), dates from no earlier than the mid fourth century B.C. Owing partly to its archaic Chinese and partly to its elliptical style of deliberate mystification, which continually emphasizes the ineffability of the Tao, this brief classic is notoriously subject to wildly variant interpretations: it is, Arthur Waley writes in* Three Ways of Thought in Ancient China *(1939), "an occultist kaleidoscope, a magic void that the reader can fill in with what images he will." As with any book, but far more than most, to read it, as Welch observes, "is an act of creation." Ever since the first English version by John Chalmers in 1868, translators have repeatedly filled in this empty canvas—the* Tao Te Ching *has been called the most translated book after the Bible—in ways that often leave the same passage wholly unrecognizable from one to another. (Waley's translation in* The Way and Its Power *[1935], which attempts "to discover what the book meant when it was first written," is wordier and less smooth than some that have followed, but since Waley, the foremost translator of Chinese and Japanese poetry in our century, was fully capable of graceful clarity, the very knottiness of his version may betoken fidelity to the original, which by all accounts is not clear.)*

To advance any one interpretation of the unnamable Tao, "formless yet complete," that underlies and, through its very emptiness, quietness, and nonactivity, gives rise to the "ten thousand things" of the phenomenal world would of course be to violate the book's essential spirit; but something of what it is can be suggested by what it manifestly is not. It is not, to begin with, a personal deity, like Enlil, Indra, or Yahweh; the incessant process of creation and change arising from and returning to it is a purely spontaneous one. Thus, though the book is remarkable for its nearly total lack of reference to the richly varied particularity of nature found so profusely in some Hebrew Psalms and hymns of the Rig Veda, *to say nothing of Kālidāsa's poems—almost no plants, birds, or animals, sun, moon, or stars, only "Heaven and Earth," water and valleys—the Tao, in contrast to Confucius's Way of the ancients, becomes something very like "the Order of Nature," in Needham's words, "which brought all things into existence and governs their every action, . . . controlling the orderly processes of change. . . . If there was one idea which the Taoist philosophers stressed more than any other," Needham writes, "it was the unity of Nature, and the eternity and uncreatedness of the Tao." Perhaps, since those who know do not speak, and those who speak do not know, it is better to say no more.*

4 The Way is like an empty vessel
That yet may be drawn from
Without ever needing to be filled.
It is bottomless; the very progenitor of all things in the world.
In it all sharpness is blunted,
All tangles untied,
All glare tempered,
All dust smoothed.
It is like a deep pool that never dries.
Was it too the child of something else? We cannot tell.
But as a substanceless image[2]
it existed before the Ancestor.[3]

6 The Valley Spirit never dies.
It is named the Mysterious Female.
And the Doorway of the Mysterious Female
Is the base from which Heaven and Earth sprang.
It is there within us all the while;
Draw upon it as you will, it never runs dry.

[2]A *hsiang*, an image such as the mental images that float before us when we think. (Waley)
[3]The Ancestor in question is almost certainly the Yellow Ancestor who separated Earth from Heaven and so destroyed the Primal Unity, for which he is frequently censured in *Chuang Tzu*. (Waley)

16 Push far enough towards the Void,
Hold fast enough to Quietness,
And of the ten thousand things none but can be worked on by you.
I have beheld them, whither they go back.
See, all things howsoever they flourish
Return to the root from which they grew.
This return to the root is called Quietness;
Quietness is called submission to Fate;
What has submitted to Fate has become part of the always-so.
To know the always-so is to be Illumined;
Not to know it, means to go blindly to disaster.
He who knows the always-so has room in him for everything;
He who has room in him for everything is without prejudice.
To be without prejudice is to be kingly;
To be kingly is to be of heaven;
To be of heaven is to be in Tao.
Tao is forever and he that possesses it,
Though his body ceases, is not destroyed.

25 There was something formless yet complete,
That existed before heaven and earth;
Without sound, without substance,
Dependent on nothing, unchanging,
All pervading, unfailing.
One may think of it as the mother of all things under heaven.
Its true name we do not know.
"Way" is the by-name that we give it.
Were I forced to say to what class of things it belongs I should call it
Great (*ta*).
Now *ta* also means passing on,
And passing on means going Far Away,
And going far away means returning. . . .[4]

The ways of men are conditioned by those of earth. The ways of earth, by those of heaven. The ways of heaven by those of Tao, and the ways of Tao by the Self-so.[5]

40 In Tao the only motion is returning;
The only useful quality, weakness.
For though all creatures under heaven are the products of Being,
Being itself is the product of Not-being.

[4]Returning to "what was there at the Beginning." (Waley)
[5]The "unconditioned"; the "what-is-so-of-itself." (Waley) Some translate this term *(tzu-jan [ziran])* as "nature."

43 What is of all things most yielding
Can overwhelm that which is of all things most hard.
Being substanceless it can enter even where there is no space;
That is how I know the value of action that is actionless.
But that there can be teaching without words,
Value in action that is actionless,
Few indeed can understand.

52 That which was the beginning of all things under heaven
We may speak of as the "mother" of all things.
He who apprehends the mother[6]
Thereby knows the sons.[7]
And he who has known the sons
Will hold all the tighter to the mother,
And to the end of his days suffer no harm:
"Block the passages, shut the doors,
And till the end your strength shall not fail.
Open up the passages, increase your doings,
And till your last day no help shall come to you."
As good sight means seeing what is very small
So strength means holding on to what is weak.
He who having used the outer-light[8] can return to the inner-light
Is thereby preserved from all harm.
This is called resorting to the always-so.

56 Those who know do not speak;
Those who speak do not know.
Block the passages,
Shut the doors,
Let all sharpness be blunted,
All tangles untied,
All glare tempered.
All dust smoothed.
This is called the mysterious leveling.
He who has achieved it cannot either be drawn into friendship or repelled,
Cannot be benefited, cannot be harmed,
Cannot either be raised or humbled,
And for that very reason is highest of all creatures under heaven.

78 Nothing under heaven is softer or more yielding than water; but when it attacks things hard and resistant there is not one of them that can prevail. For they can find no way of altering it. That the yielding conquers the resistant and the soft conquers the hard is a fact known by all men, yet utilized by none. . . .

[6]Tao, the One, the Whole. (Waley)
[7]The Many, the universe. (Waley)
[8]This corresponds to "knowing the sons." *Ming* ("inner light") is self-knowledge. (Waley)

The Torch of Chaos and Doubt

Chuang-tzu [Zhuangzi],
translated by Burton Watson

Chuang Chou [Zhuang Zhou], or Chuang Tzu [Zhuang Zi], according to the "Records of the Historian" by Ssu-ma Ch'ien [Sima Qian] (145?–85? B.C.), was "an official in the lacquer garden" of Meng in the fourth century B.C., who wrote a work of 100,000 words or more, "mostly in the nature of fable"; he was very probably a historical person, but these few facts (if they are facts) are all that we know about him outside the book that bears his name. How much if any of this book, the Chuang-tzu [Zhuangzi], *he himself wrote is impossible to say, but large parts of it are certainly later, probably incorporating material from widely different periods. As Burton Watson explains in* Complete Works of Chuang Tzu *(1968), the book is mentioned in the first century B.C. as being in fifty-two chapters; the edition we have is essentially that compiled, with commentary, by Kuo Hsiang [Guo Xiang] (died A.D. 312), who discarded much that he considered spurious and divided the rest into thirty-three chapters, calling 1–7 the "inner chapters," 8–22 the "outer chapters," and 23–33 the "miscellaneous chapters."*

The "inner" and parts of the "outer" chapters in this arrangement contain some of the most vividly imaginative, paradoxically brilliant, irreverently provocative prose in Chinese. Through seemingly disconnected, sometimes bewildering, even hilarious stories, they portray an unforgettable gallery of Taoist sages and would-be sages, from Lao Tzu and Chuang Tzu to the Yellow Emperor and Confucius, all giving expression to the central theme, as Watson says, of freedom *from the inherited conventions of the social world—above all from the stifling Confucian virtues of benevolence and righteousness and all other constraining rules and regulations. They are uncompromising advocates, in Greek terms, of* physis *over* nomos, *or nature over law. Chaos and doubt become their joyful beacon, as intuition becomes their mode of perception and paradox and contradiction their manner of speaking; they know and cheerfully affirm the relativity of all merely human values in a world where deer would flee in terror from the beautiful Lady Li, and Chuang Tzu cannot be sure whether he is himself or a butterfly.*

Like Confucius of the Analects, *the Taoist sage too looks to the past, but not to a time of idealized virtue under the Duke of Chou or the Yellow Emperor: to a mythical time, rather, of total unity—the perfect harmony of yin and yang, or even the primal void of nonbeing—before the artificial distinctions of civilization divided one man from another and each from himself. For the sage, however, who lives a life of "unvarying spontaneity" in the "jumble of wonder and mystery" all round him, dragging his tail in the mud like a tortoise and intuiting what fish enjoy, that ancient time is now, and the world has not lost its simplicity. "At ease in the illimitable," he understands that Tao is everywhere, even, Chuang Tzu says, "in the piss and the shit." Later Taoists would increasingly devote themselves to seeking everlasting life (like the fabulous Kuang Ch'eng [Guang Cheng] in Chapter 11) through bodily hygiene, or through mushrooms and alchemical elixirs; but here the true sages, like Master Lai in Chapter 6 and Chuang Tzu in Chapter 18, are those who apprehend that death is a process of change embodying the eternal alternation of yin and yang, and can therefore accept it with equanimity, or with singing and pounding of tubs. "Heaven and earth were born at the same time as I was, and the ten thousand things are one with me": this is the immortality that comes from affirming inseparable unity with the world as a whole, from highest to lowest, in life as in death—the immortality of those who intuitively know that Chuang Chou and the butterfly are, of course, one, insofar as each is unmistakably and exuberantly itself.*

(2) The understanding of the men of ancient times went a long way. How far did it go? To the point where some of them believed that things have never existed—so far, to the end, where nothing can be added. Those at the next stage thought that things exist but recognized no boundaries among them. Those at the next stage thought there were boundaries but recognized no right and wrong. Because right and wrong appeared, the Way was injured, and because the Way was injured, love became complete. But do such things as completion and injury really exist, or do they not? . . .

The torch of chaos and doubt—this is what the sage steers by.[9] So he does not use things but relegates all to the constant. This is what it means to use clarity.

There is a beginning. There is a not yet beginning to be a beginning. There is a not yet beginning to be a not yet beginning to be a beginning. There is being. There is nonbeing. There is a not yet beginning to be nonbeing. There is a not yet beginning to be a not yet beginning to be nonbeing. Suddenly there is nonbeing. But I do not know, when it comes to nonbeing, which is really being and which is nonbeing. Now I have just said something. But I don't know whether what I have said has really said something or whether it hasn't said something.

There is nothing in the world bigger than the tip of an autumn hair, and Mount T'ai is tiny. No one has lived longer than a dead child, and P'eng-tsu[10] died young. Heaven and earth were born at the same time I was, and the ten thousand things are one with me.

[Wang Ni asks Nieh Ch'üeh:] "If a man sleeps in a damp place, his back aches and he ends up half paralyzed, but is this true of a loach? If he lives in a tree, he is terrified and shakes with fright, but is this true of a monkey? Of these three creatures, then, which one knows the proper place to live? Men eat the flesh of grass-fed and grain-fed animals, deer eat grass, centipedes find snakes tasty, and hawks and falcons relish mice. Of these four, which knows how food ought to taste? Monkeys pair with monkeys, deer go out with deer, and fish play around with fish. Men claim that Mao-ch'iang and Lady Li were beautiful, but if fish saw them they would dive to the bottom of the stream, if birds saw them they would fly away, and if deer saw them they would break into a run. Of these four, which knows how to fix the standard of beauty for the world? The way I see it, the rules of benevolence and righteousness and the paths of right and wrong are all hopelessly snarled and jumbled. How could I know anything about such discriminations?"

Nieh Ch'üeh said, "If you don't know what is profitable or harmful, then does the Perfect Man likewise know nothing of such things?"

Wang Ni replied, "The Perfect Man is godlike. Though the great swamps blaze, they cannot burn him; though the great rivers freeze, they cannot chill him; though swift lightning splits the hills and howling gales shake the sea, they cannot frighten him. A

[9]He accepts things as they are, though to the ordinary person attempting to establish values they appear chaotic and doubtful and in need of clarification. (Watson)
[10]The "Chinese Methuselah."

man like this rides the clouds and mist, straddles the sun and moon, and wanders beyond the four seas. Even life and death have no effect on him, much less the rules of profit and loss!"

Once Chuang Chou dreamt he was a butterfly, a butterfly flitting and fluttering around, happy with himself and doing as he pleased. He didn't know he was Chuang Chou. Suddenly he woke up and there he was, solid and unmistakable Chuang Chou. But he didn't know if he was Chuang Chou who had dreamt he was a butterfly, or a butterfly dreaming he was Chuang Chou. Between Chuang Chou and a butterfly there must be *some* distinction! This is called the Transformation of Things.

(6) The Way has its reality and its signs but is without action or form. You can hand it down but you cannot receive it; you can get it but you cannot see it. It is its own source, its own root. Before Heaven and earth existed it was there, firm from ancient times. It gave spirituality to the spirits and to God; it gave birth to Heaven and earth. It exists beyond the highest point, and yet you cannot call it lofty; it exists beneath the limit of the six directions, and yet you cannot call it deep. It was born before Heaven and earth, and yet you cannot say it has been there for long; it is earlier than the earliest time, and yet you cannot call it old.

Suddenly Master Lai grew ill. Gasping and wheezing, he lay at the point of death. His wife and children gathered round in a circle and began to cry. Master Li, who had come to ask how he was, said, "Shoo! Get back! Don't disturb the process of change!"

Then he leaned against the doorway and talked to Master Lai. "How marvelous the Creator is! What is he going to make of you next? Where is he going to send you? Will he make you into a rat's liver? Will he make you into a bug's arms?"

Master Lai said, "A child, obeying his father and mother, goes wherever he is told, east or west, south or north. And the yin and yang—how much more are they to a man than father or mother! Now that they have brought me to the verge of death, if I should refuse to obey them, how perverse I would be! What fault is it of theirs? The Great Clod burdens me with form, labors me with life, eases me in old age, and rests me in death. So if I think well of my life, for the same reason I must think well of my death. When a skilled smith is casting metal, if the metal should leap up and say, 'I insist upon being made into a Mo-yeh!'[11] he would surely regard it as very inauspicious metal indeed. Now, having had the audacity to take on human form once, if I should say, 'I don't want to be anything but a man! Nothing but a man!', the Creator would surely regard me as a most inauspicious sort of person. So now I think of heaven and earth as a great furnace, and the Creator as a skilled smith. Where could he send me that would not be all right? I will go off to sleep peacefully, and then with a start I will wake up."

(11) The Yellow Emperor had ruled as Son of Heaven for nineteen years and his commands were heeded throughout the world, when he heard that Master Kuang Ch'eng was living on top of the Mountain of Emptiness and Identity. He therefore went to visit him. "I have heard that you, Sir, have mastered the Perfect Way. May I venture to ask about the essence of the Perfect Way?" he said. "I would like to get hold of the essence of Heaven and earth and use it to aid the five grains and to nourish the common

[11]A famous sword of King Ho-lü (r. 514–496 B.C.) of Wu. (Watson)

people. I would also like to control the yin and yang in order to insure the growth of all living things. How may this be done?"

Master Kuang Ch'eng said, "What you say you want to learn about pertains to the true substance of things, but what you say you want to control pertains to things in their divided state.[12] Ever since you began to govern the world, rain falls before the cloud vapors have even gathered, the plants and trees shed their leaves before they have even turned yellow, and the light of the sun and moon grows more and more sickly. Shallow and vapid, with the mind of a prattling knave—what good would it do to tell *you* about the Perfect Way!"

The Yellow Emperor withdrew, gave up his throne, built a solitary hut, spread a mat of white rushes, and lived for three months in retirement. Then he went once more to request an interview. Master Kuang Ch'eng was lying with his face to the south.[13] The Yellow Emperor, approaching in humble manner, crept forward on his knees, bowed his head twice and said, "I have heard that you, Sir, have mastered the Perfect Way. I venture to ask about the governing of the body. What should I do in order to live a long life?"

Master Kuang Ch'eng sat up with a start. "Excellent, this question of yours! Come, I will tell you about the Perfect Way. The essence of the Perfect Way is deep and darkly shrouded; the extreme of the Perfect Way is mysterious and hushed in silence. Let there be no seeing, no hearing; enfold the spirit in quietude and the body will right itself. Be still, be pure, do not labor your body, do not churn up your essence, and then you can live a long life. When the eye does not see, the ear does not hear, and the mind does not know, then your spirit will protect the body, and the body will enjoy long life. Be cautious of what is within you; block off what is outside you, for much knowledge will do you harm. Then I will lead you up above the Great Brilliance, to the source of the perfect Yang; I will guide you through the Dark and Mysterious Gate, to the source of the perfect Yin. Heaven and earth have their controllers, the yin and yang their storehouses. You have only to take care and guard your own body; these other things will of themselves grow sturdy. As for myself, I guard this unity, abide in this harmony, and therefore I have kept myself alive for twelve hundred years, and never has my body suffered any decay."

The Yellow Emperor bowed thrice and said, "Master Kuang Ch'eng, you have been as a Heaven to me!" . . .

(12) In the Great Beginning, there was nonbeing; there was no being, no name. Out of it arose One; there was One, but it had no form. Things got hold of it and came to life, and it was called Virtue.[14] Before things had forms, they had their allotments; these were of many kinds, but not cut off from one another, and they were called fates. Out of the flow and flux, things were born, and as they grew they developed distinctive shapes; these were called forms. The forms and bodies held within them spirits, each with its own characteristics and limitations, and this was called the inborn nature. If the nature is trained, you may return to Virtue, and Virtue at its highest peak is identical with the Beginning. Being identical, you will be empty; being empty, you will be great. You may

[12] I.e., the yin and yang, being two, already represent a departure from the primal unity of the Way. What Master Kuang Ch'eng is objecting to, of course, is the fact that the Yellow Emperor wishes to "control" them. (Watson)

[13] The Chinese ruler, when acting as sovereign, faces south. Master Ch'eng, by assuming the same position, indicates his spiritual supremacy. (Watson)

[14] Virtue *(te)* in the sense of "intrinsic power" (not "morality"), as in *Tao Te Ching.*

join in the cheeping and chirping and, when you have joined in the cheeping and chirping, you may join with Heaven and earth. Your joining is wild and confused, as though you were stupid, as though you were demented. This is called Dark Virtue. Rude and unwitting, you take part in the Great Submission.

(13) Confucius went to deposit his works with the royal house of Chou. Tzu-lu advised him, saying, "I have heard that the Keeper of the Royal Archives is one Lao Tan, now retired and living at home. If you wish to deposit your works, you might try going to see him about it."

"Excellent!" said Confucius, and went to see Lao Tan, but Lao Tan would not give permission. Thereupon Confucius unwrapped his Twelve Classics and began expounding them. Halfway through the exposition, Lao Tan said, "This will take forever! Just let me hear the gist of the thing!"

"The gist of it," said Confucius, "is benevolence and righteousness."

"May I ask if benevolence and righteousness belong to the inborn nature of man?" said Lao Tan.

"Of course," said Confucius. "If the gentleman lacks benevolence, he will get nowhere; if he lacks righteousness, he cannot even stay alive. Benevolence and righteousness are truly the inborn nature of man. What else could they be?"

Lao Tan said, "May I ask your definition of benevolence and righteousness?"

Confucius said, "To be glad and joyful in mind; to embrace universal love and be without partisanship—this is the true form of benevolence and righteousness."

Lao Tan said, "Hmm—close—except for the last part. 'Universal love'—that's a rather nebulous ideal, isn't it? And to be without partisanship is already a kind of partisanship. Do you want to keep the world from losing its simplicity? Heaven and earth hold fast to their constant ways, the sun and moon to their brightness, the stars and planets to their ranks, the birds and beasts to their flocks, the trees and shrubs to their stands. You have only to go along with Virtue in your actions, to follow the Way in your journey, and already you will be there. Why these flags of benevolence and righteousness so bravely upraised, as though you were beating a drum and searching for a lost child? Ah, you will bring confusion to the nature of man!"

(14) Confucius said to Lao Tan, "I have been studying the Six Classics—the Odes, the Documents, the Ritual, the Music, the Changes, and the Spring to Autumn, for what I would call a long time, and I know their contents through and through. But I have been around to seventy-two different rulers with them, expounding the ways of the former kings and making clear the path trod by the dukes of Chou and Shao, and yet not a single ruler has found anything to excite his interest. How difficult it is to persuade others, how difficult to make clear the Way!"

Lao Tzu said, "It's lucky you didn't meet with a ruler who would try to govern the world as you say. The Six Classics are the old worn-out paths of the former kings—they are not the thing which walked the path. What you are expounding are simply these paths. Paths are made by shoes that walk them, they are by no means the shoes themselves!

"The white fish hawk has only to stare unblinking at its mate for fertilization to occur. With insects, the male cries on the wind above, the female cries on the wind below, and there is fertilization. The creature called the *lei* is both male and female and so it can fertilize itself. Inborn nature cannot be changed, fate cannot be altered, time cannot be stopped, the Way cannot be obstructed. Get hold of the Way and there's nothing that can't be done; lose it and there's nothing that *can* be done!"

Confucius stayed home for three months and then came to see Lao Tan once again. "I've got it," he said. "The magpie hatches its young, the fish spit out their milt, the slim-waisted wasp has its stages of transformation, and when baby brother is born, big brother howls. For a long time now I have not been taking my place as a man along with the process of change. And if I do not take my own place as a man along with the process of change, how can I hope to change other men?"

Lao Tzu said, "Good, Ch'iu—now you've got it!"

(15) To repair to the thickets and ponds, living idly in the wilderness, angling for fish in solitary places, inaction his only concern—such is the life favored by the scholar of the rivers and seas, the man who withdraws from the world, the unhurried idler. To pant, to puff, to hail, to sip, to spit out the old breath and draw in the new, practicing bear-hangings and bird-stretchings, longevity his only concern—such is the life favored by the scholar who practices Induction, the man who nourishes his body, who hopes to live to be as old as P'eng-tsu.

But to attain loftiness without constraining the will; to achieve moral training without benevolence and righteousness, good order without accomplishments and fame, leisure without rivers and seas, long life without Induction; to lose everything and yet possess everything, at ease in the illimitable, where all good things come to attend—this is the Way of Heaven and earth, the Virtue of the sage. So it is said, Limpidity, silence, emptiness, inaction—these are the level of Heaven and earth, the substance of the Way and its Virtue. . . .

(16) The men of old dwelt in the midst of crudity and chaos; side by side with the rest of the world, they attained simplicity and silence there. At that time the yin and yang were harmonious and still, ghosts and spirits worked no mischief, the four seasons kept to their proper order, the ten thousand things knew no injury, and living creatures were free from premature death. Although men had knowledge, they did not use it. This was called the Perfect Unity. At this time, no one made a move to do anything, and there was unvarying spontaneity.

The time came, however, when Virtue began to dwindle and decline, and then Sui Jen and Fu Hsi stepped forward to take charge of the world. As a result there was compliance, but no longer any unity. Virtue continued to dwindle and decline, and then Shen Nung and the Yellow Emperor stepped forward to take charge of the world. As a result, there was security, but no longer any compliance. Virtue continued to dwindle and decline, and then Yao and Shun stepped forward to take charge of the world.[15] They set about in various fashions to order and transform the world, and in doing so defiled purity and shattered implicity. The Way was pulled apart for the sake of goodness; Virtue was imperiled or the sake of conduct. After this, inborn nature was abandoned and minds were set free to roam, mind joining with mind in understanding; there was knowledge, but it could not bring stability to the world. After this, "culture" was added on, and

[15]All these figures are mythical rulers or culture heroes; Sui Jen and Shen Nung are the discoverers of fire and agriculture respectively. (Watson)

"breadth" was piled on top. "Culture" destroyed the substantial, "breadth" drowned the mind, and after this the people began to be confused and disordered. They had no way to revert to the true form of their inborn nature or to return once more to the Beginning.

(17) Once, when Chuang Tzu was fishing in the P'u River, the king of Ch'u sent two officials to go and announce to him: "I would like to trouble you with the administration of my realm."

Chuang Tzu held on to the fishing pole and, without turning his head, said, "I have heard that there is a sacred tortoise in Ch'u that has been dead for three thousand years. The king keeps it wrapped in cloth and boxed, and stores it in the ancestral temple. Now would this tortoise rather be dead and have its bones left behind and honored? Or would it rather be alive and dragging its tail in the mud?"

"It would rather be alive and dragging its tail in the mud," said the two officials.

Chuang Tzu said, "Go away! I'll drag my tail in the mud!"

Chuang Tzu and Hui Tzu were strolling along the dam of the Hao River when Chuang Tzu said, "See how the minnows come out and dart around where they please! That's what fish really enjoy!"

Hui Tzu said, "You're not a fish—how do you know what fish enjoy?"

Chuang Tzu said, "You're not I, so how do you know I don't know what fish enjoy?"

Hui Tzu said, "I'm not you, so I certainly don't know what you know. On the other hand, you're certainly not a fish—so that still proves you don't know what fish enjoy!"

Chuang Tzu said, "Let's go back to your original question, please. You asked me *how* I know what fish enjoy—so you already knew I knew it when you asked the question. I know it by standing here beside the Hao."

(18) Chuang Tzu's wife died. When Hui Tzu went to convey his condolences, he found Chuang Tzu sitting with his legs sprawled out, pounding on a tub and singing. "You lived with her, she brought up your children and grew old," said Hui Tzu. "It should be enough simply not to weep at her death. But pounding on a tub and singing—this is going too far, isn't it?"

Chuang Tzu said, "You're wrong. When she first died, do you think I didn't grieve like anyone else? But I looked back to her beginning and the time before she was born. Not only the time before she was born, but the time before she had a body. Not only the time before she had a body, but the time before she had a spirit. In the midst of the jumble of wonder and mystery a change took place and she had a spirit. Another change and she had a body. Another change and she was born. Now there's been another change and she's dead. It's just like the progression of the four seasons, spring, summer, fall, winter.

"Now she's going to lie down peacefully in a vast room. If I were to follow after her bawling and sobbing, it would show that I don't understand anything about fate. So I stopped."

(20) Confucius was besieged between Ch'en and Ts'ai, and for seven days he ate no cooked food. T'ai-kung Jen went to offer his sympathy. "It looks as if you're going to die," he said.

"It does indeed."

"Do you hate the thought of dying?"

"I certainly do!"

Jen said, "Then let me try telling you about a way to keep from dying. . . . The straight-trunked tree is the first to be felled; the well of sweet water is the first to run dry. And you, now—you show off your wisdom in order to astound the ignorant, work at your good conduct in order to distinguish yourself from the disreputable, going around bright and shining as though you were carrying the sun and moon in your hand! That's why you can't escape! . . . The Perfect Man wants no repute. Why then do you delight in it so?"

"Excellent!" exclaimed Confucius. Then he said good-bye to his friends and associates, dismissed his disciples, and retired to the great swamp, wearing furs and coarse cloth and living on acorns and chestnuts. He could walk among the animals without alarming their herds, walk among the birds without alarming their flocks. If even the birds and beasts did not resent him, how much less would men!

Chuang Tzu put on his robe of coarse cloth with the patches on it, tied his shoes with hemp to keep them from falling apart, and went to call upon the king of Wei. "My goodness, Sir, you certainly are in distress!" said the king of Wei.

Chuang Tzu said, "I am poor, but I am not in distress! When a man possesses the Way and its Virtue but cannot put them into practice, then he is in distress. When his clothes are shabby and his shoes worn through, then he is poor, but he is not in distress. This is what they call being born at the wrong time. Has your Majesty never observed the bounding monkeys? If they can reach the tall cedars, the catalpas, or the camphor trees, they will swing and sway from their limbs, frolic and lord it in their midst, and even the famous archers Yi or P'eng Meng could not take accurate aim at them. But when they find themselves among prickly mulberries, brambles, hawthorns, or spiny citrons, they must move with caution, glancing from side to side, quivering and quaking with fear. It is not that their bones and sinews have suddenly become stiff and lost their suppleness. It is simply that the monkeys find themselves in a difficult and disadvantageous position where they cannot exercise their abilities to the full. And now if I should live under a benighted ruler and among traitorous ministers and still hope to escape distress, what hope would there be of doing so? Pi Kan had his heart cut out—there is the proof of the matter!"

(22) Master Tung-kuo asked Chuang Tzu, "This thing called the Way—where does it exist?"

Chuang Tzu said, "There's no place it doesn't exist."

"Come," said Master Tung-kuo, "you must be more specific."

"It is in the ant."

"As low a thing as that?"

"It is in the panic grass."

"But that's lower still!"

"It is in the tiles and shards."

"How can it be so low?"

"It is in the piss and shit!"

Master Tung-kuo made no reply.

Chuang Tzu said, "Sir, your questions simply don't get at the substance of the matter. When Inspector Huo asked the superintendent of the market how to test the fatness of a pig by pressing it with the foot, he was told that the lower down on the pig you press, the nearer you come to the truth. But you must not expect to find the Way in any particular place—there is no thing that escapes its presence! Such is the Perfect Way, and so too are the truly great words. 'Complete,' 'universal,' 'all-inclusive'—these three are different words with the same meaning. All point to a single reality.

"Why don't you try wandering with me to the Palace of Not-Even-Anything—identity and concord will be the basis of our discussions and they will never come to an end, never reach exhaustion. Why not join with me in inaction, in tranquil quietude, in hushed purity, in harmony and leisure? Already my will is vacant and blank. I go nowhere and don't know how far I've gotten. I go and come and don't know where to stop. I've already been there and back, and I don't know when the journey is done. I ramble and relax in unbordered vastness; Great Knowledge enters in, and I don't know where it will ever end. . . ."

Permutations of Heaven and Man

Changing Visions of the Natural and the Human World

Unlike Tao, which was never personified, T'ien [Tian] *("Heaven" or "sky") may once have designated a celestial divinity. Needham is among those who consider the Chinese character "undoubtedly an anthropomorphic graph (presumably of a deity) in its most ancient form," and accept H. C. Creel's supposition that rulers of the Chou substituted T'ien for the deified Shang Ti of the dynasty they overthrew. Be that as it may, Confucius, centuries later, clearly thought of Heaven, Creel remarks in* Confucius and the Chinese Way *(1949), "as an impersonal ethical force, a cosmic counterpart of the ethical sense in man, a guarantee that somehow there is sympathy with man's sense of right in the very nature of the universe." Thereafter, in Chou philosophy, as David L. Hall and Roger T. Ames observe in* Thinking Through Confucius *(1987), "there is a gradual tendency toward the depersonalization of* t'ien, *first in the relatively early identification of the will of* t'ien *with popular consensus, and further in a gradual redefining of* t'ien *as a designation for the regular pattern discernible in the unfolding processes of existence." Thus a word for sky that may once have meant "Heaven" as a providential power, or even in some sense "God," comes increasingly, without wholly losing its ethical component, to be more or less synonymous with what we call "nature," just as "Heaven and Earth" can designate the universe as a whole.*

The three passages that follow illustrate three conceptions of Heaven in relation to the human world. The first (from Sources of Chinese Tradition *[1960]), though last to be written, codifies the immemorial Chinese faith in interdependence of natural world and human king. "Between the ruler of mankind and the powers of nature," Fitzgerald writes, "there existed a close and vital relationship. The fortunes of men depended upon the balance of forces, beneficent as long as they acted in harmony, but destructive once that balance was deranged." It was this ancient belief that the First Emperor of the Ch'in (221–207 B.C.), after forcefully uniting the fractured feudal kingdoms of the late Chou and assaulting—indeed, attempting to extirpate—Confucian and other ideas not supportive of his authoritarian rule, no doubt hoped to reinforce or restore when itinerant scholars under the patronage of his prime minister, Lü Pu-wei [Lü Buwei], compiled a work—***Spring and Autumn of Mr. Lü***—reaffirming age-old practices and taboos, above all the divinely instituted monarch's role in regulating the heavens. Its opening section is an* *almanac in twelve chapters, each describing governmental functions appropriate to a particular month; our selection concerns the first month of spring. (Popular almanacs embodying a medley of ancient belief, astrological lore, and folk superstition have continued to be reprinted up to the present; see Martin Palmer's* T'ung Shu: The Ancient Chinese Almanac *[1986].)*

The next two selections are from Basic Writings of Mo Tzu, Hsün Tzu, and Han Fei Tzu *(1967), translated by Burton Watson.* ***Mo Ti [Mo Di], or Mo Tzu [Mo Zi],*** *who lived in the fifth century B.C., is best known as the apostle of universal, or indiscriminate, love, in opposition to the social and cultural distinctions propounded by Confucius before him, and by Mencius soon afterward; for a time, the ideas of his disciplined school (despite being expressed in a style that Waley, in* Three Ways of Thought, *calls "feeble, repetitive, . . . heavy, unimaginative and unentertaining, devoid of a single passage that could possibly be said to have wit, beauty or force") rivaled those of the Confucians. His conception of a Heaven that "desires righteousness" and "loves the people," arranging both the natural and the political order for their benefit, is a moralized version of traditional quasi-anthropomorphic beliefs. Far more original are the ideas of* ***Hsün Tzu [Xun Zi]*** *(late fourth to mid third century B.C.), who, by contending that human nature is evil and must be trained to goodness against its natural bent, became the leader of the "realist" school of Confucianism. As a Confucian, he is primarily concerned with society, and objects that Chuang Tzu "was obsessed by thoughts of Heaven [i.e., Nature]*[16] *and did not understand the importance of man." But in the forceful passage that concludes our selections, he sets forth a conception of Heaven radically opposed to that of Mo Tzu, or of traditional religious belief: a Heaven that acts "for no particular reason" and has no connection with, or responsibility for, order or disorder in human affairs. This Heaven is far more Nature than God, and the attitude of the sage who neither fears nor attempts to propitiate its power, seeking only to understand "phenomena which can be regularly expected," is already, in potential, more scientist's than seer's.*

Sacred Duties of the Son of Heaven

Lü-shih ch'un-ch'iu [Lüshi chunqiu] *or* Spring and Autumn of Mr. Lü

In this month, on a favorable day, the Son of Heaven shall pray to the Lord-on-High for abundant harvests. Then, selecting a lucky day, he shall himself bear a plowshare and handle in his carriage, attended by the charioteer and the man-at-arms and, leading the chief ministers, feudal princes, and officials, shall personally plow the Field of God. The Son of Heaven shall plow three furrows, the three chief ministers five, the feudal princes and officials nine. On their return, they shall assemble in the Great Hall where the emperor shall take a chalice and offer it to each of them, saying: "This is wine in recompense for your labors."

In this month the vital force of Heaven descends, the vital force of earth arises; Heaven and earth are in harmony and the grass and trees begin to burgeon.

The ruler shall order the work of the fields to begin. He shall order the inspectors of the fields to reside in the lands having an eastern exposure, to repair the borders and boundaries of the fields, to inspect the paths and irrigation ditches, to examine closely the mounts and hills, the slopes and heights and the plains and valleys to determine what lands are good and where the five grains should be sown, and they shall instruct and direct the people. This they must do in person. When the work of the fields has been well

[16] The bracketed gloss is part of Watson's translation.

begun, with the irrigation ditches traced out correctly beforehand, there will be no confusion later.

In this month, the Chief Director of Music shall be ordered to open school and train the students in dancing.

The rules for sacrifices shall be reviewed and orders given for offerings to the spirits of the mountains, forests, rivers, and lakes, but for these sacrifices no female creature may be used.

It shall be forbidden to cut down trees, to destroy nests, to kill young insects, the young yet in the womb or new born, or fledgling birds. All young of animals and eggs shall be spared.

Multitudes of people shall not be summoned for any service, nor shall any construction be done on walls or fortifications.

All bones and corpses of those who have died by the wayside shall be buried.

In this month it is forbidden to take up arms. He who takes up arms will surely call down Heaven's wrath. Taking up arms means that one may not initiate hostilities, though if attacked he may defend himself.

In all things one must not violate the way of Heaven, nor destroy the principles of earth, nor bring confusion to the laws of man.

If in the first month of spring the ruler carries out proceedings proper to summer, then the wind and rain will not come in season, the grass and trees will soon wither and dry up, and the nations will be in great fear.

If he carries out the proceedings proper to autumn, then a great pestilence will strike the people, violent winds and torrential rains will come in abundance, and the weeds of orach and fescue, darnel and southernwood will spring up together.

If he carries out the proceedings of winter, the rains and floods will cause great damage, frost and snow will wreak havoc, and the first seeds sown will not sprout.

The Beneficence of Heaven, *translated by Burton Watson*

Mo-tzu [Mozi]

26 How do I know that Heaven desires righteousness and hates unrighteousness? In the world, where there is righteousness there is life; where there is no righteousness there is death. Where there is righteousness there is wealth; where there is no righteousness there is poverty. Where there is righteousness there is order; where there is no righteousness there is disorder. Now Heaven desires life and hates death, desires wealth and hates poverty, desires order and hates disorder. So I know that Heaven desires righteousness and hates unrighteousness. . . .

Moreover, I know for the following reason that Heaven loves the people generously: It sets forth one after another the sun and moon, the stars and constellations to lighten and lead them; it orders the four seasons, spring, fall, winter, and summer, to regulate their lives; it sends down snow and frost, rain and dew, to nourish the five grains, hemp, and silk, so that the people may enjoy the benefit of them. It lays out the mountains and rivers, the ravines and valley streams, and makes known all affairs so as to ascertain the good or evil of the people. It establishes kings and lords to reward the worthy and punish the wicked, to gather together metal and wood, birds and beasts, and to see to the cultivation of the five grains, hemp, and silk, so that the people may have enough food and clothing. From ancient times to the present this has always been so.

The Indifference of Heaven,
translated by Burton Watson
Hsün-tzu [Xunzi]

17 Heaven's ways are constant. It does not prevail because of a sage like Yao; it does not cease to prevail because of a tyrant like Chieh. Respond to it with good government, and good fortune will result; respond to it with disorder, and misfortune will result. If you encourage agriculture and are frugal in expenditures, then Heaven cannot make you poor. If you provide the people with the goods they need and demand their labor only at the proper time, then Heaven cannot afflict you with illness. If you practice the Way and are not of two minds, then Heaven cannot bring you misfortune. Flood or drought cannot make your people starve, extremes of heat or cold cannot make them fall ill, and strange and uncanny occurrences cannot cause them harm. But if you neglect agriculture and spend lavishly, then Heaven cannot make you rich. If you are careless in your provisions and slow to act, then Heaven cannot make you whole. If you turn your back upon the Way and act rashly, then Heaven cannot give you good fortune. Your people will starve even when there are no floods or droughts; they will fall ill even before heat or cold come to oppress them; they will suffer harm even when no strange or uncanny happenings occur. The seasons will visit you as they do a well-ordered age, but you will suffer misfortunes that a well-ordered age does not know. Yet you must not curse Heaven, for it is merely the natural result of your own actions. Therefore, he who can distinguish between the activities of Heaven and those of mankind is worthy to be called the highest type of man.

To bring to completion without acting, to obtain without seeking—this is the work of Heaven. Thus, although the sage has deep understanding, he does not attempt to exercise it upon the work of Heaven; though he has great talent, he does not attempt to apply it to the work of Heaven; though he has keen perception, he does not attempt to use it on the work of Heaven. Hence it is said that he does not compete with Heaven's work. Heaven has its seasons; earth has its riches; man has his government. Hence man may form a triad with the other two. But if he sets aside that which allows him to form a triad with the other two and longs for what they have, then he is deluded. The ranks of stars move in progression, the sun and moon shine in turn, the four seasons succeed each other in good order, the yin and yang go through their great transformations, and the wind and rain pass over the whole land. All things obtain what is congenial to them and come to life, receive what is nourishing to them and grow to completion. One does not see the process taking place, but sees only the results. Thus it is called godlike. All men understand that the process has reached completion, but none understands the formless forces that bring it about. Hence it is called the accomplishment of Heaven. Only the sage does not seek to understand Heaven. . . .

When he turns his thoughts to Heaven, he seeks to understand only those phenomena which can be regularly expected. When he turns his thoughts to earth, he seeks to understand only those aspects that can be taken advantage of. When he turns his thoughts to the four seasons, he seeks to understand only the changes that will affect his undertakings. When he turns his thoughts to the yin and yang, he seeks to understand only the modulations which call for some action on his part. The experts may study Heaven; the ruler himself should concentrate on the Way.

Are order and disorder due to the heavens? I reply, the sun and moon, the stars and constellations revolved in the same way in the time of Yü as in the time of Chieh. Yü achieved order; Chieh brought disorder. Hence order and disorder are not due to the heavens. . . .

When stars fall or trees make strange sounds, all the people in the country are terrified and go about asking, "Why has this happened?" For no special reason, I reply. It is simply that, with the changes of Heaven and earth and the mutations of the yin and yang, such things once in a while occur. You may wonder at them but you must not fear them. The sun and moon are subject to eclipses, wind and rain do not always come at the proper season, and strange stars occasionally appear. There has never been an age that was without such occurrences. If the ruler is enlightened and his government just, then there is no harm done even if they all occur at the same time. But if the ruler is benighted and his government ill-run, then it will be no benefit to him even if they never occur at all. Stars that fall, trees that give out strange sounds—such things occur once in a while with the changes of Heaven and earth and the mutations of the yin and yang. You may wonder at them, but do not fear them. . . .

You pray for rain and it rains. Why? For no particular reason, I say. It is just as though you had not prayed for rain and it rained anyway. The sun and moon undergo an eclipse and you try to save them;[17] a drought occurs and you pray for rain; you consult the arts of divination before making a decision on some important matter. But it is not as though you could hope to accomplish anything by such ceremonies. They are done merely for ornament. Hence the gentleman regards them as ornaments, but the common people regard them as supernatural. He who considers them ornaments is fortunate; he who considers them supernatural is unfortunate. . . .

Hence if you set aside what belongs to man and long for what belongs to Heaven, you mistake the nature of things.

The Harmony of the Many

Speculations on the Cosmos

Apart from the paradoxical formulations of the Taoist classics, most of the extant Chou philosophers—not only Confucius, Mencius, and Hsün Tzu but also their rivals of the Mohist school and the Legalist school (whose principal thinker, Han Fei Tzu [Han Fei Zi], identified right with the interests of the State)—focused mainly on social and political questions and largely eschewed speculation on the cosmos, convinced that to set aside what belongs to man and long for what belongs to Heaven was to mistake the nature of things. In this they resembled what we know of ancient Chinese religion, which was also more concerned with achieving abundant harvests through ritual and sacrifice than exploring the origin and nature of the universe through myth.

But traces of ancient mythic cosmogonies do survive, notably in the cryptic T'ien Wen [Tian Wen], *or* Heavenly Questions, *a poem of uncertain date (possibly fourth century* B.C.*) included in the great collection of early poetry* Ch'u Tz'u [Chu Ci] *(songs of the southern state of Ch'u), compiled in the mid second century* A.D. *The first section of this poem begins with a series of cosmic questions concerning the origin of the universe and the differentiation of its components to which*

[17]According to *Tso chuan*, Duke Wen 15th year, when an eclipse occurs, the king should beat a drum at the altar of the soil and the feudal lords should beat drums in their courts in order to drive it away. (Watson)

no answers are given or even clearly implied—though perhaps, as Anne Birrell writes in Chinese Mythology: An Introduction *(1993), they "were meant to prompt a long-remembered knowledge and were designed to stir the mind, to entertain, and to serve as a sort of catechism concerning the most basic beliefs and sacred truths of the community of Ch'u which engendered them." Whatever its provenance, the poem thus raises cosmological questions generally slighted by the pragmatic and ritualistic religion of early China.*

Although Confucius too, in the Analects, *forswears metaphysical questions, the inclusion among the Confucian classics of the* I Ching [Yi Jing], *or* Book of Changes, *an ancient manual of divination by patterns of yarrow stalks, gave sanction to later Confucian scholars, through a series of appendices added in the Ch'in or early Han, for speculation on the interacting forces of Heaven and Earth (the first two of the eight basic trigrams of which its sixty-four hexagrams are composed) from which the cosmos had arisen. Through such interpretations, elaborated over the centuries, the* I Ching *became a bridge linking the seemingly opposed but complementary outlooks of Confucians and Taoists. Another was the school (in Ssu-ma T'an's sixfold classification) of Yin-Yang, which was perhaps less a distinct philosophical school than an elaboration of underlying tendencies common to all; for the eternal interaction of Yin—the passive, negative, female principle of darkness—and Yang—the active, positive, male principle of light—runs throughout Chinese poetry and thought, as much in the Taoist* Chuang-tzu *as in Confucian interpretations of the* I Ching, *where it becomes the fundamental source of change itself.*

Most distinctive of this putative school was the theory of the Five Elements (or Agents), which would also become a common inheritance of much Chinese philosophy. In the quasi-legendary Confucian classic the Shu Ching [Shu Jing], *or* Book of History, *King Wu, a founder of the Chou dynasty in about the eleventh century* B.C., *seeks advice from the Prince of Chi on the "constant norms" by which harmony may be achieved between human beings and the surrounding world, and learns that "in ancient times Kun dammed up the inundating waters, and brought disorder into the arrangement of the five agents. God was aroused to anger, and did not give him the Great Plan with its nine categories, whereupon the constant norms were ruined." After Kun's execution, Heaven bestowed on his son Yü the Great Plan, "and the constant norms were thereby set forth in their due order." The first of the Great Plan's nine categories, the Prince of Chi explains, was "the five agents: water, fire, wood, metal, and earth. Water is the power to soak and descend; fire, to blaze and ascend; wood, to be crooked and straight; metal, to be malleable; and earth, to take seeds and yield crops. That which soaks and descends produces saltiness; that which blazes and ascends produces bitterness; that which is crooked or straight produces sourness; that which is malleable produces acridity; that which takes seeds and yields crops produces sweetness." This is the earliest mention of the influential theory later developed—according to the brief account in the* Shih Chi [Shi Ji], *or* Records of the Historian, *of Ssu-ma Ch'ien (Sima Qian), son of Ssu-ma T'an—by the Chou scholar Tsou Yen [Zou Yan] (340–260* B.C.*?): "He claimed that from the time of the separation of heaven and earth onward, the five powers [of the five agents] had been in the process of mutual production and mutual overcoming, and that each temporal reign was in exact correspondence with one of the powers, like parts of a tally." During the Former, or Western, Han (206* B.C.*–*A.D. *8), correspondences of the five agents—with seasons, rulers, animals, grains, organs, numbers, colors, tastes, smells, directions, virtues, planets, etc.—were developed in elaborate detail. Sometime around 135* B.C. *Tung Chung-shu [Dong Zhongshu] thus summarized the central teaching of the school: "The vital forces of Heaven and earth join to form a unity, divide to become the yin and yang, separate into the four seasons, and range themselves into the five agents. . . . In the order of their succession they give birth to one another, while in a different order they overcome each other. Therefore in ruling, if one violates this order,* *there will be chaos, but if one follows it, all will be well governed." Similar correspondences were found for the hexagrams of the* I Ching *as the various schools, like Yin and Yang and the five agents themselves, evolved through continual interaction and change.*

Taoism in particular—as it absorbed not only philosophical speculations from the Yin-Yang and Five-Agents school and I Ching *commentaries, but also the search for immortality through hygiene, alchemy, and voyages to the magic island of P'eng-lai [Penglai]—was transformed to the point that Fung Yu-lan considers the teachings of Taoist philosophy and Taoist religion contradictory: "Taoism as a philosophy teaches the doctrine of following nature, while Taoism as a religion," with its denial of death, "teaches the doctrine of working* against *nature." Yet this religion also "has the spirit of science," which consists not simply (as Fung asserts) in "conquering" nature, as if it were an alien force, but in endeavoring to understand and make use of its powers: and as Needham remarks, Confucian rationalism proved less favorable than Taoist mysticism to the progress of science. Different as philosophical and popular Taoism might be, both viewed the Tao as an "Order of Nature," a* logos *underlying and engendering the multiplicity of the phenomenal world; and a proto-scientific spirit found expression not only in alchemical experimentation but also in bold speculations, mainly of Taoist inspiration, concerning the origin and nature of the cosmos.*

Thus our first selection, from the ***Huai-nan Tzu [Huainan Zi]****, a Taoist-inspired work of the Former Han, dating from about 120 B.C., describes the generation of the world by spontaneous but orderly process from an amorphous "Great Beginning"; its concepts, adopted by Han Confucianists and later by the Japanese, resemble surviving traces of early cosmogonic myths in the* Heavenly Questions *and a few other sources in the absence (as Birrell notes of these myths in contrast to those of the Judeo-Christian and other traditions) "of a creator and lack of any necessity for a divine will or benevolent intelligence to ordain the act of creation." The second selection, from* ***Chang Heng [Zhang Heng]*** *(A.D. 78–137), a scholar of the Latter, or Eastern, Han (A.D. 25–220), expounds an "ecliptical" theory of the universe—one of several in this intellectually fertile age—demonstrating, by its scrupulous precision, how far cosmology had now parted from myth. During the long and troubled period of the Three Kingdoms and Six Dynasties (220–589) between the disintegration of the Han and the beginning of the Sui, the thinkers grouped together as Neo-Taoists reinterpreted both Taoist and Confucian classical texts and located the absolute reality of Tao in nonbeing itself, out of which, as* ***Kuo Hsiang [Guo Xiang]****, who died in 312, writes in his* Commentary on Chuang Tzu *in our third selection, all things are spontaneously produced by a purely natural process: "Heaven and earth and the myriad things change and transform into something new every day and so proceed with time." In his* Introduction to Landscape Painting, *which follows,* ***Tsung Ping [Zong Bing]*** *(375–443) expresses the rapture of "responding to the call of the wilderness" through spiritual oneness with mountains and rivers.*

These centuries of turmoil were also the time when the various schools of (mainly) Mahāyāna Buddhism were spreading throughout China, reaching the acme of their influence in the T'ang [Tang] dynasty (618–907), the great age of Chinese poetry. For the most part, Chinese Buddhism remained, as Needham says, a "world-denying ascetic faith" hostile to speculation on the natural world which it repudiated; thus our fifth selection, from the Platform Scripture *attributed to the* ***Sixth Patriarch, Hui-neng*** *(638–713), and possibly recorded by his pupil Fa-hai or by a later follower of another pupil, Shen-hui, is typically Buddhist in affirming that freedom lies in nonattachment to the external world. Yet Hui-neng was the foremost exponent of the school of Buddhism called Ch'an [Chan]—derived from Sanskrit* dhyāna *("meditation" or "trance"), and passed on to Japan as Zen—and in particular of its Southern school of "sudden enlightenment." In their love of paradox, their entrancement with the nonbeing at the heart of existence,* *and their delight in solitary communion with nature from which enlightenment might suddenly come, the wild men, poets, and hermits of Ch'an Buddhism were all but indistinguishable from Taoist counterparts whose view of the external world as "real and no illusion" appeared, as Needham stresses, so opposite to their own.*

Out of all these swirling currents arose the syncretic Neo-Confucianism of the Sung [Song] dynasty (960–1279), a "joint Confucian-Taoist reaction to Buddhism" which Needham considers "the greatest of Chinese philosophical schools," culminating in the synthesis of Neo-Taoist cosmology and Confucian ethics by Chu Hsi [Zhu Xi] (1130–1200), "the greatest of all Chinese thinkers." Our last three selections are not from Chu Hsi, however, but from three earlier Sung Neo-Confucianists whose cosmological speculations, extending those of their Neo-Taoist predecessors, are especially important for the conception of nature they articulate. The sixth selection is from ***Chou Tun-yi [Zhou Dunyi]*** *(1017–73), who attempted, under the influence of Taoism, to revitalize Confucian cosmology by his concept of the "Great Ultimate" out of which the harmonious order of the myriad things of the phenomenal universe arises. In the seventh selection,* ***Shao Yung [Shao Yong]*** *(1011–77), somewhat like the Greek Pythagoreans long before, conceptualizes differentiation of the "Great Ultimate" in numerical terms: "Spirit engenders number, number engenders form, and form engenders material objects." Finally,* ***Chang Tsai [Zhang Zai]*** *(1021–77), in opposition to Buddhist rejection of an illusory phenomenal world, asserts that the central Confucian cosmological concept of* ch'i [qi], *here translated "material-force," is the primal substance, of which even the "Great Vacuity" from which all things arise and to which they return is composed; ch'i engenders, through ceaseless interaction of yin and yang, the "Great Harmony" of the ever-changing configurations of the concrete world. Successors to these three seminal thinkers, the brothers Ch'eng Hao [Cheng Hao] (1032–85) and Ch'eng Yi [Cheng Yi] (1033–1107), and Chu Hsi himself, would further develop such ideas and integrate them with Confucian ethical thought, but the foundations of a highly sophisticated natural philosophy were already firmly in place. By no accident, as Needham remarks, the Sung dynasty "saw the greatest flowering of indigenous Chinese science," far in advance of anything to which the contemporaneous Christian West could lay claim.*

All these selections are from Sources of Chinese Tradition *(1960), ed. Wm. Theodore de Bary, Wing-tsit Chan, and Burton Watson; the reader should consult this comprehensive anthology, along with* A Source Book in Chinese Philosophy *(1963), trans. Wing-tsit Chan, for more extensive passages from these and other major Chinese thinkers of these centuries. See also volume two of Joseph Needham's* Science and Civilisation in China; *Fung Yu-lan's two-volume* History of Chinese Philosophy *(English trans. 1937, 1952) and* Short History of Chinese Philosophy *(1949); and John B. Henderson,* The Development and Decline of Chinese Cosmology *(1984).*

Genesis from the Great Beginning

Huai-nan Tzu [Huainan Zi]

Before heaven and earth had taken form all was vague and amorphous. Therefore it was called the Great Beginning. The Great Beginning produced emptiness and emptiness produced the universe. The universe produced material-force[18] which had limits. That which was clear and light drifted up to become heaven, while that which was heavy and

[18] The word *ch'i* translated . . . as material-force or vital force, in order to emphasize its dynamic character, plays an important part in Chinese cosmological and metaphysical thought. At times it means the spirit or breath of life in living creatures, at other times the air or ether filling the sky and surrounding the universe, while in some contexts it denotes the basic substance of all creation. (*Sources*)

turbid solidified to become earth. It was very easy for the pure, fine material to come together but extremely difficult for the heavy, turbid material to solidify. Therefore heaven was completed first and earth assumed shape after. The combined essences of heaven and earth became the yin and yang, the concentrated essences of the yin and yang became the four seasons, and the scattered essences of the four seasons became the myriad creatures of the world. After a long time the hot force of the accumulated yang produced fire and the essence of the fire force became the sun; the cold force of the accumulated yin became water and the essence of the water force became the moon. The essence of the excess force of the sun and moon became the stars and planets. Heaven received the sun, moon, and stars while earth received water and soil. . . .

When heaven and earth were joined in emptiness and all was unwrought simplicity, then without having been created, things came into being. This was the Great Oneness. All things issued from this oneness but all became different, being divided into the various species of fish, birds, and beasts. . . . Therefore while a thing moves it is called living, and when it dies it is said to be exhausted. All are creatures. They are not the uncreated creator of things, for the creator of things is not among things. If we examine the Great Beginning of antiquity we find that man was born out of nonbeing to assume form in being. Having form, he is governed by things. But he who can return to that from which he was born and become as though formless is called a "true man." The true man is he who has never become separated from the Great Oneness.

A Theory of Cosmic Structure, by Chang Heng [Zhang Heng]

Hun-t'ien-i [Huntianyi], in Ching-tien chi-lin [Jingdian jilin], *27:1a–b, by Hung I-hsüan [Hong Yixuan]*

Heaven is like an egg, and the earth is like the yolk of the egg. Alone it dwells inside. Heaven is great and earth is small. Inside and outside of heaven there is water. Heaven wraps around the earth as the shell encloses the yolk. Heaven and earth are borne up and stand upon their vital force, floating upon the water. The circumference of heaven is $365\frac{1}{4}$ degrees. This is divided in half, so that $182\frac{5}{8}$ degrees are arched above the earth, while $182\frac{5}{8}$ degrees are cupped under the earth. Therefore there are 28 heavenly constellations, half of them visible and half invisible. Its two extremes are called the south and north poles. The north pole is the apex of heaven, but is elevated 36 degrees above the true north of the earth. The north pole is the axis, hence 72 degrees of this upper sphere are constantly visible. The south pole is the other apex of heaven, but it also is deflected 36 degrees below the true south of the earth. It is the axis for the lower half, hence 72 degrees of the lower half are constantly hidden. The two poles are 182+ degrees apart. Heaven turns about the earth like a cart wheel, revolving constantly without stop. Its form is complete and encircling, hence it is called the "encircling heaven" *(hun t'ien [hun tian])*.

Nature and Nonexistence, by Kuo Hsiang [Guo Xiang]

Commentary on Chuang Tzu

The music of nature is not an entity existing outside of things. The different apertures, the pipes and flutes and the like, in combination with all living beings, together constitute nature. Since nonexistence is nonbeing, it cannot produce being. Before being itself is produced, it cannot produce other beings. Then by whom are things produced? They spontaneously produce themselves, that is all. By this is not meant that there is an "I" to produce. The "I" cannot produce things and things cannot produce the "I." The "I" is self-existent. Because it is so by itself, we call it natural. Everything is what it is by nature, not through taking any action. Therefore [Chuang Tzu] speaks in terms of nature. The term nature [literally Heaven] is used to explain that things are what they are spontaneously, and not to mean the blue sky. But someone says that the music of nature makes all things serve or obey it. Now, nature cannot even possess itself. How can it possess things? Nature is the general name for all things. [Sec. 2, 1:21a]

Not only is it impossible for nonbeing to be changed into being. It is also impossible for being to become nonbeing. Therefore, although being as a substance undergoes infinite changes and transformations, it cannot in any instance become nonbeing. . . . What came into existence before there were things? If I say yin and yang came first, then since yin and yang are themselves entities, what came before them? Suppose I say nature came first. But nature is only things being themselves. Suppose I say perfect Tao came first. But perfect Tao is perfect nonbeing. Since it is nonbeing, how can it come before anything else? Then what came before it? There must be another thing, and so on *ad infinitum*. We must understand that things are what they are spontaneously and not caused by something else. [Sec. 22, 7:54b–55b]

Everything is natural and does not know why it is so. The more things differ in corporeal form, the more they are alike in being natural. . . . Heaven and earth and the myriad things change and transform into something new every day and so proceed with time. What causes them? They do so spontaneously. . . . What we call things are all that they are by themselves; they did not cause each other to become so. Let us leave them alone and the principle of being will be perfectly realized. The ten thousand things are in ten thousand different conditions, and move forward and backward differently, as though there were a True Lord to make them so. But if we search for evidences for such a True Lord, we fail to find any. We should understand that things are all natural and not caused by something else. [Sec. 2, 1:22b–23a]

The universe is the general name for all things. They are the reality of the universe while nature is their norm. Being natural means to exist spontaneously without having to take any action. Therefore the fabulous *p'eng* bird can soar high and the quail can fly low, the cedrela can live for a long time and the mushroom for a short time. They are capable of doing so not because of their taking any action but because of their being natural. [Sec. 1, 1:8b]

Landscapes of the Spirit, by Tsung Ping [Zong Bing]

Introduction to Landscape Painting

(Li-tai ming-hua chi [Lidai minghua ji]), *6:3b–4b*

Having embraced Tao the sage responds harmoniously to things. Having purified his mind, the worthy man enjoys forms. As to landscapes, they exist in material substance and soar into the realm of the spirit. Therefore men like the Yellow Emperor, Yao, Confucius, Kuang-ch'eng, Ta-wei, Hsü Yu, and the brothers of Ku-chu [Po I and Shu Ch'i] insisted on traveling among the mountains of K'ung-t'ung, Chü-tz'u, Miao-ku, Chi-shou, and Ta-meng. These are also called the delights of the man of humanity and the man of wisdom.[19] Now the sage, by the exercise of his spirit, follows Tao as his standard,

[19]See the passage previously quoted (introduction to Chapter 5 above) from Confucius's *Analects* VI.21: "The wise man delights in water, the Good man [or "the man of humanity," *jên*] delights in mountains."

while the worthy man understands this. Mountains and rivers in their form pay homage to Tao, and the man of humanity delights in them. Do not the sage and mountains and rivers have much in common? . . .

And so I live in leisure and nurture my vital power. I drain clean the wine-cup, play the lute, lay down the picture of scenery, face it in silence, and, while seated, travel beyond the four borders of the land, never leaving the realm where nature exerts her influence, and alone responding to the call of wilderness. Here the cliffs and peaks seem to rise to soaring heights, and groves in the midst of clouds are dense and extend to the vanishing point. Sages and virtuous men of far antiquity come back to live in my imagination and all interesting things come together in my spirit and in my thoughts. What else need I do? I gratify my spirit, that is all. What is there that is more important than gratifying the spirit?

Buddhist Freedom Through Nonattachment, by Hui-neng
The Platform Scripture of the Sixth Patriarch

Good friends, in my system, from the very beginning, whether in the sudden enlightenment or gradual enlightenment tradition, absence of thought has been instituted as the main doctrine, absence of phenomena as the substance, and nonattachment as the foundation. What is meant by absence of phenomena? Absence of phenomena means to be free from phenomena when in contact with them. Absence of thought means not to be carried away by thought in the process of thought. Nonattachment is man's original nature. [In its ordinary process,] thought moves forward without a halt; past, present, and future thoughts continue as an unbroken stream. But if we can cut off this stream by an instant of thought, the Dharma-Body will be separated from the physical body, and at no time will a single thought be attached to any dharma. If one single instant of thought is attached to anything, then every thought will be attached. That will be bondage. But if in regard to all dharmas, no thought is attached to anything, that means freedom. This is the reason why nonattachment is taken as the foundation.

Good friends, to be free from all phenomena means absence of phenomena. Only if we can be free from phenomena will the reality of nature be pure. This is the reason why absence of phenomena is taken as the substance. . . . [Sec. 18]

This being the case, in this system, what is meant by sitting in meditation? To sit means to obtain absolute freedom and not to allow any thought to be caused by external objects. To meditate means to realize the imperturbability of one's original nature. What is meant by meditation and calmness? Meditation means to be free from all phenomena and calmness means to be internally unperturbed. If one is externally attached to phenomena, the inner mind will at once be disturbed, but if one is externally free from phenomena, the inner nature will not be perturbed. The original nature is by itself pure and calm. It is only because of causal conditions that it comes into contact with external objects, and the contact leads to perturbation. There will be calmness when one is free from external objects and is not perturbed. Meditation is achieved when one is externally free from phenomena and calmness is achieved when one is internally unperturbed. Meditation and calmness mean that externally meditation is attained and internally calmness is achieved. [Sec. 19]

The Great Ultimate,
by Chou Tun-yi [Zhou Dunyi], or Chou Lien-ch'i [Zhou Lianqi]
T'ai-chi-t'u shuo [Taijitu shuo]

The Non-ultimate! And also the Great Ultimate *(T'ai-chi [Taiji])*. The Great Ultimate through movement generates the yang. When its activity reaches its limit, it becomes tranquil. Through tranquillity, the Great Ultimate generates the yin. When tranquillity reaches its limit, activity begins again. Thus movement and tranquillity alternate and become the root of each other, giving rise to the distinction of yin and yang, and these two modes are thus established.

By the transformation of yang and its union with yin, the five agents of water, fire, wood, metal, and earth arise. When these five material-forces *(ch'i [chi])* are distributed in harmonious order, the four seasons run their course.

The five agents constitute one system of yin and yang, and yin and yang constitute one Great Ultimate. The Great Ultimate is fundamentally the Non-ultimate. The five agents arise, each with its specific nature.

When the reality of the Non-ultimate and the essence of yin and yang and the five agents come into mysterious union, integration ensues. The heavenly principle *(ch'ien [qian])* constitutes the male element, and the earthly principle *(k'un [kun])* constitutes the female element. The interaction of these two material forces engenders and transforms the myriad things. The myriad things produce and reproduce, resulting in an unending transformation.

A Universe Engendered by Number, by Shao Yung [Shao Yong]
Huang-chi ching-shih shu [Huangji jingshi shu],
The Great Principles Governing the World

As the Great Ultimate becomes differentiated, the two primary modes appear. The yang descends and interacts with the yin, and yin rises to interact with yang, and consequently the four secondary forms are constituted. Yin and yang interact and generate the four secondary forms of Heaven; the element of weakness and the element of strength interact and generate the four secondary forms of earth; and consequently the eight trigrams are completed. The eight trigrams intermingle and generate the myriad things. Therefore the One is differentiated into two, two into four, four into eight, eight into sixteen, sixteen into thirty-two, and thirty-two into sixty-four. Thus "in the successive division of yin and yang and the mutual operation of strength and weakness, the six positions [of the lines in each hexagram] in the *Book of Changes* form an orderly pattern."[20] Ten is divided into 100, 1,000, and 10,000. This is similar to the fact that the root generates the trunk; the trunk, branches; and the branches, leaves. The greater the division, the smaller the result, and the finer the division, the more complex. Taken as a unit, it is One. Taken as diffused development, it is the many. Hence the heavenly principle divides, the earthly principle unites; the *chen* hexagram [symbol of development] augments, and the *sun* hexagram [symbol of bending] diminishes. Augmentation leads to division, division leads to diminution, and diminution leads to unity. [7A:24b]

The Great Ultimate is One. It produces the two [yin and yang] without engaging in

[20]*Book of Changes*, Shuo-kua 2. (*Sources*)

activity. The two constitute spirit. Spirit engenders number, number engenders form, and form engenders material objects. [8B:23a]

Forms and numbers in the universe can be calculated, but their wonderful operations cannot be fathomed. The universe can be fully investigated through principles but not through corporeal forms. How can it be fully investigated through external observation? [8A:16b]

Great Harmony, by Chang Tsai [Zhang Zai], or Chang Heng-ch'ü [Zhang Hengqu]

Cheng-meng [Zhengmeng], *I, in* Chang Heng-ch'ü chi [Zhang Henqu ji], *2:3b–10b*

Although material-force in the universe integrates and disintegrates, and attracts and repulses in a hundred ways, nevertheless the principle (*li*) according to which it operates has an order and is unerring.

The Great Vacuity of necessity consists of material-force. Material-force of necessity integrates to become the myriad things. Things of necessity disintegrate and return to the Great Vacuity. Appearance and disappearance following this cycle are all a matter of necessity. When, in the midst [of this universal operation] the sage fulfills the Way to the utmost, and identifies himself [with the universal processes of appearance and disappearance] without partiality, his spirit is preserved in the highest degree. Those [the Buddhists] who believe in annihilation expect departure without returning, and those [the Taoists] who cling to everlasting life and are attached to existence expect things not to change. While they differ, they are the same in failing to understand the Way. Whether integrated or disintegrated, my body remains the same. One is qualified to discuss the nature of man when one realizes that death is not annihilation.

When it is understood that Vacuity, Emptiness, is nothing but material-force, then existence and nonexistence, the hidden and the manifest, spirit and external transformation, and human nature and destiny, are all one and not a duality. He who apprehends integration and disintegration, appearance and disappearance, form and absence of form, and can trace them to their source, penetrates the secret of change. . . .

As the Great Vacuity, material-force is extensive and vague. Yet it ascends, descends, and moves in all ways without ever ceasing. . . . That which floats upward is the yang that is clear, while that which sinks to the bottom is the yin that is turbid. As a result of their contact and influence and of their integration and disintegration, winds and rains, snow and frost, come into being. Whether it be the countless variety of things in their changing configurations or the mountains and rivers in their fixed forms, the dregs of wine or the ashes of fire, there is nothing [in which the principle] is not revealed. . . .

Material-force moves and flows in all directions and in all manners. Its two elements unite and give rise to concrete stuff. Thus the great variety of things and human beings is produced. In their ceaseless successions the two elements of yin and yang constitute the great principles of the universe.

6

Cottage and Cosmos: Chinese Poetry of Nature and Freedom

Among countless themes in the richly varied three-thousand-year tradition of Chinese poetry, the symbiosis of human being and natural world—whether of rugged mountains or homely farmstead—is central. As in Sung landscape paintings, a human figure almost always inhabits the scene, for man and nature can hardly be thought of apart: they are not contraries but interacting components of a cosmic continuum, a perpetual interplay of yin and yang, that engenders the ten thousand things. "Silently enjoying isolation," the poet may live for a time in retirement, attuned to the passing seasons: "Thus imitating cosmic changes," Lu Yün writes in the third century A.D.*, "My cottage becomes a universe." But though he choose, like Po Chü-i [Bo Juyi] six hundred years later, a place "unfrequented by men," he is typically bound to that human world as well, a half-recluse, as Po Chü-i calls himself, who is simultaneously hermit and politician:*

In the morning I work at a Government
office-desk;
In the evening I become a dweller in the
Sacred Hills.

The rustic cottage (often more villa than hut) in which he achieves communion with the encompassing world is itself, after all, a human artifact: the two worlds, as always, are interdependent.

So abundant is short poetry in China (the Chinese, unlike Mesopotamians, Indians, and Greeks, left no epics) in the two millennia ending with the Sung dynasty, and so varied its treatments of the human and natural worlds, that any selection is arbitrary; dozens of other poets might have been included, along with hundreds of other poems by those who are. Yet even this brief sample suggests the wealth of these poems. The Book of Poetry *or, in Waley's version,* Book of Songs, *gave definitive written form in the time of Confucius to popular ballads probably composed and transmitted orally centuries before. All but a few of the 305 poems in this collection are short (between eight and sixty lines), though the longest is twice that length. Ranging in theme from courtship and marriage to farming, warfare, and dynastic legends, they are generally simple in style, with the refrains characteristic of folk poetry in many parts of the world. Once elevated to the status of Confucian classic, these poems were preserved for the didactic ends to which they could be put, through allegorical interpretations wholly alien to their original spirit. Images of nature—wild and domesticated plants and animals, birds and trees, sun, moon, and stars—run profusely throughout (in contrast to their near total absence from the* Tao Te Ching*), not as principal subjects of any poem but as implicit points of reference for the central human theme. "Often nature seems to work with man and share in his moods. But in certain moments of horror and anguish," Waley writes, "it is the indifference of nature that adds a last touch to the poignancy of human suffering. . . . Again and again it is as contrasts and not comparisons that the things of nature appear in early Chinese songs. The lotus is in the pool and the pine-tree on the hill. Nature goes its accustomed way, only man changes; such is the burden of many songs."*

In addition to fertility gods of the grain propitiated by royal rites, "The deities worshipped by shamans in Ancient China," David Hawkes writes in "The Supernatural in

Chinese Poetry" (1961), "were the gods and goddesses of rivers, mountains, stars, rain, wind, clouds, the sun, the moon, and so on," with whom shamans or mediums in the southern state of Ch'u communicated through spirit possession and ecstatic flight of the soul. This important strain of Chinese poetry, starkly different from that of the Book of Poetry, *is represented here by two of the "Nine Songs" included (like the* Heavenly Questions*) in the second major collection of Chinese poetry, the* Ch'u Tz'u [Chu Ci], *compiled in the second century A.D.; nature in these visionary poems, probably written for shamanistic performance, is clad in supernatural garb. In contrast, poems from the Han dynasty (206 B.C.–A.D. 220) and the troubled centuries (A.D. 220–618) between the Han and the T'ang are more subdued in tone and direct in style, expressing through personal statement, popular legend, or whimsical description a variety of human feelings ranging from love and longing to wonder at Taoist immortals mounting the purple mists or contentment with rural life of "Nature and Freedom."*

Of all these poets, the recluse T'ao Ch'ien [Tao Qian], for whom "a heart that is distant creates a wilderness around it," most profoundly influenced the great eighth-century poets of the High T'ang, including Wang Wei, Li Po [Li Bo], and Tu Fu [Du Fu]. Unlike court poets earlier in this cosmopolitan dynasty, the High T'ang poet, Stephen Owen writes in The Great Age of Chinese Poetry: The High T'ang *(1981), was fascinated by "lack of decorum—both in stylistic convention and thematically in the exaggerated gestures of the eccentric," taking interest not in polished society but in lower social orders to whose "aristocracy of the spirit" he endeavored to give expression in "plain and straightforward" (though highly artistic) language descending from that of T'ao Ch'ien, "the perfect model" of individuality and freedom. Each of these gifted and unmistakably individual poets—writing at a time when rebellion by An Lu-shan shook the empire to its foundations at the cost of perhaps ten million lives—was involved in the social and political life of his age, but each—the "Buddhist" Wang Wei, the "Taoist" Li Po, and the "Confucian" Tu Fu—was also intensely attracted by the recluse's communion with a more harmonious natural world. Whatever their differences, Chinese poets, as James J. Y. Liu remarks in* The Art of Chinese Poetry *(1962), "were more often eclectic than systematic, exhibiting Confucian, Taoist, and Buddhist leanings at the same time, without endeavoring to integrate them into one consistent system." Po Chü-i later gave expression, as we have seen, to the contrary callings of office desk and Sacred Hills, while his possible contemporary (for his dates are fittingly undetermined) Han-shan, or "Cold Mountain," recklessly spurned the first for the second.*

Finally, in the highly cultured Sung dynasty—both before the fall of its first capital to the Manchurian Chin [Jin] brought an end to the Northern Sung (960–1125) and in the remaining century and a half before the shrunken Southern Sung (1127–1279) fell to the Mongols of Kublai Khan—another great poetic age followed, deeply affected by the philosophical currents culminating in the Neo-Confucian synthesis. In it, Kōjiro Yoshikawa suggests in An Introduction to Sung Poetry *(1962; trans. 1967), "sublimation or transcendence of sorrow [is] brought about through a broader and more diversified vision" than that of the intense and impassioned T'ang poets. Sung poetry, much of which was printed centuries before printing reached Europe, gives greater prominence (unlike Sung landscape painting) to society than to nature. Yet here too—as in the* Book of Poetry *two millennia before—nature imagery is central to expression of human emotion: a constant of Chinese poetry, immensely varied though it otherwise is, throughout the ages.*

Even so rapid a summary gives some hint of the variation within this constant, the multeity within this unity, ranging from poignant contrasts between natural permanence and human change in the Book of Poetry *to ecstatic communion of human and divine in the* Ch'u Tz'u, *and from T'ao Ch'ien's rustic escape from "the bars of a cage," via the passionate tensions of officeholding T'ang hermits, to the serener (but far from complacent) poetry of the Sung. "The Chinese term for 'Nature'," Liu writes, "is* tzu-jan [ziran]," *the ultimate principle of the "Self-so" conditioning Tao itself (according to chapter 25 of the* Tao Te Ching*). If so, "nature poetry" seems a contradiction in terms, for like Tao this nature must forever be nameless and incommunicable. In this conception, unity of nature "is violated by conscious distinctions," Owen writes in* Traditional Chinese Poetry and Poetics: Omen of the World *(1985), "and the instrument of such distinctions is the Word. . . . A nature beyond language and all distinctions is the ultimate Other, removed from human comprehension. To seek it leads to negation, blankness, and silence."*

But our English word Nature embraces not only the essential "suchness" of things but also the infinite diversity of phenomena generated by it: not only tzu-jan *and Tao but also* T'ien *and Yin-Yang and the ten thousand things spontaneously arising from their interaction—and here the poets of China have been far from silent. For in* this *conception Nature and man are continuous and one. The world, Owen writes, "is a vast, fluctuating omenscape, and the poet is the omen-reader" of a world whose immanent order* (li) *he perceives through particular experience. "It is not a metaphorical Truth but an immanent truth that can be known only through its empirical manifestations"; the poet's function "is to see the order in the world, the pattern behind its infinite division." Certain "metaphysical" Chinese theories conceive of literature as a manifestation of Tao, "the principle of the universe immanent in the totality of all being," Pauline Yu writes in* The Poetry of Wang Wei *(1980): a doctrine postulating "fundamental correspondence . . . between the patterns* (wen) *and workings of the universe and those of human culture, such as writing (also* wen*)." In this sense, Chinese poets of every age, even when "drawing nature . . . into the world of human activities" (as Yoshikawa says of the Sung), are truly poets of Nature.*

Readers of Chinese poetry in English should be aware that it is formally far more complex than translations suggest. Chinese grammar, Liu writes, "is fluid, not architectural. Whereas in a highly inflected language such as Latin, words are solid bricks with which to build complicated edifices of periods and paragraphs, in Chinese they are chemical elements which form new compounds with great ease," often leaving the same cryptically succinct line open to multiple interpretations. In the dominant poetic form, the shih [shi] *(pronounced roughly "shrr"), each line contains a fixed number of syllables, or characters; and regular rhyme, usually of every second line, runs through a poem. In the* Book of Poetry, *lines were most often of four syllables. In what came to be called "old poetry"* (ku-shih [gushi]) *of the Han and later, lines of five or seven syllables were most common in poems that could still be of any length. A far more stringent style, called "regulated poetry"* (lü-shih [lüshi]), *gained prevalence in the T'ang, especially after composition of it became (in 680) part of civil service examinations. Such a poem consisted of eight lines with a single rhyme, strict patterns of tonal variation, and, in its middle two couplets, word-for-word antithetical parallelism; a "curtailed" variant was the four-line quatrain. (Such parallelism, Owen suggests, was "the formal linguistic manifestation of the structure of the natural world.") Other verse forms were important but less frequent.*

English translations give little sense of this verbal density and formal elegance, or of the rich fabric of allusion woven into almost all later poems. (Even rhyme has seldom proved successful in rendering Chinese verse; for one exception, see Vikram Seth's translations of Wang Wei, Li Po, and Tu Fu in Three Chinese Poets *[1992].) Yet ever since Ezra Pound's* Cathay *in 1915 and Arthur Waley's*

A Hundred and Seventy Chinese Poems *in 1919, the task has engaged major talents whose efforts have exerted a significant influence on modern poetry. The versions that follow were made by only six translators. One, Arthur Waley, was equally scholar and poet; his main formal innovation, followed by many successors, was to make the number of* stressed *syllables in an English line equal to the total number of syllables in the corresponding Chinese line. Two others, David Hawkes and Burton Watson, are scholars of exceptional poetic talent; Watson has been a prolific translator, like Waley, of both Chinese and Japanese. The others, Witter Bynner, Kenneth Rexroth, and Gary Snyder, are poets whose renditions are in some cases less literally accurate—though no readable version, of course, could be that! (In the first line of Li Po's "T'ien-mu Mountain Ascended in a Dream," the place seafarers tell of is not Bynner's Japan but—in Hawkes's translation from "The Supernatural in Chinese Poetry"—"Yingzhou, the Immortal Isle" of Taoist legend; or was Japan among the islands that might have given rise to such stories?) In addition to the volumes from which the following selections are taken—Waley's* Book of Songs *(1937) and* Chinese Poems *(1946, incorporating poems from four earlier volumes); Hawkes's* Songs of the South *(1959; 2nd ed. 1985); Watson's* Columbia Book of Chinese Poetry *(1984); Bynner's* The Jade Mountain: A Chinese Anthology *(1929); Rexroth's* One Hundred Poems from the Chinese *(no date); and Snyder's* Riprap & Cold Mountain Poems *(1965)—interested readers may wish to consult other general collections including* The White Pony: An Anthology of Chinese Verse *(1947), ed. Robert Payne;* Anthology of Chinese Literature *(1965), ed. Cyril Birch; and* Sunflower Splendor: Three Thousand Years of Chinese Poetry *(1975), ed. Wu-chi Liu and Irving Yucheng Lo.*

Four Poems from *The Book of Songs*, translated by Arthur Waley

The Shih Ching [Shi Jing], *or* Book of Poetry, *as previously noted, was compiled in the sixth century* B.C. *from oral folk songs going back several centuries before, and became a central part of the classic Confucian canon. In the first three poems given here, natural images highlight the intensity of the human experiences of love and seduction; the fourth describes a vivid agricultural scene. Numbers after each poem are those of Waley's translation and of Mao's Chinese text.*

A moon rising white
Is the beauty of my lovely one.
Ah, the tenderness, the grace!
Heart's pain consumes me.

A moon rising bright
Is the fairness of my lovely one.
Ah, the gentle softness!
Heart's pain wounds me.

A moon rising in splendor
Is the beauty of my lovely one.
Ah, the delicate yielding!
Heart's pain torments me.

—Waley 32; Mao 143

Swoop flies that falcon;
Dense that northern wood.
Not yet have I seen my lord;
Sore grieves my heart.
What will it be like, what like?
I am sure many will forget me.

On the hill is a clump of oaks
And in the lowlands, the piebald-tree.
Not yet have I seen my lord;
My grief I cannot cure.
What will it be like, what like?
I am sure many will forget me.

On the hill is a clump of plum-trees;
And on the lowlands, planted pear-trees.
Not yet have I seen my lord;
With grief I am dazed.
What will it be like, what like?
I am sure many will forget me.[1]

—Waley 80; Mao 132

In the wilds there is a dead doe;
With white rushes we cover her.
There was a lady longing for the spring;
A fair knight seduced her.

In the wood there is a clump of oaks,
And in the wilds a dead deer
With white rushes well bound;
There was a lady fair as jade.

"Heigh, not so hasty, not so rough;
Heigh, do not touch my handkerchief.[2]
Take care, or the dog will bark."

—Waley 63; Mao 23

[1] Waley comments on this poem: "The theme of the comparisons is that everything in nature goes its wonted way and is in its proper place; but I am embarking on a new, unimaginable existence."
[2] Which was worn at the girdle. (Waley)

They clear away the grass, the trees;
Their ploughs open up the ground.
In a thousand pairs they tug at weeds and roots,
Along the low grounds, along the ridges.
There is the master and his eldest son,
There the headman and overseer.
They mark out, they plough.
Deep the food-baskets that are brought;
Dainty are the wives,
The men press close to them.
And now with shares so sharp
They set to work upon the southern acre.
They sow the many sorts of grain,
The seeds that hold moist life.
How the blade shoots up,
How sleek, the grown plant;
Very sleek, the young grain!
Band on band, the weeders ply their task.
Now they reap, all in due order;
Close-packed are their stooks—
Myriads, many myriads and millions,
To make wine, make sweet liquor,
As offering to ancestor and ancestress,
For fulfilment of all the rites.
"When sweet the fragrance of offering,
Glory shall come to the fatherland.
When pungent the scent,
The blessed elders are at rest."
Not only here is it like this,
Not only now is it so.
From long ago it has been thus.

—Waley 157; Mao 290

Two of the "Nine Songs" from *Songs of the South*, translated by David Hawkes

The Ch'u Tz'u [Chu Ci], *or* Songs of Ch'u *(a partly barbarian state, in ancient times, of the Yangtze valley in what was then southern China) was compiled in the second century* A.D. *from far older materials ascribed to the first Chinese poet known by name, Ch'u Yüan (Qu Yuan), who was probably the author of at least the first and longest poem in the collection, "Li Sao" ("On Encountering Trouble"). Shamanistic influences are strong throughout, and the "Nine Songs" (actually eleven), in particular, seem to be texts of elaborate shamanistic performances. Some are apparently the shaman's erotic pleas to a god or goddess, some the divinity's own words, some dialogues between the two; in the absence of stage directions, precise division of parts is hard to determine. Hawkes conjectures that they were composed at the Ch'u court of Shou-ch'un, who reigned from 241 to 233* B.C.*, near the end of the Warring States period. The god first addressed* *in the songs, T'ai Yi, "the Great Unity," was mentioned a century later (by the great historian Ssu-ma Ch'ien) as an object of worship in the reign of the Han emperor Wu Ti [Wu Di] (140–87 B.C.), a noted patron of shamans. In the following two songs, Ssu Ming [Si Ming] "was the God of Fate, the bestower of life and death," determining how long a person would live: "The manipulation of* yin *and* yang *on whose balance a man's life depends was his special concern." Like T'ai Yi, he was also an astral god, identified with a star in Ursa Major. The distinction between "greater" and "lesser," Hawkes suggests, may reflect different aspects of the same god as worshipped by shamans of two states, Chin [Jin] and Ch'u, or may simply indicate that the second song is shorter than the first. The sublime celestial and luxuriant floral imagery of these songs (presumably heightened by use of "holy hemp," or cannabis) suggests an ecstatic dimension of nature worship closer to the Indian Rig Veda than to Confucian or indeed Taoist outlooks. Interested readers should also consult Arthur Waley's* The Nine Songs: A Study of Shamanism in Ancient China *(1955).*

V. The Greater Master of Fate (Da ssu ming [Da si ming])

Open wide the door of heaven!
On a black cloud I ride in splendor,
Bidding the whirlwind drive before me,
Causing the rainstorm to lay the dust.

In sweeping circles my lord is descending:
"Let me follow you over the Kong-sang mountain!
See, the teeming peoples of the Nine Lands:
The span of their lives is in your hand!"

Flying aloft, he soars serenely,
Riding the pure vapor, guiding *yin* and *yang*.
Speedily, lord, I will go with you,
Conducting High God to the height of heaven.

My cloud-coat hangs in billowing folds;
My jade girdle-pendants dangle low:
A *yin* and a *yang*, a *yin* and a *yang*:
None of the common folk know what I am doing.

I have plucked the glistening flower of the Holy Hemp
To give to one who lives far away.
Old age already has crept upon me:
I am no longer near him, fast growing a stranger.

He drives his dragon chariot with thunder of wheels;
High up he rides, careering heavenwards.
But I stand where I am, twisting a spray of cassia:
The longing for him pains my heart.

It pains my heart, but what can I do?
If we only could stay as we were, unchanging!
But all man's life is fated;
Its meeting and partings not his to arrange.

VI. The Lesser Master of Fate (Shao ssu ming [Shao si ming])

The autumn orchid and the deer-parsley
Grow in a carpet below the hall;
The leaves of green and the pure white flowers
Assail me with their wafted fragrance.

The autumn orchids bloom luxuriant,
With leaves of green and purple stems.
All the hall is filled with lovely women,
But his eyes swiftly sought me out from the rest.

Without a word he came in to me, without a word he left me:
He rode off on the whirlwind with cloud-banners flying.
No sorrow is greater than the parting of the living;
No happiness is greater than making new friendships.

Wearing a lotus coat with melilotus girdle,
Quickly he came and as quickly departed.
At night he will lodge in the High God's precincts.
"Whom are you waiting for at the cloud's edge?"

I will wash my hair with you in the Pool of Heaven;
You shall dry your hair on the Bank of Sunlight.
I watch for the Fair One, but he does not come.
Wildly I shout my song into the wind.

With peacock canopy and kingfisher banner,
He mounts the ninefold heaven and grasps the Broom Star;
He brandishes his long sword, protecting young and old:
"You only, Fragrant One, are worthy to be judge over men."

Five Poems from the Han and Chin

Both in the Han dynasty that followed unification of China under the harsh and short-lived Ch'in [Qin] (226–207 B.C.) and lasted—with a brief interregnum between A.D. 8 and 25—from 202 B.C. to A.D. 220, and in the chaotic succession of dynasties before China was reunited under the Sui (589–618) and its successor, the T'ang, Chinese poetry continued in many forms. The Han was an age of imperial expansion and—especially after adoption of Confucianism and the examination system based on it—of cultural consolidation. No single poet or collection of poems from this time (in which the loosely constructed descriptive genre of the fu, *influenced by the longer poems of the* Ch'u Tz'u *and sometimes mingling verse and prose, became common) had the importance of the* Ch'u Tz'u *or* Book of Poetry. *Among the finest poems of the period are the anonymous* **Nineteen Old Poems of the Han,** *written (probably in the first to second centuries A.D.) in the five-character line that would become a model for much shih poetry of later centuries. In the two poems here included, floral and astral images (in the legend of the Herdboy and Weaving Maiden of the night sky) give form to feelings of separation and longing. The whimsical* fu *"The Wangsun"—which Waley glosses as "a kind of small, tailless ape (?)" and dates around* A.D. *130—by* ***Wang Yen-shou [Wang Yanshou]*** *is remarkable for close observation and detailed description of animal life by a poet equally capable (in a poem translated as "The Nightmare") of the wildest flights of fancy. The next two poems are by poets writing in the time of the beleaguered Western Chin [Jin] dynasty (265–316), which after abandoning its northern domains to barbarian invaders fled south to become the shrunken Eastern Chin, one of the "Six Dynasties" of the period of disunion (317–589) that followed.* ***Lu Yün [Lu Yun]*** *(262–303) was the younger brother of Lu Chi [Lu Ji], author of the critical poem* Wen fu, *or "Essay on Literature"; unprotected by "retirement beyond the World," the brothers were entangled in a power struggle, convicted of treason, and executed with Lu Chi's two sons. Finally,* ***Kuo P'u [Guo Pu]*** *(276–324), in this troubled age of strife and division when the Confucian consensus of the Han was being called into question by new religious currents, portrays a "man of quiet retirement" whose spirit, amid the tranquil landscape that surrounds him, soars to the heights inhabited by Taoist immortals. For other poems from this period—including many love poems of a kind rare in later Chinese poetry—see Anne Birrell's translation of* New Songs from a Jade Terrace *(1982), an anthology compiled in about 545.*

TWO POEMS FROM *NINETEEN OLD POEMS OF THE HAN*, TRANSLATED BY ARTHUR WALEY

Crossing the river I pluck the lotus flowers;
In the orchid-swamps are many fragrant herbs.
I gather them, but who shall I send them to?
My love is living in lands far away.
I turn and look towards my own country;
The long road stretches on for ever.
The same heart, yet a different dwelling:
Always fretting, till we are grown old!

Far away twinkles the Herd-boy star;
Brightly shines the Lady of the Han River.[3]
Slender, slender she plies her white fingers;
Click, click go the wheels of her spinning loom.
At the end of the day she has not finished her task;
Her bitter tears fall like streaming rain.
The Han River runs shallow and clear;
Set between them, how short a space!
But the river water will not let them pass,
Gazing at each other but never able to speak.

[3]The "Han River" is the Milky Way. Burton Watson, in a note to his translation, explains that this poem "concerns the legend of the Herdboy and the Weaving Maiden, constellations that correspond roughly to Aquila [Herdboy] and Vega and the Lyre [Weaving Maiden]. The Weaving Maiden, daughter of the Emperor of Heaven and an expert at weaving, married the Herdboy, but after her marriage she neglected her weaving. To punish her, her father placed the couple on opposite sides of the River of Heaven. . . . They are permitted to meet only once a year, on the night of the seventh day of the seventh lunar month, when sympathetic magpies form a bridge for them over the stream of stars."

The Wangsun,
by Wang Yen-shou [Wang Yanshou], son of Wang I [Wang Yi],
translated by Arthur Waley

Sublime was he, stupendous in invention,
Who planned the miracles of earth and sky.
Wondrous the power that charged
Small things with secret beauty, moving in them all.
See now the wangsun, crafty creature, mean of size,
Uncouth of form; the wrinkled face
Of an aged man; the body of a little child.
See how in turn he blinks and blenches with an air
Pathetically puzzled, dimly gazes
Under tired lids, through languid lashes
Looks tragic and hollow-eyed, rumples his brow,
Scatters this way and that
An insolent, astonished glare;
Sniffs and snorts, snuffs and sneezes,
Snicks and cocks his knowing little ears!
Now like a dotard mouths and chews
Or hoots and hisses through his pouted lips;
Shows gnashing teeth, grates and grinds ill-temperedly,
Gobbles and puffs and scolds.
And every now and then,
Down to his belly, from the larder that he keeps
In either cheek, he sends
Little consignments lowered cautiously.
Sometimes he squats
Like a puppy on its haunches, or hare-like humps
An arching back;
Smirks and wheedles with ingratiating sweetness;
Or suddenly takes to whining, surly snarling;
Then, like a ravening tiger roars.

He lives in thick forests, deep among the hills,
Or houses in the clefts of sharp, precipitous rocks;
Alert and agile is his nature, nimble are his wits;
Swift are his contortions,
Apt to every need,
Whether he climb tall tree-stems of a hundred feet,
Or sways on the shuddering shoulder of a long bough.
Before him, the dark gullies of unfathomable streams;
Behind, the silent hollows of the lonely hills.
Twigs and tendrils are his rocking-chairs,
On rungs of rotting wood he trips
Up perilous places; sometimes, leap after leap,
Like lightning flits through the woods.

Sometimes he saunters with a sad, forsaken air;
Then suddenly peeps round
Beaming with satisfaction. Up he springs,
Leaps and prances, whoops, and scampers on his way.
Up cliffs he scrambles, up pointed rocks,
Dances on shale that shifts or twigs that snap,
Suddenly swerves and lightly passes. . . .
Oh, what tongue could unravel
The tale of all his tricks?

Alas, one trait
With the human tribe he shares; their sweet's his sweet,
Their bitter is his bitter. Off sugar from the vat
Of brewer's dregs he loves to sup.
So men put wine where he will pass.
How he races to the bowl!
How nimbly licks and swills!
Now he staggers, feels dazed and foolish,
Darkness falls upon his eyes. . . .
He sleeps and knows no more.
Up steal the trappers, catch him by the mane,
Then to a string or ribbon tie him, lead him home;
Tether him to the stable or lock him into the yard;
Where faces all day long
Gaze, gape, gasp at him and will not go away.

The Valley Wind,
by Lu Yün [Lu Yun], translated by Arthur Waley

Living in retirement beyond the World,
Silently enjoying isolation,
I pull the rope of my door tighter
And bind firmly this cracked jar.[4]
My spirit is tuned to the Spring season;
At the fall of the year there is autumn in my heart.
Thus imitating cosmic changes
My cottage becomes a Universe.

Poem on the Wandering Immortal,
by Kuo P'u [Guo Pu], translated by Burton Watson

Kingfishers frolic among the orchid blossoms,
each form and hue lending freshness to the others.
Green creepers twine over the tall grove,
their leafy darkness shadowing the whole hill.
And in the midst, a man of quiet retirement
softly whistles, strokes the clear lute strings,

[4]That serves as a window. (Waley)

frees his thoughts to soar beyond the blue,
munches flower stamens, dips from a waterfall.
When Red Pine appears, roaming on high,
this man rides a stork, mounting the purple mists,
his left hand holding Floating Hill's sleeve,
his right hand patting Vast Cliff on the shoulder.[5]
Let me ask those short-lived mayflies,
what could they know of the years of the tortoise and crane?

Four Poems and a Preface by T'ao Ch'ien [Tao Qian], or T'ao Yüan-ming [Tao Yuanming], translated by Arthur Waley and Burton Watson

T'ao Ch'ien [Tao Qian] (365–427), also called by his "courtesy name" T'ao Yüan-ming, lived in southern China (modern Kiangsi) during the turbulent Six Dynasties. After efforts to secure a satisfactory official post failed, he retired to the farming life of his rural "hut in a zone of human habitation," where he wrote the poems for which he is best known. His relatively plain and straightforward style, and his contemplative personal ruminations on day-to-day existence, though little esteemed in his own age, had enormous influence on poets of the High T'ang; no Chinese poet before him had celebrated the pleasures of country life (and of wine) so fully. "There is an overall ambiguity in his poetry," however, Watson remarks— "exclamations upon the beauties of nature and the freedom and peace of rustic life, set uneasily alongside confessions of loneliness, frustration, and fear, particularly fear of death." In a somewhat fanciful autobiographical sketch, "The Gentleman of the Five Willow Trees," translated by James Robert Hightower in The Poetry of T'ao Ch'ien *(1970), he writes: "He was of a placid disposition and rarely spoke. He had no envy of fame or fortune. . . . He could not drink without emptying his cup, and always ended up drunk, after which he would retire, unconcerned about what might come. He lived alone in a bare little hut which gave no adequate shelter against rain and sun. His short coat was torn and patched, his cooking pots were frequently empty, but he was unperturbed. He used to write poems for his own amusement, and in them can be seen something of what he thought. He had no concern for worldly success, and so he ended his days." The following poems are translated by Waley; the famous preface to his poem on the Peach Blossom Spring, a dreamlike vision of a more harmonious way of life elsewhere, or nowhere (the meaning of "utopia"), which many later poets including Wang Wei and Li Po would respond to, is translated by Watson.*

Shady, shady the wood in front of the Hall;
At midsummer full of calm shadows.
The south wind follows summer's train;
With its eddying puffs it blows open my coat.
I am free from ties and can live a life of retirement.
When I rise from sleep, I play with books and lute.

[5] The Master of the Red Pine (Sung), Floating Hill (Fu-ch'iu), and Vast Cliff (Hung-yai) are Taoist Immortals of ancient times. (After Watson)

The lettuce in the garden still grows moist;
Of last year's grain there is always plenty left.
Self-support should maintain its strict limits;
More than enough is not what I want.
I grind millet and make good wine;
When the wine is heated, I pour it out for myself.
My little children are playing at my side,
Learning to talk, they babble unformed sounds.
These things have made me happy again
And I forget my lost cap of office.
Distant, distant I gaze at the white clouds;
With a deep yearning I think of the Sages of Antiquity.

—trans. Waley

I built my hut in a zone of human habitation,
Yet near me there sounds no noise of horse or coach.
Would you know how that is possible?
A heart that is distant creates a wilderness round it.
I pluck chrysanthemums under the eastern hedge,
Then gaze long at the distant summer hills.
The mountain air is fresh at the dusk of day;
The flying birds two by two return.
In these things there lies a deep meaning;
Yet when we would express it, words suddenly fail us.

—trans. Waley

Returning to the Fields,
translated by Arthur Waley

When I was young, I was out of tune with the herd;
My only love was for the hills and mountains.
Unwitting I fell into the Web of the World's dust
And was not free until my thirtieth year.
The migrant bird longs for the old wood;
The fish in the tank thinks of its native pool.
I had rescued from wildness a patch of the Southern Moor
And, still rustic, I returned to field and garden.
My ground covers no more than ten acres;
My thatched cottage has eight or nine rooms.
Elms and willows cluster by the eaves;
Peach trees and plum trees grow before the Hall.
Hazy, hazy the distant hamlets of men;
Steady the smoke that hangs over cottage roofs.
A dog barks somewhere in the deep lanes,
A cock crows at the top of the mulberry tree.

At gate and courtyard—no murmur of the World's dust;
In the empty rooms—leisure and deep stillness.
Long I lived checked by the bars of a cage;
Now I have turned again to Nature and Freedom.

Reading the Book of Hills and Seas,
translated by Arthur Waley

In the month of June the grass grows high
And round my cottage thick-leaved branches sway.
There is not a bird but delights the place where it rests;
And I too—love my thatched cottage.
I have done my plowing;
I have sown my seed.
Again I have time to sit and read my books.
In the narrow lane there are no deep ruts;
Often my friends' carriages turn back.
In high spirits I pour out my spring wine
And pluck the lettuce growing in my garden.
A gentle rain comes stealing up from the east
And a sweet wind bears it company.
My thoughts float idly over the story of the king of Chou,
My eyes wander over the pictures of Hills and Seas.
At a single glance I survey the whole Universe.
He will never be happy, whom such pleasures fail to please!

Preface to the "Poem on the Peach Blossom Spring,"
translated by Burton Watson

[The following preface is one of the most famous and influential passages in all of early Chinese prose. The poem itself, which simply repeats the account given in the preface, is of secondary interest and has not been translated. (Watson)]

During the T'ai-yüan era (376–397) of the Chin dynasty, there was a man of Wu-ling who caught fish for a living. Once he was making his way up a valley stream and had lost track of how far he had gone when he suddenly came upon a forest of peach trees in bloom. For several hundred paces on either bank of the stream there were no other trees to be seen, but fragrant grasses, fresh and beautiful, and falling petals whirling all around.

The fisherman, astonished at such a sight, pushed ahead, hoping to see what lay beyond the forest. Where the forest ended there was a spring that fed the stream, and beyond that a hill. The hill had a small opening in it, from which there seemed to come a gleam of light. Abandoning his boat, the fisherman went through the opening. At first it was very narrow, with barely room for a person to pass, but after he had gone twenty or thirty paces, it suddenly opened out and he could see clearly.

A plain stretched before him, broad and flat, with houses and sheds dotting it, and rich fields, pretty ponds, and mulberry and bamboo around them. Paths ran north and south, east and west across the fields, and chickens and dogs could be heard from farm to farm. The men and women who passed back and forth in the midst, sowing and tilling the fields, were all dressed just like any other people, and from white-haired elders to youngsters with their hair unbound, everyone seemed carefree and happy.

The people, seeing the fisherman, were greatly startled and asked where he had come from. When he had answered all their questions, they invited him to return with them to their home, where they set out wine and killed a chicken to prepare a meal.

As soon as the others in the village heard of his arrival, they all came to greet him. They told him that some generations in the past their people had fled from the troubled times of the Ch'in dynasty (221–207 B.C.) and had come with their wives and children and fellow villagers to this faraway place. They had never ventured out into the world again, and hence in time had come to be completely cut off from other people. They asked him what dynasty was ruling at present—they had not even heard of the Han dynasty, to say nothing of the Wei and Chin dynasties that succeeded it. The fisherman replied to each of their questions to the best of his knowledge, and everyone sighed with wonder.

The other villagers invited the fisherman to visit their homes as well, each setting out wine and food for him. Thus he remained for several days before taking his leave. One of the villagers said to him, "I trust you won't tell the people on the outside about this."

After the fisherman had made his way out of the place, he found his boat and followed the route he had taken earlier, taking care to note the places that he passed. When he reached the prefectural town, he went to call on the governor and reported what had happened. The governor immediately dispatched men to go with him to look for the place, but though he tried to locate the spots that he had taken note of earlier, in the end he became confused and could not find the way again.

Liu Tzu-chi of Nan-yang, a gentleman-recluse of lofty ideals, heard the story and began delightedly making plans to go there, but before he could carry them out, he fell sick and died. Since then there have been no more "seekers of the ford."[6]

Five Poems by Five T'ang Poets, translated by Witter Bynner

So rich a poetic period was the T'ang dynasty (618–907) that not even the five major poets of the following sections—Wang Wei, Li Po, Tu Fu, Po Chü-i, and Han-shan—can fully suggest its scope. This section, therefore, represents, by a single poem of each, five other important though less well known poets of the age: Chang Hsü [Zhang Xu] (early eighth century), whose "Peach-Blossom River" is one of numerous poems of this period reflecting the abiding appeal of T'ao Ch'ien's vision; Meng Hao-jan [Meng Haoran] (689–740); Ch'ang Chien [Chang Jian] (mid eighth century); Liu Chang-ch'ing [Liu Jangching] (710?–785?); and Ch'iu Wei [Qiu Wei] (694?–789). Meng Hao-jan was an honored friend of Wang Wei and Li Po; little is known about most of the others. Throughout the five poems in this section runs a longing for the simple life of the recluse—a longing going back to Chuang Tzu and other sages a thousand years before, but

[6]An allusion to *Analects* XVIII, 6, in which Confucius sends one of his disciples to inquire about a fording place across a river. Here, of course, the phrase refers to seekers of the utopian land of the Peach Blossom Spring. (Watson)

intensely heightened in this age of conflict between Confucian duty and the appeal of the Taoist or Buddhist hermit's tranquil withdrawal from the world of servile officialdom and courtly decadence centered in the capital of Ch'ang-an.

Peach-Blossom River,
by Chang Hsü [Zhang Xu]

A bridge flies away through a wild mist,
Yet here are the rocks and the fisherman's boat.
Oh, if only this river of floating peach-petals
Might lead me at last to the mythical cave!

At the Mountain-Lodge of the Buddhist Priest Ye,
Waiting in Vain for My Friend Ting,
by Meng Hao-jan [Meng Haoran]

Now that the sun has set beyond the western range,
Valley after valley is shadowy and dim . . .
And now through pine-trees come the moon and the chill of evening,
And my ears feel pure with the sound of wind and water . . .
Nearly all the woodsmen have reached home,
Birds have settled on their perches in the quiet mist . . .
And still—because you promised—I am waiting for you, waiting,
Playing my lonely lute under a wayside vine.

A Buddhist Retreat Behind Broken-Mountain Temple,
by Ch'ang Chien [Chang Jian]

In the pure morning, near the old temple,
Where early sunlight points the tree-tops,
My path has wound, through a sheltered hollow
Of boughs and flowers, to a Buddhist retreat.
Here birds are alive with mountain-light,
And the mind of man touches peace in a pool,
And a thousand sounds are quieted
By the breathing of a temple-bell.

While Visiting on the South Stream the Taoist Priest Ch'ang,
by Liu Chang-ch'ing [Liu Jangching]

Walking along a little path,
I find a footprint on the moss,
A white cloud low on the quiet lake,
Grasses that sweeten an idle door,
A pine grown greener with the rain,
A brook that comes from a mountain source—
And, mingling with Truth among the flowers,
I have forgotten what to say.

After Missing the Recluse on the Western Mountain,
by Ch'iu Wei [Qiu Wei]

To your hermitage here on the top of the mountain
I have climbed, without stopping, these ten miles.
I have knocked at your door, and no one answered;
I have peeped into your room, at your seat beside the table.
Perhaps you are out riding in your canopied chair,
Or fishing, more likely, in some autumn pool.
Sorry though I am to be missing you,
You have become my meditation—
The beauty of your grasses, fresh with rain,
And close beside your window the music of your pines.
I take into my being all that I see and hear,
Soothing my senses, quieting my heart;
And though there be neither host nor guest,
Have I not reasoned a visit complete?
. . . After enough, I have gone down the mountain.
Why should I wait for you any longer?

Seven Poems by Wang Wei, translated by Witter Bynner

Wang Wei (699–759) was equally famous in his time as poet and painter of the natural world; all of his paintings, however, like almost all others from the T'ang, have perished. He was a youthful prodigy from a wealthy family, who wrote his poem on Peach Blossom River (inspired of course by T'ao Ch'ien) in his teens. Caught up in An Lu-shan's rebellion, he was either imprisoned or forced to serve in the rebel army and was accused of sedition after its defeat, but pardoned; thereafter he divided his life between the capital and retirement in the hills. Far from being simply a recluse, he was, Owen remarks, "one of the most social and urbane of T'ang poets." Indeed, "he continually returned to court and never resigned from service," Marsha Wagner observes in her biography Wang Wei *(1981), and "in nature he led the luxurious life of a well-to-do country gentleman, not an ascetic mountain hermit or a simple peasant farmer." He was a devout Buddhist throughout his life, and his poetry is pervaded, Watson writes, by a Buddhist "calm affirmation": "He does not assiduously seek out the wild or picturesque element in the natural landscape, as nature poets often do, nor does he avert his eyes from the evidences of human habitation or activity. He registers the scenes about him just as they appear to him, the human along with the non-human components, his very impartiality a gauge, one feels, of his level of enlightenment."*

A Song of Peach-Blossom River
Written to Music

A fisherman is drifting, enjoying the spring mountains,
And the peach-trees on both banks lead him to an ancient source.
Watching the fresh-colored trees, he never thinks of distance

Till he comes to the end of the blue stream and suddenly—strange men!
It's a cave—with a mouth so narrow that he has to crawl through;
But then it opens wide again on a broad and level path—
And far beyond he faces clouds crowning a reach of trees,
And thousands of houses shadowed round with flowers and bamboos. . . .
Woodsmen tell him their names in the ancient speech of Han;
And clothes of the Ch'in Dynasty are worn by all these people
Living on the uplands, above the Wu-ling River,
On farms and in gardens that are like a world apart,
Their dwellings at peace under pines in the clear moon,
Until sunrise fills the low sky with crowing and barking.
. . . At news of a stranger the people all assemble,
And each of them invites him home and asks him where he was born.
Alleys and paths are cleared for him of petals in the morning,
And fishermen and farmers bring him their loads at dusk. . . .
They had left the world long ago, they had come here seeking refuge;
They have lived like angels ever since, blessedly far away,
No one in the cave knowing anything outside,
Outsiders viewing only empty mountains and thick clouds.
. . . The fisherman, unaware of his great good fortune,
Begins to think of country, home, of worldly ties,
Finds his way out of the cave again, past mountains and past rivers,
Intending some time to return, when he has told his kin.
He studies every step he takes, fixes it well in mind,
And forgets that cliffs and peaks may vary their appearance.
. . . It is certain that to enter through the deepness of the mountain,
A green river leads you, into a misty wood.
But now, with spring-floods everywhere and floating peach-petals—
Which is the way to go, to find that hidden source?

In a Retreat Among Bamboos

Leaning alone in the close bamboos,
I am playing my lute and humming a song
Too softly for anyone to hear—
Except my comrade, the bright moon.

A Message from My Lodge at Wang-Ch'üan

To P'ai Ti

The mountains are cold and blue now
And the autumn waters have run all day.
By my thatch door, leaning on my staff,
I listen to cicadas in the evening wind.
Sunset lingers at the ferry,
Supper-smoke floats up from the houses.
. . . Oh, when shall I pledge the great Hermit again
And sing a wild poem at Five Willows?

Answering Vice-Prefect Chang

As the years go by, give me but peace,
Freedom from ten thousand matters.
I ask myself and always answer:
What can be better than coming home?
A wind from the pine-trees blows my sash,
And my lute is bright with the mountain moon.
You ask me about good and evil fortune? . . .
Hark, on the lake there's a fisherman singing!

My Retreat at Mount Chung-nan

My heart in middle age found the Way,
And I came to dwell at the foot of this mountain.
When the spirit moves, I wander alone
Amid beauty that is all for me. . . .
I will walk till the water checks my path,
Then sit and watch the rising clouds—
And some day meet an old wood-cutter
And talk and laugh and never return.

A Green Stream

I have sailed the River of Yellow Flowers,
Borne by the channel of a green stream,
Rounding ten thousand turns through the mountains
On a journey of less than thirty miles. . . .
Rapids hum over heaped rocks;
But where light grows dim in the thick pines,
The surface of an inlet sways with nut-horns
And weeds are lush along the banks.
. . . Down in my heart I have always been as pure
As this limpid water is. . . .
Oh, to remain on a broad flat rock
And to cast a fishing-line forever!

A Farm-House on the Wei River

In the slant of the sun on the country-side,
Cattle and sheep trail home along the lane;
And a rugged old man in a thatch door
Leans on a staff and thinks of his son, the herd-boy.
There are whirring pheasants, full wheat-ears,
Silk-worms asleep, pared mulberry-leaves.
And the farmers, returning with hoes on their shoulders,

Hail one another familiarly.
. . . No wonder I long for the simple life
And am sighing the old song, *Oh, to go Back Again!*

Four Poems by Li Po [Li Bo], or Li Pai [Li Bai], translated by Witter Bynner

Li Po [Li Bo], or Li Pai [Li Bai] (701–762), grew up in Szechwan and unlike most T'ang poets never held a government position for long; much of his life was spent wandering or in exile. He was a colorful and romantic figure, exultantly celebrating life and wine, whose poems give vivid expression, Watson writes, to "a tireless search for spiritual freedom and communion with nature, a lively imagination and a deep sensitivity to the beauties of language." He was recognized immediately (unlike Tu Fu) as a great poet, and legends quickly grew up around him, fostered no doubt by his own exuberant fantasies; thus he was said to have died by falling drunk from a boat while trying to grasp the reflection of the moon. "The evanescence of the world tormented him, drove him to frenzy," Robert Payne writes; "he would dam the water and make an everlasting flower of imperishable metal if he could, and yet he knew that the sheer beauty of the world lay in its evanescence." In "Conversation in the Mountains," as translated in Payne's The White Pony, *Li Po writes:*

If you were to ask me why I dwell among green mountains,
I should laugh silently; my soul is serene.
The peach blossom follows the moving water;
There is another heaven and earth beyond the world of men.

According to traditions reported by Hawkes, he "believed in and practiced alchemy and claimed to be an earthly representative of the Taoist divinities." Be that as it may, the transformative spirit of Taoism runs throughout his poems. Nowhere is this more evident than in "T'ien-mu Mountain Ascended in a Dream," where Taoist soul-flight combines with the allure of nature in the wild to lead the poet, disdaining office, to withdraw to the mountains where he can be free.

In the Quiet Night

So bright a gleam on the foot of my bed—
Could there have been a frost already?
Lifting myself to look, I found that it was moonlight.
Sinking back again, I thought suddenly of home.

Down Chung-nan Mountain to the Kind Pillow and Bowl of Hu Ssŭ

Down the blue mountain in the evening,
Moonlight was my homeward escort.
Looking back, I saw my path
Lie in levels of deep shadow . . .
I was passing the farm-house of a friend,
When his children called from a gate of thorn
And led me twining through jade bamboos

Where green vines caught and held my clothes.
And I was glad of a chance to rest
And glad of a chance to drink with my friend. . . .
We sang to the tune of the wind in the pines;
And we finished our songs as the stars went down,
When, I being drunk and my friend more than happy,
Between us we forgot the world.

Drinking Alone with the Moon

From a pot of wine among the flowers
I drank alone. There was no one with me—
Till, raising my cup, I asked the bright moon
To bring me my shadow and make us three.
Alas, the moon was unable to drink
And my shadow tagged me vacantly;
But still for a while I had these friends
To cheer me through the end of spring. . . .
I sang. The moon encouraged me.
I danced. My shadow tumbled after.
As long as I knew, we were boon companions.
And then I was drunk, and we lost one another.
. . . Shall goodwill ever be secure?
I watch the long road of the River of Stars.

T'ien-mu Mountain Ascended in a Dream

A seafaring visitor will talk about Japan,
Which waters and mists conceal beyond approach;
But Yüeh people talk about Heavenly Mother Mountain,
Still seen through its varying deepnesses of cloud.
In a straight line to heaven, its summit enters heaven,
Tops the five Holy Peaks, and casts a shadow through China
With the hundred-mile length of the Heavenly Terrace Range,
Which, just at this point, begins turning southeast.
. . . My heart and my dreams are in Wu and Yüeh
And they cross Mirror Lake all night in the moon.
And the moon lights my shadow
And me to Yien River—
With the hermitage of Hsieh still there
And the monkeys calling clearly over ripples of green water.
I wear his pegged boots
Up a ladder of blue cloud,
Sunny ocean half-way,
Holy cock-crow in space,
Myriad peaks and more valleys and nowhere a road.
Flowers lure me, rocks ease me. Day suddenly ends.
Bears, dragons, tempestuous on mountain and river,

Startle the forest and make the heights tremble.
Clouds darken with darkness of rain,
Streams pale with pallor of mist.
The Gods of Thunder and Lightning
Shatter the whole range.
The stone gate breaks asunder
Venting in the pit of heaven,
An impenetrable shadow.
. . . But now the sun and moon illumine a gold and silver terrace,
And, clad in rainbow garments, riding on the wind,
Come the queens of all the clouds, descending one by one,
With tigers for their lute-players and phoenixes for dancers.
Row upon row, like fields of hemp, range the fairy figures. . . .
I move, my soul goes flying,
I wake with a long sigh,
My pillow and my matting
Are the lost clouds I was in.
. . . And this is the way it always is with human joy:
Ten thousand things run for ever like water toward the east.
And so I take my leave of you, not knowing for how long.
. . . But let me, on my green slope, raise a white deer
And ride to you, great mountain, when I have need of you.
Oh, how can I gravely bow and scrape to men of high rank and men of high office
Who never will suffer being shown an honest-hearted face?

Nine Poems by Tu Fu [Du Fu], translated by Witter Bynner

Tu Fu [Du Fu] (712–70) was born of an eminent family; a noted Confucian scholar was an ancestor and a leading poet of the early T'ang a grandfather. Even so, he failed his official examinations and was not awarded a post until he was forty-three. This was in 755, when the court at Ch'ang-an ("Eternal Peace")—already disrupted by the influence exerted over Emperor Hsüan-tsung [Xuanzong] (who reigned from 713 to 755 and was known as Ming Huang, "the Bright Emperor") by his concubine Yang Kuei-fei [Yang Guifei]—was shattered by the revolt of An Lu-shan, which resulted in the Emperor's abdication in favor of his son, the execution of Yang Kuei-fei, and the death of millions before An Lu-shan was defeated and a shaky peace restored. Tu Fu fled north, and after suppression of the revolt was awarded a provincial post, which he lost in 759 for criticism of the Emperor. He "spent the remainder of his life," Watson writes, "in restless wanderings," writing many of his poems in a roomy "thatched hut" in Szechwan; while returning to the capital, he died. His dense and innovative poetry combines technical mastery—above all, of difficult "regulated" verse—with deep feeling; although he met (and wrote poems for) Li Po, he was not considered a major poet in his own lifetime. "As in the case of so many artists whose work is experimental and forward-looking," Watson remarks, "it remained for posterity to recognize the full extent of his genius," as Po Chü-i and others of a later generation did; he has since been widely regarded, to quote the subtitle of William Hung's biography Tu Fu *(1952), as "China's greatest poet." His impact has also been belatedly felt, despite neglect by both Pound and*

Waley, by Western readers; Kenneth Rexroth thinks him "the greatest non-epic, non-dramatic poet who has survived in any language."

Given his strong Confucian obligation to comment on human suffering, Tu Fu is not usually viewed (as Wang Wei sometimes too simply is) as a "nature poet." His greatest works, in Watson's view, "are at once a lament upon the appalling sorrows that he saw around him, and a reproach to those who, through folly or ignorance, were to some degree responsible for the creation of so much misery." Yet as the very title of "Autumn Meditations," translated by A. C. Graham in Poems of the Late T'ang *(1965) and regarded by Owen as possibly "the greatest poems in the Chinese language," suggests, nature imagery is no less central to the "Confucian" Tu Fu than the "Buddhist" Wang Wei or the "Taoist" Li Po. Tu Fu "sees nature not as retreat or drama," Vikram Seth writes, "but as an emotional or moral entity set in juxtaposition to human life and human events, whether in sympathy or antipathy. The noble cypress that is not uprooted by violent storms, the flowers that insist on returning in spring to a devastated war-stricken country—these appear to him to be intimately tied through either consciousness or heedlessness to human vicissitudes and griefs." When Rexroth writes, "I am sure he has made me a better man," he writes in the spirit that Tu Fu would have wished, beyond all else, to convey.*

A Spring View

Though a country be sundered, hills and rivers endure;
And spring comes green again to trees and grasses
Where petals have been shed like tears
And lonely birds have sung their grief.
. . . After the war-fires of three months,
One message from home is worth a ton of gold.
. . . I stroke my white hair. It has grown too thin
To hold the hairpins any more.

A Night Abroad

A light wind is rippling at the grassy shore. . . .
Through the night, to my motionless tall mast,
The stars lean down from open space,
And the moon comes running up the river.
. . . If only my art might bring me fame
And free my sick old age from office!—
Flitting, flitting, what am I like
But a sand-snipe in the wide, wide world!

A Hearty Welcome

To Vice-Prefect Ts'uei

North of me, south of me, spring is in flood,
Day after day I have seen only gulls . . .
My path is full of petals—I have swept it for no others.
My thatch gate has been closed—but opens now for you.
It's a long way to the market, I can offer you little—

Yet here in my cottage there is old wine for our cups.
Shall we summon my elderly neighbor to join us,
Call him through the fence, and pour the jar dry?

A View of the Wilderness

Snow is white on the westward mountains and on three fortified towns,
And waters in this southern lake flash on a long bridge.
But wind and dust from sea to sea bar me from my brothers;
And I cannot help crying, I am so far away.
I have nothing to expect now but the ills of old age.
I am of less use to my country than a grain of dust.
I ride out to the edge of town. I watch on the horizon,
Day after day, the chaos of the world.

A Long Climb

In a sharp gale from the wide sky apes are whimpering,
Birds are flying homeward over the clear lake and white sand,
Leaves are dropping down like the spray of a waterfall,
While I watch the long river always rolling on.
I have come three thousand miles away. Sad now with autumn
And with my hundred years of woe, I climb the height alone.
Ill fortune has laid a bitter frost on my temples,
Heart-ache and weariness are a thick dust in my wine.

From an Upper Story

Flowers, as high as my window, hurt the heart of a wanderer
For I see, from this high vantage, sadness everywhere.
The Silken River, bright with spring, floats between earth and heaven
Like a line of cloud by the Jade Peak, between ancient days and now.
. . . Though the State is established for a while as firm as the North Star
And bandits dare not venture from the western hills,
Yet sorry in the twilight for the woes of a long-vanished Emperor,
I am singing the song his Premier sang when still unestranged from the mountain.

Night in the Watch-Tower

While winter daylight shortens in the elemental scale
And snow and frost whiten the cold-circling night,
Stark sounds the fifth-watch with a challenge of drum and bugle.
. . . The stars and the River of Heaven pulse over the three mountains;
I hear women in the distance, wailing after the battle;
I see barbarian fishermen and woodcutters in the dawn.
. . . Sleeping-Dragon, Plunging-Horse, are no generals now, they are dust—
Hush for a moment, O tumult of the world.

A View of T'ai-shan

What shall I say of the Great Peak?—
The ancient dukedoms are everywhere green,
Inspired and stirred by the breath of creation,
With the Twin Forces balancing day and night.
. . . I bare my breast toward opening clouds,
I strain my sight after birds flying home.
When shall I reach the top and hold
All mountains in a single glance?

A Song of an Old Cypress

Beside the Temple of the Great Premier stands an ancient cypress
With a trunk of green bronze and a root of stone.
The girth of its white bark would be the reach of forty men
And its tip of kingfisher-blue is two thousand feet in heaven.
Dating from the days of a great ruler's great statesman,
Their very tree is loved now and honored by the people.
Clouds come to it from far away, from the Wu cliffs,
And the cold moon glistens on its peak of snow.
. . . East of the Silk Pavilion yesterday I found
The ancient ruler and wise statesman both worshipped in one temple,
Whose tree, with curious branches, ages the whole landscape
In spite of the fresh colors of the windows and the doors.
And so firm is the deep root, so established underground,
That its lone lofty boughs can dare the weight of winds,
Its only protection the Heavenly Power,
Its only endurance the art of its Creator.
. . . When beams are required to restore a great house,
Though oxen sway ten thousand heads, they cannot move a mountain.
Though a tree writes no memorial, yet people understand
That not unless they fell it can use be made of it. . . .
Its bitter heart may be tenanted now by black and white ants,
But its odorous leaves were once the nest of phoenixes and pheasants.
. . . Let wise and hopeful men harbor no complaint.
The greater the timber, the tougher it is to use.

Six Poems by Po Chü-i [Bo Juyi], or Pai Chü-i [Bai Juyi], translated by Arthur Waley

Po Chü-i [Bo Juyi], or Pai Chü-i [Bai Juyi] (772–846), a prolific poet ever since his early years, wrote nearly 2,800 poems, which he took care to arrange and publish himself—poems expressing, as we have seen, the division felt by many in his age between official duties and longing for a secluded life in accord with the rhythms of the natural world. His family was of modest means, and Po, after passing his examinations at twenty-nine, held a number of positions throughout his career, including governorships of Hangchow and Soochow. Such an existence, at a time when the weakened T'ang dynasty continued to be disrupted by internal rebellions and barbarian incursions, entailed continual mobility, since officials were not allowed to serve long in any one place, and Po like many contemporaries often laments his separation from friends. In a letter of 815 to his fellow poet Yüan Chen [Yuan Zhen], as translated in Waley's Life and Times of Po Chü-i *(1949), he voiced the deep sense of Confucian responsibility that helped make his poems popular with the common people in his own time and later: "the duty of literature is to be of service to the writer's generation; that of poetry to influence public affairs." In the next year, however, after being demoted for opposition to government policies and sent to a remote rural area south of the Yangtze, he determined to build a small house on nearby Mount Lu in which to find refuge from official burdens.*

In a short prose sketch translated by Watson in Four Huts: Asian Writings on the Simple Life *(1994) as "Record of the Thatched Hall on Mount Lu," he describes his inner joy and peace as "I gaze up at the mountains, bend down to listen to the spring, look around at the trees and bamboos, the clouds and rocks, busy with them every minute from sunup to evening." A similar delight, often mingled with melancholy, in solitude and in nature runs throughout his poems, reflecting the meditative Buddhism of the Ch'an Sudden Enlightenment school to which he was converted at about this time. To some critics his poetry, with its deliberate simplicity and commonplace themes, has seemed pedestrian, lacking the flair of Li Po and the depth of Tu Fu. Yet both in China and abroad—especially in Japan, where he is known by his "courtesy name" Po Lot'ien [Bo Lotian]—his influence has been immense; in Lady Murasaki's* Tale of Genji, *as Waley (who translated it, and translated Po Chü-i more often than any other poet) notes, "the numerous references to Chinese poetry are all to poems either by Po or by his friends Yüan Chen and Liu Yü-hsi." Surely one reason for this continuing impact is the deep "compassion for human suffering" that finds expression even when his subject is the intense beauty he finds in the natural world: in the beasts and the trees, the moon, the cranes, and above all in the mountains where he plays with the pebbles of the stream, and madly sings.*

Going Alone to Spend a Night at the Hsien-yu Temple

A.D. 806

The crane from the shore standing at the top of the steps,
The moon on the pool seen at the open door;
Where these are, I made my lodging-place
And for two nights could not turn away.
I am glad I chanced on a place so lonely and still
With no companion to drag me early home.
Now that I have tasted the joy of being alone,
I will never again come with a friend at my side.

The Beginning of Summer

A.D. 815

At the rise of summer a hundred beasts and trees
Join in gladness that the Season bids them thrive.
Stags and does frolic in the deep woods;
Snakes and insects are pleased by the rank grass.

Wingèd birds love the thick leaves;
Scaly fish enjoy the fresh weeds.
But to one place Summer forgot to come;
I alone am left like a withered straw . . .
In solitude, banished to the world's end;
Flesh and bone all in distant ways.
From my native place no tidings come;
Rebel troops flood the land with war.
Sullen grief, in the end, what will it bring?
I am only wearing my own heart away.
Better far to let both body and mind
Blindly yield to the fate that Heaven made.
Hsün-yang abounds in good wine;
I will fill my cup and never let it be dry.
On P'ên River fish are cheap as mud;
Early and late I will eat them, boiled and fried.
With morning rice at the temple under the hill,
And evening wine at the island in the lake . . .
Why should my thoughts turn to my native land?
For in this place one could well end one's age.

Madly Singing in the Mountains

There is no one among men that has not a special failing;
And my failing consists in writing verses.
I have broken away from the thousand ties of life;
But this infirmity still remains behind.
Each time that I look at a fine landscape,
Each time that I meet a loved friend,
I raise my voice and recite a stanza of poetry
And marvel as though a God had crossed my path.
Ever since the day I was banished to Hsün-yang
Half my time I have lived among the hills.
And often, when I have finished a new poem,
Alone I climb the road to the Eastern Rock.
I lean my body on the banks of white Stone;
I pull down with my hands a green cassia branch.
My mad singing startles the valleys and hills;
The apes and birds all come to peep.
Fearing to become a laughing-stock to the world,
I choose a place that is unfrequented by men.

Visiting the Hsi-lin Temple

A.D. 817

I dismount from my horse at the Hsi-lin Temple;
I hurry forward, speeding with light cane.
In the morning I work at a Government office-desk;

In the evening I become a dweller in the Sacred Hills.
In the second month to the north of K'uang-lu
The ice breaks and the snow begins to melt.
On the southern plantation the tea-plant thrusts its sprouts;
Through the northern crevice the veins of the spring ooze.

This year there is war in An-hui,
In every place soldiers are rushing to arms.
Men of learning have been summoned to the Council Board;
Men of action are marching to the battle-line.
Only I, who have no talents at all,
Am left in the mountains to play with the pebbles of the stream.

Pruning Trees

Trees growing—right in front of my window;
The trees are high and the leaves grow thick.
Sad alas! the distant mountain view,
Obscured by this, dimly shows between.
One morning I took knife and axe;
With my own hand I lopped the branches off.
Ten thousand leaves fell about my head;
A thousand hills came before my eyes.
Suddenly, as when clouds or mists break
And straight through, the blue sky appears.
Again, like the face of a friend one has loved
Seen at last after an age of parting.
First there came a gentle wind blowing;
One by one the birds flew back to the tree.
To ease my mind I gazed to the South-East;
As my eyes wandered, my thoughts went far away.
Of men there is none that has not some preference;
Of things there is none but mixes good with ill.
It was not that I did not love the tender branches;
But better still—to see the green hills!

The Cranes

A.D. 830

The western wind has blown but a few days;
Yet the first leaf already flies from the bough.
On the drying paths I walk in my thin shoes;
In the first cold I have donned my quilted coat.
Through shallow ditches the floods are clearing away;
Through sparse bamboos trickles a slanting light.
In the early dusk, down an alley of green moss,
The garden-boy is leading the cranes home.

Ten Poems by Han-shan ("Cold Mountain"), translated by Gary Snyder

Han-shan, whose name means "Cold Mountain," was included in no standard T'ang anthology; given the uncertainty of his dates (probably late eighth to early ninth century), we cannot be quite sure he was *a T'ang poet—or even that a single poet wrote the poems attributed to him. If so, it appears, Burton Watson writes in introducing his selection in* The Columbia Book of Chinese Poetry, *"that he was a scholar-farmer who later in life retired to a place called Cold Mountain . . . in the T'ien-t'ai mountains, a range stretching along the seacoast in northeastern Chekiang Province." Other T'ang poets expressed a tension between social obligation and longing for freedom in nature; Han-shan, in the spirit of Ch'an Buddhism, relinquished the first for the second, and legends of the madly laughing mountain hermit soon clustered round his name as a favorite subject of stories and paintings in both China and Japan. Thus a probably apocryphal preface, attributed to Liu Ch'iu-yin [Liu Qiuyin], governor of T'ai Prefecture—included in the translation of 24 poems by Bollingen Prize poet Gary Snyder—says: "He looked like a tramp. His body and face were old and beat. Yet in every word he breathed was a meaning in line with the subtle principles of things, if only you thought of it deeply." When the governor, suffering from headache, sought out Han-shan and his companion Shih-te in a temple's kitchen in the mountains, he found them "facing the fire, laughing loudly. I made a bow. The two shouted* HO! *at me. They struck their hands together—Ha Ha!—great laughter. . . . The two men grabbed hands and ran out of the temple. I cried, 'Catch them'—but they quickly ran away. Han-shan returned to Cold Mountain"—where he has remained, perhaps, still madly laughing, till today. His "vivid delineations of the natural world," Watson remarks in his introduction to* Cold Mountain: 100 Poems by the T'ang Poet Han-shan *(1962), "are at the same time allegories of spiritual questing and attainment," so that Cold Mountain—as Waley, another of his translators, observes in a passage cited by Watson— "is often the name of a state of mind rather than a locality. It is on this conception, as well as on that of the 'hidden treasure,' the Buddha who is to be sought not somewhere outside us, but 'at home' in the heart, that the mysticism of the poems is based." Snyder's translations, done under the guidance of Professor Chen Shih-hsiang at the University of California, Berkeley, were first published in* Evergreen Review *in 1958 and reprinted in* Riprap and Cold Mountain Poems *in 1965.*

The path to Han-shan's place is laughable,
A path, but no sign of cart or horse.
Converging gorges—hard to trace their twists
Jumbled cliffs—unbelievably rugged.
A thousand grasses bend with dew,
A hill of pines hums in the wind,
And now I've lost the shortcut home,
Body asking shadow, how do you keep up?

In a tangle of cliffs I chose a place—
Bird-paths, but no trails for men.
What's beyond the yard?

White clouds clinging to vague rocks.
Now I've lived here—how many years—
Again and again, spring and winter pass.
Go tell families with silverware and cars
"What's the use of all that noise and money?"

In the mountains it's cold.
Always been cold, not just this year.
Jagged scarps forever snowed in
Woods in the dark ravines spitting mist.
Grass is still sprouting at the end of June,
Leaves begin to fall in early August.
And here am I, high on mountains,
Peering and peering, but I can't even see the sky.

Men ask the way to Cold Mountain
Cold Mountain: there's no through trail.
In summer, ice doesn't melt
The rising sun blurs in swirling fog.
How did I make it?
My heart's not the same as yours.
If your heart was like mine
You'd get it and be right here.

I settled at Cold Mountain long ago,
Already it seems like years and years.
Freely drifting, I prowl the woods and streams
And linger watching things themselves.
Men don't get this far into the mountains,
White clouds gather and billow.
Thin grass does for a mattress,
The blue sky makes a good quilt.
Happy with a stone underhead
Let heaven and earth go about their changes.

Clambering up the Cold Mountain path,
The Cold Mountain trail goes on and on:
The long gorge choked with scree and boulders,
The wide creek, the mist-blurred grass.
The moss is slippery, though there's been no rain
The pine sings, but there's no wind.
Who can leap the world's ties
And sit with me among the white clouds?

Rough and dark—the Cold Mountain trail,
Sharp cobbles—the icy creek bank.
Yammering, chirping—always birds
Bleak, alone, not even a lone hiker.
Whip, whip—the wind slaps my face
Whirled and tumbled—snow piles on my back.
Morning after morning I don't see the sun
Year after year, not a sign of spring.

In my first thirty years of life
I roamed hundreds and thousands of miles.
Walked by rivers through deep green grass
Entered cities of boiling red dust.
Tried drugs, but couldn't make Immortal;
Read books and wrote poems on history.
Today I'm back at Cold Mountain:
I'll sleep by the creek and purify my ears.

If I hide out at Cold Mountain
Living off mountain plants and berries—
All my lifetime, why worry?
One follows his karma through.
Days and months slip by like water,
Time is like sparks knocked off flint.
Go ahead and let the world change—
I'm happy to sit among these cliffs.

When men see Han-shan
They all say he's crazy
And not much to look at—
Dressed in rags and hides.
They don't get what I say
& I don't talk their language.
All I can say to those I meet:
"Try to make it to Cold Mountain."

Three Poets of the Sung [Song] Dynasty

The T'ang dynasty, weakened by rebellions and barbarian pressures on its borders, collapsed in 907, to be followed by the chaotic half-century of the "Five Dynasties" before the founders of the Sung [Song], in 960, initiated a long period of internal cohesion. Unity was continually threatened, however, by invasions that resulted in the loss of northern China in 1125 and the fall of the Southern Sung to the Mongols in 1279. The Sung was a great age of landscape painting, scholarship and Neo-Confucian thought, science and technology (from the invention of gunpowder to widespread use of printing), and literature in various forms, including poetry characterized by a more philosophical, even optimistic attitude to life than in the still more turbulent T'ang. The following selections represent three important poets for whom nature imagery remains indispensable to expression of human feeling. The first, ***Su Tung-p'o [Su Dongpo]*** *(1037–1101), with the "courtesy name" of Su Shih [Su Shi], fell from favor for criticizing the government and was exiled several times; his varied poems express a Confucian ideal of social responsibility but also an equanimity reflecting Taoist and Ch'an Buddhist outlooks. Like Wang Wei before him, he was a noted painter; the poem here included (translated, like the two poems of Lu Yu, by Watson in* The Columbia Book of Chinese Poetry*) portrays nature in the mirror of painting, and affirms that the peach blossoms of Wu-ling thought by T'ao Ch'ien to belong to a mythical world and by Li Po to suggest "another heaven and earth beyond the world of men" are among us here in that human world. Our second poet,* ***Li Ch'ing Chao [Li Ching Zhao]*** *(1084?–1151?), called by Rexroth in* One Hundred Poems from the Chinese *"China's greatest poetess, of any period," came from an educated family and lived a happy life until forced to abandon her home to the victorious Chin who seized northern China. Several among her handful of surviving poems—mainly written in the* tz'u [ci] *form, fitting lines of different lengths to popular melodies, a form widespread in the Sung—celebrate the joys of married love and the "boundless beauty" of the world, especially "the profound lasciviousness of spring." Finally,* ***Lu Yu*** *(1125–1210), the foremost poet of the Southern Sung and author of some ten thousand poems, led a life of many frustrations, including divorce forced upon him by his mother and disfavor at court because of his staunch advocacy of war to regain the territories lost to the northern barbarians. The title of Watson's volume of his selected poems and prose,* The Old Man Who Does As He Pleases *(1977), translates the literary name Fang-weng that Lu Yu adopted in his later years.*

Written on a Painting Entitled "Misty Yangtze and Folded Hills" in the Collection of Wang Ting-Kuo, by Su Tung-p'o [Su Dongpo], translated by Burton Watson

Above the river, heavy on the heart, thousandfold hills:
layers of green floating in the sky like mist.
Mountains? clouds? too far away to tell
till clouds part, mist scatters, on mountains that remain.
Then I see, in gorge cliffs, black-green clefts
where a hundred waterfalls leap from thc sky,
threading woods, tangling rocks, lost and seen again,
falling to valley mouths to feed swift streams.
Where the river broadens, mountains part, foothill forests end,
a small bridge, a country store set against the slope:
now and then travelers pass beyond tall trees;
a fishing boat—one speck where the river swallows the sky.
Tell me, where did you get this painting
sketched with these clean and certain strokes?
I didn't know the world had such places—
I'll go at once and buy some land!
But perhaps you've never seen those hidden spots

near Wu-ch'ang and Fan-k'ou, where I lived five years—[7]
Spring wind shook the river and sky was everywhere;
evening clouds rolled back the rain on gentle mountains;
from scarlet maples, crows flapped down to keep the boatman company;
from tall pines, snow tumbled, startling his drunken sleep.
The peach flowers, the stream are in the world of men!
Wu-ling is not for immortals only.
Rivers, hills, clean and empty: I live in city dust,
and though roads go there, they're not for me.
I give back your picture and sigh three sighs;
my hill friends will soon be sending poems to call me home.

TWO POEMS BY LI CH'ING CHAO [LI QING ZHAO],
TRANSLATED BY KENNETH REXROTH

Two Springs

Spring has come to the Pass.
Once more the new grass is kingfisher green.
The pink buds of the peach trees
Are still unopened little balls.
The clouds are milk white jade
Bordered and spotted with green jade.
No dust stirs.
In a dream that was too easy to read,
I have already drained and broken
The cup of Spring.
Flower shadows lie heavy
On the translucent curtains.
The full, transparent moon
Rises in the orange twilight.
Three times in two years
My lord has gone away to the East.
Today he returns.
And my joy is already
Greater than the Spring.

Mist

In my narrow room, I throw
Wide the window, and let in
The profound lasciviousness
Of Spring. Confused shadows
Flicker on the half drawn curtains.
Hidden in the pavilion, wordlessly,

[7]Places south of the Yangtze opposite Huang-chou, where the poet lived in exile. In the four lines that follow, he recalls these places during the four seasons of the year. (Watson)

I strum the rose jade harp.
Far away a rocky cliff
Falls from the mountain in the
Early twilight. A gentle breeze
Blows the mist like a shadow
Across my curtain. O bright pods
Of the pepper plant, you do not
Need to bow and beg pardon.
I know you cannot hold back
The passing day.

TWO POEMS BY LU YU,
TRANSLATED BY BURTON WATSON

Idle Thoughts

Thatch gate works all right but I never open it,
afraid people walking might scuff the green moss.
Fine days bit by bit convince me spring's on the way;
fair winds come now and then, wrapped up with market sounds.
Studying the Classics, my wife asks about words she doesn't know;
tasting the wine, my son pours till the cup overflows.
If only I could get a little garden, half an acre wide—
I'd have yellow plums and green damsons growing all at once!

The Stone on the Hilltop

Autumn wind: ten thousand trees wither;
spring rain: a hundred grasses grow.
Is this really some plan of the Creator,
this flowering and fading, each season that comes?
Only the stone there on the hilltop,
its months and years too many to count,
knows nothing of the four-season round,
wearing its constant colors unchanged.
The old man has lived all his life in these hills;
though his legs fail him, he still clambers up,
now and then strokes the rock and sighs three sighs:
how can I make myself stony like you?

7

Japan: A World of Dewdrops, and Yet . . .

Buddhism was introduced to Japan from Korea in the sixth century A.D., more than a century after Chinese writing, and has been inseparable from Japanese civilization throughout its history. "It is impossible to understand the literature of premodern Japan," Donald Keene stresses in Seeds in the Heart *(1993), "without at least a modicum of knowledge of Buddhism." Yet at first glance this great religion of world-denial and world-transcendence, of liberation through extinction of desire from the sufferings of an illusory world, might seem no less ill-suited than Chinese writing initially was to the very different culture of the Japanese—characterized, as many have remarked, by affirmation of and aesthetic delight in the beauties of nature in the here-and-now. From the first, Joseph Kitagawa writes in* On Understanding Japanese Religion *(1987), they have adhered to the monistic outlook of the indigenous Shintō religion, which "took it for granted that the natural world was the original world; that is, they did not look for another order of meaning behind the phenomenal, natural world—at least until they came under the influence of Sino-Korean civilization and Buddhism." A modicum of knowledge of Shintō, therefore, would also seem necessary to understanding premodern Japan.*

Buddhism did not displace ancient Japanese "nature worship," as Christianity displaced Greco-Roman and Oriental paganism in the struggles of the late Roman Empire, but instead merged with it in a largely harmonious synthesis that continued for a millennium and has by no means vanished today. Only after the introduction of Buddhism was native Japanese religion given a name, ironically in Chinese. The countless gods, spirits, or numinous presences worshipped in manifestations of the natural world with which they were often identified—trees, birds, animals, streams, stones, and above all mountains, as well as extraordinary human beings—were called kami, *and the indigenous tribal religion, whose beliefs had never been systematized, was called* Shintō, *"the way" (Chinese* tō, *or* tao*) "of the gods," or kami (*shin, *or* shen*), as distinguished from* Butsudō, *the way of the Buddha. The attitudes of this primitive Shintō seem, to many Japanese, to lie at the heart of their affirmative view of the human and natural world, a view predicated (Matsumoto Shigeru writes in introducing the essays of* Japanese Religion *[1972]) "on the belief that this is the only world for man." Not that no other worlds existed—early myths tell of worlds above and below our own, and belief in a world from which ghosts or spirits might sometimes return is central to the shamanistic practices chronicled by Carmen Blacker in* The Catalpa Bow *(1975)—but the focus remained on this one. "The idea of absolute transcendence or negation of this-worldly values," Matsumoto states, "is conspicuously absent," for both upper and lower worlds, Katō observes, made it "possible for a mortal man from this Earth to live there for a time and then return. In this sense both were extensions of the Earth."*

Human nature, though its pollutions needed to be periodically cleansed, was fundamentally good, and the surrounding world, though needing propitiation through kotodama, *the magical power of words embodied in ritual prayer, was beneficent. "At the core of all Shintō ceremonial," Sansom writes, "is the idea of purity, and at the core of all Shintō beliefs is the idea of fertility," with which festivals like that of the harvest were vitally con-*

cerned. In the words of one invocation from Donald Philippi's Norito: A Translation of the Ancient Japanese Ritual Prayers (1990),

May the latter grain to be harvested
With foam dripping from the elbows,
Pulled hither with mud adhering to both thighs—
May this grain be prospered by you, oh Sovereign Deity,
In ears many hands long—
Then the first fruits in both liquor and stalks
Will be set up, a thousand ears, eight thousand ears,
And piled high like a long mountain range,
And will be presented in the autumn festival.

In one central aspect, then, early Shintō, like ritualistic practices of other ancient peoples, was an agricultural religion promoting increase of the crops; as such, it was fundamentally seasonal. "Ancestral kami were thought of as residing in different places at different seasons. During the winter," Miyake Hitoshi writes in Japanese Religion, "they stayed in the mountains, symbols of the other world, and were regarded as mountain deities (yama no kami). But in the spring they returned to their homes and became kami of the rice fields (ta no kami), gods of fertility and productivity, remaining until fall to watch over the work of the farm during this critical season." Thus, although ancient Shintō was not, Kitagawa cautions, simply "nature worship," reverence for natural process was basic to its affirmation "of the sacrality of the total world." And human beings were inseparable from this totality, since man and nature, Ueda Kenji writes in Japanese Religion, were "common offspring of the kami who brought Japan into existence." For such a faith, "the essence and value of existence is to be discovered not in an absolute, a priori rational principle (Greek, logos), nor again in a universal norm or law (Sanskrit, dharma), but in the possibilities inherent in concrete forms of existence. Accordingly, Shinto is a religion of the relative, . . . committed to reality in the endless process of becoming."

Over the centuries, however, the word Shintō, while always referring to indigenous Japanese practice and belief, took on different meanings. In early Shintō, Blacker suggests, "a kami had no shape of his own, his occasional visionary appearances being temporary disguises only" as he manifested himself (or perhaps as it manifested itself), after descending from a sacred mountain, in such a provisional abode as a rock-seat, pine tree, or boulder. At later stages the kami might be identified with mountain or tree itself, or take human form like the gods of the first two extant Japanese books (composed almost entirely in Chinese characters), the Kojiki of 712 and the Nihon Shoki (or Nihongi) of 720, which combine myth and legend with quasi-historical traditions. The Shintō portrayed in these first written records, describing genesis of the world from chaos and creation of the gods and Japanese islands from sexual union of Izanagi and Izanami, is complicated both by probable Chinese influence and by clear intention of validating the divine imperial line of Yamato, from which all subsequent Japanese emperors were said to descend. Izanagi, having escaped the gruesome specter of his spouse Izanami (who had died from giving birth to the fire god) in Yomi, the land of the dead, and purified himself of pollution, engendered multiple gods from his bodily parts and clothing; the sun goddess Amaterasu, born from his left eye, was ancestress to the emperors. Whatever external influences or ulterior motives entered into its composition, however, the opening words of the Kojiki (in Philippi's translation of 1968) give forceful expression to an origin myth that has become deeply Japanese:

I, Yasumarö, do say:

When the primeval matter had congealed but breath and form had not yet appeared, there were no names and no action. Who can know its form?

However, when heaven and earth

were first divided, the three deities became the first of all creation. The Male [Yang] and Female [Yin] here began, and the two spirits were the ancestors of all creation. . . .

Thus, though the primeval beginnings be distant and dim, yet by the ancient teachings do we know the time when the lands were conceived and the islands born; though the origins be vague and indistinct, yet by relying upon the sages of antiquity do we perceive the age when the deities were born and men were made to stand. . . .

Here, as in Chinese myths and Neo-Taoist cosmologies (like the Huai-nan Tzu, *Chapter 5 above) that no doubt lay behind it, "there is no first agent creating and ordering existence," as Ian Hideo Levy remarks in* Hitomaro and the Birth of Japanese Lyricism *(1984): "things are not 'made', but become, 'come into being' . . ." This conception of spontaneous process—central to Taoism—is equally central to the Japanese outlook; therefore the indigenous religion could evolve, without loss of identity, in harmony with Buddhism, which gradually adopted the native kami as avatars of the Buddha, or as bodhisattvas. Only after the Meiji Restoration of 1868 did the Japanese government, in reaction against the state-sponsored Buddhism of the Tokugawa shoguns, establish an official "State Shintō" based on emperor worship—while prohibiting as superstitions many of the folk beliefs that had lain at the heart of the ancient faith.*

If the Buddhism that reached Japan at the dawn of its history—more than a millennium after Buddhism's origins in India and half a millennium after its spread to China—was to achieve harmony with the Shintō outlook toward the phenomenal world so fundamental to Japanese mentality, it must clearly have undergone major changes. Japanese Buddhism, in contrast to that of India or indeed China, was at first an exclusively aristocratic religion which (Tamaru Noriyoshi writes in Japanese Religion*) "only gradually made its way into the ranks of people of low degree," among whom local forms of Shintō long continued with little regard for the imported religion of the court. Six schools of Buddhism were recognized in the Nara period (710–94). In the Heian (794–1185), two others of major importance were introduced from China during the early ninth century: Tendai (Chinese T'ien-t'ai), founded by Saichō (also known by his posthumous title, Dengyō Daishi), and Shingon (Chinese Chen-yen), founded by Kūkai (Kōbō Daishi). Both, but especially Shingon, drew upon esoteric lore and magical incantations of Tantric Buddhism; Tendai, centering on the Lotus Sūtra, became the major influence, through its great monastery of Mount Hiei, on Buddhist movements of later times. Not until disintegration of the Heian, and the troubled early medieval, or Kamakura, period (1185–1333), did new forms of Buddhism strike deep roots among the people. The Jōdo or "Pure Land" sect of Hōnen Shōnin, and its offshoot, the Jōdo Shinshū or "True Pure Land Sect" of Shinran, both stressed salvation in the Western Paradise or Pure Land of Amida (Sanskrit Amitābha) Buddha through meditation on (or repetition of) his name; and the Hokke or Lotus sect, generally called by the name of its founder, Nichiren, believed that incantation of the Lotus Sūtra was the sole path to salvation. In these same years (late twelfth to mid thirteenth centuries) Eisai and, a generation later, Dōgen returned from China to found the two most important Zen sects, Rinzai and Sōtō respectively, which gained their main following not among the people but among the new warrior class of the samurai.*

A Buddhism centered in a court society, like that of the Nara and Heian, easily fit the refined aestheticism of aristocratic taste; it was consequently not so much suffering—the first of the Buddha's Noble Truths—as regretful sadness that characterized its attitude toward an ephemeral yet hauntingly beautiful human and natural world. The "strong current of enjoyment of the beauties of nature" that pervades the Man'yōshū *of the mid eighth century, as it does almost all later Japanese poetry, "seems already a trifle artificial at times," Sansom suggests in* A History of Japan to 1334 *(1958), "but it does betoken a naturally aesthetic inclination—anima naturaliter poetica," rooted in a cultivated blend of Shintō nature worship with Taoist thought and poetry of the Six Dynasties and (at this early period) an occasional dash of Buddhist sentiment. After the immensely influential first imperial anthology of poetry, the* Kokinshū, *in 905, the aesthetic Heian attitude to nature was firmly established; the "dominance of cherry blossoms as a Spring image" in subsequent poetry (and eventually in popular culture as well), for example, was rooted, Helen Craig McCullough remarks in* Brocade by Night *(1985), in the* Kokinshū*'s cluster of poems on a subject only sporadically treated in the* Man'yōshū. *The unparalleled prevalence of women in early Japanese literature—at a time when educated men wrote mainly in Chinese—no doubt contributed to this refined regretfulness. Nowhere is this more evident than in Murasaki Shikibu's eleventh-century* Tale of Genji, *where* Kokinshū *and* Kokinshū*-like poems (often two or three to a page) crystallize the characters' feelings, and autumn, in life as in poetry, is "always the melancholy season." In sum, the indigenous Japanese religion, Ivan Morris writes in* The World of the Shining Prince *(1964), "combined with the influence of Buddhism and Taoism to inculcate the sense that man is an integral part of the natural world . . ."*

Although very little Nara and Heian literature in Japanese is explicitly Buddhist, then, its aesthetic sensitivity to natural beauty easily accorded with an urbane aristocratic religion; later monks and priests from Saigyō in the twelfth century to Ryōkan in the nineteenth found this sensitivity fully compatible with a far deeper commitment to Buddhism as they knew and lived it. For despite the apparent disparity between world-denial and world-affirmation, there was much in Buddhism, as it made its long journey from India through China and Korea to Japan, that lent itself to melancholy appreciation of the world's fleeting beauty. The attitude toward nature later associated with Zen, for example, drew nourishment, Heinrich Dumoulin writes in A History of Zen Buddhism *(1959), from "the cosmotheistic world-view" of the* Avatamsaka Sūtras *sacred to the Kegon school of Nara Buddhism, which affirmed "the identity of the absolute state in* nirvāna *and the relative phenomenal world in* samsāra." *The Lotus Sūtra, above all, dignified the phenomenal world through such images as that of the Buddha's dharma as fertilizing rain-cloud. This sūtra—in Chinese translations from Sanskrit, for no Japanese translations were made until the twentieth century—was most strongly influential in Japan, especially on the Tendai and Nichiren sects. In the Lotus, William LaFleur writes in* The Karma of Words: Buddhism and the Literary Arts in Medieval Japan *(1983), we find a "move to affirm the complete reality of the world of concrete phenomena in spite of the fact that they are impermanent"—or even* because *of that fact, since change and impermanence are the very nature of reality in this world.*

Nowhere was this reaffirmation of phenomenal nature, already prominent in Chinese T'ien-t'ai and Japanese Tendai Buddhism, so fully realized as in Zen; and Zen—through its disciplined following among upper-class samurai to whom it gave the training of meditation and the focused minimalism of rock gardens, tea ceremonies, and Nō plays—more profoundly affected poetry and painting than any other sect. The Sung dynasty Chinese Ch'an or Zen master Ch'ing-yüan [Qingyuan] (1067–1120), in a famous saying quoted by LaFleur, summarized this central progression of Buddhist thought in a few sentences: "Before I had studied Zen for thirty years, I saw mountains as mountains and waters as waters. When I arrived at a more intimate knowledge, I came to the point where I saw that mountains are not mountains and waters are not waters. But now that I have got the very substance I am at rest. For it is just that I see mountains once again as mountains, and waters once again as waters." If the Buddha-nature pervades all things, all things, in and of themselves, are of the nature of Buddha; thus in the Zen philosophy of Dōgen, as paraphrased by Nakamura Hajime in Ways of Thinking of Eastern Peoples *(English trans. 1964), "the ever-changing, incessant temporal flux is identified with ultimate being itself." In Dumoulin's words: "If nature is the body of the Buddha, the Buddha-body is constantly becoming. He who would depict nature from within must comprehend this endless becoming." At this point Butsudō is one with Shintō, for by synthesis of the ancient Japanese (or pre-Japanese) "notion of a cosmos permeated by* kami *(sacred) nature with the Chinese Buddhist emphasis on the phenomenal world as the locus of soteriology and the Taoist notion of 'naturalness'* (tzu-jan)," *Kitagawa declares, Japanese Buddhism, like Shintō, "affirms the sacrality of the world of nature." Through all the changes of seasons and centuries, from the* Man'yōshū *to Ryōkan more than a thousand years later, this fundamental affirmation remains a constant amid incessant flux.*

Many poems in the sections that follow are taken from anthologies of Japanese poetry of different periods, beginning with Arthur Waley's Japanese Poetry: the 'Uta' *(1919). (The translations in Waley's slim volume were "chiefly intended to facilitate the study of the Japanese text," but Waley can transform even a literal "trot" into poetry.) Others include* The Penguin Book of Japanese Verse *(1964), trans. Geoffrey Bownas and Anthony Thwaite; Donald Keene, ed.,* Anthology of Japanese Literature from the Earliest Era to the Mid-Nineteenth Century *(1955); Hiroaki Sato and Burton Watson, eds.,* From the Country of Eight Islands: An Anthology of Japanese Poetry *(1981); Kenneth Rexroth,* One Hundred Poems from the Japanese *(no date); and Steven D. Carter, trans.,* Traditional Japanese Poetry: An Anthology *(1991). General studies include Robert H. Brower and Earl Miner,* Japanese Court Poetry *(1961), and Earl Miner,* An Introduction to Japanese Court Poetry *(1968), and three literary histories: Donald Keene's* Seeds in the Heart: Japanese Literature from Earliest Times to the Late Sixteenth Century *(1993) and* World Within Walls: Japanese Literature of the Pre-Modern Era 1600–1867 *(1976); the first two volumes of Shuichi Katō's* History of Japanese Literature *(English trans. 1979 and 1983); and volumes 1–3 of Jin'ichi Konishi's* A History of Japanese Literature *(1984–1991).*

Yamato Dawn

Eight Poems from the *Kojiki* and *Nihon Shoki*, translated by Donald Philippi

The earliest poems in Japanese, gathered and translated by Donald Philippi in This Wine of Peace, This Wine of Laughter: A Complete Anthology of Japan's Earliest Songs *(1968), are mainly those recorded in the mytho-historical chronicles the* Kojiki *of 712 and the* Nihon Shoki *of 720. The poems—written in Chinese characters used, like the pictures of a rebus, for their phonetic values—are interspersed in the narrative of events, even when this (in the* Nihon Shoki*) is in Chinese. These chronicles were published in the intellectually and poetically fertile early years of the Nara period (710–94), named after the new capital of the kingdom of Yamato, which had grown up over the centuries around the imperial clan, thus becoming the nucleus of* *the Japanese state; many poems no doubt embodied far older folk traditions. For Waley, "Of the two hundred and thirty-five poems contained in these two chronicles, not one is of any value as literature"; but although their style is unpolished and their verse often irregular, the best have a vigorous energy conspicuously missing from later court poetry. In the poems that follow—the first and third from the* Kojiki, *the second from both chronicles, the rest from the* Nihon Shoki—*basic human emotions of love and grief are intensified by association or contrast with a directly experienced natural world not of dewdrops and cherry blossoms but of morning sun, wild ducks, and luxuriant bamboo.*

As soon as the sun
   Hides behind the verdant mountains,
Then jet-black
   Night will come.
Smiling resplendently
   Like the morning sun,
With your arms
   White as a rope of *taku* fibers,
   You will embrace
My breast, thrilling with youth,
   Soft as the light snow;
We shall embrace and entwine our bodies.
Your jewel-like hands
   Will twine with mine,
And, your legs outstretched,
   You will lie and sleep.
Therefore, my lord,
   Do not yearn.
Oh god
   Ya-chi-hoko!

These are
The words,
The words handed down.

The song sung by Ho-wori-no-mikoto in longing for his sea-wife who had returned to her watery homeland.

As long as I have life,
I shall never forget
My beloved, with whom I slept
On an island, where wild ducks,
Birds of the offing, came to land.

As Emperor Yūryaku was going to visit the Princess Waka-Kusaka-be (who became Empress in 457 A.D.) at her home in Kusaka, the Princess sent word that it was inauspicious for the Emperor to progress with his back to the sun, and promised that she would set out immediately to go and serve him. Then, as he was about to return to his palace, he stood atop the mountain incline and sang this song, which he sent by messenger to the Princess.

In the valleys
Here and there
Between the mountains this side
Of Kusakabe
And the rush-matting
Heguri mountain—
There stands a thriving
Wide-leaved oak tree.
At its foot
Grows entwined bamboo;
At its sides
Grows luxuriant bamboo.
Entwined bamboo:
We did not sleep entwined;
Luxuriant bamboo:
We did not sleep luxuriantly.
But later we will sleep twined—
Ah that beloved spouse of mine!

Soga-no-Miyatsuko-hime, consort of the Prince Imperial (later Emperor Tenchi), died of a broken heart in 648 A.D. after her father had been killed as the result of slander. The Prince grieved deeply at her death. Nonaka-no-Kawara-no-Fuhito-Maro came forward and presented these songs.

On the mountain stream
Is a pair of mandarin ducks—
Well matched like them
Were my beloved and I—
But who has led her away?

Second elegy by Nonaka-no-Kawara-no-Fuhito-Maro. The Prince praised the songs highly and, giving a *koto* to the scribe, had him sing the songs. He also rewarded him richly with gifts.

On each stem
Flowers are blooming;
But why
Will my dear beloved
Never bloom forth again?

In 658 A.D., Prince Takeru, the grandson of Empress Saimei, died at the age of eight and was interred in a tomb above the Imaki valley. The Empress, beside herself with grief, commanded that he be interred along with her after her own death and made three songs which she sang from time to time as she lamented bitterly.

Above the hill
At Imaki—

If even a cloud
Would only appear,
Then what should I grieve?

I did not think of him
As being a mere child, young
Like the young grass
By the river bank, where they track
The wounded deer.

Like the foaming waters
Of the Asuka river,
Moving on ceaselessly:
Without pause
Does my mind dwell on him.

What Could Surpass Today?

Fifteen Poems from the *Man'yōshū*

The first collection of poetry by Japanese poets was the Kaifūsō *("Fond Recollections of Poetry"), an anthology of* kanshi, *or poems in Chinese, most of them imitative and conventional, compiled in 751. But at some time in the next few decades appeared the greatest of all collections of Japanese poetry, the* Man'yōshū *("Collection of a Myriad Leaves" or "Collection for a Myriad Ages"). In contrast to the poems of the* Kojiki *and* Nihon Shogi, *those of the* Man'yōshū *are highly accomplished in style and technique, while compared to the imperial anthologies that would follow, this one embraces poems varying widely in subject matter and poetic form. Some of these may date back several hundred years, but most were composed between the mid seventh and mid eighth centuries; some may be ancient folk songs, and some are attributed to peasants or frontier guards, but the collection as a whole, as Sansom observes, is "the product of a small, cultivated, aristocratic society" that would govern Japanese taste for a thousand years to come. Many poems are indeed ascribed to emperors, empresses, princes, and princesses, and many others are eulogies of the imperial house.*

More than four thousand of the roughly 4,500 poems in the collection are in the form—at first generally called waka *or simply* uta, *"poem" or "song," but later more often known as* tanka*—that had made its appearance already in the* Kojiki *and* Nihon Shoki, *and would dominate Japanese poetry for most of the next millennium; this consists of thirty-one syllables arranged in five lines, 5-7-5-7-7. But the most distinctive poetic form of the* Man'yōshū *was the loosely constructed longer poem (of as many as 150 lines), the* chōka, *which alternated lines of five and seven syllables, ending in a couplet (unrhymed, like nearly all Japanese poetry) of two seven-syllable lines; one or more "envoys" in the* waka *form sometimes followed. Poems of such length on a variety of subjects, from public tributes to expressions of personal love or loss, permitted far fuller and more varied treatment of different themes than the* waka, *and the subject mat-*

ter—like the vocabulary—of the Man'yōshū is much less limited and stereotyped than that of later collections. Perhaps its main distinction, Keene suggests in Seeds in the Heart, "is that—unlike the poems in later collections that may be marvels of mood and suggestion but are too brief to state much—the Man'yōshū poems are usually direct in their statements of the poet's emotions." Over half concern love between men and women, and both on this subject and others many "display an outlook on nature"—as the introduction to the Nippon Gakujutsu Shinkōkai (Japan Society for the Promotion of Scientific Research) translation of 1940 affirms—"which excels the later anthologies of Japan in scope and in depth of sympathy," demonstrating an intimacy with nature "difficult to find . . . in the contemporary lyric poetry of any other country in the world."

The first of our selections, by **Emperor Jomei** (593–641), expresses, in a tone of wonder rare in later times, the "exultant joy" (in Keene's phrase) of viewing from a mountaintop—probably a religious rite of spring in anticipation of the harvests of autumn—his land of Yamato. In the second poem, **Princess Nukada** (the first of many women poets in this and later anthologies, in sharp contrast to the near total domination of classical Chinese poetry by men) responds to the command of Emperor Tenji (626–71) that his Prime Minister Fujiwara Kamatari judge between the claims of spring and autumn by stating the case for both and deciding—as if because of the regret that attends it—in favor of autumn.

The greatest poet of the Man'yōshū (and perhaps of all Japanese literature) is **Kakinomoto Hitomaro**, an official of low rank in the late seventh to early eighth century about whose life almost nothing is known. His chōka, both public and private, are the masterpieces of the collection, and among the finest are the two here included, to his wife before and after her death: nowhere is the deep connection of love between man and woman and intimacy with nature more movingly expressed. As Keene observes of his poetry in general, "contrast between eternal nature and the transience of man and his works gives poignancy to his observations and universality to his sorrow." **Yamabe Akahito** (early eighth century) was long esteemed second only to Hitomaro, especially for his waka on subjects from nature. Among his longer poems, the "Distant View of Mount Fuji" gives expression to a sense of sublimity rare in later poets, and the poem on Yoshinu (or Yoshino) palace is memorable above all for the two intensely lyrical envoys that crystallize the mysterious beauty of the scene.

Yamanoue Okura (660?–733?) took part in an embassy to China in 701 and thereafter held several important positions; the social concerns of his poems on the sufferings of the poor (especially his powerful "Dialogue on Poverty") are nearly unique in Japanese poetry and have gained him especially high regard in our century. The poem here included, "The Impermanence of Human Life," gives voice to this central poetic theme through immediate and helpless experience of the stream of time relentlessly sweeping away all human things subject, as only the firmest rocks seem not to be, to its power. **Takahashi Mushimaro**'s poem conveys the physical exertion and spiritual exultation of climbing Mount Tsukuba and seeing its peak "in shining clarity" with the intense delight of living fully in the present moment.

Ōtomo Yakamochi (718–85) held a wide range of government and military positions in a career that knew violent swings of fortune. He edited the later books of the Man'yōshū (if not the collection as a whole), in which he included some five hundred of his own poems. The three waka of the year 753 included here represent, to the Nippon Gakujutsu Shinkōkai translators, "the fine lyricism that characterized the poetry of the closing years of the Nara period"; as in many Chinese poems, human feeling is mirrored by the surrounding natural world. To him were addressed all twenty-nine love poems of **Lady Kasa** (late eighth century) in the Man'yōshū; here, too, natural phenomena, from dew on the grass to thunderous ocean waves, crystallize the *poet's love and anguish. Finally, two anonymous* waka, *one purportedly by a frontier guard, give further expression to the deep connection between human love and communion with nature central to much of Japanese poetry.*

In addition to the Nippon Gakujutsu Shinkōkai translation of the entire Man'yōshū, *readers should consult that of Ian Hideo Levy,* The Ten Thousand Leaves *(1981).*

Climbing Mount Kagu, by Emperor Jomei, translated by Geoffrey Bownas and Anthony Thwaite

In the land of Yamato
The mountains cluster;
But the best of all mountains
Is Kagu, dropped from heaven.
I climbed, and stood, and viewed my lands.
Over the broad water
Seagulls hover.
Beautiful, my country,
My Yamato,
Island of the dragonfly.[1]

Poem Written on the Occasion of Emperor Tenji's Ordering Fujiwara Kamatari to Judge between the Claims of Spring and Autumn, by Princess Nukada, translated by Bownas and Thwaite

When the spring comes
After winter's confining,
The birds that did not sing
Come out and sing;
The flowers that were closed
Come out and bloom.
But the mountain trees grow dense—
We cannot reach to pick the flowers:
The weed-grasses are thick—
We cannot see the flowers we pick.

We see the leaves
On an autumn hill;
We pick the red leaves,
Admiring and praising;
We leave the green ones,
Sighing and grieving.
There lies my regret:
Autumn hills for me.

[1]Pillow-word to Yamato. A dragonfly touches its tail with its mouth, thus forming a shape not unlike the circle of hills that ring Yamato. (Bownas and Thwaite) A pillow-word *(makura kotoba)* is a word conventionally associated with and qualifying another, somewhat like the kennings of Old English and Old Norse poetry.

TWO POEMS BY KAKINOMOTO HITOMARO,
TRANSLATED BY BOWNAS AND THWAITE

On Leaving His Wife

The thick sea-pine
Grows on the rocks
In the sea of Iwami
Off the cape of Kara.
The sea-tangle clings
To the rocky beach.
Like the sea-tangle
She bent and clung to me,
My wife, my love; deep
As the deep sea-pine
Was my love for her.
Yet the nights are few
When we have slept together.
The creeping ivy parts,
And we have parted too.
My heart aches when I think
Of her, but when I look
Back, the yellow leaves
Of the mountain flutter and hide
Her distant waving sleeve.
As the moon through a wide rift
Peeps, then hides in the clouds,
My wife is hidden, and I
Grieve. The sun is low.
And I, a strong man—
Or so I thought—make wet
My heavy sleeves with tears.
My glossy steed goes fast,
And far as the clouds I've come
From my wife, from my home.
You yellow leaves that cover
The autumn mountain, cease
Your falling for a while,
For I would see my love.

I loved her like the leaves,
The lush leaves of spring
That weighed the branches of the willows
Standing on the jutting bank
Where we two walked together
While she was of this world.
My life was built on her;

But man cannot flout
The laws of this world.
To the wide fields where the heat haze shimmers,
Hidden in a white cloud,
White as white mulberry scarf,
She soared like the morning bird
Hidden from our world like the setting sun.
The child she left as token
Whimpers, begs for food; but always
Finding nothing that I might give,
Like birds that gather rice-heads in their beaks,
I pick him up and clasp him in my arms.
By the pillows where we lay,
My wife and I, as one,
The daylight I pass lonely till the dusk,
The black night I lie sighing till the dawn.
I grieve, yet know no remedy:
I pine, yet have no way to meet her.
The one I love, men say,
Is in the hills of Hagai,
So I labor my way there,
Smashing rock-roots in my path,
Yet get no joy from it.
For, as I knew her in this world,
I find not the dimmest trace.

Envoys

1

The autumn moon
We saw last year
Shines again: but she
Who was with me then
The years separate for ever.

2

On the road to Fusuma
In the Hikite Hills,
I dug my love's grave.
I trudge the mountain path
And think: "Am I living still?"

TWO POEMS BY YAMABE AKAHITO,
NIPPON GAKUJUTSU SHINKŌKAI TRANSLATION

On a Distant View of Mount Fuji

Ever since heaven and earth were parted,
It has towered lofty, noble, divine,
Mount Fuji in Suruga!
When we look up to the plains of heaven,

The light of the sky-traversing sun is shaded,
The gleam of the shining moon is not seen,
White clouds dare not cross it,
And for ever it snows.
We shall tell of it from mouth to mouth,
O the lofty mountain of Fuji!

Envoy

When going forth I look far from the shore of Tago,
How white and glittering is
The lofty Peak of Fuji,
Crowned with snows!

Here in a beautiful dell where the river runs,
The Yoshinu Palace, the high abode
Of our Sovereign reigning in peace,
Stands engirdled, fold on fold,
By green mountain walls.
In spring the flowers bend with the boughs;
With autumn's coming the mist rises and floats over all.
Ever prosperous like those mountains
And continuously as this river flows,
Will the lords and ladies of the court
Come hither.

Envoys

1

Oh, the voices of the birds
That sing so noisily in the tree-tops
Of the Kisa Mountain of Yoshinu,
Breaking the silence of the vale!

2

Now the jet-black night deepens;
And on the beautiful river beach,
Where grow the *hisagi*-trees,
The sanderlings cry ceaselessly.

The Impermanence of Human Life,
by Yamanoue Okura,
translated by Bownas and Thwaite

We are helpless in this world.
The years and months slip past
Like a swift stream, which grasps and drags us down.
A hundred pains pursue us, one by one.
Girls, with their wrists clasped round
With Chinese jewels, join hands
And play their youth away.

But time cannot be stopped,
And when their youth is gone
Their jet-black hair, black as a fish's bowels,
Turns white, like a hard frost.
On their sun-browned, glowing faces,
Wrinkles are etched—by whom?

Boys, with their swords at their waists,
Clutching the hunting bow,
Mount their chestnut horses
On saddles linen-spun,
And ride on in their pride.
But is the world eternal?
He pushes back the door
Where a girl sleeps within,
Gropes to her side and lies
Arm on her jewel arm.
But how few are those nights
Before, with stick at waist,
He goes shunned and detested—
The old are always so.
We grudge life moving on
But we have no redress.
I would become as those
Firm rocks that see no change.
But I am a man in time
And time must have no stop.

When Lord Ōtomo, the Revenue Officer, Climbed Mount Tsukuba, by Takahashi Mushimaro, translated by Bownas and Thwaite

My lord came to survey
The peaks of Tsukuba,
The mountain of black clouds,
In our province of Hitachi.
In the hot summer sun,
The sweat ran down, we panted,
Hauled ourselves up by roots,
Climbed on, our breathing heavy.
Thus we reached the peak
And looked about us, where
The God of the western peak
Revealed his realm below,
The Goddess of the eastern peak
Displayed her magic power.
The crags of Tsukuba's peak,
Shrouded in mist and rain
That always hover there,

Flashed in the brilliant light:
The beauties of our land
That always lay obscured
The gods that moment showed
In shining clarity.
And in our grateful joy
We stripped away our clothes,
Ran and jumped and played
As if we were at home.

Envoys

1

The spring grass bent and swayed.
With summer, it grows rank.
And yet this summer day
Is happier even than spring.

2

What could surpass today?
The day when my father first
Came to Tsukuba's peak?
Even that day grows pale.

THREE *WAKA* BY ŌTOMO YAKAMOCHI, NIPPON GAKUJUTSU SHINKŌKAI TRANSLATION

[Composed extempore, on the twenty-third of the second month of the fifth year of Tempyō-Shōhō (753)]

Over the spring field trails the mist,
And lonely is my heart;
Then in this fading light of evening
A warbler sings.

Through the little bamboo bush
Close to my chamber,
The wind blows faintly rustling
In this evening dusk.

[Composed on the twenty-fifth day]

In the tranquil sun of spring
A lark soars singing;
Sad is my burdened heart,
Thoughtful and alone.

[In the languid rays of the spring sun, a lark is singing. This mood of melancholy cannot be removed except by poetry: hence I have composed this poem in order to dispel my gloom.]

TWO *WAKA* TO ŌTOMO YAKAMOCHI,
BY LADY KASA,
NIPPON GAKUJUTSU SHINKŌKAI TRANSLATION

In the loneliness of my heart
I feel as if I should perish
Like the pale dew-drop
Upon the grass of my garden
In the gathering shades of twilight.

Oh how steadily I love you—
You who awe me
Like the thunderous waves
That lash the sea-coast of Isé!

TWO ANONYMOUS *WAKA*

Shall we make love
Indoors
On this night when the moon has begun to shine
Over the rushes
Of Inami Moor?

—trans. Waley

Poem of a Frontier Guard

I will think of you, love,
On evenings when the grey mist
Rises above the rushes,
And chill sounds the voice
Of the wild ducks crying.

—Nippon Gakujutsu Shinkōkai translation

The World in Thirty-One Syllables

Twenty *Waka* from the Imperial Anthologies, translated by Geoffrey Bownas and Anthony Thwaite, and by Arthur Waley

The Man'yōshū *was highly honored in later days, but its direct influence was limited. Its written language became nearly indecipherable after the evolution of more flexible* kana *syllabaries in the ninth century, and its most distinctive poetic form, the* chōka *(which without the discipline* of a Hitomaro became mere prose), fell into disuse. Most important of all was the change in poetic sensibility that followed transfer of the capital, for unknown reasons, from Nara to Heian-kyō, the newly built city of "Peace and Tranquility" that gave its name to the Heian period (794–1185) and would remain, as Kyōto, the imperial capital for over a thousand years. In the court life of this age, aristocratic refinement was pervasive; distinctions of rank and manners were of consuming import, and direct expression of emotion increasingly gave way to oblique suggestion and allusion. In Heian diaries and novels, such as Sei Shōnagon's Pillow Book and Murasaki Shikibu's Tale of Genji from the early eleventh century, subtle intricacies of manners and feeling characteristic of this society, especially as viewed by its women, were developed at length and in detail. But it was almost exclusively in the thirty-one syllable waka, now and for centuries to come, that poets found they could crystallize, in a single evocative image, their elusive response to a beautiful but ephemeral world.

In the ninth century, several imperially commissioned anthologies of poems written in Chinese had been compiled. Then, in about 905, the first imperial collection of Japanese poetry, the Kokin Wakashū ("Collection of Waka Old and New"), or Kokinshū, appeared, followed over the next three centuries by seven others culminating in the Shin Kokinshū ("New Kokinshū") of about 1205; still others, of decreasing importance, came later. The Kokinshū, not the Man'yōshū, became the model for Japanese poetry thereafter, and inclusion of waka in an imperial anthology was a mark of prestige at court. In the Japanese preface to the Kokinshū, as translated by Keene, Ki no Tsurayuki proclaims that "Japanese poetry has its seeds in the human heart and burgeons into many different kinds of leaves of words," crystallizing experiences of the world such as poets of the past felt "when they saw blossoms fall on a spring morning, or heard the leaves fall on an autumn evening; when they grieved over the new snow and ripples reflected with each passing year by their looking glasses; when they were startled, seeing dew on the grass or foam on the water, by the brevity of life . . ." As Keene remarks, such images, all expressing regret at the passage of time, rarely permit "the rough edges of emotions to pierce the elegant surface" or allow passions like joy or anguish to intrude on the poet's delicate sadness. The images of the Kokinshū—especially of the seasonal poems in the first six of its twenty books—were endlessly repeated in this and later collections, whose cultivated poetic diction, dense literary allusions, intricate word-play, and restricted vocabulary (of some two thousand words) provided a paradigm for Japanese poetry down to nearly the present.

Yet despite the limitations of such a tradition, the best of these polished waka, as in the brief selection that follows, distil in their thirty-one syllables a concentrated sensibility to the beauties of a world continually passing away that is unsurpassed in any poetry of the world. Some of the finest poems in the Kokinshū are by anonymous earlier poets; among named poets in this and the later collections, one of the most moving, through the intensity of directly expressed emotions in contrast to the studied obliqueness of much waka poetry, was **Ono no Komachi** of the mid ninth century, around whose unhappy love poems grew up a profusion of legends later immortalized in the Nō plays of medieval Japan. **Ono no Yoshiki** lived in the late ninth century, dying in 902; **Lady Ise**, in the mid tenth century. **Fujiwara Shunzei** (1114–1204) was an important critic who compiled the seventh imperial anthology in about 1188. His son **Fujiwara Teika** (1162–1241) compiled the eighth and greatest, the Shin Kokinshū, in about 1205, as well as a less distinguished successor thirty years later; he was famed for a variety of writings, including his sequences of a hundred waka. But perhaps the most memorable of these poets is **Saigyō** (1118–90). In a troubled time, when the seemingly secure Heian world was beginning to disintegrate from the feudal rivalries that would dominate the coming age, he gave up a promising career in his early twenties to become a priest of the Shingon sect and spend his life as a wandering hermit. His poems, Keene observes, express both the Buddhist sense of the impermanence of all things, *called* mujō, *and a sense—no less important to Saigyō, who stressed the unity of Buddhism and Shintō—of beauty in the old, the faded, and the forlorn. As Steven Carter writes, he articulated "a strong awareness of the frailty of human existence, of the constant change that is the way of the universe, and of man's paradoxical attraction to his own frail, changeable existence through his perception of natural beauty."*

Except for the four poems indicated as Waley's, translations in this section are from Bownas and Thwaite's Penguin Book of Japanese Verse. *In addition to other anthologies and studies mentioned above, see Helen Craig McCullough,* Kokin Wakashū: The First Imperial Anthology of Japanese Poetry *(1985), and her study,* Brocade by Night: "Kokin Wakashū" and the Court Style in Japanese Classical Poetry *(1985); Laurel R. Rodd,* Kokinshū: A Collection of Poems Ancient and Modern *(1984);* The Ink Dark Moon: Love Poems by Ono no Komachi and Izumi Shikibu *(1988), trans. Jane Hirshfield with Mariko Aratani;* Fujiwara Teika's Superior Poems of Our Time: A Thirteenth-Century Poetic Treatise and Sequence *(1967), trans. Robert H. Brower and Earl Miner;* Mirror for the Moon: A Selection of Poems by Saigyō *(1978), trans. William R. LaFleur; and Saigyō,* Poems of a Mountain Home *(1991), trans. Burton Watson.*

FIVE ANONYMOUS *WAKA* FROM THE *KOKINSHŪ*

In this world is there
One thing constant?
Yesterday's depths
In Asuka River
Today are but shallows.

Beating their wings
Against the white clouds,
You can count each one
Of the wild geese flying:
Moon, an autumn night.

When a thousand birds
Twitter in spring
All things are renewed:
I alone grow old.

In the spring haze
Dim, disappearing,
The wild geese are calling
Above autumn's mist.

Even for the space of a flash
Of lightning
That flashes over the corn-ears
Of an autumn field,—
Can I forget you?

—trans. Waley

FOUR *WAKA* BY WOMEN POETS

A thing which fades
With no outward sign—
Is the flower
Of the heart of man
In this world!

—Ono no Komachi, trans. Waley

The lustre of the flowers
Has faded and passed,
While on idle things
I have spent my body
In the world's long rains.

—Ono no Komachi

My love
Is like the grasses
Hidden in the deep mountain:
Though its abundance increases,
There is none that knows.

—Ono no Yoshiki, trans. Waley

Forsaking the mists
That rise in the spring,
Wild geese fly off.
They have learned to live
In a land without flowers.[2]

—Lady Ise

[2]Wild geese migrate north with the early spring, thus showing a completely un-Japanese disregard for the cherry-blossom. (Bownas and Thwaite)

TWO *WAKA* BY FUJIWARA SHUNZEI

In Autumn, Lodging at a Temple Near His Wife's Grave

Even at midnight,
When I come so rarely,
The sad wind through the pines:
Must she hear it always
Beneath the moss?

Oh, this world of ours—
There is no way out!
With my heart in torment
I sought the mountain depths,
But even there the stag cries.

TWO *WAKA* BY FUJIWARA TEIKA

This spring night
The floating bridge of my dream
Fell apart:
Swirling away from the peak,
Dawn clouds in the eastern sky.

As far as the eye can see,
No cherry-blossom,
No crimson leaf:
A thatched hut by a lagoon,
This autumn evening.

SEVEN *WAKA* BY SAIGYŌ

Trailing on the wind,
The smoke from Mount Fuji
Melts into the sky.
So too my thoughts—
Unknown their resting-place.

Is it a shower of rain?
I thought as I listened
From my bed just awake.
But it was falling leaves
Which could not stand the wind.

On Mount Yoshino
I shall change my route
From last year's broken-branch trail,
And in parts yet unseen
Seek the cherry-flowers.

The winds of spring
Scattered the flowers
As I dreamt my dream.
Now I awaken,
My heart is disturbed.

The cry of the crickets,
As the nights grow chill
And autumn advances,
Grows weak and more distant.

Every single thing
Changes and is changing
Always in this world.
Yet with the same light
The moon goes on shining.

Since I am convinced
That Reality is in no way
Real,
How am I to admit
That dreams are dreams?

—trans. Waley

Living in the Mountains

Eight Zen Poems of the Five Mountains, translated by David Pollack

Throughout the Nara and Heian periods, the imperial court and Buddhist religion had been inseparably linked. Yet although individual poems and even whole sections of the Man'yōshū *and the imperial anthologies were devoted to Buddhist subjects, the dominant outlook of these collections can hardly be called Buddhist in any explicit sense. (Buddhist writings, like those of Kūkai in the ninth century and of course the sūtras themselves, which were left untranslated, were principally in Chinese.) The autumnal melancholy of poets contemplating a transitory world reflects a vaguely "Buddhist" aura, but their aristocratic savoring of ephemeral pleasures was at least as much an aesthetic as a religious attitude; and the delight in nature—above all in mountains—that we find from Emperor Jomei to Priest Saigyō owes as much to Shintō or Taoist as to Buddhist strains of the syncretic court religion. Beginning in the turbulent late Heian, however, and continuing through much of the early medieval or Kamakura period (1185–1333) and its successor, the late medieval or Muromachi (1333–1600)—times of continual warfare among samurai in the service of powerful feudal lords and rival shoguns who had usurped the powers of the emperor in Kyōto—new forms of Buddhism stressing not the transient pleasures of this world but the promise of salvation in the paradise or "Pure Land" of Amida Buddha took hold among the people.*

Zen, on the other hand, as heir both to the half-Taoist Ch'an Buddhism of China and to earlier Japanese Shintō and Buddhist traditions—above all of Kūkai's Shingon sect, which stressed the importance of art in religious experience—appealed not to the populace but to upper-class samurai and exerted immense influence on poetry and painting. The Zen temples were known collectively as Gozan, "Five Mountains" ("mountain" being a designation for temple, and "five" denoting their conventional number in both Kyōto and Kamakura, not an actual total). In these, and in more remote temples and hermitages, Zen monks and priests could practice the meditation that gave their sect its name in relative isolation from a violent world. Zen came to be associated, David Pollack remarks in Zen Poems of the Five Mountains *(1985), from which the following selections are taken, "with certain landscape scenes in nature—reeds and geese, for example, or orchids and rocks, sunset over a village, fishermen on a bay," so that "poems ostensibly describing a landscape are in fact demonstrations of how an enlightened mind views the world. . . . Thus developed the tradition of writing a poem to demonstrate one's enlightenment." The following poems, mostly written in Chinese, express both Zen predilection for seclusion from a world of turmoil and Zen irreverence for social and religious institutions that impede human communion with the natural world.*

Of the poets here represented, ***Kōhō Kennichi*** *(1241–1316), a son of Emperor Gosaga, was a Zen poet who uncharacteristically wrote not Chinese poems but* waka, *many of which were included in imperial anthologies; he was the teacher of* ***Musō Soseki*** *(1275–1351), who wrote both* waka, *such as the first two poems given here, and Chinese poems, such as the third. Of aristocratic birth, Musō played an important part in gaining the support of the Ashikaga shoguns of the Muromachi period for the Zen sect.* ***Kokan Shiren*** *(1278–1345) wrote an important history of Buddhism in Japan.* ***Betsugen Enshi*** *(1294–1364), after ten years in China, spent twenty years in small temples along the Japanese coast, refusing the enticements of more prestigious positions.* ***Tesshū Tokusai*** *(died 1366), who was both painter and poet, returned from China to study with Musō Soseki just before Musō's death. About* ***Seiin Shunshō*** *(1358–1422) little is known; many of his poems were meant to be written on paintings they describe. Interested readers should also consult* Poems of the Five Mountains *(1977), translated by Marian Ury.*

The white clouds
On the mountain-tops
Poke halfway into this thatched hut
I had thought too cramped
Even for myself.

—Kōhō Kennichi

THREE POEMS BY MUSŌ SOSEKI

I Hid Away at the Keizan Mountain Temple in Mino

[Though so deep in the mountains there was not even a real road of any sort to the spot, much to my annoyance people kept calling to study Zen with me.]

It would be merciful of people
Not to come calling and disturb
The loneliness of these mountains
To which I have returned
From the sorrows of the world.

Abandoning the Hermitage I Was Living in That I Had Built in Shimizu in Mino

How many times
Have I left abandoned,
Living hidden away like this,
A temporary dwelling built
In an uncertain world?

Written While Visiting

A drifter all my life, I never saved a thing:
Clouds in the mountains, moon in the creek for carpets,
East to west I trod this narrow track for nothing—
It wasn't in the dwellings along the way.

Impromptu Poem,
by Kokan Shiren

From a spider's web hangs an empty cicada shell,
Twisting and turning this way and that in the breeze;
While it was alive one heard only its pleasant song—
Who would have thought that, dead, it could dance like this?

Rhyming with the Priest Ts'ao-an's Poem "Living in the Mountains," by Betsugen Enshi

I

In this wretched, thatched-roof hut I have my secret fun:
Lounge around, sleep, whatever suits my fancy,
And laugh at preachers who come from all around
With their sneaky spiels, fervent bellows, wild gesticulations.

2
All the hundred thousand Buddhas are just dust in my eyes,
Those folks up there in heaven—they're no neighbors of mine;
I keep my fire going with kindling that smokes, flares up, goes out—
How can I leave any here for those to come after?

Living in the Mountains,
by Tesshū Tokusai

1
To shake off the dust of human ambition
I sit on moss in Zen robes of stillness,
While through the window, in the setting sun of late autumn,
Falling leaves whirl and drop to the stone dais.
2
Old cedars and ancient cypresses impale rosy mists,
Through huge boulders and hanging vines a small path winds;
Even monkeys and cranes won't come to a mountain *this* desolate—
Only the wind-borne cassia pods that fill my thatched hut.

River Plum Tree in the Twelfth Month,
by Seiin Shunshō

I don't like the plum trees at the Shogun's residence: red year after year,
They boast too much of wealth and beauty and drinking in spring breezes;
Better this single tree at the foot of a barren hill,
That opens before the year's end amidst the ice and snow.

Buddhist Retreat from a Troubled World

An Account of My Hut, by Kamo no Chōmei, translated by Donald Keene

The language of Japanese Buddhist writings—with a few exceptions, like the popular setsuwa *(anecdotes or tales) of such collections as the twelfth-century* Konjaku Monogatari, *or* Tales of Times Now Past*—long remained Chinese, even for most Zen poets. Prose written in Japanese, on the other hand—including the great diaries and novels of the Heian period, and historical accounts like the thirteenth-century* Tale of the Heike*—was mainly devoted to secular subjects such as intrigues and love affairs of the imperial court or warfare between contending rivals for power. This division was never, of course, absolute; especially after the breakup of the Heian amid religious upheavals and feudal strife, and the shift of power from emperors in Kyōto to shoguns in Kamakura, the old distinctions of subject matter, though not abandoned, became less stringent. Nowhere is this clearer than in the work of Kamo no Chōmei (1153–1216), a priest devoted to Amida Buddha, whose brief* Hōjōki*—translated in Donald Keene's* Anthology of Japanese Literature (1955) *as* An Account of My Hut*—describes his retreat from a world devastated* *by fire, wind, famine, earthquake, and war to monastic renunciation, in his fiftieth year, and seclusion in a ten-foot-square hut he built in the mountains.*

The emphasis in Chōmei's opening section on impermanence (mujō) *in the midst of unremitting flux motivates his renunciation, yet in his mountain abode and in the surrounding natural world he takes solace in this very quality: "Only in a hut built for the moment can one live without fears," finding companionship not only with pheasants and deer but also with the dying embers of the fire in a world where impermanence is inseparable from life. And though his book is perhaps, in Keene's words, "the most perfect work of literature composed in Japanese under the strong influence of Buddhism," Chōmei had by no means left the social world and its aesthetic culture wholly behind; he was a musician and poet (anthologized in imperial collections) from a family of Shintō priests, and his little book teems with literary allusions to earlier Chinese and Japanese poetry. Indeed, however much it is rooted in lived experience, the book as a whole belongs to the loosely articulated genre of* zuihitsu *(occasional essays "following the brush" wherever it leads), whose supreme earlier example was Sei Shōnagon's* Pillow Book*; more specifically, Chōmei's account is indebted to the tenth-century Buddhist priest Yoshinige no Yasutane, whose brief* Record of the Pond Pavilion, *written in Chinese, was itself based on Po Chü-i's "Record of the Thatched Hall on Mount Lu." (All three essays, along with Bashō's still shorter "Record of the Hut of the Phantom Dwelling," are collected in* Four Huts: Asian Writings on the Simple Life *[1994], trans. Burton Watson.) In the end, however, it is the direct experience of deeply felt communion with a natural world beautiful in its very transience that makes Chōmei's little book both an elegant literary composition and, like Thoreau's* Walden *more than six hundred years later, a living classic.*

The flow of the river is ceaseless and its water is never the same. The bubbles that float in the pools, now vanishing, now forming, are not of long duration: so in the world are man and his dwellings. It might be imagined that the houses, great and small, which vie roof against proud roof in the capital [of Kyōto] remain unchanged from one generation to the next, but when we examine whether this is true, how few are the houses that were there of old. Some were burnt last year and only since rebuilt; great houses have crumbled into hovels, and those who dwell in them have fallen no less. The city is the same, the people are as numerous as ever, but of those I used to know, a bare one or two in twenty remain. They die in the morning, they are born in the evening, like foam on the water.

Whence does he come, where does he go, man that is born and dies? We know not. For whose benefit does he torment himself in building houses that last but a moment, for what reason is his eye delighted by them? This too we do not know. Which will be the first to go, the master or his dwelling? One might just as well ask this of the dew on the morning-glory. The dew may fall and the flower remain—remain, only to be withered by the morning sun. The flower may fade before the dew evaporates, but though it does not evaporate, it waits not the evening.

Renunciation of the World

. . . For over thirty years I had tormented myself by putting up with all the things of this unhappy world. During this time each stroke of misfortune had naturally made me realize the fragility of my life. In my fiftieth year, then, I became a priest and turned my back on the world. Not having any family, I had no ties that would make abandoning the world difficult. I had no rank or stipend—what was there for me to cling to? How many years had I vainly spent among the cloud-covered hills of Ohara?

The Hut Ten Feet Square

Now that I have reached the age of sixty, and my life seems about to evaporate like the dew, I have fashioned a lodging for the last leaves of my years. It is a hut where, perhaps, a traveler might spend a single night; it is like the cocoon spun by an aged silkworm. This hut is not even a hundredth the size of the cottage where I spent my middle years.

Before I was aware, I had become heavy with years, and with each remove my dwelling grew smaller. The present hut is of no ordinary appearance. It is a bare ten feet square and less than seven feet high. I did not choose this particular spot rather than another, and I built my house without consulting any diviners. I laid a foundation and roughly thatched a roof. I fastened hinges to the joints of the beams, the easier to move elsewhere should anything displease me. What difficulty would there be in changing my dwelling? A bare two carts would suffice to carry off the whole house, and except for the carter's fee there would be no expenses at all.

Since first I hid my traces here in the heart of Mount Hino, I have added a lean-to on the south and a porch of bamboo. On the west I have built a shelf for holy water, and inside the hut, along the west wall, I have installed an image of Amida. The light of the setting sun shines between its eyebrows. On the doors of the reliquary I have hung pictures of Fugen and Fudō.[3] Above the sliding door that faces north I have built a little shelf on which I keep three or four black leather baskets that contain books of poetry and music and extracts from the sacred writings. Beside them stand a folding koto and a lute.

Along the east wall I have spread long fern fronds and mats of straw which serve as my bed for the night. I have cut open a window in the eastern wall, and beneath it have made a desk. Near my pillow is a square brazier in which I burn brushwood. To the north of the hut I have staked out a small plot of land which I have enclosed with a rough fence and made into a garden. I grow many species of herbs there.

This is what my temporary hut is like. I shall now attempt to describe its surroundings. To the south there is a bamboo pipe which empties water into the rock pool I have laid. The woods come close to my house, and it is thus a simple matter for me to gather brushwood. The mountain is named Toyama. Creeping vines block the trails and the valleys are overgrown, but to the west is a clearing, and my surroundings thus do not leave me without spiritual comfort. In the spring I see waves of wisteria like purple clouds, bright in the west.[4] In the summer I hear the cuckoo call, promising to guide me on the road of death. In the autumn the voice of the evening insects fills my ears with a sound of lamentation for this cracked husk of a world. In winter I look with deep emotion on the snow, piling up and melting away like sins and hindrances to salvation. . . .

On mornings when I feel myself short-lived as the white wake behind a boat, I go to the banks of the river and, gazing at the boats plying to and fro, compose verses in the style of the Priest Mansei. Or if of an evening the wind in the maples rustles the leaves, I recall the river at Jinyō, and play the lute in the manner of Minamoto no Tsunenobu.[5] If still my mood does not desert me, I often tune my lute to the echoes in the pines, and play the "Song of the Autumn Wind," or pluck the notes of the "Melody of the Flowing

[3]Fugen . . . is the highest of the bodhisattvas. Fudō Myōō . . . is the chief of the Guardian Kings. (Keene)
[4]The west is the direction of Paradise and it was thus auspicious that it should have been clear in that direction. The purple cloud is the one on which Amida Buddha descends to guide the believer to the Western Paradise. (Keene)
[5]Reference to the famous "Lute Song" *(P'i-p'a Chi)* by Po Chü-i. Minamoto no Tsunenobu (1016–97) was a famous musician and poet. (Keene)

Stream," modulating the pitch to the sound of the water. I am but an indifferent performer, but I do not play to please others. Alone I play, alone I sing, and this brings joy to my heart. . . .

Sometimes I pick flowering reeds or the wild pear, or fill my basket with berries and cress. Sometimes I go to the rice fields at the foot of the mountain and weave wreaths of the fallen ears. Or, when the weather is fine, I climb the peak and look out toward Kyoto, my old home, far, far away. The view has no owner and nothing can interfere with my enjoyment.

When I feel energetic and ready for an ambitious journey, I follow along the peaks to worship at the Iwama or Ishiyama Temple. Or I push through the fields of Awazu to pay my respects to the remains of Semimaru's hut, and cross the Tanagami River to visit the tomb of Sarumaru.[6] On the way back, according to the season, I admire the cherry blossoms or the autumn leaves, pick fern-shoots or fruit, both to offer to the Buddha and to use in my house.

If the evening is still, in the moonlight that fills the window I long for old friends or wet my sleeve with tears at the cries of the monkeys. Fireflies in the grass thickets might be mistaken for fishing-lights off the island of Maki; the dawn rains sound like autumn storms blowing through the leaves. And when I hear the pheasants' cries, I wonder if they call their father or their mother; when the wild deer of the mountain approach me unafraid, I realize how far I am from the world. And when sometimes, as is the wont of old age, I waken in the middle of the night, I stir up the buried embers and make them companions in solitude.

It is not an awesome mountain, but its scenery gives me endless pleasure regardless of the season, even when I listen in wonder to the hooting of the owls. How much more even would the sights mean to someone of deeper thought and knowledge!

When I first began to live here I thought it would be for just a little while, but five years have already passed. My temporary retreat has become rather old as such houses go: withered leaves lie deep by the eaves and moss has spread over the floor. When, as chance has had it, news has come to me from the capital, I have learned how many of the great and mighty have died since I withdrew to this mountain. And how to reckon the numbers of lesser folk? How many houses have been destroyed by the numerous conflagrations? Only in a hut built for the moment can one live without fears. It is very small, but it holds a bed where I may lie at night and a seat for me in the day; it lacks nothing as a place for me to dwell. The hermit crab chooses to live in little shells because it well knows the size of its body. The osprey stays on deserted shores because it fears human beings. I am like them. Knowing myself and the world, I have no ambitions and do not mix in the world. I seek only tranquillity; I rejoice in the absence of grief. . . .

A man's friends esteem him for his wealth and show the greatest affection for those who do them favors. They do not necessarily have love for persons who bear them warm friendship or who are of an honest disposition. It is better to have as friends music and the sights of nature. . . .

My clothing and food are as simple as my lodgings. I cover my nakedness with whatever clothes woven of wisteria fiber and quilts of hempen cloth come to hand, and I eke out my life with berries of the fields and nuts from the trees on the peaks. I need not feel ashamed of my appearance, for I do not mix in society and the very scantiness of the food gives it additional savor, simple though it is.

I do not prescribe my way of life to men enjoying happiness and wealth, but have

[6]Semimaru was a poet of the Heian Period who lived in a hut near the Barrier of Ausakayama. . . . Sarumaru-dayū was an early Heian poet, but nothing is known about him. (Keene)

related my experiences merely to show the differences between my former and present life. Ever since I fled the world and became a priest, I have known neither hatred nor fear. I leave my span of days for Heaven to determine, neither clinging to life nor begrudging its end. My body is like a drifting cloud—I ask for nothing, I want nothing. My greatest joy is a quiet nap; my only desire for this life is to see the beauty of the seasons. . . .

Now the moon of my life sinks in the sky and is close to the edge of the mountain. Soon I must head into the darkness of the Three Ways:[7] why should I thus drone on about myself? The essence of the Buddha's teaching to man is that we must not have attachment for any object. It is a sin for me now to love my little hut, and my attachment to its solitude may also be a hindrance to salvation. Why should I waste more precious time in relating such trifling pleasures?

Buddhist Delight in Idleness

Essays in Idleness, by Kenkō, translated by Donald Keene

Kenkō (1283–1352?) was the Buddhist name of Urabe Kaneyoshi (or Yoshida Kaneyoshi). His family, descended from Shintō priests, had turned to Buddhism, and one of his brothers was a priest. After serving at court, where his father and another brother were officials, he became a priest at age thirty in the Tendai temple of Mount Hiei, from which he soon returned to Kyōto. He was a waka *poet in a long if depleted tradition, but is now known solely for* Tsurezuregusa, *or* Essays in Idleness, *as Keene entitles his 1967 translation. This book of 243 loosely connected reflections on whatever came into his head, probably composed between 1330 and 1332, "is now almost universally accepted," Keene writes, "as one of two Japanese masterpieces of the zuihitsu genre, along with* The Pillow Book of Sei Shōnagon.*" Unlike Buddhist recluses from Saigyō to Chōmei, Kenkō "was too much involved with this world to renounce it for a hermitage," remaining attached to the court life from which he was never far for long. According to a no doubt apocryphal tradition, he wrote his thoughts down on scraps of paper which he pasted to his walls, and which others collected after his death. Nowhere are Buddhist awareness of the world's impermanence and aesthetic pleasure in its beauty—always heightened by associations with poetry—more fused than in Kenkō, who contemplated moon, dew, and cherry blossoms no less as a refined connoisseur of* waka *than as a priest, and took profound delight in the very impermanence of a charmingly illusory phenomenal world.*

What a strange, demented feeling it gives me when I realize I have spent whole days before this inkstone, with nothing better to do,[8] jotting down at random whatever nonsensical thoughts have entered my head. . . .

7

If man were never to fade away like the dews of Adashino, never to vanish like the smoke over Toribeyama, but lingered on forever in the world, how things would lose their

[7]The three paths in the afterworld leading to different types of hells. (Keene)
[8]*Tsurezure naru mama ni*, translated here as "with nothing better to do," opens the introductory section and gives rise to the title, *Tsurezuregusa*. (Keene)

power to move us![9] The most precious thing in life is its uncertainty. Consider living creatures—none lives so long as man. The May fly waits not for the evening, the summer cicada knows neither spring nor autumn. What a wonderfully unhurried feeling it is to live even a single year in perfect serenity! If that is not enough for you, you might live a thousand years and still feel it was but a single night's dream. We cannot live forever in this world; why should we wait for ugliness to overtake us? The longer man lives, the more shame he endures. To die, at the latest, before one reaches forty, is the least unattractive. Once a man passes that age, he desires (with no sense of shame over his appearance) to mingle in the company of others. In his sunset years he dotes on his grandchildren, and prays for long life so that he may see them prosper. His preoccupation with worldly desires grows ever deeper, and gradually he loses all sensitivity to the beauty of things, a lamentable state of affairs.

11

About the tenth month I had the occasion to visit a village beyond the place called Kurusono.[10] I made my way far down a mosscovered path until I reached a lonely-looking hut. Not a sound could be heard, except for the dripping of a water pipe buried in fallen leaves. Sprays of chrysanthemum and red maple leaves had been carelessly arranged on the holy-water shelf. Evidently somebody was living here. Moved, I was thinking, "One can live even in such a place," when I noticed in the garden beyond a great tangerine tree, its branches bent with fruit, that had been enclosed by a forbidding fence. Rather disillusioned, I thought now, "If only the tree had not been there!"

13

The pleasantest of all diversions is to sit alone under the lamp, a book spread out before you, and to make friends with people of a distant past you have never known. The books I would choose are the moving volumes of *Wen Hsüan*,[11] the collected works of Po Chü-i, the sayings of Lao Tzu, and the chapters of Chuang Tzu. Among works by scholars of this country, those written long ago are often quite interesting.

19

The changing of the seasons is deeply moving in its every manifestation. People seem to agree that autumn is the best season to appreciate the beauty of things. That may well be true, but the sights of spring are even more exhilarating. The cries of the birds gradually take on a peculiarly springlike quality, and in the gentle sunlight the bushes begin to sprout along the fences. Then, as spring deepens, mists spread over the landscape and the cherry blossoms seem ready to open, only for steady rains and winds to cause them to scatter precipitously. The heart is subject to incessant pangs of emotion as the young leaves are growing out.

Orange blossoms are famous for evoking memories, but the fragrance of plum blossoms above all makes us return to the past and remember nostalgically long-ago events. Nor can we overlook the clean loveliness of the *yamabuki*[12] or the uncertain beauty of wisteria, and so many other compelling sights.

Someone once remarked, "In summer, when the Feast of Anointing the Buddha and the Kamo Festival come around, and the young leaves on the treetops grow thick and

[9]This phrase, *mono no aware*, Keene notes, means more literally "the pity of things." Adashino was and Toribeyama still is the name of a graveyard in or near Kyōto: "The word *adashi* (impermanence), contained in the place name, accounted for the frequent use of Adashino in poetry as a symbol of impermanence."

[10]A village east of Kyōto. (Keene)

[11]A collection of poetry compiled by Prince Chao Ming of Liang (501–31). It was known as *Monzen* in Japan and exercised great influence. (Keene)

[12]Sometimes translated as kerria roses, a yellow flower. (Keene)

cool, our sensitivity to the touching beauty of the world and our longing for absent friends grow stronger." Indeed, this is so. When, in the fifth month, the irises bloom and the rice seedlings are transplanted, can anyone remain untroubled by the drumming of the water rails? Then, in the sixth month, you can see the whiteness of moonflowers glowing over wretched hovels, and the smouldering of mosquito incense is affecting too. The purification rites of the sixth month are also engrossing.

The celebration of *Tanabata* is charming.[13] Then, as the nights gradually become cold and the wild geese cry, the under leaves of the *hagi*[14] turn yellow, and men harvest and dry the first crop of rice. So many moving sights come together, in autumn especially. And how unforgettable is the morning after an equinoctial storm!—As I go on I realize that these sights have long since been enumerated in *The Tale of Genji* and *The Pillow Book*, but I make no pretense of trying to avoid saying the same things again. If I fail to say what lies on my mind it gives me a feeling of flatulence; I shall therefore give my brush free rein. Mine is a foolish diversion, but these pages are meant to be torn up, and no one is likely to see them. . . .

20

A certain hermit once said, "There is one thing that even I, who have no worldly entanglements, would be sorry to give up, the beauty of the sky." I can understand why he should have felt that way.

21

Looking at the moon is always diverting, no matter what the circumstances. A certain man once said, "Surely nothing is so delightful as the moon," but another man rejoined, "The dew moves me even more." How amusing that they should have argued the point! What could fail to be affecting in its proper season? This is obviously true of the moon and cherry blossoms. The wind seems to have a special power to move men's hearts.

Regardless of the season, however, a clear-flowing stream breaking against rocks makes a splendid sight. I remember how touched I was when I read the Chinese poem, "The Yüan and Hsiang flow ever east, night and day alike; they never stop an instant to soothe the grieving man." Chi K'ang also has the lines, "The heart rejoices to visit mountains and lakes and see the birds and fish." Nothing gives so much pleasure as to wander to some spot far from the world, where the water and vegetation are unsullied.

25

The world is as unstable as the pools and shallows of Asuka River.[15] Times change and disappear: joy and sorrow come and go; a place that once thrived turns into an uninhabited moor; a house may remain unaltered, but its occupants will have changed. The peach and the damson trees in the garden say nothing—with whom is one to reminisce about the past? I feel this sense of impermanence even more sharply when I see the remains of a house which long ago, before I knew it, must have been imposing. . . .

137

Are we to look at cherry blossoms only in full bloom, the moon only when it is cloudless? To long for the moon while looking on the rain, to lower the blinds and be unaware of

[13]At this point the description shifts to autumn. *Tanabata*, a feast celebrated on the seventh night of the seventh month, commemorated the annual meeting of two stars. (Keene)

[14]The *lespedeza bicolor*, a lavender or white flower traditionally associated with the season of autumn rains. The turning of the colors of its under leaves is often mentioned in poetry as a sign of approaching winter. (Keene)

[15]The Asuka River, a stream near Nara, figures prominently in Japanese poetry. Reference is here made to the anonymous poem in *Kokinshū* [the first poem in "The World in Thirty-One Syllables," above], "In this world what is constant? In the Asuka River yesterday's pools are today's shallows." (Keene) The phrases that follow this first sentence, Keene notes, are borrowed from the Japanese preface to *Kokinshū*.

the passing of the spring—these are even more deeply moving. Branches about to blossom or gardens strewn with faded flowers are worthier of our admiration. Are poems written on such themes as "Going to view the cherry blossoms only to find they had scattered" or "On being prevented from visiting the blossoms" inferior to those on "Seeing the blossoms"? People commonly regret that the cherry blosssoms scatter or that the moon sinks in the sky, and this is natural; but only an exceptionally insensitive man would say, "This branch and that branch have lost their blossoms. There is nothing worth seeing now."

In all things, it is the beginnings and ends that are interesting. Does the love between men and women refer only to the moments when they are in each other's arms? The man who grieves over a love affair broken off before it was fulfilled, who bewails empty vows, who spends long autumn nights alone, who lets his thoughts wander to distant skies, who yearns for the past in a dilapidated house—such a man truly knows what love means.

The moon that appears close to dawn after we have long waited for it moves us more profoundly than the full moon shining cloudless over a thousand leagues. And how incomparably lovely is the moon, almost greenish in its light, when seen through the tops of the cedars deep in the mountains, or when it hides for a moment behind clustering clouds during a sudden shower! The sparkle on hickory or white-oak leaves seemingly wet with moonlight strikes one to the heart. One suddenly misses the capital, longing for a friend who could share the moment.

And are we to look at the moon and the cherry blossoms with our eyes alone? How much more evocative and pleasing it is to think about the spring without stirring from the house, to dream of the moonlit night though we remain in our room! . . .

243

When I turned eight years old I asked my father, "What sort of thing is a Buddha?" My father said, "A Buddha is what a man becomes." I asked then, "How does a man become a Buddha?" My father replied, "By following the teachings of Buddha." "Then, who taught the Buddha to teach?" He again replied, "He followed the teachings of the Buddha before him." I asked again, "What kind of Buddha was the first Buddha who began to teach?" At this my father laughed and answered, "I suppose he fell from the sky or else he sprang up out of the earth."

My father told other people, "He drove me into a corner, and I was stuck for an answer." But he was amused.

A Strong Desire to Wander

The Narrow Road to the Deep North, by Matsuo Bashō, translated by Nobuyuki Yuasa

Matsuo Bashō (1644–94), the foremost haiku *poet of his or any other time (see next section), was equally an outstanding writer of travel journals intermingling prose and verse. Born in Ueno, a provinicial town founded only sixty years before, of a low-ranking samurai family, he began at an early age to compose* haikai *verse in the modern style; after moving to Edo in 1672, and to its suburb of Fukagawa eight years later—where he took his pen name from the* bashō *in his garden, a plantain tree known for its delicate leaves—he soon gathered admiring disciples around him. Kyōto remained the old capital of the imperial court, looking back to the glories of ancient* waka, *whereas Edo (later renamed Tokyo) was an upstart city of merchants and plebeians, the capital of the Tokugawa shoguns who had brought the chaotic strife of medieval times to an end. Under an autocratic regime in which Japan was virtually sealed off from contact with the outside world, and foreign ideas, including Christianity, were proscribed, the peaceful Tokugawa or Edo period (1600–1868) was inaugurated. In Edo, especially during the lively Genroku phase (1688–1704), a new urban and "bourgeois" literature came into being, commemorating in the puppet plays of Chikamatsu the loves and sorrows, and in the novels of Saikaku the pleasures, not of refined aristocrats but of shopkeepers and geishas.*

Bashō inhabited this "floating world" of Edo but, with his "strong desire to wander," never long remained. Five of his brief works—translated by Nobuyuki Yuasa in The Narrow Road to the Deep North and Other Travel Sketches *(1966)—are records of travel, mainly on foot but sometimes on horseback or by boat, to various parts of Japan; a sixth is the diary of his retreat to a small house outside Kyōto; and a seventh (translated by Burton Watson in* Four Huts*) is an account of his stay in a "phantom" hut south of Lake Biwa. "I'm just a mountain dweller, sleepy by nature," he writes, with characteristic self-mockery, "who has turned his footsteps to the steep slopes and sits here in the empty hills catching lice and smashing them." The best of his journals is* Oku no Hosomichi (The Narrow Road of Oku *or, in the translation from which our selections come,* The Narrow Road to the Deep North). *"Oku" in the title refers to a road leading north to the province of that name, but also means "within"; as Keene remarks in* World Within Walls, *"the title thus suggests a narrow road—perhaps Bashō's art—leading to the inner depths of poetry." Written in the elliptical* haibun *style, the prose equivalent of* haikai *poetry—"scribbled down," he professes in the earlier* Manuscript in My Knapsack, *"without any semblance of order" like "the ramblings of a drunkard or the mutterings of a man talking in his sleep"—*The Narrow Road *describes major stages in Bashō's journey of 1689–91 to the north: both natural scenes (often chosen because famous poets such as Saigyō had written about them) and the poetic associations they evoke.*

His prose teems with allusions to earlier Chinese and Japanese poetry, and his visits to famous sights typically culminate in one or more haiku crystallizing the effect of what he has seen; but far from contradicting the genuineness of his responses to nature, these poetic overtones deepen his perceptions and differentiate them from those of the animals. "One and the same thing runs through the waka of Saigyō, the renga ["linked verse"] of Sōgi, the tea ceremony of Rikyū," he writes in Manuscript in My Knapsack: *"What is common to all these arts is their following nature and making a friend of the four seasons. Nothing the artist sees but is flowers, nothing he thinks of but is the moon . . . . I say, free yourselves from the barbarian, remove yourself from the birds and beasts; follow nature and return to nature!" As a student of Zen, he had no interest in merely external veracity; indeed, the publication in 1943 of the diary of his traveling companion Sora revealed that Bashō had taken considerable liberties in rearranging the facts of his journey. But for all his poetic allusions and literary embellishments, his attitude was by no means only aesthetic: his "belief in poetry was religious," Keene stresses, "and his bonds with Saigyō and other poets of the past were expressions of his faith." Far from merely imitating his illustrious predecessors (as* waka *poets had self-consciously done for centuries), throughout his journey both to the deep north and into his innermost self he followed his own prescription: "Do not seek to follow in the footsteps of the men of old; seek what they sought!"*

Days and months are travelers of eternity. So are the years that pass by. Those who steer a boat across the sea or drive a horse over the earth till they succumb to the weight of years, spend every minute of their lives traveling. There are a great number of ancients, too, who died on the road. I myself have been tempted for a long time by the cloud-moving wind—filled with a strong desire to wander.

It was only towards the end of last autumn that I returned from rambling along the coast. I barely had time to sweep the cobwebs from my broken house on the River Sumida before the New Year, but no sooner had the spring mist begun to rise over the field than I wanted to be on the road again to cross the barrier-gate of Shirakawa in due time. The gods seemed to have possessed my soul and turned it inside out, and roadside images seemed to invite me from every corner, so that it was impossible for me to stay idle at home. Even while I was getting ready, mending my torn trousers, tying a new strap to my hat, and applying *moxa* to my legs to strengthen them, I was already dreaming of the full moon rising over the islands of Matsushima. Finally, I sold my house, moving to the cottage of Sampū for a temporary stay. . . .

It was early on the morning of March the twenty-seventh that I took to the road. There was darkness lingering in the sky, and the moon was still visible, though gradually thinning away. The faint shadow of Mount Fuji and the cherry blossoms of Ueno and Yanaka were bidding me a last farewell. My friends had got together the night before, and they all came with me on the boat to keep me company for the first few miles. When we got off the boat at Senju, however, the thought of the three thousand miles before me suddenly filled my heart, and neither the houses of the town nor the faces of my friends could be seen by my tearful eyes except as a vision.

The passing spring,
Birds mourn,
Fishes weep
With tearful eyes.

With this poem to commemorate my departure, I walked forth on my journey, but lingering thoughts made my steps heavy. My friends stood in a line and waved good-bye as long as they could see my back . . . .

After climbing two hundred yards or so from the shrine, I came to a waterfall, which came pouring out of a hollow in the ridge and tumbled down into the dark green pool below in a huge leap of several hundred feet. The rocks of the waterfall were so carved out that we could see it from behind, though hidden ourselves in a craggy cave. Hence its nickname, See-from-behind.

Silent a while in a cave
I watched a waterfall,
For the first of
The summer observances. . . .

I went to see the willow tree which Saigyō celebrated in his poem, when he wrote, "Spreading its shade over a crystal stream."[16] I found it near the village of Ashino on the bank of a rice-field. I had been wondering in my mind where this tree was situated, for the ruler of this province had repeatedly talked to me about it, but this day, for the first time in my life, I had an opportunity to rest my worn-out legs under its shade.

[16]Bashō is referring to the following poem by Saigyō in *Shin Kokin Shū*. "Under this solitary willow / Spreading its grateful shade / Over the crystal stream, / Let us rest our tired legs / On our way to the North." (Yuasa)

When the girls had planted
A square of paddy-field,
I stepped out of
The shade of a willow tree.

After many days of solitary wandering, I came at last to the barrier-gate of Shirakawa, which marks the entrance to the northern regions. Here, for the first time, my mind was able to gain a certain balance and composure, no longer a victim to pestering anxiety, so it was with a mild sense of detachment that I thought about the ancient traveler who had passed through this gate with a burning desire to write home. . . .

Pushing towards the north, I crossed the River Abukuma, and walked between the high mountains of Aizu on the left and the three villages of Iwaki, Sōma, and Miharu on the right, which were divided from the villages of Hitachi and Shimotsuke districts by a range of low mountains. I stopped at the Shadow Pond, so called because it was thought to reflect the exact shadow of any object that approached its shore. It was a cloudy day, however, and nothing but the grey sky was reflected in the pond. I called on the Poet Tōkyū at the post town of Sukagawa, and spent a few days at his house. He asked me how I had fared at the gate of Shirakawa. I had to tell him that I had not been able to make as many poems as I wanted, partly because I had been absorbed in the wonders of the surrounding countryside and the recollections of ancient poets. . . .

My heart leaped with joy when I saw the celebrated pine tree of Takekuma, its twin trunks shaped exactly as described by the ancient poets. I was immediately reminded of the Priest Nōin[17] who had grieved to find upon his second visit this same tree cut and thrown into the River Natori as bridge-piles by the newly-appointed governor of the province. This tree had been planted, cut, and replanted several times in the past, but just when I came to see it myself, it was in its original shape after a lapse of perhaps a thousand years, the most beautiful shape one could possibly think of for a pine tree. The Poet Kyohaku[18] wrote as follows at the time of my departure to express his good wishes for my journey.

Don't forget to show my master
The famous pine of Takekuma,
Late cherry blossoms
Of the far north.

The following poem I wrote was, therefore, a reply.

Three months after we saw
Cherry blossoms together
I came to see the glorious
Twin trunks of the pine. . . .

Much praise had already been lavished upon the wonders of the islands of Matsushima. Yet if further praise is possible, I would like to say that here is the most beautiful spot in the whole country of Japan, and that the beauty of these islands is not in the least inferior to the beauty of Lake Dōtei or Lake Seiko in China. The islands are situated in a bay about three miles wide in every direction and open to the sea through a narrow mouth on the south-east side. Just as the River Sekkō in China is made full at each swell

[17]Nōin Hoshi (988–?) was one of the outstanding poets of the Heian period. He is believed to have had a deep influence on Saigyō. (Yuasa)
[18]Kyohyaku (?–1696) was one of Bashō's disciples in Edo. (Yuasa)

of the tide, so is this bay filled with the brimming water of the ocean, and innumerable islands are scattered over it from one end to the other. Tall islands point it to the sky and level ones prostrate themselves before the surges of water. Islands are piled above islands, and islands are joined to islands, so that they look exactly like parents caressing their children or walking with them arm in arm. The pines are of the freshest green, and their branches are curved in exquisite lines, bent by the wind constantly blowing through them. Indeed, the beauty of the entire scene can only be compared to the most divinely endowed of feminine countenances, for who else could have created such beauty but the great god of nature himself? My pen strove in vain to equal this superb creation of divine artifice.

Ojima Island where I landed was in reality a peninsula projecting far out into the sea. This was the place where the Priest Ungo had once retired, and the rock on which he used to sit for meditation was still there. I noticed a number of tiny cottages scattered among pine trees and pale blue threads of smoke rising from them. I wondered what kind of people were living in those isolated houses, and was approaching one of them with a strange sense of yearning, when, as if to interrupt me, the moon rose glittering over the darkened sea, completing the full transformation to a night-time scene. I lodged in an inn overlooking the bay, and went to bed in my upstairs room with all the windows open. As I lay there in the midst of the roaring wind and driving clouds, I felt myself to be in a world totally different from the one I was accustomed to. My companion Sora wrote:

> Clear voiced cuckoo,
> Even you will need
> The silver wings of a crane
> To span the islands of Matsushima.

I myself tried to fall asleep, suppressing the surge of emotion from within, but my excitement was simply too great. I finally took out my notebook from my bag and read the poems given me by my friends at the time of my departure—a Chinese poem by Sodō, a *waka* by Hara Anteki, *haiku* by Sampū and Dakushi, all about the islands of Matsushima. . . .

According to the gate-keeper, there was a huge body of mountains obstructing my way to the province of Dewa, and the road was terribly uncertain. So I decided to hire a guide. The gate-keeper was kind enough to find me a young man of tremendous physique, who walked in front of me with a curved sword strapped at his waist and a stick of oak gripped firmly in his hand. I myself followed him, afraid of what might happen on the way. What the gate-keeper had told me turned out to be true. The mountains were so thickly covered with foliage and the air underneath was so hushed that I felt as if I were groping my way in the dead of night. There was not even the cry of a single bird to be heard, and the wind seemed to breathe out black soot through every rift in the hanging clouds. I pushed my way through thick undergrowth of bamboo, crossing many streams and stumbling over hidden rocks, till at last I arrived at the village of Mogami after much shedding of cold sweat. My guide congratulated me by saying that I was indeed fortunate to have crossed the mountains in safety, for accidents of some sort had always happened on his past trips. I thanked him sincerely and parted from him. However, fear lingered in my mind some time after that. . . .

The River Mogami rises in the high mountains of the far north, and its upper course runs through the province of Yamagata. There are many dangerous spots along this river, such as Speckled Stones and Eagle Rapids, but it finally empties itself into the sea at Sakata, after washing the north edge of Mount Itajaki. As I descended this river in a boat, I felt as if the mountains on both sides were ready to fall down upon me, for the boat was a tiny one—the kind that the farmers had used for carrying sheaves of rice in old times—and the trees were heavily laden with foliage. I saw the Cascade of the Silver Threads sparkling through the green leaves and the temple called Sennindō standing close to the shore. The river was swollen to the brim, and the boat was in constant peril.

> Gathering all the rains
> Of May,
> The River Mogami rushes down
> In one violent stream. . . .

I climbed Mount Gassan on the eighth. I tied around my neck a sacred rope made of white paper and covered my head with a hood made of bleached cotton, and set off with my guide on a long march of eight miles to the top of the mountain. I walked through mists and clouds, breathing the thin air of high altitudes and stepping on slippery ice and snow, till at last through a gateway of clouds, as it seemed, to the very paths of the sun and moon, I reached the summit, completely out of breath and nearly frozen to death. Presently the sun went down and the moon rose glistening in the sky. I spread some leaves on the ground and went to sleep, resting my head on pliant bamboo branches. When, on the following morning, the sun rose again and dispersed the clouds, I went down towards Mount Yudono. . . .

I had seen since my departure innumerable examples of natural beauty which land and water, mountains and rivers, had produced in one accord, and yet in no way could I suppress the great urge I had in my mind to see the miraculous beauty of Kisagata, a lagoon situated to the north-east of Sakata. I followed a narrow trail for about ten miles, climbing steep hills, descending to rocky shores, or pushing through sandy beaches, but just about the time the dim sun was nearing the horizon a strong wind rose from the sea, blowing up fine grains of sand, and rain, too, began to spread a grey film of cloud across the sky, so that even Mount Chōkai was made invisible. I walked in this state of semi-blindness, picturing all sorts of views to myself, till at last I put up at a fisherman's hut, convinced that if there was so much beauty in the dark rain, much more was promised by fair weather.

A clear sky and brilliant sun greeted my eyes on the following morning, and I sailed across the lagoon in an open boat. I first stopped at a tiny island named after the Priest Nōin to have a look at his retreat where he had stayed for three years, and then landed on the opposite shore where there was the aged cherry tree which Saigyō honored by writing "sailing over the waves of blossoms."[19] . . . Although little more than a mile in width, this lagoon is not in the least inferior to Matsushima in charm and grace. There is, however, a remarkable difference between the two. Matsushima is a cheerful laughing beauty, while the charm of Kisagata is in the beauty of its weeping countenance. It is not only lonely but also penitent, as it were, for some unknown evil. Indeed it has a striking resemblance to the expression of a troubled mind.

> A flowering silk tree
> In the sleepy rain of Kisagata
> Reminds me of Lady Seishi
> In sorrowful lament. . . .[20]

[19]Bashō is referring to the following poem by Saigyō: "Buried in the waves / So that it seems / Fishermen's boats are sailing / Over the waves of blossoms— / A cherry tree at Kisagata." (Yuasa)
[20]Lady Seishi or Hsi-shih is known for her melancholy beauty. She was sent to Fu-chai of the Wu dynasty by Kou-chien of the Yüeh dynasty as a gift. (Yuasa)

It was again a fine day on the sixteenth. I went to the Colored Beach to pick up some pink shells. I sailed the distance of seven miles in a boat and arrived at the beach in no time, aided by a favorable wind. A man by the name of Tenya accompanied me, with servants, food, drinks and everything else he could think of that we might need for our excursion. The beach was dotted with a number of fishermen's cottages and a tiny temple. As I sat in the temple, drinking warm tea and *sake*, I was overwhelmed by the loneliness of the evening scene.

Lonelier I thought
Than the Suma beach—
The closing of autumn
On the sea before me. . . .

On September the sixth . . . I left for the Ise Shrine, though the fatigue of the long journey was still with me, for I wanted to see the dedication of a new shrine there. As I stepped into a boat, I wrote:

As firmly cemented clam-shells
Fall apart in autumn,
So I must take to the road again,
Farewell, my friends.

The World in Seventeen Syllables

Nineteen *Haiku*, translated by Geoffrey Bownas and Anthony Thwaite

Half a millennium after the Kokinshū, *the aristocratic* waka *tradition had become so ingrown that guardians of its purity claimed access to esoteric meanings to which only a few were privy. Yet this central tradition of Japanese poetry held within it the seeds of its own renewal. Poets confined within its thirty-one syllables might compose sequences of up to a hundred loosely related poems, either by themselves, like Fujiwara Teika, or with other poets. Such collaboration, called* renga, *or "linked verse," which soon evolved elaborate rules, became the dominant poetic form of the late medieval or Muromachi period. Its foremost practitioner was Sōgi (1421–1502), a Rinzai Zen priest of humble birth who journeyed, like Saigyō before and Bashō after him, to different parts of Japan; he was co-author of the* rengas *"Three Poets at Minase" in 1488 and "Three Poets at Yuyama" in 1491. A* renga*'s links were between a* waka*'s initial three lines of 5-7-5 syllables (the* hokku*), composed by one poet, and its remaining two-line verse of 7-7 syllables, by another; this in turn was linked with a new three-line verse by a third poet, and so forth, throughout a sequence, in general, of a hundred verses. (To the Mexican poet Octavio Paz—co-author with three fellow poets of* Renga: A Chain of Poems *[1971] composed in four languages—this mode of composition suggested "aspiration toward a collective poetry" similar to that of twentieth-century modernists: though* renga *"was governed by rules as strict as those of etiquette," he writes, "its object was not to put a brake on personal spontaneity, but to open up a free space so that the genius of each one could manifest itself without doing harm either to others or oneself.")*

Further emancipation from classical constraints came with the increasing popularity of haikai, *comic poems freed from the* waka*'s—or* tanka*'s, as the thirty-one syllable form came to be* *known, since* waka *properly referred to any poem in Japanese—narrow restrictions of both subject matter and vocabulary. Such poems date back to the first two imperial anthologies but became prevalent only in Tokugawa times, when comic linked poetry* (haikai no renga) *addressed contemporary themes in colloquial language forbidden to the classical poets. (As Keene remarks, "The haikai poets were not merely allowed to use nontraditional vocabulary but absolutely required to.") One pioneer of this new style in the late Muromachi was the Shintō priest* ***Arakida Moritake*** *(1473–1549), whose gently humorous poems influenced* haikai *poets of the next century. Our first two selections show how* hokku *extracted from longer sequences can stand by themselves as seventeen-syllable* haiku*—a name familiar in Anglo-American poetry since Ezra Pound.*

The great master of haiku*—which continued until the nineteenth century to be called* hokku, *and to be considered, despite their effective independence, as opening verses to which others might be added—was* ***Bashō****, who transformed a seemingly trivial verse form into a supremely evocative poetic genre. The* haiku and haikai *sequences of Bashō and his followers observe many conventions, such as inclusion in each verse of a word at least obliquely suggesting a season; their language, in contrast to the faded diction of many latter-day* tanka, *is fresh and immediate, and often achieves a startling or humorous effect. But the hallmark of Bashō's* haiku, *which crystallize a world in their seventeen-syllable microcosm, is juxtaposition—as in his famous poem on a frog's splash disrupting the silence of an old pond—of the timeless and the momentary: "Only by suggesting the age of the pond, its unchanging nature," Keene remarks, "is the momentary life of the frog evoked. . . . He believed that the smallest flower or insect if properly seen and understood could suggest all of creation, and each had its reason for existence." Thus the very poetic sensibility that separates man from beast paradoxically permits him to merge with a natural world to whose innermost essence—the silence or emptiness at the core of being—Bashō gives incomparably concise expression, following his own counsel to "learn about pines from the pines, study about bamboos from the bamboos."*

Other poets in the selection that follows are Bashō's pupils ***Hattori Ransetsu*** *(1654–1707) and* ***Enomoto (Takarai) Kikaku*** *(1661–1707);* ***Uejima Onitsura*** *(1661–1738), Bashō's younger contemporary but not a member of his school, who stressed* makoto *or truth to lived experience and "believed that each element of creation had its own nature," in Keene's words, "and the task of the poet was to understand and distinguish it";* ***Yosa Buson*** *(1716–83), leader of the "haikai revival" of the 1740s to 1780s, for whom the direct observation central to Bashō "was of less significance," Keene remarks, "than the blending of literary tradition and the poet's private emotions"; Buson's younger contemporaries* ***Miura Chora*** *(1729–80) and* ***Katō Gyōdai (or Kyōtai)*** *(1732–92); and* ***Kobayashi Issa*** *(1763–1827), a farmer's son whose poems are known for plainness of language and sympathy for other living things, from sparrows and frogs to spiders and fleas. From the deceptively brief and seemingly simple verses of these poets, intently contemplated with heart and mind, whole worlds can arise.*

On renga *and* haikai *poetry see also Earl Miner,* Japanese Linked Poetry *(1979); Steven D. Carter,* Three Poets at Yuyama *(1983); Harold G. Henderson,* An Introduction to Haiku *(1958); Makoto Ueda,* Matsuo Bashō *(1970); Earl Miner and Hiroko Odagiri,* The Monkey's Straw Raincoat and Other Poetry of the Bashō School *(1981); Lewis Mackenzie,* The Autumn Wind: A Selection from the Poems of Issa *(1957); and Nobuyuki Yuasa,* The Year of My Life: A Translation of Issa's Oraga Haru *(1960).*

TWO *HOKKU* BY ARAKIDA MORITAKE

Fallen flower I see
Returning to its branch—
Ah! a butterfly.

As the morning glory
Today appears
My span of life.

SIX *HAIKU* BY MATSUO BASHŌ

An old pond
A frog jumps in—
Sound of water.

The sea dark,
The call of the teal
Dimly white.

On a bare branch
A rook roosts:
Autumn dusk.

Summer grasses—
All that remains
Of soldiers' visions.[21]

Spring:
A hill without a name
Veiled in morning mist.

A flash of lightning:
Into the gloom
Goes the heron's cry.

[21] Written at Takadate, the "Castle on the Height," where Yoshitsune, a Minamoto general, and his faithful followers were killed by the armies of his jealous brother. (Bownas and Thwaite)

TWO *HAIKU* BY PUPILS OF BASHŌ

Painting pines
On the blue sky,
The moon tonight.

—Hattori Ransetsu

Harvest moon:
On the bamboo mat
Pine-tree shadows.

—Enomoto Kikaku

THREE *HAIKU* BY UEJIMA ONITSURA

They bloom and then
We look and then they
Fall and then . . .

Trout leaping:
On the river-bed
Clouds floating.

Green cornfield:
A skylark soaring,
There—swooping.

THREE *HAIKU*

Mosquito-buzz
Whenever honeysuckle
Petals fall.

—Yosa Buson

Peering at the stars
Through the gaps between branches,
The lonely willow.

—Miura Chora

Mournful wind:
Night after night
The moon wanes.

—Katō Gyōdai

THREE *HAIKU* BY KOBAYASHI ISSA

The world of dew is
A world of dew . . . and yet,
And yet . . .

Far-off mountain peaks
Reflected in its eyes:
The dragonfly.

For fleas, also, the night
Must be so very long,
So very lonely.

Dewdrops on a Lotus Leaf

Six Zen Poems of Ryōkan, translated by John Stevens

Ryōkan (1758–1831) was the son of a village headman, Shintō priest, and haikai *poet on the coast of Japan; at seventeen he entered a local Zen temple and at twenty-one became a Sōtō Zen monk. After attaining enlightenment and after wandering for several years, he settled in a one-room hut belonging to a temple on Mount Kugami; here, mixing with villagers and playing with their children, he supported himself by begging. In the Zen tradition, many of his poems are in Chinese (with great disregard for formal regularity). Most, however, he wrote in an archaic Japanese modeled on the* Man'yoshū, *reviving the* chōka *after a thousand years' neglect and making use of ancient poetic diction such as* makura kotoba *("pillow-words"), as well as writing* waka *and haiku. What he most esteemed in the* Man'yōshū, *Burton Watson suggests in his translation* Ryōkan: Zen Monk-Poet of Japan *(1977), "and tried hardest to capture in his own work is its spirit of simplicity and openness, its freedom from literary artifice and intellectualization," attempting "to record the actual scenes and experiences of his own life."*

Estimations of so unconventional a poet have varied widely, but in general Ryōkan's reputation—like that of his favorite Chinese poet and model, Han-shan—has risen with time. He teaches oneness with nature by "giving myself up to the whim of the wind," by letting be rather than holding on, and by following nature's course through the Zen (and Taoist) wisdom of "not knowing." He thus connects back with the beginnings not only of Japanese poetry in the Man'yōshū *but also of Taoist insight in the* Tao Te Ching *and in Chuang Tzu's dream of the butterfly he could not distinguish from himself (Chapter 5 above). The Zen nun Teishin (forty years his junior, with whom he became close in his final years) published in about 1835 a posthumous collection of his Japanese poems, later greatly augmented by discovery of others preserved by local people. John Stevens has adapted the collection's title for his translations, from which the following selections are taken, from both Chinese and Japanese:* Dewdrops on a Lotus Leaf: Zen Poems of Ryōkan *(1993).*

If someone asks
My abode
I reply:
"The east edge of
The Milky Way."

Like a drifting cloud,
Bound by nothing:
I just let go
Giving myself up
To the whim of the wind.

In my garden
I raised bush clover,
suzuki grass.
violets, dandelions,
flowery silk trees,
banana plants, morning glories,
boneset, asters,
spiderwort, daylilies:
Morning and evening,
Cherishing them all,
Watering, nourishing,
Protecting them from the sun.
Everyone said my plants
Were at their best.
But on the twenty-fifth of May,
At sunset,
A violent wind
Howled madly,
Battering and rending my plants;
Rain poured down,
Pounding the vines and flowers
Into the earth.
It was so painful
But as the work of the wind
I have to let it be . . .

The flower invites the butterfly with no-mind;
The butterfly visits the flower with no-mind.
The flower opens, the butterfly comes;
The butterfly comes, the flower opens.
I don't know others,
Others don't know me.
By not-knowing we follow nature's course.

For Children Killed in a Smallpox Epidemic

When spring arrives
From every tree tip
Flowers will bloom,
But those children
Who fell with last autumn's leaves
Will never return.

My legacy—
What will it be?
Flowers in spring,
The cuckoo in summer,
And the crimson maples
Of autumn . . .

I must go there today—
Tomorrow the plum blossoms
Will scatter.

FOUR

Coming-to-Be and Passing-Away: Classical Greece

The *experience* of nature—of some human relationship with a surrounding world of plants and animals wild and domestic, of rivers and mountains, winds and rain, and of heavenly bodies majestically encircling us all—is of course universal, immensely varied though its forms have been, in every society from tribal hunter-gatherer to modern megalopolitan or suburban. Though veiled by skyscrapers and smog, violated by litter and waste, or confined to window boxes and well-weeded lawns, nature, we know or dimly remember, is *there*, and will not go away; it may indeed, as it long antedated our advent, long survive our departure. But our *concept* of nature as an autonomous process, a dynamic self-regulating order pervading and governing all aspects of the phenomenal world, had its origin in a particular place and time. This concept (historically unique to the West, whatever its analogues in Chinese and other cultures have been) was essential to the development of ancient natural philosophy and its offspring, modern natural science. G. E. R. Lloyd is thus justified in provocatively entitling an essay of 1989, included in *Methods and Problems in Greek Science* (1991), "The Invention of Nature."

Not that nature could be "invented" out of whole cloth. The materials had long been at hand, and every invention (as the Latin derivation of our word should remind us) is a discovery, a "coming upon" what was already there, if unnoticed. But the concept of *physis* that originated in ancient Greece—specifically in the Ionian city of Milétus (Milêtos)[1] in the sixth century B.C.—marked a major revolution in thought. Most of the pre-Socratic philosophers (so called because Socrates effected a second major shift by concerning himself not with nature but with ethics and definitions) boldly speculated on the origin and composition—the *physis*—of a world continually in the process of change, and hence were known to their successors as "natural philosophers" *(physikoi* or *physiólogoi)*. Even those who denied the reality of this world, and of change itself, were said to have written *peri physeôs*, "on nature," which thus remained, in their very denial, a central concern. To these same thinkers, by no accident, were ascribed the beginnings of what we now consider science in such fields as astronomy, geography and geology, biology, mathematics, acoustics, and harmonics. Nature, in their conception, was susceptible not only to daring speculation but also to careful observation and even, in some of its aspects, to precise quantification.

The Greek word *physis*, which takes on such prominence in these seminal thinkers, apparently derives from *phyein*, "to grow"; thus "at the heart of the ancient conception of *physis*," as Charles Kahn writes in *Anaximander and the Origins of Greek Cosmology* (1960), "the etymological link with growth and vital development is preserved." The principal meanings of this developing concept in Greek philosophical and scientific thought are the subject of William A. Heidel's classic paper of 1910, "*Peri Physeôs.* A study of the Conception of Nature Among the Pre-Socratics," which distinguishes three basic meanings: *physis* as process; as the beginning of a process; and as the result of a process. The very word suggests the continual change and development implicit in growth; "nature" for most Greek philosophers is accordingly not a static entity but a life-giving and life-sustaining process. Of especial interest to the earliest thinkers was

[1]Some Greek names are here given in traditional Latinized spelling followed in parentheses by English transliteration of the Greek. Accents are occasionally added as an aid to pronunciation.

the origin or beginning *(arkhê)* of this process, whether conceived as a material "first principle" out of which the universe spontaneously arises or (by some later philosophers) as the creative power that brings it into being. Nature is thus the organic process—an inherently *orderly* process subject to its own intrinsic law or necessity *(anankê)*, not to arbitrary or external forces acting upon it—by which one primary substance (such as water or air) takes many shapes and again reverts to one, or by which a plurality of substances interact to produce the continually changing shapes of the phenomenal world.

"The main sense of Nature was, however," Heidel writes, this product itself: "the sum of things as constituted by the elements and the cosmic laws and processes." Out of the orderly process of nature arises an orderly result, the *kosmos* so named for the very order it manifests: a being both living and divine, of which the human being, a microcosm of the cosmos as a whole, is inseparably part. For these thinkers—unlike the Sophists of the fifth century, who introduced a sharp distinction between *physis* and human convention, or *nomos*—there can be "no unbridgeable gulf between man and nature," Kahn remarks, "and the larger order of the universe is one in which the 'little cosmos' has its natural place. For the course of man's life keeps step with the seasons and with the ceaseless pacing of the stars, while the elemental forces dwell, as he does, within an ordered community that holds their antagonism in check." This correspondence between an orderly (thus at least a theoretically comprehensible) world and the human mind finds expression in another pre-Socratic concept, especially important in Heraclítus (Hêrakleitos) of Ephesus, the *logos*, signifying at once "speech," "thought," or "reason" and "measure," "proportion," or "ratio": "the eternal structure of the world," Kahn writes in *The Art and Thought of Heraclítus* (1979), "as it manifests itself in discourse." Though hidden, this structure underlies and repeatedly counteracts the apparent amorphousness of a world forever in flux.

But the fitting together or attunement *(harmonia)* of parts constituting the cosmic natural order, as Heraclitus and later Empédocles stress, is a "harmony" entailing continual tension between opposites: here again, cosmic and human correspond. Through incessant cooperation and strife, attraction and repulsion, of conflicting forces acting on the elemental materials of nature arise the continual processes of coming-to-be or generation *(genesis)* and passing-away or corruption *(phthora)* central to a world of perpetual change or Becoming. The status and interrelationship of these complementary processes would remain a central philosophical problem for Plato and Aristotle, who strove in very different ways to reconcile the transitory and the eternal.

Such in brief are some central concerns of Greek natural philosophy in the great creative period between Thales of Miletus in the sixth century B.C. and Aristotle in the fourth. Yet revolutionary as the concept of nature was, it was not born from nothing. Many scientific "discoveries" attributed to the pre-Socratics—prediction of eclipses, identity of morning and evening star, "Pythagorean" theorem—drew upon Babylonian astronomical and mathematical lore going back a millennium or more, though only recently revealed to the upstart Greeks, who soon went far beyond it. Still more fundamentally, major elements of the new concept had important antecedents in older Greek (and even "primitive") myth, as F. M. Cornford, under the influence of Sir James George Frazer's *The Golden Bough* and Jane Ellen Harrison's *Prolegomena to the Study of Greek Religion*, stressed (somewhat overzealously) in *From Religion to Philosophy* (1912). If in the natural philosophers, as Heidel remarks, "*logos peri physeôs* [discourse about nature] succeeds *mythos peri theôn* [myth about gods]," the distinction marks a continuity also, for myth about gods never wholly vanished from pre-Socratic (or indeed Platonic) *logos.* Even the most ruthlessly logical of thinkers, Parménides, portrays himself as a poet oracularly speaking truth revealed by a goddess.

Not, of course, that the gods of Thales, Parmenides, or Plato were the Olympians of Homer—those all-too-human deities whose uncontrolled desires and willful actions Xenóphanes and Plato scathingly condemned. Yet Plato, through the mouth of Socrates *(Cratylus* 397C–D), suggests that "the sun, moon, earth, stars, and heaven, which are still the gods of many barbarians, were the only gods known to the aboriginal Greeks," and these hypothetically pre-Olympian nature divinities perhaps presage the divinized nature of the early pre-Socratics. Nor were the Olympians themselves, in Homer but especially in "Homeric Hymns" such as those to Dêmêtêr or Pan, though humanized in outward form, wholly remote from the natural forces which they continued to embody. Like those forces they were subject, as Zeus in the *Iliad* knows, to necessity, albeit from above and beyond them. (The Sumerian and Vedic gods, too, personified natural forces, yet neither gave rise to a concept of nature like that of the Greeks.) Aristotle recognized in the poet Hesiod a precursor of the natural philosophers; the account of spontaneous origin of the world out of chaos in the *Theogony* strikingly anticipates others that followed. As Werner Jaeger affirms in *Paideia* (1939; 2nd ed. 1945), with perhaps a trace of hyperbole, "There is no discontinuity between Ionian natural philosophy and the Homeric epics. The history of Greek thought is an organic unity, whole and complete."

If the gods of Greek poetry personify quasi-natural forces, its world is a prototype of the philosophers' cosmos. Nowhere is the splendid order of a world embracing man and nature alike, from city to farm and from the stars above to the encompassing ocean, more majestically depicted than on the shield of Achilles in the *Iliad.* Yet this harmony, as in Heraclitus, arises from conflict, both in scenes on the shield and in the surrounding world of incessant war. The annual farm life of Hesiod's *Works and Days* is a cosmos ruled by the rising and setting of seasonal constellations, but this world too is subject to unending strife; and in the *Bacchae* of Eurípides the social order upheld by the Theban king is torn asunder by his attempted denial of the elemental forces embodied in Dionysus. The world of nature—and of human beings in contact and tension with nature—portrayed in classical Greek poetry is thus no less a dynamically ordered world than that of the natural philosophers, and no less a world where perpetual coming-to-be and passing-away, in human life as in the world of which it is part, are of compelling concern. For, as Glaucus declares to Diomêdês on the battlefield of the *Iliad* (VI.146–49):

> *Like leaves' are mankind's fleeting generations.*
> *Some the wind scatters to the ground, but others*
> *grow from new-budding trees when spring returns:*
> *so too men's generations grow and die.*

The repeated verb "grow" *(phyei)* is the root from which *physis* would spring. The Greek philosophy of nature, revolutionary though its consequences would be, might almost be called an extended gloss on these lines.

8

Poets from Homer to Aristophanes

There could be no category of "nature poetry" in classical Greece, for the natural world could never be conceived apart from the human. The actions and passions of mortal beings in relation to the immortal gods are at the center of Greek poetry: the wrath of Achilles, the homecoming of Odysseus, the loves and wars of the lyric poets, the victories of Pindar's athletes, the reversals of tragedy and the revels of comedy. Yet though set on the battlefield or at sea, in stadium, law court, or palace, this poetry repeatedly reveals an extraordinary intimacy with a natural world that is never its explicit theme; its very lack of self-conscious awareness is perhaps the key to the intimacy. As Schiller would write many centuries later, when nature (he thought) had become an object *as it could never have been to the Greeks, "They felt naturally; we feel the natural."*

The Iliad *is a poem of war taking place on the plain of windy Troy in between the Achaean encampment by the loud-roaring, fruitless sea and the beleaguered city of Priam; except for occasional moments when natural forces like the river Scamander or the fire personified by the god Hephaestus take on mythic proportions as they join in the human combat, the world of nature could hardly seem farther away. And yet, it is never far. Humanized though they are, the Homeric gods occasionally reveal—like Hephaestus when he rages against Scamander, fire against water; or like the Nereid Thetis, Achilles' mother, when she rises like mist from the sea—their affinity with the elements, whose permanence the recurrent epithets of wine-dark sea and rosy-fingered dawn repeatedly contrast with the bloodshed and slaughter of war. Most of all, it is the similes that introduce the wealth and diversity of the natural world into this caldron of battle by comparing or contrasting the actions of men with, for example, waves, clouds, flooded rivers, forest fires, thunderstorms, snowflakes, oak or olive trees, grain, shooting stars, eagles, vultures, sheep, deer, lions, boars, bulls, runaway horses, wasps, bees, flies, locusts, and even mosquitoes. As Paolo Vivante comments in* The Homeric Imagination *(1970), Homer's "deep sympathy for the life of nature as a whole" finds expression in similes that stress not objects but events, "not so much a static pattern as the life of nature" to which later thinkers would give the name* physis.

Only in such similes, and in the extended portrayal on the shield of Achilles of urban and rural life in tense harmony with its natural surroundings, do moments of seemingly peaceful communion with the encompassing world find sporadic expression amid the tumultuous violence of the Iliad. *In the more "domestic"* Odyssey *such scenes are more frequent, but here the disruption of the human world by the aftermath of the siege and sack of Troy and by the anarchic misrule of Penelope's suitors in Ithaca produces a* disharmony *with nature as well, so that those—like the swineherd Eumaeus and Odysseus's father Laërtês—who live close to the soil have no satisfaction in doing so. Odysseus finds occasional comfort in closeness to nature, as when he shelters himself from the howling storm that has swamped his raft by crawling beneath the intertwined branches of a hybrid wild and domestic olive bush, or clings to a fig tree above the whirlpool of Charybdis, or slowly recognizes the coast of his rocky homeland of Ithaca after a twenty-year absence. Only the eternally bountiful fairyland of the Phaeacians ruled by King Alcínoüs, or the primitive pastoral of the savagely anarchic Cyclopes, however, seems to promise a harmony with nature that might* be without conflict, and Odysseus rejects both as illusory, just as he rejects the nymph Calypso's offer of an immortality that would remove him from life and thus from nature forever.

If the ever-present world of nature makes an explicit appearance only peripherally in the Homeric epics, it provides the constant and essential backdrop for the farmer's unending struggle to gain his livelihood in Hesiod's Works and Days. Once long ago a golden race of human beings had lived "remote from toil and care," but those days are gone forever, and in this age of iron, hard work to make the grudging soil yield its fruits is the lot of mortal men. Hesiod's poem unforgettably evokes the peasant's constant toil: his struggles with the land, the weather, and a good-for-nothing brother; his crusty temper, sententious values, and inveterate prejudices; and always, of course, the cycle of the seasons for plowing, planting, and harvesting, as marked by the rising and setting of the constellations, the coming of the rains, the cry of the crane or the cuckoo, and the cicadas' song. Hesiod has no abstract word for nature, no philosophical concept of physis; but the thing itself, in all its concrete actuality, is inescapably and continually part of his life.

Whatever their ultimate origins may have been, no major Greek god could of course be reduced to mere embodiment of a natural force. Far too many diverse components entered into the making of each, and in Homeric epic any elemental powers the Olympians continue to wield seem thoroughly subordinated, most of the time, to their conspicuously fallible humanity. In some of the "Homeric Hymns," on the other hand, the forces of nature always latent beneath the anthropomorphic persona manifest themselves much more directly—as when the long-haired youth kidnapped by pirates sits "impassively smiling with dark eyes" as the bonds fall away from him, then makes the masts sprout with grape vines and the decks run with wine as he transforms himself into a lion and reveals himself as "loud-roaring Dionysus." The close connection between the rape of Perséphonê and Dêmêtêr's withholding of the grain, on which earthly life depends, until her daughter is restored to the upper world for two-thirds of each year is a seasonal myth distantly related to those of ancient Mesopotamia and Anatolia and closely connected, in this hymn, with the Eleusinian Mystery religion that promised renewal of life to its initiates in the world beyond. And the hymns to Pan and to Earth, Mother of All, address powers inseparably identified with the vitality and bounty of nature itself.

The principal themes of the Greek lyric poets, in the pitifully few fragments of their verse that survive, include war and politics, companionship and, above all, love; but here, as in the epic, the world of nature repeatedly makes its appearance in images that place the human concerns within the larger framework of the surrounding nonhuman world. Alcman's haunting fragment "Mountain summits and deep ravines are sleeping" and Sappho's "Stars around the beautiful brightly shining moon" are among the few (of more than a line or two) devoted, as we now have them, entirely to description or evocation of a natural scene. More often, natural imagery is used to make a vivid comparison, as when Sappho likens erotic desire to a whirlwind; or to set a contrasting scene, as when Alcaeus finds shelter from a wintry gale through wine and a blazing fire—or to do both, as when Ibycus, after evoking the orderly growth of quince trees and vines in springtime, compares his own raging desire to a thunderstorm driven by the howling north wind. These poets, as Jaeger says of Alcaeus and his companions, "did not see nature as an objective or aesthetic spectacle, like the shepherd in Homer who gazes with joy from the mountain peak upon the splendor of the midnight stars; they felt that the changes of sky and season, the succession of light and darkness, calm and storm, winter frost and the cheering breath of spring, reflected the changing emotions of the human soul, and that earth and sky echoed and strengthened their cries of love and grief."

In tragedy, too, nature generally remains ostensibly on the sidelines, manifesting itself

—above all in splendid choral odes—through images evocative of the never distant world beyond, as the conflict of values among men (and gods) takes place at center stage, often in front of a palace or temple, tent or tomb. In a number of tragedies, however, the setting is far removed from city or camp, and the imagery of nature is vividly present, as in the Caucasian mountain crags of the Promêtheus Bound *of Aeschylus; on the rocky island of Lemnos to which Philoctétes bids farewell after ten years of suffering in Sóphocles' play; or in the grove of the Euménides in Sophocles'* Oedipus at Colonus, *where the chorus of old Athenians, welcoming the exiled stranger to their city, sing the glories of a maternal land blessed by the horse and olive, gifts of Poseidon and Athena. And in the* Bacchae *of Euripides, Dionýsus himself, god of wine and ecstatic oneness with nature, is hymned by his wildly fervent worshippers—and takes terrible vengeance on those who attempt to deny his undeniable power. Nature, we once again remember, is not invariably benevolent or gentle.*

Finally, two choruses from Aristóphanes suggest that in comedy, where excess was the norm and members of the chorus might appear sporting oversized red leather phalluses, or fantastically decked out as wasps, clouds, birds, or frogs, nature stands ready, as in the Birds, *to reassert its ancient dominion (albeit in burlesque) and gaily evict its impotently overcivilized opponents.*

Cosmos, Orchard, and Cave

Three Homeric Visions of the Natural Life, translated by Robert M. Torrance

Greek poetry as we now have it begins with its greatest masterpieces, the Iliad *and the* Odyssey *of Homer, probably composed in the Ionian islands or on the adjacent coast of Asia Minor in the mid to late eighth century* B.C. *Whether Homer (whose name means "hostage") was one poet, as tradition has always held, or two is a fruitless question; each poem is clearly a supreme artistic unity forged from materials handed down by generations of singers.*

The Bronze Age Mycenaean civilization—so called for the city of Mycénae once ruled, the epics relate, by King Agamémnon, leader of the Greek or Achaean expedition against Troy—that provided the subject matter for the epics, and thus for most Greek myths, had its dim beginnings as far back as the third millennium B.C.; *it reached its height after the middle of the second millennium under the stimulus of the older "Minoan" civilization of Crete with which the Mycenaeans traded, and which they may have conquered. Heinrich Schliemann's excavations in the late nineteenth century of both Troy and Mycenae proved that these cities, long thought fabulous, were very real. The archeological evidence, though inconclusive, is compatible with the traditional story that Troy (which guarded the entrance to the Dardanelles near the coast of Asia Minor) was sacked and burned, possibly in the thirteenth century, and that this triumph of Mycenaean Greece was followed, within a few generations, by disastrous wars among its cities, and their final destruction by iron-age Dorian Greek invaders called, in myth, the sons of Héracles, or Hêrakleidai. During the "Dark Ages" that followed, roughly from the eleventh to the eighth century, stories of these great events must have taken shape in orally composed songs that culminated in the two Homeric epics. The researches of Milman Parry, Albert Lord, and others in the twentieth century have demonstrated that poems of this length—the* Iliad *is over 15,000 lines—could be composed and recited by illiterate poets making use of oral formulas inherited from a long epic tradition. Their fortunate preservation may have been due to the reintroduction of writing—the only writing known to Mycenaean Greece, the "Linear B" script adapted from the Minoans, had wholly vanished—somewhere around the time of their composition.*

The selections that follow are three very different portrayals of what might be called human life in accord with nature in these two epics. The first, from Book XVIII of the Iliad, *describes the wondrous shield forged by the divine smith Hephaestus for Achilles (Akhilleus), at the request of his mother Thetis, just before his return to the battlefield from which he withdrew in anger at the beginning of the poem. The description, whose varied urban and rural, peaceful and violent scenes are in striking contrast to the singleness and intensity of war that dominates most of the epic, are a comprehensive emblem of human life encompassed by nature. On its inner circle are fashioned earth, sea, and sky with the sun and stars—that is, the cosmos as a whole. Here, as James M. Redfield writes in* Nature and Culture in the *Iliad (1975; 2nd ed. 1994), "we are shown nature, in the absence of man, as a realm of order and significance." The long description of two cities, one at peace and one at war, on the second ring portrays the multiple activities of the human world. On the third ring, representing "the four agricultural seasons: plowing, harvest, vintage, and the fallow time when the cattle are driven into the fields to fertilize them for the plowing to come," Redfield continues, "we see man* with *nature and the cycle of man's productive activity." In the round dance on the fourth ring, the society portrayed in the two cities as "a structure of cooperation and conflict" now "appears as pure* communitas," *overcoming distinctions through the absorption of natural process into the rhythms of human life. "And as the fourth ring echoes the second, with a difference, so also the fifth ring echoes the first," Redfield remarks; "here is the ocean stream, which, as it runs around the whole world, so also runs around this picture of the whole world. The inner two rings portray nature and culture as meaningful structure; the outer two portray culture and nature as pure act or pure process," while the middle ring, representing the seasons, shows the fusion of the natural and the human. Thus the shield moves, in a symmetrical construction typical of Homeric style, "from nature to culture to productivity, which is the inclusion of nature in culture; it then moves back through culture to nature."*

The two passages from the Odyssey, *on the other hand, represent much more partial, indeed illusory, visions of the "natural" life, both of which the hero Odysseus leaves behind and implicitly repudiates by his return to Ithaca. The first evokes the enchanting fairytale world of the Phaeacians through the ever-abundant orchard of King Alcínoüs, father of the lovely Nausicaa; Odysseus wonderingly admires it, but passes on. The second, a brief description of the one-eyed Cyclôpês whom Odysseus foolishly visits on his homeward travels, savage cavemen for whom "everything grows unplanted and unplowed," might seem an early example of romantic primitivism. Later poets such as Theócritus (Chapter 11 below) would sentimentalize Homer's cannibalistic Cyclops Polyphémus as a lovelorn shepherd; Odysseus inebriates him, bores out his eye, and (unwisely) defies him. A "natural" life of this kind has no appeal to a civilized Greek who knows nature too well to deny its often harsh and dangerous reality.*

The Shield of Achilles
Iliad *XVIII*

First he forged an enormous sturdy shield,
embellished on all sides, and round it cast
a shining triple rim, with silver strap. 480
Five folds composed the shield, on which he wrought
countless intricate works of artful cunning.
On it he fashioned earth and sky and sea,

the never wearied sun, the moon when full,
and all the signs that crown the starry sky: 485
Pleíades, Hýades, and strong Oríon,
and last the She-Bear, also called the Wagon,
who circles one fixed place, looks toward Oríon,
and never drops, like others, into Ocean.
On it he fabricated two fair cities 490
of mortal men. In one were feasts and weddings,
brides ushered from their rooms by blazing torchlight
all through the town to hymeneal singing,
and youthful dancers whirling, while among them
flutes and lyres clamorously played, and women 495
stood at their outer doorways marveling.
The people, gathered in the marketplace,
heard two men angrily dispute the blood-price
for murder: one man vowed full restitution
publicly, but the other scorned his offer. 500
Both sought a judgment from an arbitrator
as people urged them on, supporting both.
Heralds restrained the people while the elders
sat in a sacred circle on smoothed stones,
grasping the loudly shouting heralds' staffs. 505
To them the men rushed, pleading each in turn,
and placed two golden talents[1] in between
to give whoever spoke the fairest verdict.
Around the other city lay two armies
sparkling with weapons. These disputed whether 510
to waste the city or divide in half
the wealth its splendid citadel contained.
Meanwhile, within, unyielding men prepared
an ambush. While dear wives and little children
stood guard with agèd men upon the walls, 515
they set forth led by Ares and Athena
both made of gold and wearing golden raiment,
immense and beautifully armed, like gods
conspicuous from afar; the men were smaller.
Reaching the designated place of ambush, 520
a river where all animals came to water,
they sat, enveloped by bright shining bronze.
Two of the people's scouts sat far apart
watching for sheep and cattle with curved horns
which soon appeared, attended by two herdsmen 525
happily piping—and suspecting nothing.
Those in wait saw them, sprang out, instantly
cut off the cattle herds and lovely flocks

[1]A Greek measure of weight and (like the British pound sterling) of money; its value, though varying greatly, was high.

of glistening white sheep, then killed the shepherds.
But when the armies seated in assembly 530
heard the loud din of cattle, they at once
mounted swift horses, charged, and overtook them.
They stood and battled by the river banks
hurling brass-pointed spears at one another.
Strife and Tumult were with them; baleful Death 535
clutched one man newly injured, one unwounded,
and dragged a corpse through slaughter by the foot,
her garment dripping red with human blood.
These specters joined in battle like the living,
pulling dead men away from one another. 540
On it he put a soft and fertile plowland
broad and thrice tilled, where many plowmen wheeled
their teams about and drove them back and forward
till, when they reached the boundary of the plowland,
a man arrived and handed them a flagon 545
of sweet wine, and they turned again to plowing,
eager to reach the deep field's boundary line.
Behind them earth resembled blackened furrows
though made of gold, so wondrously he forged it.
On it he put a king's domain, where reapers 550
were mowing, clutching sickles in their hands.
Some swaths fell densely to the ground in windrows,
others, sheaf-binders tied with ropes of straw.
Three binders stood nearby; behind them, children
gathered and carried sheaves, incessantly 555
passing them onward, while the king, in silence,
holding his staff, stood next to them rejoicing.
Heralds beneath a far-off oak, preparing
a banquet, cleaned a slaughtered ox while women
scattered white barley for the men to dine on. 560
On it he put a lovely golden vineyard
bursting with clustered grapes of darker color
set off by silver poles thrust in among them.
Round it he drew a ditch of dark blue steel,
then a tin fence: one path alone gave access 565
by which men came to gather in the vintage.
Young men and women innocent of heart
carried away sweet fruit in woven baskets,
and in their midst a boy melodiously
plucked his sharp lyre and sang the Linus-song[2] 570
in lilting tones, while all together danced
singing and shouting as they skipped behind him.
On it he made a herd of straight-horned oxen.

[2]The "Linus" (Greek Linos), to which this is the first extant reference, was usually a lament sung for a dead god or hero, perhaps originally a god of vegetation (here of the vintage) who died and was reborn each year. In a later myth, Linus was killed by Apollo for boasting he could sing as well as the god.

Fashioned of gold and tin, the lowing cattle
hastened forth from their dungy stall to pasture 575
beside a roaring river's quivering reeds.
Herdsmen of gold paraded with the oxen,
four of them, followed by nine speedy dogs.
Among the foremost oxen two fierce lions
held a loud bellowing bull and dragged him off 580
lowing in pain as dogs and men pursued him.
The lions, ripping up the thick oxhide,
gorged themselves on black blood and guts, as herdsmen
frantically urged their swift dogs on against them.
The dogs, afraid of biting at the lions, 585
bayed at close quarters, keeping clear of harm.
Next the famed limping blacksmith made a spacious
pasture for white sheep in a lovely valley,
and resting places: covered huts and sheepfolds.
Next the famed limping smith elaborated 590
a dance floor such as Daedalus[3] once made
in Knossos for long-braided Ariadne.
There youths and eagerly desired young maidens
danced holding one another by the wrist.
The girls wore fine-spun linen gowns, the boys 595
thin tunics brushed with glistening olive oil;
the girls were crowned with lovely wreaths, the boys
wore golden daggers from their silver belts.
Sometimes with understanding feet they ran
lightly, as when a potter sits and twirls 600
the running wheel held firmly in his palms;
sometimes they ran in rows to one another.
Around the lovely dance floor stood a crowd 603
delighting; in their midst two acrobats[4] 605
struck up the song and nimbly whirled among them.
Lastly he put the mighty river Ocean
around the well-wrought shield's encircling rim.

The Orchard of Alcinous
Odyssey *VII*

Outside the courtyard near the gates, an orchard
four acres large was girded by a fence.
There tall trees grow luxuriantly, pears

[3]Daedalus (Daídalos), an Athenian exile, was the fabled craftsman of King Minos of Knossos, in Crete; among his creations, besides the dancing floor, were the cow into which Queen Pasíphaë climbed to be impregnated by a bull, and the labyrinth in which the child of that illicit union, the Minotaur, was imprisoned until slain by Thêseus of Athens with the help of Minos's daughter Ariádnê, whom he later abandoned. Daedalus escaped from Knossos by making wax wings for himself and his son Ícarus (Ikaros), who was killed when he flew too near the sun and fell into the sea.
[4]Here, as elsewhere in these translations, irregular numbering indicates omission of lines considered spurious in the original.

and pomegranates, apples bright with fruit, 115
delicious figs, and flowering olive trees.
Their fruit can never fail nor ever perish,
winter or summer, all year long, for always
Zephyr[5] blows, nurturing some and ripening others.
Apple on apple, pear on pear matures, 120
fig upon fig, and grape on clustered grape.
A vineyard rich in fruit is rooted there:
part is spread out on level ground to dry
in sunlight, part is harvested at vintage,
and part they trample. Unripe grapes in front 125
shed flowers while the others slowly darken.
By the last row, well-ordered garden plots
grow bountiful fresh greens of every kind.
Two springs flow, one dispersed throughout the garden,
the other gushing from the courtyard, near 130
the towering house where townsfolk come for water.
These gifts the gods gave to Alcinous' line.
Long-suffering Odysseus stood admiring.
But when his heart had done with admiration
he swiftly crossed the threshold to the house. 135

The Cave of the Cyclopes
Odyssey *IX*

We reached the land of arrogantly lawless
Cyclopes, who rely upon the gods,
never plant seeds by hand, and never plow.
Everything grows unplanted and unplowed:
barley and wheat and richly clustered vines 110
of wine grapes fertilized by Zeus's rains.
They make no plans or laws in marketplaces
but dwell upon the peaks of lofty mountains
in hollow caves, and each dispenses laws
for wives and children, unconcerned for others. 115

Chaos, Cosmos, and Farm

Hesiod on Strife and Work, translated by Robert M. Torrance

Hesiod (Hêsíodos), the first poet of Greece, and perhaps of the world, who emerges as a distinct personality, is generally thought to have composed his poems somewhat later than Homer, probably in the late eighth to mid seventh century B.C.*, in Ascra, a village of Boeotia (Boiôtia) in*

[5] The mild West Wind.

mainland Greece. Concerning his life, he tells us (in H. G. Evelyn-White's translation of Works and Days *633–40) that "your father and mine, foolish Perses, used to sail on shipboard because he lacked sufficient livelihood. And one day he came to this very place crossing over a great stretch of sea; he left Aeolian Cyme [Kymê, on the coast of Asia Minor] and fled, not from riches and substance, but from wretched poverty which Zeus lays upon men, and he settled near Helicon in a miserable hamlet, Ascra, which is bad in winter, sultry in summer, and good at no time." In another passage (650–59), he says he has never "sailed by ship over the wide sea, but only to Euboea from Aulis . . . Then I crossed over to Chalcis, to the games of wise Amphidamas where the sons of the great-hearted hero proclaimed and appointed prizes. And there I boast that I gained the victory with a song and carried off a handled tripod which I dedicated to the Muses of Helicon, in the place where they first set me in the way of clear song." From this episode arose the legend of a song contest between Homer and Hesiod on the island of Dêlos, birthplace of Apollo. In another legend, Hesiod fled to Oenoë in Locris to escape an oracle and was killed by his hosts for seducing their sister, after which his body, cast out to sea, was brought to shore by dolphins.*

A number of apocryphal poems were attributed to Hesiod (as to Homer), but only three works survive in reasonably complete form, the Shield of Heracles *(an imitation of the Homeric shield of Achilles), the* Theógony, *and the* Works and Days. *Whether or not the latter two poems are by the same author,*[6] *both are conventionally known as Hesiod's; they employ many of the oral formulas familiar from Homer, but since the subject matter of the* Works and Days *in particular is by no means traditional, it is possible that Hesiod wrote them down or dictated them to a scribe. The* Theogony *or* Generation of the Gods *recounts the genealogy and deeds of the gods; it remains one of our primary sources for knowledge of Greek mythology. Among the most striking characteristics of its account of the origin of Earth and the universe from the primordial yawning gap of Chaos is the absence—in contrast to origin myths of many other cultures—of any deliberate act of creation. Its emphasis, instead, is on spontaneous process or generation, with or without sexual union, as expressed by forms of* genesthai *("become," "come-to-be," "be born," "originate," "arise," "bring forth," "give birth to"), the verb from whose root is derived the noun* genesis, *"generation" or "coming-to-be"; for this reason Aristotle could legitimately view Hesiod as a predecessor of the* physikoi, *or natural philosophers.*

The Works and Days, *in contrast, is a very personal poem relating the poet's struggles with his indolent brother Perses over the land their father bequeathed them, and with the hard Boeotian soil that only grudgingly yields a meager livelihood to the hard-working farmer. Strife, both of the good and the bad kind, is as central to this world as it will be to that of Heraclitus or Empedocles. Much of the early part of the poem (the "Works") consists of practical and moral apophthegms (trust no one, especially a woman!) and exhortations ("work, work, work!") expressing*

[6]Prevalent scholarly opinion considers the *Theogony* the earlier poem both on tenuous linguistic grounds and because its poetic style—often descending to mere catalogues—seems far less accomplished than that of the *Works and Days*. At the beginning of the *Theogony* the poet invokes the Muses of Mount Helicon (Hélikôn) and declares, in Evelyn-White's translation of lines 22–24, that "one day they taught Hesiod glorious song while he was shepherding his lambs under holy Helicon, and this word first the goddesses said to me . . ." If these lines are assumed to be authentic, and "Hesiod" and "me" to be the same person, it is indeed hard to believe that the ruder *Theogony* was not the earlier work. But if, as Evelyn-White reads these lines, the poet is here recounting "his own inspiration by the same Muses who once taught Hesiod glorious song," it might follow that "the *Works and Days* is the oldest, as it is the most original, of the Hesiodic poems," and that the *Theogony* was written by a later and lesser admirer of Hesiod. A passage in the *Guide to Greece* (IX.31) by Pausanias in the second century A.D. (in Peter Levi's translation [1971; rev. 1979]) supports this view: "The Boiotians living around Helikon hand down a tradition that Hesiod wrote nothing but the *Works and Days;* and even from that they take away the 'Prelude to the Muses,' saying that the lines about Strife are the beginning of the poem. They showed me at the spring [of Hippokrênê] a sheet of lead which time had injured, with the *Works and Days* engraved on it."

the peasant's decidedly unromantic view of life, and of myths (Promêtheus, Pandôra, the five successive "races" of man) told to illustrate a moral; the latter portion (the "Days") is a kind of farmer's almanac of the seasons and the work appropriate to each. Out of this unique mixture of disparate components emerges an astonishingly vivid and realistic picture of the poet-peasant Hesiod himself, of the hard-scrabble life of subsistence farming in his place and time (and most other places and times), and of the encompassing world of crops and animals, sun and circling stars with which the peasant's life is inseparably connected. The selections included here from the Works and Days *comprise about 470 of the poem's roughly 825 lines, translated from H. G. Evelyn-White's Loeb Classical Library text of* Hesiod, The Homeric Hymns, and Homerica *(1914; rev. 1936), with occasional use of the text and commentary of M. L. West (1968).*

First of All, Chaos
Theogony

Daughters of Zeus, hail! Grant enchanting song
to praise the holy race of everlasting 105
immortals, born of Earth and starry Sky
and murky Night, and nursed by salt Sea waters.
Tell me how first the gods and Earth and rivers
were born, the boundless wildly tossing sea,
wide heaven high above, and sparkling stars; 110
how gods born from them, fountainheads of all
good gifts, divided up their wealth and honors
and occupied the heights of steep Olympus.
From the beginning tell me this, Olympian
Muses, recounting all that came to be. 115
First of all, Chaos came to be, and then
wide-bosomed Earth, foundation of all things, 117
dim Tártarus, deep beneath much-traveled earth, 119
and Love, the fairest of immortal gods, 120
who slackens joints, demolishing wise counsel
and reason in the hearts of gods and mortals.
From Chaos, Érebus and black Night arose;
Aether and Day from Night, who intermingled
in love with Erebus and gave them birth. 125
Earth first brought forth her equal, starry Sky,
to fold himself around her everywhere
and be the blessèd gods' abode forever;
then brought forth mountain chains to be the haunts
of lovely nymphs inhabiting their valleys. 130
Without sweet love's delights she bore the fruitless
rabidly foaming Sea: then went to bed
with Sky and soon bore deeply swirling Ocean,
Coeus, Crius, Hypérion, Iápetus,
Theia, Rhea, Themis, Mnemósyne, 135
Phoebe the golden crowned, and lovely Tethys.
Last, crafty Kronos, youngest child and most
dreadful, was born: he loathed his thriving father!

A Race of Iron
Works and Days

Pierian Muses, come, and celebrate
your father Zeus in high rhapsodic song!
Through him are mortals spoken or unspoken,
famed or unfamed, as mighty Zeus decrees.
Easily he exalts, and then brings low, 5
humbles the great and magnifies the little,
straightens the bent and shrivels up the proud—
high-thundering Zeus, inhabitant of heaven.
Seeing and hearing, heed me, passing judgment
rightly: for, Perses, I would tell the truth! 10

More than one breed of Strife was on the earth:
two are. One wins the praise of all who know her,
the other, blame: each is unlike the other.
The first arouses evil war and cruel
conflict: no one may love, but grudgingly 15
heeding the gods, all pay grim Strife due honor.
The other one is black Night's eldest daughter
whom Kronos' son Zeus, dwelling in the aether,
rooted in earth and made more kind to mortals.
Even the idler, whom she stirs to work, 20
eagerly works when he beholds another's
riches, how zealously he plows and plants
and sets up house. Thus neighbor vies with neighbor,
striving for wealth: *this* Strife is good for mortals.
Potter begrudges potter, builder builder, 25
beggarman beggar, and each bard another.

Perses, store up these matters in your heart.
Let ruinous Strife not turn you from your work,
snooping around in marketplace disputes.
Disputes and marketplaces matter little 30
to one without abundant livelihood
in store, the gift of earth, Demeter's grain.
Surfeit yourself on that before disputing
over another's goods: you'll never get
a second chance to do so! Let us settle 35
our quarrel with true justice, Zeus's blessing.
We had divided our estate already
when you embezzled most of it, cajoling
gift-guzzling magistrates, litigious fools
ignorant how much half exceeds the whole 40
or how much asphodel and mallow[7] nourish.

The gods conceal their livelihood from mortals.
Otherwise one day's work would easily

[7]The humble meal of poor farmers.

meet a year's needs, and leave you free of work
to hang your steering oar above the fireplace 45
while fields, once plowed by ox and mule, lie wasted.
But Zeus concealed it, angered in his heart
because Prometheus craftily deceived him:
therefore he planned dire misery for mortals.
He hid fire; but Iápetus' brave son[8] 50
stole it again for mortals from wise Zeus
the Thunderer, in a hollow fennel stalk.
Angrily cloud-assembling Zeus said to him:
"Son of Iapetus, passing all in cunning:
stealing fire and defrauding me delights you, 55
ruining both yourself and men to come!
To them, in recompense for fire, I'll send
an evil they will cherish—to their sorrow."
Thus the father of men and gods spoke, laughing.
He bade renowned Hephaestus quickly mingle 60
water and earth, endowed with human voice
and strength, then mold a lovely maiden's form
like an immortal goddess; bade Athena
teach her to spin and weave elaborate webs
while golden Aphrodite shed upon her 65
troubling allure and enervating cares;
then bade the great Guide Hermes Argeïphóntes[9]
give her a doglike mind and thievish nature.
He spoke, and they obeyed Zeus son of Kronos.
At once the famous cripple molded clay 70
to look, as Zeus willed, like a modest maiden;
grey-eyed Athena girded and adorned her;
queenly Persuasion and the immortal Graces
draped golden necklaces; and richly braided
Hours crowned her with garlands of spring flowers. 75
Guiding Argeïphontes then endowed her 77
with lies, deceitful wiles, a thievish nature
as thundering Zeus willed; then the godly herald
injected her with speech and named this woman 80
Pandora,[10] since all gods upon Olympus
gave her a gift to plague hard-working men.

[8]Promêtheus ("Forethought"), along with his brothers Atlas, Menoetius, and the "scatter-brained" Epimêtheus ("Afterthought"), was the son of Iápetus—one of the Titans born from the union of Earth (Gaia) and Sky (Ouranos)—and of Clýmene, daughter of Iapetus's brother Ocean (Okéanos). Some have speculated that the name Iapetus might be distantly related to that of the biblical Japhet, son of Noah. For another version of the stories of Prometheus and Pandora, see *Theogony* 507–616.

[9]Probably "Slayer of Argus," an epithet of Hermes.

[10]"Pandôra" means "all gifted." In some early versions of the myth she may have been an earth goddess, like Demeter; in others, the wife of Prometheus or Epimetheus and mother of Deucalion or Pyrrha, the first human beings; in one fragment attributed to Hesiod, she is Deucalion's daughter. In the somewhat confusing account of the *Works and Days*, Pandora is not, as she is often called, the first woman, since other mortals happily existed before her. See M. L. West's note on this line in his commentary.

This hopelessly ensnaring trick once finished,
father Zeus bade his runner Argeïphontes
swiftly convey the gift to Epimetheus 85
who, quite forgetting what Prometheus told him—
"Take no gift from Olympian Zeus, but give it
back, lest some evil come to mortals"—took it,
and only later understood its evil.
Before this, tribes of humans lived on earth 90
free of misfortune and oppressive toil
and harsh diseases bringing death to men: 92
all these the woman, opening her jar,[11] 94
scattered abroad, contriving grief for mortals. 95
Hope alone stayed securely in her home
under the rim inside the jar, not flying
out, but prevented by the lid from leaving. 98
Numberless plagues descend on humankind, 100
for earth and sea alike are full of evils;
diseases unremittingly befall
men day and night, inflicting hardship on them
in silence, for wise Zeus denied them speech.
The plan of Zeus can never be evaded. 105
If you wish, I shall tell another story
briefly and well—retain it in your heart!—
how men originally were like the gods.
First, the immortals dwelling on Olympus
fashioned a golden race of human beings 110
under the reign of Kronos, king of heaven.
They lived like gods, with hearts immune from care,
remote from toil and trouble, ignorant
of vile old age; with never-failing limbs
they danced and feasted, far from every evil, 115
dying as though subdued by sleep. All blessings
were theirs, for fertile fields, uncultivated,
brought forth abundant crops, and people lived
peacefully off the land, with untold bounty,
rich in flocks and beloved by all the gods. 120
But after earth had covered up this race—
daemons now[12] by the will of mighty Zeus,
kindly earth-spirits guarding mortal men 123
and bringing wealth, by sovereign right, to many— 126

[11]Probably a jar containing the "gifts" the gods gave Pandora in lines 81–82. Since the Renaissance, because of a mistaken translation by Erasmus, the jar has been known as a box; see Dora and Erwin Panofsky, *Pandora's Box: The Changing Aspects of a Mythical Symbol* (1956; rev. 1962).
[12]The Greek word *daimôn*, often roughly synonymous with *theos* (god), designated the divine power embodied in various supernatural beings, especially those associated with the earth, and more widely still the power of destiny sometimes taking the form of a personal guardian spirit attending a man or woman throughout life. In Christian times these and all other pagan divinities became identified as "demons" or devils.

those who dwell on Olympus then created
a second, far inferior silver race
unlike the gold in either mind or body.
Each child grew up a hundred years at home 130
foolishly playing by its mother's side,
but when it reached the prime of young adulthood
would live a short time only, through sheer folly
coming to grief, unable to refrain
from reckless violence, nor would they serve 135
the gods and sacrifice on holy altars
as every righteous nation must. Zeus then
angrily did away with them for failing
to pay due honor to the Olympian gods.
After earth covered up this race as well— 140
men call them blessèd spirits of the world
below, and give them secondary honors—
yet a third race of mortals father Zeus
made out of bronze, unequal to the silver,
from ash trees, terrible and strong. They cherished 145
Ares' lugubrious works and violent deeds
and ate no bread, hardhearted, half-formed men
of adamant. Invincibly strong hands
grew from their shoulders down their sturdy limbs.
Bronze were their weapons, bronze their houses, bronze 150
their implements; black iron was unknown.
Conquered by their own hands they soon departed
down to chill Hades' moldering abode,
leaving no name. Black death, fierce though they were,
captured them as bright sunlight dimmed behind them. 155
After earth covered up this race as well,
Kronos' son Zeus created yet a fourth
on the rich land, more noble and more righteous,
a godlike hero race called demigods
who dwelt upon the boundless earth before us. 160
Some perished in fierce wars and fearful battles
fighting in seven-gated Thebes, the city
of Cadmus, over Oedipus's flocks
or sailing on the sea's unfathomed depths
to Troy for Helen of the lovely hair. 165
There death enshrouded some in final darkness,
but Zeus gave others livelihoods and dwellings
apart from mortals at the ends of earth, 168
where, free from care and sorrow, they inhabit 170
isles of the blest beside deep-swirling Ocean:
fortunate heroes whom the bounteous soil
thrice a year blesses with delicious fruit.

Oh, that I had no part among the fifth race
but might have died first or not yet been born! 175
Now is a race of iron that never ceases
toiling and suffering by day, and dying
at nightfall; heavy cares the gods shall give them!
Yet even here some good shall temper evil:
Zeus shall destroy this race of mortal men 180
when newborn heads shall manifest grey hairs.
Fathers shall challenge children, children fathers,
guests betray hosts, and comrades one another;
brothers shall not, as in the past, be friends;
men shall dishonor swiftly aging parents 185
with harsh abusive words, hardheartedly
scorning the gods, and never recompensing
parents, now old, who reared them. Thinking right
is might, they shall despoil each other's cities.
The good and righteous man who keeps his oath 190
shall be dishonored, but the evildoer
be praised: might shall be right, and reverence
shall cease; the evil man shall harm his better,
bearing false witness, under oath, against him.
Envy, foul-mouthed, malevolent, and scowling, 195
shall walk alongside every wretched mortal.
Then Shame and Nemesis, their lovely forms
wrapped in white robes, shall spurn earth's spacious paths,
forsaking men to join the company
of gods high on Olympus, leaving mortals 200
harrowing pains and no defense from evil. . . .

Always remember my advice to you
and work, illustrious Perses, so that Hunger
may shun you, and augustly crowned Demeter 300
love you and fill your granary with plenty:
for Hunger is the lazy man's companion!
Mortals and gods alike disdain the idle
loafer, in character like stingless drones
that dissipate the work of honeybees, 305
eating scotfree. Set your affairs in order
and granaries will burst with ripe provisions.
By working, men abound in flocks and treasure
and win great favor from the immortal gods. 309
Work is no shame, but idleness is shameful. 311
If you work hard, the indolent will envy
your growing wealth, for wealth brings fame and glory.
Whatever life you lead, hard work is best,
so turn your fantasies from others' riches 315
and work to earn your living as I bid you.

Shame without good accompanies the pauper,
Shame, which both damages and profits men:
Shame attends poverty, assurance wealth. . . . 319

Let wages promised to a friend be plenty. 370
Smile on your brother—but secure a witness,
for trust and distrust ruin men alike.
No slinky woman should deceive your judgment,
wheedling and coaxing while she eyes your barn.
Whoever heeds a woman heeds a fraud! 375
Let there be one son only to replenish
his father's house and cultivate the homestead:
plan to die old, should you beget a second!
Yet Zeus might easily lavish wealth on more,
since more, with greater cares, bring greater bounty. 380
Thus, if your heart yearns inwardly for riches,
listen to what I say, and work, work, work!
When once the Pleiades, Atlas' daughters, rise,[13]
begin your harvest; when they set, your plowing.
Hidden for forty days and forty nights, 385
they once again, with each revolving year,
shine forth when sickle-honing season comes.
This is the law for dwellers in the plains
or near the sea, or in lush mountain glens
far from the tossing waves, where fertile soil 390
abounds: strip bare the land to sow, strip bare
to plow, strip bare to reap, if you intend
to glean Demeter's seasonable fruits
and see them grow—lest afterwards you beg
at other people's doors, receiving nothing, 395
as you have done from me: but I shall give you
no more, mete out no more! O foolish Perses,
work at the work the gods ordained for man,
lest, with your wife and children, plunged in sorrow,
you scrabble after food from hardened neighbors. 400
Twice or three times you may succeed: but vex them
further, and pleas will be of no avail,
your eloquence unheeded. So I bid you,
plan to repay your debts and banish hunger!
First, get a house; an ox to plow; a woman— 405
a slave-girl, not a wife—to drive your oxen;
then put your household goods in decent order
and never beg from others, emptyhanded
as seasons pass and labors come to nothing.
Leave no task till tomorrow or beyond it. 410

[13]In early May, in Hesiod's time, beginning to set in late October or early November. For roughly forty nights between late March and early May these seven stars in the constellation Taurus, important signals of planting and harvest time in much of the ancient world, were beneath the horizon.

Procrastinators never fill their barns,
nor lazybones. Work thrives on diligence,
but ruin overpowers good-for-nothings.
When the strong piercing sun's oppressive heat
begins to slacken, and almighty Zeus 415
sends down autumnal rains, and human skin
feels lighter—for the dog-star Sirius then[14]
passes above the heads of wretched mortals
briefly by day, but longer every night—
trees chopped when leaves are falling down, and buds 420
done sprouting, will be least devoured by worms.
Remember, this is timber-cutting season.
Carve out a three-foot mortar, then a pestle
three cubits long; an axle of seven feet
does best: if eight feet long, lop off a mallet. 425
Hew two-foot wheels for a wagon three feet long.[15]
Cut many bent poles, searching mountainsides
and fields to find a holm oak for a plow tree,
and bring it home—the strongest kind for oxen
to plow with, once Athena's workman pegs it 430
fast to the stock and joins it to the pole.
Best make two plows, one from a single piece,
one jointed, manufacturing both at home:
if one breaks, yoke your oxen to the other.
Laurel and elm poles, stocks of oak, and holm-oak 435
plow trees are least worm-eaten. Get two oxen,
bulls nine years old, whose energetic strength
is undiminished; these are best for working.
They will not fight amid the furrows, smashing
the plow and bringing all your work to nothing. 440
A sturdy man of forty who has eaten
double a quartered loaf should walk behind them,
minding his job and driving straight his furrow,
too old to gape at comrades, caring only
for work. No younger man will scatter seed 445
better, or be more careful not to waste it:
more youthful men find comrades more distracting.
Listen intently for the crane's shrill voice[16]
crying above you from the clouds each year:
she signals plowing time and marks the season 450
of winter rains; men without oxen dread her.

[14]In September or October. In summer Sirius, the brightest fixed star in the night sky (in the constellation Canis Major), shines more briefly at night and was thought to add its intensity to the searing heat of the sun during the day. It was sometimes called the "dog star" in Greek, as in English, though Hesiod here writes simply "the star Sirius," or "the Scorching star," which some later commentators identified with the sun itself.
[15]Here as elsewhere the measurements, proportions, and precise meanings of some terms are very obscure in the Greek; the English translation ranges from an approximation to a wild guess.
[16]Around mid-November.

Then fatten up horned oxen in their stalls.
Easy to say: "Yoke oxen to my wagon!"
Easy to hear: "They've other work to do!"
The fancied rich man thinks his wagon built, 455
fool! not to know one wagon needs a hundred
planks: so take care to lay them up beforehand.
As soon as plowing time arrives for mortals,
rouse yourself, with your slaves, and hasten forth
to plow in season, be it wet or dry, 460
striving each dawn to make your fields abundant.
Plow in the spring; heaped furrows will reward you
in summer. Sow when fallow soil turns light:
fallow protects from dearth and safeguards children.
Pray, when you grasp the handle of the plow 465
for the first time and strike the backs of oxen
straining against the yoke, that Chthonian Zeus[17]
and chaste Demeter bless with ripened bounty
Demeter's sacred grain. Then bid a slave-boy
follow, and with his mattock vex the birds 470
by burying seeds. Good management is best
for mortal men; bad management worst of all!
Thus ripened ears will bow with heaviness,
provided Zeus looks favorably upon them,
and you will sweep your granaries of cobwebs, 475
reveling in the livelihood you've garnered,
and reach bright spring in comfort, never looking
to others: they will stand in need of you!
If you delay your plowing to the solstice[18]
you'll gather skimpy handfuls squatting down, 480
dusty and joyless as you wrap your sheaves
and bring one basket home: few will applaud you!
Yet aegis-bearing Zeus's mind may change,
and mortals cannot easily discern it.
Though you plow late, this remedy may save you: 485
when first the cuckoo from the oak cries "cuckoo!"[19]
gladdening men throughout the boundless earth,
if on the third day Zeus sends rainfall, rising
up to an ox's hoof but not above it,
late plowers may yet overtake the early. 490
Keep this in mind, and never fail to notice
bright spring's arrival, nor the rainy season's.
Pass by the smithy's warm receptacle
in wintertime, when cold keeps men from working
outside (though busy men improve their homes), 495
lest cruel winter's penury reduce you
to rubbing swollen feet with shriveled hands.

[17]Zeus in his aspect of god of the earth. [18]The winter solstice, in December. [19]In March.

The loiterer, who makes his living only
on empty hopes, fills up his heart with evil.
No good hope can accompany a pauper 500
who dawdles when his livelihood's uncertain.
  While it is still midsummer, tell your slaves,
"Summer won't last forever: build the barns!"
Avoid Lenaeon's[20] evil days—fit only
for skinning oxen—and the murderous frosts 505
that come when Boreas howls across the land,
whipping the sea to frenzy as he crosses
horse-breeding Thrace, while earth and forest moan.
On many a towering oak and massive pine
he falls, in mountain glens, and hurls them crashing 510
to bounteous earth while all the vast forest roars.
Beasts, covering their privates with their tails,
shiver, though thickly furred: for Boreas' cold
penetrates even the shaggiest among them,
blows through the ox's hide, which cannot bar him, 515
and through the long-haired goat's. The woolly sheep's
thick fleece alone his icy blast can never
pierce, though it bends old people into hoops.
Nor does it pierce the soft skin of a girl
who keeps her mother company indoors, 520
innocent still of golden Aphrodite,
and, having washed her smooth skin and anointed
herself with oil, lies down inside her bedroom
in winter, when the boneless one[21] is munching
his foot deep in his dank and fireless home. 525
For the sun shows no pasture to repair to,
but, circling over dark-skinned peoples' cities
and lands, shines tardily on all Hellenes.
Then horned and unhorned forest dwellers flee,
teeth chattering wretchedly, through wooded groves, 530
driven by one thought only—how to find
shelter in some entangled undergrowth
or rocky cave as, like a three-legged person
with broken back and bowed head looking earthward,
they wander seeking refuge from white snow. 535
  Shield your skin then by wearing, as I bid you,
an ankle-grazing tunic underneath
a soft cloak, weaving thick weft on thin warp.
Wrap these around you, and your hair will stay
in place, not stand up over all your body. 540
Onto your feet secure tight-fitting boots
made from a slaughtered ox and lined with felt.

[20]A Greek month corresponding to late January and early February. Boreas, in the following lines, is the North Wind.
[21]The octopus.

And when the frosty season comes, stitch skin
of newborn kids with ox-thongs, to protect
your back from rain, then pull down on your head 545
a conical felt cap to keep your ears dry.
For dawn is chill once Boreas plummets down
and mist spreads from the starry sky at dawn
to fertilize the fields of blessèd mortals.
Sucked up from rivers flowing on forever 550
and lifted skyward by tempestuous winds,
it sometimes turns to evening rain and sometimes
to wind, when Thracian Boreas whips dense clouds.
Get home in time, accomplishing your work
before dark clouds wrap everything around you, 555
drenching your skin and soaking through your clothes.
Avoid this month, hardest of all the year,
wintermost, hard on sheep and hard on men.
Feed oxen half, but give your man-slave more
provisions, for industrious nights are long. 560
Keep this in mind until the year has ended,
days equal nights in length, and Earth the mother
of all brings forth her varied fruits again.

When Zeus completes the sixtieth winter day
after the solstice, then the star Arcturus, 565
abandoning the sacred stream of Ocean,
first rises,[22] shining brilliantly at twilight.
Pandíon's sharp-voiced child, the swallow, follows,
arriving just as spring begins to flower.
Prune vines before she comes; this way is best. 570

When the house-carrier[23] flees the Pleiades
and climbs your plants, no longer dig your vineyard
but sharpen sickles and rouse up your slaves.
Flee restful shade and do not sleep till daylight
at harvest-time when sun dries out your skin. 575
Get busy then, and bring the harvest home,
rising at dawn to make your living sure.
For dawn takes up a third of human work;
dawn benefits both travelers and workers:
dawn, whose appearance rouses many men 580
to journey forth, and yokes up many oxen.

Cicadas, when the artichoke is blooming,[24]
pour shrill song down continually from the trees,
scraping their wings in weary summertime.
Then goats are fattest and the vintage best, 585

[22]From February to March.
[23]The snail, around mid-May.
[24]In June. The cicada (Greek *tettix*, sometimes translated as grasshopper, or cricket) will reappear frequently in Greek literature; see the selections from Alcaeus below, and from Plato's *Phaedrus* in Chapter 9, and Theocritus, Meleager, *Daphnis and Chloe*, and the anonymous "Godlike Cicada" in Chapter 11.

women most lecherous, but men most feeble,
for Sirius enervates both head and knees,
baking the skin bone-dry. So in that season
give me a shady rock, a wine of Biblis,
a clotted curd-cake, milk from goats drained empty, 590
meat from a grazing cow that never calved
and newborn kids; then let me, cooled by shade
and satisfied by food, drink shimmering wine
and, turning round to greet refreshing Zephyr,
thrice from perpetual springs pour forth libations 595
of sparkling water—and a fourth of wine.
Exhort your slaves, when strong Oríon first
rises, to thresh Demeter's sacred grain
on a well-rounded, breezy threshing-floor.
Measure and store it well in jars, and after 600
carefully stacking all your household goods,
hire a man with no home, and find a childless
serving girl: one with children causes trouble.
Feed your sharp-toothed dog well, so men who sleep
by day will never steal away your goods. 605
Provide yourself with ample stores of fodder
and straw for mules and oxen. Then, at last,
let weary slaves rest and unyoke your oxen.
When Sirius and Oríon reach mid-heaven
and rosy-fingered Dawn beholds Arcturus, 610
Perses, bring home the clustered grapes you've gathered.
Ten days expose them to the sun; five more
cover them; on the sixth, press into vessels
jubilant Dionysus' gifts. And then,
when Pleiades, Hyades, and strong Oríon 615
begin to set,[25] remember plowing season.
Thus a full year drops down beneath the earth . . . .

All-Mother Earth and Her Children

The "Homeric Hymns," translated by Robert M. Torrance

The "Homeric Hymns" are a collection of thirty-three or thirty-four invocations of various gods, ranging in length from four or five to 580 lines (the "Hymn to Hermes"). Composed in the dactylic hexameter meter of Homer and Hesiod, with frequent oral formulas, they were ascribed to Homer but appear to have been the creation of anonymous poets over several centuries. The first extant reference, in the History *(III.104) of Thucydides (Thoukydidês), toward the end of the fifth century* B.C., *calls the Hymn to (Delian) Apollo a* prooimion, *or "prelude"; some were*

[25]In late October and early November.

possibly intended as preliminaries to recitations from the Homeric epics. The portrait of the gods in some hymns is significantly different, however, from Homer's. If no major god can be simply reduced to a personified natural force, none, on the other hand, is wholly lacking in this dimension. Whatever their other functions, Zeus remains a god of thunder and lightning, Poseidon of tempest and earthquake, Hermes of phallic worship, and Artemis the mistress of wild animals; even the deities most symbolic of Hellenic culture, Apollo (a god of disease no less than of song) and Athena (a goddess of war no less than of reason), continued to be associated, respectively, with wolves and with snakes. But whereas in Homer's humanized Olympians such dimensions are often vestigial, the gods of some Homeric Hymns more overtly express forces prominent both in primitive "nature worship" and in cults that later developed into the Mystery religions of the Greek city-states.

This conjunction of old and new is especially evident in the ***Hymn to Dêmêtêr****, probably composed in the late seventh to early sixth century. Demeter, whose name the Greeks interpreted to mean "Earth Mother," was an ancient goddess of the grain and the fruits of the earth, worshipped since Mycenaean times. The myth of her frantic search for her daughter Korê ("the Maiden"), or Persephonê, raped by Hadês (Aïdôneus), or Pluto (Ploutôn), god of the underworld, and restored only after Demeter withholds the grain on which life depends, parallels seasonal fertility myths of the Sumerians, Canaanites, Hittites, and other peoples of Mesopotamia, Syria, and Anatolia. As told in the Homeric Hymn (our first extant account), however, this ancient nature myth is specifically connected with foundation of the Mysteries of Eleusis outside Athens and erection of the temple of Demeter in which its rituals took place. (The first three-fifths of the poem, omitted here, recount not only the rape of Persephone and her mother's grief but also Demeter's arrival at King Celeus's city of Eleusis in the guise of an old woman, and her anger when her effort to make the young prince Dêmophoôn immortal by placing him in the fire each night is disrupted by his dismayed mother Metaneira, whom Demeter, revealing her true form, denounces for her ill-timed intrusion.) Helene P. Foley, in her interpretive edition of* The Homeric Hymn to Demeter *(1994), remarks that "the Hymn puts female experience at the center of the narrative by giving the privileged place to the point of view of the divine mother and daughter in their shared catastrophe," and notes that the female quest "is defined by issues relating to marriage and fertility," not (as in the male quest) to war and kingship; by withholding her life-giving power, Demeter wins new honors* (timai) *for herself and her daughter in the world above as in the world below. The female experience is closely related, indeed, to the nature myth: "Demeter, as a goddess of the grain, and Korê, who is often associated in myth [though not in this Hymn] with the grain itself, have, like mortal women in Greek cult, a special symbolic relation to and power over nature. The promise of happiness in the Mysteries was almost certainly linked with the natural cycle itself, with its endless and necessary alternations between procreation and death." Whatever unspeakable secret was revealed at the culmination of the Mysteries surely entailed (as in other agricultural festivals of Demeter and Dionysus) some promise of a human counterpart to the annual rebirth of the grain.*

In the ***Hymn to Dionysus****, possibly dating from the seventh century, the ancient Greek god of the vine, whose worship had been renewed by ecstatic cults thought to have been imported from Phrygia in Asia Minor, eerily demonstrates his awesome powers of fertility and transformation—and the danger of denying them. The* ***Hymn to Pan****, probably from the late sixth to mid fifth century, celebrates the goat-footed god whose name (perhaps derived from a root meaning "shepherd") the Greeks identified with their word for "all"; no "nature god" could be more inclusive. Here too was a god both old and new. In his* Histories *(II.145), Herodotus calls Pan, with Dionysus and Heracles, the youngest of the gods, but adds that in Egypt (from which he thought all Greek gods came) he was one of the ancient "eight gods" who existed before the twelve Olym-*

pians. His worship, as described in Philippe Borgeaud's The Cult of Pan in Ancient Greece *(1979), spread from the rugged mountains of Arcadia to Attica and other parts of Greece, where he was venerated not in temples but in caves, one of which was located on the Acropolis of Athens. Like his father Hermes, he was phallic god, patron of shepherds, and master of wild animals. In him, as in nature itself, god and beast merge, and he brings both joyous laughter, as at the end of this Hymn, and the irrational fear we still call "panic." Finally, the brief* ***Hymn to Earth Mother of All****, of uncertain date, extols the "primeval nurturer of all that lives," the bounteous source from which nature and the gods alike derive.*

The Hymns in this selection are translated from H. G. Evelyn-White's Loeb Classical Library text, with reference to T. W. Allen, W. R. Halliday, and E. E. Sikes, eds., The Homeric Hymns *(2nd ed., 1936), and N. J. Richardson, ed.,* The Homeric Hymn to Demeter *(1974). For a complete translation, see Thelma Sargent,* The Homeric Hymns *(1973).*

FROM *To Demeter*
Hymn 13

But golden-haired Demeter
sat far apart from all the blessèd gods,
consumed with yearning for her deep-girt daughter.
Throughout the bounteous earth she made the year 305
dreadful and doglike for mankind. No seed
sprang from the ground, for flower-crowned Demeter
concealed them; oxen plowed the fields in vain,
and thick white barley strewed the earth for nothing.
She would have utterly destroyed mankind 310
with cruel famine, robbing the Olympian
gods of their offerings and sacrifices
had Zeus's mind not apprehended all.
First he sent gold-winged Iris down to summon
fair-haired Demeter of alluring form. 315
He spoke, and she obeyed Zeus, son of cloud-wrapped
Kronos, and crossed mid-heaven with swift feet.
Reaching the fragrant city of Eleusis
she found dark-robed Demeter in her temple
and spoke, addressing her with wingèd words: 320
"Demeter, all-perceiving father Zeus
calls you to join the everlasting gods.
Come, let the word of Zeus not go unheeded."
She spoke imploringly—to no avail.
Once more the father sent the ever-blessèd 325
eternal gods all forth: one after one
they wooed her, offering many lovely gifts
and privileges among the immortal gods.
Yet none could make her mind forget her anger,
so stubbornly did she refuse their pleas. 330
For she had sworn to set foot on Olympus
never again, nor let the soil bear fruit,
until her eyes beheld her lovely daughter.

When Zeus, far-seeing thunderer, heard this,
he bade gold-wanded Argeïphontes[26] down 335
to Erebus, to urge with soothing words
that Hades send Persephone from darkness
to light among the gods, so that her mother,
seeing her, might relinquish bitter anger.
Hermes at once obeyed and left Olympus, 340
rushing down to the hidden world below.
He found its king at home inside his palace
seated in bed beside a reluctant wife
sullenly yearning for her far-off mother,
who hatched dire schemes against the blessèd gods. 345
Strong Argeïphontes stood nearby and said:
"Hades, dark-haired commander of the dead,
father Zeus bade me bring Persephone back
from Erebus to her kin, so that her mother,
seeing her, may relinquish strife and anger 350
against the immortals. She is direly scheming
to obliterate the feeble tribe of earthlings,
hiding seed underground, and wreck the honors
due to the gods, from whom fierce anger keeps her
apart, inhabiting her rocky stronghold, 355
Eleusis, deep inside her fragrant temple."
He spoke; Aïdoneus, lord of those below,
smiled darkly and obeyed king Zeus's bidding,
strongly exhorting wise Persephone:
"Return, Persephone, to your dark-robed mother 360
with gentleness and kindness in your heart,
and do not feel forlorn beyond all others.
Among the gods I am no ill-matched husband—
brother of father Zeus! You shall be queen
of everything that lives when you are with me, 365
honored most highly of the immortal gods.
Eternal punishment will come to all
who wrong you by withholding sacrifices
and rich gifts to appease your sovereign power."
He spoke, and proud Persephone sprang up 370
bursting with joy: but stealthily he gave her
delicious pomegranate seed to eat,
contriving that she would not spend her days
always with reverent dark-robed Demeter.
Aïdoneus, lord of multitudes, rigged up 375
his golden chariot to undying horses;
she mounted; then strong Argeïphontes seized
reins and whip in his hand and promptly drove
out from the hall with gladly flying horses.

[26]An epithet, probably meaning "slayer of Argus," of Hermes, messenger of the gods.

Over long roads they sped, and neither sea 380
nor river waters, grassy vales nor mountains
checked the onrush of these immortal horses
cleaving deep air all round them in their flight.
He stopped where flower-crowned Demeter waited
before her fragrant temple: when she saw them 385
she ran like a maenad down thick-wooded mountains.[27]
Persephone, when she saw her mother coming,
sprang down and left the chariot and horses
and threw her arms around her mother's neck.
But as she held her darling child, Demeter 390
suddenly trembled with a dark foreboding
of trickery, ceased hugging her, and asked:
"Child, did you taste of any food down there?
Speak out, hide nothing, so we both may know.
If not, though you have been with hateful Hades, 395
you shall return to me and cloud-wrapped Kronos'
son Zeus and be revered by all the immortals.
But if you tasted, you shall dwell again
deep underground one third of every year,
spending two thirds with me and the other gods, 400
and when earth blooms with various fragrant flowers
in spring, you shall return again from misty
darkness, a miracle to gods and humans.
How did the mighty Host of Throngs deceive you?"
Lovely Persephone in turn replied: 405
"All that I tell you, mother, shall be true.
When lucky Hermes came, swift messenger
from father Zeus and the other heavenly gods,
bidding me leave dark Erebus so that you
might see me and relinquish strife and anger, 410
I leapt for joy: but stealthily he gave me
delicious food of pomegranate seed,
compelling me, against my will, to eat.
How he tore me away and bore me off
deep down below, through father Zeus' dark plan, 415
I shall relate, in answer to your question.
All of us, in one lovely meadow land—
Leucippe, Phaeno, Electra, and Ianthe,
Melite, Iache, Rhodeia, Callirhoë,
Tyche, Melobosis, flowerlike Ocyrhoë, 420
Chryseia, Ianeira, Acaste and Admete,
Rhodope, Pluto and beautiful Calypso,
Styx and Urania, lovely Galaxaure,
warlike Pallas, and Artemis lover of arrows—

[27] Translation of the next 15 lines or so is based on highly conjectural reconstruction of a manuscript full of major gaps. A maenad is a frenzied female worshipper of Dionysus.

were playing, as we gathered lovely flowers: 425
soft crocus mixed with hyacinth and iris,
rosebuds and lilies wondrous to behold,
narcissus covering the ground like crocus.
I plucked them joyously: then under me
earth gaped, out sprang the mighty Host of Throngs 430
and bore me against my will deep down below
in his golden chariot, shrieking loud and shrill.
All that I say is true, much though it pains me."

Thus all day long, with minds in unison
they comforted each other's hearts and spirits, 435
entwined in love, their spirits free of sorrow,
giving reciprocal joy to one another.
Brightly scarfed lady Hecate[28] then drew near
and warmly greeted chaste Demeter's daughter,
whose close companion she remained thereafter. 440

All-seeing thunderous Zeus sent thick-haired Rhea
to summon back dark-robed Demeter then
to join her heavenly kin, and promised any
honors she wished among the immortal gods,
ordaining that her daughter spend one third 445
of each revolving year in misty darkness,
two with her mother and the other gods.
He spoke: the goddess heeded Zeus's message,
hastening down from pinnacled Olympus
to Rharus, fertile nurturer of grain 450
once—now no longer fertile, but abandoned,
leafless and bare, since shapely-legged Demeter
hid its white barley underground—but soon
destined to sprout ears tumbling down like hair
in bounteous springtime, burdening rich furrows 455
with grain heaped on the ground or sheaved in bundles.
There she first lighted from the barren aether,
and joyously did each behold the other.
Then bright-scarfed Rhea thus addressed Demeter:

"Come, child: all-seeing Zeus the thunderer 460
summons you to your heavenly kin and pledges
honor among the immortals shall be yours,
ordaining that your daughter spend a third
of each revolving year in misty darkness,
two thirds with you and all the other gods. 465
This he ordains shall be, and nods approval.
Come, child, obey: do not be unrelenting
in anger at dark-clouded Kronos' son,
but make crops bountiful again for mortals."

[28]Hékatê, a mysterious goddess of the underworld and of crossroads, often associated with witches and with ghosts.

She spoke, and flower-crowned Demeter heeded, 470
instantly made rich soil abound with crops,
loaded the wide earth down with leaves and flowers,
then went and showed the kingly magistrates—
Triptólemus and Diocles the horseman,
strong Eumolpus and Celeus, mighty leader— 475
how to conduct her sacred rites, and told them—
Triptólemus, Polyxeinus, and Diocles—
awesome mysteries none may violate
or speak: for reverence of the gods forbids it.
Happy the human who beholds these things! 480
But one who is uninitiate will never
share in their blessings down in misty darkness.
After the dazzling goddess taught them all,
they went to Olympus, where the gods were gathered;
there beside thunder-hurling Zeus they dwell, 485
awesomely venerable. Truly blessed
of humans whom they generously love!
Soon to their homes will come a welcome guest,
Plutus,[29] who brings abundant wealth to mortals.
Come, mistress of Eleusis' fragrant land, 490
of sea-girt Paros and of rocky Antron,
giver of gifts and seasons, Deo, queen,
you and your lovely child Persephone:
kindly reward my song with pleasing bounty
and I shall remember you in song again. 495

To Dionysus
Hymn 7

Of Dionysus, far-famed Sémele's son,
listen how once near jutting promontories
beside the barren sea he walked, resembling
a flowering youth, his long hair floating darkly
downward, a purple robe flung round his sturdy 5
shoulders. Soon afterward, Tyrsenian pirates
sailing their stout ship on the wine-dark sea
came swiftly, led by evil fate. They saw him,
signaled each other, leapt out, seized upon him,
and put him jubilantly aboard their ship— 10
thinking he was the son of kings descended
from Zeus—and sought to bind him with strong vines,
which could not hold him. As they dropped away
from hands and feet he sat impassively

[29]Ploutos, the god of wealth; not to be confused with Pluto (Ploutôn), the god of the underworld, nor of course with Pluto (Ploutô, feminine), Persephone's companion in line 422 above.

smiling with dark eyes. Then at once the helmsman, 15
seeing this, cried aloud to his companions:
"Madmen! What mighty god have you attempted
to capture? No stout ship can ever hold him!
He must be Zeus, or silver-bowed Apollo,
or else Poseidon: like no mortal man 20
he looks, but like the gods upon Olympus.
Come, set him free at once on the dark shore
and lay no hand upon him lest, in anger,
he stir wild winds up to tempestuous frenzy."
Disdainfully the captain answered him: 25
"Madman, observe the favoring breeze, hoist sail,
haul on the sheets: leave *him* for men to handle!
He's bound for Egypt, I suppose, or Cyprus,
the Hyperboreans,[30] or beyond. At length
he'll tell us all about his friends, possessions, 30
and brothers, since fate puts him in our hands."
Speaking thus, up he hoisted mast and sail.
Wind bellied out the sail, the crew hauled sheets
taut: and soon marvelous deeds were done among them.
First sweetly fragrant wine ran murmuring through 35
the swift black ship, ambrosial fumes arose,
astonishing the wonder-stricken sailors.
Then all at once a vine spread from the sail-top
this way and that, thick-clustered grapes hung downward
and round the mast dark ivy tendrils twisted, 40
bursting with flowers and voluptuous berries;
garlands bedecked the oarlocks. Seeing this
the pirates finally bade the helmsman land,
but now the god turned to a savage lion
high on the prow, and fiercely roared; amidships 45
he made a shaggy bear miraculously
rear up with ravening jaws, while from the fo'c's'le
the lion glared hideously. The crew stampeded
astern and crowded terrified around
the sober helmsman. Then the lion sprang 50
and seized the captain: seeing this, all leapt
overboard to escape their evil fate,
and turned to dolphins. On the gladdened helmsman
the god showed mercy, held him back, and told him:
"Courage, good man! for you have pleased my heart. 55
I am loud-roaring Dionysus, he
whom Cadmus' child Sémele bore for love of Zeus."
Hail, lovely Sémele's son: one who forgets you
never again will modulate sweet song!

[30]A legendary people thought to live in the far north.

To Pan
Hymn 19

Tell me now, Muse, of Hermes' cherished son,
goatfooted horned cacophonous wanderer
through forest glades, accompanied by nymphs
dancing atop vertiginous precipices
shouting PAN! to the disheveled brighthaired god 5
of shepherds, lord of snowenveloped peaks,
towering mountaintops, and granite crags.
He wanders here and there through bushy thickets
drawn by the lure of softly murmuring streams
or presses on past overhanging cliffs, 10
scaling steep pinnacles to view his flocks.
Often he runs through shimmering high mountains,
often darts over hillsides slaying beasts
fixed by his piercing eyes. At eventide,
after the hunt, he modulates his reed, 15
making music so sweet not even the bird
who pours forth honeyed threnodies through the leaves
of flowery spring could match its melody.
Mellifluous mountain nymphs crowd close about him,
nimble of foot, beside the stream's dark waters, 20
and sing while Echo wails across the mountains.
Dancing beside or stealing in among them,
the deity, shouldering a bloodbespattered
lynx's pelt, glories in their high-pitched song
amid soft fields of fragrant hyacinth 25
and crocus interspersed among the grasses.
They sing of blissful gods high on Olympus,
telling, beyond all other tales, how Hermes,
swift messenger of the gods, came speeding down
to Arcady of many streams, rich mother 30
of flocks, his shrine as guardian of Cylléne.
There, though a god, he tended shaggy flocks
under a man, for sheer desire consumed him
to merge in love with Dryops' braided daughter:
a marriage soon accomplished. In her chambers 35
she bore a monstrous looking son to Hermes,
goatfooted horned cacophonous softly laughing,
one glimpse of whose grotesque fullbearded face
sent his nurse flying forth in utter terror.
But Hermes quickly lifted him in his hands, 40
welcomed him with a god's prodigious joy,
and took him to the immortals' bright abode
wrapped warmly in the skins of mountain hares,
then, setting him by Zeus and all the immortals,
showed them his boy; and all the immortal gods 45

were glad at heart, above all Dionysus.
They named him *Pan* for bringing joy to *all*.
    Hail to you, lord: grant favor to my song
and I shall remember you in song again.

To Earth Mother of All
Hymn 30

All-Mother deeply grounded Earth I sing,
primeval nurturer of all that lives!
Whatever roams the land or swims the waters,
or soars above, draws sustenance from her bounty.
Mortals through you abound in crops and children, 5
Lady, for you both grant and snatch away
the gift of life, and fortunate is the man
you heap with honors! All he asks is his:
his fertile farmland overflows, his pastures
flourish with herds, his homestead teems with plenty. 10
Such men benevolently govern cities
of lovely women; wealth and bliss attend them;
their sons exult in ever-fresh delights
while choirs of daughters garlanded with blossoms
skip gleefully through meadows soft with flowers. 15
    Hail, Mother of gods, wife of the starry Heaven!
Kindly reward my song with pleasing bounty
and I shall remember you in song again.

Earthly Seasons and Human Desires

Fragments of Greek Lyric Poets, translated by Robert M. Torrance

Except for the victory odes of Pindar (Píndaros), no collection of poems from the great age of Greek lyric poetry between the seventh and fifth centuries B.C. *has survived intact. What we have are fragments—sometimes of a word or two, sometimes of a line or a stanza, but very rarely an entire poem—quoted for stylistic or philological reasons by later grammarians or rhetoricians, or reconstructed from mangled scraps of papyrus discovered in the deserts of Egypt. The following nine selections from four poets are a fragment of these fragments chosen to illustrate a few ways in which natural imagery becomes a central component of poems primarily devoted to other subjects, from mythological narrations to descriptions of the poet's personal circumstances and emotions.*

Alcman *was a choral poet of Sparta (possibly a native of Lydia in Asia Minor), probably in the mid to late seventh century; his fragment "Mountain summits and deep ravines are sleeping" is a haunting evocation of the nocturnal tranquility of nature.* ***Alcaeus (Alkaios)****, descendant of a noble family on the island of Lesbos, took active part in the turbulent politics of the late seventh* *to early sixth century; some of his fragments are political diatribes, others mythological episodes, others personal professions or drinking songs. Natural images such as the wintry gale of the first selection enter vividly into many of his poems; the second selection, "Wet your gullet with wine," recasts in lyric meter a passage from Hesiod's* Works and Days *(582–88).* ***Sapphô*** *of Lesbos, a contemporary of Alcaeus, was the greatest woman poet and one of the greatest lyric poets of ancient Greece, famed above all for passionate love poems to the close-knit* thiasos *or circle of girls and women whom she gathered around her; the three fragments give some indication of ways in which sharply delineated natural images provide an "objective correlative" (in T. S. Eliot's phrase) for her intense emotions. Finally, the selection from* ***Íbycus (Ibykos)****, a poet of the late sixth century who migrated from Rhêgion (modern Reggio di Calabria) on the toe of Italy to the Aegean island of Samos, provides a clear contrast, as David A. Campbell notes in his edition of* Greek Lyric Poetry *(1967), "between the seasonal regularity of nature and the ever-present love of Ibycus which knows no seasons," and "between the tranquillity of nature and Ibycus' unresting love, harsh in its onslaught."*

The translations, which approximate the varied meters of the originals, are made from Campbell's edition; numbers of fragments are those of E. Diehl, Anthologia Lyrica Graeca, *and of D. Page,* Poetae Melici Graeci *or E. Lobel and D. Page,* Poetarum Lesbiorum Fragmenta. *Many translations of Greek lyrics are available, including Willis Barnstone's* Greek Lyric Poetry *(1962; 2nd ed. 1967); for a general introduction, see C. M. Bowra,* Greek Lyric Poetry from Alcman to Simonides *(1936; 2nd ed. 1961).*

TWO FRAGMENTS OF ALCMAN

Mountain summits and deep ravines are sleeping,
jutting crags and rocky chasms,
all the creeping inhabitants nurtured by the black earth,
mountain predators, legions of honeybees,
and monstrous beasts in the depths of the purple sea:
sleeping too are the birds'
tribes with elongated wings.

—Diehl No. 58; Page No. 89

Three seasons he[31] created: summer
and winter, and the third was autumn,
and then a fourth one, spring, when earth
bursts into blossom, and no man
can eat too much.

—Diehl No. 56; Page No. 20

THREE FRAGMENTS OF ALCAEUS

Zeus sends rain howling down with a wintry gale
from heaven; flowing waters are frozen fast . . .

[31]Presumably Zeus.

Ignore the ice-cold weather and pile a log
upon the fire, distributing honeyed wine
unstintingly, beneath your tired head
casually tossing a downy cushion.

—Diehl No. 90; Lobel and Page No. 338

Wet your gullet with wine
now, for the Dog
Star is revolving round,
harsh the season becomes,
everything droops,
parched with oppressive heat,

sweetly cicadas sound
down from the leaves,
pouring, from underneath
scraped wings, piercingly shrill
showers of song
after the flaming bright

summer [comes] . . .

artichokes are in bloom,
women are most
pestilent with desire
now when men have become
languid and frail,
Sirius having scorched

brains and knees . . .

—Diehl No. 94; Lobel and Page No. 347

What birds do I behold
flying from far
Ocean, remotest bound
earth knows, mottled wild ducks
stretching their splotched
necks and extended wings?

—Diehl No. 135; Lobel and Page No. 345

THREE FRAGMENTS OF SAPPHO

Stars around the beautiful brightly shining
moon conceal their glimmering apparition
when she most illuminates all the earth with
radiant fullness.

—Diehl No. 4; Lobel and Page No. 34

Desire overwhelmed my heart
like a whirlwind plummeting down upon mountain oaks.
—Diehl No. 50; Lobel and Page No. 47

Hesperus,[32] you, bringing back
everything the bright Dawn had scattered,
bring back the sheep, bring the goats,
and bring back the child to its mother.
—Diehl No. 120; Lobel and Page No. 104

"In springtime Cydonian quince . . . ," by Ibycus

In springtime Cydonian quince
flourishes, watered by plentiful rills
flowing from streams by the undefiled
garden of Maidens, and vine-blossoms sprout
up from beneath the flowering grape vines'
shadowy branches—
but my desire never
peacefully sleeps, whatever the season:
rather, like Thracian Boreas blazing
up with lightning and frenziedly rushing
forth from the Cyprian goddess[33] with searing
madness, somber and shameless,
overpowering all resistance, it
shatters my heart to the core.
—Diehl No. 6; Page No. 286

Desert Island and Flowering City

Two Passages from Sophocles

Sóphocles (Sophoklês, 496–406 B.C.) wrote some 123 plays, of which seven survive. He won his first victory in the tragic contests of the Theater of Dionysus over his great predecessor Aeschylus (Aískhylos) before he was thirty, but his surviving plays were probably all written after the age of fifty. The last two, Philoctétes (Philoktêtês) *and* Oedipus at Colónus, *Sophocles wrote when approaching his death at ninety, as Athens, in whose political and military life he had played an active part, was nearing final defeat in the long and bitter Peloponnesian War against Sparta.* Philoctetes, *produced in 409, is set on the rocky island of Lemnos where the archer Philoctetes had been abandoned ten years before because a snakebite had incurably wounded his foot. Odysseus and the young son of the dead Achilles, Neoptólemus, have been sent to bring him back after a prophecy revealed that Troy could not be taken without him; during the play Neoptolemus repudiates Odysseus's cynical deceptions and learns to act in accord with his own noble nature by reveal-*

[32]The evening star (Latin Vesper). [33]Aphrodítê, born on the island of Cyprus (Kypros).

ing the truth to Philoctetes, who is persuaded to return by an epiphany of the deified hero Heracles. His final speech is a poignant farewell to the barren island on which he had suffered so long but which he has learned to love through his very sufferings.

Oedipus at Colonus, *Sophocles' last play, was produced by his grandson and namesake in 401, after Sophocles' death and his city's defeat. In it Oedipus, former King of Thebes, self-blinded and exiled after his terrible discovery that he had unwittingly killed his father and married his mother, arrives—an old beggar guided by his daughters (and half-sisters) Antígonê and Ismênê—at the sacred grove of the Euménidês (the "Kindly Ones," as the Erínyês or Furies were euphemistically called) in the deme of Colonus, Sophocles' birthplace a mile and a half from the Acropolis of Athens. Those who banished him seek his return (as those who deserted Philoctetes sought his) when they learn that he will bring not pollution but benefit to the place that receives him. It is not Thebes, however, but the Athens of Thêseus that recognizes his value and welcomes him to its land, into which he is mysteriously absorbed and to which he brings the blessing of a* daimôn, *or deified hero. In our selection, a* stásimon *or choral ode (lines 668–719) sung near the middle of the play, the chorus rhapsodically celebrates the beauty and glory of Colonus and Athens in verses splendidly rendered by Robert Fitzgerald. The translations are from* Philoctetes and the Women of Trachis *(1966), translated by Robert M. Torrance, and* The Oedipus Cycle *(1949), translated by Dudley Fitts and Robert Fitzgerald.*

O Land of Sea-Circled Lemnos, Farewell!,
translated by Robert M. Torrance
Philoctetes

Now as I leave I will call on my island.
Farewell to the chamber that shared in my vigil,
and the nymphs of the meadows, nymphs of the streams,
and the masculine roar of the sea-swept coast. 1455
Often my head has been damp with the blowing
of southerly winds, though deep in my cave;
and often the distant mountain of Hermes
has heard my voice and answered to me
with echoing groans in my tempest of sorrow. 1460
But O my streams and my Lycian spring,
I am leaving you now, I am leaving at last,
though I had thought I would never depart.
O land of sea-circled Lemnos, farewell!
Do not begrudge me a fair voyage now 1465
to whatever place great Destiny calls,
and my friends' advice, and the almighty god
who has brought these things to fulfillment.

Athens, Land of Olive and Horse,
translated by Robert Fitzgerald
Oedipus at Colonus

The land of running horses, fair
Colonus takes a guest;
He shall not seek another home,

For this, in all the earth and air,
Is most secure and loveliest.

In the god's untrodden vale
Where leaves and berries throng,
And wine-dark ivy climbs the bough,
The sweet, sojourning nightingale
Murmurs all night long.

No sun nor wind may enter there
Nor the winter's rain;
but ever through the shadow goes
Dionysus reveler,
Immortal maenads in his train.

Here with drops of heaven's dews
At daybreak all the year,
The clusters of narcissus bloom,
Time-hallowed garlands for the brows
Of those great ladies whom we fear.

The crocus like a little sun
Blooms with its yellow ray;
The river's fountains are awake,
And his nomadic streams that run
Unthinned forever, and never stay;

But like perpetual lovers move
On the maternal land.
And here the choiring Muses come,
And the divinity of love
With gold reins in her hand.

And our land has a thing unknown
On Asia's sounding coast
Or in the sea-surrounded west
Where Pelops' kin hold sway:
The olive, fertile and self-sown,
The terror of our enemies
That no hand tames or tears away—
The blessed tree that never dies!—
But it will mock the swordsman in his rage.

Ah, how it flourishes in every field,
Most beautifully here!
The grey-leafed tree, the children's nourisher!
No young man nor one partnered by his age
Knows how to root it out nor make
Barren its yield;
For Zeus Protector of the Shoot has sage

Eyes that forever are awake,
And Pallas watches with her sea-grey eyes.

Last and grandest praise I sing
To Athens, nurse of men,
For her great pride and for the splendor
Destiny has conferred on her.
Land from which fine horses spring!
Land where foals are beautiful!
Land of the sea and the sea-farer,
Enthroned by her pure littoral
By Cronus' briny son in ancient time.

That lord, Poseidon, must I praise again,
Who found our horsemen fit
For first bestowal of the curb and bit,
To discipline the stallion in his prime;
And strokes to which our oarsmen sing,
Well-fitted, oak and men,
Whose long sea-oars in wondrous rhyme
Flash from the salt foam, following
The track of winds on waters virginal.

Wine, Honey, and Blood

Two Passages from *The Bacchae*, by Euripides, translated by Robert M. Torrance

Eurípidês (ca. 485–ca. 406 B.C.), the third of the great Athenian tragic dramatists, after Aeschylus and Sophocles, wrote some ninety-two plays, of which nineteen survive. Among the last was the Bacchae (Bakkhai), *written when he was nearing eighty and produced in Athens, after his death at the court of King Archelaus in Macedonia, in about 405. As his plays repeatedly show, Euripides was deeply affected by the rationalistic and skeptical currents of his time, including the sophistic rhetoric of Prôtágoras and Gorgias, the philosophical speculations of Anaxágoras (Chapter 9 below), and Socrates' questioning of received assumptions. In the* Bacchae, *however, it is the ecstatic popular worship of Dionysus—which seems to have gained an increasingly fervid following, especially among women, in the troubled years of the Peloponnesian War—that dominates the tragedy when Pentheus, King of Thebes, unwisely attempts to oppose it.*

This is a play, E. R. Dodds emphasizes in his edition (1944) of the Greek text, "about an historical event—the introduction into Hellas of a new religion"—though not an entirely new religion, and certainly not a new god. Dionýsos—also known by many other names such as Bacchus (Bakkhos), Iacchus (Iakkhos), Bromios ("the Roarer"), and Euius (from the ecstatic cry "euoi")—had been worshipped in Greece for many centuries, and was far more prominent in popular festivals (including those associated with the origins of tragic drama) than in Homeric epic; among the Olympians he remained an outsider. Far from being confined to wine, "his domain is, in Plutarch's words," Dodds writes, "the whole of the hygra physis *['moist nature']—*

not only the liquid fire in the grape, but the sap thrusting in a young tree, the blood pounding in the veins of a young animal, all the mysterious and uncontrollable tides that ebb and flow in the life of nature. Our oldest witness, Homer, nowhere explicitly refers to him as a wine god; and it may well be that his association with certain wild plants, such as the fir and the ivy, and with certain wild animals, is in fact older than his association with the vine." His worship had long been linked with bands of ecstatic female followers called bakkhai *("bacchants," or followers of Bacchus) or* maínades *("maenads," or madwomen); the one extended reference in Homer (*Iliad *VI.130–40) anticipates the* Bacchae *in telling how the Thracian King Lycurgus (Lykourgos) was punished by Zeus for chasing "the nurses of maddened Dionysus," who dropped their Bacchic wands in flight.*

The orgiastic "new religion" of fifth-century Greece portrayed by Euripides—a religion possibly influenced by the newly introduced worship of the Phrygian "Great Mother" Cýbele (Kybelê)—was thus the continuation or revival of a much older one, and Euripides places his story in a distant mythical past: for Pentheus is the grandson, by his mother Agáve (Agauê), of Cadmus (Kadmos), founder of the ancient city of Thebes. Dionysus, too, in the birth myth retold in this play, is a grandson of Cadmus by another of the old king's daughters, Sémelê, who when pregnant by Zeus was tricked by Hêra into asking to see her lover in his true shape and incinerated by his thunderbolt; Zeus then sewed up the unborn Dionysus in his thigh until he was ready to be born. Dionysus thus partakes of human, natural, and divine in one, and contradiction is essential to him: "a most terrible god, but most gentle to human beings"— if they accept his divinity (861). As Geoffrey S. Kirk writes in introducing his annotated translation (1970), "he is god of joy, peace, and festivity on the one hand, of terror and the dead on the other. These are two sides of fertility; for fertility comes from the earth, but the earth is the place where the dead go." His maenads worship him by dancing in the mountains, clad in dappled fawnskins, while waving his thyrsus (a wand wreathed with ivy and vine leaves and topped with a pine cone) to torchlight and kettledrum. Their wild ecstasy culminates in the savage rites known as sparagmos *and* ômophagia, *the "tearing apart" and "eating raw" of a living animal, even a bull. We should be wary, as Kirk and others caution, of taking this picture too literally. Yet at the very least it gives an unforgettably vivid picture of a wildly ecstatic, joyous yet fearful union with nature—a union free, for the duration of ecstasy, from the repressions of civilization.*

Our first selection is the parodos *of the play, the passage chanted by the chorus as they enter the theater and then dance, turning first in one direction (strophê), then back in the other (antístrophê), or stand in place (epode); the irregular line numbering follows the printed Greek text. It is a hymn of praise to the god of untamed nature who brings joyous release to his devoted worshippers, subduing and channeling their bacchic frenzy, as Dodds suggests, "to the service of religion." Not only the women of Thebes but also old Cadmus and the prophet Teiresias join in the revels. Pentheus scornfully repudiates what he pruriently thinks is their orgiastic lechery; and when a long-haired effeminate stranger (in fact Dionysus) is brought to him as his "quarry," Pentheus ignores his warnings and attempts (like the pirates in the Homeric Hymn) to enchain the god, who frees himself by shattering the palace amid earthquake and flames. Not even the Theban herdsman's report (our second selection) of the miracles performed by Pentheus's mother Agave and her comrades under the influence of the god can dissuade the king from his folly. In the rest of the play he himself becomes the quarry of his own enslaving madness, dressing himself, at the stranger's advice, as a woman and setting forth to Mount Cithaeron to catch the maenads in their lewd activities—only to be seized when they uproot the fir tree he has climbed and, mistaking him for a wild beast, tear him limb from limb; in a scene of climactic horror, his mother enters with her son's head impaled on her thyrsus. Pentheus meets this terrible fate as a result of attempting to deny the elemental power of nature beyond and within him; his punishment, Dodds remarks, "is the sudden complete collapse of the inward dykes when the elemental breaks through perforce and civilization vanishes."*

Readers should consult, besides Dodds's commentary to his Greek edition (on which this translation is based) and Kirk's to his English translation, Dodds's The Greeks and the Irrational *(1951); Walter F. Otto's* Dionysus: Myth and Cult *(English trans. 1965); Marcel Détienne's* Dionysos at Large *(1986); and Wole Soyinka's twentieth-century version,* The Bacchae of Euripides: A Communion Rite *(1973).*

I Shall Hymn Dionysus

Parodos

From the land of Asia
I have crossed over sacred Tmolus[34] and hasten 65
to perform Bromius' sweet toil,
his unlaborious labor,
crying "*Euoi*, O Bacchus!"
Who is in the street? Who is outside?
Withdraw to your houses in hushed veneration,
each keeping reverent silence: 70

for by custom as always
I shall hymn Dionysus.

[Strophe 1]

Oh,
blest who, happily knowing
sacraments of the gods,
leads a life of devotion
celebrating in spirit; 75
on mountaintops serves Bacchus
with sacred purifications;
practices holy rituals
of Cybele, Great Mother;
brandishes the thyrsus 80
and, garlanded with ivy,
worships Dionysus.
On bacchae, on bacchae!
Come, and bring Dionysus
Bromius, god and god's son, 85
down from Phrygian mountains
home to Hellas's broad streets—
Bromius the roaring

[Antistrophe 1]

whom
Sémele, his pregnant
mother prematurely

[34]Tmôlos was a mountain of Lydia in Asia Minor, on which the fabled river of gold Pactólus (Paktôlos) arose.

bore in pangs of childbirth 90
forced by Zeus' swift lightning,
losing her life when stricken
by a thunderbolt from heaven:
at once Zeus son of Kronos
wrapped him in dark recesses, 95
hiding him in his own thigh,
and sewed him in with golden
pins, kept secret from Hera.
When the Fates decreed the moment,
he gave birth to a bull-horned 100
god, crowning him with garlands
of snakes: wherefore the maenads
twine in their flowing tresses
a beast-nurtured booty.

[Strophe 2]

O nurse of Sémele, Thebes, come 105
now, crown yourself with ivy:
abound, abound with greening
bryony's lovely berries,
dedicating yourselves to Bacchic frenzies
with twigs of oak or fir-tree, 110
then fringe your garment of dappled
fawnskin around with tufted
white wool, and make yourselves holy with violent
thyrsuses: soon all the land will be dancing
when Bromius leads his revelers 115
to the mountain, up to the mountain,
where thronging women await him
far from their looms and shuttles,
stung mad by Dionysus.

[Antistrophe 2]

O secret lair of Curetes,[35] 120
O sacred dwelling places
of Crete that Zeus was born in,
where three-helmed Corybantes
created this hide-covered kettle-drum
for me, down in their caverns, 125
then, in wild Bacchic worship,
mixed it with sweetly inviting
Phrygian flute sounds and placed it in mother Rhea's

[35]The Kourêtes were mythical creatures of the island of Crete who protected the infant Zeus, son of Krônos and Rhea (Rheia), by clattering their shields and cymbals to drown out his crying when he was hidden in a cave to prevent him from being devoured by his father, to whom Rhea had given a stone in swaddling clothes to swallow in Zeus's place. They were sometimes confused or identified with the Corybantes, frenzied priests of Cybele (Kybelê), the "Great Goddess" of Phrygia in Asia Minor, whose ecstatic worship was introduced to Athens in the fifth century B.C., and who was herself sometimes identified with the Titan Rhea.

hands to beat time for the joyous cries of the bacchae:
from her, from the mother goddess, 130
ecstatic satyrs received it
and made it part of biennial
dances in which Dionysus
joins, and greatly rejoices.

[Epode]

Welcome is he on the mountains who falls down among the running 135
bands of revelers onto the ground,
wears the sacred fawnskin garment while hunting
slain goats' blood in the joy of eating live flesh, and yearning
for Phrygian or Lydian mountains: and the leader is Bromius, *euoi!* 140
The ground is flowing with milk, with wine, and with honeybees' nectar.
Like Syrian frankincense's
smoke, the Bacchic god, raising 145
the fiery flame of his pine torch,
sends it flaring up from his thyrsus,
and with running and dancing
rouses the laggard stragglers
reviving them all with his shouting,
then flings his luxuriant locks to the breezes, 150
roaring *thus*, amid cries of exultance:
"On bacchae, on!
On bacchae, on!
Pride of gold-running Tmolus,
sing praising of Dionysus 155
to the kettledrum's deep roaring,
with shouts of *euoi!* exalt the god Euius
amid Phrygian howling and battle-cries
as the holy melodious flute 160
roars out holy tunes to accompany frenzied
women up to the mountain, the mountain!" Rejoicing 165
then like a foal alongside its grazing
mother, each nimble bacchant moves her legs with frolicsome leaping.

Some God Must Surely Be Among Them

Messenger's speech

The grazing cattle herds were just beginning
to amble toward high pastures, and the warming
rays of the sun to spread across the land,
when I beheld three bands of women revelers. 680
Autónoë led one of them, your mother
Agáve led the second, Ino the third.
All were asleep to rest their weary bodies,
some propping up their backs against a fir-tree,
others resting their heads on fallen oak leaves, 685

casually, yes, but modestly, and not—
as *you* say—hunting singly through the forest
for Cyprian pleasures, drunk with wine and flute-song.
 Your mother, when she heard the cattle lowing
stood up and cried aloud "*olololú!*" 690
amid the bacchae, rousing them from slumber.
Then, shaking off the luxury of sleep,
they leapt straight up, amazingly well ordered,
young and old both, and maidens yet unmarried.
First they shook hair cascading down their shoulders, 695
then fastened up their fawnskins—those whose knots
had come undone—while writhing serpents, girdling
dappled pelts, licked their cheeks with flicking tongues.
Some cradled in their arms young fawns or savage
wolf cubs and suckled them with warm white milk— 700
mothers with swollen breasts who left their newborn
babies behind—all garlanded with wreaths
of ivy, oak, and flowering bryony.
One, picking up her thyrsus, struck a rock
out of which dewy streams of water gushed; 705
another hurled her staff against the ground,
whence the god sent a wine-spring bubbling up;
and anyone who wanted white drink, needed
only to scrape the ground with fingertips,
and milk came spurting; from their ivy-clustered 710
thyrsuses dribbled down sweet streams of honey.
This god whom you revile, had you been present
and seen this, you would supplicate with prayer!
 We met together then, cowherds and shepherds,
and vied with one another in exchanging 715
stories of what strange marvels they were doing.
One, who had been to town and had a way
with words, said to us all: "You who inhabit
majestic mountain plains, what say we flush
Agave, Pentheus' mother, from her revels 720
and thus oblige our king?" To us it seemed
he had spoken wisely, so we lay in ambush,
hiding in leafy bushes. When the time came,
shaking the thyrsus, they began the revels
crying in unison "Iacchus! Bromius, 725
son of Zeus!" All the mountain, and mountain beasts,
joined the wild dance that left no thing unshaken.
 It happened that Agave darted near me,
so out I jumped, in my desire to catch her,
abandoning the lair that kept me hidden. 730
She howled aloud, "O my swift running hounds,
these men are hunting us, but follow me:
follow me, armed with thyrsuses in hand!"

We ran for dear life and avoided being
torn apart by the bacchae, who assaulted 735
the grazing cattle with bare hands, unarmed.
You could have seen one tear an unmilked, lowing
heifer asunder with her own two hands,
while others lacerated cows to ribbons—
could have seen ribs or cloven hooves hurled madly 740
backward and forward till they dangled dripping
down from the fir-trees, clotted red with blood.
Bulls, violent till then with savage fury
up to the horns, now stumbled to the ground
dragged by innumerable maidens' hands: 745
the garment of their flesh shred into gobbets
sooner than you could blink your kingly eyes.
Off they sped then like birds upraised in flight
over the rolling plains Asopus waters[36]
to fertilize rich crops of Theban grain. 750
On Hysia and Erythra, in the foothills
of Mount Cithaeron, they at once descended
like enemies, and turned these villages
inside out, snatching children from their homes.
Whatever they draped across their shoulders stayed 755
without being tied or falling to the ground,
even bronze and iron utensils; from their curls
fire blazed, but did not burn them. Angered townsmen
took arms against the plundering bacchants:
a fearful marvel to behold, my lord! 760
The sharp points of their weapons drew no blood,
but maenads, wielding only thyrsuses,
wounded and routed them. If women did this
to men, some god must surely be among them! 765
Then they retraced their steps to where they started—
the very springs the god made spurt for them—
and washed away the blood, while serpents' tongues
licked from their skin the driblets on their cheeks.
Whoever he may be, O master, welcome
this deity to our city. He is great 770
for many things, but most of all, I hear,
for giving men the vine that ends all troubles.
Without wine there could be no Aphrodite,
nor anything delighting humankind.

[36]The river Asôpos flowed between Thebes (Thêbai) and the mountain range of Cithaeron (Kithairôn) to the south.

The Cosmic and the Comic

Two Choruses of Aristophanes

Aristóphanês was born in the mid fifth century B.C. *and died in about 385. His eleven extant plays (out of more than forty) include the only surviving examples of Greek "Old Comedy," which combined scathing satire, ribald farce, exuberant fantasy, and soaring lyrics. Comedy originated, Aristotle tells us, in phallic processions, and the climactic* kômos *("revels") to which it owes its name is often explicitly a celebration of life and fertility. Its frequent animal choruses, with their oversized phalluses, are a graphic reminder of its close adhesion to nature. In the* Clouds, *however—produced in 423 but revised after it won third (and last) prize in the theatrical competition—nature takes more ethereal form in the chorus of cloud-maidens, whose lovely opening song (the* parodos, *here given in Oscar Wilde's free translation) contrasts with the pompous flummery of meteoric speculation and sophistic chicanery pilloried by the young Aristophanes, with total disregard for fairness or accuracy, in the person of Socrates.*

The Birds, *produced in 414, is perhaps Aristophanes' masterpiece. The long section here included, in B. B. Rogers's rollicking translation of 1920 (with line numbers corresponding to the Greek original), is from the first* parabasis *of the play—an interlude in which the chorus directly addresses the audience, often in the poet's own name. Here the Birds, having been persuaded by two disgruntled Athenian self-exiles to reassert their ancient dominion over the world, ecstatically recount (in verses that fantastically combine pre-Socratic speculation with Hesiodic and Orphic myth) the birth of Love from the wind-egg of Chaos—and of the Birds, followed by Earth, Sea, Sky, and the gods, from Love. Burlesque though it is, this panegyric is at the same time a fervent celebration of the dominion of nature that preceded and will long survive the upstart human inhabitants of cities at war, unless they can learn to utilize nature not by opposing but by joining it, donning wings as citizens, along with the birds, of a new Cloudcuckooland.*

For further reading, see Rogers's translations of Aristophanes in the Loeb Classical Library and William Arrowsmith's of The Clouds *(1962) and* The Birds *(1961); Francis M. Cornford,* The Origin of Attic Comedy *(1914); Cedric H. Whitman,* Aristophanes and the Comic Hero *(1964); and Robert M. Torrance,* The Comic Hero *(1978).*

Song of the Clouds,
translated by Oscar Wilde

Parodos of the Clouds, *275–90, 298–313*

Cloud-maidens that float on forever,
  Dew-sprinkled, fleet bodies, and fair,
Let us rise from our Sire's loud river,
  Great Ocean, and soar through the air
To the peaks of the pine-covered mountains where the pines hang as tresses of hair!
Let us seek the watch-towers undaunted,
  Where the well-watered cornfields abound,
And through murmurs of rivers nymph-haunted
  The songs of the sea-waves resound;
And the sun in the sky never wearies of spreading his radiance around!
    Let us cut off the haze

Of the mists from our band,
Till with far-seeing gaze
We may look on the land!

Cloud-maidens that bring the rain-shower,
To the Pallas-loved land let us wing.
To the land of stout heroes and Power,
Where Kekrops was hero and king,[37]
Where honor and silence is given
To the mysteries that none may declare,
Where the gifts to the high gods in heaven
When the house of the gods is laid bare,
Where are lofty-roofed temples and statues well-carven and fair;
Where are feasts to the happy immortals
When the sacred procession draws near,
Where garlands make bright the bright portals
At all seasons and months of the year;
And when Spring days are here,
Then we tread to the wine-god a measure
In Bacchanal dance and in pleasure,
Mid the contests of sweet-singing choirs,
And the crash of loud lyres!

The Birds' Celebration of Love,
translated by Benjamin Bickley Rogers
First Parabasis of the Birds

O darling! O tawny-throat!
Love, whom I love the best,
Dearer than all the rest,
Playmate and partner in
All my soft lays,
Thou art come! Thou art come! 680
Thou hast dawned on my gaze,
I have heard thy sweet note,
Nightingále! Nightingále!
Thou from thy flute   Softly sounding canst bring
Music to suit   With our songs of the Spring:
Begin then I pray
Our own anapestic address[38] to essay.

Ye men who are dimly existing below, who perish and fade as the leaf,
Pale, woebegone, shadowlike, spiritless folk, life feeble and wingless and brief,
Frail castings in clay, who are gone in a day, like a dream full of sorrow and sighing,
Come listen with care to the Birds of the air, the ageless, the deathless, who flying
In the joy and the freshness of Ether, are wont to muse upon wisdom undying.

[37]Kekrops was the mythical first king of Athens, the city loved by the goddess Pallas Athena.
[38]The passage that follows is mainly in anapests (two short syllables followed by a long), a rhythm suited to military marches.

We will tell you of things transcendental; of Springs and of Rivers the mighty
upheaval; 690
The nature of Birds; and the birth of the Gods: and of Chaos and Darkness primeval.
When this ye shall know, let old Prodicus go,[39] and be hanged without hope of
reprieval.
There was Chaos at first, and Darkness, and Night, and Tartarus vasty and dismal;
But the Earth was not there, nor the Sky, nor the Air, till at length in the bosom
abysmal
Of Darkness an egg, from the whirlwind conceived, was laid by the sable-plumed
Night.
And out of that egg, as the Seasons revolved, sprang Love, the entrancing, the bright,
Love brilliant and bold with his pinions of gold, like a whirlwind, refulgent and
sparkling!
Love hatched us, commingling in Tartarus wide, with Chaos, the murky, the darkling,
And brought us above, as the firstlings of love, and first to the light we ascended.
There was never a race of Immortals at all till Love had the universe blended; 700
Then all things commingling together in love, there arose the fair Earth, and the Sky,
And the limitless Sea; and the race of the Gods, the Blessed, who never shall die.
So we than the Blessed are older by far; and abundance of proof is existing
That we are the children of Love, for we fly, unfortunate lovers assisting.
And many a man who has found, to his cost, that his powers of persuasion have failed,
And his loves have abjured him for ever, again by the power of the Birds has prevailed;
For the gift of a quail, or a Porphyry rail, or a Persian, or goose, will regain them.
And the chiefest of blessings ye mortals enjoy, by the help of the Birds ye obtain them.
'Tis from us that the signs of the Seasons in turn, Spring, Winter, and Autumn are
known.
When to Libya the crane flies clanging again, it is time for the seed to be sown, 710
And the skipper may hang up his rudder awhile, and sleep after all his exertions,
And Orestes[40] may weave him a wrap to be warm when he's out on his thievish
excursions.
Then cometh the kite, with its hovering flight, of the advent of Spring to tell,
And the Spring sheep-shearing begins; and next, your woollen attire you sell,
And buy you a lighter and daintier garb, when you note the return of the swallow.
Thus your Ammon, Dodona, and Delphi are we;[41] we are also your Phoebus Apollo.
For whatever you do, if a trade you pursue, or goods in the market are buying,
Or the wedding attend of a neighbor and friend, first you look to the Birds and their
flying.
And whene'er you of omen or augury speak, *'tis a bird* you are always repeating;
A Rumor's a bird, and a sneeze is a bird, and so is a word or a meeting, 720
A servant's a bird, and an ass is a bird. It must therefore assuredly follow
That the birds are to you (I protest it is true) your prophetic divining Apollo.

Then take us for Gods, as is proper and fit,
And Muses Prophetic ye'll have at your call

[39]Prôdikos of Keos was a contemporary sophist and rhetorician.
[40]Nickname of a robber known for stealing travelers' clothes, possibly by pretending to be mad like the Oréstês of myth.
[41]Ammon was god of the Egyptian city of Thebes; his oracle at Siwa rivaled those of Zeus at Dodôna and of Apollo at Delphi.

Spring, winter, and summer, and autumn and all.
And we won't run away from your worship, and sit
Up above in the clouds, very stately and grand,
Like Zeus in his tempers: but always at hand
Health and wealth we'll bestow, as the formula runs,
ON YOURSELVES, AND YOUR SONS, AND THE SONS OF YOUR SONS; 730
And happiness, plenty, and peace shall belong
To you all; and the revel, the dance, and the song,
And laughter, and youth, and the milk of the birds
We'll supply, and we'll never forsake you.
Ye'll be quite overburdened with pleasures and joys,
So happy and blest we will make you.

O woodland Muse,
tío tio, tío tiotinx,
Of varied plume, with whose dear aid
On the mountain top, and the sylvan glade, 740
tío tio, tío tiotinx,
I, sitting up aloft on a leafy ash, full oft,
tío tio, tío tiotinx,
Pour forth a warbling note from my little tawny throat,
Pour festive choral dances to the mountain mother's praise,
And to Pan the holy music of his own immortal lays;
totótotótotótotótotinx,
Whence Phrynichus of old,[42]
Sipping the fruit of our ambrosial lay,
Bore, like a bee, the honied store away, 750
His own sweet songs to mold.
tío, tío, tío, tío, tíotinx. . . .

9

Philosophers, Physiologues, and Physicians

Greek philosophy and its offshoot, science, undoubtedly had deep roots in mythical thought, for myth is among other things a way of thinking, through stories, about the origin and nature of the world. Thus the origin myth of Hesiod's Theogony *anticipated pre-Socratic speculation on the genesis of the cosmos, to which Plato, in the* Timaeus, *opposed his own "probable myth" of creation; and from the Homeric Hymn to Demeter to Euripides'* Bacchae *myths explored the relation of human beings to the divinely natural world around them. Yet philosophy was at the same time radically different in that it postulated a dynamic reality—the reality of* physis, *or nature—with a structure corresponding to that of the human mind (since man is inherently part of nature) but by no means simply*

[42]Phrýnikos was an early Athenian tragic poet.

anthropomorphic. The spontaneously self-regulated order of nature, though susceptible of understanding by human reason, is wholly independent of human—or divine—will.

Since no complete work by any thinker of this seminal period of Western natural philosophy survives, our only sources are testimonies of later writers describing their views and, most importantly, fragments of presumably direct quotations incorporated in these testimonies. Among major sources are Plato and above all Aristotle in the fourth century B.C., *and a host of later "doxographers" (recorders of opinions), all drawing directly or indirectly on the lost* Physikôn Doxai (Opinions of the Natural Philosophers) *of Aristotle's successor Theophrastus. Among the most valuable are Diógenês Laërtius (third century* A.D.?), *author of* Lives of the Philosophers; *the Christian polemicists Clement of Alexandria and Hippólytus of Rome (author of a* Refutation of All Heresies) *in the second to third centuries* A.D.; *the Neoplatonists Porphyry (Porphyrios) in the third and Proclus (Proklos) in the fifth; and Simplicius (Simplikios), a commentator on Aristotle, in the sixth—a thousand years after the pre-Socratics whose words he did much to preserve.*[1]

The beginnings of natural philosophy in the late seventh to mid sixth century B.C. *are closely associated with the southern Ionian city of Milétus (Milêtos) on the coast of Asia Minor, the region in which the Homeric epics had taken shape a century or more before. First as an independent* polis, *or city-state, then as a protectorate under the Lydian King Croesus (Kroisos), who subdued the Ionian cities, and under the Persian kings who conquered him, Miletus was a vigorous maritime center and perhaps the largest and wealthiest Greek city until it led an Ionian revolt against the Persian empire and was captured and devastated in 494. The first great Milesian thinker was Thalês, one of the "Seven Wise Men" of ancient Greece, traditionally dated from about 636 to 546. Said to be of Phoenician descent, he was prominent in the political life of his city, like a number of later pre-Socratic philosophers, and many stories grew up around his name. Thus Plato (*Theaetetus *174A) tells how a Thracian servant girl mocked him for falling into a well when looking up at the stars, saying "that he was eager to know things in the sky, but what was behind him or next to his feet escaped him." Aristotle, on the other hand (*Politics *1259a), stresses his practical acumen, reporting that through his astronomical observations Thales was able to predict a large olive crop and outbid others on buying olive presses, which he rented out on his own terms at a large profit, demonstrating "that it is easy for philosophers to be rich if they want, but this is not what they aspire to." He was famed for his knowledge of many fields, including astronomy, mathematics, engineering, and geography; thus he was said by Herodotus (I.74) to*

[1]The standard modern edition of the pre-Socratics is *Die Fragmente der Vorsokratiker*, ed. Hermann Diels (1903), as revised by Walther Kranz; my translations are mainly based on the three-volume seventh edition of 1954. For each thinker, Diels-Kranz includes Greek texts of ancient testimonies in Section A ("Life and Doctrine") and Greek texts with German translations of presumed direct quotations in Section B ("Fragments"); "DK" numbers below are from Section B. For English translations of the fragments (but not testimonies), see Kathleen Freeman, *Ancilla to the Pre-Socratic Philosophers* (n.d.). The fullest—but far from complete—English-language editions of Greek texts with translation and commentary are G. S. Kirk and J. E. Raven, *The Presocratic Philosophers* (1957), and M. Schofield's substantially revised second edition (1983) of Kirk and Raven. Among many studies are Friedrich Nietzsche's early but posthumously published *Philosophy in the Tragic Age of the Greeks* (1903); Eduard Zeller, *A History of Greek Philosophy from the Earliest Period to the Time of Socrates* (Eng. trans. 1881); F. M. Cornford, *From Religion to Philosophy: A Study in the Origins of Western Speculation* (1912); John Burnet, *Early Greek Philosophy* (4th ed. 1930); William A. Heidel, *Selected Papers* (1980); Werner Jaeger, *The Theology of the Early Greek Philosophers* (1937); Kathleen Freeman, *The Pre-Socratic Philosophers* (1946; 2nd ed. 1949); W. K. C. Guthrie, *A History of Greek Philosophy*, vol. I (1962), vol. II (1965); Alexander P. D. Mourelatos, ed., *The Pre-Socratics: A Collection of Critical Essays* (1974); and David J. Furley and R. E. Allen, eds., *Studies in Presocratic Philosophy*, vol. I (1970), vol. II (1975).

have predicted the year of a solar eclipse that took place in 585 during a battle between Lydians and Medes, and by Diogenes Laërtius (I.23, 27) to have determined the times of the solstices and measured the height of the Egyptian pyramids by their shadow. Above all, he was known as the founder of natural philosophy: "the first, according to tradition," Simplicius writes in concurrence with Theophrastus, "to have revealed the investigation of nature (tôn peri physeôs historian) *to the Greeks; although many others came before him, . . . he differed from them so much as to obscure them all." Whether or not he left any writings behind, no fragment in his own words has survived. He is said to have thought that the earth floated like a log on water, that the magnet possessed a soul because it moved iron, and that soul was diffused through the universe—"whence perhaps Thales also thought," Aristotle writes (*On the Soul *411a), "that all things are full of gods." Most significantly, he theorized that the primal element—later called the* arkhê, *or "beginning"—of all things was water, so that through its multiple transformations, as Aristotle describes his doctrine (*Metaphysics *983b), the underlying substance persists and nothing, despite appearances, wholly comes to be or passes away. Whatever influence older cosmogonic myths may have had in shaping his ideas, Thales was thus, in effect, the inventor of* physis.

Thales' successor Anaximander (Anaximandros, ca. 610–540) was also famed for scientific accomplishments; he was said to have made the first map of earth and sea, and also a celestial globe. He believed in a plurality of (possibly successive) worlds, and speculated that the first living creatures were born in moisture, and that man was originally similar to a fish. "As one examines the remains of this early period," Charles Kahn writes in Anaximander and the Origins of Greek Cosmology, *"it is Anaximander who emerges more and more clearly as the central figure in sixth-century thought. It is, in all probability, his work which laid down the lines along which ancient science was to develop and his mind which gave the Greek philosophy of nature its characteristic stamp. . . . What the system of Anaximander represents for us is nothing less than the advent, in the West at any rate, of a rational outlook on the natural world." To Kahn, Anaximander's view of the earth as suspended in equipoise at the center of the cosmos presages "a purely geometric approach to astronomy" freed from celestial myth. His* arkhê *or first principle was not a material substance, like water or air, but the* apeiron, *the "boundless" or "indefinite," which he thought divine; in his dynamic universe, only the limitless substratum of all things is changeless. "That which truly* is *. . . cannot possess definite characteristics," Nietzsche writes, "or it would come-to-be and pass away like all the other things. In order that coming-to-be shall not cease, primal being must be indefinite . . . insuring thereby eternity and the unimpeded course of coming-to-be."*

In the Milesian view, as Kahn reconstructs it, whatever the arkhê *may be, elemental opposites such as hot and cold, dry and damp, are continually being separated out from one another and merging together again in a process of perpetual change: the process of nature. "What little we know of Anaximander's cosmogony suggests that Hot and Cold, the basic constituents of the heavens, appeared as the eldest progeny of the Boundless," from whose interaction other opposites arose and continue to arise. The one probable quotation from Anaximander's treatise on nature (possibly the earliest work of Greek prose) is cited in the commentary of Simplicius at the end of antiquity; in Kahn's translation, and with Kahn's hypothetical quotation marks, it reads:*

> Anaximander . . . declared the Boundless to be principle and element of existing things, having been the first to introduce this very term of "principle" *[arkhê]*; he says that "it is neither water nor any other of the so-called elements, but some different, boundless nature *[physin apeiron]*, from

which all the heavens arise and the *kosmoi* [ordered worlds] within them; out of those things whence is the generation [or coming-to-be, *genesis*] for existing things, into these again does their destruction [or passing-away, *phthora*] take place, according to what must needs be; for they make amends and give reparation to one another for their offense, according to the ordinance of time," speaking of them thus in rather poetical terms. . . .

In Kahn's interpretation, it is the elemental opposites that arise from the apeiron *that "are one another's source of generation, just as they are the mutual cause of death. . . . The wet is generated from the dry, the light from the darkness. But the birth of such a thing involves the death of its reciprocal, and this loss must eventually be repaired by a backward swing of the pendulum. . . . The words and imagery of the fragment indicate above all that the exchange of birth and death is sure, remorseless, inescapable, like the justice which the gods send upon guilty men. It will come at last, when its hour is full. For Necessity enforces the ordinance which Time lays down." This ineluctable necessity is the divinely established order of the cosmos itself, for the "Decalogue" of Anaximander's impersonal monotheism, Kahn writes, "is the Law of Nature," whose "revelation is to be read in the ever-turning cycles of the sun, the moon, the planets, and the stellar sphere."*

Anaxímenês (mid sixth century), said to have been Anaximander's pupil, thought the primal substance was air, or vapor. According to Theophrastus, as cited by Simplicius, he held that air, becoming finer, turns to fire, and becoming thicker, to wind, then to cloud, water, earth, and stone in a process of eternal motion. Anaximenes' air, like Thales' water, remains in substance eternally the same beneath the continual transformations it undergoes. From it (Hippolytus writes) "things that are coming-to-be and that have come-to-be and that shall be, and gods and things divine, come-to-be, and other things from its offspring," through perpetually alternating rarefaction and condensation, the former producing heat and the latter cold. Thus within a generation or two the Milesian philosophers had revolutionized Greek and thereby Western thought, introducing problems and concepts central to natural philosophy for centuries and millennia to come: the concepts—in most cases, probably also the words—of physis *or nature itself; of a* kosmos *ordered by internal necessity or law* (anankê)*; of an abiding first substance* (arkhê) *underlying the diverse and changing phenomena of the world; and of transformations of this primal substance by interaction among the opposites arising from it in a continual process of phenomenal generation and corruption, coming-to-be and passing-away. Under the impact of ancient but newly discovered Babylonian observations, these thinkers also undertook bold astronomical and cosmological speculations on which their successors would build—speculations characterized, in Anaximander at least (as Kahn interprets him), by their "geometric character" and the revolutionary belief "that the universe was governed by mathematical ratios, and by ratios of a simple kind."*

This mathematical dimension of the Greek concept of nature, which would play a central role in later Platonic philosophy as well as in mathematics, geometry, and mathematical astronomy, has been associated since ancient times with Pythágoras and the Pythagoreans. About Pythagoras we know next to nothing, since he left no writings and surrounded himself with a cult of silence. Practically all we can say with any confidence about this mysterious figure is that he was an Ionian Greek who lived in the sixth century B.C. *and emigrated from his native Samos in the Aegean to the southern Italian colony of Croton (Krotôn) and later—after a political upheaval in Croton ended his domination of its government—to nearby Metapontum, where he died. He built up an influential secret society of followers closely knit by shared rituals and taboos (notably against eating meat and beans); several early writers, including Heraclítus and Xenóphañes, who scorned him, testify to his reputedly wide learning* (polymathiê) *and belief in reincarnation or metempsychosis, a central tenet of Pythagorean religion. Herodotus (II.123) relates that "the Egyptians were the first to tell the story that the human soul is immortal, and when the body dies it enters into another animal being born, and when it has gone through all those of land, sea, and air, it again enters into a human body at birth, and its cycle is completed in three thousand years. There are some Greeks who have adopted this story, some earlier and some later, as if it were their own; I know their names but do not write them." Xenophanes, a near contemporary of Pythagoras, was less reticent and far less respectful:*

They say Pythagoras expressed compassion,
seeing a puppy beaten, in this fashion:
"Oh, cease your blows! This soul I recognize:
It is a friend's; I knew him from his cries!"

(DK 7)

Porphyry much later summarized Pythagoras' reputed views: that "the soul is immortal, then that it changes into other kinds of animals, and that things happen again in cycles and nothing is simply new, and that all living things should be regarded as akin." The ultimate source of these beliefs is less likely to have been Egypt, as Herodotus thought, than India, where Jainism and Buddhism were arising at approximately the same time (Chapter 4 above).

Such beliefs remind us of mystical, even "shamanistic" elements that remained, along with mythological survivals, important dimensions of pre-Socratic thought (especially in the western Greek colonies of southern Italy known as "Greater Greece" or Magna Graecia)—and indeed of Plato's. But the "Pythagorean" doctrines that would be most crucial to later natural philosophy and science were those concerned with mathematics and music: the "discovery" of the "Pythagorean" theorem long known to the Babylonians; the duality of limit (associated with odd numbers and the good) and the unlimited (associated with even numbers and the bad), and the many permutations somehow generated by their interactions; the theory that all things are composed of numbers, and that the soul itself is an attunement or harmony; the definition of numerical ratios determining the intervals of the musical scale; the astronomical system (attributed to Philoláus in the late fifth century) positing a "central fire" around which the earth revolves along with sun, moon, planets, fixed stars, and an invisible "counter earth"; and belief that "music of the spheres" is created by revolution of the heavenly bodies but is inaudible because we have been accustomed to it from birth.

This medley of numerology and mathematics had an enormous and often fruitful impact on philosophical and scientific thought in antiquity, and beyond; since there is no evidence for ascribing any of these ideas (most of which are first mentioned by Aristotle) to Pythagoras, some probably took shape later under Platonic influence. Scholars have generally subscribed to Heidel's conclusion of 1940 that the Pythagorean role in Greek mathematics has been "much exaggerated"; thus Kahn discounts "the originality of Pythagoras as a figment—or at least an exaggeration—of the Hellenistic imagination," and Walter Burkert's painstaking analysis in Lore and Science in Ancient Pythagoreanism *(1962) demonstrates that much which later tradition attributed to Pythagoras was—as Aristotle knew—"Platonism and not Pythagoreanism."*

Whatever their provenance and history, however, some rudimentary "Pythagorean" numerical notions must have helped stimulate Plato's thought, becoming, in the form they eventually took (above all in Plato's Timaeus*), a key component of the Greek conception of nature. If the Milesians of eastern Greece had developed the seminal concepts of nature and of a living cosmos governed by its own necessity, and of coming-to-be and passing-away as orderly transformations of opposites engendered by an enduring* *primal substance, the Pythagoreans of the West came to be associated with concepts of mathematical proportion—already anticipated by Anaximander—no less important to later natural philosophy and science. The primary elements from which Greek thought would develop were ready, at this incipient stage, to be tested and challenged—as they surely would be—by equally brilliant thinkers soon to follow.*

Mythos and Logos

Fragments of the Pre-Socratic Philosophers, translated by Robert M. Torrance

Xenophanes of Colophon

Like Pythagoras, Xenóphanês was an Ionian emigrant to the west; born perhaps around 570 B.C.*, he left Kolophôn in Asia Minor, possibly after the Persian conquest in 545, and lived a long life in Sicily and southern Italy. He was not a systematic thinker but an iconoclastic rationalist whose verses mock the anthropomorphic gods of Homer and Hesiod and proclaim belief in (though not certain knowledge of) a single god "unlike men in mind or body." Clearly stimulated by Milesian speculation (tradition made him a pupil of Anaximander), he advanced bold scientific theories with unusual attention to empirical data. "Xenophanes thinks a mixture of earth with the sea is taking place," Hippolytus writes, "and that in time it is dissolved by its moisture. He claims to have proofs such as these: that seashells are found inland and in the mountains, and in the quarries of Syracuse he says an impression of a fish and of seaweed has been found, and in Paros an impression of a bay leaf deep in the rock, and in Malta flat forms of all marine life. These he says came into being when all things were covered with mud long ago, and the impression dried out in the mud; and that all human beings are destroyed whenever earth is carried down into the sea and becomes mud; then coming-to-be begins again, and this becomes the foundation for all worlds." Yet the Eleatic Stranger in Plato's* Sophist *(242D) makes Xenophanes a precursor of the Eleatic* mythos *(which arose in opposition to Milesian natural philosophy) "that what we call all things are really one," and Aristotle (*Metaphysics *986b) even reports that "Parmenides is said to have been his pupil."*

DK 11 — Homer and Hesiod ascribed to gods
all that brings infamy and shame to men:
adultery, theft, and tricking one another.

DK 14 — But mortal men think gods are born like them,
have clothing, speech, and bodies like their own.

DK 16 — The Ethiopians say their gods are snub-nosed
and dark, the Thracians, blue-eyed and red-headed.

DK 15 — If horses, cows, and lions all had hands
to draw with, making works of art, like men,
horses would draw their gods to look like horses,
cattle like cattle, each one representing
bodily forms exactly like their own.

DK 34 — No man knows certainly, or ever will,
about the gods and everything I speak of;
for even should he chance to speak the truth,
he would not know it: semblance governs all.

DK 18 — The gods have not revealed all things to start with,
but men by seeking find out more in time.

DK 35 — Let our conjectures seem to be like truth.

DK 23 — One god exists, among both gods and mortals
mightiest, unlike men in mind or body.

DK 24 — Wholly he sees, thinks wholly, hears things wholly.

DK 26 — Always in one place he remains, unmoving,
nor inappropriately shifts position

DK 25 — but effortlessly moves all things by thinking.

DK 29 — All that is born and grows is earth and water.

DK 27 — All comes from earth and ends in earth again.

DK 28 — Here by our feet is seen earth's upper limit
touching the air, but going down forever.

DK 30 — Sea is the source of water and of wind,
for neither would strong winds come blowing forth
out of the clouds without the mighty sea,
nor river streams, nor rain down from the aether,
but clouds and winds and rivers all alike
the mighty sea begets.

DK 32 — What they call Iris[2] is a cloud by nature,
purple and red and yellow to behold.

Heraclitus of Ephesus

Heraclítus (Hêrakleitos) was born, in the mid to late sixth century, of aristocratic family in the Ionian city of Éphesus in Asia Minor, which was spared by the Persians when Miletus was destroyed. Known to tradition as the weeping philosopher (in contrast to the laughing philosopher Demócritus, for whose temper Montaigne would express a preference two thousand years later), he was famed for his darkly cryptic style and arrogant contempt of his fellow men. "Learning many things does not teach intelligence," he writes (DK 40), "or it would have taught Hesiod and Pythagoras and again Xenophanes and [the Milesian historian] Hecataeus." Many legends soon grew up around him: "In the end," Diogenes Laërtius reports (IX.3), "becoming a misanthrope and withdrawing from society, he dwelt in the mountains eating grass and plants," and died at sixty when he shut himself up in a stall hoping to cure himself of dropsy by the heat from cow manure. His book On Nature, *which he deposited in the Temple of Artemis, was divided, Diogenes says (IX.5), into three discourses—on the universe, politics, and theology—and was*

[2]The goddess of the rainbow.

written obscurely so that only the powerful would have access to it and the populace could not easily despise it.

Though writing in prose, he "is not merely a philosopher but a poet, and one who chose to speak in terms of prophecy," as Kahn remarks in The Art and Thought of Heraclitus *(1979), and his "more than Delphic delight in paradox, enigma, and equivocation" leaves his riddling pronouncements susceptible to endless interpretation. His name is most often associated with universal flux, and though the words "all things flow"* (panta rhei) *do not appear among his fragments, it is this doctrine that Plato's Socrates stresses in* Cratylus 402A *(a dialogue named for the disciple of Heraclitus said by Aristotle to have deeply influenced the youthful Plato): "Heraclitus says somewhere that all things are in motion and nothing stays still, and comparing existing things to a river stream says that you could not step twice into the same river." Still more important, however, is the complementary doctrine of the* logos, *the proportion, order, or law underlying and uniting the seemingly disparate phenomena of a universe in flux and providing a measure* (metron) *common to all, though as difficult to understand as the logos or discourse of Heraclitus himself. The ever-living (thus ever-changing) fire of the cosmos is "kindled in measures and quenched in measures," so that nature—though it "loves to hide," and reveals itself only in the "unapparent attunement" or harmony* (harmonia) *of opposites continually at war yet continually transformed into one another—is potentially comprehensible by the human soul that seeks the cosmic logos within itself, knowing that its depths can never be fully plumbed. Though contributing little or nothing to the scientific achievements of the Milesians, Heraclitus deepened their conception of nature by exploring the hidden unity of world and mind that permits us to formulate conceptions corresponding to realities both within and beyond us.*

Therefore one must follow what is common [to all]; but although the logos is common, most people live as though they had a private understanding. (DK 2)

Listening not to me but to the logos it is wise to agree that all things are one. (DK 50)

Those things of which there is seeing, hearing, and learning, these I prefer. (DK 55)

Evil witnesses are eyes and ears for men if they have barbarian souls. (DK 107)

Nature loves to hide. (DK 123)

If one does not expect the unexpected one will not find it out, since it is not to be found out and impassable. (DK 18)

It is not possible to step twice into the same river. (DK 91)

Into the same rivers we both step and do not step, we are and are not. (DK 49a)

The path up and down is one and the same. (DK 60)

Hesiod is the teacher of most; they are convinced he knows most, who did not recognize day and night, for they are one. (DK 57)

Immortals mortals, mortals immortals, living each other's death, dying each other's life. (DK 62)

Fire lives the death of earth and air lives the death of fire, water lives the death of air, earth of water. (DK 76)

This cosmos, the same for all, no one of the gods or men made, but it ever was and is and shall be an ever-living fire, kindled in measures and quenched in measures. (DK 30)

Things taken together are whole and not whole, brought together brought apart, in tune out of tune; from all things one and from one all things. (DK 10)

God is day night, winter summer, war peace, satiety famine; he alters just as fire mingling with spices is named after the fragrance of each. (DK 67)

To god all things are beautiful and good and just, but men have supposed some things unjust and some just. (DK 102)

An unapparent harmony is stronger than an apparent one. (DK 54)

They do not understand how what differs with itself is in agreement: a harmonious tension as of the bow and the lyre. (DK 51)

It is necessary to know that war is common [to all things], and justice is strife, and all things come to be by strife and necessity. (DK 80)

War is father of all things, king of all things, and has shown some to be gods some human, made some slaves some free. (DK 53)

For souls to become water is death, for water to become earth is death; from earth water comes to be, from water soul. (DK 36)

A dry soul is wisest and best. (DK 118)

The limits of soul you would not discover, though you went tramping every road; so deep is its logos. (DK 45)

The soul's logos is self-increasing. (DK 115)

I sought after myself. (DK 101)

Character for man is destiny. (DK 119)

Parmenides of Elea

Parménidês of Elea (a colony founded in 540 B.C. on the Tyrrhenian coast of southern Italy) appears from Plato's Parmenides *to have been born around 515; according to Plato's account, he visited Athens with Zeno (Zênôn) when he was about 65, and there met the young Socrates. Later called both a pupil (but not follower) of Xenophanes and a Pythagorean, he was the first philosopher native to Magna Graecia, and founder of the influential Eleatic school whose main adherents were his pupils Zeno (author of famous paradoxes upholding the impossibility of motion) and Melissus. He is said, like Thales and Pythagoras before and Empedocles after him, to have played an important part in his city's political life, legislating (according to Plato's successor Speusippus) for its citizens. Diogenes Laërtius (IX.21) declares him the first to conceive of the earth as a sphere situated in the center of the universe (Anaximander had thought it a shallow cylinder) and perhaps the first (IX.23) to detect the identity of morning and evening star (though this had been known to the Babylonians for centuries).*

But immeasurably his most important work is his great philosophical poem On Nature, *as it came (like almost all pre-Socratic writings) to be called, major portions of which survive, largely in quotations by Simplicius a thousand years later. Written in epic hexameters said to have been addressed to Zeno, the poem begins in* mythos *with a proem (of which the closing lines are here included) narrating how Parmenides, a "knowing man," was borne aloft in a chariot drawn by "wise steeds" to the gates of Night and Day, which the daughters of the Sun persuaded Justice to open, admitting him to the presence of an unnamed goddess who promises to teach him all things, "both the unshaken heart of rounded Truth* [alêtheiê] *and men's opinions* [doxai]*"; the remainder of the poem is divided between the "Way of Truth" and the "Way of Opinion" (or Belief). Of these two, the first and far more influential part relentlessly applies a newly rigorous philosophical* logos, *or logic, to demolition of the very foundations on which the Milesian view of nature as a dynamic and self-transformative reality had been based. The only valid path of thought, Parmenides proclaims, is that Being* IS, *one and indivisible, ungenerated and undying, eternal and unchanging, continuous and immobile; non-being* IS NOT, *for it is "unthinkable and nameless." Since Being is one, it cannot become two or many: "Therefore plurality, becoming, change, motion," as Cornford writes in* Plato and Parmenides *(1939), "are in some sense unreal." The coming-to-be and passing-away which the Milesians thought to explain by transformations of a single underlying substance are inconceivable, and therefore illusions—to which the second and more fragmentary part of the poem is nevertheless devoted, much as Buddha in the* Shūrangama Sūtra *(Chapter 4 above) addresses the illusion of change in the phenomenal world.*

Thus Parmenides unleashed a radical assault both on the Milesian concept of nature and on the Heraclitean doctrines of flux and harmonized opposites: for if "Heraclitus was the prophet of a Logos *which could only be expressed in seeming contradictions," Cornford remarks, "Parmenides is the prophet of a logic which will tolerate no semblance of contradiction." Yet, as Karl Reinhardt astutely observes,*[3] *"he, the great revolutionary, the uncompromising gainsayer to all speculation in the mode of the* physikoi, *became, nonetheless, a more fruitful influence on natural philosophy than any of its adherents." Because it raised questions that no future thinker could ignore, "the Parmenidean attack on generation and corruption," as Kahn asserts (in* Anaximander*), "dominates the entire development of natural philosophy in the fifth century," and through elevation of logic above the deceptions of sense it deeply influenced Plato and thus all subsequent Western thought.*

The selections below (with lines numbered consecutively for convenient reference) comprise a large part of Parmenides' surviving fragments translated not literally—the language is often obscure; for scholarly translations see Kirk, Raven, and Schofield, and Leonardo Tarán's Parmenides *[1965]—but into English verse; for contrary to opinions that Parmenides is an uninspired writer of "clodhopping hexameters" who ought to have philosophized in prose, "every line he wrote pulsates," as Jaeger affirms in* Paideia, *"with his ardent faith in the newly discovered powers of pure reason. . . . Parmenides was a natural poet, because he was carried away by his conviction that he must preach his discovery, the discovery which he believed to be in part at least a revelation of the truth."*

From the PROEM

DK 1

. . . Kindly the goddess welcomed me, and taking
my right hand in her own, she thus addressed me:
"Young comrade of immortal charioteers,
you whom these mares have borne to my abode,
greetings! No evil fate has made you travel 5
this path, far from the trodden ways of men,

[3] In an excerpt from *Parmenides und die Geschichte der griechischen Philosophie* (1916) translated in Mourelatos, ed., *The Pre-Socratics.*

but right and Justice. You must learn all things,
both the unshaken heart of rounded Truth
and men's opinions, lacking true conviction.
Nonetheless, you shall learn how things that seem — 10
to be pervade the universe with semblance."

From THE WAY OF TRUTH

DK 2 — Come, I shall tell you—hear my words and heed them!—
what paths are open for investigation:
either IT IS and cannot ever NOT BE,
the path of firm conviction, serving Truth, — 15
or IT IS NOT and must NOT BE forever—
a pathway utterly beyond conceiving.
For what IS NOT could never possibly
DK 3 — be known or said: to think and be are one.

DK 6 — What can be said and thought must BE, for Being — 20
can BE, but Nothing cannot: ponder this!
That path I first forbid you from pursuing:
then this, on which poor mortals, knowing nothing,
wander two-headed (helplessness directing
errant thoughts in their breasts) as they are swept — 25
away, deaf, blind, dazed, undecided hordes
who think to be and not to be the same
and not the same: this path leads only backward!

DK 8 — Never shall this prevail, that what IS NOT
IS: keep your thoughts from following that path! — 30
Let worldly ways not force your blinded eye,
echoing ear, or tongue to roam this path,
but with your reason judge the thorny proof
spoken by me.
One path alone is left
to tell of: that IT IS. On this are many — 35
signs that what IS is unborn and immortal,
sound of limb, imperturbable, unending.
It never was, nor shall be; all now IS
one indivisibly. Why seek its birth?
Or whence it grew? I shall not let you say — 40
or think: "from what IS NOT"; no speech or thought
can make it NOT BE. What could make it grow,
now or before, if it began from nothing?
Thus it must either wholly BE, or not.
Trustworthy faith lets nothing but itself[4] — 45
rise from Non-Being; therefore Justice never
unchains it to be born or pass away

[4]Some, following Cornford, have construed these lines to read "lets nothing else but Being . . ." The precise meaning is in either case imperfectly clear.

but holds it fast, still leaving this decision:
IS it, or not? But it has been decided
that one path is unthinkable and nameless— 50
no true path—and the other real and valid.
How could what IS then die? How come to be?
It IS not if it came to be, or shall:
thus birth dies out, and death is past conceiving.
Nor can this homogeneous whole be sundered, 55
nor greater density obstruct cohesion,
nor lesser: everything is full of being.
All is continuous; being clings to being.
Motionless now within its binding limits,
beginningless and endless, it has banished 60
birth and destruction through true understanding.
Unmoved and changeless in itself, it stays
firmly at rest, for strong Necessity
binds and contains it by encircling limit
since being cannot rightly be unbounded . . . 65

From THE WAY OF SEEMING

DK 8

Here my trustworthy discourse on the truth
I end: now listen to my artfully
dissembling words concerning men's opinions.
Men have made up their minds to name two forms
but gone astray in naming even one, 70
judging them opposites and giving each
a different sign: to one, aetherial flame,
weightless and mild, alike all round, but unlike
its no less self-sufficient opposite,
dark night, corporeal and densely massive. 75
I shall describe its seeming dispensation
so that no mortal judgment may outstrip you.

DK 10

You shall perceive the nature of the aether
and all its constellations, the destructive
works of the sun's bright torch and whence they came, 80
learn the works and the nature of the wandering
circular moon, and know the sky around it
from which it grew, and how Necessity
bound it within the limits of the stars.

Empedocles of Akragas

Empédocles (Empedoklês) was a native of Ákragas (Roman Agrigentum), a major Greek city on the southern coast of Sicily founded around 580 B.C.; he probably lived from around 493 to 433. Of aristocratic birth, he was a fervid democrat and skilled orator (called by Aristotle the inventor of rhetoric) and was said to have refused the kingship of his city, ruled for much of the previous century by tyrants. Legend attributed important public works to him, including diver-

sion of a pestilential river, and he was famed as a physician and even miracle-worker—a belief he promoted in his Purifications, *which he recited at Olympia, apparently during exile from Akragas. The most famous legend in a life that inspired poetic elaboration from antiquity to modern times (notably in Friedrich Hölderlin's* Empedokles *and Matthew Arnold's* Empedocles on Etna) *tells of his death by leaping into the fiery crater of Mount Etna in order to show that he had become (as he proclaims himself in the* Purifications*) a god.*

He is variously reputed to have been a pupil of Parmenides, the Pythagoreans, and even Xenophanes; in his philosophical poem to his pupil Pausanias, predictably known as On Nature, *Parmenides' influence is obvious. In the substantial fragments (of uncertain order) surviving from this poem, Empedocles reaffirms Parmenides' tenet that the universe is unborn and undying, but not his repudiation of sensory perception and change. If it is impossible for a single underlying substance or* arkhê *to generate plurality, change can be explained, Empedocles proposes, by incessant interaction of four eternal principles—fire, air, water, and earth—which he called "roots"* (rhizômata) *and which others later called "elements"* (stoikheia, *literally "letters" of the alphabet): "only a mix and constant intermixture," in his new conception, "is this which human beings know as nature." As H. F. Cherniss writes,*[5] *"Generation and destruction are expressly declared to be impossible, merely conventions of human language; and all the multitudinous differences in the world are simply the external appearances of the various combinations of the identical roots." Two forces, Love or Amity* (Philotês) *and Strife or Hatred* (Neikos), *govern the "continual interchange" of the roots, alternately binding them into unity and splitting them up again in a cosmic cycle. In some interpretations of the confusingly scattered fragments, this cycle progresses from perfect harmony of the two contrary forces in a primordial "Sphere," to a chaotic vortex in which Strife scatters the roots in randomly combined "undifferentiated shapes," to blissful dominance of Love in a golden age destined not to last because it cannot assimilate Strife, which has begun to dominate the present age. (What the outcome will be after Strife's ascendance—whether return to the Sphere from which the cycle began, or terminal dispersion of the roots—remains disputable in this cosmology, as it does in modern accounts of the Big Bang.)*

In Empedocles' second and probably later poem, the Katharmoi *or* Purifications, *the prominence of Pythagorean dimensions—including metempsychosis, ritual purification, condemnation of animal sacrifice, and an overall tone of exalted religiosity—is so striking that some have postulated an intervening "conversion" experience. Yet the difference is of degree, not kind:* mythos *is prominent in the earlier poem, where the roots are given names of gods (Zeus, Hera, Aidoneus, and Nestis), and Love is identified with Aphrodite, or Cypris; nor is it always clear to which poem some fragments belong. Empedocles is perhaps best viewed, as by Dodds in* The Greeks and the Irrational, *as "a very old type of personality, the shaman who combines the still undifferentiated functions of magician and naturalist, poet and philosopher, preacher, healer, and public counsellor," for whom* mythos *and* logos *are complementary forms of expression. As with Parmenides, selections from Empedocles are here translated into verse—for despite Aristotle's dismissal* (Poetics *1447b), he has been widely acclaimed as a poet—and lines numbered consecutively. See also William E. Leonard's verse translation,* The Fragments of Empedocles *(1908), and M. R. Wright's in prose, with scholarly commentary,* Empedocles: The Extant Fragments *(1981). Besides works already cited, see Helle Lambridis's* Empedocles: A Philosophical Investigation *(1976) for a basic introduction, and D. O'Brien's* Empedocles' Cosmic Cycle *(1969) for detailed scholarly analysis of this central theme.*

[5]In "The Characteristics and Effects of Presocratic Philosophy" (1948), included in Furley and Allen, eds., *Studies in Presocratic Philosophy*, vol. I.

From On Nature

DK 1
Pausanias, listen, son of wise Ánchitês!

DK 2
Narrow means are distributed through our bodies,
and hordes of pressing troubles stifle thought.
Seeing but little of their life while living,
then swiftly dead, men fly away like smoke 5
only believing what each randomly
comes upon, thinking he has found the whole.
Differently must men see and hear these things,
and understand: you, shunning misconception,
shall learn whatever human wit can compass. 10

DK 3
Deflecting folly from my tongue, O gods,
draw from my pious lips a lucid spring!
And, much-remembering white-armed maiden Muse,
you I entreat: drive down, so far as mortals
may know, the chariot of Reverence. . . . 15

DK 11
Foolish are they who lack farsighted thoughts,
thinking what never was can come to be,
or anything die out and be destroyed.

DK 12
Impossible that something come from nothing;
unthinkable that anything should perish! 20
Wherever placed, it always shall remain.

DK 16
[Of Love and Strife:]
As once they were, they shall be, nor, I think,
shall endless time be emptied of these two.

DK 6
First hearken to the four roots of all things:
dazzling Zeus, Hera bringing life, Aidoneus, 25
and Nestis, watering mortal springs with tears.

DK 8
This too I tell you: no one mortal thing
gives nature birth or causes its destruction:
only a mix and constant intermixture
is this which human beings know as nature. 30

DK 17
Twofold my tale: first, out of many one
grew up, then grew apart from one to many.
Double the birth and fall of mortal beings:
union of all creates one, then destroys it,
breeding another from its dissolution. 35
None ceases this continual interchange,
all merging into unity through Love,
then splitting up through hateful Strife again.
Thus growing out of many into one
and, as they grow apart again, to many, 40

they come to be, and have no long duration;
but by their never-ceasing interchange
unmoved through every cycle they remain.
  Come, hear my words, for wisdom comes from learning.
Again, to clarify my meaning further, 45
my tale is twofold: first, from many one
grew up, then grew apart from one to many,
fire, water, earth, and air's unmeasured height,
dread Strife, their segregated counterbalance,
and Love, inextricably mixed among them. 50
Her with your mind observe, not with dazed eyes!
In mortal limbs she is thought to be inborn;
friendly thoughts and beneficence she fosters,
earning the names of Joy and Aphrodite.
Mortal men never see her as she whirls 55
among them: heed my undeceiving story!
  Though these are equal, and of equal age,
each has its proper character and function,
in turn prevailing as time circles round.
Besides these, nothing comes to be, or ceases, 60
for, if they died off, nothing more would be.
What could increase this All? Whence could it come?
Where could it perish, since no place is empty?
No, these alone exist and, intermingling,
change forms and evermore remain the same. 65

DK 20

This may be clearly seen in human bodies.
Sometimes, when life is in its flowering prime,
all bodily limbs combine in one through love;
sometimes, divided by destructive conflicts,
each wanders by the shore of life alone. 70
Likewise with shrubs, fish in their watery houses,
beasts in their mountain lairs, and wingèd birds.

DK 21

Come, witness proofs of what I said before,
if form was insufficient to my matter.
See the warm sun shine brightly everywhere; 75
heavenly bodies steeped in its dazzling rays;
rain permeating all with chilly darkness;
things firmly grounded flowing forth from earth.
In Anger, all are disparate and divided,
in Love, made one by mutual desire. 80
From these, all things that were and are and shall be
spring into being: trees and men and women,
animals, birds, fish nurtured in the water,
and the long-living gods, supreme in honor.
Only these things exist and, intermingling, 85
transform their shapes, so much does mixture change them.

DK 22

All these—the beaming sun, earth, sky, and sea—
are bound in unity from component parts
scattered throughout the world of mortal beings.
Likewise, all things most suitable to mixture 90
are made alike through love by Aphrodite.
But those that differ most, through mutual hatred,
in origin, mixture, and in forms of molding,
wholly repudiate combining, prompted
by Strife, to whom they owe their generation. 95

DK 23

As painters, decorating votive tablets—
artisans skilled and cunning in their craft—
take many-colored pigments in their hands,
mix them in harmony, some more, some less,
and fashion semblances of all things from them, 100
manufacturing trees and men and women,
animals, birds, fish nurtured in the water,
and the long-living gods, supreme in honor:
be not deceived, or fancy mortal beings'
numberless forms have any other source. 105
Be sure of this, since from a god you heard it!

DK 35

I shall retrace the pathway of the song
I told before, connecting words together.
When Strife attains the whirlpool's lowest depth
and from its vortex Love comes into being, 110
all things unite in one, not suddenly,
but voluntarily, from all directions;
and as they mingled, countless mortal tribes
poured forth while others, unmixed, jostled with them,
held back as hovering Strife—not blamelessly 115
withdrawn yet to the circle's outer limits—
clung to some parts but drew apart from others.
But always, as it ran away, there entered
the mild divinity of blameless Love,
till what had learned to be immortal swiftly 120
grew mortal, unmixed mixed, exchanging paths;
and as they mingled, countless mortal tribes
of every shape poured forth, a stunning marvel!

DK 38

Come, I shall tell you whence the sun and other
bodies we now behold came into being, 125
earth and the sea of many waves, moist air,
and Titan Aether who encircles all.

DK 57

Out of [earth] sprang forth many neckless faces,
arms wandered naked, unattached to shoulders,
and eyes deprived of foreheads roamed alone. 130

DK 59

As one divinity increasingly
mixed with the other, chance brought these together,
and many more were constantly engendered.

DK 61

Many grew up with doubled face and breast,
cows born with human faces, and again 135
men with cows' heads, and creatures mingling men's
and women's parts, equipped with shadowy members.

DK 62

Learn how fire, separating out, draws shoots
of woeful men and women forth from night.
Not ignorant, nor aimless, is my story! 140
Undifferentiated shapes of earth first rose,
partaking both of water and of heat;
striving to reach its like, fire elevated
parts not displaying yet the lovely limbs,
voice, or the member suitable to men. 145

DK 98

Earth, anchoring at last in Cypris' harbors,
blended with these in roughly equal parts—
Hephaestus, moisture, and all-shining aether—
with some a little more, with others less:
thence blood and other forms of flesh arose. 150

DK 103

Thus by the will of Chance all things are conscious.

DK 106

Human intelligence grows toward what meets it.

DK 109

By earth we know of earth, by water, water,
aether by aether, fire by raging fire,
fondness by fondness, strife by grievous strife. 155

DK 110

For if you firmly contemplate these [truths]
and guard them faithfully with pure endeavor,
throughout life they will always stay beside you
bestowing boundless profit by augmenting
qualities proper to each person's nature. 160
But if you reach for other, worthless things
such as assail and blunt men's thoughts by thousands,
these, as time circles round, will soon forsake you,
longing to find their own dear kind again:
for all partake of thought and understanding. 165

DK 111

All drugs preventing illness and old age
I shall reveal; for you alone I do this!
You shall restrain the impetuous force of winds
swooping to earth and devastating crops,
and, if you wish, restore reviving breezes; 170
after dark rain, create a drying season
for men, then after summer's drought, invoke
rivers to nurture trees down from the aether
and up from Hades bring a thriving dead man.

From the PURIFICATIONS

DK 112

Friends on the heights of yellow Akragas'
great citadel, beneficently minded,
harboring strangers, innocent of evil,
greetings: I mingle with you now, no longer
human, but honored as a deathless god 5
suitably crowned with ribbons and fresh flowers.
All those whose prosperous cities I approach,
women and men alike, revere me; thousands
follow, beseeching what road leads to riches,
some clamoring for prophecies while others, 10
pierced by incessant torments, beg to learn
remedies for diseases of all kinds.

DK 115

This oracle Necessity and the gods
decreed of old and sealed with binding oaths:
when any daemon destined to long life 15
contaminates his hands with bloody crimes
or negligently falsifies an oath,
for thrice ten thousand years, remote from blessings
he wanders, born in countless mortal forms,
stumbling from one hard pathway to another. 20
Winds from the aether blow him toward the sea,
sea spits him onto land, earth into blazing
rays of the sun, sun back to swirling aether:
each takes him from the last, but all abhor him.
I too am now an exile from the gods, 25
a wanderer trusting none but raving Strife.

DK 124

Woe to the wretched race of cursèd mortals:
out of such strife and moaning you were born!

DK 117

Already I have been boy, girl, and bush,
bird, and dumb fish cavorting in the sea. 30

DK 128

Ares was not their god, nor din of battle,
nor Zeus the king, nor Kronos nor Poseidon,
but Cypris was their queen. . . .
Her they appeased with sacred images
of animals, and intricate perfumes, 35
then sacrificed pure myrrh and frankincense,
pouring libations of sweet yellow honey.
No altars dripped with gore of slaughtered bulls,
but this, of all defilements, seemed the greatest:
to rip the life from sturdy limbs, and eat them. 40

DK 130

All things were tame and well-disposed to humans,
both animals and birds, aglow with friendship.

Anaxagoras of Clazomenae

Anaxágoras of Clazómenae (Klazomenai), off the coast of Asia Minor, lived from about 500 to about 428 B.C., and was one of the last great Ionian natural philosophers. He spent many years, probably from about 480 to 450, in Athens, which until then, like other cities of mainland Greece, had played no important part in Greek philosophy. Through his acquaintance with Anaxagoras, Péricles (Periklês), the Athenian leader, acquired "high-mindedness," Socrates says in Plato's Phaedrus *(270A), and by study of the heavens* (meteôrologia) *learned also of mind* (nous) *and the art of words. The friendship proved dangerous, however, when Pericles' enemies brought Anaxagoras to trial—charged either with impiety for his materialistic view that the sun is a red-hot mass of metal or with treasonous sympathy for the Persians—and was either fined and exiled or condemned to death in absentia after fleeing to Lampsacus on the Asian shore of the Hellespont, where he died. He too followed Parmenides in affirming that nothing comes-to-be or passes-away, and like the younger Empedocles saw what is called coming-to-be as a mixing together and passing-away as a sorting out. But his solution to the dilemma that plurality could not arise from unity was to postulate an indefinite number of infinitely divisible "seeds" or elements (one kind for every existing substance), once mingled indiscriminately together and each still containing traces of all others in different proportions. Only mind or reason* (nous), *which initiates the rotational process of cosmic sorting, "is infinite and self-ruling, and is mixed with no thing, but is alone by itself," moving all other things.*

All things were together, infinite in number and in smallness. . . . (DK 1)

Air and aether are separated out from the multiplicity surrounding them, and what surrounds them is infinite in number. (DK 2)

. . . Before these things were separated out, while all things were together, no color was distinct, for the mixing of all things prevented this, of moist and dry and of hot and cold and of bright and dark, and there was much earth in the mixture and seeds infinite in number and not at all like one another. For none of the other things, either, is like any other. And this being so, we must believe that all things were present in the whole. (DK 4)

And since parts of great and small are equal in number, so in everything there would be everything; nor is it possible to exist apart, but everything has a portion of everything. . . . (DK 6)

Things in the one cosmos are not separated off from one another by an axe, neither hot from cold nor cold from hot. (DK 8)

In all things there is a portion of everything except mind, and in some there is mind also. (DK 11)

Other things have a portion of everything, but mind is infinite and self-ruling, and is mixed with no thing, but is alone by itself. For if it were not by itself, but were mixed with anything else, it would share in all things, if it were mixed with anything; for in all things there is a portion of everything, as I said before. And the things mixed [with mind] would have prevented it from ruling over any thing in the same way as it can, being alone by itself. For it is the finest of all things and the purest, and it has complete knowledge about everything, and greatest strength; and all things, both greater and smaller, that have life [or soul], mind rules. And mind ruled the entire rotation, so that it began rotating in the beginning. And it first began rotating from a small place, but now it rotates more widely and will rotate more widely still. And things that were mixed together and separated out and divided, all these mind knew. And whatever things were to be, and were, that are not now, and all things that are now, and whatever shall be, mind arranged them all, and also this rotation of the stars and the sun and the moon, and the air and aether that were separated out. This rotation made them separate out. And dense is separated out from rare and hot from cold and bright from dark and dry from moist. There are many portions of many things. Nothing is altogether separated out or divided one from the other except mind. Mind is all alike, both greater and less. Nothing else is like anything else, but each single thing is and was most plainly those things of which it contains the most. (DK 12)

And when mind initiated motion, from all that was moved it was separated out, and all that mind moved was divided, and as things moved and divided, the rotation greatly increased the dividing. (DK 13)

The dense and moist and cold and dark came together here, where [earth] now is, and the rare and hot and dry went out to the furthest part of aether. (DK 15)

The Greeks do not think correctly about coming-to-be and passing-away, for no thing comes-to-be or passes-away, but is mixed together or separated out from existing things. And thus they would be correct to call coming-to-be "mixing together" and passing-away "separating out." (DK 17)

The sun gives the moon its brightness. (DK 18)

Through feebleness [of our sense perceptions] we are unable to judge truth. (DK 21)

Democritus of Abdera

*Atomism originated, according to Aristotle (*De Generatione *425a), with Leucíppus (Leukippos), who in the fifth century* B.C. *advanced a theory that he thought accorded with sense perception and would not, like that of Parmenides, abolish either coming-to-be and passing-away or motion and multiplicity. This theory, developed mainly by Demócritus (Dêmokritos, ca. 460–ca. 360?) of Abdêra on the coast of Thrace, held that being is not, as Parmenides insisted, continuous and unbroken, thereby precluding motion, but consists of both matter and void—the nonbeing whose possibility Parmenides vehemently denied. (In Democritus's words, "No less than aught is there naught.") Through this infinite void in which countless worlds are interspersed, imperceptibly tiny and imperishable atoms (*atomos *means "uncuttable" or "indivisible") are eternally falling, alike in substance but differing in size and shape. From their random collisions matter is formed, so that "coming-to-be" results from their entanglements and "passing-away" from their reboundings. Atoms and void alone truly are by nature, all else exists provisionally and by convention. Only fragments of some seventy works attributed to Democritus survive, and many of these are ethical proverbs for which this "laughing philosopher" of good cheer was renowned: "A life without holidays," he remarked (DK 230), "is like a long road without an inn." Other sayings proclaim the uncertainty of knowledge stressed by this undogmatic thinker for whom, Aristotle says (*Metaphysics *1009b), "either there is no truth or to us at least it is not evident." Full exposition of the atomic theory which Epicurus (Epikouros) adopted from Democritus would be transmitted—via a single manuscript—only in the great Latin philosophical* *poem of Lucretius,* De Rerum Natura (On the Nature of Things, *Chapter 10 below), from the first century B.C. The materialistic atomism of Democritus, in which pre-Socratic natural philosophy culminates and which, more than any other of its speculations, anticipates modern physical science, evidently won little following throughout most of antiquity.*

Sweet exists by convention, bitter by convention, hot by convention, cold by convention, color by convention, but atoms and void truly exist . . . . We perceive nothing accurately in reality, but [only] as it changes, according to the disposition of our body and of things that strike into and press against it. (DK 9)

There are two forms of knowledge, one legitimate, one bastard, and all these belong to the bastard form: sight, hearing, smell, taste, touch. The legitimate form is separate from this. . . . (DK 11)

Nature and instruction are akin. For instruction refashions the man, and by refashioning him forms his nature. (DK 33)

Man is a little cosmos. (DK 34)

We really know nothing; truth is in the abyss. (DK 117)

We are pupils of the animals in the greatest matters; of the spider in spinning and mending, of the swallow in building houses, and of the songbirds, the swan and the nightingale, in singing, by imitation. (DK 154)

No less than aught is there naught. (DK 156)

Do not endeavor to understand everything, lest you become ignorant of everything. (DK 169)

Chance is munificent but unstable; nature, however, is self-sufficient. Therefore it is victorious over the greater [promise of] hope. (DK 176)

More people become good through practice than by nature. (DK 242)

To a wise man all the earth is open, for the whole cosmos is the native land of a good soul. (DK 247)

Alike Divine and Human

Selections from the "Hippocratic Corpus," translated by Robert M. Torrance

Hippócrates (Hippokratês), the most famous of ancient physicians, was a contemporary of Socrates from the Dorian island of Cos (Kôs) off Asia Minor; little is known of his life, though late sources say he was born in 460 B.C. and died in 357, at the age of 104. From a remark in Plato's Phaedrus *(270C), on which Socrates approvingly expands—"Hippocrates the Asclepiad says that the nature even of the body can only be understood as a whole"—it appears that Hippocrates held an organic and holistic conception of the interrelation of bodily parts to the whole. So great was his renown that the various texts of the "Hippocratic Corpus" were attributed to him in antiquity, though scholars now consider them all anonymous (with the possible exception of* The Nature of Man, *ascribed to Hippocrates' son-in-law Pólybus), dating most of them in the century between about 430 and 330, and some even later. The treatises differ greatly among themselves, some more popular, others more technical, some more empirical, others more theoretical; they contain extended case studies and evidence of close observation and even experimentation, but also much "bluff and dogmatism" (in G. E. R. Lloyd's phrase) that would help perpetuate false opinions and unexamined assumptions.*

Yet what is most significant in the best of these writings—including the excerpts translated below from volumes I, II, and IV (1923–31) of the Loeb Classical Library Hippocrates, *ed. W. H. S. Jones—is their firm rejection not only of magical and superstitious beliefs but also of speculations not confirmed by evidence, and their affirmation of a conception of* physis, *governed by discernible laws of cause and effect, that made more concrete the revolutionary ideas of the pre-Socratic* physiologoi *and extended them from cosmic speculation to human practice. As Lloyd notes in* Magic, Reason and Experience *(1979), "the deliberate investigation of how particular kinds of natural phenomena occur only begins with the philosophers," but becomes a fundamental procedure in some medical writers; the author of* On the Sacred Disease, *in particular, "has a conception of nature, and a view of what constitutes a causal explanation, that rule out supernatural intervention in diseases." That the observations of these early physicians were imperfect, and the conclusions of their "art" (or "science"; the Greek* tekhnê *might be translated either way) sometimes mistaken, was inevitable; the treatment prescribed for the "sacred disease" (epilepsy) in later parts of that treatise could only, as Lloyd remarks, have been ineffective. Nor were their doctrines entirely free of the philosophical abstractions they somewhat ungratefully scorned. The immensely influential theory of the four bodily fluids or "humors" (as they were later called, from Latin* humor*)—blood, phlegm, yellow bile, and black bile—is clearly analogous to Milesian polarities of hot and cold, dry and moist, and to Empedocles' four elements. Yet their insistence that no disease is more "sacred" (hence more susceptible to religious cures or superstitious quackery) than any other—"since all are alike divine and human"—and their sober awareness both of the possibilities of their art and of its fallibility and limitations mark an impressive stage in the progress of rational Greek science. In the attention many of these writers give both to the physician's ethical responsibility to his patients (as embodied in the Hippocratic oath) and to prevention of disease through a temperate regimen of diet and exercise, their works seem closer to us than many of far more recent times. For further selections, see G. E. R. Lloyd, ed.,* Hippocratic Writings *(1978).*

FROM On Ancient Medicine

1 All who have tried to speak or write about medicine by laying down some postulate for their argument—heat or cold or moisture or dryness or whatever else they wish—thus narrowing down the first principle of causation for human diseases and death to the same one or two for all, are obviously mistaken, on the evidence of their own words, in many ways. They are especially blameworthy because their mistake concerns an art of which all make use in matters of greatest importance and for which good craftsmen and practitioners are most honored. Some indeed are incompetent, but others very different; yet this could not be the case if there were no such thing as medicine and nothing were investigated or discovered by it, for all would be equally inexperienced and ignorant in it, and patients would be treated entirely by chance. But this is not so; rather,

just as in all other arts practitioners differ greatly from one another in skill and in knowledge, so they do in medicine. For this reason I think it has no need of any empty postulate, as invisible or unresolvable matters do, in which anyone trying to say something might find it necessary to form a postulate—about things in the sky, for example, or under the earth. If anyone said he had learned how these things are, it would not be clear either to the speaker or his listeners whether this was true or not; for there is no way to test it and know for sure.

2 But for medicine all this has long been feasible; it has discovered both a first principle and a method by which many valid discoveries have been made over a long period of time, and by which other things will be discovered if one capably sets out to seek them with knowledge of what has been discovered already. Anyone who rejects and casts all this aside, trying to conduct his search by another method and in another fashion, and says he has discovered something, has been and is being deceived. For this is impossible. . . .

20 Some doctors and sophists say it would be impossible for anyone to know medicine who does not know what man is: without learning this he will not be able to treat patients properly. But this subject belongs to philosophy—things such as Empedocles and others have written about nature: what man is from the beginning, how he first came to be, and of what elements he was constructed. But I think that everything sophists or doctors have said or written about nature belongs less to the medical than to the literary art. I also think that sure knowledge of nature can come from nothing but medicine, and that one can gain this knowledge only by rightly understanding medicine: until then, one will fall far short of accurately mastering this inquiry into what man is, and through what causes he comes to be, and other such things.

FROM On the Sacred Disease

1 As for the disease called "sacred," it seems to me no more divine or sacred than other diseases, but has a natural cause; men have thought it something divine out of inexperience and astonishment that it has no resemblance to others. Through incapacity to understand it they perpetuate its supposed divinity, which they then undermine through facile cures by purifications and incantations. But if it is thought divine because it is astonishing, there will be not one sacred disease but many, since I will show that others, which no one thinks sacred, are no less astonishing and portentous. Thus quotidian, tertian, and quartan fevers seem to me no less sacred and divine in origin than this disease, yet no one is astonished at them. Thus too I see men go mad and become delirious for no apparent cause, and do many strange things, and I know of many who moan and scream in their sleep, others who choke, or leap up and run outside, delirious until they wake up. Then they are as healthy and sensible as before, though pale and weak; and this happens not once but often. There are many other such maladies of every sort, but it would take too long to discuss them all.

2 In my opinion those who first called this disease sacred were men like our present-day magicians and purifiers and charlatans and quacks who pretend to be extremely pious and to have superior knowledge. In their helplessness at having nothing beneficial to

offer they cloaked and shielded themselves with the divine, and to conceal their total ignorance they called this affliction sacred; then by making up plausible stories they established a treatment that saved their own reputations . . .

3 Therefore I think that those who attempt to cure these diseases in this way consider them neither sacred nor divine; for if they can be removed by such treatments as purifications, why should attacks on men not originate from devices of a similar kind? If so, nothing divine is the cause, but something human. Anyone who is able to dispel such an affliction by magical purifications could also bring it on by other such devices, and by this argument its divinity is undermined. By saying and contriving such things they pretend to superior knowledge, and deceive men by prescribing cleansings and purifications and by a great deal of talk about divinities and spirits. Yet in my opinion their talk is not of piety, as they think, but of impiety, as if the gods did not exist, and their "pious" and "divine," as I shall demonstrate, are impious and unholy. . . .

5 In my view, this disease is no more divine than others, but has the same nature and cause as other diseases. It is also no less curable than others, unless through long passage of time it has become so ingrained as to be too powerful for the medicines prescribed for it. Its origin, like that of other diseases, is hereditary. . . .

17 Men ought to know that from nothing else but the brain do our pleasures, joys, laughter, and amusements come, and our sorrows, troubles, cares, and tears. Through it, in particular, we think, see, and hear, and distinguish ugly from beautiful, bad from good, pleasant from unpleasant, judging some of these things by convention, others by our perceptions of what is expedient. By its means, too, we go mad or delirious, and terrifying specters present themselves to us, whether by night or by day; from it come insomnia and wandering thoughts, absentmindedness and unaccustomed memory lapses. These afflictions all come when an unhealthy brain becomes hotter than is natural to it, or colder, or moister, or drier, or suffers some other unnatural affliction to which it is not accustomed. Moistness gives rise to madness; for when it is unnaturally moist, it is necessarily perturbed, and when it is perturbed neither sight nor hearing is firm, but we see and hear now one thing, now another, and the tongue speaks about things that are seen and heard at any given time. Whenever the brain is unperturbed, however, a man remains in his senses. . . .

21 This so-called sacred disease arises from the same causes as others do, from things that come and go, and from cold, and the sun, and the shifting and ever restless winds. These things are divine, so that there is no need to distinguish this disease as diviner than others, since all are alike divine and human. Each has a nature and faculty of its own, and none is beyond help or treatment; most are curable by the same things as caused them. One thing nourishes one thing, another another, but sometimes also harms it. The physician must know how to discern the season for restoring nourishment and growth to one thing and for depriving and harming another. For in this disease as in all others it is imperative not to increase the illness but to wear it down by applying to each disease what is most contrary to it, not what it is accustomed to: from the accustomed it flourishes and grows, but from the contrary it weakens and wastes away. Whoever understands a regimen for bringing about dryness and moisture, heat and cold, in men, could cure this disease also, if he distinguished the seasons for profitable treatment, with no need for purifications and magic.

FROM On the Medical Art

3 . . . Concerning medicine, the subject of this discussion, I shall give an exposition, first defining what I think it is. In general, it is doing away with the troubles of the sick and blunting the intensity of their diseases while not undertaking to treat patients already mastered by their diseases, knowing that medicine is powerless here. The rest of my discussion will show that medicine accomplishes these things and is continually able to do so. . . .

8 There are some who fault medicine because physicians will not undertake to treat patients already mastered by their diseases, saying they treat only cases that would be cured by themselves but will not touch those in need of most help, whereas if medicine were an art, it should cure all alike. If those who say these things blamed physicians for not treating their own delirium their blame would be more appropriate. If anyone expects that an art can perform what does not belong to art, or nature what is unnatural, his ignorance is more akin to madness than to lack of learning. For we can only be craftsmen of things we master through the tools of nature or of art, and nowhere else. Whenever a man suffers from an ill stronger than the tools medicine has at its disposal, then, he must not expect it to be vanquished by medicine. . . .

FROM On the Nature of Man

1 Anyone used to hearing people talk about "human nature" beyond what is proper to medicine will not find the present discussion of interest. For I do not claim that man is all air, or fire, or water, or earth, or anything else that is not manifestly part of man. Those who wish to talk of such things are free to do so, but I do not think they know what they are talking about. They all employ the same theory, but do not say the same things; they tack the same conclusion onto their theory and say that what Is is some One Thing, and this One Thing is the All, but cannot agree on its name. One declares that this One and All is air, another fire, another water, another earth, and each tacks onto his discussion arguments and proofs, which come to nothing. When people make use of the same theory but do not say the same things, they clearly know nothing about it. . . .

4 A man's body contains blood and phlegm and yellow bile and black bile; these constitute the nature of his body, and because of these he feels pain or is healthy. He is healthiest when they are well proportioned in strength and quantity, and thoroughly blended; he feels pain when one is deficient or excessive, or is isolated in his body and not mixed in with the others. For when one is isolated and stands by itself, not only must the part it leaves become diseased, but the part where it gathers and overflows swells up and causes pain and distress. When the body discharges a superfluity, the drainage causes pain. But if the drainage and transfer of a fluid separating out from others takes place internally, it necessarily causes a double pain, as I have said already: in both the part it leaves and the part in which it accumulates.

5 I said I would demonstrate that the human constituents I have defined are always the same, as both custom and nature confirm; these are blood, phlegm, yellow bile, and black bile. To begin with, their accustomed names differ, and none has the same name as the others; moreover, their natural forms differ, since phlegm is not at all similar to blood, nor blood to bile, nor bile to phlegm. How could they be similar, when their colors

do not look alike to the eye, nor do they feel alike to the hand? For they are not equal in heat, cold, dryness, or moisture. It is impossible, when forms and faculties are so dissimilar, that they should be identical, any more than fire and water are identical. To understand that all are not identical, but that each has its own faculty and nature, consider the following: give someone medicine to bring up phlegm, he vomits phlegm; give him medicine to bring up bile, he vomits bile. Similarly, black bile is purged by medicine to bring up black bile; and if a man's body is cut open, blood will flow from him. All these things will happen every day and night of both winter and summer until the man can no longer inhale or exhale, or until one of these congenital constituents is lacking. For how could the things I have spoken of not be congenital? It is obvious that a human being always contains them all so long as he lives; moreover, he is born from a human being and nursed by a human being who contains all these things that I have discussed and demonstrated.

7 . . . A man's body, then, always contains all these things, but as the seasons revolve, they increase and decrease in turn, each according to its nature. Every year participates in the hot, the cold, the dry, and the moist, none of which could survive for a moment apart from the totality of things in the universe: if one should fail, all would vanish, for all are necessarily constituted and nourished by one another. Just so, if any one of these congenital constituents of a human being should fail, the human being could not live. Sometimes in a year winter prevails, sometimes spring, sometimes summer, sometimes autumn: just so in a human being, sometimes phlegm prevails, sometimes blood, sometimes bile, first yellow, then what is called black. . . .

8 . . . The physician must therefore treat diseases knowing that each prevails in the body according to the season most suited to its nature. . . .

9 . . . In sum, the physician should combat the characteristics of diseases—the forms they take in different seasons and at different times of life—slackening what is tight and tightening what is slack; for in this way the affliction would be most alleviated, and this, in my view, is a cure.

The Best Kind of Work and of Knowledge

The *Oeconomicus* of Xenophon, translated by Rev. J. S. Watson and Robert M. Torrance

Among the bold theorists and meticulous observers of the natural world in the seminal age of Greek thought between Thales and Aristotle, Xénophôn (ca. 428?–ca. 354 B.C.?) is something of a parenthesis, for his outlook is that of a farmer, soldier, and hunter—even, Edith Hamilton suggests in her popular book The Greek Way *(1930; 2nd ed. 1943), of an "ordinary Athenian gentleman," albeit of quite extraordinary accomplishments. Everything in his early experience indicates that this seeming ordinariness was at least in part deliberately cultivated in reaction to the turbulent events of his youth, when this wealthy aristocrat associated, during the late years of the Peloponnesian War and the oligarchic revolution and democratic restoration that followed the defeat of Athens, with the dissident young men around the chameleonic and subversive Socrates. In 401, three years after his city's surrender, he joined the mercenary Greek army of Cyrus the Younger of Persia (whose cooperation with the Spartan general Lysander had been instrumental to Peloponnesian victory over Athens) in his campaign against his brother, Artaxerxes II, for the Persian throne; after the defeat and death of Cyrus at Cunaxa, and the treacherous slaughter of the Greek generals under a flag of truce, Xenophon (by his own account in his stirring* Anabasis*) helped rally the Greeks and lead them on their long and perilous march through unknown country to the sea. In the years that followed (including the year of Socrates' execution, 399), Xenophon campaigned against the Persians in Asia Minor under a series of Spartan generals, and was exiled from Athens. He accompanied the Spartan king Agesiláus back to Greece in about 395, and was rewarded by the Spartans with an estate near Olympia, where he lived for over twenty years. During his subsequent stay around Corinth, his exile was rescinded, and he returned in about 366 to Athens, where he died some twelve years later.*

Among his many works are the Hellenica, *a history of Greece from 411 to 362, which takes up where the unfinished history of Thucydides broke off; the* Cyropaedia, *or* Education of Cyrus, *a "historical novel" (as the Russian critic Mikhael Bakhtin has called it) portraying the sixth-century Persian king Cyrus the Great as an ideal ruler; and several works devoted to the memory of Socrates, including the* Memorabilia *or* Memoirs *in four books, and two shorter works paralleling those of Plato by the same title, an* Apology *or* Defense of Socrates *and a* Symposium *or* Banquet. *The Socrates of these works, in contrast to Plato's, as Hamilton remarks, "was a soberly thinking man, distinguished for common sense, and . . . what he chiefly does for his young friends is to give them practical advice on how to manage their affairs." Practicality is indeed the hallmark of this down-to-earth, conservative, somewhat pedestrian ("Eternal truths," Hamilton notes, "were not in his line"), but engagingly congenial writer; among his treatises are several on politics (the* Constitution of Sparta *and the* Hieron, *a dialogue on tyranny), one on horsemanship, one on hunting* (Cynegéticus), *and one on cavalry command* (Hipparchus).

In the Oeconómicus (Oikonomikos), *or "Household Manager," the narrator reports a conversation between Socrates and Critóbulus (Kritoboulos)—a farmer's son attracted to the pleasures of city life—on household management* (oikonomia, *from which "economy" derives), and especially farm management, as an art* (tekhnê) *or form of knowledge. Throughout most of history, as opposed to prehistory, human contact with the "natural" world of other living things has taken place predominantly on or around the farmlands where the immense majority of the population has lived. Xenophon's book—standing between Hesiod's hardscrabble account of farm drudgery (Chapter 8 above) and the idealized portrayals of country life by Horace, Virgil, and Claudian (Chapter 11 below), and later by such poets as Jonson (Chapter 17) and Herrick (Chapter 19)—is an important testimonial to the solid yet comfortable virtues of life on a prosperous aristocratic estate; among Xenophon's major successors in this regard were the Roman agricultural writers Cato, Varro, and Columella (Chapter 12). In the first of the following excerpts, Socrates calls the farmer's life worthy not only of a "gentleman" (a pallid English equivalent for the Greek* kalos kagathos, *a "beautiful and good" man), but of a great king, exemplified by the younger Cyrus. Socrates then tells Critobulus how he himself, when young, had learned of farming from Ischómachus (Iskhomakhos), whom he met in the city—for not even Xenophon's Socrates is a countryman!—and who by his questioning had convinced Socrates that he already possessed the knowledge of farming of which he had thought himself ignorant. As Socrates would thereafter preeminently demonstrate, questioning itself can teach what is already somehow known.*

Thus for Xenophon, as Leo Strauss observes in Xenophon's Socratic Discourse: An Interpretation of the Oeconomicus *(1970), "Socrates' later view of teaching and learning is the outcome of his meditation on a thought first suggested to him by the practice of the perfect gentle-*

man par excellence," the gentleman farmer. Indeed, in Strauss's view, "the Oeconomicus *describes Socrates' famous turning away from his earlier pursuit," the useless study of things above alluded to in Plato's* Phaedo *and satirized in Aristophanes'* Clouds, *"which brought him the reputation of being an idle talker and a man who measures the air, . . . toward the study of only the human things and the things useful to human beings"—in particular, of the most useful thing of all, the proper management of a farm, which is nothing less than a paradigm for the just rule of a state. The translation is based on that of the Rev. J. S. Watson in* Xenophon's Minor Works *(1877), substantially revised, for the sake both of greater fidelity to the Greek and of more concise and idiomatic expression, with reference to the text and translation of Xenophon's* Oeconomicus: A Social and Historical Commentary, *ed. Sarah B. Pomeroy (1994), and to the translation by Carnes Lord in Strauss.*

IV.4–5, 13–18, 20–25 "In what sorts of employment, then, Socrates, would you recommend us to engage?" asked Critobulus.

"Ought we to be ashamed," replied Socrates, "to imitate the king of the Persians? For they say that he considers the art of farming, and that of war, to be among the most honorable and necessary occupations, and has serious regard for both of them."

Critobulus, on hearing this, said, "Do you then, believe, Socrates, that farming is among the king of the Persians' concerns?"

". . . In whatever provinces he resides," Socrates said, "and wherever he travels, he provides for gardens, such as those called *paradeisoi*,[6] stocked with everything good and valuable that the soil will produce; and in these gardens he himself spends the greatest part of his time, whenever the season of the year does not prevent him."

"By Zeus, then, Socrates," Critobulus said, "he must be concerned that the pleasure gardens in which he spends time be supplied as beautifully as possible with trees and all other choice things that the earth produces."

"Some relate, too, Critobulus," Socrates said, ". . . that Cyrus [the Great], who was certainly the most illustrious of their kings, once remarked to those who had been called to receive rewards [for distinction in war and in farming] that he himself might justly receive both sorts of presents; for he was the best, he said, both in cultivating land and in defending the land that he cultivated."

"Cyrus," said Critobulus, "if he made this observation, took no less pride, then, Socrates, in making his land fertile, and in keeping it in order, than in his ability in war."

"Yes, by Zeus," said Socrates, "Cyrus [the Younger], if he had lived, would surely have proved an excellent ruler . . . It is this Cyrus who is said to have paid Lysander[7] many marks of civility when he came with presents from the allies (as Lysander once remarked to a friend of his at Megara), and to have shown him (as Lysander related) his pleasure

[6]Xenophon is the first Greek writer (here and in *Hellenica* IV.1.15) to use the word *paradeisos*, of Persian origin, from which our word "paradise" derives, meaning "park" or "pleasure garden." According to Pomeroy, in a note to this passage, "Xenophon not only created the first park on the Greek mainland," on his estate in Scillus (*Anabasis* V.3.7–11), "but through his descriptions he also contributed to the creation of parks in the Hellenistic and Roman worlds. In these parks the plantings were often in rows as straight as military formations . . ." In the Septuagint (the Greek translation of the Hebrew Bible made in or after the third century B.C. for the Jewish community of Alexandria), the word *paradeisos* is used of the garden of Eden in Gen. 2:8 ("And the LORD God planted a *garden* in Eden"). See also Sir Thomas Browne's reflections in *The Garden of Cyrus* in Chapter 19 below.

[7]A Spartan commander of the Peloponnesian fleet toward the end of the Peloponnesian War. Xenophon in the *Hellenica* (I.5.1 ff.) describes his effort, in a meeting of 407, to obtain from Cyrus Persian assistance in the war against Athens, but does not mention the incident narrated here.

garden at Sardis. When Lysander expressed admiration of it, observing how fine the trees were, how regularly they were planted, how straight their rows were, and how elegantly all the rows formed angles with one another, while many sweet odors accompanied the two as they walked about—admiring all this, he said: 'I marvel at all these things for their beauty, Cyrus, but am much more amazed at the man who measured out and arranged each of them for you.' On hearing this, Cyrus was delighted, and said: 'All these things, Lysander, I measured out and arranged myself; and some of them,' he added, 'I planted with my own hands.' And Lysander, looking at him and contemplating the beauty of the robes he had on, and smelling the perfume that wafted from them, and the splendor of the necklaces, bracelets, and other ornaments which he wore, said that he had said: 'What are you telling me, Cyrus? Did you plant some of these with your own hands?' And Cyrus had replied: 'Do you marvel at this, Lysander? I swear to you by Mithras that, whenever I am healthy, I never dine until I have worked up a sweat by some military or agricultural activity, or some competitive enterprise.' And Lysander said that on hearing this he took Cyrus's right hand and said: 'Cyrus, you seem to me justly happy, for you are happy while being a good man.'

V.1–12, 17 "This I relate to you, Critobulus," said Socrates, "to show that not even those wholly blessed by fortune can do without farming, for its pursuit seems a means both of enjoyment and of increasing their resources; and it is also an exercise for the body, accomplishing everything that befits a free man. To begin with, the earth yields to those who cultivate it the food on which men live, and also produces things from which they receive gratification. . . . Yet, though it offers blessings in abundance, it does not allow us to receive them idly, but requires us to acclimatize ourselves to endure the cold of winter and the heat of summer. To those whom it exercises in manual labor, it gives increased strength; and even in those who only oversee its cultivation, it produces a manly vigor by making them rise early in the morning, and forcing them to move about energetically; for in the country, as in the city, the most important matters are always done at a particular season. Again, if a man wishes to defend his city as a horseman, farming is most suited for feeding a horse, or if one is a foot soldier, it keeps the body robust; and it also affords some incitement to exertion in hunting, providing an easy supply of nourishment for the dogs and at the same time supporting wild animals. The horses and dogs, moreover, which are maintained by farming, benefit the farm in turn: the horse by carrying his master early in the morning to the scene of his labors, and furnishing him the means of returning late; the dogs, by preventing wild beasts from damaging the crops or the livestock and by providing security even in the most solitary places.

"The earth also stimulates farmers, in some degree, to defend their country in arms, since it produces its crops exposed to all, for the strongest to take possession of them. What art, moreover, makes men more fit for running, throwing, and jumping than farming does? What art offers men more gratification for their labor? Or welcomes its practitioner with greater pleasure, inviting whoever comes along to take what he needs? Or receives strangers with richer hospitality? Where is there greater occasion for passing the winter amid plenty of fires, and warm baths, than on the farm? Or where can we spend a summer more agreeably, by streams, amid breezes, and under shade, than in the fields? What offers more pleasing first-fruits to the gods, or richer banquets on festive days? What is more welcome to servants, more delightful to a wife, more appealing to children, or more gratifying to friends? I would be amazed if any free person has ever possessed anything more pleasant than this, or discovered any pursuit more attractive, or more conducive to making a living, than farming is.

"The earth, being a goddess, also teaches justice to those who can learn; for those who treat her best she recompenses with the most numerous benefits. . . . Well did he speak who said that farming is mother and nurse of the other arts; for when farming flourishes, all other arts also prosper, but when the earth is forced to lie barren, the other arts are almost extinguished, both on land and at sea. . . ."

VI.2, 8–10 "Suppose," said Socrates, "we first go back over those points on which we agreed in our discussion, then try, if we can, to reach agreement on the remaining matters? . . . We decided that for a gentleman the best occupation and knowledge is farming, by which men procure livelihood. For we thought this occupation was easiest to learn and most pleasant to practice; made men's bodies most beautiful and vigorous; and most provided their souls with leisure to be concerned with their friends and cities. We further thought that farming emboldens those who pursue it by producing and sustaining necessities beyond the city walls. Therefore this way of life, we thought, is most esteemed by cities, as making citizens most virtuous and best disposed to the common good . . ."

XI.14–18 "'I am accustomed, Socrates,' said Ischomachus, 'to rise from my bed at an hour when I am likely to find at home anyone I need to see. If I have any business in the city, I use the occasion to take a walk there. Or if I have no need to be in the city, my servant takes my horse into the fields, and perhaps, by the walk along the road into the country, I get more benefit than I would if I were to walk under a covered colonnade. When I reach the open fields, whether the laborers are planting trees, or turning up the soil, or sowing or bringing in the crops, I observe how everything is going on, and set them right if I think of any way of improving what is being done. After this, I generally mount my horse and perform equestrian exercises as similar as possible to those required in war, avoiding neither cross roads, slopes, ditches, nor streams, though taking care not to lame my horse by doing so. When this is over, the servant gives the horse a roll, and then takes him home, carrying into town whatever we need from the country; and I return home, sometimes walking, sometimes running, and scrape off my perspiration. Next I take my morning meal, Socrates, eating just so much as to pass the day neither empty nor overfull.'

'By Hera, Ischomachus,' I said, 'you do all this in a way that I approve; for to occupy yourself, at the same time, in pursuits to improve your health and strength, in exercises suited to war, and in concern for your fortune, seems to me in the highest degree admirable. . . .'"

XIX.14–15 "'By Zeus,' I said, 'I am ignorant of none of the things you have mentioned, Ischomachus; but I am again puzzled how it was, when you asked me in a general way, a while ago, whether I understood planting, that I said no; for I thought I had nothing to say about how planting should be done. But now that you have set about questioning me on each particular, my answers agree with what you, who are called a clever farmer, say. Is questioning then teaching, Ischomachus?' I asked. 'For I have just now recognized,' I went on, 'the way you questioned me about each particular: by leading me through things I understand, and showing me things I thought I did not understand, but which are similar to those that I do, you persuade me, I think, that I understand these too.' . . ."

A Nature of Wondrous Beauty

The Dialogues of Plato, translated by Benjamin Jowett

No philosophy would appear to be farther from that of the pre-Socratic natural philosophers than Plato's, which reacted against it almost as strongly as against the deceptive rhetoric of the contemporary Sophists, yet was deeply influenced by it. Born of a wealthy family, Plato (Platôn, ca. 429–347 B.C.) spent his youth in a democratic Athens being led to ruin, he thought, by its demagogic leaders. After defeat of his city by Sparta (whose code of disciplined conduct he admired), and after condemnation of Socrates (Sôkratês) by the Athenians in 399 on charges of introducing new gods and corrupting the young, Plato passed the remainder of his life—apart from a period of travel and a series of futile efforts to transform the younger Dionysius of Syracuse (in Sicily) from tyrant to philosopher-king—as leader of the Academy which he founded around 385 in a grove outside Athens, and which remained a center of philosophical education in the ancient world until it was closed by the Emperor Justinian in A.D. 529.

Socrates, the principal influence on Plato's early years and central interlocutor of all but his late dialogues, had turned away, we are told in the Phaedo (Phaidôn), *from early fascination with natural and cosmological questions (such as Aristophanes travestied in the* Clouds*) out of disappointment at finding that not even Anaxagoras' cosmic* nous *or "mind" could explain why things should be as they are. From idle preoccupation with matters beyond human knowledge he turned to ethical questions concerning the good, declaring in his* Apology, *or Defense before the Athenian court, "that I have nothing to do with physical speculations"—a statement confirmed by Xénophôn's* Memorabilia *(I.11). It is this Socrates, for whom "the men who dwell in the city are my teachers, and not the trees or the country," yet who responds to the fragrance of flowers and the chorus of cicadas, and offers humble prayer to the god Pan, that Plato engagingly depicts in the* Phaedrus (Phaidros). *Here and elsewhere Plato himself betrays an acute susceptibility to the beauties of the natural world, for even in resisting the seductions of poetry he knows the force of poetic inspiration and remains (as he was known to be in antiquity) a poet. In poems attributed to him—preserved in the* Palatine Anthology *of the tenth century A.D.—he evokes both the heavenly bodies, in an epigram (VII.670) translated by Percy Bysshe Shelley—*

Thou wert the morning star among the living,
Ere thy fair light had fled—
Now, having died, thou art as Hesperus, giving
New splendor to the dead—

and, in another poem (IX.823), the vibrant world over which Pan presides:

Hushed be the hill where dryads throng, the rocks
whence fountains tumble, hushed the bleating flocks:
playing melodiously, the great god Pan
trills with moist lips his reeds' segmented span,
while round about, with nimbly flashing feet
tree-nymphs and water-nymphs take up the beat.

But under the influence both of Eleatic logic and of Pythagorean formalism Plato turned far more radically than Socrates away from investigation of the natural world perceived by the senses toward contemplation of an eternal, supersensible or "intelligible" world of immutable forms (eidea or ideai) apprehensible only to the trained philosophic mind: a world of which the phenomenal or "sensible" world of coming-to-be and passing-away is only—in the famous image of Socrates' parable of the cave in Republic VII—a shadow.

The very concept of physis becomes a negative one for Plato to the extent that the natural philosophers (especially the atomists) identify it with a mechanical universe ruled by chance, and the Sophists with fulfillment of selfish desires in defiance of human reason and divine law. The Sophists' elevation of physis (reduced to brute power) over nomos (law, convention, or custom), most crassly by Cállicles in the Gorgias, and their definition of justice, as by Thrasýmachus in Republic I, as the interest of the stronger, aroused Plato's determined opposition throughout his career. Sophistic relativism, in alliance with materialistic naturalism, reduced the gods themselves to mere creatures of human convention, thereby undermining, as the Athenian Stranger warns in the Laws (the longest and possibly last of Plato's dialogues), religion and justice. Yet Plato by no means repudiates nature, as some later thinkers, in revulsion from the deceptions of the sensory world, would do. On the contrary, Socrates at the end of the Gorgias counters a false conception of natural good—that of the sophistical cook who pleases small boys by stuffing them with sweets—with the unpopular conception of the philosophical physician who administers medicines to restore by art the true good of nature, which is health. And in the Symposium, Socrates' instruction in the "mysteries of love" by Diotíma, priestess of Mantineia, began (he tells his fellow banqueters at Ágathôn's dinner party) with the example of mortal nature "seeking as far as possible," through procreation, "to be everlasting and immortal." Thus physis is the starting point for ascent from bodily love to love of the soul and finally to "a nature of wondrous beauty" that in its immutable wholeness seems both outgrowth and negation of nature as coming-to-be and passing-away—an eternal physis transcending the growth in which it began and with which it remains, through participation of the temporal in the eternal, connected.

The relation of being to becoming is a central problem addressed in the Timaeus (Timaios), which Kahn calls Plato's "attempt to counteract the materialistic tendencies of Ionian science—as embodied above all in the atomic system—by resuming and reinterpreting the achievements of the entire tradition." Plato, unlike Parmenides, does not deny existence to the physical cosmos of generation and corruption, though he too believes (as the Pythagorean Timaeus affirms in this dialogue) that judgments about it can only be opinions, not—as in judgments concerning absolute being—truths accessible to reason. Any explanation of the cosmos must therefore, like Parmenides', be at best a "likely story" or "probable myth" (eikós mythos); and such is the tale Timaeus tells, forswearing even the possibility of exact knowledge. Yet this mythic account, approximative and uncertain as it must be, is rooted, as Cornford observes in Plato's Cosmology (1937), in "the truth, firmly believed by Plato, that the world is not solely the outcome of blind chance or necessity, but shows the working of a divine intelligence." Hence Plato "introduced, for the first time in Greek philosophy, the alternative scheme of creation by a divine artificer, according to which the world is like a work of art designed with a purpose." What this artificer or "Demiurge" (Dêmiourgos) creates, however, is a divinely living cosmos—nature itself—not through blindly colliding bodies ("for the world is the fairest of creations and he is the best of causes") but on the paradigm of eternal being, in which it participates through the agency of the soul, the first and "best of things created."

Thus, "declared enemy and would-be persecutor of the physiologoi though Plato certainly was," as Gregory Vlastos writes in Plato's Universe (1975), "he nonetheless accepted their discovery of the cosmos," and highly valued the very natural world whose deficiencies he continually sought to transcend. Nor does he deny that in the imperfect world of creation necessity as well as mind has its role, for in addition to the eternal and the created world, Timaeus postulates a third or "intermediate nature," eternal space (khôra), "the receptacle and, in a sense, the nurse, of all *generation," whose given limits impose a partial necessity by restricting the Demiurge's capacity to model Becoming on Being. The cosmology Timaeus advances in later parts of the dialogue, not included here, would exert enormous influence for two thousand years—the* Timaeus *was the only dialogue of Plato widely known to the Latin Middle Ages—inspiring Kepler as late as the seventeenth century. This universe is constructed of geometric shapes, based on the triangle, which combine into solid figures that are the first principle or* arkhê *from which the four visible elements are formed. Over the gate of Plato's Academy were the words "Let no one ignorant of geometry enter here," and in* Republic *VII Socrates declares that geometrical knowledge is of the eternal, not the transient; the* Timaeus *is the dialogue that perhaps comes closest to expounding a theory of "ideal numbers" such as Plato may have taught—see Burkert's discussion in* Lore and Science in Ancient Pythagoreanism*—in the "unwritten doctrines" of which Aristotle speaks* (Physics *209b). And if geometry is one means by which reason approaches knowledge of Being, astronomy, too, the Athenian Stranger declares in the final book of the* Laws, *can be valuable to an aspiring ruler, provided that he perceives in the stars not lifeless bodies but the eternal mind that governs created things. Despite his repudiation of Ionian natural philosophy, then, and his turn toward contemplation of changeless Being, there are indeed, as Lloyd affirms in* Methods and Problems in Greek Science, *"many features of Plato's position that allow and even encourage the inquiry into nature, construed as an inquiry into the intelligible ordered world that the challenging phenomena presuppose."*

The translations by Benjamin Jowett have been lightly revised.

FROM Phaedrus

229–30, 279

Socrates. Let us turn aside and go by the Ilissus; we will sit down at some quiet spot.

Phaedrus. I am fortunate in not having my sandals, and as you never have any, I think that we may go along the brook and cool our feet in the water; this will be the easiest way, and at midday and in the summer is far from being unpleasant.

Socrates. Lead on, and look out for a place in which we can sit down.

Phaedrus. Do you see the tallest plane-tree in the distance?

Socrates. Yes.

Phaedrus. There are shade and gentle breezes, and grass on which we may either sit or lie down.

Socrates. Move forward.

Phaedrus. I should like to know, Socrates, whether the place is not somewhere near here at which Boreas is said to have carried off Orithyia from the banks of the Ilissus?

Socrates. Such is the tradition.

Phaedrus. And is this the exact spot? The little stream is delightfully clear and bright; I can fancy that there might be maidens playing near.

Socrates. I believe that the spot is not exactly here, but about a quarter of a mile lower down, where you cross to the temple of Artemis, and there is, I think, some sort of an altar of Boreas at the place.

Phaedrus. I have never noticed it; but I beseech you to tell me, Socrates, do you believe this tale?

Socrates. The wise are doubtful, and I should not be singular if, like them, I too doubted. I might have a rational explanation that Orithyia was playing with Pharmacia, when a northern gust carried her over the neighboring rocks; and this being the manner of her death, she was said to have been carried away by Boreas. There is a discrepancy, however, about the locality; according to another version of the story she was taken from Areopagus, and not from this place. Now I quite acknowledge that these allegories are very nice, but he is not to be envied who has to invent them; much labor and ingenuity will be required of him; and when he has once begun, he must go on and rehabilitate Hippocentaurs and chimeras dire. Gorgons and winged steeds flow in apace, and numberless other inconceivable and portentous natures. And if he is skeptical about them, and would fain reduce them one after another to the rules of probability, this sort of crude philosophy will take up a great deal of time. Now I have no leisure for such inquiries; shall I tell you why? I must first know myself, as the Delphic inscription says; to be curious about that which is not my concern, while I am still in ignorance of my own self, would be ridiculous. And therefore I bid farewell to all this; the common opinion is enough for me. For, as I was saying, I want to know not about this, but about myself: am I a monster more complicated and swollen with passion than the serpent Typho, or a creature of a gentler and simpler sort, to whom Nature has given a diviner and lowlier destiny? But let me ask you, friend: have we not reached the plane-tree to which you were conducting us?

Phaedrus. Yes, this is the tree.

Socrates. By Hera, a fair resting place, full of summer sounds and scents. Here is this lofty and spreading plane-tree, and the agnus castus high and clustering, in the fullest blossom and the greatest fragrance; and the stream which flows beneath the plane-tree is deliciously cold to the feet. Judging from the ornaments and images, this must be a spot sacred to Achelous and the Nymphs. How delightful is the breeze—so very sweet; and there is a sound in the air shrill and summerlike which makes answer to the chorus of the cicadas. But the greatest charm of all is the grass, like a pillow gently sloping to the head. My dear Phaedrus, you have been an admirable guide.

Phaedrus. What an incomprehensible being you are, Socrates: when you are in the country, as you say, you really are like some stranger who is led about by a guide. Do you ever cross the border? I rather think that you never venture even outside the gates.

Socrates. Very true, my good friend; and I hope that you will excuse me when you hear the reason, which is, that I am a lover of knowledge, and the men who dwell in the city are my teachers, and not the trees or the country. Though I do indeed believe that you have found a spell with which to draw me out of the city into the country, like a hungry cow before whom a bough or a bunch of fruit is waved. For only hold up before me in like manner a book, and you may lead me all around Attica, and over the wide world. And now having arrived, I intend to lie down, and you may choose any posture in which you can read best. Begin . . .

*

Phaedrus. . . . And now, as the heat is abated, let us depart.

Socrates. Should we not offer up a prayer first to the local deities?

Phaedrus. By all means.

Socrates. Beloved Pan, and all ye other gods who haunt this place, give me beauty in the inward soul; and may the outward and inward man be at one. May I reckon the wise to be the wealthy, and may I have such a quantity of gold as a temperate man and he only can bear and carry.—Anything more? The prayer, I think, is enough for me.

Phaedrus. Ask the same for me, for friends should have all things in common.

Socrates. Let us go.

FROM Gorgias

482–84, 491–92, 521–22

Callicles. . . . For the truth is, Socrates, that you, who pretend to be engaged in the pursuit of truth, are appealing now to the popular and vulgar notions of right, which are not natural, but only conventional. Convention and nature are generally at variance with one another: and hence, if a person is too modest to say what he thinks, he is compelled to contradict himself; and you, in your ingenuity perceiving the advantage to be gained thereby, slyly ask of him who is arguing conventionally a question which is to be determined by the rule of nature; and if he is talking of the rule of nature, you slip away to custom: as, for instance, you did in this very discussion about doing and suffering injustice. When Polus was speaking of the conventionally dishonorable, you assailed him from the point of view of nature; for by the rule of nature, to suffer injustice is the greater disgrace because the greater evil; but conventionally, to do evil is the more disgraceful. For the suffering of injustice is not the part of a man, but of a slave, who indeed had better die than live; since when he is wronged and trampled upon, he is unable to help himself, or any other about whom he cares. The reason, as I conceive, is that the makers of laws are the majority who are weak; and they make laws and distribute praises and censures with a view to themselves and to their own interests; and they terrify the stronger sort of men, and those who are able to get the better of them, in order that they may not get the better of them; and they say that dishonesty is shameful and unjust; meaning, by the word injustice, the desire of a man to have more than his neighbors; for knowing their own inferiority, I suspect that they are too glad of equality. And therefore the endeavor to have more than the many is conventionally said to be shameful and unjust, and is called injustice, whereas nature herself intimates that it is just for the better to have more than the worse, the more powerful than the weaker; and in many ways she shows, among men as well as among animals, and indeed among whole cities and races, that justice consists in the superior ruling over and having more than the inferior. For on what principle of justice did Xerxes invade Hellas, or his father the Scythians—not to speak of numberless other examples? Nay, but these are the men who act according to nature; yes, by Heaven, and according to the law of nature: not, perhaps, according to that artificial law, which we invent and impose upon our fellows, of whom we take the best and strongest from their youth upwards, and tame them like young lions, charming them with the sound of the voice, and saying to them that they must be content with equality, and that the equal is the honorable and just. But if there were a man who had sufficient force, he would shake off and break through, and escape from all this; he would trample under foot all our formulas and spells and charms, and all our laws which are against nature: the slave would rise in rebellion and be lord over us, and the light of natural justice would shine forth. . . . And this is true, as you may ascertain, if you will leave philosophy and go on to higher things: for philosophy, Socrates, if pursued in moderation and at the proper age, is an elegant accomplishment, but too much philosophy is the ruin of human life. . . .

For how can a man be happy who is the servant of anything? On the contrary, I plainly assert that he who would truly live ought to allow his desires to wax to the uttermost, and not to chastise them; but when they have grown to their greatest he should have courage and intelligence to minister to them and to satisfy all his longings. And this I affirm to be natural justice and nobility. To this however the many cannot attain; and they blame the strong man because they are ashamed of their own weakness, which they desire to conceal, and hence they say that intemperance is base. As I have remarked already, they enslave the nobler natures, and being unable to satisfy their pleasures, they praise temperance and justice out of their own cowardice. For if a man had been originally the son of a king, or had a nature capable of acquiring an empire or a tyranny or sovereignty, what could be more truly base or evil than temperance—to a man like him, I say, who might freely be enjoying every good, and has no one to stand in his way, and yet has admitted custom and reason and the opinion of other men to be lords over him?—must not he be in a miserable plight whom the reputation of justice and temperance hinders from giving more to his friends than to his enemies, even though he be a ruler in his city? Nay, Socrates, for you profess to be a votary of the truth, and the truth is this: that luxury and intemperance and license, if they be provided with means, are virtue and happiness—all the rest is a mere bauble, agreements contrary to nature, foolish talk of men, worth nothing.

Socrates. There is a noble freedom, Callicles, in your way of approaching the argument; for what you say is what the rest of the world think, but do not like to say. . . .

I would indeed be a fool, Callicles, if I did not know that in the Athenian State any man may suffer anything. . . . Nor shall I be surprised if I am put to death. . . . I shall be tried just as a physician would be tried in a court of little boys at the indictment of the cook. What would he reply under such circumstances, if someone were to accuse him, saying, "O my boys, many evil things has this man done to you: he is the death of you, especially of the younger ones among you, cutting and burning and starving and suffocating you, until you know not what to do; he gives you the bitterest potions, and compels you to hunger and thirst. How unlike the variety of meats and sweets on which I feasted you!" What do you suppose that the physician would be able to reply when he found himself in such a predicament? If he told the truth he could only say, "All these evil things, my boys, I did for your health," and then would there not be a clamor among a jury like that? How they would cry out! . . . And I too shall be treated in the same way, as I well know, if I am brought before the court. . . . And therefore there is no telling what may happen to me.

FROM Apology of Socrates

19

I will begin at the beginning and ask what is the accusation which has given rise to the slander of me, and in fact has encouraged Meletus to bring this charge against me. Well, what do the slanderers say? They shall be my prosecutors, and I will sum up their words in an affidavit: "Socrates is an evildoer, and a curious person, who searches into things under the earth and in heaven, and he makes the worse appear the better cause; and he teaches the aforesaid doctrines to others." Such is the nature of the accusation: it is just what you have yourselves seen in the comedy of Aristophanes, who has introduced a man whom he calls Socrates, going about and saying that he walks on air, and talking a great deal of nonsense concerning matters of which I do not pretend to know either much or little—not that I mean to speak disparagingly of anyone who is a student of natural philosophy. I should be very sorry if Meletus could bring so grave a charge against me. But the simple truth is, O Athenians, that I have nothing to do with physical speculations. Very many of those here present are witnesses to the truth of this, and to them I appeal. Speak then, you who have heard me, and tell your neighbors whether any of you have ever known me to hold forth in few words or in many upon such matters. . . . You hear their answer. And from what they say of this part of the charge you will be able to judge of the truth of the rest. . . .

FROM Phaedo

96–98

When I was young, Cebes, I had a prodigious desire to know that department of philosophy which is called the investigation of nature; to know the causes of things, and why a thing is and is created or destroyed appeared to me to be a lofty profession; and I was always agitating myself with the consideration of questions such as these: Is the growth of animals the result of some decay which the hot and cold principle contracts, as some have said? Is blood the element with which we think, or air, or fire? or perhaps nothing of the kind—but the brain may be the originating power of the perceptions of hearing and sight and smell, and memory and opinion may come from them, and science may be based on memory and opinion when they have attained fixity. And then I went on to examine the corruption of them, and then to the things of heaven and earth, and at last I concluded myself to be utterly and absolutely incapable of these inquiries, as I will satisfactorily prove to you. For I was fascinated by them to such a degree that my eyes grew blind to things which I had seemed to myself, and also to others, to know quite well; I forgot what I had before thought self-evident truths; for example, such a fact as that the growth of man is the result of eating and drinking; for when by the digestion of food flesh is added to flesh and bone to bone, and whenever there is an aggregation of congenial elements, the lesser bulk becomes larger and the small man great. Was that not a reasonable notion? . . .

Then I heard someone reading, as he said, from a book of Anaxagoras, that mind was the disposer and cause of all, and I was delighted at this notion, which appeared quite admirable, and I said to myself: If mind is the disposer, mind will dispose all for the best, and put each particular in the best place; and I argued that if anyone desired to find out the cause of the generation or destruction or existence of anything, he must find out what state of being or doing or suffering was best for that thing, and therefore a man had only to consider the best for himself and others, and then he would also know the worse, since the same science comprehended both. And I rejoiced to think that I had found in Anaxagoras a teacher of the causes of existence such as I desired, and I imagined that he would tell me first whether the earth is flat or round; and whichever was true, he would proceed to explain the cause and the necessity of this being so, and then he would teach me the nature of the best and show that this was best; and if he said that the earth was in the center, he would further explain that this position was the best, and I should be satisfied with the explanation given, and not want any other sort of cause. And I thought that I would then go on and ask him about the sun and moon and stars, and that he would explain to me their comparative swiftness, and their returnings and various states, active and passive, and how all of them were for the best. For I could not imagine that when he spoke of mind as the disposer of them, he would give any other account of their being as they are, except that this was best; and I thought that when he had explained to me in detail the cause of each and the cause of all, he would go on to explain to me what was best for each and what was good for all. These hopes I would not have sold for a large sum of money, and I seized the books and read them as fast as I could in my eagerness to know the better and the worse.

What expectations I had formed, and how grievously was I disappointed! As I proceeded, I found my philosopher altogether forsaking mind or any other principle of order, but having recourse to air, and aether, and water, and other eccentricities. I might compare him to a person who began by maintaining generally that mind is the cause of the actions of Socrates, but who, when he endeavored to explain the causes of my several actions in detail, went on to show that I sit here because my body is made up of bones and muscles; and the bones, as he would say, are hard and have joints which divide them, and the muscles are elastic, and they cover the bones, which have also a covering or environment of flesh and skin which contains them; and as the bones are lifted at their joints by the contraction or relaxation of the muscles, I am able to bend my limbs, and this is why I am sitting here in a curved posture—that is what he would say; and he would have a similar explanation of my talking to you, which he would attribute to sound, and air, and hearing, and he would assign ten thousand other causes of the same sort, forgetting to mention the true cause, which is, that the Athenians have thought fit to condemn me, and accordingly I have thought it better and more right to remain here and undergo my sentence; for I am inclined to think that these muscles and bone of mine would have gone off long ago to Megara or Boeotia—by the dog, they would, if they had been moved only by their own idea of what was best, and if, instead of playing truant and running away, I had not chosen the better and nobler part, of enduring any punishment which the state inflicts. There is surely a strange confusion of causes and conditions in all this. It may be said, indeed, that without bones and muscles and the other parts of the body I cannot execute my purposes. But to say that I do as I do because of them, and that this is the way in which mind acts, and not from the choice of the best, is a very careless and idle mode of speaking. I wonder that they cannot distinguish the cause from the condition, which the many, feeling about in the dark, are always mistaking and misnaming. And thus one man makes a vortex all round and steadies the earth by the heaven; another gives the air as a support to the earth, which is a sort of broad trough. Any power which in arranging them as they are arranges them for the best never enters into their minds; and instead of finding any superior strength in it, they rather expect to discover another Atlas of the world who is stronger and more everlasting and more containing than the good: of the obligatory and containing power of the good they think nothing; and yet this is the principle which I would fain learn if anyone would teach me. . . .

FROM Symposium

207–11

All this she [Diotíma, priestess of Mantineia] taught me at various times when she spoke of love. And I remember her once saying to me, "What is the cause, Socrates, of love, and the attendant desire? Do you not see how all animals, birds, as well as beasts, in their desire of procreation, are in agony when they take the infection of love, which begins with the desire of union; whereto is added the care of offspring, on whose behalf the weakest are ready to battle against the strongest even to the uttermost, and to die for them, and will let themselves be tormented with hunger or suffer anything in order to maintain their young? Man may be supposed to act thus from reason; but why should animals have these passionate feelings? Can you tell me why?" Again I replied that I did not know. She said to me: "And do you expect ever to become a master in the art of love, if you do not know this?" "But I have told you already, Diotima, that my ignorance is the reason why I come to you; for I am conscious that I want a teacher; tell me then the cause of this and of the other mysteries of love." "Marvel not," she said, "if you believe that love is of the immortal, as we have several times acknowledged; for here again, and on the same principle too, the mortal nature is seeking as far as possible to be everlasting and immortal: and this is only to be attained by generation, because generation always leaves behind a new existence in the place of the old. . . .

"Those who are pregnant in the body only, betake themselves to women and beget children—this is the character of their love; their offspring, as they hope, will preserve their memory and give them the blessedness and immortality which they desire in the future. But souls which are pregnant—for there certainly are men who are more creative in their souls than in their bodies—conceive that which is proper for the soul to conceive or contain. And what are these conceptions?—wisdom and virtue in general. And such creators are poets and all artists who are deserving of the name inventor. But the greatest and fairest sort of wisdom by far is that which is concerned with the ordering of states and families, and which is called temperance and justice. And he who in youth has the seed of these implanted in him and is himself inspired, when he comes to maturity desires to beget and generate. . . .

"He who has been instructed thus far in the things of love, and who has learned to see the beautiful in due order and succession, when he comes forward to the end will suddenly perceive a nature of wondrous beauty (and this, Socrates, is the final cause of all our former toils)—a nature which in the first place is everlasting, not growing and decaying, or waxing and waning; secondly, not fair in one point of view and foul in another, or at one time or in one relation or at one place fair, at another time or in another relation or at another place foul, as if fair to some and foul to others or in the likeness of a face or hands or any other part of the bodily frame, or in any form of speech or knowledge, or existing in any other being, as for example, in an animal, or in heaven, or in earth, or in any other place; but beauty absolute, separate, simple, and everlasting, which without diminution and without increase, or any change, is imparted to the ever-growing and perishing beauties of all other things. He who ascends from these under the influence of true love, and begins to perceive that beauty, is not far from the end. And the true order of going, or being led by another, to the things of love, is to begin from the beauties of earth and mount upwards for the sake of that other beauty, using these as steps only, and from one going on to two, and from two to all fair forms, and from fair forms to fair practices, and from fair practices to fair notions, until from fair notions he arrives at the notion of absolute beauty, and at last knows what the essence of beauty is. This, my dear Socrates," said the stranger of Mantineia, "is that life above all others which man should live, in the contemplation of absolute beauty. . . ."

FROM *Timaeus*

27–34, 36–37, 47–52

Timaeus. First then, in my judgment, we must make a distinction and ask, What is that which always is and has no becoming; and what is that which is always becoming and never is? That which is apprehended by intelligence and reason is always in the same state; but that which is conceived by opinion with the help of sensation and without reason, is always in a process of becoming and perishing and never really is. Now everything that becomes or is created must of necessity be created by some cause, for without a cause nothing can be created. The work of the creator, whenever he looks to the unchangeable and fashions the form and nature of his work after an unchangeable pattern, must necessarily be made fair and perfect; but when he looks to the created only, and uses a created pattern, it is not fair or perfect. Was the heaven then or the world, whether called by this or any other more appropriate name, . . . always in existence and without beginning? or created, and had it a beginning? Created, I reply, being visible and tangible and having a body, and therefore sensible; and all sensible things are apprehended by opinion and sense and are in a process of creation and created. Now that which is created must, as we affirm, of necessity be created by a cause. But the father and maker of all this universe is past finding out; and even if we found him, to tell of him to all men would be impossible. And there is still a question to be asked about him: Which of the patterns did the artificer have in view when he made the world—the pattern of the unchangeable, or of that which is created? If the world be indeed fair and the artificer good, it is manifest that he must have looked to that which is eternal; but if what cannot be said without blasphemy is true, then to the created pattern. Everyone will see that he must have looked to the eternal; for the world is the fairest of creations and he is the best of causes. And having been created in this way, the world has been framed in the likeness of that which is apprehended by reason and mind and is unchangeable, and must therefore of necessity, if this is admitted, be a copy of something. Now it is all-important that the beginning of everything should be according to nature. And in speaking of the copy and the original we may assume that words are akin to the matter which they describe; when they relate to the lasting and permanent and intelligible, they ought to be lasting and unalterable, and, as far as their nature allows, irrefutable and immovable—nothing less. But when they express only the copy or likeness and not the eternal things themselves, they need only be likely and analogous to the real words. As being is to becoming, so is truth to belief. If then, Socrates, amid the many opinions about the gods and the generation of the universe, we are not able to give notions which are altogether and in every respect exact and consistent with one another, do not be surprised. Enough, if we adduce probabilities as likely as any others; for we must remember that I who am the speaker, and you who are the judges, are only mortal men, and we ought to accept the tale which is probable and inquire no further. . . .

Let me tell you then why the creator made this world of generation. He was good, and the good can never have any jealousy of anything. And being free from jealousy he desired that all things should be as like himself as they could be. This is in the truest sense the origin of creation and of the world, as we shall do well to believe on the testimony of wise men: God desired that all things should be good and nothing bad, so far as this was attainable. Wherefore also finding the whole visible sphere not at rest, but moving in an irregular and disorderly fashion, out of disorder he brought order, considering that this was in every way better than the other. Now the deeds of the best could never be or have been other than the fairest; and the creator, reflecting on the things which are by nature visible, found that no unintelligent creature taken as a whole was fairer than the intelligent taken as a whole; and that intelligence could not be present in anything which was devoid of soul. For which reason, when he was framing the universe, he put intelligence in soul, and soul in body, that he might be the creator of a work which was by nature fairest and best. Wherefore, using the language of probability, we may say that the world became a living creature truly endowed with soul and intelligence by the providence of God. . . .

God in the beginning of creation made the body of the universe to consist of fire and earth. But two things cannot be rightly put together without a third; there must be some bond of union between them. . . . God placed water and air in the mean between fire and earth, and made them to have the same proportion so far as was possible (as fire is to air so is air to water, and as air is to water so is water to earth); and thus he bound and put together a visible and tangible heaven. . . . And he gave to the world the figure which was suitable and also natural. Now to the animal which was to comprehend all animals, that figure was suitable which comprehends within itself all other figures. Wherefore he made the world in the form of a globe, round as from a lathe, having its extremes in every direction equidistant from the center, the most perfect and the most like itself of all figures; for he considered that the like is infinitely fairer than the unlike. . . . And in the center he put the soul, which he diffused throughout the body, making it also to be the exterior environment of it; and he made the universe a circle moving in a circle, one and solitary, yet by reason of its excellence able to converse with itself, and needing no other friendship or acquaintance. Having these purposes in view he created the world a blessed god. . . .

Now when the Creator had framed the soul according to his will, he formed within her the corporeal universe, and brought the two together, and united them center to center. The soul, interfused everywhere from the center to the circumference of heaven, of which also she is the external envelopment, herself turning in herself, began a divine beginning of never-ceasing and rational life enduring throughout all time. The body of heaven is visible, but the soul is invisible, and partakes of reason and harmony, and being made by the best of intellectual and everlasting natures, is the best of things created. . . .

Thus far in what we have been saying, with small exception, the works of intelligence have been set forth; and now we must place by the side of them in our discourse the things which come into being through necessity—for the creation is mixed, being made up of necessity and mind. Mind, the ruling power, persuaded necessity to bring the greater part of created things to perfection, and thus and after this manner in the beginning, when the influence of reason got the better of necessity, the universe was created. But if a person will truly tell of the way in which the work was accomplished, he must include the other influence of the variable cause as well. Wherefore, we must return again and find another suitable beginning, as about the former matters, so also about these. To which end we must consider the nature of fire, and water, and air, and earth, such as they were prior to the creation of the heaven, and what was happening to them in this previous state; for no one has as yet explained the manner of their generation, but we speak of fire and the rest of them, whatever they mean, as though men knew their natures, and we maintain them to be the first principles and letters or elements of the whole, when they cannot reasonably be compared by a man of any sense even to syllables or first compounds. . . .

This new beginning of our discussion of the universe requires a fuller division than the former; for then we made two classes, now a third must be revealed. The two sufficed for the former discussion: one, which we assumed, was a pattern intelligible and always the same; and the second was only the imitation of the pattern, generated and visible. There is also a third kind which we did not distinguish at the time, conceiving that the two would be enough. But now the argument seems to require that we should set forth in words another kind, which is difficult of explanation and dimly seen. What nature are we to attribute to this new kind of being? We reply, that it is the receptacle, and in a sense the nurse, of all generation. . . .

For the present we have only to conceive of three natures: first, that which is in process of generation; secondly, that in which the generation takes place; and thirdly, that of which the thing generated is a resemblance. And we may liken the receiving principle to a mother, and the source or spring to a father, and the intermediate nature to a child; and may remark further, that if the model is to take every variety of form, then the matter in which the model is fashioned will not be duly prepared, unless it is formless, and free from the impress of any of those shapes which it is hereafter to receive from without. For if the matter were like any of the supervening forms, then whenever any opposite or entirely different nature was stamped upon its surface, it would take the impression badly, because it would intrude its own shape. . . . In the same way that which is to receive perpetually and through its whole extent the resemblances of all eternal beings ought to be devoid of any particular form. Wherefore, the mother and receptacle of all created and visible and in any way sensible things is not to be termed earth, or air, or fire, or water, or any of their compounds or any of the elements from which these are derived, but is an invisible and formless being which receives all things and in some mysterious way partakes of the intelligible, and is most incomprehensible. In saying this we shall not be far wrong; as far, however, as we can attain to a knowledge of her from the previous considerations, we may truly say that fire is that part of her nature which from time to time is inflamed, and water that which is moistened, and that the mother substance becomes earth and air, insofar as she receives the impressions of them. . . .

Thus I state my view: If mind and true opinion are two distinct classes, then I say that there certainly are these self-existent ideas unperceived by sense, and apprehended only by the mind; if, however, as some say, true opinion differs in no respect from mind, then everything that we perceive through the body is to be regarded as most real and certain. But we must affirm them to be distinct, for they have a distinct origin and are of a different nature; the one is implanted in us by instruction, the other by persuasion; the one is always accompanied by true reason, the other is without reason; lastly, every man may be said to share in true opinion, but mind is the attribute of the gods and of very few men. Wherefore also we must acknowledge that there is one kind of being which is always the same, uncreated and indestructible, never receiving anything into itself from without, nor itself going out to any other, but invisible and imperceptible by any sense, and of which the contemplation is granted to intelligence only. And there is another nature of the same name with it, and like to it, perceived by sense, created, always in motion, becoming in place and again vanishing out of place, which is apprehended by opinion and sense. And there is a third nature, which is space, and is eternal, and admits not of destruction and provides a home for all created things, and is apprehended without the help of sense, by a kind of spurious reason, and is hardly real; which we beholding as in a dream, say of all existence that it must of necessity be in some place and occupy a space, but that what is neither in heaven nor in earth has no existence. Of these and other things of the same kind, relating to the true and waking reality of nature, we have only this dreamlike sense, and we are unable to cast off sleep and determine the truth about them. . . .

FROM Laws
X.889–92; XII.966–68

Athenian Stranger. They [philosophers] say that the greatest and fairest things are the work of nature and of chance, the lesser of art, which, receiving from nature the greater and primeval creations, molds and fashions all those lesser works which are generally termed artificial.

Cleinias. How is that?

Athenian Stranger. I will explain my meaning still more clearly. They say that fire and water, and earth and air, all exist by nature and chance, and none of them by art, and that as to the bodies which come next in order—earth, and sun, and moon, and stars—they have been created by means of these absolutely inanimate existences. The elements are severally moved by chance and some inherent force according to certain affinities among them—of hot with cold, or of dry with moist, or of soft with hard, and according to all the other accidental admixtures of opposites which have been formed by necessity. After this fashion and in this manner the whole heaven has been created, and all that is in the heaven, as well as animals and all plants, and all the seasons come from these elements, not by the action of mind, as they say, or of any God, or from art, but as I was saying, by nature and chance only. Art sprang up afterwards and out of these, mortal and of mortal birth, and produced in play certain images and very partial imitations of the truth, having an affinity to one another, such as music and painting create and their companion arts. And there are other arts which have a serious purpose, and these cooperate with nature, such, for example, as medicine, and husbandry, and gymnastic. And they say that politics cooperate with nature, but in a lesser degree, and have more of art; also that legislation is entirely a work of art, and is based on assumptions which are not true.

Cleinias. What do you mean?

Athenian Stranger. In the first place, my dear friend, these people would say that the Gods exist not by nature, but by art, and by the laws of states, which are different in different places, according to the agreement of those who make them; and that the honorable is one thing by nature and another thing by law, and that the principles of justice have no existence at all in nature, but that mankind are always disputing about them and altering them; and that the alterations which are made by art and by law have no basis in nature, but have authority for the moment and at the time when they are made. These, my friends, are the sayings of wise men, poets and prose writers, which find a way into the minds of youth. They are told by them that the highest right is might, and in this way the young fall into impieties, under the idea that the Gods are not such as the law bids them imagine; and hence arise factions, these philosophers inviting them to lead a true life according to nature, that is, to live in real dominion over others, and not in legal subjection to them.

Cleinias. What a dreadful picture, Stranger, have you given, and how great is the injury which is thus inflicted on young men to the ruin both of states and families!

Athenian Stranger. True, Cleinias; but then what should the lawgiver do when this evil is of long standing? . . .

Cleinias. Why, Stranger, if such persuasion be at all possible, then a legislator who has anything in him ought never to weary of persuading men; he ought to leave nothing unsaid in support of the ancient opinion that there are Gods, and of all those other truths which you were just now mentioning; he ought to support the law and also art, and acknowledge that both alike exist by nature, and no less than nature, if they are the creations of mind in accordance with right reason, as you appear to me to maintain, and I am disposed to agree with you in thinking. . . .

Athenian Stranger. Well, then, tell me, Cleinias—for I must ask you to be my partner—does not he who talks in this way conceive fire and water and earth and air to be the first elements of all things? these he calls nature, and out of these he supposes the soul to be formed afterwards; and this is not a mere conjecture of ours about his meaning, but is what he really means. . . . Then I suppose that I must repeat the singular argument of those who manufacture the soul according to their own impious notions; they affirm that which is the first cause of the generation and destruction of all things, to be not first, but last, and that which is last to be first, and hence they have fallen into error about the nature of the Gods.

Cleinias. Still I do not understand you.

Athenian Stranger. Nearly all of them, my friends, seem to be ignorant of the nature and power of the soul, especially in what relates to her origin: they do not know that she is among the first of things, and before all bodies, and is the chief author of their changes and transpositions. And if this is true, and if the soul is older than the body, must not the things which are akin to the soul be of necessity prior to those which appertain to the body?

Cleinias. Certainly.

Athenian Stranger. Then thought and attention and mind and art and law will be prior to that which is hard and soft and heavy and light; and the great and primitive works and actions will be works of art; they will be the first, and after them will come nature and works of nature, which however is a wrong term for men to apply to them; these will follow, and will be under the government of art and mind.

Cleinias. But why is the word "nature" wrong?

Athenian Stranger. Because those who use the term mean to say that nature is the first creative power; but if the soul turn out to be the primeval element, and not fire or air, then in the truest sense and beyond other things the soul may be said to exist by nature; and this would be true if you proved that the soul is older than the body, but not otherwise.

Cleinias. You are quite right. . . .

*

Athenian Stranger. If a man look upon the world not lightly or ignorantly, there was never anyone so godless who did not experience an effect opposite to that which the many imagine. For they think that those who handle these matters by the help of astronomy, and the accompanying arts of demonstration, may become godless, because they see, as far as they can see, things happening by necessity, and not by an intelligent will accomplishing good.

Cleinias. But what is the fact?

Athenian Stranger. Just the opposite, as I said, of the opinion which once prevailed among men, that the sun and stars are without soul. Even in those days men wondered about them, and that which is now ascertained was then conjectured by some who had a more exact knowledge of them—that if they had been things without soul, and had no mind, they could never have moved with numerical exactness so wonderful; and even at that time some ventured to hazard the conjecture that mind was the orderer of the universe. But these same persons again mistaking the nature of the soul, which they conceived to be younger and not older than the body, once more overturned the world, or rather, I should say, themselves; for the bodies which they saw moving in heaven all appeared to be full of stones, and earth, and many other lifeless substances, and to these they assigned the causes of all things. Such studies gave rise to much atheism and perplexity, and the poets took occasion to be abusive, comparing the philosophers to she-dogs uttering vain howlings, and talking other nonsense of the same sort. But now, as I said, the case is reversed.

Cleinias. How so?

Athenian Stranger. No man can be a true worshipper of the Gods who does not know these two principles—that the soul is the eldest of all things which are born, and is immortal and rules over all bodies; moreover, as I have now said several times, he who has not contemplated the mind of nature which is said to exist in the stars, and gone through the previous training, and seen the connection of music with these things, and harmonized them all with laws and institutions, is not able to give a reason of such things as have a reason. And he who is unable to acquire this in addition to the ordinary virtues of a citizen, can hardly be a good ruler of a whole state; but he should be the subordinate of other rulers. . . .

The Perfection of Coming-to-Be

Selected Writings of Aristotle

The great creative period of Greek natural philosophy culminates in Aristotle (Aristótelês), for whom—as he writes in the Parts of Animals*—"every realm of nature is marvellous," from highest to lowest: "for each and all will reveal to us something natural and something beautiful." Indeed, despite the immense importance of Aristotle's writings on logic, politics, ethics, rhetoric, and poetics, the bulk of his voluminous work (from which only a small sample can be included here) is devoted to questions concerning nature; and about a third of the total, including his longest book, the* History of Animals (Historia Animalium),[8] *records his extensive biological investigations—a subject slighted by all philosophers before (and most after) him.*

Born in the northern city of Stagíros or Stagíra in 384 B.C.*, Aristotle may have spent his early years at the Macedonian court at Pella and developed an interest in biology through his father Nicómachus (Nikomakhos), physician to King Amyntas II. At seventeen he entered Plato's Academy in Athens, where he remained for some twenty years, until Plato's death around 347. Thereafter he lived in Assos (Asia Minor) and in Mytilênê (on the island of Lesbos), where he appears to have conducted many of his zoological inquiries. In about 342 King Philip summoned him to tutor his young son, the future Alexander the Great. In 335 he returned to Athens and founded the Lycéum (Lykeion), from whose covered walkway* (perípatos) *his school came to be known as the "peripatetic." Apart from the later Museum of Alexandria, Lloyd remarks in* Magic, Reason and Experience, *the Lyceum would remain a center for research in the natural sciences (to say nothing of other fields) "unsurpassed in the whole of antiquity." After the death of Alexander in 323, and the outbreak of anti-Macedonian sentiment that followed in Athens, Aristotle was charged (as Anaxagoras and Socrates had been) with impiety and prudently withdrew rather than let the Athenians "sin twice against philosophy." He died in Chalcis (Khalkis) in 322.*

The dialogues published by Aristotle during and after his years at the Academy, of which only fragments survive, were much admired for their style and were immensely influential throughout antiquity; for more than two centuries these were the principal, if not the only, writings

[8]Titles in parentheses are those of Latin translations of Aristotle, by which his works are often known. Thus the book titled in Greek *Peri Geneseôs kai Phthoras* is widely called *De Generatione et Corruptione* (or simply *De Generatione*), and translated "On Generation and Corruption," "On Generation and Destruction," or "On Coming-to-Be and Passing-Away." The title translated into Latin as *Historia Animalium* (and back into English as *History of Animals*) means more literally "Investigation of Animals" or "of Living Things."

known to the outside world. Aristotle's extant works—apart from an account of the Constitution of Athens, unearthed in Egypt in 1890—are apparently expanded lecture notes or memoranda intended for students of the Lyceum; published in the late first century B.C., *some (notably the* Metaphysics*) apparently combine writings from different periods of Aristotle's career. Their terse style and uncertain arrangement inevitably result in obscurities and inconsistencies, yet together they constitute incalculably the most important body of theoretical speculation and empirical research in natural philosophy before the modern age.*

Profoundly influenced though Aristotle was by Plato, he reacted against what he considered the abstract and mathematical tendencies of Plato in his later years (and of Plato's successor, Speusippus). To him as to Plato, form was a critical philosophical concept, but he explicitly rejected any separation (khôrismos) *between an intelligible world of changeless forms and the sensible world of corruptible matter such as Plato, in his view, postulated. For Aristotle, form and matter are inseparable: matter* (hylê) *is merely a potentiality* (dynamis) *given actuality* (energeia) *through the form* (eidos) *that brings it to fulfillment. Thus the natural world of matter, as actualized in the form that is its "entelechy" (the fulfillment of its end, or* telos), *is the* only *world, and the cosmos is an indivisible whole. Physics—*ta physiká, *"things pertaining to nature"—was in his view, W. D. Ross writes in* Aristotle *(1923; 5th ed. 1953), "the study not of form alone nor of matter alone but of informed matter or of inmattered form." Especially in the first four books of the* Physics, *known as his books* peri physeôs, *or "on nature"—defined as an inherent "source or cause of being moved or at rest"—Aristotle is not only differentiating himself from Plato but also, as John Herman Randall suggests in* Aristotle (1960), *"trying to reinstate, reconstruct, and defend the ancient Ionian conception of 'Nature,'* physis, *and of natural career or process, against the critics who had discredited it, Parmenides and the Eleatics, whose criticism had culminated in the mechanistic views of Empedocles and the atomists." In opposition to their repudiation of "nature" in the sense of process—in his poem "On Nature," Aristotle objects* (De Generatione *333b), Empedocles "says nothing about nature"!—and to their assertion that instead of "genuine coming into being" there is "only a mixing and unmixing of elements which themselves do not change," Aristotle is insisting, Randall asserts, "that the world displays real geneses, real comings into being, with a fundamental unity and continuity, a basic temporal pattern or structure." This structured process, moreover, is by no means random, like that of the atomists; for Aristotle repeatedly affirms that nature does nothing in vain and works toward the best possible result. For him, no less than for the Milesians whose work he brings to completion, the cosmos is a living being with its own internal goals, and nature, he writes in "On Prophecy in Sleep" (in* Parva Naturalia *463b), though not divine, is infused with divinity: it is* daimonia, *if not* theia.

Aristotle indeed sharply differentiates, in what seems a quasi-Platonic duality, between our world and the incorruptible heavens, moving eternally in perfect circular patterns and composed not of the four sublunary elements but of a "fifth element," the aether. Yet though the ungenerated heavenly bodies are "excellent beyond compare and divine" (Parts of Animals *644b), the heaven "in which we believe the divine to reside" is explicitly called (*On the Heavens [De Caelo] *278b) a "natural body"* (sôma physikon). *Thus* both *the perishable and imperishable are natural substances* (Parts of Animals *644b), and both—though in different degrees—are living and divine; the true relation between them (as between body and soul, and between all matter and its perfecting form) is of potential to actual* (On the Heavens *311a). To Aristotle, the cosmos as a whole—the earth no less than the heavens—though finite, is eternal and ungenerated, and thus has no* arkhê *in the sense of origin. But it is of course on earth that the constituents and processes of nature are most in evidence, and much of Aristotle's natural philosophy—in the* Physics *and*

Metaphysics,[9] *the* De Generatione, *and the biological writings including* On the Soul (De Anima)*—is devoted to their examination.*

No concept is more central to understanding the natural world than that of its "causes": the Greek aitiai *means more nearly, in Ross's words, "conditions necessary but not separably sufficient to account for the existence of a thing." Aristotle defines these differently in different places, but his discussions in the* Physics *and* Metaphysics *(of which the passage below from* Metaphysics Δ *[V] is the most concise) distinguish four, not all of which always pertain: the "material cause," or matter of which a thing is composed; the "efficient cause" acting upon it to bring about a change (our usual understanding of "cause"); the "formal cause," or form that the matter takes as a result; and the "final cause," or end toward which the process is directed. (Elsewhere the last two, or even the last three, are grouped together in contradistinction to the material cause.) Any adequate conception of nature requires attention to the interaction of these causes. This was lacking, Aristotle contends in the first book of the* Metaphysics, *in his predecessors; definition of the causes is thus his principal contribution to the conception of* physis.

As for the matter of which nature is composed, for Aristotle (unlike Empedocles) the four elements are not indissoluble, since they variously combine the contraries of hot and cold, dry and moist, formed from a common "substratum" (hypokeimenon) *of "prime matter" underlying all things. "We maintain," Aristotle writes* (Meteorologica *339a–b), "that fire, air, water and earth are transformable one into another, and that each is potentially latent in the others, as is true of all other things that have a single common substratum underlying them into which they can in the last resort be resolved." Matter, Ross writes, "is a purely relative term—relative to form," for "in the realm of nature, the elements, which are the determinate product of prime matter and the primary contrarieties hot and cold, dry and fluid, are matter relatively to their simple compounds the tissues; these again are matter relatively to the organs, and these are matter relatively to the living body. Prime matter . . . never exists apart; the elements are the simplest physical things, and within them the distinction of matter and form can only be made by an abstraction of thought."*

There is thus a continuum in the physical world, just as in the biological there is a "scale of ascent" from inanimate to animate and plant to animal, culminating in man, the microcosm whose upper and lower parts correspond to those of the universe (History of Animals *494a), and who combines the nutritive soul of plants and the sentient soul of animals with the rational soul whose matter is the divine* pneuma, *or spirit, analogous to the ethereal element of the stars* (Generation of Animals *736b). Such expressions, of which there are many, of an anthropocentric teleology linking man with the divine were sources of great influence in antiquity and the Middle Ages, but also of later repudiation of Aristotle's thought as antipathetic to modern science. But as Randall reminds us, "'final causes' and 'natural ends' are in no sense whatever to be taken as 'purposes': they involve no conscious intent," and in general—traces of divine pneuma apart—express a functionalism compatible with much modern biology, which unlike physics acknowledges the importance of teleological explanations. Aristotle's teleology is "not external but internal"* (Parts of Animals *641b); his necessity not absolute but "hypothetical" or conditional.*

The exhaustive collection of data in his biological writings testifies to his meticulous observation and (despite the theoretical bias of Greek philosophy and science) to impressive instances of experimentation and even dissection. Mistakes were inevitable, but the detailed accuracy of his

[9]*Metaphysika* ("After Physics") was the title given by Aristotle's early editors to the highly composite work they placed after his *Physics (Physika);* its contents pertain at least as much to natural philosophy as to what we have come to call "metaphysics" but Aristotle called "first philosophy." The fourteen books of the *Metaphysics,* which may have been written at very different times, and have been differently divided, are generally designated by Greek letters; the books from which our selections come are A or Alpha (known also by the Roman numeral I), Δ or Delta (V), and Z or Zeta (VII).

inductions and classifications was unsurpassed by any biologist for more than two thousand years. "He recognized, for example," Ross writes, "the mammalian character of the cetaceans—a fact which was overlooked by all other writers till the sixteenth century. . . . He detected a remarkable feature in the copulation of cephalopods, which was not rediscovered till the nineteenth century." Nor did the upward scale of life imply a world without conflict. Aristotle recognizes an almost Malthusian "war against each other among all animals that occupy the same places and get their living from the same things" (History of Animals *608b), and remarks on the adaptive power of nature, which "will often quite spontaneously take some part that is common to all animals and press it into service for some specialized purpose"* (Parts of Animals *662a). For such reasons as these, the philosophy rejected by Bacon and Newton could forcefully appeal to Darwin, who wrote to William Ogle, Victorian translator of the* Parts of Animals, *that "Linnaeus and Cuvier have been my two gods, though in very different ways, but they were mere schoolboys to old Aristotle." Nor is it only in biology that Aristotle's natural philosophy seems strikingly modern: in an age of quantum mechanics, his emphasis on form over matter and on self-transformative processes over mechanistic laws is "far closer to present-day physical theory," Randall contends, "than are the ideas of the nineteenth century." Above all, in proclaiming that "the closest approximation to eternal being" lies not in some ungenerated and incorruptible realm but in perpetual coming-to-be, in which the true perfection of the universe lies, Aristotle decisively affirms the supreme dignity and value of nature itself.*

Selections from Aristotle's works are from the Oxford translations as reprinted in Richard McKeon, ed., The Basic Works of Aristotle *(1941).*

FROM Physics,

translated by R. P. Hardie and R. K. Gaye

II.1 (192b–193b); II.8 (198b–199b)

II.1 Of things that exist, some exist by nature, some from other causes. "By nature" the animals and their parts exist, and the plants and the simple bodies (earth, fire, air, water)—for we say that these and the like exist "by nature."

All the things mentioned present a feature in which they differ from things which are *not* constituted by nature. Each of them has *within itself* a principle of motion and of stationariness (in respect of place, or of growth and decrease, or by way of alteration). On the other hand, a bed and a coat and anything else of that sort, *qua* receiving these designations—i.e. in so far as they are products of art—have no innate impulse to change. But in so far as they happen to be composed of stone or of earth or of a mixture of the two, they *do* have such an impulse, and just to that extent—which seems to indicate that *nature is a source or cause of being moved and of being at rest in that to which it belongs primarily,* in virtue of itself and not in virtue of a concomitant attribute.

I say "not in virtue of a concomitant attribute," because (for instance) a man who is a doctor might cure himself. Nevertheless it is not in so far as he is a patient that he possesses the art of medicine: it merely has happened that the same man is doctor and patient—and that is why these attributes are not always found together. So it is with all other artificial products. None of them has in itself the source of its own production. But while in some cases (for instance houses and the other products of manual labor) that principle is in something else external to the thing, in others—those which may cause a change in themselves in virtue of a concomitant attribute—it lies in the things themselves (but not in virtue of what they are).

"Nature" then is what has been stated. Things "have a nature" which have a principle of this kind. Each of them is a substance; for it is a subject, and nature always implies a subject in which it inheres.

The term "according to nature" is applied to all these things and also to the attributes which belong to them in virtue of what they are, for instance the property of fire to be carried upwards—which is not a "nature" nor "has a nature" but is "by nature" or "according to nature."

What nature is, then, and the meaning of the terms "by nature" and "according to nature," has been stated. *That* nature exists, it would be absurd to try to prove; for it is obvious that there are many things of this kind, and to prove what is obvious by what is not is the mark of a man who is unable to distinguish what is self-evident from what is not. (This state of mind is clearly possible. A man blind from birth might reason about colors. Presumably therefore such persons must be talking about words without any thought to correspond.)

Some identify the nature or substance of a natural object with that immediate constituent of it which taken by itself is without arrangement, e.g. the wood is the "nature" of the bed, and the bronze the "nature" of the statue.

As an indication of this Antiphon points out that if you planted a bed and the rotting wood acquired the power of sending up a shoot, it would not be a bed that would come up, but *wood*—which shows that the arrangement in accordance with the rules of the art is merely an incidental attribute, whereas the real nature is the other, which, further, persists continuously through the process of making.

But if the material of each of these objects has itself the same relation to something else, say bronze (or gold) to water, bones (or wood) to earth and so on, *that* (they say) would be their nature and essence. Consequently some assert earth, others fire or air or water or some or all of these, to be the nature of the things that are. For whatever any one of them supposed to have this character—whether one thing or more than one thing—this or these he declared to be the whole of substance, all else being its affections, states, or dispositions. Every such thing they held to be eternal (for it could not pass into anything else), but other things to come into being and cease to be times without number.

This then is one account of "nature," namely that it is the immediate material substratum of things which have in themselves a principle of motion or change.

Another account is that "nature" is the shape or form which is specified in the definition of the thing.

For the word "nature" is applied to what is according to nature and the natural in the same way as "art" is applied to what is artistic or a work of art. We should not say in the latter case that there is anything artistic about a thing, if it is a bed only potentially, not yet having the form of a bed; nor should we call it a work of art. The same is true of natural compounds. What is potentially flesh or bone has not yet its own "nature," and does not exist "by nature," until it receives the form specified in the definition, which we name in defining what flesh or bone is. Thus in the second sense of "nature" it would be the shape or form (not separable except in statement) of things which have in themselves a source of motion. (The combination of the two, e.g. man, is not "nature" but "by nature" or "natural.")

The form indeed is "nature" rather than the matter; for a thing is more properly said to be what it is when it has attained to fulfilment than when it exists potentially. Again man is born from man, but not bed from bed. That is why people say that the figure is not the nature of a bed, but the wood is—if the bed sprouted, not a bed but wood would come up. But even if the figure *is* art, then on the same principle the shape of man is his nature. For man is born from man.

We also speak of a thing's nature as being exhibited in the process of growth by which its nature is attained. The "nature" in this sense is not like "doctoring," which leads not to the art of doctoring but to health. Doctoring must start from the art, not lead to it. But it is not in this way that nature (in the one sense) is related to nature (in the other). What grows *qua* growing grows from something into something. Into what then does it grow? Not into that from which it arose but into that to which it tends. The shape then is nature. . . .

II.8 . . . A difficulty presents itself: why should not nature work, not for the sake of something, nor because it is better so, but just as the sky rains, not in order to make the corn grow, but of necessity? What is drawn up must cool, and what has been cooled must become water and descend, the result of this being that the corn grows. Similarly if a man's crop is spoiled on the threshing-floor, the rain did not fall for the sake of this—in order that the crop might be spoiled—but that result just followed. Why then should it not be the same with the parts in nature, e.g. that our teeth should come up *of necessity*—the front teeth sharp, fitted for tearing, the molars broad and useful for grinding down the food—since they did not arise for this end, but it was merely a coincident result; and so with all other parts in which we suppose that there is a purpose? Wherever then all the parts came about just what they would have been if they had come to be for an end, such things survived, being organized spontaneously in a fitting way; whereas those which grew otherwise perished and continue to perish, as Empedocles says his "man-faced ox progeny" did.

Such are the arguments (and others of the kind) which may cause difficulty on this point. Yet it is impossible that this should be the true view. For teeth and all other natural things either invariably or normally come about in a given way; but of not one of the results of chance or spontaneity is this true. We do not ascribe to chance or mere coincidence the frequency of rain in winter, but frequent rain in summer we do; nor heat in the dog-days, but only if we have it in winter. If then, it is agreed that things are either the result of coincidence or for an end, and these cannot be the result of coincidence or spontaneity, it follows that they must be for an end; and that such things are all due to nature even the champions of the theory which is before us would agree. Therefore action for an end is present in things which come to be and are by nature.

Further, where a series has a completion, all the preceding steps are for the sake of that. Now surely as in intelligent action, so in nature; and as in nature, so it is in each action, if nothing interferes. Now intelligent action is for the sake of an end; therefore the nature of things also is so. Thus if a house, e.g., had been a thing made by nature, it would have been made in the same way as it is now by art; and if things made by nature were made also by art, they would come to be in the same way as by nature. Each step then in the series is for the sake of the next; and generally art partly completes what nature cannot bring to a finish, and partly imitates her. If, therefore, artificial products are for the sake of an end, so clearly also are natural products. The relation of the later to the earlier terms of the series is the same in both.

This is most obvious in the animals other than man: they make things neither by art nor after inquiry and deliberation. Wherefore people discuss whether it is by intelligence or by some other faculty that these creatures work—spiders, ants, and the like. By gradual advance in this direction we come to see clearly that in plants too that is produced which is conducive to the end—leaves, e.g., grow to provide shade for the fruit. If then it is both by nature and for an end that the swallow makes its nest and the spider its web, and plants grow leaves for the sake of the fruit and send their roots down (not up) for the sake of nourishment, it is plain that this kind of cause is operative in things which come to be and are by nature. And since "nature" means two things, the matter and the form, of which the latter is the end, and since all the rest is for the sake of the end, the form must be the cause in the sense of "that for the sake of which."

Now mistakes come to pass even in the operations of art: the grammarian makes a mistake in writing and the doctor pours out the wrong dose. Hence clearly mistakes are possible in the operations of nature also. If then in art there are cases in which what is rightly produced serves a purpose, and if where mistakes occur there was a purpose in what was attempted, only it was not attained, so must it be also in natural products, and monstrosities will be failures in the purposive effort. . . .

FROM Metaphysics,
translated by W. D. Ross
Δ (V).2 (1013a); A (I).3–6 (983b–987b); A (I).9 (990a–991a, 992a–992b); Δ (V).4 (1014b–1015a); Z (VII).7 (1032a)

Δ (V).2 "Cause" means (1) [the "material cause":] that from which, as immanent material, a thing comes into being, e.g. the bronze is the cause of the statue and the silver of the saucer, and so are the classes which include these. (2) [The "formal cause":] The form or pattern, i.e. the definition of the essence, and the classes which include this (e.g. the ratio 2:1 and number in general are causes of the octave), and the parts included in the definition. (3) [The "efficient cause":] That from which the change or the resting from change first begins; e.g. the adviser is a cause of the action, and the father a cause of the child, and in general the maker a cause of the thing made and the change-producing of the changing. (4) [The "final cause":] The end, i.e. that for the sake of which a thing is; e.g. health is the cause of walking. For "Why does one walk?" we say; "that one may be healthy"; and in speaking thus we think we have given the cause. . . .

A (I).3 . . . Of the first philosophers, then, most thought the principles which were of the nature of matter were the only principles of all things. That of which all things that are consist, the first from which they come to be, the last into which they are resolved (the substance remaining, but changing in its modifications), this they say is the element and this the principle of things, and therefore they think nothing is either generated or destroyed, since this sort of entity is always conserved, as we say Socrates neither comes to be absolutely when he comes to be beautiful or musical, nor ceases to be when he loses these characteristics, because the substratum, Socrates himself, remains. Just so they say nothing else comes to be or ceases to be; for there must be some entity—either one or more than one—from which all other things come to be, it being conserved.

Yet they do not all agree as to the number and the nature of these principles. Thales, the founder of this type of philosophy, says the principle is water (for which reason he declared that the earth rests on water), getting the notion perhaps from seeing that the nutriment of all things is moist, and that heat itself is generated from the moist and kept alive by it (and that from which they come to be is a principle of all things). He got his notion from this fact, and from the fact that the seeds of all things have a moist nature, and that water is the origin of the nature of moist things.

Some[10] think that even the ancients who lived long before the present generation, and first framed accounts of the gods, had a similar view of nature; for they made Ocean and Tethys the parents of creation, and described the oath of the gods as being by water,[11] to which they gave the name of Styx; for what is oldest is most honorable, and the most honorable thing is that by which one swears. It may perhaps be uncertain whether this opinion about nature is primitive and ancient, but Thales at any rate is said to have declared himself thus about the first cause. Hippo no one would think fit to include among these thinkers, because of the paltriness of his thought.

Anaximenes and Diogenes make air prior to water, and the most primary of the simple bodies, while Hippasus of Metapontium and Heraclitus of Ephesus say this of fire, and Empedocles says it of the four elements (adding a fourth—earth—to those which have been named); for these, he says, always remain and do not come to be, except that they come to be more or fewer, being aggregated into one and segregated out of one.

Anaxagoras of Clazomenae, who, though older than Empedocles, was later in his philosophical activity,[12] says the principles are infinite in number; for he says almost all the things that are made of parts like themselves, in the manner of water or fire, are generated and destroyed in this way, only by aggregation and segregation, and are not in any other sense generated or destroyed, but remain eternally.

From these facts one might think that the only cause is the so-called material cause; but as men thus advanced, the very facts opened the way for them and joined in forcing them to investigate the subject. However true it may be that all generation and destruction proceed from some one or (for that matter) from more elements, why does this happen and what is the cause? For at least the substratum itself does not make itself change; e.g. neither the wood nor the bronze causes the change of either of them, nor does the wood manufacture a bed and the bronze a statue, but something else is the cause of the change. And to seek this is to seek the second ["efficient"] cause, as *we* should say—that from which comes the beginning of the movement. Now those who at the very beginning set themselves to this kind of inquiry, and said the substratum was one,[13] were not at all dissatisfied with themselves; but some at least of those who maintain it to be one[14]—as though defeated by this search for the second cause—say the one and nature as a whole is unchangeable not only in respect of generation and destruction (for this is a primitive belief, and all agreed in it), but also of all other change; and this view is peculiar to them. Of those who said the universe was one, then, none succeeded in discovering a cause of this sort, except perhaps Parmenides, and he only inasmuch as he supposes that there is not only one but also in some sense two causes. But for those who make more elements[15] it is more possible to state the second cause, e.g. for those who make hot and cold, or fire and earth, the elements; for they treat fire as having a nature which fits it to move things, and water and earth and such things they treat in the contrary way.

When these men and the principles of this kind had had their day, as the latter were found inadequate to generate the nature of things men were again forced by the truth itself, as we said, to inquire into the next kind of cause [the "efficient"]. For it is not likely either that fire or earth or any such element should be the reason why things manifest

[10] Probably Plato (*Cratylus* 402B, *Theaetetus* 152E, 162D, 180C).

[11] See *Iliad* XIV.201, 246; II.755; XIV.271; XV.37.

[12] Kahn, in *Anaximander*, disputes this translation, taking Aristotle's phrase to mean "Anaxagoras comes before Empedocles in time, but after him in his philosophical achievements."

[13] Thales, Anaximenes, and Heraclitus.

[14] Parmenides and the Eleatics.

[15] Probably Empedocles.

goodness and beauty both in their being and in their coming to be, or that those thinkers should have supposed it was; nor again could it be right to entrust so great a matter to spontaneity and chance. When one man[16] said, then, that reason was present—as in animals, so throughout nature—as the cause of order and of all arrangement, he seemed like a sober man in contrast with the random talk of his predecessors. We know that Anaxagoras certainly adopted these views, but Hermotimus of Clazomenae is credited with expressing them earlier. Those who thought thus stated that there is a principle of things which is at the same time the cause of beauty, and that sort of cause from which things acquire movement.

A (I).4 One might suspect that Hesiod was the first to look for such a thing—or some one else who put love or desire among existing things as a principle, as Parmenides, too, does; for he, in constructing the genesis of the universe, says:—

> Love first of all the Gods she planned.

And Hesiod says:—

> First of all things was chaos made, and then
> Broad-breasted earth, . . .
> And love, 'mid all the gods pre-eminent,

which implies that among existing things there must be from the first a cause which will move things and bring them together. How these thinkers should be arranged with regard to priority of discovery let us be allowed to decide later; but since the contraries of the various forms of good were also perceived to be present in nature—not only order and the beautiful, but also disorder and the ugly, and bad things in greater number than good, and ignoble things than beautiful—therefore another thinker introduced friendship and strife, each of the two the cause of one of these two sets of qualities. For if we were to follow out the view of Empedocles, and interpret it according to its meaning and not to its lisping expression, we should find that friendship is the cause of good things, and strife of bad. Therefore, if we said that Empedocles in a sense both mentions, and is the first to mention, the bad and the good as principles, we should perhaps be right, since the cause of all goods is the good itself.

These thinkers, as we say, evidently grasped, and to this extent, two of the causes which we distinguished in our work on nature[17]—the matter ["material cause"] and the source of the movement ["efficient cause"]—vaguely, however, and with no clearness, but as untrained men behave in fights; for they go round their opponents and often strike fine blows, but they do not fight on scientific principles, and so too these thinkers do not seem to know what they say; for it is evident that, as a rule, they make no use of their causes except to a small extent. For Anaxagoras uses reason as a *deus ex machina* for the making of the world, and when he is at a loss to tell from what cause something necessarily is, then he drags reason in, but in all other cases ascribes events to anything rather than to reason. And Empedocles, though he uses the causes to a greater extent than this, neither does so sufficiently nor attains consistency in their use. At least, in many cases he makes love segregate things, and strife aggregate them. For whenever the universe is dissolved into its elements by strife, fire is aggregated into one, and so is each of the other

[16]Anaxagoras.
[17]*Physics* II.3, 7.

elements; but whenever again under the influence of love they come together into one, the parts must again be segregated out of each element.

Empedocles, then, in contrast with his predecessors, was the first to introduce the dividing of this cause, not positing one source of movement, but different and contrary sources. Again, he was the first to speak of four material elements; yet he does not *use* four, but treats them as two only; he treats fire by itself, and its opposites—earth, air, and water—as one kind of thing. We may learn this by study of his verses.

This philosopher, then, as we say, has spoken of the principles in this way, and made them of this number. Leucippus and his associate Democritus say that the full and the empty are the elements, calling the one being and the other non-being—the full and solid being being, the empty non-being (whence they say being no more is than non-being, because the solid no more is than the empty); and they make these the material causes of things. And as those who make the underlying substance one, generate all other things by its modifications, supposing the rare and the dense to be the sources of the modifications, in the same way these philosophers say the differences in the elements are the causes of all other qualities. These differences, they say, are three—shape and order and position. . . . The question of movement—whence or how it is to belong to things—these thinkers, like the others, lazily neglected.

Regarding the two causes, then, as we say, the inquiry seems to have been pushed thus far by the early philosophers.

A (I).5 Contemporaneously with these philosophers and before them, the so-called Pythagoreans, who were the first to take up mathematics, not only advanced this study, but also having been brought up in it they thought its principles were the principles of all things. Since of these principles numbers are by nature the first, and in numbers they seemed to see many resemblances to the things that exist and come into being—more than in fire and earth and water (such and such a modification of numbers being justice, another being soul and reason, another being opportunity—and similarly almost all other things being numerically expressible); since, again, they saw that the modifications and the ratios of the musical scales were expressible in numbers;—since, then, all other things seemed in their whole nature to be modeled on numbers, and numbers seemed to be the first things in the whole of nature, they supposed the elements of numbers to be the elements of all things, and the whole heaven to be a musical scale and a number. And all the properties of numbers and scales which they could show to agree with the attributes and parts and the whole arrangement of the heavens, they collected and fitted into their scheme; and if there was a gap anywhere, they readily made additions so as to make their whole theory coherent. E.g. as the number 10 is thought to be perfect and to comprise the whole nature of numbers, they say that the bodies which move through the heavens are ten, but as the visible bodies are only nine, to meet this they invent a tenth—the "counter-earth." We have discussed these matters more exactly elsewhere.[18]

But the object of our review is that we may learn from these philosophers also what they suppose to be the principles and how these fall under the causes we have named. Evidently, then, these thinkers also consider that number is the principle both as matter for things and as forming both their modifications and their permanent states, and hold that the elements of number are the even and the odd, and that of these the latter is limited, and the former unlimited; and that the One proceeds from both of these (for it is

[18]*On the Heavens (De Caelo)* II.13.

both even and odd), and number from the One; and that the whole heaven, as has been said, is numbers. . . .

From these facts we may sufficiently perceive the meaning of the ancients who said the elements of nature were more than one; but there are some who spoke of the universe as if it were one entity, though they were not all alike either in the excellence of their statement or in its conformity to the facts of nature. The discussion of them is in no way appropriate to our present investigation of causes, for they do not, like some of the natural philosophers, assume being to be one and yet generate it out of the one as out of matter, but they speak in another way; those others add change, since they generate the universe, but these thinkers say the universe is unchangeable. Yet *this* much is germane to the present inquiry: Parmenides seems to fasten on that which is one in definition, Melissus on that which is one in matter, for which reason the former says that it is limited, the latter that it is unlimited; while Xenophanes, the first of these partisans of the One (for Parmenides is said to have been his pupil), gave no clear statement, nor does he seem to have grasped the nature of either of these causes, but with reference to the whole material universe he says the One is God. Now these thinkers, as we said, must be neglected for the purposes of the present inquiry—two of them entirely, as being a little too naïve, viz. Xenophanes and Melissus; but Parmenides seems in places to speak with more insight. For, claiming that, besides the existent, nothing non-existent exists, he thinks that of necessity one thing exists, viz. the existent and nothing else (on this we have spoken more clearly in our work on nature),[19] but being forced to follow the observed facts, and supposing the existence of that which is one in definition, but more than one according to our sensations, he now posits two causes and two principles, calling them hot and cold, i.e. fire and earth; and of these he ranges the hot with the existent, and the other with the non-existent. . . .

A (I).6 After the systems we have named came the philosophy of Plato, which in most respects followed these thinkers, but had peculiarities that distinguished it from the philosophy of the Italians.[20] For, having in his youth first become familiar with Cratylus and with the Heraclitean doctrines (that all sensible things are ever in a state of flux and there is no knowledge about them), these views he held even in later years. Socrates, however, was busying himself about ethical matters and neglecting the world of nature as a whole but seeking the universal in these ethical matters, and fixed thought for the first time on definitions; Plato accepted his teaching, but held that the problem applied not to sensible things but to entities of another kind—for this reason, that the common definition could not be a definition of any sensible thing, as they were always changing. Things of this other sort, then, he called Ideas, and sensible things, he said, were all named after these, and in virtue of a relation to these; for the many existed by participation in the Ideas that have the same name as they. Only the name "participation" was new; for the Pythagoreans say that things exist by "imitation" of numbers, and Plato says they exist by participation, changing the name. But what the participation or the imitation of the Forms could be they left an open question.

Further, besides sensible things and Forms he says there are the objects of mathematics, which occupy an intermediate position, differing from sensible things in being eternal and unchangeable, from Forms in that there are many alike, while the Form itself is in each case unique.

[19] *Physics* I.3.
[20] The Pythagoreans and Eleatics, from southern Italy.

Since the Forms were the causes of all other things, he thought their elements were the elements of all things. As matter, the great and the small were principles; as essential reality, the One; for from the great and the small, by participation in the One, come the Numbers. . . .

A (I).9 Let us leave the Pythagoreans for the present; for it is enough to have touched on them as much as we have done. But as for those who posit the Ideas as causes, firstly, in seeking to grasp the causes of the things around us, they introduced others equal in number to these, as if a man who wanted to count things thought he would not be able to do it while they were few, but tried to count them when he had added to their number. For the Forms are practically equal to—or not fewer than—the things, in trying to explain which these thinkers proceeded from them to the Forms. For to each thing there answers an entity which has the same name and exists apart from the substances, and so also in the case of all other groups there is a one over many, whether the many are in this world or are eternal.

Further, of the ways in which we prove that the Forms exist, none is convincing; for from some no inference necessarily follows, and from some arise Forms even of things of which we think there are no Forms. . . .

Above all one might discuss the question what on earth the Forms contribute to sensible things, either to those that are eternal or to those that come into being and cease to be. For they cause neither movement nor any change in them. But again they help in no wise either towards the knowledge of the other things (for they are not even the substance of these, else they would have been in them), or towards their being, if they are not *in* the particulars which share in them; though if they were, they might be thought to be causes, as white causes whiteness in a white object by entering into its composition. But this argument, which first Anaxagoras and later Eudoxus and certain others used, is very easily upset; for it is not difficult to collect many insuperable objections to such a view.

But, further, all other things cannot come from the Forms in any of the usual senses of "from." And to say that they are patterns and the other things share in them is to use empty words and poetical metaphors. . . . Again, the Forms are patterns not only of sensible things, but of Forms themselves also; i.e. the genus, as genus of various species, will be so; therefore the same thing will be pattern and copy. . . .

Nor have the Forms any connection with what we see to be the cause in the case of the arts, that for whose sake both all mind and the whole of nature are operative—with this ["final"] cause which we assert to be one of the first principles; but mathematics has come to be identical with philosophy for modern thinkers, though they say that it should be studied for the sake of other things. . . .[21] The whole study of nature has been annihilated. . . .

Δ (V).4 "Nature" means (1) the genesis of growing things—the meaning which would be suggested if one were to pronounce the *y* in *physis* long.[22] (2) That immanent part of a growing thing, from which its growth first proceeds. (3) The source from which the primary movement in each natural object is present in it in virtue of its own essence. Those things are said to grow which derive increase from something else by contact and either by organic unity, or by organic adhesion as in the case of embryos. Organic unity differs from contact; for in the latter case there need not be anything besides the contact,

[21]See Plato, *Republic* VII.531D, 533 B–E.

[22]*Phyesthai* ("to grow"), from which *physis* derives, has a long *y* (upsilon) in most of its forms.

but in organic unities there is something identical in both parts, which makes them grow together instead of merely touching, and be one in respect of continuity and quantity, though not of quality.—(4) "Nature" means the primary material of which any natural object consists or out of which it is made, which is relatively unshaped and cannot be changed from its own potency, as e.g. bronze is said to be the nature of a statue and of bronze utensils, and wood the nature of wooden things; and so in all other cases; for when a product is made out of these materials, the first matter is preserved throughout. For it is in this way that people call the elements of natural objects also their nature, some naming fire, others earth, others air, others water, others something else of the sort, and some naming more than one of these, and others all of them.—(5) "Nature" means the *essence* of natural objects, as with those who say the nature is the primary mode of composition, or as Empedocles says:—

Nothing that has a nature,
But only mixing and parting of the mixed,
And nature is but a name given them by men.

Hence as regards the things that are or come to be by nature, though that *from which* they naturally come to be or are is already present, we say they have not their nature yet, unless they have their form or shape. That which comprises both of these[23] exists *by* nature, e.g. the animals and their parts; and not only is the first matter nature . . . , but also the form or essence, which is the end of the process of becoming.—(6) By an extension of meaning from this sense of "nature" every essence in general has come to be called a "nature," because the nature of a thing is one kind of essence.

From what has been said, then, it is plain that nature in the primary and strict sense is the essence of things which have in themselves, as such, a source of movement; for the matter is called the nature because it is qualified to receive this, and processes of becoming and growing are called nature because they are movements proceeding from this. And nature in this sense is the source of the movement of natural objects, being present in them somehow, either potentially or in complete reality.

Z (VII).7 Of things that come to be, some come to be by nature, some by art, some spontaneously. Now everything that comes to be comes to be by the agency of something and from something and comes to be something. And the something which I say it comes to be may be found in any category; it may come to be either a "this" of some size or of some quality or somewhere.

Now natural comings to be are the comings to be of those things which come to be by nature; and that out of which they come to be is what we call matter; and that by which they come to be is something which exists naturally; and the something which they come to be is a man or a plant or one of the things of this kind, which we say are substances if anything is—all things produced either by nature or by art have matter; for each of them is capable both of being and of not being, and this capacity is the matter in each—and, in general, both that from which they are produced is nature, and the type according to which they are produced is nature (for that which is produced, e.g. a plant or an animal, has a nature), and so is that by which they are produced—the so-called "formal" nature, which is specifically the same (though this is in another individual); for man begets man. . . .

[23]Matter and form.

FROM On Generation and Corruption (On Coming-to-Be and Passing-Away),
translated by Harold H. Joachim
II.3–4 (330b–331b); II. 10 (336b)

II.3 . . . The "simple" bodies, since they are four, fall into two pairs which belong to the two regions, each to each: for Fire and Air are forms of the body moving towards the "limit," while Earth and Water are forms of the body which moves towards the "center." Fire and Earth, moreover, are extremes and purest: Water and Air, on the contrary, are intermediates and more like blends. And, further, the members of either pair are contrary to those of the other, Water being contrary to Fire and Earth to Air; for the qualities constituting Water and Earth are contrary to those that constitute Fire and Air. Nevertheless, since they are four, each of them is characterized *par excellence* by a single quality: Earth by dry rather than by cold, Water by cold rather than by moist, Air by moist rather than by hot, and Fire by hot rather than by dry.

II.4 It has been established before that the coming-to-be of the "simple" bodies is reciprocal. At the same time, it is manifest, even on the evidence of perception, that they *do* come-to-be: for otherwise there would not have been "alteration," since "alteration" is change in respect to the qualities of the objects of touch. Consequently, we must explain (1) what is the manner of their reciprocal transformation, and (2) whether every one of them can come-to-be out of every one—or whether some can do so, but not others.

Now it is evident that all of them are by nature such as to change into one another: for coming-to-be is a change into contraries and out of contraries, and the "elements" all involve a contrariety in their mutual relations because their distinctive qualities are contrary. For in some of them *both* qualities are contrary—e.g. in Fire and Water, the first of these being dry and hot, and the second moist and cold: while in others *one* of the qualities (though only one) is contrary—e.g. in Air and Water, the first being moist and hot, and the second moist and cold. It is evident, therefore, if we consider them in general, that every one is by nature such as to come-to-be out of every one: and when we come to consider them severally, it is not difficult to see the manner in which their transformation is effected. For, though all will result from all, both the speed and the facility of their conversion will differ in degree.

Thus (1) the process of conversion will be quick between those which have interchangeable "complementary factors," but slow between those which have none. The reason is that it is easier for a single thing to change than for many. Air, e.g., will result from Fire if a single quality changes: for Fire, as we saw, is hot and dry, while Air is hot and moist, so that there will be Air if the dry be overcome by the moist. Again, Water will result from Air if the hot be overcome by the cold: for Air, as we saw, is hot and moist while Water is cold and moist, so that, if the hot changes, there will be Water. So too, in the same manner, Earth will result from Water and Fire from Earth, since the two "elements" in both these couples have interchangeable "complementary factors." For Water is moist and cold while Earth is cold and dry—so that, if the moist be overcome, there will be Earth: and again, since Fire is dry and hot while Earth is cold and dry, Fire will result from Earth, if the cold pass-away.

It is evident, therefore, that the coming-to-be of the "simple" bodies will be cyclical; and that this cyclical method of transformation is the easiest, because the *consecutive* "elements" contain interchangeable "complementary factors."[24] On the other hand (2) the transformation of Fire into Water and of Air into Earth, and again of Water and Earth into Fire and Air respectively, though possible, is more difficult, because it involves the change of more qualities. . . .

II.10 Coming-to-be and passing-away will, as we have said, always be continuous, and will never fail. . . . And this continuity has a sufficient reason on our theory. For in all things, as we affirm, Nature always strives after "the better." Now "being" . . . is better than "not-being": but not all things can possess "being," since they are too far removed from the "originative source." God therefore adopted the remaining alternative, and fulfilled the perfection of the universe by making coming-to-be uninterrupted: for the greatest possible coherence would thus be secured to existence, because that "coming-to-be should itself come-to-be perpetually" is the closest approximation to eternal being. . . .

FROM History of Animals,
translated by D'Arcy Wentworth Thompson
VIII.1 (588b)

VIII.1 . . . Nature proceeds little by little from things lifeless to animal life in such a way that it is impossible to determine the exact line of demarcation, nor on which side thereof an intermediate form should lie. Thus, next after lifeless things in the upward scale comes the plant, and of plants one will differ from another as to its amount of apparent vitality; and, in a word, the whole genus of plants, whilst it is devoid of life as compared with an animal, is endowed with life as compared with other corporeal entities. Indeed, as we just remarked, there is observed in plants a continuous scale of ascent towards the animal. So, in the sea, there are certain objects concerning which one would be at a loss to determine whether they be animal or vegetable. For instance, certain of these objects are fairly rooted, and in several cases perish if detached; thus the pinna is rooted to a particular spot, and the solen (or razor-shell) cannot survive withdrawal from its burrow. Indeed, broadly speaking, the entire genus of testaceans have a resemblance to vegetables, if they be contrasted with such animals as are capable of progression.

In regard to sensibility, some animals give no indication whatsoever of it, whilst others indicate it but indistinctly. Further, the substance of some of these intermediate creatures is fleshlike, as is the case with the so-called tethya (or ascidians) and the acalephae (or sea-anemones); but the sponge is in every respect like a vegetable. And so throughout the entire animal scale there is a graduated differentiation in amount of vitality and in capacity for motion.

FROM On the Parts of Animals,
translated by William Ogle
I.1 (641b–642a); I.5 (644b–645a)

I.1 . . . everything that Nature makes is means to an end. For just as human creations are the products of art, so living objects are manifestly the products of an analogous cause or principle, not external but internal, derived like the hot and the cold from the environ-

[24]Aristotle has shown that, by the conversion of a single quality in each case, Fire is transformed into Air, Air into Water, Water into Earth, and Earth into Fire. This is a *cycle* of transformations. Moreover, the "elements" have been taken in their natural consecutive series, according to their order in the Cosmos. (Joachim)

ing universe. And that the heaven, if it had an origin, was evolved and is maintained by such a cause, there is therefore even more reason to believe, than that mortal animals so originated. For order and definiteness are much more plainly manifest in the celestial bodies than in our own frame; while change and chance are characteristic of the perishable things of earth. Yet there are some who, while they allow that every animal exists and was generated by nature, nevertheless hold that the heaven was constructed to be what it is by chance and spontaneity; the heaven in which not the faintest sign of haphazard or of disorder is discernible! Again, whenever there is plainly some final end, to which a motion tends should nothing stand in the way, we always say that such final end is the aim or purpose of the motion; and from this it is evident that there must be a something or other really existing, corresponding to what we call by the name of Nature. For a given germ does not give rise to any chance living being, nor spring from any chance one; but each germ springs from a definite parent and gives rise to a definite progeny. And thus it is the germ that is the ruling influence and fabricator of the offspring. . . . Moreover, the seed is potentially that which will spring from it, and the relation of potentiality to actuality we know.

There are then two causes, namely, necessity and the final end. For many things are produced, simply as the results of . . . what may be called hypothetical necessity. For instance, we say that food is necessary; because an animal cannot possibly do without it. . . . Here is another example of it. If a piece of wood is to be split with an axe, the axe must of necessity be hard; and, if hard, must of necessity be made of bronze or iron. Now exactly in the same way the body, which like the axe is an instrument—for both the body as a whole and its several parts individually have definite operations for which they are made—just in the same way, I say, the body, if it is to do its work, must of necessity be of such and such a character, and made of such and such materials.

It is plain then that there are two modes of causation, and that both of these must, so far as possible, be taken into account in explaining the works of nature, or that at any rate an attempt must be made to include them both; and that those who fail in this tell us in reality nothing about nature. For primary cause [*arkhê*] constitutes the nature of an animal much more than does its matter. There are indeed passages in which even Empedocles hits upon this, and following the guidance of fact, finds himself constrained to speak of the ratio [*logos*] as constituting the essence and real nature of things. Such, for instance, is the case when he explains what is a bone. For he does not merely describe its material, and say it is this one element, or those two or three elements, or a compound of all the elements, but states the ratio of their combinations. As with a bone, so manifestly is it with the flesh and all other similar parts.

The reason why our predecessors failed in hitting upon this method of treatment was, that they were not in possession of the notion of essence, nor of any definition of substance. The first who came near it was Democritus. . . . In the time of Socrates a nearer approach was made to the method. But at this period men gave up inquiring into the works of nature, and philosophers diverted their attention to political science and to the virtues which benefit mankind. . . .

I.5 Of things constituted by nature some are ungenerated, imperishable, and eternal, while others are subject to generation and decay. The former are excellent beyond compare and divine, but less accessible to knowledge. The evidence that might throw light on them, and on the problems which we long to solve respecting them, is furnished but scantily by sensation; whereas respecting perishable plants and animals we have abundant information, living as we do in their midst, and ample data may be collected concerning all their various kinds, if only we are willing to take sufficient pains. Both departments, however, have their special charm. The scanty conceptions to which we can attain of celestial things give us, from their excellence, more pleasure than all our knowledge of the world in which we live; just as a half glimpse of persons that we love is more delightful than a leisurely view of other things, whatever their number and dimensions. On the other hand, in certitude and in completeness our knowledge of terrestrial things has the advantage. Moreover, their greater nearness and affinity to us balances somewhat the loftier interest of the heavenly things that are the objects of the higher philosophy. Having already treated of the celestial world, as far as our conjectures could reach, we proceed to treat of animals, without omitting, to the best of our ability, any member of the kingdom, however ignoble. For if some have no graces to charm the sense, yet even these, by disclosing to intellectual perception the artistic spirit that designed them, give immense pleasure to all who can trace links of causation, and are inclined to philosophy. Indeed, it would be strange if mimic representations of them were attractive, because they disclose the mimetic skill of the painter or sculptor, and the original realities themselves were not more interesting, to all at any rate who have eyes to discern the reasons that determined their formation. We therefore must not recoil with childish aversion from the examination of the humbler animals. Every realm of nature is marvellous: and as Heraclitus, when the strangers who came to visit him found him warming himself at the furnace in the kitchen and hesitated to go in, is reported to have bidden them not to be afraid to enter, as even in that kitchen divinities were present, so we should venture on the study of every kind of animal without distaste; for each and all will reveal to us something natural and something beautiful. Absence of haphazard and conduciveness of everything to an end are to be found in Nature's works in the highest degree, and the resultant end of her generations and combinations is a form of the beautiful.

If any person thinks the examination of the rest of the animal kingdom an unworthy task, he must hold in like disesteem the study of man. For no one can look at the primordia of the human frame—blood, flesh, bones, vessels, and the like—without much repugnance. Moreover, when any one of the parts or structures, be it which it may, is under discussion, it must not be supposed that it is its material composition to which attention is being directed or which is the object of the discussion, but the relation of such part to the total form. Similarly, the true object of architecture is not bricks, mortar, or timber, but the house; and so the principal object of natural philosophy is not the material elements, but their composition, and the totality of the form, independently of which they have no existence. . . .

Spontaneous and Cultivated Nature

Explanations of Plants, by Theophrastus, translated by Robert M. Torrance

Theophrastus was a native of the island of Lesbos, born (according to Diogenes Laërtius) around 370 B.C. and dying around 285, at the age of 85. Whether as a fellow pupil in Plato's Academy or in Asia Minor or Lesbos, he became one of Aristotle's closest colleagues and succeeded him as head of the Lyceum in 322. Best known in modern times for his Characters, *thirty brief sketches of human types similar to those portrayed in the comedy of his time—tradition made the comic poet Menander his pupil—he was a prolific researcher and writer. His lost* Opinions of the Natural Philosophers, *of which only the section "On the Senses" survives, is the ultimate source, after Aristotle, of most of our knowledge of pre-Socratic natural philosophy. Theophrastus appears to have been a thinker of little originality, extending and modifying Aristotle's ideas rather than advancing new concepts. His longest surviving works, the* Enquiry into Plants *and* De Causis Plantarum *("Explanations of Plants"), both available in the Loeb Classical Library, are counterparts of Aristotle's investigations of animals, distinguished for their painstaking empirical observations. The following passages are important for their affirmation, in the spirit of Aristotle, that art and cultivation, far from being necessarily opposed to nature, can work together with it and foster its processes.*

I.16 ... One might perhaps bring up a difficulty common to plants in general, whether to study their nature from those that grow spontaneously, or by cultivation, and which of these is natural. (Similar to this, or rather part of it, is whether to study their nature from wild or domesticated varieties.)

For nature contains its principles in itself—we are speaking of the natural, as in plants that grow spontaneously—whereas plants externally aided, especially by art, have a different principle. Nor should it be thought natural for animals to be molded or forced to small or large size or to any overall shape. Nature always strives for the best, and on this there is general agreeement.

But cultivation does the same; for nature simultaneously finds fulfillment when it obtains through art what it happens to lack, such as abundant food of the right kind and removal of impediments and hindrances—things which their native habitats, where it is best to study their natures, obviously provide to each plant. But these provide only such external aids as air, wind, soil, and food, whereas farming also alters and rearranges. If a plant's nature needs external aids to growth, it will also welcome the latter kind as suitable to it; and it will reasonably demand and seek these things, especially since it is dependent on them and has its principles in them. A strange and seemingly unnatural thing happens to plants that grow spontaneously: sprouting from seed, they deteriorate and completely change kind. This is not natural, for it is natural that each produce its own kind.

The difficulties, then, are roughly speaking such as these.

V.1 Some things occur spontaneously in trees and plants, others through provision and cultivation.

Each of these has its subdivision. Some spontaneous occurrences are natural, some unnatural, for this happens in plants as well as animals—if, for example, a plant does not sprout or bear fruit properly, or at the normal season, or from its usual parts, or something else of this kind, all these things are unnatural. Art and cultivation, on the other hand, can either work together with nature to produce good and plentiful fruit or can aim at specialized and unusual fruits—by growing a seedless grape-cluster, for example, or black and white grapes from the same twig and on the same cluster, and any number of other such things.

FIVE

Oneness and Separation: The Hellenistic World and Rome

Chapter 11: In the Footsteps of Saturn: Poets of Pasture, Farmland, and Forest *441*

The fertile centuries of archaic and classical Greece between Homer and Aristotle, roughly from the eighth to the fourth century B.C., saw the beginnings and some of the greatest achievements of Western epic, lyric, and drama, sculpture and architecture, political institutions and ideas (notably democracy) originating in the *polis*, historical writing and geography, mathematics and astronomy—and philosophy. Among philosophy's seminal concepts were those of dynamic nature or *physis*, of an ordered universe or *kosmos*, and of the reason or *logos* by which it is ordered: concepts on which physics and cosmology (which still bear their names) are founded. However much some of these accomplishments may have owed to the civilizations of Egypt, Babylonia, and Persia, they remain astonishing testimonies to one of the most creative moments in the history of human thought.

The subsequent age, embracing the establishment of powerful Hellenistic monarchies by the successors of Alexander the Great, followed by the relentless spread and eventual collapse, in the fifth and sixth centuries A.D., of the Roman Empire in the West, was in many ways a time of consolidation and diffusion of these momentous achievements. It spread first (somewhat ephemerally) to the conquered peoples of the East, as far as India, and then (far more permanently) to those who conquered the Greeks from the West, as captive Greece—in the Roman poet Horace's famous phrase—took her fierce victor captive and brought the arts of civilization to Italy, and thence, in time, to northern and western Europe. Yet despite the immense historical importance of this conservative function, these were by no means centuries of stagnation in the Mediterranean world, or of parasitic repetition of a glorious past. On the contrary—even apart from the dissident impulses of post-exilic Judaism and early Christianity, among others—Greco-Roman civilization in this tumultuous era not only assimilated but also continued, in countless ways, to adapt and transform the riches of its classical inheritance. Both during the Hellenistic age—properly dated from the death of Alexander in 323 to the victory of Octavian over Antony and Cleopatra (last of the Ptolemies of Egypt) at the battle of Actium in 31 B.C., though sometimes loosely used of the whole period between classical Greece and the Byzantine Empire that succeeded the fall of Rome—and under Roman rule, the Greek-speaking peoples of the *oikouménê*, or "inhabited world," as they called the territories of the eastern Mediterranean dominated by their culture and language, continued to develop new philosophies and to make important discoveries in mathematics, astronomy, and medicine; these would profoundly affect later civilizations of both the Arab East and Christian West. Meanwhile, the Romans, though making few if any original contributions to science or philosophy, not only absorbed what their Greek teachers handed down, but gave vigorous expression, above all through their great poets (at a time when Greek literature had largely succumbed to its overshadowing past), to the hopes and tensions of a moment acutely conscious of following a more creative age—one which might or might not yet be reborn—when the divinely natural and the human worlds had seemed, at least in retrospect, more nearly one.

For both Plato and Aristotle (as for Socrates, Empedocles, Pythagoras, Thales, and others before them), philosophy had been closely tied up with the Greek city-state, and the theoretical life of the mind had been inseparable from practical, and indeed political, concerns. Contemplation of the immutable forms might be the Platonic philosopher's highest goal, yet (like a Buddhist bodhisattva) he must always return from the light above to the cave where others dwell; by no accident, Plato's two longest dialogues were the *Republic* and the *Laws*. But after the subjugation of Greece by the Macedonian phalanxes of Philip the Great and his son Alexander, and after the powerful monarchies of Alexander's successors—principally the Ptolemies in Egypt, the Seleucids in Syria and Mesopotamia, and the Antigonids in Macedonia—began to combat one another for hegemony over the Greek-speaking world, the precarious autonomy of all but a few among the old Greek city-states (notably Rhodes) soon became little more than illusion. Partly in reaction, no doubt, to the loss of political freedom, the three major philosophical schools of the new age sought freedom through detachment from the failed world of the city-state: Epicureanism through withdrawal from political life; Stoicism through involvement in the larger citizenship of the world, the *kosmópolis;* and Scepticism through questioning of all dogmatic or positive beliefs.

In many basic respects the two major Hellenistic systems of thought (Scepticism repudiated all systems) would view themselves as opposites, Epicureanism emphasizing pleasure—or at least absence of pain—as the goal of human life in a universe composed of atoms randomly colliding in the void, Stoicism exalting rationality and denial of all emotion in a world divinely ordered by fate in accord with the providence of God. But for both, an understanding of nature, conceived mainly in material terms inherited from the pre-Socratics (even the Stoics, while drawing heavily on Plato and Aristotle, rejected their idealism and thought of the active principle of the world as a Heraclitean fire), was a precondition of ethics: the good life for both was life according to nature, and suffering or evil arose through separation from nature by misunderstanding of its fundamental conditions. Many centuries later, when the Neoplatonism of Plotinus, the third great philosophical system of the Greco-Roman world, renewed and intensified the idealistic tendencies of Platonic philosophy, it significantly viewed nature not (like the Gnostics of its age) as an irremediable evil categorically opposed to the divine but as an emanation of the ultimate One that shared in the contemplation of its origin; through such contemplation all animate life might aspire to return to the unity from which it had been divided. To this extent, even Neoplatonism might be called in part what Epicureanism and Stoicism had been more fully, a philosophy of life in accord with nature.

The increasing urbanization of an age dominated by great heterogeneous imperial capitals like Alexandria, Antioch, and Rome helped to create, on a more immediate level, a widespread sense of increasing separation from a putatively more natural condition associated temporally with a bygone pastoral or agricultural existence, or an idealized mythological past in the lost Golden Age of Kronos or Saturn, and spatially with a simple farmstead far from the ever-encroaching megalopolis—from what Horace, in another famous line, called "the smoke and wealth and tumult of Rome." Only such a condition, could it but be attained or revived (in imagination if not in reality), might possibly restore primordial oneness with nature and overcome the separation from it which now afflicted humankind. The one important new genre of poetry created by the poets of Alexandria—who were very nearly the last major poets of ancient Greece—under the early Ptolemies in the third century B.C. was the pastoral or bucolic idyll of Theocritus. It celebrated, in a highly artificial (and sometimes humorous) style, the occupations and loves of idealized shepherds and cowherds living a rustic existence as remote as possible from the intrigues of city or court. This genre would have a long and influential future, growing ever more distant from any actual acquaintance with the life it purported to extol.

The great Roman poets of the first centuries B.C. and A.D., who inherited the shop-worn conventions of Alexandrian poetry, renewed them through acquaintance with the more vigorous traditions of earlier Greek epic, lyric, and philosophical poetry, and with the conflicts and troubles of their own time. Lucretius and Manilius, in their impassioned didactic poems, express respectively the Epicurean and Stoic visions of nature and human life, while Horace and Virgil evoke the longing for a more natural state of existence—increasingly associated with agriculture and the nostalgically remembered simplicity of village life—that will re-echo through centuries of European poetry. Toward the end of antiquity the joyous exaltation of spring that resonates in Lucretius, Virgil, and Horace reappears with renewed erotic intensity (and with a new sense of foreboding) in the Latin *Vigil of Venus.* And the nostalgia for a lost communion with birds, lambs, and flowers takes popular form in Greek prose romances such as *Daphnis and Chloe,* whose hero and heroine inhabit a pastoral world so impossibly innocent that the act of lovemaking eludes them: shepherds of the Neverland who embody deep-seated romantic longings for a "nature" too pure to be true. In their various ways such works all express a troubled human need for reunion with a condition that threatens—perhaps because it had never existed—to be lost forever.

The philosophers of this age, then, continued to be almost exclusively Greeks, even though many doctrines of both Epicureans and Stoics have survived most fully in expositions by Roman writers such as Lucretius and Cicero; the greatest poets, on the other hand, after Theocritus, were overwhelmingly Romans. The concerns to which both philosophers and poets gave expression were the subject also of reflections by various prose writers of the time, including Roman moralists and historians troubled by decline of the old agricultural virtues; Greek and Roman encyclopedists cataloguing marvels of nature often indistinguishable from fable; and major Greek scientists such as the astronomer Ptolemy and the physician Galen exploring, near the end of Western antiquity, problems raised by the pre-Socratics and Hippocratics near its beginning. One of the most important of these late Greco-Roman writers, Plutarch, draws in his wide-ranging essays upon philosophy and science, poetry and history, of Greeks and Romans alike to reflect on the disturbing paradox that human beings may be less happy, less virtuous, and even less rational than animals depicted as being more fully in accord with nature than humans can ever be. It is fittingly Plutarch who relates the haunting story, in his essay on the obsolescence of oracles in a pagan world now passing forever, of the sailors who mysteriously heard a voice proclaiming, during the reign of Tiberius Caesar, that the great god Pan, the very embodiment of inseparable oneness with nature, is dead.

For general treatments of Hellenistic and Greco-Roman history and literature, see W. W. Tarn, *Hellenistic Civilisation* (3rd ed. 1951); F. E. Peters, *The Harvest of Hellenism* (1970); Peter Green, *Alexander to Actium: The Historical Evolution of the Hellenistic Age* (1990); and Albrecht Dihle, *Greek and Latin Literature of the Roman Empire* (1989; Eng. trans. 1994).

10

Chance, Fate, and Contemplation: Greco-Roman Philosophies of Nature

Although the philosophical thought of the Hellenistic and early Greco-Roman periods was dominated by Epicureanism, Stoicism, and Scepticism, which rapidly gained adherents among the educated classes of the Greek- (and soon of the Latin-) speaking world, the earlier philosophies of Plato and Aristotle continued to be of immense importance. Both Plato's Academy, under Speusippus, Xenócratês, and their successors, and Aristotle's Lycéum under Theophrastus, Strato, and theirs, remained major centers of philosophical and scientific learning in Athens, training teachers who promulgated their views throughout the oikoumene. *Their doctrines were by no means static, especially within the Academy, where Speusippus, its head from 347 to 339 B.C., stressed dialectic and mathematics but rejected—like those who followed him—the Platonic theory of Forms. After Arcesilaus, head of the Academy in the mid third century, emphasized not the positive theories of Plato but the suspension of judgment characteristic of the early Socratic dialogues, what came to be called the Middle or Late Academy was effectively allied with the Scepticism of Pyrrho (ca. 365/360–275/270 B.C.) against the dogmatic systems of Epicureans and Stoics. Changes in Peripatetic doctrine were less pronounced, but with its focus on continuing research, the Lyceum too resisted doctrinaire fixity. Thus Theophrastus, its scholarch or head from 322 to ca. 285, though mainly following Aristotle, moderated his teleological emphasis, and his successor Strato, scholarch from ca. 285 to 269, contradicted Aristotle (and the Stoics) by affirming the existence of void within the cosmos.*

Nor was the influence of Plato and Aristotle limited to formal adherents of the Academy or Lyceum. On the contrary, their views (and those of their precursors and contemporaries) were to a great extent the common property of the age. Thus the scientific and philological research conducted under the auspices of the great Alexandrian Museum (Mouseion) *founded by Ptolemy I, and the investigations of generations of physicians in Alexandria and elsewhere, whatever their philosophical allegiance, continued the exploratory spirit of Aristotle and of the Hippocratics. In the same way, great geometers of the age—Euclid, Archimêdês, Apollonius of Pergê, and others—advanced in practice the mathematical apprehension of reality advocated by Plato and the Pythagoreans. Moreover, just as the atomic theory of Leucippus and Democritus became the basis of Epicurean physics, many concepts of the pre-Socratics, and of both Aristotle (including interchangeability of the four elements) and Plato (including the Demiurge of the* Timaeus*), were adapted to the developing system of the Stoics.*

Neo-Pythagoreanism, and a host of hybrid eclecticisms under the Roman Empire, drew heavily on Plato and Aristotle alike. Their continued vitality was especially evident after the third century A.D. In the first great Neoplatonist, Plotínus (A.D. 205–269/70), not only are Plato's dialogues, along with many Stoic doctrines, a pervasive influence, but also Aristotle's Metaphysics, *his disciple Porphyry remarked, "is condensed in its entirety." By the end of antiquity neither Stoics nor Epicureans were to be found in the Greek-speaking world, where the Eastern "Roman Empire" centered in Constantinople would outlast that of the West through the* near-millennium of Byzantine history. Christian theology was suffused with transmutations of Platonic doctrine, and in philosophy Neoplatonism reigned unchallenged; among its last representatives were the great sixth-century commentators on Aristotle, the pagan Simplicius (who preserved much of what now survives from the pre-Socratics more than a thousand years before) and the Christian John Philoponus. Thus the partial eclipse of Plato and Aristotle by the Hellenistic philosophies proved to be only temporary, and only apparent; it would not be Epicurus or Zeno who would help shape the Islamic and Christian conceptions of the world for centuries to come.

Stoicism and Epicureanism were above all ethical systems teaching men and women—Greek and barbarian, free and slave—how to attain tranquility of soul in a troubled time when the city-state, for many, had ceased to provide an adequate structure of meaning. At the beginning of the Hellenistic age, in consequence, it was less the politically oriented ethics of Plato and Aristotle that promised accessible models for new ways of life than the dissident practices of a diverse group of Socrates' fourth-century followers. Among these were Aristippus of Cyrene (Kyrênê), founder of the Cyrenaic school, which taught that the end of life was pleasure, and Antísthenês of Athens, whose follower, Diógenês of Sinôpê on the Black Sea, was the leading exponent of the Cynic (kynikos) school—so named after Diogenes' nickname of kyôn, "the dog"—which despised pleasure and taught that the end of life was virtue attained by a strenuously abstemious existence in accord with nature. For Aristippus and the Cyrenaics, pleasure was instinctive or natural, and therefore good. For the Cynics, on the contrary, what was natural was the ascetic pursuit of virtue in defiance of social convention, for only in stripping away the extraneous trappings of culture could man be free. Aristippus therefore flaunted the luxurious perquisites bestowed upon him by his patron Dionysius, tyrant of Syracuse, while Diogenes, according to the many legends that gathered about him, lived in a pot or "tub" outside Athens, ate and even masturbated in public—"if only it were that easy," he says in Diogenes Laërtius (VI.69), "to stop hunger by rubbing the belly!"—and, when asked by Alexander the Great what favor he wished to request, told the king to stand out of his sunlight. The followers of Socrates, who of course included Plato also, were clearly a varied group, and no less varied were the meanings they attached to the flexible concept of nature.

For the Epicureans, too, pleasure was the end of human life, and they were therefore often associated, despite fundamental differences, with the Cyrenaics. Aristippus and his school rejected the study of physics, Diogenes Laërtius tells us (II.92), because of its incomprehensibility; Epicurus (Epikouros, 341–270), in contrast, again according to Diogenes Laërtius (X.27), wrote thirty-seven of his three hundred or so books or scrolls—far more than on any other topic—"On Nature." Not that Epicurus partook—like the Ionian natural philosophers, Aristotle, or indeed Plato—in a spirit of scientific investigation; far from it. But he believed that only a true understanding of nature could allow human beings to free themselves from superstitious terrors of the gods and of death. They could thus attain the liberating condition of ataraxia, "imperturbability," the lasting pleasure of the soul, defined by absence of mental pain, that was the goal of each human being. (Epicurus and his followers in the school commonly called the Garden, after their gathering place in Athens, were known for unusual sobriety and moderation; they held that pleasure was natural, but the pleasures they sought were far from those of a libertine hedonism.) To achieve this understanding, Epicurus adopted, without acknowledgment, the atomic theory of matter from the pre-Socratics Leucippus (whose very existence he denied) and Democritus. He then incorporated a few important innovations, apparently of his own, notably by postulating an occasional unpredictable "swerve" in the course of atoms otherwise fall-

ing forever in parallel lines through endless space. From these swerves arose the collisions of atoms that repeatedly brought about the creation and dissolution of worlds—and also, perhaps, the freedom of the human will. Epicurus's cosmos was thus composed of atoms hurtling at random through the infinite void, a universe whose irrelevant gods remained eternally aloof from human affairs; and freedom consisted in serene understanding that in such a world death, like all else, was a natural occurrence by which the atoms of each human soul would return to the universe whence they came and be recombined in random patterns throughout eternity. Life in accord with nature, then, presupposed clearsighted recognition of the sovereignty of chance in a material world whose processes, though forever the same, were forever unpredictable in outcome. In this recognition was the emancipation from which alone tranquility of soul, and therefore happiness, could arise.

Stoicism, which originated in the same place (Athens) and at the same time (the end of the fourth and beginning of the third century B.C.) as Epicureanism, had much in common with it, since both developed comprehensive philosophical systems that stressed—in contrast to the earlier "Socratics"—the necessity of understanding nature in order to pursue the ethical life that presupposed harmony with it. Both, far more than the Academy or the Lyceum, developed followings among various classes and ethnic groups of the cosmopolitan Hellenistic world (the concept of the "cosmopolite," or "citizen of the world," derives from the Cynics and Stoics) and propounded doctrinal orthodoxies that gave their philosophies a quasi-religious dimension. But whereas the Epicurean system was all but fully formed by its founder, whom Lucretius and others revered as very nearly a god, and became a sect or even a cult, the Stoic system changed and developed throughout its early centuries, proving far more adaptable to its time than that of its more exclusive rival. A few of Epicurus's many writings—three letters and a collection of sayings—survive in their entirety, all preserved in the Life *of Diogenes Laërtius. In the case of the early Stoics, however, nothing but scattered fragments remain. Some of these are from works by the Stoics' founder, Zeno (Zênôn, 335–263), a Hellenized Phoenician of Citium (Kition) in Cyprus, and his successors, Cleanthes (331–232) of Assos in northwestern Asia Minor and Chrysippus (ca. 280–207) of Soli in southeastern Asia Minor; others are by the "middle Stoics" who significantly moderated their doctrines and brought Stoicism to Rome, Panaetius (ca. 185–109) of Rhodes and Posidonius (ca. 135–ca. 50) of Apaméa in Syria. Only from the Roman imperial period have complete works, mainly of moral philosophy, survived—by the Spanish-Roman moralist Seneca (ca. 2 B.C.–65 A.D.), the Greek freedman Epictetus (Epiktêtos, ca. A.D. 55–ca. 135), and the Roman Emperor Marcus Aurelius (121–180), who wrote his* Meditations *in Greek.*

Because it was both a changing philosophy and one whose seminal writings, especially on nature, have almost entirely vanished, apart from discussions of them in Cicero (first century B.C.), Diogenes Laërtius (early third century A.D.?), and others, the Stoic system must be reconstructed with cautious awareness that no summary can embrace the range of its variations. Zeno, who founded the school—known as the Stoa after the stoa poikilê *or "painted colonnade" in Athens (sometimes known as "the Porch" in English) where he held his discussions in public, in contrast to the private gatherings in Epicurus's Garden—began as a pupil of the Cynic Crates, and connections between Cynicism and early Stoicism were many. Above all, both philosophies held that the object of life was a self-denying virtue attained through deliberately living in accord with nature, which for both was an embodiment not of passion or pleasure but, in the Platonic phrase, of "right reason." For the Stoic, Diogenes Laërtius writes (VII.85–86), self-preservation is a natural instinct of all animals, but pleasure is only a secondary by-product. In an animal, "Nature's rule is to fol-*

low the direction of impulse. But when reason by way of a more perfect leadership has been bestowed on the beings we call rational," Diogenes explains in R. D. Hicks's Loeb Classical Library translation (1925), "for them life according to reason rightly becomes the natural life." Therefore, for the human being, virtue (Greek aretê, Latin virtus), reason (Greek logos, Latin ratio), and nature (Greek physis, Latin natura) are one and the same.

As they became increasingly influential, the Stoics greatly tempered the scandalous nonconformity of Cynic behavior. Yet from Zeno to Chrysippus traces of earlier antinomianism remained, as in their supposed advocacy of community of women, incest, and even cannibalism: "natural" activities from which the respectable middle and later Stoics hastened to distance themselves. So uncompromising was Stoic rationalism that "passion or emotion [pathos]," Diogenes writes (VII.110), "is defined by Zeno as an unnatural or irrational movement in the soul," and although "rational" emotions of joy, caution, and wishing were approved, their irrational and thus unnatural counterparts, pleasure, fear, and desire, along with grief, pity, and most other emotions, were sternly condemned. The Peripatetics, Seneca pithily observes (Epistles CXVI.1), moderate the emotions, the Stoics expel them. Through his godlike apathy (apatheia), or absence of emotion—the negative virtue corresponding to the Epicurean goal of ataraxia—the Stoic wise man, or saint, whatever his external condition, attained a freedom greater than that of kings, believing, in Diogenes' words (VII.122–23), that "the wise are infallible, not being liable to error." From this lofty eminence the Stoic elect, like their Puritan counterparts of a later age, "are not pitiful and make no allowance for anyone; they never relax the penalties fixed by the laws, since indulgence and pity and even equitable consideration are marks of a weak mind." Such a "natural" condition, totally removed from that of the animals—human "nature" having become a normative, and no longer a descriptive, construct—was, perhaps fortunately, one which very few human beings could hope to achieve.

Yet understanding of human nature, for the Stoic as for the Epicurean, required understanding of the nature of the universe, since between microcosm and macrocosm the Stoics, like others before them, saw a close correspondence. Their elaborate cosmology drew heavily on the pre-Socratics (notably Heraclitus, on whom Cleanthes wrote four books), the Hippocratics, Plato, and Aristotle, whose views the early Stoics, especially Zeno and Chrysippus, welded into an integrated physical system of their own. Zeno's universe, F. E. Peters writes in The Harvest of Hellenism, "was essentially that of Aristotle cleansed of its last traces of transcendentalism and pushed even further back toward the dynamism of pre-Parmenidean philosophers." The cosmos is a being at once living and divine. God, nature, and reason (the logos) are one, sometimes identified with the cosmos as a whole, but more often with its active principle of nature. This principle might be characterized as "an artistically working fire, going on its way to create," in Zeno's phrase (Diogenes Laërtius VII.156): a pure form of the material element of fire (equated with aether but not considered, as by Aristotle, a fifth element pertaining to a separate superlunar sphere) in which the cosmos begins and to which it periodically returns as the four elements again blend into the fiery destruction and renewal of ekpyrôsis. Or it might be identified with the pervasive pneuma (spirit or breath), a mixture of fire and air binding together the passive matter of the cosmos (as the pneuma of the medical writers binds together the body) in a continuum of incessant Heraclitean tension or harmony. Both the active and the passive components consist of matter, yet "it is misleading," A. A. Long cautions in Hellenistic Philosophy (2nd ed., 1986), "to describe the Stoics as 'materialists.' Bodies, in the Stoic system, are compounds of 'matter' and 'mind' (God or logos). Mind is not something other than *body but a necessary constituent of it, the 'reason' in matter. The Stoics are better described as vitalists."*

God, as nature, fire, or spirit, is the demiurge or craftsman continually re-creating the cosmos and ruling over it by reason or providence (pronoia), *so that everything in it, rightly understood, is purposeful and good. Far more than in Aristotle, teleology extends to all aspects of existence in what is truly, despite all appearances, the best of all possible worlds. This supremely rational god could be given the name (as in Cleanthes' hymn) of Zeus, and by extensive allegorical reinterpretation the Stoics portrayed their philosophy as compatible with traditional Greco-Roman religion. God's ineluctable plan or providence is experienced as destiny or fate (Greek* heimarménê, *Latin* fatum*), and the wise man is free not by resisting but by acknowledging and consciously cooperating with the fate or reason by which God or "governing nature" rules all things. The Stoics could thus incorporate into their system both inherited forms of divination and the newer astrology—most fully expounded in Manilius's great poem of the first century A.D.—widely accepted, after the Hellenistic period, by virtually all schools except the Epicureans. Both the universe and society—in which Stoics, unlike Epicureans, were encouraged to take an active role—were thus ruled by law in accord with divine and human reason, and therefore with nature.* Nomos *and* physis *could no longer be opposed, as by the Sophists of an earlier generation, since for the Stoics, as Peters writes, "the two were synonymous.* Nomos *was, in the first instance, God's will immanent and operative in the universe"; in the human sphere, "it represented the conversion of his providence into a moral imperative." Nature was "the cosmic rhythm that provided a paradigm for man's activity," and it was in accord with nature in this comprehensive sense, and not merely with "nature" as the expulsion of human emotion, that the great Stoics—most movingly, perhaps, the Emperor Marcus Aurelius—earnestly endeavored to live.*

Stoicism thus provided an integrated system of physics and ethics that appealed to a wide spectrum of Greco-Roman society unwilling to accept either the more uncompromisingly materialistic naturalism of Epicurus or the irrationalism and mysticism of the various religious currents that repeatedly swept over their world. It was by far the most widely disseminated philosophy of its age, and through it some of the central concepts of Greek thought, including logos *(the Word) and* pneuma *(the Holy Spirit), entered into Christianity; the Neoplatonism of Plotínus and his successors, too, was deeply indebted to the Stoic as well as to the Platonic and Aristotelian legacies. For further reading, see Whitney J. Oates, ed.,* The Stoic and Epicurean Philosophers *(1940); A. A. Long and D. N. Sedley, eds.,* The Hellenistic Philosophers, *2 vols. (1987); Emile Bréhier,* History of Philosophy: The Hellenistic and Roman Age *(1931; English trans. 1965); A. A. Long,* Hellenistic Philosophy: Stoics, Epicureans, Sceptics *(2nd ed., 1986); Martha C. Nussbaum,* The Therapy of Desire: Theory and Practice in Hellenistic Ethics *(1994); J. M. Rist,* Stoic Philosophy *(1969); and S. Sambursky,* Physics of the Stoics *(1959) and* The Physical World of Late Antiquity *(1961).*

A World of Infinite Beginnings

The Extant Remains of Epicurus, translated by Cyril Bailey

Epicurus (Epikouros) was born an Athenian citizen on the island of Samos in 341 B.C.*; after returning to Athens at the age of 18, he studied at Colophon on the coast of Asia Minor under Nausíphanês, from whom he learned of the atomism of Democritus. Thereafter, in his early thirties, both at Mitylênê on Lesbos and at Lámpsacus on the Hellespont, he began to gather around him the group of devoted followers that would continue to grow throughout his life. In 307 or 306, at about 35, he returned to Athens and purchased both a house and the garden where he and his circle held their philosophical conversations, and which would give its name to his school. Here he spent almost the entire remainder of his life, dying in 270.*

For Epicurus, as for other philosophers of his age, the goal of philosophy was to promote the good and happy life, the life of imperturbable serenity, freedom, and self-sufficiency (autarkeia), *"the greatest of all riches." "Vain is the word of a philosopher," one of his fragments declares, "which does not heal any suffering of man." Athens and other Greek cities had by now become pawns in the power politics of the Hellenistic monarchs, and Epicurus counseled his disciples to "release ourselves from the prison of affairs and politics" (Vatican LVIII). But self-sufficiency and withdrawal from political life—"Live unknown" was his motto—by no means meant solitude; Epicurus and his tight-knit sect that included women (some of them courtesans) along with men, and slaves along with citizens, highly valued friendship. "Intimacy without difference, amiability, courtesy, and fellow-feeling rather than cultivation of the self: these formed the premises," Thomas G. Rosenmeyer writes in* The Green Cabinet *(1969), "on which living in the Garden was founded," and Epicurus, he notes, was significantly "the first philosopher to use the medium of the epistle to formulate his position." Indeed, of the writings that survive from Epicurus's voluminous output, apart from miscellaneous fragments quoted by other authors, three are letters to friends giving short summaries of his thoughts on physics (to Herodotus), "meteorology" or the study of things above (to Pythocles), and ethics (to Menoeceus); and two are collections of aphorisms, the "Principal Doctrines," included with the three letters in Diogenes Laërtius's* Life of Epicurus, *and the "Vatican Sayings" discovered in a fourteenth-century Vatican manuscript in 1888.*

The master may have considered the study of nature "a necessary evil," as J. M. Rist suggests in Epicurus, *but to it he dedicated enormous time and attention, believing (he writes to an unknown correspondent in a fragment of another letter) that "in so far as you are in difficulties, it is because you forget nature; for you create for yourself unlimited fears and desires" which a true understanding of nature could dispel. The "Letter to Herodotus" sets forth a brief epitome of his natural philosophy, including the basic doctrines that nothing comes from nothing, that only bodies and space truly exist, that the boundless and eternal universe contains an infinite number of indivisible and unalterable atoms differing only in shape, weight, and size, bringing about by repeated collisions and separations the creation and dissolution of infinite worlds. The "Letter to Pythocles" (thought by some to have been written by a pupil) repudiates all attribution of natural phenomena to the actions of gods, who are blessedly unconcerned with mortal affairs—"If God listened to the prayers of men," a fragment declares, "all men would quickly have perished: for they are for ever praying for evil against one another."But it also reveals how far Epicu-*

rus is from seeking precise scientific knowledge of nature, since he suggests that alternative and incompatible explanations of phenomena such as eclipses, thunder, and planetary motions may be equally valid: what is important is not to attain certainty in such details but to renounce the superstitions that subjugate men to false terror of the gods and of imagined punishments after death.

Even so, it is the constancy and regularity of the fundamental processes of nature that Epicurus, like Democritus and the Milesian naturalists before him, most emphasizes: "Nothing new happens in the universe," a fragment affirms, "if you consider the infinite time past." The cosmos is no more ordered by mere accident than by arbitrarily willful divinities. In Vatican Saying XLVII Epicurus's principal disciple, Metrodôrus of Lampsacus (ca. 330–ca. 277), repudiates the dominion of Chance or Fortune (Greek Tykhê, *Latin* Fortuna*), widely worshipped throughout the Hellenistic world—a goddess who crystallized, as Peter Green observes in* Alexander to Actium, *"the negative, unknown, random, and, thus, arguably unpredictable element of existence" no less tyrannically than the most implacable fate, to which she was only theoretically opposed. Yet the random dimension introduced by the unpredictable swerve of the atoms was crucial. This was an ordered, though never a deterministic, universe, for within strict limits chance was inherently part of its order, and therein lay the momentous possibilities of indeterminacy, change, and freedom. In this insight—which appears to anticipate both Darwinian natural selection by random mutation and the uncertainty principle of twentieth-century quantum theory—Epicurus's conception of nature, as expounded most fully by Lucretius, seems astonishingly modern.*

Quotations are from Epicurus: The Extant Remains, *ed. Cyril Bailey (1926). See also* The Philosophy of Epicurus, *trans. George K. Strodach (1963);* Letters, Principal Doctrines, and Vatican Sayings, *trans. Russel M. Geer (1964); and J. M. Rist,* Epicurus: An Introduction *(1972).*

FROM *Letter to Herodotus*

. . . Having made these points clear, we must now consider things imperceptible to the senses. First of all, that nothing is created out of that which does not exist: for if it were, everything would be created out of everything with no need of seeds. And again, if that which disappears were destroyed into that which did not exist, all things would have perished, since that into which they were dissolved would not exist. Furthermore, the universe always was such as it is now, and always will be the same. For there is nothing into which it changes: for outside the universe there is nothing which could come into it and bring about the change.

Moreover, the universe is bodies and space: for that bodies exist, sense itself witnesses in the experience of all men, and in accordance with the evidence of sense we must of necessity judge of the imperceptible by reasoning, as I have already said. And if there were not that which we term void and place and intangible existence, bodies would have nowhere to exist and nothing through which to move, as they are seen to move. And besides these two nothing can even be thought of either by conception or on the analogy of things conceivable such as could be grasped as whole existences and not spoken of as the accidents or properties of such existences. Furthermore, among bodies some are compounds, and others those of which compounds are formed. And these latter are indivisible and unalterable (if, that is, all things are not to be destroyed into the non-existent, but something permanent is to remain behind at the dissolution of compounds): they are completely solid in nature, and can by no means be dissolved in any part. So it must needs be that the first-beginnings are indivisible corporeal existences.

Moreover, the universe is boundless. For that which is bounded has an extreme point: and the extreme point is seen against something else. So that as it has no extreme point, it has no limit; and as it has no limit, it must be boundless and not bounded. Furthermore, the infinite is boundless both in the number of the bodies and in the extent of the void. For if on the one hand the void were boundless, and the bodies limited in number, the bodies could not stay anywhere, but would be carried about and scattered through the infinite void, not having other bodies to support them and keep them in place by means of collisions. But if, on the other hand, the void were limited, the infinite bodies would not have room wherein to take their place.

Besides this the indivisible and solid bodies, out of which too the compounds are created and into which they are dissolved, have an incomprehensible number of varieties in shape: for it is not possible that such great varieties of things should arise from the same atomic shapes, if they are limited in number. And so in each shape the atoms are quite infinite in number, but their differences of shape are not quite infinite, but only incomprehensible in number.

And the atoms move continuously for all time, some of them falling straight down, others swerving, and others recoiling from their collisions. And of the latter, some are borne on, separating to a long distance from one another, while others again recoil and recoil, whenever they chance to be checked by the interlacing with others, or else shut in by atoms interlaced around them. For on the one hand the nature of the void which separates each atom by itself brings this about, as it is not able to afford resistance, and on the other hand the hardness which belongs to the atoms makes them recoil after collision to as great a distance as the interlacing permits separation after the collision. And these motions have no beginning, since the atoms and the void are the cause.

These brief sayings, if all these points are borne in mind, afford a sufficient outline for our understanding of the nature of existing things.

Furthermore, there are infinite worlds both like and unlike this world of ours. For the atoms being infinite in number, as was proved already, are borne on far out into space. For those atoms, which are of such nature that a world could be created out of them, have not been used up either on one world or on a limited number of worlds, nor again on all the worlds which are alike, or on those which are different from these. So that there nowhere exists an obstacle to the infinite number of worlds. . . .

Moreover, we must suppose that the atoms do not possess any of the qualities belonging to perceptible things, except shape, weight, and size, and all that necessarily goes with shape. For every quality changes; but the atoms do not change at all, since there must needs be something which remains solid and indissoluble at the dissolution of compounds, which can cause changes; not changes into the non-existent or from the non-existent, but changes effected by the shifting of position of some particles, and by the addition or departure of others. For this reason it is essential that the bodies which shift their position should be imperishable and should not possess the nature of what changes, but parts and configuration of their own. For thus much must needs remain constant. For even in things perceptible to us which change their shape by the withdrawal of matter it is seen that shape remains to them, whereas the qualities do not remain in the changing object, in the way in which shape is left behind, but are lost from the entire body. Now these particles which are left behind are sufficient to cause the differences in compound bodies, since it is essential that some things should be left behind and not be destroyed into the non-existent.

Moreover, we must not either suppose that every size exists among the atoms, in order that the evidence of phenomena may not contradict us, but we must suppose that there are some variations of size. For if this be the case, we can give a better account of what occurs in our feelings and sensations. But the existence of atoms of every size is not required to explain the differences of qualities in things, and at the same time some atoms would be bound to come within our ken and be visible; but this is never seen to be the case, nor is it possible to imagine how an atom could become visible. . . .

Furthermore, the motions of the heavenly bodies and their turnings and eclipses and risings and settings and kindred phenomena to these, must not be thought to be due to any being who controls and ordains or has ordained them and at the same time enjoys perfect bliss together with immortality (for trouble and care and anger and kindness are not consistent with a life of blessedness, but these things come to pass where there is weakness and fear and dependence on neighbors). Nor again must we believe that they, which are but fire agglomerated in a mass, possess blessedness, and voluntarily take upon themselves these movements. But we must preserve their full majestic significance in all expressions which we apply to such conceptions, in order that there may not arise out of them opinions contrary to this notion of majesty. Otherwise this very contradiction will cause the greatest disturbance in men's souls. Therefore we must believe that it is due to the original inclusion of matter in such agglomerations during the birth-process of the world that this law of regular succession is also brought about.

Furthermore, we must believe that to discover accurately the cause of the most essential facts is the function of the science of nature, and that blessedness for us in the knowledge of celestial phenomena lies in this and in the understanding of the nature of the existences seen in these celestial phenomena, and of all else that is akin to the exact knowledge requisite to our happiness: in knowing too that what occurs in several ways or is capable of being otherwise has no place here, but that nothing which suggests doubt or alarm can be included at all in that which is naturally immortal and blessed. Now this we can ascertain by our mind is absolutely the case. But what falls within the investigation of risings and settings and turnings and eclipses, and all that is akin to this, is no longer of any value for the happiness which knowledge brings, but persons who have perceived all this, but yet do not know what are the natures of these things and what are the essential causes, are still in fear, just as if they did not know these things at all: indeed, their fear may be even greater, since the wonder which arises out of the observation of these things cannot discover any solution or realize the regulation of the essentials. . . .

And besides all these matters in general we must grasp this point, that the principal disturbance in the minds of men arises because they think that these celestial bodies are blessed and immortal, and yet have wills and actions and motives inconsistent with these attributes; and because they are always expecting or imagining some everlasting misery, such as is depicted in legends, or even fear the loss of feeling in death as though it would concern them themselves; and, again, because they are brought to this pass not by reasoned opinion, but rather by some irrational presentiment, and therefore, as they do not know the limits of pain, they suffer a disturbance equally great or even more extensive than if they had reached this belief by opinion. But peace of mind is being delivered from all this, and having a constant memory of the general and most essential principles. . . .

FROM *Letter to Pythocles*

. . . A world is a circumscribed portion of sky, containing heavenly bodies and an earth and all the heavenly phenomena, whose dissolution will cause all within it to fall into confusion: it is a piece cut off from the infinite and ends in a boundary either rare or dense, either revolving or stationary: its outline may be spherical or three-cornered, or any kind of shape. For all such conditions are possible, seeing that no phenomenon is evidence against this in our world, in which it is not possible to perceive an ending. And that such worlds are infinite in number we can be sure, and also that such a world may come into being both inside another world and in an interworld, by which we mean a space between worlds; it will be in a place with much void, and not in a large empty space quite void, as some say: this occurs when seeds of the right kind have rushed in from a single world or interworld, or from several: little by little they make junctions and articulations, and cause changes of position to another place, as it may happen, and produce irrigations of the appropriate matter until the period of completion and stability, which lasts as long as the underlying foundations are capable of receiving additions. For it is not merely necessary for a gathering of atoms to take place, nor indeed for a whirl and nothing more to be set in motion, as is supposed, by necessity, in an empty space in which it is possible for a world to come into being, nor can the world go on increasing until it collides with another world, as one of the so-called physical philosophers says. For this is a contradiction of phenomena. . . .

The eclipse of sun and moon may take place both owing to their extinction, as we see this effect is produced on earth, or again by the interposition of some other bodies, either the earth or some unseen body or something else of this sort. And in this way we must consider together the causes that suit with one another and realize that it is not impossible that some should coincide at the same time. Next the regularity of the periods of the heavenly bodies must be understood in the same way as such regularity is seen in some of the events that happen on earth. And do not let the divine nature be introduced at any point into these considerations, but let it be preserved free from burdensome duties and in entire blessedness. For if this principle is not observed, the whole discussion of causes of celestial phenomena is in vain, as it has already been for certain persons who have not clung to the method of possible explanations, but have fallen back on the useless course of thinking that things could only happen in one way, and of rejecting all other ways in harmony with what is possible, being driven thus to what is inconceivable and being unable to compare earthly phenomena, which we must accept as indications. . . .

Thunder may be produced by the rushing about of wind in the hollows of the clouds, as happens in vessels on earth, or by the reverberation of fire filled with wind inside them, or by the rending and tearing of clouds, or by the friction and bursting of clouds when they have been congealed in a form like ice: phenomena demand that we should say that this department of celestial events, just like them all, may be caused in several ways. . . .

That some of the stars should wander in their course, if indeed it is the case that their movements are such, while others do not move in this manner, may be due to the reason that from the first as they moved in their circles they were so constrained by necessity that some of them move along the same regular orbit, and others along one which is associated with certain irregularities: or it may be that among the regions to which they are carried in some places there are regular tracts of air which urge them on successively in the same direction and provide flame for them regularly, while in other places the tracts are irregular, so that the aberrations which we observe result. But to assign a single cause for these occurrences, when phenomena demand several explanations, is madness, and is quite wrongly practised by persons who are partisans of the foolish notions of astrology, by which they give futile explanations of the causes of certain occurrences and all the time do not by any means free the divine nature from the burden of responsibilities. . . .

All these things, Pythocles, you must bear in mind; for thus you will escape in most things from superstition and will be enabled to understand what is akin to them. And most of all give yourself up to the study of the beginnings and of infinity and of the things akin to them, and also of the criteria of truth and of the feelings, and of the purpose for which we reason out these things. For these points when they are thoroughly studied will most easily enable you to understand the causes of the details. But those who have not throughly taken these things to heart could not rightly study them in themselves, nor have they made their own the reason for observing them.

FROM *Principal Doctrines*

XI. If we were not troubled by our suspicions of the phenomena of the sky and about death, fearing that it concerns us, and also by our failure to grasp the limits of pains and desires, we should have no need of natural science.

XII. A man cannot dispel his fear about the most important matters if he does not know what is the nature of the universe but suspects the truth of some mythical story. So that without natural science it is not possible to attain our pleasures unalloyed.

XIII. There is no profit in securing protection in relation to men, if things above and things beneath the earth and indeed all in the boundless universe remain matters of suspicion.

XV. The wealth demanded by nature is both limited and easily procured; that demanded by idle imaginings stretches on to infinity.

XXXI. The justice which arises from nature is a pledge of mutual advantage to restrain men from harming one another and save them from being harmed.

FROM *Vatican Sayings*

IX. Necessity is an evil, but there is no necessity to live under the control of necessity.

XXI. We must not violate nature, but obey her; and we shall obey her if we fulfil the necessary desires and also the physical, if they bring no harm to us, but sternly reject the harmful.

XXIX. In investigating nature I would prefer to speak openly and like an oracle to give answers serviceable to all mankind, even though no one should understand me, rather than to conform to popular opinions and so win the praise freely scattered by the mob.

XXXI. Against all else it is possible to provide security, but as against death all of us mortals alike dwell in an unfortified city.

XXXVII. Nature is weak towards evil, not towards good: because it is saved by pleasures, but destroyed by pains.

XLI. We must laugh and philosophize at the same time and do our household duties and employ our other faculties, and never cease proclaiming the sayings of the true philosophy.

XLV. The study of nature does not make men productive of boasting or bragging nor apt to display that culture *[paideia]* which is the object of rivalry with the many, but high-spirited and self-sufficient, taking pride in the good things of their own minds and not of their circumstances.

XLVII. [From Metrodorus:] I have anticipated you, Fortune *[Tykhê]*, and entrenched myself against all your secret attacks. And we will not give ourselves up as captives to you or to any other circumstance; but when it is time for us to go, spitting contempt on life and on those who here vainly cling to it, we will leave life crying aloud in a glorious triumph-song that we have lived well.

If Nature Suddenly Should Find a Voice

On the Nature of Things, by Lucretius,
translated by Robert M. Torrance

Titus Lucretius Carus was born between 99 and 94 and died between 55 and 50 B.C. *Apart from these uncertain dates, almost nothing is known about him except that he was the author of the poem* De Rerum Natura (On the Nature of Things, *or simply* On Nature), *in which he expounds, in six books of 7,415 lines, the Epicurean view of the world and man. Cicero highly valued (and may have edited) the poem, which Lucretius dedicated to Gaius Memmius, son-in-law of the dictator Sulla and patron of the poet Catullus, and Virgil paid moving tribute to this happy predecessor who "plumbed things' hidden causes, / trampling fears of inexorable fate / underfoot, and loud-roaring Ácheron"* (Georgics *II.490–92, Chapter 11 below). But by the late first century* A.D. *the rhetorician Quintilian, though quoting and commending his poem, found Lucretius's archaic style "difficult," and thought him surpassed by his more polished successors. As Epicureanism—always a dissident among ancient philosophical schools because of its materialism and rejection of divine providence and teleology—fell out of favor, so apparently did its great poet. St. Jerome relates the legend (on which Tennyson's "Lucretius" is based) that he was driven mad by a love potion and committed suicide; his poem (like those of Catullus) survived from the ancient world in a single manuscript.*

As it stands, the apparently uncompleted De Rerum Natura *is both the fullest exposition (given the loss of most of Epicurus's writings) of the Epicurean philosophy and perhaps the most powerfully reasoned and splendidly written long philosophical poem in world literature. For Lucretius went back beyond more tepid Hellenistic models of didactic poetry such as the astronomical poem of Aratus of Soli (ca. 315–ca. 240* B.C.*), the* Phaenomena, *to find inspiration for his prophetic undertaking in the seminal pre-Socratic poems* On Nature *of Parmenides and, above all, of Empedocles four centuries before him (Chapter 9 above). In the selections here translated (604 lines, barely more than eight percent of the whole), the poet begins his first book with a glorious invocation of Venus as mythic embodiment of the generative force that permeates nature, "striking soft love into the breast of all." He declares his conviction that only knowledge of "the ordered ways of nature" can disperse the "terrifying mental darkness" of men enslaved by fear of death and the gods, and states his first principle, that nothing can come from nothing or, conversely, vanish into nothing (the law of the conservation of matter). The atoms of which all things are composed are immortal, and therefore nothing in nature is ever wholly destroyed, but everything is continually being renewed. "From heavenly seed each one of us is born," through repeated impregnation of Mother Earth by the sky, he writes in Book II, and death only changes the shapes and colors, not the underlying components, of things. The universe is infinite, and "other worlds must exist in other regions." The theme of Book III is that "Death is a matter, then, of no concern," since the mind is mortal, and no one can suffer who* is *not; if Nature herself could find a voice, as she does in this poem, she would reprimand mortals for complaining of death and clinging to life, when reason declares that nothing can shorten "the countless aeons following extinction,"* *which are the same for all. In Book V, after one of several passages in the poem lauding Epicurus as a god and the deliverer of mankind, Lucretius relates the creation (and foretells the end) of the world through the natural processes of continual change that govern it, and describes the progression of human beings from primitive savagery to civilization: a description that emphasizes, in contrast to so many pictures of a lost Golden Age, both the gains and the losses of this long evolution.*

Never before, and perhaps never since, has rational faith in human liberation through understanding of nature found such sustained and elevated poetic expression. For this reason, "Lucretius, more than any other man," George Santayana writes in Three Philosophical Poets *(1910), "is the poet of nature. Of course, being an ancient, he is not particularly a poet of landscape. He runs deeper than that; he is a poet of the source of landscape, a poet of matter. . . . One breath of lavish creation, one iron law of change, runs through the whole, making all things kin in their inmost elements and in their last end. Here is the touch of nature indeed, her largeness and eternity. Here is the true echo of the life of matter." At the same time, much of the poem's profundity lies in the latent contradictions that underlie Lucretius's earnest pronouncements. He exhorts us to banish fear of "immortal death" in verses that continually reveal how deeply ingrained that fear is in him, as in us all. And, as Martha Nussbaum perceptively observes in* The Therapy of Desire, *the passionate yearning that he attributes to Epicurus "to burst through the narrow confines of the gates of nature" shows that the true Epicurean, by learning nature's boundaries, is able to move beyond them, so that Epicurean reflection becomes "an assault on the secrets of nature, an assertion of the human into the realm of the gods." This tension between nature and transcendence, which Nussbaum finds characteristic, in some degree, of Stoicism and Scepticism as well as of Epicureanism, nowhere finds such intense expression as in the poet of impassioned reason who aspires to emulate the godlike liberator of mankind from terror of the gods by verses "interpreting the nature of all things."*

The translation is mainly based on the text of De Rerum Natura *edited by William Ellery Leonard and Stanley Barney Smith (1942) and on the Loeb Classical Library text and translation by W. H. D. Rouse (1924). For complete verse translations, see those of William Ellery Leonard (1916), Frank O. Copley (1977), and Anthony M. Esolen (1995).*

I

Mother of Rome, delight of men and gods,
nurturing Venus who beneath the circling
heavenly signs suffuse ship-laden seas
and fruitful lands: conceived through you alone
all living things rise up to see the sun. 5
You, goddess, you fierce winds and clouded skies
flee, when you near: for you the wondrous earth
quickens sweet flowers, oceans laugh for you,
and tranquil skies grow radiant with light.
For once spring manifests its smiling face, 10
freeing the generative breeze to frolic,
birds of the air, hearts smitten by your power,
goddess, first herald you and your arrival;
then frenzied herds skip through exultant pastures
and swim swift rivers, captured by your charm, 15
avid to follow your alluring call.
Through seas and mountains and rapacious floods,

the leafy homes of birds, and verdant meadows,
striking soft love into the breasts of all,
you make them wildly propagate their kind! 20
Since you alone rule over all of nature,
nor without you can anything arise
toward dazzling light in joyousness and beauty,
assist me, I implore, to write these verses
interpreting the nature of all things . . . 25

This terrifying mental darkness, then,
not sunbeams nor bright arrows of the day
can scatter, but the ordered ways of nature.
From this first principle let us begin:
nothing from nothing can divinely come. 150
Fearfulness holds all mortal men in bondage
when they observe, on earth and in the heavens,
events whose cause no reason can explain,
and think divinities have made them happen.
Thus when we see no thing can be created 155
from nothing, we shall better understand
our object: both the source of all creation
and how it comes to pass without the gods. . . .

Furthermore, nature once again resolves 215
things to their elements, not into nothing.
For if all parts of anything could perish,
things could be suddenly snatched from our sight
and fly apart, with no external force
needed to nullify the bond between them. 220
Yet now, because its seeds are everlasting,
though force may shatter something with a blow
or penetrate within and undermine it,
nature lets nothing wholly be destroyed.
Besides, if age consumes and utterly 225
destroys all matter worn away by time,
from whence does Venus bring each living species
back to the light of life, or wondrous earth
furnish the food to nourish each in turn?
Whence do abundant springs and distant rivers 230
supply the sea? or aether feed the stars?
For everything endowed with mortal body
passage of endless time must wear away;
but if the sum of things persists, renewed
by bodies unimpaired through endless time, 235
immortal must their nature surely be:
therefore none can return again to nothing.
One force would liquidate all things alike
did not eternal matter hold them fast
entwined by more or less entangling bonds. 240

One touch would be enough to bring destruction
and any force could dissipate the union
of things not made from everlasting bodies.
But since the bonds connecting elements
are different, and their matter is eternal, 245
things will remain intact until sufficient
force is applied to tear apart their texture.
No thing returns to nothing; all, when sundered,
return to their material elements.
Thus raindrops disappear when father Aether 250
hurls them into the groin of mother Earth,
but glistening crops surge upward, budding branches
turn green, and trees grow heavy with their offspring.
Hence humankind and animals are nourished;
hence cities blossom joyfully with boys 255
and young birds serenade in leafy forests;
hence fattened flocks refurbish weary bodies
in happy pastures, spurting milky streams
from swollen udders; hence their sportive newborns
frolic on wobbly limbs through slender grass, 260
inebriated by deep drafts of milk.
Things do not wholly perish, then, that seem to,
since nature, making one thing from another,
lets none be born but by another's death. . . .

II From heavenly seed each one of us is born.
All have one sire, by whom our nurturing mother
Earth, pregnant with his fertilizing moisture,
gives birth to glistening crops and joyous trees,
the human race, and every animal, 995
providing them with food to feed their bodies,
live a sweet life, and propagate their kind:
hence she deservedly is named our mother.
What first arose from earth returns to earth,
and what descended from the realms of aether, 1000
heavenly regions soon again reclaim.
Nor does death, in destroying things, demolish
their elements, but dissipates their union
and then rejoins them, so that everything
changes its shape and color, first acquiring 1005
sensation, then forsaking it again.
Thus you may understand the great importance
of how primordial bodies join together
through give-and-take of interacting movements,
and not confuse eternal building blocks 1010
with what we see adrift upon the surface
of things, no sooner born than perishing . . .

Look at the sheer bright color of the sky, 1030
the roaming constellations it embraces,
the moon, the dazzlingly resplendent sun:
if these now first revealed themselves to mortals
astonished at their sudden apparition,
what could be called more marvelous than they 1035
or would have seemed less credible before?
Nothing: yet now satiety of seeing
this miracle so wearies us that no one
deigns to look up and see the shimmering sky!
Cease to be fearful, then, of new ideas, 1040
banishing rationality, but weigh them
judiciously, and if you find them true,
surrender—or, if false, prepare for battle.
For since space infinitely spreads beyond
the world's walls, reason yearns to penetrate 1045
that distant realm through which imagination
pursues its own emancipated flight.

First, nowhere in this boundless universe,
left or right, high or low, does any limit
confine it: thus I teach, thus things themselves 1050
proclaim, and deepest nature manifests.
By no means can it be thought probable,
since boundless space lies empty everywhere
and seeds incomprehensible in number
fly about driven by eternal motion, 1055
that only this one world has been created
or bodies so profuse accomplish nothing,
considering that nature made this world
when elements spontaneously colliding
drifted at random aimlessly together 1060
and coalesced, when suddenly united,
giving rise everywhere to mighty things—
land, sea and sky, and every living creature.
In consequence, you must acknowledge, elsewhere
assemblages of matter must exist 1065
like this which aether ardently embraces.

Besides, when matter congregates in plenty,
and space is all around, and nothing hinders,
things must assuredly come to completion.
If the supply of seeds is now beyond 1070
all counting in the time since life began,
and natural force is able, as before,
to join the seeds of things in other places
just as it joined them here, it clearly follows
other worlds must exist in other regions, 1075
and different races, both of men and beasts.

III Death is a matter, then, of no concern, 830
once mind is understood as being mortal.
Just as, before our birth, we felt no anguish
when Carthaginian warriors, massed for battle,
convulsed and terrified a world resounding
with savage tumult underneath high heaven, 835
casting in doubt beneath whose domination
all human things on land and sea would fall,
so, when we shall no longer be, when body
and soul, whose union constitutes us, sever,
nothing will then be able to affect us 840
who shall no longer be, or stir our senses,
though earth be intermixed with sea and sky.
Even suppose the faculty of mind
and soul, when sundered from the body, feels—
still this is nothing to us, since our being 845
consists in unity of mind and body.
If time, when we are dead, should reassemble
and organize our matter as it was,
giving us back the light of life again,
even this would be no concern of ours 850
once recollection of ourselves was severed.
Nothing we were before now matters to us
nor can *their* agonies affect us now.
For when you contemplate unmeasured time's
bygone expanse, how manifold the motions 855
of matter are, you too will be convinced
the very seeds of which we now consist
frequently must have been arranged as now,
though memory no longer can recall it:
for in between, life having stopped, its motions 860
scattered afar, dispersing all sensation.
No one can suffer any future pain
unless, when injury befalls, he then
exists: since death releases him, absolving
any, whom troubles might assail, from being, 865
nothing, assuredly, in death is fearful.
For none who *is* not suffers (whether born
or not, no matter) once immortal death
has spirited his mortal life away. . . .

If Nature suddenly should find a voice,
thus might she reprimand someone among us:
"Mortal, why agitate yourself, indulging
in sick laments, tears, and complaints of death?
For if your former life has satisfied you, 935
and all your blessings have not seeped away—
stored in a perforated jar—and vanished,

why not now, like a banqueter fed full
of life, poor fool, retire and savor peace?
But if your happiness has dissipated, 940
and life repels you, why prolong it further,
to see it cease again and sadly perish?
Why not bring life and sorrow to an end?
For nothing more can I devise or fashion
to please you: all is always still the same. 945
Even should body be unwithered, limbs
unweakened, everything remains the same,
though you survive uncounted centuries,
and all the more if you should never die"—
what could we answer, but that Nature enters 950
just charges, and the case she pleads is true? . . .

Everyone flees himself (though by its nature
self clutches inescapably) and hates
himself, sick with disease he cannot fathom. 1070
Otherwise, each would liberate his mind
to contemplate the nature of all things,
since not one hour, but eternity
is here in doubt: in what condition mortals
will pass the time remaining after death. 1075
What turbulent, malign desire for living
dogs us with insecurity and peril?
The terminus of mortal life is certain,
death inescapable: for all must perish.
Ceaselessly moving in a single place, 1080
we fabricate no new delights by living:
though what we crave is paramount when absent,
gaining it, we desire another object,
panting incessantly with thirst for life.
None knows what fate futurity conceals, 1085
what chance will bring, or what result awaits us;
nor, by protracting life, can we diminish
the span of death, or undertake to shorten
the countless aeons following extinction.
However many centuries you live, 1090
eternal death lies equally before you,
and no less long will nonexistence blanket
him who relinquishes his life today
than one who perished months or years before. . . .

V Who, by the might of intellect, could fashion
song equaling discoveries so splendid?
Who, by mere words, is competent to praise
the merits of the man whose intellect

bequeathed to us the treasures he attained? 5
No one, I think, of mortal body born.
For, speaking as the majesty of things
now known demands, a god, famed Memmius,
a god he was who first revealed the truth
of life, called wisdom now, and skilfully 10
out of tempestuous waves and shadowy darkness
steered life to radiant tranquility. . . .

Treading his footsteps, I now emulate 55
his reasoning, and teach by what accord
all things were made and must abide forever,
impotent to annul the laws of time:
first how the composition of the mind
is found to be an outgrowth of the body, 60
unable to endure intact for ages
(yet visions often cheat the mind in dreams,
making us fantasize a man now dead);
further, how reason clearly demonstrates
the world consists of perishable body 65
and must in consequence have once been born;
how this assembled multitude of matter
created earth, sky, sea, the constellations,
sun, and round moon; which animals arose
then from the earth, and which were never born; 70
how human beings first communicated
in varied tongues by giving names to things;
and how gods infiltrated mortal hearts
with terror, sanctifying far and wide
their temples, lakes, groves, images, and altars. 75
I shall explain, moreover, by what power
nature controls the course of sun and moon,
lest we believe they randomly meander
year after year between the earth and heaven
nurturing crops and animals as beckoned, 80
or think their orbitings divinely planned.
Even knowing the gods live free of care,
nonetheless men may wonder in what fashion
things come about—especially in matters
glimpsed in ethereal regions overhead— 85
and thus may turn to ancient superstitions,
thinking their cruel overlords almighty
in wretched ignorance of what can be
and what cannot: how everything is finite
in strength, and cannot overstep its limits. 90
Next—to delay no more with promises—
contemplate first of all sea, land, and sky:
this triple nature, Memmius—three bodies,

three different forms, three interwoven textures—
one day shall end, when, after many years, 95
this massive world-machine shall crash to ruin.
Nor do I overlook how new and strange
future collapse of sky and earth must seem,
how difficult to prove by argument;
as happens, when you speak of unaccustomed 100
matters impossible to visualize
or grasp: the road by which conviction leads
straight to the temple of the human mind.
Yet I shall speak. The facts themselves will give
my words conviction: soon you shall behold 105
earthquakes convulsing all the universe.
May guiding fortune keep this fate afar,
and reason, not reality, exhibit
worlds crashing down with terrifying din. . . .

First, since the mass and moisture of the earth, 235
the gently blowing breeze and steaming heat
seen to compose this universal sum
consist of a body that is born and dies,
all the world's nature must be thought the same;
for when we see the parts and limbs of beings 240
composed of bodies that are born and die,
we postulate these beings also perish
once born. Therefore, when I observe the mighty
parts and limbs of the world regenerated,
heaven and earth, I know, had some beginning 245
and will hereafter sometime be destroyed. . . .

Further, since I have clarified how things
occur throughout the world's vast blue expanses,
and made it known what force originated
the varied movements of the sun and moon, 775
and how their suddenly obstructed light
envelops the astonished earth in shadow
and seems to wink, yet soon with open eye
contemplates regions shimmering with brightness:
now to the childhood of the world I turn, 780
when soft fields, giving birth, first brought to light
shoots trusted to the vacillating winds.

In the beginning, earth brought forth the grasses
glistening green around the hills and plains,
and flowering meadows burst with dazzling green; 785
among the various trees a mighty struggle
of growth began, a race into the air.
As feathers, hair, and bristles start to cover
limbs of four-footed beasts and wings of birds,

so the new earth first generated grasses 790
and thickets, then the countless breeds of creatures
rising in many ways by various means.
For animals cannot have dropped from heaven
or crawled from salty waters onto land.
Therefore the earth deserves the name of mother, 795
since everything was once created by her.
Even now many animals arise
from earth, formed by the rain and warming sun:
no wonder, then, that many more, and greater,
uprose when earth and air alike were young. 800
The variegated race of wingèd birds
hatched first from eggs relinquished in the spring,
just as cicadas now desert their slender
husks in the summer, seeking nourishment.
Earth then first brought forth mortal generations. 805
For heat and moisture overflowed the fields,
so that in many suitable locations
wombs, penetrating earth with roots, would grow,
which infants, with the ripening of time,
burst, fleeing moisture and desiring air; 810
nature provided apertures of earth,
forcing them to discharge from open veins
a milklike fluid, just as women now,
having delivered, swell with warm sweet milk,
a surge of food directed to the breasts. 815
Earth fed her children, warmth provided clothing,
grass the abundance of a downy bed.
The infant world produced no freezing cold,
no searing heat, no overpowering winds,
for strength increases with advancing years. 820
Therefore again and yet again earth merits
a mother's name, for she alone created
the human race, producing in due time
each animal that courses lofty mountains
along with varied shapes of soaring birds. 825
Since parturition could not last forever,
she halted, like a woman worn by age.
For time transmutes the nature of the world;
one state of things must pass into another;
nothing remains the same, but always changes 830
transformed by nature and compelled to alter.
One thing decays and languishes with age,
another rises out of what we scorn.
Time therefore transubstantiates the nature
of all; one state of earth succeeds another, 835
passing from barrenness again to plenty. . . .

That human race, because hard earth had made it, 925
was fittingly much harder in the fields,
strengthened by solider and larger bones
and mighty sinews fastening their flesh,
so that no heat or cold, no unknown food
or bodily flaw could easily subdue them. 930
Through many cycles of the circling sun
they passed their lives like migratory beasts.
No sturdy moderator of the plow
existed, none to work the fields with iron,
plant seedlings in the ground, or with a sickle 935
lop dying branches off the tops of trees.
What sun and rainfall gave, what earth itself
created, was enough to please their hearts.
Among the acorn-laden oaks they often
rested themselves; strawberry trees that now 940
ripen in wintertime with crimson color
earth nurtured more abundantly and larger.
Bounteous foods the flowering infant world
brought forth—hard, but enough for wretched mortals.
Rivers and streams invited them to drink, 945
as now swift waters rushing from high mountains
loud and far summon hordes of thirsting beasts.
Next in their wanderings they occupied
sylvan shrines of the nymphs, who indicated
smooth brooks splashing wet rocks with running waters, 950
splashing wet rocks, bespattering green moss,
gushing and overflowing level meadows.
Nor did they have the skill to handle fire
or use the skins of animals for clothing,
but dwelt in woods and mountain caves and forests, 955
hiding their rough limbs in the underbrush
to circumvent the blows of wind and rain.
They could not contemplate the common good
or rule themselves by custom or by law,
but what good fortune brought, each seized upon 960
learning to thrive each by himself alone.
Venus joined lovers' bodies in the woods,
either united by some shared desire,
by the man's violent lust, or by a bribe
of acorns, berries, or delicious pears. 965
Solely relying on strong hands and feet,
they hunted tribes of savage animals
with flying stones and blows from heavy clubs,
slaying those not concealed in hidden places,
and then, like bristling hogs, threw down their savage 970
limbs to the ground when nighttime overtook them,

wrapping themselves around with leaves and branches.
Nor did they roam the fields to seek the sun,
fearfully wailing through the darkened night,[1]
but rested quietly, interred by sleep, 975
till the sun's rosy torch inflamed the sky.
Since early childhood they had been accustomed
always to see light alternate with shadows,
and thus could not have ever been astonished
or feared eternal night would overpower 980
the world, and sunlight be withdrawn forever.
Far more disturbing, tribes of savage beasts
often endangered their precarious rest.
Driven from home, they fled their rocky caves
when foaming boars or mighty lions appeared, 985
and, panic-stricken in the dead of night,
yielded to savage guests their leafy beds.
Not much more frequently than now did mortals
leave evanescent life's sweet light behind.
Each was more likely then to be the living 990
food of wild beasts, torn into tiny shreds
filling the woods and mountaintops with groans
as living tombs engorged their living flesh;
while those whom flight preserved with mangled bodies
covered repulsive sores with trembling hands 995
thereafter, calling horribly on Orcus[2]
till agonies released them from their life,
with none to give assistance to their wounds.
But one day did not butcher many thousands
under the colors, nor the raging sea 1000
smash on the rocks both ships and mariners.
Then all for nothing, all in vain the waters
uprose and howled, or calmed their idle threats,
nor could the placid sea's deceptive guile
or laughing waves lure anyone to ruin. 1005
Infamous navigation then lay hidden.
Formerly, lack of food brought fainting limbs
to death, but now abundance overwhelms them.
Then men unknowingly would often poison
themselves, who artfully now poison others. 1010
After acquiring cabins, skins, and fire,
and sanctioning the marriage of one woman
to each man, and the children they engendered,
then human beings first began to soften.
For fire made shivering bodies less resistant 1015

[1]There seems to have been some ancient belief that the early men thought the sun, once set, would never return; it is alluded to by Manilius i.66 and Statius *Thebaid* iv.282. (Munro) Perhaps Lucretius's picture was suggested by the common custom of beating drums at an eclipse. (Rouse)
[2]A god of the dead, often identified with Greek Hades.

to cold beneath the canopy of heaven,
Venus curtailed their strength, and wheedling children
easily broke their parents' haughty temper.
Then neighbors first cemented friendships, eager
neither to do nor suffer any wrong, 1020
and sought protection for their wives and children,
indicating by babbling sounds and gestures
that all the powerless deserve compassion.
Though concord could not utterly prevail,
many—nay, most—observed their covenant, 1025
otherwise human beings would have perished
long since, and left no progeny behind them. . . .

What motive propagated deities
throughout great nations, filling towns with altars,
and founded customary sacrifices
whose flourishing performance in high places
inculcates reverential fear in mortals, 1165
driving men everywhere to elevate
shrines thronged with celebrating worshippers—
all this my words can readily explain.
For mortal generations long ago
beheld divinities of wondrous stature 1170
when wide awake, and even more in dreams.
To these they thus attributed sensation,
because to all appearances they moved
and spoke according to their strength and beauty;
thus they endowed them with eternal life 1175
because their shapes persistently remained,
and thought they could not easily be vanquished,
with strength so great, by any force on earth.
Therefore they thought their happiness surpassed
all others', since no fear of death disturbed them, 1180
and saw the many marvelous achievements
they did in dreams while suffering no sorrows.
Likewise they watched the orderly progression
of heavenly signs and seasons of the year,
and could not understand their explanation. 1185
Thus they took refuge in ascribing all
to gods whose dispensation ruled the world.
In heaven they placed the gods' eternal dwellings,
seeing the moon and night revolve through heaven:
moon, day, and night, the sober constellations, 1190
night-wandering flames and torches of the sky,
clouds, sun and rain, snow, winds and hail, and lightning's
immensely menacing, mighty, murmurous roar.

Unhappy human beings, who ascribed
to gods such deeds, such bitterness and anger! 1195

What groans they generated for themselves,
what wounds for us, what sufferings hereafter!
No piety is shown by turning often,
veiled, toward a stone, approaching every altar,
falling prostrate upon the ground and stretching 1200
palms to the gods' shrines, nor by smearing altars
with blood of quadrupeds, vow after vow—
but by regarding all with tranquil mind.
For when we contemplate the universe's
heavenly realms, the aether's sparkling stars, 1205
remembering the paths of sun and moon,
then, in hearts crushed by other tribulations,
this trouble too begins to rear its head:
whether perchance the gods' enormous power
controls the varied movements of bright stars. 1210
For reason's impotence disturbs the mind:
Did the world have a day of origin?
Is there some limitation to the world's
endurance of incessant agitation?
Or can it glide through time's unbounded regions 1215
divinely gifted with eternal health,
spurning the potency of endless time?
Whose mind does not contract with apprehension,
whose limbs not crawl with terror of the gods
when thunderbolts incinerate the shaken 1220
earth, and reverberations rend the sky?
Do not whole nations tremble, haughty kings
huddle themselves, transfixed by fear of gods,
lest for some heinous crime or haughty saying
the time of retribution be at hand? 1225
And when supremely furious winds at sea
batter a fleet's commander on the waters
along with elephants and mighty legions,
does he not supplicate the gods in panic
to pacify the winds and send sweet breezes— 1230
vainly, since even so the furious tempest
dashes him often on the shoals of death!
To such a point some hidden force grinds down
humanity, and tramples fiercely shining
fasces and axes[3] with derisive laughter. 1235
Thus, when earth totters underneath our feet
and shaken cities fall, or threaten falling,
what wonder mortals hold themselves in scorn,
relinquishing all great and wondrous powers
over this world to gods who govern all. . . . 1240

[3]Carried by the lictors as symbols of office. (Rouse)

Sowing and grafting took their paradigm
from Nature, prime creatrix of all things:
berries and acorns dropping from the trees
gave rise in time to swarming undergrowth,
suggesting shoots could be combined with branches 1365
and seedlings planted in the fertile ground.
Then they tried various means of cultivating
their gardens, learning how the earth could soften
wild fruits through tenderly indulgent care.
Daily they made the woods ascend the mountain 1370
further, surrendering the dales to farming,
thus gaining pools, streams, crops, meads, joyous vineyards
on hills and plains, and making blue-green rows
of olive trees extend as boundary-markers
between them, over mounds and plains and valleys, 1375
just as you now see everything distinct
with varied charms, partitioned by sweet orchards
and planted all around with happy trees.

Imitating by mouth sweet melodies
of birds long antedated serenading 1380
men's ears by repetition of smooth songs.
Rustling through hollow reeds, warm zephyrs first
taught country folk to blow on hollow stalks;
then gradually they learned the plaintive tunes
pipes lavish at the touch of players' fingers— 1385
pipes found amid untrodden woods and forests
and solitary haunts where shepherds loiter.
These quieted their mind and gave delight 1390
after their fill of food, when song is pleasant.
Often, therefore, relaxing on soft meadows,
next to a stream beneath a tree's high branches,
they gratified their bodies at small cost,
especially when weather smiled upon them 1395
and springtime brightened up green plants with flowers.
Then was the time for jesting, talk, and laughter,
for then the rustic muse was in her prime;
then they draped head and shoulders round with garlands
of leaves and flowers in their sportive mood, 1400
marching forth out of step and tramping hard
down on their mother earth with hardened feet:
whence mirth arose and joyous peals of laughter
at wondrous novelties that so engaged them.
This was their recompense for sleep when waking: 1405
to modulate their variegated songs,
running pursed lips across connected reeds.
Hence watchmen even now preserve tradition
and learn to master every sort of rhythm,

yet take no greater satisfaction from it 1410
than the indigenous woodland folk of yore.
For what is near—unless we've tasted something
sweeter before—impresses and prevails
till something else then recognized as better
destroys our preference for ancient things. 1415
Thus acorns grew repugnant, couches strewn
with grass and leaves were utterly abandoned.
Garments of fur fell into disrepute,
though I imagine these aroused such envy
once that the first to brandish one was ambushed 1420
by murderers who ripped his bloody garment
to pieces none could ever wear again.
Furs formerly, now gold and purple vex
men's life with cares and weary it with war—
in which the greater fault, I think, is ours. 1425
For without furs cold agonized the naked
children of earth; but lacking purple robes
fretted with golden filigree harms no one
so long as some plebeian garment cloaks us.
Thus all in vain humanity forever 1430
labors, consuming time with empty cares,
doubtless because we recognize no limit
to having, or how far true pleasure grows:
little by little this has overwhelmed
life and stirred up the mighty storms of war. 1435
But sun and moon, those sentinels that light
the great revolving temple of the sky,
certify that the seasons circle round
and all occurs in orderly progression.
Men were already cramped in sturdy towers 1440
and land distributed for cultivation,
the sea already flowering with sails,
auxiliaries and allies under treaty,
when poets first commemorated deeds
in song, soon after letters were invented. 1445
Therefore our age can only know what happened
before if reasoning detects its traces.
Sailing and farming, fortification, laws,
weapons, roads, clothes, all other such attainments,
prizes, and every luxury of life, 1450
pictures and poems, artfully shaped statues—
all this, progressing step by step, men mastered
through use and through experimenting mind.
So by degrees time manifests all things
and reason elevates them toward the light: 1455
for one by one men saw the various arts
grow famous, and attain their culmination. . . .

Epicurean and Stoic Views of Nature

On the Nature of the Gods, by Cicero, translated by H. Rackham

Marcus Tullius Cicero (106–43 B.C.) was the greatest of Roman orators in the turbulent decades that saw the dissolution of the Roman Republic, and one of the masters of Latin prose. Born in Arpinum, he was educated in Rome, Athens, and Rhodes, where he heard the Stoic philosopher Posidonius. In Rome he held a variety of public offices culminating in his consulship of 63 B.C., during which he defeated the conspiracy of Catiline. As the rivalry between Caesar and Pompey mounted, Cicero finally threw his lot in with Pompey and the Republic, but after Pompey's defeat at Pharsalus in 48, he was pardoned by the victor. Following Caesar's assassination in 44, Cicero delivered a series of orations, the Philippics, *denouncing Mark Antony, who attained the assent of his fellow triumvirs Octavian (the future Augustus) and Lepidus to condemn Cicero to death. Overtaken as he attempted to flee, Cicero was beheaded, and his head displayed in the Roman forum.*

In addition to his many orations, his rhetorical works, and his letters, he composed a series of philosophical essays and dialogues setting forth his own views and those of others on a variety of questions ranging from politics (the Republic *and the* Laws*) to a series of works on moral, epistemological, and theological subjects written between 45 and 44 during the years of Caesar's dictatorship; these included* On Duties (De Officiis), On Ends (De Finibus), Tusculan Disputations, *and* On the Nature of the Gods (De Natura Deorum). *Cicero was not an original thinker, and his philosophical position tended to be somewhat eclectic and inconsistent; he generally identified with the sceptical reservations of the Academic philosophy, but frequently inclined (despite criticism of various doctrines) toward sympathy with the Stoics. In the passages quoted below, from the Loeb Classical Library edition of* De Natura Deorum *(1933), he represents the views on nature and cosmology both of the Epicureans, through their spokesman Gaius Velleius—who scoffs at the Stoics for holding that the world was created by a "prying busybody of a god" employing "bellows and anvils," and belittles the doctrine of fate as a belief for "ignorant old women"—and of the Stoics, through Lucilius Balbus, who gives a lucid explanation of the view of nature as teleological process and as "the sustaining and governing principle of the world," identified with god as its eternal artificer or craftsman. Many of these views clearly derive from Plato and Aristotle, including the view that the highest of arts is, in Balbus's phrase, "Nature's art," by which the living world is perpetually created and sustained in accord with purpose and reason.*

[Gaius Velleius upholds the Epicurean against the Stoic view of nature (I.53–56).]

We for our part deem happiness to consist in tranquility of mind and entire exemption from all duties. For he who taught us all the rest has also taught us that the world was made by nature, without needing an artificer to construct it, and that the act of creation, which according to you cannot be performed without divine skill, is so easy, that nature will create, is creating and has created worlds without number. You on the contrary cannot see how nature can achieve all this without the aid of some intelligence, and so, like the tragic poets, being unable to bring the plot of your drama to a *dénouement*, you have recourse to a god; whose intervention you assuredly would not require if you would but contemplate the measureless and boundless extent of space that stretches in every direction, into which when the mind projects and propels itself, it journeys onward far and wide without ever sighting any margin or ultimate point where it can stop. Well then, in this immensity of length and breadth and height there flits an infinite quantity of atoms innumerable, which though separated by void yet cohere together, and taking hold each of another form unions wherefrom are created those shapes and forms of things which you think cannot be created without the aid of bellows and anvils, and so have saddled us with an eternal master, whom day and night we are to fear; for who would not fear a prying busybody of a god, who foresees and thinks of and notices all things, and deems that everything is his concern? An outcome of this theology was first of all your doctrine of Necessity or Fate, *heimarmenê*, as you termed it, the theory that every event is the result of an eternal truth and an unbroken sequence of causation. But what value can be assigned to a philosophy which thinks that everything happens by fate? it is a belief for old women, and ignorant old women at that. And next follows your doctrine of *mantikê*, or Divination, which would so steep us in superstition, if we consented to listen to you, that we should be the devotees of soothsayers, augurs, oraclemongers, seers and interpreters of dreams. But Epicurus has set us free from superstitious terrors and delivered us out of captivity, so that we have no fear of beings who, we know, create no trouble for themselves and seek to cause none to others, while we worship with pious reverence the transcendent majesty of nature.

[Lucilius Balbus expounds the Stoic conceptions of god and nature (II.33–36, 45–47, 57–58, 81, 85, 87).]

Again, if we wish to proceed from the first rudimentary orders of being to the last and most perfect, we shall necessarily arrive in the end at deity. We notice the sustaining power of nature first in the members of the vegetable kingdom, toward which her bounty was limited to providing for their preservation by means of the faculties of nurture and growth. Upon the animals she bestowed sensation and motion, and an appetite or impulse to approach things wholesome and retire from things harmful. For man she amplified her gift by the addition of reason, whereby the appetites might be controlled, and alternately indulged and held in check. But the fourth and highest grade is that of beings born by nature good and wise, and endowed from the outset with the innate attributes of right reason and consistency; this must be held to be above the level of man: it is the attribute of god, that is, of the world, which must needs possess that perfect and absolute reason of which I spoke. Again, it is undeniable that every organic whole must have an ultimate ideal of perfection. As in vines or in cattle we see that, unless obstructed by some force, nature progresses on a certain path of her own to her goal of full development, and as in painting, architecture and the other arts and crafts there is an ideal of perfect workmanship, even so and far more in the world of nature as a whole there must be a process towards completeness and perfection. The various limited modes of being may encounter many external obstacles to hinder their perfect realization, but there can be nothing that can frustrate nature as a whole, since she embraces and contains within herself all modes of being. Hence it follows that there must exist this fourth and highest grade, unassailable by any external force. Now this is the grade on which universal nature stands; and since she is of such a character as to be superior to all things and incapable of frustration by any, it follows of necessity that the world is an intelligent being, and indeed also a wise being. . . .

It remains for us to consider the qualities of the divine nature; and on this subject nothing is more difficult than to divert the eye of the mind from following the practice of bodily sight. This difficulty has caused both uneducated people generally and those philosophers who resemble the uneducated to be unable to conceive of the immortal gods without setting before themselves the forms of men: a shallow mode of thought which Cotta has exposed and which therefore calls for no discussion from me. But assuming that we have a definite and preconceived idea of a deity as, first, a living being, and secondly, a being unsurpassed in excellence by anything else in the whole of nature, I cannot see anything that satisfies this preconception or idea of ours more fully than, first, the judgement that this world, which must necessarily be the most excellent of all things, is itself a living being and a god. Let Epicurus jest at this notion as he will—and he is a person who jokes with difficulty, and has but the slightest smack of his native Attic wit,—let him protest his inability to conceive of god as a round and rotating body. Nevertheless he will never dislodge me from one belief which even he himself accepts: he holds that gods exist, on the ground that there must necessarily be some mode of being of outstanding and supreme excellence; now clearly nothing can be more excellent than the world. Nor can it be doubted that a living being endowed with sensation, reason and intelligence must excel a being devoid of those attributes; hence it follows that the world is a living being and possesses sensation, intelligence and reason; and this argument leads to the conclusion that the world is god. . . .

Now Zeno gives this definition of nature: "nature (he says) is a craftsmanlike fire, proceeding methodically to the work of generation." For he holds that the special function of an art or craft is to create and generate, and that what in the processes of our arts is done by the hand is done with far more skilful craftsmanship by nature, that is, as I said, by that "craftsmanlike" fire which is the teacher of the other arts. And on this theory, while each department of nature is "craftsmanlike," in the sense of having a method or path marked out for it to follow, the nature of the world itself, which encloses and contains all things in its embrace, is styled by Zeno not merely "craftsmanlike" but actually "a craftsman," whose foresight plans out the work to serve its use and purpose in every detail. And as the other natural substances are generated, reared and sustained each by its own seeds, so the world-nature experiences all those motions of the will, those impulses of conation and desire, that the Greeks call *hormae*, and follows these up with the appropriate actions in the same way as do we ourselves, who experience emotions and sensations. Such being the nature of the world-mind, it can therefore correctly be designated as prudence or providence (for in Greek it is termed *pronoia*); and this providence is chiefly directed and concentrated upon three objects, namely to secure for the world, first, the structure best fitted for survival; next, absolute completeness; but chiefly, consummate beauty and embellishment of every kind. . . .

Next I have to show that all things are under the sway of nature and are carried on by her in the most excellent manner. But first I must briefly explain the meaning of the term "nature" itself, to make our doctrine easily intelligible. Some persons define nature as a non-rational force that causes necessary motions in material bodies; others as a rational and ordered force, proceeding by method and plainly displaying the means that she takes to produce each result and the end at which she aims, and possessed of a skill that no handiwork of artist or craftsman can rival or reproduce. For a seed, they point out, has such potency that, tiny though it is in size, nevertheless if it falls into some substance that conceives and enfolds it, and obtains suitable material to foster its nurture and growth, it fashions and produces the various creatures after their kinds, some designed merely to absorb nourishment through their roots, and others capable of motion, sensation, appetition and reproduction of their species. Some thinkers again denote by the term "nature" the whole of existence—for example Epicurus, who divides the nature of all existing things into atoms, void, and the attributes of these. When we on the other hand speak of nature as the sustaining and governing principle of the world, we do not mean that the world is like a clod of earth or lump of stone or something else of that sort, which possesses only the natural principle of cohesion, but like a tree or an animal, displaying no haphazard structure, but order and a certain semblance of design.

But if the plants fixed and rooted in the earth owe their life and vigor to nature's art, surely the earth herself must be sustained by the same power, inasmuch as when impregnated with seeds she brings forth from her womb all things in profusion, nourishes their roots in her bosom and causes them to grow, and herself in turn is nourished by the upper and outer elements. Her exhalations moreover give nourishment to the air, the ether and all the heavenly bodies. Thus if earth is upheld and invigorated by nature, the same principle must hold good of the rest of the world, for plants are rooted in the earth, animals are sustained by breathing air, and the air itself is our partner in seeing, hearing and uttering sounds, since none of these actions can be performed without its aid; nay, it even moves as we move, for wherever we go or move our limbs, it seems as it were to give place and retire before us. And those things which travel towards the center of the earth which is its lowest point, those which move from the center upwards, and those which rotate in circles round the center, constitute the one continuous nature of the world. Again the continuum of the world's nature is constituted by the cyclic transmutations of the four kinds of matter. For earth turns into water, water into air, air into aether, and then the process is reversed, and aether becomes air, air water, and water earth, the lowest of the four. Thus the parts of the world are held in union by the constant passage up and down, to and fro, of these four elements of which all things are composed. And this world-structure must either be everlasting in this same form in which we see it or at all events extremely durable, and destined to endure for an almost immeasurably protracted period of time. Whichever alternative be true, the inference follows that the world is governed by nature. . . .

But if the structure of the world in all its parts is such that it could not have been better whether in point of utility or beauty, let us consider whether this is the result of chance, or whether on the contrary the parts of the world are in such a condition that they could not possibly have cohered together if they were not controlled by intelligence and by divine providence. If then the products of nature are better than those of art, and if art produces nothing without reason, nature too cannot be deemed to be without reason. When you see a statue or a painting, you recognize the exercise of art; when you observe from a distance the course of a ship, you do not hesitate to assume that its motion is guided by reason and by art; when you look at a sun-dial or a water-clock, you infer that it tells the time by art and not by chance; how then can it be consistent to suppose that the world, which includes both the works of art in question, the craftsmen who made them, and everything else besides, can be devoid of purpose and of reason? . . .

God of Nature and Universal Law

Hymn to Zeus, by Cleanthes, translated by Robert M. Torrance

Cleanthês, said to have lived from 331 to 232 B.C., was born in Assos, a coastal town in the Troad of northwestern Asia Minor. He arrived in Athens, according to Diogenes Laërtius, as a boxer with only four drachmas, and became the pupil of Zeno, whom he succeeded as head of the Stoic school from Zeno's death in 263 until his own thirty years later, when he was succeeded by Chrysippus. His contributions to early Stoic doctrine appear to have been less substantial than theirs: "He had industry but no natural aptitude for physics," Diogenes relates (VII.170–71), "and was extraordinarily slow," called "the ass" by his fellow pupils because "he alone was strong enough to carry the load of Zeno." The religious bent that would contribute prominently to the spread of Stoicism is especially striking in his "Hymn to Zeus," one of the few complete writings that survive from the early Stoa, as preserved by the anthologist Ioannes Stobaeus in about the fifth century A.D. Here Cleanthes adapts to the purposes of Stoic philosophy a form descending from the Homeric hymns, for his key terms (each appearing twice or more) are physis *("nature"),* logos *(here translated "plan"), and* nomos *("law")—twice called* koinos nomos *("common" or "universal" law, an adjective also applied to* logos*). This Zeus thus embodies the rational law of nature by which the ordered world or cosmos (another key term) is ruled, and which the wise—the poem implies—perceive as fate, and only the wicked foolishly attempt to resist. Such a Zeus is far indeed from the capricious adulterer of popular myth. Yet we should remember that already in Homer (to say nothing of Aeschylus), Zeus, though himself subject to* moira, *had been the just dispenser of the fates allotted to mortals. The translation is based on the text in Hans von Arnim, ed.,* Stoicorum Veterum Fragmenta *(1902), vol. I, #537.*

Famed among gods by many names, almighty
Zeus, nature's sovereign, governing all by law,
greetings! For rightly mortals may address you,
since, born from you, to them alone of creatures
crawling the earth, your echo's trace remains: 5
by which I shall forever hymn your power!
This whole world, wheeling round the earth, submits,
obediently following your lead
as, with unconquered hands, you wield the ready
forked, fiery, ever-living thunderbolt, 10
by whose stroke nature's deeds are all accomplished.
By it you guide the common, all-pervading
plan interfused with lights both great and small;
by it you reign supremely over all!
No deed on earth is done without you, god— 15
in heaven's aethereal vault, nor on the sea—
save by the wicked in their ignorance.
What is irregular you even out,
order discord, and find the unloved lovely.

Thus good and evil you have joined in one, 20
so that one everlasting plan rules all,
which wicked men shun, in their misery
always craving more goods, and forfeiting
nobility of life by foolishly
flouting god's common law—both deaf and blind. 25
Madly each rushes to his own destruction,
some striving zealously for reputation,
some groping unrestrainedly for profit,
others for sensual pleasures of the flesh.
But all encounter evil everywhere, 30
finding the opposite of what they strive for.
Bountiful Zeus, whose lightning cleaves black clouds,
save mortals from their wretched ignorance:
scatter it from their souls, and help them reach
knowledge by which you govern all with justice. 35
Thus honored, we in turn shall honor you,
hymning your deeds forever, as men should,
since both for men and gods this prize is highest:
justly to hymn your universal law.

Potencies of Heaven

The *Astronomica* of Manilius, translated by Robert M. Torrance

About Marcus Manilius nothing is known except that he was the author of the Astronomica, *a didactic poem of five books and 4,258 lines, composed in Latin under the Emperors Augustus and Tiberius in the first century* A.D., *setting forth an astrological view of the cosmos. Whether because of its recondite subject or depreciation of its poetic quality, the poem has been little read and is indeed hardly readable in its entirety. "I began to read Manilius's* Astronomicon," *the young Goethe writes in his* Ephemerides *of 1770 (as quoted in G. P. Goold's introduction to the Loeb Classical Library edition of Manilius [1977]), "and soon had to put it down: no matter how much this philosophical poet festoons his work with lofty thoughts, he cannot redeem the barrenness of his subject."*

Yet barren and mechanically deterministic as the subject can seem, for the Stoics, to whose views Manilius gives powerful expression, astrology was widely considered not a superstitious but a rational and even a scientific *view of nature, in contrast to popular conceptions of the universe as subject to the whims of Chance or of arbitrary divinities. Fate or destiny was the expression of a divinely ordered natural world subject to just and inexorable law, and astrology (which was nearly synonymous with astronomy) was thought to be the surest, and the loftiest, means by which human beings might understand fate. In the 299 lines here translated Manilius, in Book I, states his elevated theme, "to draw divine arts down from heaven through song," addresses his poem to Caesar (either Augustus, adopted son of Julius, or Tiberius, adopted son of Augustus), and declares that only through knowledge given by the gods through astrology can the "potencies of heaven" be revealed. He then briefly traces the progress of human skill and reason, culminating in its ascent of heaven and grasp of the natural explanations that liberate mortals from false belief in marvels. After surveying rival cosmologies of Hesiod and the pre-Socratics, he sets forth an Aristotelian-Stoic view of the cosmos created by interaction of the four elements and suspended by the equilibrium of forces holding it in balance: an eternally godlike form "identically the same at every point." In Book II he reaffirms the originality of his poetic enterprise as he sings of a universe "guided by reason" of the god who rules through nature, which alone gives "sacred vision / to kindred minds directed toward herself / through this great work"—that is, through astrology, the laborious "mastery of fate's laws." The concluding passage of Book IV (from which the last 25 lines are here omitted) celebrates the mind's capacity "to penetrate the universe's heart," as it can because human beings (in a belief going back at least to the Pythagoreans, if not to their shamanistic antecedents) originated from the heavens: "What wonder mortals / can know of heaven, since heaven is within them?"*

Thus this astrological poem is dedicated to the proposition, as Manilius will declare in IV.932, that "reason conquers all" (ratio omnia vincit). *Nor is this neglected poet of reason poetically negligible. Though little read, his text (whose difficulty is suggested by the frequent omission and rearrangement of its lines) attracted three of the great editors of classical philology, Julius Caesar Scaliger in the sixteenth century, Richard Bentley in the eighteenth, and the poet A. E. Housman in the twentieth. At his best, Manilius is worthy of their efforts, for through passionate devotion to his momentous theme he is capable of rising to poetic heights comparable to those of his great Epicurean opposite, Lucretius. Fittingly, as Goold notes, the same Goethe who discarded Manilius in 1770 inscribed in the visitors' book at the summit of the Brocken (on which the Walpurgis Night in* Faust *takes place), after climbing it on September 4, 1784, the lines (II.115–16):*

> Quis coelum posset nisi coeli munere nosse,
> Et reperire Deum, nisi qui pars ipse Deorum est?
> ("Who could know heaven save by heaven's gift
> and find god, save by being part divine?")

The translation is based on the text (with English prose translation) of Goold's edition, with reference to Housman's text in M. Manilii Astronomica: Editio Minor *(1932).*

I

To draw divine arts down from heaven through songs
telling of stars, fate's confidants, by which
celestial reason varies human fortunes,
I strive: the first to agitate by singing
the windswept woods of Helicon's green summit,[4] 5
spreading strange mysteries till now concealed.
Caesar, first father of your fatherland,
who rule a willing world by august laws—
a god deserving heaven, like your father—
you are my song's great strength and inspiration! 10
Heaven now favors those who probe its secrets,
numbering regions of the sky through song.
Peace alone makes it possible to soar
joyfully through the air and boundless sky,
learning the movements of the stars and planets. 15

[4]Hélikôn was a mountain range of Boeotia in Greece, whose Muses Hesiod (Chapter 8 above) invoked in the opening lines of the *Theogony*; from it flowed the springs of Aganippê and Hippokrênê, which were sacred to the Muses and thought to inspire poets who drank from them. See also below, II.50–52.

Yet far intenser pleasure comes from learning
the mighty universe's inmost being,
telling in rhythms measured by Apollo
how heaven's signs rule every earthly creature.
Two altars flame before me; at two temples — 20
I pray, enveloped by a twofold passion:
for song, and substance. Strict laws bind the poet,
round whom the world's vast sphere reverberates
beyond what even unbound words could compass.

Through the gods' favor earth first came to know — 25
the universe: for who, if they concealed it,
could steal away all-ruling heaven's secrets?
Who, with mere human intellect, could wish
to seem a god without the gods' assent,
revealing orbits round the earth, and stars — 32
hurtling obediently through the void?
Author of this great mystery, Cyllenian — 30
god:[5] through you deeper knowledge of the sky
and constellations comes—names, paths, and powers— — 34
to magnify and consecrate not only
appearances but potencies of heaven,
and manifest God's majesty to mortals . . .

Skill had not yet brought science into being,
and rustic farmers left the earth untilled;
gold lay untouched in mountainous recesses, — 75
and unmolested seas hid pristine worlds.
None hazarded his life to waves, nor pleaded
with winds; what each man knew appeared sufficient.
But after long time sharpened human wits,
and work instructed wretched men, and hardship — 80
exacted vigilance of every mortal,
all vied to consummate their varied tasks,
gladly contributing to the common good
what sage experience found by trial and error.
Their barbarous speech was disciplined by laws, — 85
and crops arose from cultivated wastelands;
the roving sailor entered seas uncharted,
opening trade with unfamiliar lands.
Time then devised the arts of war and peace,
for practice generates one art from others. — 90
Men (so tradition tells us) mastered birdsong,
animal divination, charms for serpents,
summoning shades and rousing Ácheron,[6]
and turning day to night, and night to daytime.

[5]Mercury (Hermês), said to have been born on Cyllene (Kyllênê), the highest mountain of the Peloponnesus (on the frontiers of Arcadia and Achaea), is here invoked as the founder of astrology.
[6]Ácheron was a river of the dead.

Taught by experience, skill conquered all. 95
Reason submitted to no limitation
until, ascending heaven, it grasped the deepest
nature and cause of everything that is.
Why clouds are shattered by such tumult, winter
snowfalls are softer than the hail of summer, 100
volcanoes blaze and solid earth is shaken,
why rain pours, and what causes winds to blow,
it saw—emancipating men from marvels
by wresting mighty thunderbolts from Jove,
assigning noise to winds, and fire to clouds. 105
Having ascribed to earthly things true causes,
reason went on to learn the universe's
neighboring mass, by mind embracing heaven,
and specified the forms and names of signs,
discovering what destiny governed movements 110
made by celestial will and disposition
of constellations varying human fates.

This work, till now unsanctified by song,
rises before me now: may fortune speed
my task, and may a long and peaceful lifetime 115
let me achieve this monumental labor
with equal care to great and little things!
And since my song aspires to heaven's heights,
from which fate's ordinance descends to earth,
first must I sing of nature's true appearance 120
and show the mighty universe's likeness.[7]
Some think it traces back its seeds to nothing
and had no origin, but always was
and shall be, lacking birth and death alike;
or chaos, giving birth, first separated 125
mixed elements, and darkness, having borne
this shimmering world, fled to infernal shadows;
or nature's indivisible beginnings
vanish and reappear, unchanged, forever—
a sum of almost nothing—endlessly 130
fashioning heaven and earth from lifeless matter;
or flickering fire composed this work, creating
eyes that now permeate the body of heaven
and lightning-bolts that rumble through the sky;
or water, swallowing the very fire 135
by which it perishes, engendered heaven;
or neither earth, fire, air, nor water knows
a father, but compose one four-limbed god

[7]The following 25 lines set forth competing theories of the origin and composition of the universe (see Chapters 8 and 9 above), including its birth from chaos (Hesiod); the atomistic theory (Leucippus, Democritus, Epicurus); and its composition from fire (Heraclitus), water (Thales), or the four elements (Empedocles). Beginning in line 147, Manilius expounds the synthetic Stoic cosmology of his time.

that built the universe, precluding further
questions, since they themselves created all, 140
mingling cold things with hot, and wet with dry,
airy with solid, in harmonious discord
fitly uniting generative powers
in elements from which all things are born.
Men will dispute forever what will always 145
lie hidden, far beyond both man and god.
However it began, the world's appearance
is clear, its bodily order firmly fixed.
Wingèd fire soared aloft to realms of aether
and, compassing the starry heaven's rooftops, 150
walled nature off with ramparts made of flame.
Next air sank down, becoming tenuous breezes
seeping between the earth's interstices. 153
Third was the undulating billows' place, 155
tumbling forth as the widespread sea was born
through which the waters exhale tenuous vapors
that first inseminate, then feed the air
whose breath stokes fires beneath the neighboring stars. 154
Last, earth sank downward, molded by its weight 159
into a ball, while mud and shifting sands
gathered, and tenuous liquid slowly surfaced.
As segregated moisture formed fresh waters,
the more the filtered seas built up firm land,
the more they settled down in hollow valleys;
mountains rose from the waters, then a world 165
leapt up, surrounded by the boundless ocean.
Stable it stands, since all the firmament 168
is equidistant from it; having fallen
from every side, its core can fall no further. 170
Unless earth's equilibrium kept it balanced, 173
Phoebus' car could not pass beyond the sunset[8]
as stars appeared, nor ever rise again, 175
nor the moon journey on submerged in darkness;
nor, having shone as Hesperos, then crossing
the sky, could Lucifer blaze forth at dawn.[9]
Now, since earth hangs suspended in mid-air,
not plunged in nether depths, all paths are open 180
for skies to pass beneath, and rise again.
That stars should reappear by chance, or heaven
be born so often, suns die daily, then
be resurrected, passes all belief,
since heavenly signs remain unchanged for ages; 185
the selfsame Phoebus rises in the east;

[8]Phoebus Apollo was the god who drove the chariot of the sun.
[9]"Venus is not seen as both morning-star [Lucifer; Greek Phósphoros] and evening-star [Vesper; Greek Hésperos] within the same day, though this fanciful notion was popular with the ancient poets." (Goold)

the moon recurrently goes through its phases;
Nature pursues the paths she made herself,
not stumbling like an amateur; the day's
eternal light marks out identical 190
times now in one, now in another region;
and travelers find both sun's and heaven's risings
and settings further off, as they approach them.
Nor should the nature of this dangling world
surprise you, since the firmament itself 195
hangs without footsteps planted on the ground,
as manifested by its rapid movement.
Poised aloft, Phoebus nimbly steers his car
now here, now there, observing heavenly limits,
while moon and stars revolve through empty space, 200
and earth hangs—mimicking celestial laws—
deep in a cavern hollowed from mid-air
equally far from each extreme of heaven:
not flattened out, but shaped into a sphere
raised and compressed alike in all directions. 205
Thus appears nature; thus the universe,
arching above, gives stars their pointed shapes.
Circular as the sun's orb is the moon's,
seeking light for its fully rounded body
when slanting fires illuminate parts only. 210
This godlike form abides eternally,
nowhere containing either start or finish,
identically the same at every point.
So earth is rounded in the shape of heaven,
lowest, but therefore centermost, of all. . . . 167

II . . . Things of all kinds the learnèd sisters sang:[10]
well worn is every path to Helicon. 50
Turbid are waters flowing from its fountains,
nor can crowds jostling round them drink their fill.
Let us seek meadows fresh with dewy grasses,
streams rehearsing murmurs in deep caves
whose waters neither beaks of birds have tasted 55
nor Phoebus grazed with his aethereal fire.
I speak words borrowed from no bard—my own
creation, not a theft—and soar toward heaven
alone, in my own ship traverse the waves.
For I shall sing god, nature's silent-minded 60
monarch, who permeates sky, land, and sea,
ruling this mighty mass by just accord:
sing how the universe lives by agreement

[10]"The learnèd sisters" are the Muses. For Helicon, see note on I.5.

guided by reason, since one single spirit
dwells in all parts and, flitting everywhere, 65
nurtures the world, and molds one living creature.
Unless this whole machine—composed of kindred
limbs, and obedient to one overlord—
persisted, and providence controlled the world,
earth would not stand unmoved, nor stars revolve, 70
and heaven would go astray, or stop and stiffen;
nor would the constellations hold their courses,
night alternately chase and flee the sunlight,
rains feed the earth or winds the aether, seas
feed pregnant clouds, or rivers nourish seas 75
and seas the streams; nor would the sum of things
be always equal, by parental justice
forbidding waves to vanish, land to dwindle,
and heaven to shrink or swell beyond due measure.
Motion maintains this stable work: thus ordered, 80
the world continues, following its master.
This god, and all-controlling reason, guided
all earthly creatures by their heavenly signs,
making their influences felt from far
away, establishing the lives and fates 85
of nations, and the characters of men. . . .

Who can doubt, then, that man is linked with heaven? 105
Nature pre-eminently gave him language,
broad knowledge, swift intelligence: in him
god, searching for himself, descends and dwells.
Pass over arts whose mastery arouses
envy for gifts beyond our human state: 110
who could know heaven save by heaven's gift 115
and find god, save by being part divine?
Who, by his narrow intellect, could compass
this vaulted infinite, the dancing stars,
the flaming roof of heaven, the everlasting 119
combat of planets with the constellations, 121
if Nature had not given sacred vision
to kindred minds directed toward herself
through this great work,[11] or if from heaven nothing 125
called us to heaven's sacred fellowship? 127
Who could deny the impiety of dragging
a self-imprisoned universe to earth?
But what is manifest to all, arousing 130
trustworthy confidence through undeceived
and undeceiving reason, needs no proof.
Follow this rightly trusted path, and all
predictions faithfully shall come to pass:

[11]The science of astrology.

for who denies what fortune brings about 135
or challenges fate's overwhelming ballot?
This to the stars would I exalt, with holy
breath singing neither in the crowd nor for it:
but, all alone, as though I freely guided
my chariot around with none to cross me 140
or drive along beside me as my comrade,
for heaven I sing—astonishing the stars
as worlds delight in my prophetic songs—
and for that smallest company on earth,
those to whom stars reveal their sacred motions.
Vast is the crowd that treasures gold and riches, 145
power and pomp, soft luxury and ease,
and pleasant feelings roused by soothing music
far less laboriously learned than fate.
Yet mastery of fate's laws is likewise fated. . . .

IV What does it profit subtle minds to search
the gleaming universe, if fearful thoughts
demolish hope and bar the gates of heaven?
"See," someone says, "how deeply nature lies
buried beyond men's sight and contemplation, 870
nor does it help that destiny rules all
when destiny can by no means be fathomed."
What good is self-reproach, depriving men
of blessings god himself has not begrudged them,
and eyes that nature planted in the mind? 875
We see the heavens: why not heaven's gifts?
The human mind can leave its proper place[12]
and penetrate the universe's heart;
from its own seeds construct that mighty mass;
lead heaven's child through regions whence he came;
follow the farthest sea; and plunge beneath 880
the hanging earth, inhabiting all the world.
Now nature lies revealed: we see it all; 883
apprehend heaven, looking on the parent
whose parts we are; and reach the very stars. 885
Who doubts a god is dwelling in our breasts,
and souls return to heaven, whence they came?
Just as the universe, composed of every
element—lofty fire, air, earth, and sea—
houses a mind pervading all it governs, 890
so with us, bodies of terrestrial makeup
and lifeblood house a mind that governs all

[12]The italicized words translate a Latin line editorially interpolated to supply an apparent lacuna in Manilius's text.

and constitutes the man. What wonder mortals
can know of heaven, since heaven is within them,
and each exemplifies a god in little? 895
Can we believe men born from anything
but heaven? All the animals lie prostrate
on earth, submerged in water, or suspended
in air: all prize sleep, food, and strength of limb,
and, lacking thought, relinquish language also. 900
One breed, all-governing man alone, examines
the world, has powerful speech, broad intellect,
and varied arts. Withdrawing into cities,
he tamed the earth for crops, caught animals,
built roads at sea, and stood with towering head 905
upraised: victoriously lifting starry
eyes to the stars, he contemplates Olympus
and questions Jove; content with no mere figures
of gods, he probes the heavens' womb for kindred
beings, and seeks himself among the stars. . . . 910

The Integrity of the Whole

The *Meditations* of Marcus Aurelius, translated by George Long (revised)

Edward Gibbon, who saw human history as "little more than the register of the crimes, follies, and misfortunes of mankind," declared in Chapter III of The Decline and Fall of the Roman Empire *(1776) that "if a man were called to fix the period in the history of the world, during which the condition of the human race was most happy and prosperous, he would, without hesitation, name that which elapsed from the death of Domitian to the accession of Commodus," that is, the reigns of Nerva (A.D. 96–98), Trajan (98–117), Hadrian (117–38), Antoninus Pius (138–61), and Marcus Aurelius (161–80). No doubt many crimes, follies, and misfortunes are cloaked by such a judgment. Yet given the cruelties and depravities of predecessors such as Caligula, Nero, and Domitian and successors such as Commodus (son of Marcus Aurelius), Caracalla, and Elagabalus, and the calamities that ensued, idealization of this era, when the Roman Empire reached its maximum expansion and still maintained a precarious bulwark against the invaders who continually menaced its peace, is easily understandable. It was a time not of original thought or great literature but of an ancient pagan civilization that sustained, in decline, a high level of culture and even, in the astronomer Ptolemy and the physician Galen (Chapter 12 below), of scientific achievement, while the new Christian culture, despite sporadic persecutions (including one by Marcus), grew and prospered.*

Marcus Aurelius (121–80), surnamed Antoninus after his adoption by Antoninus Pius, received an excellent education under the Hellenizing emperor Hadrian. He then, after the death of Antoninus Pius, ruled the Empire—at first jointly with his stepbrother—from 161 until his death twenty years later. Much of his reign was spent in the field, fighting the central European Germanic tribes of the Marcomanni and Quadi, both between 170 and 174 and again from 178 until his death, not far from modern Vienna. Yet what this commander of armies and ruler of men is mainly remembered for is the slim volume, written in Greek and entitled To Himself, *but generally known as the* Meditations, *in which he sets forth his random thoughts, many of them recorded during military campaigns, on human life in a world that often seems without meaning. "There is nothing new," he writes (VII.1), in words reminiscent of the Hebrew Preacher: "all things are both familiar and short-lived," and all is determined by fate. "Whatever may happen to you was prepared for you from all eternity; and the implication of causes was from eternity spinning the thread of your being, and of what is incident to it" (X.5). But if Marcus, like other Stoics of the Roman Empire, is primarily a moralist, as such he is deeply concerned with the close interrelation of the human and the natural worlds. Imbued though he is with Stoic cosmology, his views on nature (as on human life) are tolerant and undogmatic, showing openness to Platonic and even Epicurean ideas; for in this world, what can be known with certainty? Marcus's world, like that of Heraclitus, is one of constant flux, and the object of life is to harmonize the individual self with the continuity of the cosmos as a living being, possessing a single soul, in which all things are "implicated with one another." To do so is to transcend personal smallness by participation in the greater world to which, through our common nature, we belong; "nothing is injurious to the part," Marcus believes (X.6), "if it is for the advantage of the whole." Marcus is thus a man of action for whom the contemplative life is essential, since only philosophy can teach us to wait for death with a cheerful mind. He makes no pretense to original speculation, yet through his deeply felt meditations he brings to life the Stoic view of the world more fully and movingly than any other writer. Two of his many sayings, both included here, summarize what is most essential in that view: "Nothing according to nature is evil" (II.17) and "To a rational animal the same act is according to nature and to reason" (VII.11).*

The translation is slightly revised from that of George Long (1862), with reference to the Loeb Classical Library text edited by C. R. Haines (1916).

II 9 This you must always bear in mind, what is the nature of the whole, and what is my nature, and how this is related to that, and what kind of a part it is of what kind of whole; and that there is no one who hinders you from always doing and saying the things which are according to the nature of which you are a part.

17 Of human life the time is a point, the substance is in flux, and the perception dull, and the composition of the whole body subject to putrefaction, and the soul a whirl, and fortune hard to divine, and fame a thing devoid of judgment. And, in a word, everything which belongs to the body is a river, and what belongs to the soul a dream and vapor, and life is a warfare and a stranger's sojourn, and after-fame is oblivion. What then can guide a man? One thing and only one, philosophy. But this consists in keeping the daemon within free from violence and unharmed, superior to pains and pleasures, doing nothing without a purpose, nor yet falsely and with hypocrisy, not feeling the need of another man's doing or not doing anything; and, besides, accepting all that happens, and all that is allotted, as coming from wherever it is that he himself came; and, finally, waiting for death with a cheerful mind, as being nothing but a dissolution of the elements of which every living being is compounded. But if the elements themselves are not harmed by each continually changing into another, why should a man have any apprehension about the change and dissolution of all the elements? For this is according to nature, and nothing according to nature is evil.

This in Carnuntum.[13]

[13]A town on the Danube on the northern frontier of the Roman province of Pannonia (roughly corresponding to modern Slovenia and parts of Austria and Hungary), from which Marcus Aurelius waged his long wars against the invading Germanic tribe of the Marcomanni.

IV 23 Everything harmonizes with me, which is in harmony with you, O universe. Nothing for me is too early nor too late, which is in due time for you. Everything is fruit to me which your seasons bring, O nature: from you are all things, in you are all things, to you all things return. The poet says, "Dear city of Cecrops,"[14] and will you not say, "O dear city of Zeus"?

29 If he is a stranger to the universe who does not know what is in it, no less is he a stranger who does not know what is happening in it. He is an exile, who flees from civic reason; he is blind, who shuts the eyes of the understanding; he is poor, who has need of another, and has not from himself all things useful for life. He is an abscess on the universe who withdraws and separates himself from the reason of our common nature by being displeased with the things that happen, for the same nature produces this, that produced you; he is a piece cut off from the community, who tears his own soul from the soul of reasonable beings, which is one.

40 Constantly regard the universe as one living being, having one substance and one soul; and observe how all things have reference to one perception, the perception of this one living being; and how all things act with one movement; and how all things are joint causes of all things that exist; observe too the continuous spinning of the thread and the contexture of the web.

48 Think continually how many physicians are dead after often contracting their eyebrows over the sick; and how many astrologers after predicting with great pretensions the deaths of others; and how many philosophers after endless discourses on death or immortality; how many heroes after killing thousands; and how many tyrants who have used their power over men's lives with terrible insolence as if they were immortal; and how many cities are entirely dead, so to speak—Helice and Pompeii and Herculaneum,[15] and others innumerable. Add to the reckoning all whom you have known, one after another. One man after burying another has been laid out dead, and another buries him: and all this in a short time. To conclude, always observe how ephemeral and worthless human things are, and what was yesterday a little mucus tomorrow will be a mummy or ashes. Pass then through this little space of time conformably to nature, and end your journey in content, just as an olive falls off when it is ripe, blessing nature who produced it, and thanking the tree on which it grew.

V 8 Just as we understand it when Asclepius[16] is said to have prescribed to this man horse-exercise, or bathing in cold water or going without shoes; so we must understand it when it is said that the nature of the universe prescribed to this man disease or mutilation or loss or anything else of the kind. For in the first case "prescribed" means something like this: he prescribed this treatment for this man as a thing suitable to his health; and in the second case it means that what happens to every man is fixed in a manner suitable to his destiny *[heimarmenê]*. For this is what we mean when we say that things are suitable to us, as the workmen say squared stones in walls or the pyramids are suitable when they fit them to one another in some kind of connection. For there is altogether one fitness *[harmonia]*. And as the universe is made up out of all bodies to be such a body as it is, so out of all existing causes destiny is made up to be such a cause as it is. Even

[14]Cecrops was the mythical first king of Athens; the line is probably a fragment from Aristophanes.
[15]Hélikê was a town of Achaia on the bay of Corinth destroyed by a flood (see Ovid, *Metamorphoses* XV.293–94); Pompeii and Herculaneum were destroyed by the eruption of Vesuvius in A.D. 79.
[16]The Greek god of healing (Roman Aesculapius).

those who are completely ignorant understand what I mean, for they say destiny brought this to someone: this then was brought, and this was prescribed to him. Let us then accept these things like those that Asclepius prescribes. Many among his prescriptions are disagreeable as a matter of course, but we accept them in the hope of health. Let the perfecting and accomplishment of things which the common nature judges to be good be judged by you in the same way as your health. And so accept everything which happens, even if it seems disagreeable, because it leads to the health of the universe and to the prosperity and felicity of Zeus; for he would not have brought on any man what he has brought, if it were not useful for the whole. Nor does the nature of anything, whatever it may be, cause anything unsuitable to what is directed by it. For two reasons then it is right to be content with what happens to you: first, because it was done for you and prescribed for you, and had some kind of reference to you, spun from the beginning out of the most ancient causes; and secondly, because even what comes separately to every man is, to the power that administers the universe, a cause of felicity and perfection, indeed of its very continuation. For the integrity of the whole is mutilated if you cut off anything whatever from the conjunction and the continuity of either the parts or the causes. And you do cut something off, as far as it is in your power, when you are dissatisfied and cast anything aside.

VI 10 [The universe is] either confusion and entanglement and dispersion, or unity and order and providence. If then it is the former, why do I desire to tarry in a fortuitous combination of things, and such a disorder? And why do I care about anything else than how I shall at last become earth? And why am I disturbed, for the dispersion of my elements will happen whatever I do. But if the other supposition is true, I venerate, and I am firm, and I trust in him who governs.

15 Some things are hurrying into existence, and others are hurrying out of it; and of that which is coming into existence part is already extinguished. Motions and changes are continually renewing the world, just as the uninterrupted course of time is always renewing the infinite duration of ages. In this flowing stream then, on which there is no abiding, what is there of the things which hurry by on which a man would set a high price? It would be just as if a man should fall in love with one of the sparrows which fly by, but has already passed out of sight. Something like this is the life of every man, like the exhalation of the blood and the respiration of the air. For just like drawing in the air and giving it back, which we do every moment, so is it to give back your whole respiratory power to the place from which you received it at your birth, yesterday or the day before.

38 Consider frequently the connection of all things in the universe and their relation to one another. For in a sense all things are implicated with one another, and all in this way are friendly to one another; for one thing comes in order after another by virtue of the active movement and intermingling and unity of substance.

VII 9 All things are implicated with one another, and the bond is holy; and there is hardly anything unconnected with any other thing. For things have been co-ordinated, and they combine to form the same universe. For there is one universe made up of all things, and one God who pervades all things, and one substance, and one law, one common reason in all intelligent animals, and one truth—if indeed there is also one perfection of all animals that are of the same stock and participate in the same reason.

10 Everything material soon disappears in the substance of the whole; and every cause is very soon taken back into the universal reason; and the memory of everything is soon overwhelmed in time.

11 To a rational animal the same act is according to nature and to reason.

25 Nature, which governs the whole, will soon change all things which you see, and out of their substance will make other things, and again other things from the substance of them, in order that the world may be ever new.

47 Look round at the course of the stars, as if you were going around with them; and constantly consider the changes of the elements into one another; for such thoughts purge away the filth of earthly life.

IX 19 All things are changing, and you yourself are in continuous mutation and, in a sense, in continuous destruction, and the whole universe too.

28 The periodic movements of the universe are the same, up and down from age to age. And either the universal intelligence puts itself in motion for every separate effect (and if this is so, be content with the result of its activity); or it puts itself in motion once, and everything else comes, as it were, by way of sequence; or indivisible elements are the origin of all things. In a word, if there is a god, all is well; and if chance rules, do not yourself act by chance.

Soon the earth will cover us all; then the earth, too, will change, and the things also which result from change will continue to change forever, and these again forever. For if a man reflects upon the changes and transformations which follow one another like wave after wave, and their rapidity, he will despise everything which is perishable.

X 6 Whether [the universe is composed of random] atoms or [of a rational] nature, let it first be established that I am a part of the whole which is governed by nature; next, I am in a manner intimately related to the parts which are of the same kind with myself. For remembering this, inasmuch as I am a part, I shall be discontented with none of the things which are assigned to me out of the whole; for nothing is injurious to the part, if it is for the advantage of the whole. For the whole contains nothing which is not for its advantage; and all natures indeed have this common principle, but the nature of the universe has this principle besides, that it cannot be compelled even by any external cause to generate anything harmful to itself. By remembering, then, that I am a part of such a whole, I shall be content with everything that happens. And inasmuch as I am in a sense intimately related to the parts which are of the same kind with myself, I shall do nothing unsocial, but I shall rather direct myself to the things which are of the same kind with myself, and I shall turn all my efforts to the common interest, and divert them from the contrary. Now, if these things are done so, life must flow on happily, just as you may observe that the life of a citizen is happy, who continues a course of action which is advantageous to his fellow-citizens, and is content with whatever the state may assign to him.

14 To nature, who gives and takes back all, the man who is instructed and modest says, "Give what you will; take back what you will." And he says this not proudly, but with simple obedience and good will toward her.

15 Little of life remains to you. Live as on a mountain. For it makes no difference whether a man lives there or here, if he lives everywhere as if in the city of the world. Let men see, let them know a real man who lives according to nature. If they cannot endure him, let them kill him. For that is better than to live thus as men do.

XI 10 Nature is never inferior to art, for the arts imitate the nature of things. But if this is so, that nature which is the most perfect and the most comprehensive of all natures cannot fall short of the skill of art. Now all arts do the inferior things for the sake of the superior; therefore the universal nature does so too. And here, indeed, is the origin of justice, and in justice the other virtues have their foundation: for justice will not be observed, if we either care for things indifferent, or are easily deceived and careless and changeable.

XII 21 Consider that before long you will be nobody and nowhere, nor will any of the things exist which you now see, nor any of those who are now living. For all things are formed by nature to change and be turned and to perish in order that other things in continuous succession may exist.

The Wondrous Beauty of Multiplicity from Oneness

The *Enneads* of Plotinus, translated by A. H. Armstrong

*Plotínus (*A.D. *205–269/70), the first and greatest of philosophers now called Neoplatonists, was probably born in Egypt; for eleven years after the age of 27 he studied in Alexandria under the Platonist or Neo-Pythagorean Ammonius Saccas, who like Socrates wrote nothing. In 242/43 Plotinus, hoping to learn more about Eastern philosophy, joined the Persian expedition of Gordian III (who was murdered by his own troops). Then, at age 40, he settled in Rome, where he taught for most of the remaining twenty-five years of his life. The most important of his disciples, Malchus of Tyre (232/33–ca. 305), who took the Greek name Porphyry (Porphyrios), wrote a brief life of his master after his death; and in around* A.D. *300 he edited fifty-four treatises, written in Greek during Plotinus's last fifteen years, in six books of nine each, which he called the* Enneads *(from Greek* ennea, *"nine").*

"Though not systematic in intention," E. R. Dodds remarks in the Oxford Classical Dictionary, *"the* Enneads *form a more complete body of philosophical teaching than any other which has come down to us from antiquity outside the Aristotelian corpus." For Plotinus, as for his great model Plato, reality lies not in the "sensible" but in the "intelligible world." Indeed, the ultimate reality, "which he identifies with the 'One' of the* Parmenides *and the 'Good' of the* Republic*" of Plato, "is strictly insusceptible of any predicate or description," and can be glimpsed only in rare moments of ecstatic* enôsis, *or mystical unification, such as Plotinus, according to Porphyry, experienced four times; this wholly impersonal One is Plotinus's God. Through "overflow" from the One arise the second and third "hypóstases" of the intelligible world, respectively Intellect* (nous)*—the realm of the Platonic Forms—and the World-Soul* (psykhê)*; from the latter in turn arise the illusory manifestations of Nature* (physis) *and Matter* (hylê)*. This descent by degrees from the One to Matter "is marked by increasing individuation and diminishing unity." In this "fearful state" of plurality and separation, J. M. Rist writes in* Plotinus *(1967), "the soul is in chains, a prisoner shut up in a bodily cave." But because it participates not only in Nature but also, at least potentially, in Mind, the human soul is capable through contemplation (as in Plato's* Symposium*) of reascending toward its source.*

Nor are Nature and Matter merely evil (any more than they had been for Plato); on the contrary, Plotinus vigorously rebutted the Christian Gnostics of his day who postulated a categor-

ical opposition between the natural and the divine. Everything here below derives from those upper regions, he writes in Ennead *V.8, and what is there pure is adulterated here. Yet the source of both is the same: for if the intelligible world "was not transcendently beautiful with an overwhelming beauty, what would be more beautiful than this visible universe?" Thus Plotinus's very idealism affirms the value of the dreamlike but wondrously beautiful multiplicity of nature, which (he half playfully writes in III.8), having come to exist through contemplation by the World-Soul—as the World-Soul did through contemplation by Intellect—is "at rest in contemplation of the vision of itself." Through the* logos *that descends from the World-Soul and rationally informs the very silence of its quasi-contemplative activity,* physis *remains, as it was for the pre-Socratics, a dynamically formative process.* Physis *falls infinitely short, to be sure, of the One that is the inexhaustible source of all creation, yet it shares, however partially, in the contemplation that connects the sensible to the intelligible world and makes it possible for a human being, who partakes of both, to give back in death (as Porphyry reported Plotinus's last words) "the Divine in myself to the Divine in the All." The greatness of the One can be seen only "by the things which exist after it and through it"; those who condemn the visible universe, then, blame it only for not being the still more beautiful intelligible world.*

The translations are from volumes III and V of the Loeb Classical Library Plotinus *(7 vols., 1966–88). See also* The Enneads, *trans. Stephen MacKenna (4th ed., rev. by B. S. Page, 1969), and, besides Dodds's account and Rist's book, Emile Bréhier,* The Philosophy of Plotinus *(1928; Eng. trans. 1958), and John N. Deck,* Nature, Contemplation, and the One: A Study in the Philosophy of Plotinus *(1967).*

FROM *On Nature and Contemplation and the One*
III.8

1 Suppose we said, playing at first before we set out to be serious, that all things aspire to contemplation, and direct their gaze to this end—not only rational but irrational living things, and the power of growth *[physis]* in plants, and the earth which brings them forth—and that all attain to it as far as possible for them in their natural state, but different things contemplate and attain their end in different ways, some truly, and some only having an imitation and image of this true end—could anyone endure the oddity of this line of thought? Well, as this discussion has arisen among ourselves, there will be no risk in playing with our own ideas. Then are we now contemplating as we play? Yes, we and all who play are doing this, or at any rate this is what they aspire to as they play. And it is likely that, whether a child or a man is playing or being serious, one plays and the other is serious for the sake of contemplation, and every action is a serious effort towards contemplation . . . But we will discuss this later: but now let us talk about the earth itself, and trees, and plants in general, and ask what their contemplation is, and how we can relate what the earth makes and produces to its activity of contemplation, and how nature, which people say has no power of forming mental images[17] or reasoning, has contemplation in itself and makes what it makes by contemplation, which it does not have.

2 Well, then, it is clear, I suppose, to everyone that there are no hands here or feet, and no instrument either acquired or of natural growth, but there is need of matter on

[17]The Stoics used the terms *physis aphantastos* ["nature with no power of forming mental images"] and *noera physis* ["intellectual nature"] to distinguish between "nature" in the sense of the Aristotelian growth-principle and in their own sense of the all-pervading divine reason. (Armstrong)

which nature can work and which it forms. But we must also exclude levering from the operation of nature. For what kind of thrusting or levering can produce this rich variety of colors and shapes of every kind?[18] . . . In fact, of course, nature must be a form, and not composed of matter and form; for why should it need hot or cold matter? For matter which underlies it and is worked on by it comes to it bringing this [heat or cold] or rather becomes of this quality (though it has no quality itself) by being given form by a rational principle. For it is not fire which has to come to matter in order that it may become fire, but a forming principle *[logos]*; and this is a strong indication that in animals and plants the forming principles are the makers and nature is a forming principle; and this is a strong indication that in animals and plants the forming principles are the makers and nature is a forming principle, which makes another principle, its own product, which gives something to the substrate *[hypokeimenon]*, but stays unmoved itself. This forming principle, then, which operates in the visible shape, is the last, and is dead and no longer able to make another, but that which has life is the brother of that which makes the shape, and has the same power itself, and makes in that which comes into being.

3 How then, when it makes, and makes in this way, can it attain to any sort of contemplation? If it stays unmoved as it makes, and stays in itself, and is a forming principle, it must itself be contemplation. . . . But nature possesses, and just because it possesses, it also makes. Making, for it, means being what it is, and its making power is coextensive with what it is. But it is contemplation and object of contemplation, for it is a rational principle. So by being contemplation and object of contemplation and rational principle, it makes in so far as it is these things. So its making has been revealed to us as contemplation, for it is a result of contemplation, and the contemplation stays unchanged and does not do anything else but makes by being contemplation.

4 And if anyone were to ask nature why it makes, if it cared to hear and answer the questioner it would say: "You ought not to ask, but to understand in silence, you, too, just as I am silent and not in the habit of talking. Understand what, then? That what comes into being is what I see in my silence, an object of contemplation which comes to be naturally, and that I, originating from this sort of contemplation have a contemplative nature. And my act of contemplation makes what it contemplates, as the geometers draw their figures while they contemplate. But I do not draw, but as I contemplate, the lines which bound bodies come to be as if they fell from my contemplation. What happens to me is what happens to my mother and the beings that generated me,[19] for they, too, derive from contemplation, and it is no action of theirs which brings about my birth; they are greater rational principles, and as they contemplate themselves I come to be."

What does this mean? That what is called nature is a soul, the offspring of a prior soul with a stronger life; that it quietly holds contemplation in itself, not directed upwards or even downwards, but at rest in what it is, in its own repose and a kind of self-perception,

[18] It is part of Plotinus's consistent effort to eliminate materialistic and spatial conceptions from our ideas of spiritual existence and activity that he insists frequently that soul and nature are not to be thought of as forming the material world with hands and tools and machines. He seems to have in mind the sort of crude Epicurean criticism of Plato which we find in [the discourse by Velleius in] Cicero *De Natura Deorum* I.8.19: *quae molitio? quae ferramenta? qui vectes? quae machinae?* ["What method of engineering was employed (in building the structure of the universe)? What tools and levers and derricks?"—trans. H. Rackham]. (Armstrong)

[19] "My mother" = the higher soul; "the beings that generated me" = the *logoi* in soul which are the immediate expressions of the Forms in Intellect. (Armstrong)

and in this consciousness and self-perception it sees what comes after it, as far as it can, and seeks no longer, but has accomplished a vision of splendor and delight. If anyone wants to attribute to it understanding or perception, it will not be the understanding or perception we speak of in other beings; it will be like comparing the consciousness of someone fast asleep to the consciousness of someone awake. Nature is at rest in contemplation of the vision of itself, a vision which comes to it from its abiding in and with itself and being itself a vision; and its contemplation is silent but somewhat blurred. For there is another, clearer for sight, and nature is the image of another contemplation. For this reason what is produced by it is weak in every way, because a weak contemplation produces a weak object. Men, too, when their power of contemplation weakens, make action a shadow of contemplation and reasoning. Because contemplation is not enough for them, since their souls are weak and they are not able to grasp the vision sufficiently, and therefore are not filled with it, but still long to see it, they are carried into action, so as to see what they cannot see with their intellect. When they make something, then, it is because they want to see their object themselves and also because they want others to be aware of it and contemplate it, when their project is realized in practice as well as possible. Everywhere we shall find that making and action are either a weakening or a consequence of contemplation; a weakening, if the doer or maker had nothing in view beyond the thing done, a consequence if he had another prior object of contemplation better than what he made. For who, if he is able to contemplate what is truly real will deliberately go after its image? . . .

10 What is [the one] then? The productive power *[dynamis]* of all things; if it did not exist, neither would all things, nor would Intellect be the first and universal life. But what is above life is cause of life; for the activity of life, which is all things, is not first, but itself flows out, so to speak, as if from a spring. For think of a spring which has no other origin, but gives the whole of itself to rivers, and is not used up by the rivers but remains itself at rest, but the rivers that rise from it, before each of them flows in a different direction, remain for a while all together, though each of them knows, in a way, the direction in which it is going to let its stream flow; or of the life of a huge plant, which goes through the whole of it while its origin remains and is not dispersed over the whole, since it is, as it were, firmly settled in the root. So this origin gives to the plant its whole life in its multiplicity, but remains itself not multiple but the origin of the multiple life. And this is no wonder. Or, yes, it is a wonder how the multiplicity of life came from what is not multiplicity, and the multiplicity would not have existed, if what was not multiplicity had not existed before the multiplicity. For the origin is not divided up into the All, for if it were divided up it would destroy the All too; and the All could not any more come into being if the origin did not remain by itself, different from it. Therefore, too, we go back everywhere to *one*. And in each and every thing there is some *one* to which you will trace it back, and this in every case to the *one* before it, which is not simply one, until we come to the simply one; but this cannot be traced back to something else. But if we take the *one* of the plant—this is its abiding origin—and the *one* of the animal and the *one* of the soul and the *one* of the universe, we are taking in each case what is most powerful and really valuable in it; but if we take the *one* of the beings which truly exist, their origin and spring and productive power, shall we lose faith and think of it as nothing? It is certainly none of the things of which it is origin; it is of such a kind, though nothing can be predicated of it, not being, not substance, not life, as to be above all of these things. But if you grasp it by taking away being from it, you will be filled with wonder. And, throwing

yourself upon it and coming to rest within it, understand it more and more intimately, knowing it by intuition and seeing its greatness by the things which exist after it and through it.

FROM *On the Intelligible Beauty*
V.8

1 . . . Everything which is extended departs from itself: if it is bodily strength, it grows less strong, if heat, less hot, if power in general, less powerful, if beauty, less beautiful. Every original maker must be in itself stronger than that which it makes; it is not lack of music which makes a man musical, but music, and music in the world of sense is made by the music prior to this world. But if anyone despises the arts because they produce their works by imitating nature, we must tell him, first, that natural things are imitations too. Then he must know that the arts do not simply imitate what they see, but they run back up to the forming principles from which nature derives; then also that they do a great deal by themselves, and, since they possess beauty, they make up what is defective in things. For Pheidias[20] too did not make his Zeus from any model perceived by the senses, but understood what Zeus would look like if he wanted to make himself visible.

2 But let us leave the arts; and let us contemplate those things whose works they are said to imitate, which come into existence naturally as beauties and are so called, all the rational and irrational living creatures and especially those among them which have succeeded since the craftsman who formed them dominated the matter and gave it the form he wished. . . . But certainly nature which produces such beautiful works is far before them in beauty, but we, because we are not accustomed to see any of the things within and do not know them, pursue the external and do not know that it is that within which moves us: as if someone looking at his image and not knowing where it came from should pursue it. . . .

3 There is therefore in nature a rational forming principle *[logos]* which is the archetype of the beauty in body, and the rational principle in soul is more beautiful than that in nature, and is also the source of that in nature. It is clearest in a nobly good soul and is already advanced in beauty: for by adorning the soul and giving it light from a greater light which is primarily beauty it makes us deduce by its very presence in the soul what that before it is like, which is no longer in anything else but in itself. For this reason it is not an expressed forming principle at all, but is the maker of the first forming principle which is the beauty present in the matter which is soul; but this [primary principle of beauty] is Intellect *[nous]*, always and not just sometimes Intellect, because it does not come to itself from outside. . . .

5 Some wisdom makes all the things which have come into being, whether they are products of art or nature, and everywhere it is a wisdom which is in charge of their making. But if anyone does really make according to wisdom itself, let us grant that the arts are like this. But the craftsman goes back again to the wisdom of nature, according to which he has come into existence, a wisdom which is no longer composed of theorems, but is one thing as a whole, not the wisdom made into one out of many components, but

[20]The famous Athenian sculptor of the fifth century B.C., whose statue of Zeus at Olympia came to be known as one of the seven wonders of the ancient world.

rather resolved into multiplicity from one. If then one is going to make this the first, that is enough: for it no longer comes from another and is not in another. But if people are going to say that the rational forming principle is in nature, but its origin is nature, from where shall we say that nature has it—is it perhaps from that other? If it is from itself, we shall stop there; but if they are going on to Intellect, we must see at this point if Intellect generated wisdom; and if they assent to this, from where did it get it? If from itself, this is impossible unless it is wisdom itself. The true wisdom, then, is substance, and the true substance is wisdom; and the worth of substance comes from wisdom, and it is because it comes from wisdom that it is true substance. Therefore all the substances which do not possess wisdom, because they have become substance on account of some wisdom but do not possess wisdom in themselves, are not true substances. One must not then suppose that the gods or the "exceedingly blessed spectators" in the higher world contemplate propositions, but all the Forms we speak about are beautiful images in that world, of the kind which someone imagined to exist in the soul of the wise man, images not painted but real.[21] This is why the ancients said that the Ideas were realities and substances. . . .

7 This All, if we agree that its being and its being what it is come to it from another, are we to think that its maker conceived earth in his own mind, with its necessary place in the center, and then water and its place upon earth, and then the other things in their order up to heaven, then all living things, each with the sort of shapes which they have now, and their particular internal organs and outward parts, and then when he had them all arranged in his mind proceeded to his work? Planning of this sort is quite impossible—for where could the ideas of all these things come from to one who had never seen them? And if he received them from someone else he could not carry them out as craftsmen *[dêmiourgoi]* do now, using their hands and tools; for hands and feet come later. The only possibility that remains, then, is that all things exist in something else, and, since there is nothing between, because of their closeness to something else in the realm of real being something like an imprint and image of that other suddenly appears, either by its direct action or through the assistance of soul—this makes no difference for the present discussion—or of a particular soul.[22] All that is here below comes from there, and exists in greater beauty there: for here it is adulterated, but there it is pure. All this universe is held fast by forms from beginning to end: matter first of all by the forms of the elements, and then other forms upon these, and then again others; so that it is difficult to find the matter hidden under so many forms. Then matter, too, is a sort of ultimate form;[23] so

[21]The "images in the soul of the wise man" . . . come from the speech of Alcibiades in praise of Socrates at the end of the *Symposium*, where he speaks of the wonderful images which are concealed within his Silenus-like outside: Plato *Symposium* 215B1–3 and 216E6–217A1. The form of reference ("someone imagined") is curious for a reference by Plotinus to a Platonic passage, but Plotinus is probably attributing the imagination of the Silenus-figure containing divine images to Alcibiades himself rather than to Plato. It brings out excellently that the Forms in Intellect are concrete living realities, not mental abstractions like propositions, a point on which Plotinus is much concerned to insist in this treatise. (Armstrong)

[22]The insistence on the immediate and intimate relationship of the intelligible and sensible universes and the comparative unimportance of the mediation of soul should be noted. Soul in Plotinus never has a world of its own intermediate between the intelligible and sensible worlds; it belongs to both worlds, and is normally thought of as linking them; but here it seems to be hardly necessary even as a link. (Armstrong)

[23]This passing remark, which is very difficult to reconcile with Plotinus's normal view of *hylê* [matter] as the principle of evil (there are no evil Forms in Plotinus), is the nearest he ever comes to a totally positive valuation of matter. (Armstrong)

this universe is all form, and all the things in it are forms; for its archetype is form; the making is done without noise and fuss, since that which makes is all real being and form. . . .

8 Who, then, will not call beautiful that which is beautiful primarily, and as a whole, and everywhere as a whole when no parts fail by falling short in beauty? . . . For this reason Plato, wishing to indicate this by reference to something which is clearer relatively to ourselves, represents the Craftsman *[dêmiourgos]* approving his completed work, wishing to show by this how delightful is the beauty of the model, which is the Idea.[24] For whenever someone admires a thing modeled on something else, he directs his admiration to that on which the thing is modeled. But if he does not know what is happening to him, that is no wonder: since lovers also, and in general all the admirers of beauty here below, do not know that this is because of the intelligible beauty: for it is because of the intelligible beauty. Plato deliberately makes it clear that he refers the "was delighted" to the model by the words which follow: for he says, "he was delighted, and wanted to make it still more like its model," showing what the beauty of the model is like by saying that what originates from it is itself, too, beautiful because it is an image of the intelligible beauty: for, if that was not transcendently beautiful with an overwhelming beauty, what would be more beautiful than this visible universe? Those who blame it, then, do not do so rightly, except perhaps in so far as it is not the intelligible world.

II

In the Footsteps of Saturn: Poets of Pasture, Farmland, and Forest

Nature, for the ancient Greeks and Romans, whether designating primarily the vital principle of growth or the totality of its products, the cosmos itself, could never be thought of as separate from the human nor reduced to mere landscape or scenery. Life in accord with nature, in this larger sense, was the goal, as we have seen, of the Epicurean and Stoic philosophers and philosophical poets of the Hellenistic and Greco-Roman age. Understanding and acceptance of nature were the means to human freedom: a freedom accorded to Epicureans by random deviations that prevented the universe or man from being wholly predictable, and to Stoics by the conscious collaboration with fate that we still call "stoical." For the Neoplatonist, on the other hand, human freedom came from following to its source in the intelligible world the contemplative activity imperfectly adumbrated in nature. Yet in all three schools acceptance of nature was incomplete and entailed deep contradictions. To the Neoplatonist, of course, the contradiction was overt, since the beauty of the natural world that could *lead, through contemplation of it, to the higher realities of Soul, Intellect, and the One might more often be an ensnaring delusion that impeded that ascent and resulted in not freedom but enslavement. The largely unacknowledged dissidence in the*

[24]The reference is to Plato *Timaeus* 3C77–D1. But there is nothing in Plato to suggest the interpretation given by Plotinus here, which is wholly based on his own doctrine that all perfect activity is contemplation, and that creation or action should be the spontaneous reflex of contemplation . . . (Armstrong)

Epicurean and Stoic views was more insidious. Although neither conceded any alternative (save illusion or error) to life in accord with nature, which was the wise man's goal and his promise of liberation, both were forever striving, in the very passion with which they pursued imperturbability or apathy, against the boundaries whose necessity they proclaimed. In all three philosophies, indeed, the longing for oneness—with nature or what lay beyond—was an index of the disquieting deprivation with which men felt their separation from it.

To this longing, and deprivation, poets of the age (including of course Lucretius and Manilius in the previous chapter) gave intense and varied expression. Because they are poets, the nature they evoke is, not surprisingly, more concrete than that of the philosophers, whether embodied in pastures, forests, or farms, goats or birds, cicadas or flowers, or in vividly evoked mythical incarnations of immemorial natural powers. These incarnations include Pan and Bacchus, chaste Diana and the fecund Venus who quickens living things in the ferment of spring—and primeval Saturn, who ruled a happier world before disunity between human beings and the rest of nature was known. The separation is already apparent in the characteristic new genre of Hellenistic poetry, the bucolic (from Greek boukolos, *"cowherd") or pastoral (from Latin* pastor, *"shepherd") idyll of Theocritus. By celebrating the herdsman's simple life, the genre might appear to deny the separation that is nevertheless implicit, both in its extreme remoteness from the urban cosmopolis in reaction to which it arose, and in the artificiality of language and treatment that so greatly distance it from the (highly domesticated) nature that is its ostensible setting. "The artificiality of the pastoral framework lies partly in the one-sidedness, the peacefulness and tameness of the natural world that it pictures. It is a world constructed for men of the city," Charles Segal writes in "Nature and the World of Man in Greek Literature" (1963), "a world wherein man leaves the limits of his urban environment, not to be exposed to the elemental and uncontrollable, but to receive a measured dose of pleasurable rusticity made agreeable by the grace and refinement of the poetic form. Ultimately then he is merely receiving another product of his own urbane civilization in altered dress." The nostalgic pastoral longing for an inaccessibly "natural" life will remain a strong undercurrent throughout much Hellenistic and Roman poetry of different genres and moods, finding direct expression again in the prose romances of late antiquity.*

In some poems, especially earlier in this period, any disjunction there may be between human and extra-human worlds remains implicit. Meleager, writing around 100 B.C., can celebrate the vibrant beauty of spring with a freshness reminiscent of the earliest Greek lyrics (and without Ibycus's anguished contrast between tranquil nature and furious desire), yet in another lyric can represent Pan as deserting the mountains for the city. Catullus's joyous celebration, in the mid first century B.C., of the laughing waves of his native lake or the solemnity of Diana, queen of mountains and forests, contrasts, in his work as a whole, with his tormented love-hate for Lesbia. In the great poets of the Augustan period, at the end of this century and the beginning of the next, however, the disjunction becomes a recurrent and central theme. In Horace the cyclical return of grasses and leaves in the spring of each year accentuates the brevity of human life as "dust and a shadow"; and yearning for a simpler country existence continually jostles against the turmoil of the city. Virgil's loving evocation of the farmer's beatitude ("Oh happy farmers, did they only know / their blessings!") is tinged with elegiac sadness for a way of life endangered if not extinct: the Golden Age of Saturn, which he sometimes associates—in contrast to Hesiod long before and Ovid shortly after him—not with the pre-agrarian but with the agricultural life. In the anonymous late Latin Vigil of Venus, *a lushly erotic rhapsody of springtime and love is shadowed by the poet's doubt that spring will come to him, and by apparent regret for a pagan world now passing or past. To these late centuries belong also Claudian's variation on Horace's and Virgil's theme of the ennobling country life, and three anonymous works of latter-day Greek celebration of nature: the pastoral romance of* Daphnis and Chloe; *a whimsical Anacreontic poem in praise of the Stoic cicada; and an inscription from the temple of Asclepius at Epidaurus celebrating, well into Christian times, that youngest and oldest of nature gods, goat-footed Pan.*

Cowherding Muses

The *Idylls* of Theocritus, translated by Robert M. Torrance

Theócritus (Theokritos) lived from around 300 to around 260 B.C. Born in Syracuse, on Sicily, he spent part of his life on the Aegean island of Cos before settling in Alexandria under Ptolemy II; here he became, with the erudite Callímachus and his rival Apollonius of Rhodes (author of the epic Argonautica), *a leader of the new Hellenistic poetry. He was the inventor of the bucolic idyll, in which herdsmen of Sicily or Cos, far from the refined corruptions of the city, talk and sing of their loves and pastoral pursuits. The new genre was immensely influential in later antiquity, when Moschus and Bion in Greek, and Virgil, Calpurnius Siculus, and Nemesianus in Latin composed increasingly artificial poems that would find countless imitators in Renaissance Europe and after. Yet although the bucolic idylls of Theocritus—not all his idylls are bucolic—are set in natural surroundings amid flocks and herds, and are spoken by simple herdsmen in relatively plain if euphonious language in the broad Doric dialect of Sicily, and although they abound with rustic local color and with rural deities, notably Pan, it is remarkable how much the pastoral, as Thomas Rosenmeyer observes in* The Green Cabinet: Theocritus and the European Pastoral Lyric *(1969), "eschews descriptions of nature. . . . The Greek pastoral does not attempt to solemnize man's power over nature, or his response to nature's guidance, nor does it, overtly, assert any bond with nature. . . . The ease of the rural scene, with its tree, its greensward, and its brook, is a dramatic convenience," whose main objective is the Epicurean condition of tranquility or liberation.*

Only in the first of our selections, the closing verses of Idyll VII—in which the narrator Simíchidas (thought to be a stand-in for Theocritus himself) rests in the shade with his companions after singing of love on the day of a harvest festival to Demeter on Cos—does the poet indulge in extended description of nature, expanding on Hesiod's famous vision of leisure in Works and Days *588–96 (Chapter 8 above). The second selection, Idyll IX, is the third of three singing matches (after Idylls VI and VIII). It is a very slight poem whose authenticity editors have questioned—Gow rejects it as "hackwork" for its "inherent badness"—yet it not only exhibits the Theocritean conjunction of elegant symmetry and artfully naïve expression but also takes for its subject, with a lightness of touch that may have eluded sober editors, the rival claims of two kinds of rustic life, the cowherd's as sung by Daphnis and the shepherd's or goatherd's as praised by Menalcas. Idyll XI, on the other hand, is one of Theocritus's most accomplished: here the Cyclops Polyphêmos, the brutal savage of Homer's* Odyssey, *is portrayed, in a delicious combination of sentiment and humor, as a lovesick swain, the grotesquely inept yet pathetically moving adolescent suitor of the disdainful sea nymph Galatéa, whom he vainly woos by inventorying the attractions of the cowherd's simple life.*

The translations are based on the text, translation, and commentary of A. S. F. Gow (2 vols., 2nd ed., 1952).

Opulent Summer's Harvest
Idyll VII

But Eúcritus and I, with fair Amyntas,
strolling toward Phrasidámus' farm, lay down
awhile on beds piled deep with fragrant rushes,
joyfully reveling in fresh-cut vine-leaves.
Many a poplar waved above our heads, 135
and many an elm; nearby, the sacred water
gurgled down, murmuring, from the Nymphs' own cave.
On shadowy boughs above, sunburnt cicadas
kept up their chattering toil, while far away,
hid in thick bramble thorns, a tree-frog muttered. 140
Larks and goldfinches sang, the ring-dove cooed,
tawny honeybees flitted round the fountains.
Everything smelled of opulent summer's harvest.
Pears at our feet and apples by our sides
rolled in profusion, as the pliant saplings, 145
laden with plums, bent downward toward the ground.
Wine-jars we stripped of stoppers four years old.
Castalian Nymphs inhabiting steep Parnassus,[1]
did old man Chiron pour a bowl so fine
for Heracles in Pholus' rocky cavern? 150
Did that strong shepherd dwelling near Anápus,
Polyphémus, who pummeled ships with mountains,
dance round his sheepfold drunk with sweeter nectar,
Nymphs, than you then decanted near the altar
of Threshing-Floor Demeter?
Once more let me 155
plant my fan on her heaped-up grain, while she
smiles, holding sheaves and poppies in her hands.

Cowherd Versus Shepherd: A Singing Match
Idyll IX

Sing me a cowherd's song: begin now, Daphnis,
Daphnis, begin your song; Menalcas, follow!

[1]Castalia was a fountain on Mount Parnassus (overlooking Delphi), sacred to the Muses and Apollo. Pholus was a Centaur who entertained Heracles in his expedition against the Erymanthian boar, but refused him wine; in the version to which Theocritus alludes, the Centaur Chiron (Kheirôn) apparently poured the wine for him (although in other versions Heracles takes it for himself). Anápus is a river of Sicily near Syracuse.

Place calves beneath cows, bulls among the sterile
heifers, and let them graze on leaves together,
not scorn the herd. Sing me a cowherd's song 5
from your side; then from yours, Menalcas, answer.

Daphnis

"Sweet is the sound of calf or heifer lowing,
sweet is the cowherd's piping, sweet my singing!
By the cool stream my bed lies; heaped upon it
are skins of calves hurled by a stiff southwester 10
down from a cliff while browsing there on berries.
Hence I pay no more heed to scorching summer
than lovers heed their parents' words of warning."

Thus Daphnis sang to me; and thus Menalcas:

Menalcas

"Aetna, my mother, I too dwell in lovely 15
caves in the hollow rock; all I imagine
in dreams, I have: ewes and she-goats aplenty,
whose fleeces near my feet and head lie scattered.
Sausage cooks on oak fires, and roasted chestnuts
in winter—winter, which I scorn as greatly 20
as toothless men do nuts, when cakes are handy."

Clapping my hands, I gave them both a gift:
Daphnis, a staff grown in my father's fields
(and grown so straight no carpenter could fault it);
the other one, a spiral shell I spotted 25
atop Icarian rocks, and ate its flesh
split five ways: how he trumpeted upon it!

Cowherding Muses, greetings to you! Tell me
the song I sang those herdsmen in their presence.
Never shall blisters grow upon my tongue![2] 30

"Cicada loves cicada, ant loves ant,
and hawk loves hawk: I love the Muse and singing!
Let my house overflow with song, for neither
sleep, sudden spring, nor flowers to honeybees
are sweeter: so I love the Muses! Those 35
they favor, Circe's philter never injured!"[3]

The Song of Polyphemus
Idyll XI

"White Galatéa, whiter than curdled milk,
gentler than lamb and friskier than heifer, 20

[2]Probably a reference to the superstition that blisters would grow on a liar's tongue, though others cite other folk beliefs.
[3]Circe's (Kirkê's) philter, as related in Book X of Homer's *Odyssey*, turned men to swine.

smoother than unripe grapes—why scorn your lover?
Hither you come when sweet sleep overcomes me,
hence you depart when sweet sleep lets me go,
swift as a ewe that flees a grizzled wolf!
I fell in love with you, sweet maid, when first 25
you came to gather hyacinths with my mother[4]
on the hillside; I showed you both the way.
To stop, from that time forward, once I saw you,
I could not—yet you reckon this as nothing!
I know, delightful maid, why you are fleeing: 30
because one shaggy eyebrow spans my forehead
from ear to ear in one unbroken line
over one eye above my flaring nostrils.
Such I am—yet I herd a thousand cattle,
savoring milk deliciously drawn from them. 35
Cheese I have both in summertime and autumn,
and winter too, in baskets brimming over.
No Cyclops pipes more skilfully than I,
singing of you, sweet apple, and myself
often, deep in the night. Eleven collared 40
fawns, and four bear cubs, I am tending for you.
Come to me, then, and you shall lack for nothing:
leave the grey sea to roll against the shore,
and pass your nights more pleasantly in my cave
beside me. Slender cypresses and laurels 45
grow here, dark ivy, and grape-clustered vines;
deep-wooded Aetna sends me freezing waters
tumbling down like ambrosia from white snow.
Who could prefer the salty waves to these?
But if I seem too shaggy for you still, 50
oak logs and unextinguished embers have I;
and I would even let you burn my soul
and my one eye, beyond all things the sweetest.
Oh that I had been born a finny creature,
to dive beneath the waves and kiss your hand 55
(should you forbid your lips), and offer you
white lilies and bright crimson-petaled poppies:
but one sprouts up in summer, one in winter,
and so I cannot bring you both together.
As soon as any stranger's ship comes calling, 60
darling, I'll learn immediately to swim,
and see why dwelling in the deep delights you.
Come, Galatea, come—and then forget
(as I who sit here do) your homeward journey!
Join me in shepherding and milking flocks, 65
and fixing cheese by mixing in sharp rennet.

[4]Polyphêmos's mother was the sea nymph Thoôsa, daughter of the sea deity Phorkys; his father was Poseídôn, god of the sea.

Only my mother wrongs me, and I blame her!
Never did she commend me in your presence,
though day by day I waste away before her.
I'll say my head—and both my feet—are throbbing, 70
and make her suffer, just as I am suffering!
O Cyclops, Cyclops, have you lost your mind?
Go weave your baskets, gather twigs, and bring them
home to your lambs, for this would be far wiser!
Milk what is yours, renouncing her who flees you! 75
Surely you'll find a fairer Galatea!
Many girls ask to play with me at night,
giggling when I accept their invitations:
on land, at least, I seem to be somebody!"

Thus Polyphemus shepherded his love 80
with song, enjoying ease no gold could buy him.

Bard of the Springtime

Three Poems by Meleager, translated by Robert M. Torrance

Meleager (Meleagros), who was active sometime around 100 B.C., was born in Syria, lived much of his life in Tyre, and died in Cos. As a Cynic philosopher he wrote Menippean satires (now lost) mingling verse and prose, and he edited the Garland, *the first major anthology of Greek poetry from the beginnings to his own time. About a hundred of his poems, mostly epigrams on love, are preserved in the* Palatine Anthology *compiled in the tenth century A.D. The following poems (of which the first is in hexameters, the second and third in elegiac couplets) are translated from volumes II (1917) and III (1915) of the Loeb Classical Library edition of* The Greek Anthology, *ed. W. R. Paton.*

How Can a Bard Stay Silent?

Palatine Anthology *IX.363*

Winter storms having vanished from the sky,
flowery springtime's purple season smiles.
Dark earth is garlanded with verdant grass
and plants are burgeoning with leafy tresses.
Drinking the soft dew of the nurturing dawn, 5
meadows laugh as the rosebuds burst in bloom.
Shepherds rejoice at piping on the mountains
while goatherds take delight in dappled kids.
Already sailors navigate wide seas
as harmless zephyrs belly out their sails. 10
Already, garlanding their heads with ivy,

men shout *Euoi!* to clustered Dionysus.
Honeybees born from carcasses of oxen[5]
tend to their lovely works, and congregating
on hives build honeycombs of fresh white wax. 15
Races of birds all sing their high-pitched notes:
swallows on roofs, kingfishers on the waves,
swans on riverbanks, nightingales in the forest.
If foliage rejoices, earth is blooming, 20
sailors set sail and Dionysus dances,
birds make music and honeybees give birth—
how can a bard stay silent in the springtime?

Noisy Cicada

Palatine Anthology *VII.196*

Noisy cicada, drunk with drops of dew,
wastelands explode with melody through you
strumming, on leaf-tips, with your sawtoothed shin
tunes on a cithara of swarthy skin.
Make the tree-nymphs rejoice, friend, if you can, 5
piping new songs responsively to Pan,
so that a noonday nap may set me free
from Love, stretched underneath this shadowy tree.

With Goats No More

Palatine Anthology *VII.535*

With goats no more will I, goat-footed Pan,
inhabit crags unfit for beast or man!
Daphnis is dead, who set my heart afire:
what can I find in mountains to desire?
Give me the city! Hunt the savage boar 5
whoever will, that life is Pan's no more!

What Could Be Happier?

Three Poems of Catullus, translated by Robert M. Torrance

Gaius Valerius Catullus (ca. 84–ca. 54 B.C.) was born in Verona, in the north Italian province of Cisalpine Gaul ("Gaul on this side of the Alps"), but appears to have spent most of his short life in Rome. His book—which survived into the Middle Ages in a single manuscript—of 116 poems in varied meters is the only extant collection written by the young group known, from

[5]Bees were long thought to be spontaneously generated from putrefying flesh. See, e.g., Virgil's *Georgics* IV.554–58.

*Cicero's letters, as "new poets"—*neôteroi, *or* poetae novi. *The poems range in subject and style from obscenely witty abuse and scathing invective to passionate eroticism and bitter self-incrimination, and from the direct simplicity of everyday speech to the erudite ingenuity of the longer mythological poems. Catullus (who won the epithet of* doctus, *"learned") was deeply influenced by the polished formality of the Alexandrians—one of his mythological poems is translated from Callimachus—but drew also on the robust if technically far cruder traditions of Latin poetry of the third to second centuries (and on its great Greek antecedents), both in the high epic-tragic line of Ennius and the low comic line of Plautus and Terence; the latter especially contributed, Kenneth Quinn argues in* The Catullan Revolution *(1959; rev. 1969), to "the racy directness of speech" that Catullus adapted to his formally sophisticated new poetry. But what Catullus most significantly introduced, above all to the poems describing the rhapsody and revulsion of his love for "Lesbia," was a passionate personal involvement with no precedent in Roman, Alexandrian, or indeed in any Greek lyric poetry since Sappho (whom Catullus also translated). Nature was not a central theme of this urbane young poet, yet the following three poems—the first two expressing his ardent joy in returning home in springtime to his native place from foreign travels, the third a solemn hymn to Diana, virgin goddess of mountains and forests—evoke with unforgettable freshness emotions of happy or even worshipful oneness with the natural world that sharply contrast with the tormented divisions of his love poems. The translations are based on the texts and commentaries of Elmer Truesdell Merrill (1893) and C. J. Fordyce (1961).*

Departure in Springtime

46

Now fresh spring is reviving balmy breezes;
now the furor of equinoctial heavens
softens under the soothing breath of Zephyr.
Leave the Phrygian fields behind, Catullus,
leave behind you Nicaea's fertile farmland, 5
flying toward the resplendent towns of Asia.[6]
Now my tremulous spirit yearns to wander,
now my jubilant feet regain new vigor.
Farewell, cherished companions of my travels!
You, who started from distant homes together, 10
far-flung various paths are now returning.

Returning Home

31

Sirmio, jewel of almost-isles and islands[7]—
all that both Neptunes, fresh and salt alike,
raise from the sparkling lakes and open sea—
how willingly and gladly I behold you,
scarcely believing Thynia and the Bithynian 5

[6] Catullus is about to return to Italy from Nicaea in Bithynia—a region of northwestern Asia Minor bordering (Lesser) Phrygia to the south—passing en route through the famous cities of Asia Minor.

[7] Catullus writes of his return from Bithynia (or Thynia) to Sirmio (now Sirmione), a peninsula—Latin *paene insula*, "almost-island"—at the southern end of Benacus (Lago di Garda), near his birthplace of Verona. The lake is called "Lydian" because the Etruscans who once inhabited the region were said to have been natives of Lydia in Asia Minor.

plains are behind me, now I safely see you!
What could be happier than casting off
all tribulations burdening a mind
weary of foreign toils, and coming home
to rest upon the couch so long desired? 10
This is full recompense for many labors.
Fair Sirmio, rejoice to greet your master:
rejoice you too, waves of the Lydian lake,
laughing aloud all laughter lodged within you!

Hymn to Diana

34

Diana's faithful followers
are we, chaste girls and boys together:
come let us, faithful boys and girls,
together sing Diana!

O thou, Latona's magnificent 5
child by almighty Jupiter,[8]
thou whom thy mother bore beneath
the olive tree of Delos

to be the queen of lofty mountains
and forest lands forever green, 10
of secretly sequestered glens
and rivers loudly roaring:

Juno Lucína thou art named
by mothers giving birth in labor;
Trívia thou art named, and Moon 15
reflecting light from others.[9]

Measuring in thy monthly course,
goddess, the circuit of the year,
with fruitful blessings thou dost fill
the farmer's rustic cottage. 20

Hallowed be thou by any name
that pleases thee, and as of old
thou wert accustomed, keep the race
of Romulus in safety![10]

[8]Artemis (here identified with Roman Diana) was in Greek myth the daughter of Zeus (Jupiter) and Lêtô (Latóna), who gave birth to her and her twin brother Apollo on the island of Dêlos.
[9]In addition to her roles as mistress of wild animals and forests, and goddess of fertility, Diana was identified as a goddess of childbirth with Juno Lucína (as Artemis was with the Greek birth goddess Ilithyia); as a goddess of the underworld with Hécatê or "Trivia" (goddess of the "three ways," that is, of the crossroads); and as a moon goddess with Luna (Greek Selênê).
[10]Rómulus was, of course, with his twin brother Remus, the mythical founder of Rome.

The Sum of Today

Six Poems by Horace

Quintus Horatius Flaccus (65–8 B.C.) was born in the southern Italian town of Venusia. His father, a freed slave and auctioneer, promoted his education in Rome and Athens, where Horace joined the army of Brutus and took part in the battle of Philippi in which Brutus and Cassius were defeated in 42 B.C. by Caesar Octavian (the future Augustus) and Mark Antony. Back in Rome, Virgil introduced Horace in about 39 to Octavian's powerful minister Maecenas, the great patron of artists and poets, from whom he later received the Sabine farm near Tibur (modern Tivoli) celebrated in many poems. Two books of satires— saturae *or "medleys," which Horace also called* sermones, *or "chats," because of their colloquial tone and rambling form—and a book of sometimes caustic epodes were published between about 35 and 30, followed by three books of odes in about 23, and later by two books of epistles, the "Art of Poetry," and a fourth book of odes.*

Whether in the low style inspired by the "pedestrian Muse" of his satires or in the metrically varied odes Horace was master not only of style and technique but also of tonal nuances ranging from self-deprecating humor to poignant celebration of life's fleeting joys. Love of rural scenes and the simple life they promote in contrast to the turmoil and falsity of Rome is a central theme to which Horace gave classic expression that would profoundly influence Roman poets and their successors for centuries to come. His younger contemporary Albius Tibullus, to whom Horace probably addressed an epistle (I.iv), begins his first elegy with a very Horatian evocation of rustic seclusion—

> Let other men amass what yellow gold
> acres of cultivated land can hold,
> trembling with terror when the foe appears
> and Martian trumpets rouse their sleeping ears:
> from me may poverty, and quiet life
> lit by perpetual hearthfires, banish strife!
> Let me plant timely vines throughout my land,
> and sturdy fruit-trees with a farmer's hand,
> as Hope unfailingly supplies rich crops
> and wine cascading down in endless drops—

and nostalgic eulogies of ancient country virtues surface even in the scathing satirist Juvenal in the late first century A.D., whose Tiburtine farm in Satire XI echoes Horace's a century before. But Horace's encomium is always infused with multiple perspectives through ironical awareness of contradiction: contradiction between the fervid praise of humble poverty in Epode 2 (here translated by the great seventeenth-century poet John Dryden) and the revelation of its speaker's hypocrisy; between the unreconciled pulls of city and country in Satire II.vi, where the man, unlike the mouse, cannot wholly abandon the city; and, most fundamentally, between eternal return of the seasons and the ineluctable brevity of human life. These contradictions, far from undermining the joys of Bandusian spring, village festival, or the stately seasons, make them intensely real. Both country and city, nature and art, have their claims—yet these can never be equal (this country-loving urban poet writes in Epistle I.x), since in the midst of the city "you may drive out Nature with a pitchfork, yet she will always hurry back":

Naturam expelles furca, tamen usque recurret.

My translations are based on Loeb Classical Library editions of the Satires, *ed. H. R. Fairclough (1929), and (approximating the original meters) of the* Odes, *ed. C. E. Bennett (reprinted 1952).*

Beatus Ille,
translated by John Dryden
Epode 2

"How happy in his low degree,
How rich in humble poverty, is he,
Who leads a quiet country life;
Discharg'd of business, void of strife,
And from the griping scrivener free! 5
(Thus, ere the seeds of vice were sown,
  Liv'd men in better ages born,
Who plow'd with oxen of their own
  Their small paternal field of corn.)
Nor trumpets summon him to war, 10
  Nor drums disturb his morning sleep,
Nor knows he merchants' gainful care,
  Nor fears the dangers of the deep.
The clamors of contentious law,
  And court and state, he wisely shuns, 15
Nor brib'd with hopes, nor dar'd with awe,
  To servile salutations runs;
But either to the clasping vine
  Does the supporting poplar wed,
Or with his pruning-hook disjoin 20
  Unbearing branches from their head,
  And grafts more happy in their stead;
Or, climbing to a hilly steep,
  He views his herds in vales afar,
Or shears his overburden'd sheep, 25
  Or mead for cooling drink prepares,
  Of virgin honey in the jars.
Or, in the now declining year,
  When bounteous autumn rears his head,
He joys to pull the ripen'd pear, 30
  And clust'ring grapes with purple spread.
The fairest of his fruit he serves,
  Priapus,[11] thy rewards:
Sylvanus too his part deserves,
  Whose care the fences guards. 35
Sometimes beneath an ancient oak

[11]A phallic god of fertility, sometimes called the son of Dionysus (Bacchus) and Aphrodite (Venus), and considered the protector of gardens, in which his grotesque statue was placed. Silvánus (from *silva*, "forest") was a Latin god of fields and forests sometimes identified with Faunus or the Greek Pan.

Or on the matted grass he lies:
No god of sleep he need invoke;
The stream, that o'er the pebbles flies,
With gentle slumber crowns his eyes. 40
The wind, that whistles thro' the sprays,
Maintains the consort of the song;
And hidden birds, with native lays,
The golden sleep prolong.
But when the blast of winter blows, 45
And hoary frost inverts the year,
Into the naked woods he goes,
And seeks the tusky boar to rear,
With well-mouth'd hounds and pointed spear;
Or spreads his subtile nets from sight, 50
With twinkling glasses, to betray
The larks that in the meshes light,
Or makes the fearful hare his prey.
Amidst his harmless easy joys
No anxious care invades his health, 55
Nor love his peace of mind destroys,
Nor wicked avarice of wealth.
But if a chaste and pleasing wife,
To ease the business of his life,
Divides with him his household care, 60
Such as the Sabine matrons were,
Such as the swift Apulian's bride,
Sunburnt and swarthy tho' she be,
Will fire for winter nights provide,
And without noise will oversee 65
His children and his family;
And order all things till he come,
Sweaty and overlabor'd, home;
If she in pens his flocks will fold,
And then produce her dairy store, 70
With wine to drive away the cold,
And unbought dainties of the poor;
Not oysters of the Lucrine lake
My sober appetite would wish,
Nor turbet, or the foreign fish 75
That rolling tempests overtake,
And hither waft the costly dish.
Not heathpout, or the rarer bird
Which Phasis or Ionia yields,
More pleasing morsels would afford 80
Than the fat olives of my fields;
Than shards or mallows for the pot,
That keep the loosen'd body sound,
Or than the lamb, that falls by lot

To the just guardian of my ground. 85
Amidst these feasts of happy swains,
The jolly shepherd smiles to see
His flock returning from the plains;
The farmer is as pleas'd as he
To view his oxen, sweating smoke, 90
Bear on their necks the loosen'd yoke:
To look upon his menial crew,
That sit around his cheerful hearth,
And bodies spent in toil renew
With wholesome food and country mirth." 95
This Alfius said within himself,[12]
Resolv'd to leave the wicked town,
And live retir'd upon his own.
He call'd his money in;
But the prevailing love of pelf 100
Quickly reviv'd his former self,
And put it out again.

Of Mice and Men,
translated by Robert M. Torrance
Satire II.vi

For this I prayed:
a modest plot of land,
a house with flowing waters close at hand,
a garden near a wood.
On every score
the gods surpassed my wish.
I ask no more,
Maia's son,[13] than deserving such rich treasures. 5
If I have never by dishonest measures
swollen my wealth, and shall not by neglect
foolishly waste it, praying . . .
"Just connect
that little corner, rounding out my farm!
A buried treasure trove would do no harm, 10
surely: a plowman found enough to buy
the field he plowed before, so why not I,
by Hercules?"
. . . if I am satisfied:
fatten my flocks, I pray—and all, beside
my mind, befriending me, as in the past! 15

Here in my hillside citadel at last,
far from the town, what could my lowly Muse

[12]I have substituted Horace's original "Alfius" for Dryden's "Morecraft" as the moneylender's name and, for the sake of clarity, rewritten line 101 below from Dryden's "Soon split him on the former shelf."
[13]Mercury (Greek Hermes), son of Jupiter (Zeus) and Maia, was considered the god of luck.

more fitly praise? Here no ambition skews
my peace, no leaden autumn blowing ill.

Father of Dawn—or Janus, if you will— 20
to whom, god granting, origins belong
in human life, initiate my song!

At Rome you rushed me off, before I knew it—
"Hurry, or someone else will beat you to it!"—
to act as guarantor: though tempests blow 25
and snowstorms swirl around me, I must go!
My case once made, I navigate the town,
jostling the crowd and trampling laggards down.
"Imbecile! Watch yourself!" some angry lout
screams: "Is there nothing you can think about 30
but barging toward Maecenas's retreat?"
That, I confess, is truly honey-sweet.
But on the gloomy Esquiline,[14] I find
countless concerns of others vex my mind:
"Roscius needs to meet you near the forum 35
early tomorrow!"
"Clerks require a quorum,
Quintus, so don't forget: come back today!"
"Get me Maecenas' signature, I pray!"
"I'll try," you say: and he insists, "You can!"

Eight years, almost, have passed since I began 40
first to be counted as Maecenas' friend—
one at least sometimes chosen to attend
his party on a journey, and confide
gems of this kind:
"What time is it?"
"Which side
will win the match?"
"Beware of frost this year": 45
chitchat distilled into a leaky ear.
Each hour and each day that passes by,
envy surrounds me: if beside him, I
attend the games, "O Fortune's friend!" they cry.
Suppose some chilling rumor hits the streets: 50
everyone who encounters me entreats,
"Friend, do your contacts on Olympus give you
news of the Dacians?"
"None."
"May god forgive you,
always the jester!"
"None at all, I swear!"

[14]The Esquiline Hill, outside the Servian wall of Rome, was the site of a cemetery mainly devoted to criminals and paupers.

"Come now: in Sicily . . . Italy . . . or *where* 55
will Caesar give his troops their promised prize?"
I swear I'm ignorant, but in their eyes
never was man so furtive in his ways.

Amid the misery of wasted days,
I pray:
My country home, when shall I look 60
upon you, and relax with some old book
in idleness, oblivious of troubles?
When shall I feast on beans (Pythagoras' doubles)[15]
and vegetables well greased with bacon fat,
banquets befitting gods—at which I chat 65
with friends at home, bequeathing saucy slaves
the scraps? Whatever any drinker craves,
he has: be it a flagon, or a cup,
paying no heed to rules, he drinks it up,
guzzled with zest or sipped in moderation. 70
The topic of our dinner conversation
is not what riches buy (a house? a poet?)
but what concerns us so, that *not* to know it
would harm us: whether wealth, or virtue, makes
men happy; love, or interest, most partakes 75
in friendship: and what constitutes the good.
Now and then Cervius, our neighbor—should
a guest, in praising wealth, ignore its woes—
relates an old wives' tale.
The story goes:
"Once on a time, into his simple house 80
a country mouse received a city mouse,
his friend: though frugal of his meager stores,
he opened wide his soul's unstinting doors
and shared his shriveled peas and oaten grits,
hoping, with raisins and half-nibbled bits 85
of lard, his feast would overcome (forsooth!)
abstention by his friend's disdainful tooth,
while he, the scrimping host, outstretched on straw,
left the best morsels for his guest to gnaw.
Finally, City Mouse said: 'Friend, what pride 90
can living on this bushy ridge provide?
Forests lack all redeeming city features!
Let's hit the road together! Earthly creatures
all equally are mortal, great or small,
and death eventually will come to all: 95
therefore, remembering your life's brief span,
wee mouse, enjoy its pleasures while you can!'

[15]Pythagoras was said to have prohibited the eating of both meat (believing the souls of men could be reincarnated in animals) and beans, which Horace here humorously calls his relatives ("doubles").

Struck by his words, the country mouse departed,
led by his friend, then trudged the road they'd started
together, creeping toward the town at night. 100

Shades were already veiling heaven's light
when, in a splendid mansion, both set paw
stealthily. Strewn on ivory beds they saw
gleaming red coverlets; not far away
baskets contained rich scraps from yesterday. 105
Plumping the country mouse on purple covers
outstretched, his host industriously hovers
around, serves endless dishes, never tarries,
and, like a slave, tastes everything he carries.
Country Mouse, reveling in untold favors 110
of fortune, plays the pampered guest and savors
unending bliss, when **CRASH!** both bolt from bed
as doors bang open, scamper off half dead
with terror, as the elevated house
echoes with barking dogs.
Said Country Mouse, 115
"No life for me: I'm off! My wooded cave
shelters and gives me everything I crave!"

To Sestius: On the Coming of Spring,
translated by Robert M. Torrance
Ode I.iv

Winter's severity soon dissolves with the welcome change to springtime,
and ships long high and dry are hoisted seaward;
cattle loiter no longer in stalls, nor the plowman by the fireside,
and frosty white no longer blankets meadows.

Venus the Cytherean[16] already is leading forth her dancers 5
beneath the dangling moon, and lovely Graces
arm in arm with the Nymphs tread earth with alternating footsteps
while Vulcan stokes the Cyclops' mighty furnace.

Now is the time to circle glistening locks with verdant myrtle
or flowers born of earth's emancipation; 10
now in shadowy groves it is time to sacrifice to Faunus
a lamb or kid, according to his wishes.

Pallid Death comes pounding alike on cottages of paupers
and castles: O blest Sestius, the trifling
span of our mortal life prohibits inaugurating long hopes. 15
Soon Night and phantom spirits shall oppress you

[16]Aphrodite (Roman Venus) was called the Cytheréan after one of the places where she was said to have been born from the sea, the island of Cýthera, off the Peloponnesus in Greece, just as she was called the Cyprian after her other birthplace, Cyprus.

down in Pluto's narrow abode: once there, you shall not ever
again by rolling dice command the banquet,
never marvel at Lycidas' beauty, which inflames young men now
with passion soon to warm the hearts of virgins. 20

To the Bandusian Spring,
translated by Robert M. Torrance
Ode III.xiii

O Bandusian spring, splendidly crystalline,
fitly honored with wine not without flowers too,
I shall bring you a firstling
kid tomorrow, whose budding brow

hints incipient horns destined to love and war— 5
all for nothing! This young scion of sportive flocks
soon will color your frigid
waters crimson with gushing blood.

You the savagely fierce heat of the blazing Dog
Star can never affect:[17] cooling benevolence 10
bullocks weary of plowing
find, and wandering flocks, in you.

Soon you too shall become famous among all springs
through me, singing the oak planted above the stone
grotto whence your loquacious 15
waters bubblingly tumble down.

To Faunus,
translated by Robert M. Torrance
Ode III.xviii

Faunus, avid lover of Nymphs who flee you,
through my territories and sunny pastures
gently come and, when you depart, on nurslings
shower your blessing

if before year's ending a tender kid falls 5
sacrifice; bowls (intimate friends of Venus)
overflow with wine; and familial altars
billow with incense.

Sportive flocks, when nones of December greet you,[18]
skip cavorting over the grassy meadows; 10
village folk, along with their shambling oxen,
festively saunter;

[17]On the Dog Star (Latin *canicula*), or Sirius, see note on Hesiod, *Works and Days* 417 (Chapter 8 above).
[18]The nones (nine days before the ides in the Roman calendar) of December, or December 5, was apparently the date of a rural festival of Faunus.

lambs are unafraid of the wolf among them;
forests shed rich foliage in your honor;
farmers gaily trample with triple tread their 15
miserly farmland.

To Torquatus: On the Return of Spring,
translated by Robert M. Torrance
Ode IV.vii

Snows have vanished and grass is already returning to meadows,
leaves returning to trees;
earth is transforming herself and rivulets, no longer swollen,
softly traversing their banks.

Now with her Nymphs and twin sisters the Grace audaciously ventures 5
naked to enter the choir:
"hope for nothing immortal," the years and the hours that ravish
radiant daylight proclaim.

Zephyrs soften the cold, then springtime is trampled by summer,
destined to perish in turn 10
once the fruit-bearing autumn has showered its harvest, and winter
lifelessly circles around.

Yet the mutable moons soon repair their celestial losses:
we, having made our descent
down to where pious Aenéas, rich Tullus, and Ancus are waiting, 15
dust and a shadow shall be.

Who can say if the heavenly gods will be adding tomorrow's
time to the sum of today?
Only gifts bestowed on your heart shall escape the awaiting
hands of your covetous heir. 20

Once you perish, Torquátus, and Minos pronounces upon you
judgment beyond all appeal,
neither family, eloquent speech, nor righteousness ever
more will restore you again:

even prudish Hippólytus chaste Diana will never 25
free from the shadows below,
nor can Theseus pry Piríthoüs loose from Lethaean
fetters enchaining his friend.[19]

[19]Hippólytus was the son of Thêseus and the Amazon Hippólyta. In the myth recounted in Euripides' *Hippolytus*, Aphrodite seeks vengeance on him for his devotion to the virgin goddess Artemis (Roman Diana). After Hippolytus's stepmother, Phaedra, whose advances he has rejected, kills herself and unjustly accuses him of incestuous desires in a letter she has left behind, Theseus calls on his own father, Poseidon, for revenge, and Hippolytus is dragged to death by his horses. Piríthoüs the Lapith took part with Theseus in battles against the Centaurs and Amazons; together they descended to the underworld (perhaps to bring back Persephone as Pirithous's wife), but though Theseus, in some versions, was rescued by Heracles, Pirithous remained attached to the stone seat on which he had sat and from which he could never arise. "Lethaean" refers to Lêthê, the underworld river of forgetfulness.

What Makes the Fields Rejoice

Selections from Virgil, translated by Robert M. Torrance

Publius Vergilius Maro (70–19 B.C.) was born in the country district of Andes near Mantua, not far from Catullus's Verona, in what was then (until incorporated into Italy in 42 B.C.) the province of Cisalpine Gaul. His father appears to have been a landowning farmer. Virgil was educated at Cremona and Mediolanum (Milan), then probably studied rhetoric in Rome and Epicurean philosophy in Naples. Deeply influenced in his youth by the neôteroi *or "new poets" of Catullus's circle, and by their Alexandrian precursors, he took Theocritus as the chief model for his first important work, the* Bucolics *or, as they are generally known,* Eclogues *(from Greek* ekloga, *"selection"), ten pastoral idylls, probably written between about 45 and 37. These hauntingly musical poems, with their wistful longing for escape from the troubled present to a dreamlike world of pastoral indolence, or* dolce far niente, *are set in the landscape of an imaginary "Arcadia," the legendary home of Pan. This place is characterized, as Bruno Snell writes in* The Discovery of the Mind *(1948; Eng. trans. 1953), by "tenderness and warmth and delicacy of feeling . . . which suffuses everything with its glow," yet in which contemporary events—including the assassination of Julius Caesar and the expropriation by Octavian of Virgil's family farm for distribution to war veterans in the aftermath of the civil wars with Brutus and Cassius—continually intrude. In this highly civilized poetic refuge from harsh reality, nature is far less prominent than art.*

During or shortly after the composition of the Eclogues, *Virgil received the patronage of Maecenas, the confidant of Octavian, who in 27 B.C. (four years after his defeat of Antony and Cleopatra at Actium) was given the title of Augustus. The* Georgics, *an agricultural poem in four books, was finished by 29, when Virgil and Maecenas (according to ancient* Lives *of Virgil) read it aloud to the future Augustus for four consecutive days; the* Aeneid, *an epic in twelve books written, at the emperor's behest, on the wanderings of Aeneas, founder of the Roman race and of the Julian line of the Caesars, was written in the last decade of his life. Virgil died in the port of Brundisium (modern Brindisi) after a voyage to Greece, leaving the* Aeneid *unfinished, and was buried at Naples; his final days are the subject of Hermann Broch's novel* The Death of Virgil *(1945).*

John Dryden, who translated both the Georgics *and the* Aeneid *into rhymed couplets in the late seventeenth century, considered the* Georgics *"the best Poem of the best Poet," and many since have thought it his most nearly perfect work. It is ostensibly a didactic poem on farming that took its subject from Hesiod's* Works and Days *and its title from the lost* Georgica *of the second-century Hellenistic poet Nicander; it drew for its agricultural and meteorological lore on numerous other writers, including Aristotle, Theophrastus, Aratus (whose immensely popular astronomical and meteorological poem the* Phaenomena *Cicero had translated, and Germanicus Caesar, adoptive son of Tiberius, would later again translate, into Latin), and the learned Roman Varro, whose* On Agriculture *is excerpted in Chapter 12 below. But insofar as the poem is didactic, its true aim is not to provide a handbook for farmers but, as W. Y. Sellar writes in* The Roman Poets of the Augustan Age: Virgil *(1876; 3rd ed. 1908), "to revive and extend the love of the land, and to restore the fading ideal of a life of virtue and happiness, passed in the labors of a country life." Beyond this, the poem is a philosophical and a deeply religious celebration of the rewards and the joys—in its world the very fields rejoice—of life in accord with nature, experienced not as an abstract force or concept but in its inexhaustible particulars.*

The farmer's life is one of ceaseless struggle with nature, to be sure, for ever since the Golden Age when acorns and berries provided food without labor, Jove has imposed "obdurate toil" on men, through which they have gained a livelihood from the sweat of their brow. Yet Virgil knew that Aratus had included oxen and the plow in his brief portrayal of the Age of Kronos, when Justice had dwelt among men. In his own nostalgic celebration of the farmers' unrecognized happiness in Book II, their way of life becomes the true remnant of the reign of Saturn for whose return he had fervently hoped in the messianic fourth Eclogue. That condition of dignified primitive simplicity was embodied also, in Book VIII of the Aeneid, *by King Evander of Pallantéum, an Arcadian very different from the melodious shepherds of the* Eclogues, *who welcomes Aeneas to his simple home on the very site where centuries later (though neither can know it) the marble palaces of Rome will arise.*

Nor is life according to nature only a moral norm, for the world of the Georgics *is vibrantly suffused with a power reminiscent of the nurturing Venus of Lucretius's poem; far more than Aratus or Varro, or even Hesiod, it is Lucretius whose dynamic presence is felt throughout. "The influence, direct and indirect, exercised by Lucretius on the thought, composition, and even the diction of the* Georgics *was perhaps stronger than that ever exercised, before or since," Sellar declares, "by one great poet on the work of another." Yet in several ways Virgil's treatment of the natural world differs profoundly from that of his predecessor. His is not a universe ruled by the mechanical forces of Epicurean philosophy, but more resembles the cosmos of Plato or the Stoics in being subject to Providence, embodied by Jupiter in the heavens and by Augustus on earth. (Virgil's praise of the emperor who brought peace to Italy after generations of savage civil wars goes far beyond mere flattery and recalls the ancient Mesopotamian, and more recent Hellenistic, associations of the divine king with the cycles of the natural world.) And as he himself affirms in Book II, the majestic cosmic themes of Lucretius are less suited to his temper than humble rural scenes and flowing brooks, woods and rivers—or the everyday country life embodied by the old Corycian he portrays in Book IV who in his mind, despite outward poverty, "matched the wealthiest kings." Joy in oneness with the natural world, and love of all it contains, from the movements of the heavens and the burgeoning of spring to the tilling of the fields and the ceaseless labors of the bees, is the keynote of a poem permeated throughout by rhapsodic celebration of a beauty that is forever passing away even as we behold it:*

Sed fugit interea, fugit inreparabile tempus,
singula dum capti circumvectamur amore. (III.284–85)
("Meanwhile, as time irreparably flees onward,
we linger round each detail, held by love.")

Modern verse translations of the Georgics *include those of Smith Palmer Bovie (1956) and L. P. Wilkinson (1982). For further reading (in addition to Sellar's excellent study), see Brooks Otis,* Virgil: A Study in Civilized Poetry *(1964) and L. P. Wilkinson,* The Georgics of Virgil: A Critical Survey *(1969). The translations below are based on the texts of the* Bucolics and Georgics *(1894) and of the* Aeneid *(vol. 2, 1900), ed. T. E. Page, with consultation of the Loeb Classical Library* Virgil, *ed. H. R. Fairclough (2 vols., rev. ed. 1934–35).*

FROM THE Georgics

I What makes the fields rejoice, beneath what star
earth should be plowed, Maecenas, and the vine
wedded to elms; what tending oxen need,

what care the flocks, what skill the thrifty bees—
here I begin my song. O most resplendent 5
lights of the universe, who guide the year;
Liber and bounteous Ceres, by whose favor
earth changed Chaonian acorns for rich grain,
commingling wine with Achelóüs' waters;[20]
and Fauns, gods never far removed from farmers— 10
step to the fore, Fauns, leading Dryad maidens!—
your gifts I sing: you, too, for whom the trident-
stricken earth first brought forth the neighing steed,
Neptune; and you[21] who guard the groves where thrice-
a-hundred snow-white steers crop Cea's thickets; 15
shepherd-god Pan, deserting native forests
and glades: if Maenalus is precious to you,
lord of Tegéa,[22] favor us; Minerva,
inventress of the olive; youth who gave us
the plow;[23] Silvánus, brandishing the cypress; 20
all gods and goddesses who tend the fields,
you who spontaneously engender fruits
and you who irrigate the crops with rains—
most of all, Caesar, you, whom some assembly
of gods shall shortly welcome: whether guarding 25
cities and lands so caringly that all
the world proclaims you lord of seasons, author
of bounty, crowned with myrtle by your mother:[24]
whether as god of boundless seas the sailors
worship you only, farthest Thule serves you, 30
and billowy Tethys begs you wed her daughters:[25]
or whether, as a newborn star amid
slow months, you fill the expanding space where claws
grasp after Virgo[26] (since the blazing Scorpion

[20]Liber, an ancient Italian fertility god of vine and fields, was identified with Greek Bacchus or Dionysus, as Ceres, goddess of the grain, was with Greek Demeter. Chaonia was a region of Epírus (Epeiros) once inhabited by Pelasgians, the mythical first people of Greece, where the oracle of Dôdôna, with its sacred oak trees, was located. Achelóüs (Akhelôos) was the largest river of Greece.

[21]Aristaeus, son of Apollo and Cyrênê, was worshipped as a god of fields and pastures; in *Georgics* IV, Virgil tells how he recovered bees destroyed by Nymphs avenging Orpheus's wife Eurydicê, who died of a snakebite when fleeing from Aristaeus. In the previous lines Virgil refers to the creation of the horse by Neptune (Poseidon) in competition with Minerva (Athena), after whom the city of Athens was named when her gift, the olive, was judged the more valuable.

[22]Maenalus, a mountain located near the city of Tegéa in Arcadia, was a favorite haunt of the god Pan.

[23]Triptólemus, son of Celeus, king of Eleusis, was a favorite of Ceres (Demeter) and a hero said to have made many discoveries in agriculture, including the plow. Silvánus was a Latin god of fields and forests sometimes identified with Faunus or Pan.

[24]Venus was the mother of Aeneas and thus, through Aeneas's son Iulus, of the Julian *gens* (clan) to which Julius Caesar and his adopted son Octavian (later Caesar Augustus) belonged.

[25]Thule, an island some have identified as Iceland, Greenland, or the Shetlands, was considered the northernmost point of the world and was thus called *ultima Thule*. Tethys, daughter of Ouranus and Gaia (Sky and Earth), was the wife of Océanus and mother of the Océánides, sea nymphs whom she is here pictured as enticing Caesar, with the dowry of her waves, to wed.

[26]Virgil imagines the deified Caesar Augustus as a new star in the constellation Libra (the scales, symbolizing justice) between Virgo and Scorpio.

for you retracts its arms and opens heaven!): 35
whatever you shall be—and may you never
grimly desire to reign in Tártarus,
though Greece may marvel at Elysian fields
and Proserpine ignore her mother's summons—
grant a smooth course to my bold enterprise; 40
pity both me and farmers gone astray,
and from this moment forward, hear our prayer!

When spring is new, and icy streams are flowing
from snow-capped peaks, and clods commence to crumble,
let the bull then begin to groan beneath 45
the yoke, the plowshare sparkle in its furrow.
Only that field can grant the greedy farmer's
prayer that has twice encountered frost and sunlight:
granaries burst with harvests garnered from it!
Yet before plowing unknown plains, be sure 50
to learn the winds and changing ways of heaven,
each region's character and cultivation,
and what each climate yields, and each refuses.
Here grain, there grapes grow more abundantly,
elsewhere young trees and grasses spring to life 55
unbidden. Tmolus sends us fragrant saffron,
India tusks, Sabaeans frankincense,
nude Chálybês iron, Pontus pungent oils,
Epírus triumphs by Olympian mares.
Nature has laid eternal covenants 60
and laws on certain places, since the time
Deucálion struck with stones an empty world:
whence men, a stony race, arose.[27] Come then,
where soil is rich, let sturdy oxen turn it
from the year's opening months, and dusty summer 65
bake with maturing suns the fallen clods;
where it is poor, suffice it, when Arcturus
rises, to lift it up with shallow furrows—
thus weeds may neither choke the joyful grain
nor scant rain drain away from barren sand. 70
By alternation let your fields lie fallow
once mowed, and sluggish plains be scurf-encrusted;
or, when the stars have changed, sow yellow spelt
in plots first purified of beans, that revel
in quivering pods; fruits of the slender vetch; 75

[27]Deucalion, son of Prometheus, and his wife Pyrrha, according to Greek myth, were the only mortals saved by Zeus, because of their piety, when he destroyed all other humans. Surviving the flood in a ship they had built on the advice of Prometheus, they landed on Mount Parnassus and were told by the goddess Themis, daughter of Earth and mother of Prometheus, that they could restore the human race by throwing the bones of their mother behind them. Realizing that what was meant by their mother's bones was the stones of the earth, mother of all, they threw these, and from them sprang up men and women. Ovid tells the story in the first book of the *Metamorphoses.*

and bitter lupine's brittly rattling stems.
Flax or oat crops, or poppies redolent
of Lethe's slumber, soon exhaust the soil,
but changing crops is easy: do not shrink
from fattening dry soil with rich manure 80
or dirtying depleted fields with ashes.
Thus alternation lets the land lie idle,
nor will the unplowed earth neglect to thank you.
Often it pays to kindle barren fields
and burn light stubble up with crackling flames: 85
whether earth thence derives some hidden strength
from fertile nutrients, or fire bakes off
all taint, and harmful moisture sweats away;
or heat enlarges paths, and opens up
blocked pores, releasing sap to tender shoots, 90
or dries it out and narrows gaping veins,
so that no seeping rain, no searing sun
may waste it, nor the North-wind's piercing cold.
Much does he benefit the land who hoes
dull clods, and breaks them up with wicker harrows— 95
on him from high Olympus Ceres smiles—
much, too, who wheels around and cuts again
crosswise through ridges that his plow created,
zealous to tame the land and master fields.
Pray for wet summer days and sunny winters, 100
farmers: the dust of winter gladdens crops
and fields; no better tillage Mysia boasts,
nor Gárgarus, proud of abounding harvests.[28]
What shall I say of him who, flinging seed,
grapples with soil and levels mounds of sand, 105
guides an obedient river toward his crops
and, when scorched topsoil withers dying sprouts,
charms from a slope crisscrossed by flowing channels
waters that gurgle past smooth stones and hoarsely
murmur, cascading down through thirsty fields? 110
Or him who rescues heavy stalks from drooping
by grazing down luxuriating crops
once the shoot tops the furrow? Or of him
who drains away marsh pools with soaking sands,
especially in the changeful months when rivers 115
flow over, coating everything with mud
and filling ditches whence warm vapors rise?
Nor, though the laboring of men and oxen
so often turns the soil, are impudent geese,
Strymonian cranes,[29] or bitter-fibered endive 120

[28]Mysia, a region of western Asia Minor, and Gárgarus, a town and mountain in the nearby Troad, were famed for fertility.
[29]Strymon was a river between Macedonia and Thrace around which cranes were said to congregate in summer.

harmless, or shadows. Father Jove has made
the path of farming hard, first rousing fields
by art, and sharpening men's minds with cares,
lest torpid lethargy weigh down his kingdom.
Before Jove,[30] none subdued the land by farming: 125
even to mark a field, or subdivide it,
was not allowed; all things were common; earth
unstintingly provided what none asked.
From Jupiter black vipers took vile venom,
wolves learned to plunder, seas to surge in fury; 130
he drained the leaves of honey, hid the fire,
and dammed up overflowing streams of wine.
Thought and experience then gradually
hammered out arts, discovered grain in furrows,
and liberated fire from veins of flint. 135
Then hollowed logs first navigated rivers,
and sailors numbered stars and gave them names—
Pleíades, Hýades, Lycáon's shining
she-bear[31]—and men first undertook to hunt
with traps and bird-lime, circling glades with dogs. 140
One man assaults wide streams with casting-nets,
plumbing the depths, another trolls the seas.
Then iron's rigor, then the saw-blade's shrillness
(formerly men divided wood with wedges)
came, and the varied arts. Toil conquered all: 145
obdurate toil, and unrelenting need.

Ceres instructed men to turn the earth
with iron, when acorns and wild berries failed
Dodona's sacred wood, denying food.
Grain was afflicted when destructive blight 150
gnawed at its stems, while sluggish thistles bristled
up from the fields. Crops die as prickly bushes
spring up: burs, thorns, and, in the midst of glistening
grain, fruitless weeds and barren oats take over.
Therefore, unless you hoe assiduously, 155
frighten birds off with shouting, check with sickles
shade darkening the land, and pray for rain,
vainly, alas, will you behold your neighbor's
huge stores, while shaking oaks to slake your hunger. . . .

II

Thus far the stars and tillage of the fields,
now, Bacchus, you I sing—and with you forest
saplings, and slowly growing olive trees.

[30]The age of Saturn (Kronos), before he was overthrown by his son Jove or Jupiter (Zeus), was considered the golden age, especially in Italy where Saturn was said to have taken refuge.
[31]Juno (Hêra), discovering Jupiter's affair with Lycáon's daughter Callísto, changed her to a she-bear, and Jupiter placed her in the sky as the constellation Arctos, or the Bear.

Lenaean father, come![32] Here all proclaims
your bounty, fields are heavy with the vine's 5
autumnal harvesting, vats froth with vintage.
Lenaean father, come! Strip off your sandals:
plunge naked ankles in new must with mine!
Varied are Nature's ways of making trees.
Some, with no human guidance, by themselves 10
spring up, and cover plains and winding rivers
far and wide: supple osier, limber broom,
poplar, and willow, pale with silvery leaves.
Some rise from scattered seed: the towering chestnut,
and oaks, commanders of the forest, spreading 15
toward Jove leaves thought oracular by the Greeks.
Dense undergrowth sprouts up from leaves of others,
cherries and elms; Parnassian laurels rise
from springs beneath their mother's mighty shade.
These methods Nature gave; by these, all species 20
of trees and shrubs and sacred forests burgeon.
Others, experience derived by practice.
One implants twigs torn from their mother's tender
body in furrows, while another buries
stems, cut across or sharpened at the bottom. 25
Some trees await strips sliced in arching segments
from shoots pressed down alive in native soil;
some need no root, and pruners confidently
implant their topmost pinnacles in the earth:
from severed trunks (miraculous to tell!) 30
out of dry wood an olive root may thrust,
and often one tree's branches harmlessly
turn to another's: transformed pears grow grafted
apples, and plum trees blush with cornel berries.
Rise up, then, farmers: learn the cultivation 35
proper to each, and tame wild fruits by tillage;
leave no lands idle! Sow all Ísmarus
with vines, and clothe Taburnus with the olive.[33]
And you, to whom the greatest part of glory
belongs, Maecenas, share with me the toil 40
I've undertaken, sail the open sea!
Not everything my verse aspires to compass—
not though I had a hundred tongues and mouths,
a voice of iron! Come now, skirt the shoreline:
land is nearby! I shall not long detain you 45
with rambling fictive song, or lengthy preludes.
Trees that spontaneously reach shores of light
rise without fruit, exulting in the strength
Nature supplies the soil. Yet even these—

[32]Lenaeus (from Greek *lênos*, "winepress") was a name of Bacchus.
[33]Ísmarus was a mountain of Thrace, famed for wine; Taburnus a mountain of Campania, in Italy, known for olives.

grafted, or planted in a well-worked trench— 50
cast off their savage spirit, and with constant
tillage submit to any arts you teach them. . . .

Listen to no authority who tells you 315
to stir the rigid soil when Boreas blows.
Winter congeals the land with frost, and blocks
seeds from affixing frozen roots in earth.
Vines are best planted when, in blushing spring,
the white stork, enemy of snakes, arrives; 320
or autumn's first chill, when the blazing sun
nears winter, and already summer passes.
Spring benefits the leaves and all the forest
as earth demands revivifying seeds;
in fertile rains almighty father Aether 325
plunges into his wife's glad lap and, mixing
with her strong body, nourishes all offspring.
Then pathless woods resound with singing birds,
and herds at intervals petition Venus.
Bountiful fields give birth as Zephyr's breezes 330
loosen their bosom; moisture overflows,
and grasses safely venture to encounter
new suns; vine-tendrils fear no rising South-winds
or rains hurled down from furious northern skies,
but burst with buds, unfolding countless leaves. 335
Even such days, I might suppose, illumined
the infant world's beginnings, such a course
they followed: *that* was spring; the mighty globe
kept spring, and East-winds tempered wintry blasts,
when the first cattle drank the light, and men's 340
iron race reared its head from rocky fields,
savage beasts filled the woods, and stars the sky.
Nor could things delicate endure this hardship
unless long respites separated cold
from heat, and heaven's kindness spared the earth. . . . 345

Oh happy farmers, did they only know
their blessings! Far from clashing arms, for them
righteous Earth pours forth easy nourishment. 460
No lofty mansion, from proud halls and portals,
vomits vast waves of visitors at dawn;
nor do they gape at tortoise-inlaid doors,
clothes tricked with gold, or Ephyreian bronzes;
their white wool no Assyrian poisons color; 465
no spice corrupts their virgin olive oil.
Carefree and innocent of fraud they live,
rich in uncounted treasures. Quiet farmsteads,
caverns and living lakes, cool vales of Tempe,[34]

[34]The vale of Tempê in Thessaly, through which the river Penéüs (Peneios) reached the sea, was famed for tranquil beauty.

lowing of kine, soft sleep beneath the trees 470
are theirs, with forest glades and haunts of game,
a youth inured to toil and needing little,
sacred rites, honored parents: Justice, fleeing
earth, left her final vestiges among them.[35]
Me, first of all, may Muses sweetly welcome, 475
whose sacraments, pierced by enormous love
I keep, and show me heaven's paths, the stars,
the sun's eclipses and the moon's great labors;
whence the earth trembles, by what force high seas
swell, bursting barriers, then again subside; 480
why suns in winter plunge precipitously
in Ocean, and what slows the lingering nights.
But if the frigid blood around my heart
bars me from reaching those domains of nature,
may rural scenes and flowing brooks delight me, 485
inglorious in my love of woods and rivers.
Oh, for Sperchéüs' plains! Taÿgetus,
where Spartan virgins revel! Set me down,
sheltered by cooling shade, in Haemus' valley![36]
Happy the man who plumbed things' hidden causes, 490
trampling fears of inexorable fate
underfoot, and loud-roaring Ácheron![37]
Fortunate, too, who knows the rustic gods,
Pan, old Silvánus, and the sister Nymphs.
No popular acclaim, no royal purple 495
bends him, no strife inciting rival brothers,
no Dacian from confederated Danube,[38]
matters of Rome, or kingdoms doomed to perish,
pity for poverty, or envied wealth.
Fruits his own branches or consenting fields 500
spontaneously bear, he plucks, ignoring
stern laws, mad forum, archives of the people.
Others with oars vex unknown seas, take up
the sword, or throng the courts and gates of kings;
one devastates a city and its gods 505
to drink from jewels and sleep on Tyrian purple;
another covets hoards of buried gold.

[35]Astraea, daughter (according to some) of Jupiter and Themis, was a goddess of justice who lived among men in the Golden Age; when their wickedness increased, she was the last of the gods to withdraw to heaven, becoming (as Aratus had related in the *Phaenomena*) the constellation Virgo. Virgil here suggests that farmers alone retained traces of her, and thus of the Golden Age itself.

[36]Sperchéüs (Sperkheios) is a river of Thessaly; Taÿgetus is the mountain overlooking Sparta, where girls celebrated Bacchic rites; Haemus was a lofty mountain range (the Balkan Mountains of modern Bulgaria) on the northern border of ancient Thrace.

[37]Ácheron was a river of the underworld where souls of the dead gathered. These lines almost certainly refer to Lucretius.

[38]Dacia (roughly modern Romania), north of the Danube, was a warlike country whose people Augustus feared would ally against him; it remained a source of unease until conquered by Trajan and made a Roman province in A.D. 101–6.

The Rostra dazzles one;[39] redoubled plaudits
of plebs and elders transport yet another
open-mouthed. Reveling in brothers' blood, 510
they change sweet hearth and home for foreign exile,
seeking their country under alien skies.
Clearing the earth with crooked plow, the farmer
meanwhile tends to his annual work, sustaining
country, grandchildren, herds, and worthy oxen. 515
Immediately the year abounds with fruits,
offspring of flocks, and sheaves of Ceres' grain,
burdening rows and bursting barns with produce.
Winter comes; Sicyon's olives glut the presses;
swine return gorged with acorns; woods breed berries; 520
autumn brings forth its varied yield, while mellow
vintage spreads out to dry on sun-drenched rocks.
Meanwhile, dear children hang upon his kisses;
his modest home preserves its virtue; cattle
droop swollen udders; on the joyful meadow 525
fattened kids lock their horns in playful battle.
Stretched on the grass on holidays, while comrades
around the fire wreathe bowls, he pours libations
and calls on you, Lenaeus, placing targets
on elms for games with wingèd javelins 530
by strong-limbed shepherds stripping bare to wrestle.
Such a life ancient Sabines once observed,[40]
such Remus and his brother: thus Etruscans,
surely, grew strong, and Rome incomparable,
circling her seven hills within one wall. 535
Before the reign of Dicte's king,[41] before
an impious people dined on slaughtered oxen,
such a life golden Saturn lived on earth—
long before trumpet blasts had yet been heard, 540
or swords resounding on hard-tempered anvils.
Now, having raced across immense expanses,
let us release our smoking steeds from harness.

IV Truly, were I not furling sail, and urging
landward my prow, my labors nearly ended,
how cultivation ornaments rich gardens
too might I sing—twice-blooming Paestum's rose-beds;

[39]The Rostra ("beaks") was the speaker's platform in the Roman Forum, made from prows of ships captured from the Volscians of Antium, south of Rome, in 338 B.C. The "plaudits" are from supporters of the plebeian and senatorial classes in the theater.
[40]The Sabines, neighbors of the Romans, joined them in a common citizenship, according to ancient tradition, after the "rape of the Sabine women" in the time of Romulus; they were sturdy farmers known for bravery and love of freedom.
[41]Jupiter (Zeus) was brought up on Mount Dicte in Crete; in the golden age of Kronos (Saturn), before his reign, no animals were sacrificed to the gods.

how the endive delights in drinking streams, 120
green banks in parsley; how the cucumber, twining
through grass, expands its paunch—nor leave in silence
late daffodils or curly-stemmed acanthus,
pale ivy, and myrtle hugging shores with love.
For I remember, under Oebalia's towers[42] 125
where black Galaesus waters yellow fields,
seeing an old Corycian tend his few
acres of unclaimed land, too poor for plowing,
unsuitable for flocks, nor fit for Bacchus:
yet, as he planted herbs among the brambles, 130
interspersing white lilies, vervain, poppies,
in mind he matched the wealthiest kings, and coming
home late, spread unbought banquets on his table.
First in the spring to pluck the rose, and apples
in autumn: when bleak winter still was splitting 135
rocks with its cold and curbing streams with ice,
he would be clipping hyacinth-blooms already,
chiding late summertime and tardy Zephyrs.
Thus was he first to cultivate great swarms
of pregnant bees and gather frothing honey 140
from squeezed combs; lindens and sap-rich pines were his,
and all the fruits that clothed his bounteous trees
in early bloom still ripened in the autumn.
Even grown elms he planted out in rows,
hard pear trees, thorns already bearing plums, 145
and plane trees sheltering drinkers in their shade.
Cut off by narrow space, however, I
omit what others after me may mention. . . .

FROM THE *Aeneid*
Book VIII

Having performed the sacred rites, all turned
back to the town. The king, besieged by age,
walked with Aeneas and his son beside him,
easing their way with varied conversation.
Aeneas marvels, turning eager eyes 310
around him, charmed by spots where joyfully
he learns the monuments of men of yore.
Then King Evander, founder of Rome's fortress:
"These woods once native fauns and nymphs possessed,

[42]Here "Oebalia" is apparently Tarentum (modern Taranto), a Greek colony of Calabria in southern Italy founded by Spartans, whose ancestors had been governed by King Oebalus. The Galaesus was a river flowing into the Gulf of Tarentum. The old man of the following lines, called only "Corycius," was probably a native of Corycus in Cilicia (southeastern Asia Minor); the ancient commentator Servius, writing in the fourth century A.D., claims that this old man was one of the Cilicians settled in Calabria by Pompey after his defeat of the Cilician pirates in 67 B.C.

and sturdy men born from hard oaken trunks,[43] 315
uncultured, ignorant of yoking oxen,
acquiring wealth, or husbanding their gains.
Branches, and rugged hunting, gave them food.
Saturn first came from heavenly Olympus,
banished from realms Jove's weapons wrested from him. 320
He culled an independent race dispersed
through lofty mountains, gave them laws, and named them
Latins, since here security was *latent*.
Under his reign occurred the golden age
men tell about: so peacefully he ruled, 325
till, bit by bit, a drabber, lesser time
crept in, and rabid war, and covetous gain.
Then came Ausonian bands, Sicanian tribes,
and frequently Saturnia changed its name;[44]
next kings, and savagely gigantic Thybris 330
came, for whom river Tiber afterward
was named, her true name, Álbula, being lost.
Myself, omnipotent fortune and unbending
fate drove across the furthest seas from home,
and placed me here, warned dreadfully by my mother, 335
the nymph Carmentis,[45] and the god Apollo."
So speaking, he advances, pointing out
the altar and Carmental gate—as Romans
still call them—honoring the nymph Carmentis,
the first who truly prophesied Aeneas' 340
sons would be great, and Pallantéüm splendid;
then shows the vast grove Romulus made a refuge
near a chill cave, the Lúpercal, named after
Lycaean Pan in old Arcadian fashion.[46]
He bids the wood of sacred Argilétum 345
bear witness to his tale of Argus' death;
thence to Tarpeia's home and the Capitol
he leads—gold now, but bristling once with thickets.

[43]See *Odyssey* XIX.162–63, where Penelope says to Odysseus, disguised as a beggar: "Tell me the race from which you come, for you are not from an oak, as in ancient tales, or a rock."

[44]With the arrival of other peoples Saturn's land took new names, such as Ausonia, Hesperia, Oenotria, and Italia.

[45]Carmentis, or Carmenta (from *carmen*, "song" or "spell"), was one of the Caménae (prophetic fountain nymphs sometimes identified with the Greek Muses), with a temple on the Capitoline Hill and altars near the Carmental Gate; she is said to have warned Evander, her son by Mercury (Hermes), to leave Arcadia and transport his city of Pallantéüm to Italy. See Ovid's *Fasti* I.467–542. Apollo was the god of the oracle of Delphi.

[46]The Lupercal was a cave on the Palatine Hill, sacred to Lupercus (from *lupus*, "wolf"), a deity sometimes identified with Faunus or the Arcadian Pan (whose epithet "Lycaeus" comes from Greek *lykos*, "wolf"). From this cave, in which the she-wolf was said to have nursed the infant Romulus and Remus, the ancient fertility festival of the Lupercalia started forth each February 15. See Ovid's *Fasti* II.381–452, and Sir James George Frazer's appendix to the Loeb Classics edition. The Argilétum was a district near the Forum, here said to be named for a certain Argus, who died as a guest of Evander. The Tarpeian rock on the Capitoline was a steep cliff (named after the traitress Tarpeia) from which criminals were thrown to their deaths.

That sanctuary even then struck terror
in rustics trembling at its rock and forest. 350
"This wood," he says, "atop this leafy hill,
some unknown god inhabits; my Arcadians
think they have often seen Jove's right hand brandish
his stormy aegis, summoning the clouds.
Here, in two towns with devastated walls, 355
you see the monuments of men of old.
This fortress, father Janus built; that, Saturn:
one named Janiculum, and one Saturnia."[47]
Speaking thus, they approached impoverished
Evander's house and gazed on cattle lowing 360
over the forum, over lush Carínae.[48]
Reaching his home, he said: "This threshold mighty
Alcídes crossed; this mansion bade him welcome.
Dare disdain wealth, my guest, and make yourself
worthy of gods by not refusing hardship." 365
He spoke, and underneath his narrow roof
led tall Aeneas, spreading him a bed
of strewn leaves covered by a Libyan bearskin.
With dusky wings night falls, embracing earth.

Into New Shapes

The *Metamorphoses* of Ovid, translated by Robert M. Torrance

Publius Ovidius Naso (43 B.C.–A.D. 17?) was born in the central Italian town of Sulmo and studied rhetoric in Rome; though not of Maecenas's circle, he was acquainted in his younger years with Horace, Propertius, and Tibullus, and later became the leading poet of his age. In A.D. 8 he was banished by Augustus, for unknown but apparently flagrant indiscretions, to Tomis on the Pontus (Black Sea), where despite repeated appeals to Augustus and his successor Tiberius, he died in much-lamented exile.

The cosmopolitan Ovid, in most of his many extant poems—mainly love elegies in a light and sometimes a frivolously witty vein (Amores, Ars Amatoria, Remedia Amoris), *or complaints from his wretched place of banishment* (Tristia, Epistulae ex Ponto) *—is anything but a poet of nature; the city was very much his home turf. Yet his fascination with mythological and religious explanations of origins (a central theme of Callimachus and other Alexandrians before him), and with processes of change perpetually shaping and reshaping the world, is evident in the uncompleted* Fasti *(a poetical calendar of Roman festive days) and above all in the fifteen books of*

[47]The citadel of Janus (an Italian deity of beginnings) was on the Janículum Hill across the Tiber from the Capitoline, which was sacred to Saturn and Jupiter.
[48]Carínae was a fashionable district on the Esquiline. Alcídês (below) was a name of Hercules, grandson of Alcaeus, to whom Evander (as he had told Aeneas in lines 184–279) had been host after Hercules slew the giant Cacus, son of Vulcan, who had stolen his oxen. See also Ovid's *Fasti* I.543–86.

his masterpiece, the Metamorphoses, *narrating in fluid epic hexameters widely varied stories of mythical transformations. The opening lines of the first book, which follow, give a vivid account of the creation of the world that draws upon Hesiod, Empedocles, and Stoic cosmogony but welds the different components into a very Ovidian story that allows for uncertainty as to which god ("or some better nature") sorted out the confused elements of primeval chaos, and whether man was created by a divine artificer or formed spontaneously from aethereal seeds. Like his account of creation, Ovid's description of the Golden Age, when "spring was eternal," and of the iron age of fraud and deception—when earth, once common to all, was parceled out and violated—was enormously influential in later times.*

The translation is based on the text in the Loeb Classical Library edition of Frank Justus Miller (1916).

FROM *Book I*

Into new shapes my mind inclines to tell
of bodies changed . . .
Gods who likewise transformed them,
favor my enterprise! and lead my song
from the world's origin to the present time.

Before land, sea, and overarching sky, 5
nature displayed one face to all the world,
chaos: a rude and undigested mass,
nothing but lifeless bulk and warring seeds
of ill-matched elements heaped up together.
No Titan then supplied the world with light, 10
no Phoebe yet renewed her slender horns,
nor did earth hang in equilibrium,
encompassed by the air, nor Amphitríte's
long arms entwine the margins of the land.[49]
Earth, sea, and atmosphere existed; yet 15
earth was unstable, waves unswimmable,
air without light; no form remained the same;
things clashed with one another; in a single
body, cold battled hot, wet battled dry,
soft fought with hard, and weightless things with heavy. 20
God—or some better nature—stilled this strife,
cutting off land from sky and sea from land,
and severing dense air from radiant sky.
Once he had sorted out things blindly jumbled,
he tied their strands harmoniously in place: 25
the fiery, weightless thrust of vaulted aether
leapt up, ascending to the topmost height;
next in both lightness and in place is air;
heavier earth dragged down large elements,

[49]The Titan is the Sun (Greek Hêlios); Phoebe (another name of Diana or Artemis, sister of Phoebus Apollo) is goddess of the moon; Amphitrítê, a Néreid or Océanid, as wife of Neptune (Poseidon), is goddess of the sea.

sinking from its own weight; encircling moisture 30
took last place, and embraced the solid globe.
Whichever god it was, he thus divided
that formless heap, distributing its members,
first modeling the mighty ball of earth—
created equidistant everywhere— 35
then ordering seas to spread and swell with rushing
winds and encompass shores around the earth.
Springs and enormous lakes and pools he added,
hemming swift rivers in with sloping banks;
in various places these are partly swallowed 40
by earth and partly reach the sea, united
with freer tides, pounding not banks but shores.
He bids the plains expand, the valleys sink,
forests grow leaves, and rocky mountains rise.
Just as two zones divide the heavens both 45
to right and left—a fifth between is hotter—
so, with no less a number, god distinguished
this center, stamping earth with equal regions.
Heat makes the mid-zone uninhabitable;
snow enwraps two; two more he placed between, 50
tempering each by mixing frost and flame.
Air overhangs them—heavier than fire
no less than earth is heavier than water.
There he commanded mists and clouds to dwell,
and thunder frightening men's minds, and winds 55
creating lightning bolts and freezing cold.
To these the fabricator of the world
allotted air unequally: even so,
each in his separate domain, they nearly
mangle the world through brotherly dissension! 60
Eurus retired to Nabataean dawn
and Persian mountains lit by morning sunlight;
evening, and shores warmed by the western sun,
are Zephyr's neighbors; fearful Boreas
assails the Scythian north; while Auster drenches 65
the antipodes with ceaseless mist and rain.[50]
Above them all he placed the radiant aether,
weightless, and purified of earthly dregs.
Scarcely had he thus given all things limits,
when constellations, long oppressed by darkness, 70
suddenly effervesced throughout the sky.
Lest any region be deprived of life,
stars and divine forms occupied the heavens,

[50]Ovid names four of the principal Greek and Roman winds: Eurus from the (south)east; Zephyr(us) from the west; Boreas from the north; and Auster (Greek Notus) from the south(west). The Nabataeans were Arabians; the Scythians a loosely defined nomadic people of regions around the Black and Caspian seas.

waters became the shiny fishes' homes,
earth received beasts, and fluttering air the birds. 75
A creature loftier and more capable
of thought, to dominate the rest, was lacking
till man was born, fashioned from godlike seed
by the artificer of a better world—
or else the young earth, recently divided 80
from aether, harbored seeds of kindred sky.
Mixing this earth with rain, Iápetus' son[51]
molded an effigy of the omnipotent gods;
and though all other animals gaze earthward,
to man he gave a face sublimely lifted, 85
bidding him elevate it toward the stars.
Thus earth, till now inchoate and unformed,
sported the unfamiliar shapes of men.
Gold was that primal age, which freely practiced
goodness and virtue, uncompelled by law. 90
No fearful penalties, no menacing words
were fixed in bronze, no judge intimidated
the suppliant throng, for none required a judge.
No pine had yet been felled on native mountains
to float on sparkling seas to foreign lands, 95
and mortals knew no shores except their own;
no steep moats yet encircled city walls;
no curving horns, no trumpets of straight bronze,
no swords, no helmets: nations then pursued
sweet ease, securely free of men at arms. 100
Earth of her own accord, inviolate
by hoe, untouched by plow, produced all things;
and men, content with foods none cultivated,
gathered strawberries growing on the mountains,
cherries, blackberries stuck among the brambles, 105
and acorns dropping down from Jove's wide tree.
Spring was eternal; gentle zephyrs warmly
caressed bright flowers springing up unseeded;
soon the unplowed earth burst with teeming grain
as white ears blanketed fields left unfallowed; 110
rivers flowed now with nectar, now with milk,
and from green ilexes dripped yellow honey.
Once shadowy Tărtarus had swallowed Saturn,
a silver race soon followed under Jove,
baser than gold but worthier than bronze. 115
Jupiter now curtailed the span of springtime,
making each year traverse four seasons: winter,
summer, unequal autumn, fleeting spring.
Then the parched air first glowed with searing heat,

[51]Prometheus. (See note on Hesiod's *Works and Days*, line 50, in Chapter 8 above.)

and icicles hung down, congealed by winds; 120
then the inhabitants of caves, of bushes,
and branches bound with bark, first entered houses;
then in long furrows Ceres' seeds were first
buried, and oxen groaned beneath the plow.
After this came a third race, made of bronze, 125
fiercer, and more inclined to savage warfare,
yet not depraved; last came the race of iron.
Into this age of baser vein all evil
broke, as faith, modesty, and truth departed,
leaving fraud and deception in their place, 130
tricks, violence, and impious love of gain.
Sails were unfurled to winds that sailors scarcely
yet knew, and keels that lately crowned the summits
of mountains, vaulted untried waves in triumph.
Ground previously shared like air and sunlight, 135
careful surveyors marked with drawn-out lines.
Demanding more than crops and nourishment
of bounteous earth, men pierced her very bowels,
whence wealth sequestered deep in Stygian shadows
was excavated, prodding men to crime. 140
Now baneful iron and still more baneful gold
appeared; now war appears, which fights with both,
brandishing clangorous arms with bloody hands.
Men live on plunder; neither son-in-law
nor guest is safe; few brothers please each other. 145
Husbands plot wives' destruction, wives their husbands';
stepmothers dreadfully mix deadly poisons;
sons prematurely plot their fathers' life spans;
piety fails, and—last among heavenly beings—
virgin Astraea left lands stained by slaughter.[52] 150

The Springtime of Love

The *Vigil of Venus (Pervigilium Veneris)*, translated by Robert M. Torrance

In the "Silver Age" of Roman poetry in the first century A.D., *poems pertaining to natural themes survive from writers such as Petronius, Statius, and the pastoral poet Calpurnius Siculus (who praised the rule of Nero as the Golden Age), but few of their verses are of great originality. In the third century Nemesianus could do little more than emulate the eclogues of Virgil and Calpurnius and the hunting poems* (Cynegetica) *of his Augustan predecessor Grattius and the second-century Greek poet Oppian. In contrast, the anonymous* Pervigilium Veneris, *written sometime between the second and fifth centuries in vigorous trochaics that anticipate the accen-*

[52]See note to Virgil's *Georgics*, Book II, line 474.

tual meters of the Latin Middle Ages, both recaptures spring's erotic renewal of the world, celebrated long before by Lucretius and Virgil, and expresses, in its conclusion, the melancholy twilight of this rhapsodic pagan oneness with a vitally living cosmos. With its intermingling of colloquial simplicity and baroque elaboration, and its haunting refrain, the poem—the earliest manuscript of whose corrupt text dates from the seventh to eighth century—"gives at once the impression," Robert Schilling writes in his edition, La Veillée de Vénus *(1961), "of the new and of the déjà vu: of classical inspiration with an astonishing spontaneity." It has fascinated readers ever since its rediscovery in the sixteenth century. The narrator of Walter Pater's novel* Marius the Epicurean *(1885) describes it as "a kind of nuptial hymn, which, taking its start from the thought of nature as the universal mother, celebrated the preliminary pairing and mating together of all fresh things, in the hot and genial springtime—the immemorial nuptials of the soul of spring itself and the brown earth . . ." And among the fragments "shored against my ruins" in the closing lines of T. S. Eliot's* The Waste Land *(1922) are the words* "Quando fiam uti chelidon" *("when shall I be like the swallow?") from the sad final lines of this otherwise rapturous poem. The translation is based on Schilling's text.*

Those who never loved, tomorrow they and lovers too shall love!

Spring is youthfulness and singing, springtime is the birth of love;
spring, when lovers come together; spring, the mating time of birds,
when the woods let down their tresses under matrimonial rains.
She who couples loves tomorrow, in among the shady trees 5
interweaves from sprays of myrtle cottages of living green:
for tomorrow fair Dione[53] promulgates laws throned on high.

Those who never loved, tomorrow they and lovers too shall love!

Then the Sea, from bloody droplets falling down in balls of foam
in among bipedal horses prancing in cerulean herds, 10
fashioned undulant Dione out of matrimonial rains.[54]

Those who never loved, tomorrow they and lovers too shall love!

She herself with jeweled blossoms paints the year in purple hues,
then compresses budding nipples, springing up with Zephyr's breath,
into ever-swelling clusters; these she irrigates with streams 15
fallen from nocturnal moisture sprinkling all the world with dew.
Out of these amassed deposits quivering teardrops sparkle forth,
each precipitated globule forming as it filters down.
See, the little flower petals manifest their crimson shame.
Then this condensation, drizzling out of cloudless starry skies, 20
liberates the virgin nipples sheathed in filmy robes at dawn.

[53]Diônê (a feminine form of Zeus) was perhaps the original consort of Zeus in Greek mythology, before Hera displaced her; she was worshipped at the ancient oracle of Dôdôna in northern Greece, and later at Athens, and was sometimes identified (as in *Iliad* V. 370) as the mother of Aphrodite (Roman Venus) or even, as she is throughout the present poem, with Venus herself.

[54]In the myth told by Hesiod (*Theogony* 176–206), Aphrodite was born not (as in Homer) from the union of Zeus and Dione but when Kronos (Roman Saturn), at the instigation of his mother Gaia (Earth), castrated his father Ouranos (Sky), and cast his genitals into the sea; from its foam Aphrodite, or Venus, was born. The present lines, with their "bipedal horses" (two-finned sea-horses), etc., are a fanciful variation on that story.

She herself commands the virgin roses, moist with dawn, to wed,
fashioned from the blood of Cypris, from the kiss of Cupid's mouth,
coruscating gems and flickering flames and purple-blazing sun.
Unashamed, the bride tomorrow from its virgin knot shall free 25
all the crimson blood sequestered underneath her fiery garb.

Those who never loved, tomorrow they and lovers too shall love!

She herself, the goddess, orders nymphs to leave the myrtle grove;
with them is a boy companion—yet could anyone believe
Love was truly on vacation if his arrows went along? 30
Come, nymphs: Love has laid his weapons down and takes a holiday!
Unarmed he has been commanded, then, and naked, to proceed,
lest his bow and arrows injure, or his searing fires consume.
Nonetheless, nymphs, tread with caution: Cupid is a lovely boy,
still the same, though lacking weapons, still, though naked, wholly Love! 35

Those who never loved, tomorrow they and lovers too shall love!

Venus sends attendant virgins to you, bashful like the bride.
(This alone we ask: O virgin born on Delos,[55] now withdraw
lest the slaughtering of creatures desecrate with blood the wood
where green shadows now are flitting over flowers freshly blown.) 58
She herself desires to ask you: might your chastity now bend? 40
might you now attend her vigil? might a virgin find it fit?
Three nights long you may have noticed bands of festive dancers come
passing through the thronged assemblies in among the forest glades,
in among the crowns of flowers, in among the myrtle huts.
Ceres comes along with Bacchus, not without the poets' god.[56] 45
Let the long night vigilantly stay awake with ceaseless song:
let Dione rule the forest: Delian maiden, now withdraw!

Those who never loved, tomorrow they and lovers too shall love!

Flowery Hybla[57] has the goddess named to be her judgment seat:
there, surrounded by her Graces, let her promulgate her laws. 50
Shower countless flowers, Hybla—strew the bounty of the year!
Hybla, clothe yourself in flowers burgeoning on Aetna's plain!
Country maidens will be coming here, and mountain maidens too,
all who live in groves and forests, all inhabiting the springs:
each of these has Cupid's mother ordered to be seated near, 55
bidding every maid abandon confidence in naked Love.

Those who never loved, tomorrow they and lovers too shall love!

[55]Diana (Greek Artemis) is asked to depart, during Venus's festival, from the woods that are her sacred precinct so that she may not witness the shedding of blood within them by sacrificial slaughter (and perhaps by deflowering of the virgin bride). I have followed J. W. Mackail's hypothetical transposition of line 58, clearly out of place in the manuscripts, to this passage.
[56]Ceres (Greek Dêmêtêr) is goddess of the grain; Bacchus (Greek Dionysos), god of wine; "the poets' god," Phoebus Apollo.
[57]A city on the slopes of Mount Etna in Sicily famed for its honey.

On the morrow primal Aether celebrated nuptial bonds.
Fathering year-long creation by the fertile clouds of spring, 60
from the bridegroom rain has fallen deep into his lady's groin,
intermingling with her body, out of which all things are born.
She herself, the procreatress, governs, by pervasive breath
with impenetrable power, blood and intellect within.
Over lands, across the heavens, through the oceans down below, 65
she, upon her sprouting pathway, takes her fecundating course,
bidding all the mighty cosmos learn the mysteries of birth.

Those who never loved, tomorrow they and lovers too shall love!

She it was who guided Trojan offspring to Italian shores:[58]
to her son she gave the princess from Laurentum as his bride, 70
giving Mars a modest virgin taken from a sacred shrine;
then she wedded Sabine women to the sons of Romulus,
whence the Ramnes and Quirítes and—continuing the race
born of Romulus the father—Caesar, to complete his line.

Those who never loved, tomorrow they and lovers too shall love! 75

Pleasure fecundates the meadows blossoming at Venus' touch;
Love himself, Dione's offspring, people call a rustic child:
she, when fields convulsed in labor, welcomed him upon her breast,
rearing him herself with tender kisses from sweet flower cups.

Those who never loved, tomorrow they and lovers too shall love! 80

Look, already rutting bullocks brush their flanks against the broom,
each united to another through the marriage covenant;
see, beneath the shadows, bleating ewes accompanied by rams;
hear the singing birds the goddess never lets be silent long.
Now loquacious swans are hoarsely trumpeting across the pools; 85
underneath a shady poplar Tereus's lady sings
so melodiously you fancy tunes of love are spilling forth—
not a sister's lamentations over her barbaric lord.[59]

She is singing, we are silent: when will springtime come to me?
When shall I be like the swallow, so my silence too may cease? 90

[58]Venus, in the myth recounted in Virgil's *Aeneid*, was the mother (by the Trojan hero Anchíses) of Aenéas, who after the destruction of Troy came to Italy to found what would later be Rome. Lauréntum was a town in Latium (the district around Rome) ruled by Latínus, whose daughter Lavínia became Aeneas's wife. Later, the war god Mars raped the Vestal Virgin Rhea Sylvia (or Ilia), daughter of King Numitor of Alba; she gave birth to Romulus and Remus but was buried alive for violating the laws of chastity imposed by Vesta (Greek Hestia), goddess of the hearth. Livy's *History of Rome* tells how the Romans, under Romulus, seized and married the young Sabine women of nearby central Italy, whom they had lured to a festival, thus uniting the two peoples. The Ramnes were one of the oldest Patrician tribes of Rome; "Quirítes" was the name given to the citizens of Rome after they admitted the Sabines of the city of Cures as citizens. Julius Caesar was considered a descendant of Aeneas's son Iulus.
[59]In Greek myth, the Thracian king Têreus, after marrying Procnê, daughter of King Pandíon of Attica, concealed her and said she was dead in order to marry her sister Philomêla, whose tongue he cut out. Philomela revealed the truth to Procne, who, after killing her son Itys and serving him for dinner to Tereus, fled with her sister; when Tereus overtook them, they prayed to be changed into birds, and Procne was changed to a nightingale and Philomela to a swallow (or vice versa in other versions).

I have lost the muse by silence, now that Phoebus turns away:
even thus Amýclae perished once its voice had fallen dumb.[60]

Those who never loved, tomorrow they and lovers too shall love!

The Longest Voyage

Carmina Minora 20, by Claudian, translated by Abraham Cowley

Claudius Claudianus, last of the major classical Latin poets, was a Greek-speaking Alexandrian who came to Italy in the late fourth century. His facility in extolling his patrons in Latin verse won him fame and fortune as a court poet, in Rome and Mediolanum (Milan), to the young western Emperor Honorius (reigned 393–423) and his powerful minister and general, the Vandal Stilicho, victor in several battles over the Visigoth Alaric. (Alaric was destined, however, after Claudian's death—probably in the opening years of the fifth century—and the execution of Stilicho by Honorius, to sack Rome in 410.) To Stilicho's extravagant praise much of Claudian's surviving poetry is devoted. Very different from his grandiose political and mythological poetry is the eulogy, in elegiac couplets, of an old man of the district around Verona; it echoes the mood of Horace's second Epode ("Beatus Ille"), without its ironical ending, and Virgil's verses in the fourth Georgic *on the old Corycian gardener. The translation is by the seventeenth-century poet Abraham Cowley (1618–67).*

Happy the man who his whole time doth bound
Within the enclosure of his little ground.
Happy the man whom the same humble place,
The hereditary cottage of his race,
From his first rising infancy has known, 5
And by degrees sees gently bending down
With natural propension to that earth
Which both preserved his life and gave him birth.
Him no false distant lights by Fortune set
Could ever into foolish wanderings get. 10
He never dangers either saw or feared;
The dreadful storms at sea he never heard,
He never heard the shrill alarms of war,
Or the worse noises of the lawyers' Bar.
No change of consuls marks to him the year; 15
The change of seasons is his calender.
The cold and heat Winter and Summer shows,
Autumn by fruits, and Spring by flowers he knows.

[60]Concerning Amýclae, a Pythagorean town of southern Italy, Lemprière writes in his *Classical Dictionary*: "Once a report prevailed in Amyclae that enemies were coming to storm it; upon which the inhabitants made a law that forbade such a report to be credited, and when the enemy really arrived, no one mentioned it, or took up arms in his own defense, and the town was easily taken. From this circumstance the epithet of *tacitae* ['silent'] has been given to Amyclae."

He measures time by landmarks, and has found
For the whole day the Dial of his ground. 20
A neighboring wood born with himself he sees,
And loves his old contemporary trees.
He's only heard of near Verona's name,
And knows it, like the Indies, but by fame:
Does with a like concernment notice take 25
Of the Red Sea and of Benacus Lake.[61]
Thus health and strength he to a third age enjoys,
And sees a long posterity of boys.
About the spacious world let others roam,
The Voyage Life is longest made at home. 30

Birds, Bees, Lambs, and Lovers

Daphnis and Chloe, by Longus, translated by George Thornley

Though written in prose, the Greek romances of late antiquity continue a romantic vein that found poetic expression in Euripidean tragedies such as the Ion, *and in the New Comedy of Menander and others in the fourth century* B.C.*; their improbable plots of lovers happily reunited after hair-raising adventures are sometimes combined with pastoral themes descended from Theocritus and the Alexandrians. Nowhere are such themes more prominent, or more beautifully evoked, than in the brief romance of* Daphnis and Chloe, *attributed to a certain Longus and dated anywhere between the second and sixth centuries* A.D.*; and nowhere has flagrant artifice been so unabashedly used to paint an impossibly idealized yet irresistibly charming portrayal of the "natural" existence of young lovers in innocent communion with the living world around them. So discerning a critic as Goethe called the book (in conversations with Eckermann of March 1831, the year before he died) "a masterpiece, which I have often read and admired; in which Understanding, Art, and Taste, appear at their highest point, and beside which the good Virgil retreats somewhat into the background. . . . The book is so beautiful that, amid the bad circumstances in which we live, we cannot retain the impression we receive from it, but are astonished anew every time we read it" by "a perfection, and a delicacy of feeling, which cannot be excelled." Here once again we meet the cicada or grasshopper (Greek* tettix*) that has played so prominent a role in Greek poetry and prose from Hesiod and Alcaeus through Plato to Theocritus and Meleager. The delightful translation (slightly revised) is that of George Thornley, published in 1657 as* Daphnis and Chloe: A Most Sweet and Pleasant Pastorall Romance for Young Ladies.

I.9 . . . It was the beginning of spring, and all the flowers of the lawns, meadows, valleys, and hills were now blowing; all was fresh, and green, and odorous. The bee's humming from the flowers, the bird's warbling from the groves, the lamb's skipping on

[61]Benacus (modern Lago di Garda, not far from Verona in northern Italy) was the lake of laughing waves celebrated by Catullus in his poem of homecoming to his "almost-isle" of Sirmio.

the hills, were pleasant to the ear and eye. And now, when such a fragrancy had filled those blessed and happy fields, both the old men and the young would imitate the pleasant things they heard and saw; and hearing how the birds did chant it, they began to carol too; and seeing how the lambs skipped, tripped their light and nimble measures; then to emulate the bees, they fall to cull the fairest flowers, some of which in toysome sport they cast in one another's bosoms, and of some, plaited garlands for the nymphs—and, always keeping near together, had and did all things in common. For Daphnis often gathered in the straggling sheep, and Chloe often drove the bolder venturous goats from the crags and precipices; and sometimes to one of them the care of both the flocks was left, while the other did intend some pretty knack or toysome play. For all their sport were sports of children and of shepherds. Chloe, scudding up and down, and here and there picking up the stalks of grass, would make in pleats a trap to catch a grasshopper, and be so wholly bent on that that she was careless of her flocks. Daphnis, on the other side, having cut the slender reeds and bored the quills, or intervals, between the joints, and with his soft wax joined and fitted one to another, took no care but to practice or devise some tune, even from morning to the twilight. Their wine, and their milk, and whatever was brought from home to the fields, they had still in common. And a man might sooner see all the cattle separate from one another than he should see Chloe and Daphnis asunder. . . .

I.23–26 For now the cooler spring was ended, and the summer was ended, and the autumn was come on, and all things were got to their highest flourishing acme and vigor: the trees with their fruits, the fields with standing corn. Sweet then was the singing of the grasshoppers; sweet was the odor of the fruits; and not unpleasant, the very bleating of the sheep. A man would have thought that the very rivers, by their gentle gliding away, did sing; and that the softer gales of wind did play and whistle on the pines; that the cattle, as languishing with love, lay down and slumbered on the ground; and that the sun, as a lover of beauty unveiled, did strive to undress and turn the rurals all naked. By all these was Daphnis inflamed; and therefore often he goes to the rivers and brooks, there to bathe and cool himself, and often he drinks of the clear rills, as thinking by that to quench his scorching inward fever. When Chloe had spent much time, because the flies were importunate and vexatious, milking the sheep and the goats and curdling and pressing it into smaller cheeses, she washed herself and crowned her head with pine boughs; and when she had girt her kid-skin about her, she took a piggin, and with wine and milk she made a sillabub[62] for her dear Daphnis and herself. When it grew towards noon, they fell to their fascination, or catching of one another, by their eyes. For Chloe, seeing Daphnis naked, thought she had fallen on the most sweet and florid beauty, and therefore could not choose but melt, as being not able to find in him the least moment to dislike or blame. Daphnis again, if he saw Chloe in her kid-skin and her pine coronet give him the sillabub to drink, thought he saw one of the nymphs, the fairest of the holy cave. Therefore taking off her pine, he would put it on his own head; and when he had kissed it over and over, set it upon hers again. And Chloe, when he was naked and bathing, would take up his garment, and when she kissed it, put it on upon herself. Sometimes they flung apples at one another, sometimes they dressed and distinguished one another's hair into curious trammels and locks. And Chloe likened Daphnis's hair to the myrtle, because it was black; Daphnis again, because her face was white and ruddy, compared it to the fairest apple. He taught her too to play on the pipe, and always when she

[62]A beverage of wine mixed with milk.

began to blow would catch the pipe away from her lips, and run it presently over with his: he seemed to teach her when she was off, but with that specious pretext, by the pipe he kissed Chloe. But it happened, when he played on his pipe at noon, and the cattle took shade, that Chloe fell unawares asleep. Daphnis observed it, and laid down his pipe, and without any shame or fear was bold to view her all over, and every limb, insatiably; and withal spoke softly thus:

"What sweet eyes are those that sleep! How sweetly breathes that rosy mouth! The apples smell not like to it, nor the flowery lawns and thickets. But I am afraid to kiss her. For her kiss stings to my heart, and makes me mad, like new honey. Besides, I fear lest a kiss should chance to wake her. O ye prating grasshoppers, ye make a noise to break her sleep! And the goats beside are fighting, and they clatter with their horns. Ye wolves, worse dastards than the foxes, come and ravish them away." While he was muttering this passion, a grasshopper that fled from a swallow took sanctuary in Chloe's bosom, and the pursuer could not take her; but her wing, by reason of her close pursuit, flapped the girl upon the cheek; but she, not knowing what was done, cried out, and started from her sleep. But when she saw the swallow flying near by, and Daphnis laughing at her fear, she began to give it over, and rub her eyes that yet were sleeping. The grasshopper sang out of her bosom, as if her suppliant were now giving thanks for the protection. Therefore Chloe again squeaked out; but Daphnis could not hold laughing, nor pass the opportunity to put his hand into her bosom and draw forth the grasshopper, which still did sing even in his hand. When Chloe saw it, she was pleased, and put it in her bosom again, and it prattled all the way. . . .

Singers Unsilenced

Two Anonymous Late Greek Poems, translated by Robert M. Torrance

Pagan poetry written in Greek during the Roman period ranges from brief epigrams to the gigantic fifth-century epic of Nonnus, the Dionysiaca, *in forty-eight books; except for scattered shorter poems preserved, along with others from classical and Hellenistic times, in the tenth-century* Palatine Anthology, *few have great interest for modern readers. The following two poems nevertheless attest to the continuation in later centuries of favorite themes descending from early Greek poetry. The first is from the collection known as the* Anacreontea, *after Anacreon, the famed author of lighthearted drinking songs and love poems in the sixth century* B.C.*; these poems of uncertain date, appended to the* Palatine Anthology, *were probably composed during the Roman period. The poem on the cicada, which Goethe translated into German, celebrates—with its humorous parallel to the passionless Stoic philosopher—this humble musician whom we have met before in Hesiod and Alcaeus, Plato's* Phaedrus, *Theocritus, Meleager, and* Daphnis and Chloe. *(See also Abraham Cowley's version, "The Grasshopper," in Chapter 19 below.) And the poem "To Pan," inscribed on the shrine of Asclepius, the god of healing, at his great sanctuary of Epidaurus, in the third to fourth century* A.D. *(though possibly dating from the previous century), is a final tribute to the god of nature whose very name was thought to mean "All."*

The translations are based on texts in volumes 2 (1988) and 5 (1993), respectively, of the Loeb Classical Library Greek Lyrics, *ed. David A. Campbell.*

The Godlike Cicada

Anacreontea, *34*

Cicada, blest are you!
Sipping a little dew,
atop a tree you sing
blissfully as a king!
Whatever you survey 5
throughout the fields today
is yours, and through the trees.
Farmers you greatly please
by doing harm to none.
Honored by everyone, 10
you pleasantly foretell
summer; the Muses well
love you—and Phoebus, bringing
the gift of shrill-toned singing.
Age cannot do you wrong, 15
wise earth-born lover of song.
Bloodless, passionless too,
how like the gods are you!

To Pan

Inscription from the shrine of Asclepius at Epidaurus

Of Pan the leader of Nymphs, now
I sing, the lover of Naiads,
adornment of golden dancers
and lord of sonorous music,
whose god-given siren-singing 5
pours from his well-tuned reed-pipes,
as, nimbly stepping in rhythm,
he leaps from shadowy caverns
spinning his *all*-shaped body,
gracious in face and in movement, 10
conspicuously blond-bearded.
To star-enclustered Olympus
his *all*-melodious sound comes,
shedding immortal music
on Olympian gods assembled. 15
All earth and sea are commingled
in oneness, thanks to you only,
    for you are the bulwark of *all* things,
oh iê Pan, Pan!

12

Precarious Communion: Reflections on the Human and Natural Worlds

In addition to the philosophers and poets of the previous two chapters, many prose writers of Greco-Roman times—ranging from agricultural writers to historians, moral essayists, encyclopedists, astronomers, and physicians—reflected in varied ways on the relation between the human and natural worlds, often with a troubled awareness that the communion between them celebrated by earlier poets and wistfully remembered by later ones was now imperiled. No longer does man unquestioningly belong (if he ever had) to the encompassing natural world from which he now finds himself uneasily divided. This chapter, by gathering together some of the most important of these writers, both Greek and Roman, suggests not only the centrality of such concerns for men of varied professions and backgrounds but also the diversity of outlooks to which their writings gave expression.

Virtue for the sternly traditional Roman was inseparably associated with the agricultural way of life traditionally embodied by the sturdy Sabine and Latin farmers of old who, like the Cincinnatus of legend, wished for nothing more, after taking up arms against the enemies of the Republic, than to return to the plow. Yet already in Cato the Elder's time, after the Second Punic War, to say nothing of Varro's or Columella's in the late Republic and early Empire, this way of life, ideally memorialized in the happy farmers of Virgil's Georgics, *was vanishing or had long vanished. When Tacitus, toward the turn of the second century A.D., sought examples of uncorrupted virtue, he found them not (of course) in the city, nor in the Italian countryside, where the small independent farmer was largely a sentimental memory, but among the Germanic tribes that would eventually bring the mighty Roman Empire to ruin. Even in an age that found itself so far removed as that of Nero from the nature in accord with which the Stoic endeavored to live, Seneca could find recompense in contemplation of its elevating mysteries. Still, these remained difficult to penetrate, and the thought that "nothing is difficult for nature" could be disconcerting, given her vast potentiality for apocalyptic destruction.*

The great age of Greek scientific discovery and speculation, beginning with the pre-Socratics of the sixth to fifth centuries B.C. and continuing into the Hellenistic age, had left a rich legacy for writers of the Greco-Roman period to assimilate and continue, and some of the most influential Greek scientists of antiquity would carry out their studies under the Empire. Yet the encyclopedists Pliny the Elder in Latin and Aelian in Greek—both of whom, drawing uncritically on literary sources from their own and earlier times, place helpfully reliable factual information indiscriminately side by side with unsubstantiated marvels, and consider both equally the products of nature—remind us how far from the spirit of scientific inquiry even those who professed to be students of nature could be. In contrast, the meticulous observations and careful calculations of Ptolemy in the second century A.D. resulted in a consolidation of the geocentric system of earlier Greek astronomy that would remain the standard of Greco-Roman, and then of Arabic and European, science until the time of Copernicus and Vesalius well over a thousand years later. This was true also of the medical theories of Ptolemy's younger contemporary, Galen. In these thinkers, the spirit of ancient Greek natural philosophy came to a fitting culmination.

The chapter concludes with three excerpts from Plutarch, whose essays, written around the turn of the second century and known as the Moralia, *reflect not only on moral questions but also on widely varied aspects of the relation between the human and natural worlds in a time of great ferment and change.*

An Important and Noble Art

Three Roman Writers on Agriculture

In addition to Virgil's Georgics *in verse, three Roman prose treatises on agriculture—all known as* De Re Rustica *or* De Agricultura*—survive nearly intact. The first, the earliest extant work of Latin prose, is a short, down-to-earth book of practical instructions recorded with little evident art by* ***Marcus Porcius Cato*** *(234–139 B.C.), known as Cato the Censor or Cato the Elder, who was born of plebeian stock at Tusculum near Rome and fought in the Second Punic War. Thereafter he became an intransigent opponent both of Carthage's continued existence and of the corrupting influence of the newfangled Greek culture, and in the office of Censor made himself a stern guardian of what were already the old and threatened virtues of the peasant farmer-soldier. In the selections that follow, Cato declares the farmer's to be the most respectable of callings and gives a few pieces of homespun advice somewhat reminiscent of Hesiod's five centuries before him.*

The second treatise is a far more literary work written, in dialogue form, by ***Marcus Terentius Varro*** *(116–27 B.C.), famed as the most learned Roman of his time for his Menippean satires and dozens of books (of which only* On Agriculture *and parts of* On the Latin Language *survive) on almost every recognized domain of human knowledge. In the excerpts given here, his spokesmen affirm the nobility of agriculture, trace its origin and progress (in accord with the scheme of the Greek philosopher Dicaearchus) from the earlier "state of nature" and pastoral stage, and declare country life much more ancient than that of the city, "since it was divine nature which gave us the country, and man's skill that built the cities."*

Finally, the far longer work (in twelve books, as opposed to Varro's three) of ***Lucius Junius Moderatus Columella****, a Spaniard from Gades (Cádiz) writing around A.D. 60–65, is the most comprehensive of these treatises, combining literary skill—his tenth book, on gardens, is written in hexameter verse, taking up Virgil's invitation (*Georgics *IV.147–48) for others to mention what he had omitted—with practical information and instruction. In the following passages from his preface to Book I, he insists that the decline of agriculture in his time is due not to depletion of the inexhaustible earth but to shameful abandonment, for licentious urban pleasures, of the ancient discipline that had made Rome great.*

The translations are from the Loeb Classical Library Cato and Varro *(rev. ed. 1935) and* Columella, *vol. 1 (1940).*

FROM On Agriculture, *by Cato,*
translated by William Davis Hooper; revised by Harrison Boyd Ash

It is true that to obtain money by trade is sometimes more profitable, were it not so hazardous; and likewise money-lending, if it were as honorable. Our ancestors held this view and embodied it in their laws, which required that the thief be mulcted double and the usurer fourfold; how much less desirable a citizen they considered the usurer than the thief, one may judge from this. And when they would praise a worthy man their praise took this form: "good husbandman," "good farmer"; one so praised was thought to have received the greatest commendation. The trader I consider to be an energetic man, and one bent on making money; but, as I said above, it is a dangerous career and one subject to disaster. On the other hand, it is from the farming class that the bravest men and the sturdiest soldiers come, their calling is most highly respected, their livelihood is most assured and is looked on with the least hostility, and those who are engaged in that pursuit are least inclined to be disaffected. And now, to come back to my subject, the above will serve as an introduction to what I have undertaken.

I When you are thinking of acquiring a farm, keep in mind these points: that you be not over-eager in buying nor spare your pains in examining, and that you consider it not sufficient to go over it once. However often you go, a good piece of land will please you more at each visit. Notice how the neighbors keep up their places; if the district is good, they should be well kept. Go in and keep your eyes open, so that you may be able to find your way out. It should have a good climate, not subject to storms; the soil should be good, and naturally strong. If possible, it should lie at the foot of a mountain and face south; the situation should be healthful, there should be a good supply of laborers, it should be well watered, and near it there should be a flourishing town, or the sea, or a navigable stream, or a good and much traveled road. It should lie among those farms which do not often change owners; where those who have sold farms are sorry to have done so. It should be well furnished with buildings. Do not be hasty in despising the methods of management adopted by others. It will be better to purchase from an owner who is a good farmer and a good builder. When you reach the steading, observe whether there are numerous oil presses and wine vats; if there are not, you may infer that the amount of the yield is in proportion. The farm should be one of no great equipment, but should be well situated. See that it be equipped as economically as possible, and that the land be not extravagant. Remember that a farm is like a man—however great the income, if there is extravagance but little is left. If you ask me what is the best kind of farm, I should say: a hundred iugera[1] of land, comprising all sorts of soils, and in a good situation; a vineyard comes first if it produces bountifully wine of good quality; second, a watered garden; third, an osier-bed; fourth, an oliveyard; fifth, a meadow; sixth, grain land;[2] seventh, a wood lot; eighth, an arbustum [orchard]; ninth, a mast grove. . . .

XXXIX When the weather is bad and no other work can be done, clear out manure for the compost heap; clean thoroughly the ox stalls, sheep pens, barnyard, and farmstead; and mend wine-jars with lead, or hoop them with thoroughly dried oak wood. If you mend it carefully, or hoop it tightly, closing the cracks with cement and pitching it thoroughly, you can make any jar serve as a wine-jar. Make a cement for a wine-jar as follows: Take one pound of wax, one pound of resin, and two-thirds of a pound of sulphur, and mix in a new vessel. Add pulverized gypsum sufficient to make it of the consistency of a plaster, and mend the jar with it. To make the color uniform after mending,

[1]A iugerum is approximately two-thirds of an acre.

[2]During the long and destructive Second Punic War against Hannibal, as Hooper notes, Italian farmers had been conscripted into the army and their lands devastated. As a result, "grain farming was no longer profitable, and it had become the custom to import grain from Sicily and Africa," where large farms were worked by slaves. "The new Rome that emerged from this horrible war centered around a nobility of wealth and was in a state of demoralization. Such a condition naturally caused the cultivation of grain to be less important than that of the vine, the olive, domestic vegetables, or the rearing of cattle." The mast grove, below, furnished feed for cattle.

mix two parts of crude chalk and one of lime, form into small bricks, bake in the oven, pulverize, and apply to the jar.

In rainy weather try to find something to do indoors. Clean up rather than be idle. Remember that even though work stops, expenses run on none the less.

FROM On Agriculture, *by Varro,* *translated by William Davis Hooper; revised by Harrison Boyd Ash*

I.3 "Well then," said Agrasius, "since we have decided the nature of the subjects which are to be excluded from agriculture, tell us whether the knowledge of those things used in agriculture is an art or not, and trace its course from starting-point to goal." Glancing at Scrofa, Stolo said: "You are our superior in age, in position, and in knowledge, so you ought to speak." And he, nothing loath, began: "In the first place, it is not only an art but an important and noble art. It is, as well, a science, which teaches what crops are to be planted in each kind of soil, and what operations are to be carried on, in order that the land may regularly produce the largest crops.

I.4 "Its elements are the same as those which Ennius[3] says are the elements of the universe—water, earth, air, and fire. You should have some knowledge of these before you cast your seed, which is the first step in all production. Equipped with this knowledge, the farmer should aim at two goals, profit and pleasure; the object of the first is material return, and of the second enjoyment. The profitable plays a more important role than the pleasurable; and yet for the most part the methods of cultivation which improve the aspect of the land, such as the planting of fruit and olive trees in rows, make it not only more profitable but also more saleable, and add to the value of the estate. . . ."

II.1 . . . I began: "As it is a necessity of nature that people and flocks have always existed (whether there was an original generating principle of animals, as Thales of Miletus and Zeno of Citium thought, or, on the contrary, as was the view of Pythagoras of Samos and of Aristotle of Stagira, there was no point of beginning for them), it is a necessity that from the remotest antiquity of human life they have come down, as Dicaearchus teaches,[4] step by step to our age, and that the most distant stage was that state of nature in which man lived on those products which the virgin earth brought forth of her own accord; they descended from this stage into the second, the pastoral, in which they gathered for their use acorns, arbutus berries, mulberries, and other fruits by plucking them from wild and uncultivated trees and bushes, and likewise caught, shut up, and tamed such wild animals as they could for the like advantage. There is good reason to suppose that, of these, sheep were first taken, both because they are useful and because they are tractable; for these are naturally most placid and most adapted to the life of man. For to his food they brought milk and cheese, and to his body wool and skins for clothing. Then by a third stage man came from the pastoral life to that of the tiller of the soil; in this they retained much of the former two stages, and after reaching it they went far before reaching our stage. Even now there are several species of wild animals in various places:

[3]Quintus Ennius (239–169 B.C.), though Greek by birth, was revered as "the father of Roman poetry"; only fragments of his epic on the history of Rome, the *Annales*, and of his other poems now survive. Varro's citation is from Ennius's *Epicharmus*, a Latin translation of a Greek poem on nature falsely attributed to Epicharmus of Sicily.

[4]Zeno of Citium, in Cyprus (335–263 B.C.), was the founder of the Stoic school. Dicaearchus of Messana, in Sicily (late fourth century B.C.), was a disciple of Aristotle and a contemporary of Theophrastus.

as of sheep in Phrygia, where numerous flocks are seen. . . . As to swine, everybody knows—except those who think that wild boars ought not to be called swine. There are even now many quite wild cattle in Dardania, Maedica, and Thrace; wild asses in Phrygia and Lycaonia, and wild horses at several points in Hither Spain.

"The origin is as I have given it; the dignity, as I shall now show. Of the ancients the most illustrious were all shepherds, as appears in both Greek and Latin literature, and in the ancient poets, who call some men 'rich in flocks,' others 'rich in sheep,' others 'rich in herds'; and they have related that on account of their costliness some sheep actually had fleeces of gold—as at Argos the one which Atreus complains that Thyestes stole from him; or as in the realm of Aeëtes in Colchis, the ram in search of whose golden fleece the Argonauts of royal blood are said to have fared forth; or as among the Hesperides in Libya, from which Hercules brought from Africa to Greece golden *mala*,[5] which is the ancient manner of naming goats and sheep. . . . But if the flock had not been held in high honor among the ancients, the astronomers, in laying out the heavens, would not have called by their names the signs of the zodiac; they not only did not hesitate to give such names, but many of them begin their enumeration of the twelve signs with the names of the Ram and the Bull, placing them ahead of Apollo and Hercules. . . ."

III. 1 "Though there are traditionally two ways in which men live—one in the country, the other in the city—there is clearly no doubt, Pinnius, that these differ not merely in the matter of place but also in the time at which each had its beginning. Country life is much more ancient—I mean the time when people lived on the land and had no cities. For tradition has it that the oldest of all cities is a Greek one, Thebes in Boeotia, founded by King Ogygus; while the oldest on Roman territory is Rome, founded by King Romulus. For we may now say, with regard to this, with more accuracy than when Ennius wrote:

> 'Seven hundred years are there, a little more or less,
> Since glorious Rome was founded, with augury august.'

Thebes, however, which is said to have been founded before the deluge which takes its name from Ogygus, is some 2,100 years old. If, now, you compare this span of time with that early day when fields were first tilled, and men lived in huts and dugouts, and did not know what a wall or a gate was, farmers antedate city people by an enormous number of years. And no marvel, since it was divine nature which gave us the country, and man's skill that built the cities; since all arts are said to have been discovered in Greece within a thousand years, while there never was a time when there were not fields on earth that could be tilled. And not only is the tilling of the fields more ancient—it is more noble. It was therefore not without reason that our ancestors tried to entice their citizens back from the city to the country; for in time of peace they were fed by the country Romans, and in time of war aided by them. It was also not without reason that they called the same earth 'mother' and 'Ceres,' and thought that those who tilled her lived a pious and useful life, and that they were the only survivors of King Saturnus.[6] And it is in accordance

[5]*Mala* in Latin means "apples"; *mêla* in Greek can mean either "apples" or "sheep" (or "goats"). Varro thus suggests that the fabled golden apples of the Hespérides (which Hercules, in the eleventh of his twelve labors, obtained for Eurystheus, king of Argos and Mycenae) were really sheep or goats. The expedition of the Argonauts to obtain the golden fleece from King Aeëtes of Colchis (east of the Black Sea) was led by Jason, who obtained the fleece with the help of Aeëtes' daughter, Medéa.
[6]See Virgil's *Georgics*, II.473–74, in Chapter 11 above.

with this that the sacred rites in honor of Ceres are beyond all others called 'Initiations.' . . . At first, because of their poverty, people practiced agriculture, as a rule, without distinction, the descendants of the shepherds both planting and grazing on the same land; later, as these flocks grew, they made a division, with the result that some were called farmers and others herdsmen. . . ."

FROM On Agriculture, *by Columella, translated by Harrison Boyd Ash*

I. PREFACE Again and again I hear leading men of our state condemning now the unfruitfulness of the soil, now the inclemency of the climate for some seasons past, as harmful to crops; and some I hear reconciling the aforesaid complaints, as if on well-founded reasoning, on the ground that, in their opinion, the soil was worn out and exhausted by the over-production of earlier days and can no longer furnish sustenance to mortals with its old-time benevolence. Such reasons, Publius Silvinus, I am convinced are far from the truth; for it is a sin to suppose that Nature, endowed with perennial fertility by the creator of the universe, is affected with barrenness as though with some disease; and it is unbecoming to a man of good judgment to believe that Earth, to whose lot was assigned a divine and everlasting youth, and who is called the common mother of all things—because she has always brought forth all things and is destined to bring them forth continuously—has grown old in mortal fashion. And, furthermore, I do not believe that such misfortunes come upon us as a result of the fury of the elements, but rather because of our own fault; for the matter of husbandry, which all the best of our ancestors had treated with the best of care, we have delivered over to all the worst of our slaves, as if to a hangman for punishment. . . .

When I observe these things, reviewing in my mind and reflecting upon the shameful unanimity with which rural discipline has been abandoned and passed out of use, I am fearful lest it may be disgraceful and, in a sense, degrading or dishonorable to men of free birth. But when I am reminded by the records of many writers that it was a matter of pride with our forefathers to give their attention to farming, from which pursuit came Quinctius Cincinnatus, summoned from the plow to the dictatorship to be the deliverer of a beleaguered consul and his army, and then, again laying down the power which he relinquished after victory more hastily than he had assumed it for command, to return to the same bullocks and his small ancestral inheritance of four *iugera*; from which pursuit came also Gaius Fabricius and Curius Dentatus, the one after his rout of Pyrrhus from the confines of Italy, the other after his conquest of the Sabines, tilling the captured land which they had received in the distribution of seven *iugera* to a man, with an energy not inferior to the bravery in arms with which they had gained it;[7] and, not unseasonably to run through individual cases at this time, when I observe that so many other renowned captains of Roman stock were invariably distinguished in this twofold pursuit of either defending or tilling their ancestral or acquired estates, I understand that yesterday's morals and strenuous manner of living are out of tune with

[7]According to tradition, Cincinnatus was called from the plow to the dictatorship in 458 B.C., to save the Roman army besieged by the Aequians on Mt. Algidus. He delivered the consul Minucius and his army, resigned the dictatorship, and returned to his little farm after holding the office only sixteen days. *Cf.* Livy, III.26–29. Gaius Fabricius was consul in 282 and 278 B.C.; his noble conduct toward Pyrrhus, king of Epirus (in Greece), led to the evacuation of Italy by that king. Curius Dentatus was consul in 290 and 275 B.C. Famous for his frugality and his conquests over the Samnites, Sabines, Lucanians, and Pyrrhus, he retired to his farm, refusing all share in the booty. (Ash)

our present extravagance and devotion to pleasure. For, even as Marcus Varro complained in the days of our grandfathers, all of us who are heads of families have quit the sickle and the plow and have crept within the city-walls; and we ply our hands [by applauding] in the circuses and theaters rather than in the grainfields and vineyards; and we gaze in astonished admiration at the posturings of effeminate males, because they counterfeit by their womanish motions a sex which nature has denied to men, and deceive the eyes of the spectators. And presently, then, that we may come to our gluttonous feasts in proper fettle, we steam out our daily indigestion in sweat-baths, and by drying out the moisture of our bodies we arouse a thirst; we spend our nights in licentiousness and drunkenness, our days in gaming or in sleeping, and account ourselves blessed by fortune in that "we behold neither the rising of the sun nor its setting."[8] The consequence is that ill health attends so slothful a manner of living; for the bodies of our young men are so flabby and enervated that death seems likely to make no change in them.

But, by heaven, that true stock of Romulus, practiced in constant hunting and no less in toiling in the fields, was distinguished by the greatest physical strength and, hardened by the labors of peace, easily endured the hardships of war when occasion demanded, and always esteemed the common people of the country more highly than those of the city. For as those who kept within the confines of the country houses were accounted more slothful than those who tilled the ground outside, so those who spent their time idly within the walls, in the shelter of the city, were looked upon as more sluggish than those who tilled the fields or supervised the labors of the tillers. It is evident, too, that their market-day gatherings were employed for this purpose—that city affairs might be transacted on every ninth day only and country affairs on the other days. For in those times, as we have previously remarked, the leading men of the state used to pass their time in the fields and were summoned from their farms to the senate when advice on matters of state was wanted; as a result of which those who summoned them were called *viatores* or "road-men." And so long as this custom was preserved, with a most persevering enthusiasm for tilling their lands, those old Sabine *Quirites*[9] and our Roman forefathers, even though exposed to fire and sword, and despite the devastation of their crops by hostile forays, still laid by a greater store of crops than do we, who, with the sufferance of long-continued peace, might have extended the practice of agriculture.

So, then, in "this Latium and Saturnian land,"[10] where the gods had taught their offspring of the fruits of the fields, we let contracts at auction for the importation of grain from our provinces beyond the sea, that we may not suffer hunger; and we lay up our stores of wine from the Cyclades Islands and from the districts of Baetica[11] and Gaul. Nor is it to be wondered at, seeing that the common notion is now generally entertained and established that farming is a mean employment and a business which has no need of direction or of precept. But for my part, when I review the magnitude of the entire subject, like the immensity of some great body, or the minuteness of its several parts, as so many separate members, I am afraid that my last day may overtake me before I can comprehend the entire subject of rural discipline. . . .

[8]Cato, in Seneca's *Epistles* 122.2. (Ash)
[9]"Citizens." See *Pervigilium Veneris* line 73, and note to line 69 (Chapter 11 above).
[10]The authorship of this phrase is attributed to Ennius. (Ash)
[11]The Cyclades are [Aegean islands] in Greece. Baetica was a district of southern Spain (modern Andalusia), where Columella was born in the town of Gades (Cádiz). (Ash)

The Stern Virtues of Barbarian Life

Germania, by Tacitus, translated by Alfred John Church and William Jackson Brodribb

Publius (or Gaius) Cornelius Tacitus, greatest historian of the Roman Empire, was born around A.D. *56 and probably died after 115. He belonged to the upper-class society of his age, having in 77 married the daughter of Gnaeus Julius Agricola, a leading general under Vespasian and his sons Titus and Domitian. The younger Pliny wrote to him (in Betty Radice's translation of Letter VII.20), "I am delighted to think that if posterity takes any interest in us, the tale will everywhere be told of the harmony, frankness, and loyalty of our lifelong relationship." Three short works survive almost in their entirety, the* Agricola *(a laudatory biography of his father-in-law, especially as legate in Britain) and the* Germania *(an account of the peoples of Germany, most of whom lay outside of—and continually threatened—the Empire), both published in 98, and the* Dialogue on Oratory, *probably composed soon afterward. His masterpieces, the* Histories *and above all the* Annals, *which between them covered most of the first century* A.D. *(from the death of Augustus in 14 to the murder of the tyrannical Domitian in 96), are a devastatingly mordant indictment of the corruptions of absolute power; they survived, in fragmentary form, in a single manuscript.*

Like other Roman writers, Tacitus yearned for spiritual or imaginative refuge from an often murderous civilization, especially before Nerva and Trajan, following the terror of Domitian's final years, introduced (Tacitus wrote at the beginning of his Histories*) "the rare happiness of times, when we may think what we please, and express what we think." He found alternatives not in rustic retreat but in poetry—where "the soul withdraws herself to abodes of purity and innocence, and enjoys her holy resting-place" not in "present money-getting and blood-stained eloquence" but in "the happy golden age" of the poets (*Dialogue *12)—or in a ruder society, like that of the Germans, still uncorrupted by the terrible excesses of his own. He by no means sentimentalizes these Germans as noble savages free of faults, yet his admiring account of their stern moral virtues stands in stark contrast to his indictment of the depravities of Tiberius, Nero, or Domitian.*

The translation is from The Complete Works of Tacitus, *ed. Moses Hadas (1942).*

16 It is well known that the nations of Germany have no cities, and that they do not even tolerate closely contiguous dwellings. They live scattered and apart, just as a spring, a meadow, or a wood has attracted them. Their villages they do not arrange in our fashion, with the buildings connected and joined together, but every person surrounds his dwelling with an open space, either as a precaution against the disasters of fire, or because they do not know how to build. No use is made by them of stone or tile; they employ timber for all purposes, rude masses without ornament or attractiveness. Some parts of their buildings they stain more carefully with a clay so clear and bright that it resembles painting, or a colored design. They are wont also to dig out subterranean caves, and pile on them great heaps of dung, as a shelter from winter and as a receptacle for the year's produce, for by such places they mitigate the rigor of the cold. And should an enemy approach, he lays waste the open country, while what is hidden and buried is either not known to exist, or else escapes him from the very fact that it has to be searched for.

17 They all wrap themselves in a cloak which is fastened with a clasp, or, if this is not forthcoming, with a thorn, leaving the rest of their persons bare. They pass whole days on the hearth by the fire. The wealthiest are distinguished by a dress which is not flowing, like that of the Sarmatae and Parthi,[12] but is tight, and exhibits each limb. They also wear the skins of wild beasts; the tribes on the Rhine and Danube in a careless fashion, those of the interior with more elegance, as not obtaining other clothes by commerce. These select certain animals, the hides of which they strip off and vary them with the spotted skins of beasts, the produce of the outer ocean, and of seas unknown to us. The women have the same dress as the men, except that they generally wrap themselves in linen garments, which they embroider with purple, and do not lengthen out the upper part of their clothing into sleeves. The upper and lower arm is thus bare, and the nearest part of the bosom is also exposed.

18 Their marriage code, however, is strict, and indeed no part of their manners is more praiseworthy. Almost alone among barbarians they are content with one wife, except a very few among them, and these not from sensuality, but because their noble birth procures for them many offers of alliance. The wife does not bring a dower to the husband, but the husband to the wife. The parents and relatives are present, and pass judgment on the marriage-gifts, gifts not meant to suit a woman's taste, nor such as a bride would deck herself with, but oxen, a caparisoned steed, a shield, a lance, and a sword. With these presents the wife is espoused, and she herself in her turn brings her husband a gift of arms. This they count their strongest bond of union, these their sacred mysteries, these their gods of marriage. Lest the woman should think herself to stand apart from aspirations after noble deeds and from the perils of war, she is reminded by the ceremony which inaugurates marriage that she is her husband's partner in toil and danger, destined to suffer and to dare with him alike both in peace and in war. The yoked oxen, the harnessed steed, the gift of arms, proclaim this fact. She must live and die with the feeling that she is receiving what she must hand down to her children neither tarnished nor depreciated, what future daughters-in-law may receive, and may be so passed on to her grandchildren.

19 Thus with their virtue protected they live uncorrupted by the allurements of public shows or the stimulant of feastings. Clandestine correspondence is equally unknown to men and women. Very rare for so numerous a population is adultery, the punishment for which is prompt, and in the husband's power. Having cut off the hair of the adulteress and stripped her naked, he expels her from the house in the presence of her kinsfolk, and then flogs her through the whole village. The loss of chastity meets with no indulgence; neither beauty, youth, nor wealth will procure the culprit a husband. No one in Germany laughs at vice, nor do they call it the fashion to corrupt and to be corrupted. Still better is the condition of those states in which only maidens are given in marriage, and where the hopes and expectations of a bride are then finally terminated. They receive one husband, as having one body and one life, that they may have no thoughts beyond, no further-reaching desires, that they may love not so much the husband as the married state. To limit the number of their children or to destroy any of their subsequent offspring is accounted infamous, and good habits are here more effectual than good laws elsewhere. . . .

[12]The Sarmatae were a nomadic people of eastern Europe; the Parthi the ruling people of Persia.

22 On waking from sleep, which they generally prolong to a late hour of the day, they take a bath, oftenest of warm water, which suits a country where winter is the longest of the seasons. After their bath they take their meal, each having a separate seat and table of his own. Then they go armed to business, or no less often to their festal meetings. To pass an entire day and night in drinking disgraces no one. Their quarrels, as might be expected with intoxicated people, are seldom fought out with mere abuse, but commonly with wounds and bloodshed. Yet it is at their feasts that they generally consult on the reconciliation of enemies, on the forming of matrimonial alliances, on the choice of chiefs, finally even on peace and war, for they think that at no time is the mind more open to simplicity of purpose or more warmed to noble aspirations. A race without either natural or acquired cunning, they disclose their hidden thoughts in the freedom of the festivity. Thus the sentiments of all having been discovered and laid bare, the discussion is renewed on the following day, and from each occasion its own peculiar advantage is derived. They deliberate when they have no power to dissemble; they resolve when error is impossible. . . .

26 Of lending money on interest and increasing it by compound interest they know nothing—a more effectual safeguard than if it were prohibited.

Land proportioned to the number of inhabitants is occupied by the whole community in turn, and afterwards divided among them according to rank. A wide expanse of plains makes the partition easy. They till fresh fields every year, and they have still more land than enough; with the richness and extent of their soil, they do not laboriously exert themselves in planting orchards, inclosing meadows, and watering gardens. Corn is the only produce required from the earth; hence even the year itself is not divided by them into as many seasons as with us. Winter, spring, and summer have both a meaning and a name; the name and blessings of autumn are alike unknown.

27 In their funerals there is no pomp; they simply observe the custom of burning the bodies of illustrious men with certain kinds of wood. They do not heap garments or spices on the funeral pile. The arms of the dead man and in some cases his horse are consigned to the fire. A turf mound forms the tomb. Monuments with their lofty elaborate splendor they reject as oppressive to the dead. Tears and lamentations they soon dismiss; grief and sorrow but slowly. It is thought becoming for women to bewail, for men to remember the dead.

Such on the whole is the account which I have received of the origin and manners of the entire German people. . . .

Nothing Is Difficult for Nature

Natural Questions, by Seneca,
translated by Thomas H. Corcoran

Lucius Annaeus Seneca—sometimes called Seneca the Younger to distinguish him from his father, a famed rhetorician of the same name—was born in Corduba (Cordova), Spain, near the beginning of the first century A.D.*, and studied in Rome, where he made his reputation as a leading orator and writer. Between 41 and 49 he was exiled to Corsica on charges by Messalina, wife of the feeble new Emperor Claudius, of adultery with Julia Livilla, sister of the recently* *murdered Emperor Gaius (Caligula); after Messalina's murder, he was recalled by Claudius's new wife Agrippina (sister of Gaius and Julia) to act as tutor to her young son by a previous marriage, Nero. When Nero succeeded Claudius (whose murder Agrippina may have arranged) in 54, Seneca and Sextus Afranius Burrus effectively acted as co-regents to give the Empire a brief period of good rule. But after Nero (with or without their knowledge) arranged his mother's murder, the increasingly depraved emperor turned against them too. In 62 Burrus died, probably poisoned. Seneca retired to private life and writing until he was suspected by Nero of plotting against him and commanded, in 65, to commit suicide; his end is vividly described by Tacitus (*Annals *XV.60–64).*

*Such was the background against which this Stoic philosopher wrote the moral essays and letters which, along with some highly oratorical tragedies and a Menippean satire, make up the bulk of his writings. Seneca was by no means an original thinker, and the seeming abyss between his rhetorical moralizing and his actions has disturbed many readers, but through his pithy style and probing intellect he exerted immense influence for many centuries. Ethical questions were central to him, but his interest extended to nature as well; thus he notes, in *Moral Epistle *CXXI, that "whatever art transmits is uncertain and unequal; what nature distributes turns out alike." The *Naturales Quaestiones*—here quoted from the two-volume Loeb Classical Library translation (1971–72)—examine a variety of natural phenomena with an emphasis not on scientific investigation but on moral lessons to be derived from them through their elevating effect on the mind. Yet in Seneca the natural like the human world tends to catastrophe. In the year of his forced retirement, Pompeii was shattered by an earthquake (prelude to the volcanic eruption that destroyed it in 79), and this less than perfectly apathetic Stoic seems obsessed by the troubling realization that nature's omnipotence extends to cosmic ruin, since "any deviation from the existing state of the universe is enough for the destruction of mankind."*

The Elevating Study of Nature

I. PREFACE . . . I, for one, am very grateful to nature, not just when I view it in that aspect which is obvious to everybody but when I have penetrated its mysteries; when I learn what the stuff of the universe is, who its author or custodian is, what god is; whether he keeps entirely to himself or whether he sometimes considers us; whether he creates something each day or has created it only once; whether he is a part of the universe or is the universe. . . .

If I had not been admitted to these studies it would not have been worth while to have been born. . . . After all, man is a contemptible thing unless he rises above his human concerns. . . . I do not see why a man should feel pleased who is simply less sick than others in the hospital. . . .

That special virtue which we seek is magnificent, not because to be free of evil is in itself so marvelous but because it unchains the mind, prepares it for the realization of heavenly things, and makes it worthy to enter into an association with god. The mind possesses the full and complete benefit of its human existence only when it spurns all evil, seeks the lofty and the deep, and enters the innermost secrets of nature. Then as the mind wanders among the very stars it delights in laughing at the mosaic floors of the rich and at the whole earth with all its gold. . . . When the mind contacts those regions it is nurtured, grows, and returns to its origin just as though freed from its chains. As proof of its divinity it has this: divine things cause it pleasure, and it dwells among them not as being alien things but things of its own nature. Serenely it looks upon the rising and setting of the stars and the diverse orbits of bodies precisely balanced with one another. The mind observes where each star first shows its light to earth, where its culmination, the highest altitude of its course, lies and how far it descends. As a curious spectator the mind separates details and investigates them. Why not do this? It knows that these things pertain to itself. Then it despises the limitation of its former dwelling place. . . .

What, then, is the difference between our nature and the nature of god? In ourselves the better part is the mind, in god there is no part other than the mind. He is entirely reason. None the less, meanwhile, a great error possesses mortals: men believe that this universe, than which nothing is more beautiful or better ordered or more consistent in plan, is an accident, revolving by chance, and thus tossed about in lightning bolts, clouds, storms, and all the other things by which the earth and its vicinity are kept in turmoil. Nor does this nonsense exist among only the common people; it also infects those who say they have knowledge. There are some men who conclude that they themselves have a mind, indeed a provident one, evaluating situations, both their own and other peoples'; but the universe, in which we also exist, they presume is lacking in plan and either moves along in some haphazard way or else nature does not know what it is doing. . . .

III. Preface On this point it will help us to study nature. In the first place we will get away from sordid matters. Second, we will free the mind—and we need one that is sound and great—from the body. Third, the subtlety of thought exercised on the mysteries of nature will be no less successful in dealing with plain problems. Moreover, nothing is plainer than those salutary lessons we are taught against our own iniquities and follies which we condemn but do not renounce. . . .

The Destructive Powers and the Mysteries of Nature

III.27 But this subject reminds me to wonder how a great part of the earth will be covered over by water when the fated day of the deluge comes.[13] Will it be by the force of the ocean and the rising of the outer sea against us or will heavy rains fall without ceasing and persistent winter eliminate summer and hurl the full force of water down from burst clouds? Or will the earth pour out rivers far and wide and open new springs? Or will there be no single cause for such a catastrophe but rather all principles working together; at the same time the rains will descend, the rivers rise, the seas rush violently from their places, and all things in a united effort will apply themselves to the destruction of the human race? And so it will be. Nothing is difficult for nature, especially when she rushes to destroy herself. At the beginning of things she uses her strength sparingly and apportions herself out in imperceptible increases. For destruction she comes suddenly with all her violence. A long time is needed so that a child, once conceived, may come to be born. The tender infant is reared only with great toil. The frail body finally develops only with diligent nurture. But how with no effort it is all undone! It takes an age to establish cities, an hour to destroy them. A forest grows for a long time, becomes ashes in a moment. Great safeguards may exist and all things may be flourishing, but quickly and suddenly they all fall apart.

Any deviation by nature from the existing state of the universe is enough for the destruction of mankind. So, when that destined time comes the fates put into motion many

[13]Seneca's treatment of the universal catastrophe as if occurring in his time may be caused by the idea that it recurs; and thus can be thought of as a permanent feature of the universe. Hence his present tenses. (Corcoran)

causes at the same time. For according to some thinkers, among them Fabianus,[14] such a great change does not occur without a shattering of the universe.

At first, excessive rain falls. There is no sunshine, the sky is gloomy with clouds, and there is continuous mist, and from the moisture a thick fog which no winds will ever dry out. Next, there is a blight on the crops, a withering of fields of standing grain as it grows without fruit. Then all things sown by hand rot and swamp plants spring up in all the fields. Next the stronger plants also feel the blight. For, indeed, trees fall when their roots are loosened, and vines and all shrubs are not held by the soil, which becomes soft and fluid. No longer does the ground sustain grazing land and water-loving pasturage. There is suffering from famine, and recourse is had to the diet of ancient times: food is shaken down wherever there is an ilex or an oak or where any tree on a hillside stands, still held firm by tightly joined rocks. Houses sag and drip, and when the moisture penetrates deeply the foundations settle and all the ground becomes marshy. Props are tried for collapsing houses, uselessly; for every foundation is set in slipping and muddy soil. Nothing is stable.

After the clouds have massed more and more, and the accumulated snows of centuries have melted, a torrent which has rolled down from the highest mountains carries off forests that are unable to cling fast and tears boulders free from their loosened structures and rolls them along, washes away villas and carries down sheep and owners intermixed. The smaller houses are plucked up by the torrent which carries them off as it passes. Finally the torrent is diverted violently against larger dwellings and drags along cities and peoples who are forced back to their city walls and uncertain whether they should complain of a cave-in or a shipwreck. In such a way the disaster comes, which both crushes and submerges them at the same time. Eventually, as the torrent passes along it is increased by the absorption into itself of several other torrents and spreads out in scattered devastation on the plain. Finally, it pours out in all directions, loaded with the vast stuff of nations. . . .

VII.30 How many animals we have learned about for the first time in this age; how many are not known even now! Many things that are unknown to us the people of a coming age will know. Many discoveries are reserved for ages still to come, when memory of us will have been effaced. Our universe is a sorry little affair unless it has in it something for every age to investigate. Some sacred things are not revealed once and for all. Eleusis keeps in reserve something to show to those who revisit there.[15] Nature does not reveal her mysteries once and for all. We believe that we are her initiates but we are only hanging around the forecourt. Those secrets are not open to all indiscriminately. They are withdrawn and closed up in the inner sanctum. This age will glimpse one of the secrets; the age which comes after us will glimpse another. . . .

VII.32 There is no interest in philosophy. Accordingly, so little is found out from those subjects which the ancients left partially investigated that many things which were discovered are being forgotten. But, by Hercules, if we applied ourselves to this with all our might—if youth soberly applied itself to it, if the elders taught it and the younger generation learned it—we would scarcely reach to the bottom where truth is located, which we now seek on the surface of the earth and with slack effort.

[14]Servius Flavius Papirius Fabianus lived in the first century A.D. and wrote *Naturalium Causarum Libri [Books of Natural Causes]*. (Corcoran)

[15]Eleusis, near Athens, was the site of the most sacred mystery rites of ancient Greece, those devoted to Demeter and Persephone (Roman Ceres and Proserpina).

No Statement About Her Incredible

The *Natural History* of Pliny the Elder

Gaius Plinius Secundus—called Pliny the Elder to distinguish him from his nephew and adopted son, the letter writer Gaius Plinius Caecilius Secundus, or Pliny the Younger—was born in Comum (modern Como) in northern Italy in 23/24 A.D. and educated in Rome. He spent twelve years in military service and held various offices under Vespasian and Titus. His only surviving work is the immense Naturalis Historia *in thirty-seven books: "a learned and comprehensive work," the younger Pliny aptly writes (Letter III.5, in Betty Radice's translation), "as full of variety as nature itself." In this same letter the younger Pliny goes on to describe his uncle's indefatigable work habits and amazing powers of concentration, recalling "how he scolded me for walking" in Rome, instead of being carried in a chair, "for he thought any time wasted that was not devoted to work." His inquisitiveness indeed led to his death. When stationed at Misenum near Naples in 79, in charge of the fleet, he observed huge clouds of smoke from the eruption of Vesuvius that destroyed Pompeii and Herculaneum, and determined—as the younger Pliny, who remained at a safe distance, describes it in a letter (VI.16) to Tacitus—to investigate, and then to help rescue those trapped near the shore, "describing each new movement and phase of the portent to be noted down exactly as he observed them," until he fell dead from suffocation by the sulfurous fumes.*

In his vast Natural History, *however—which embraces books on the universe,the geography of the earth, anthropology, animals, plants, medicine (over a third of the whole), and metals and stones (including their use in art and architecture)—it is not personal observation that dominates but massive citation of innumerable literary sources. The elder Pliny's work is the most imposing extant monument of what William H. Stahl, writing of the erudite Varro in* Roman Science *(1962), calls "the classical Roman attitude which survived among Latin writers of the Middle Ages: that the man who could assemble the greatest number of authorities on a subject was himself the most reliable authority." He voraciously read everything and seldom questioned the statements, however improbable, of his authorities. One result is that in his collection, as in many of the older Greek and Roman writers he cites (to say nothing of Aelian and the medieval encyclopedists [Chapter 13 below] who would follow), detailed and useful information of every kind is mingled higgledy-piggledy with the most absurd fantasies. All are ascribed to the workings of Nature, who thereby becomes less the rationally guided process of the Milesian natural philosophers, Aristotle, or the Stoics, than a miracle worker capable of creating virtually anything in virtually any way. This, too, would be a conception of nature that later times would inherit from the Greeks and Romans, largely through Pliny.*

"Pliny's science oscillates between the intent to recognize an order in nature and the recording of what is extraordinary or unique," Italo Calvino writes in "Man, the Sky, and the Elephant: On Pliny's Natural History*" (in Daniel Halpern, ed.,* On Nature *[1986]): "and the second aspect of it always wins out. Nature is eternal and sacred and harmonious, but it leaves a wide margin for the emergence of inexplicable prodigious phenomena." This sprawling book, whose credulity is sometimes held in check but whose curiosity is wholly insatiable, is animated "by admiration for everything that exists and respect for the infinite variety of things." Within its world, where man is next of kin to the elephant, nature is "external to man, but not to be separated from what is most intrinsic to his mind—the alphabet of dreams, the code book of the imagination, without which there is neither thought nor reason." Already in the first century A.D., Pliny thus*

stands more than halfway betwen Aristotle's empirical biology and the fabulous bestiaries of the Middle Ages (Chapter 13 below) that would begin to take shape not long after his time.

The translations are from the Loeb Classical Library edition of Pliny's Natural History *in ten volumes (1938–62).*

The Madness of Looking Beyond This World, translated by H. Rackham
II.i

The world and this—whatever other name men have chosen to designate the sky whose vaulted roof encircles the universe, is fitly believed to be a deity, eternal, immeasurable, a being that never began to exist and never will perish. What is outside it does not concern men to explore and is not within the grasp of the human mind to guess. It is sacred, eternal, immeasurable, wholly within the whole, nay rather itself the whole, finite and resembling the infinite, certain of all things and resembling the uncertain, holding in its embrace all things that are without and within, at once the work of nature and nature herself.

That certain persons have studied, and have dared to publish, its dimensions, is mere madness; and again that others,[16] taking or receiving occasion from the former, have taught the existence of a countless number of worlds, involving the belief in as many systems of nature, or, if a single nature embraces all the worlds, nevertheless the same number of suns, moons and other unmeasurable and innumerable heavenly bodies, as already in a single world; just as if owing to our craving for some End the same problem would not always encounter us at the termination of this process of thought, or as if, assuming it possible to attribute this infinity of nature to the artificer of the universe, that same property would not be easier to understand in a single world, especially one that is so vast a structure. It is madness, downright madness, to go out of that world, and to investigate what lies outside it just as if the whole of what is within it were already clearly known; as though, forsooth, the measure of anything could be taken by him that knows not the measure of himself, or as if the mind of man could see things that the world itself does not contain.

Man the Stepchild of Nature, translated by H. Rackham
VII.Preface

The above is a description of the world, and of the lands, races, seas, important rivers, islands and cities that it contains.

The nature of the animals also contained in it is not less important than the study of almost any other department, albeit here too the human mind is not capable of exploring the whole field.

The first place will rightly be assigned to man, for whose sake great Nature appears to have created all other things—though she asks a cruel price for all her generous gifts, making it hardly possible to judge whether she has been more a kind parent to man or more a harsh stepmother. First of all, man alone of all animals she drapes with borrowed resources. On all the rest in various wise she bestows coverings—shells, bark, spines, hides, fur, bristles, hair, down, feathers, scales, fleeces; even the trunks of trees she has

[16]The founders of the atomic theory, Leucippus and Democritus [and Epicurus]. (Rackham)

protected against cold and heat by bark, sometimes in two layers: but man alone on the day of his birth she casts away naked on the naked ground, to burst at once into wailing and weeping, and none other among all the animals is more prone to tears, and that immediately at the very beginning of life; whereas, I vow, the much-talked-of smile of infancy even at the earliest is bestowed on no child less than six weeks old. This initiation into the light is followed by a period of bondage such as befalls not even the animals bred in our midst, fettering all his limbs; and thus when successfully born he lies with hands and feet in shackles, weeping—the animal that is to lord it over all the rest, and he initiates his life with punishment because of one fault only, the offense of being born. Alas the madness of those who think that from these beginnings they were bred to proud estate!

His earliest promise of strength and first grant of time makes him like a four-footed animal. When does man begin to walk? when to speak? when is his mouth firm enough to take food? how long does his skull throb, a mark of his being the weakest among all animals? Then his diseases, and all the cures contrived against his ills—these cures also subsequently defeated by new disorders! And the fact that all other creatures are aware of their own nature, some using speed, others swift flight, others swimming, whereas man alone knows nothing save by education—neither how to speak nor how to walk nor how to eat; in short the only thing he can do by natural instinct is to weep! Consequently there have been many who believed that it were best not to be born, or to be put away as soon as possible. On man alone of living creatures is bestowed grief, on him alone luxury, and that in countless forms and reaching every separate part of his frame; he alone has ambition, avarice, immeasurable appetite for life, superstition, anxiety about burial and even about what will happen after he is no more. No creature's life is more precarious, none has a greater lust for all enjoyments, a more confused timidity, a fiercer rage. In fine, all other living creatures pass their time worthily among their own species: we see them herd together and stand firm against other kinds of animals—fierce lions do not fight among themselves, the serpent's bite attacks not serpents, even the monsters of the sea and the fishes are only cruel against different species; whereas to man, I vow, most of his evils come from his fellow-man.

If One Is Willing to Believe It,
translated by H. Rackham
VII.ii.15–30

. . . Beyond the Nasamones and adjacent to them Calliphanes records the Machlyes, who are Androgyni[17] and perform the function of either sex alternately. Aristotle adds that their left breast is that of a man and their right breast that of a woman. Isogonus and Nymphodorus report that there are families in the same part of Africa that practice sorcery, whose praises cause meadows to dry up, trees to wither and infants to perish. Isogonus adds that there are people of the same kind among the Triballi and the Illyrians, who also bewitch with a glance and who kill those they stare at for a longer time, especially with a look of anger, and that their evil eye is most felt by adults; and that what is more remarkable is that they have two pupils in each eye. Apollonides also reports women of this kind in Scythia, who are called the Bitiae, and Phylarchus also the Thibii tribe and many others of the same nature in Pontus, whose distinguishing marks he records as being a double pupil in one eye and the likeness of a horse in the other, and he also says that they are incapable of drowning, even when weighed down with clothing. Damon

[17]Androgynes, or hermaphrodites, were said to combine the sexual organs of male and female.

records a tribe not unlike these in Ethiopia, the Pharmaces, whose sweat relieves of diseases bodies touched by it. Also among ourselves Cicero states that the glance of all women who have double pupils is injurious everywhere. In fact when nature implanted in man the wild beasts' habit of devouring human flesh, she also thought fit to implant poisons in the whole of the body, and with some persons in the eyes as well, so that there should be no evil anywhere that was not present in man.

There are a few families in the Faliscan territory, not far from the city of Rome, named the Hirpi, which at the yearly sacrifice to Apollo performed on Mount Soracte walk over a charred pile of logs without being scorched, and who consequently enjoy exemption under a perpetual decree of the senate from military service and all other burdens. Some people are born with parts of the body possessing special remarkable properties, for instance King Pyrrhus in the great toe of his right foot, to touch which was a cure for inflammation of the spleen; it is recorded that at his cremation it proved impossible to burn the toe with the rest of the body, and it was stored in a chest in a temple.

India and parts of Ethiopia especially teem with marvels. The biggest animals grow in India: for instance Indian dogs are bigger than any others. Indeed the trees are said to be so lofty that it is not possible to shoot an arrow over them, and the richness of the soil, temperate climate and abundance of springs bring it about that, if one is willing to believe it, squadrons of cavalry are able to shelter beneath a single fig-tree; while it is said that reeds are of such height that sometimes a single section between two knots will make a canoe that will carry three people. It is known that many of the inhabitants are more than seven feet six inches high, never spit, do not suffer from headache or toothache or pain in the eyes, and very rarely have a pain in any other part of the body—so hardy are they made by the temperate heat of the sun; and that the sages of their race, whom they call Gymnosophists,[18] stay standing from sunrise to sunset, gazing at the sun with eyes unmoving, and continue all day long standing first on one foot and then on the other in the glowing sand. Megasthenes states that on the mountain named Nulus there are people with their feet turned backwards and with eight toes on each foot, while on many of the mountains there is a tribe of human beings with dogs' heads, who wear a covering of wild beasts' skins, whose speech is a bark and who live on the produce of hunting and fowling, for which they use their nails as weapons; he says that they numbered more than 120,000 when he published his work. Ctesias writes that also among a certain race of India the women bear children only once in their life-time, and the children begin to turn grey directly after birth; he also describes a tribe of men called the Monocoli who have only one leg, and who move in jumps with surprising speed; the same are called the Umbrella-foot tribe, because in the hotter weather they lie on their backs on the ground and protect themselves with the shadow of their feet; and that they are not far away from the Cave-dwellers; and again westward from these there are some people without necks, having their eyes in their shoulders. There are also satyrs in the mountains in the east of India (it is called the district of the Catarcludi); this is an extremely swift animal, sometimes going on all fours, and sometimes standing upright as they run, like human beings; because of their speed only the old ones or the sick are caught. Tauron gives the name of Choromandae to a forest tribe that has no speech but a horrible scream, hairy bodies, keen grey eyes and the teeth of a dog. Eudoxus says that in the south of India men have feet eighteen inches long and the women such small feet that they are called

[18]Stories of the "Gymnosophists," or "naked philosophers," of India may have originated with reports of the ascetic Jains (Chapter 4 above), some of whom renounced even clothing.

Sparrow-feet. Megasthenes tells of a race among the Nomads of India that has only holes in the place of nostrils, like snakes, and bandy-legged; they are called the Sciritae. At the extreme boundary of India to the East, near the source of the Ganges, he puts the Astomi tribe, that has no mouth and a body hairy all over; they dress in cottonwool and live only on the air they breathe and the scent they inhale through their nostrils; they have no food or drink except the different odors of the roots and flowers and wild apples, which they carry with them on their longer journeys so as not to lack a supply of scent; he says they can easily be killed by a rather stronger odor than usual. Beyond these in the most outlying mountain region we are told of the Three-span men and Pygmies, who do not exceed three spans, *i.e.* twenty-seven inches, in height; the climate is healthy and always spring-like, as it is protected on the north by a range of mountains; this tribe Homer[19] has also recorded as being beset by cranes. It is reported that in springtime their entire band, mounted on the backs of rams and she-goats and armed with arrows, goes in a body down to the sea and eats the cranes' eggs and chickens, and that this outing occupies three months; and that otherwise they could not protect themselves against the flocks of cranes that would grow up; and that their houses are made of mud and feathers and egg-shells. Aristotle says that the Pygmies live in caves, but in the rest of his statement about them he agrees with the other authorities. The Indian race of Cyrni according to Isigonus live to 140; and he holds that the same is true of the Long-lived Ethiopians, the Chinese and the inhabitants of Mount Athos [in Greece]—in the last case because of their diet of snakes' flesh, which causes their head and clothes to be free from creatures harmful to the body. Onesicritus says that in the parts of India where there are no shadows there are men five cubits and two spans high,[20] and people live a hundred and thirty years, and do not grow old but die middle-aged. Crates of Pergamum tells of Indians who exceed a hundred years, whom he calls Gymnetae, though many call them Long-livers. Ctesias says that a tribe among them called the Pandae, dwelling in the mountain valleys, live two hundred years, and have white hair in their youth that grows black in old age; whereas others do not exceed forty years, this tribe adjoining the Long-livers, whose women bear children only once. Agatharchides records this as well, and also that they live on locusts, and are very swift-footed. Clitarchus gave them the name of Mandi; and Megasthenes also assigns them three hundred villages, and says that the women bear children at the age of seven and old age comes at forty. Artemidorus says that on the Island of Ceylon the people live very long lives without any loss of bodily activity. Duris says that some Indians have union with wild animals and the offspring is of mixed race and half animal; that among the Calingi, a tribe of the same part of India, women conceive at the age of five and do not live more than eight years, and that in another part men are born with a hairy tail and extremely swift, while others are entirely covered by their ears. . . .

The Greatest of Nature's Works,
translated by W. H. S. Jones
XXXII.i.1–3

The course of my subject has brought me to the greatest of Nature's works, and I am actually met by such an unsought and overwhelming proof of hidden power that inquiry should really be pursued no further, and nothing equal or similar can be found, Nature surpassing herself, and that in numberless ways. For what is more violent than sea, winds,

[19]*Iliad* III.6. [20]About eight feet. (Rackham)

whirlwinds, and storms? By what greater skill of man has Nature been aided in any part of herself than by sails and oars? Let there be added to these the indescribable force of tidal ebb and flow, the whole sea being turned into a river. All these, however, although acting in the same direction, are checked by a single specimen of the sucking fish,[21] a very small fish. Gales may blow and storms may rage; this fish rules their fury, restrains their mighty strength, and brings vessels to a stop, a thing no cables can do, nor yet anchors of unmanageable weight that have been cast. It checks their attacks and tames the madness of the Universe with no toil of its own, not by resistance, or in any way except by adhesion. This little creature suffices in the face of all these forces to prevent vessels from moving. But armored fleets bear aloft on their decks a rampart of towers, so that fighting may take place even at sea as from the walls of a fortress. How futile a creature is man, seeing that those rams, armed for striking with bronze and iron, can be checked and held fast by a little fish six inches long! . . .

Nature's Grandeur in the Narrowest Limits,
translated by D. E. Eichholz
XXXVII.i; lxxviii.205

In order that the work that I have undertaken may be complete, it remains for me to discuss gemstones. Here Nature's grandeur is gathered together within the narrowest limits; and in no domain of hers evokes more wonder in the minds of many who set such store by the variety, the colors, the texture and the elegance of gems that they think it a crime to tamper with certain kinds by engraving them as signets, although this is the prime reason for their use; while some they consider to be beyond price and to defy evaluation in terms of human wealth. Hence very many people find that a single gemstone alone is enough to provide them with a supreme and perfect aesthetic experience of the wonders of Nature. . . .

Hail, Nature, mother of all creation, and mindful that I alone of the men of Rome have praised thee in all thy manifestations, be gracious unto me.

Fabulous Though They Be

On the Characteristics of Animals, by Aelian, translated by A. F. Scholfield

Claudius Aelianus was born around 170 in Praeneste (modern Palestrina), southeast of Rome, and was over sixty when he died; though a Roman, who spent most of his life in Rome, he wrote only in Attic Greek. He was a member of the influential cultural circle that Julia Domna, wife of Septimius Severus (emperor from 193 to 211), gathered around herself. The group also included the poet Oppian (author of Greek poems on hunting and fishing), the physician Galen, and Flavius Philostratus, a leading Greek rhetorician of the so-called Second Sophistic, whose Life *of Apollonius of Tyana portrayed this Pythagorean miracle worker as a sort of pagan counterpart to Jesus. Aelian's extant works are the seventeen books* On the Characteristics of Animals, *often known by its Latin title as* De Natura Animalium, *from which the following selec-*

[21] *Echenais* or *echeneis*, the suckerfish or remora.

tions are taken from the three-volume Loeb Classical Library edition (1958–59); the twenty-four Country Letters *describing rural life; and the fourteen books of the* Miscellaneous History, *combining historical and mythological anecdotes.*

In the disorganized miscellany of the Characteristics of Animals, *facts and fables gleaned from earlier and contemporary Greek writers to illustrate the habits of the animal world are set down side by side in "a search after the picturesque, the startling, even the miraculous," in which, Scholfield notes, "mythology, mariners' yarns, vulgar superstitions, the ascertained facts of nature—all serve to adorn a tale and, on occasion, point a moral" of a vaguely Stoic kind. The tenuous order and discipline Pliny had imposed on his material has almost vanished, and Nature has become a wonder-working magician (or is she a charlatan?) no less than the fabulous Apollonius of Philostratus.*

Two Beasts of India
IV.21, 27

There is in India a wild beast, powerful, daring, as big as the largest lion, of a red color like cinnabar, shaggy like a dog, and in the language of India it is called *Martichoras.*[22] Its face however is not that of a wild beast but of a man, and it has three rows of teeth set in its upper jaw and three in the lower; these are exceedingly sharp and larger than the fangs of a hound. Its ears also resemble a man's, except that they are larger and shaggy; its eyes are blue-grey and they too are like a man's, but its feet and claws, you must know, are those of a lion. To the end of its tail is attached the sting of a scorpion, and this might be over a cubit in length; and the tail has stings at intervals on either side. But the tip of the tail gives a fatal sting to anyone who encounters it, and death is immediate. If one pursues the beast it lets fly its stings, like arrows, sideways, and it can shoot a great distance; and when it discharges its stings straight ahead it bends its tail back; if however it shoots in a backward direction, as the Sacae do,[23] then it stretches its tail to its full extent. Any creature that the missile hits it kills; the elephant alone it does not kill. These stings which it shoots are a foot long and the thickness of a bulrush. Now Ctesias asserts (and he says that the Indians confirm his words) that in the places where those stings have been let fly others spring up, so that this evil produces a crop. And according to the same writer the Mantichore for choice devours human beings; indeed it will slaughter a great number; and it lies in wait not for a single man but would set upon two or even three men, and alone overcomes even that number. All other animals it defeats: the lion alone it can never bring down. That this creature takes special delight in gorging human flesh its very name testifies, for in the Greek language it means *man-eater*, and its name is derived from its activities. Like the stag it is extremely swift.

Now the Indians hunt the young of these animals while they are still without stings in their tails, which they then crush with a stone to prevent them from growing stings. The sound of their voice is as near as possible that of a trumpet.

Ctesias declares that he has actually seen this animal in Persia (it had been brought from India as a present to the Persian King)—if Ctesias is to be regarded as a sufficient authority on such matters. At any rate after hearing of the peculiarities of this animal, one must pay heed to the historian of Cnidos. . . .[24]

[22]The English form is *mantichore*. The word is derived from the Persian *mardkora* = "man-slayer"; perhaps a man-eating tiger. (Scholfield)
[23]Iranian nomads inhabiting the country SE of the Sea of Aral between the rivers Jaxartes and Oxus. They contributed a contingent to the Persian army. (Scholfield)
[24]Ctesias.

I have heard that the Indian animal the Gryphon is a quadruped like a lion; that it has claws of enormous strength and that they resemble those of a lion. Men commonly report that it is winged and that the feathers along its back are black, and those on its front are red, while the actual wings are neither but are white. And Ctesias records that its neck is variegated wih feathers of a dark blue; that it has a beak like an eagle's, and a head too, just as artists portray it in pictures and sculpture. Its eyes, he says, are like fire. It builds its lair among the mountains, and although it is not possible to capture the full-grown animal, they do take the young ones. And the people of Bactria, who are neighbors of the Indians, say that the Gryphons guard the gold in those parts; that they dig it up and build their nests with it, and that the Indians carry off any that falls from them. The Indians however deny that they guard the aforesaid gold, for the Gryphons have no need of it (and if that is what they say, then I at any rate think that they speak the truth), but that they themselves come to collect the gold, while the Gryphons fearing for their young ones fight with the invaders. They engage too with other beasts and overcome them without difficulty, but they will not face the lion or the elephant. . . .

Pupils of All-Wise Nature
VI.57–59

It seems after all that Spiders are not only dexterous weavers after the manner of Athena the Worker and the goddess of the Loom, but that they are by nature clever at geometry. Thus, they keep to the center and fix with the utmost precision the circle with its boundary based upon it, and have no need of Euclid, for they sit at the very middle and lie in wait for their prey. And they are, as you might say, most excellent weavers and adept at repairing their web. And any thread that you may chance to break of their skilled and delicate workmanship they repair and render sound and whole again.

The Phoenix knows how to reckon five hundred years without the aid of arithmetic, for it is a pupil of all-wise Nature, so that it has no need of fingers or anything else to aid it in the understanding of numbers. The purpose of this knowlege and the need for it are matters of common report. But hardly a soul among the Egyptians knows when the five-hundred-year period is completed; only a very few know, and they belong to the priestly order. But in fact the priests have difficulty in agreeing on these points, and banter one another and maintain that it is not now but at some date later than when it was due that the divine bird will arrive. Meantime while they are vainly squabbling, the bird miraculously guesses the period by signs and appears. And the priests are obliged to give way and confess that they devote their time "to putting the sun to rest with their talk"; but they do not know as much as birds. . . .

If even animals know how to reason deductively, understand dialectic, and how to choose one thing in preference to another, we shall be justified in asserting that in all subjects Nature is an instructress without a rival. . . .

One Tale Too Many
XII.3

The Egyptians assert (though they are far from convincing me), they assert, I say, that in the days of the far-famed Bocchoris[25] a Lamb was born with eight feet and two tails, and that it spoke. They say also that this Lamb had two heads and four horns. It is right

[25]Bócchoris was a king of Egypt in perhaps the ninth century B.C. (Scholfield)

to forgive Homer who bestows speech upon Xanthus the horse [*Iliad* XIX.404], for Homer is a poet. And Alcman could not be censured for imitating Homer in such matters, for the first venture of Homer is a plea sufficient to justify forgiveness. But how can one pay any regard to Egyptians who exaggerate like this? However, fabulous though they be, I have related the peculiarities of this lamb.

The Simplicity of Heavenly Things

The *Almagest* of Ptolemy, translated by G. J. Toomer

Claudius Ptolemaeus, the last great astronomer of the ancient world, lived from about 100 to 175 (his celestial observations are datable between 127 and 148), in Alexandria. His many works—some complete, some fragmentary; some extant in Greek, others in Arabic or Latin translations—include half a dozen or so on astronomy, the Tetrabiblos *on astrology, and others on geography, music, optics, and philosophy. The* Mathematical Treatise, *known since the Middle Ages by the hybrid title of* Almagest, *"The Greatest [Treatise]" (from Arabic* al *and Greek* megistê*), consummated astronomical theories that began with the pre-Socratics, and remained the definitive guide to the heavens until displaced by the heliocentric theory of Copernicus nearly fifteen hundred years later.*

Five hundred years before Ptolemy, Eudoxus of Cnidus (ca. 390–ca. 340 B.C.) had made the first major effort to "save the appearances," or "save the phenomena," of circular planetary movements by postulating concentric spheres whose complex rotations could explain the apparent irregularities—the "stations" and "retrogressions"—of the planets. His geocentric theory, which Aristotle modified and refined, was the basis for Aratus's third-century poem the Phaenomena. *In the mid second century B.C., Hipparchus of Nicaea, building on the work of his third-century predecessor, the great geometer Apollonius of Pergê, advanced a theory combining eccentrics (center points of circular motions displaced from the earth) and epicycles (circles on circles) to "save" the motions of sun and moon in accord with the best available observations. (Meanwhile, Aristarchus of Samos, in the early third century, had proposed a heliocentric theory anticipating Copernicus, but this was widely rejected not only on religious grounds—the Stoic Cleanthes was among those who assailed it—but also because leading mathematicians, including Archimedes, failed to be convinced by its calculations.)*

But Hipparchus, whose only surviving work is a critique of Eudoxus and Aratus, proposed no planetary theory because he lacked sufficient data. As so far as we know, Ptolemy—who throughout the Almagest *generously credits predecessors, whose contributions are now mainly known because of him—was the first to work out the complex mathematics that allowed him to reconcile the best available observational data with an eccentric-epicyclic theory of uniform circular motions capable of explaining planetary movements. Of Aristotle's three divisions of theoretical philosophy—theology, physics, and mathematics—only mathematics (including astronomy), Ptolemy argues, can provide "sure and unshakeable knowledge" pertaining both to eternity (the realm of theology) and to movement (the realm of physics), since the heavenly bodies are both eternal and in motion. Therefore mathematical astronomy, whose apparent complexity reveals an underlying simplicity not to be judged by earthly standards, can elevate its devotees to a Platonic "love of the contemplation of the eternal and unchanging." Here the Greek concep-*

*tion of nature governed by its own laws, a conception at once scientific and religious, finds one of its finest expressions. In the words of an epigram attributed to Ptolemy (*Palatine Anthology *IX.577):*

Mortal I know I am: and yet when I
ponder the stars revolving through the sky,
I touch the earth no more, at Zeus's side
surfeiting on ambrosia far and wide.

Selections are from Ptolemy's Almagest, *trans. Toomer (1984). See also M. R. Cohen and I. E. Drabkin, eds.,* A Source Book in Greek Science *(1948), and G. E. R. Lloyd,* Greek Science After Aristotle *(1973).*

I.1 The true philosophers, Syrus, were, I think, quite right to distinguish the theoretical part of philosophy from the practical. For even if practical philosophy, before it *is* practical, turns out to be theoretical, nevertheless one can see that there is a great difference between the two: in the first place, it is possible for many people to possess some of the moral virtues even without being taught, whereas it is impossible to achieve theoretical understanding of the universe without instruction; furthermore, one derives most benefit in the first case [practical philosophy] from continuous practice in actual affairs, but in the other [theoretical philosophy] from making progress in the theory. Hence we thought it fitting to guide our actions (under the impulse of our actual ideas) in such a way as never to forget, even in ordinary affairs, to strive for a noble and disciplined disposition, but to devote most of our time to intellectual matters, in order to teach theories, which are so many and beautiful, and especially those to which the epithet "mathematical" is particularly applied. For Aristotle divides theoretical philosophy too, very fittingly, into three primary categories, physics, mathematics and theology. For everything that exists is composed of matter, form and motion; none of these can be observed in its substratum by itself, without the others: they can only be imagined. Now the first cause of the first motion of the universe, if one considers it simply, can be thought of as an invisible and motionless deity; the division [of theoretical philosophy] concerned with investigating this [can be called] "theology," since this kind of activity, somewhere up in the highest reaches of the universe, can only be imagined, and is completely separated from perceptible reality. The division which investigates material and ever-moving nature, and which concerns itself with "white," "hot," "sweet," "soft" and suchlike qualities one may call "physics"; such an order of being is situated (for the most part) amongst corruptible bodies and below the lunar sphere. That division which determines the nature involved in forms and motion from place to place, and which serves to investigate shape, number, size, and place, time and suchlike, one may define as "mathematics." Its subject-matter falls as it were in the middle between the other two, since, firstly, it can be conceived of both with and without the aid of the senses, and, secondly, it is an attribute of all existing things without exception, both mortal and immortal: for those things which are perpetually changing in their inseparable form, it changes with them, while for eternal things which have an aethereal nature, it keeps their unchanging form unchanged.

From all this we concluded: that the first two divisions of theoretical philosophy should rather be called guesswork than knowledge, theology because of its completely invisible and ungraspable nature, physics because of the unstable and unclear nature of matter; hence there is no hope that philosophers will ever be agreed about them; and that only mathematics can provide sure and unshakeable knowledge to its devotees, provided one approaches it rigorously.[26] For its kind of proof proceeds by indisputable methods, namely arithmetic and geometry. Hence we were drawn to the investigation of that part of theoretical philosophy, as far as we were able to the whole of it, but especially to the theory concerning divine and heavenly things. For that alone is devoted to the investigation of the eternally unchanging. For that reason it too can be eternal and unchanging (which is a proper attribute of knowledge) in its own domain, which is neither unclear nor disorderly. Furthermore it can work in the domains of the other [two divisions of theoretical philosophy] no less than they do. For this is the best science to help theology along its way, since it is the only one which can make a good guess at [the nature of] that activity which is unmoved and separated; [it can do this because] it is familiar with the attributes of those beings[27] which are on the one hand perceptible, moving and being moved, but on the other hand eternal and unchanging, [I mean the attributes] having to do wih motions and the arrangements of motions. As for physics, mathematics can make a significant contribution. For almost every peculiar attribute of material nature becomes apparent from the peculiarities of its motion from place to place. [Thus one can distinguish] the corruptible from the incorruptible by [whether it undergoes] motion in a straight line or in a circle, and heavy from light, and passive from active, by [whether it moves] towards the center or away from the center. With regard to virtuous conduct in practical actions and character, this science, above all things, could make men see clearly; from the constancy, order, symmetry and calm which are associated with the divine, it makes its followers lovers of this divine beauty, accustoming them and reforming their natures, as it were, to a similar spiritual state.

It is this love of the contemplation of the eternal and unchanging which we constantly strive to increase, by studying those parts of these sciences which have already been mastered by those who approached them in a genuine spirit of inquiry, and by ourselves attempting to contribute as much advancement as has been made possible by the additional time between those people and ourselves. We shall try to note down everything which we think we have discovered up to the present time; we shall do this as concisely as possible and in a manner which can be followed by those who have already made some progress in the field. For the sake of completeness in our treatment we shall set out everything useful for the theory of the heavens in the proper order, but to avoid undue length we shall merely recount what has been adequately established by the ancients. However, those topics which have not been dealt with [by our predecessors] at all, or not as usefully as they might have been, will be discussed at length, to the best of our ability. . . .

IX.2 . . . Now it is our purpose to demonstrate for the five planets, just as we did for the sun and moon, that all their apparent anomalies can be represented by uniform circular motions, since these are proper to the nature of divine beings, while disorder and non-uniformity are alien [to such beings]. Then it is right that we should think success in such a purpose a great thing, and truly the proper end of the mathematical part of theoretical philosophy. But, on many grounds, we must think that it is difficult, and that there is good reason why no one before us has yet succeeded in it. . . .

XIII.2 . . . Now let no one, considering the complicated nature of our devices, judge such hypotheses to be over-elaborated. For it is not appropriate to compare human [con-

[26]In this exaltation of mathematics above the other two divisions of philosophy Ptolemy parts company with Aristotle, for whom theology was the most noble pursuit of the human mind. (Toomer) So, too, for Plato, philosophical apperception of the ideal Forms revealed a still higher truth than that of mathematics.

[27]The heavenly bodies. (Toomer)

structions] with divine, nor to form one's beliefs about such great things on the basis of very dissimilar analogies. For what [could one compare] more dissimilar than the eternal and unchanging with the ever-changing, or that which can be hindered by anything with that which cannot be hindered even by itself?[28] Rather, one should try, as far as possible, to fit the simpler hypotheses to the heavenly motions, but if this does not succeed, [one should apply hypotheses] which do fit. For provided that each of the phenomena is duly saved by the hypotheses, why should anyone think it strange that such complications can characterize the motions of the heavens when their nature is such as to afford no hindrance, but of a kind to yield and give way to the natural motions of each part, even if [the motions] are opposed to one another? Thus, quite simply, all the elements can easily pass through and be seen through all other elements, and this ease of transit applies not only to the individual circles, but to the spheres themselves and the axes of revolution. We see that in the models constructed on earth the fitting together of these [elements] to represent the different motions is laborious, and difficult to achieve in such a way that the motions do not hinder each other, while in the heavens no obstruction whatever is caused by such combinations. Rather, we should not judge "simplicity" in heavenly things from what appears to be simple on earth, especially when the same thing is not equally simple for all even here. For if we were to judge by those criteria, nothing that occurs in the heavens would appear simple, not even the unchanging nature of the first motion, since this very quality of eternal unchangingness is for us not [merely] difficult, but completely impossible. Instead [we should judge "simplicity"] from the unchangingness of the nature of things in the heaven and their motions. In this way all [motions] will appear simple, and more so than what is thought "simple" on earth, since one can conceive of no labor or difficulty attached to their revolutions.

The Geometrical Forethought of Bees

The *Collection* of Pappus of Alexandria, translated by T. L. Heath

Pappus of Alexandria, in the late third to early fourth century A.D., *wrote commentaries on Euclid and geographical treatises; his one extant work is the* Synagôgê, *or* Collection, *a compilation preserving many data invaluable for the history of Greek mathematics. The following passage from the Preface to Book V shows what admiration a rationalistic Greek could have for the capacities of even the most humble animals. Many other writers, including Aristotle, Pliny, Seneca, and the poets Nicander (who in the second century* B.C. *wrote a lost didactic poem on farming) and Virgil, had marveled at these social insects, but Pappus's comments are especially notable in that they come from a mathematician and suggest, once again, that for the Greeks the "highest" and "lowest" realms of nature are closely united. The translation is from volume 2 of T. L. Heath's* A History of Greek Mathematics *(1921).*

It is of course to men that God has given the best and most perfect notion of wisdom in general and of mathematical science in particular, but a partial share in these things he allotted to some of the unreasoning animals as well. To men, as being endowed with reason, he vouchsafed that they should do everything in the light of reason and demon-

[28]I.e. the substance of the heavenly bodies, the "fifth essence." (Toomer)

stration, but to the other animals, while denying them reason, he granted that each of them should, by virtue of a certain natural instinct, obtain just so much as is needful to support life. This instinct can be observed to exist in very many other species of living creatures, but most of all in bees. In the first place their orderliness and their submission to the queens who rule in their state are truly admirable, but much more admirable still is their emulation, the cleanliness they observe in the gathering of honey, and the forethought and housewifely care they devote to its custody. Presumably because they know themselves to be entrusted with the task of bringing from the gods to the accomplished portion of mankind a share of ambrosia in this form, they do not think it proper to pour it carelessly on ground or wood or any other ugly and irregular material; but, first collecting the sweets of the most beautiful flowers which grow on the earth, they make from them, for the reception of the honey, the vessels which we call honeycombs, (with cells) all equal, similar and contiguous to one another, and hexagonal in form. And that they have contrived this by virtue of a certain geometrical forethought we may infer in this way. They would necessarily think that the figures must be such as to be contiguous to one another, that is to say, to have their sides common in order that no foreign matter could enter the interstices between them and so defile the purity of their produce. Now only three rectilineal figures would satisfy the condition, I mean regular figures which are equilateral and equiangular; for the bees would have none of the figures which are not uniform. . . . There being then three figures capable by themselves of exactly filling up the space about the same point, the bees by reason of their instinctive wisdom chose for the construction of the honeycomb the figure which has the most angles, because they conceived that it would contain more honey than either of the two others.

Bees, then, know just this fact which is of service to themselves, that the hexagon is greater than the square and the triangle, and will hold more honey for the same expenditure of material used in constructing the different figures. . . .

Rational and Empirical Medicine

On Medicine, by Celsus, translated by W. G. Spencer

Almost nothing is known about Aulus Cornelius Celsus, who under Tiberius (14–37 A.D.) compiled an encyclopedic work embracing agriculture, medicine, warfare, rhetoric, philosophy, and jurisprudence. Of this vast undertaking, only the eight books De Medicina *survived from antiquity, and until their rediscovery in the Renaissance, when Celsus's style became a model for humanists, they appear to have had little influence. Celsus is mentioned only in passing by the elder Pliny, and is depreciated by the rhetorician Quintilian as a man "of mediocre talent." Yet on the basis of what survives, he appears to have had a gift for organization and a capacity for discrimination far superior to that of Pliny or later encyclopedists; his work is now regarded, William Stahl writes in* Roman Science *(1962), "on the basis of its form and content, as the most satisfactory book in the entire collection of extant Latin scientific writings and, by some critics, as the most orderly and best arranged single treatise in the entire range of ancient scientific literature, Latin or Greek."*

The following passage from the Preface to Book One (quoted from volume I of the Loeb Classi-

cal Library edition of 1935) succinctly describes the major schools that dominated Greek medical thinking between the age of Hippocrates and that of Galen: the rationalists or "dogmatists," emphasizing theory (as derived, however, from practice, including dissection); the empiricists, stressing the variant experience of each particular case and rejecting abstract theories and hidden causes; and the "methodists," seeking a narrower and more mechanically applied method for the treatment of given diseases. These controversies, central to the history of medicine for centuries, have broader importance as well, since the opposing positions of rationalists and empiricists suggest the views of nature upheld by the "dogmatic" schools of philosophy (Epicurean and especially Stoic) on the one hand and by the Sceptics on the other. The rationalists were among the foremost to attempt applying the theory of nature to medical practice, whereas the empiricists embraced the influential view that the workings of nature are too complex to be comprehended by imposition of theoretical paradigms that inherently distort her infinitely variable processes. The two views were not, as Celsus sensibly remarks in proposing an intermediate position, incompatible.

They, then, who profess a reasoned theory of medicine propound as requisites, first, a knowledge of hidden causes involving diseases, next, of evident causes, after these of natural actions also, and lastly of the internal parts.

They term hidden, the causes concerning which inquiry is made into the principles composing our bodies, what makes for and what against health. For they believe it impossible for one who is ignorant of the origin of diseases to learn how to treat them suitably. They say that it does not admit of doubt that there is need for differences in treatment, if, as certain of the professors of philosophy have stated, some excess, or some deficiency, among the four elements, creates adverse health; or, if all the fault is in the humors, as was the view of Herophilus;[29] or in the breath, according to Hippocrates; or if blood is transfused into those blood-vessels which are fitted for pneuma, and excites inflammation which the Greeks term *phlegmónê*, and that inflammation effects such a disturbance as there is in fever, which was taught by Erasistratus; or if little bodies by being brought to a standstill in passing through invisible pores block the passage, as Asclepiades contended—his will be the right way of treatment, who has not failed to see the primary origin of the cause. They do not deny that experience is necessary; but they say it is impossible to arrive at what should be done unless through some course of reasoning. . . .

But they call evident those causes, concerning which they inquire, as to whether heat or cold, hunger or surfeit, or such like, has brought about the commencement of the disease; for they say that he will be the one to counter the malady who is not ignorant of its origin.

Further, they term "natural" actions of the body, those by which we draw in and emit breath, take in and digest food and drink, as also those actions through which food and drink are distributed into every part of the members. . . . Moreover, as pains, and also various kinds of diseases, arise in the more internal parts, they hold that no one can apply remedies for these who is ignorant about the parts themselves; hence it becomes neces-

[29]Heróphilus of Chálcedon was a leading rationalist physician of Alexandria in the first half of the third century B.C.; he based his theories on the four "humors" (blood, phlegm, choler or yellow bile, and black bile) expounded, for example, in the Hippocratic treatise "On the Nature of Man" (Chapter 9 above). The theory of "breath" (Latin *spiritus*, translating Greek *pneuma*) in the clauses that follow meant, Spencer suggests, "breath with the addition of some vital spirit — almost the Greek equivalent to oxygen." Erasístratus of Ceos was a physician of Alexandria in and after the time of Herophilus in the third century B.C. Asclepiádes of Prusa in Bithynia was a physician in Rome in the first century B.C., who subscribed to the atomic theory of Epicurus.

sary to lay open the bodies of the dead and to scrutinize their viscera and intestines. They hold that Herophilus and Erasistratus did this in the best way by far, when they laid open men whilst alive—criminals received out of prison from the kings—and whilst these were still breathing, observed parts which beforehand nature had concealed, their position, color, shape, size, arrangement, hardness, softness, smoothness, relation, processes and depressions of each, and whether any part is inserted into or is received into another. For when pain occurs internally, neither is it possible for one to learn what hurts the patient, unless he has acquainted himself with the position of each organ or intestine; nor can a diseased portion of the body be treated by one who does not know what that portion is. When a man's viscera are exposed in a wound, he who is ignorant of the color of a part in health may be unable to recognize which part is intact, and which part damaged; thus he cannot even relieve the damaged part. External remedies too can be applied more aptly by one acquainted with the position, shape and size of the internal organs, and like reasonings hold good in all the instances mentioned above. Nor is it, as most people say, cruel that in the execution of criminals, and but a few of them, we should seek remedies for innocent people of all future ages.

On the other hand, those who take the name of Empirici from their experience, do indeed accept evident causes as necessary; but they contend that inquiry about obscure causes and natural actions is superfluous, because nature is not to be comprehended. That nature cannot be comprehended is in fact patent, they say, from the disagreement among those who discuss such matters; for on this question there is no agreement, either among professors of philosophy or among actual medical practitioners. Why, then, should anyone believe rather in Hippocrates than in Herophilus, why in him rather than in Asclepiades? If one wants to be guided by reasoning, they go on, the reasoning of all of them can appear not improbable; if by method of treatment, all of them have restored sick folk to health: therefore one ought not to derogate from anyone's credit, either in argument or in authority. Even philosophers would have become the greatest of medical practitioners, if reasoning from theory could have made them so; as it is, they have words in plenty, and no knowledge of healing at all. They also say that the methods of practice differ according to the nature of localities, and that one method is required in Rome, another in Egypt, another in Gaul; but that if the causes which produce diseases were everywhere the same, the same remedies should be used everywhere; that often, too, the causes are apparent, as, for example, of ophthalmia, or of wounds, yet such causes do not disclose the treatment: that if the evident cause does not supply the knowledge, much less can a cause which is in doubt yield it. Since, therefore, the cause is as uncertain as it is incomprehensible, protection is to be sought rather from the ascertained and explored, as in all the rest of the Arts, that is, from what experience has taught in the actual course of treatment: for even a farmer, or a pilot, is made not by disputation but by practice. That such speculations are not pertinent to the Art of Medicine may be learned from the fact that men may hold different opinions on these matters, yet conduct their patients to recovery all the same. This has happened, not because they deduced lines of healing from obscure causes, nor from the natural actions, concerning which different opinions were held, but from experiences of what had previously succeeded. Even in its beginnings, they add, the Art of Medicine was not deduced from such questionings, but from experience; for of the sick who were without doctors, some in the first days of illness, longing for food, took it forthwith; others, owing to distaste, abstained; and the illness was more alleviated in those who abstained. Again, some partook of food whilst actually under the fever, some a little before, others after its remission, and it went best with those who did so after the fever had ended; and similarly some at the beginning adopted at once a rather full diet, others a scanty one, and those were made worse who had eaten plentifully. When this and the like happened day after day, careful men noted what generally answered the better, and then began to prescribe the same for their patients. Thus sprang up the Art of Medicine, which, from the frequent recovery of some and the death of others, distinguished between the pernicious and the salutary.

It was afterwards, they proceed, when the remedies had already been discovered, that men began to discuss the reasons for them: the Art of Medicine was not a discovery following upon reasoning, but after the discovery of the remedy, the reason for it was sought out. They ask, too, does reasoning teach the same as experience? If the same, it was needless; if something else, then it was even opposed to it: nevertheless, at first remedies had to be explored with the greatest care; now, however, they have been explored already; there were neither new sorts of diseases to be found out, nor was a novel remedy wanted. For even if there happened nowadays some unknown form of malady, nevertheless the practitioner had not to theorize over obscure matters, but straightway would see to which disease it came nearest, then would make trial of remedies similar to those which have succeeded often in a kindred affection, and so through its similarities find help; that was not to say that a practitioner had no need to take counsel, and that an irrational animal was capable of exhibiting this art, but that these conjectures about concealed matters are of no concern because it does not matter what produces the disease but what relieves it; nor does it matter how digestion takes place, but what is best digested, whether concoction comes about from this cause or that, and whether the process is concoction or merely distribution. We had no need to inquire in what way we breathe, but what relieves labored breathing; not what may move the blood-vessels, but what the various kinds of movements signify. All this was to be learnt through experiences; and in all theorizing over a subject it is possible to argue on either side, and so cleverness and fluency may get the best of it; it is not, however, by eloquence but by remedies that diseases are treated. A man of few words who learns by practice to discern well, would make an altogether better practitioner than he who, unpracticed, overcultivates his tongue.

Now the matters just referred to they deem to be superfluous; but what remains, cruel as well, to cut into the belly and chest of men whilst still alive, and to impose upon the Art which presides over human safety someone's death, and that too in the most atrocious way. Especially is this true when, of things which are sought for with so much violence, some can be learnt not at all, others can be learnt even without a crime. . . . It follows, therefore, that the medical man just plays the cut-throat, not that he learns what our viscera are like when we are alive. If, however, there be anything to be observed whilst a man is still breathing, chance often presents it to the view of those treating him. . . . Thus, they say, an observant practitioner learns to recognize site, position, arrangement, shape and such like, not when slaughtering, but whilst striving for health; and he learns in the course of a work of mercy, what others would come to know by means of dire cruelty. . . .

Since all these questions have been discussed often by practitioners, in many volumes and in large and contentious disputations, and the discussion continues, it remains to add such views as may seem nearest the truth. These are neither wholly in accord with one opinion or another, nor exceedingly at variance with both, but hold a sort of intermediate place between diverse sentiments, a thing which may be observed in most controversies when men seek impartially for truth, as in the present case. . . . Although, therefore, many things, which are not strictly pertinent to the Arts as such, are yet helpful by stimulating the minds of those who practice them, so also this contemplation of the nature of things, although it does not make a practitioner, yet renders him more apt and perfected in the Art of Medicine. And it is probable that Hippocrates, Erasistratus and certain others, who were not content to busy themselves over fevers and ulcerations, but also to some extent searched into the nature of things, did not by this become practitioners, but by this became better practitioners. . . .

Further, certain practitioners of our time, following, as they would have it appear, the authority of Themison,[30] contend that there is no cause whatever, the knowledge of which has any bearing on treatment: they hold that it is sufficient to observe certain general characteristics of diseases; that of these there are three classes, one a constriction, another a flux, the third a mixture. . . . They hold that the Art of Medicine consists of such observations; which they define as a sort of way, which they name *méthodos*, and maintain that medicine should examine those characteristics which diseases have in common. They do not want to be classed with reasoners from theory, nor with those who look to experience only; for in so naming themselves Methodici, they dissent from the former because they are unwilling that the Art should consist in conjecture about hidden things, and from the latter because they think that in the observation of experience there is little of an Art of Medicine. . . . But disciples of Themison, if they hold their precepts to be of constant validity, are reasoners even more than anybody else. . . .

Therefore, to return to what I myself propound, I am of the opinion that the Art of Medicine ought to be rational, but to draw instruction from evident causes, all obscure ones being rejected from the practice of the Art, although not from the practitioner's study. But to lay open the bodies of men whilst still alive is as cruel as it is needless; that of the dead is a necessity for learners, who should know positions and relations, which the dead body exhibits better than does a living and wounded man. As for the remainder, which can only be learnt from the living, actual practice will demonstrate it in the course of treating the wounded in a somewhat slower yet much milder way.

Nature's Work of Art

On the Usefulness of the Parts of the Body, by Galen, translated by Margaret Tallmadge May

Galen (Galênos, A.D. 129?–99), the most renowned Greek physician since Hippocrates, was born at Pérgamum, in Asia Minor, which before Rome absorbed it had long been an independent kingdom proud of its artistic and intellectual heritage. In his voluminous extant writings (some 20,000 pages in Greek; still others survive in Arabic or Latin translations), he often speaks of his own life and character, formed on the model of his temperate father in contrast, as he portrays her, to his shrewish mother. (Galen, like many others from long before his time up to nearly our own, justified his belief in the irrationality of women on purportedly scientific grounds; we should never forget that "nature" has been used as a rationalization for countless prejudices through the ages.) At Pergamum and in Athens and Alexandria, he studied philosophy under Stoics, Platonists, Peripatetics, and Epicureans, and he always affirmed that the true physician must be

[30]Thémison of Laodicéa was a pupil of Asclepiades in Rome at the time of Augustus (late first century B.C. to early first century A.D.).

a philosopher. From 162 to 166 and again from 169 until his death, Galen practiced medicine in Rome, where he became physician to the Emperor Marcus Aurelius and his son Commodus.

In addition to his own contributions to anatomy and physiology, based on careful dissection and close firsthand observation, Galen upheld a flexible rationalism, or "eclectic dogmatism," affirming the fundamental compatability of reason and experience exemplified, in his view, by both Plato and Hippocrates. His aim, Albrecht Dihle writes, "was to keep Hippocratic medicine alive by means of a modernization, with a comprehensive inclusion of recent advances in medical knowledge." So successful was he that not until the sixteenth century did medicine advance significantly beyond the knowledge he so comprehensively expounded. From Plato, Aristotle, and the Stoics, he derived the strong teleological emphasis in his writings, insisting repeatedly that there is a purpose (we might say a function) for every bodily part, and indeed for every event in the heavens or earth. Nature, he believed, in accord with the Greek tradition descending from Thales and Anaximander, is ruled by its own coherent laws, not—as the Jews and Christians, whose views he vigorously opposed, contended—by a transcendent deity capable of suspending those laws at will and creating not simply something from nothing but anything from anything.

In the following selections from one of his major works, On the Usefulness of the Parts of the Body, *in the two-volume translation by Margaret Tallmadge May (1968), Galen acknowledges that some things (like the commingling of bodies in the mythical Centaur) are "impossible for Nature"—a realization that eluded Pliny and Aelian, along with many others—but endeavors repeatedly to demonstrate the perfection of "Nature's art" even when working on the humblest and most perishable materials. The praise of nature that permeates Galen's writing becomes, as he calls it, "a true hymn to our Creator" in the spirit of Hippocrates, Plato, and Aristotle, for whom "every realm of nature is marvellous."*

II.3 Now let me once for all make this general statement to apply to my whole treatise so as not to be forced to say the same thing repeatedly: I am now explaining the structures actually to be seen in dissection, and no one before me has done this with any accuracy. Hence, if anyone wishes to observe the works of Nature, he should put his trust not in books on anatomy but in his own eyes and either come to me, or consult one of my associates, or alone by himself industriously practise exercises in dissection; but so long as he only reads, he will be more likely to believe all the earlier anatomists because there are so many of them. . . .

III.1 Man is the only one of all the animals to have been provided with hands, instruments suitable for an intelligent animal; likewise, of animals that go afoot he alone was made biped and erect, for the reason that he had hands. . . . But why, then, was he not given four legs and hands as well, like the centaur? The reason is that, in the first place, a commingling of such widely different bodies was impossible for Nature. . . .

III.10 Is it, then, only in the parts I have mentioned that Nature has arranged everything justly for the hand and foot, making the necessary analogies and differences, and has she been so careless in constructing the skin as to stretch beneath the foot one that is insensitive, loose-textured, thin, and soft? Surely if you examine the skin in dissection, even if you are one of those who through ignorance of Nature's works accuse her of lack of skill, I think you will repent with shame and change your opinion for the better, agreeing with Hippocrates, who is continually singing the praises of Nature's righteousness and the foresight she displays in the creation of animals. Do you think it is profitless for the skin of the palm of the hand and the sole of the foot to grow fast to the underlying parts? Or are you totally ignorant that it is so intimately attached to the tendons beneath that it cannot be stripped off like the skin over all the rest of the body? Or do you know this and still think that it would be better if the sole of the foot were covered with a skin loose-textured and easily movable? If you say that such a skin would be better, I suppose that, instead of close-fitting sandals bound tightly all around, you would prefer those that are loose and slip in every direction; for so you may assert your cleverness in everything and not scruple to cry down even what is clearly known to all men. Or do you grant that an artificial sandal must certainly be bound to the foot all around if it is to fulfill its purpose, but not that Nature's sandal itself has a much greater need to be bound on, held firmly, and closely united to the parts under which it has been placed? A man would be a veritable Coroebus[31] if he failed to marvel at works of Nature such as these and even presumed to censure them.

It is time now for you, my reader, to consider which chorus you will join, the one that gathers round Plato, Hippocrates, and the others who admire the works of Nature, or the one made up of those who blame her because she has not arranged to have the superfluities discharged through the feet. Anyone who dares to say these things to me has been spoiled by luxury to such an extent that he considers it a hardship to rise from his bed when he voids, thinking that man would be better constructed if he could simply extend his foot and discharge the excrement through it. How do you suppose such a man feels and acts in private? How wantonly he uses all the openings of his body! How he maltreats and ruins the noblest qualities of his soul, crippling and blinding that godlike faculty by which alone Nature enables a man to behold the truth, and allowing his worst and most bestial faculty to grow huge, strong, and insatiable of lawless pleasures and to hold him in a wicked servitude! But if I should speak further of such fatted cattle, right-thinking men would justly censure me and say that I was desecrating the sacred discourse which I am composing as a true hymn to our Creator. And I consider that I am really showing him reverence not when I offer him unnumbered hecatombs of bulls and burn incense of cassia worth ten thousand talents, but when I myself first learn to know his wisdom, power, and goodness, and then make them known to others. I regard it as proof of perfect goodness that one should will to order everything in the best possible way, not grudging benefits to any creature, and therefore we must praise him as good. But to have discovered how everything should best be ordered is the height of wisdom, and to have accomplished his will in all things is proof of his invincible power.

Then do not wonder so greatly at the beautiful arrangement of the sun, moon, and the whole chorus of stars, and do not be so struck with amazement at the size of them, their beauty, ceaseless motion, and ordered revolutions that things here on earth will seem trivial and disorganized in comparison; for here too you will find displayed the same wisdom, power, and foresight. Consider well the material of which a thing is made, and cherish no idle hope that you could put together from the catemenia and semen an animal that would be deathless, exempt from pain, endowed with never-ending motion, and as radiantly beautiful as the sun. You should rather estimate the art of the creator of all things just as you judge the art of Phidias. Now perhaps you are struck with admiration of the decoration covering the image of Zeus at Olympia, its gleaming ivory, its massy gold, and the great size of the whole statue, and if you saw such a statue made of clay, you would perhaps turn away in contempt. Not so, however, the man who is an artist and able to recognize the art employed in the work; no, he commends Phidias equally, even if he sees him working in cheap wood, common stone, wax, or clay. For the unculti-

[31]A Phrygian, son of Mygdon and Anaximene, who came too late to the aid of Priam at Troy and hence gave rise to the proverb, "More stupid than Coroebus." (May)

vated man sees beauty in material, whereas it is the art itself that seems beautiful to the artist. Come, then, let us make you skillful in Nature's art so that we may call you no longer an uncultivated person, but a natural philosopher instead. Disregard differences of material and look only at the naked art itself, keeping in mind when you inspect the structure of the eye and the foot that the one is an instrument of vision and the other of locomotion. If you think it proper for the eyes to be made of material like the sun's or for the feet to be pure gold instead of bones and skin, you are forgetting the substance of which you have been formed. Bear it in mind and reflect whether your substance is celestial light or slime of the earth, if you will permit me to give such a name to the mother's blood flowing into the uterus. Then, just as you would never demand an ivory statue of Phidias if you had given him clay, so in the same way, when blood is the material you give, you would never obtain the bright and beautiful body of the sun or moon. For they are divine and celestial and we are mere figures of clay, but in both cases the art of the Creator is equally great.

Who will deny that the foot is a small, ignoble part of an animal? And we know full well that the sun is grand and the most beautiful thing in the whole universe. But observe where in the whole universe was the proper place for the sun, and where in the animal the foot had to be placed. In the universe the sun had to be set in the midst of the planets, and in the animal the foot must occupy the lowest position. How can we be sure of this? By assuming a different location for them and seeing what would follow. If you put the sun lower down where the moon is now, everything here would be consumed by fire, and if you put it higher, near Pyroeis or Phaëthon,[32] no part of the earth would be habitable because of the cold. The size and character of the sun are qualities inherent in its nature, but its particular position in the universe is the work of One who has arranged it so. For you could find no better place in the whole universe for a body of the size and character of the sun, and in the body of an animal you could find no better place for the foot than the one it occupies. You should observe that the same skill has been employed in locating both sun and foot. (I am intentionally comparing the noblest of the stars with the lowliest member of the animal body.) What is more insignificant than the heel? Nothing. But it could not be better located in any other place. What is nobler than the sun? Nothing. But neither could the sun be better located anywhere else in the whole universe.

What is the grandest and most beautiful of created things? The universe, as everyone admits. But the Ancients,[33] well-versed in Nature, say that an animal is, so to speak, a little universe, and you will find the same wisdom displayed by the Creator in both his works. Then show me, you say, a sun in the body of an animal. What a thing to ask! Are you willing to have the sun formed from the substance of blood, so prone to putrefy and so filthy? Wretched fellow, you are mad! This, and not failure to make offerings and burn incense, is true sacrilege. I will not, indeed, show you the sun in the body of an animal, but I will show you the eye, a very brilliant instrument, resembling the sun as closely as is possible [for a part located] in the body of an animal. I will explain the position of the eye, its size, shape, and all its other qualities, and I will show that everything about it is so beautifully ordered that it could not possibly be improved; but I will do this farther on in my discourse. . . .

X.9 Justly then, in my opinion, it is something to wonder at that the sophists, who have not as yet been able to discover or explain the works of Nature, still accuse her of a

[32]Respectively, the fiery one (Mars) and Jupiter. (May)

[33]See Democritus, fragment 34, in Chapter 9 above, and Aristotle, *Physics* VIII.2 (252b).

lack of skill. For I think they should be expected to show that it was better that the eyes should not be provided with lids, or that if they were, the lids should be without motion, or that if the lids did move, the motion should be involuntary, or that if it was voluntary, the muscles should be arranged thus and so. But they have reached such heights of cleverness that although the eyelids obviously move, they do not understand how it happens and do not discover any other movement! And they are so senseless as not yet to admit that the One forming and framing so many wonderful parts is a Craftsman. Furthermore, if there should be a discussion among artisans as to how best a house, a door, a little couch, or something of the sort should be constructed to serve the usefulness for which it was formed, and if, while the others were at a loss, there was one who knew how, he would justly be admired and considered a clever craftsman. And shall not we who, not to mention being incapable of planning the works of Nature, are not even able to understand them when we see them already formed—shall not we admire them more than man-made creations? . . .

X.14 . . . Not only here but also in many other places in these commentaries, if it depended on me, I would omit demonstrations requiring astronomy, geometry, music, or any other logical discipline, lest my books should be held in utter detestation by physicians. For truly on countless occasions throughout my life I have had this experience: persons for a time talk pleasantly with me because of my work among the sick, in which they think me very well trained, but when they learn later on that I am also trained in mathematics, they avoid me for the most part and are no longer at all glad to be with me. Accordingly, I am always wary of touching on such subjects, and in this case it is only in obedience to the command of a divinity, as I have said, that I have used the theorems of geometry. . . .

XI.14 Since I have nowhere mentioned earlier in my discourse that out of her abundance Nature sometimes also aims at beauty of form and that this too must be recognized by those studying Nature, I have thought it most proper to speak of it now. Well, then, the hair of the beard not only protects the cheeks but also serves to ornament them; for a man seems more stately, especially as he grows older, if he has everywhere a good covering of hair. . . . On the other hand, for woman, the rest of whose body is always soft and hairless like a child's, the bareness of the face would not be inappropriate, and besides, this animal does not have an august character as the male has and so does not need an august form. For I have already shown many times, indeed throughout the work, that Nature makes for the body a form appropriate to the character of the soul. And the female sex does not need any special covering as protection against the cold, since for the most part women stay within doors, yet they do need long hair on their heads for both protection and ornament, and this need they share with men. . . .

As regards the eyelashes and eyebrows, however, if you either added or subtracted anything, you would destroy their usefulness. For the former are set like a palisade before the open eyes so that no small bodies may fall into them, and the latter must provide shelter like a wall and be the first to receive all that flows down from the head. If, then, you made them shorter or thinner than they should be, you would to that extent impair their usefulness; for whatever they formerly kept out would be allowed by the eyelashes to fall into the eyes and by the eyebrows to flow into them. But again, if you made them longer or thicker, they would no longer be a palisade and a wall for the eyes, but coverings very like a prison, hiding and darkening the pupils, which ought least of all instruments to be obscured.

Has, then, our Creator commanded only these hairs to preserve always the same length, and do the hairs preserve it as they have been ordered either because they fear the injunction of their Lord, or reverence the God who commands it, or themselves believe it better to do so? Is this the way in which Moses reasons about Nature (and it is a better way than Epicurus')? Yet it is best for us to adopt neither, but, continuing to derive the principle of generation from the Creator in all things generable, as Moses does, to add to this the material principle. For our Creator has made these hairs feel the necessity of preserving always an even length for the reason that this was the better thing. And since he had decided that it was necessary to make them so, he spread beneath some of them [the eyelashes] a hard body like cartilage [the tarsus] and under the others [the eyebrows] a hard skin united to cartilage by means of the brows. Now it was not enough merely to will that they should be so; for even if he wished to make a rock into a man all of a sudden, it would be impossible. And this is the point at which my teaching and that of Plato and the other Greeks who have treated correctly of natural principles differs from that of Moses. For him it suffices for God to have willed material to be arranged and straightway it was arranged, because Moses believed everything to be possible to God, even if he should wish to make a horse or beef out of ashes. We, however, do not feel this to be true, saying rather that some things are naturally impossible and that God does not attempt these at all but chooses from among the possible what is best to be done.

Accordingly, when it was better that the length and number of the hairs of the eyelids should be always the same, we do not say that he willed it and straightway they were made so; for though he willed it times without number, they would never become such as they are if they grew out from soft skin. As for the other requirement, they could not possibly stand erect unless they were implanted in something hard. We say, then, that God is the cause of two things, namely, the choice of the better in what is being made and the selection of material. Since the hairs of the eyelids must stand erect and at the same time remain always of the same length and number, he implanted them firmly in a cartilaginous body. If he had implanted them in a soft, fleshy substance, he would be worse than either Moses or some wretched general who planted his wall or palisade in a swamp. That the hairs of the eyebrows too are kept always the same depends on the same selection of material. For just as grass and plants coming up from damp, rich soil grow very tall, whereas those that come from dry, rocky soil remain without increase, small and hard, so in the same way, I suppose, the hair coming from soft, moist parts has a very good growth, like the hair of the head, armpits, and pudenda, whereas that from the hard, dry parts does not grow and remains short. Thus, like herbs and plants, hair has a twofold generation, stemming in part from the providence of the Creator and in part from the nature of the place.

XIV.6 Now just as mankind is the most perfect of all animals, so within mankind the man is more perfect than the woman, and the reason for his perfection is his excess of heat, for heat is Nature's primary instrument. Hence in those animals that have less of it, her workmanship is necessarily more imperfect, and so it is no wonder that the female is less perfect than the male by as much as she is colder than he. In fact, just as the mole has imperfect eyes, though certainly not so imperfect as they are in those animals that do not have any trace of them at all, so too the woman is less perfect than the man in respect to the generative parts. For the parts were formed within her when she was still a fetus, but could not because of the defect in the heat emerge and project on the outside, and this, though making the animal itself that was being formed less perfect than one that is complete in all respects, provided no small advantage *(khreia)* for the race; for there needs must be a female. Indeed, you ought not to think that our Creator would purposely make half the whole race imperfect and, as it were, mutilated, unless there was to be some great advantage in such a mutilation.

XVII.1 Who would not straightway conclude that some intelligence possessed of marvelous power was walking the earth and penetrating its every part? For you see everywhere that the animals produced all have a marvelous structure. And yet, what part of the universe is more ignoble than the earth? Nevertheless, even here there appears to be some intelligence reaching us from the bodies above, and anyone seeing these is at once forced to admire the beauty of their substance, first and foremost that of the sun, after the sun that of the moon, and then of the stars.

It is reasonable to suppose that the intelligence dwelling in them is as much better and more perfect than that in earthly bodies as their bodily substance is the purer. For when in mud and slime, in marshes, and in rotting plants and fruits animals are engendered[34] which yet bear a marvelous indication of the intelligence constructing them, what must we think of the bodies above? But you can see the nature of the intelligence in man himself when you consider Plato, Aristotle, Hipparchus, Archimedes, and many others like them. When a surpassing intelligence comes into being in such slime—for what else would one call a thing composed of fleshes, blood, phlegm, and yellow and black bile?—how great must we consider the pre-eminence of the intelligence in the sun, moon, and stars? As I meditate on these matters, it seems to me that a certain not inconsiderable intelligence pervades even the very air surrounding us; certainly the air could not partake of the sun's light without receiving its power too. I am sure that you will also regard all these things in the same way when you examine carefully and justly the skill displayed in animals, unless, as I have said, you are prevented by some doctrine which you have rashly posited about the elements of the universe. Thus, when anyone looking at the facts with an open mind sees that in such a slime of fleshes and juices there is yet an indwelling intelligence and sees too the structure of any animal whatsoever—for they all give evidence of a wise Creator—he will understand the excellence of the intelligence in the heavens.

Then a work on the usefulness of the parts, which at first seemed to him a thing of scant importance, will be reckoned truly to be the source of a perfect theology, which is a thing far greater and far nobler than all of medicine. Hence such a work is serviceable not only for the physician, but much more so for the philosopher who is eager to gain an understanding of the whole of Nature. And I think that all men of whatever nation or degree who honor the gods should be inititated into this work, which is by no means like the mysteries of Eleusis and Samothrace. For feeble are the proofs that these give of what they strive to teach, but the proofs of Nature are plain to be seen in all animals.

In fact, you must not suppose that such skill as I have been explaining in this book is displayed in man alone; on the contrary, any other animal you may care to dissect will show you as well both the wisdom and skill of the Creator, and the smaller the animal the greater the wonder it will excite, just as when craftsmen carve something on small objects. There are such craftsmen even now, one of whom recently carved on a signet

[34]The doctrine of spontaneous generation was taken for granted in antiquity and persisted unquestioned till the seventeenth century, when Francesco Redi led the attack on it. In the eighteenth century it was the subject of a prolonged controversy between Lazzaro Spallanzani and John T. Needham, the latter defending and the former seeking new grounds for discrediting it. It was Louis Pasteur who dealt the final blows that destroyed it. (May)

ring Phaëthon drawn by four horses, each with its bit, mouth, and front teeth so small that I did not see them at all except by turning the marvel around under a bright light, and even then, like many others, I did not see all the parts. But when anyone was able to see any of them clearly, he agreed that they were in perfect proportion. For instance, we had difficulty in even counting the sixteen limbs of the four horses, but to those who could see them the parts of each one appeared marvelously articulated. Yet not one of these displays more perfect workmanship than the leg of a flea, and besides, the whole leg of the living, nourished, growing flea is pervaded by skill. It is reasonable, however, that the wisdom and power of the skillful Creator of the flea should be sufficiently great to preserve it, increase its size, and nourish it without difficulty. At all events, if the Creator's skill is such when displayed incidentally, as one might say, in insignificant animals, how great must we consider his wisdom and power when displayed in animals of some importance!

The Death of Pan

Selections from the *Moralia* of Plutarch

Plutarch (Ploutarkhos) was born in the Boeotian town of Chaeronéa in mainland Greece sometime before 50 A.D.; though he traveled to Athens, Alexandria, and Rome, and became widely known through his writings, he spent most of his later life in his native place and probably died there sometime after 120. He is renowned both for his Parallel Lives, *in which he placed biographies of famous Greeks and Romans (for example, Alexander the Great and Julius Caesar) side by side and drew comparisons between them, and for the literary and philosophical essays and dialogues known by the comprehensive title of* Moralia.

Plutarch was in love with the glorious heroic and intellectual achievements of classical Greece, though he knew they had long been in decline; awareness of that disparity permeates his writings. At a time when new religions were sweeping through the Greco-Roman world, he was a priest of Delphi despite the obsolescence of its oracles; at a time when Stoicism and Epicureanism were the prevalent schools of philosophy, he criticized both and maintained his strong adherence to a flexible and eclectic Platonism. The word "nature" resonates throughout his works. Yet the excerpts that follow repeatedly betray a tension between the longing for oneness with nature (which Plutarch closely associates with the gods of the classical past) and the experience of separation from, or of very precarious communion with, it.

The first selection is from the "scientific" portion of Concerning the Face Which Appears in the Orb of the Moon, *in volume 12 (1957) of the Loeb Classical Library edition. This dialogue is strangely divided between assertion of the earthly nature of the moon (in contrast to traditional assertions of its aethereal purity) and a concluding eschatological myth portraying the moon as the abode of souls who have left their bodies after death on earth. Plutarch's spokesman Lamprias here refutes the Aristotelian-Stoic belief that each element or part of the whole has an inherently "natural" place; instead, no position or motion is "unconditionally natural," but is natural relative to the purpose it serves, both in the body and in the cosmos. As in Plato's world, not mechanical necessity but "the rational principle" is in control. How shakily this is true, if at all, becomes clear in the next selection, the (somewhat abridged) dialogue generally known as* Gryllus *("Grunter"), from the same Loeb Classical Library volume; for here one of the shipmates whom Circe has transformed to a swine utterly rejects the proverbially wise Odysseus's offer to restore his human condition. The "conservative" Plutarch could hardly have questioned more radically the rationalistic assumptions of classical Greek culture, or raised more troublingly the question of which condition, Odysseus's or Grunter's, is more "natural" for a human being. Finally, in a brief passage from* The Obsolescence of Oracles, *in volume 5 (1936) of the Loeb Classical Library edition, one of the interlocutors, Philip, tells a haunting story of the death of the god who more than any other embodied the Greek sense of both communion with nature and awestruck panic in its presence. If Pan could be pronounced dead, the pagan world might indeed be near its end.*

The Natural Not Unconditioned,
translated by Harold Cherniss

Concerning the Face Which Appears in the Orb of the Moon *(927D–928D)*

14 What is more, if we are finally to throw off the habits and opinions that have held our minds in thrall and fearlessly to say what really appears to be the case, no part of a whole all by itself seems to have any order, position, or motion of its own which could be called unconditionally "natural." On the contrary, each and every such part, whenever its motion is usefully and properly accommodated to that for the sake of which the part has come to be and which is the purpose of its growth or production, and whenever it acts or is affected or disposed so that it contributes to the preservation or beauty or function of that thing, then, I believe, it has its "natural" position and motion and disposition. In man, at any rate, who is the result of "natural" process if any being is, the heavy and earthy parts are above, chiefly in the region of the head, and the hot and fiery parts are in the middle regions; some of the teeth grow from above and some from below, and neither set is "contrary to nature"; and it cannot be said that the fire which flashes in the eyes above is "natural," whereas that in the bowels and heart is "contrary to nature," but each has been assigned its proper and useful station. Observe, as Empedocles says, the nature of

Tritons and tortoises with hides of stone

and of all testaceans,

You will see earth there established over flesh;

and the stony matter does not oppress or crush the constitution on which it is superimposed, nor on the other hand does the heat by reason of lightness fly off to the upper region and escape, but they have been somehow intermingled and organically combined in accordance with the nature of each.

15 Such is probably the case with the cosmos too, if it really is a living being: in many places it has earth and in many fire and water and breath as the result not of forcible expulsion but of rational arrangement. After all, the eye has its present position in the body not because it was extruded thither as a result of its lightness, and the heart is in the chest not because its heaviness has caused it to slip and fall thither but because it was better that each of them should be so located. Let us not then believe with regard to the parts of the cosmos either that earth is situated here because its weight has caused it to subside or that the sun, as Metrodorus of Chios once thought, was extruded into the upper regions like an inflated skin by reason of its lightness or that the other stars got into their present positions because they tipped the balance, as it were, at different weights. On the contrary, the rational principle is in control; and that is why the stars

revolve fixed like "radiant eyes" in the countenance of the universe, the sun in the heart's capacity transmits and disperses out of himself heat and light as it were blood and breath, and earth and sea "naturally" serve the cosmos to the ends that bowels and bladder do an animal. The moon, situate between sun and earth as the liver or another of the soft viscera is between heart and bowels, transmits hither the warmth from above and sends upward the exhalations from our region, refining them in herself by a kind of concoction and purification. It is not clear to us whether her earthiness and solidity have any use suitable to other ends also. Nevertheless, in everything the better has control of the necessary. Well, what probability can we thus conceive in the statements of the Stoics? They say that the luminous and tenuous part of the ether by reason of its subtility became sky and the part which was condensed or compressed became stars, and that of these the most sluggish and turbid is the moon. Yet all the same anyone can see that the moon has not been separated from the ether but that there is still a large amount of it about her in which she moves and much of it beneath her in which . . . comets whirl. So it is not the inclinations consequent upon weight and lightness that have circumscribed the precincts of each of the bodies, but their arrangement is the result of a different principle. . . .

The Virtues of Beasts,
translated by William C. Helmbold
Gryllus, *or* Beasts Are Rational

2 *Gryllus.* Hello, Odysseus.

Odysseus. And you, too, Gryllus, for heaven's sake!

Gryllus. What do you want to ask?

Odysseus. Since I am aware that you have been men, I feel sorry for all of you in your present plight; yet it is only natural that I should be more concerned for those of you who were Greeks before you fell into this misfortune. So now I have asked Circe to remove the spell from any Greek who chooses and restore him to his original shape and let him go back home with us.

Gryllus. Stop, Odysseus! Not a word more! You see, we don't any of us think much of you either, for evidently it was a farce, that talk of your cleverness and your fame as one whose intelligence surpassed all the rest—a man who boggles at the simple matter of changing from worse to better because he hasn't considered the matter. For just as children dread the doctor's doses and run from lessons, the very things that, by changing them from invalids and fools, will make them healthier and wiser, just so you have shied away from the change from one shape to another. At this very moment you are not only living in fear and trembling as a companion of Circe, frightened that she may, before you know it, turn you into a pig or a wolf, but you are also trying to persuade us, who live in an abundance of good things, to abandon them, and with them the lady who provides them, and sail away with you, when we have again become men, the most unfortunate of creatures!

Odysseus. To me, Gryllus, you seem to have lost not only your shape, but your intelligence also under the influence of that drug. You have become infected with strange and completely perverted notions. Or was it rather an inclination to swinishness that conjured you into this shape?

Gryllus. Neither of these, king of the Cephallenians.[35] But if it is your pleasure to dis-

[35]After Homer, *Iliad*, ii.631; *Odyssey*, xxiv.378; or, taking the pun [on Greek *képhalos*, "brain"], "King of Brains," "Mastermind." (Helmbold)

cuss the matter instead of hurling abuse, I shall quickly make you see that we are right to prefer our present life in place of the former one, now that we have tried both.
Odysseus. Go on. I should like to hear from you.

3 *Gryllus.* And I, in that case, to instruct you. Let us begin with the virtues, which, we note, inspire you with pride; for you rate yourselves as far superior to animals in justice and wisdom and courage and all the rest of them. But answer me this, wisest of men! Once I heard you telling Circe about the land of the Cyclopes, that though it is not plowed at all nor does anyone sow there, yet it is naturally so fertile and fecund that it produces spontaneously every kind of crops. Do you, then, rate this land higher than rugged, goat-pasturing Ithaca, which barely yields the tiller a meager, churlish, trifling crop after great efforts and much toil? And see that you don't lose your temper and give me a patriotic answer that isn't what you really believe.
Odysseus. I have no need to lie; for though I love and cherish my native soil more, the other wins my approval and admiration.
Gryllus. Then this, we shall say, is the situation: the wisest of men thinks fit to commend and approve one thing while he loves and prefers another. Now I assume that your answer applies to the spiritual field also, for the situation is the same as with the land: that spiritual soil is better which produces a harvest of virtue as a spontaneous crop without toil.
Odysseus. Yes, this too you may assume.
Gryllus. At this moment, then, you are conceding the point that the soul of beasts has a greater natural capacity and perfection for the generation of virtue; for without command or instruction, "unsown and unplowed," as it were, it naturally brings forth and develops such virtue as is proper in each case.
Odysseus. And what sort of virtue, Gryllus, is ever found in beasts?

4 *Gryllus.* Ask rather what sort of virtue is not found in them more than in the wisest of men? Take first, if you please, courage, in which you take great pride, not even pretending to blush when you are called "valiant" and "sacker of cities." Yet you, you villain, are the man who by tricks and frauds have led astray men who knew only a straightforward, noble style of war and were unversed in deceit and lies; while on your freedom from scruple you confer the name of the virtue that is least compatible with such nefariousness. Wild beasts, however, you will observe, are guileless and artless in their struggles, whether against one another or against you, and conduct their battles with unmistakably naked courage under the impulse of genuine valor. No edict summons them, nor do they fear a writ of desertion. No, it is their nature to flee subjection; with a stout heart they maintain an indomitable spirit to the very end. Nor are they conquered even when physically overpowered; they never give up in their hearts, even while perishing in the fray. In many cases, when beasts are dying, their valor withdraws together with the fighting spirit to some point where it is concentrated in one member and resists the slayer with convulsive movements and fierce anger until, like a fire, it is completely extinguished and departs.

Beasts never beg or sue for pity or acknowledge defeat: lion is never slave to lion, or horse to horse through cowardice, as man is to man when he unprotestingly accepts the name whose root is cowardice.[36] And when men have subdued beasts by snares and tricks,

[36]"Slavery" (*douleia*) as though derived from "cowardice" (*deilia*). (Helmbold)

such of them as are full grown refuse food and endure the pangs of thirst until they induce and embrace death in place of slavery. But nestlings and cubs, which by reason of age are tender and docile, are offered many beguiling allurements and enticements that act as drugs. These give them a taste for unnatural pleasures and modes of life, and in time make them spiritless to the point where they accept and submit to their so-called "taming," which is really an emasculation of their fighting spirit.

These facts make it perfectly obvious that bravery is an innate characteristic of beasts, while in human beings an independent spirit is actually contrary to nature. The point that best proves this, gentle Odysseus, is the fact that in beasts valor is naturally equal in both sexes and the female is in no way inferior to the male. She takes her part both in the struggle for existence and in the defense of her brood. . . . Surely from what has been said it is perfectly obvious that men have no natural claim to courage; if they did, women would have just as great a portion of valor. It follows that your practice of courage is brought about by legal compulsion, which is neither voluntary nor intentional, but in subservience to custom and censure and molded by extraneous beliefs and arguments. When you face toils and dangers, you do so not because you are courageous, but because you are more afraid of some alternative. . . . And the reason is that the spirit of anger is, as it were, the tempering or the cutting edge of courage. Now beasts use this undiluted in their contests, whereas you men have it mixed with calculation, as wine with water, so that it is displaced in the presence of danger and fails you when you need it most. Some of you even declare that anger should not enter at all into fighting, but be dismissed in order to make use of sober calculation; their contention is correct so far as self-preservation goes, but is disgracefully false as regards valorous defense. For surely it is absurd for you to find fault with Nature because she did not equip your bodies with natural stings, or place fighting tusks among your teeth, or give you nails like curved claws, while you yourselves remove or curb the emotional instrument that Nature has given.

5 *Odysseus.* Bless me, Gryllus, you must once have been a very clever sophist, one may judge, since even as things are, and speaking from your swinishness, you can attack the subject with such fervent ardor. But why have you failed to discuss temperance, the next in order?

Gryllus. Because I thought that you would first wish to take exception to what I have said. But you are eager to hear about temperance since you are the husband of a model of chastity and believe that you yourself have given a proof of self-control by rejecting the embraces of Circe.[37] And in this you are no more continent than any of the beasts; for neither do they desire to consort with their betters, but pursue both pleasure and love with mates of like species. . . . As for the chastity of Penelope, the cawing of countless crows will pour laughter and contempt upon it; for every crow, if her mate dies, remains a widow, not merely for a short time, but for nine generations of men. It follows that your fair Penelope is nine times inferior in chastity to any crow you please.

6 Now since you are not unaware that I am a sophist, let me marshal my arguments in some order by defining temperance and analyzing the desires according to their kinds. Temperance, then, is a curtailment and an ordering of the desires that eliminate those that are extraneous or superfluous and discipline in modest and timely fashion those that are essential. You can, of course, observe countless differences in the desires . . . and the

[37]In the *Odyssey* (X.472–74) it was Odysseus's shipmates who had to remind him, after a year on Circe's island, to "remember your fatherland" and think of returning home.

desire to eat and drink is at once natural and essential, while the pleasures of love, which, though they find their origin in nature, yet may be forgone and discarded without much inconvenience, have been called natural but not essential. But there are desires of another kind, neither essential nor natural, that are imported in a deluge from without as a result of your inane illusions and because you lack true culture. So great is their multitude that the natural desires are, every one of them, all but overwhelmed, as though an alien rabble were overpowering the native citizenry. But beasts have souls completely inaccessible and closed to these adventitious passions and live their lives as free from empty illusions as though they dwelt far from the sea.[38] They fall short in the matter of delicate and luxurious living, but solidly protect their sobriety and the better regulation of their desires since those that dwell within them are neither numerous nor alien. . . . None, then, of such adventitious desires has a place in our souls; our life for the most part is controlled by the essential desires and pleasures. As for those that are non-essential, but merely natural, we resort to them without either irregularity or excess. . . .

7 . . . Even men themselves acknowledge that beasts have a better claim to temperance and the non-violation of nature in their pleasures. Not even Nature, with Law for her ally, can keep within bounds the unchastened vice of your hearts; but as though swept by the current of their lusts beyond the barrier at many points, men do such deeds as wantonly outrage Nature, upset her order, and confuse her distinctions. For men have, in fact, attempted to consort with goats and sows and mares, and women have gone mad with lust for male beasts. From such unions your Minotaurs and Aegipans, and, I suppose, your Sphinxes and Centaurs have arisen.[39] Yet it is through hunger that dogs have occasionally eaten a man; and birds have tasted of human flesh through necessity; but no beast has ever attempted a human body for lustful reasons. But the beasts I have mentioned and many others have been victims of the violent and lawless lusts of man.

8 Though men are so vile and incontinent where the desires I have spoken of are concerned, they can be proved to be even more so in the case of essential desires, being here far inferior to animals in temperance. These are the desires for food and drink, in which we beasts always take our pleasure along with some sort of utility; whereas you, in your pursuit of pleasure rather than natural nourishment, are punished by many serious ailments which, welling up from one single source, the surfeit of your bodies, fill you with all manner of flatulence that is difficult to purge. In the first place, each species of animal has one single food proper to it, grass or some root or fruit. Those that are carnivorous resort to no other kind of nourishment, nor do they deprive those weaker than themselves of sustenance; but the lion lets the deer, and the wolf lets the sheep, feed in its natural pasture. But man in his pleasures is led astray by gluttony to everything edible; he tries and tastes everything as if he had not yet come to recognize what is suitable and proper for him; alone of all creatures he is omnivorous.

In the first place his eating of flesh is caused by no lack of means or methods, for he can always in season harvest and garner and gather in such a succession of plants and grains as will all but tire him out with their abundance; but driven on by luxurious de-

[38] The sea is the symbol of mischievous foreign influence. (Helmbold)

[39] The Minotaur, half bull and half man, was born of the illicit union of Pasiphaë, wife of King Minos of Knossos, with a bull; Aegipan ("Goat-Pan") was a name given to goat-footed Pan; the Sphinx, in Greece, was usually portrayed as a wingèd lion with a woman's breasts and upper body; the Centaur was half horse and half man.

sires and satiety with merely essential nourishment, he pursues illicit food, made unclean by the slaughter of beasts; and he does this in a much more cruel way than the most savage beasts of prey. Blood and gore and raw flesh are the proper diet of kite and wolf and snake; to man they are an appetizer. Then, too, man makes use of every kind of food and does not, like beasts, abstain from most kinds and consequently make war on a few only that he must have for food. In a word, nothing that flies or swims or moves on land has escaped your so-called civilized and hospitable tables.

9 Well, then. It is admitted that you use animals as appetizers to sweeten your fare. . . . Animal intelligence, on the contrary, allows no room for useless and pointless arts; and in the case of essential ones, we do not make one man with constant study cling to one department of knowledge and rivet him jealously to that; nor do we receive our arts as alien products or pay to be taught them. Our intelligence produces them on the spot unaided, as its own congenital and legitimate skills. I have heard that in Egypt everyone is a physician; and in the case of beasts each one is not only his own specialist in medicine, but also in the providing of food, in warfare and hunting as well as in self-defense and music, in so far as any kind of animal has a natural gift for it. From whom have we swine learned, when we are sick, to resort to rivers to catch crabs? Who taught tortoises to devour marjoram after eating the snake?[40] And who instructed Cretan goats, when they are pierced by an arrow, to look for dittany, after eating which the arrowhead falls out? For if you speak the truth and say that Nature is their teacher, you are elevating the intelligence of animals to the most sovereign and wisest of first principles. If you do not think that it should be called either reason or intelligence, it is high time for you to cast about for some fairer and even more honorable term to describe it, since certainly the faculty that it brings to bear in action is better and more remarkable. It is no uninstructed or untrained faculty, but rather self-taught and self-sufficient—and not for lack of strength. It is just because of the health and completeness of its native virtue that it is indifferent to the contributions to its intelligence supplied by the lore of others. . . . If you are doubtful that we can learn arts, then let me tell you that we can even teach them. When partridges are making their escape, they accustom their fledglings to hide by falling on their backs and holding a lump of earth over themselves with their claws. You can observe storks on the roof, the adults showing the art of flying to the young as they make their trial flights. Nightingales set the example for their young to sing; while nestlings that are caught young and brought up by human care are poorer singers, as though they had left the care of their teacher too early. . . . And since I have entered into this new body of mine, I marvel at those arguments by which the sophists[41] brought me to consider all creatures except man irrational and senseless.

10 *Odysseus.* So now, Gryllus, you are transformed. Do you attribute reason even to the sheep and the ass?

Gryllus. From even these, dearest Odysseus, it is perfectly possible to gather that animals have a natural endowment of reason and intellect. For just as one tree is not more nor less inanimate than another, but they are all in the same state of insensibility, since none is endowed with soul, in the same way one animal would not be thought to be more

[40]See Aelian, *On the Characteristics of Animals* III.5, trans. A. F. Scholfield: "If a Tortoise eats part of a snake and thereafter some marjoram, it becomes immune from the poison which was bound to be quite fatal to it."

[41]Probably the Stoics are meant (by anachronism). (Helmbold)

sluggish or indocile mentally than another if they did not all possess reason and intellect to some degree—though some have a greater or less proportion than others. Please note that cases of dullness and stupidity in some animals are demonstrated by the cleverness and sharpness of others—as when you compare an ass and a sheep with a fox or a wolf or a bee. It is like comparing Polyphemus to you or that dunce Coroebus to your grandfather Autolycus.[42] I scarcely believe that there is such a spread between one animal and another as there is between man and man in the matter of judgment and reasoning and memory.
Odysseus. But consider, Gryllus: is it not a fearful piece of violence to grant reason to creatures that have no inherent knowledge of God?
Gryllus. Then shall we deny, Odysseus, that so wise and remarkable a man as you had Sisyphus for a father?[43]

Great Pan Is Dead,
translated by Frank Cole Babbitt
The Obsolescence of Oracles *(419B–E)*

17 . . . The father of Aemilianus the orator, to whom some of you have listened, was Epitherses, who lived in our town and was my teacher in grammar. He said that once upon a time in making a voyage to Italy he embarked on a ship carrying freight and many passengers. It was already evening when, near the Echinades Islands, the wind dropped, and the ship drifted near Paxi. Almost everybody was awake, and a good many had not finished their after-dinner wine. Suddenly from the island of Paxi was heard the voice of someone loudly calling Thamus, so that all were amazed. Thamus was an Egyptian pilot, not known by name even to many on board. Twice he was called and made no reply, but the third time he answered; and the caller, raising his voice, said, "When you come opposite to Palodes, announce that Great Pan is dead." On hearing this, all, said Epitherses, were astounded and reasoned among themselves whether it were better to carry out the order or to refuse to meddle and let the matter go. Under the circumstances Thamus made up his mind that if there should be a breeze, he would sail past and keep quiet, but with no wind and a smooth sea about the place he would announce what he had heard. So, when he came opposite Palodes, and there was neither wind nor wave, Thamus from the stern, looking toward the land, said the words as he had heard them: "Great Pan is dead." Even before he had finished there was a great cry of lamentation, not of one person, but of many, mingled with exclamations of amazement. As many persons were on the vessel, the story was soon spread abroad in Rome, and Thamus was sent for by Tiberius Caesar. Tiberius became so convinced of the truth of the story that he caused an inquiry and investigation to be made about Pan; and the scholars, who were numerous at his court, conjectured that he was the son born of Hermes and Penelope.[44]

[42]Coroebus was proverbially so stupid that he tried to count the waves of the sea. Autólycus (see *Odyssey* XIX.394ff.) surpassed all men "in thefts and perjury," a gift of Hermes. (Helmbold)
[43]Homer's crafty Odysseus was the son of Laërtês and of Anticléa (daughter of Autólycus), but as his reputation later declined, he was sometimes called the son of the cruel and crafty Sísyphus (son of Autolycus), who for his sins was condemned in the underworld to roll a boulder up a hill from the top of which it repeatedly fell again. The dialogue abruptly ends at this point.
[44]Herodotus (*Histories* II.145) reports that the Egyptians considered Pan one of the oldest of the gods, but the Greeks thought him one of the youngest, calling him the son of Hermes and of Penelope (rather than of Dryops's daughter, as in the Homeric Hymn to Pan in Chapter 8 above).

SIX

The Vicar's Dominion: Late Antiquity and the Middle Ages

Most classical and later Greek and Roman writers saw nature as a beneficent or at least a neutral power, participating in if not identical with the divine. Platonists and those most influenced by them indeed repudiated Milesian and Epicurean conceptions of an autonomous or mechanical nature, a realm of necessity governed by no laws but its own, and posited an "intelligible" world beyond the "sensible" phenomena of the world of change and becoming. Yet the natural world, far from being corrupt, had in Plato's myth been created by a demiurge who governed by reason a cosmos that continued to partake of the eternal forms from which it could never be severed. Thus even those for whom the Good, the Beautiful, and the One infinitely transcended the world of nature saw that world as wondrously beautiful and good.

For others in late antiquity and the Latin Middle Ages—Jews, Christians, and followers of various Hermetic, Gnostic, and Manichaean traditions—the dualistic opposition between the divine and the natural latent in Platonism (Plotinus "seemed ashamed," Porphyry tells us, "of being in his body") was intensely troubling. Gnostics radically depreciated the physical world and sought salvation by eliminating attachment to it; many believed the Creator of such a world (sometimes identified with the Hebrew God) had himself been evil and the source of evil, which only purification of earthly desire and dedication to a categorically different divinity (sometimes identified with Christ) could overcome. Thus for the second-century Christian Gnostic Valentinus, the physical universe, Peter Brown writes in *The Body and Society* (1988), "was a mistake that must be rectified." Similar views surfaced from time to time in Western Christendom, notably among the Cathari ("Pure") of the Middle Ages, whose sect a crusade of the early thirteenth century undertook to annihilate by fire and the sword.

But if the Church zealously banished Gnostic and Manichaean heresies, its own beliefs and practices—especially after St. Anthony in the early fourth century devoted himself to ascetic renunciation in the Egyptian desert—frequently encouraged the loathing of the body and *contemptus mundi* (contempt of the world) advocated by many monastic writers for centuries to come. The majority of Christians could choose, as Paul counseled the Corinthians, to marry rather than burn with passion, but this compromised choice was not Paul's own nor that of the saints who would follow. Of the two ways of Christian life defined in the fourth century by Bishop Eusebius of Caesarea in his *Proof of the Gospel* (as quoted by Brown), it was not the second—the humble way of marriage and child-rearing, farming and trade—that was most revered, but the first: "above nature, and beyond common human living . . . it admits not marriage, child-bearing, property nor the possession of wealth. . . . Like some celestial beings, these gaze down upon human life, performing the duty of a priesthood to Almighty God for the whole race." The laity might be tolerantly indulged in their traffic with the world, but the clergy, according to this persistent (if often violated) Christian ideal, should fervently renounce its fleshly corruptions. As the future Pope Innocent III wrote in his bleak treatise *On the Misery of Man* in the late twelfth century, the "food of worms" and "mass of putridity" that is physical man pollutes his soul with "the infection of sin" and "filth of iniquity" through the "foul stench of wantonness" in sexual union. Throughout medieval Christianity runs, not far below the surface, a strong undercurrent of quasi-Gnostic aversion to the natural world in general and the human body in particular.

Even so, the Church *did* evolve a view of the natural world very far from the dualistic antinomies of the Gnostics and Manichaeans. The reverent celebration of the beauty and goodness of the LORD's creation that pervades the Hebrew Bible (adopted as the Christian Old Testament) was a potent legacy, reaffirmed by major Jewish thinkers from Philo of Alexandria to Moses Maimonides, that continually worked against ascetic repudiation. Moreover, the fundamental Christian doctrines of the Incarnation, by which God took human flesh, and resurrection not of the soul or spirit alone—the "pneumatic man" of the Gnostics—but of the body scorned by them, affirmed the essential dignity of man's physical dimension and, by extension, the universe of which he is part. "Do you not know that your body is a temple of the Holy Spirit within you, which you have from God?" St. Paul asks (1 Corinthians 6:19–20): "So glorify God in your body." Far from standing opposed to the natural world, like the fastidiously untainted deity of the Gnostics, Paul's God is manifest in it. "Ever since the creation of the world," the Epistle to the Romans states (1:20), "his invisible nature, namely, his eternal power and deity, has been clearly perceived in the things that have been made." This belief would underlie countless allegorizations throughout the Middle Ages—both by learned thinkers and poets and in popular bestiaries and fables—of the world as the book in which could be read the handwriting of God. Nor could the doctrine of the Trinity admit any possibility of an evil creator God in the Christian universe: whether as Father, Son, or Holy Spirit, God was good, and beside him there was no other. The benevolence of God, as stern Father or Good Shepherd, extended to humankind as it did to the lilies of the field; and Solomon in all his glory, Jesus had said (Matthew 6:29), "was not arrayed like one of these."

As Christian doctrine developed during the patristic period of the early Christian centuries and during the Middle Ages, then, a highly complex attitude toward nature evolved. The very concept of nature as a realm ruled by laws of its own derived from the pagan Greek philosophers or their Roman transmitters—in the Hebrew Bible the heavens and earth were the handiwork of God, not powers in their own right—yet in a universe governed by an omnipotent God, this realm could not be autonomous, or truly divine. "Everywhere in mediaeval philosophy the natural order leans on a supernatural order" above and beyond it, Etienne Gilson writes in *The Spirit of Mediaeval Philosophy* (1936): "And how, if the case stands thus, do we dare to talk of nature at all in a Christian philosophy?"

To this question, thinkers and poets of the Middle Ages responded in varied ways, for any relation between God's will and nature's impersonal laws could only be problematic. Some writers, reaffirming with St. Ambrose that all depends on God, dismissed natural philosophy as unprofitable speculation. Others, deeply influenced by Platonic philosophy, like Origen and Gregory of Nyssa in the Greek East (and Augustine in the Latin West), saw the paramount reality of physical phenomena in the spiritual realm, and sought to discover the truth of the world through moral and allegorical interpretation not only of biblical texts but also of the natural world. "The books of nature and Scripture thus provided two kinds of knowledge and two modes of interpreting experience," Richard Hamilton Green remarks:[1] "one kind was appropriate to the divine revelation

[1]In his article "Alan of Lille's *De Planctu Naturae*," in *Speculum* 31 (1956).

in the sacramental world of natural phenomena and was properly exercised by reason; the other was addressed to the revelation of Christ in Scripture and was finally the object not of reason but of faith. But for the mediaeval theologian the two were not really separable." Counteracting the *contemptus mundi* of both Gnostic heretic and monkish ascetic, in "a world filled with references, reminders and overtones of Divinity, manifestations of God in things," Umberto Eco observes in *Art and Beauty in the Middle Ages* (1959; Eng. trans. 1986), "even at its most dreadful, nature appeared to the symbolical imagination to be a kind of alphabet through which God spoke to men and revealed the order in things, the blessings of the supernatural, how to conduct oneself in the midst of this divine order and how to win heaven."

Thus despite the revulsion from the flesh and the world expressed by certain Christian writers, much medieval thought and literature throbs with the vitality and joy of a world ruled by the benevolent Natura whom some would even portray, in a daring metaphor, as a "goddess." It is scarcely too much to say that "nature poems"—poems devoted mainly to describing scenic landscapes and evoking the impressions they create—begin, or at least begin to be prominent, in Western literature in late antiquity and the Middle Ages. Lyrics on gardens (including that of Eden) and flowers, forests and meadows, birds and the seasons of summer and spring, adorn even the "Dark Ages" before and after the Carolingian revival of the eighth to ninth centuries; they proliferate with explosive energy after the twelfth-century Renaissance in the vibrant celebrations of life of the Latin *Carmina Burana* and their vernacular counterparts. Between the collapse of ancient and the rise of late medieval urban civilizations, as Marc Bloch remarks in *Feudal Society* (1940; Eng. trans. 1961), men and women were necessarily "close to nature" in a Europe that was largely untamed forest and wilderness; both wild and cultivated nature were always close at hand.

Throughout much of the period between the decline of the ancient schools and the rise of universities in the twelfth to thirteenth centuries, and even later, "learned" accounts of the natural world drew far less on direct experience than on secondhand citation of the limited number of classical works that survived in monasteries and cathedral schools. Many writers seemed far more concerned to cite an ancient authority on nature than to observe the world with their own eyes. Even so, medieval man's relationship with nature, especially in the High Middle Ages, entailed "a discovery of himself in the external world," as A. J. Gurevich suggests in *Categories of Medieval Culture* (1972; Eng. trans. 1985), "combined with a perception of the cosmos as subject. . . . No clear boundaries separated man from the world: finding in the world an extension of himself, he discovers in himself an analogue of the universe. The one mirrors the other." And if man—in an ancient conception that gained new currency in the Middle Ages—was a microcosm intimately connected with the universe reflected in him, Nature, as Chaucer (echoing Alan of Lille and the *Roman de la rose*) described her in *The Parliament of Fowls*, could be conceived as "the vicaire of the almyghty Lord": not supreme or wholly autonomous, indeed (only the triune God could be that), yet exercising an authority over the physical universe analogous to that which the Pope, as vicar of Christ, held in the universal Church. And this, in the Middle Ages, was no small dominion.

13

Created Creatrix: Medieval Conceptions of Nature

The root meaning of physis, *as we have seen, was "growth," and Greek concepts of nature—in Hesiod, in the Milesian naturalists, in the atomists, and even in the overflow of successive hypostases from the Plotinian One—characteristically depicted the origin of the world as a spontaneous process, not as the product of an external agent. Yet Plato had introduced in the* Timaeus*—the one Platonic dialogue widely available, in the partial translation and commentary of Calcidius (or Chalcidius), to the Latin Middle Ages—his "likely story" of the formation of the cosmos by a rational demiurge out of preexisting matter.*

In other traditions, above all the Judeo-Christian but also various Gnostic and Hermetic accounts, creation of the universe by a good or evil god, sometimes ex nihilo *(out of nothing), was an article of faith or dogma. The Aristotelian or Epicurean believed in the eternity of the world, in which coming-to-be and passing-away (or their appearance), however conceived, were continual processes in a cosmos that had not begun and would never end. In contrast, both Jew and Christian held that the universe—and matter, and time itself—began with the Word of an almighty God and would end (for the Christian at least) when God willed, not in the cyclical fires of a recurrent Stoic* ekpyrôsis, *but in an apocalyptic final Judgment. Once these other traditions had absorbed, as to some degree all did, Aristotelian, Stoic, and above all Platonic conceptions of nature, questions concerning creation and the respective roles of God's will and of nature's laws in creation were central to efforts to reconcile such fundamentally different visions of the world as the Judeo-Christian and the Greek. Not only major philosophers of the present chapter, from Philo of Alexandria to the Scholastic theologians thirteen hundred years later, but also important poets in the chapters that follow occupied themselves with articulating the part played by created nature in the creation of the world.*

Questions concerning creation related to larger questions concerning the value of the natural as opposed to the supernatural world, and of the body as opposed to the spirit. Such questions critically concerned even thinkers principally occupied with "otherwordly" matters of faith and revelation. The most influential of them—such as Augustine, affirming that "every nature, insofar as it is a nature, is good," or Boëthius, discussing questions of nature with Lady Philosophy while awaiting death in his prison cell—strove to integrate seemingly disparate conceptions of nature and God, reason and faith, into a new Christian philosophy.

Through important Christian thinkers of the Greek East, such as Origen, Gregory of Nyssa, and Pseudo-Dionysius the Areopagite, as well as through Augustine and Boëthius in the West, Platonic and Neoplatonic conceptions entered into the mainstream of Christian thought: the one major Western thinker between the sixth and twelfth centuries, John the Scot (Johannes Scotus Eriugena) in the ninth century, translated both Gregory and Dionysius from the Greek and made the division of nature—in a sense so comprehensive as to include even God—the central question of his major work, the Periphyseon. *Though his highly original thought was several times condemned as heretical, it appears to have exerted significant influence on thinkers and poets (such as Bernard Silvestris and Alan of Lille in Chapter 15) loosely associated with the twelfth-century "School of Chartres." Among Scholastic theologians of the thirteenth century, Albert the Great devoted much atten-*

tion to natural inquiries, and Roger Bacon boldly declared that "all things must be verified by experience."

Alongside the subtle formulations of these and other thinkers of late antiquity and the Middle Ages, the crude (if often delightful) zoology of popular bestiaries like the Greek Physiologus, *with its simplistic allegorizations of animals real and imagined, and the largely derivative medleys of fact and fancy crammed into encyclopedic collections like those of Bartholomew the Englishman or Brunetto Latini seem strangely out of place. Yet here too relations between the natural and the supernatural are continually being addressed. It is a long way from John the Scot's affirmation that everything in the world of generation is a theophany of divine goodness to the solemn assurance of the bestiaries that the unicorn, elephant, and pelican are living symbols of Christ. Yet these statements belong, after all, to the same world.*

On the Creation and Eternity of the World

The Writings of Philo of Alexandria

Philo of Alexandria, or Philo Judaeus ("Philo the Jew," ca. 30/20 B.C.–ca. A.D. 45/50), undertook, in his voluminous writings, to establish truths of philosophy through allegorical reading of biblical scriptures. Alexandria had for centuries been, and would for centuries remain, the center of a thriving Jewish community (which Philo claims was as large as a million), for whom Greek had replaced Aramaic and Hebrew as the spoken and written language. It was here that the Greek translation of the Hebrew Bible, the Septuagint, had been made (according to legend, as far back as the third century B.C.), and here that Philo absorbed the thought of Aristotle, the Stoics, the neo-Pythagoreans, and above all Plato, and evolved a philosophy that in important respects anticipated Neoplatonic thought and Christian exegesis.

Christians rather than Jews were mainly responsible for preserving his works, which had a great impact on Clement of Alexandria and Origen in the second to third centuries, and on many others. "Philo became widely known as one of the greatest scholars, not only among our own people [Christians] but also among those brought up as pagans," Eusebius wrote in his Ecclesiastical History *(in G. A. Williamson's translation of 1965): "The constant and conscientious labor that he bestowed on theological and traditional studies is plain for all to see, while . . . in his enthusiasm for the systems of Plato and Pythagoras he surpassed all his contemporaries." Yet for Philo, as Emil Schürer observes in his* History of the Jewish People in the Time of Jesus Christ *(2nd ed. 1886–90), "Not in Plato, Pythagoras and Zeno, but above all in the writings of Moses, is to be found the deepest and most perfect instruction concerning things divine and human. In them was already comprised all that was good and true, which the Greek philosophers subsequently taught."*

In the first of the following treatises, he upholds the Mosaic account of the creation of the world against those who contend it has existed forever, but affirms that the universe, though created, is indestructible. In the second treatise, "On the Eternity of the World" (the authenticity of which has been questioned), he appears to entertain, instead of the Platonic (and Mosaic) account, the Aristotelian view of an "uncreated and indestructible" world. Philo's overall thought is characterized by a sharp dualism between God and the world (bridged by intermediate beings, including the divine Logos), by a Stoical ethic of renunciation of sensual desire, and by insistence that evil originates not from God but from matter. Yet the outlook reflected in his gnôsis *anticipates more that of Plotinus than that of the Gnostics whom Plotinus would oppose; for throughout his writings runs a celebration (in the spirit of both Hebrew psalms and Greek myths) of the fertility of the earth as a bounteous mother, of nature as a living force desiring "the conservation of the All," and of man "as the borderland between mortal and immortal nature." By demonstrating that the Mosaic Law (Torah) is at one with nature, "Philo is emphasizing the unity of all realms of human experience," Samuel Sandmel writes in* Philo of Alexandria: An Introduction *(1979), "against those views which stress the sense in which* physis *and* nomos *can be opposites and hence an obstacle to unity"; his "main accomplishment" is thus that "he has blended* physis *and Torah so thoroughly that in his thought they are inextricably bound together."*

The translations are from volumes I (1929) and IX (1941) respectively of the Loeb Classical Library edition. For further reading, see also Harry A. Wolfson, Philo, *2 vols. (1947).*

FROM *On the Account of the World's Creation Given by Moses, translated by F. H. Colson and G. H. Whitaker*

7–11; 42–44; 82; 133; 135; 145–46

II There are some people who, having the world in admiration rather than the Maker of the world, pronounce it to be without beginning and everlasting, while with impious falsehood they postulate in God a vast inactivity; whereas we ought on the contrary to be astonished at His powers as Maker and Father, and not to assign to the world a disproportionate majesty. Moses, both because he had attained the very summit of philosophy, and because he had been divinely instructed in the greater and most essential part of Nature's lore, could not fail to recognize that the universal must consist of two parts, one part active Cause and the other passive object; and that the active Cause is the perfectly pure and unsullied Mind of the universe, transcending virtue, transcending knowledge, transcending the good itself and the beautiful itself; while the passive part is in itself incapable of life and motion, but, when set in motion and shaped and quickened by Mind, changes into the most perfect masterpiece, namely this world. Those who assert that this world is unoriginate unconsciously eliminate that which of all incentives to piety is the most beneficial and the most indispensable, namely providence. For it stands to reason that what has been brought into existence should be cared for by its Father and Maker. For, as we know, it is a father's aim in regard of his offspring and an artificer's in regard of his handiwork to preserve them, and by every means to fend off from them aught that may entail loss or harm. He keenly desires to provide for them in every way all that is beneficial and to their advantage: but between that which has never been brought into being and one who is not its Maker no such tie is formed. It is a worthless and baleful doctrine, setting up anarchy in the well-ordered realm of the world, leaving it without protector, arbitrator, or judge, without anyone whose office it is to administer and direct all its affairs. . . .

XIII Now in the original creation of all things, as I have said already, God caused all shrubs and plants to spring out of the earth perfect, having fruits not unripe but at their prime, to be perfectly ready for the immediate use and enjoyment of the animals that were forthwith to come into being. God then enjoins the earth to give birth to all these, and the earth, as though it had been long pregnant and in travail, brings forth all kinds of things sown, all kinds of trees, and countless kinds of fruits besides. But not only were the several fruits nourishment for animals, but also a provision for the perpetual reproduction of their kind, containing within them the seed-substances. Hidden and imperceptible in these substances are the principles or nuclei of all things. As the seasons go round these become open and manifest. For God willed that Nature should run a course that brings it back to its starting-point, endowing the species with immortality, and making them sharers of eternal existence. For the sake of this He both led on the beginning speedily towards the end, and made the end to retrace its way to the beginning. For it is the case both that the fruit comes out of the plants, as an end out of a beginning, and that out of the fruit again, containing as it does the seed in itself, there comes the plant, a beginning out of an end. . . .

XXVII . . . God, being minded to unite in intimate and loving fellowship the beginning and end of created things, made heaven the beginning and man the end, the one the most perfect of imperishable objects of sense, the other the noblest of things earth-born and perishable, being, in very truth, a miniature heaven. He bears about within himself, like holy images, endowments of nature that correspond to the constellations. He has capacities for science and art, for knowledge, and for the noble lore of the several virtues. For since the corruptible and the incorruptible are by nature contrary the one to the other, God assigned the fairest of each sort to the beginning and the end, heaven (as I have said) to the beginning, and man to the end. . . .

XLV . . . Nature has bestowed on every mother as a most essential endowment teeming breasts, thus preparing in advance food for the child that is to be born. The earth also, as we all know, is a mother, for which reason the earliest men thought fit to call her "Demeter," combining the name of "mother" with that of "earth"; for, as Plato says,[1] earth does not imitate woman, but woman earth. Poets quite rightly are in the habit of calling earth "All-mother," and "Fruit-bearer," and "Pandora" or "Give-all," inasmuch as she is the originating cause of existence and continuance in existence to all animals and plants alike. Fitly therefore on earth also, most ancient and most fertile of mothers, did Nature bestow, by way of breasts, streams of rivers and springs, to the end that both the plants might be watered and all animals might have abundance to drink. . . .

XLVI . . . It says, however, that the formation of the individual man, the object of sense, is a composite one made up of earthly substance and of Divine breath: for it says that the body was made through the Artificer taking clay and moulding out of it a human form, but that the soul was originated from nothing created whatever, but from the Father and Ruler of all: for that which He breathed in was nothing else than a Divine breath that migrated hither from that blissful and happy existence for the benefit of our race, to the end that, even if it is mortal in respect of its visible part, it may in respect of the part that is invisible be rendered immortal. Hence it may with propriety be said that man is the borderland between mortal and immortal nature, partaking of each so far as is needful, and that he was created at once mortal and immortal, mortal in respect of the body, but in respect of the mind immortal. . . .

LI Of the beauty of the first-made man in each part of his being, in soul and body, we have now said what falls perhaps far short of the reality but yet what for our powers was possible. It could not but be that his descendants, partaking as they did in the original form in which he was formed, should preserve marks, though faint ones, of their kinship with their first father. Now what is this kinship? Every man, in respect of his

[1] *Menexenus* 238A. (Colson and Whitaker)

mind, is allied to the divine Reason, having come into being as a copy or fragment or ray of that blessed nature, but in the structure of his body he is allied to all the world, for he is compounded of the same things, earth, water, air, and fire, each of the elements having contributed the share that falls to each, to complete a material absolutely sufficient in itself for the Creator to take in order to fashion this visible image. . . .

FROM *On the Eternity of the World, translated by F. H. Colson*
7–11; 35–38; 59; 62–64; 75

III Three views have been put forward on the question before us. Some assert that the world is eternal, uncreated and imperishable. Some on the contrary say that it is created and destructible. Others draw from both of these. From the latter they take the idea of the created, from the former that of the indestructible and so have laid down a composite doctrine to the effect that the world is created and indestructible. Democritus with Epicurus and the great mass of Stoic philosophers maintain the creation and destruction of the world but in different ways. The two first named postulate many worlds, the origin of which they ascribe to the mutual impacts and interlacings of atoms and its destruction to the counterblows and collisions sustained by the bodies so formed. The Stoics admit one world only; God is the cause of its creation but not of its destruction. This is due to the force of the ever-active fire which exists in things and in the course of long cycles of time resolves everything into itself and out of it is constructed a reborn world according to the design of its architect. According to these the world may be called from one point of view an eternal, from another a perishable world; thought of as a world reconstructed it is perishable, thought of as subject to the conflagration it is everlasting through the ceaseless rebirths and cycles which render it immortal. But Aristotle surely showed a pious and religious spirit when in opposition to this view he said that the world was uncreated and indestructible and denounced the shocking atheism of those who stated the contrary and held that there was no difference between hand-made idols and that great visible God who embraces the sun and moon and the pantheon as it may be truly called of the fixed and wandering stars. He is reported to have said in bitter mockery that in the past he had feared for his house lest it should be overthrown by violent winds or terrific storms or lapse of time or neglect of proper care. But now he lived under the fear of a greater menace from the theorists who would destroy the whole world. . . .

VII . . . Another point which must be clear to everyone is this. Nature in each case strives to maintain and conserve the thing of which it is the nature and if it were possible to render it immortal. Tree nature acts so in trees, animal nature in each kind of animal, but the nature of any particular part is necessarily too feeble to carry it into a perpetual existence. For privation or scorching or chilling or the vast multitude of other circumstances which ordinarily affect it descend to shake it violently and loosen and finally break the bond which holds it together, though if no such external force were lying ready to attack it, so far as itself was concerned, it would preserve all things small or great proof against age. The nature of the world then must necessarily desire the conservation of the All. For it is not inferior to the nature of particular parts that it should take to its heels and leave its post and try to manufacture sickness rather than health, destruction rather than complete preservation, since

High o'er them all she rears her head and brows
Easy to recognize though all are fair.[2]

But if this is true the world will not be susceptible to destruction. Why so? Because the nature which holds it together fortified by its great fund of strength is invincible and prevails over everything which could injure it. And so Plato says well:[3] "For nothing went out from it nor entered it from anywhere. For there was nothing. For by design it was created to supply its own sustenance by its own wasting and have all its actions and passions in itself and by itself. For its framer deemed that were it self-sufficing it would be far better than if it required aught else." . . .

XI . . . But there is no swerving in the nature of the universe, for that nature is supreme above all and so steadfast are its decisions once taken that it keeps immutable the limits fixed from the beginning. . . . A clear proof that the earth retains its vigor continually and perpetually at its height is its vegetation, for purified either by the overflow of rivers, as they say is the case in Egypt, or by the annual rains, it takes a respite and relaxation from the weary toil of bearing fruit, and then after this interval of rest recuperates its native force till it reaches its full strength and then begins again to bear fruits like the old and supplies in abundance to each kind of living creature such food as they need.

XII And therefore it seems to me that the poets did not do amiss in giving her the name of Pandora, because she gives all things that bring benefit and pleasurable enjoyment not to some only but to all creatures endowed with conscious life. Suppose one soaring aloft on wings when spring has reached its height were to survey the uplands and the lowlands, he would see the lowlands verdant with herbage, producing pasturage and grass fodder and barley and wheat and numerous other forms of grain, some sown by the farmer, others provided self-grown by the season of the year. He would see the uplands overshadowed with the branches and foliage which deck the trees and filled with a vast quantity of fruits, not merely those which serve for food, but also those which prove to be a cure for troubles. For the fruit of the olive heals the weariness of the body and that of the vine if drunk in moderation relaxes the violence of sorrow in the soul. Further he would perceive the sweet fragrance of the exhalations wafted from the flowers and the multitudinous varieties of their colors diversified by superhuman skill. Again looking away from the cultivated vegetation, he would survey poplars, cedars, pines, firs, tall towering oaks and the other deep, unbroken forests of wild trees which overshadow the vast expanse of the huge mountains and the wide stretch of deep soil which lies at their feet. Seeing all this he will recognize that the ever-youthful earth still has the indomitable and unwearying vigor of its prime. . . .

XV . . . Again if there was no everlasting form of nature to be seen, those who propound the destruction of the world might seem to have a good excuse for their iniquity, since they had no example of perpetual existence before them. But since according to the best professors of natural philosophy, fate has no beginning or end, being a chain connecting the causes of each event in unfailing continuity without a gap or break, why should we not also declare that the nature of the world or cosmic system is age-long, since

[2] *Odyssey* 6:107–108, said of Artemis and her nymphs. (Colson)
[3] *Timaeus* 33c. (Colson)

it is order of the disordered, adjustment of the unadjusted, concord of the discordant, unification of the discrepant, appearing as cohesion in wood and stone, growth in crops and trees, conscious life in all animals, mind and reason in men and the perfection of virtue in the good? And if the nature of the world is uncreated and indestructible, clearly the world also is the same, held together as it is by the might of an eternal bond. . . .

Downward-Inclining Nature's Lover

The *Hermetica*, by "Hermes Trismegistus," translated by Robert M. Torrance

The writings grouped together as Hermetica*—seventeen or so discourses of the Greek* Corpus Hermeticum; *three parts of the Latin* Asclepius, *translated from a lost Greek original (but wrongly ascribed in the Middle Ages to Apuleius, author of* The Golden Ass*); and various fragments collected by Stobaeus in about the fifth century* A.D., *along with a hodgepodge of works on astrology, alchemy, and magic—though probably written in the second to third century* A.D., *were attributed to the Egyptian god Thoth, inventor of writing, whom the Greeks called "thrice-great Hermes" (Hermes Trismegistus): an attribution first challenged by Isaac Casaubon in 1614. The treatises are permeated by Platonic, Pythagorean, and Stoic concepts (some possibly transmitted by Philo), and by a mystical religiosity typical of their troubled time. Though of little originality, these writings, through their pretended antiquity and exalted spiritual tone, were greatly influential in the Latin Middle Ages (through the* Asclepius*) and in the Renaissance, when Marsilio Ficino translated the Greek texts into Latin.*

In the Greek Poimandrês *(first treatise of the* Corpus Hermeticum*), the highest god, androgynous Mind* (Nous), *gives birth by the Word* (Logos) *to Mind the maker (the* dêmiourgos *of Plato's* Timaeus*), who combines with the Word to bring forth the irrational animals. Man* (Anthrôpos), *in contrast, was born directly from Mind the androgynous father, who fell in love with his offspring; but, enamored of the image of "downward-inclining Nature" (composed of mere matter,* hylê*), Man mingled with her; hence he alone of all animals is twofold, "mortal in body, immortal in the essential man." The* Asclepius *praises man as a great marvel insofar as he "scorns the merely human part of his nature, trusting in the divinity of its other part": only in contemplation of his heavenly dimension and contempt for everything connected with matter and body can he become truly human. In this fundamentally Gnostic dualism, "everything that comes from matter, thus all the visible world, is bad," A.-J. Festugière writes in* Hermétisme et Mystique Païenne *(1967); "and one can say as much of all forms of gnosis, pagan or Christian." Even so, through many Hermetic writings runs an optimistic current more akin to Neoplatonism. "In this conception of the spiritual life, the sensible world leads to the intelligible," Festugière comments, whereas in Gnosticism proper "the sensible world, being bad, leads away from the intelligible." For in the divine, the third treatise of the* Corpus *proclaims, "Nature too has her being."*

The translation is based on Corpus Hermeticum, *vol. 1, ed. A.D. Nock, with French translation by A.-J. Festugière (1938; 2nd ed. 1960). See also* Hermetica, *in four volumes, edited with introduction and translation by Walter Scott (1924);* Hermetica, *translated by Brian P. Copenhaver (1992); and Festugière's four-volume study* La Révélation d'Hermès Trismégiste *(1942–53; 2nd ed. 1983).*

FROM Poimandres

I.9 But Mind the god, being androgynous, existing as life and light, gave birth by the Word to a second Mind, the maker, who, as god of fire and spirit *[pneuma]*, made seven managers[4] surrounding the sensible world with their circles, and their management is called fate *[heimarmenê]*.

I.10 At once the Word of God leapt from the downward-inclining elements toward the pure product of Nature,[5] and was united with Mind the maker (for it was of the same substance); and the downward-inclining elements of nature were left without reason [or "without the Word"], so as to be mere matter.

I.11 But Mind the maker, along with the Word, surrounding the circles and spinning them round in a whirl, made his creations revolve and let them go on revolving from an indeterminate beginning toward a limitless end: for it starts where it ends. And this rotation, as Mind willed, brought forth from the downward-inclining elements animals without reason [or "without the word"], for they no longer retained reason; the air brought forth birds, and the water swimming things. The earth and water had separated from one another as Mind willed, and earth had brought forth from herself four-footed animals and creeping things, beasts wild and tame.

I.12 Mind the father of all, being life and light, gave birth to Man in his own likeness, with whom he fell in love as his own child; for Man was very beautiful, retaining the image of his father; indeed God fell in love with his own form, and handed over all his creations to him. . . .

I.14 . . . and [Man] showed downward-inclining Nature the beautiful form of God, and when she saw him [Man] who had within himself the insatiate beauty and all the activity of the managers, with the form of God, she smiled with love, as having seen the reflection of Man's most beautiful form on the water and his shadow on the earth. And he, having seen in her a form similar to himself on the water, cherished it and wished to dwell there; his wish was immediately enacted, and he inhabited unreasoning form. And Nature, receiving her loved one, entwined him all round and mingled with him; for they were in love.

I.15 And therefore unlike all animals on earth man is twofold, mortal in body, immortal in the essential man. For though immortal and having power over all things, he suffers the mortal condition of submission to fate. Though raised above the conjuncture *[harmonia]* [of the spheres], he is enslaved to it; [though] androgynous because he comes from an androgynous father, and sleepless because from a sleepless father, he is vanquished [i.e., by sexual desire and by sleep, and to that extent subject to Nature]. . . .

III.4 . . . For the divine is the whole cosmic mixture renewed by Nature; for in the divine, Nature too has her being.

FROM Asclepius

I.6 For these reasons, Asclepius, man is a great marvel, an animal to be revered and honored. For he passes over into the nature of a god as if he himself were a god; he is

[4] The seven planets.
[5] I. e., heaven, or the highest sphere of heaven. (Scott)

acquainted with the race of daemons, since he knows that he arose from the same source; he scorns the merely human part of his nature, trusting in the divinity of its other part. Oh, how much more happily blended is the nature of man! United with the gods by his kindred divinity, he scorns in himself the part which makes him earthly; all other creatures, with which he knows he is linked by the disposition of heaven, he binds to himself by the ties of affection; he looks upward to heaven. Thus placed in a fortunate intermediate position, he cherishes all below him, and is himself cherished by those above him. He cultivates the earth, mingles with the elements by the swiftness and descends to the depths of the sea by the sharpness of his mind. All things are allowed him: heaven does not seem too high, for by his sagacious thoughts he measures it as if it were near by. No mist of the air obscures his mental exertion; no density of the earth obstructs his work; no depth of the water blurs his downward gaze. He is all things in one; he is everywhere the same. . . .

I.7 . . . —Why then was it necessary, Trismegistus, that man be placed in the [material] world, and not live in supreme happiness in that part where God is?

—You are right to ask, Asclepius, and I pray that God may give me the means to answer you. . . . [10] . . . His images are two: the world and man. Hence it comes about that, since man is a single construct, through that part in which he is divine by means of the higher elements, as it were, of soul and intellect, spirit and reason, he seems able to ascend to heaven, but through the worldly part, consisting of fire [and earth], water and air, he remains a mortal on earth, so as not to neglect and abandon all things entrusted to his care. Thus humanity, though partly divine, was created mortal in its other part, that of the body. [11] . . . Whatever earthly things he possesses to gratify his bodily desire are alien from all parts of him that are kindred with the divine; hence they are rightly called possessions, since they were not born with us but began to be possessed by us later. Thus all things of this kind are alien to man, even the body, so that we should scorn both the objects of our appetite and the source from which this vice of appetite derives. For according to the view to which strenuous reason leads me, man ought to be [considered] human only to the extent that through contemplation of divinity he scorns and despises the mortal part joined to him by his need to take care of the lower world. . . .

God Distinct from Creation

The *Divine Institutes*, by Lactantius, translated by Bernard McGinn

Lucius Caelius (or Caecilius) Firmianus, called Lactantius (ca. 240–ca. 320), was born in Roman North Africa. He converted to Christianity sometime around the beginning of the fourth century, and lost his position as a teacher of rhetoric in Nicomedia (a major city of Bithynia in northwest Asia Minor) around 303, when the Emperor Diocletian's persecution of the Christians began. In old age he became tutor, in about 317, to the Christian Emperor Constantine's eldest son, Crispus. His major work, the Divinae Institutiones, *written roughly between 303 and 313, was an apology for Christianity and a refutation of its opponents aimed primarily at educated pagans; the many classical Latin writers whom he cites strikingly include Lucretius, and in his fifth book he argues that the true Golden Age of justice described by the poets was the biblical Eden. In this "first synthetic treatise on theology by a Western [i.e., Latin] writer," E. K. Rand writes in* Founders of the Middle Ages *(1928), Lactantius "seasons his apologetics with philosophy and tops it with a Ciceronian style" for which he was greatly admired from the time of St. Jerome through the Renaissance. Though by no means an original thinker, Lactantius played an important part in establishing the principle, Rand remarks, that "Christian faith . . . may, or rather must, draw freely for its sustenance on the thought, the poetry, and the inspiration of the past." In several places (notably II.8), he considers problems relating to God's creation of the world; in the following passages from Book VII, he censures the error of the Stoics in including both God and world, creator and created, under "nature," for in the Christian view the incorruptible God cannot be identical with the corruptible elements of the world. Man is "the only divine and heavenly animal," because he alone separates himself from the perishable world through contemplation of God.*

The translation is from Apocalyptic Spirituality *(1979), ed. Bernard McGinn; see also* The Divine Institutes, Books I–VII, *translated by Sister Mary Francis McDonald (1964).*

VII. 3 Since we are speaking of the errors of the philosophers, the Stoics divide nature into two parts, the one that makes, the other that offers itself for formation. In the first there is power to understand, in the latter matter. The one can do nothing without the other. How can it be the same being that both forms and is formed? If someone said that the potter is the same as the clay or the clay the same as the potter, would he not seem clearly crazy? The Stoics include two completely different things, God and the world, the Maker and the work, under the single name of "nature." They say that the one can do nothing without the other, as though nature were God mixed in with the world. Sometimes they so confuse things that God himself is the mind of the world and the world the body of God, as if the world and God came into being at the same time and God did not make the world. At other times they themselves admit creation when they announce that the world was made for the sake of man, and that God, as a divine and eternal mind separated and free of a body, is able to exist without the world if he wished. Since they were not able to understand his power and majesty, they mixed him up with the world, that is, with his work. Hence the passage in Vergil:

> Mind spread through all the limbs
> impels the entire mass and mingles
> itself in the vast body.
>
> —*Aeneid* VI.726–727

Where then is what they say was both made by divine providence and is ruled by it? If he made the world, he existed without it. If he rules it, he does so not as mind rules the body, but as a master rules a house, a driver a chariot, that is, as not mixed up with the things they rule. If all the things we see are parts of God, because they lack perception he lacks perception also. He is mortal because the parts are mortal.

I can count how often lands shaken by sudden earthquakes have split open or sunk abruptly, how often cities and islands, submerged by waves, have gone to the bottom, how often swamps have swallowed up fruitful plains, rivers and lakes have dried out, and mountains have either cracked and fallen or been leveled with the plains. Hidden unknown fire has consumed many regions and the foundations of many mountains. It is not enough that God does not spare his own members; man is also allowed to act against

God's body! Large bodies of water are formed, mountains are cut down and the inner bowels of the earth are dug out to find wealth. Is plowing possible without tearing the divine body? We who violate the members of God are criminals and evildoers. Does God allow his body to be abused and himself weaken it or permit man to do so? Or perhaps that divine understanding that is mingled with the world and all its parts has abandoned the earth's outer surface and buried itself in the depths lest it feel some pain from continuous wounding. If this is vain and absurd, the Stoics are just as lacking in sense as the things of this world. They have not understood that the divine spirit is diffused everywhere and holds all things together, but not in such a way that the incorrupt God himself is mingled with the solid corruptible elements. . . .

VII. 9 . . . Can anyone have considered the nature of other living beings, which Almighty God's providence created humbled with bent bodies and prostrate upon the earth, and not have understood that they have no relation with heaven? Can he not understand that man is the only divine and heavenly animal, he whose body is raised up from the ground and whose countenance is held high? In his upright state he seeks his origin. Contemptuous of the lowliness of earth, he stretches out to the heights because he understands he is to seek his supreme good on high. Mindful of the creation by which God made him special, does he not look toward his Maker? Trismegistus very rightly called this looking "contemplation." It does not exist in animals. Because the wisdom given to man alone is nothing else than knowledge of God, it is clear that the soul does not perish or dissolve. It lasts forever because it seeks and loves the eternal God. By the prompting of its nature it understands where it has come from and where it is going.

That man alone makes use of the heavenly element is no small argument for immortality. The universe consists of these two elements, fire and water, mutually opposed and at odds with each other. The one belongs to heaven, the other to earth. The other living things, because they are earthly and mortal, use the earthly and heavy element; man alone has the use of fire, the light, lofty, and heavenly element. Heavy things press down to death, light things rise up to life, because life is above, death below. Just as light cannot exist without fire, so life cannot be without light. Fire is the element of life and light; hence it appears that man who uses it is immortal in his condition, because he is at home with that which causes life. . . .

All Things Depend on His Will

The Six Days of Creation (Hexameron, I.6), by St. Ambrose, translated by John J. Savage

Ambrosius—the future St. Ambrose (ca. 339–97)—was born at Augusta Treverorum (Trèves) on the Moselle in north-central Gaul of an aristocratic Roman family, and educated at Rome. Choosing a legal career, he was made consul and governor of Aemilia-Liguria, with headquarters at Mediolanum (Milan). When the bishop of Milan, an adherent of the Arian heresy, died, Ambrose was named to succeed him by popular acclamation even before he had been baptized. Throughout his remaining years he was a master administrator and tireless defender of Catholic orthodoxy against Arianism, Judaism, paganism, and other "errors," making the Church a for-

midable power in the Roman state and even daring to excommunicate the Emperor Theodosius I for ordering a massacre at Thessalonica in 390. Milan, Rand writes, "became a model of discipline for the whole Latin world."

But Ambrose was by no means merely an administrator. By his personal example and his sermons and writings, he deeply influenced others, like the young Augustine. He was the first great composer of Latin hymns, and a scholar who drew heavily not only on Judeo-Christian Greek thinkers, such as Philo, Clement of Alexandria, and Origen, but also on pagans, including Plato, Aristotle, the Stoics, Galen, and Plotinus, to say nothing of Latin writers such as Cicero, Lucretius, Virgil (especially of the Georgics*), and Horace. He was the first Latin theologian to make wide use of the allegorical methods of exegesis developed in the Greek East: "As Cicero translated Greek thought into Roman," Rand observes, "so Ambrose translated it into Christian."*

His homilies on the six days of Creation, the Hexameron *(or* Hexaëmeron*), based on a previous Greek work of that title by St. Basil, were probably delivered in 387, and became an influential model for later Latin works on the subject. "Following in the footsteps of his great model," John J. Savage writes in introducing his translation of Ambrose's* Hexameron, Paradise, and Cain and Abel *(1961), from which the following selections are taken, "Ambrose has made these sermons into a series of Christian and humanistic observations on nature and man in their relations to their Creator, who formed them out of no pre-existing material." In these passages, while showing an acquaintance with the natural philosophy of the Greeks and Romans, he depreciates their speculations as not "fruitful for eternal life." Yet the work as a whole, as Rand notes, "is full of observation of the facts of the natural world," including "minute descriptions of quails and storks and swallows, of bees and crickets, of trees and their modes of reproduction, of evaporation and the action of rain, of human anatomy and physiology," which show how deep a sympathy with the natural world he retained despite his dismissal of the investigation of nature's laws.*

20 In the beginning of time, therefore, God created heaven and earth. Time proceeds from this world, not before the world. And the day is a division of time, not its beginning. . . .

In fact, with heaven and earth were created those four elements from which are generated everything in the world. The elements are four in number: heaven, fire, water, and earth—elements which are found mingled in all things. You may find fire also in earth, for it frequently arises from stones and iron; you may find it also in the heavens, since it may take fire and the skies may gleam with brilliant stars. In the heavens, too, we can perceive the presence of water, which is either above the heavens or from that high position falls frequently to earth in heavy rainstorms.

We can in many ways demonstrate this, if we observe that these elements are of advantage in the building of a church. But, since it is not profitable to be concerned with this, let us rather turn our attention to those matters which may be fruitful for eternal life.

21 It is sufficient, therefore, to set forth what we find in the writings of Isaias concerning the nature and the substance of the heavens. In modest and familiar language he described the nature of the heavens when he said that God "hath fixed the heavens like smoke" [Isa. 51:6], desiring to declare it to be not of solid but of subtle nature. . . .

22 On the nature and position of the earth there should be no need to enter into discussion at this point with respect to what is to come. It is sufficient for our information

to state what the text of the Holy Scriptures establishes, namely, that "he hangeth the earth upon nothing" [Job 26:7]. . . .

Let others hold approvingly that the earth never will fall, because it keeps its position in the midst of the world in accordance with nature. They maintain that it is from necessity that the earth remains in its place and is not inclined in another direction, as long as it does not move contrary to nature but in accordance with it. Let them take occasion to magnify the excellence of their divine Artist and eternal Craftsman. What artist is not indebted to Him? "Who gave to women the knowledge of weaving or the understanding of embroidery?"[6] However, I who am unable to comprehend the excellence of His majesty and His art do not entrust myself to theoretical weights and measures. Rather, I believe that all things depend on His will, which is the foundation of the universe and because of which the world endures up to the present. . . .

23 Why should I enumerate the theories which philosophers in their discussions have woven concerning the nature and composition of the substance of the heavens? . . .

24 . . . Accordingly, let us leave these men to their contentions, men who contradict themselves by their mutual disputes. Sufficient for our salvation is not disputatious controversy but doctrine—not the cleverness of argumentation, but fidelity of the mind—that we may serve, not a creature, but our Creator, who is God, blessed for all ages.

All Nature Is Good

St. Augustine on Nature and the Flesh

Aurelius Augustinus (354–430) was born at Thagaste in Roman Numidia (now Algeria), and taught rhetoric in Carthage, Rome, and Milan. Though his mother, St. Monica, was a Christian, only slowly and tortuously did Augustine, as he describes his spiritual journey in the Confessions *(written between about 397 and 400), grope by way of Manichaean heresy and pagan philosophy toward conversion to the Christian faith. He became its foremost defender and exponent in the Latin West, both in his own time, as Bishop of Hippo in North Africa, and for centuries thereafter. Although he never fully mastered Greek and lacked the learning of Ambrose or Jerome, he absorbed the Neoplatonic heritage of the Greek East and transmitted it in a form thoroughly adapted to the temper of the West. Perhaps his very lack of Greek helped liberate him from the "anxiety of influence" and make him the first great original Christian thinker of Latin antiquity.*

The following selections from various writings throughout his career give expression to a wide range of attitudes toward nature and the body. The first, from the late Enchiridion *("Handbook")* on Faith, Hope, and Love, *written in 421–22, sets forth a position similar to that of Ambrose: "it is not necessary to probe into the nature of things," as the Greek* physici *did, for it suffices for the Christian to believe in the goodness of the Creator. In other works, however, Augustine struggles to attain a balanced view of created nature. In general, his early works appear to express a more optimistic attitude toward the world: thus in* On Free Choice of the Will (De libero arbitrio), *written in 388, when rejection of Manichaean dualism and conversion to Catholicism were still recent (he had been baptized the year before), he affirms that "every nature,*

[6] Cf. Job 38:36 (Septuagint). (Savage)

insofar as it is a nature, is good." But by the time of On Nature and Grace, *written in 415 to combat the heresy of the British theologian Pelagius, who by emphasizing the freedom of human beings to choose good appeared to deemphasize the unmeritable gift of divine grace, Augustine's thinking—like the crumbling Roman Empire, in which the Visigoths had sacked Rome and the Vandals were threatening North Africa—had visibly darkened. Man's nature, created faultless, now needs a physician to heal the malady of original sin committed by free will and passed on from Adam to his descendants in an inherently corrupted world.*

Yet this more pessimistic vision by no means implies repudiation of the natural world; in his late Retractationes *Augustine writes that in* Nature and Grace *"I defend grace, not indeed as in opposition to nature, but as that which liberates and controls nature." (The "optimistic" view espoused by Pelagius, who declared that "since perfection is possible for man, it is obligatory," seemed, Peter Brown writes in* Augustine of Hippo: A Biography *[1967], "to establish an icy puritanism as the sole law of the Christian community. Paradoxically, therefore, it is Augustine, with his harsh emphasis on baptism as the only way to salvation, who appears as the advocate of moral tolerance: for within the exclusive fold of the Catholic church he could find room tor a whole spectrum of human failings.") And in* The City of God (De civitate Dei), *written between about 413 and 427 to rebut the charge that Alaric's sack of Rome in 410 was divine retribution for abandonment of the pagan gods, Augustine reaffirms that natural beauty is a "requisite part of this world," since all natures (and flesh itself) are good in their place: even Platonists, though not detesting the body as an evil, like the Manichaeans, err in ascribing to it "the whole viciousness of human life."*

Selections from On Free Choice of the Will *are from the Benjamin/Hackstaff translation of 1964; all others are from* Basic Writings of Saint Augustine, *2 vols., ed. Whitney J. Oates (1948).*

FROM The Enchiridion on Faith, Hope, and Love, *translated by J. F. Shaw*

IX. WHAT WE ARE TO BELIEVE. IN REGARD TO NATURE IT IS NOT NECESSARY FOR THE CHRISTIAN TO KNOW MORE THAN THAT THE GOODNESS OF THE CREATOR IS THE CAUSE OF ALL THINGS When, then, the question is asked what we are to believe in regard to religion, it is not necessary to probe into the nature of things, as was done by those whom the Greeks call *physici;* nor need we be in alarm lest the Christian should be ignorant of the force and number of the elements—the motion, and order, and eclipses of the heavenly bodies; the form of the heavens; the species and the natures of animals, plants, stones, fountains, rivers, mountains; about chronology and distances; the signs of coming storms; and a thousand other things which those philosophers either have found out, or think they have found out. For even these men themselves, endowed though they are with so much genius, burning with zeal, abounding in leisure, tracking some things by the aid of human conjecture, searching into others with the aids of history and experience, have not found out all things; and even their boasted discoveries are oftener mere guesses than certain knowledge. It is enough for the Christian to believe that the only cause of all created things, whether heavenly or earthly, whether visible or invisible, is the goodness of the Creator, the one true God; and that nothing exists but Himself that does not derive its existence from Him; and that He is the Trinity—to wit, the Father, and the Son begotten of the Father, and the Holy Spirit proceeding from the same Father, but one and the same Spirit of Father and Son.

FROM On Free Choice of the Will, *translated by Anna S. Benjamin and L. H. Hackstaff*

III.XIII. TO BLAME THE FAULT OF A CREATURE IS TO PRAISE ITS ESSENTIAL NATURE
Every nature which can become less good is good, and every nature becomes less good when it is corrupted. Now, either corruption does not harm the nature and it is not corrupted, or if the nature is corrupted, corruption causes harm; and corruption, if it causes harm, takes something away from the good of the nature and makes the nature less good. Now, if corruption completely removes all good from the nature, what remains cannot be corrupted since there will not exist any good to be removed and harmed by corruption. Moreover, what corruption cannot harm is not corrupted, and the nature which is not corrupted is incorruptible. This line of reasoning leads to the absurd conclusion that a nature will be incorruptible after corruption has occurred. Therefore, the truth is that every nature, insofar as it is a nature, is good; since if it is incorruptible, it is better than a corruptible nature, and if it is corruptible, it is without doubt good because when it is corrupted, it becomes less good. Every nature, moreover, is either corruptible or incorruptible. Every nature, therefore, is good. . . .

FROM On Nature and Grace, *translated by P. Holmes*

III. NATURE WAS CREATED SOUND AND WHOLE; IT WAS AFTERWARDS CORRUPTED BY SIN
Man's nature, indeed, was created at first faultless and without any sin; but that nature of man in which every one is born from Adam, now wants the Physician, because it is not sound. All good qualities, no doubt, which it still possesses in its make, life, senses, intellect, it has of the Most High God, its Creator and Maker. But the flaw, which darkens and weakens all those natural goods, so that it has need of illumination and healing, it has not contracted from its blameless Creator—but from that original sin, which it committed by free will. Accordingly, criminal nature has its part in most righteous punishment. For, if we are now newly created in Christ, we were, for all that, children of wrath, even as others, "but God, who is rich in mercy, for His great love wherewith He loved us, even when we were dead in sins, hath quickened us together with Christ, by whose grace we were saved" [Romans 3:24].

XXXIX. PELAGIUS GLORIFIES GOD AS CREATOR AT THE EXPENSE OF GOD AS SAVIOR
Beyond this, however, although he flatters himself that he vindicates the cause of God by defending nature, he forgets that by predicating soundness of the said nature, he rejects the Physician's mercy. He, however, who created him is also his Savior. We ought not, therefore, so to magnify the Creator as to be compelled to say, nay, rather as to be convicted of saying, that the Savior is superfluous. Man's nature indeed we may honor with worthy praise, and attribute the praise to the Creator's glory; but at the same time, while we show our gratitude to Him for having created us, let us not be ungrateful to Him for healing us. Our sins which He heals we must undoubtedly attribute not to God's operation, but to the wilfulness of man, and submit them to *His* righteous punishment; as, however, we acknowledge that it was in our power that they should not be committed, so let us confess that it lies in His mercy rather than in our own power that they should be healed. . . .

FROM The City of God,
translated by Marcus Dods

XII.4. OF THE NATURE OF IRRATIONAL AND LIFELESS CREATURES, WHICH IN THEIR OWN KIND AND ORDER DO NOT MAR THE BEAUTY OF THE UNIVERSE But it is ridiculous to condemn the faults of beasts and trees, and other such mortal and mutable things as are void of intelligence, sensation, or life, even though these faults should destroy their corruptible nature; for these creatures received, at their Creator's will, an existence fitting them, by passing away and giving place to others, to secure that lowest form of beauty, the beauty of seasons, which in its own place is a requisite part of this world. For things earthly were neither to be made equal to things heavenly, nor were they, though inferior, to be quite omitted from the universe. Since, then, in those situations where such things are appropriate, some perish to make way for others that are born in their room, and the less succumb to the greater, and the things that are overcome are transformed into the quality of those that have the mastery, this is the appointed order of things transitory. Of this order the beauty does not strike us, because by our mortal frailty we are so involved in a part of it, that we cannot perceive the whole, in which these fragments that offend us are harmonized with the most accurate fitness and beauty. And therefore, where we are not so well able to perceive the wisdom of the Creator, we are very properly enjoined to believe it, lest in the vanity of human rashness we presume to find any fault with the work of so great an Artificer. . . .

XII.5. THAT IN ALL NATURES, OF EVERY KIND AND RANK, GOD IS GLORIFIED
All natures, then, inasmuch as they are, and have therefore a rank and species of their own, and a kind of internal harmony, are certainly good. And when they are in the places assigned to them by the order of their nature, they preserve such being as they have received. And those things which have not received everlasting being, are altered for better or for worse, so as to suit the wants and motions of those things to which the Creator's law has made them subservient; and thus they tend in the divine providence to that end which is embraced in the general scheme of the government of the universe. So that, though the corruption of transitory and perishable things brings them to utter destruction, it does not prevent their producing that which was designed to be their result. And this being so, God, who supremely is, and who therefore created every being which has not supreme existence (for that which was made of nothing could not be equal to Him, and indeed could not be at all had He not made it), is not to be found fault with on account of the creature's faults, but is to be praised in view of the natures He has made. . . .

XIV.5. THAT THE OPINION OF THE PLATONISTS REGARDING THE NATURE OF BODY AND SOUL IS NOT SO CENSURABLE AS THAT OF THE MANICHAEANS, BUT THAT EVEN IT IS OBJECTIONABLE, BECAUSE IT ASCRIBES THE ORIGIN OF VICES TO THE NATURE OF THE FLESH
There is no need, therefore, that in our sins and vices we accuse the nature of the flesh to the injury of the Creator, for in its own kind and degree the flesh is good; but to desert the Creator good, and live according to the created good, is not good, whether a man choose to live according to the flesh, or according to the soul, or according to the whole human nature, which is composed of flesh and soul, and which is therefore spoken of either by the name flesh alone, or by the name soul alone. For he who extols the nature

of the soul as the chief good, and condemns the nature of the flesh as if it were evil, assuredly is fleshly both in his love of the soul and hatred of the flesh; for these his feelings arise from human fancy, not from divine truth. The Platonists, indeed, are not so foolish as, with the Manichaeans, to detest our present bodies as an evil nature; for they attribute all the elements of which this visible and tangible world is compacted, with all their qualities, to God their Creator. Nevertheless, from the death-infected members and earthly construction of the body they believe the soul is so affected, that there are thus originated in it the diseases of desires, and fears, and joy, and sorrow, under which four perturbations, as Cicero calls them, or passions, as most prefer to name them with the Greeks, is included the whole viciousness of human life. . . . So that even they themselves acknowledge that the soul is not only moved to desire, fear, joy, sorrow, by the flesh, but that it can also be agitated with these emotions at its own instance.

A Certain Affinity with the Divine

The (Great) Catechetical Oration (5), by St. Gregory of Nyssa, translated by Henry Bettenson

St. Gregory of Nyssa (ca. 335–ca. 395), along with his elder brother St. Basil and sister St. Macrina, and their friend St. Gregory of Nazianzus, was one of the four "Cappadocians" (so called from their native province of Cappadocia in eastern Asia Minor) who dominated Greek Christian thought in the fourth century, opposing the Arian heresy (which denied the divinity and eternity of Christ) and promoting the spread of ascetic monasticism. Deeply influenced, via Philo, Plotinus, and Origen, by Platonic and Neoplatonic thought, he takes a less somber view than his younger Western contemporary Augustine of human nature and divine grace. Gregory conceives them, according to Werner Jaeger in Early Christianity and Greek Paideia *(1961), "as the cooperation of the divine Spirit with the effort of man himself." For Gregory, Christian paideia ("education" or "culture"), in Jaeger's view, "reaches its conclusion in the final restoration of the perfect status of God's original creation. Here again appears his basic belief in the essential goodness of man and of the whole world, which God in the beginning created good." As God's creation, the cosmos (like that of Plato's demiurge in the* Timaeus*) is an expression of order and beauty, and Gregory—who took an interest, like the other Cappadocians, in Greek mathematics and medicine—can write, in two passages from his treatise* Contra Eunomium *cited by Jaroslav Pelikan in* Christianity and Classical Culture *(1993), "Investigate the work of nature!" (1.388) and even "The divine will become nature" (2.232). The following selection from* The (Great) Catechetical Oration, *a compendium of Christian teaching composed in about 385 for the benefit of teachers, gives expression to this positive view of human nature, whose immortal element can unite man with the divine not in opposition to the world of nature (as in Gnostic belief) but within the framework of that world.*

The translation is from The Later Christian Fathers *(1970), ed. Henry Bettenson.*

If therefore man came into being for this purpose, to share in the good things of God, he must inevitably be created with the capacity of enjoying those goods. The eye comes to partake of the light in virtue of the beam which is naturally implanted in it, attracting what is akin to it through its innate capacity. In the same way it was necessary that a certain affinity with the divine should be mingled with the nature of man, so that by means of this correspondence it might have an impulse towards what is congenial to it. Irrational creatures, to whom the water or the air has been allotted as their environment, must be given a constitution adapted to their manner of life, so that because of the particular organization of their bodies each finds its own congenial and kindred element, the one kind in the air, the other in the water. Similarly, it was necessary that man, who came into being in order to enjoy the good things of God, should have something in his nature akin to that in which he is to share. Therefore he has been equipped with life and reason and wisdom and all the qualities appropriate to God, so that through each of those he might have a desire for what is congenial to him. Now since one of the good things pertaining to the divine nature is eternity, it was absolutely necessary that the organization of our nature should not be deprived of this attribute, but should contain an immortal element, so that by reason of his innate capacity man might recognize the transcendent and be seized with a desire for the divine eternity. The narrative of the creation in fact demonstrated this, in a comprehensive expression, by one phrase: the statement that man came into being "after the image of God." . . . For if necessity in any way ruled the life of man, the "image" would have been falsified in that particular, since it would have become remote from its original by this lack of resemblance. How could a nature which was subjugated and enslaved to any kind of necessity be called an "image" of the nature of the King? Surely that which resembles the divine in every respect must inevitably possess in its nature the principle of self-determination and freedom, so that participation in good becomes the reward of virtue. . . . It is not God who is responsible for the present evils, since he has constituted your nature so as to be uncontrolled and free. The responsibility is with the perverse will which has chosen the worse rather than the better.

There Is No Evil Nature

The Divine Names, by Pseudo-Dionysius the Areopagite, translated by Colm Luibheid

*When St. Paul preached the resurrection of the dead on the Areopagus of Athens, some mocked, but "some men joined him and believed, among them Dionysius the Areopagite" (Acts 17:34). To this early convert of the Apostle were subsequently ascribed a series of Greek writings—*The Divine Names; The Celestial Hierarchy; The Ecclesiastical Hierarchy; The Mystical Theology; *and ten letters—probably composed around* A.D. *500; tradition later identified their author with yet another Dionysius, the patron saint of France, St. Denys. The Pseudo-Dionysian corpus, with its synthesis of Christian doctrine and Neoplatonism—including that of Proclus (ca. 410–85), whom E. R. Dodds in the* Oxford Classical Dictionary *calls "the last great systematizer of the Greek philosophical inheritance"—was immensely influential throughout the Middle Ages not only in defining the angelic orders that mediate between man and God but also in setting forth a negative theology of progressive "unknowing" leading to union with a God definable only by what He is not. Strikingly, however, this mystical spirituality does not denigrate but affirms, even celebrates, the body and the physical world; evil, as the following passages assert* *against the Gnostic view, is not inherent in matter, which indeed "has a share in the cosmos, in beauty and form." God may be indescribable and unkowable, but, as Gordon Leff writes in* Medieval Thought *(1958), "we know Him through the different natures He has created: all nature is therefore evidence of God and all creatures are moved by love of Him."* The Divine Names, *in Eco's words, "is a work which describes the universe as an inexhaustible irradiation of beauty, a grandiose expression of the ubiquity of the First Beauty, a dazzling cascade of splendors."*

The translations are from Pseudo-Dionysius: The Complete Works *(1987).*

IV.26 Evil is not an inherent part of nature as a whole. If all the laws of nature derive from the universal system of nature, no contrary will be found there. It is only in the realm of particulars that something is said to be natural or unnatural. With regard to what is unnatural, it can be so in one respect and not so in another. Evil in the domain of nature is against nature, a deficiency of what should be there in nature. Thus, there is no evil nature, for this is evil to nature. Rather, evil lies in the inability of things to reach their natural peak of perfection.

IV.27 And there is no evil in our bodies, for ugliness and disease are a defect in form and a lack of due order. What is here is not pure evil but a lesser beauty. If beauty, form, and order could be destroyed completely, the body itself would disappear.

It is also obvious that the body is not the cause of evil in the soul. Evil does not require a body to be nearby, as is clear in the case of demons. Evil in minds, in souls, and in bodies is a weakness and a defect in the condition of their natural virtues.

IV.28 There is no truth in the common assertion that evil is inherent in matter *qua* matter, since matter too has a share in the cosmos, in beauty and form. If matter lacked these, if it were inherently deficient in quality and form, if it lacked even the capacity to be affected, how could it produce anything?

Surely matter cannot be evil. If it has being in no way at all, then it is neither good nor evil. If it has some kind of being then it must derive from the Good, since every being owes its origin to the Good. Hence Good produces evil, because evil coming from Good is good, or else the Good is itself produced by evil and is therefore evil because of its source. Or, once again, it may be that there are two sources. But if so these must in turn be derived from some anterior source.

If it is said that matter is a necessity for the fulfillment of the whole cosmos, how can matter be evil? Evil and necessity are two different things. How can the Good bring something into being from evil? And how can that be evil which needs the Good, for evil surely flees the nature of the Good? How indeed could matter produce and sustain nature if it is evil? Evil *qua* evil cannot produce and cannot sustain anything, cannot make or preserve anything.

But if it is asserted that matter does not cause the evil in souls but that it pulls them down, how can this be true? For many souls have their gaze directed toward the Good and how could this happen if matter completely drags them down to evil? Hence the evil in souls does not owe its origin to matter but comes from disorder and error. . . .

Nature's Appetite for Subsistence

Boëthius on Nature, translated by H. F. Stewart, E. K. Rand, and S. J. Tester

Anicius Manlius Severinus Boëthius (ca. 480–ca. 524) was born of a noble Roman family and rose to great influence under the Ostrogoth King Theodoric, successor in the West to the Roman emperors who had ceased to reign after 476. He was later arrested on suspicion of treason, imprisoned, and executed; during his imprisonment he wrote The Consolation of Philosophy, *a dialogue composed in alternating prose and verse passages (the genre known as "Menippean satire," or* prosimetrum*) between himself and Philosophy, who appears to him in personified form.*

More than any other Western writer, Boëthius—in Rand's judgment "the most thoroughgoing philosopher, and, with the exception of St. Augustine, the most original philosopher that Rome had ever produced"—marks the transition between antiquity and the Middle Ages. His ambition, at a time when knowledge of Greek was rapidly vanishing in the barbarized West, had been to translate all of Aristotle and Plato into Latin, and to write commentaries showing that their philosophies were essentially one. All that he accomplished of this great project was to translate and comment on Aristotle's logical works (and on the Neoplatonist Porphyry's Isagôgê*). Yet these, along with Calcidius's partial translation and commentary on Plato's* Timaeus, *were to remain, for centuries to follow, almost the only works of Greek classical philosophy directly accessible to the Latin Middle Ages. His other works include treatises on music, astronomy, and arithmetic, and five theological tractates, among which are* On the Trinity *and* Contra Eutychen [et Nestorium], *written in about 512 in opposition to heretical views on the nature and person (or persons) of Christ espoused by the Monophysite Eutyches and by Nestorius.*

In the following passages from Contra Eutychen, *Boëthius sets forth a series of definitions of nature that would deeply influence subsequent medieval philosophy; not without reason Rand calls him both "the last of the Romans" and "the first of the scholastics." In the prose passages selected from the* Consolation, *Philosophy herself affirms the beauty of creation, and asserts that nature embodies the love of all things for continuation and survival. Along with St. Augustine, Boëthius—whose English translators would include King Alfred the Great and Chaucer—was throughout the Middle Ages the most widely read and pervasively influential Christian writer of Latin antiquity.*

The translations of the Theological Tractates *by H. F. Stewart, E. K. Rand, and S. J. Tester, and of* The Consolation of Philosophy *by Tester, are from the Loeb Classical Library edition (1918; new ed. 1973); see also V. E. Watts's translation of* The Consolation of Philosophy *(1969), and Edmund Reiss's introductory* Boethius *(1982).*

FROM Contra Eutychen

I Nature, then, may be predicated either of bodies alone or of substances alone, that is, of corporeals and incorporeals, or of all things which are said to exist in any way at all. Since, then, nature can be predicated in three ways, it must obviously be defined in three ways. For if you choose to predicate nature of all things, a definition will be given of such a kind as to be able to include all things that are. It will accordingly be something of this kind: "Nature belongs to those things which, since they exist, can in some way be apprehended by the intellect." This definition, then, includes the definition of both accidents and substances, for they all can be apprehended by the intellect. But I add "in some way" because God and matter cannot be apprehended by the intellect, be it never so whole and perfect, but still they are apprehended in some way through the removal of other things. The reason we add the words, "since they exist," is that even the word "nothing" itself signifies something, though not nature. For it signifies, indeed, not that something is, but rather non-existence; but every nature exists. And if we choose to predicate nature of all things, the definition will be as we have given it above.

But if nature is predicated of substances alone, we shall, since all substances are either corporeal or incorporeal, give to nature signifying substances a definition of the following kind: "Nature is either that which can act or that which can be acted upon." On the one hand, be acted upon and act, as all corporeals and the soul of corporeals; for the soul acts and is acted upon in the body and by means of the body. On the other hand, only act, as God and other divine substances.

Here, then, you have the definition of that signification of nature which is only applied to substances. This definition comprises also the definition of substance. For if the word nature indicates substance, when we have described nature we have also given a description of substance. But if we neglect incorporeal substances and confine the name nature to corporeal substances so that they alone appear to possess the nature of substance—which is the view of Aristotle and the adherents both of his and various other schools—we shall define nature as those do who have posited nature as not existing except in bodies. Now, its definition is as follows: "Nature is the principle of movement, *per se* and not accidental." I said "principle of movement" because every body has its proper movement, as fire upwards, earth downwards. Again, that I propose that nature is "the principle of movement *per se* and not accidental" is so expressed because a wooden bed is necessarily borne downward and is not carried downward by accident. For it is drawn downward by weight and heaviness because it is of wood, *i. e.* an earthly material. For it falls downwards not because it is a bed, but because it is earth, that is, because it has happened of earth that it should be a bed; hence we call it wood in virtue of its nature, but bed in virtue of the art that shaped it.

Nature has, further, another signification according to which we speak of the different nature of gold and silver, wishing thereby to indicate the special property of things; this signification of nature will be defined as follows: "Nature is the specific difference that gives form to anything." Thus, although nature is predicated or defined in so many ways, both Catholics and Nestorius hold that there are in Christ two natures according to our last definition, but the same differences cannot apply to God and man.

FROM The Consolation of Philosophy

II.v "Does the beauty of the countryside delight you? As why should it not? It is a beautiful part of the whole creation, which is beautiful. So we sometimes take pleasure in the calm aspect of the sea, and so also we admire the sky with its stars and the moon and the sun. Does any of these things belong to you? Dare you boast of the splendor of any of them? Are *you* adorned with flowers in spring? Is it *your* plenteousness which grows big with summer fruits? Why are you captivated by empty pleasures, why embrace external goods as though they were your own? Fortune will never make yours what nature has made otherwise. The fruits of the earth are surely intended for the sustenance of living things. But if you want to satisfy your needs, which is enough for nature, there is no need to ask fortune for abundance. For nature is content with few things and small: if you want to overlay that satisfaction with superfluity, then what you add will be either unpleasant or positively harmful. . . ."

III.xi "Then you know," she said, "that everything that is, endures and subsists so long as it is one, and perishes and is destroyed as soon as it ceases to be one?"

"How is that?"

"For example, in living things," she replied, "while the body and soul come together and remain as one, the result is called a living thing; but when this unity is dissolved by the separation of the two, clearly it perishes and is no longer a living thing. And the body itself, so long as by the conjunction of its members it remains in one form, is seen as a human shape; but if the parts, being separated and scattered, tear apart the unity of the body, it ceases to be what it was. In the same way it will be obvious beyond doubt to anyone surveying other examples that each thing subsists so long as it is one, but when it ceases to be one, it perishes."

"If I consider many more things," I said, "it seems not in the least different."

"Then is there anything," she asked, "that so far as it acts naturally, abandons the appetite for subsistence and desires to come to its own corruption and destruction?"

"If I consider living things," I answered, "which have some natural ability to want or not to want a thing, I find nothing which with no forces working from outside is such as to cast aside the effort to remain alive, and hasten voluntarily to its own destruction. For every animal strives to guard its own safety and avoids death and destruction. But what I should think of plants and trees, or of things altogether without life, I am very much in doubt."

"But there is nothing that you could be in doubt about in their case either, since you perceive first that plants and trees grow in places suitable to them, where, so far as their nature permits, they are able to avoid withering swiftly and perishing. For some spring up in the fields, others on mountains; others marshes bring forth, and others cling to stones, while the barren sands are productive of others which would wither if one tried to transplant them into other places. But nature gives to each what is fitting for it, and labors to prevent their dying for as long as they can endure. Have you not noticed that they all, with, as it were, their mouths buried in the ground, draw nourishment through their roots and diffuse strength through their pith and bark? Have you not noticed that all that is softest, like the pith, is hidden always in an inside place, covered without by some woody hardness, and lastly the bark is set as a defense against the inclemency of the weather, as able to bear its ill-usage? Again, how great indeed is nature's care that all are propagated by the multiplication of seed! Who does not know that they are all as it were a kind of mechanism not only for enduring for a time, but also from one generation to another as if to last for ever? And do not all those things which are believed to be without life in a similar way desire each what is fitting for itself? For why else does their lightness bear flames upwards, or its weight press earth downwards, except because these directions and motions are fitting for each? And further, whatever is suited to any thing preserves that thing, whatever it is; just as those things injurious to it destroy it. Again, those things which are hard, like stones, cling most tenaciously to their parts and resist easy dissolution; but those things which are flowing, as air or water, yield easily it is true to forces dividing them, but the parts so divided swiftly flow together as one again; while fire shuns all division.

Nor are we now dealing with the voluntary motions of the intelligent soul, but with the exertion of nature, such as when we digest food we have taken in without any conscious thought, or when we draw breath in our sleep without knowing it. For not even in living things does the love of survival proceed from the acts of will of the soul, but from natural principles. For often for compelling reasons the will embraces death, which nature fears and avoids, and on the other hand, though nature always desires it, the will sometimes restrains that act of generation by which alone the perpetuation of mortal things is assured. So this love of self proceeds not from a motion of the soul but from an exertion of nature; for providence has given to her creatures this most important cause of enduring, that by their nature they desire to endure so far as they can. Therefore there is nothing that could in any way make you doubt that all things that are seek naturally the continuance of their own survival, and avoid destruction."

"I confess," I said, "that now I see without any doubt what previously seemed doubtful." . . .

Encompassing All and Encompassed by Nothing

The *Periphyseon*, or *On the Division of Nature*, by John the Scot (Johannes Scotus Eriugena), translated by Myra L. Uhlfelder

"Scot" originally meant "Irishman"; the name Johannes Scotus (or Scottus) Eriugena (or Erigena) indicates that John the Scot (ca. 810–ca. 877) was born in Ireland, whose monasteries had for centuries—during the Dark Ages following the disintegration of the Roman Empire in the West—played a crucial part in preserving and transmitting remnants of classical learning (see Chapter 14 below). In about 847 he was invited to the court of Charlemagne's grandson Charles the Bald, who made him head of his palace school, possibly at Laon in northern France. He took a leading part in theological disputes of his time, and as one of the few Western scholars with a command of Greek, translated all of Dionysius the Areopagite as well as some works of Gregory of Nyssa and Maximus the Confessor, thus making central texts of Christian Platonism available to the Latin West. Above all, he was the one major thinker of the short-lived Carolingian revival (soon ended by a second Dark Age of Viking invasions), and therefore the only important Western philosopher in the half-millennium between Boëthius in the early sixth and Anselm in the late eleventh century—truly, as Henry Bett affirms in Johannes Scotus Erigena: A Study in Mediæval Philosophy *(1925), "the loneliest figure in the history of European thought."*

His masterpiece, the Periphyseon, *or* De Divisione Naturae (On the Division of Nature), *probably written between about 860 and 866, is a dialogue in five books between a teacher (Nutritor, "Nourisher," or Magister, "Master") and his student (Alumnus, "Pupil," or Discipulus, "Disciple") which sets forth, under the inspiration of Greek Neoplatonism, the fullest Christian philosophy of nature in the Middle Ages. The central argument develops the fourfold division of nature stated at the beginning of Book I. As summarized in Emile Bréhier's* History of Philosophy: The Middle Ages and the Renaissance *(1931; Eng. trans. 1965): "First comes nature that creates and is not created, or God as the principle of things; then comes nature that is created and creates, or the Word that is engendered by the principle and that produces the sensible world; then comes nature that is created and does not create, or the sensible world; last comes nature that is neither created nor creative, or God as the supreme end in whom the motion of things in search of perfection is terminated. But beneath these differences," Bréhier observes, "we detect an essential unity. . . . The first division, with God as the principle, is identical to the fourth, with God as the end; the second, with the Word as the creator, is identical to the third, the created world; and finally the second and third, which together constitute the totality of created beings, are shown through redemption to be identical to the fourth."*

Thus God himself, no less than his creation, is first and last among the finally indivisible divisions of nature; in John's words: "Since Nature, the Creator of the whole universe, is infinite, It is confined by no limits above or below. It encompasses everything Itself, and is encompassed by nothing." All phenomena of a natural world created from nothing are "theophanies" of God, for in John's view it is impossible, as Bett remarks, to know God as he is, except negatively and by way of contradiction: "He is known to be only through the things He has created," and ultimately God and creation, John writes, are "one and the same." Such a conception has strong elements of both mysticism and pantheism, and John's views were officially condemned as heretical both in his own time and in the thirteenth century. Yet as Bett writes, "If by pantheism we mean (as we ought to do) that the actuality of the universe is taken as one and all, and identified with the Absolute—that the totality of things is taken for God—then Erigena is emphatically not a pantheist. If, on the other hand, any doctrine which holds the essential unity of the Creator and the creation is identified with pantheism, then indeed Erigena is a pantheist, for his doctrine is absolute monism. But it is either confusion of thought or abuse of language to call this pantheism."

John's complex, abstruse, and sometimes seemingly self-contradictory thought—of which the following selections from Myra L. Uhlfelder's partial translation (1976) can give only a glimmering—is open to multiple interpretations; thus Dermot Moran, in The Philosophy of John Scottus Eriugena *(1989), gives a radically idealist reading in which "the whole of nature, which includes God, proceeds or externalizes itself in its multifarious forms through the operation of the human mind," since "the hierarchical order of nature is in fact a product of mind, and is absorbed and transcended by the mind of the spiritually liberated person." But whatever specific interpretation is given, for John the Scot as for Pseudo-Dionysius before him, nature bespeaks the glory of God. For both thinkers, as Eco remarks, "the world was a great theophany, manifesting God through its primordial and eternal causes, and manifesting these causes in its sensuous beauties. . . . God's creative power, marvellous and ineffable, was at work in every creature; thus did He manifest and reveal Himself, though He can be known only in secret and is ultimately incomprehensible."*

For a complete translation of the first three books (with facing Latin text), see I. P. Sheldon-Williams's three-volume edition, Iohannis Scotti Eriugenae Periphyseon *(1968–81).*

I.1 *Teacher.* The division of nature seems to me to admit of four species through four differentiae. The first is the division into what creates and is not created; the second into what is created and creates; the third, into what is created and does not create; the fourth, into what neither creates nor is created. Of these four, two pairs consist of opposites. The third is the opposite of the first, the fourth of the second. But the fourth is among the things which are impossible, and its differentia is in its inability to be. Does such a division seem to you correct?

Student. It surely does, but would you please go over it to clarify the opposition of the species just mentioned?

Teacher. Unless I'm mistaken, you see the opposition of the third species to the first.

The first creates and is not created, and its opposite is that which is created and does not create. Likewise the opposition of the second to the fourth, since the second is created and creates; the fourth, which neither creates nor is created, is contrary to it in every respect.

Student. I see that clearly, but I am quite perplexed about the fourth species which you added. As for the other three, I should not venture to have any misgivings; for I judge that the first is understood in the Cause of all things which have and all which do not have being, the second in the primordial causes, the third in those things known by generation in time and place. I see, therefore, that we must have a more detailed discussion about the individual species. . . .

III.1 *Student.* . . . Since the first book is about the Nature that creates everything and is created by none, and is understood in reference to God alone; and the second, by a reasonable progression, deals with that which both is created and creates and is recognized in the first beginnings of things; surely the third would, according to proper sequence, take as its subject the third nature which is created but does not create. But before we turn to an explanation of this part of nature, I should like to know why you chose to posit as the first part of the universe itself that Nature which is removed from the universe of all natures by Its excellence and Its infinity. The universe is filled with the numbers of its forms and parts, and hence does not proceed to infinity. It is limited above and below by its own boundaries. Beginning from intellectual creation, consisting of angels, or, to go higher, from the primordial causes, above which true reason finds nothing higher except God alone, it descends through the natural orders of intelligible and celestial essences and the visible things which comprise this world, and is borne to the lowermost order of all creation, which is filled with bodies and the growth and decay of bodies, and with the various disappearances and successions through the coming together of the universal elements into particulars and their dissolution again into universals. Since Nature, the Creator of the whole universe, is infinite, It is confined by no limits above or below. It encompasses everything Itself, and is encompassed by nothing. No wonder, since It may not be encompassed even by Itself because It cannot be encompassed at all. How It can be embraced at all by Itself, not to say by anything else, either in something limited or in something supernaturally definable, eludes the understanding. Unless perhaps one should say that It encompasses Itself solely in realizing that it cannot be encompassed; It comprehends Itself in realizing that It cannot be comprehended; It understands Itself in realizing that It cannot be understood in anything because It surpasses all that is and can be. Since this is the case and no true philosopher rashly counters these arguments, I do not clearly see why It is established by you among the divisions of the universe.

Teacher. I should by no means set It among the divisions of the created universe; but I judged for many reasons, not just one, that It should be set among the divisions of the universe itself, which is comprehended by the single designation of universal nature. By that name *Nature*, not only the created universe but also its Creator is usually signified. In fact, the first and greatest division of universal nature is into the Creator of the founded universe and the nature created in that founded universe itself; for surely that natural division is uniformly preserved to infinity in all universals. . . .

III.3–4 *Teacher.* . . . Thus one can understand very readily that participation is simply the derivation from higher essence to the essence following after it, and the distribution from that which first has being to what follows, so that it may be. We can demonstrate this point by examples taken from nature. [4] A whole river flows from its source as point of origin, and through its channel the water, after rising in the source, pours out constantly and without stopping throughout the river's whole length, however great. So Divine Goodness, Essence, Life, Wisdom, and all things which are in the Source of all things, first flow down into the primordial causes and give them being; then, in some ineffable way, they course down through the primordial causes into their effects through the appropriate orders of the universe, always flowing down through the higher to the lower; and, through hidden natural passages by a carefully concealed path, they return to their source. That is the origin of all good, all essence, all life, all reason, all wisdom, every genus, every species, all beauty, all order, all unity, all equality, all difference, all place, all time, all that has being, all that has no being, all that is understood, all that is sensed, and all that surpasses sense and intellect. The changeless motion and the simple multiplication and the inexhaustible diffusion from Itself, in Itself, toward Itself of the highest, threefold, and only true Goodness in Itself is the Cause of—or rather is—everything. If the Intellect of all things is all things and It alone understands all things, then It alone is all things since It is the only power of knowledge *(gnostica virtus)* which recognized all things before they had being. It did not recognize all things outside Itself, because nothing is outside It but It contains everything. It encompasses everything, and nothing within It truly has being except Itself, because It alone truly has being. The other things which are said to have being are Its theophanies, which also truly subsist in It. So God is all that truly has being, for He makes everything and is made in everything, as Saint Dionysius the Areopagite says. Everything understood and sensed is merely the appearance of the Non-appearing, the manifestation of the Hidden, the affirmation of the Denied, the comprehension of the Incomprehensible, the expression of the Ineffable, the approach to the Inaccessible, the understanding of the Unintelligible, the body of the Incorporeal, the essence of the Superessential, the form of the Formless, the measure of the Immeasurable, the number of the Unnumberable, the weight of the Weightless, the turning into flesh of the Spiritual, the visibility of the Invisible, the setting in place of the Unplaceable, the setting in time of the Timeless, the definition of the Infinite, the circumscription of the Uncircumscribed, and the other things which are reflected about and perceived by pure intellect, but which cannot be grasped within the limits of memory and which elude mental insight. . . . There are many other marvelous and ineffable thoughts which can be framed about the nature made in God's image. These examples suffice, however, to give an idea of the ineffable diffusion of Divine Goodness through all things from the highest down—i.e., through the universe created by It. This ineffable diffusion makes everything and is made in everything and is everything. . . .

III.5 *Teacher.* Concerning formless matter, which the Greeks call *hyle*, no one versed in sacred scripture who rightly considers the creation of natures doubts that it was created by the Creator of everything both causally among causal things and proportionately among the effects of causes. He who made the world from formless matter also made formless matter from absolutely nothing, since the Creator of the world made from formless matter is one and the same as the Creator of matter previously made from absolutely nothing; for all things with being, whether formless or formed, proceed from a single First Principle. The universe has been created from One just as all numbers are created from the monad and all lines emerge from the center. On this point in particular one can refute the error of secular philosophers who have dared to deal with the making of this world. They have said that formless matter is coeternal with God, and that God undertook His constructive work using it, as though it subsisted outside Him and were coeternal with Him. They considered it unfitting for formless matter to have been created by God. How, they ask, blinded by the mists of their own false reasoning, would the formless come into being from the Form of all things; the variable and changeable from the Immovable and altogether inherently Invariable; that which is subject to various accidents from Him who has no accident; that which receives intervals of places, times, and quantities from what is not extended through spaces of times and places? How could what is receptive of different qualities and shapes be made by what is subject to no quality; the corruptible by the incorruptible; the compound by the simple; and so on? We, on the other hand, perceiving the truth of Scripture and following in the footsteps of its godly interpreters, believe by faith and observe, insofar as our intellect permits, that everything—the formlessness of things, the forms and everything in them, whether essential or accidental—has been created by the single Cause of all. . . .

III.15 *Teacher.* . . . All things which are seen to arise in the order of the ages in times and places through generation were made eternally at once and together in God's Word. For we must not believe that they have just begun to be made at the time when they are seen arising in the world. They always had being substantially in God's Word, and their rising and setting in the order of times and places by generation—i.e., by their taking on of accidents—always was in God's Word, where things which are to be have already been made. Divine Wisdom encompasses times, and all things which arise in time in the nature of things have prior and eternal subsistence in It. It is Itself the measureless Measure of all things, the numberless Number, and the Weight (i.e., Order) without weight. It is Itself time and age; It is past, present, and future. The Greeks call it *Epekeina* ["Beyond"] because It creates and circumscribes all times in Itself, while It is above all times in Its eternity, preceding, encompassing, and limiting all intervals. No one can give an account even about the things which we see produced every year in the course of nature in the order of seasons. Who, for example, in reflection upon the force of seeds and how they burst forth through the numbers of places and times and the various species of animals, shrubs, and grasses, would presume to say why and how it happens or would be able to discern their occasions clearly, and would not at once proclaim that they must all be assigned to divine laws which surpass all sense and intellect? And one should not conjecture why and how things are this way and not obliged to fill the order of times another way; and why, though from invisible causes once established in the force of seeds, they proceed into sensible forms not at once but at intervals of times and places. One should not conjecture, I say, as though they could not be made otherwise if that seemed best to the Divine Will, which is bound by no law. Often, in fact, many things happen contrary to the usual course of nature to show us that Divine Providence can govern all things not only in one way, but in infinitely many. If, then, the government of the universe by divine laws is not known to any intellect, by what rational or intellectual creature can the eternal creation of that same universe in God's Word be clearly seen? But none of the faithful should be unaware that all things are at once eternal and made in God's Word, even though he does not understand how the eternal are made and the made are eternal; for this is known only to the Word, in which they are both made and eternal. . . .

III.17 *Teacher.* We should not therefore understand God and creation as two different things, but as one and the same. For creation subsists in God, and God is created in creation in a remarkable and ineffable way, manifesting Himself and, though invisible, making Himself visible, and, though incomprehensible, making Himself comprehensible, and, though hidden, revealing Himself, and, though unknown, making Himself known; though lacking form and species, endowing Himself with form and species; though superessential, making Himself essential; though supernatural, making Himself natural; though simple, making Himself compound; though free from accidents, making Himself subject to accidents and an accident; and, though infinite, making Himself finite; though uncircumscribed, making Himself circumscribed; though above time, making Himself temporal; though above place, making Himself local; though creating everything, making Himself created in everything. The Maker of all, made in all, begins to be eternal and, though motionless, moves into everything and becomes all things in all things. Nor am I talking about the incarnation of the Word and Its becoming Man, but about the ineffable condescension of the Highest Good, which is Unity and Trinity, to things with being in order that they may have being or rather that It Itself may be in everything from the highest down, always eternal, always made by Itself in Itself, eternal by Itself, made in Itself; and while It is eternal, It does not cease to be made; and though made, It does not cease to be eternal, and makes Itself from Itself. It has no need of other matter besides Itself, in which It makes Itself. . . .

IV.5 *Teacher.* . . . I affirm that man is one and the same rational soul joined to a human body in some ineffable way; and that man himself, by a marvelous and intelligible division, in the part in which he was made in the Creator's image and likeness, does not participate at all in the nature of animals and is absolutely free from it; but in the part in which he does share the nature of animals, he was produced in the universal genus of animals from the earth, i.e., from the common nature of all. . . .

Of course, the division of all creation is fivefold. It is corporeal, vital, sensitive, rational, or intellectual. All of these aspects are contained in man in every respect. The lowest part of his nature is body; then comes the life contained in seed which governs the body and which is under the dominion of sense; next reason, which rules over the parts of nature beneath itself; and mind holds the highest place of all. Thus human nature as a whole, insofar as it has a share with animals, is rightly considered an animal. Moreover, it has a share with them insofar as it is a body, and a life governing the body, and sense and a memory of sensible things that deals with *phantasiae.* But insofar as it is a participant in Divine and Celestial Essence, it is not an animal, but participates in Celestial Essence by reason, intellect, and the memory of eternal things. There, consequently, it is wholly devoid of the nature of animal. . . .

V.3 *Teacher.* . . . No one instructed by the study of wisdom can doubt, however, that nature, generated in times and places and encompassed by the other accidents, will perish at an interval predetermined by the Creator of all things. By these and similar motions and returns of the sensible world both as a whole and in its parts, and by the firmly fixed restoration to the very beginning from which motions had proceeded . . . , what is mystically intimated to us except the return of our nature to its beginning, from which it was made, in and through which it moves, and to which it always strives to return? Generally in all men, whether they are perfect or imperfect, pure or defiled, renewed in Christ and so aware of the truth or held fast in the shades of ignorance in the old man, there is a single, same natural striving for being, well-being, and being perpetually; and, as St. Augustine briefly summarizes it, of living blessedly and of avoiding wretchedness. That motion of living and subsisting happily comes from Him who is always and in a state of well-being and is present in all. And if every natural motion necessarily refrains from ceasing or resting until it reaches the end which it seeks, what can inhibit, check, or stop the necessary motion of human nature from being able to arrive at the goal of its natural striving? . . .

The Best Proof for Design

The Guide for the Perplexed, by Moses Maimonides, translated by M. Friedländer

Of the vast territories once ruled by the Roman Empire, those of the Latin West remained (with important exceptions) an intellectual backwater for much of the Dark Ages between about 500 and 1100. Literacy was almost entirely limited to monks, few of whom could read Greek, and only a handful of translations by Boëthius and Calcidius made even a tiny fraction of the rich legacy of Greek philosophy and science accessible to the Latin world. In the East, however, Greek writings were far more widely available both in the original (in the powerful if continually shrinking empire that we call the Byzantine, but which, though centered in Constantinople, was known to itself and its neighbors as "Rome") and in translations into Armenian, Syriac, Hebrew, and above all Arabic. Especially under the caliphate of Harun-al-Rashid (785–809), extensive translations were made not only from Aristotle but also from Archimedes, Euclid, Ptolemy, the Hippocratic Corpus, and Galen. In Islamic philosophy Aristotle was always the major authority, but his influence initially took a heavily neoplatonized form; indeed, writings by Plotinus and Proclus were wrongly attributed to him. Major Islamic philosophers from Alkindi in ninth-century Baghdad to Alfarabi in tenth-century Damascus, the eleventh-century Persian physician Avicenna (Ibn Sina), and Averroës (Ibn Rushd) in twelfth-century Spain dealt in different ways with problems of reconciling Greek-inspired philosophy with the faith of the Prophet. Averroës (1126–78), who became the preeminent commentator on Aristotle for Jews and Christians who followed him, pared away the Neoplatonic accretions and gave priority to philosophy over faith.

Medieval Jewish philosophy was in turn deeply influenced by Arabic thought. Its most important exponent, Moses ben Maimon, or Maimonides, was born in Cordova, Spain, in 1135 and moved to Morocco, Palestine, and Egypt, where he died in Cairo in 1204. He was both a rabbi and a physician, who strove throughout his writings, including commentaries on the Talmud and his major philosophical work, The Guide for the Perplexed *(published in Arabic as* Dalalat al-haïrin *in about 1190 and translated into Hebrew as* Moreh Nebukhim*), to conjoin faith, as revealed in the Bible, and reason, as embodied by Aristotle—though always affirming the primacy of revelation. In the following selections from Book II in the second edition of M. Friedländer's translation (1904), Maimonides remarks that Aristotle's denial that the products of nature are due to chance does not, like God's deliberate creation of the world from nothing, imply design, such as the motions of the heavenly spheres attest, and affirms the central Jewish (and indeed Muslim and Christian) doctrine that "the existence or non-existence of things depends solely on the will of God and not on fixed laws." The role of nature thus remained, for all of Aristotle's prestige, strictly circumscribed and distinctly secondary.*

II.xix . . . The best proof for design in the Universe I find in the different motions of the spheres, and in the fixed position of the stars in the spheres. For this reason you find all the prophets point to the spheres and stars when they want to prove that there must exist a Divine Being. Thus Abraham reflected on the stars, as is well known; Isaiah (40:26) exhorts to learn from them the existence of God, and says, "Lift up your eyes on high, and behold who hath created these things?" Jeremiah [calls God] "The Maker of the heavens"; Abraham calls Him "The God of the heavens" (Gen. 24:7); [Moses,] the chief of the Prophets, uses the phrase explained by us (Part I, chap. lxx), "He who rideth on the heavens" (Deut. 33:26). The proof taken from the heavens is convincing; for the variety of things in the sublunary world, though their substance is one and the same, can be explained as the work of the influences of the spheres, or the result of the variety in the position of the substance in relation to the spheres, as has been shown by Aristotle. But who has determined the variety in the spheres and the stars, if not the Will of God? . . . We have thus been brought to examine two questions:—(1) Is it necessary to assume that the variety of the things in the Universe is the result of Design, and not of fixed laws of Nature, or is it not necessary? (2) Assuming that all this is the result of Design, does it follow that it has been created after not having existed, or does *Creatio ex nihilo* [creation from nothing] not follow, and has the Being which has determined all this done always so? Some of those who believe in the Eternity of the Universe hold the last opinion. I will now begin the examination of these two questions, and explain them as much as necessary in the following chapters.

II.xx According to Aristotle, none of the products of Nature are due to chance. His proof is this: That which is due to chance does not reappear constantly nor frequently, but all products of Nature reappear either constantly or at least frequently. . . . It is therefore clear that Aristotle believes and proves that things in real existence are not accidental; they cannot be accidental, because they are essential, i.e., there is a cause which necessitates that they should be in their actual condition, and on account of that cause they are just as they in reality are. This has been proved, and it is the opinion of Aristotle. But I do not think that, according to Aristotle, the rejection of the spontaneous origin of things implies the admission of Design and Will. For as it is impossible to reconcile two opposites, so it is impossible to reconcile the two theories, that of necessary existence by causality, and that of Creation by the desire and will of a Creator. For the necessary existence assumed by Aristotle must be understood in this sense, that for everything that is not the product of work there must be a certain cause that produces it with its properties; for this cause there is another cause, and for the second a third, and so on. The series of causes ends with the Prime Cause, from which everything derives existence, since it is impossible that the series should continue *ad infinitum.* He nevertheless does not mean to say that the existence of the Universe is the necessary product of the Creator, i.e., the Prime Cause, in the same manner as the shadow is caused by a body, or heat by fire, or light by the sun. Only those who do not comprehend his words attribute such ideas to him. . . . The notion of design and determination applies only to things not yet in existence, when there is still the possibility of their being in accordance with the design or not. I do not know whether the modern Aristotelians understood his words to imply that the existence of the Universe presupposes some cause in the sense of design and determination, or whether, in opposition to him, they assumed design and determination, in the belief that this does not conflict with the theory of the Eternity of the Universe. . . .

II.XXVII We have already stated that the belief in the Creation is a fundamental principle of our religion; but we do not consider it a principle of our faith that the Universe will again be reduced to nothing. It is not contrary to the tenets of our religion to assume that the Universe will continue to exist for ever. It might be objected that everything produced is subject to destruction, as has been shown; consequently the Universe, having had a beginning, must come to an end. This axiom cannot be applied according to our views. We do not hold that the Universe came into existence, like all things in Nature, as the result of the laws of Nature. . . . According to our theory, taught in Scripture, the existence or non-existence of things depends solely on the will of God and not on fixed laws, and, therefore, it does not follow that God must destroy the Universe after having created it from nothing. It depends on His will. He may, according to His desire, or according to the decree of His wisdom, either destroy it, or allow it to exist, and it is therefore possible that He will preserve the Universe for ever, and let it exist permanently as He Himself exists. It is well known that our Sages never said that the throne of glory will perish, although they assumed that it has been created. . . . There remains only the question as to what the prophets and our Sages say on this point; whether they affirm that the world will certainly come to an end, or not. Most people amongst us believe that such statements have been made, and that the world will at one time be destroyed. I will show you that this is not the case; and that, on the contrary, many passages in the Bible speak of the permanent existence of the Universe. Those passages which, in the literal sense, would indicate the destruction of the Universe, are undoubtedly to be understood in a figurative sense, as will be shown. If, however, those who follow the literal sense of the Scriptural texts reject our view, and assume that the ultimate certain destruction of the Universe is part of their faith, they are at liberty to do so. But we must tell them that the belief in the destruction is not necessarily implied in the belief in the Creation; they believe it because they trust the writer, who used a figurative expression, which they take literally. Their faith, however, does not suffer by it.

Superlunary Nature's Sublunary Works

The *Didascalion*, by Hugh of St. Victor, translated by Jerome Taylor

Hugh of St. Victor, born in the late eleventh century (whether in France, Flanders, or Saxony remains uncertain), in about 1115 entered the Augustinian house of canons in the abbey of St. Victor in Paris, founded in 1113, and died there in 1172. As head of its monastery school and author of important works of mystical and religious philosophy (including a commentary on Pseudo-Dionysius the Areopagite's Celestial Hierarchy*), he played a leading role in making St. Victor a center of orthodox theology in the twelfth century. Already in the early years of the century a few leading scholars like Adelard of Bath—who boasts in his* Natural Questions *of learning from his Arab masters by the guidance of reason while others slavishly follow the halter of authority—had begun to rediscover and transmit the wealth of philosophical and scientific texts that would transform the intellectual map of Europe. Adelard himself translated Euclid's* Elements *from the Arabic, and by the end of the century much of Aristotle, Ptolemy, and Galen (whether translated from Arabic or Greek or both), along with Avicenna and Averroës, would be available.*

Along with the new prominence of Aristotle, and the beginnings of Scholastic philosophy in St. Anselm (ca. 1033–1109), Peter Abelard (or Abailard, 1079–1142), and Peter Lombard (ca. 1100–60), the early twelfth century also saw a strong revival of interest in Platonic cosmology as transmitted from late antiquity in Calcidius's translation of and commentary on the early sections of the Timaeus; *among the writers loosely associated with the Neoplatonic "School of Chartres" were Thierry of Chartres, William of Conches, and Gilbert de la Porrée. (On this subject, see M.-D. Chenu,* Nature, Man, and Society in the Twelfth Century *[1957; Eng. trans. 1968]; J. M. Parent,* La Doctrine de la création dans l'école de Chartres *[1938]; and the introductions to Bernard Silvestris and Alan of Lille in Chapter 15 below.) In his* Didascalion, *composed in the late 1120s and subtitled, in Jerome Taylor's translation of 1961, "A Medieval Guide to the Arts," Hugh was intensely aware of these different currents jostling with one another in the ferment of an age aptly called by Charles Homer Haskins "the Renaissance of the twelfth century." As he undertook to write a counterpart for his own times to earlier guides to the liberal arts by such ancient writers as Cassiodorus and Martianus Capella, his aim, Taylor writes in his excellent introduction, was "to select and define all the areas of knowledge important to man and to demonstrate not only that these areas are essentially integrated among themselves, but that in their integrity they are necessary to man for the attainment of his human perfection and his divine destiny."*

But Hugh, though drawing on Calcidius and Boëthius, on Macrobius's commentary on Cicero's Dream of Scipio *and the Hermetic* Asclepius, *and on the commentaries of Thierry of Chartres and William of Conches, along with Aristotle and many others, was determined to interpret them in accord with an orthodox Christian view. This view he based above all on St. Augustine (whose* De Doctrina Christiana *was among his major inspirations), in contrast to the taint of heterodoxy associated with the Chartrian Platonists—at least indirectly influenced by John the Scot—and with Abelard. Unlike those who appear to elevate Plato to near-divine authority, Hugh repeatedly remarks, Taylor notes, "upon the disparity between the cosmogony of the* Timaeus *and the truth taught by Christian authors according to the true faith": thus "the* Didascalion *is true to Augustine, as to its own time, in adhering to a conception of philosophy as coterminous with all knowledge whatever, including revealed or scriptural knowledge, and as consisting essentially in man's search for reunion with the God by and for whom he was made and without whom he can have neither peace nor profit in any earthly pursuit."*

In the following chapters from Book I (accompanied by selections from Taylor's scholarly notes), Hugh gives great prominence to nature in its various meanings, but is also concerned to distinguish clearly among the eternal, the perpetual, and the temporal; the superlunary and the sublunary; and the works of God, of nature, and of human art. For Hugh, Eco writes (citing his De Tribus Diebus [On the Three Days]*), "the earth was 'like a book written by the finger of God'* (quasi quidam liber scriptus digito Dei). . . . *The pleasures of sound and sight, of smell and touch, bring us face to face with the beauty of the world, so that we may see in it the reflection of God." Thus we rise from knowledge of the natural world through the eye of the flesh to self-knowledge through the eye of reason to the final stage, in this Platonic-Augustinian ascent, of knowledge of God through the eye of contemplation. Rather than usurping the sovereignty of God, nature thus continually points upward and back to the invisible hand of its Creator.*

FROM BOOK I

Chapter Six: Concerning the Three "Manners" of Things

Among things there are some which have neither beginning nor end, and these are named eternal; there are others which have a beginning but are terminated by no end, and these are called perpetual; and there are yet others which have both beginning and end, and these are temporal.[7]

In the first category we place that in which the very being *(esse)* and "that which is" *(id quod est)* are not separate, that is, in which cause and effect are not different from one another, and which draws its subsistence not from a principle distinct from it but from its very self. Such alone is the Begetter and Artificer of nature.

But that type of thing in which the very being *(esse)* and "that which is" *(id quod est)* are separate, that is, which has come into being from a principle distinct from it, and which, in order that it might begin to be, flowed into actuality out of a preceding cause—this type of being, I say, is nature, which includes the whole world, and it is divided into two parts:[8] it is that certain being which, in acquiring existence from its primordial causes,[9] came forth into actuality not as moved thereto by anything itself in motion, but solely by the decision of the divine will, and, once in existence, stood immutable, free from all destruction or change (of this type are the substances of things, called by the Greeks *ousiai*) and it is all the bodies of the superlunary world, which, from their knowing no change, have also been called divine.

The third type of things consists of those which have both beginning and end and which come into being not of their own power but as works of nature. These come to be upon the earth, in the sublunary world, by the movement of an artifacting fire[10] which descends with a certain power to beget all sensible objects.

Now, of the second sort it is said, "Nothing in the world perishes," for no essence suffers destruction. Not the essences but the forms of things pass away. When the form is said to pass away, this is to be understood as meaning not that some existing thing is believed to have perished altogether and lost its being, but rather that it has undergone change, perhaps in one of the following ways: that things once joined are now apart, or once apart are now joined; or that things once standing here now move there; or that things which now are only "have-beens" once subsisted; in all these instances the being of the things suffers no loss. Of the third sort it is said, "All things which have arisen fall, and all which have grown decline": for all the works of nature, as they have a beginning, so have they an end. Of the second sort, again, it is said, "Nothing comes from nothing, and into nothingness can nothing revert," from the fact that all of nature has both a primordial cause and a perpetual subsistence. And of the third sort, once more, "That

[7]These distinctions among "eternal," "perpetual," and "temporal" occur in William of Conches' commentary on the *De Consolatione philosophiae* [of Boethius] III.m.ix (see [J.-M.] Parent, *Doctrine de la création* [*dans l'école de Chartres*, 1938]). (Taylor)

[8]"Nature" and "world" in this passage are coextensive with the whole of creation, not merely with the physical universe. Nature's two parts are (1) the incorporeal, invisible, rational creation or created wisdom (angels and human souls), made in the likeness not of a single exemplar in the Mind of God, but of the entire Mind of God with all the exemplary causes it contains . . . ; and (2) the corporeal, visible creation, which, as initially created (primordial matter) was changeless because unformed, and out of which God made the immutable bodies of the superlunary world and the changing bodies of the sublunary world. . . . Note that though John the Scot divides "nature" into two parts (*De divisione naturae* III.1 . . . : "The first and greatest division of all nature is into that Nature which creates the established universe and that nature which is created *in* the established universe") and teaches that the former part "encompasses all things" . . . , the resemblance to Hugh is only superficial. The parts of John's "nature" are Creator and creation; Hugh's "nature," in the present context, that is, is like Augustine's, entirely created. (Taylor)

[9]The primordial causes are the uncreated exemplars of creation subsisting in the divine Wisdom, or Mind. (Taylor)

[10]For the famous Stoic definition of nature as "an artistically working fire, going on its way to create" (Diogenes Laërtius VII.156), usually attributed to Zeno, see introduction to Chapter 10 above, and the selection in that chapter from Cicero's *On the Nature of the Gods*.

which before was nothing returns again thereto": for just as every work of nature flows temporarily into actuality out of its hidden cause, so when its actuality has temporarily been destroyed, that work will return again to the place from which it came.

Chapter Seven: Concerning the Superlunary and Sublunary World

Because of these facts, astronomers *(mathematici)* have divided the world into two parts: into that, namely, which stretches above the sphere of the moon and that which lies below it. The superlunary world, because in it all things stand fixed by primordial law, they called "nature," while the sublunary world they called "the work of nature," that is, the work of the superior world, because the varieties of all animate beings which live below by the infusion of life-giving spirit, take their infused nutriment through invisible emanations from above, not only that by being born they may grow but also that by being nourished they may continue in existence. Likewise they called that superior world "time" because of the course and movement of the heavenly bodies in it, and the inferior world they called "temporal" because it is moved in accordance with the movements of the superior. Again, the superlunary, from the perpetual tranquility of its light and stillness, they called *elysium*, while the sublunary, from the instability and confusion of things in flux, they called the underworld or *infernum.*

Into these things we have digressed somewhat more broadly in order to explain how man, in that part in which he partakes of change, is likewise subject to necessity, whereas in that in which he is immortal, he is related to divinity. From this it can be inferred, as said above, that the intention of all human actions is resolved in a common objective: either to restore in us the likeness of the divine image or to take thought for the necessity of this life, which, the more easily it can suffer harm from those things which work to its disadvantage, the more does it require to be cherished and conserved. . . .

Chapter Nine: Concerning the Three Works

"Now there are three works—the work of God, the work of nature, and the work of the artificer, who imitates nature."[11] The work of God is to create that which was not, whence we read, "In the beginning God created heaven and earth"; the work of nature is to bring forth into actuality that which lay hidden, whence we read, "Let the earth bring forth the green herb," etc.; the work of the artificer is to put together things disjoined or to disjoin those put together, whence we read, "They sewed themselves

[11]Adapted from Chalcidius's commentary on the *Timaeus*. . . . Hugh makes radical changes in the meaning of his source. In Chalcidius, the *opera Dei* [works of God] are ascribed to the *summus Deus* [highest God], who founds all things according to the exemplary causes existing in his eternal Providence *(pronoia)* or Mind *(nous)*, and promulgates the law or "fate" of natural things in created intelligences subsisting visibly in the fixed stars of the outer *ignis (aplanê)* and invisibly in the *daemones* or blessed angels of the *aether*. The *opera naturae* [works of nature] are the province of the *anima mundi* [world soul], a *secunda mens* [second mind] and the substantial projection of God's law of "fate"; ruling all things according to their proper natures, the *anima mundi* may be called their "law." The *opera artificis* [works of the artificer]—human arts, disciplines, and the things effected with their aid—are regular and fruitful because, imitating nature, they too are ruled by the law, idea, and order of the *anima mundi*. . . . Hugh accepts the Chalcidian idea of the exemplary causes . . . and their communication to the angelic intellect . . . , rejecting, however, the identification of angels with the heavenly bodies. . . . Among the *opera Dei* he includes the ordering and disposition of the movements of all things, which, in Chalcidius, were *opera naturae*; the *anima mundi*, in which certain of his contemporaries saw reference to the Holy Spirit . . . , he altogether rejects, limiting the *opera naturae* to the sun's superintendence of growth and decay in terrestrial life. . . . The *opera artificis imitantis naturam* [works of the artificer, who imitates nature] are associated not, as in Chalcidius, with the *anima mundi*, but with man's efforts, through the mechanical arts, to supply both the internal and external needs of his body. . . . (Taylor)

aprons." For the earth cannot create the heaven, nor can man, who is powerless to add a mere span to his stature, bring forth the green herb.

Among these works, the human work, because it is not nature but only imitative of nature, is fitly called mechanical, that is adulterate, just as a skeleton key is called a "mechanical" key.[12] How the work of the artificer in each case imitates nature is a long and difficult matter to pursue in detail. . . . But it is not without reason that while each living thing is born equipped with its own natural armor, man alone is brought forth naked and unarmed. For it is fitting that nature should provide a plan for those beings which do not know how to care for themselves, but that from nature's example, a better chance for trying things should be provided to man when he comes to devise for himself by his own reasoning those things naturally given to all other animals. Indeed, man's reason shines forth much more brilliantly in inventing these very things than ever it would have had man naturally possessed them. . . .

Chapter Ten: What "Nature" Is

But since we have already spoken so many times of "nature," it seems that the meaning of this word ought not to be passed over in complete silence, even though as Tully[13] says, "Nature is difficult to define." Nor, because we are unable to say of it all we might wish, ought we to maintain silence about what we can say.

Men of former times, we find, have said a great deal concerning "nature," but nothing so complete that no more should seem to remain to be said. So far as I am able to conclude from their remarks, they were accustomed to use this word in three special senses, giving each its own definition.[14]

In the first place, they wished by this word to signify that archetypal Exemplar of all things which exists in the divine Mind, according to the idea of which all things have been formed; and they said that nature was the primordial cause of each thing, whence each takes not only its being *(esse)* but its "being such or such a thing" *(talis esse)* as well. To the word in this sense they assigned the following definition: "Nature is that which gives to each thing its being."

In the second place they said that "nature" meant each thing's peculiar being *(proprium esse)*, and to "nature" in this sense they assigned this next definition: "The peculiar difference giving form to each thing is called its nature."[15] It is with this meaning in mind that we are accustomed to say, "It is the nature of all heavy objects to tend toward the earth, of light ones to rise, of fire to burn, and of water to wet."

The third definition is this: "Nature is an artificer fire coming forth from a certain power to beget sensible objects." For physicists tell us that all things are procreated from heat and moisture. Therefore Vergil calls Oceanus "father," and Valerius Soranus, in a certain verse which treats Jove as a symbol of aethereal fire, says:

> Jupiter omnipotent, author of things as of kings,
> Of all true gods the father and womb in one!

[12]Hugh [mistakenly] associates "mechanical" with the Greek *moikhos*, Latin *moechus*, adulterer, rather than with *mêkhanê*, machine [or device]. . . . (Taylor)

[13][Marcus Tullius] Cicero, *De Inventione* I.xxiv.34. (Taylor)

[14]The passage which follows is a good example of the generalizing and schematizing habit of Hugh's mind. With his reduction of numerous current definitions of "nature" to three basic senses, cf. the more diffuse survey in John of Salisbury *Metalogicon* I.viii. . . . (Taylor) See Taylor's notes for detailed discussion of some of Hugh's principal sources.

[15]Adapted from Boethius *Contra Eutychen*.i . . . (Taylor)

Signs Divinely Bestowed

The Mind's Road to God, by St. Bonaventura, translated by George Boas

The early twelfth century, which saw both the revival of Neoplatonism in the "School of Chartres" and the ascent of Aristotelian rationalism, embodied by the Sic et non *of Abelard, that would culminate in thirteenth-century Scholasticism, was also a great age of mysticism, exemplified above all by Abelard's great opponent, St. Bernard of Clairvaux (1090–1153). But rationalism and mysticism, as Hugh of St. Victor attests, were by no means exclusive, and in the leading Franciscan Scholastic theologian, the "Seraphic Doctor" St. Bonaventura (1221–74), they were once again harmoniously combined. Born Giovanni di Fidanza near Viterbo in Tuscany, he studied at the great center of Scholasticism, the University of Paris, where he both held the Franciscan chair of theology established by his teacher, the Englishman Alexander of Hales (at the same time as St. Thomas Aquinas was holding the Dominican chair), and became general of the Franciscan order and a cardinal.*

His later years were mainly devoted to administrative duties of his order, but his writings include a major commentary on the Sentences *of Peter Lombard and a brief treatise,* Itinerarium mentis in Deum, *from which the following selections are given in George Boas's translation,* The Mind's Road to God *(1953). He endeavored to assimilate Aristotle to orthodox Augustinianism, and in this, as in his mysticism (which also reflects the influence of St. Bernard), he is the successor of Hugh in the previous century. "In everything," Bréhier writes, "he searches for expressions, images, vestiges, shadows of the nature of God. The solutions to the most technical questions that he disputes with St. Thomas are provided by the vast symbolism that makes him consider nature, along with the Bible, as a book whose divine meaning must be deciphered. . . . Nature and the soul reach an understanding only when turned toward God: then nature stands as evidence of divine attributes, and the soul unites us to God through its essential function of love."*

FROM *Chapter Two: Of the Reflection of God in His Traces in the Sensible World*

3 Therefore man, who is called a "microcosm," has five senses like five doors, through which enters into his soul the cognition of all that is in the sensible world. . . .

7 . . . These all, however, are traces in which we can see the reflection of our God. . . .

10 . . . Since, therefore, all things are beautiful and in some way delightful, and beauty and delight do not exist apart from proportion, and proportion is primarily in number, it needs must be that all things are rhythmical (*numerosa*). And for this reason number is the outstanding exemplar in the mind of the Maker, and in things it is the outstanding trace leading to wisdom. . . . It causes Him to be known in all corporeal and sensible things while we apprehend the rhythmical, delight in rhythmical proportions, and through the laws of rhythmical proportions judge irrefragably.

11 From these . . . initial steps by which we are led to seeing God in His traces, as if we had two wings falling to our feet, we can determine that all creatures of this sensible world lead the mind of the one contemplating and attaining wisdom to the eternal God;

for they are shadows, echoes, and pictures, the traces, simulacra, and reflections of that First Principle most powerful, wisest, and best; of that light and plenitude; of that art productive, exemplifying, and ordering, given to us for looking upon God. They are signs divinely bestowed which, I say, are exemplars or rather exemplifications set before our yet untrained minds, limited to sensible things, so that through the sensibles which they see they may be carried forward to the intelligibles which they do not see, as if by signs to the signified.

12 The creatures of this sensible world signify the invisible things of God [Rom. 1:20], partly because God is of all creation the origin, exemplar, and end, and because every effect is the sign of its cause, the exemplification of the exemplar, and the way to the end to which it leads; partly from its proper representation; partly from prophetic prefiguration; partly from angelic operation; partly from further ordination. For every creature is by nature a sort of picture and likeness of that eternal wisdom, but especially that which in the book of Scripture is elevated by the spirit of prophecy to the prefiguration of spiritual things. But more does the eternal wisdom appear in those creatures in whose likeness God wished to appear in angelic ministry. And most specially does it appear in those which He wished to institute for the purpose of signifying which are not only signs according to their common name but also Sacraments.

13 From all this it follows that the invisible things of God are clearly seen, from the creation of the world, being understood by the things that are made . . .

Nature as Instrument of Intelligence

The *Book of Six Principles*, by St. Albert the Great, translated by J. P. Mullally

St. Albert the Great, or Albertus Magnus (1206–80), the Dominican "Universal Doctor," was born in Bollstädt of a noble Swabian family and taught at Hildesheim, Freiburg, Regensburg, Strasbourg, and Cologne before becoming doctor of theology at the University of Paris in 1245; he later taught again at Cologne from 1248 to 1260 and from 1270 until his death. He was the first Scholastic to assimilate the full influence of Aristotle, on all of whose available works he wrote extensive commentaries; but his scientific interest in nature allowed him to go beyond mere explication of "the Philosopher" and to make firsthand observations in a number of his works, such as De mineralibus, De vegetabilibus, *and the later books of* De animalibus.

In Albert's view, which deeply influenced that of his greatest pupil, St. Thomas Aquinas, "we can attain to God only through the sensible world," as Bréhier writes, "through a cosmological proof proceeding from effect to cause, and not through an ontological proof" (deducing God as the ultimate cause of the goodness in individual things), such as St. Anselm had proposed. "By contemplating the world we can probably infer the existence of God, but we cannot even know with rational certainty whether the world had a beginning in time"; for this we must turn not to reason but to revelation alone. In the passage that follows—as translated in Herman Shapiro, ed., Medieval Philosophy *(1964)—Albert considers an aspect of the central Scholastic problem of universals and argues that the universal essence common to different things seems to be a product neither of art (which is accidental, not substantial) nor of nature (which generates the particular, not the universal). His resolution of this dilemma is that "nature operates in the operations or actions of art" (as evidenced in beehives or swallows' nests) insofar as it is an instrument of divine intelligence, which is the subject not of physics but of "first philosophy," that is, of metaphysics or theology.*

I.5 ... The general or universal essence of those things which are in many, predicated of them, cannot exist by an operation of art, because it is impossible for such an essence to exist by an operation of art, by virtue of the fact that every form of art is an accident and accidental. But such a form is substantial and belongs essentially *[per se]* to those things whose form it is. But still such a universal form does not seem to be caused by nature, for the reason that those things which are produced by the power of nature, seem to be caused by an act of creating and generating and they derive [their] origin from the creating or generating. But the end of creating or generating is always the particular and not the universal; therefore it is not by a work of nature, as it seems. . . .

However, speculating subtly in this way we discover that nature operates in the operations or actions of art, by virtue of the fact that even in the case of animals, deprived of reason, we discover nature managing works of art, as for example in the case of the hives of the bees and the nests of swallows. And nature causes this, insofar as it is a work and an instrument of intelligence. Certainly in this way we discover the first creator of all creatures operating in the creature, and we perceive this from the very act of the creator. For we see, that he has stabilized the whole of nature according to a certain number and proportion, which proportion does not exist except for the wisdom of the creator. For intelligence, insofar as it is intelligence, does not possess as a quality that intelligence works in nature proportionate to this or that nature, rather it possesses this quality insofar as it is divine. For "this" depends upon the wisdom of the one who disposes and produces and orders the whole in its proper proportions. But what we have said up to this point suffices for an explanation of what was stated. For these matters belong to another and a more profound investigation; for they belong to the part of First philosophy, which is concerned with the universal and primary causes.

God at Work in the Operations of Nature

On the Power of God, Question III, Article VII ("Does God Work in Operations of Nature?"), by St. Thomas Aquinas, translated by the English Dominican Fathers

St. Thomas Aquinas (ca. 1225–74), known as the "Dumb Ox" in his own time and as the "Angelic Doctor" to posterity, was born near Naples of a noble family of Aquino and educated at the Benedictine school at Montecassino and later at Naples. He joined the Dominican order in 1244 and studied at the University of Paris from 1245 to 1248 under Albert the Great, whom he followed to Cologne; after returning to Paris in 1252, he became professor of theology. From 1259 to 1269, when again called to Paris, he taught in Italy, returning to Naples in 1272. His written output, ranging from commentaries on Aristotle, Peter Lombard, Boëthius, and the Bible to his two great "summas," the Summa contra gentiles *and the massive but uncompleted* Summa theologica *(or* theologiae*), was prodigious. Though controversial in his own time, when Aris-*

totle's works were suspect and some of Thomas's propositions were posthumously condemned in 1277, he has since been recognized as the greatest of Scholastic theologians and perhaps the greatest Christian thinker since St. Augustine; in 1879 a papal encyclical declared his works the basis of the official philosophy of the Catholic Church.

Like Aristotle and Albert before him, Thomas held that all rational knowledge begins by way of the senses. Though vigorously opposing the "Averroist" exaltation of reason at the expense of faith (associated with his opponent Siger of Brabant), he himself, in contrast to more conservatively orthodox Augustinians, sharply distinguished between their appropriate domains, declaring that certain beliefs—notably God's creation of the world—were not susceptible to rational demonstration but must be held by faith on the basis of revelation. In Thomas's view, the disparity between divine and natural causes is only apparent, for "one and the same effect can proceed simultaneously," as Etienne Gilson writes in The Christian Philosophy of St. Thomas Aquinas *(5th ed. 1948; Eng. trans. 1957), "from two different causes: God and the natural agent which produces it. . . . Far from encroaching upon the Creator's privileges, the perfections attributed to second causes can only increase His glory, since He is their first cause and since this is a new occasion for glorifying him. It is because there is causality in nature that we can go back step by step to the first cause, God. . . . As soon as we realize the significance of this principle, all shadow of antinomy between God's perfection and that of created beings disappears."*

On the Power of God (Quaestiones disputatae de potentia Dei), *from which the following selection is taken, as translated by the English Dominican Fathers (1932), was part of a series of "disputed questions" composed as public debates during the years between 1256 and 1272. In his typical Scholastic method, Thomas takes up the question "whether God works in the operations of nature" and examines considerations that might lead to a negative answer before stating his own affirmative conclusion, his reasons for it, and his responses to the objections previously made.*

The seventh point of inquiry is whether God works in the operations of nature: and apparently the answer should be in the negative.

1 Nature neither fails in necessary things nor abounds in the superfluous. Now the action of nature requires nothing more than an active force in the agent, and passivity in the recipient. Therefore there is no need for the divine power to operate in things.

2 It may be replied that the active force of nature depends in its operation on the operation of God.—On the contrary as the operation of created nature depends on the divine operation, so the operation of an elemental body depends on the operation of a heavenly body: because the heavenly body stands in relation to the elemental body, as a first to a second cause. Now no one maintains that the heavenly body operates in every action of an elemental body. Therefore we must not say that God operates in every operation of nature.

3 If God operates in every operation of nature God's operation and nature are either one and the same operation or they are distinct. They are not one and the same: since unity of operation proves unity of nature: wherefore as in Christ there are two natures [i.e., divine and human], so also are there two operations: and it is clear that God's nature and man's are not the same. Nor can they be two distinct operations: because distinct operations cannot seemingly terminate in one and the same product, since movements and operations are diversified by their terms. Therefore it is altogether impossible that God operate in nature. . . .

16 Given a cause whose action suffices, it is superfluous to require the action of another cause. Now it is clear that if God operates in nature and will, his action is sufficient, since *God's works are perfect* (Deut. 22:4). Therefore all action of nature and will would be superfluous. But nothing in nature is superfluous, and consequently neither nature nor will would do anything, and God alone would act. This, however, is absurd: therefore it is also absurd to state that God operates in nature and will.

On the contrary it is written (Isa. 26:12): *Lord, thou hast wrought all our works in us.*

Moreover, even as art presupposes nature, so does nature presuppose God. Now nature operates in the operations of art: since art does not work without the concurrence of nature: thus fire softens iron so as to render it malleable under the stroke of the smith. Therefore God also operates in the operation of nature. . . .

Further, nothing can act except what exists. Now nature cannot exist except through God's action, for it would fall into nothingness were it not preserved in being by the action of the divine power, as Augustine states. . . . Therefore nature cannot act unless God act also.

Again, God's power is in every natural thing, since he is in all things by his essence, his presence and his power. Now it cannot be admitted that God's power forasmuch as it is in things is not operative: and consequently it operates as being in nature. And it cannot be said to operate something besides what nature operates, since evidently there is but one operation. Therefore God works in every operation of nature.

I answer that we must admit without any qualification that God operates in the operations of nature and will. Some, however, through failing to understand this aright fell into error, and ascribed to God every operation of nature in the sense that nature does nothing at all by its own power. . . .

Consequently we may say that God works in everything forasmuch as everything needs his power in order that it may act: whereas it cannot properly be said that the heaven always works in an elemental body, although the latter acts by its power. Therefore God is the cause of everything's action inasmuch as he gives everything the power to act, and preserves it in being and applies it to action, and inasmuch as by his power every other power acts. And if we add to this that God is his own power, and that he is in all things not as part of their essence but as upholding them in their being, we shall conclude that he acts in every agent immediately, without prejudice to the action of the will and of nature.

Reply to the First Objection. The active and passive powers of a natural thing suffice for action in their own order: yet the divine power is required for the reason given above.

Reply to the Second Objection. Although the action of the forces of nature may be said to depend on God in the same way as that of an elemental body depends on the heavenly body, the comparison does not apply in every respect.

Reply to the Third Objection. In that operation whereby God operates by moving nature, nature itself does not operate: and even the operation of nature is also the operation of the divine power, just as the operation of an instrument is effected by the power of the principal agent. Nor does this prevent nature and God from operating to the same effect, on account of the order between God and nature. . . .

Reply to the Sixteenth Objection. God acts perfectly as first cause: but the operation of nature as second cause is also necessary. Nevertheless God can produce the natural effect even without nature: but he wishes to act by means of nature in order to preserve order in things.

The Necessity of Experience

Opus Majus, by Roger Bacon, translated by Robert Belle Burke

Roger Bacon (ca. 1214–ca. 1294) studied at Oxford, where he was no doubt influenced by the scientific interests of Robert Grosseteste (1175–1253), and after about 1240 at Paris, mastering Hebrew, Greek, and possibly Arabic. He became a Franciscan in about 1247, and in Paris attracted the interest of Guy Fulcodi, who in 1265 was elected Pope Clement IV and, in the following year, requested a copy of Bacon's writings. Bacon hastily composed his Opus Majus (Greater Work) *and sent it, along with two shorter works based on it, the* Opus Minus (Lesser Work) *and* Opus Tertium (Third Work), *to the Pope, who died, however, in 1268, probably without having read them.*

In the Opus Majus, *Bacon enumerates causes of error, commends a philosophy completed by Christian faith, urges the study of languages, examines the importance of mathematics (including music and astrology), and reflects on optics and the nature of light (a central concern of Grosseteste) before affirming, at the beginning of Part Six, the necessity of "experimental science," or knowledge attained through direct experience and not through abstract reason alone. It is this emphasis that has caused Bacon to be considered a medieval precursor of modern science, yet his reputation for centuries to come (as in Robert Greene's play of 1594,* Friar Bacon and Friar Bungay*) was as a quasi-Faustian magician and alchemist of awesomely mysterious powers: a reputation often attached to early modern science itself. "When Bacon speaks of experimental science," Bréhier writes, "he is thinking of a secret, traditional science that consists in the investigation of occult forces and of the power that knowledge of these forces confers on the expert. The universe of the* experts *is essentially the universe described by Plotinus: a set of interpenetrating forces, enchantment, magic words, forces which have emanated from the stars and to which people are unwittingly subject." Bacon's view of nature thus looks backward as much as forward but was, in any case, disturbingly unlike the orthodoxies of his age. In 1277 he was apparently condemned (and perhaps imprisoned) by the General of the Franciscan Order for "suspect novelties" and "dangerous doctrines."*

The following selections are taken from volume II of Robert Belle Burke's translation (1928).

FROM *Part Six, Chapter I*

Having laid down fundamental principles of the wisdom of the Latins so far as they are found in language, mathematics, and optics, I now wish to unfold the principles of experimental science, since without experience nothing can be sufficiently known. For there are two modes of acquiring knowledge, namely, by reasoning and experience. Reasoning draws a conclusion and makes us grant the conclusion, but does not make the conclusion certain, nor does it remove doubt so that the mind may rest on the intuition of truth, unless the mind discovers it by the path of experience; since many have the arguments relating to what can be known, but because they lack experience they neglect the arguments, and neither avoid what is harmful nor follow what is good. For if a man who has never seen fire should prove by adequate reasoning that fire burns and injures things and destroys them, his mind would not be satisfied thereby, nor would he avoid fire, until he placed his hand or some combustible substance in the fire, so that he might prove by experience that which reasoning taught. But when he has had actual experience of combustion his mind is made certain and rests in the full light of truth. Therefore reasoning does not suffice, but experience does.

This is also evident in mathematics, where proof is most convincing. But the mind of one who has the most convincing proof in regard to the equilateral triangle will never cleave to the conclusion without experience, nor will he heed it, but will disregard it until experience is offered him by the intersection of two circles, from either intersection of which two lines may be drawn to the extremities of the given line; but then the man accepts the conclusion without any question. Aristotle's statement, then, that proof is reasoning that causes us to know is to be understood with the proviso that the proof is accompanied by its appropriate experience, and is not to be understood of the bare proof. His statement also in the first book of the Metaphysics that those who understand the reason and the cause are wiser than those who have empiric knowledge of a fact, is spoken of such as know only the bare truth without the cause. But I am here speaking of the man who knows the reason and the cause through experience. These men are perfect in their wisdom, as Aristotle maintains in the sixth book of the Ethics, whose simple statements must be accepted as if they offered proof, as he states in the same place.

He therefore who wishes to rejoice without doubt in regard to the truths underlying phenomena must know how to devote himself to experiment. For authors write many statements, and people believe them through reasoning which they formulate without experience. Their reasoning is wholly false. For it is generally believed that the diamond cannot be broken except by goat's blood, and philosophers and theologians misuse this idea. But fracture by means of blood of this kind has never been verified, although the effort has been made; and without that blood it can be broken easily. For I have seen this with my own eyes, and this is necessary, because gems cannot be carved except by fragments of this stone. Similarly it is generally believed that the castors employed by physicians are the testicles of the male animal. But this is not true, because the beaver has these under its breast, and both the male and female produce testicles of this kind. Besides these castors the male beaver has its testicles in their natural place; and therefore what is subjoined is a dreadful lie, namely, that when the hunters pursue the beaver, he himself knowing what they are seeking cuts out with his teeth these glands.[16] Moreover, it is generally believed that hot water freezes more quickly than cold water in vessels, and the argument in support of this is advanced that contrary is excited by contrary, just like enemies meeting each other. But it is certain that cold water freezes more quickly for any one who makes the experiment. People attribute this to Aristotle in the second book of the Meteorologics; but he certainly does not make this statement, but he does make one like it, by which they have been deceived, namely, that if cold water and hot water are poured on a cold place, as upon ice, the hot water freezes more quickly, and this is true. But if hot water and cold are placed in two vessels, the cold will freeze more quickly. Therefore all things must be verified by experience. . . .

[16]See the first selection from the Greek *Physiologus* in the section immediately following, and Bernard Silvestris's *Cosmographia* I.3.229–30 in Chapter 15 below.

The Auguries of Nature

Selections from Two Medieval Bestiaries

From Philo of Alexandria in the first century to Roger Bacon in the thirteenth, Jewish and Christian thinkers had variously explored the problematic relationships between God's will and nature's laws: between supernatural and natural, spirit and flesh, faith and reason, authority and experience in a world created by God out of nothing and always subject to Him, yet governed from day to day by the regular processes of nature. For most men and women of the Middle Ages, however, the principal written sources of information concerning the natural world were the popular bestiaries (along with their botanical and mineralogical counterparts, the herbals and lapidaries) and the more ostensibly learned encyclopedias; and in these the intellectual sophistication of the philosophers and theologians is notably absent.

The archetype of all bestiaries, both Latin and vernacular, was the Greek Physiologus, *an anonymous or pseudonymous work (Physiologus means "naturalist") of uncertain provenance that took shape in perhaps the second or third century* A.D., *or even earlier, possibly in Alexandria, where Clement and Origen were adapting the allegorical methods of the Stoics and Philo to Christian exegesis. In the form that its forty-eight or forty-nine chapters take in most manuscripts, the* Physiologus, *Florence McCullough writes in* Mediaeval Latin and French Bestiaries *(1960), "is a compilation of pseudo-science in which the fantastic descriptions of real and imaginary animals, birds, and even stones were used to illustrate points of Christian dogma and morals" through allegorization of qualities they were said to embody; it is possible that Christian allegories were added to an originally pagan work. At a fairly early date, perhaps by about* A.D. *400, the Greek* Physiologus *was translated into Latin, and variant if similar versions—in which the moral allegories tend to be even more obtrusive than in the Greek—proliferated over the following centuries.*

Beginning in about the twelfth century, more elaborate Latin bestiaries began to appear in both prose and verse; these were still based on the Physiologus *but incorporated materials (as in the second and third selections from the Latin bestiary quoted below) from other sources, including the encyclopedic tradition. At the same time, vernacular versions—dating back to the fragmentary Old English verse* Physiologus *of about the ninth century—became increasingly popular. Numerous prose and verse renditions survive in Old French, Middle English, and other languages of medieval Europe, some of them blending, along with Aesopian fables, into more secular traditions that culminated in the irreverent beast epic of Reynard the Fox. To modern readers these tales will seem at best quaint, at worst ludicrously simplistic; but their enormous popularity throughout the Middle Ages suggests the extent to which they provided a readily accessible guide to the "forest of symbols" (in Baudelaire's phrase) that comprised the natural world and pointed, when rightly interpreted, to the spiritual world beyond it.*

Selections from the Greek Physiologus *in James Carlill's translation are from William Rose, ed.,* The Epic of the Beast *(n.d.); those from the Latin bestiary are from T. H. White, ed.,* The Book of Beasts *(1954). See also Michael J. Curley's translation of the Latin* Physiologus *(1979) and, in addition to McCulloch's study, Louis Charbonneau-Lannay's* The Bestiary of Christ *(1940; Eng. trans. 1991).*

FROM THE GREEK *PHYSIOLOGUS*,
TRANSLATED BY JAMES CARLILL

The Beaver

There is a beast called the Beaver, who is very gentle and quiet; but his organs of reproduction are very useful to him in the protection of his body; for, when he is pursued by the hunter and is about to be captured, he bites them off and casts them back to the hunter; and afterwards when the beaver encounters any other hunter he throws himself on his back, and, when the hunter perceives that he is mutilated, he leaves him alone.

So do thou, oh man! give back to the hunter, the Devil, that which belongs to him, such as unchastity, adultery, greediness. Cut away all such and give them to the Devil and he will let thee go, and thou shalt say: "My soul is as a bird escaped from the net of the fowler." (Psalm 124:7)

The Unicorn

"Thou hast exalted my horn," said the Psalmist, "like the horn of the Unicorn." (Psalm 92:11)

Physiologus relates of the Unicorn that it has the following attribute. It is a small beast like a goat; but it is very wary and the hunter cannot approach it because it possesses great cunning. It has a horn in the middle of its head. Let us now relate how it is caught. They send to it a pure virgin all robed. And the Unicorn springs into the lap of the maiden, and she subdues him, and he follows her; and so she leads him to the King's palace.

By this we see that the Unicorn is the image of our Savior, the horn of salvation raised for us in the house of our father David. The heavenly powers could not of themselves accomplish the work, but he had to become flesh and to dwell in the body of the true virgin Mary.

There is a further attribute of the Unicorn. In the places where he dwells there is a great lake, and to this lake all beasts resort to drink. But before they assemble themselves comes the snake, and casts her poison on the water. And the beasts when they observe the poison dare not drink, but stand aside and wait for the Unicorn. He comes and goes straight into the lake, and marks the sign of the Cross with his horn; and thereupon the poison becomes harmless, and all those beasts drink.

The Antholops

There is a beast called the Antholops. He is very crafty, so that the hunter cannot approach him. He has on his head great horns in the shape of a saw, so that he can saw through the trunks of the highest trees reaching up to heaven and throw them down to the ground. When he is thirsty, he goes to the Euphrates and drinks. Near that river grow sweet-brooms with branches bending over. The Antholops begins to play with the sweet-brooms, and entangles his horns in the branches; and presently he finds himself held fast and twists round and cries out loudly, because he wants to get away but cannot. When the hunter hears this he knows the Antholops is caught, and comes and kills it.

And thou, oh man! thou hast both horns, the Old and the New Testaments, with which thou canst attack thine enemies, which are unchastity, adultery, greed, and other passions. Remember that they are like the sweet-broom: entangle not thyself in them lest the evil hunter be able to kill thee.

The Elephant

There is a beast called the Elephant. The male has no desire for offspring. When, therefore, the female wishes to bear, she resorts to the far East near to Paradise. In that part grows a tree called mandragora [mandrake]. To this tree comes the male Elephant with the female, and she eats first of the tree and gives of the tree to him, and plays with him until he also partakes of it. And when the time draws near for the birth, she goes into a lake until the water reaches her breast. And so at length is born the young, just above the water, and forthwith goes to the breast of the mother and sucks. But meantime the male Elephant keeps a strict watch against snakes, for the snake is an enemy of the Elephant. When he finds a snake, he tramples on it, and kills it.

The Elephant has the following attribute. When he falls down, he is unable to rise again, for his legs have no joints. But how comes he ever to fall? In this way. When he wants to sleep, he leans against a certain tree, and so sleeps. Now the Indian, who knows the sleeping place of the Elephant, goes there, and saws the tree partly through. The Elephant comes now to lean thereon as he is accustomed, and, as soon as he comes close to the tree, it gives way, and he falls with it to the ground.

Now after he has fallen he cannot rise again. He begins therefore to weep and cry aloud. Another Elephant hears his cry, and comes to help him; but he cannot raise the fallen one. Thereupon they both lament and cry aloud; and twelve more Elephants now come to help, but they are not able to raise the fallen one. Thereupon they all cry out. Last of all comes the little Elephant, and lays his trunk round the fallen Elephant and lifts him up from the ground.

The nature of the little Elephant is such that, if you burn his hair or his bones in any place whatever, that place is for ever free from Devils or snakes, nor will ever any evil thing be found there.

The pair of Elephants is like to Adam and Eve. Adam and his wife, as long as they lived in the plenty of Paradise, were innocent of all carnal desire; but, when the woman had eaten of the tree, the potent mandragora, and given it to him, then they fell to evil passions, and she bare Cain over the miry waters (Genesis 4:1). As David said: "Save me, O God, for the waters rise to my soul." (Psalm 69:2)

When now the great Elephant, which is the Law, was come, he could not raise the fallen one. Thereupon came the twelve Elephants, namely the prophets, but these could not raise him. At the last came the true Elephant, Christ the Lord, and raised the fallen one from the earth. For the first of all was the smallest of all. He humbled himself and took the form of a servant that he might redeem all. (Philippians 2:7)

The Peacock

The Peacock is the most gaily colored of all birds. He is beautiful of color and lordly in plumage. When he passes by, he looks at himself and rejoices much over himself. He shakes himself, turns a somersault, and looks proudly around. But, when he glances at his feet, he screams wildly, for his feet are not suitable to his beautiful appearance.

And thou too, wise man, when thou regardest thy pomp and thy possessions dost delight thyself and rejoice and feel proud; but, when thou lookest at thy feet, that is thy sins, then cry aloud and lament to God, and despise thy sins as the Peacock his feet, so that thou mayest appear right in the presence of thy bridegroom.

Well spake Physiologus concerning the Peacock.

The Phoenix

Our Lord Jesus Christ said: "I have power to lay down my life and I have power to take it again" (John 10:18). And the Jews were angry at his saying.

Now there is a bird in India called Phoenix. And at the end of five hundred years he comes to the trees of Lebanon, and fills his wings with pleasant odors, and he makes known his return to the priest of Heliopolis in the month Nisan or Adar (that is Phamenoti or Pharmuti). And the priest, when he hears the tidings, comes there and fills up the altar with wood of vines. And the bird comes to Heliopolis laden with odors of pleasant spices, and settles on the altar, and kindles a fire and burns himself. And on the following morning the priest searches through the ashes on the altar and finds therein a small worm. And on the second day, behold! he achieves feathers and becomes as a young bird. And on the third day they find him even as before, the Phoenix, and he salutes the priest, and flies away and returns to his old dwelling-place.

If now this bird has the power to slay himself and come to life again, how should reasonable men complain of our Lord Jesus Christ when He said: "I have power to lay down my life and to take it again."

For the Phoenix takes on itself the image of our Lord, when, coming down from heaven, he brought with him both wings full of pleasant odors, the excellent heavenly words, so that as we stretch out our hands in prayer we become filled with the pleasant scent of his mercy.

Well spake Physiologus of the Phoenix.

FROM A TWELFTH-CENTURY LATIN BESTIARY, TRANSLATED BY T. H. WHITE

The Pelican

Pelicanus the Pelican is a bird which lives in the solitude of the River Nile, whence it takes its name. The point is that, in Greek, Egypt is called *Can*opos. The Pelican is excessively devoted to its children. But when these have been born and begin to grow up, they flap their parents in the face with their wings, and the parents, striking back, kill them. Three days afterward the mother pierces her breast, opens her side, and lays herself across her young, pouring out her blood over the dead bodies. This brings them to life again.

In the same way, Our Lord Jesus Christ, who is the originator and maker of all created things, begets us and calls us into being out of nothing. We, on the contrary, strike him in the face. As the prophet Isaiah says: "I have borne children, and exalted them and truly they have scorned me." We have struck him in the face by devoting ourselves to the creation rather than the creator.

That was why he ascended into the height of the cross, and, his side having been pierced, there came from it blood and water for our salvation and eternal life.

The Compassion of Fishes

What human love can compare with the compassion of fishes? For us the kisses of the mouth are enough. For them it is not too much to open their whole insides for the reception of their children, to bring them out again unharmed, to reanimate them as it were with the warmth of their own heat, to revive them with their own spirit, and for two to live in one body. So they go on, until they have either cheered the babies up or defended them from dangers by interposing their own bulk.

What person, on seeing these things and being able to prove them, must not give pride of place to the wonderful piety of fishes? Who is not amazed and astonished that Nature should exhibit in them what she does not exhibit in men?

For lots of human mothers, driven to distrust by malevolent hatreds, kill their own offspring; others of bad repute, as we read, have actually eaten up their young, so that the mother has become the grave of her human children. The inside of the mamma-fish, on the contrary, is a wall with ramparts to the babes, in which she preserves the child undisturbed in her innermost organs.

Well, different species of fish have different customs. Some produce eggs, others bring forth formed and living young. Those who lay eggs do not build nests as birds do, nor do they engage in the lengthy labor of hatching eggs, nor do they trouble to feed them. The egg is simply dropped, because Water like a pleasant wet-nurse brings it up by her own nature, in her bosom as it were, and vivifies it into an animal by rapid incubation. Animated by a touch of its parent the Ocean, the egg finally falls apart and a fish comes out.

How pure and unsullied is the succession among fishes, consequently, where none is cross-bred except with a member of its own race! They are indeed ignorant of adulterous contacts with strange fish, unlike those of us who go in for adultery.

Before the intervention of men, the great clans of donkeys and horses continued uninterruptedly among themselves. But now, on the contrary, donkeys are crossed with horses, bastardizing nature. This is certainly a greater sin than mere fornication, because it is committed contrary to nature: it injures natural affinity, apart from the injury in respect of the person.

You, O Man, manage these things like a pander to adulterous horses and you think the new animal valuable more because it is counterfeit than because it is genuine. You mix up strange species; you mingle opposite semen and frequently collect together the unwilling victims to a forbidden coitus, and you call this husbandry!

Because you are not able to do this with men themselves, inasmuch as human offspring cannot mix with contrary species, you think of another thing to do. You take away what a man is born with and remove his manhood, and, cutting off part of his body, you destroy sex to create an eunuch. So, what Nature did not produce among men, your shamelessness produces! What would good Mother Water say to all this?

Man, you have taught children the renunciation of their fathers, you have taught them separations, hatreds and enmity. Now learn what the relationship of parents and children could be. Fishes do not seek to live without Water, to be separated from the fellowship of their parents, to bc parted from the nourishment of the mother. It is their nature that, if separated from the sea, they die immediately.

The Wise Sea-Urchin

Echinus the Sea-Urchin is said to be a poor, paltry and contemptible animal. He is very frequently the herald of a coming tempest to sailors, or the announcer of a calm. When he senses a storm of wind he seizes a stout stone and carries it as ballast, or drags it as an anchor, so as not to be tossed about by the waves. Thus he is saved, not by his own strength, but held firm by an outside help and by the weight which he carries. Sailors snatch eagerly at this information as a sign of the coming disturbance and take care that no hurricane shall suddenly find them unprepared.

What mathematician, what astrologer, or what Chaldaean can understand the course of the stars or the movement and signs of the heavens so well? By what natural quality does the sea-urchin comprehend what is taught among us by learned men? Who was the interpreter to it of so great an augury?

Men often see the disorder of the atmosphere and are deceived—for the clouds frequently disperse without a storm. Echinus is not deceived, the signs never escape Echinus. There is so much science in this one poor animal that it foretells the future. Since there is nothing more in it than this one bit of wisdom, we must believe that it is through the tenderness of God to all things that the urchin also gets his function of prescience. Moreover, if God makes lovely the grass of the field so that we marvel; if he feeds the birds and provides food for the ravens, whose young are truly turned toward the Lord; if he gave women the knowledge of weaving and does not leave even the spider destitute of that wisdom, who now minutely and skilfully hangs his roomy webs in the doorways; if God himself gives courage to the horse and unharnesses fear from his neck—so that he leaps about on the plains and is pleasing to kings as he gallops—that horse who detects war from a distance by the smell, and is excited by the sound of the trumpet; and if there are so many unreasoning things and others of no account, such as herbs, such as the lilies which are filled with the ordering of their own knowledge: can we doubt then that he also assigns to Echinus the service of this foresight? God leaves nothing unexplored, nothing unnoticed. He who feeds all things sees all things. He completes all things in wisdom. As it is written: "He makes everything with knowledge."

And thus, if he does not neglect poor, blind Echinus, if he takes care of him and trains him in the signs of the future, will he not carefully consider your things too, O Man? Indeed, he truly takes care of you when his divine wisdom is called upon, saying: "If he has regard to the birds of the air, if he feeds them, are you not more than they? If God adorns the grass of the field, which today is and tomorrow is cast into the fire, how much more will he consider you, O ye of little faith?"

The Nature of the Things of the World

Selections from Two Medieval Encyclopedias

The medieval encyclopedic tradition, though drawing heavily on earlier sources such as Pliny the Elder (Chapter 12 above), dates back to the Etymologies *of St. Isidore of Seville (ca. 560–636), whose wide-ranging if shallow and derivative learning and sometimes bizarre etymologies were immensely influential throughout the Middle Ages. (See the selections in Ernest Brehaut,* An Encyclopedist of the Dark Ages: Isidore of Seville *[1912].) "The twenty books of the* Etymologies *which stood for the sum of human knowledge during many centuries," Haskins writes in* The Renaissance of the Twelfth Century *(1927), "covered the seven liberal arts, medicine and law, the church and the alphabet, man and animals (the longest book is on animals), the earth and the universe, political as well as physical geography, architecture and surveying, agricultural and military sciences, ships and household utensils, and the practical arts in general. . . . It was popular because it was compendious and succinct in an age which wanted all its learning in tabloid form, because it was bookish, because it was credulous, because it made use of allegorical and mystical interpretation."*

For long thereafter, encyclopedists both in Latin and later in the vernaculars echoed or simply paraphrased or quoted Pliny, Isidore, and one another in compilations that made no claim to originality or literary distinction. At first their sources remained meager. As Haskins observes, "Those indefatigable compilers, Isidore and Bede, knowing no Greek, were perforce limited to Latin sources, and their science, thin and barren and often fantastic, carried on a bare modicum of ancient learning to the mediaeval world." But later encyclopedias, especially of the thirteenth century—like Bartholomew the Englishman's De proprietatibus rerum (On the Properties of Things) *and Brunetto Latini's* Li Livres dou Tresor (The Book of the Treasure), *from which the following selections are taken, or Vincent of Beauvais's* Speculum naturale (Mirror of Nature), *or Rabbi Gershon ben Shlomoh of Arles's Hebrew* Shaar ha-Shamayim *(trans. F. S. Bodenheimer as* The Gate of Heaven *[1953])—made use not only of these oft-pillaged sources (and of the ubiquitous* Physiologus *and bestiaries). They utilized also the newly rediscovered learning of Aristotle, Galen, and Ptolemy in Greek, and of the Arabs and recent Latin writers such as Adelard of Bath, William of Conches, and Albert the Great. These works thus encapsulate, with little discrimination, the beliefs of their age, ranging from popular superstition to Scholastic speculation.*

Bartholomew the Englishman (Bartholomaeus Anglicus) wrote his Latin encyclopedia toward the mid thirteenth century; the English translation, made in 1397 by John Trevisa, chaplain to Thomas of Berkeley, is a model of Middle English prose, here modernized by Robert Steele in Mediæval Lore from Bartholomew Anglicus *(n.d.). Brunetto Latini, whom Dante remembered as the master who "taught me how man makes himself eternal," yet placed among those damned for sodomy in Canto 15 of the* Inferno, *was a Florentine diplomat of the mid to late thirteenth century who spent part of his life in exile in France. He composed his allegorical poem the* Tesoretto, *a dream vision influenced by Alan of Lille and the* Roman de la rose *(Chapter 15 below), in which Nature and Philosophy discourse together, in Italian, but wrote his encyclopedic* Book of the Treasure *in French. The selections here given in the Barrette/Baldwin translation of 1993 provide a much-simplified synopsis of advanced views of nature current in his time.*

FROM *ON THE PROPERTIES OF THINGS*, BY BARTHOLOMEW THE ENGLISHMAN, TRANSLATED BY JOHN TREVISA (REVISED BY ROBERT STEELE)

A Universe of Matter and Form

Matter and form are principles of all bodily things; and privation of matter and form is naught else but destruction of all things. And the more subtle and high matter is in kind, the more able it is to receive form and shape. And the more thick and earthly it is, the more feeble is it to receive impression, printing of forms and of shapes. And matter is principle and beginning of distinction, and of diversity, and of multiplying, and of things that are [en]gendered. For the thing that gendereth and the thing that is gendered are not diverse but touching matter. And therefore where a thing is gendered without matter, the thing that gendereth, and the thing that is gendered, are all one in substance and in kind: as it fareth of the persons in the Trinity. Of form is diversity, by the which one thing is diverse from another, and some form is essential, and some accidental. Essential form is that which cometh into matter, and maketh it perfect; and accordeth therewith to the perfection of some thing. And when form is had, then the thing hath its being, and when form is destroyed nothing of the substance of the thing is found. And form accidental is not the perfection of things, nor giveth them being. But each form accidental needeth a form substantial. And each form is more simple and more actual and noble than matter. And so the form asketh that shall be printed in the matter, the matter ought to be disposed and also arrayed. For if fire shall be made of matter of earth, it needeth that the matter of earth be made subtle and pur[ifi]ed and more simple. Form maketh matter known. Matter is cause that we see things that are made, and so nothing is more common and general than matter. And natheless [nevertheless] nothing is more unknown than is matter; for matter is never seen without form, nor form may not be seen in deed, but joined to matter.

Elements are simple, and the least particles of a body that is compound. And it is called least touching us, for it is not perceived by wits [senses] of feeling. For it is the least part and last in undoing of the body, as it is first in composition. And is called simple, not for [because] an element is simple without any composition, but for it hath no parts that compound it, that be diverse in kind and in number as some medlied [mixed] bodies have: as it fareth [happens] in metals of the which some parts be diverse; for some part is air, and some is earth. But each part of fire is fire, and so of others. Elements are four, and so there are four qualities of elements, of the which every body is composed and made as of matter. The four elements are Earth, Water, Fire, and Air, of the which each hath his proper qualities. Four be called the first and principal qualities, that is, hot, cold, dry, and moist: they are called the first qualities because they slide first from the elements into the things that be made of elements. Two of these qualities are called Active—heat and coldness. The others are dry and wetness and are called Passive.

The Whale

It is said that the whale hath great plenty of sperm, and after that he gendereth, superfluity thereof fleeteth [floats] above the water; and if it be gathered and dried it turneth to the substance of amber. And in age, for greatness of body, on his ridge powder and earth is gathered, and so digged together that herbs and small trees and bushes grow thereon, so that that great fish seemeth an island. And if shipmen come unwarily thereby, unneth [barely] they [e]scape without peril. For he throweth as much water out of his mouth upon the ship, that he overturneth it sometime or drowneth it.

Also he is so fat that when he is smitten with fishers' darts he feeleth not the wound, but it passeth throughout the fatness. But when the inner fish is wounded, then is he most easily taken. For he may not suffer the bitterness of the salt water, and therefore he draweth to the shoreward. And also he is so huge in quantity, that when he is taken, all the country is better for the taking. Also he loveth his whelps with a wonder love, and leadeth them about in the sea long time. And if it happeth that his whelps be let [hindered] with heaps of gravel, and by default of water, he taketh much water in his mouth, and throweth upon them, and delivereth them in that wise out of peril, and bringeth them again into the deep sea. And for to defend them he putteth himself against all things that he meeteth if it be noyful [harmful] to them, and setteth them always between himself and the sun on the more secure side. And when strong tempest ariseth, while his whelps are tender and young, he swalloweth them up into his own womb. And when the tempest is gone and fair weather come, then he casteth them up whole and sound.

Also Jorath saith, that against the whale fighteth a fish of serpent's kind, and is venomous as a crocodile. And then other fish come to the whale's tail, and if the whale be overcome the other fish die. And if the venomous fish may not overcome the whale, then he throweth out of his jaws into the water a fumous [cloudy] smell most stinking. And the whale throweth out of his mouth a sweet smelling smoke, and putteth off the stinking smell, and defendeth and saveth himself and his in that manner wise.

Satyrs

Satyrs be somewhat like men, and have crooked nose and horns in the forehead, and like to goats in their feet. Saint Anthony saw such a one in the wilderness, as it is said, and he asked what he was, and he answered Anthony, and said: "I am deadly, and one of them that dwelleth in the wilderness." These wonderful beasts be divers[e]: for some of them be called Cyno[ce]phali, for they have heads as hounds, and seem by the working, beasts rather than men, and some be called Cyclops, and have that name, for one of them hath but one eye, and that in the middle of the forehead, and some be all headless and noseless, and their eyen be in the shoulders, and some have plain faces without nostrils, and the nether lips of them stretch so, that they hele [cover] therewith their faces when they be in the heat of the sun: and some of them have closed mouths, in their breasts only one hole, and breathe and suck as it were with pipes and veins, and these be accounted tongueless, and use signs and becks [gestures] instead of speaking. Also in Scythia be some with so great and large ears, that they spread their ears and cover all their bodies with them, and these be called Panchios. . . . And other be in Ethiopia, and each of them have only one foot so great and so large, that they beshadow themselves with the foot when they lie gaping on the ground in strong heat of the sun; and yet they be so swift, that they be likened to hounds in swiftness of running, and therefore among the Greeks they be called Cynopodes. And some have the soles of their feet turned backward behind the legs, and in each foot eight toes, and such go about and stare [stay] in the desert of Lybia.

Mermaids

The mermaid is a sea beast wonderly shapen, and draweth shipmen to peril by sweetness of song. The Gloss on Is[aiah]. xiii. saith that sirens are serpents with crests.[17] And some men say, that they are fishes of the sea in likeness of women. Some men feign that there are three Sirens some-deal [in part] maidens, and some-deal fowls with claws and wings, and one of them singeth with voice, and another with a pipe, and the third with an harp, and they please so shipmen, with likeness of song, that they draw them to peril and to shipbreach [shipwreck], but the sooth [truth] is, that they were strong [w]hores, that drew men that passed by them to poverty and to mischief. And Physiologus saith it is a beast of the sea, wonderly shapen as a maid from the navel upward and a fish from the navel downward, and this wonderful beast is glad and merry in tempest, and sad and heavy in fair weather. With sweetness of song this beast maketh shipmen to sleep, and when she seeth that they are asleep, she goeth into the ship, and ravisheth which she may take with her, and bringeth him into a dry place, and maketh him first lie by her, and if he will not or may not, then she slayeth him and eateth his flesh. Of such wonderful beasts it is written in the great Alexander's story.[18]

[17]In St. Jerome's Latin Vulgate translation of Isaiah 13:22, the Hebrew word translated "dragons" in the English Authorized (King James) Version and "jackals" in the Revised Standard Version is "sirenes."
[18]The popular and often fabulous romances of Alexander the Great, in which Alexander encounters the Sirens, were widely read in the medieval West, where Homer's *Odyssey* was unknown.

FROM BOOK I OF *THE BOOK OF THE TREASURE* (*LI LIVRES DOU TRESOR*), BY BRUNETTO LATINI, TRANSLATED BY PAUL BARRETTE AND SPURGEON BALDWIN

Chapter 8: The Role of Nature

Now you have heard three ways in which God made all things.[19] The fourth way was when he had finished making all things, he ordained the nature of all things individually, and gave them a way in which they should be born and die, and the strength and characteristics and nature of each one. 2. You should know that all things which have a beginning—that is to say, those which were made out of some thing—will have an end; but those which were created out of nothing will have no end. It is the role of nature, made by the sovereign Father, to govern this fourth category. He is creator, she was created; He is without beginning and she has a beginning; He is commander, she must obey; He will have no end and she will end, as will all her works; He is all powerful, and she has no power beyond that which God has given her; He knows all things past and present and future, and she knows only what God reveals to her; He orders the world, she follows his order. Thus we can understand that each thing is subject to its nature; nevertheless He who made everything can adjust and change the course of its nature by divine miracle, as He did in the Glorious Virgin Mary, who conceived the Son of God without carnal knowledge, and was pure and virgin before and after; and He himself arose from the dead. These other divine miracles are not at all against nature, 3. and if anyone says that God ordained a certain course for nature and then went against that course and changed his first plan, and that if he changes his plan he is not everlasting, I would say to him that nature has no role in the things which God retains in his power, and that God always had in his will the birth and passion and resurrection of his Son, just as it happened.

Chapter 99: How Nature Works in the Elements and Other Things

Here the narrative states that the principal material to be treated in these books is the nature of the things of the world, which are made up of four complexions, namely hot and cold, dry and moist, of which all things are composed. 2. But the four elements, which are like the foundation of the world, are composed of these four complexions, for fire is hot and dry while water is cold and moist, air is hot and moist, earth is cold and dry. Of similar composition are the bodies of men and beasts and all animals, for there are four humors in them; choler is hot and dry; phlegm is cold and moist, melancholy is cold and dry, and blood is hot and moist. The year itself is divided into four seasons which are similarly composed, for spring is hot and moist, and summer is hot and dry, autumn is cold and dry, and winter is cold and moist. Thus may you know that fire and choler and summer are made up of one complexion, and water, phlegm and winter are of another, but air and blood and spring are tempered with more than one nature, and for this reason they are of a better makeup than all the others. Their opposites are earth, melancholy and autumn, and for this reason they have a very bad nature. 3. Now it is easy to understand how it is the function of nature to harmonize discordant matters and make unequal things equal in such a way that all diversity returns to unity, and it adjusts them and assembles them in one body and substance or into something else which is continually

[19]The first was the image (or "World Archetype") in God's mind; the second, the formless matter *(hyle)* and other primal beings, like light and the angels, created out of nothing; the third, the world and its creatures created from this matter.

reborn in the world, either through plants or seeds, whereby some produce an egg which is filled with a living being; others produce fleshly shapes, as the narrative will detail hereafter, in the appropriate time and place. 4. Through these words it is clear that nature is to God as a hammer is to a blacksmith, who at one time forms a spear, at another a helmet, or a nail, or a needle, or one thing or another according to what the blacksmith wishes, and just as there is one way of forming a needle, similarly nature works in the stars differently than in plants, and differently in men and beasts and other animals.

14

The Bounty of Earth: Poems in Praise of Creation

Only rarely does a philosopher or encyclopedist of the Middle Ages (or of any age) convey the vivid immediacy *of the natural world, even when commending the goodness of nature or the beauty of God's creation. This task has of course, since time immemorial, been primarily the poet's. Medieval poetry—from its precursors in late antiquity to the Carolingian revival, and especially after the renewal of the twelfth century—gives intense lyric expression not only to pious devotion to God and the Virgin but also to human intimacy with the encompassing world of meadow and forest, birds and flowers: the created world that recalls, in its fallen state, the Earthly Paradise. St. Bernard himself, in his introductory letter to Aelred's* Speculum Caritatis (Mirror of Charity), *remarked that monks might learn more in the woods and fields than from books. "When the monks and priests of the liturgical age spoke of nature," Georges Duby writes in* The Age of Cathedrals *(1976; Eng. trans. 1981), "they referred to the abstract idea of perfection beyond the reach of the senses. For them, nature was the conceptual form which revealed the substance of God, not the transient fictitious aspects which sight, hearing, and smell could perceive. Not the hazardous and disordered appearances of the world, but what had been the Garden of Paradise for Adam before his sin; a universe of peace, measure, and virtue, obeying divine reason and escaping the upheavals and decadence later introduced along with the powers of desire and death."*

This vision of nature deeply affected Christian poets too, but for them it was no abstract idea but a luminous presence spread all around them and manifest in the luxuriance of spring or summer, in the cry of the cuckoo or the song of the nightingale. Like Lucretius, Horace, Virgil, and the Vigil of Venus *(Chapter 11 above), medieval poets tirelessly celebrate the glories of an earth rejuvenated by the coming of spring, as Chaucer does in the Prologue to the Canterbury Tales (1–12):*

Whan that Aprill with his shoures soote
The droghte of March hath perced to the
roote,
And bathed every veyne in swich licour
Of which vertu engendred is the flour;
Whan Zephirus eek with his sweete breeth
Inspired hath in every holt and heeth
The tendre croppes, and the yonge sonne
Hath in the Ram his halve cours yronne,
And smale foweles maken melodye
That slepen al the nyght with open ye
(So priketh hem Nature in hir corages);
Thanne longen folk to goon on pilgrimages.

("When April with his sweet showers has pierced to the root the drought of March, and bathed every vein of sap with moisture, by whose potency the flower is engendered; when Zephyr too has inspired with his sweet breath

the tender sprouts in every holt and heath, and the young sun has run his half course in the Ram, and small birds that sleep all night with open eye make melody—so Nature pierces them in their hearts—then people long to go on pilgrimages.")

Certain pagan poems of antiquity anticipate portrayals of nature typical of the Middle Ages. Thus the Greek Orphic Hymn to Nature was among the first poems to personify and even deify Physis, much as later poets (such as Bernard Silvestris, Alan of Lille, and Jean de Meun in Chapter 15 below) would personify Natura—"this noble goddesse Nature" of Chaucer's Parliament of Fowls. *Some late Roman poems, like those of Tiberianus and Asmenius, in which a landscape is "painted" (a frequent word) purely for its own effect and not as part of a larger composition, are "nature poems" of a kind all but unknown in earlier classical antiquity. Not, of course, that verbal painting of natural scenes and celebration of their restorative powers were entirely new; Homer's description of the orchard of Alcinous in* Odyssey *VII and Hesiod's of momentary respite from the heat of summer in* Works and Days *588–96 (Chapter 8 above) are famed examples, but both are essentially digressions from the serious work at hand. Such a scene of peaceful seclusion in garden or meadow, especially when incorporated in pastoral poems like Theocritus's seventh idyll (Chapter 11 above), came to be known as a* locus amoenus *("pleasant place," or "pleasance"), and with increasing elaboration played a central role in much medieval and Renaissance poetry.*

*Imagery of the Golden Age, going back to Hesiod and Aratus in Greek but most familiar to the medieval West through Virgil and Ovid (Chapter 11), entered into such idealized landscapes. Ovid's description of the lovely grove near the pool of Pergus from which Pluto snatched Proserpina (*Metamorphoses *V.388–96) is an example:*

Woods crown its waters all around, and soothe
Phoebus' blows with their canopy of leaves:
branches give coolness; moist ground, purple flowers;
there spring is everlasting. In this grove
Prósperpine, picking violets or white lilies,
played, and with girlish eagerness filled basket
and bosom, striving to outdo her playmates:
no sooner seen by Dis than prized and seized,
so impetuous was his love.

*Claudian similarly describes Venus's mountain home on her native island of Cyprus (*Epithalamium of Honorius and Maria, *52–68):*

Hoarfrost dares not enwrap it, winds shrink back
from pounding, clouds from darkening this realm
of Venus' idle pleasures. Harsher seasons
banished, eternal spring's indulgence reigns.
Down to a plain the mountain slopes; a golden
hedge girds its meadows with a yellow wall.
Uxorious Mulciber with this wall and towers,
they say, once bought the kisses of his wife.[1]
Fields gleam within, perpetually in flower
though tended by no hand, content with Zephyr's
husbandry; in this shady grove no birds
enter, unless their song delights the goddess:
if so, they throng the trees, if not, withdraw.
Leaves live for Venus, every tree in turn
happily loves; in mutual union, palms
bow downward, poplar sighs at poplar's stroke,
alder to alder whispers, plane to plane-tree . . .

Christian writers, as Giles Constable notes,[2] "had long applied paradisiacal imagery to the Church in general and to monasteries in particular and drawn on the topos of the locus amoenus that dictated the descriptions of natural beauty and paradise from Antiquity down to the sixteenth century," and these classical images soon merged with accounts of the biblical Eden. In particular, phrases such as "everlasting" or "eternal spring" (ver perpetuum, ver aeternum, *as*

[1]Mulciber, or Vulcan (Greek Hephaestus), the limping smith of the gods, was the husband of the less than faithful Venus (Aphrodite), known especially for her adultery with her brother Mars (Ares).
[2]"Renewal and Reform in Religious Life," in Robert L. Benson and Giles Constable, eds., *Renaissance and Renewal in the Twelfth Century* (1982).

also in Ovid's Metamorphoses *I.107 [Chapter 11]) regularly characterize the Earthly Paradise from Dracontius and Avitus in the fifth and sixth centuries to Dante in the fourteenth (Chapter 15 below), and beyond. In Dante's account, with its references both to loss of perpetual spring through the rape of Proserpine and to the vanished Golden Age of the classical poets, the connection is explicit, as it is in Honorius Agustodunensis's twelfth-century definition of Paradise (in his* Elucidarium*) as "locus amoenissimus in Oriente"—"the most pleasant of places in the East." The deliberate artifice of such scenes is striking. Indeed, ideal poetic landscapes in late antiquity and the Middle Ages are often portrayed as divine or human works of art, and are thus in effect a form of* ekphrasis *(a verbal description of a work of art), of which Homer's description of the shield of Achilles in* Iliad *XVIII (Chapter 8) was a classic instance.*

Evocations of the beauties of spring and summer, of melodious bird songs and bright-colored flowers, often tinged with reminiscences of the lost Earthly Paradise that was both the culmination and the antithesis of nature (insofar as its splendors were not subject, like all of nature since the Fall, to change, corruption, and death), are among the most prominent characteristics of much medieval poetry. It both maintained and transformed the Roman classical tradition in whose poetic meters most nonliturgical Latin verse, through the Carolingian revival and for some time later, continued to be written. But beginning with the vernacular poetry of Ireland and England, whose traditions were much less directly dependent on Roman sources, classic and indigenous influences intermingle. In the Gaelic Voyage of Bran *the Celtic paradise overshadows the biblical, and in the Anglo-Saxon* Seafarer *the harsh beauty of life on the wintry sea is more immediate than the Christian sentiments to which it gives rise. Nature is by no means always a locus amoenus; it may be a realm of snow, frost, and hail, of "high-surging streams and tumultuous salt-waves."*

In Latin poetry after about the eleventh century, much of which is written in rhymed verse, old themes regain new vigor through the wandering scholars, or "Goliards," for whom the world itself seemed renewed, like the earth in springtime. Vernacular poets in French, Spanish, and Middle English (and in Provençal, Middle High German, and other European languages) share in the sense of vital union between the human and natural worlds for which St. Francis of Assisi gave rhapsodic thanks to God in his "Song of Brother Sun," in praise of created things. But for the irreverent Goliards, unlike the pious St. Francis, nature—above all as embodied in the rebellious flesh—was often opposed to Christian piety, as the anonymous "Archpoet" of Cologne openly avers in his frankly sensual "Confession":

Res est arduissima vincere naturam,
in aspectu virginis mentem esse puram . . .
Unicuique proprium dat natura donum:
ego versus faciens bibo vinum bonum.

("It is a most arduous thing to conquer nature and keep a pure mind in a virgin's presence . . . To each of us nature gives a gift: mine is making verses and drinking good wine.")

Here was still another dimension of nature absent, at least so explicitly, from the bucolic locus amoenus, and certainly from the Earthly Paradise. It finds candid voice in the medieval pastourelle *(typically portraying a sexual encounter between a knight or scholar and a peasant girl), and more crudely in raunchy verse* fabliaux *culminating in Chaucer's bawdier stories, such as "The Miller's Tale." Nature was emphatically not to be confined within the conventional pleasance, nor within the monastic cloister. As Horace had said, though you drive her out with a pitchfork, she will always be back.*

Among general anthologies, with Latin texts and verse translations, see Mediaeval Latin Lyrics, *trans. Helen Waddell (1929), and* The Goliard Poets, *trans. George F. Whicher (1949). For an overall history, see F. J. E. Raby,* A History of Christian-Latin Poetry from the Beginnings to the Close of the Middle Ages *(1927; 2nd ed. 1953) and* A History of Secular Latin Poetry in the Middle Ages, *2 vols. (1934). On ideal landscapes and the Earthly Paradise in late antiquity and the Middle Ages, see Ernst Robert Curtius,* European Literature and the Latin Middle Ages *(1948; Eng. trans. 1953), Chapter 10 ("The Ideal Landscape"), and the opening chapters of A. Bartlett Giamatti,* The Earthly Paradise and the Renaissance Epic *(1966).*

Various Thy Essence

Orphic Hymn IX [X]: "To Nature," translated by Thomas Taylor

The date of the eighty-seven Orphic Hymns, *composed in Greek hexameters in the general style of the far earlier "Homeric Hymns" (Chapter 8 above), is unknown; no clear reference to them is made before the twelfth century, and a tentative date in the late third century A.D. can be only, as Apostolos N. Athanassakis acknowledges in the introduction to his translation (1977), "as good a guess as any." A number of indications, including specification of incense to be burnt with each hymn, appear to connect them with a pagan mystery cult that claimed descent from the ancient Orphic and Dionysiac religions. First translated into Latin by the great Florentine humanist Marsilio Ficino (who left them unpublished) in the fifteenth century, they were of course unavailable to the Latin Middle Ages.*

Yet the personification of Physis in the hymn "To Nature" is among the earliest examples of a poetic practice—found less elaborately in such classical Latin poets as Statius and Claudian—that would have a long legacy in the Middle Ages. (See Curtius's Chapter 6, "The Goddess Natura," in European Literature and the Latin Middle Ages, *and George D. Economou,* The Goddess Natura in Medieval Literature *[1972].) The translation of this thirty-line hymn into English rhymed couplets is from* The Mystical Hymns of Orpheus *(1824), an amended second edition of* The Mystical Initiations; or, Hymns of Orpheus *(1787), translated by Thomas Taylor (1758–1835), as reprinted in* Thomas Taylor the Platonist: Selected Writings, *ed. Kathleen Raine and George Mill Harper (1969); it is numbered not IX but X in the text of Wilhelm Quandt's* Orphei Hymni *(1941; 2nd ed. 1955), as reprinted along with Athanassakis's literal translation.*

Taylor, whose translations of these hymns and of Plotinus and Porphyry, and commentaries on Neoplatonic philosophy and Greek mystery religions, greatly influenced English and American Romantics and W. B. Yeats, remarks in his introductory note to this hymn as follows:

> Nature, according to the theologists, as related by Proclus, in [his commentary on Plato's] Tim[aeus]. p. iv, is the last of the demiurgic causes of this sensible world, and the boundary of the latitude of incorporeal essences: and is full of reasons and powers, by which she governs the universe, every where connecting parts with their wholes. . . . Now the reason why the epithets of much-mechanic [Taylor's translation of the traditional epithet *polymêkhanos*, "of many devices," which Homer applied to Odysseus], all-artist, connecting, all-wise, providence &c. are given to nature, which evince her agreement with Minerva [Greek Athena], is be-

cause that Goddess, according to the Orphic theology, fabricated the variegated veil of nature, from that wisdom and virtue of which she is the presiding divinity. And Proclus informs us, that she connects all the parts of the universe together: containing in herself intellectual life, by which she illuminates the whole, and unifying powers by which she superintends all the opposing natures of the world. Nature, therefore, from her connecting, and unifying power, and from her plenitude of seminal reasons, has an evident agreement with Minerva; whose divine arts according to the Orphic theology, reduce whatever in the universe is discordant and different, into union and consent.

THE FUMIGATION FROM AROMATICS

Nature, all parent, ancient, and divine,
O much-mechanic mother, art is thine;
Heav'nly, abundant, venerable queen,
In ev'ry part of thy dominions seen.
Untam'd, all-taming, ever splendid light, 5
All-ruling, honor'd, and supremely bright.
Immortal, first-born, ever still the same,
Nocturnal, starry, shining, glorious dame.
Thy feet's still traces in a circling course,
By thee are turn'd, with unremitting force. 10
Pure ornament of all the pow'rs divine,
Finite and infinite alike you shine;[3]
To all things common and in all things known,
Yet incommunicable and alone.
Without a father of thy wondrous frame, 15
Thyself the father whence thy essence came.
All-flourishing, connecting, mingling soul,
Leader and ruler of this mighty whole.
Life-bearer, all-sustaining, various nam'd,
And for commanding grace and beauty fam'd. 20
Justice, supreme in might, whose general sway
The waters of the restless deep obey.
Aetherial, earthly, for the pious glad,
Sweet to the good, but bitter to the bad.
All-wise, all-bounteous, provident, divine, 25
A rich increase of nutriment is thine;
Father of all, great nurse, and mother kind,
Abundant, blessed, all-spermatic mind:
Mature, impetuous, from whose fertile seeds
And plastic hand, this changing scene proceeds. 30
All-parent pow'r, to mortal eyes unseen,
External, moving, all-sagacious queen.

[3]Philolaus [of Croton in Magna Graecia, the Pythagorean philosopher of the fifth century B.C. said to have influenced Plato's *Timaeus*] according to Demetrius (in [Diogenes] Laert[ius VIII.85]) published a discourse concerning Nature, of which this is the beginning: . . . "Nature, and the whole world, and whatever it contains, are aptly connected together from infinites and finites [or unlimited and limiting elements]." (Taylor)

By thee the world, whose parts in rapid flow,[4]
Like swift descending streams, no respite know,
On an eternal hinge, with steady course 35
Is whirl'd, with matchless, unremitting force.
Thron'd on a circling car, thy mighty hand
Holds and directs the reins of wide command.
Various thy essence, honor'd, and the best,
Of judgement too, the general end and test. 40
Intrepid, fatal, all-subduing dame,
Life-everlasting, Parca,[5] breathing flame.
Immortal, Providence, the world is thine,
And thou art all things, architect divine.
O blessed Goddess, hear thy suppliant's pray'r, 45
And make my future life, thy constant care;
Give plenteous seasons, and sufficient wealth,
And crown my days with lasting peace and health.

Joys of Farmstead, River, and Garden

Four Latin Poets of the First Four Centuries A.D.

Of Gaius (or Titus) **Petronius,** *Nero's "arbiter of elegance," best known for his racy fragmentary novel of imperial Roman decadence, the* Satyricon, *Tacitus writes* (Annals *16.18–19, trans. Church and Brodribb): "His days he passed in sleep, his nights in the business and pleasures of life. . . . And indeed his talk and his doings, the freer they were and the more show of carelessness they exhibited, were the better liked, for their look of natural simplicity." Forced to commit suicide, like Seneca (Chapter 12 above), by Nero in A.D. 66, "he did not fling away life with precipitate haste, but having made an incision in his veins and then, according to his humor, bound them up, he again opened them, while he conversed with his friends . . . as they repeated, not thoughts on the immortality of the soul or on the theories of philosophy, but light poetry and playful verses. . . . He dined, indulged himself in sleep, that death, though forced on him, might have a natural appearance." If the handful of short poems ascribed to him are by the same author, they represent a different side of his genius. The poem here included, "Parvula securo tegitur mihi culmine sedes," though belonging chronologically to an earlier period and sharing in the Horatian spirit of refined "natural simplicity," anticipates medieval poems in its portrayal of the mod-*

[4]Since the world has an extended and composite essence, and is on this account continually separated from itself, it can alone be connected by a certain indivisible virtue infused from the divine unity. Again, since from a natural appetite, it is ever orderly moved towards good, the nature of such an appetite and motion must originate from a divine intellect and goodness. But since, from its material imperfection, it cannot receive the whole of divine infinity at once, but in a manner accommodated to its temporal nature: it can only derive it gradually and partially, as it were by drops, in a momentary succession. So that the corporeal world is in a continual state of flowing and formation, but never possesses real being; and is like the image of a lofty tree seen in a rapid torrent, which has the appearance of a tree without the reality; and which seems to endure perpetually the same, yet is continually renewed by the continual renovation of the stream. (Taylor)

[5]Taylor uses the Latin "Parca" as his personified equivalent for the Greek *aisa* (fate, destiny).

est pleasures of a small house with orchard and garden; for this reason, no doubt, it is included in Helen Waddell's Mediaeval Latin Lyrics *(1929).*

The other three poets wrote some three centuries later. Little is known about ***Tiberianus****; the poem here translated, "Amnis ibat inter arva valle fusus frigida" (text in* The Penguin Book of Latin Verse, *ed. Frederick Brittain [1962]), is one of the first self-contained "nature poems" in European literature—although nature, by being set off by itself, has become a painted landscape. The popular meter (the fifteen-syllable trochaic tetrameter catalectic) is that of the* Vigil of Venus *(Chapter 11), which some have attributed, on no other basis, to Tiberianus.*

Decimus Magnus ***Ausonius*** *(ca. 310–395) was born at Bordeaux (Burdigala) in Roman Gaul, where he taught grammar and rhetoric until summoned by the Emperor Valentinian I to tutor his son Gratian. He became governor of Gaul in 378 and consul in 379, but returned to private life in his native region after the Emperor Gratian was murdered in 383 by rebellious troops loyal to the usurper Maximus (himself overthrown by Theodosius in 388). He wrote much of his poetry—collected in a two-volume Loeb Classical Library edition, ed. Hugh G. Evelyn-White (1919, 1921)—in his old age; his former pupil, Meropius Pontius Paulinus, became bishop of Nola and an important poet in his own right. His poetic output was large and varied, but for the most part contrived and uninspired. Gibbon, in a footnote to Chapter XXVII of* The Decline and Fall of the Roman Empire, *declared that "the poetical fame of Ausonius condemns the taste of his age." According to Evelyn-White, "From first to last his verse is barren of ideas: not a gleam of insight or of broad human sympathy, no passion, no revolt: his attitude towards life is a mechanical and complacent acceptance of things as they are"—a sterility significantly offset, however, by "a distinct appreciation for the beauties of nature without reference to the comfort and gratification which they may afford to mankind." Of the poems here given, the first is Helen Waddell's translation of a fifty-line elegy, "De rosis nascentibus," attributed to Ausonius but now thought to be by an unknown poet, and consigned to Evelyn-White's Appendix. The second and third passages are from Ausonius's finest poem,* Mosella, *a descriptive account of the river Moselle, its fish and surrounding sights, in 483 hexameters written after Ausonius accompanied Valentinian and Gratian on an expedition against the Germans in 368–69. The first passage (lines 50–74) describes the weeds and pebbles in the clear river's bed; the second, in Waddell's translation, is a brief moment (lines 191–95) of evening serenity. Here again was a new kind of poem, the scenic travelogue, that would have a long progeny: "naturae mirabor opus," the poet exclaims (line 51), "I shall marvel at the work of nature."*

Of ***Asmenius****, probably a native of Gaul who lived around 400, nothing more is known; his iambic poem in praise of gardens (text in* Penguin Book*) speaks for itself.*

Small House and Quiet Roof Tree, by Petronius,
translated by Helen Waddell

Small house and quiet roof tree, shadowing elm,
Grapes on the vine and cherries ripening,
Red apples in the orchard, Pallas' tree
Breaking with olives, and well-watered earth,
And fields of kale and heavy creeping mallows 5
And poppies that will surely bring me sleep.
And if I go a-snaring for the birds
Or timid deer, or angling the shy trout,
'Tis all the guile that my poor fields will know.

Go now, yea, go, and sell your life, swift life, 10
For golden feasts. If the end waits me too,
I pray it find me here, and here shall ask
The reckoning from me of the vanished hours.

A Woodland Scene,
by Tiberianus, translated by Robert M. Torrance

Winding through the fields a river flowed through coolly shadowed vales,
glistened laughingly with pebbles past bright flower-painted lawns.
High above, the gentle breezes, whispering seductive songs,
stirred through dark green laurel branches overhanging myrtle shrubs.
Underneath them, tender grasses flowered lushly in their prime: 5
all the earth was red where crocus blossomed, white where lilies sprang,
everywhere the fragrant forest smelled of violets in bloom.
Sprinkled in among the gracious sparkling jeweled gifts of spring,
bloomed the regal queen of odors, morning-star of radiant hues,
first of flowers, fair Dione's[6] flaming golden-petaled rose. 10
Forest branches soaked with dewdrops drenched dank grasses down below,
rivulets profusely welling upward murmured here and there,
caves were overgrown with tangled moss and myrtle deep within
dark recesses whence their bubbling waters issued sparkling bright.
Every bird among the shadows raised triumphant songs of spring, 15
intermingled with soft warblings unimaginably sweet.
Murmurs from the babbling rivers sang in harmony with leaves
roused to tunefully melodious rustling by the zephyr's muse.
Thus whoever crossed this meadow, glorious with scent and song,
reveled joyously in river, bird, breeze, flower, grove, and shade. 20

FROM THE POEMS OF AUSONIUS

On Newblown Roses, attributed to Ausonius,
translated by Helen Waddell

Spring, and the sharpness of the golden dawn.
Before the sun was up a cooler breeze
Had blown, in promise of a day of heat,
And I was walking in my formal garden,
To freshen me, before the day grew old. 5

I saw the hoar frost stiff on the bent grasses,
Sitting in fat globes on the cabbage leaves,
And all my Paestum roses[7] laughing at me,
Dew-drenched, and in the East the morning star,
And here and there a dewdrop glistening white, 10
That soon must perish in the early sun.

[6]Venus's. See note to "Pervigilium Veneris" (Chapter 11 above), line 7.
[7]Perhaps a reference to Virgil's *Georgics* IV.119 (Chapter 11 above).

Think you, did Dawn steal color from the roses,
Or was it new-born day that stained the rose?
To each one dew, one crimson, and one morning,
To star and rose, their lady Venus one. 15
Mayhap one fragrance, but the sweet of Dawn
Drifts through the sky, and closer breathes the rose.

A moment dies: this bud that was new-born
Has burgeoned even fold on even fold;
This still is green, with her close cap of leaves, 20
This shows a red stain on her tender sheath,
This the first crimson of the loosened bud;

And now she thinks to unwind her coverings,
And lo! the glory of the radiant chalice,
Scatt'ring the close seeds of her golden heart. 25
One moment, all on fire and crimson glowing,
All pallid now and bare and desolate.
I marvelled at the flying rape of time;
But now a rose was born: that rose is old.
Even as I speak the crimson petals float 30
Down drifting, and the crimsoned earth is bright.

So many lovely things, so rare, so young,
A day begat them, and a day will end.
O Earth, to give a flower so brief a grace!
As long as a day is long, so long the life of a rose. 35
The golden sun at morning sees her born,
And late at eve returning finds her old.
Yet wise is she, that hath so soon to die,
And lives her life in some succeeding rose.
O maid, while youth is with the rose and thee, 40
Pluck thou the rose: life is as swift for thee.

Beneath Exultant Waters,
translated by Robert M. Torrance
The Moselle

But I, contemptuous of wealth and honors, 50
shall sing of nature's work, not riotous waste
reveling in the loss of children's fortune!
Here firm sands overspread well-watered shores
and footsteps leave no trace to be remembered.
Through your smooth surface glassy depths gleam bright, 55
river concealing nothing! As the gentle
breeze is transparent to our fluid vision,
and calm winds pose no obstacle to seeing,
so, penetrating deep, we see things sunk
far below, and your inner shrine laid bare 60

whenever your smooth-gliding limpid waters
reveal shapes shimmering in your azure light:
furrowed sand rippled by soft-flowing currents
and grasses quivering on your bed of green.
Under their native fountains swaying plants 65
bend, buffeted by your waters; lurking pebbles
glisten, and gravel decorates green moss.
As Britons, when the ebbing tide exposes
all Caledonia's coast,[8] behold green seaweed,
coral bright red, and white pearls, seeds of shellfish, 70
men's trinkets; and beneath the opulent waves
strings of false jewels counterfeit our fashions:
so beneath calm Moselle's exultant waters
weeds of all hues expose commingled pebbles.

Evening on the Moselle,
translated by Helen Waddell
The Moselle

What color are they now, thy quiet waters?
The evening star has brought the evening light,
And filled the river with the green hillside;
The hill-tops waver in the rippling water,
Trembles the absent vine and swells the grape 5
In thy clear crystal.

In Praise of Gardens, by Asmenius,
translated by Robert M. Torrance

Come, Muses, children of almighty Jove,
let us proclaim the praise of fertile gardens.
Gardens provide the body healthy foods,
furnishing varied fruits to gardeners,
fresh vegetables, and many kinds of herbs, 5
glistening grapes, and produce from the trees.
Gardens abound in infinite delights
and joys accompanied by countless uses.
Crystalline waters murmuringly lap them,
and branching rivulets irrigate their crops. 10
Flowers with many-colored buds shine brightly,
adorning all the earth with jeweled glory.
Soft-humming bees buzz gratefully around them,
sipping at flowers moistened with fresh dew.
Fruitful vines burden down fast-wedded elms, 15
shading the reeds their tendrils intertwine.
Darkly shadowing trees provide asylum

[8]Caledonia was the Roman name for Scotland.

from blazing sunlight with their tangled hair.
Melodious birds pour forth their garrulous chatter,
soothing the ear perpetually with song. 20
Gardens delight, divert, support, and nourish,
alleviating melancholy spirits,
reinvigorate limbs, enchant tired eyes,
recompense toil with still intenser pleasure,
and give their gardeners joys of every kind! 25

Sickened World and Happiest Garden

Three Christian Visions of Nature, translated by Robert M. Torrance

Claudian was a pagan, the religions of Tiberianus and Asmenius are unknown, and Ausonius was at most an external Christian, distressed by his pupil Paulinus's fervor. But beginning with St. Paulinus himself, and above all with Aurelius ***Prudentius*** *Clemens (348–ca. 405), Christian poetry, hitherto largely confined to hymns like those of St. Ambrose (Chapter 13 above), laid claim to its rich classical poetic inheritance in the twilight years of the Western Roman Empire. Prudentius was born in Roman Spain, possibly at Caesaraugusta (Saragossa), and appears to have spent much of his life there. His varied and original poems, including the allegorical* Psychomachia (Battle of the Soul), *mainly are devoted less to this world than to the world of the spirit: the rewards of martyrdom, the divinity of Christ, the truths of the Bible. But vivid awareness of the natural world enters his poetry from time to time, as in the following passage from the second of two books* Contra Orationem Symmachi (Against the Oration of Symmachus). *The work is a polemic against paganism written in the opening years of the fifth century in response to an oration delivered in 384 by the pagan orator Quintus Aurelius Symmachus (who had died in the meantime); Symmachus had advocated restoration of state funding for pagan rites, a motion denied by Valentinian II. Prudentius argues (as St. Augustine would soon argue in* The City of God*) that the troubles of the world—in this case, of the natural world—cannot, as pagans charged, be laid to the Christians, but are inherent in nature itself. His parallel between world and body continues the pagan homology of macrocosm and microcosm that would persist through the Middle Ages and the Renaissance, but with emphasis now on their shared imperfection.*

Our second poet, Blossius Aemilius ***Dracontius****, lived in Carthage in the late fifth century. The Germanic Vandals, along with their allies, the Alans, had invaded Spain in 409 and North Africa in 429; in 439 their king, Gaiseric, occupied Carthage and declared his kingdom independent of the Roman Empire. Dracontius, a lawyer, was imprisoned by a later Vandal king, Gunthamund, for the crime of eulogizing the Roman emperor. While in prison he composed his major poem,* De Laudibus Dei (On the Praises of God), *in three books of hexameters; his description of the Earthly Paradise as a realm of perfected nature is from the first book.*

Finally, Alcimus Ecdicius ***Avítus****, about whom little is known, was bishop of Vienne (Roman Vienna) in Gaul from about 490 to 518; his account of the Earthly Paradise is from Book I of his five-book hexameter epic,* De Spiritalis Historiae Gestis (On the Deeds of Spiritual History).

For text and translation of Prudentius, see volume 2 of the Loeb Classical Library edition of H. J. Thomson (1953); of Dracontius, Liber I Dracontii de Laudibus Dei, *ed. James F. Irwin (1942). The passage from Avitus is in* The Penguin Book of Latin Verse; *see also* The Fall of Man: De Spiritalis Historiae Gestis Libri I–III, *ed. Daniel J. Nodes (1985).*

A Sickened World, by Prudentius
Contra Orationem Symmachi, *Book II*

Long-tottering elements disintegrate,
losing their properties, and often hurtle
lawlessly to unseasonable outcomes. 975
Gluttonous rust, engendered by malignant
vapors, now gnaws the crops; now arctic winds,
following balmy vernal zephyrs, scorch
stalks, blemishing their heads with soot-black stains;
or else, when swelling shoots spring up from tender 980
seedlings, relentlessly harsh frost destroys them,
letting no slender fibers pierce the earth:
forced above ground by penetrating ice,
their mangled roots lie bare and unprotected.
Double-edged spikes sprout up from arid soil, 985
and prickly thorns from inundating moisture.
Shortfall or superfluity of weather
thus plagues the earth and wounds a sickened world.
Just so malfunctions of our body often
lead to defects and plunge us in disorder, 990
sapping our limbs through lack of governance.
For the world's constitution and the body's
is one; one nature underlies them both.
Born out of nothing, nothing soon awaits them,
tottering from disease, or bowed with age. 995
Nature, ordained to end, must lack perfection.

The Earthly Paradise, by Dracontius
De Laudibus Dei, *Book I*

A place there is from which four rivers flow, 180
painted with jeweled turf and deathless flowers,
full of sweet-scented, never-fading herbs,
happiest garden in God's universe.
Fruit, disregarding seasons, grows year-long;
earth flowers with perennial spring forever, 185
decking arboreal choirs with pleasant garb;
a wall of densely interwoven branches
shades it, and fruit hangs down from every tree
or drops throughout the meadow. No hot sun
burns with its rays, no windy blast convulses; 190

no whirlwind, reinforced by storms, there rages;
no stringent ice subdues, no pounding hail
batters, no fields turn white with chilling frost.
Tranquil winds generated by a gentle
breeze from bright fountains in the garden blow there, 195
stirring the treetops: when their gentle breath
rustles the leaves, immobile shadows quiver,
and dangling fruits sway in the rippling foliage.
Perpetual spring there tempers welcome breezes,
helping fruit ripen on uninjured branches. 200
Bees need not manufacture waxen cells:
aethereal nectar drips from trees like honey,
hanging from leaves in tempting future cupfuls;
medicine of life-giving health hangs down,
with everything this artful picture fashions. 205

The Earthly Paradise, by Avitus

De Spiritalis Historiae Gestis, *Book I*

East of the Indies, where the world begins,
where earth and sky are said to meet together,
a grove stands, inaccessible to mortals,
on heights surrounded by eternal walls 215
whence the primeval criminal was banished:
now heavenly ministers occupy the sacred
land from whose bliss the guilty pair were driven.
There with the alternating seasons comes
no wintry cold dispersed by summer suns. 220
Though the revolving year brings searing heat,
and fields turn white with thickening frost and snow,
there temperate climes preserve eternal spring;
rough winds are absent, under sunny skies
clouds yield to unremittingly clear weather. 225
The nature of the place requires no rain,
for grasses thrive contentedly on dew.
The ground is always green, earth's smiling face
radiates warmth; plants grow on hills forever,
and leaves on trees: for though they burst in flower 230
often, swift sap invigorates their buds.
All that here takes a year to reach fruition,
month after month there ripens fully grown.
Lilies unwithered by the sun shine brightly,
inviolate violets preserve their blushing 235
freshness, suffused by everlasting grace.
Since winter and fierce summer heat are absent,
autumn yields yearlong fruit, and springtime, flowers.

The Conflict of Spring and Winter

Latin Poets of the Dark Ages

The last Roman emperor of the West, Romulus Augustulus, was deposed in 476 by the Germanic king Odoacer, himself defeated and murdered in 493 by the Ostrogoth Theodoric, under whom Boëthius (Chapter 13 above) served and died. But although 476 is the traditional end of the Western Roman Empire, much of Italy and North Africa, and parts of Spain, came again under imperial rule during the sixth century after reconquest by the Eastern Emperor Justinian (ruled 527–565) and his general Belisarius; only with the fall of northern Italy to the Germanic Lombards in 568 did the reality of Roman Empire in the West come to an end, though the fiction would linger on for a thousand years. The Dark Ages of Western Europe, roughly from the late sixth to the late eleventh century, was a time of almost continuous upheaval during which the political, cultural, and educational institutions of the ancient world had broken down and no others with any lasting stability, apart from the Church itself, had yet arisen.

Only in the monasteries—especially in regions outside of or peripheral to the Roman sphere, notably Ireland and England—and later in a small number of palace and cathedral schools did literacy and education maintain a fragile connection between the classical tradition and a medieval civilization struggling to be born. Another such region was Spain, before the Moorish (Muslim Berber) conquest by Tarik in 711 put an end to the partly Romanized Visigoth rule. Here in the seventh century Isidore of Seville (Chapter 13 above) composed his Etymologies, *and here* **Eugenius of Toledo** *(ca. 600–658), a Visigoth who became archbishop of Toledo, wrote, in classical elegiac couplets, his rhapsodic praise of the nightingale, whose melodious song redounds to the glory of Christ. (For the text see* The Penguin Book of Latin Verse.*) The poems that follow are from the century of the Carolingian revival under Charles the Great, or Charlemagne (king of the Franks from 768 and emperor of the West from 800 to 814), and his lineal succcessors, Louis the Pious (emperor 814–40) and Charles the Bald (king of the West Franks from 843 and emperor from 875 to 877). The period was increasingly marked by fratricidal warfare among Louis's sons, by the rise of local feudal barons, and by regression to the chaos of the Dark Ages under the impact of the Norse (or Viking) invasions beginning in northern France around 843.*

The most important writer of the early Carolingian period was **Alcuin**, *or Albinus (died 804), an English prelate of noble Northumbrian stock whom Charles invited in about 782 to set up a school at his court in Aachen (Aix-la-Chapelle). There Alcuin established a system of elementary study of the seven liberal arts, consisting of the trivium (grammar, logic, and rhetoric) and quadrivium (arithmetic, geometry, astronomy, and music), as formulated by such late Roman writers as Cassiodorus and Martianus Capella. In addition to extensive prose writings, Alcuin wrote numerous workmanlike poems in classical meters. The first included below is Waddell's freely abridged translation of a twenty-eight-line poem in elegiac couplets, "Quae te dextra mihi rapuit, luscinia, ruscis" (poem 61 in vol. 1 of Ernst Dümmler's* Poetae Latini Aevi Carolini *[1881]), lamenting the loss of his nightingale. The second, "Conflictus Veris et Hiemis" (poem 58), thought by some scholars to be of Irish authorship, is a pastoral eclogue in which two shepherds preside over a debate between Winter and Spring, contesting whether "the shepherd's friend, the cuckoo" (a harbinger of spring) shall come or not.*

Walafrid Strabo *(ca. 809–49) was a German educated at the monasteries of Reichenau* *and Fulda who became tutor to the future emperor Charles the Bald. As abbot of Reichenau he wrote his poem* De Cultura Hortorum (On the Culture of Gardens), *also known as* Hortulus (Little Garden), *describing in 444 hexameters some twenty-three herbs and flowers in his monastery garden and their medicinal powers (for the text see Dümmler, vol. 2), ending with the roses that symbolize the blood of the martyrs and the lilies that symbolize their heavenly reward. The dedication to Grimold, Chancellor to Louis the Pious and Abbot of St. Gall, concludes the poem.*

Wandalbert*, a monk of Prüm in Lorraine born in 813, was the author of a martyrology and of a poem in 366 hexameters,* De Mensium Duodecim Nominibus Signis Culturis Aerisque Qualitatibus (On the Names, Signs, Times of Planting, and Qualities of Weather of the Twelve Months), *in which he vividly describes the rural activities of each month in his native Rhineland. As L. P. Wilkinson remarks in* The Georgics of Virgil *(1969), from which this translation is taken, the influence of Virgil's* Georgics *is prominent. (The selection is from "December" [lines 336–66]; the entire text is given in Dümmler's second volume.)*

Finally, the anonymous lyric ***"The Sadness of the Woods Is Bright"*** *("Vestiunt silvae tenera merorem") is one of the tenth-century "Cambridge Songs," possibly from the Rhineland, that survive in two corrupt manuscripts, one of Verona, one of Cambridge by way of Canterbury. Written in classical Sapphics (a popular medieval meter), its vivid evocation of bird song in summer sharply contrasts with the conventional allegory of the final stanza—which may have saved it, Waddell conjectures, from the attempted erasure inflicted on more unmistakably secular poems of the Cambridge manuscript.*

To the Nightingale, by Eugenius of Toledo,
translated by Robert M. Torrance

Philomel, yours the voice imparting song
by which I praise you in my rustic tongue;
Philomel, yours the voice no lilting lyres
can match, melodious beyond all choirs;
Philomel, yours the voice that can allay 5
anxious hearts, driving seeds of care away!
In flowery fields you dwell, in grass delight,
tending to fledglings unprepared for flight.
See how the bushes tunefully resound
while leafy forests harmonize all round. 10
Swans and chattering swallows, in my view,
and parrots, famed for talking, yield to you.
No bird can imitate your highs and lows,
such honeyed sweetness from your murmuring flows.
Sing, then, with vibrant tongue, each tremulous note, 15
and pour forth liquid melody from your throat!
Gorge with sweet sounds our avid listening,
and do not cease, oh, do not cease to sing!
Christ, to thee be all glory and all praise,
blessing thy servants in such pleasant ways! 20

TWO POEMS BY ALCUIN,
TRANSLATED BY HELEN WADDELL

Written for His Lost Nightingale

Whoever stole you from that bush of broom,
 I think he envied me my happiness,
O little nightingale, for many a time
 You lightened my sad heart from its distress,
 And flooded my whole soul with melody. 5
And I would have the other birds all come,
 And sing along with me thy threnody.
So brown and dim that little body was,
 But none could scorn thy singing. In that throat
That tiny throat, what depth of harmony, 10
 And all night long ringing thy changing note.
 What marvel if the cherubim in heaven
Continually do praise him, when to thee,
 O small and happy, such a grace was given?

The Strife between Winter and Spring, attributed to Alcuin

From the high mountains the shepherds come together,
Gathered in the spring light under branching trees,
Come to sing songs, Daphnis, old Palemon,
All making ready to sing the cuckoo's praises.
Thither came Spring, girdled with a garland, 5
Thither came Winter, with his shaggy hair.
Great strife between them on the cuckoo's singing.

Spring I would that he were there,
 Cuckoo!
Of all winged things most dear, 10
To every roof the most beloved guest.
Bright-billed, good songs he sings.

Winter Let him not come,
 Cuckoo!
Stay on in the dark cavern where he sleeps, 15
For Hunger is the company he brings.

Spring I would that he were here,
 Cuckoo!
Gay buds come with him, and the frost is gone,
Cuckoo, the age-long comrade of the sun. 20
The days are longer and the light serene.

Winter Let him not come,
Cuckoo!
For toil comes with him and he wakens wars,
Breaks blessed quiet and disturbs the world, 25
And sea and earth alike sets travailing.

Spring And what are you that throw your blame on him?
That huddle sluggish in your half-lit caves
After your feasts of Venus, bouts of Bacchus?

Winter Riches are mine and joy of reveling, 30
And sweet is sleep, the fire on the hearth stone.
Nothing of these he knows, and does his treasons.

Spring Nay, but he brings the flowers in his bright bill,
And he brings honey, nests are built for him.
The sea is quiet for his journeying, 35
Young ones begotten, and the fields are green.

Winter I like not these things which are joy to you.
I like to count the gold heaped in my chests;
And feast, and then to sleep, and then to sleep.

Spring And who, thou slug-a-bed, got thee thy wealth? 40
And who would pile thee any wealth at all,
If spring and summer did not toil for thee?

Winter Thou speakest truth; indeed they toil for me.
They are my slaves, and under my dominion.
As servants for their lord, they sweat for me. 45

Spring No lord, but poor and beggarly and proud.
Thou couldst not feed thyself a single day
But for his charity who comes, who comes!
Cuckoo!

Then old Palemon spake from his high seat, 50
And Daphnis, and the crowd of faithful shepherds.
'Have done, have done, Winter, spendthrift and foul,
And let the shepherd's friend, the cuckoo, come.
And may the happy buds break on our hills,
Green be our grazing, peace in the plowed fields, 55
Green branches give their shadow to tired men.
The goats come to the milking, udders full,
The birds call to the sun, each one his note.
Wherefore, O cuckoo, come, O cuckoo, come!
For thou art Love himself, the dearest guest, 60
And all things wait thee, sea and earth and sky.
All hail, beloved: through all ages, hail!'

To Grimold, Abbot of St. Gall, with His Book "Of Gardening," by Walafrid Strabo, translated by Helen Waddell

A very paltry gift, of no account,
My father, for a scholar like to thee,
But Strabo sends it to thee with his heart.
So might you sit in the small garden close
In the green darkness of the apple trees 5
Just where the peach tree casts its broken shade,
And they would gather you its shining fruit
With the soft down upon it; all your boys,
Your little laughing boys, your happy school,
And bring huge apples clasped in their two hands. 10
Something the book may have of use to thee.
Read it, my father, prune it of its faults,
And strengthen with thy praise what pleases thee.
And may God give thee in thy hands the green
Unwithering palm of everlasting life. 15

December's Tasks, by Wandelbert, translated by L. P. Wilkinson

On the Names . . . of the Twelve Months

Now too, when winter's parching cold has gripped
The land and night is longest, farmers seize
The excuse of shorter daylight to enjoy
Rest, and at length forgetful of their toil
Give up the aching burden of their limbs 5
To sleep's perfusive balm. Yet even so
The month of rain has tasks that are its own,
Nor does the season with its icy blasts
And snows that veil the face of earth compel
Total cessation. Even when the fields 10
With windy showers are sodden, farmers then
Should furrow with the plow the lumpish clods
Where barley later or luxuriant beans
They hope to sow and rear; and foul manure
Can then be tossed about over the land. 15
But when the soil lies rigid and inert,
Free to get on with much indoors at home
The numbing cold of winter they relieve.
Hence came experienced skill to catch in nets
The varied sea-birds, or with fire and noise 20
To cheat birds of the field, or set up snares
To catch them; hence in rivers full of fish
To lower screens of wicker openwork

They learnt; and to the banks, where gently flows
The stream in shallows, closely fitted lurk 25
Bundles with nets to catch an easy prey.
Swine in this month now sleek from acorn-feed,
Full swoll'n with pendulous belly to be seen,
Men slaughter, and hang up the chines in smoke
Well saturated first in salty brine. 30

Such tasks through twelve revolving months they ply,
Such ways of life the Gallic fields preserve,
Which briefly noting, reader, in my song,
I, Wandalbert, recorded, by a friend
Pressed and encouraged, while I passed my days 35
Along the pleasant margins of the Rhine
Crowned by the ancient ramparts of Cologne.

The Sadness of the Woods Is Bright (Anonymous),
translated by Helen Waddell

The sadness of the wood is bright
With young green sprays, the apple trees
Are laden, in their nests high overhead
  Wood pigeons croon.

The doves make moan, deep throated sings the thrush, 5
The blackbirds flute their ancient melody;
The sparrow twitters, making his small jests
  High underneath the elm.

The nightingale sings happy in the leaves,
Pouring out on the winds far carrying 10
Her solemn melody: the sudden hawk
  Quavers in the high air.

The eagle takes his flight against the sun;
High overhead the lark trills in the sky,
Down dropping from her height and changing note, 15
  She touches earth.

Swift darting swallows utter their low cry;
The jackdaw jargons, and clear cries the quail;
And so in every spot some bird is singing
  A summer song. 20

Yet none among the birds is like the bee,
Who is the very type of chastity,
Save she who bore the burden that was Christ
  In her inviolate womb.

Forest Huts and Mysterious Isles

Lyric Poems and Voyages of Medieval Ireland

Ireland, which the Romans called Hibernia, was never part of the Empire, and between its conversion to Christianity (by St. Palladius and the great British missionary St. Patrick) in the mid fifth century and the devastating Viking raids of the eighth, it enjoyed the "golden age" of its culture when most of Western Europe was plunged into chaos. Orthodox in doctrine, the "Celtic Church" differed from others in being more monastic than episcopal in its organization. "Independence and individualism were the keynotes of Celtic monastic life," J. F. Webb writes in the introduction to his translation of Lives of the Saints *(1965), from which our selections from* The Voyage of St. Brendan *are taken: "each house was autonomous, most of them tracing their creation to some holy founder in the sixth century whose rule remained the sole authority for the community. There was no dependence of house on house, no Rule of St. Benedict, no liturgical uniformity." In this Age of the Saints, Ireland was known for its forest-dwelling hermits, for its devotion to Latin learning, which it did much to preserve, and for its far-traveled missionaries such as St. Brendan, St. Columba, and St. Columban in the sixth and seventh centuries, who propagated both faith and learning in distant lands. In much of Europe, Ireland was known "as the one haven of rest in a turbulent world overrun by hordes of barbarians," Kuno Meyer remarks in* Selections from Ancient Irish Poetry *(1911), and "as the great seminary of Christian and classical learning"; as late as the ninth century, John the Scot (Chapter 14 above) brought to the court of Charles the Bald both a knowledge of Greek nearly unknown to the Franks and the most original philosophical mind of his age.*

During these same centuries, the Irish were creating what Nora Chadwick, in The Celts *(1969), reminds us was "the oldest vernacular literature north of the Alps." In a Christian Ireland divided into clans and ruled by petty (and often warring) kings, an Ireland where the old heroic ethos survived in prose cycles relating the deeds of Cuchulainn, Fergus, and other great warriors, the official bard attached to a king had retained much of the prestige of his druid heritage, and now he was joined by monks and hermits who not only copied Latin manuscripts but also composed lyric poems in their native Irish Gaelic. The earliest manuscripts date from about the twelfth century, but some poems appear to go back to the eighth or even to the sixth century. Our first poem,* **"The Mystery"** *(from Kathleen Hoagland, ed.,* 1000 Years of Irish Poetry *[1953]), in which the identification of divinity with every aspect of nature strangely parallels Arjuna's vision in the* Bhagavad Gītā *(Chapter 4 above), was attributed to the legendary bard Amergin, supposed to have lived centuries before Christ. Love of nature, which reflects the glory of God's creation, permeates this poetry. "To seek out and watch and love Nature, in its tiniest phenomena as in its grandest," Meyer writes, "was given to no people so early and so fully as to the Celt. . . . It is a characteristic of these poems that in none of them do we get an elaborate or sustained description of any scene or scenery, but rather a succession of pictures and images which the poet, like an impressionist, calls up before us by light and skilful touches. Like the Japanese, the Celts were always quick to take an artistic hint; they avoid the obvious and the commonplace; the half-said thing to them is dearest." The second and third poems* (**"Summer Has Come"** *and* **"Song of Summer"**)*, from Meyer's* Selections*; the fourth* (**"The Open-Air Scriptorium"**)*, a ninth-century poem from Sean O'Faolain, ed.,* The Silver Branch *(1938); and the fifth (a humorous parallel between* **"The Philologian and His Cat"**)*, from Philip Schuyler Allen, ed.,*

The Romanesque Lyric *(1928), richly illustrate varied aspects of the medieval Irish poet's closeness to nature.*

Two longer works tell of fabulous voyages in which the spiritual and the natural mysteriously intermingle. **The Voyage of Bran Son of Febal to the Land of the Living**, *translated by Meyer (1895), is a mythical* imram *(voyage), dating to the seventh or eighth century. It tells of Bran's summons, after sweet music, by "a woman in strange raiment" who sings of a distant isle where nature and art, silver twigs and blossoming trees, combine to bring everlasting joy. She vanishes with her silver apple branch, and Bran sets out. After hearing from Manannan son of Ler of the "wood of beautiful fruit" with golden leaves, which is the sea on which he is sailing—and of the future coming of Christ—Bran sails to the Land of Women, where many years seem one and no savor is wanting. Like classical portrayals of the Golden Age and Christian descriptions of the Earthly Paradise, which may have influenced it, this largely pagan saga gives expression to perennial dreams of natural bounty exempt from decay and death.*

A similar spirit pervades the Latin **Voyage of St. Brendan** (Navigatio Brendani), *an immensely popular account, probably written in the ninth century but influenced by earlier mythical* imrama, *of a fabled missionary voyage (565–73) of the historical saint. Here the miracles retain a Celtic hue, even though the birds that sing to Brendan are now messengers of the Christian God.*

FIVE OLD IRISH POEMS

The Mystery, attributed to Amergin, translated by Douglas Hyde

I am the wind which breathes upon the sea,
I am the wave of the ocean,
I am the murmur of the billows,
I am the ox of the seven combats,
I am the vulture upon the rocks, 5
I am a beam of the sun,
I am the fairest of plants,
I am a wild boar in valor,
I am a salmon in the water,
I am a lake in the plain, 10
I am a word of science,
I am the point of the lance of battle,
I am the God who created in the head the fire.
Who is it who throws light into the meeting on the mountains?
Who announces the ages of the moon? 15
Who teaches the place where couches the sun?
(If not I)

Summer Has Come,
translated by Kuno Meyer

Summer has come, healthy and free,
Whence the brown wood is bent to the ground:
The slender nimble deer leap,
And the path of seals is smooth.

The cuckoo sings gentle music, 5
Whence there is smooth peaceful calm:
Gentle birds skip upon the hill,
And swift grey stags.

Heat has laid hold of the rest of the deer—
The lovely cry of curly packs! 10
The white extent of the strand smiles,
There the swift sea is roused.

A sound of playful breezes in the tops
Of a black oakwood is Drum Daill,
The noble hornless herd runs, 15
To whom Cuan-wood is a shelter.

Green bursts out on every herb,
The top of the green oakwood is bushy,
Summer has come, winter has gone,
Twisted hollies wound the hound. 20

The blackbird sings a loud strain,
To him the live wood is a heritage,
The sad angry sea is fallen asleep,
The speckled salmon leaps.

The sun smiles over every land,— 25
A parting for me from the brood of cares.
Hounds bark, stags tryst,
Ravens flourish, summer has come!

Song of Summer,
translated by Kuno Meyer

Summer-time, season supreme!
Splendid is color then.
Blackbirds sing a full lay
If there be a slender shaft of day.

The dust-colored cuckoo calls aloud: 5
Welcome, splendid summer!
The bitterness of bad weather is past,
The boughs of the wood are a thicket.

Panic startles the heart of the deer,
The smooth sea runs apace— 10
Season when ocean sinks asleep,
Blossom covers the world.

Bees with puny strength carry
A goodly burden, the harvest of blossoms;
Up the mountain-side kine take with them mud, 15
The ant makes a rich meal.

The harp of the forest sounds music,
The sail gathers—perfect peace;
Color has settled on every height,
Haze on the lake of full waters. 20

The corncrake, a strenuous bard, discourses,
The lofty cold waterfall sings
A welcome to the warm pool—
The talk of the rushes has come.

Light swallows dart aloft, 25
Loud melody encircles the hill,
The soft rich mast buds,
The stuttering quagmire prattles.

The peat-bog is as the raven's coat,
The loud cuckoo bids welcome, 30
The speckled fish leaps—
Strong is the bound of the swift warrior.

Man flourishes, the maiden buds
In her fair strong pride.
Perfect each forest from top to ground, 35
Perfect each great stately plain.

Delightful is the season's splendor,
Rough winter has gone:
Every fruitful wood shines white,
A joyous peace is summer. 40

A flock of birds settles
In the midst of meadows,
The green field rustles,
Wherein is a brawling white stream.

A wild longing is on you to race horses, 45
The ranked host is ranged around:
A bright shaft has been shot into the land,
So that the water-flag is gold beneath it.

A timorous, tiny, persistent little fellow
Sings at the top of his voice, 50
The lark sings clear tidings:
Surpassing summer-time of delicate hues!

The Open-Air Scriptorium,
translated by Robin Flower

Over my head the forest wall
Rises; the ousel sings to me;
Above my booklet lined for words
The woodland birds shake out their glee.

There's the blithe cuckoo chanting clear 5
In mantle grey from bough to bough;
God keep me still! For here I write
His gospel bright in great woods now.

The Philologian and His Cat,
translated by Howard Mumford Jones

Pangur is proof the arts of cats
And men are in alliance;
His mind is set on catching rats,
And mine on snaring science.

I make my book, the world forgot, 5
A kind of endless class-time;
My hobby Pangur envies not—
He likes more childish pastime.

When we're at home time quickly flies—
Around us no one bustles; 10
Untiringly we exercise
Our intellectual muscles.

Caught in his diplomatic net,
A mouse jumps down his gullet;
And sometimes I can half-way get 15
A problem when I mull it.

He watches with his shining eye
The wall that guards his earnings;
As for my eyesight—well, I try
To match my stare with learning's. 20

His joy is in his lightning leap;
Me—I'm a mental wizard;
My claws are sunk in problems deep,
His, in a mouse's gizzard.

As comrades we admit we shine, 25
For each observes his station;
He practices his special line,
And I, my avocation.

Our rivalry you'll find is nice,
If in the scale you weigh us: 30
Each day Pangur goes hunting mice,
I bring forth light from chaos.

FROM The Voyage of Bran Son of Febal to the Land of the Living, *translated by Kuno Meyer*

. . . This is the beginning of the story. One day, in the neighborhood of his stronghold, Bran went about alone, when he heard music behind him. As often as he looked back, 'twas still behind him the music was. At last he fell asleep at the music, such was its sweetness. When he awoke from his sleep, he saw close by him a branch of silver with white blossoms, nor was it easy to distinguish its bloom from that branch. Then Bran took the branch in his hand to his royal house. When the hosts were in the royal house, they saw a woman in strange raiment on the floor of the house. 'Twas then she sang the fifty[9] quatrains to Bran, while the host heard her, and all beheld the woman.

And she said:

A branch of the apple-tree from Emain
I bring, like those one knows;
Twigs of white silver are on it,
Crystal brows with blossoms.

There is a distant isle, 5
Around which sea-horses[10] glisten:
A fair course against the white-swelling surge,—
Four feet uphold it.

A delight of the eyes, a glorious range,
Is the plain on which the hosts hold games: 10
Coracle contends against chariot
In southern White-Silver Plain.

Feet of white bronze under it
Glittering through beautiful ages.
Lovely land throughout the world's age, 15
On which the many blossoms drop.

An ancient tree there is with blossoms,
On which birds call to the Hours.[11]
'Tis in harmony it is their wont
To call together every Hour. 20

Splendors of every color glisten
Throughout the gentle-voiced plains.
Joy is known, ranked around music,
In southern Silver-Cloud Plain.

Unknown is wailing or treachery 25
In the familiar cultivated land,
There is nothing rough or harsh,
But sweet music striking on the ear.

[9]All the MSS contain only twenty-eight quatrains. (Meyer) In the translation that follows, I have incorporated from Meyer's footnotes the English equivalents for some of the Irish proper names; thus in line 12, "White-Silver Plain" for Mag Findargat, and in line 24, "Silver-Cloud Plain" for Mag Argatnnél.
[10]A *kenning* for "crested sea-waves." (Meyer)
[11]The canonical hours, an allusion to church music. (Meyer)

Without grief, without sorrow, without death,
Without any sickness, without debility, 30
That is the sign of Emain—
Uncommon is an equal marvel.

A beauty of a wondrous land,
Whose aspects are lovely,
Whose view is a fair country, 35
Incomparable is its haze. . . .

Many-shaped Emne[12] by the sea, 65
Whether it be near, whether it be far,
In which there are many thousands of motley women,
Which the clear sea encircles. . . .

There will come happiness with health
To the land against which laughter peals,
Into Imchiuin at every season 75
Will come everlasting joy.

It is a day of lasting weather
That showers silver on the lands,
A pure-white cliff on the range of the sea,
Which from the sun receives its heat. 80

The host race along Mag Mon,[13]
A beautiful game, not feeble,
In the variegated land over a mass of beauty.
They look for neither decay nor death.

Listening to music at night, 85
And going to Ildathach,[14]
A variegated land, splendor on a diadem of beauty,
Whence the white cloud glistens. . . .

Thereupon the woman went from them, while they knew not whither she went. And she took her branch with her. The branch sprang from Bran's hand into the hand of the woman, nor was there strength in Bran's hand to hold the branch.

Then on the morrow Bran went upon the sea. The number of his men was three companies of nine. One of his foster-brothers and mates was set over each of the three companies of nine. When he had been at sea two days and two nights, he saw a man in a chariot coming towards him over the sea. That man also sang thirty[15] other quatrains to him, and made himself known to him, and said that he was Manannan son of Ler, and said that it was upon him to go to Ireland after long ages, and that a son would be born to him, even Mongan son of Fiachna—that was the name which would be upon him.

So he sang these thirty quatrains to him:

[12]Here and in line 111 below the nominative Emne is used instead of Emain. (Meyer)
[13]"Plain of Sports." (Meyer)
[14]"Many-colored Land." (Meyer)
[15]The MSS again contain only twenty-eight quatrains. (Meyer)

Bran deems it a marvelous beauty
In his coracle across the clear sea:
While to me in my chariot from afar
It is a flowery plain on which he rides about. . . .

Sea-horses glisten in summer
As far as Bran has stretched his glance:
Rivers pour forth a stream of honey 15
In the land of Manannan son of Ler.

The sheen of the main, on which thou art,
The white hue of the sea, on which thou rowest about,
Yellow and azure are spread out,
It is land, and is not rough. 20

Speckled salmon leap from the womb
Of the white sea, on which thou lookest:
They are calves, they are colored lambs
With friendliness, without mutual slaughter. . . .

Along the top of a wood has swum
Thy coracle across ridges,
There is a wood of beautiful fruit
Under the prow of thy little skiff. 40

A wood with blossom and fruit,
On which is the vine's veritable fragrance,
A wood without decay, without defect,
On which are leaves of golden hue.

We are from the beginning of creation 45
Without old age, without consummation of earth,[16]
Hence we expect not that there should be frailty;
The sin has not come to us. . . .

A noble salvation will come 61
From the King who has created us,
A white law will come over seas;
Besides being God, He will be man. . . .

Steadily then let Bran row,
Not far to the Land of Women, 110
Emne with many hues of hospitality
Thou wilt reach before the setting of the sun.

. . . It was not long thereafter when they reached the Land of Women. They saw the leader of the women at the port. Said the chief of the women: "Come hither on land, O Bran son of Febal! Welcome is thy advent!" Bran did not venture to go on shore. The woman throws a ball of thread to Bran straight over his face. Bran put his hand on the ball, which stuck to his palm. The thread of the ball was in the woman's hand, and she

[16]I.e., of the grave. (Meyer)

pulled the coracle towards the port. Thereupon they went into a large house, in which was a bed for every couple, even thrice nine beds. The food that was put on every dish vanished not from them. It seemed a year to them that they were there,—it chanced to be many years. No savor was wanting to them. . . .

FROM The Voyage of St. Brendan,
translated by J. F. Webb
Section 11

They rowed towards the island on which they had previously made a three days' stay. They climbed its summit, which faces westwards across the sea, and from there they espied another island close at hand. It was grassy, covered with flowers, full of glades, and separated from the island they were on by only a narrow strait. They sailed round it, looking for a harbor, and put in at the mouth of a stream on the southern shore. This stream was about as wide as the coracle. The monks disembarked, and Brendan instructed them to fix ropes to the sides of the coracle and pull it, with himself on board, as hard as they could, against the current. He was conveyed about a mile upstream to the source. "Our Lord Jesus Christ," he said, "has led us to a place in which to stay and celebrate his Resurrection." Then he added: "I think that, even if we had brought no supplies at all, this spring would provide us with all the nourishment we need."

Beyond the spring, on higher ground, there was an exceptionally tall tree growing, with a trunk of colossal girth. This tree was full of pure white birds; so thickly had they settled on it that there was hardly a branch, or even a leaf, to be seen. Brendan wondered why so vast a number of birds should have flocked together. So keenly did he long to unravel the mystery that he threw himself on his knees in tears and prayed silently: "O God, to whom nothing is unknown and who can bring to light every hidden fact, you see how anxious I am. I beseech your infinite majesty to deign to make known to me, a sinner, this secret design of yours which I see before me. I presume to ask, not because of any merit or dignity of my own, but solely on account of your boundless clemency."

He sat down in the boat and one of the birds flew down from the tree towards him. The flapping of its wings sounded like a bell. It settled on the prow, spread out its wings as a sign of joy, and looked placidly at Brendan. He realized at once that God had paid heed to his prayer. "If you are God's messenger," he said to the bird, "tell me where these birds come from and why they are gathered together."

"We are fallen angels," the bird replied, "part of the host which was banished from Heaven through the sin of man's ancient foe. Our sin lay in approving the sin of Lucifer; when he and his band fell, we fell with them. Our God is faithful and just and, by His great justice, we were placed here. Thanks to His mercy, we suffer no torment: our only punishment is to have no part in the vision of His glory which those who stand before His throne in Heaven enjoy. Like the other messengers of God, we wander through the air, over the bowl of Heaven, and upon the earth, but on Sundays and holy days we take on this physical form and tarry here to sing the praises of our creator. You and your companions have completed one year of your journey; six more years remain. Every year you will celebrate Easter in the same place as you are going to spend it today, and at the end of your travels you will achieve your heart's desire—you will find the Land of Promise of the Saints." With that the bird flew away from the prow of the boat and rejoined the flock.

When it was almost time for vespers, the birds all began to sing in unison: "*Thou, O*

God, art praised in Sion: and unto thee shall the vow be performed in Jerusalem . . . ," beating their wings against their sides, and they continued singing the verse antiphonally for a whole hour. To the man of God and his companions the rhythm of the melody combined with the sound of their beating wings seemed as sweet and moving as a plaintive song of lament. . . .

Blessed Abodes and Times of Hard Travail

Three Old English Poems

Celtic Britain, twice raided by Julius Caesar (in 55 and 54 B.C.), was invaded in A.D. 43 under Claudius, and most of the island south of the Scottish highlands was in time made part of the Empire. By the mid fifth century Roman control had disintegrated, and by the end of the century the indigenous (and Christianized) peoples—led, according to legend, by King Arthur—had been pushed back to the borderlands of Wales, Cornwall, and Brittany (across the Channel) by invading north-Germanic tribes, the Angles, Saxons, and Jutes. Subsequent invasions by Vikings, or "Danes," beginning in the eighth century, were resisted in the ninth and tenth centuries by Alfred the Great, king of Wessex from 871 to 899, and his successors. But by 1016 the Danish King Canute ruled all England, and fifty years later William the Conqueror of Normandy brought an end to the Anglo-Saxon period at the battle of Hastings. Converted to Christianity by Pope Gregory the Great's missionary, St. Augustine of Canterbury, at the end of the sixth century, England, like Ireland, became a center of learning in Dark Age Europe with the Venerable Bede (ca. 673–735), Alcuin, and King Alfred himself.

As in Ireland, vernacular literature also developed far earlier and more vigorously than on the continent. Heroic poetry in the four-beat alliterative line grew out of an oral tradition that culminated in the epic of Beowulf. *But religious poetry is no less prominent and important. Our first poem, the earliest in Old English, is the* **"Hymn of Caedmon,"** *a seventh-century poet whose "delightful and moving poetry in his own English language" (according to Bede's* Ecclesiastical History, *or* History of the English Church and People, *completed in 731, as translated by Leo Sherley-Price [1955]) "stirred the hearts of many folk to despise the world and aspire to heavenly things." As Bede tells the story, Caedmon had withdrawn from a feast because he had no talent for singing and had fallen asleep. A man appeared in a dream, called him by name, and bade him "Sing about the Creation of all things"; then "Caedmon immediately began to sing verses in praise of God the Creator that he had never heard before." Bede gives the hymn in Latin, but the Old English text is transcribed in various manuscripts of his* History *and included in its vernacular translation. Brief as it is, it splendidly evokes "the world in its beauty" as created by God. (The translation is from A. S. Cook and C. B. Tinker, eds.,* Select Translations from Old English Poetry *[1926].)*

The second selection consists of 146 (out of 677) lines from **The Phoenix** *(traditionally if wrongly ascribed to the ninth-century religious poet Cynewulf), preserved in the late-tenth-century* Exeter Book, *a poetic miscellany probably transcribed from a lost ninth-century manuscript. (The translation is from J. Duncan Spaeth,* Old English Poetry *[1921]; for the text see N. F. Blake, ed.,* The Phoenix *[1964; 2nd ed. 1990].) The phoenix—a Greek name for the fabulous* bennu *bird of Egyptian Heliopolis, site of the temple of the Sun-god Ra—is first mentioned, in a fragment of Hesiod quoted by Plutarch* (Obsolescence of Oracles *415c), as outliving nine ravens. Herodotus, in the fifth century B.C., narrates the story as follows (II.73, trans.*

*de Sélincourt): "Another sacred bird is the phoenix; I have not seen a phoenix myself, except in paintings, for it is very rare and visits the country (so at least they say in Heliopolis) only at intervals of 500 years, on the occasion of the death of the parent bird. To judge by the paintings, its plumage is partly golden, partly red, and in shape and size it is exactly like an eagle. There is a story about the phoenix; it brings its parent in a lump of myrrh all the way from Arabia and buries the body in the temple of the Sun. To perform this feat, the bird first shapes some myrrh into a sort of egg as big as it finds, by testing, that it can carry; then it hollows the lump out, puts its father inside and smears some more myrrh over the hole. The egg-shaped lump is then just the same weight as it was originally. Finally it is carried by the bird to the temple of the Sun in Egypt. Such, at least, is the story." The tale was repeated, with variations but seldom with Herodotus's skepticism, by many classical authors—including Ovid (*Metamorphoses* XV.391–407); Pliny (X.2), who also withheld assent to its truth; Aelian, for whom the phoenix was "a pupil of all-wise Nature" (Chapter 12 above); and Claudian (*Carmina Minora* 27)—and by countless medieval successors, to say nothing of writers from Arabia to China. Though it has surely never existed on earth, it was as much a part of the "natural world" for the Middle Ages as the eagle or elephant.*

*The implications of the ancient myth as a symbol of Christian resurrection were evident as early as St. Clement of Rome in the late first century A.D.; in the Greek *Physiologus* (Chapter 13 above), identification with the resurrected Christ is explicit. The Old English poet's main source for the first part of his poem was a 170-line Latin poem in elegiac couplets ascribed to Lactantius, the "De Ave Phoenice." (For text and translation, see Mary Cletus Fitzpatrick's edition [1933].) Lactantius's poem too portrays the land of the phoenix as a paradise free from disease and death, in language reminiscent of classical descriptions of the Golden Age and their Christian counterparts; but the English poem's account of the "happy land far in the east" abounding in "ever fresh and fragrant fruit" is far richer and more detailed, and its later parts associate rebirth of the phoenix not only with bodily resurrection but also with fruitfulness of the land—"the germ of life in the corn," when the sun awakens "the wealth of the world."*

Finally, the first 67 lines of the 124-line **Seafarer** *(also from the* Exeter Book*) vividly portray a life of hardship on the wintry seas that has its own stern rewards both in allaying the hunger of longing for its austere beauty and in preparing through suffering for the far greater "joy in the Lord" that will follow "this dead life on loan here on land"; the untranslated remainder of the poem is an expansion on these Christian sentiments. The text used for the translation is that of James W. Bright's* Anglo-Saxon Reader, *rev. James R. Hulbert (1935), in the normalized orthography of Francis P. Magoun's* The Anglo-Saxon Poems *(2nd printing, 1961).*

The Hymn of Caedmon,
translated by Albert S. Cook
Bede's Ecclesiastical History *IV.24*

Now must we hymn the Master of heaven,
The might of the Maker, the deeds of the Father,
The thought of His heart. He, Lord everlasting,
Established of old the source of all wonders:
Creator all-holy, He hung the bright heaven, 5
A roof high upreared, o'er the children of men;
The King of mankind then created for mortals
The world in its beauty, the earth spread before them,
He, Lord everlasting, omnipotent God.

FROM The Phoenix, *attributed to Cynewulf, translated by J. Duncan Spaeth*

Lo, I have heard of a happy land
Far in the East, of a fair country,
Happier, fairer, than earth-folk know.
Far remote the mighty Creator
Planted this realm, where few may reach it; 5
Sinful mortals seek it in vain.
Blest are those fields, abloom with the fragrance
Of all sweet odors that earth exhales.
Peerless the island, peerless her maker,
Glorious the Lord who laid her foundations. . . . 10

'Tis a region calm of sunny groves
Woodlands glad, whose wondrous trees
Stand fair and fresh in unfading hues, 35
Goodly and green at God's behest.
Ever the same, summer and winter,
In living green those groves are clad,
Laden with fruit. No leaf shall waste
No branch be blackened with blast of lightning 40
Till doomsday come. When the deluge swept
With might of waters the world of men,
And the flood o'erwhelmed the whole of earth,
This isle withstood the storm of billows
Serene and steadfast 'mid raging seas 45
Spotless and pure by the power of God.
Thus blest it abides till the bale-fire come,
The day of doom when death's dark chambers,
Abodes of shade, shall be broken asunder. . . .
No sleet or snow assails that isle; 60
No pelting rains pour from the clouds,
Lashed by the gale; but living streams
Wondrously gush from woodland springs.
Lapping the earth with limpid ripples.
Each month of the year in the midmost grove 65
The winsome waters well sea-cold
From the mossy turf; at the time appointed
Wind through the wood in wandering streams.
For God decreed that the joy of waters
Should twelve times play through that land of plenty. 70
Thick hangs the fruit in the forest-glades;
The shining clusters never decay,
The holy burden of the bending trees.
No withered blooms are wafted down;
No leaves are shed; but laden boughs 75
Of bounteous ever-bearing trees

Yield ever-fresh and fragrant fruit.
Green are the groves on the grassy sward
Decked and adorned by the deed of God,
In beauty unwasting. Through the woodlands bright 80
A holy fragrance floats and hovers.
Changeless through ages the isle shall remain,
Till He that uplifted the land at the first
Shall end his wisdom's ancient work.

A glorious bird guardeth this grove, 85
Noble in flight, Phoenix by name.
Alone in the land he liveth, a hermit,
Proudly dwelleth, proof against death,
In this wood of delight, while the world endures.
'Tis said he watches the way of the sun, 90
Eager to greet the candle of God,
The gleaming gem, and joyously waits
Till the day-star come at dawn from the east,
Shining bright o'er the billowy sea,
First of lights by the Father created, 95
Glorious sign of God. When the stars are gone,
Dipped in the waves of the western sea,
Or hid in the dawn, and dusky night
Darkling departs, then poised for flight
The strong-winged Phoenix scans the ocean, 100
Sky and wave, and waits the time
When the glorious light shall glide from the east
And radiant rise o'er the rounding sea.
This peerless bird abides by the fountain,
Haunting ever the hallowed streams. . . . 105

Tall in the grove a great tree towers,
Firmly rooted 'neath heaven's roof,
Named from the bird, and known as the Phoenix.
The Maker of man, the mighty Creator, 175
Hath granted a glorious growth to this tree.
I have heard that it passes in height by far
The tallest tree that towers on earth;
Its foliage fair shall flourish and thrive;
Blight shall not touch it, its branches shall wave, 180
Winsome and green while the world endures.

When winds are laid and weather is calm,
The lamp of heaven shines holy and pure;
Clouds are scattered and skies are clear;
The mighty surge of the sea is stilled; 185
Storms are asleep and warm in the south
Gleams the sun and gladdens the world.
Then begins the bird to build in the branches,

To furnish his nest for his hour of need,
When his spirit's fervor shall urge him to change 190
The years of his age, restoring his youth,
And renewing his life. From near and far
He gathers together the goodliest herbs;
Blossoms and leaves he brings from the wood;
Fills with fragrance his forest-abode; 195
Culls each sweet that the King of glory,
The Father, created o'er earth's wide realm,
To charm and delight the children of men. . . .

When the beautiful nest is burnt to a cinder,
And body and bones of the bird are crumbled,
In the waning glow of the whitening embers
A ball is found, in the bed of ashes 230
Rolled together, round like an apple;
Out of it comes a curious creature,
Wondrous in hue, as though it were hatched,
Shining bright, from the shell of an egg.
It grows in the shade to the shape of an eaglet, 235
A nestling fair, then further increases,
Lustily thriving, larger still,
Equaling soon an eagle in size.
At length he is fledged with feathers gay,
Bright as of old with beauteous plumes, 240
His body renewed by the birth of fire,
Taint of evil all taken away.
Like as when men in the month of harvest
Gather for food the fruits of the earth;
Garner their crops 'gainst coming of winter; 245
Shelter and shield them from showers and storms,
Laying in stores and living in plenty,
While roaring winter rages amain,
And covers the fields with coat of snow;
Out of those winter-stores, wealth abounding 250
Shall come through the germ of life in the corn,
Cleanly sown as a seed in the spring.
When the sun returns, the token of life,
And his warm rays waken the wealth of the world,
Sprouteth afresh each fruit of the earth, 255
Each in its own kind quickened and kindled
To brighten the field. So the Phoenix old
After many years his youth renews;
Is girt again with a garment of flesh. . . .

When the Savior Christ on the souls of the blest
Shines from on high, toward heaven's gate 590
They mount, like beautiful birds, to meet him;
Glad is the song and glorious the shape

Of the spirits-elect in that land of joy,
Where envy and malice no more shall touch them:
For ever and ever from evil free, 595
They live in peace, appareled in light,
Girt with glory, by God defended,
Like the Phoenix wondrous. The works of each
Sun-like gleam and glow in splendor,
Bright before the face of the Lord, 600
In clear abodes of blessed calm. . . .

FROM The Seafarer,
translated by Robert M. Torrance

Of myself I am able   to utter a true song,
to say of my journeys   how often I suffered
toilsome days   and times of hard travail,
bitterly bearing   grief in my breast:
made trial of countless   troubles on shipboard, 5
fierce-pounding waves,   when frequently posted
on anxious night-watches,   awake at the bowsprit
bumping past cliffs.   By fierce cold afflicted,
often my feet   were frozen by frost,
fettered by cold: sorrowful care sighed 10
hot round my heart,   and hunger within
tore my sea-weary mind.
A man cannot know
when living a carefree   life on the land
how, wretched and weary,   I sailed for a winter
the ice-cold sea,   the path of the exile, 15
cut off from kinsmen,
hung with icicles;   hail fell in showers.
There I heard nothing   but heavy seas pounding,
icy waves swirling.   Sometimes a swan's song
gave me gladness: the cry of a gannet, 20
the sound of sea-birds   instead of men's laughter,
the singing of mews   instead of mead-drinking.
There storms beat on stone cliffs,   there the tern answered them,
icy-feathered;   the eagle screamed often,
dewy-feathered;   no friend or protector 25
could have consoled   my desolate spirit.

One who enjoys life   little believes—
finding in cities   few of life's sorrows,
proud and wine-sodden—   how often I wearily
had to pass over   the pathway of ocean. 30
Night-shadows darkened,   snow from the north dropped,

frost held the ground fast,   hail fell to earth,
coldest of grains.
                        Thus my thoughts clamor,
importune my heart   to explore for itself
the high-surging streams   and tumultuous salt-waves;   35
my mind's desire ever   urgently presses
my spirit to venture   forth now and visit
some faraway homeland   of foreign inhabitants.
For no man on earth   is so haughtily minded,
so giving of gifts,   so young and aggressive,   40
so daring in feats,   to his lord so faithful,
as not to feel sorrow   about his seafaring,
concern what the Lord   now wishes to send him.
He thinks not of harps,   of ring-giving in halls,
of delight in a woman   or hope in this world,   45
nor of anything other   than upsurging waves,
yet never stops yearning   in his strife with the waters.

Groves are in blossom,   towns growing beautiful,
fields turning fair,   refreshing the world;
everything urges   a man who is eager   50
at heart to go journeying   just as he chooses,
faring afar off   over the flood-ways.
Just so the cuckoo's   sad voice advises:
summertime's ward sings,   foreboding sorrow,
bitter in breast-trove.   None who is blessed   55
by fortune surmises   what some men suffer
following tracks   of far-distant exile!

Thus my heart roams now,   leaving my rib-cage—
mind and soul   in the midst of the sea-flood—
over the whale's realm   it ranges afar   60
through all the earth's quarters,   then comes back to me,
hungry with longing: the lone-flier howls,
irresistibly whetting   my heart for the whale-path
over wide sea-lanes.
                        Thus joy in the Lord
to me is more living   than this dead life   65
on loan here on land: for I do not believe
riches of earth   will be lasting forever. . . .

Applaudamus Igitur Rerum Novitati

Latin Lyrics of the High Middle Ages

Latin poetry of the High Middle Ages, beginning in the late eleventh century, was continuous in important ways with that of the Carolingian revival, just as the latter had been continuous with the poetry of late antiquity. Like the poets of that earlier time, and like the theologians of their own, the largely anonymous lyric poets of the age recognized no national boundaries. Latin remained the lingua franca of these footloose clerics, but their Latin was far more colloquial than that of the self-consciously classicizing Carolingians. Although poetry continued to be written in classical meters, as by Bernard Silvestris and Alan of Lille (Chapter 15 below), rhymed accentual stanzas, closely akin to emerging vernacular verse forms, now dominated lyric poetry. "Thus the Latin poetry of the twelfth century," Haskins remarks, "was far more than a mere revival of ancient modes and subjects; it was a manifold expression of the vigorous and many-sided life of the age, an age of romance as well as an age of religion. This very variety, however, is a sign of the impending decline of Latin."

The linguistic and prosodic innovation reflects the widespread sense of renewal—intellectual, physical, and poetic—that marks this period. The celebration of newness—"applaudamus igitur rerum novitati" ("let us therefore applaud the newness of things"), the poet of our fourth poem sings—finds voice in an explosion of paeans to spring. No doubt there is a risk of exaggerating the novelty, yet the claim can plausibly be made, as by Gurevich, that " 'rehabilitation' of the world and of nature does not really begin until the twelfth century." At this time, "interest in the study and the elucidation of nature grew steadily. But this interest was not in nature as such*—nature is not independent, but a creation of God whose glory it declares." Under the impact of the Platonic and Aristotelian revivals, nature becomes more sharply delineated in the poetry and art of the time. "The depiction of foliage and fruit in Gothic sculpture is so exact," Haskins notes, quoting Emile Mâle, "that modern naturalists have identified a large number of the originals among the flora of modern France: plantain, arum, buttercup, fern, clover, coladine, hepatica, columbine, cress, parsley, strawberry, ivy, snapdragon, the leaf of the oak, and the flower of the broom—spring flowers and buds for the most part, so that 'all the spring delights in the Middle Ages live again in the work' of these artists of an epoch often considered indifferent to natural beauty."*

In the poems, it is not so much precise depiction as emotional response to the natural world that is paramount. The first and earliest, ***"Softly the West Wind Blows"*** *("Levis exsurgit zephirus"; text in* Penguin Book of Latin Verse *and Waddell), probably from the eleventh century, is transitional. As Gerhardt B. Ladner observes,*[17] *it "is full of that vernal imagery which was later to serve as a favorite and characteristic expression of the consciousness of natural renewal." In it, "the speaker, a woman, contrasts her sad state of mind with the peaceful life of the animals in spring, with the joyful songs of the birds among the blossoming trees."*

The next four poems— ***"The Earth Lies Open-Breasted"*** *("Terra iam pandit gremium"),* ***"Now Are the Meadows Laughing"*** *("Iamiam rident prata"), and* ***"Happily Returning"*** *("Laetabundus rediit"; texts in Waddell), and the humorous* ***Pastoral*** *("Exiit diliculo"; text in* Penguin Book*), in which the peasant girl propositions the scholar rather than the reverse—are all from the most famous manuscript of goliardic poetry, the* Carmina Burana. *This collection,*

[17]"Terms and Ideas of Renewal," in Benson and Constable, eds., *Renaissance and Renewal in the Twelfth Century* (1982).

named for the monastery of Benedictbeuern in Upper Bavaria where it was copied in the thirteenth century, in Waddell's words "seems to have lived a kind of stowaway existence, hidden to save it from the censor's gall. . . . These poets are young," she writes, "as Keats and Shelley and Swinburne were never young, with the youth of wavering branches and running water. They do not look before and after, they make light of frozen thawings and of ruined springs."

Finally, **"Down from the Branches Fall the Leaves"** *("De ramis cadunt folia"; text in Waddell and* Penguin Book*), from a thirteenth-century manuscript, ends these incomparable songs of spring with a rare love song of winter.*

Softly the West Wind Blows,
translated by Robert M. Torrance

Softly the west wind blows,
warmly the sunshine glows;
earth is already showing
breasts sweetly overflowing.

Clad in bright crimson, spring 5
with jewels is glistening,
sprinkling the ground with flowers,
with leaves, the blossoming bowers.

Quadrupeds strew their lairs,
sweet birds build nests by pairs 10
while flowery boughs are ringing
with their exultant singing.

I see it all so clear!
How piercingly I hear!
Yet all this paradise 15
burdens my heart with sighs.

Pondering thus, alone
and pale, I softly moan;
rising from bended knee
I neither hear nor see. 20

You, at least, need not wait,
fair Spring, to celebrate
flower and grass and leaf,
leaving *me* to my grief.

The Earth Lies Open-Breasted,
translated by Helen Waddell

The earth lies open-breasted
In gentleness of spring,
Who lay so close and frozen
In winter's blustering.
The northern winds are quiet, 5

The west wind winnowing,
In all this sweet renewing
How shall a man not sing?

Now go the young men singing,
And singing every bird, 10
Harder is he than iron
Whom Beauty hath not stirred.
And colder than the rocks is he
Who is not set on fire,
When cloudless are our spirits, 15
Serene and still the air.

Behold, all things are springing
With life come from the dead,
The cold that wrought for evil
Is routed now and fled. 20
The lovely earth hath brought to birth
All flowers, all fragrancy.
Cato himself would soften
At such sweet instancy.

The woods are green with branches 25
And sweet with nightingales,
With gold and blue and scarlet
All flowered are the dales.
Sweet it is to wander
In a place of trees, 30
Sweeter to pluck roses
And the fleur-de-lys,
But dalliance with a lovely lass
Far surpasseth these.

And yet when all men's spirits 35
Are dreaming on delight,
My heart is heavy in me,
And troubled at her sight.
If she for whom I travail
Should still be cold to me, 40
The birds sing unavailing,
'Tis winter still for me.

Now Are the Meadows Laughing,
translated by Robert M. Torrance

Now are the meadows laughing,
and now the virgins play
merrily, while the laughing earth
joins in their holiday.
Now is summertime here, 5
brightening all the world with flowery cheer.

Forests again are turning
  green, and bushes sprout,
fierce winter having vanished:
  look happily about, 10
youths, and rejoice in flowers
as love allures you to the virgins' bowers.

Let us make war together
  with Venus at our side,
banishing all sadness, 15
  hearts tenderly allied:
and may soft whispers, bright
faces, and hope and love bring us delight.

Happily Returning,
translated by Robert M. Torrance

Happily returning
sings the feathered choir;
spring with joyous yearning
sets young hearts afire,
bringing new delight; 5
everything turns bright,
dazzled by Phoebus' rays,
redolent with spring;
Flora, newly flowering,
renews her sparkling gaze. 10

Clammy winter shrinks afar
from brightly smiling Jove;
ever higher climbs Sun's car
through clear skies above,
melting frosts away: 15
warming day by day,
balmy climes return.
Just as cold departs,
Venus within our hearts
begins again to burn. 20

Dryads now are basking
under shady trees,
Oreads relaxing
gloriously at ease;
Satyrs, with merry dance, 25
choiring songs, advance
through Tempe's lovely vales;
Philomel, caroling
along, recalls a spring
sweeter to nightingales. 30

Freed from exile and dearth,
much-desired summertime,
painting the lap of earth,
now renovates his prime;
snugly housed in thickets, 35
softly chirping crickets
rhapsodically sing,
till, with jubilant sound
tunefully whistling round,
forests joyfully ring. 40

Come, let us celebrate
things delightfully new!
Loved by a loving mate,
all our prayers will come true
through Venus's great might, 45
whose altars flicker bright
with fragrant flowers' savors.
But miserably lost
is one who, freed from frost,
despondently still labors! 50

Pastoral,
translated by Robert Torrance

With dawn's first ray
a buxom hick
starts on her way
with flock and stick.

Sheep with her staff 5
this country lass
drives, heifer, calf,
goat, kid, and ass.

Stretched at his ease
a scholar lay. 10
"Sir, would it please
you come and play?"

Down from the Branches Fall the Leaves,
translated by Helen Waddell

Down from the branches fall the leaves,
A wanness comes on all the trees,
The summer's done;
And into his last house in heaven
Now goes the sun. 5

Sharp frost destroys the tender sprays,
Birds are a-cold in these short days.
    The nightingale
Is grieving that the fire of heaven
    Is now grown pale. 10

The swollen river rushes on
Past meadows whence the green has gone,
    The golden sun
Has fled our world. Snow falls by day,
The nights are numb. 15

About me all the world is stark,
And I am burning; in my heart
    There is a fire,
A living flame in me, the maid
    Of my desire. 20

Her kisses, fuel of my fire,
Her tender touches, flaming higher.
    The light of light
Dwells in her eyes: divinity
    Is in her sight. 25

Greek fire can be extinguishèd
By bitter wine; my fire is fed
    On other meat.
Yea, even the bitterness of love
    Is bitter-sweet. 30

In Praise of Created Things

The Mirror of Perfection, Section XII: Of His Love for Creatures and of Creatures for Him, translated by Robert Steele (prose) and Robert M. Torrance (verse)

The wandering goliards, with their ebullient celebration of wine, women, and song, and their audacious mockery of conventional pieties, represent one pole of the medieval love of nature. "Voluptatis avidus magis quam salutis," the Archpoet of Cologne brashly declares, "mortuus in anima, curam gero cutis": "more greedy for pleasure than for salvation, dead in my soul, I take care of my skin!" St. Francis (1182–1226), founder of the first order of wandering friars, embodies the other, in which love of nature is not in conflict with devotion to God but is among its most intense expressions. Francis, most revered of medieval saints, was born Giovanni di Bernardone in the Umbrian town of Assisi, but was nicknamed Francesco (the Frenchman) because of his rich merchant father's travels to France. As a boy he dreamed of martial glory, but after imprisonment and illness had ended his career as a soldier, he turned toward the service of God. He exchanged clothes with a beggar (his legend relates) on a pilgrimage to Rome; in 1208, taking

Christ's charge to leave everything for His sake as a personal call, he founded the band of followers that became the Franciscan order, to which Innocent III gave his blessing in 1210; in 1212 St. Clare founded a similar order, the Poor Clares, for nuns. After travels to Dalmatia, France, and Spain, he set out in 1219 on a pilgrimage to Egypt (where he attempted to convert the Sultan) and the Holy Land, returning to Italy to help resolve dissensions in his order. In 1224 he is said to have received the stigmata, imprinting the wounds of Christ's passion on his own body; in 1228, two years after his death, he was canonized by Pope Gregory IX. His example of personal humility, love of poverty (he was known as "il poverello," the little pauper), and joyous celebration of God rapidly gained him followers throughout Europe, and his order, along with that founded a few years later by St. Dominic of Spain, dominated much of the spiritual and intellectual life of the High Middle Ages.

Early lives were written by Thomas of Celano and St. Bonaventura. In about 1318 the anonymous Speculum Perfectionis (Mirror of Perfection) *was composed in Latin by adherents of the "Spirituals," the more uncompromising of his followers. Somewhat later, Brother Ugolino di Monte Santa Maria composed the* Actus Beati Francisci (Acts of the Blessed Francis), *whose anonymous Italian translation, the* Fioretti (Little Flowers), *spread further still the legend of this saint who preached to the birds and fishes and even converted the fierce wolf of Gubbio. The following selections from the* Speculum *amply attest to Francis's ardent devotion to the fellow creatures of God's world, his brothers and sisters, for whose creation, in his "Song of Brother Sun," or "Canticle of the Creatures"—whether or not he actually composed it, in his final illness, as we now have it—he gives joyous praise to God.*

The translation of the prose passages by Robert Steele is from The Little Flowers and the Life of St. Francis, with The Mirror of Perfection, *ed. T. Okey (1910); the Italian text of the Song is from* The Penguin Book of Italian Verse, *ed. George Kay (1958).*

Chapter 113: And Firstly, of the Love Which He Especially Had for the Birds Which Are Called Larks

Blessed Francis, wholly wrapped up in the love of God, discerned perfectly the goodness of God not only in his own soul, now adorned with the perfection of virtue, but in every creature. On account of which he had a singular and intimate love of creatures, especially of those in which was figured anything pertaining to God or the Order. Whence above all other birds he loved a certain little bird which is called the lark, or by the people, the cowled lark. And he used to say of it, "Sister Lark hath a cowl like a Religious; and she is a humble bird, because she goes willingly by the road to find there any food. And if she comes upon it in foulness, she draws it out and eats it. But flying she praises God very sweetly like a good Religious, despising earthly things, whose conversation is always in the heavens, and whose intent is always to the praise of God. Her clothes are like to the earth (that is her feathers), and she gives an example to Religious that they should not have delicate and odored garments, but vile in price and color, as earth is viler than the other elements." And because he perceived this in them, he looked on them most willingly. Therefore it pleased the Lord, that these most holy little birds should show some sign of affection towards him in the hour of his death. For late in the Sabbath day, after vespers, before the night in which he passed away to the Lord, a great multitude of that kind of birds called larks came on the roof of the house where he was lying; and flying about, made a wheel like a circle round the roof, and sweetly singing, seemed likewise to praise the LORD.

Chapter 118: Of His Love for Water and Stones and Wood and Flowers

After fire, he most singularly loved water, by which is figured holy penitence and tribulation, whereby the filth of the soul is washed away, and because the first ablution of the soul is by the water of baptism. Whence, when he washed his hands, he used to choose such a place that the water which fell should not be trodden by his feet; when he would walk over stones, moreover, he used to walk with great fear and reverence, for the love of Him Who is called "The Rock."[18] Whence when he used to say that verse of the Psalm [40:2], *Thou didst exalt me on a rock*, he used to say out of great reverence and devotion, "Under the foot of the rock hast thou exalted me." He used also to say to the friar who made ready the wood for the fire, that he should never cut down a whole tree; but so that always some part of a tree should remain whole for the love of Him Who did work out our salvation on the wood of the cross. Likewise he used to say to the friar who did the garden, not to till the whole ground for pot-herbs; but to leave some part of it to produce green herbs, which in their time should produce flowers for the friars, for the love of Him Who is called the "flower of the field" and "the lily of the valley." Nay, he used to say to that brother gardener that he ought always to make a fair pleasaunce in some part of the garden; setting and planting there all sweet-smelling herbs and all herbs which bring forth fair flowers, that in their time they might call them that looked upon those herbs and flowers to the praise of God. For every creature cries aloud, "God made me for thee, O man!" Whence we who were with him used to see him rejoice, within and without, as it were, in all things created; so that touching or seeing them his spirit seemed to be not on earth but in heaven. And by reason of the many consolations which he used to have in things created, a little before his death he composed certain Praises of the Lord for His creatures, to incite the hearts of those who should hear them to the praise of God, and that the Lord Himself might be praised by men in His creatures.

Chapter 119: How He Used to Commend the Sun and Fire Above All Other Created Things

Above all other creatures wanting reason, he loved the sun and fire with most affection. For he was wont to say, "In the morning when the sun rises, every man ought to praise God, Who created it for our use, because through it our eyes are enlightened by day. Then in the even when it becomes night, every man ought to give praise on account of Brother Fire, by which our eyes are enlightened by night; for we be all as it were blind, and the Lord by these two, our brothers, doth enlighten our eyes. And therefore we ought specially to praise the Creator Himself for these and the other creatures which we daily use." The which he himself always did to the day of his death, nay, when he was struck down with great infirmity he begun to sing the Praises of the Lord which he had made concerning created things, and afterwards he made his fellows sing, so that in considering the praise of the Lord, he might forget the bitterness of his pains and infirmities. And because he deemed and said that the sun is fairer than other created things, and is more often likened to our Lord, and that in Scripture the Lord Himself is called "the Sun of Righteousness" [Malachi 4:2], therefore giving that name to those Praises which he had made of the creatures of the Lord, what time the Lord did certify him of His kingdom, he called them "The Song of Brother Sun."

[18]St. Peter (Matthew 16:18: "You are Peter, and on this rock [Greek *petra*] I will build my church").

Chapter 120: This Is the Praise of Created Things, Which He Made When the Lord Certified Him of His Kingdom

Most High, omnipotent, good Lord,
yours are the praise, the glory and the honor and every blessing.
To you alone, Most High, they belong,
and no man is worthy to mention you.

Be praised, my Lord, with all your creatures, 5
especially master brother sun,
who makes it day, and through him you give us light;
and he is fair and radiant with great splendor;
from you, Most High, he takes his signification.

Be praised, my Lord, for sister moon and the stars; 10
in heaven you have made them clear, precious, and beautiful.

Be praised, my Lord, for brother wind,
and for air both cloudy and clear, and in every weather,
through which you give sustenance to your creatures.

Be praised, my Lord, for sister water, 15
who is very useful and humble and precious and chaste.

Be praised, my Lord, for brother fire,
by whom you illumine the night,
and he is beautiful and joyous and vigorous and strong.

Be praised, my Lord, for sister our mother earth, 20
who sustains and governs us,
and brings forth various fruits, with colored flowers and grass.

Be praised, my Lord, for those who for love of you forgive,
and who endure sickness and tribulation.
Blessed are they who will endure in peace, 25
for by you, Most High, they shall be crowned.

Be praised, my Lord, for sister our bodily death,
from whom no living man can escape.
Woe to those who shall die in mortal sin;
blessed those who shall be found in accord with your most holy will, 30
for the second death shall do them no harm.

Give praise and blessings and thanks to my Lord,
and serve him with great humility.

Le Temps a laissié son manteau

Medieval French Lyrics

With a few important exceptions, notably Ireland and Anglo-Saxon England, Latin had been the only language written *in most of Europe for much of the Middle Ages, and those who wrote and read it were almost all clerics. Beginning in late-eleventh-century Provence (in what is now southern France), however, and flourishing there throughout the twelfth and into the thirteenth century—roughly when the Albigensian heresy was at its height—the songs of the troubadours commending "courtly love" of a high-born lady gave rise to immensely influential forms of vernacular poetry that spread to northern France, Germany, and beyond. But whereas Latin lyrics of the period were highly colloquial, even racy, and were often anonymous products of impoverished wandering scholars, the troubadours were prominent members of aristocratic circles. Their poems were almost always—after the earliest Provençal poet, Guillaume IX, count of Poitiers and duke of Aquitaine, who indulged a nobleman's libertine bawdry—highly refined and artificial. Time and again the beauties of nature—above all, the birds and flowers of spring—appear in these poems and in their French and German successors, but mainly to set the scene for a poem celebrating love not of nature but of the poet's lady.*

The troubadour Bernard de Ventadorn (ca. 1150–ca. 1180; texts in Hill and Bergin, eds., Anthology of the Provençal Troubadours *[1941]) thus begins one song ("Can l'erba fresch' e-lh folha par"):*

When the fresh grass and leaves appear
and on the branch the flower blooms,
and, with voice lifted high and clear,
the nightingale his song resumes,
joy in these and myself I take,
but mostly for my lady's sake:
circled by joys as by a wall,
I find this joy surpasses all!

Another ("Can vei la lauzeta mover") begins as follows:

When in the sun's bright rays I see
the lark rejoicing in her wings,
letting herself obliviously
plummet, her heart so sweetly sings,
alas! such envy have I felt
of all whose raptures I behold,
surely, it seems, my heart must melt
at once with yearnings unconsoled!

In similar fashion begin many poems of the trouvères of northern France, such as the following love song of the late-twelfth-century Châtelain de Coucy ("Commencement de douce saison belle"; text in vol. 1 of The Penguin Book of French Verse, *ed. Brian Woledge [1961]):*

When the beginning of the lovely time
  of newborn spring I see,
and memory of love renews my rhyme
  (love I shall never flee!),
and the thrush warbles tunefully,

and waters through bright gravel chime
with coruscating glee,
then they bring back to me
her whom my heart now yearningly
desires, and shall, till life no longer be.

So do many songs by German love poets (minnesinger), *such as the following by Walther von der Vogelweide (ca. 1170–1230) ("Sô die bluomen ûz dem grase dringent"; text in Leonard Forster, ed.,* The Penguin Book of German Verse *[1957; rev. 1959]):*

When through the grass bright flowers sprang
laughingly in the sun's bright glow
early one morningtime in May,
and all the little songbirds sang
the sweetest melodies they know:
what joy could equal such a day?
Halfway to heaven this must be!
Yet I shall tell what equally
delights my eyes, declaring how
one thing has often pleased me better,
and would still, could I see her now.

Such settings, like the poems they introduce, have both a wonderful initial freshness and, before long, the monotony of overused convention; nature's beauty, like the lady's, stales from repetition and from lack of specificity or discrimination.

In other medieval French poems, nature is more than merely introductory. Of the selections that follow, the first three are variations of the popular reverdie, *defined by the* Princeton Encyclopedia of Poetry and Poetics *as a "dance poem which celebrates the coming of spring, the new green of the woods and fields, the singing of the birds, and the time of love." These anonymous popular songs appear to retain traces of "pagan" beliefs surviving since early times in the French countryside. The first,* ***"In April at Eastertide"*** *("En avril au tens pascour"; text in* Penguin Book of French Verse*), is unusual in its flowingly nonstanzaic verse form. The second,* ***"In the Sweet New Time of May"*** *("En mai au douz tens novel"), with its lighthearted refrain, and the third,* ***"Would You Wish Me Sing to You"*** *("Volez vos que je vos chant"), with its air of dreamlike fantasy, are from Claude Colleer Abbott,* Early Mediaeval French Lyrics *(1932), which contains both text and translation. The anonymous* pastourelle, ***"When Flowers Are Born in Dewy May,"*** *also from Abbott, exemplifies this popular medieval genre of male sexual fantasy (which no doubt often reflected brute reality) of love in the haystack. Here the knight's desires are fulfilled; in many other instances—in French, Provençal, and Italian, and in the* serranillas, *or mountain songs, of Spain—they are soundly rebuffed.*

Finally, this section ends with two polished poems by the aristocrat Charles d'Orléans (1349–1465), the rondeau ***"The Return of Spring"*** *("Now Time throws off his cloak again": "Le temps a laissié son manteau") and the* ballade ***"Lovely Springtime"*** *("Bien moustrez, printemps gracieux"; texts in* Penguin Book*). Even when the terrible Hundred Years' War was devastating France (Charles d'Orléans was captured at Agincourt), even when "the feeling of general insecurity . . . was further aggravated," as Johan Huizinga writes in* The Waning of the Middle Ages *(1924), "by the obsession of the coming end of the world, and by the fear of hell, of sorcerers, and of devils," a member of the royal family could still delight in the "splendors of earth" annually renewed by the jocund company of Spring's attendants.*

THREE ANONYMOUS *REVERDIES*

In April, at Eastertide,
translated by Robert M. Torrance

In April, at Eastertide,
when flowers bloom in the grass
and larks at daybreak far and wide
sing joyous tunes none can surpass
in celebration of spring, 5
I rose when morning was young
and heard a small bird sweetly sing,
perched on a tree, in its own tongue.
Then, raising up my eyes
before its song was past, 10
I saw, to my surprise,
birds coming thick and fast,
the oriole,
the nightingale,
the chaffinch, 15
and the merlin,
God! and so many other birds
whose names I know not of,
all sitting in that tree
and singing high above. 20
So I went beneath its flowering boughs
to learn of the joys of love.
I saw the god of love there ride
quietly over the mead,
and, hearing him call, went to his side 25
and became his squire indeed.
His horse was made of pleasures,
his saddle of haughty wiles,
his shield in equal measures
of kisses and of smiles; 30
his hauberk bore traces
of impassioned embraces,
and with flowers all bright
was his helmet alight:
God! his lance is of courtesy, 35
of gladiolus his sword,
his hose of sweet gentility,
of jays' beaks his spurs.
The birds were all singing one air,
though no human minstrel was anywhere! 40

In the Sweet New Time of May,
translated by Claude Colleer Abbott

In the sweet new time of May
When the meads get green and gay,
I heard from out a leafy tree
A nightingale sing merrily,
Saderala don don doon! 5
How great a boon
To sleep beside the briar tree.

And as I was pensive grown,
By the tree I sat me down,
A little time I slumbering slept 10
In the sweet singing he kept.
Saderala don don doon!
How great a boon
To sleep beside the briar tree.

When I wakened in the shade, 15
To the little bird I prayed
He might give such joy to me
That I for it should merrier be.
Saderala don don doon!
How great a boon 20
To sleep beside the briar tree.

And when I was stood upright
In my lute I found delight,
And I made the small bird wing
Before me in the mead, and sing. 25
Saderala don don doon!
How great a boon
To sleep beside the briar tree.

The nightingale he said I had
Within a little made him mad. 30
'Twas for him such bitter dole
That a villein heard the whole.
Saderala don don doon!
How great a boon
To sleep beside the briar tree. 35

Would You Wish Me Sing to You,
translated by Claude Colleer Abbott

Would you wish me sing to you
Song of love no villein knew,
        Song of many charms?
For a knight this singing made
Lying neath an olive's shade   5
        In his lady's arms.

Linen was her small camise,
White with ermine her pelisse,
        She had a silken gown.
Tiger lilies were her hose,   10
Flowers o' may her little shoes,
        Fitted tightly on.

For a girdle, tender leaves,
When the weather rained, grew green,
        Buttoned up with gold.   15
Cords of flowers swung above
Her wallet shapen all for love,
        And Love the giver bold.

On a mule she rode along,
And the mule was silver shod,   20
        Saddle gold inlaid;
On the crupper right behind,
Three rose trees stood up in line
        For to give her shade.

She went riding through the mead;   25
All the knights who met her steed
        Bowed with courtly state.
"Lady fair, whence are you sped?"
"I am the boast of France," she said,
        "Of renowned estate.   30

My father is the nightingale
Who sings within the bosky dale
        On the tallest tree.
The mermaiden my mother is,
She who sings her melodies   35
        In the deep salt sea."

"Lady, blessed was your birth,
Parentage of famous worth
        And renowned estate.
Would that God our Father dear   40
Gave you for to be my peer
        And my wedded mate."

When Flowers Are Born in Dewy May (Anonymous pastourelle*),*
translated by Claude Colleer Abbott

When flowers are born in dewy May
And the rose is sweet, at break of day,
In every tree
Birds frolic joy
So merrily 5
When I hear their employ
Nothing on earth would hold
Me back from love, the bold.

Alluring fair this maiden is
With her green eyes and laughing lips. 10
Blessed is he
Begot her so.
O, how she taketh me!
I seat me low.
"O now, my pretty maid, 15
Give me your love," I prayed.

"Knight of Champagne," she said to me,
"'Twill be long time ere you win me.
The courteous son
Of dame Marie, 20
Robin, I love
Who kirtles me.
And never does he let
Me want fine chapelet."

But when my prayer was nothing worth 25
Straightway I bedded her on earth,
Twitched up her dress
And burned the more
So white her flesh;
Taught her love's lore. 30
Nor did she say me nay
But did desire such play.

When with the shepherdess my need
Was fed, then I bestrode my steed,
And she did say 35
"By our Lady
Sir knight, I pray
Forget not me.
Always your love am I,
Often pass by." 40

TWO POEMS OF CHARLES D'ORLÉANS

The Return of Spring (Rondeau), *translated by Henry Wadsworth Longfellow*

Now Time throws off his cloak again
Of ermined frost, and wind, and rain,
And clothes him in the embroidery
Of glittering sun and clear blue sky.
With beast and bird the forest rings, 5
Each in his jargon cries or sings;
And Time throws off his cloak again
Of ermined frost, and wind, and rain.

River, and fount, and tinkling brook
Wear in their dainty livery 10
Drops of silver jewelry;
In new-made suit they merry look;
And Time throws off his cloak again
Of ermined frost, and wind, and rain.

Lovely Springtime (Ballade), *translated by Robert M. Torrance*

Lovely springtime, you demonstrate
great mastery in the work you do,
for winter burdens with its weight
hearts that delight to welcome you.
Once you come dancing into view, 5
instantly he is forced to flee
along with his malignant crew
far from your jocund company.

In winter, fields and trees grow old,
their beards becoming white with snow; 10
so dank is he, so vile and cold,
that, huddled by the hearthfire's glow
and fearing out of doors to go,
like birds in moulting time are we:
yet everything revives anew 15
thanks to your jocund company.

Wintry clouds darken dreary days,
muffling the sun in leaden skies;
now—to almighty God the praise!—
with your arrival darkness flies, 20
splendors of earth once more arise:
winter has labored fruitlessly,
for with the year's rebirth he dies,
banned from your jocund company.

Soft and Mellow Notes

Poems from Middle English and Spanish

Anglo-Norman England after the conquest of 1066 was for several centuries, in its literature, a virtual appendage of France. Only after the language we now call Middle English had evolved, in its various dialects, from the fusion of Norman French and Old English does poetry in the native tongue begin to be written again, in the thirteenth century, culminating, in the fourteenth, in major works such as the allegorical Piers Plowman, *the Arthurian* Sir Gawain and the Green Knight *(Chapter 15 below), and Chaucer's* Troilus and Criseyde *and* The Canterbury Tales.

The three anonymous lyrics that follow suggest the freshness and variety of poetic expression in the emergent English language. The first, ***"Somer Is I-Comen In"*** *(MS Harley 978), from around 1300, in the text of Robert D. Stevick, ed.,* One Hundred Middle English Lyrics *(1964), is an exultant announcement of the coming of spring ("somer" could mean spring as well as summer) as heralded not only by the cuckoo's song but by the rutting (and even the farting!) of barnyard and woodland animals. The second,* ***"Spring Comes to Town"*** *("Lenten ys come with love to toune"; translated from the text of MS Harley 2253 in G. L. Brook, ed.,* The Harley Lyrics *[4th ed. 1968]), from the early fourteenth century, likewise gives new life to the poetry of springtime by its vividness of detail; here even "worms woo each other underground"! The third,* ***"Western Wind,"*** *probably from the fifteenth century, though printed only at the beginning of the sixteenth, unites the wind and rain, Christ and the poet's love in four lines of passionate intensity.*

In Spain, the energies of the Christian peoples were devoted to the centuries-long reconquest of the peninsula from the Moors (which ended only in 1492)—a feat celebrated in the epic Poem of the Cid *and in many ballads of the* Romancero. ***"The Praise of Spring"*** *(text and translation in Eleanor L. Turnbull, ed.,* Ten Centuries of Spanish Poetry *[1955]) is from* Milagros de nuestra Señora *(*Miracles of Our Lady, *stanzas 2–7) by the first Spanish poet known by name, Gonzalo de Berceo, in the late twelfth to mid thirteenth century. It inaugurates a poem dedicated to the Virgin with yet another evocation of the world revitalized by spring. Finally, the poignant anonymous lyric* ***"In the Wind Are Murmuring"*** *("Con el viento murmuran"), dating from perhaps the fifteenth or sixteenth century, portrays a young girl's dreamlike vision of life restored by the murmuring of the leaves.*

THREE ANONYMOUS MIDDLE ENGLISH POEMS

Somer Is I-Comen In

Somer is i-comen in,
Loude syng cuckow!
Groweth seed and bloweth[19] meed
And springeth the wode[20] now.
Syng cuckow! 5

[19]"Blows," "blooms." [20]"Wood," "forest."

Ewe bleteth after lamb,
Loweth after calve cow;
Bullock sterteth,[21] bukke[22] farteth,—
Myrie syng cuckow!
Cuckow! Cuckow! 10
Wel syngest thou cuckow:

Ne swik[23] thou nevere now.
Syng cuckow, now, syng cuckow!
Syng cuckow, syng cuckow, now!

Spring Comes to Town,
translated by Robert M. Torrance

To town with Love Spring's come along,
with blossoms and with birds' sweet song
that bliss and rapture brings:
daisies abloom throughout the dales,
tuneful refrains of nightingales, 5
melodies each bird sings.
Ever the song-thrush warbles so;
far away now is winter's woe;
fragrantly woodruff springs.
Wondrously many songbirds these, 10
chirping with such delightful ease
that all the forest rings.

The rose puts on her rosiness;
leaves brighten up the trees they dress,
growing with all their will. 15
The moon is shining radiantly,
the lily is a joy to see,
fennel, and sweet chervil.
Wantonly now the wild drakes woo;
everywhere beasts their mates pursue, 20
as streams flow deep and still.
Passionate men lament their woes:
I know that I am one of those,
for love has served me ill.

Moonbeams proliferate their light; 25
the sun shines gloriously bright
when songbirds cast their spell.
Fresh dewdrops dampen vale and hill;
from beasts clandestine murmurs spill
that ardent yearnings tell. 30

[21]"Starts," "leaps." [22]"Buck," "stag." [23]"Cease," "stop."

Worms woo each other underground.
Women preen and parade around,
all pleases them so well:
if none will take delight in me,
forgoing pleasure, I shall flee 35
and in the forest dwell.

Western Wind

Western wind, when will thou blow,
The small rain down can rain?
Christ, if my love were in my arms,
And I in my bed again!

TWO MEDIEVAL SPANISH POEMS

The Praise of Spring, by Gonzalo de Berceo, translated by Henry Wadsworth Longfellow

I, Gonzalo de Berceo, in the gentle summertide,
Wending upon a pilgrimage, came to a meadow's side;
All green was it and beautiful, with flowers far and wide,—
A pleasant spot, I ween, wherein the traveler might abide.

Flowers with the sweetest odors filled all the sunny air, 5
And not alone refreshed the sense, but stole the mind from care;
On every side a fountain gushed, whose waters pure and fair,
Ice-cold beneath the summer sun, but warm in winter were.

There on the thick and shadowy trees, amid the foliage green,
Were the fig and the pomegranate, the pear and apple seen; 10
And other fruits of various kinds, the tufted leaves between,
None were unpleasant to the taste and none decayed, I ween.

The verdure of the meadow green, the odor of the flowers,
The grateful shadows of the trees, tempered with fragrant showers,
Refreshed me in the burning heat of the sultry noontide hours: 15
Oh, one might live upon the balm and fragrance of those bowers!

Ne'er had I found on earth a spot that had such power to please,
Such shadows from the summer sun, such odors on the breeze:
I threw my mantle on the ground, that I might rest at ease,
And stretched upon the greensward lay in the shadow of the trees. 20

There soft reclining in the shade, all cares beside me flung,
I heard the soft and mellow notes that through the woodland rung:
Ear never listened to a strain, for instrument or tongue,
So mellow and harmonious as the songs above me sung.

In the Wind Are Murmuring (Anonymous),
translated by Robert M. Torrance

In the wind are murmuring,
mother, the leaves;
in their shade, to the sound of
their rustling, I sleep.

A mild wind is blowing 5
delightfully soft
and dreamily moving
the ship of my thoughts;
such bliss has it brought
that it seems to have given 10
premonitions of heaven
enticingly near:
in their shade, to the sound of
their rustling, I sleep.

On occasion, whenever 15
I wake among flowers,
I scarcely remember
my numberless sorrows,
soon wholly forgotten
as I peacefully doze, 20
and life is restored
by the murmuring leaves:
in their shade, to the sound of
their rustling, I sleep.

15

From the Workshop of Art: Major Poets of the Twelfth Through Fourteenth Centuries

In the six poets of the High Middle Ages represented in this chapter, many central themes of the preceding two find culminating expression. Bernard Silvestris and Alan of Lille, in the twelfth century, are often loosely associated with the Platonic or Neoplatonic revival of the "School of Chartres." Between the Dark Ages—when monasteries and a few palace schools were almost the sole educational institutions—and the rise of universities to preeminence in the late twelfth century, the cathedral schools of northern France were the principal centers of learning in Europe. "Chartres is unique among cathedral schools in having had two *masters of international standing separated by a century: Fulbert in the early eleventh century and Bernard in the early twelfth," R. W. Southern writes in*

"Humanism and the School of Chartres" (in Medieval Humanism and Other Studies *[1970]): "No other cathedral school can show so much."*

Renewal of Platonic cosmology, and with it renewal of the study of nature, found expression in the School of Chartres, mainly through commentary on Calcidius's translation of Plato's Timaeus *and on Boëthius, whose* Consolation of Philosophy *contains in Meter 9 of Book III a poetic epitome of the* Timaeus *as interpreted by the Neoplatonist Proclus. Prominent among such commentaries were the lectures and writings of Bernard of Chartres, whom the English scholar John of Salisbury called "the foremost Platonist of our time" (*Metalogicon *IV.35), Thierry of Chartres (possibly Bernard's brother), William of Conches, and Gilbert de la Porrée. Though this Platonic tendency may not have been so exclusively associated as many have assumed, Southern argues, with Chartres as opposed to other cathedral schools such as Paris, Tours, or Orléans, it was a central intellectual current of the age, only slowly displaced by Scholastic philosophy under the stimulus of newly translated texts of Aristotle and the Arabs. "To me," Peter Dronke writes in "New Approaches to the School of Chartres" (1971, in* Intellectuals and Poets in Medieval Europe *[1992]), responding to Southern's doubts about the extent of its role, the school of Chartres "has always stood for what is freshest in thought, richest and most adventurous in learning, in northern Europe in the earlier twelfth century."*

Most important for our purposes is the emphasis of these writers on the created world as a "theophany" or revelation of God—to use the term of John the Scot (Chapter 13 above), whose influence was felt anew through a digest of his Periphyseon, *the* Clavis physicae *of Honorius Augustodunensis. "More than certain of their contemporaries, the Chartrians had a feeling for the intrinsic value of nature. They recognized an activity proper to it," J. M. Parent writes in* La Doctrine de la création dans l'école de Chartres *(1938), "for which they paid homage to a God operating in it and associating it with the realization of his plans. They also thought that nature could be studied in itself without immediate reference to its author." In consequence, Eco observes, "the work of God was the* kosmos*—the all-encompassing order, opposite of primeval chaos. . . . Nature, in the Chartrian metaphysics, was not merely an allegorical personification but an active force which presided at the birth and the becoming of things." This emphasis immensely influenced, besides those immediately associated with Chartres, others such as Hugh of St. Victor (Chapter 13) and especially Bernard Silvestris of Tours (who dedicated his* Cosmographia *to Thierry of Chartres) and Alan of Lille.*

The hegemony of Aristotle with the victory of various forms of Scholasticism at the universities of Paris and Oxford in the thirteenth century found no comparable reflection in Latin poetry, both because Latin was rapidly giving way to the vernaculars as the primary language of poetic expression and because the argumentative form of Scholasticism was resolutely anti-literary: not until Dante in the fourteenth century would the vision of St. Thomas Aquinas (Chapter 13) be triumphantly transmuted into poetry. But many philosophical concerns of the age find place in Guillaume de Lorris's Roman de la rose (Romance of the Rose), *and especially in Jean de Meun's sprawlingly encyclopedic continuation, where Nature herself—as in Bernard Silvestris and Alan of Lille before—is a central character, and the relation of nature to God and man a central concern. No medieval poem, apart from Dante's* Divine Comedy, *survived in so many manuscripts, or had so varied and lasting an influence both in France and beyond. Chaucer was among Guillaume's English translators, and from Jean de Meun's poem—and Jean's immediate source, Alan of Lille's* Complaint of Nature *—he took the noble goddess who presides over his* Parliament of Fowls *as "the vicaire of the almyghty Lord":*

And right as Aleyn, in the Pleynt of Kynde,
Devyseth Nature of aray and face,
In swich aray men myghte hire there fynde.
(316–18)

In such ways, even during the Scholastic ascendancy of the thirteenth and fourteenth centuries poets continued, in their portrayal of nature, to reflect the legacy of twelfth-century Chartrian Platonism and, indirectly, of John the Scot, Boëthius, and Dionysius the Areopagite (Chapter 13), of Calcidius, of Plotinus, and of Plato himself.

Chaucer's omission from the six major poets of this chapter is due not to absence of nature from his writings but to the pervasiveness of its presence, which can only rarely be quoted in isolation from the context of his richly human comedy. The characters of many Canterbury Tales*—like those of the* Libro de buen amor (Book of Good Love) *of Juan Ruiz, archpriest of Hita in Spain, several generations before—are firmly rooted in the physical world, whether of town or country. The opening words of the Wife of Bath's prologue,*

Experience, though noon auctoritee
Were in this world, were right ynogh for me,

give robust voice to a new affirmation of experience (decidedly including sexual experience) in conflict with received authority—an affirmation far earthier than those of her philosophical English predecessors Adelard of Bath or Roger Bacon (Chapter 13). And if the natural world is overshadowed in Chaucer, Juan Ruiz, and François Villon by the human, in Dante's Comedy, *where it is strictly subordinated to the divine, it again repeatedly makes its presence felt. Vivid imagery brings life and growth into the changeless regions of hell and heaven, much as the similes of Homer's* Iliad *vary the brutal monotony of the battlefield with the teeming multiplicity of encompassing nature. Atop Mount Purgatory (the one realm of the afterlife that is not eternal), Dante enters the Earthly Paradise first described by Christian poets nearly a thousand years before, and gives a classic account of nature in its formerly uncorrupted perfection.*

While Dante brings the High Middle Ages to ripe fulfillment, the last two poets of this chapter look both forward to a new age and back to a primeval past. Petrarch, in his Latin letters and Italian poems, anticipates both the polished humanism and the restless striving of the European Renaissance that began in Italy a century later and claimed Petrarch as one of its founders. Yet the poet who climbs Mount Ventoux "from desire to see its conspicuous height" learns from St. Augustine's Confessions, *after reaching the summit, the vanity of earthly aspirations, and the lover who wanders, as if in anticipation of Rousseau or Werther, "di pensier in pensier, di monte in monte" ("from thought to thought, from mountaintop to mountain"), will end his love poems to Laura, the* Canzoniere, *with a penitent hymn to the Virgin Mary: much as he foreshadows the Renaissance, Petrarch firmly belongs to the Middle Ages. And though the Green Knight of the anonymous Middle English* Sir Gawain and the Green Knight *is no simple vegetation spirit or tree god out of Frazer's* Golden Bough, *this mysterious figure embodies qualities that link him not only with medieval canons of hospitality and courtesy but also with more primitive, even half-savage conduct, and with the periodic renewal of nature's green world on which continuation of human life depends. In the Green Knight, primordial nature and cultured humanity are so interfused as to make distinction between them impossible.*

Simple Nature's Varied Forms

The *Cosmographia*, by Bernard Silvestris

Next to nothing is known about Bernardus Silvestris of Tours. His writings include the Cosmographia *or* De Mundi Universitate (On the Universe of the World), *a work in two books (of roughly 65 pages) mingling prose and verse, composed between about 1143 and 1148; a poem, the* Mathematicus, *of nearly 900 lines in elegiac couplets; and parts of a didactic* Experimentarius *translated from the Arabic. He is also probably the author of an allegorical* Commentary on the First Six Books of Virgil's *Aeneid; a partial commentary on Martianus Capella's* The Marriage of Mercury and Philology*; a lost commentary on Plato's* Timaeus*; and two shorter poems. (For details, see the introduction to Peter Dronke's edition of the* Cosmographia *[1978], which replaces that of Carl Sigmund Barach and Johann Wrobel [1876].)*

The Cosmographia, *deeply influenced by Plato's* Timaeus *(with a "decisive impulse," Dronke believes, from John the Scot), "is possibly the most complex literary product of the early twelfth century," Brian Stock writes in* Myth and Science in the Twelfth Century: A Study of Bernard Silvester *(1972): an "audacious" attempt "to rewrite the myth of the creation of the world and man." Our first selection is the introductory prose Summary (translated by Winthrop Wetherbee in* The *Cosmographia* of Bernardus Silvestris *[1973]), which states the central (if incomplete) argument of this complex myth. The process of creation begins in the first book ("Megacosmos") when Nature herself complains to Noys—divine "mind," transmuted from the Platonic-Plotinian masculine monosyllable* nous *to a bisyllabic feminine goddess—about the chaotic formlessness of matter (Greek* Hyle *or Latin* Silva*). Noys in response separates the four elements and creates the angels and stars, the earth with its mountains and rivers, and the multitudinous living creatures. The second book ("Microcosmos") tells of the creation of man by Nature with the help of Urania and Physis. Thus this "cosmological epic," Dronke writes in* Fabula: Explorations into the Uses of Myth in Medieval Platonism *(1974), reaches its climax in the creation of man brought about "by a triad of goddesses: Urania, the principle of celestial generation, Physis, the principle of physical generation, and Natura, who partakes of both these realms, the principle of that coalescence of the celestial and the physical which effects the handiwork [of] creation. The three are assigned their roles by Noys, the transcendent goddess Mind. For all three principles are needed for the making of man: Urania must fashion the human soul, Physis the human body; Natura must conjoin these."*

Of the assumptions concerning creation of the world that medieval Christian Platonists drew from the Timaeus, *three are fundamental, Wetherbee writes in* Platonism and Poetry in the Twelfth Century: The Literary Influence of the School of Chartres *(1972): "that the visible universe is a unified whole, a 'cosmos'; that it is the copy of an ideal exemplar; and that its creation was the expression of the goodness of its creator. The ideal exemplar,* archetypus mundi, *was identified by the Chartrians with the divine wisdom, while the abiding goodness of the creator was seen as expressed by Plato's World Soul, which thus assumed a providential as well as an organizational function." Bernard's mythic vision is extraordinarily dynamic. "The universe of the* Cosmographia, *and the life of man within it," Wetherbee observes,*[1] *"are charged with conflict and presided over by forces which convey strong suggestions of necessity. The intractability*

[1] "Philosophy, Cosmology, and the Twelfth-Century Renaissance," in Peter Dronke, ed., *A History of Twelfth-Century Philosophy* (1988).

of Silva, the primal chaos, powerfully dramatized in the opening scene of the poem, has its counterpart in the unruliness of human passion and appetite. Like the life of the universe at large, human experience is an interplay of rational and irrational forces, and the tension between them as man seeks to maintain right relations with the cosmic order becomes at times a heroic, almost tragic theme."

Bernard's "celebration of creativity and sexuality as divine manifestations in the universe" is well illustrated here by passages translated from the 482-line elegiac poem (I.iii) on the wonders of the newly created world, with its curious intermingling of direct observation and classical erudition. Dronke, in his introduction, sees within it parallels to—even, despite "scanty manuscript tradition," possible influence by—Lucretius, as well as Manilius (Chapter 10 above). How far the Platonizing assumptions and mythic vision of Bernard and the School of Chartres were compatible with orthodox interpretation of Genesis remained in doubt. Hugh of St. Victor, and later St. Thomas Aquinas (Chapter 13), among others, would attempt to formulate more acceptably orthodox positions; and John the Scot's Periphyseon *would be condemned by Pope Honorius III in 1225. At the very least, the difference in emphasis is striking. In Genesis, Dronke remarks, "God said, 'Let there be light: and there was light'—the effect follows the cause with the inevitability of a deduction. But here in the* Cosmographia *all is open and uncertain: Silva can terrify God by her disfigured looks, Natura is not sure how Noys will react, the infant World does not know his fate. It is a scene full of risks, full of surprises, where anything may go awry. Even in the* Timaeus *there is not this sense of the unpredictable in the shaping of the cosmos. This is Bernard's own poetic contribution, and a sign of how little his* fabula *can be reduced to the sources he knew."*

Summary of the Work,
translated by Winthrop Wetherbee

In the first book of this work, which is called *Megacosmos*, or "the Greater Universe," Nature, as if in tears, makes complaint to Noys, or Divine Providence, about the confused state of the primal matter, or Hyle, and pleads that the universe be more beautifully wrought. Noys, moved by her prayers, assents willingly to her appeal, and straightway separates the four elements from one another. She sets the nine hierarchies of angels in the heavens; fixes the stars in the firmament; arranges the signs of the Zodiac and sets the seven planetary orbs in motion beneath them; sets the four cardinal winds in mutual opposition. Then follows the creation of living creatures and an account of the position of earth at the center of things. Then famous mountains are described, followed by the characteristics of animal life. Next are the famous rivers, followed by the characteristics of trees. Then the varieties of scents and spices are described. Next the kinds of vegetables, the characteristics of grains, and then the powers of herbs. Then the kinds of swimming creatures, followed by the race of birds. Then the source of life in animate creatures is discussed. Thus in the first book is described the ordered disposition of the elements.

In the second book, which is called *Microcosmos*, or "the Lesser Universe," Noys speaks to Nature, glories in the refinement of the universe, and promises to create man as the completion of her work. Accordingly she orders Nature to search carefully for Urania, who is queen of the stars, and Physis, who is deeply versed in the nature of earthly life. Nature obeys her instructress at once, and after searching for Urania through all the celestial spheres, finds her at last, gazing in wonder at the stars. Since the cause of Nature's journey is already known to her, Urania promises to join her, in her task and in her journey. Then the two set out, and after having passed through the circles of the planets and forewarned themselves of their several influences, they at last discover Physis, dwelling in the very bosom of the flourishing earth amid the odors of spices, attended by her two daughters, Theory and Practice. They explain why they have come. Suddenly Noys is present there, and having made her will known to them she assigns to the three powers three kinds of speculative knowledge, and urges them to the creation of man. Physis then forms man out of the remainder of the four elements and, beginning with the head and working limb by limb, completes her work with the feet.

The Wonders of the Newly Created World,
translated by Robert M. Torrance
I.iii

Narrowly jointed mountain zones allow
small space for soil, and devastate the plow. 200
Wolves howl in shrubs, in deserts lions roar,
rocks shelter serpents, forests hide the boar.
Genus divides in species, simple nature
takes varied form in each particular creature. . . .

Bones bulwark elephants, the camel swells 205
with humps, horns grace the foreheads of gazelles.
Stags gird themselves to run; with knees held high
does raise up slender legs, prepared to fly.
Lions rely on courage, bears on claws,
tigers on fangs, and boars on tusky maws. 210
Sheep in soft fleecy warmth luxuriate,
while sack-cloth canopies she-goat and mate.
High-hearted is the horse; the donkey bears
a sluggish spirit bowed by heavy ears.[2]
Bloodthirsty wolves and leopards howl for prey 215
deep in the woods and mountains where they stay.
Bulls have a mightier spirit, but the small
fox in his narrow frame knows most of all.
Oxen are born to serve; afraid of stronger
beasts, skittish rabbits grow ears ever longer. 220
Escaping mountainward, the wild ass trims
uses and obligations from his limbs.[3]
Friendly by native instinct or long use
dogs endure savagery of men's abuse.
Lynxes come forth to hide from human sight 225
the wondrous fountain of their liquid light.[4]

[2]I have adapted these two lines from C. S. Lewis's translation in *The Allegory of Love* (1936).
[3]Isidore, *Etymologiae* 12.1.39, reports that male *onagri* [wild asses], each of whom presides over a herd of females, are wont to eat the testicles of newborn males, so that mothers, fearing this fate for their offspring, hide them away in deserted places. (Wetherbee)
[4]Pliny, *Nat. Hist.* 8.38.137, reports that the urine of the lynx crystallizes into a precious gem; having learned that this gem is coveted by man, he is very secretive about where he urinates. Cp. Isidore, *Etymologiae* 12.2.20. . . . (Wetherbee)

Degenerate nature's manlike ape comes after,
rousing, by hideous counterfeit, loud laughter.
The beaver willingly takes drastic measures
to leave the greedy enemy his treasures.[5] 230
Marten and squirrel creep forth to be worn
by potentates whom beaver pelts adorn,
and costlier-smelling sables, turning coats,
pick pockets as they line plump princes' throats. . . .

Nearer to dawn and Eurus[6] lies a place
encompassed round by flowery earth's embrace,
on which the rising sun shines soft and warm
with youthful flames incapable of harm. 320
There temperate rays from heaven's clement powers
impregnate earth with varied grains and flowers.
One corner nurtures scents, breeds species, holds
all riches and delights the world enfolds.
Tall gálbanum and ginger in this clime 325
prosper, and with valerian grows sweet thyme:
acanthus, honored by perpetual bloom,
and nard, whose ointment radiates perfume.
Crocus near purple hyacinth grows pale,
and over cassia mace strives to prevail. 330
Among these happy woods a sinuous rill
strays in whatever winding course it will.
Gurgling through pebbles, roaring past tall trees,
water flows on with unobstructed ease.
The primal man, in this lush painted spot, 335
dwelt as a transient guest—but knew it not.
This grove did nature make with loving care;
else, tangled wilderness rose everywhere. . . .

Though birds flit round their aerial domain,
many in kindred waters yet remain:
white gulls that fly before the lunar tides 445
rising, then follow when their surge subsides;
thick-feathered bitterns, herons long of shank,
fish-glutted divers, ducks absurdly frank;
and swans that, having sensed with every breath
life's dangers, die with songs defying death. 450
Most ascend skyward: phoenix, all alone
resuscitated wholly on her own;
the king of birds who seized the Bacchic boy
to be a gift for Jove's nocturnal joy;[7]

[5]On the beaver's reputed act of protective self-castration, see the passage from the Greek *Physiologus* in Chapter 13 above. Among many possible sources for this legend, Wetherbee cites Isidore 12.2.21.
[6]The (south-)east wind. Bernard is of course here describing the Earthly Paradise.
[7]Jupiter (or Jove; Greek Zeus) took the form of an eagle ("the king of birds") to carry off the beautiful boy Ganymede to be his cupbearer and catamite—a word ultimately derived from "Ganymede."

falcon and hawk, whose predatory way 455
makes every other bird their helpless prey;
crane, who devises figures as he flies
forth from Strymonian waters through the skies;[8]
Juno's bird,[9] sportive nature's work; the white
dove that unsparingly her breast may bite; 460
Philomel, who laments her wrongs anew
each spring, and Procne, stained with bloody hue;[10]
both cocks, the stay-at-home, and roving race
named Phasis[11] for Medea's native place;
turtledove, loving truly; bobtailed quail; 465
wise thrush who finds provision without fail;
partridge, who might have lived more, learning less;[12]
lark, welcoming each day with joyousness;
hiphopping sparrow; crow that sees ahead;
magpie with twofold color overspread; 470
voracious vulture, ravening kite—foul blots!;
ostrich who cultivates deserted spots;
finch, sweetly singing love's delicious passion;
parrot who prattles on in human fashion;
oracular raven who too soon forgets 475
nests of young offspring left without regrets;
kingfisher, woodpecker, guarding shores and woods;
goose, loving lakes' wide-open neighborhoods;
owl, whom the friendly sunlight fiercely scourges;
and screech-owl, singing lamentable dirges. 480
Once having found these forms, the feathered race
diversified in body, mind, and place.

Creation's Book and Nature's Complaints

Selections from the Writings of Alan of Lille

Alan of Lille, or Alanus de Insulis, though later famed as "doctor universalis," was little mentioned during his lifetime. He was probably born at Lille in Flanders sometime between 1116 and 1128. He taught at Paris and later at Montpellier, entered the abbey of Cîteaux as a monk late in life, and died in 1203. His numerous theological writings and sermons included an Art of Preaching *and a manual for confessors, but he is best known for his Latin poetry.*

[8]On Strymonian cranes, see Virgil's *Georgics* I.120, and note, in Chapter 11 above.
[9]The peacock.
[10]On Philomel, the nightingale, and her sister Procne, the swallow, see note on *The Vigil of Venus* 88, in Chapter 11 above.
[11]The pheasant took its name from the river Phasis, which flowed into the Euxine (Black Sea) in Colchis, the mythical kingdom of Aeëtes and his daughter Medéa, to which Jason and the Argonauts sailed in search of the Golden Fleece.
[12]Perdix, in Greek myth, was so skillful in the use of the saw, chisel, and compasses he invented that his uncle Daedalus jealously threw him from the temple of Athena on the acropolis; Athena caught him in his fall and transformed him into the partridge named after him.

All of Creation, Like a Book *("Omnis mundi creatura"; text in* Penguin Book of Latin Verse*), one of his two rhymed lyrics, is a classic expression of the brevity of "our human life's brief spring." The remaining selections are from Alan's two long poems.* **The Complaint of Nature** (De Planctu Naturae), *variously dated between about 1160 and 1182, is—like Boëthius's* Consolation of Philosophy, *Martianus Capella's* Marriage of Mercury and Philology, *and Bernard Silvestris's* Cosmographia*—a* prosimetrum, *or Menippean satire. Passages in prose and in meter alternate for roughly a hundred pages. "I turn from laughter to tears, from joy to grief, from merriment to lament, from jests to wailing," the first meter begins (in Sheridan's prose translation), "when I see that the essential decrees of Nature are denied a hearing, while large numbers are shipwrecked and lost because . . . Venus wars wih Venus and changes 'he's' into 'she's' and with her witchcraft unmans man." As the poet mourns the vices of sodomy and other forms of illicit sexuality, Nature herself appears to him; her garments, adorned with innumerable creatures and flowers, are described at length. Then the poet describes in Prose 2 Nature's attendants, and in Meter 3 evokes the coming of spring in classical elegiac couplets. In Prose 3, Nature reminds the poet that she had formed man "so that in him, as in a mirror of the universe itself, Nature's lineaments might be there to see." Yet sensuality has corrupted man's mind; and Nature's powers, great though they are, are nothing compared with God's, for "He is the creator, I was created." In the Sapphics of Meter 4, the poet prays to Nature, who governs the heavens and rejuvenates earth, and asks the reason for her sorrowful tears. The remainder of the poem will explain that man alone, of all created things, does not obey her laws. Overwhelmed by her tasks as the vicar of God, she had instructed Venus as her sub-vicar, along with her husband Hymenaeus and son Cupid, to regulate the generation of humanity. But Venus had committed adultery and given birth to Cupid's perverse opposite, Jocus (Sport), whose dominance has perverted mankind's natural instincts. After a saddened Hymenaeus enters, followed by a train of Virtues, Genius pronounces anathema on all who have sinned against Nature. The vision fades, and the poet awakens.*

Our final selection is from Alan's later and longer **Anticlaudianus** de Antirufino, *a poem in nine books (and a verse prologue) of 4,352 Latin hexameters probably written after 1180. Its title indicates that this epic on a wholly good man was meant to be the antithesis of the ancient poet Claudian's portrayal, in* Against Rufinus, *of a completely evil man. In Book I, Nature calls upon the Virtues to assist her in creating a perfect man. The passage translated here describes the Earthly Paradise and the palace where Nature dwells not as an Eden in the distant past but a realm forever potentially present. The rest of the poem will describe how Prudence, prompted by Reason, ascends with the aid of Theology and Faith to heaven and enters the palace of God, who creates a beautiful soul for which Nature fashions a body. Aided by the Virtues, the New Man defeats the Vices stirred up by the Fury Allecto, and brings peace and harmony to the earth.*

This resolutely orthodox author thus continually emphasizes subordination of both Nature to God and natural desire to Christian virtues. Yet in the prominence he gives to Nature's creative power, of which he writes with a passion rare in his otherwise often monotonous (though immensely influential) verses, Alan continues the tradition of Chartrian Platonism exemplified by the cosmogonic mythmaking of Bernard Silvestris. "Without the pervasive inspiration of Bernard in language, dramatis personae, *and conceptions," as Dronke affirms in introducing the* Cosmographia, *Alan's two major poetic works would be inconceivable. Latin texts of both poems are included in Thomas Wright, ed.,* The Anglo-Latin Satirical Poets and Epigrammatists of the Twelfth Century *(1872); the* Anticlaudianus *has been separately edited by R. Bossuat (1955). See also the translations and commentaries by James J. Sheridan of both* Anticlaudianus *(1973) and* The Plaint of Nature *(1980); Douglas M. Moffat's translation of* The Complaint of Nature *(1908); and G. R. Evans's study* Alan of Lille *(1983).*

All of Creation, Like a Book,
translated by Robert M. Torrance

All of creation, like a book,
pictures to us, with every look,
our accurate reflection,

so that all human life and death,
and fate that governs every breath, 5
are mirrored to perfection.

The rose depicts our true condition,
glossing with learned erudition
how life is rightly read:

flowers that bloom with early dawn, 10
deflowered as the day goes on,
are all by evening shed.

Thus, as each flower breathes its last,
pallor possesses it so fast
it dies when scarcely born: 15

youthful and old at once, the rose,
combining maid and matron, blows
and withers with the morn.

Therefore our human life's brief spring
with the first flush of dawn takes wing 20
awhile, when youth is bright,

yet evening quickly drives away
the dawn, dispersing every ray
of life's remaining light.

The glory of its youthful prime 25
is decimated soon by time,
which flows, but never stands:

bud turns to mud, to weeds the flower,
to ashes man, who yields all power
to death's triumphant hands. 30

FROM The Complaint of Nature,
translated by James J. Sheridan (prose) and Robert M. Torrance (verse)

From PROSE 2 [NATURE'S ATTENDANTS] . . . At the arrival of the above-mentioned maiden, you would think that all the elements were having a celebration by having, so to speak, their native powers renewed. The firmament, as though lighting the maiden's path with its candles, bade its stars to shine with more than their wonted radiance. As a result the daylight itself seemed to be astonished at this great boldness on their part, when it saw them appear in its sight with too great a show of insolence. Phoebus, too, assuming an unusually gladsome countenance, poured forth all the riches of his light to greet the maiden's arrival. His sister, too,[13] whom he had robbed of her ornaments of brilliance, had her mantle of delight restored her and was ordered to go forth to meet the approaching maiden.

The air, wiping off the clouds with their tear-laden appearance, smiled on the maiden's steps with the benediction of a fair countenance. This had at first been ruffled by the mad blast of angry Aquilo,[14] but now found rest on the propitious bosom of Favonius. The birds, by some inspiration of nature, played a pleasing game with their wings and in their faces showed reverence for the maiden. Juno, who for a long time had scorned the playful touches of Jupiter, was so intoxicated with joy that, with the uninterrupted fore-play of her eyes, she kept inviting her husband to the enticements of love.

The sea, that beforehand had been raging with tossing waves, now keeping a solemn holiday in honor of the maiden's arrival, promised unbroken calm and quiet. For Aeolus had chained the storm-winds in his prisons to prevent them from stirring up a greater than civil war in the maiden's presence. The fish, swimming on the surface of the water, in so far as the slowness of their natural senses allowed, proclaimed the arrival of the maiden with a certain festive gaiety. Thetis, too, marrying Nereus, decided to conceive a second Achilles.[15] Maidens, whose beauty would not only rob man of his reason but would also force the heavenly beings to forget their divinity, emerged from the streambeds and like tributaries presented their queen with little gifts of aromatic nectar. When the maiden had graciously received these, she showed the maidens her love for them by the clasp of long embraces and oft-repeated kisses.

Earth, for long stripped of its ornaments by plundering Winter, acquired a purple garment of flowers from the bounteous spirit of Spring, lest, dishonored in ragged attire, she might present an improper appearance to the eyes of the tender maid.

Spring, too, skilled craftsman in the weaver's art, wishing to show greater happiness in applauding the maiden's approach, wove garments for trees, which let down their foliage-tresses and bent over in a type of adoration, like a genuflection, offered prayers to the young maiden. Maidens,[16] emerging from the trees, by the light of their beauty enriched the riches of the actual day. In small vessels of cedarwood they brought spices made from types of herbs that have given their names to the spices: by offering these as though in payment of returns due from them, they purchased her favor by their gifts. Nymphs of the dell,[17] their laps filled with flowers, at times with roses colored the royal chariot the rich red of blood, at times with petals of white flowers made it shine like a lily. Flora, in lavish mood, presented the maiden with the cotton night-gown she had woven for her husband to earn his embraces. Proserpine, disdaining the marital bed of the lord of Tartarus,[18] returned to her home in the upper world, refusing to be cheated of a face-to-face meeting with her mistress. The land animals, instructed by some force or other in their nature, learned of the maiden's presence and were indulging in some playful frolic. Thus everything in the universe, swarming forth to pay court to the maiden, in wondrous contest toiled to win her favor.

[13]Diana (Greek Artemis or Phoebe) as moon goddess, sister of Apollo or Phoebus as god of the sun.
[14]Aquilo is the stormy North Wind, Favonius the gentle West Wind.
[15]Achilles was the son of the sea nymph Thetis and the mortal hero Peleus; in Alan's vision, Thetis has married the sea god Nereus and plans to give birth (long after Peleus's and Achilles' deaths) to a "second Achilles." The "maidens" of the following sentence are Naiades, or water nymphs.
[16]Hamadryades, tree nymphs.
[17]Napaeae, nymphs of the wooded vales.
[18]Pluto, the husband of Proserpine (Greek Persephone) in the underworld, here called Tartarus.

Meter 3 [The Coming of Spring]

Flowery Zephyr ended winter's lease,
quelling pugnacious Boreas with mild peace,
then rained down privet in a hail of flowers
coating with snow-white petals fields and bowers;
Spring, like a youthful fuller, now reclaims 5
fields, kindling flowers with her purple flames.
Leaves have returned to winter-shaven trees,
renewing branches stripped by winter's freeze.
This was the time when, cheered by Dryads, Spring
scatters the bounteous wealth her graces bring; 10
when, with the vigorous strength of recent birth,
emergent blossoms spring from mother Earth
and shimmering violets, cradled on the ground,
suck with young avid mouths the air all round:
then rose-starred earth self-confidently vies, 15
chockfull of constellations, with the skies;
banners of almond-trees prognosticate
summer, while blossoms trumpet spring's estate;
the budding vine, fast to the elm-tree grown
in wedlock, fancies children on her own. 20
Sunlight's bright candle outlaws winter's shade,
making frigidity a renegade,
yet winter's phantom lurks in many a wood
where recently its brooding shadows stood.
Juno now tenders dewy breasts, the first 25
to pacify her flower-nurslings' thirst.
This was the time when Phoebus' strength arouses
dead grasses, raised from subterranean houses,
while warm sunbeams suffuse the world with grace
and dry the tears of winter from its face, 30
entrusting flower-buds to air's broad womb
lest icy cold destroy their infant bloom.
Phoebus shines on a world that winter's blight
darkens, and greets it with exultant light;
youth finds decrepit age bereft of joy, 35
and old man world begins to be a boy;
Phoebus impoverishes the nighttime's length
as pygmy days reclaim gigantic strength;
Phrixus' flock, paying tribute to the one—
Phoebus—who gave them shelter, hail the sun;[19] 40

[19]Phrixus was son of Athamas [king of Thebes] and Nephele and brother of Helle. Athamas later married Ino. She became jealous of Phrixus and Helle. When the grain (which she had roasted) failed to grow, she bribed the messengers sent to Delphi to say that the sacrifice of Phrixus (and possibly of Helle) was required. The children escaped on a golden-fleeced ram given by Hermes [Mercury]. The ram carried Phrixus to Colchis. Helle fell into the strait and it came to be called Hellespont: Ovid, *Met.* 11.195; Hyginus *Fab.* 1–4. (Sheridan) The golden fleece presented by Phrixus to King Aeëtes of Colchis was later the prize sought by Jason and the Argonauts. On Philomena (or Philomela), the nightingale, in the following lines, see note to *The Vigil of Venus*, line 88 (Chapter 11 above).

and Philoména celebrates the spring's
rites with the ode her honey-sweet mouth sings:
full-throatedly her vocal chords she plays
to solemnize the god on festive days.
The lark with sweet sounds imitates the lyre 45
while flying up to Jove, her heavenly sire.
Silvery splendor cloaks rambunctious streams,
illumining their waves with dazzling gleams.
Down from their source, see, babbling brooklets leap—
a murmuring prologue flowing on toward sleep— 50
while, sparkling splendidly, the source invites
exhausted men to drink its cool delights.

From PROSE 3 [NATURE'S POWERS AND LIMITATIONS] Despite the fresh youth of the charming season, by no applause from created nature could the maiden be cheered and moderate her above-mentioned grief. But, lowering the chariot to earth and hallowing the ground with her footprints, she approached me with modest gait. When I saw this kinswoman of mine close at hand, I fell upon my face and stricken with mental stupor, I fainted; completely buried in the delirium of a trance, with the powers of my senses impeded, I was neither alive nor dead and being neither, was afflicted with a state between the two. The maiden, kindly raising me up, strengthened my reeling feet with the comforting aid of her sustaining hands. Entwining me in an embrace and sweetening my lips with chaste kisses, she cured me of my illness of stupor by the medicine of her honey-sweet discourse.

When she realized that I had been brought back to myself, she fashioned for me, by the image of a real voice, mental concepts and brought forth audibly what one might call archetypal words that had been preconceived ideally. She said:

"Alas, what blindness of ignorance, what delirium of mind, what impairment of sense, what weakness of reason, have cast a cloud over your intellect, driven your reason into exile, dulled the power of your senses, forced sickness of mind on you, so that your mind is not only robbed of an intimate knowledge of your foster-mother but also that at my first rising the star of your judgment is forced to set as though stricken by some monstrous and unheard-of appearance? Why do you force the knowledge of me to leave your memory and go abroad, you in whom my gifts proclaim me who have blessed you with the right bounteous gifts of so many favors; who, acting by an established covenant as the deputy of God, the creator, have from your earliest years established the appointed course of your life; who of old brought your material body into real existence from the mixed substance of primordial matter; who, in pity for your ill-favored appearance that was, so to speak, haranguing me continually, stamped you with the stamp of human species and with the improved dress of form brought dignity to that species when it was bereft of adornments of shape? In dealing with that species, I made arrangements for various work-companies of members to serve the body and gave orders that the senses keep vigil there like sentries, so to speak, in the state of the body, so that sighting in advance enemies from outside, they might protect the body against external harassment and thus the entire material body, adorned with the noble purple vestments of nature, might proceed to a marriage alliance and be joined in a more acceptable union with spirit as a husband and that the husband might not be disgusted by the baseness of his partner and repudiate the espousals. I also gave the spirit in you distinctive powers and capacities, lest, being poorer than the body, it might envy its successes. I allotted to the spirit the native faculty and power of hunting down subtle matters in the chase for knowledge and of retaining them when it apprehends them. I stamped it also with the stamp of ratiocination to separate, with the winnowing-fan of judgment, vain falsehood from important truth. Through me, too, the power of recollection becomes your handmaiden and lays up a treasure, a noble record of knowledge, in the strong-box of her memory. I have blessed both parts of you, then, with these endowments so that neither should bemoan its own poverty or complain of the other's riches.

"But just as the above-mentioned marriage was solemnized by my consent, so, too, at my discretion this marital union will be annulled. Moreover, my bounteous power does not shine forth in you alone individually but also universally in all things. For I am the one who formed the nature of man according to the exemplar and likeness of the structure of the universe so that in him, as in a mirror of the universe itself, Nature's lineaments might be there to see. For just as concord in discord, unity in plurality, harmony in disharmony, agreement in disagreement of the four elements unite the parts of the structure of the royal palace of the universe, so too, similarity in dissimilarity, equality in inequality, like in unlike, identity in diversity of four combinations[20] bind together the house of the human body. Moreover, the same qualities that come between the elements as intermediaries establish a lasting peace between the four humors. Just as any army of planets fights against the accepted revolution of the heavens by going in a different direction,[21] so in man there is found to be continual hostility between sensuousness and reason. For the movement of reason, springing from a heavenly origin, escaping the destruction of things on earth, in its process of thought turns back again to the heavens. On the other hand, the movements of sensuality, going planet-like in opposition to the fixed sky of reason, with twisted course slip down to the destruction of earthly things. The latter, then, draws man's mind down to the destruction arising from vice so that he may fall, the former invites him to come to the source of virtue so that he may rise; the one, corrupting man, changes him into a beast, the other has the power to transform man into a god; one illuminates the dark night of the mind with the light of contemplation, the other removes the light of the mind by the dark night of concupiscence; one enables man to hold converse with angels, the other drives him to wanton with brute beasts; one shows the man in exile how to get back to his fatherland, the other forces the one in his fatherland to go into exile. . . .

"However, see how the universe, with Proteus-like succession of changing seasons, now plays in the childhood of Spring, now grows up in the youth of Summer, now ripens in the manhood of Autumn, now grows hoary in the old age of Winter. Comparable changes of season and the same variations alter man's life. When the dawn of man's life comes up, man's early Spring morning is beginning. As he completes longer laps in the course of his life, man reaches the Summer-noon of his youth; when with longer life he has completed what may be called the ninth hour of his time, man passes into the manhood of Autumn. And when his day sinks to the West and old age gives notice of life's evening, the Winter's cold forces man's head to turn white with the hoar frost of old age. In all these things the effects of my power shine forth to an extent greater than words can express. . . .

[20]The four humors. See "On the Nature of Man" from the Hippocratic Corpus, in Chapter 9 above.
[21]Reference here is to the retrograde motion of the planets. . . . Thus, for example, . . . from Earth, Mars seems to slow down as Earth approaches it, stop momentarily as Earth passes it, reverse its course as Earth leaves it behind, and finally resume its regular motion. (Sheridan)

"But, lest by thus first canvassing my own power, I seem to be arrogantly detracting from the power of God, I most definitely declare that I am but the humble disciple of the Master on High. For in my operations I have not the power to follow closely in the footprints of God in His operations, but with sighs of longing, so to speak, gaze on His work from afar. His operation is simple, mine is multiple; His work is complete, mine is defective; His work is the object of admiration, mine is subject to alteration. He is ungeneratable, I was generated; He is the creator, I was created; He is the creator of my work, I am the work of the Creator; He creates from nothing, I beg the material for my work from someone; He works by His own divinity, I work in His name; He, by His will alone, bids things come into existence, my work is but a sign of the work of God. You can realize that in comparison with God's power, my power is powerless; you can know that my efficiency is deficiency; you can decide that my activity is worthless. Consult the authoritative teaching of theology on whose trustworthiness you should base your assent rather than on the strength of my arguments. According to its reliable testimony man is born by my work, he is reborn by the power of God; through me he is called from non-being into being, through Him he is led from being to higher being; by me man is born for death, by Him he is reborn for life. But my professional services are set aside in the mystery of this second birth. . . . But, although my accomplishment fail in comparison with God's power, it, nevertheless, has the advantage when placed side by side with man's power. Thus on the table of comparison, so to speak, we can find three degrees of power and they are termed the *superlative* power of God, the *comparative* power of Nature, and the *positive* power of man.

"Without one ounce of questioning, all these things bestow on you an intimate knowledge of me. To speak more intimately still, I am Nature who, by the gift of my condescension, have made you a sharer in my presence here and have deigned to bless you with my conversation."

While Nature was revealing aspects of her nature to me in these words and by her instruction, as by an opening key, was unlocking for me the door of her knowledge, the cloudlet of stupor was drifting away from the confines of my mind. By the final instruction, as by some healing potion, the stomach of my mind, as if nauseated, spewed forth all the dregs of phantasy. When I came completely back to myself after my mind's trip abroad, I fell down at Nature's feet and marked them with the imprint of many a kiss to take the place of formal greeting. Then straightening up and standing erect, with humbly bowed head, I poured out for her, as for a divine majesty, a verbal libation of good wishes. . . . I went on to a question in the following words.

Meter 4 [Prayer to Nature]

Child of God above, universal mother,
binding all the world in unbroken linkage,
earthly gem, bright mirror of fallen mortals,
star of the morning,

peace, love, virtue, guidance and sovereign power, 5
order, law, end, origin, light and pathway,
life, praise, splendor, beauty and form united,
rule of the cosmos:

moderating firmly with reins the world's course,
everything you join in a knot of concord, 10
stably wedding heaven and earth together
peacefully bonded:

contemplating pure intellect's ideas,[22]
every single species of things you fashion,
cloaking things with form, then with thumb re-form their 15
outward apparel:

heaven heeds your wishes and air attends you,
earth pays homage, waves of the sea revere you,
every creature, owning your world-dominion,
offers you tribute: 20

binding day and night in their alternations,
day you radiate with the sun's bright candle,
putting clouds of nighttime to sleep by brilliant
moonlight's reflection:

overlaying heaven with sparkling starlight, 25
bright you make the throne of our native aether,
ornamenting night with the constellations'
heavenly armies:

altering, with Protean variations,
heaven's changing face, you endow with veering 30
flocks of birds[23] the airy expanse you always
lawfully govern:

by your nod the world is rejuvenated,
trees coiffured with burgeoning leafy branches;
earth, adorned with flowery ostentation, 35
flaunts her apparel:

howling seas you pacify, then upheave them,
syncopate the course of the ocean's fury,
lest its surging waters entomb forever
sunlight's dominion: 40

give me now, I pray you, the explanation
why you visit earth and abandon heaven?
why you offer creatures of earth your priceless
deity's presence?

why your face is moistened with raining teardrops? 45
what your tearful features are prophesying?
Surely tears abundantly voice some hidden
sorrow within you.

[22]"The ideas of pure *noys*," a term adopted from Bernard Silvestris as the counterpart of Greek *nous*, "mind" or "intellect," transmitted to the Middle Ages via Latin writers such as Macrobius and John the Scot from the Neoplatonists and ultimately from earlier Greek thinkers from Anaxagoras to Aristotle.
[23]Reading *aviumque vulgus* for *animumque vulgus* (or *aridumque vulgus*) of the MSS.

The Earthly Paradise,
translated by Robert M. Torrance
Anticlaudianus, *Book I*

Far from our clime, a distant place there is 55
that smiles at all the turbulence of our world,
uniting in one place all others' powers
and making good what every other lacks.
Whatever Nature's bounteous hand can do,
whatever gifts she lavishes, are here. 60
The earth, pubescent with soft down of flowers,
studded with stars, ablaze with purple roses,
strives to depict a heaven here below.
No newborn flowers lose their beauty here,
dying once born, nor is the rose a maiden 65
at dawn, a hag at night, but with unaging
features delights in spring's eternal youth.
No winter nips, no summer sears this flower,
no frenzied Boreas rages here, no Notus[24]
hurls lightning bolts, no stinging hailstones pommel. 70
Whatever feasts the eyes, intoxicates
the ears, beguiles taste, captivates the nostrils
and soothes the touch, this place of places holds.
Untroubled by the plowshare, it produces
whatever fights diseases and restores 75
health by preventing pestilent afflictions.
Bearing not common but miraculous offspring,
without external husbandry, the earth,
content with Nature's hand and Zephyr's favor,
gives birth, and revels in its wealth of children. 80

Rejoicing in fresh flowers, green with foliage,
ungnawed by rot, unslashed by angry axes,
unleveled, and unstrewn with cluttered branches,
a wall-like wood encompasses the garden.
Winter, that nips the tender youth of flowers, 85
here plunders none, and shears no leafy trees.
Whatever tree cannot pay nature's tribute
of seed and fruit, must suffer banishment.
Vying by better fruit to purchase Nature's
favor, and to excel the gifts of others, 90
every tree thinks continually of childbirth.

The sirens of the grove, spring's minstrels, birds
gather here, practicing their honeyed songs
all round, and play the organ of their throats,
mimicking lyres in song, while listeners 95
drink the sweet food the sound affords their ears.

[24]Boreas is the cold north wind, Notus the rainy south(west) wind. Zephyr (ten lines below) is the warm west wind.

Right in the midst, by beatific weeping
the ground sheds tears that generate a fountain
sobbing perpetually sweet drafts of water.
The silvery stream, returning to its pure 100
element's laws, and casting off all foreign
dregs, again sparkles with its native brightness.
Its drafts, inebriating pregnant Earth's
lap, stimulate a mother's prayers for childbirth.
Earth not unwillingly bestows like drafts 105
on trees, and prompts desire of offspring in them.
At the grove's center surges toward the sky,
kissing the clouds, a lofty mountain plain.
Here is the house of Nature built—if thus
one may denominate what numinously 110
surpassess starry homes of gods above,
scorning comparison with palaces.
Remote from our abodes, this happier hall
pierces the air, suspended on high columns,
flashing with starlike gems, ablaze with gold, 115
and equally adorned with silver's splendor.
Materials inferior to these,
or subject to decay, here have no rights.
A graceful picture here portrays men's manners,
faithfully cleaving to its proper function 120
lest the thing painted deviate from truth.
Strange miracles of painting! What is nothing
comes into being: painting apes the true,
sporting with new arts, and converts things' shadows
to things themselves, and every lie to truth. . . . 125

Settling each matter with deep insight, Nature
rules this domain with laws she providently
fashions, and promulgates throughout the world.
Causes and seeds of all she scrutinizes: 190
sees who redeemed old chaos with a fairer
aspect, when, mourning its own tumult, matter
solicited a more appealing form;
who, with trustworthy bond, curbed civil wars
and brothers' strife, giving the elements 195
the kiss of peace, and binding them with number;
who beholds earthquakes with a steadfast mind—
the roar of thunder, wrath of ocean, battles
of winds—and skilfully contains the weather's
assaults; why Winter, sorrowing at hoarfrost, 200
grieves, Springtime smiles, and, after Summer rages,
Autumn pours forth a mighty flood of riches;

why land stays still, streams flow, air circulates,
and fire soars, keeping faith with all the others;
why running water, daring not break faith, 205
covenants with the earth to hold one course. . . .

Nature's Passing Hour

The Romance of the Rose, by Guillaume de Lorris and Jean de Meun, translated by F. S. Ellis (revised)

The Roman de la rose, *an allegorical poem of 21,780 lines, "had a greater influence on medieval European literature," Heather M. Arden writes in her introductory study* The Romance of the Rose *(1987), "than any work after Boethius's* Consolation of Philosophy." *By no accident these were the two major works that Chaucer took part in translating. Less than a fifth of this immense poem (the first 4,058 lines) was written by its first author, Guillaume de Lorris, between about 1225 and 1230. At some time after his death, another author from the same region of the Loire valley around Orléans, Jean de Meun (or Meung), also called Jean Chopinel (or Clopinel), took up the unfinished poem and between about 1267 and 1278 wrote its continuation in the same meter—the octosyllabic rhymed couplet of poetry ranging from bawdy* fabliaux *to courtly romances—but in a strikingly different range of tones and styles.*

Guillaume's story is more unified in its evocation of the Lover's emotions and the obstacles he must surmount; Jean's gives free rein to apparent digressions of enormous length, and embraces a vivid cast of characters along with encyclopedic learning. The six selections that follow represent only a few of the poem's multifarious concerns. In **The Joys of Spring***—the only passage from Guillaume de Lorris—the poet describes his dream in the amorous month of May five years before. This sets the scene for his entry into the paradisiacal garden of Delight (Deduit), where he will see the Rose—first glimpsed in the fountain of Narcissus—that will be the object of his desires and frustrations. His approaches, despite instruction both by the God of Love and by Reason, are twice repulsed; and Fair Welcome (Bel Acueil), after permitting him at Venus's urging to kiss the Rose, is imprisoned by Jealousy.*

The long disquisition of Reason with which Jean's continuation begins fails to discourage the Lover, but he gladly hears a worldly-wise Friend (Amis) discoursing on ways to retain, by hook or crook, a woman's love. Our second passage, Friend's account of "earth's simpler ways" in **The Golden Age,** *comes immediately after his caustic denunciation of women—"Far different once was man's estate, / But now the world's degenerate"—and precedes the stridently antifeminist tirade of a Jealous Husband. After this, the God of Love rebukes the Lover for having momentarily listened to Reason, and with the help of False Seeming (Faus Semblant) and other worthies prepares to assault Jealousy's castle and liberate Fair Welcome. Having slain the doorkeeper, they employ the Old Woman (La Vieille) as a go-between; she delivers a cynical lesson in deceitful ways to Fair Welcome, denouncing the perfidy of men much as the Jealous Husband had railed at women's lechery. Our third passage,* **The Old Woman on Nature and Freedom,** *sets forth her credo that Nature "made each for all and all for each," so that woman's natural freedom is realized through sexual promiscuity but sacrificed (like a monk's in the cloister) in marriage. Thereafter, another attempt by the Lover on the Rose is fended off by Aloofness (Dangiers),*

Shame, and Fear, who reimprison Fair Welcome and repel the assaults of the God of Love and his vassals.

At this point, in our fourth passage, **Nature's Workshop,** Nature herself appears, sitting at the forge where she tirelessly labors, in abhorrence of death, at regeneration of life: works created by art (including Jean's poem) can do no more than copy Nature's living models. There follows **Nature's Confession to Genius,** of some three thousand lines, from which our fifth selection is taken. Here Nature relates how God, having created the world from nothing, made her his constable and vicar, so that everything on earth obeys her laws—with the one exception, she laments, of man. Man's much-abused reason, a "subtle and mysterious thing" given by God, lies outside her domain. Her confessor, the suave Genius, then goes to the camp of the God of Love and delivers a thousand-line sermon, which contains our final selection, **Genius's Exhortation to Follow Nature.** He then describes the "parc du champ joli" (park of the lovely meadow), where those who obey his command to hammer and plow at the work of procreation will live in eternal bliss as white lambs of Christ. Having ignited all women in the world with a flaming candle, Genius departs. The poem ends when Venus leads an assault on the castle and gives the Rose to the Lover, who plucks her after forcing his staff into the opening in a reliquary shaped like a beautiful woman: "Thus I had the crimson rose; then it was day, and I awoke."

Despite its pervasive influence and many imitations, Jean de Meun's poem was denounced as immoral and misogynistic by Jean Gerson, chancellor of the University of Paris, and by the poet Christine Pisan in the "quarrel of the Rose" at the start of the fifteenth century; it has been subject to widely different interpretations ever since. Because of the outspoken advocacy of natural (and especially sexual) impulse expressed by some of its characters, Huizinga declared it "impossible to imagine a more deliberate defiance of the Christian ideal," and Alan Gunn, in The Mirror of Love: A Reinterpretation of "The Romance of the Rose" (1952), thought the poem "a challenge to chastity, the most exalted in the medieval hierarchy of virtues." But others, notably John V. Fleming in The Roman de la Rose: A Study in Allegory and Iconography (1969) and Rosemond Tuve in Allegorical Imagery (1966), have stressed the irony of such "challenges." For them, Reason's discourse is the key to the poem; Amis and La Vieille are satirized as immoral and vulgar; and Nature and Genius are comic figures whose pronouncements cannot be taken at face value. Amis fatuously "thinks he can happily traipse through the terrestrial paradise clothed in his ill-fitting post-lapsarian human nature plucking fruits with impunity," Fleming writes; La Vieille's defense of women's liberation "is satire, not socialist realism"; and Genius, who embodies "natural concupiscence," is "a hypocritical friar of diabolically evil character." "For men to behave like beasts was not generally considered natural in the Middle Ages," Fleming observes. ". . . Bestial lechery was said to be particularly inappropriate to men and even 'against human nature.' . . . Indeed, indiscriminate abandon to the works of Venus, excused by La Vieille on natural grounds, was frequently said to denature (denaturare) man."

It is surely correct to see Genius—who promises heaven to those who indulge their unrestrained sexual desires—as "the satiric portrait of a demagogic preacher," as Arden suggests, "who has taken an idea (procreation has a function in the natural order) to an extreme (procreation is the only good), a monomaniac who has some familiarity with current theological disputes and much skill in rhetorical manipulation." His preachments can no more be taken seriously than those of the archpriest Bernart the Ass, who proclaims in a late branch of the parodic beast epic, the Roman de Renart:

Fucking seems suitable to me.
Therefore fucking shall never be
Forbidden, I say to all of you.

For fucking the cunt was split in two . . .
And those who follow my advice
Will find true bliss in paradise.[25]

At the same time, the values upheld by Nature—though certainly partial, and subordinate to higher values beyond her ken—clearly have a positive place in the poem; some passages, as C. S. Lewis affirms in The Allegory of Love *(1936), are "nothing less than a triumphal hymn in honor of generation and of Nature's beauty and energy at large." As Fleming acknowledges, "While Jean could realize a modest comic potential in Lady Nature's prolix femininity, he does not cease to respect her greatly and to see in her, damaged as she is, the majesty of the divine Creator."*

In this she both resembles and differs from the Nature of Alan of Lille. For as Arden writes, "Whereas unnatural sex stood for all the vices in the Plaint, *procreative sex is a metaphor for all the virtues in the* Rose." *Nature is, in Tuve's words, "not wrong but incomplete"; she resembles far less her fraudulent confessor Genius, or Bernart the Ass, than she does Chaucer's frequently married and equally irrepressible Wife of Bath:*

God bade us for to wexe and multiplye:
That gentil text kan I wel understonde.

The translation by F. S. Ellis (3 vols., 1900) has been revised to eliminate some obtrusive archaisms and to bring a few lines slightly closer to the original. See also the prose translation by Charles Dahlberg (1971). The marginal line numbers are those of Ellis's translation. For comparison with the Old French, approximate line numbers in Ernest Langlois's five-volume edition (1920–24) are given at the head of each selection; these differ only slightly from those of Félix Lecoy's more recent three-volume edition (1966–74), whose text is thirty lines shorter.

The Joys of Spring, by Guillaume de Lorris

Langlois 45–83

Five years, or more, have passed away, 45
I think, since in the amorous May
I dreamed this dream: O month of joy
When everything, without alloy,
Delights in life; when bush and brake
Again their vernal raiment take 50
Out of cold storage, where it lay
Neglected many a dreary day,
As woods and thickets don bright green,
Casting off winter's arid mien.
The lovely earth once more grows vain, 55
And, cheered by balmy dews and rain,
Forgets the indigent estate
In which cold winter made her wait;
For pride awakens new desire
To deck herself in bright attire, 60
And therefore does she fashion quaint

[25]Translated by Robert M. Torrance in *The Comic Hero* (1978) from branch XVII (873–76, 911–12) in vol. 2 of the edition of Ernest Martin (1885).

Lovely habiliments, and paint
Each with an iridescent hue,
Green herb, and flowers white, red, blue:
Sporting robes of such brilliant sheen, 65
Earth takes delight in being seen.
The merry birds that silence kept
While all the world through winter slept,
And wild winds roared, and skies were grey
With rain, burst forth to welcome May 70
With lusty notes, and let sweet song
Trumpet their joy that winter's wrong
Has vanished now, when gladly reigns
Sweet springtime over earth's domains.
Now nightingales with earnest voice 75
Constantly make delightful noise,
While larks and parrots stay awake
Rejoicing in the songs they make.
Responsive to such sweetness, soon
Young hearts throb to the amorous tune 80
Enrapturing the lovely spring.
Oh dull the soul that caroling
Of birds cannot delight when they
Sing piteously the songs of May!

The Golden Age, by Jean de Meun
Langlois 8355–8409

How pleasant were earth's simpler ways
In our progenitors' first days!
Old legends tell us how the fires
Of love burned bright amongst our sires; 8770
No man called this or that his own,
And lust and rapine were unknown.
During that glorious golden age
Simplicity was all the rage,
And none affected robes of state 8775
Or craved for overdelicate
Spiced meats, but simple woodland fruits,
Beech mast, or nuts, or wholesome roots
Out of the earth all needs supplied.
Fish and flesh then were left aside 8780
As needless; then through coverts wild
Men sought kind Nature's store of mild
And bloodless food; the untamed vine
Gave berries though men knew not wine,
And apples, pears and mulberries, 8785
Rich plums and chestnuts, beans and peas
They ate, and mushrooms from the field;

For valleys, plains, and heights would yield
Due sustenance from day to day.
From ears of corn they rubbed away 8790
The chaff betwixt their palms; they sipped
The brown bees' store which plenteous dripped
From ancient oak-tree boles; they drank
Clear water kneeling on the bank
Of crystal streams, nor added spice 8795
To delicacies of lofty price,
Nor longed for wine aged in the tun,
With weary treading hardly won.
They had no need to smash the earth,
Which, under God, unplowed gave birth. 8800
Thus amply fed, they had no wish
For salmon, pike, or any fish.
Sheepskins they wore against the harm
Of winter's cold, or thence wove warm
And simple garments, which no dye 8805
Fashioned from grain or herb came nigh.
Green broom or rushes roofed their cots,
Or else they hid in hillside grots
Fenced in with boughs, or hollow oaks
Gave shelter from rude winter's strokes. . . . 8810
And when at night they sought their rest, 8815
No beds of down their bodies pressed;
They strewed the ground with fragrant leaves,
Or moss in heaps, or fresh cut sheaves
Of grass or reeds, and heaven's sweet air
Was ever soft and gentle there, 8820
In one unvarying time of spring,
While tuneful birds made morning ring
With the sweet latin of their lay
That greeted every dawning day.

The Old Woman on Nature and Freedom, by Jean de Meun

Langlois 13875–14038

Women as free as men are born;
Only men's law has wrongly torn
Their charter, and that freedom riven
Away that was by Nature given. 14590
For Nature is not such a fool
To order by unbending rule
Margot to love but Robichon,
Nor that for him the only one
Should be Marie or fair Perrette, 14595
Jane, Agnes, or sweet Mariette,

But she, dear son, I scarce need teach,
Made each for all and all for each,
And every one for all alike,
Just as the taste and fancy strike. 14600
So that (although the marriage laws
Might interject a moment's pause),
To satisfy Dame Nature's call,
To which they hearken one and all,
And strifes and murders to avoid, 14605
To which their lusts might be decoyed,
Ever have women, foul and fair,
Whether the name of maid they bear,
Or wife, done all within their power
To win back freedom as their dower. . . . 14610

If any man beheld a dame,
In primal days, who lit a flame 14620
Within his heart, he thought no wrong
To ravish her, till one more strong
Should rob him of her, or he changed
His mind and somewhat further ranged.
But hence arose contention great, 14625
And homes were oft left desolate;
And so by wise men were laid down
Rules now as laws of marriage known. . . .

So each bright youth to servitude 14715
Goes who adopts the monkish hood,
For neither cowl, nor broad-brimmed hat,
Nor cloister gown, can smother that
Which Nature in his heart did plant,
And, unfulfilled, still leaves a want. 14720
He's worse than dead, for all his life
Is racked and torn by mental strife,
Or else with broken spirit he
Feigns virtue by necessity.
Dame Nature does not lie, but still 14725
His mind with bitter thoughts does fill
Of freedom lost. Horace this thing
Says well, 'tis worth remembering:
"Though any man should seize a fork,
To drive dame Nature from her work, 14730
Beat her, and chase her out of doors,
She'll quick return to pay old scores."[26]
What matters it? Do what you will,
Each living creature must fulfill
Its nature, and although you drive 14735
It far, it will return and thrive.

[26]See Horace, Epistle I.x.24 (quoted in the introduction to Horace in Chapter 11 above).

Nature despises violence,
And makes of man-made law small sense:
Thus Venus finds a good excuse
When from her trammels she breaks loose; 14740
And so with damsels, I allow,
Who chafe beneath the marriage-vow:
Nature it is who draws them still
Toward freedom, both for good and ill.
So strong is she, men seek to rein 14745
And curb her power all in vain.

Nature's Workshop, by Jean de Meun
Langlois 15893–16232

Nature, who tends to everything
Beneath vast heaven's circling ring,
Into her workshop entered straight,
Where she both early works and late, 16650
To forge such objects as may be
Used for the continuity
Of life; for she designs things so
That never shall any species know
Death's might, but as one creature dies 16655
Forthwith another may arise
To fill its place. In vain does death
With hurrying footsteps spend his breath. . . .
But even though a thousand died,
Nature another does provide.
And in this way does everything
That dies, through Nature once more spring
To life anew. Beneath the moon 16765
Whatever fails shall late or soon
Revive if only one remain
From whence the race new life may gain,
For Nature, pitiful and good,
Abhors and hates Death's envious mood, 16770
Who ruthlessly would mar and break
The fairest thing her skill can make,
And since no thing more fair can be,
Her own form she industriously
Stamps on her works, as men who mint 16775
New coins, put on them their imprint,
And form and color give to each.
Such workmanship Art strives to reach
By copying Nature's models, though
Products so perfect none can show. 16780
Art, falling on his knees before
Dame Nature, humbly does implore,

Beseech, and earnestly require
In suppliant form, that she inspire
His heart, if but in small degree, 16785
How he may copy carefully
Her handiwork, and reproduce
Its form, for ornament and use,
Acknowledging inferior far
His works to be than Nature's are. 16790
Each method Art does closely watch
And painfully essay to catch
Of Nature's working, as an ape
His gestures after man's does shape;
But vainly, vainly, Art may try 16795
To come near Nature's mastery.
To nothing that man's hand makes live
Can he her touch supernal give.
For Art, though he no labor shirk
To imitate great Nature's work, 16800
And set his hand to every kind
Of thing he may around him find,
Of whatsoever sort it be,
Painting and decking curiously
(And none of all the arts men leave 16805
Untried, but paint, dye, carve, and weave)
Warriors on chargers roan or white,
Caparisoned in colors bright,
Purple and yellow, green and blue,
And many another varied hue; 16810
Fair birds that pipe on branches green,
And fish in crystal waters seen,
And all the savage beasts that roam
In forest haunts, their native home;
And flowers and herbs in sunny glades, 16815
Which merry youths and gladsome maids
Go forth in pleasant days of spring
To gather in their wandering:
Tame birds, and beasts all unafraid
And games and dances in the shade, 16820
And noble dames in vesture fair,
In metal, wax, or wood with care
Portrayed, as they in life might stand,
And lovers clasping hand in hand:
But never on plank, cloth, or wall, 16825
Can subtlest art, whate'er befall,
Make Nature's figures live and move,
Or speak, or feel joy, grief, or love. . . .

Far better were it I had died
Than venture to indulge such pride
As think that I could comprehend, 17005
For all the pains I might expend,
Fair Nature's glorious paradise—
Beyond all words—past thought of price;
Nor though aloft my thoughts had flown,
That I should dare to write thereon; 17010
No—in my spirit so am I
Abashed, that fear my tongue does tie,
And so does shame my being steep,
That silence it behoves me keep;
For even as more and more I think 17015
Of Nature's loveliness, I shrink
From lauding, in my faltering phrase,
Her perfect works and wondrous ways.

Nature's Confession to Genius, by Jean de Meun
Langlois 16729–16800; 18967–19088

When God, whose beauties know no bound, 17545
First made this beautiful earth's round,
Whose wondrous form had always been
In providential thought foreseen:
How everything at last should be
In time, from all eternity 17550
(For from himself he had evolved
All that should be at last resolved,
Since, high or low though he might look,
Nought was yet writ in Nature's book
From which he could example take, 17555
For heaven and earth were yet to make:
Sun, moon and stars, and air and sea,
Vast chaos of immensity,
And all from nought he did create—
That God who is himself innate— 17560
Though to this work nought else did move
His will, but all-abounding love;
Perfect and pure, past envious strife,
Exists he, fount and spring of life),
Athwart infinity of space 17565
He made the world and fixed its place;
Out of a rude unshapely mass
To form and order all did pass
Beneath his will; the parts estranged
In perfect symmetry he ranged, 17570

And marked the proper bound to each
Division, whereto it should reach;
And formed all things in circles so
That each one should its function know. . . .
The lighter made he up to fly,
The heavier in the center lie, 17580
With intermediates between.
All this by God has ordered been
Aright, both as to time and space.
And when he had through bounteous grace
Disposed his creatures here and there 17585
With perfect knowledge, love, and care,
He then to honor me did deign,
Appointing me his chamberlain. . . .
His chamberlain! nay, constable 17595
Made me, and vicar general, . . .
Though but for his good grace am I
Unworthy of such dignity. 17600
God honors me as fit to hold
Within my hands the chain of gold[27]
Which the four elements enlace,
As all bow down before my face.
To me he granted all the things 17605
Within those interlinking rings,
Commanding me to watch their fate,
And all their forms perpetuate.
Everything must my laws obey,
Following where I point the way, 17610
Forgetting or omitting ne'er
Closely by my commands to fare
Through time to all eternity.
All as it was shall ever be
Observed, wherever shines the sun, 17615
By all my creatures—save by one. . . .

Nor do I of the elements
Complain, which work out my intents,
Blending together as it were 19795
The revolutions of the air.
Beneath the moon all living creatures,
I know, display corruption's features,
And never creature yet so well
Nourished itself, but that it fell 19800

[27]See Macrobius, *Commentary on the Dream of Scipio* I.xiv.15, trans. William Harris Stahl (1952): "Accordingly, since Mind emanates from the Supreme God and Soul from Mind, and Mind, indeed, forms and suffuses all below with life, . . . the close observer will find that from the Supreme God even to the bottommost dregs of the universe there is one tie, binding at every link and never broken. This is the golden chain of Homer [*Iliad* VIII.19] which, he tells us, God ordered to hang down from the sky to the earth." Jean de Meun puts Nature in charge of this Neoplatonic chain of being.

To death as Nature does direct
By ineluctable effect.
This is a rule so absolute
And fixed, that vain is all pursuit
Of means by which to change its course; 19805
Never does it abate its force.
Nor will I of the plants complain,
Whose loveliness to all is plain.
Faithfully following my laws
As the primordial spring and cause 19810
Of life, they duly send forth roots,
And boles and branches, flowers and fruits,
And then successive being give
To others, once they cease to live.
Nor with the birds or scaly fish 19815
Do I find fault; my every wish
Fulfill they with abundant care,
Proving what love to me they bear.
Each one I find a worthy scholar,
And all set shoulder to my collar; 19820
According to their kind and use,
All breed, engender, and produce
So that their lineage shall not die,
Which pleases me exceedingly.
Nor of the beasts whose heads are prone 19825
To look toward earth do I make moan,
For none against me do rebel,
But love my yoke, and serve me well:
To me they cling, and as I bid,
They act, as once their forebears did. 19830
Right merry festivals they keep,
When males upon the females leap,
Engendering, in their lustihood,
At any time they feel is good.
Thereof but small debate beasts make, 19835
But simply love for love's sweet sake;
What this desires, will that afford
With kind and debonair accord;
And with the blessings I provide,
All vow their hearts are satisfied. 19840
The smallest creatures men despise,
Beetles, ants, gnats, and butterflies,
And worms that from corruption come,
Finding in carcasses a home,
And snakes and adders (in whom lurk 19845
Poisons) delight to do my work.
Man only, unto whom I've given
Freely of all things under heaven,

Man, whom alone I form with face
Lifted to seek God's dwelling place,[28] 19850
Man, whom alone from earth's dull sod
I make in image of his God—
My last and fairest work—'tis he
Alone insults and angers me.
Yet not a thing in all his frame, 19855
Saving what through my bounty came,
No bodily trait, nor any member
Is worth a single lump of amber;
Even, indeed, his very soul
He owes to me, upon the whole, 19860
Except one part.
For from my hand,
Which exercises firm command,
Man has three powers of body and soul—
Truly might I affirm his whole
Existence he derives from me, 19865
The power to live, to feel, to be,
And would the wretch prove good and wise,
A glorious field before him lies,
For God's great love to him has given
Whatever lies beneath the heaven, 19870
That he may at his will employ
Them all, nor need thereof be coy.
His origin comes from the bones
Of mother Earth, Deucalion's stones;[29]
With thriving herbs he lives and deals, 19875
And with mute living beasts he feels.
Nay more, in understanding he
May with God's angels equal be.
What more of man then can I say?
Whatever he intends, he may. 19880
A small world in himself is he,
Yet worse than wolves behaves to me.
Man's intellect I recognize
As something that beyond me lies,
A subtle and mysterious thing 19885
Which was not of my fostering.
Whate'er is mortal count I mine,
But have no hand in things divine.
And Plato shows distinctly what

[28]See Ovid, *Metamorphoses* I.84–86, in Chapter 11 above.
[29]See note on Virgil's *Georgics* I.63 in Chapter 11 above. Here the French text says only, "Il a son estre avec les pierres" ("He has his being with the stones"), and no reference to the myth told by Virgil (and Ovid, among others) need be implied. Langlois cites a sermon attributed to Alan of Lille, which declares that man "has some resemblance with every creature: with stones in being; with trees and plants in living; with beasts in sensing; with spirits in reasoning."

Belongs to me, and what does not: 19890
When speaking of the gods that ne'er
Shall suffer death, he says: They were
By their creator ordered so,
That death they shall not undergo,
But subject to his will must be 19895
Their gift of immortality.
All Nature's works, all that draws breath,
Great Plato says, are doomed to death;
In God's sight vile are they, and must,
Their part once played, return to dust. 19900
Nature before the Almighty power
Of God has but a passing hour;
He as a lightning flash does see
Time past, time present, time to be.
Great Emperor is he, lord, and king; 19905
Unto the gods he says: You spring
From me as father. This well know
All learned men who read Plato;
The words he wrote when Greece was young
Go thus in our rude Frankish tongue: 19910
"O gods, your God am I, creator,
Father, and prime originator
Of all your being; every feature
You own proclaims each one my creature.
Nature but made you mortal; I 19915
Alone give immortality."[30]

Genius's Exhortation to Follow Nature,
by Jean de Meun
Langlois 19689–19735

In Nature's service now be quick
As squirrels amid the branches thick,
Swift as the wind, or merry bird
To love by happy springtime stirred.
Plenary pardon I bestow 20545
For all you do where'er you go.
In following Nature's high behest
Be diligent, and only rest
That work you may once more begin
When morrow dawns, new joys to win, 20550
Wage Nature's war ere stiff and cold
Your limbs become—worn, weak, and old. . . .

[30]See Plato's *Timaeus* 41A–B, which in Calcidius's translation (in *Plato Latinus*, ed. Raymond Klibansky, vol. IV: *Timaeus*, ed. J. H. Waszink [1962]) reads: "opera siquidem vos mea, dissolubilia natura, me tamen ita volente indissolubilia" ("you are indeed my works, dissoluble by nature, yet indissoluble if I wish it so").

To work, my masters, then, to work,
Seek not Dame Nature's laws to shirk;
Unless in labor you engage
With right good will, your lineage 20560
Must perish. Quickly seize the plow
With ready hands, and gladly bow
Your backs in manner of the sail
That bellies to the ruffling gale.
Plow-handles let your sturdy hands 20565
Grip, and across the fallow lands
Drive the bright coulter, while the share
Plays its due part, and then with care
Scatter around the precious grain;
In faith, 'twill render back again 20570
In autumn-tide a manifold
Rich harvesting of bearded gold,
Which stored within fair barns may keep
The wolf afar while winter's sleep
Enwraps the world. The human race, 20575
If labor fails, from off the face
Of earth must perish; nought can live
Unless with heart and soul men give
Themselves to work, and earnest will
Possesses them all gaps to fill 20580
Wrought in their ranks from day to day
By death, whose scythe knows no delay.
For as by Nature's laws men die,
So 'tis her will that they supply
Successors who may carry on 20585
The same good work that they have done
With unremitting ardor, and
With sons and daughters store the land
Which God created for men's use:
That done, you well may have excuse 20590
From hard laborious toil to rest.

Human Nature's Generative Nest

Purgatorio, Canto XXVIII, by Dante,
translated by Robert M. Torrance

Dante Alighieri (1265–1321) was born in Florence, where he studied philosophy and held office as a councilman and prior until factional disputes led to his exile in 1302; thenceforth he wandered from city to city under the patronage of various rulers, including Cangrande della Scala of Verona and Guido da Polenta of Ravenna, where he died and is buried. The formative experi-

ence of his spiritual and poetic life was his encounter, in a Florentine church in 1274, with Beatrice Portinari, who died in 1290; in The New Life (La Vita nuova), *he narrates his exalted love for her in prose intermixed with poems in the "sweet new style"* (dolce stil novo). *Beatrice would later become his guide through the heavenly Paradise in the masterpiece written during his exile, the* Divine Comedy, *comprising the* Inferno, Purgatorio, *and* Paradiso.

His description of the Earthly Paradise in Canto XXVIII of the Purgatorio *(here translated from Charles S. Singleton's edition of 1973) echoes—and encompasses—earlier accounts both of the pagan Golden Age and of the biblical Eden such as we have met in poets from Hesiod (Chapter 8) to Virgil and Ovid (Chapter 11), and from Dracontius and Avitus (Chapter 14) to Bernard Silvestris, Alan of Lille, and Jean de Meun in the present chapter. Having journeyed under Virgil's guidance through hell and emerged from the center of earth to the opposite hemisphere, Dante had laboriously climbed the purgatorial mountain and been joined partway up by a second companion, the Roman poet Statius (whom legend made a Christian), released from centuries of purgation to complete his climb toward heaven. In Canto XXVII he had passed through a wall of fire purging the sin of lust, drawn onward by Virgil's reminder that Beatrice lay beyond, and had dreamt of Leah, for whom Jacob had toiled in order to win her sister Rachel. "She with seeing, I with doing am satisfied," Leah says, linking herself and Rachel respectively with the active and the contemplative life. He then had risen at dawn and had climbed to the top step of the mountain, where Virgil had spoken his final words to him: "Free, upright, and whole is your will, and it would be wrong not to act according to its pleasure; wherefore I crown and miter you over yourself."*

Now, at the beginning of Canto XXVIII, eager and waiting no more, he enters a fragrant wood and sees on the other side of a stream (which he will learn is Lethe, the river that obliterates the memory of sin) a solitary maiden singing and plucking flowers; not until Canto XXXIII will he learn that her name is Matelda. She tells Dante that he and his companions are in the garden of Eden, and answers his puzzled questions with a discourse on the meteorology and botany of this seedbed of life. Location of the Earthly Paradise had been a much-discussed question in the Middle Ages. (See Chapter IX, "A Lament for Eden," in Singleton's Journey to Beatrice *[1958], and Chapter XI, "Il Mito dell'Eden" [1922], in Bruno Nardi,* Saggi di Filosofia dantesca *[1967].) Most often it was placed, as by Avitus and Bernard Silvestris, in the East; in the fourteenth century John Mandeville, in his* Travels *(trans. C. W. R. D. Moseley, 1983), located it east of the kingdom of Prester John, adding: "Of Paradise I cannot speak properly, for I have not been there; and that I regret." But in Dante it lies at the summit of Purgatory, which rises out of the sea in the uninhabited southern hemisphere of the globe directly across from the central point of the northern hemisphere, Jerusalem. It represents for Dante the potential recovery, insofar as possible for sinful man, of the once natural (though supernaturally given) state of human innocence before the Fall. But it is not his final goal, for above the earthly rises the heavenly Paradise, through which Dante will journey under the tutelage of Beatrice in the* Paradiso. *"Twofold, therefore, are the ends which unerring Providence has ordained for man," Dante wrote in* De Monarchia *(trans. Herbert W. Schneider as* On World-Government *[2nd ed. 1957]): "the bliss of this life, which consists in the functioning of his own powers, and which is typified by the earthly Paradise; and the bliss of eternal life, which consists in the enjoyment of that divine vision to which he cannot attain by his own powers, except they be aided by the divine light, and this state is made intelligible by the celestial Paradise."*

In his evocation of the perpetual springtime of the Earthly Paradise, in which the Golden Age dreamed by ancient poets on Parnassus had, perhaps, its reality, this poet of the world beyond pays glorious tribute to the world of nature. Like Plotinus (Chapter 10 above), he might well have asked: if the vision of God in the heavenly paradise "was not beautiful with an overwhelming beauty, what would be more beautiful than this visible universe," as distilled in the primal beauty of Eden? Nor had this supreme poet of the supernatural neglected nature outside this privileged enclosure, for throughout his poem vividly concrete images testify to his keen observation of the phenomenal world. "When he is fully himself . . . Dante declares his originality by deliberately avoiding the proverbial associations of a given species," such as the bestiaries and encyclopedias promulgate (Chapter 13), "and giving us instead," Patrick Boyde notes in Dante Philomythes and Philosopher *(1981), "his own independent observations. He describes the snail not for its slowness but for the eerie power of withdrawing its horns into its head. Similarly, his ant is not a symbol of industry and dedication ('Go to the ant, thou sluggard'). Dante has noticed how ants will move in columns to and from the nest, and how an individual ant will sometimes 'muzzle' a fellow ant proceeding in the opposite direction as if it were seeking news and information. . . . Passages like these show the sharp eye of the naturalist who will record such humble impressions as the wet fur of an otter, fish rising to the surface of a fish-pond, the wriggling twist of an eel, a dog scratching, an ox licking its nose, or two billy-goats butting each other."*

Here in the Earthly Paradise there are indeed no eels or billy-goats, for this is a sublimated idyll of nature in its first perfection. Matelda is like the lovely shepherdess of a pastourelle *who inspires love not of the flesh but of God. She cannot be simply reduced, Singleton argues in* Journey to Beatrice, *to a symbol, like Leah in Dante's dream, of the active as opposed to the contemplative life, but "figuring as she does that perfection of human nature which man enjoyed in Eden before his fall, is presented by the poet as figuring a perfection of nature not to be enjoyed by any living man." Such temporal perfection (which Adam himself enjoyed for only six hours [*Paradiso *XXVI]) prefigures the eternal perfection of divine revelation embodied by Beatrice, who will lead Dante on from this garden to the true and everlasting "orto / de l'ortolano etterno" ("garden of the eternal gardener"). After Matelda's first apparition, John D. Sinclair remarks in his edition of the* Purgatorio *(1939; rev. ed. 1948), the tone abruptly changes, "as it seems to us, from the lyrical beauty of the first part to the didactic and scientific manner of the rest," in which Matelda painstakingly explains the causes of winds and sources of waters in Eden. Yet in this change Dante "does not fall away from his imaginative purpose but fulfills it. Not only does the formal explanation of the phenomena of the mountain give it an added verisimilitude as something that can bear to be examined and accounted for, but it brings the garden under credible heavenly influences and connects it with the world we know; it tells of the soft, steady wind caused by the moving spheres on the summit which rises clear of the exhalations and disturbances of the earth below, the seeds which float from the garden and are self-sown round the world, and the rivers flowing from a supernatural and unfailing spring."*

The Earthly Paradise as Dante portrays it, like other visions of the Golden Age that it brings to completion, is thus both the perfection of nature and also—to the extent that it is *perfect in the changelessness of perpetual spring—the antithesis of nature in the world as we know it, of which change is the essence. Yet perhaps the paradox is merely apparent: for even this seemingly changeless and perpetual nature will in the end, like the world of which the mountain of Purgatory and the garden of Eden are part, pass away and be no more. The only truly lasting Paradise lies above it, in everlasting contemplation of the love that moves the sun and the other stars.*

Eager already to explore what way
 traversed that wood, dense, living, and divine,
 which tempered for my eyes the dawning day,
waiting no more, I left the slope aside
 and slowly, slowly took the path that leads 5
 over ground smelling fragrant far and wide.

Murmuring with uninterrupted ease,
whispers of air began to stroke my brow,
inconspicuous as a gentle breeze,
causing the quivering foliage to bow 10
readily all around me toward the part
the sacred mount began to shadow now:[31]
yet not so limberly did branches arch
that birds atop them ceased at any time
the endless variations of their art 15
but with full-throated exultation chimed
among the leaves, to greet the morning hours,
the unremitting burden of such rhyme
as gathers strength amid the pine-tree bowers,
from branch to branch, upon Chiassi's shore[32] 20
when Aeolus sets free Sirocco's powers.
Into that ancient wood my slow steps bore
me soon so far, I now no longer could
discern where I had entered it before;
and where a stream meandered past,[33] I stood 25
blocked by waves bending left the grasses near
the flowering banks that intersect the wood.
All sparkling waters that are purest here
would equally seem blemished by some stain
compared with *that*, through which all shapes gleam clear, 30
dark though it flows, and ever clear remain
under perpetual overarching shade
so dense, both sun and moon there shine in vain.
With feet I stopped and with my eyes surveyed
what lay beyond the stream, and gazed upon 35
the rich variety of that burgeoning glade:
when there appeared—just as a thought may dawn
suddenly on our mind and put to flight
others, in utter wonderment withdrawn—
a solitary lady fair of sight, 40
singing, and plucking flowers one by one
that painted all the path behind her bright.
"Lady, illumined by love's warming sun,
if your delightful semblance may compel
belief, by which heart's confidence is won, 45
step forth, I pray you now, that I may tell,
here where the stream is sweetly murmuring,
what song you sing so ravishingly well.

[31]Toward the west, the direction in which the mountain casts its shade in early morning. (Singleton)
[32]The pine groves along the Adriatic near Ravenna, where Dante is buried. Chiassi (ancient Roman Classis, modern Classe), though no longer near the coast, was an ancient seaport of the Western Roman Empire. Aeolus, in the next line, was the Greek god of the winds; the sirocco is the moist south wind that blows on southern Europe from Africa.
[33]This stream, as Dante will soon discover, is Lethe, the river of forgetfulness in classical myth.

Back to half-wakened memory you bring
where and what Prόserpine was when her mother 50
lost her, and with her lost perpetual spring."[34]
As when a lady stands with feet together,
then, in her dance, turns round in stately wise
and hardly puts one foot before the other,
toward me she turned, and to my awed surmise 55
seemed, amid yellow flowers mixed with red,
a modest virgin casting down her eyes;
then all my supplicating prayers she sped
by coming nearer, till I heard the sound
her sweet voice made, and every word she said. 60
Reaching the place where grasses all around
were freshly bathed by waves of that fair river,
she raised her eyes, and thus my rapture crowned.
I think from under Venus' eyelids never
flashed so much light, beholding gashes bleed 65
with barbs shot from her son's misguided quiver.[35]
Standing across the stream, she smiled indeed,
fingering flowers grown on every side
of this high land, engendered by no seed.
Three paces only was the river wide, 70
yet Hellespont, which once King Xerxes crossed—
that bridle to recurrent human pride—
Leander no more hated then, when tossed
toward Sestos from Abýdos on the crest
of waves, than I did this, when hope seemed lost.[36] 75
"Since I am smiling, and you three are guests,"
she then began, "new to this place designed
as human nature's generative nest,
astonishment and doubt have kept you blind:
let the psalm *Delectasti* shed its light 80
to scatter mists now darkening your mind.[37]
And you who questioned me here in my sight,
tell me what more you wish to learn, for I
will answer all your inquiries aright."

[34]For the story of Ceres (Greek Demeter) and Prosérpina, or Prόserpine (Greek Perséphonê), see the Homeric Hymn to Demeter in Chapter 8 above and the lines quoted from Ovid's *Metamorphoses* (V.388–96), perhaps Dante's principal source, in the introduction to Chapter 14 above.

[35]While Venus was kissing her son Cupid, she was unintentionally wounded by him and fell madly in love with Adonis. The incident is recounted by Ovid, *Metamorphoses* X.525–532. (Singleton)

[36]Xerxes, king of Persia, crossed the Hellespont (Dardanelles) to conquer Greece in 480 B.C., but after defeat of his navies at Salamis he fled back without his armies. Herodotus tells the story in Books VII–VIII, but Dante probably knew it from the first Christian historian, Paulus Orosius, an early-fifth-century pupil of St. Augustine's (*Histories* II.x.8–10, as translated by Singleton in his note). For the story of Leander, drowned when swimming the Hellespont from Abydos to Sestos to visit his love, Hero, priestess of Venus, see Ovid's *Heroides* XVIII–XIX.

[37]Matelda refers Dante and his two companions to Psalm 91:5–6 of the Vulgate, "Quia delectasti me, Domine, in factura tua; et in operibus manuum tuarum exsultabo. Quam magnificata sunt opera tua, Domine" (Psalm 92:4–5 in RSV: "For thou, O LORD, hast made me glad by thy work; at the works of thy hands I sing for joy. How great are thy works, O LORD!"). Her smile is from her joy in God's works.

"Water, and forest sounds," I made reply, 85
"seem to be contradictory to my new
belief, which their appearances deny."[38]
Whence she: "I shall explain how all is due
to one cause, wondrous though it seem to be,
and dissipate the mist surrounding you. 90
The Highest Good, who is entirely free
to please Himself, created man for good,
pledging peace here throughout eternity.
By his own fault, man forfeited this wood;
by his own fault, changed honest mirth and play 95
for sorrow, and what recompense he could.
So that disturbances—once under way
from exhalations both on sea and land
that follow heat however far they may—
might not subject mankind to their command, 100
heavenward rose this mountain from the ground
above capriciousness forever banned.
Because the air is circling all around
in correlation with the primal sphere[39]
(if no break in its orbiting be found), 105
its movement strikes this height, serenely clear
in the emancipated living air,
and makes the dense wood echo far and near.
Plants, when thus struck, possess a force so rare
their potency impregnates every breeze, 110
which, circulating, spreads it here and there;
the other land,[40] so far as worth decrees
it fit in soil and climate, soon bears fruit,
with varied growths of varied properties.
Be not amazed, then, once this attribute 115
is understood, if yonder, by its grace,
without apparent seed some plant takes root,
knowing that here the sacred fertile place
wherein you stand is full of every seed,
with fruits of which, beyond, there is no trace. 120
The water you observe does not proceed
from vapors cold condenses on this mount,
much as a river gains, then loses speed,

[38]In *Purgatorio* XXI.43–57 (trans. Singleton), Statius explained to Dante that Purgatory "is free from every change . . . wherefore neither rain, nor hail, nor snow, nor dew, nor hoarfrost falls any higher than the short little stairway of three steps" at the gate near the bottom of the mountain. " . . . It trembles perhaps lower down, little or much, but up here, from wind that is hidden in the earth, it never trembles, I know not how." Dante asks Matelda to explain the apparent discrepancy between the "new belief" he has gained from this explanation and the streams and breezes he encounters here in the Earthly Paradise.
[39]The primum mobile, the outermost sphere of the heavens, by its turning causes not only the revolution of the fixed stars but also the movement of the atmosphere.
[40]The inhabited northern hemisphere opposite from the Mount of Purgatory. Only some of the seeds blown from the Earthly Paradise can grow there.

but issues from a sure and steady fount
open in two directions, by God's will 125
losing and gaining back the same amount.
Here on this side, meandering downhill,
all memory of sin it nullifies;
there restores memory of deeds done well.
On this side, 'Lethe,' in its other guise 130
'Eunoe' it is called;[41] until each one
is tasted, their effect cannot arise:
this one in sweetness cannot be outdone.
And notwithstanding that by now your thirst
ought to be slaked, though I have scarce begun, 135
a corollary I shall now disburse,
nor think that you will find my words less dear
if they surpass the promise made at first.
Those in whose ancient verses did appear
the golden age of unsurpassed content, 140
dreamed on Parnassus then, perhaps, of here.
Here was the root of mankind innocent;
eternal spring brought forth perpetual fruit:
by nectar, *this* the poets truly meant."
I turned that very moment to salute 145
my poets, and beheld them in that place
smiling at her last words, and standing mute:
then toward that lovely lady turned my face.

In Lofty Mountains and Wild Woods

Selections from the Letters and Poems of Petrarch, translated by Robert M. Torrance

Francesco Petrarca (1304–74), known as Petrarch in English, was born in Arezzo two years after his father Pietro—from whose nickname "Petracco" his son later took his surname—had been exiled (a few months after Dante) from his native Florence. After living with his mother at Incisa in Florentine territory and in Pisa, he moved with his family in 1312 to Carpentras in Provence, fifteen miles from Avignon. Pope Clement V had transferred the papal court to Avignon in 1309, and it remained there, despite Petrarch's pleas for its return to Rome, all his life. He studied law from 1316 to 1320 at Montpellier and then at Bologna, returning at his father's death in 1326 to Avignon. There, in a church (one of the few places an Italian woman could *be seen by a stranger), he beheld the "Laura" to whom the Italian love poems of his* Canzoniere, *or* Rime, *were written, both in her lifetime and after her death in 1348.*

[41]Unlike the traditional Lethe, river of forgetfulness, Eúnoë, the river of good remembrance, was Dante's invention.

In 1330 he entered the service of Cardinal Giovanni Colonna, and in 1341 the growing fame of his Latin poetry culminated in the proudest moment of his life, when he was crowned poet laureate (in what he mistakenly thought was an ancient Roman custom) after examination by King Robert of Naples, and made a citizen of Rome. Sojourning in various cities of Italy (he spent eight years in Milan), and returning frequently to his home in the valley of Vaucluse near Avignon, he took part in the discovery of important manuscripts, including Cicero's Letters to Atticus, *edited the first critical text of the Roman historian Livy, and promoted both a Latin style purified of medieval "barbarisms" and a "new learning" that deprecated Scholastic Aristotelianism in favor of Plato and the ancient Romans, especially Cicero and Virgil. But his love of Rome extended beyond classical to Christian writers—above all St. Augustine—and beyond ancient to modern Italy, for whose revival he fervently hoped.*

Despite his urbane sophistication (he was among the most famous men in Italy when he died), Petrarch was a lover of natural beauty who could humorously portray himself—in a Latin letter of 1362 to Francesco Bruni (translated by Hans Nachod in The Renaissance Philosophy of Man, *ed. Ernst Cassirer et al. [1948])—as "a backwoodsman who is roaming around through the lofty beech trees all alone, humming to himself some silly little tune, and—the very peak of presumption and assurance—dipping his shaky pen into his inkstand while sitting under a bitter laurel tree." We learn more about him from his own writings than about any Western writer since St. Augustine. He despised Avignon as an infernal region worse than Babylon or Tartarus, but of his beloved Vaucluse he wrote, in a letter of 1347 to Cola di Rienzo (translated in David Thompson's anthology* Petrarch: A Humanist among Princes *[1971]): "The hills cast a grateful shadow in the morning and in the evening hours; and at noon many a nook and corner of the vale gleams in the sunlight. Round about, the woods lie still and tranquil, woods in which the tracks of wild animals are far more numerous than those of men. Everywhere a deep and unbroken stillness, except for the babbling of running waters, or the lowing of the oxen browsing lazily along the banks, or the singing of birds."*

Our first selection, from Petrarch's Latin letter to Dionigi da Borgo San Sepolcro, describes his ascent, with his brother Gherardo, of Mont Ventoux in Provence in 1336; though dated April 26 of that year, the letter was almost certainly composed much later, in 1352–53. It splendidly combines the appreciation of landscape and urge for exploration typical of the Renaissance—the "indefinable longing for a distant panorama" that Jacob Burckhardt, in his Civilization of the Renaissance in Italy *(1860), attributed to Petrarch—and the allegorization of human experience characteristic of the Middle Ages. (Gherardo, who took the straight upward path, became a monk in 1342.) Our second selection, "Di pensier in pensier, di monte in monte," probably written in Parma between May 1341 and January 1342, is one of the finest "songs"* (canzoni) *intermingled with the sonnets of the* Canzoniere. *Separated by the Alps from his love, whose image is always with him, Petrarch finds what tranquility he can not among men but in the pathless woods and mountains. His intense expression of the unappeased restlessness of conflicting emotions was one principal reason for Petrarch's unrivaled influence on European poetry for over two centuries, and for his continuing appeal, two centuries later, to Jean-Jacques Rousseau (Chapter 21 below).*

For Italian texts of the poems, with English verse translations, see Petrarch, Sonnets and Songs, *trans. Anna Maria Armi (1946). Translation of the Latin letter is based on the text in vol. I of Francesco Petrarca,* Le Familiari, *ed. Ugo Dotti (1974).*

FROM *The Ascent of Mont Ventoux*
Epistolae familiares *IV.1*

Today, led solely by desire to see so famous a height, I ascended the loftiest mountain of this region, which they rightly call Ventosus, the Windy.[42] For many years I had had this journey in mind; for ever since infancy, as you know, I have been tossed about in this region by the fate that tosses human affairs, and this mountain, visible far and wide from everywhere, has been almost always in my view. At last the impulse seized me to do what I had thought of doing every day: specifically, after I encountered by chance, while rereading Roman history in Livy a few days ago, the passage where Philip, king of Macedon—he who waged war on the Roman people—ascends Mount Haemus in Thessaly, from whose summit he believed the rumor that two seas could be seen, the Adriatic and the Black.[43] Whether he was right or wrong, I have no way of knowing for certain, both because the mountain is far from our country, and because differences among writers make the matter doubtful. The cosmographer Pomponius Mela, to mention but one of many, unhesitatingly affirms its truth; Titus Livy supposes the rumor false. If experience of that mountain were as readily available to me as it was of this one, I would not long leave the question in doubt. . . .

But as I thought about a traveling companion, scarcely any of my friends, to my surprise, seemed suitable in all respects . . . At last I turned to my own household for help and disclosed my plan to my younger and only brother, whom you know well. He could have heard nothing more gladly, pleased that he would hold the place of my friend as well as my brother.

Leaving home on the appointed day, we reached Malaucène towards evening; this place is at the foot of the mountain, on the northern side. There we spent a day; today we ascended the mountain, each attended by one servant, not without great difficulty, for it is a steep and almost inaccessible rocky mass of earth. But the poet said it well:

Toil conquers all;
obdurate toil.[44]

The long day, the mild air, the vigor of our minds, the strength and dexterity of our bodies, and other things of this kind helped us on our way; only the nature of the place obstructed us. We found a shepherd of advanced years in the hollows of the mountain, who endeavored with many words to hold us back from the ascent, saying that fifty years before, in a surge of just the same youthful ardor, he had climbed to the top of the mountain, and had brought back nothing but regret and pain, his body and cloak alike ripped by rocks and thorns. Never before or after had anyone been heard of who dared do a

[42]The name of this mountain, northeast of Carpentras, near Avignon and Vaucluse in Provence, Hans Nachod notes in his translation of this letter in Cassirer et al., eds., *The Renaissance Philosophy of Man* (1948), "appears as 'Ventosus' ['Windy'] in Latin documents as early as the tenth century, though originally it had nothing to do with the strong winds blowing about that isolated peak. Its Provençal form 'Ventour' proves that it is related to the name of a deity worshiped by the pre-Roman (Ligurian) population of the Rhone Basin, a god believed to dwell on high mountains . . ."

[43]As Nachod remarks, "In his *History of Rome* (xl.21.2–22.7) Livy tells that Philip V of Macedonia went up to the top of Mount Haemus, one of the highest summits of the Great Balkans . . . , when he wanted to reconnoiter the field of future operations before the Third Macedonian War, which he was planning to fight against the Romans (181 B.C.). Since Petrarca knew the exact location of this mountain from Pliny's *Natural History* (iv.1.3 and xi.18.41), it must have been a slip of his pen that made him substitute 'Thessaly' for 'Thrace.'"

[44]Virgil, *Georgics* I.145–46, changing the past tense (*vicit*) to the present (*vincit*). See Chapter 11 above.

similar thing. While he was shouting these words, his very prohibition increased our desire, for young people's minds pay no attention to advisers. When the old man noticed that he was exerting himself in vain, he went forward with us a little way through the rocks and with his finger pointed out a steep path, giving us much advice and repeating it often after we had left him some distance behind.

We left with him any clothes and other objects that might burden us, taking only what facilitated our ascent, and joyously began to climb. But, as usually happens, swift fatigue followed great effort; not far from there we rested on a rock. Thence we advanced again, but more slowly. I in particular picked my way up the mountain at a more leisurely pace, while my brother attempted to gain the heights by a short cut along the ridge of the mountain itself; being less hardy, I inclined toward the lower regions, and when he called me back and indicated the better way, I replied that I hoped to find an easier approach on the other side, and did not shrink from a longer way by which I might go more smoothly. This excuse I lazily offered, and when the others had already gained the upper regions, I was still wandering through the valleys, since no easier access lay open in any direction: instead, the way became longer and the vain labor more burdensome. Meanwhile, overcome by loathing, I repented my puzzling error, and determined to assail the heights wholeheartedly; worn out and weary, I reached my brother, whom a long rest had refreshed, and we went on for a while at the same pace. Hardly had we left that ridge, however, when I forgot the roundabout path I had taken a while before and again headed downward, again wandered through the valleys seeking a longer and easier path, and falling into longer difficulties. Thus I doubtless delayed the trouble of climbing, but nature is not annulled by human contrivance, nor can anything corporeal reach the heights by descending. Why waste words? To my brother's amusement and my own annoyance, this happened thrice or more within a few hours. Thus frequently deluded, I sat down in a valley. There my swift thought leapt from corporeal to incorporeal things, and I addressed myself in such words as these:

"What you have so often experienced while ascending this mountain today happens both to you, be assured, and to many others making their way toward the blessed life; but this is not easily understood by men, because the body's motions are manifest, those of the mind are invisible and hidden. Truly the life we call 'blessed' is placed on a lofty height; 'Narrow is the road,'[45] as they say, that leads to it. Many hills rise up between, and we must walk 'from virtue to virtue' with determined steps; at the top is the goal of all, the end toward which our pilgrimage is ordained. All want to arrive there, but as Naso says:

Wanting is little: yearn, and you shall have.[46]

You, certainly—unless you deceive yourself in this as in many other things—not only want but yearn. What keeps you back, then? Nothing, surely, except that the way through vile earthly pleasures at first appears smoother and less obstructed . . . " You would never believe how much this thought exalted my mind and body for what remained. Would that my mind might accomplish that journey for which I sigh day and night, just as my bodily feet accomplished today's journey, overcoming all difficulties at last . . .

There is a hilltop higher than all the others . . . where at last we found rest from weari-

[45]Matthew 7:14 (Sermon on the Mount).
[46]Ovid (Publius Ovidius Naso), *Ex Ponto* III.1.35.

ness . . . The frontier of France and Spain, the ridge of the Pyrenees, cannot be seen from there, though no barrier that I know of intervenes, except the weakness of mortal sight; but the mountains of the province of Lyons were clearly visible to the right and, to the left, the sea at Marseilles, and that which pounds against Aigues Mortes, several days' journey from here. The Rhone was right under our eyes.

Having admired each detail, now savoring things of the earth, now elevating my mind to higher regions after the example of my body, I thought it fitting to look into the volume of Augustine's *Confessions* that is the gift of your kindness, and which I preserve and keep always at hand in remembrance of both author and donor:[47] a book of tiny size but of infinite sweetness. I opened it to read whatever I happened upon; for what could I find that was not pious and devout? By chance the tenth book of the work offered itself to me. My brother stood intently expecting to hear something of Augustine from my lips. I call God as my witness, and him who was with me, that where I first fixed my eyes, it was written: "And men go to admire the mountains' heights, the vast floods of the sea, the broad streams of the rivers, the embrace of the ocean, and the orbits of the stars—and forsake themselves."[48] I was stunned, I confess; and asking my brother, who wanted to hear more, not to bother me, I closed the book, angry with myself that I still admired earthly things, though I ought long since to have learned even from pagan philosophers that "nothing is admirable except the mind, beside whose greatness nothing is great."[49]

Satisfied with what I had seen of the mountain, I turned my inner eyes on myself, and from that hour no one heard me speak till we arrived at the bottom. . . .

How often, do you think, on my way down today, did I turn back and look up at the mountaintop? It seemed scarcely more than a cubit high compared to the height of human contemplation, were this not plunged in earthly filth and squalor. "If you are willing to undergo so much sweat and toil to bring the body a little nearer to heaven," I mused at every step, "what cross, what dungeon, what rack should frighten a mind that is nearing God and trampling upon the inflated peaks of insolence and on mortal fate?" And again: "How few will not swerve from this path because of fear of hardship or desire for luxury?" Truly fortunate would such a man be, if he exists anywhere! Of him I should think the poet meant to speak:

> Happy the man who plumbed things' hidden causes,
> trampling fears of inexorable fate
> underfoot, and loud-roaring Ácheron![50]

How eagerly should we toil to place underfoot not a higher spot of ground, but appetites swollen by earthly impulses!

Amid such agitations of my storm-tossed breast, inattentive to the rugged path, I returned in the depth of night to the little rustic inn from which I had left before dawn; the moon, shining all night long, offered welcome aid to the travelers. While the ser-

[47]The small-sized manuscript codex of Augustine's *Confessions*, a present from Dionigi, accompanied Petrarca wherever he went until the last year of his life, when he could no longer read its minute script and gave the book to Luigi Marsili . . . as a token of friendship. (Nachod)

[48]Augustine, *Confessions* X.8.15. Petrarch's reported experience somewhat parallels that of St. Augustine himself when, in the crisis of his conversion to Catholic Christianity, he opened the New Testament and found in the first passage on which his eyes fell (Romans 13:13–14) a message to put aside reveling and debauchery and "put on the Lord Jesus Christ" (*Confessions* VIII.12.29).

[49]Seneca, *Moral Epistles* VIII.5.

[50]Virgil, *Georgics* II.490–92 (Chapter 11 above). Petrarch points his Christian moral by quoting the lines of a pagan poet in praise of the Epicurean materialist Lucretius (Chapter 10).

vants busied themselves preparing supper, I withdrew all alone into a secluded part of the house to write you this letter extemporaneously and in haste, lest my intention to write, if I deferred it, might cool, and my feelings might alter with the change of scene. . . .

From Thought to Thought, from Mountaintop to Mountain
Canzoniere *CXXIX*

From thought to thought, from mountaintop to mountain
Love leads me on, for any well-blazed trail
is antithetic to my tranquil mood.
If, by some lonely shore, some brook or fountain,
between two hillocks lies a shady vale, 5
there the dejected soul finds quietude.
Yet, since Love makes it brood,
it laughs, it cries, it fears, it dares to hope;
and, trailing after it, a face is seen
clouded now, now serene, 10
giving each feeling momentary scope:
seeing this, one experienced in life
would say, He burns with neverending strife.

In lofty mountains and wild woods I find
some respite; any place where people dwell 15
seems in my eyes a mortal enemy.
New fancies of my lady spring to mind
at every step, converting all this hell
I feel on her account to mockery;
and should I wish to flee 20
the bittersweetness of the life I live,
I think: Perhaps some destiny less ill
Love is preparing still:
vile though you are, some pleasure might you give!
In such a state I wander, sighing now: 25
Could it then truly be? but when? and how?

Where a tall pine or hillside offers shade,
sometimes I halt, and on the nearest stone
draw with my mind the beauty of her face.
Awakening, I find teardrops have made 30
my bosom moist, and murmur all alone:
Where have you come to? from what distant place?
So long as I can trace
my first thought, steadying my rambling mind,
and gaze on her, forgetting I am here, 35
then I feel love so near,
my soul finds happiness in being blind.
So beautiful I see her everywhere:
should error last, for nothing else I care!

Often (but who believes what I shall say?) 40
vividly in clear streams and on green grass
I've seen her figure, in a smooth beech tree,
or on a white cloud, formed in such a way
Leda would swear her beauty did surpass
her child's,[51] as sunbeams force bright stars to flee. 45
Wild though the spot may be
in which I am, and desolate the shore,
so much more beautiful in thought she seems;
then, when my lovely dreams
fade before truth, I sit here as before, 50
cold, a dead stone upon a living stone,
like one who thinks, and weeps, and writes alone.

Up to where loftiest pinnacles arise
untouched by shadows of some other peak,
burning desire impels me ever on. 55
Thence I begin to measure with my eyes
my sorrows, and with flowing tears then seek
to ease the heart dank mists have sunk upon,
thinking how far I've gone
from the fair beauty of her distant face 60
that even so remains intensely near.
Soft to myself, but clear
I cry: What madness! Maybe, in that place,
one sighs that *she* is far away from you:
and with this thought my spirit breathes anew. 65

My song, beyond that alp,
where the sky shines more radiantly gay,
you shall behold me, by a running stream
where groves of laurel teem
with the cool aura of the fragrant bay:[52] 70
there is *she*, there my heart, which she bereft;
here nothing but my effigy is left.

A Marvel Beyond Denial

Sir Gawain and the Green Knight, translated by J. R. R. Tolkien

The Middle English Sir Gawain and the Green Knight *is the last of four poems—following* Pearl, Purity *(or* Cleanness*), and* Patience, *generally assumed to have been written by the same author in the last quarter of the fourteenth century—that survived in a single manuscript*

[51]Helen of Troy, daughter of Leda by Zeus.
[52]Here, as often in the *Canzionere*, Petrarch plays on the similarity of the words *l'aura* (the breeze) and *lauro* (laurel)—or, in this line, *laureto* (laurel grove)—to Laura, the name of his lady.

dated around 1400. Its anonymous author would thus appear to have been a younger contemporary of Petrarch, yet no poem could be more different from the Italian poet's mellifluous love songs than this masterpiece deliberately archaic in both its subject matter and its alliterative verse form. It is one of the latest, and one of the finest, Arthurian romances in a literary tradition that began with Geoffrey of Monmouth's Latin History of the Kings of Britain *in 1137 and continued, through Chrétien de Troyes's French romances of the late twelfth century and the many poems influenced by them—notably Gottfried von Strassburg's* Tristan *and Wolfram von Eschenbach's* Parzifal—*down to Malory's* Morte Darthur, *posthumously printed by Caxton in 1485. Nature is not a central theme of these romances, yet the scene of a hunt or a falconing expedition is often vividly painted. And in the Waste Land and the forest hermitages of the Grail legend, or the Cave of Lovers in Gottfried's* Tristan, *various aspects of the natural world take on a thematic and not merely a decorative function.*

But the English Sir Gawain, *though drawing on earlier sources, is unique. Larry D. Benson, in* Art and Tradition in Sir Gawain and the Green Knight *(1965), notes that the "beheading tale" of the first part of the poem appears in ten other surviving medieval works (two Irish, four French, two German, and two English), and there are many analogues for the testing of Gawain's chastity in the second part; but no previous poet had combined these two stories. The stanzas here included from J. R. R. Tolkien's translation (1975) relate how on New Year's Day a horseman with green hair and skin rides a green horse into Camelot and rudely challenges one of Arthur's knights to strike him with his axe and receive a return blow in a year's time. When Gawain takes up the challenge and strikes off his head, the green knight picks it up, bids Gawain meet him at the Green Chapel a year later, and rides off. The rest of the romance tells how the year goes by and the seasons change (stanza 23):*

After the season of summer with its soft breezes,
when Zephyr goes sighing through seeds and herbs,
right glad is the grass that grows in the open,
when the damp dewdrops are dripping from the leaves,
to greet a gay glance of the glistening sun.
But then Harvest hurries in, and hardens it quickly,
warns it before winter to wax to ripeness.
He drives with his drought the dust, till it rises
from the face of the land and flies up aloft;
wild wind in the welkin makes war on the sun,
the leaves loosed from the linden alight on the ground,
and all grey is the grass that green was before:
all things ripen and rot that rose up at first,
and so the year runs away in yesterdays many,
and here winter wends again, as by the way of the world
it ought,
until the Michaelmas moon
has winter's boding brought;
Sir Gawain then full soon
of his grievous journey thought . . .

Gawain sets out, on All Saints' Day, in search of the Green Chapel. Finally, on Christmas Eve, he comes upon a castle in a wild forest whose lord entertains him lavishly. On each of the three days before the New Year, however, while the lord of the castle is out hunting, his beautiful wife visits Gawain and offers herself to him. He politely turns her down, but agrees on the third day to accept the green silk girdle that she says will protect him from being wounded. On New

Year's Day he rides to the Green Chapel to receive the Green Knight's blow. He shrinks at the first two feints; on the third swing, the Green Knight nicks his neck, then reveals that he is the lord of the castle, Bercilak de Hautdesert, and has drawn blood because Gawain had broken his troth by taking the girdle. Bercilak praises Gawain for his virtue, but Gawain feels only shame and wears the green girdle back to court as a sign of his fault. There he is greeted by laughter, and all the lords and ladies of the Round Table agree to wear a green baldric in solidarity with the imperfect virtue of this very human knight. "In this Gawain, the blithe young embodiment of chivalry at its best," Laura Hibbard Loomis writes in "Gawain and the Green Knight" (1959),[53] *"goodness is made manifest and radiant, but not, as in Galahad of the Grail romances, a supernatural virtue touched by a mysterious divinity. . . . . Unlike other Arthurian heroes, he returns to Arthur's court, not in conventional glory, but in self-confessed shame. Yet . . . that shame gave him new grace, and the Round Table achieved a new nobility by its act of compassionate fellowship."*

But while the poem thus presents a tolerantly flexible lesson in virtue and honor, it is no less importantly a reminder of the deep connection between the human and the natural worlds. Though the Green Knight cannot be reduced to the "Green Man" or "May King" of popular folklore, he has unquestionable affinities with the vegetative vitality embodied by such figures, and also—as Benson notes—with the Wild Man who "seems to have had the same sort of ritual origin as the green man. In popular belief the two figures are closely linked, and in folk ritual they are interchangeable. . . . However, in literary works the two figures are quite distinct. A close association with nature is the only quality they share, and even this reveals a contrast. Spring and greenery are the natural phenomena associated with the green man; he develops from the pleasant aspects of nature, and in literature he becomes an attractive, youthful figure. The wild man seems to have developed from the sterner side of nature. Winter is the more suitable season for him, and, in folklore, he delights in storms and rides with the Wild Hunt."

The Green Knight of our poem reunites these opposites in one. His green skin, the one element in his description unparalleled in the poem's sources and analogues, "is the color of life as well as death . . . and it is associated with spring and rebirth . . . It is the ambiguity of the greenness and the relevance of its ambiguous implications to the challenger's character that maintains the balance of attractiveness and fearfulness that the combined figures of the literary green man and wild man produce," allowing the poet to emphasize "the one characteristic they have in common, their association with nature." And if Gawain is a preeminently human hero, the Green Knight embodies, as Francis Berry suggests,[54] *"something even more primary," the creative energy that underlies the perennial fertility of the natural world. "In the poem, Gawain and his 'society' humbly come to terms with the Green Knight. They had been in danger of forgetting their own* sine qua non"*—the roots that even the most golden and chivalric of cultures always have in the repeatedly resurrected greenness of nature.*

3

This king lay at Camelot at Christmas-tide,
with many a lovely lord, lieges most noble,
indeed of the Table Round all those tried brethren,
amid merriment unmatched and mirth without care.

[53]Reprinted in Donald R. Howard and Christian K. Zacher, eds., *Critical Studies of Gawain and the Green Knight* (1968).
[54]"Sir Gawayne and the Greene Knight," in Boris Ford, ed., *The Age of Chaucer* (1954).

There tourneyed many a time the trusty knights, 5
and jousted full joyously these gentle lords;
then to the court they came at carols to play.
For there the feast was unfailing full fifteen days,
with all meats and all mirth that men could devise,
such gladness and gaiety as was glorious to hear, 10
din of voices by day, and dancing by night;
all happiness at the highest in halls and in bowers
had the lords and the ladies, such as they loved most dearly.
With all the bliss of this world they abode together,
the knights most renowned after the name of Christ, 15
and the ladies most lovely that ever life enjoyed,
and he, king most courteous, who that court possessed.
For all that folk so fair did in their first estate
abide,
Under heaven the first in fame, 20
their king most high in pride;
it would now be hard to name
a troop in war so tried.

4

While New Year was yet young that yestereve had arrived,
that day double dainties on the dais were served,
when the king was there come with his courtiers to the hall,
and the chanting of the choir in the chapel had ended.
With loud clamor and cries both clerks and laymen 5
Noel announced anew, and named it full often;
then nobles ran anon with New Year gifts,
Handsels, handsels they shouted, and handed them out,
Competed for those presents in playful debate;
ladies laughed loudly, though they lost the game, 10
and he that won was not woeful, as may well be believed. . . .

5

But Arthur would not eat until all were served;
his youth made him so merry with the moods of a boy,
he liked lighthearted life, so loved he the less
either long to be lying or long to be seated:
so worked him his young blood and wayward brain. 5
And another rule moreover was his reason besides
that in pride he had appointed: it pleased him not to eat
upon festival so fair, ere he first were apprised
of some strange story or stirring adventure,
or some moving marvel that he might believe in 10
of noble men, knighthood, or new adventures. . . .

7

Now of their service I will say nothing more,
for you are all well aware that no want would there be.
Another noise that was new drew near on a sudden,
so that their lord might have leave at last to take food.
For hardly had the music but a moment ended, 5
and the first course in the court as was custom been served,
when there passed through the portals a perilous horseman,
the mightiest on middle-earth in measure of height,
from his gorge to his girdle so great and so square,
and his loins and his limbs so long and so huge, 10
that half a troll upon earth I trow that he was,
but the largest man alive at least I declare him;
and yet the seemliest for his size that could sit on a horse,
for though in back and in breast his body was grim,
both his paunch and his waist were properly slight, 15
and all his features followed his fashion so gay
in mode:
for at the hue men gaped aghast
in his face and form that showed;
as a fay-man fell he passed, 20
and green all over glowed.

8

All of green were they made, both garments and man:
a coat tight and close that clung to his sides;
a rich robe above it all arrayed within
with fur finely trimmed, shewing fair fringes
of handsome ermine gay, as his hood was also, 5
that was lifted from his locks and laid on his shoulders;
and trim hose tight-drawn of tincture alike
that clung to his calves; and clear spurs below
of bright gold on silk broideries banded most richly,
though unshod were his shanks, for shoeless he rode. 10
And verily all this vesture was of verdure clear,
both the bars on his belt, and bright stones besides
that were richly arranged in his array so fair,
set on himself and on his saddle upon silk fabrics:
it would be too hard to rehearse one half of the trifles 15
that were embroidered upon them, what with birds and with flies
in a gay glory of green, and ever gold in the midst.
The pendants of his poitrel, his proud crupper,
his molains, and all the metal to say more, were enameled,
even the stirrups that he stood in were stained of the same; 20
and his saddlebows in suit, and their sumptuous skirts,
which ever glimmered and glinted all with green jewels;
even the horse that upheld him in hue was the same,

I tell:
a green horse great and thick, 25
a stallion stiff to quell,
in broidered bridle quick:
he matched his master well.

9

Very gay was this great man guised all in green,
and the hair of his head with his horse's accorded:
fair flapping locks enfolding his shoulders,
a big beard like a bush over his breast hanging
that with the handsome hair from his head falling 5
was sharp shorn to an edge just short of his elbows,
so that half his arms under it were hid, as it were
in a king's capadoce that encloses his neck.
The mane of that mighty horse was of much the same sort,
well curled and all combed, with many curious knots 10
woven in with gold wire about the wondrous green,
ever a strand of the hair and a string of the gold;
the tail and the top-lock were twined all to match
and both bound with a band of a brilliant green;
with dear jewels bedight to the dock's ending, 15
and twisted then on top was a tight-knitted knot
on which many burnished bells of bright gold jingled.
Such a mount on middle-earth, or man to ride him,
was never beheld in that hall with eyes ere that time;
for there 20
his glance was as lightning bright,
so did all that saw him swear;
no man would have the might,
they thought, his blows to bear.

10

And yet he had not a helm, nor a hauberk either,
not a pisane, not a plate that was proper to arms;
not a shield, not a shaft, for shock or for blow,
but in his one hand he held a holly-bundle,
that is greatest in greenery when groves are leafless, 5
and an axe in the other, ugly and monstrous,
a ruthless weapon aright for one in rhyme to describe:
the head was as large and as long as an ellwand,
a branch of green steel and of beaten gold;
the bit, burnished bright and broad at the edge, 10
as well shaped for shearing as sharp razors;
the stem was a stout staff, by which sternly he gripped it,
all bound with iron about to the base of the handle,
and engraven in green in graceful patterns,

lapped round with a lanyard that was lashed to the head 15
and down the length of the haft was looped many times;
and tassels of price were tied there in plenty
to bosses of the bright green, braided most richly.
Such was he that now hastened in, the hall entering,
pressing forward to the dais—no peril he feared. 20
To none gave he greeting, gazing above them,
and the first word that he winged: "Now where is," he said,
"the governor of this gathering? For gladly I would
on the same set my sight, and with himself now talk
in town." 25
On the courtiers he cast his eye,
and rolled it up and down;
he stopped, and stared to espy
who there had most renown.

11

Then they looked for a long while, on that lord gazing;
for every man marveled what it could mean indeed
that horseman and horse such a hue should come by
as to grow green as the grass, and greener it seemed,
than green enamel on gold glowing far brighter. 5
All stared that stood there and stole up nearer,
watching him and wondering what in the world he would do.
For many marvels they had seen, but to match this nothing;
wherefore a phantom and fay-magic folk there thought it,
and so to answer little eager was any of those knights, 10
and astounded at his stern voice stone-still they sat there
in a swoooning silence through that solemn chamber,
as if all had dropped into a dream, so died their voices
away.
Not only, I deem, for dread; 15
but of some 'twas their courtly way
to allow their lord and head
to the guest his word to say.

12

Then Arthur before the high dais beheld this wonder,
and freely with fair words, for fearless was he ever,
saluted him saying: "Lord, to this lodging thou'rt welcome!
The head of this household Arthur my name is.
Alight, as thou lovest me, and linger, I pray thee; 5
and what may thy wish be in a while we shall learn."
"Nay, so help me," quoth the horseman, "He that on high is throned,
to pass any time in this place was no part of my errand.
But since thy praises, prince, so proud are uplifted,
and thy castle and courtiers are accounted the best, 10

the stoutest in steel-gear that on steeds may ride,
most eager and honorable of the earth's people,
valiant to vie with in other virtuous sports,
and here is knighthood renowned, as is noised in my ears:
'tis that has fetched me hither, by my faith, at this time. 15
You may believe by this branch that I am bearing here
that I pass as one in peace, no peril seeking.
For had I set forth to fight in fashion of war,
I have a hauberk at home, and a helm also,
a shield, and a sharp spear shining brightly, 20
and other weapons to wield too, as well I believe;
but since I crave for no combat, my clothes are softer.
Yet if thou be so bold, as abroad is published,
thou wilt grant of thy goodness the game that I ask for
by right." 25
Then Arthur answered there,
and said: "Sir, noble knight,
if battle thou seek thus bare,
thou'lt fail not here to fight."

13

"Nay, I wish for no warfare, on my word I tell thee!
Here about on these benches are but beardless children.
Were I hasped in armor on a high charger,
there is no man here to match me—their might is so feeble.
And so I crave in this court only a Christmas pastime, 5
since it is Yule and New Year, and you are young here and merry.
If any so hardy in this house here holds that he is,
if so bold be his blood or his brain be so wild,
that he stoutly dare strike one stroke for another,
then I will give him as my gift this guisarm costly, 10
this axe—'tis heavy enough—to handle as he pleases;
and I will abide the first brunt, here bare as I sit.
If any fellow be so fierce as my faith to test,
hither let him haste to me and lay hold of this weapon—
I hand it over for ever, he can have it as his own— 15
and I will stand a stroke from him, stock-still on this floor,
provided thou'lt lay down this law: that I may deliver him another.
Claim I!
And yet a respite I'll allow,
till a year and a day go by. 20
Come quick, and let's see now
if any here dare reply!" . . .

17

. . . Gawain goes to the great man with guisarm in hand, 10
and he boldly abides there—he blenched not at all.

Then next said to Gawain the knight all in green:
"Let's tell again our agreement, ere we go any further.
I'd know first, sir knight, thy name; I entreat thee
to tell it me truly, that I may trust in thy word." 15
"In good faith," quoth the good knight, "I Gawain am called
who bring thee this buffet, let be what may follow;
and at this time a twelvemonth in thy turn have another
with whatever weapon thou wilt, and in the world with none else
but me." 20
The other man answered again:
"I am passing pleased," said he,
"upon my life, Sir Gawain,
that this stroke should be struck by thee." . . .

19

The Green Knight on the ground now gets himself ready,
leaning a little with the head he lays bare the flesh,
and his locks long and lovely he lifts over his crown,
letting the naked neck as was needed appear.
His left foot on the floor before him placing, 5
Gawain gripped on his axe, gathered and raised it,
from aloft let it swiftly land where 'twas naked,
so that the sharp of his blade shivered the bones,
and sank clean through the clear fat and clove it asunder,
and the blade of the bright steel then bit into the ground. 10
The fair head to the floor fell from the shoulders,
and folk fended it with their feet as forth it went rolling;
the blood burst from the body, bright on the greenness,
and yet neither faltered nor fell the fierce man at all,
but stoutly he strode forth, still strong on his shanks, 15
and roughly he reached out among the rows that stood there,
caught up his comely head and quickly upraised it,
and then hastened to his horse, laid hold of the bridle,
stepped into stirrup-iron, and strode up aloft,
his head by the hair in his hand holding; 20
and he settled himself then in the saddle as firmly
as if unharmed by mishap, though in the hall he might wear
no head.
His trunk he twisted round, 25
that gruesome body that bled,
and many fear then found,
as soon as his speech was sped.

20

For the head in his hand he held it up straight,
towards the fairest at the table he twisted the face,
and it lifted up its eyelids and looked at them broadly,

and made such words with its mouth as may be recounted.
"See thou get ready, Gawain, to go as thou vowedst, 5
and as faithfully seek till thou find me, good sir,
as thou has promised in this place in the presence of these knights.
To the Green Chapel go thou, and get thee, I charge thee,
such a dint as thou hast dealt—indeed thou hast earned
a nimble knock in return on New Year's morning! 10
The Knight of the Green Chapel I am known to many,
so if to find me thou endeavor, thou'lt fail not to do so.
Therefore come! Or to be called a craven thou deservest."
With a rude roar and rush his reins he turned then,
and hastened out through the hall-door with his head in his hand, 15
and fire of the flint flew from the feet of his charger.
To what country he came in that court no man knew,
no more than they had learned from what land he had journeyed.
Meanwhile,
the king and Sir Gawain 20
at the Green Man laugh and smile;
yet to men had appeared, 'twas plain,
a marvel beyond denial. . . .

SEVEN

Worlds in Revolution: The Renaissance in Letters and Science

The European High Renaissance, from the mid fifteenth century in Florence to the mid seventeenth century in Spain and England, overlaps with the scientific revolution in astronomy, physics, and physiology that begins with Copernicus and Vesalius, continues with Kepler, Galileo, and Harvey, and culminates with Newton. But though the "renaissance" of letters and arts coincided in large part with the "revolution" in science during these intensely creative centuries, the words by which we label these movements suggest significantly—if perhaps exaggeratedly—different dimensions of their achievements. "Renaissance" means "rebirth"; this term for the revival of letters presumed comatose in the "Middle Ages," though most widely used since Jacob Burckhardt's *Civilization of the Renaissance in Italy* (1860), perpetuates the view held by humanists (as those who prided themselves on classical erudition were called) and by leading writers and artists of the time. The renewal of ancient glory was thought to have originated in Italy over a century before, in the age of Giotto, Dante, and Petrarch (Chapter 15 above), the forefather of humanism. Thus Giorgio Vasari, in his preface to *Lives of the Artists* (1550; revised 1568), affirms that "the attainment of perfection in the arts" in classical Rome and their subsequent "ruin" in the medieval centuries had been followed by "their restoration, or to put it better still, their rebirth" (*rinascita*) beginning with Giotto and culminating with Michelangelo.

The literary and artistic Renaissance thus looked to the ancient past for its paradigm of human accomplishment (much as its somber twin, the Protestant Reformation, looked to apostolic times for a Christianity pure of medieval corruptions), whereas the scientific revolution, its name suggests, was a radical overturning of cardinal assumptions of previous ages, looking not to the past but toward an increasingly enlightened future. Yet these closely intertwined movements were complementary rather than opposite: the Renaissance and Reformation were of revolutionary consequence, and the scientific revolution drew strength from the rebirth of antiquity in which it shared. At its weakest, to be sure, Renaissance humanism was often a narrowly pedantic exaltation of fixed models from an unsurpassable past, prizing pseudo-Ciceronian Latin prose and pseudo-Virgilian Latin hexameters above all other human achievements. But leading thinkers and writers, though inspired by rediscovery of long-lost classics unearthed by the early humanists and disseminated by the new medium of the printed book—Plato, Homer, and other Greeks unknown in the West for a thousand years, and Catullus, Lucretius, Tacitus, and much of Cicero, who had shared their oblivion—were no mere imitators, but drew upon multiple sources in their efforts both to emulate and to surpass that liberating inheritance. In doing so, they were creating the future.

Humanists might scorn "medieval darkness" (ignoring the vigorous originality of Germanic and Romance literatures and the profound continuities between their own Renaissance and earlier renewals of the classical tradition); but Rabelais, Cervantes, Spenser, Marlowe, and Shakespeare drew at least as much on the chapbooks and romances, allegories and moralities, of medieval and popular tradition as on the ancients. Above all, they paid less heed to authority than to experience. And though earlier voices, from Roger Bacon's (Chapter 13 above) to that of Chaucer's Wife of Bath, had been raised in its advocacy, only in Renaissance writers does personal experience of the world (including ourselves) become a defining theme of the age. Through individual experience stimulated by curiosity and fed by the classics arose the archetypally Renaissance "discovery of the world and of man" (in Burckhardt's phrase) to which exploration of the natural world was central. Montaigne's essays and Rabelais's account of Gargantua's humane reeducation both emphasize the necessity of learning not just from books that have drawn on experience (these are the true classics) but also on firsthand acquaintance with the great book of the world—though here, too, they were continuing a medieval tradition especially prominent since the twelfth century. As Gargantua and his comrades "walked through the meadows, or other grassy places," Rabelais writes in a passage from Chapter 23 of *Gargantua* (in J. M. Cohen's translation) that crystallizes the best new humanist learning—in which experience and reading stimulate and reinforce each other—"they examined the trees and the plants, comparing them with the descriptions in the books of such ancients as Dioscorides, Marinus, Pliny, Nicander, Macer, and Galen; and they brought back whole handfuls to the house." Renaissance herbalists such as John Gerard enriched the accounts of classical and medieval predecessors with personal observations. Above all the poets of this poetically bountiful age, whose verses teem with the flora and fauna of worlds both familiar and strange, convey the inexhaustible abundance of nature, ultimate source of all knowledge and wisdom. "The more simply we trust to Nature," Montaigne writes in "Of Experience," "the more wisely we trust to her."

The Renaissance "discovered and championed this independence of nature by means of immediate, *sensible-empirical observation*," Ernst Cassirer writes in *The Individual and the Cosmos in Renaissance Philosophy* (1927; Eng. trans. 1963). But the discovery was made by humanists and poets, painters and sculptors, before the scientists—who increasingly tended, Cassirer notes, to abandon "exact observation of nature" in a search either for occult power through magic or for knowledge of abstract laws through mathematics. It was a painter, Leonardo da Vinci, who surpassingly combined the precise observations of the herbalists, the wide-ranging curiosity of the humanists (whose pomposities he disdained), the bold speculations of the natural philosophers, and the restless aspirations of the poets of his age in his endlessly fascinating comments on nature. For Leonardo, Cassirer remarks, art "is and remains a genuine and indispensable organ for the understanding of reality itself." In him, "The creative power of the artist, the imagination that creates a 'second nature,' does not consist in his inventing the law, in his creating it *ex nihilo*; it consists in his discovery and demonstration of the law." This understanding of nature as a power subject to no laws but its own, both in the external world and in the human mind that apprehends it, was perhaps the most potent lesson the poets and artists of the Renaissance learned from the Greeks and Romans, thus turning the revival of classical letters into a force for revolutionary change. In this respect, the "philosophes" of the eighteenth century, self-proclaimed prophets of Newtonian science though they were, continued "The Appeal to Antiquity" (as Peter Gay entitles the opening book of *The Rise of Modern Paganism* [1966]) inherited from the Renaissance. Diderot even declared, in his *Salon of 1765*, "that we must study antiquity in order to learn to see nature." No Renaissance humanist could have made a more sweeping claim.

If the Renaissance discovered a future through the past, the scientific revolution owed its very existence to the revival of learning. The word "revolution" signified a turning round (as of the celestial spheres in Copernicus's *De Revolutionibus Orbium Coelestium*), or a recurrence, before it designated a radical overturning of the past. In this sense, too, the great scientific discoveries of the age were a revolution. To Ptolemy's mathematically scrupulous geocentric system (Chapter 12 above), even more than to anticipations of his heliocentric hypothesis by the Pythagoreans and Aristarchus, Copernicus owed the impetus that allowed him to simplify Ptolemy's calculations by displacing the sun to the center of the orbiting planets. Similarly, Vesalius's anatomy would have been impossible without Galen's (Chapter 12); and Harvey, though refuting central tenets of Aristotelian physiology, honored Aristotle himself, from whom he and his age, even in their revulsion from Scholasticism, had learned so much. Nor was it only on ancient scientists that their Renaissance successors drew: Copernicus cites Hermes Trismegistus; Kepler built his astronomical system on the geometrical solids of Plato's *Timaeus*; Newton devoted years of study to alchemy and biblical prophecy; and Bacon, who harshly repudiated the humanistic exaltation of classical philosophers, was influenced by the very Hermetic magicians he scorned. This was a revolution inconceivable—like any other, perhaps—without a concomitant renaissance.

Occult concepts of nature in thinkers such as Paracelsus and Giordano Bruno, John Dee and Robert Fludd, loosely grouped together as Hermetists, mystics, alchemists, or magicians, might seem to have little in common either with Renaissance humanism or with the scientific revolution. Yet the foremost Italian humanist, Marsilio Ficino, translated the *Corpus Hermeticum* and *Orphic Hymns*, and in his *Platonic Theology* propounded a mystical Christian Neoplatonism anticipated by Pseudo-Dionysius the Areopagite and John the Scot (Chapter 13 above). His younger colleague in the Platonic Academy of Florence, Pico della Mirandola, whose "Oration on the Dignity of Man" is frequently cited as a manifesto of humanism, assimilated fundamental doctrines of Hermetic, Neoplatonic, and Kabbalistic mysticism. (Less aristocratically educated occult thinkers such as Paracelsus showed small regard for most of the ancients, notably Galen—by whose ideas their own continued nevertheless to be deeply shaped.)

Any presumed opposition between Hermetic concepts and the scientific view of nature would be equally misleading. Important scientists took many of these concepts ("unscientific" as they appear in retrospect) with great seriousness, and even contributed to them, and some Hermetists and magicians were among the main defenders of the new science in their opposition—which united humanists, Hermetists, and scientists—to the rigid Aristotelianism of the universities. Thus Pico denounced astrology, Bruno proselytized for the Copernican system (which Bacon and many astronomers rejected), and Fludd welcomed Harvey's discovery of the circulation of the blood long before most of the medical community. Magic and science ("natural philosophy") were by no means distinct. Those who elaborated ancient concepts such as the correspondence of microcosm and macrocosm, and those who, by empirical observation or mathematical calculation, created the immensely powerful—if surely incomplete—vision of nature that we regard as scientific were often difficult to tell apart. Only Newton's genius, it has been said, could have isolated from Kepler's Neoplatonic astronomical treatises the three planetary laws that made possible his formulation of the law of universal gravitation—a law rejected by many scientists because of its seemingly occult affirmation of "attraction at a distance."

Virtually all poets in this age of revolutionary rebirth adhered more closely to the cosmologies of Pythagoras or Plato, Bruno or Fludd (or indeed of Aristotle and Thomas Aquinas), than to those of Copernicus or Galileo, yet few could have been untouched by the opposing views of the world that continually swirled around them, and by the intellectual ferment to which their verses gave expression. But the Renaissance celebration of nature and man coincided also with the Reformation's annihilation of both before the omnipotence of a fearful God. In the clash between these radically opposed visions, the universe itself—"this brave o'erhanging firmament, this majestical roof fretted with golden fire"—might equally appear, as to Hamlet, "a foul and pestilent congregation of vapors," and man be at once "the beauty of the world" and a "quintessence of dust." Increasingly, writers such as Donne and Pascal expressed disquiet that the new philosophy had rent apart the certainties of the old, leaving in their place the terror of infinite spaces now eternally silent. This disquiet, too, was a legacy of Renaissance, Reformation, and scientific revolution to the ages that followed.

16

Humanists, Hermetists, and Herbalists: Renaissance Prose

Renaissance views of nature reflect the immense variety of this age of transition between medieval and modern. The humanists were mainly concerned not with formulating new systems of thought but with reviving those of the ancients, several of which—notably Platonism and the Epicurean materialism of Lucretius—would have a major impact on this and the following age. The Reformers, on the other hand, turned their attention emphatically from nature (the source of corruption and sin) to grace, and from this world, except as the testing ground of faith, to the next. There was, indeed—perhaps in reaction to the excesses of Scholasticism—little formal philosophy before the seventeenth century, when Bacon and Descartes, Spinoza and Leibniz, gave expression to suppositions underlying the new science and to potential conflicts between these and traditional religious beliefs. Yet rediscovery of ancient concepts of nature and human nature, though producing no new philosophical synthesis, profoundly affected the views both of Latin humanists and of vernacular authors who absorbed their learning while rejecting their pedantries. Writers of the occult or Hermetic tradition meanwhile gave vigorous expression to older concepts of the cosmos as the creation, and of nature as the instrument, of God.

Despite the epochal intellectual changes the world was undergoing, the predominant Renaissance view of the cosmos and of man's place within it—as expounded for example in E. M. W. Tillyard's The Elizabethan World Picture *(1943)—is a lineal offshoot of the early Christian and medieval world view set forth by C. S. Lewis in* The Discarded Image *(1964). According to such a view, the hierarchical universe is linked by a "great chain of being" ascending from plants and animals to angels and God, and by elaborate correspondences between the little world of man and the greater world around him. Some Hermetic writers might, with Bruno, daringly combine these traditional conceptions with heretical new ideas, like the infinity of the universe or the multiplicity of worlds within it, anticipated by Lucretius and Nicholas Cusanus. Or they might lay claim to ominous magical powers, thus bringing suspicion and persecution upon them. But by and large their characteristic assumptions are closer to prevailing beliefs, and even prevailing superstitions—including many shared by most poets—than are those of either the eruditely elitist humanists or the radically innovative natural philosophers who were reshaping beliefs inherited from the past into those of ages to come.*

While concepts of nature were very much in flux during this age, the direct experience of nature was being stimulated both by rediscovery of the ancients and by discovery of new lands across the seas. Descriptions of plants and their medicinal properties could trace a long lineage from antiquity through the Middle Ages, but herbals of the Renaissance, exemplified by that of John Gerard, are characterized not only by book learning but also by firsthand observation of plants both domestic and exotic. The astonishing accounts by explorers of both plants and animals and "savage" peoples of the New World led Montaigne to look anew at people nearest to him, above all himself, in relation to the natural (or unnatural) world around them. That great question—what kind of human existence is most in accord with the rhythms of nature?—continued to be raised in the Renaissance by the myth of the Golden Age, by the vogue of a pastoral poetry celebrating nature as artifice, and by alternate exaltation and condemnation of "primitive" peoples including both the cannibals of Brazil and the gypsies of Europe itself. These conflicting visions obsessed Cervantes, who praised the Golden Age through a madman and exposed the idealization of shepherds and gypsies through the observations of a disillusioned dog.

The Finite Spirit of the Infinite Sphere

Of Learned Ignorance, by Nicholas Cusanus, translated by Fr. Germain Heron

Nicholas Cusanus, or Nicholas of Cusa (the Latinized form of Cues, his birthplace on the Moselle in Germany), who lived from 1401 to 1464, more than any other philosopher bridges the Middle Ages and the Renaissance. After studies at Heidelberg in 1416 and then at Padua, where he took his doctorate in canon law in 1423, he participated in the Council of Basel. In 1437, disillusioned by internal divisions in the conciliar movement, he entered the service of Pope Eugenius IV. He was made a cardinal by the humanist Pope Nicholas V in 1448 and Bishop of Brixen in 1450, and after 1458 served the humanist Pope Pius II, the former Aeneas Sylvius Piccolomini. Throughout his career as churchman and thinker, he was a voice for change. He advocated reform of the monasteries and the calendar, contributed to the humanist enterprise by discovering several plays by the Roman comic dramatist Plautus, and rejected the Scholastic synthesis of the previous centuries in favor of a Christian Neoplatonism that looked back to "Hermes Trismegistus" (known only through the Latin Asclepius*) and Augustine, Pseudo-Dionysius the Areopagite and John the Scot (Chapter 13 above), and the German mystic Meister Eckhart, and forward to the Platonism of Marsilio Ficino.*

His most important work, Of Learned Ignorance (De Docta Ignorantia), *completed in 1440, was conceived, he writes, while returning from Constantinople, where he had helped negotiate the short-lived reunion of the Eastern and Western churches. In its plan, as D. J. B. Hawkins observes in his introduction to Fr. Germain Heron's translation (1954), "the first book is concerned with the* maximum absolutum, *which is God, the second with the* maximum contractum, *the sum of limited things which is the universe, and the third with the* maximum *within the universe, that created nature in which the universe as a whole is fulfilled and which links the world with its creator. This is the human nature of the Incarnate Word: in Christ there is a personal and existential identity of the* maximum absolutum *and the* maximum contractum."

Only by the "learned ignorance" of negation and contradiction can we approach a knowledge of the unknowable Absolute Maximum, characterized by its conjunction of opposites (coincidentia oppositorum), *an infinite sphere that is also an infinite line, triangle (identified with the trinity), and circle. Only thus can we begin to understand (I.xxi) "how the Maximum is a being which is neither the same as, nor different from any other, and how all things are in it, from it and by it, because it is the circumference, diameter and center." Such a universe can have "no fixed, immovable center" (II.xi); therefore, "the center and the circumference are identical" and the world, though not infinite (in the absolute sense that only God is infinite), "cannot be conceived as finite, since there are no limits within which it is enclosed." Thus, as Walter Pagel writes in* Paracelsus *(2nd ed., 1982), "Cusanus demolished the whole concept" of the medieval cosmos "based on the ancient doctrine of a closed finite system of spheres," and to that extent paved the way for the infinite Copernican universe of Bruno and Pascal (if not of Copernicus himself).*

The following passages from the Second Book set forth Cusanus's concept—indebted to Aristotle as well as to the Christian Neoplatonists—of nature as a finite created "spirit diffused throughout the entire universe" and embracing "all things which owe their origin to movement." Only human nature, "the microcosm or world in miniature," Book III declares, is capable of being "raised to union with the maximum." Although such a union was fully possible only in Christ, the potentiality is partially present in every human being, thus making it possible to know nature, if not God. "Nature is not only the reflection of the divine being and divine force," Cassirer writes in The Individual and the Cosmos; *"rather, it becomes the book God has written with his own hand. . . . Neither subjective feeling nor mystical sentiment suffice to understand the meaning of the book of nature. Rather, it must be investigated, it must be deciphered word for word, letter for letter. The world may no longer remain a divine hieroglyph, a holy sign; instead, we must analyze and systematically interpret the sign." This was the task that both Hermetic and scientific thinkers would take up, in very different ways, in the centuries to come.*

FROM THE *Second Book, Chapter X: Spirit of the Universe*

. . . In fact, from the aptitude of matter to receive a form springs a desire: it desires a form as evil desires good and privation desires possession. Form, too, has a desire—a desire to be actualized; but since its existence cannot be absolute, for it is not its own being and it is not God, it descends in order to have a limited existence in matter. In other words, whilst matter ascends towards being actual, form descends to limit, perfect, and determine matter. Thus from the ascent and descent there arises the movement which connects the two of them. Potency and act are connected by means of this movement, for movement itself, which is the intermediary, springs from moveable matter and a formal mover.

This, then, is the finite spirit that is diffused throughout the entire universe and all its parts and to which is given the name nature. Nature, therefore, embraces, as it were, all things which owe their origin to movement. How this movement is reduced in due order and by degrees from the universal to the particular is illustrated by the following example. When I say "God exists" that pronouncement is formed by a movement and according to a definite order; first I utter the letters, then the syllables, then the words, and finally the sentence, though these orderly stages are not perceived by the ear. So, too, does movement descend by stages from the universal to the particular, where it is limited by the temporal or natural order. This movement or spirit descends from the

Spirit of God who put all things in movement by movement itself. Just as, therefore, in a speaker there is a breath or spirit which proceeds from him and is formed into a sentence, as we have already said, so God, Who is a Spirit, is He from Whom all movement descends. Truth Himself tells us: "It is not you who speak but the Spirit of your Father who speaks in you." The same has to be said of all other movements and operations.

This created spirit, then, is the spirit without which there is no unity, without which nothing can subsist. It fills the whole earth; by it the entire world and all it contains are naturally what they are connectedly, so that by means of this spirit potency is in act and act, through it, is in potency. This is the movement of a love that links all in one with the result that all things form one universe. All things, in fact, are moved individually to be precisely what they are; and they would not be another thing equally as well, for they are what they are in the best possible way. Yet, at the same time, in its own proper way each thing shares in and limits the movement of all others either mediately or immediately: the elements share in and limit the movement of the heavens and the elements and all the members share in and limit the movement of the heart, with the result that the universe is a unity. By this movement things exist in the best way possible; and the purpose of the movement is their own preservation, or the preservation of their species through the union of the sexes, for, though in the individuals the sexes are defined and separate, in nature, in which the movement exists, they are united. . . .

FROM THE *Second Book, Chapter XIII: The Divine Design in the Creation of the World and Its Constituent Parts Is Wholly Admirable*

That the vast bulk, the beauty, and the ordered adjustment of this visible world must fill us with amazement at the incomparable skill of its creator, goes without question among wise men. We have touched upon some examples of the astonishing creative skill of the divine mind. Let us now briefly add a word more, in admiration of the placing and mutual adjustment of the elements of creation.

When we measure the size and analyze the elements and study the behavior of things, we make use of the sciences of arithmetic and geometry and even of music and astronomy. Now these same sciences God employed when He made the world. With arithmetic He adjusted it into unity, with geometry He gave it a balanced design upon which depends its stability and its power of controlled movement; with music He allotted its parts that there should be no more earth in the earth than water in the water, than air in the air, or than fire in the fire, so that no element could be wholly transmuted into another; whence it comes that the physical system cannot sink into chaos. Some of one element may be transformed into another, but (for example) the air which is mingled with water can never all be changed into water, the surrounding air preventing this; it is this intransmutability that makes possible the mingling of elements. Nevertheless, God has so arranged it that there should be part transmutation of the elements; and when this takes place successively there is brought into existence a new thing that endures in being as long as the agreement of the elements remains. If the agreement is broken the new substance disappears.

God has set up the elements in an admirable order, for He created all things in number, weight, and measure. Number appertains to arithmetic, weight to music, and measure to geometry. Heaviness is kept in place by the action of lightness—the heavy earth for example is suspended in the middle by the action of fire—and lightness adheres to heaviness as does fire to the earth. In setting up these things eternal wisdom employed an indescribably accurate proportioning. The Measurer of all things foresaw that one element should demand the earlier existence of another and that water should be lighter than earth in the same proportion as air is lighter than water and as fire is lighter than air. Weight and bulk will thus coincide and the container will occupy a larger space than the thing contained. And he related one element with another in such intimacy that one must necessarily dwell in another. The earth, as Plato says, is like some vast animal whose bones are stones, whose veins are rivers, and whose hairs are the trees; and the animals that feed among those hairs of the earth are as the vermin to be found in the hair of beasts. . . .

Who could help admiring this craftsman who in spheres and stars and in the vast stellar spaces employs such skill that, with no discontinuity, achieves in the widest diversity the highest unity, in one single world so weighing and adjusting the vast bulk and position and movement of the stars, so minutely ordering the distances that lie between them, that each astral area, if it is to be, and the universe, if it is to continue, must be just as it is and in no other way. He gives to each star its own splendor, its own power to influence, its own shape, color, and heat. This heat accompanies its illumination and shares the influence it exerts over other things. And lastly, He in each star so adjusts and proportions the parts to each other that there is in each a movement of parts that secures the whole, downward to the center in heavy parts, upwards from the center in the lighter parts, together with a constant movement round the center; so that we perceive each star to move only through its orbit.

In such a high diversity of endlessly admirable things learned ignorance has taught us never to hope to penetrate to the reasons of all the works of God, but only to admire; for the Lord is great and of His greatness there is no end. He is the absolute maximum and the author and comprehender of all His works, as He is also the end of them all, for in Him are all things and outside Him is nothing. He is the beginning, the middle, and the end of all things, the center and circumference of all that is, and in all things He only is to be sought; for apart from Him all things are nothing. Possessing only Him, we possess all things, for He is all. Knowing Him we know all, for He is the truth of all things. . . .

FROM THE *Third Book, Chapter III: Human Nature and Only Human Nature Peculiarly Adapted to Be This Maximum*

. . . Now, human nature it is that is raised above all the works of God and made a little lower than the angels. It contains in itself the intellectual and the sensible natures, and therefore, embracing within itself all things, has very reasonably been dubbed by the ancients the microcosm or world in miniature. Hence is it a nature that, raised to union with the maximum, would exhibit itself as the fullest perfection of the universe and of every individual in it, so that in this humanity itself all things would achieve their highest grade. But humanity has no real existence except in the limited existence of the individual. Wherefore it would not be possible for more than one real man to rise to union with the maximum; and this man assuredly would so be man as to be God, would so be God as to be man, the perfection of all things and in all things holding the primacy. In him the smallest things of nature, the greatest and all between, would so coincide in a nature united with the absolute maximum, as to form in him the perfection of all things; and all things, in their limitation, would repose in him as in their perfection. This man's measure would also be that of the angel and of every one of the angels, as St. John says in the Apocalypse (21:17), for he would be the universal contracted entity of each creature through his union with the absolute, which is the absolute entity of all things. From him all things would receive the beginning and end of their limitation. By him who is the maximum in limitation, all things are to come forth into their limited being from the Absolute Maximum, and by means of him revert to the maximum. For he is the first beginning of their setting forth and the last end of their return. . . .

Hidden Alliances and Affinities

Heptaplus: On the Sevenfold Narration of the Six Days of Genesis, by Giovanni Pico della Mirandola, translated by Douglas Carmichael

Count Giovanni Pico della Mirandola, in the thirty-one years of his life (1463–94), was famed as a prodigy of learning and a brilliant paragon of Renaissance humanism. Born in his family's castle near Bologna, he studied at Bologna, Ferrara, and Padua, and joined the Platonic Academy of Florence in 1484. By this time, Marsilio Ficino, who had established the academy under Cosimo de' Medici in 1462 and continued it under Cosimo's son Piero and grandson Lorenzo the Magnificent, had translated the Greek Corpus Hermeticum *and the dialogues of Plato and written, besides commentaries on the* Symposium *and* Philebus, *his* Platonic Theology *in eighteen books, and was about to undertake translations of the Neoplatonists Plotinus, Porphyry, Proclus, and Pseudo-Dionysius. Pico fell heir to this legacy and took up the study of Hebrew and the Kabbala.*

In December 1486, he published in Rome 900 theses aspiring to reconcile Christianity and Platonism (united by an esoteric tradition stemming from Hermes Trismegistus and paralleled by the Mosaic "philosophy"), and invited scholars to attend a public disputation the following month. His famous Oration on the Dignity of Man, *proclaiming that man "can have that which he chooses and be that which he will," was intended as a prelude to this event. But Pope Innocent VIII canceled the disputation and charged a commission to scrutinize Pico's theses for heresy; Pico, despite recantation of thirteen theses, was briefly arrested in 1488 in Lyons. "In the nine hundred theses and in the* Apologia, *composed to defend the theses against the charges of heresy," Cassirer writes in* The Individual and the Cosmos, *"Pico defines magic as the sum of all natural wisdom and as the practical part of natural science. With this, he is only expressing a conviction common in the natural philosophy of the Renaissance. In this view, magic is nothing but the active side of the knowledge of nature." In his later years this celebrant of unfettered freedom of the will became a follower of the puritanical monk Savonarola, who would later rule Florence for a short time until he was burned at the stake.*

The following selections from the Heptaplus *of 1489—a commentary on the six days of Creation in a tradition descending from Basil and Ambrose (Chapter 13 above)—expound Pico's theory of occult correspondences between the intelligible, celestial, and sublunary worlds and the "fourth world" of man, who brings together "all the natures of the world"; it crystallizes central assumptions of the Hermetic-Neoplatonic conception of nature. "The permanent interest and value of Pico's view of nature," Paul J. Miller suggests in his introduction to* On the Dignity of Man, On Being and the One, *and* Heptaplus *(1965), from which the present translation is taken, "comes from his seeing the physical order as a translation of philosophical and religious truth. In this way, physics, philosophy, and Scripture literally say the same things in different languages."*

FROM *Second Proem to the Whole Work*

Antiquity imagined three worlds. Highest of all is that ultramundane one which theologians call the angelic and philosophers the intelligible, and of which, Plato says in the *Phaedrus* [247c], no one has worthily sung. Next to this comes the celestial world, and last of all, this sublunary one which we inhabit. This is the world of darkness; that the world of light; the heavens are compounded of light and darkness. This world is symbolized by water, a flowing and unstable substance; that by fire, for the splendor of its light and the elevation of its position; of a middle nature, the heavens are on that account called by the Hebrews *asciamaim*, as if composed of *es* and *maim*, that is, of the fire and water of which we spoke. Here there is an alternation of life and death; there, eternal life and unchanging activity; in the heavens, stability of life but change of activity and position. This world is composed of the corruptible substance of bodies; that one of the divine nature of the mind; the heavens of body, but incorruptible, and of mind, but enslaved to body. The third is moved by the second; the second is governed by the first; and there are among them many further differences which I do not propose to enumerate here, where we are skimming the surface of such things without fathoming their depths. . . .

It should above all be observed, a fact on which our purpose almost wholly depends, that these three worlds are one world, not only because they are all related by one beginning and to the same end, or because regulated by appropriate numbers they are bound together both by a certain harmonious kinship of nature and by a regular series of ranks, but because whatever is in any of the worlds is at the same time contained in each, and there is no one of them in which is not to be found whatever is in each of the others. If we have understood him rightly, I believe that this was the opinion of Anaxagoras, as expounded by the Pythagoreans and the Platonists. Truly, whatever is in the lower world is also in the higher ones, but of better stamp; likewise, whatever is in the higher ones is also seen in the lowest, but in a degenerate condition and with a nature one might call adulterated. In our world there is the elemental quality of heat, in the heavens there is a heating power, and in angelic minds there is the idea of heat. I shall speak more precisely: among us there is the fire which is an element; the sun is fire in the sky; in the ultramundane region the fire is the seraphic intellect. But see how they differ. The elemental fire burns, the celestial gives life, and the supercelestial loves. There is water in our world; there is water in the heavens, the mover and mistress of ours, namely, the moon, the vestibule of the heavens; and above the heavens, the waters are the minds of the cherubim. But see what a disparity of condition there is in the same nature: the elemental moisture quenches the heat of life; the celestial feeds it; the supercelestial understands it. . . .

Bound by the chains of concord, all these worlds exchange natures as well as names wih mutual liberality. From this principle (in case anyone has not yet understood it) flows the science of all allegorical interpretation. The early Fathers could not properly represent some things by the images of others unless trained, as I have said, in the hidden alliances and affinities of all nature. Otherwise there would be no reason why they should have represented this thing by this image, and another by another, rather than each by its opposite. But versed in all things and inspired by that Spirit which not only knows all these things but made them, they aptly symbolized the natures of one world by those which they knew corresponded to them in the other worlds. Therefore, those who wish to interpret their figures of speech and allegorical meanings correctly need the same knowledge (unless the same Spirit helps them also).

There is, moreover, besides the three that we have mentioned, a fourth world in which are found all those things that are in the rest. This is man himself, who is, as the Catholic doctors say, referred to in the Gospel by the name of every creature, since Christ gave the Gospel to be preached to men, not to brutes and angels, but nevertheless to be preached to every creature [Mark 16:15]. It is a commonplace expression in the schools that man is a lesser world, in which are seen a body compounded from the elements, and a heavenly spirit, and the vegetative soul of plants, and the sense of brutes, and reason, and the angelic mind, and the likeness of God. . . .

Therefore the first principle, which, as we have shown, is the greatest of all, is that whatever is in any of the worlds is contained in each. As the imitator of nature, Moses had to treat of each of these worlds in such a way that in the same words and in the same context he could treat equally of all. Hence there arises immediately a fourfold exposition of the whole Mosaic text, so that, in the first place, whatever is written there we interpret in relation to the angelic and invisible world, making no mention whatever of the others. In the second place, we interpret everything in relation to the celestial world; then in relation to this sublunary and corruptible one; and fourthly, in relation to the nature of man. If there is anywhere a discussion of the intelligible world, for instance, we surely can, or at least we should, interpret all the details in respect to all the others, so that just as that world contains in itself all the lower natures, so also the same passage may put us in mind of the rest of the worlds. . . .

FROM *Fifth Exposition. Chapter Six*

. . . Truly, just as God is God not only because He understands all things, but because in Himself He assembles and unites the total perfection of the true substance of things, so also man (although differently, as we shall show, else he would be not the image of God, but God) collects and joins to the completeness of his substance all the natures of the world.

We cannot say this of any other creature, angelic, heavenly, or sensible. The difference between God and man is that God contains all things in Himself as their origin, and man contains all things in himself as their center. Hence in God all things are of better stamp than in themselves, whereas in man inferior things are of nobler mark and the superior are degenerate.

Fire, water, air, and earth in the true peculiarity of their natures exist in this gross, earthly human body which we see. Besides these, there is another, spiritual body more divine than the elements, as Aristotle says, which by analogy corresponds to heaven. There is also in man the life of the plants, performing all the same functions in him as in them—nutrition, growth, and reproduction. There is the sense of the brutes, inner and outer; there is the soul, powerful in its heavenly reason; there is participation in the angelic mind. There is the truly divine, simultaneous possession of all these natures flowing together into one, so that we may exclaim with Mercury,[1] "A great miracle, O Asclepius, is man!"

Human nature can take its greatest glory in this name, because of which no created substance disdains to serve it. The earth, the elements, and the beasts wait upon it as

[1]Hermes Trismegistus; see Chapter 13 above.

its servants, the heavens labor for it, and the angelic minds look after its salvation and beatitude, if what Paul writes is true [Hebrews 1:14], that all ministering spirits are sent to minister to those who are destined for salvation as their inheritance. It ought not to seem wonderful to anyone that all creatures should love the one in whom they all recognize something of themselves, nay, even their entire selves with all their qualities.

The Greatest School for All

Selections from the Writings of Paracelsus, translated by Norbert Guterman

Of Paracelsus (1493?–1541), Marie Boas writes in The Scientific Renaissance, 1450–1630, *"Everything about him is so complex and difficult that almost any interpretation is justified. Even his name is surrounded with obscurity. By the end of his life he was known as Philippus Aureolus Theophrastus Bombastus von Hohenheim Paracelsus, though at the beginning Bombast von Hohenheim only appears." The name Paracelsus may have been adopted to signify that he surpassed Celsus, the ancient Roman medical writer (Chapter 12 above) recently rediscovered by the humanists and hailed as a model of Latin prose—which Paracelsus largely disdained in favor of his native German. Born at Einsiedeln in Switzerland of a father who appears to have been an illegitimate Swabian nobleman and a mother probably of peasant descent, he stands in sharp contrast, with his restless contempt for all authority, to the aristocratic humanisms of Cardinal Nicholas of Cusa and Count Pico della Mirandola. "I am different, let this not upset you," he writes at the beginning of the* Selected Writings, *from which our passages are taken:*[2] *". . . those who are raised in soft clothes and in women's apartments and we who are brought up among the pine-cones have trouble in understanding one another well."*

He studied medicine at several universities, including Ferrara in Italy, acquired a knowledge of metallurgy and alchemy, and acted as an army surgeon—a position of far lower standing than that of physician—in various wars as far afield as Denmark. All his life he traveled throughout Europe, mainly in German-speaking lands, seldom staying long in one place. His growing reputation led him to Salzburg and Strassburg and then, in 1527, to Basel, where he lectured in his native Swiss German—lecturing in a language other than Latin was nearly unheard of—until his unfailing capacity to make enemies drove him forth to wander again.

The Chemical Philosophy that underlay the "iatrochemistry" of Paracelsus and his disciples "was to be a universal philosophy of nature founded on new observations and indisputable philosophical precepts which conformed to religious truth. The teachings of the schools," Allen G. Debus writes in "The Chemical Debates of the Seventeenth Century,"[3] *"were rejected as moribund and impious. . . . Clearly related to this was their distrust of mathematics as an interpreter of nature." Paracelsus continually places emphasis on personal (including mystical) experiences, and the most important affinities revealed by the scattered writings that gained him a fervent following after his death are with practitioners of "natural magic," such as Cornelius Agrippa of Nettesheim and Giambattista della Porta.*

[2] *Paracelsus: Selected Writings*, ed. Jolande Jacobi (1942; English trans. by Norbert Guterman 1951, 2nd ed. 1958). References at the end of each selection are to the part, volume, and page numbers in *Paracelsus: Sämtliche Werke*, ed. Karl Sudhoff and Wilhelm Matthiessen (1922–33).

[3] In M. L. Righini Bonelli and William R. Shea, eds., *Reason, Experiment, and Mysticism in the Scientific Revolution* (1975).

But despite his contempt both for the Aristotelianism of the universities and for ancient writers in general, he is unthinkable, as Pagel writes in Paracelsus: An Introduction to Philosophical Medicine in the Era of the Renaissance *(2nd ed., 1982), "without the Hellenistic blending of Jewish, Christian, Greek and Oriental ideas and symbolism ('syncretism') as expressed in Neoplatonism, Gnosticism and Kabbala, Alchemy, Astrology and Magic"—the sources of which had been revived in his time by the humanists. Even doctrines that he explicitly assails, like that of the four elements (only partly displaced by the three alchemical principles of mercury, sulfur, and salt) or of the four humors descending through Galen (Chapter 12 above) from the Hippocratics (Chapter 9), repeatedly find a place in his writings. Nowhere is the ancient concept of correspondence between microcosm and macrocosm developed so fully, or the Hermetic-Neoplatonic-Christian insistence on man's fundamental allegiance not to his animal but to his "sidereal" nature insisted upon more strongly.*

But his faith that nature, "the greatest school for all of us," "emits a light and by its radiance she can be known" has an empirical dimension that partly offsets his mystical tendencies and his conventional prejudices, such as the belief that woman exists for the sake of the womb. Though the alchemist labors to bring nature to perfection, "the physician is only the servant of nature, not her master"; by careful attention to the signatures inscribed in the great apothecary's shop of nature he may patiently learn medicinal properties such as those that gained Paracelsus a reputation for finding cures that had escaped others. Thus for all his belief that man is "a part of nature whose end lies in heaven," Pagel writes, Paracelsus remains "first and foremost a naturalist. . . . He sets out to explore how Nature works, to discover what are the ephemeral phenomena and the rational laws by which Nature is governed." To that extent, his philosophy has much in common with the new natural science whose view of nature is in other respects so fundamentally different.

Heaven encompasses both spheres—the upper and the lower—to the end that nothing mortal and nothing transient may reach beyond them into that realm which lies outside the heaven that we see. . . . For mortal and immortal things must not touch each other, and must not dwell together. Therefore, the Great World, the macrocosm, is closed in itself in such a way that nothing can leave it, but that everything that is of it and within it remains complete and undivided. Such is the Great World. Next to it subsists the Little World, that is to say, man. He is enclosed in a skin, to the end that his blood, his flesh, and everything he is as a man may not become mixed with that Great World. . . . For one would destroy the other. Therefore man has a skin; it delimits the shape of the human body, and through it he can distinguish the two worlds from each other—the Great World and the Little World, the macrocosm and man—and can keep separate that which must not mingle. Thus the Great World remains completely undisturbed in its husk . . . and similarly man in his house, that is to say, his skin. Nothing can penetrate into him, and nothing that is in him can issue outside of him, but everything remains in its place. (I/9, 178)

Man emerged from the first matrix, the maternal womb, of the Great World. This world—formed by God's hand along with all other creatures—gave birth to man in his flesh and placed him in a transient life. For this reason man became "earthly" and "carnal"; he received his material body from earth and water. These two elements constitute the body in its transient, animal life, which man as a natural being received from divine creation. . . . In his earthly life man consists of the four elements. Water and earth, of which his body is formed, constitute the dwelling place and the physical envelope of life. And I am not referring here to that life of the soul, which springs from the breath of God

. . . but to the transient life, of the earthly kind. For we must know that man has two kinds of life—animal life and sidereal life. . . . Hence man has also an animal body and a sidereal body; and both are one, and are not separated. The relations between the two are as follows. The animal body, the body of flesh and blood, is in itself always dead. Only through the action of the sidereal body does the motion of life come into the other body. The sidereal body is fire and air; but it is also bound to the animal life of man. Thus mortal man consists of water, earth, fire, and air. (I/14, 597–98)

The world edifice is made of two parts—one tangible and perceptible, and one invisible and imperceptible. The tangible part is the body, the invisible is the Stars. The tangible part is in turn composed of three parts—sulphur, mercury, and salt; the invisible also consists of three parts—feeling, wisdom, and art. The two parts together constitute life. (I/12, 20)

The mysteries of the Great and the Little World are distinguished only by the form in which they manifest themselves; for they are only *one* thing, *one* being. Heaven and earth have been created out of nothingness, but they are composed of three things—*mercurius, sulphur,* and *sal.* . . . Of these same three things the planets and all the stars consist; and not only the stars but all bodies that grow and are born from them. And just as the Great World is thus built upon the three primordial substances, so man—the Little World—was composed of the same substances. Thus man, too, is nothing but mercury, sulphur, and salt. (I/8, 280)

The body has four kinds of taste—the sour, the sweet, the bitter, and the salty. . . . They are to be found in every creature, but only in man can they be studied. . . . Everything bitter is hot and dry, that is to say, choleric; everything sour is cold and dry, that is to say, melancholic. . . . The sweet gives rise to the phlegmatic, for everything sweet is cold and moist, even though it must not be compared to water. . . . The sanguine originates in the salty, which is hot and moist. . . . If the salty predominates in man as compared with the three others, he is sanguine; if the bitter is predominant in him, he is choleric. The sour makes him melancholic, and the sweet, if it predominates, phlegmatic. Thus the four tempers are rooted in the body of man as in garden mold. (I/1, 211–13)

Woman is like the earth and all the elements, and in this sense she must be considered a matrix; she is the tree which grows from the earth, and the child is like the fruit that is born of the tree. Just as a tree stands in the earth and belongs not only to the earth but also to the air and the water and the fire, so all the four elements are in woman—for the Great Field, the lower and the upper sphere of the world, consists of these—and in the middle of it stands the tree; woman is the image of the tree. Just as the earth, its fruits, and the elements are created for the sake of the tree and in order to sustain it, so the members of woman, all her qualities, and her whole nature exist for the sake of her matrix, her womb. (I/9, 209–10)

Everything that comes from the flesh is animal and follows an animal course; heaven has little influence on it. Only that which comes from the stars is specifically human in us; this is subject to their influence. But that which comes from the spirit, the divine part of man, has been formed in us in the likeness of God, and upon this neither earth nor heaven has any influence. (I/12, 18)

You should look upon man as a part of nature whose end lies in heaven. In the heavens you can see man, each part for itself; for man is made of heaven. And the matter out of which man was created also indicates to you the pattern after which he was formed. . . . External nature molds the shape of internal nature, and if external nature vanishes, the inner nature is also lost; for the outer is the mother of the inner. Thus man is like the image of the four elements in a mirror; if the four elements fall apart, man is destroyed. If that which faces the mirror is at rest, then the image in the mirror is at rest too. And so philosophy is nothing other than the knowledge and discovery of that which has its reflection in the mirror. And just as the image in the mirror gives no one any idea about his nature, and cannot be the object of cognition, but is only a dead image, so is man, considered in himself: nothing can be learned from him alone. For knowledge comes only from that outside being whose mirrored image he is. (I/8, 71–72)

Heaven is man, and man is heaven, and all men together are the one heaven, and heaven is nothing but one man. You must know this to understand why one place is this way and the other that way, why this is new and that is old, and why there are everywhere so many diverse things. But all this cannot be discovered by studying the heavens. . . . All that can be discovered is the distribution of their active influences. . . . We, men, have a heaven, and it lies in each of us in its entire plenitude, undivided and corresponding to each man's specificity. Thus each human life takes its own course, thus dying, death, and disease are unequally distributed, in each case according to the action of the heavens. For if the same heaven were in all of us, all men would have to be equally sick and equally healthy. But this is not so; the unity of the Great Heaven is split into our diversities by the various moments at which we are born. As soon as a child is conceived, it receives its own heaven. If all children had been born at the same moment, all of them would have had the same heaven in them, and their lives would have followed the same course. Therefore, the starry vault imprints itself on the inner heaven of a man. A miracle without equal! (I/8, 100–101)

Just as the firmament with all its constellations forms a whole in itself, likewise man in himself is a free and mighty firmament. And just as the firmament rests in itself and is not ruled by any creature, the firmament of man is not ruled by other creatures, but stands for itself and is free of all bonds. For there are two kinds of created things: heaven and the earth are of one kind, man is of the other. . . . Everything that astronomical theory has profoundly fathomed by studying the planetary aspects and the stars . . . can also be applied to the firmament of the body. (I/1, 202–3)

In nature we find a light that illumines us more than the sun and the moon. For it is so ordered that we see but half of man and all the other creatures, and therefore must explore them further. . . . Nor should we become drowned in our daily work, for whosoever seeks . . . shall find. . . . And if we follow the light of nature, we learn that there exists another half of man, and that man does not consist of blood and flesh alone . . . but also of a body that cannot be discerned by our crude eyesight. (I/9, 254–55)

Know that our world and everything we see in its compass and everything we can touch constitute only one half of the cosmos. The world we do not see is equal to ours in weight and measure, in nature and properties. From this it follows that there exists another half of man in which this invisible world operates. If we know of the two worlds, we realize that both halves are needed to constitute the whole man; for they are like two men united in one body. (I/9, 258)

Nature emits a light, and by its radiance she can be known. But in man there is still another light apart from that which is innate in nature. It is the light through which man experiences, learns, and fathoms the supernatural. Those who seek in the light of nature speak from knowledge of nature; but those who seek in the light of man speak from knowledge of super-nature. For man is more than nature; he is nature, but he is also a spirit, he is also an angel, and he has the properties of all three. If he walks in nature, he serves nature; if he walks in the spirit, he serves the spirit; if he walks with the angel, he serves the angel. The first is given to the body, the others are given to the soul, and are its jewel. (I/14, 115)

The physician comes from nature, from nature he is born; only he who receives his experience from nature is a physician, and not he who writes, speaks, and acts with his head and with ratiocinations aimed against nature and her ways. (I/6, 52)

The physician is only the servant of nature, not her master. Therefore it behooves medicine to follow the will of nature. (I/7, 150)

He who would be a good physician must find his faith in the rational light of nature, he must work with it, and not undertake anything without it. . . . For Christ would have you draw your faith from knowledge and not to live without knowledge. . . . If you desire to apply an art, let it be only in the light of nature, and not in superficial action. God has given to each man the light that was his due; so that he need not go astray. (I/1, 300)

. . . The mysteries of the firmament are revealed to the physician; to him the mysteries of nature are manifest, and he communicates them to other learned men. Thus philosophy comprises knowledge of all the organs and limbs, health and sickness. The condition of urine must be read from the outer world, the pulse must be understood in relation to the firmament, physiognomy to the stars, chiromancy to the minerals, the breath to the east and west winds, fever to earthquakes, etc.—If the physician understands things exactly and sees and recognizes all illnesses in the macrocosm outside man, and if he has a clear idea of man and his whole nature, then and only then is he a physician. Then he may approach the inside of man; then he may examine his urine, take his pulse, and understand where each thing belongs. This would not be possible without profound knowledge of the outer man, who is nothing other than heaven and earth. It would be bold and presumptuous to approach the study of man without such knowledge and to defend the sandy ground of speculation, which is more unstable than a reed in the wind. (I/8, 76)

Everything external in nature points to something internal; for nature is both inside man and outside man. An example. . . . Herbs are gathered together in an apothecary's ship and can be bought there, and in one shop more numerous and varied herbs can be found than in another; similarly there is in the world a natural order of apothecary's shops, for all the fields and meadows, all the mountains and hills are such shops. Nature has given us all of them, from which to fill our own shops. All nature is like one single apothecary's shop, covered only with the roof of heaven; and only One Being works the pestle as far as the world extends. But man has such a shop only in part, not wholly; he possesses something, not everything. For nature's apothecary's shop is greater than man's. (I/11, 195)

Nature is so careful and exact in her creations that they cannot be used without great skill; for she does not produce anything that is perfect in itself. Man must bring everything to perfection. This work of bringing things to their perfection is called "alchemy." And he is an alchemist who carries what nature grows for the use of man to its destined end. But within this art distinctions must again be made; if someone takes a sheepskin and uses it untanned as a coat, how crude and clumsy it is in comparison with the work of a furrier or clothmaker! If a man fails to perfect a thing that nature has given him, he is guilty of even greater crudeness and clumsiness, especially when man's health, body, and life are at stake. For this reason more diligence should be spent on alchemy, in order to obtain still greater results. Artisans have explored nature and its properties in order to learn to imitate her in all things, and to bring out the highest that is in her. Only in medicine has this been neglected, and therefore it has remained the crudest and clumsiest of all the arts. (I/8, 180)

Everything that man does and has to do, he should do by the light of nature. For the light of nature is nothing other than reason itself. (I/1, 306)

Only he is the enemy of nature who fancies himself wiser than nature, although she is the greatest school for all of us. (I/9, 216)

He who is guided by the ways of nature becomes well versed in the two branches of philosophy, that of heaven and that of the earth. He is blessed with such great knowledge that neither life nor death, neither health nor illness is hidden from him. (I/8, 373)

There is nothing that nature has not signed in such a way that man may discover its essence. . . . The stars have their orbits by which they are known. The same is true of man. As you can see, each herb is given the form that befits its nature; similarly, man is endowed with a form corresponding to his inner nature. And just as the form shows what a given herb is, so the human shape is a sign which indicates what a given man is. This does not refer to the name, sex, or similar characteristics, but to the qualities inherent in the man. The art of signs teaches us to give each man his true name in accordance with his innate nature. A wolf must not be called a sheep, a dove must not be called a fox; each being should be given the name that belongs to its essence. . . . Since nothing is so secret or hidden that it cannot be revealed, everything depends on the discovery of those things which manifest the hidden. . . . The nature of each man's soul accords with the design of his lineaments and arteries. The same is true of the face, which is shaped and formed according to the content of his mind and soul, and the same is again true of the proportions of the human body. For the sculptor of Nature is so artful that he does not mold the soul to fit the form, but the form to fit the soul; in other words, the shape of a man is formed in accordance with the manner of his heart. . . . Artists who make sculptures proceed no differently. . . . And the more accomplished an artist would be, the more necessary it is that he master the art of signs. . . . No artist can paint or carve, no one can produce an accomplished work, without such knowledge. . . . Only he who has some knowledge of this can be a finished artist. (I/12, 91–93)

The Infinite Unity of the Many

The Dialogues of Giordano Bruno

Even more than Paracelsus, Giordano Bruno (1548–1600) embodies the restless spirit of the Renaissance. More perhaps than in any other man of his age, the Hermetic mysticism revived by the humanists is combined in him with the revolutionary impetus of the new science, producing a ceaseless conflict of opposites. Out of this arises the insatiable aspiration to which Christopher Marlowe gives voice in Tamburlaine *(1590), for which Bruno may indeed have served as a model:*

Nature that framed us of four elements,
Warring within our breast for regiment,
Doth teach us all to have aspiring minds:
Our souls, whose faculties can comprehend
The wondrous architecture of the world,
And measure every wandering planet's course,
Still climbing after knowledge infinite,
And always moving as the restless spheres,
Wills us to wear ourselves and never rest,
Until we reach the ripest fruit of all . . .

For Bruno, unlike Marlowe's Scythian shepherd turned conqueror, this was not "the sweet fruition of an earthly crown" but that unattainable union with the infinite One toward which Bruno, in a sonnet of his own (as translated by John Addington Symonds), imagines himself ascending, "spurning the earth, soaring to heaven" through "boundless air" and refusing to curb his "daring will" by heeding Icarus's fearful end:

Dread not, I answer, that tremendous fall:
Strike through the clouds, and smile when death is near,
If death so glorious be our doom at all!

Born in Nola, he attended university at nearby Naples and joined the Dominican order, but fled when suspected of heresy in about 1576, and began a life of itinerant teaching that took him to Padua, Lyons, Geneva, Toulouse, Paris, London, Oxford, Wittenberg, Prague, and Frankfurt. In England, from 1583 to 1585, he was intimate with the aristocratic literary circle of Sir Philip Sidney and Sir Fulke Greville, participated in a debate at Oxford on the Copernican system, and wrote six Italian dialogues that expound his complex (if less than fully coherent) philosophical ideas. These include three important cosmological works, all published in 1584 with the false imprint of Venice: La Cena de le ceneri *(*The Ash Wednesday Supper, *trans. Edward A. Gosselin and Lawrence S. Lerner [1977]);* De la Causa, principio et uno (Concerning the Cause, Principle, and One, *trans. Sidney Greenberg in* The Infinite in Giordano Bruno *[1950]); and* De l'infinito universo et mondi (On the Infinite Universe and Worlds, *trans. Dorothea Waley Singer, in* Giordano Bruno: His Life and Thought *[1950]); our selections are taken from the latter two. In Frankfurt he published his main Latin works, including* De triplici minimo (On the Triple Minimum), *which expands on the Italian dialogues and sets forth his theory of an indivisible monad, or atom, the irreducibly finite minimum of which the components of the infinite universe are composed. In 1591 he traveled to Venice, where he was denounced to the Inquisition, charged with heresy, taken to Rome, and after more than eight years of imprisonment and examination by the Holy Office, burned at the stake in the Campo di Fiori in February 1600.*

Despite his bold advocacy of the Copernican system—which was not among the charges brought against him by the Inquisition—Bruno was neither a scientist nor deeply informed in science, but a mystic. Versed in magic, Hermetic and Pythagorean/Neoplatonic thought, and Kabbala (as Frances Yates maintains in Giordano Bruno and the Hermetic Tradition *[1964]), he saw in Copernican astronomy a confirmation of his occult beliefs. "The only passion that reigns in Bruno," Cassirer remarks in* The Individual and the Cosmos, *"is the passion of the self-affirmation of the Ego, heightened to titanic and heroic proportions," and in his writings "the problems of 'natural magic' take up so much room that they threaten to stifle the speculative-*

philosophical problem." Having glimpsed a macrocosmic infinity corresponding to the infinite within, he drew upon many sources in developing his conception of a universe accessible only to the intellect. "The basis of his belief was the Epicurean theory, which he derived from Lucretius, of an infinite universe with a plurality of (inhabited) worlds," Boas writes; ". . . indeed Bruno was probably the first philosopher who really comprehended the possibilities inherent in the idea of infinity. With Lucretius, Bruno blended the Platonic concept of the world-soul, and, from Nicholas of Cusa, a pantheistic concept of the relation of God and the universe." The greatest secrets of nature are revealed through apprehension of its contraries and opposites, but the supreme good to which knowledge of these secrets leads is the Neoplatonic "unity that embraces all," which is the infinite universe itself, or God.

FROM Concerning the Cause, Principle, and One, *translated by Sidney Greenberg*

From THE THIRD DIALOGUE . . . *Gervasius.* Pardon, Theophilus, do me this favor first—since I am not so experienced in philosophy—expound for me what you understand under the name of matter, and what is truly matter in natural things.
Theophilus. All those who wish to distinguish matter and consider it in itself, separate from form, recur to the analogy of Art. So it is with the Pythagoreans, the Platonists, and the Peripatetics. Take, for example, an art like that of carpentry: it has wood as a subject for all of its forms and for all of its work; just so is it with the blacksmith whose subject is iron, as the subject of the tailor is cloth. All of these arts bring into a particular matter diverse pictures, arrangements, and figures, none of which is proper and natural to matter. Therefore, nature, to which art is similar, needs to have for its operations a matter; because it is not possible that there be an agent which, when it wishes to make something, does not have that out of which it can make it; or likewise, if it wishes to work, does not have that on which to work. There is then a kind of substratum from which, with which, and in which, nature effects its operations and its work; and which is by nature endowed with so many forms that it presents for our consideration such a variety of species. And just as wood in itself has no artificial form, but can have all of them through the operations of the carpenter, in a similar way, matter, of which we speak, has no natural form by itself, and in its nature, but can have all forms through the operations of the active agent, the principle of nature. This natural matter is not, therefore, as perceptible as artificial matter is because the matter of nature has absolutely no form; but the matter of art is a thing already formed by nature, since art can only work on the surface of things formed by nature, as is the case with wood, iron, stone, wool, and others of this kind; but nature, so to speak, works from within its subject or matter, which throughout is formless. Therefore, the subjects of art are manifold, but the subject of nature is one; because the former, being diversely formed by nature, are different and variegated; the latter, being formless, is entirely indifferent, since all difference and diversity stem from the form.
Gervasius. So that the things formed by nature are the matter of art, and one thing alone that is formless is the matter of nature?
Theophilus. So it is.
Gervasius. Is it then possible that just as we can see and know clearly the substrata of the arts, we can similarly know the substratum of nature?
Theophilus. Without a doubt, but with diverse principles of cognition; for just as we do

not know colors and sounds by means of the same sense, similarly, we cannot see by means of the same eye the substratum of the arts and the substratum of nature.
Gervasius. You wish to say, then, that we see the former with the sensible eyes, and the latter with the eyes of reason.
Theophilus. Completely right.
Gervasius. May it please you then to develop this argument.
Theophilus. Willingly. That same relation and reference which the form of art has to its matter is likewise, according to due analogy, had by the form of nature to its matter. Just as in art, the forms varying to infinity (if this were possible), there always remains under all those forms the one same matter; as, for example, after the form of the tree, there is the form of the trunk; so then that of the board, that of the table, bench, stool, chest, the comb, and so forth; and throughout all the wood remains the same; so, analogously, is it in nature, where although the forms vary themselves to infinity, the one succeeding the other, the matter remains ever the same.
Gervasius. And how would you corroborate this comparison?
Theophilus. Do you not see that that which was seed becomes a herb, that that which was herb becomes corn, that that which was corn becomes bread—from bread, chyle; from chyle, blood; and from this, seed; from this, Embryo, man, corpse, earth, stone, and something else, and so on, taking in all of the natural forms?
Gervasius. I see this very easily.
Theophilus. It is necessary, therefore, that there be one same thing which in itself is not stone, not earth, not corpse, not man, not embryo, not blood, or anything else; but that which, after it was blood, became embryo, receiving the embryonic being; and after it was embryo received the human being, becoming man (himself); just so that which is formed by nature, which is the subject of art: for example, that which was a tree becomes a table, and receives the being table; that which was table, receiving the being door, becomes a door.
Gervasius. I have understood this very well. But, it seems to me, this subject of nature cannot be corporeal, nor of a certain quality; because this subject which is escaping now under one form and natural being, now under another form and being, does not show itself corporeally, as the wood or the stone, which make themselves seen as that which they are while they serve as material or subject, under whatever form they may appear.
Theophilus. Completely correct.
Gervasius. What shall I then do, when it shall come to pass that in discussing this thought with some obstinate person, he does not wish to believe that there is one matter under all the forms of nature, just as there is one under all the formations of each art? . . .
Theophilus. Then you will say to him more secretly thus: My most illustrious gentleman, or, your sacred majesty, since some things cannot become evident except through the hands and the sense of touch, and others with the ears, and still others only with the tongue, and others with the eyes—so this matter of natural things cannot become evident except through the intellect. . . .

From THE FIFTH DIALOGUE *Theophilus.* . . . First, then, I want you to note that there is one and the same scale, through which nature descends to the production of things, and the intellect ascends to the cognition of them, and that one and the other proceeds from unity to unity, passing through the multiplicity of media. . . .

Second, consider that the intellect, wishing to free itself and loosen itself from the imagination to which it is joined, besides recurring to mathematical and imaginable figures in order that it may either through those or through similarities to that understand the being and substance of things, also comes to refer the multitude and diversity of species to one and the same root; as Pythagoras, who posited the numbers as the specific principles of things, understood unity as the foundation and substance of all; [as] Plato and others, who placed the stable species in the figures, understood the point as the trunk and root of all, as substance and universal genus . . . Those, then, who say that the substantial principle is one take the substances as numbers; the others who understand the substantial principle as the point take the substances of things as figures; and all agree to set an indivisible principle. But better still and purer than Plato's is the method of Pythagoras, because unity is the cause and reason of indivisibility and punctuality, and is a principle more absolute and more conformable to the universal being. . . .

. . . Who does not see that the principle of generation and corruption is one? Is not the end of corruption the principle of generation? . . . In substance and root, therefore, love and hate, friendship and conflict, is one and the same thing. . . .

In conclusion, he who wishes to know the greatest secrets of nature should regard and contemplate the minimum and maximum of contraries and opposites. It is profound magic to know how to draw out the contrary after having found the point of union. . . .

. . . The highest good, the highest object of desire, the highest perfection, the highest beatitude, consists in the unity that embraces all. . . . We delight in a sensible, but greatest delight is in that which comprehends all the sensibles; we delight in a knowable that comprises everything knowable; we delight in the apprehensible that embraces all that which can be comprehended; we delight in a being which embraces all that which can be comprehended; we delight in a being which embraces all, but greatest delight is in that one which is the all itself. . . .

Praised be the Gods, and extolled by all the living be the infinite, the simplest, the most unified, the highest, and the most absolute cause, principle, and the one.

FROM On the Infinite Universe and Worlds,
translated by Dorothea Waley Singer (revised)

From THE THIRD DIALOGUE *Fracastoro.* I would conclude as follows. The famous and received order of the elements and of the heavenly bodies is a dream and vainest fantasy, since it can neither be verified by observation of nature nor proved by reason or argued, nor is it either convenient or possible to conceive that it exist in such fashion. But we know that there is an infinite field, a containing space which embraces and interpenetrates the whole. In it is an infinity of bodies similar to our own. No one of these more than another is in the center of the universe, for the universe is infinite and therefore without center or limit, though these appertain to each of the worlds within the universe in the way I have explained on other occasions, especially when we demonstrated that there are certain determined definite centers, namely, the suns, fiery bodies around which revolve all planets, earths, and waters, even as we see the seven wandering planets take their course around our sun. Similarly we showed that each of these stars or worlds, spinning around his own center, has the appearance of a solid and continuous world which takes by force all visible things which can become stars and whirls them around himself as the center of their universe. Thus there is not merely one world, one earth, one sun, but as many worlds as we see bright lights around us, which are neither more nor less in one heaven, one space, one containing sphere than is this our world in one containing universe, one space, or one heaven. So that the heaven, the infinitely

extending air, though part of the infinite universe, is not therefore a world or part of worlds; but is the womb, the receptacle and field within which they all move and live, grow and render effective the several acts of their vicissitudes; produce, nourish, and maintain their inhabitants and animals; and by certain dispositions and orders they minister to higher nature, changing the face of single being through countless subjects. Thus each of these worlds is a center toward which converges every one of his own parts; toward it every kindred thing tends just as the parts of this our star, even though at a certain distance, are yet brought back to their own field from all sides of the surrounding region. Therefore, since no part which flows thus outward from the great Body fails ultimately to return thereto, it happens that every such world is eternal though dissoluble; albeit, if I mistake not, the inevitability of such eternity depends on an external maintaining and provident Being and not on intrinsic power and self-sufficiency. . . .
Burchio. Then the other worlds are inhabited like our own?
Fracastoro. If not exactly as our own, and if not more nobly, at least no less inhabited and no less nobly. For it is impossible that a rational being fairly vigilant can imagine that these innumerable worlds, manifestly like our own or yet more magnificent, should be destitute of similar and even superior inhabitants . . .

From The Fifth Dialogue *Philotheo.* . . . There is in fact no *primum mobile* that draws those many bodies around ourselves as center. Rather it is our globe which causes the appearance of this happening, for reasons which Elpino will expound to you.
Albertino. Willingly I will hear him.
Philotheo. When you have heard and have well marked that such an opinion is contrary to nature, while ours is consonant with all reason, perception, and verification in nature, you will no longer say that there is an edge or limit either to the extent or to the motion of the universe; you will esteem the belief in a *primum mobile*, an uppermost and all-containing heaven, to be a vain fantasy. You will conceive rather a general womb in which are situated all worlds alike, even as this terrestrial globe in this our local space is surrounded by our atmosphere and is in no way nailed or attached to any other body, nor has any base but his own center. And if it is found that this our globe cannot be proved to be of a constitution different from the surrounding stars, since it manifests accidents no different from theirs, then should it no more than any one of them be regarded as occupying the central position of the universe, nor as being more fixed than they, nor will they appear to revolve around it rather than it around them. Whence, since such indifference on the part of nature must be inferred, so also we must infer the vanity of [imagined] deferent orbs; and we must accept the inner impulse toward motion implanted in the souls of these globes, the indifference throughout the vast space of the universe and the irrationality of conceiving any edge or external shape thereto. . . .

. . . There are then an infinity of mobile bodies and motive forces, and all of these reduce to a single passive principle and a single active principle, just as every number reduces to unity, and as infinite number coincides with unity; and just as the supreme Agent and supreme active power coincides in a single principle with the supreme potentiality, patient of all creation, as has been shown at the end of our book *Concerning the Cause, Principle, and One*. In number then, and in multitude, there is infinite possibility of motion and infinite motion. But in unity and singularity is infinite motionless motive force, an infinite motionless universe. And the infinite number and magnitude coincide with the infinite unity and simplicity in a single utterly simple and indivisible principle, which is Truth and Being. Thus there is no *primum mobile*, no order from it of second and other mobile bodies either to a last body or yet to infinity. But all mobile bodies are equally near to and equally far from the prime and universal motive power, just as (logically speaking) all species are equally related to the same kindred, and all individuals to a single species. Thus from a single infinite and universal motive force in a single infinite space there is but one infinite universal motion on which depend an infinity of mobile bodies and of motor forces, each of which is finite both as to size and power. . . .

You see further that our philosophy is by no means opposed to reason. It reduces everything to a single origin and relates everything to a single end, and makes contraries to coincide, so that there is one primal foundation both of origin and of end. From this coincidence of contraries we deduce that ultimately it is divinely right to say and to hold that contraries are within contraries, wherefore it is not difficult to compass the knowledge that each thing is within every other—which Aristotle and the other Sophists could not comprehend. . . .

Strings of the Universal Lyre

Passages from the Writings of John Dee

John Dee (1527–1608), J. L. Heilbron writes in his introduction to John Dee on Astronomy *(1978)—which contains Wayne Shumaker's Latin text and English translation of the* Propaedeumata Aphoristica*—"was a geometer, physicist, astrologer, antiquarian, hermetist, and conjurer, a mixture of mathematician and magician, of scholar and enthusiast, of schemer and dupe." His writings "show a continuous progress toward the occult and irrational," by which he became a paradigm of the Elizabethan magus and one possible model, Frances Yates has suggested, for Marlowe's Doctor Faustus, the Prospero of Shakespeare's* Tempest, *and the fraudulent "Paracelsist physician" Subtle in Jonson's* The Alchemist. *He attended St. John's College, Cambridge, in 1542, then studied mathematics at Louvain and traveled to Brussels and Paris. He was briefly imprisoned under Queen Mary in 1555, but after the accession of Elizabeth in 1558—the year when he published his* Propaedeumata, *whose 120 aphorisms set forth his cosmological and astrological concepts—he was introduced to the queen by Robert Dudley, earl of Leicester, and for two decades appears to have enjoyed the discreet patronage of these powerful allies. During these years he built up an outstanding library and collection of scientific instruments at his home at Mortlake and engaged in various geographical, mathematical, and alchemical pursuits. In 1564 he published his alchemical* Monas hieroglyphica, *and in 1570 his "Mathematicall Praeface" to Henry Billingsley's English translation of Euclid's* Elements *(photostatically reprinted, with an introduction by Allen G. Debus, in 1975). In the latter, he asserts—somewhat like Ptolemy in the* Almagest *(Chapter 12 above)—that mathematics mediates between the natural and the supernatural.*

"Frequently compared to Francis Bacon's Advancement of Learning,*" the Preface, Nicholas H. Clulee writes in* John Dee's Natural Philosophy: Between Science and Religion *(1988), "seems a manifesto of modern science more farsighted than Bacon's because Dee combined an understanding of experimental method with an emphasis on the importance of mathematics and quantification for the study of nature." Yet the Preface begins with the words "Divine Plato," and belongs to the mystical at least as much as to the scientific tradition. Its mathematical arts include not only the practical, such as geography and architecture, but also the occult, such as*

astrology, "an art mathematical, which reasonably demonstrateth the operations and effects of the natural beams of light and secret influence of the stars and planets." Its culminating art of "archemastry," which "because it proceedeth by experiences and searcheth forth the causes of conclusions by experiences . . . is named of some scientia experimentalis," *or "experimental science" (a term that Dee found in both Nicholas Cusanus and Roger Bacon), could pertain to "natural magic" as much as to natural science. In 1582 Dee met Edward Kelley, and their endeavors to conjure up angels by kabbalistic magic—experiments recorded in the diaries published in 1659 as* A True and Faithful Relation of What Passed for Many Years Between Dr. John Dee . . . and Some Spirits—*contributed to his renown as a practitioner of the black arts. Shortly after he departed for Poland with Prince Albertus Laski in 1583, a mob pillaged Mortlake and destroyed part of his library. When he returned in 1589 from Cracow and Prague, he was unable to win back the favor of Queen Elizabeth or to attain that of her successor, King James, and died impoverished and nearly alone at Mortlake.*

FROM Propaedeumata Aphoristica, *translated by Wayne Shumaker*

An "Aphoristic Introduction" to Certain Especially Important Virtues of Nature

APHORISM I As God created all things from nothing against the laws of reason and nature, so anything created can never be reduced to nothing unless this is done through the supernatural power of God and against the laws of reason and nature.

II In actual truth, wonderful changes may be produced by us in natural things if we force nature artfully by means of the principles of pyronomia.[4] I call Nature whatever has been created.

III Not only are those things to be said to exist which are plainly evident and known by their action in the natural order, but also those which, seminally present, as it were, in the hidden corners of nature, wise men can demonstrate to exist.

IV Whatever exists by action emits spherically upon the various parts of the universe rays which, in their own manner, fill the whole universe. Wherefore every place in the universe contains rays of all the things that have active existence.

VIII Whatever acts upon something else is like it in some respect; but in another way it differs utterly from that upon which it acts, or there is no action.

IX Whatever is in the universe possesses order, agreement, and similar form with something else.

X Whatever things are of the same order, or harmonious, or of similar form sometimes imitate each other of their own accord and sometimes even move toward one another; one protects and defends the other as much as it can even if, at the moment, it appears to be drawing energy from the other. By the joining of such natural things that exist separately in the universe, in their differing fashions, and by the activating of other

[4]Pyronomia is defined by Andreas Libavius as "the science of using and regulating heat and fire in one's operations" (*Alchemia*, 1597, I, xiiii, 24–25). . . . Although the "wonderful changes" would include alchemical transformations, Dee's main purpose is to acknowledge that despite the indestructibility of matter its forms can be changed. (Shumaker)

things placed somewhat higher, seminally, in nature, more wonderful things can be performed truly and naturally, without violence in faith to God or injury to the Christian religion, than any mortal might be able to believe.

XI The entire universe is like a lyre tuned by some excellent artificer, whose strings are separate species of the universal whole. Anyone who knew how to touch these dextrously and make them vibrate would draw forth marvelous harmonies. In himself, man is wholly analogous to the universal lyre.[5]

XII Just as the lyre is an arrangement of harmonious and disharmonious tones, most apt for expressing a very sweet harmony which is wonderful in its infinite variety, so the universe includes within itself parts among which a most close sympathy can be observed, but also other parts among which there is harsh dissonance and a striking antipathy. The result is that the mutual concord of the former and the strife and dissension of the latter together produce a consent of the whole and a union eminently worthy of admiration.

FROM *Mathematicall Praeface to the Elements of Geometrie of Euclid of Megara*

All things which are, and have being, are found under a triple diversity general.

For either they are deemed supernatural, natural, or of a third being. Things supernatural are immaterial, simple, indivisible, incorruptible, and unchangeable. Things natural are material, compounded, divisible, corruptible, and changeable. Things supernatural are of the mind only comprehended: things natural of the sense exterior are able to be perceived. In things natural, probability and conjecture hath place: but in things supernatural, chief demonstration and most sure science is to be had. By which properties and comparisons of these two, more easily may be described the state, condition, nature, and property of those things which we before termed of a third being: which, by a peculiar name also, are called things mathematical. For these, being (in a manner) middle between things supernatural and natural, are not so absolute and excellent as things supernatural, nor yet so base and gross as things natural, but are things immaterial and nevertheless by material things able somewhat to be signified. And though their particular images by art are aggregable and divisible, yet the general forms, notwithstanding, are constant, unchangeable, untransformable, and incorruptible. Neither of the sense can they at any time be perceived or judged. Nor yet, for all that, in the royal mind of man first conceived. But, surmounting the imperfection of conjecture, weening, and opinion, and coming short of high intellectual conception, are the mercurial fruit of dianoetical discourse, in perfect imagination subsisting. A marvelous neutrality have these things mathematical, and also a strange participation between things supernatural, immortal, intellectual, simple and indivisible, and things

[5]As is well known, the notion of a cosmic harmony goes back at least to Pythagoras (sixth century B.C.) [see Chapter 9 above], who thought that the planetary intervals were such that the movements of the planets in their orbits made ravishing, though inaudible, music. In Christian thought, its inaudibility to man was caused by the Fall. See S. K. Heninger, Jr., *Touches of Sweet Harmony* (1974), especially pp. 179–189. As the microcosm, man also is a harmonious little world patterned on the cosmos. For "touching" the strings of the universal lyre, see D. P. Walker, *Spiritual and Demonic Magic from Ficino to Campanella* (1958), I, i:2, "Ficino's Astrological Music"; also Marsilio Ficino, "De vita caelitus comparanda" (Part III of *De vita triplici*), summarized in Shumaker, *Occult Sciences*, pp. 132–133. Cornelius Agrippa explains "what sounds accord with each star" (i.e., planet) in *De occulta philosophia*, II, xxvi. (Shumaker)

natural, mortal, sensible, compounded, and divisible. Probability and sensible proof may well serve in things natural, and is commendable: in mathematical reasonings, a probable argument is nothing regarded, nor yet the testimony of sense any whit credited, but only a perfect demonstration of truths certain, necessary, and invincible, universally and necessarily concluded, is allowed as sufficient for an argument exactly and purely mathematical. . . .

The Admirable Double World Machine

The Technical, Physical and Metaphysical History of the Macrocosm and Microcosm, by Robert Fludd, translated by Patricia Tahil

*Robert Fludd (1574–1637) gives one of the last and fullest expressions to the Neoplatonic-Hermetic-Paracelsian tradition of Renaissance cosmology based on the ancient theory of correspondence between the little world of man and the greater world around him; the title and contents of his major work—*Utriusque cosmi majoris scilicet et minoris metaphysica, physica, atque technica historia (Technical, Physical and Metaphysical History of the Macrocosm and Microcosm)*—published between 1617 and 1621, reflect that foundation. Born of an aristocratic family in Shropshire, he was educated at St. John's College, Oxford; traveled in France, Spain, Italy, and Germany; took his doctorate in medicine at Christ Church, Oxford, in 1605; and despite controversy over his adherence to the "Chemical Philosophy" of the Paracelsians rather than to orthodox Galenic medicine, was admitted in 1609 to the College of Physicians in London. In 1616 and 1617 he wrote two defenses of the mysterious (and possibly nonexistent) Order of the Rosy Cross, or Rosicrucians, in whose name two anonymous works had been published in Germany, creating a sensation in much of Europe. In 1617 he published in Germany the first of four volumes of his* Utriusque cosmi historia, *accompanied by elaborately symbolical illustrations. His later years were marked by extensive controversies with others, including Johannes Kepler, Marin Mersenne, and Pierre Gassendi, and by advocacy of William Harvey's theory of the circulation of the blood (Chapter 18 below), which Fludd saw as confirmation of his own beliefs; his last major work,* Philosophia Moysaica, *was published in 1638, and translated as* Mosaicall Philosophy *in 1659.*

"Fludd's natural conservatism," Allen G. Debus writes in his introduction (1979) to Fludd's unpublished Philosophical Key *of about 1619, "was combined with a philosophy that many of his contemporaries considered to be radical in the extreme" in its rejection of orthodox Aristotelianism in the name of both Paracelsian medicine and the new philosophy of science. Yet what is perhaps most striking about the tripartite cosmology set forth in the following selections (from* Robert Fludd: Essential Readings *[1992], ed. William H. Huffman) is how similar it is, in its affirmation of a geocentric universe encircled by the* primum mobile, *and in its division between sublunary and superlunary spheres, both to the medieval "world picture" and to the cosmology that lay behind it. This was the cosmology not only of Ptolemy but above all of Aristotle, whom almost all hermetical writers after Pico had attacked—Bruno even declaring him "the stupidest of all philosophers"—but whom none could escape, if only because his ideas had been from the beginning as deeply embedded in the Neoplatonism they embraced as in the Scholasticism they spurned.*

FROM *Book One: The History of the Macrocosm*

CHAPTER ONE: ON INFINITE NATURE AND THE MAKER OF ALL THINGS Infinite nature, which is boundless Spirit, unutterable, not intelligible, outside of all imagination, beyond all essence, unnameable, known only to the heart, most wise, most merciful, FATHER, WORD, HOLY SPIRIT, the highest and only good, incomprehensible in height, the unity of all creatures, which is stronger than all power, greater than all distinction, more worthy than all praise, indivisible TRINITY, most splendid and indescribable light, in short, the divine mind, free and separate from mortal matter, glory of all, necessity, extremity and renewal: Here, I say, GOD, the highest and greatest of all, whose name was made blessed in eternity, skilfully formed the admirable machine of the entire Macrocosm, and beautifully adorned his structure. . . .

FROM *Book Two: The Structure of the Cosmos*

CHAPTER ONE: HOW THE UNIVERSE IS DIVIDED . . . We are about to deal with the double world—the Macrocosm and the Microcosm—distinguishing the first from Man, or the Microcosm, and considering the entire volume of the Primal Matter as the World, Cosmos, or Macrocosm, which the spiritual light, or spirit, of the Lord, surrounding the waters, enfolds in its circular embrace. A certain round portion of the abyss's material is divided into three different parts by Formative Nature, by arranging light and darkness in varying blends, i.e. distinguishing between the blends according to their purity and impurity. The highest of these parts is that expanse of the universe where the fiery spirit was, and it is contained in the primal light-stuff, extending from the concave limit of the Sphere of the Trinity to the convex surface of the Starry Sky. And because it contains an abundance of pattern, the matter of this part of the universe is so fine and pure that it is totally imperceptible, and cannot be seen by us; so the Philosophers called it intellectual, and spiritual in the highest degree. Now the middle part is adorned with the stars, both fixed and wandering, for it occupies the entire cavity of the Universe held between the convex outline of the Moon's orbit, and the concave outline of the *Primum Mobile*; and the matter in this region is solid compared to the higher one. Finally the lowest part of the universe is the entire expanse embraced by the concave outline of the Moon's orbit. It therefore follows that the mass of the universe is divided into two main parts, of which one is bodiless, spiritual, and very pure and fine, that is the upper one, whilst the other one is material. And this material part is subdivided again into two parts: one is rare, fine, and imperishable—that is the middle part—the other, gross, impure, and subject to corruption—this is the bottom part of the universe, which we shall call sublunary and elemental. . . .

CHAPTER FIFTEEN: WHAT THE AFORESAID ELEMENTS ARE MADE OF . . . Lastly, the spirit of the darkness in the third and lowest region of the universe, which is by far the densest and heaviest of all, succeeds the spirit of Ether in the natural order, since it is its close neighbor; and, therefore, it occupies the region of Water, and is called Water, but the fragments of this spirit that contain any portion of the light peculiar to this heaven produce minerals. For this reason the retreating spirits of darkness, disturbed from the Empyrean Sky and the highest region of all, composed the element Fire, and when thrown down out of the middle region, or Ether, they escaped and became the element

Air; and finally, when they were expelled from the lowest region of the sky, they acquired the essential nature of Water.

Now Earth is like the midden of them all, the receptacle of their surpluses, and (here I agree with the chemical philosophers), the *caput mortuum* or dung of the whole spiritual mass; next to its grossness and filth is Water, which is, as it were, the more servile part of Air, as Air is the more servile part of Fire, and Fire of the Quintessence.

Now the following is the chemical extraction of a substance: the better part is the Quintessence, which is raised to the Ether along with its soul, as we discover when we rectify spirits of wine: the grosser part of the spirit, called burning water, is distilled next, in which there are, as it were, three parts of Quintessence and one of phlegma, likened to the denser part of the spirit of darkness in the middle heaven, constituting the element Fire. After this follows that weak liquid commonly called *Aqua Vitae*, in which there are two parts of spirits or wine, and the same number of water, which is compared to the element Air, and finally phlegma is distilled, in which there is almost no spirit, but is all a watery mucus, in which the element Water is collected. Moreover, the dregs or *caput mortuum* of all these, found at the bottom of the flask, may not improperly be likened to the element Earth. The marvelous way in which Nature works, selecting and purifying the material of the elements and putting them in order, is thus clearly demonstrated. . . .

The Weight of a Tiny Bird

Selections from the Notebooks of Leonardo da Vinci, translated by Edward MacCurdy

Leonardo da Vinci (1452–1519), whose prodigal genius found expression in drawing and sculpture, architecture and music, engineering and invention, and in a few paintings unsurpassed in his own or any age, was the illegitimate and largely self-taught son of a Florentine notary and a peasant girl. Born in a small Tuscan village and apprenticed in 1466 to the Florentine painter Andrea del Verrocchio, in 1482 he moved to Milan, where he spent some sixteen years at the court of Lodovico Sforza, leaving in 1499 after Sforza's overthrow by the French, and returning to Florence in 1500. In the service of Cesare Borgia as a military engineer in 1502 he became acquainted with Machiavelli, then returned to Milan as an architect and engineer in 1506, and in 1513 was called to Rome as a painter by Pope Leo X. In about 1516 he accepted the invitation of King Francis I of France to settle at the castle of Cloux near Amboise in the Loire Valley, and there he pursued his studies undisturbed until his death.

For the arrogant humanists of his time, who imitated their predecessors rather than reading nature "by the light of experience," Leonardo had only contempt. Yet for him, unlike Paracelsus or others in the Hermetic-Neoplatonic tradition, "experience" denotes not mystical insight into truths inaccessible to the senses but empirical exploration of the processes of the natural world. In contrast to the medical and "chemical" sciences of his time, "which saw nature as an interplay of occult qualities," J. Bronowski writes in The Horizon Book of the Renaissance *(1961), "Leonardo looked at nature directly, not through the mind but through the eye. . . . He wanted no speculation about the soul of the bird, which scholars were still repeating from Pythagoras; he wanted to understand the harsh mechanics of its flight" by penetrating beneath its surface and apprehending what makes it work. To this extent he embodies the assumptions and practices of the new science. "Among his many interests," Vasari wrote a generation after his death,*[6] *"was included the study of nature; he investigated the properties of plants and then observed the motion of the heavens, the path of the moon, and the course of the sun." But despite intense interest in mathematics and important discoveries in anatomy and physiology, Leonardo, Kenneth Clarke remarks in* Leonardo da Vinci *(1939; rev. 1959), disliked "general principles" and lacked a "synthetic faculty" capable of drawing together the acute observations scattered throughout his astonishing Notebooks—written in a minute hand from right to left and never intended for publication—into a unity on which others might build: "it was only mistrust of generalizations which prevented him from anticipating many of the discoveries of later scientists, amongst them the circulation of the blood."*

He was "volatile and unstable," Vasari remarked, "for he was always setting himself to learn many things only to abandon them almost immediately." Yet in this restless drive to embrace all learning about the world is the great strength of Leonardo's ceaselessly exploratory mind, which resembles less that of a systematic scientist like Galileo than that of a humanist like Petrarch (Chapter 15 above), ever wandering "from thought to thought, from mountaintop to mountain," or Montaigne, tirelessly asking "What do I know?" Despite his scorn for humanist pretensions, Leonardo taught himself Latin and voraciously read the classics of ancient Greek medicine and mathematics in translation. Galen's On the Usefulness of the Parts of the Body *(Chapter 12 above), Marie Boas remarks, "stimulated him to further studies on bones and muscles, taught him much about anatomical fact and procedure, and interested him in physiological functioning," and the works of Archimedes, Clarke comments (following Gabriel Séailles), "were the greatest single influence on Leonardo's thought." Unlike literary humanists, however, he turned a painter's eye on the natural world, which he scrutinized with a precision all but unique in any age, whether its object was the color of leaves, the cascading of waters, the fabric of the human body, or any other phenomena of a world in which "movement is the cause of all life."*

But Leonardo's insights go far beyond the meticulous observations found in some herbalists of his time; his scrutiny of natural processes continually betrays a sympathetic identification with the objects of his interest and a deep personal yearning to incorporate the qualities—from the free flow of waters to the soaring flight of birds—that he so intently investigates. "Often when he was walking past the places where birds were sold," Vasari relates, "he would pay the price asked, take them from their cages, and let them fly off into the air, giving them back their lost freedom. In return he was so favored by nature that to whatever he turned his mind or thoughts the results were always inspired and perfect; and his lively and delightful works were incomparably graceful and realistic." "Everything interests him," Paul Valéry writes of the Leonardo he imagines in The Method of Leonardo da Vinci *(1895):*[7] *"It is of the universe that he thinks always. And he thinks of rigor. He is so made that he misses nothing of all that enters into the tangle of what exists—not a single shrub. He goes down into the depth of that which is for all men to see, but there he wanders away and studies himself."*

Many passages of the endlessly fascinating Notebooks—here excerpted from MacCurdy's two-volume translation of 1938, to which the parenthetical volume and page numbers refer—are simultaneously precise observations and moving prose poems. Thus, when describing a cavern whose darkness he shrinks from entering, stupefied by conflicting fear and desire, he leaves the

[6]In George Bull's translation of *The Lives of the Artists* (1965).
[7]As translated by Paul McGreevy in Paul Valéry, *Selected Writings* (1950).

narrative unfinished, Serge Bramly comments in Leonardo *(1988; English trans. 1991), "because he is evoking the terrible moment when he discovered in himself a window opening onto the demons and marvels of the night." Nowhere is he more truly a prophet of science than in his most poetically intuitive insights—as when he appears to anticipate the "butterfly effect" of modern chaos theory, with its emphasis on the unpredictable consequences of even the most seemingly minor event in our interconnected world, by remarking that "the earth is moved from its position by the weight of a tiny bird resting upon it."*

If indeed I have no power to quote from authors as they have, it is a far bigger and more worthy thing to read by the light of experience, which is the instructress of their masters. They strut about puffed up and pompous, decked out and adorned not with their own labors but by those of others, and they will not even allow me my own. And if they despise me who am an inventor, how much more should blame be given to themselves, who are not inventors but trumpeters and reciters of the works of others? (I.57)

Those who are inventors and interpreters between Nature and Man as compared with the reciters and trumpeters of the works of others, are to be considered simply as is an object in front of a mirror in comparison with its image when seen in the mirror, the one being something in itself, the other nothing: people whose debt to nature is small, for it seems only by chance that they wear human form, and but for this one might class them with the herds of beasts. (I.57)

I am fully aware that the fact of my not being a man of letters may cause certain arrogant persons to think that they may with reason censure me, alleging that I am a man ignorant of book-learning. Foolish folk! Do they not know that I might retort by saying, as did Marius to the Roman Patricians: "They who themselves go about adorned in the labor of others will not permit me my own?" They will say that because of my lack of book-learning, I cannot properly express what I desire to treat of. Do they not know that my subjects require for their exposition experience rather than the words of others? And since experience has been the mistress of whoever has written well, I take her as my mistress, and to her in all points make my appeal. (I.57–58)

Like an eddying wind scouring through a hollow, sandy valley, and with speeding course driving into its vortex everything that opposes its furious onset . . .

Not otherwise does the northern blast drive back with its hurricane . . .

Nor does the tempestuous sea make so loud a roaring when the northern blast beats it back in foaming waves between Scylla and Charybdis, nor Stromboli nor Mount Etna when the pent up, sulphurous fires, bursting open and rending asunder the mighty mountain by their force, are hurling through the air rocks and earth mingled together in the issuing belching flames. . . .

Nor when Etna's burning caverns vomit forth and give out again the uncontrollable element, and thrust it back to its own region in fury, driving before it whatever obstacle withstands its impetuous rage. . . .

And drawn on by my eager desire, anxious to behold the great abundance of the varied and strange forms created by the artificer Nature, having wandered for some distance among the overhanging rocks, I came to the mouth of a huge cavern before which for a time I remained stupefied, not having been aware of its existence, my back bent to an arch, my left hand clutching my knee, while with the right I made a shade for my lowered and contracted eyebrows; and I was bending continually first one way and then another in order to see whether I could discern anything inside, though this was rendered impossible by the intense darkness within. And after remaining there for a time, suddenly there were awakened within me two emotions, fear and desire, fear of the dark threatening cavern, desire to see whether there might be any marvelous thing therein. (II.472)

Among the great things which are found among us the existence of Nothing is the greatest. This dwells in time, and stretches its limbs into the past and the future, and with these takes to itself all works that are past and those that are to come, both of nature and of the animals, and possesses nothing of the indivisible present. It does not, however, extend to the essence of anything. (I.64)

. . . Nature is continually producing, and she does not change the ordinary kinds of things which she creates in the same way that from time to time the things which have been created by man are changed; and indeed man is nature's chiefest instrument, because nature is concerned only with the production of elementary things, but man from these elementary things produces an infinite number of compounds, although he has no power to create any natural thing except another like himself, that is his children. And of this the old alchemists will serve as my witnesses, who have never either by chance or deliberate experiment succeeded in creating the smallest thing which can be created by nature . . . She will completely cure you of your folly by showing you that nothing which you employ in your furnace will be numbered among the things which she employs in order to produce this gold [in the mines]. For these is there no quicksilver, no sulphur of any kind, no fire nor other heat than that of nature giving life to our world; and she will show you the veins of the gold spreading through the stone—the blue lapis lazuli, whose color is unaffected by the power of the fire.

And consider carefully this ramification of the gold, and you will see that the extremities of it are continually expanding in slow movement, transmuting into gold whatever they come in contact with; and note that therein is a living organism which it is not within your power to produce. (I.137–38)

Nothing is superfluous and nothing is lacking in any species of animal or product of nature unless the defect comes from the means which produce it. (I.156)

Every part is disposed to unite with the whole, that it may thereby escape from its own incompleteness. (I.59)

Experience the interpreter between resourceful nature and the human species teaches us that that which this nature works out among mortals constrained by necessity cannot operate in any other way than that in which reason which is its rudder teaches it to work. (I.61)

There is no result in nature without a cause; understand the cause and you will have no need of the experiment. (I.61)

You who speculate on the nature of things, I praise you not for knowing the processes which nature ordinarily effects of herself, but rejoice if so be that you know the issue of such things as your mind conceives. (I.68)

Movement is the cause of all life. (I.69)

Nature is full of infinite causes which were never set forth in experience. (I.69)

The earth is moved from its position by the weight of a tiny bird resting upon it.

The surface of the sphere of the water is moved by a tiny drop of water falling upon it. (I.70)

Every action done by nature is done in the shortest way. (I.71)

[A disputation]

Against. Why nature did not ordain that one animal should not live by the death of another.

For. Nature being capricious and taking pleasure in creating and producing a continuous succession of lives and forms because she knows that they serve to increase her terrestrial substance, is more ready and swift in creating than time is in destroying, and therefore she has ordained that many animals shall serve as food one for the other; and as this does not satisfy her desire, she sends forth frequently certain noisome and pestilential vapors and continual plagues upon the vast accumulations and herds of animals and especially upon human beings, who increase very rapidly because other animals do not feed upon them; and if the causes are taken away the results will cease.

Against. Therefore this earth seeks to lose its life while desiring continual reproduction for the reason brought forth, and demonstrated to you. Effects often resemble their causes. The animals serve as a type of the life of the world.

For. Behold now the hope and desire of going back to one's own country or returning to primal chaos, like that of the moth to the light, of the man who with perpetual longing always looks forward with joy to each new spring and each new summer, and to the new months and the new years, deeming that the things he longs for are too slow in coming; and who does not perceive that he is longing for his own destruction. But this longing is in its quintessence the spirit of the elements, which finding itself imprisoned within the life of the human body desires continually to return to its source.

And I would have you to know that this same longing is in its quintessence inherent in nature, and that man is a type of the world. (I.72–73)

In fact man does not vary from the animals except in what is accidental, and it is in this that he shows himself to be a divine thing; for where nature finishes producing its species there man begins with natural things to make with the aid of nature an infinite number of species; and as these are not necessary to those who govern themselves rightly as do the animals it is not in their disposition to seek after them. (I.116)

Given the cause nature produces the effect in the briefest manner that it can employ. (I.74)

Therefore, O students, study mathematics and do not build without foundations. (I.80)

Necessity is the mistress and guardian of nature.

Necessity is the theme and artificer of nature—the bridle, the law and the theme. (I.88)

Here nature seems in many or for many animals to have been rather a cruel step-mother than a mother, and for some not a step-mother but a compassionate mother. (I.77)

The lying interpreters of nature assert that mercury is a common factor in all the metals; they forget that nature varies its factors according to the variety of things which it desires to produce in the world. (I.293)

. . . In Art we may be said to be grandsons unto God. If poetry treats of moral philosophy, painting has to do with natural philosophy; if the one describes the workings of the mind, the other considers what the mind effects by movements of the body . . . (II.212)

How he who despises painting has no love for the philosophy in nature:

If you despise painting, which is the sole imitator of all the visible works of nature, it is certain that you will be despising a subtle invention which with philosophical and ingenious speculation takes as its theme all the various kinds of forms, airs and scenes, plants, animals, grasses and flowers, which are surrounded by light and shade. And this truly is a science and the true-born daughter of nature, since painting is the offspring of nature. But in order to speak more correctly we may call it the grandchild of nature; for all visible things derive their existence from nature, and from these same things is born painting. So therefore we may justly speak of it as the grandchild of nature and as related to God himself. (II.213)

When the poet ceases to represent in words what exists in nature, he then ceases to be the equal of the painter; for if the poet, leaving such representation, were to describe the polished and persuasive words of one whom he wishes to represent as speaking, he would be becoming an orator and be no more a poet or a painter. And if he were to describe the heavens he makes himself an astrologer, and a philosopher or theologian when speaking of the things of nature or of God. But if he returns to the representation of some definite thing he would become the equal of the painter if he could satisfy the eye with words as the painter does with brush and color, [for with these he creates] a harmony to the eye, even as music does in an instant to the ear. (II.215)

. . . Any master who let it be understood that he could himself recall all the forms and effects of nature would certainly appear to me to be endowed with great ignorance, considering that these effects are infinite and that our memory is not of so great a capacity as to suffice thereto. . . . (II.239)

Shadow partakes of the nature of universal things which are all more powerful at their beginning and grow weaker towards the end. I refer to the beginning of all forms and qualities visible or invisible, and not of things brought from small beginnings to a mighty growth by time, as a great oak would be which has its feeble beginning in a tiny acorn; though I would rather say the oak is most powerful at the spot where it is born in the ground, for there is the place of its greatest growth. Darkness, therefore, is the first stage of shadow, and light is the last. See, therefore, O painter, that you make your shadow darkest near to its cause and make the end of it become changed into light so that it seems to have no end. (II.230)

This benign nature so provides that over all the world you find something to imitate. (II.253)

The supreme misfortune is when theory outstrips performance. (II.256)

. . . The painter will produce pictures of little merit if he takes the works of others as his standard; but if he will apply himself to learn from the objects of nature he will produce good results. This we see was the case with the painters who came after the time of the

Romans, for they continually imitated each other, and from age to age their art steadily declined.

After these came Giotto the Florentine, and he—reared in mountain solitudes inhabited only by goats and such like beasts—turning straight from nature to his art, began to draw on the rocks the movements of the goats which he was tending, and so began to draw the figures of all the animals which were to be found in the country, in such a way that after much study he not only surpassed the masters of his own time but all those of many preceding centuries. After him art again declined, because all were imitating paintings already done; and so for centuries it continued to decline until such time as Tommaso the Florentine, nicknamed Masaccio, showed by the perfection of his work how those who took as their standard anything other than nature, the supreme guide of all thc masters, were wearying themselves in vain. Similarly I would say about those mathematical subjects, that those who study only the authorities and not the works of nature are in art the grandsons and not the sons of nature, which is the supreme guide of the good authorities.

Mark the supreme folly of those who censure such as learn from nature, leaving uncensured the authorities who were themselves the disciples of this same nature! (II.258)

The painter contends with and rivals nature. (II.268)

Although leaves with a smooth surface are for the most part of the same color on the right side as on the reverse, it so happens that the side exposed to the atmosphere partakes of the color of the atmosphere, and seems to partake of its color more closely in proportion as the eye is nearer to it and sees it more foreshortened. And the shadows will invariably appear darker on the right side than on the reverse, through the contrast caused by the high lights appearing against the shadow.

The underside of the leaf, although its color in itself may be the same as that of the right side, appears more beautiful; and this color is a green verging upon yellow; and this occurs when the leaf is interposed between the eye and the light which illumines it from the opposite side. Its shadows also are in the same positions as those on the opposite side.

Therefore, O painter, when you make trees near at hand, remember that when your eye is somewhat below the level of the tree you will be able to see its leaves, some on the right side and some on the reverse; and the right sides will be a deeper blue as they are seen more foreshortened, and the same leaf will sometimes show part of the right side and part of the reverse, and consequently you must make it of two colors. (II.285–86)

When the sun is in the east, the trees seen towards the east will have the light surrounding them all around their shadows, except towards the earth, unless the tree has been pruned in the previous year; and the trees in the south and in the north will be half in shadow and half in light, and more or less in shadow or in light according as they are more or less to the east or to the west.

The fact of the eye being high or low causes a variation in the shadows and lights of trees, for when the eye is above, it sees the trees with very little shadow, and when below with a great deal of shadow.

The different shades of green of plants are as varied as are their species. (II.289)

The extremities of the branches of trees if not dragged down by the weight of their fruit turn towards the sky as much as possible.

The upper sides of their leaves are turned towards the sky in order to receive nourishment from the dew that falls by night.

The sun gives spirit and life to plants, and the earth nourishes them with moisture.

In this connection I once made the experiment of leaving only one small root on a gourd and keeping this nourished with water; and the gourd brought to perfection all the fruits that it could produce, which were about sixty gourds of the long species; and I set myself diligently to consider the source of its life, and I perceived that it was the dew of the night which steeped it abundantly with its moisture through the joints of its great leaves, and thereby nourished the tree and its offspring, or rather the seeds which were to produce its offspring.

The rule as to the leaves produced on the last of the year's branches is that on twin branches they will grow in a contrary direction, that is, that the leaves in their earliest growth turn themselves round towards the branch, in such a way that the sixth leaf above grows over the sixth leaf below; and the manner of their turning is that if one turns towards its fellow on the right, the other turns to the left.

The leaf serves as a breast to nourish the branch or fruit which grows in the succeeding year. (II.292–93)

Water is that which serves the vital humor of this arid earth.

It is the cause which moves it through its veins contrary to the natural course (desire) of weighty things; it is like that which moves the humors in all kinds of living bodies, and . . .

And as the water is driven up from the lower part of the vine towards its severed stems and afterwards falls back to its roots, penetrates these and rises again anew, so from the lowest depth of the sea the water rises to the tops of the mountains, and falls down through their burst veins and returns to the sea and rises again anew. Thus up and down, in and out, unresting, now with fortuitous, and now with natural motion, now in its liberty and now constrained by its mover, it goes revolving and, after returning in force to its mover, rises again anew and then falls anew; so as one part rises the other descends.

Thus from the lowest depths of the sea the water rises up to the summits of the mountains and falls down low through the burst veins, and at the same time other water is rising: so the whole element ranges about and makes its passage many times through the rivers that fall into the sea.

At one time it becomes changed to the loftiest clouds, and afterwards it is pent up within the deep caverns of the earth.

It has nothing of itself, but moves and takes everything, as is clearly shown when it is distilled.

Thus hither and thither, up and down, it ranges, never resting at all in quietude, always flowing to help wherever the vital humor fails.

Now taking away the soil, now adding to it, here depositing logs there stones here bearing sand there mud, with nothing stable in bed or bank:

Now rushing on with headlong course, now descending in tranquility, now showing itself with fierce aspect, now appearing bright and calm, now mingling with the air in fine spray, now falling down in tempestuous rain; now changed to snow or storms of hail, now bathing the air with fine rain; so also now turning to ice and now hot; never keeping any stability; now rising aloft in thin cloud, compressing the air where it shuts it in, so that it moves through the other air after the fashion of a sponge squeezed beneath the water, when what is enclosed within it is driven out through the rest of the water. (II.100–101)

Before birds start on long journeys they wait for the winds favorable to their movements, and these favorable winds are of a different kind with different sorts of birds, because those which fly in jerks or bounds are obliged to fly against the wind, others receive the wind on one of their sides at different angles, and others receive it on both sides. But the birds that fly by jerks, such as fieldfares and other birds like these which fly in companies, have the feathers of their wings weak and poorly protected by the lesser feathers which form a covering for the larger ones. And this is why it is necessary that their flight should be against the course of the wind, for the wind closes up and presses one feather upon another and so renders their surface smooth and slippery when the air tries to penetrate it. It would be the contrary if the wind were to strike these birds towards the tail, because then it would penetrate under each feather and turn it over towards the head, and thus their flight would have a confused movement, like that of a leaf blown about in the course of the winds which goes perpetually whirling through the air continually revolving; and in addition to this their flesh would be without protection against the buffeting of the cold winds. And in order to avoid such accidents they fly against the course of the wind with a curving movement, and their bounds acquire great impetus in their descent, which is made with wings closed under the wind. And the reflex movement proceeds with wings open above the wind, which brings the bird back to the same height in the air as that from which it first descended, and so it continues time after time until it arrives at the desired spot.

The reflex movement and the falling movement vary in birds in two ways, of which one variation occurs when the reflex movement is in the same direction as the falling movement, and the second when the reflex movement is in one direction and the falling movement in another.

The bird in the falling movement closes its wings and in the reflex movement it opens them: it does this because a bird becomes heavier in proportion as it folds its wings and so much lighter as it opens its wings more.

The reflex movement is always made against the wind and the falling movement is made in the direction in which the wind is moving. (I.427)

The bird beats its wings repeatedly on one side only when it wishes to turn round while one wing is held stationary; and this it does by taking a stroke with the wing in the direction of the tail, like a man rowing in a boat with two oars, who takes many strokes on that side from which he wishes to escape, and keeps the other oar fixed. (I.455)

You will make an anatomy of the wings of a bird together with the muscles of the breast which are the movers of these wings.

And you will do the same for a man, in order to show the possibility that there is in man who desires to sustain himself amid the air by the beating of wings. (I.400)

A bird is an instrument working according to mathematical law, which instrument it is within the capacity of man to reproduce with all its movements, but not with a corresponding degree of strength, though it is deficient only in the power of maintaining equilibrium. We may therefore say that such an instrument constructed by man is lacking in nothing except the life of the bird, and this life must needs be supplied from that of man.

The life which resides in the bird's members will without doubt better conform to their needs than will that of man which is separated from them, and especially in the almost imperceptible movements which preserve equilibrium. But since we see that the bird is equipped for many obvious varieties of movements, we are able from this experience to declare that the most rudimentary of these movements will be capable of being comprehended by man's understanding; and that he will to a great extent be able to provide against the destruction of that instrument of which he has himself become the living principle and the propeller. (I.467)

The great bird will take its first flight upon the back of the great swan, filling the whole world with amazement and filling all records with its fame; and it will bring eternal glory to the nest where it was born.[8] (I.399)

Man has been called by the ancients a lesser world, and indeed the term is rightly applied, seeing that if man is compounded of earth, water, air, and fire, this body of the earth is the same; and as man has within himself bones as a stay and framework for the flesh, so the world has the rocks which are the supports of the earth; as man has within him a pool of blood wherein the lungs as he breathes expand and contract, so the body of the earth has its ocean, which also rises and falls every six hours with the breathing of the world; as from the said pool of blood proceed the veins which spread their branches through the human body, in just the same manner the ocean fills the body of the earth with an infinite number of veins of water. In this body of the earth there is lacking, however, the sinews, and these are absent because sinews are created for the purpose of movement, and as the world is perpetually stable within itself no movement ever takes place there, and in the absence of any movement the sinews are not necessary; but in all other things man and the world show a great resemblance. (II.20)

The sun does not move. (I.278)

All your discourse points to the conclusion that the earth is a star [planet] almost like the moon, and thus you will prove the majesty of our universe . . . (I.267)

The Pleasures of Country Life

Letter of December 10, 1513, to Francesco Vettori in Rome, by Niccolò Machiavelli, translated by Robert M. Torrance

Almost since the lifetime of Niccolò Machiavelli (1469–1527), his name has connoted cynical opportunism and unprincipled use of tyrannical power, and Machiavelli himself has often been portrayed in diabolical colors. In Part Three of Shakespeare's Henry VI *(III.ii), dating from about 1590, the unscrupulous Gloucester (future Richard III), plotting his treacherous rise to power at a historical moment when Machiavelli would scarcely have been born, declares:*

I can add colors to the chameleon,
Change shapes with Proteus for advantages,
And set the murderous Machiavel to school,

and in Marlowe's Jew of Malta, *from about the same year, Machiavelli in the prologue dismisses religion as "a childish toy." The hardheaded advice of his most famous, even infamous, book,* The Prince (Il Principe), *was indeed totally contrary to the high-minded political sentiments of the*

[8] In the phrase "*sopra del dosso del suo magnio cecero*," "upon the back of the great swan," Leonardo was apparently referring to Monte Ceccri, the mountain above Fiesole immediately to the south. It was from the summit or from a ridge of this mountain that he intended to make a trial of his flying machine ["the great bird"]. (MacCurdy)

Christian Middle Ages. Yet both Machiavelli's political and diplomatic career as chancellor and secretary of war to the Florentine Republic between the fall of Savonarola in 1498 and restoration of the Medici by Spanish troops in 1512, and his other writings, above all his Discourses on the First Ten Books of Titus Livius *(historian of the Roman Republic), bear witness to his lifelong dedication to republican government.*

The Prince, *written in the year after the Republic's collapse and Machiavelli's imprisonment and torture on suspicion of plotting against the Medici, expressed not his political preference but his desperate hope that a strong if necessarily ruthless leader might yet unite Italy against the foreign invaders then putting an end to what little remained of Italian independence. In his work as a whole, "Machiavelli is not specially concerned with the opportunism of individuals; the ideal before his eyes," Isaiah Berlin writes in "The Originality of Machiavelli" (1972),*[9] *"is a shining vision of Florence or of Italy; in this respect he is a typically impassioned humanist of the Renaissance, save that his ideal is not artistic or cultural but political." The letter to Vettori, translated from Machiavelli's* Opere, *ed. Mario Bonfantini (1954), vividly depicts the pleasures that this cultured, urban humanist could find—despite his amused distaste for rustic neighbors—in forced retirement to his farmstead at the very time when he was writing* The Prince *in a vain appeal to the Medici to employ his talents. He died in 1527, the year when the sack of Rome by troops of Emperor Charles V brought Italy under Austrian and Spanish subjection for centuries to come.*

. . . . I am dwelling in my country house, and ever since my recent mishaps befell me, have not spent a total of twenty days, all told, in Florence. Up until now I have been snaring thrushes with my own hands; I would rise before daybreak, spread birdlime, and set out with a bundle of cages on my back, looking like Geta when he returned from port with Amphitryon's books:[10] I would catch at least two, at most six thrushes. In this way I spent all of September; thereafter this trifling and alien diversion, to my great displeasure, ceased; what kind of life I now lead, I shall tell you. I rise in the morning with the sun and go into a wood that I am having cut, where I remain for two hours to review the work of the previous day, and to pass the time with the woodcutters, who are always disputing among themselves or with their neighbors. About this wood, I could tell you a thousand fine things that have happened to me, both with Frosino da Panzano and with others who wanted some firewood from it. . . .

Leaving the wood, I go to a spring, and from there to one of my bird-snares; I bring a book along, either Dante or Petrarch or one of the minor poets such as Tibullus, Ovid, and the like: I read about their amorous passions and loves. I remember my own; I rejoice for a while in this thought. I then proceed up the road to the inn; I speak with those who pass, asking them for news of their region; I learn various things, and note the varied tastes and different fancies of men. Meanwhile lunchtime comes, when my family and I eat such food as my poor country house and tiny patrimony permit. Having eaten, I return to the inn: here I usually meet the innkeeper, a butcher, a miller, and two bakers. With these I amuse myself all day playing cards, whence arise a thousand contentions and countless quarrels and harsh words; most of the time we fight about a penny or two, and can nevertheless be heard yelling as far away as San Casciano. Thus entangled amid these lice, I clear my brain of mold and vent the malignity of my Fate, content as I tread this road to see if she will be ashamed of the lot she has given me.

[9]Reprinted in *Against the Current: Essays in the History of Ideas* (1979), ed. Henry Hardy.
[10]An allusion to the slave of Hercules' father Amphitryon in a popular verse novella of the fifteenth century, *Geta and Birria*, based on a medieval Latin burlesque of Plautus's comedy *Amphitruo*.

With evening, I return home and enter my study; at its door, I strip off my everyday clothes, clotted with mud and mire, and don regal and curial garments; thus suitably attired I enter the ancient courts of ancient men where I am welcomed kindly by them and feed on food that is mine alone, and for which I was born, and where I am not ashamed to speak with them, and ask them the reason for their actions; and they, in their humanity, answer me; and for four hours at a time I feel no boredom, I forget every affliction, I do not fear poverty, death does not terrify me: so wholly do I identify with these men. And as Dante says that "to have heard without retaining does not make knowledge,"[11] I have noted down what I have turned to profit from their conversation, and composed a little work, *De principatibus [On Princes]*, where I delve as deeply as I can into thoughts on this subject, discussing what a principality is, of what species they are, how they are acquired, how they are maintained, why they are lost. . . .

Under the Guidance of Nature

The Praise of Folly, by Desiderius Erasmus, translated by John Wilson (revised)

Desiderius Erasmus (1466 or 1469–1536), through his many Latin writings and his editions of pagan and Christian authors and of the Greek New Testament, was the most influential humanist after Petrarch. Born at Rotterdam in Holland, educated at the school of Deventer and at Paris, and ordained as a priest, he was distinguished both for wide learning and for the tolerant moderation of spirit that vainly sought a "middle way" between the Roman Catholic Church, whose abuses he witheringly exposed yet with which he chose to remain, and the Reformation of Martin Luther, whose contentious violence he deplored. Above all he differed with Luther in his reaffirmation of the central belief of Renaissance humanism, as expounded in Pico della Mirandola's Oration on the Dignity of Man, *in the freedom and autonomy of the human will.*

Of all his writings, only The Praise of Folly (Moriae Encomium)—*which he wrote in a week in 1509 for Sir Thomas More, with whom he was staying in England, and published in 1511—has retained its popularity throughout the centuries, for in this mock-oration by Folly in praise of herself Erasmus was able not only to satirize the foolish things his foolish persona esteems but also to insinuate that in such foolishness might lie wisdom. Thus in the following passage, revised from John Wilson's translation of 1668, Folly proclaims the utter superiority of nature to art and the good fortune of those who, "living by the instinct of nature," like birds and flies, "look no further than the present," in contrast to the misery of man, who alone of all creatures endeavors to exceed nature's bounds. Surely these are positions that the earnest Christian humanist Erasmus could not have* seriously *entertained?*

. . . As therefore those arts are best that have the nearest affinity with folly, so are they most happy of all others that have least commerce with sciences and follow the guidance of Nature, who is in no way deficient, unless perhaps we endeavor to leap over those bounds she has appointed to us. Nature hates all false coloring and is ever best where she is least adulterated with art.

Well then, don't you find among the various kinds of living creatures that those thrive

[11]The words are Beatrice's in *Paradiso* V.41–42.

best that understand no more than what Nature taught them? What is more prosperous or wonderful than the bee? And though they have not the same sensory capacities as other bodies have, yet in what way has architecture ever gone beyond their building of houses? What philosopher ever founded the like republic? Whereas the horse, that comes so near man in understanding and is therefore so familiar with him, also partakes of his misery. For while he thinks it a shame to lose the race, it often happens that his breath fails him; and in the battle, while he contends for victory, he's cut down himself, and together with his rider "bites the dust"; not to mention those strong bits, sharp spurs, close stables, arms, blows, rider, and briefly, all the slavery that he willingly submits to while, imitating men of valor, he eagerly strives to be revenged of the enemy. How much more is the life of flies or birds to be wished for, who living by the instinct of nature look no further than the present—if man would but let them alone in it. And if at any time they chance to be taken, and being shut up in cages endeavor to imitate our speaking, it is strange how they degenerate from their native gaiety. So much better in every respect are the works of nature than the adulterations of art.

In like manner I can never sufficiently praise that cock, who was really Pythagoras: though only one, he had yet been everything, a philosopher, a man, a woman, a king, a private man, a fish, a horse, a frog, and, I believe too, a sponge; and at last concluded that no creature was more miserable than man, since all other creatures are content with those bounds that nature set them; only man endeavors to exceed them. And again, among men he gave precedence not to the learned or the great, but to the fool. Nor did Gryllus have less wit than Ulysses with his many counsels, when he chose rather to lie grunting in a hog sty than be exposed with the other to so many hazards.[12] Nor does Homer, that father of trifles, dissent from me, for he not only called all men "wretched and full of calamity," but often his great pattern of wisdom, Ulysses, "miserable"—Paris, Ajax, and Achilles never. And why, I pray, except that, like a cunning fellow and one that was master of his craft, Ulysses did nothing without the advice of Pallas?[13] In a word, he was too wise, and by that means ran wide of nature. As therefore among men they are least happy that study wisdom, being in this twice fools, that when they are born men, they should yet so far forget their condition as to affect the life of gods and, after the example of the giants, with their philosophical gimcracks make war upon nature: so they on the other side seem as little miserable as is possible who come nearest to beasts and never attempt anything beyond man. . . .

A Wonder to Behold

Three Descriptions of the New World

The great European expansion that began with the Crusades and the establishment of Venetian and Genoese outposts in the Levant, and found a voice in Marco Polo's account of his travels in the late thirteenth to early fourteenth centuries, became an explosion after Christopher Columbus, a Genoese sailing under the flag of Spain, crossed the Atlantic in 1492 and discovered a New World while seeking the Old, and Vasco da Gama of Portugal, beginning in 1497, sailed round the Cape of Good Hope and opened a sea route across the Indian Ocean to the East. In the two centuries

[12]See Plutarch's *Gryllus* in Chapter 12 above.
[13]Pallas Athena (Roman Minerva), the goddess of wisdom and protectress of Odysseus (Ulysses).

that followed, the English, Dutch, and French joined in the scramble to explore the lands and exploit the wealth of vast new territories opened to commerce and pillage. It was the lure of gold and silver, silks and spices, that drew the voyagers on and persuaded monarchs and trading companies to finance their risky enterprises, but from the beginning—especially in the Americas, a hemisphere whose very existence had been unsuspected—even hardheaded sailors and coldhearted conquerors sometimes interrupt the chronicles of their adventures to express wonder at the strange new world spread before them. Thus in crossing the causeway to the Aztec capital of Tenochtitlán, Bernal Díaz del Castillo writes in his account of Cortés's conquest of New Spain, "We were amazed and said that it was like the enchantments they tell of in the legend of Amadis . . . And some of our soldiers even asked whether the things that we saw were not a dream."

Not only the fabulous cities of Mexico and Peru, however, but also the exotic floras and faunas and peoples of the New World astonished both voyagers and readers, some of whom, like Montaigne and Shakespeare, learned from them to see their own world anew. The following are excerpts from three of many travelers' accounts—Columbus's letter on his first voyage, translated from the original Latin by R. H. Major (with a few minor revisions) in Select Letters of Christopher Columbus *(1847), and relations of Florida by the English captain Sir John Hawkins and of Mexico and Guatemala by the merchant Henry Hawks, as published in the great collection edited between 1598 and 1600 by Richard Hakluyt,* The Principal Navigations, Voyages, Traffiques, and Discoveries of the English Nation, *and reprinted in volumes 7 and 6, respectively, of the Everyman's Library edition of 1907. They evoke both lush forests and the imagined profusion of gold in tropical and semitropical America, both flying fish and flamingos observed in Florida and unicorns projected onto its landscape from travelers' yarns such as those of Pliny and Aelian (Chapter 12 above). Marvels so new could be seen and absorbed only by being somehow connected to the familiar and old.*

FROM *Letter on His First Voyage, by Christopher Columbus, translated by R. H. Major*

In the meantime I had learned from some Indians whom I had seized that that country [Cuba, called "Juana" by Columbus] was certainly an island; and therefore I sailed towards the east, coasting to the distance of 322 miles, which brought us to the extremity of it; from this point I saw lying eastwards another island, 54 miles distant from Juana, to which I gave the name of Hispana [Hispaniola, now comprising Haiti and the Dominican Republic]. I went thither, and steered my course eastward as I had done at Juana, even to the distance of 564 miles along the north coast. This said island of Juana is exceedingly fertile, as indeed are all the others; it is surrounded with many bays, spacious, very secure, and surpassing any that I have ever seen; numerous large and healthful rivers intersect it, and it also contains many very lofty mountains. All these islands are very beautiful, and distinguished by a diversity of scenery; they are filled with a great variety of trees of immense height, and which I believe to retain their foliage in all seasons; for when I saw them they were as verdant and luxuriant as they usually are in Spain in the month of May—some of them were blossoming, some bearing fruit, and all flourishing in the greatest perfection, according to their respective stages of growth, and the nature and quality of each; yet the islands are not so thickly wooded as to be impassable. The nightingale and various birds were singing in countless numbers, and that in November, the month in which I arrived there. There are besides in the same island of Juana seven or eight kinds of palm trees, which, like all the other trees, herbs, and fruits, considerably surpass ours in height and beauty. The pines also are very handsome, and there are very extensive fields and meadows, a variety of birds, different kinds of honey, and many sorts of metals, but no iron. In that island also which I have before said we named Hispana, there are mountains of very great size and beauty, vast plains, groves, and very fruitful fields, admirably adapted for tillage, pasture, and habitation. The convenience and excellence of the harbors in this island, and the abundance of the rivers, so indispensable to the health of man, surpass anything that would be believed by one who had not seen it. The trees, herbage, and fruits of Hispana are very different from those of Juana, and moreover it abounds in various kinds of spices, gold, and other metals. . . .

FROM *The Voyage Made by M. John Hawkins Esquire, and Afterward Knight, Captain of the Jesus of Lubek, One of Her Majesty's Ships, and General of the Salomon, and Other Two Barks Going in His Company, to the Coast of Guinea, and the Indies of Nova Hispania, Begun in An. Dom. 1564*

. . . The Floridians have pieces of unicorns' horns which they wear about their necks, whereof the Frenchmen obtained many pieces. Of those unicorns they have many; for that they do affirm it to be a beast with one horn, which coming to the river to drink, putteth the same into the water before he drinketh. Of this unicorn's horn there are of our company, that having gotten the same of the Frenchmen, brought home thereof to show. It is therefore to be presupposed that there are more commodities as well as that, which for want of time, and people sufficient to inhabit the same, can not yet come to light: but I trust God will reveal the same before it be long, to the great profit of them that shall take it in hand. Of beasts in the country besides deer, foxes, hares, polecats, conies,[14] ounces, and leopards, I am not able certainly to say: but it is thought that there are lions and tigers as well as unicorns; lions especially; if it be true that is said, of the enmity between them and the unicorns: for there is no beast but hath his enemy, as the cony the polecat, a sheep the wolf, the elephant the rhinoceros; and so of other beasts the like: insomuch, that whereas the one is, the other can not be missing. And seeing I have made mention of the beasts of this country, it shall not be from my purpose to speak also of the venomous beasts, as crocodiles, whereof there is great abundance, adders of great bigness, whereof our men killed some of a yard and a half long. Also I heard a miracle of one of these adders, upon the which a falcon seizing, the said adder did clasp her tail about her; which the French captain seeing, came to the rescue of the falcon, and took her slaying the adder; and this falcon being wild, he did reclaim her, and kept her for the space of two months, at which time for very want of meat he was fain to cast her off. On these adders the Frenchmen did feed, to no little admiration of us, and affirmed the same to be a delicate meat. And the captain of the Frenchmen saw also a serpent with three heads and four feet, of the bigness of a great spaniel, which for want of a harquebus he durst not attempt to slay. Of fish also they have in the river, pike, roach, salmon, trout, and divers other small fishes, and of great fish, some of the length of a man and longer, being of bigness accordingly, having a snout much like a sword of a yard long. There be also of sea fishes, which we saw coming along the coast flying, which are of the bigness of a smelt, the biggest sort whereof have four wings, but the other have but two: of these we saw coming out of Guinea a hundred in a company, which being chased by the gilt-heads, otherwise called the bonitos, do to avoid them the better, take their flight out of the water, but yet are they not able to fly far, because of the drying of their wings, which

[14]Rabbits.

serve them not to fly but when they are moist, and therefore when they can fly no further they fall into the water, and having wet their wings, take a new flight again. These bonitos be of bigness like a carp, and in color like a mackerel, but it is the swiftest fish in swimming that is, and followeth her prey very fiercely, not only in the water, but also out of the water: for as the flying fish taketh her flight, so doth this bonito leap after them, and taketh them sometimes above the water. There were some of those bonitos, which being galled by a fisgig,[15] did follow our ship coming out of Guinea 500 leagues. There is a sea fowl also that chaseth this flying fish as well as the bonito: for as the flying fish taketh her flight, so doth this fowl pursue to take her, which to behold is a greater pleasure than hawking, for both the flights are as pleasant, and also more often than an hundred times: for the fowl can fly no way, but one or other lighteth in her paws, the number of them are so abundant. There is an innumerable young fry of these flying fishes, which commonly keep about the ship, and are not so big as butterflies, and yet by flying do avoid the unsatiableness of the bonito. Of the bigger sort of these fishes we took many, which both night and day flew into the sails of our ship, and there was not one of them which was not worth a bonito: for being put upon a hook drabbling in the water, the bonito would leap thereat, and so was taken. Also, we took many with a white cloth made fast to a hook, which being tied so short in the water, that it might leap out and in, the greedy bonito thinking it to be a flying fish leapeth thereat, and so is deceived. We took also dolphins which are of very goodly color and proportion to behold, and no less delicate in taste. Fowls also there be many, both upon land and upon sea: but concerning them on the land I am not able to name them, because my abode was there so short. But for the fowl of the fresh rivers, these two I noted to be the chief, whereof the flamingo is one, having all red feathers, and long red legs like a hern, a neck according to the bill, red, whereof the upper neb hangeth an inch over the nether; and an egret, which is all white as the swan, with legs like to an hearnshaw,[16] and of bigness accordingly, but it hath in her tail feathers of so fine a plume, that it passeth the estridge his[17] feather. Of the sea fowl above all other not common in England, I noted the pelican, which is feigned to be the lovingest bird that is; which rather than her young should want, will spare her heart blood out of her belly:[18] but for all this lovingness she is very deformed to behold; for she is of color russet: notwithstanding in Guinea I have seen of them as white as a swan, having legs like the same, and a body like a hern, with a long neck and a thick long beak, from the nether jaw whereof down to the breast passeth a skin of such a bigness, as is able to receive a fish as big as one's thigh, and this her big throat and long bill doth make her seem so ugly. . . .

FROM *A Relation of the Commodities of Nova Hispania, and the Manners of the Inhabitants, Written by Henry Hawks, Merchant, Which Lived Five Years in the Said Country, and Drew the Same at the Request of M. Richard Hakluyt, Esquire, of Eyton in the County of Hereford, 1572*

. . . There are near about this city of Mexico many rivers and standing waters, which have in them a monstrous kind of fish [crocodiles], which is marvelous ravening and a great devourer of men and cattle. He is wont to sleep upon the dry land many times, and if there come in the meantime any man or beast and wake or disquiet him, he speedeth

[15] A kind of harpoon. (OED) [16] Heron. [17] Ostrich's.
[18] See the account of the pelican from the twelfth-century Latin bestiary in Chapter 13 above.

well if he get from him. He is like unto a serpent saving that he doth not fly, neither hath he wings. . . .

There are many great rivers and great store of fish in them, not like unto our kinds of fish. And there are marvelous great woods and as fair trees as may be seen of divers sorts, and especially fir trees, that may mast any ship that goeth upon the sea, oaks and pineapples [pines], and another tree which they call mesiquiquez [mesquite]; it beareth a fruit, like unto a peascod marvelous sweet, which the wild people gather and keep it all the year, and eat it instead of bread. . . .

In certain provinces which are called Guatemala and Soconusco there is growing great store of cacao, which is a berry like unto an almond. It is the best merchandise that is in all the Indies. The Indians make drink of it and in like manner meat[19] to eat. It goeth currently for money in any market or fair, and may buy any flesh, fish, bread, or cheese or other things.

There are many kind of fruits in the country, which are very good, as plantains, sapotes, guavas, pinas, aluacatas [avocados], tunas [prickly pears], mamios,[20] lemons, oranges, walnuts, very small and hard with little meat in them, grapes which the Spaniards brought into the country, and also wild grapes, which are of the country and are very small, quinces, peaches, figs, and but few apples and very small, and no pears: but there are melons and calabaças or gourds.

There is much honey, both of bees and also of a kind of tree which they call maguez. This honey of maguez is not so sweet as the other honey is, but it is better to be eaten only with bread than the other is; and the tree serveth for many things, as the leaves make thread to sew any kind of bags and are good to cover and thatch houses, and for divers other things.

They have in divers places of the country many hot springs of water: as above all other I have seen one in the province of Mechuacan. In a plain field without any mountain there is a spring which hath much water, and it is so hot that if a whole quarter of beef be cast into it, within an half hour it will be as well sodden as it will be over a fire in half a day. I have seen half a sheep cast in, and immediately it hath been sodden, and I have eaten part of it.

There are many hares, and some conies. There are no partridges, but abundance of quails.

They have great store of fish in the South Sea and many oysters, and very great. The people do open the oysters, and take out the meat of them, and dry it as they do any other kind of fish, and keep them all the year; and when the times serve they send them abroad into the country to sell, as all other fish. They have no salmon, nor trout, nor peal [young salmon], nor carp, tench, nor pike, in all the country.

There are in the country mighty high mountains and hills, and snow upon them. They commonly burn; and twice every day they cast out much smoke and ashes at certain open places which are in the tops of them.

There is among the wild people much manna. I have gathered of the same and have eaten it, and it is good; for the apothecaries send their servants at certain times to gather of the same for purgations, and other uses.

There are in the mountains many wild hogs, which all men may kill, and lions and

[19]"Food."

[20]Mammees: a large tree of tropical America which bears a large fruit with a yellow pulp of pleasant taste; also, the fruit of this tree. (OED)

tigers [cougars and jaguars], which tigers do much harm to men that travel in the wilderness. . . .

The wild people go naked, without anything upon them. The women wear the skin of a deer before their privities and nothing else upon all their bodies. They have no care for anything, but only from day to day for that which they have need to eat. They are big men and likewise the women. They shoot in bows, which they make of a cherry tree, and their arrows are of cane, with a sharp flintstone [obsidian] in the end of the same; they will pierce any coat of mail, and they kill deer, and cranes, and wild geese, ducks, and other fowl, and worms, and snakes, and divers other vermin, which they eat. They live very long, for I have seen men that have been an hundred years of age. They have but very little hair in their face, nor on their bodies. . . .

There remain some among the wild people that unto this day eat one another. I have seen the bones of a Spaniard that have been as clean burnished as though it had been done by men that had no other occupation. And many times people are carried away by them, but they never come again, whether they be men or women. . . .

Our Great and Powerful Mother Nature

Essays, by Michel de Montaigne,
translated by Donald M. Frame

Michel de Montaigne (1533–92) was born Michel Eyquem at the Château de Montaigne in Périgord, France, of a Catholic landowning father and a mother of Spanish and probably Jewish descent. Though he wrote his "essays" or "trials," as he called this experimental new genre, in informal conversational French at the opposite pole from the classically "correct" Latin of the humanists, for whose pedantries he had no more use than he did for their lofty Neoplatonic flights, in many ways he is the finest product of Renaissance humanism. Tutored in Latin from earliest childhood, to the extent that he spoke no other language until age six, he received an excellent humanist education at the Collège de Guyenne in Bordeaux, and his essays, permeated with admiration for heroes and philosophers known from the Lives *of Plutarch and Diogenes Laërtius, are interspersed with countless Latin quotations from thinkers and poets—Seneca and Cicero, Horace and Lucretius and Virgil—who remained models of excellence all his life.*

In the tolerant moderation of his temper, and in his aversion to the fanaticisms tearing his country apart in recurrent wars between Catholics and Huguenots (as the French Calvinists were called), he resembles the foremost humanist of his century, Erasmus. Weary of wars in which he reluctantly took part on the Catholic side, he withdrew between 1571 and 1580 to his library in a round tower of his family château, where he composed the first two books of his Essays, *published in 1580. In later years, besides serving as mayor of Bordeaux between 1581 and 1585, he wrote the generally longer essays of Book III, published in 1588 along with numerous additions—his revisions were always additions—to Books I and II; after his death at 59, an expanded edition appeared in 1595. (Donald M. Frame's English translation in* The Complete Works of Montaigne *[1957] follows French editions in indicating by markers—here omitted—changes made in different editions.)*

But Montaigne continues the humanist heritage in the value he ascribes to nature, purged of "transcendental humors" and approached not as an object of impersonal scientific investigation (in which Montaigne had small interest) but as a mother and guide by whose instruction we may hope to achieve the "great and glorious masterpiece" of living well. In his early essay (I.26) "Of the Education of Children," he commends the world as the student's book and affirms that "whoever considers as in a painting the great picture of our mother Nature in her full majesty; whoever reads such universal and constant variety in her face; whoever finds himself there, and not merely himself, but a whole kingdom, as a dot made with a very fine brush; that man alone estimates things according to their true proportions."

"Of Cannibals," which in John Florio's Elizabethan translation influenced Shakespeare's Tempest, *calls the cannibals of Brazil "barbarous" only in being "still very close to their original naturalness," which recalls the Golden Age of the poets: not plants and peoples that Nature has fostered are "wild" but those (like ourselves) "that we have changed artificially and led astray from the common order." The "romantic primitivism" of this essay ("The whole day is spent in dancing . . .") is undercut by ironical awareness that European man, with his truly savage vices, could never belong to a world where—superior though it be to Plato's* Republic*—"they don't wear breeches."*

The second excerpt, from the long "Apology for Raymond Sebond" (the Catalan author of a Natural Theology *that Montaigne had translated at his father's request, but which his "Apology" largely subverts by its pervasive skepticism), proclaims the equality or superiority of animals to human beings in rational action as well as instinct. By confining man "within the boundaries of his finite individuality," Jean Starobinski writes in* Montaigne in Motion *(1982; English trans. 1985), "Montaigne's text does not simply oppose man to the incomprehensible omnipotence of God; it also opposes him to the infinite richness of the world and of nature. . . . Facing the world, man knows that he is a finite creature in a physical infinity." Nature, in Montaigne's words, "has universally embraced all her creatures," and his object is "to bring us back and join us to the majority."*

Finally, "Of Experience" affirms the commonality of every human being in bodily functions ("Both kings and philosophers defecate, and ladies too"), disease, and death, but also in pleasures. Wary of the "singular accord" between "supercelestial thoughts" and "subterranean conduct," Montaigne bids us look earthward, affirm the oneness of body and soul, and "give Nature a chance" by following her wise guidance in the occupation of life.

FROM *I.31: Of Cannibals*

When King Pyrrhus passed over into Italy, after he had reconnoitered the formation of the army that the Romans were sending to meet him, he said: "I do not know what barbarians these are" (for so the Greeks called all foreign nations), "but the formation of this army that I see is not at all barbarous." The Greeks said as much of the army that Flaminius brought into their country, and so did Philip, seeing from a knoll the order and distribution of the Roman camp, in his kingdom, under Publius Sulpicius Galba. Thus we should beware of clinging to vulgar opinions, and judge things by reason's way, not by popular say.

I had with me for a long time a man who had lived for ten or twelve years in that other world which has been discovered in our century, in the place where Villegaignon landed,[21] and which he called Antarctic France. This discovery of a boundless country seems worthy of consideration. I don't know if I can guarantee that some other such discovery will not be made in the future, so many personages greater than ourselves hav-

[21]In Brazil, in 1557. (Frame)

ing been mistaken about this one. I am afraid we have eyes bigger than our stomachs, and more curiosity than capacity. We embrace everything, but we clasp only wind. . . .

This man I had was a simple, crude fellow—a character fit to bear true witness; for clever people observe more things and more curiously, but they interpret them; and to lend weight and conviction to their interpretation, they cannot help altering history a little. They never show you things as they are, but bend and disguise them according to the way they have seen them; and to give credence to their judgment and attract you to it, they are prone to add something to their matter, to stretch it out and amplify it. We need a man either very honest, or so simple that he has not the stuff to build up false inventions and give them plausibility; and wedded to no theory. Such was my man; and besides this, he at various times brought sailors and merchants, whom he had known on that trip, to see me. So I content myself with his information, without inquiring what the cosmographers say about it. . . .

Now, to return to my subject, I think there is nothing barbarous and savage in that nation, from what I have been told, except that each man calls barbarism whatever is not his own practice; for indeed it seems we have no other test of truth and reason than the example and pattern of the opinions and customs of the country we live in. *There* is always the perfect religion, the perfect government, the perfect and accomplished manners in all things. Those people are wild, just as we call wild the fruits that Nature has produced by herself and in her normal course; whereas really it is those that we have changed artificially and led astray from the common order, that we should rather call wild. The former retain alive and vigorous their genuine, their most useful and natural, virtues and properties, which we have debased in the latter in adapting them to gratify our corrupted taste. And yet for all that, the savor and delicacy of some uncultivated fruits of those countries is quite as excellent, even to our taste, as that of our own. It is not reasonable that art should win the place of honor over our great and powerful mother Nature. We have so overloaded the beauty and richness of her works by our inventions that we have quite smothered her. Yet wherever her purity shines forth, she wonderfully puts to shame our vain and frivolous attempts:

> Ivy comes readier without our care;
> In lonely caves the arbutus grows more fair;
> No art with artless bird song can compare.
>
> —Propertius[22]

All our efforts cannot even succeed in reproducing the nest of the tiniest little bird, its contexture, its beauty and convenience; or even the web of the puny spider. All things, says Plato, are produced by nature, by fortune, or by art; the greatest and most beautiful by one or the other of the first two, the least and most imperfect by the last.

These nations, then, seem to me barbarous in this sense, that they have been fashioned very little by the human mind, and are still very close to their original naturalness. The laws of nature still rule them, very little corrupted by ours; and they are in such a state of purity that I am sometimes vexed that they were unknown earlier, in the days when there were men able to judge them better than we. I am sorry that Lycurgus[23] and Plato did not know of them; for it seems to me that what we actually see in these nations surpasses not only all the pictures in which poets have idealized the golden age and all

[22]Sextus Propertius (late first century B.C.), Elegy I.ii.10–11, 14.
[23]The legendary lawgiver of ancient Sparta.

their inventions in imagining a happy state of man, but also the conceptions and the very desire of philosophy. They could not imagine a naturalness so pure and simple as we see by experience; nor could they believe that our society could be maintained with so little artifice and human solder. This is a nation, I should say to Plato, in which there is no sort of traffic, no knowledge of letters, no science of numbers, no name for a magistrate or for political superiority, no custom of servitude, no riches or poverty, no contracts, no successions, no partitions, no occupations but leisure ones, no care for any but common kinship, no clothes, no agriculture, no metal, no use of wine or wheat. The very words that signify lying, treachery, dissimulation, avarice, envy, belittling, pardon—unheard of. How far from this perfection would he find the republic that he imagined: *Men fresh sprung from the gods* [Seneca].[24]

These manners nature first ordained.

—Virgil[25]

For the rest, they live in a country with a very pleasant and temperate climate, so that according to my witnesses it is rare to see a sick man there; and they have assured me that they never saw one palsied, bleary-eyed, toothless, or bent with age. They are settled along the sea and shut in on the land side by great high mountains, with a stretch about a hundred leagues wide in between. They have a great abundance of fish and flesh which bear no resemblance to ours, and they eat them with no other artifice than cooking. The first man who rode a horse there, though he had had dealings with them on several other trips, so horrified them in this posture that they shot him dead with arrows before they could recognize him.

Their buildings are very long, with a capacity of two or three hundred souls; they are covered with the bark of great trees, the strips reaching to the ground at one end and supporting and leaning on one another at the top, in the manner of some of our barns, whose covering hangs down to the ground and acts as a side. They have wood so hard that they cut with it and make of it their swords and grills to cook their food. Their beds are of a cotton weave, hung from the roof like those in our ships, each man having his own; for the wives sleep apart from their husbands.

They get up with the sun, and eat immediately upon rising, to last them through the day; for they take no other meal than that one. Like some other Eastern peoples, of whom Suidas tells us, who drank apart from meals, they do not drink then; but they drink several times a day, and to capacity. Their drink is made of some root, and is of the color of our claret wines. They drink it only lukewarm. This beverage keeps only two or three days; it has a slightly sharp taste, is not at all heady, is good for the stomach, and has a laxative effect upon those who are not used to it; it is a very pleasant drink for anyone who is accustomed to it. In place of bread they use a certain white substance like preserved coriander. I have tried it; it tastes sweet and a little flat.

The whole day is spent in dancing. The younger men go to hunt animals with bows. Some of the women busy themselves meanwhile with warming their drink, which is their chief duty. Some one of the old men, in the morning before they begin to eat, preaches to the whole barnful in common, walking from one end to the other, and repeating one single sentence several times until he has completed the circuit (for the buildings are fully a hundred paces long). He recommends to them only two things:

[24]*Letters (Epistulae Morales)* XC. [25]*Georgics* II.20.

valor against the enemy and love for their wives. And they never fail to point out this obligation, as their refrain, that it is their wives who keep their drink warm and seasoned.

There may be seen in several places, including my own house, specimens of their beds, of their ropes, of their wooden swords and the bracelets with which they cover their wrists in combats, and of the big canes, open at one end, by whose sound they keep time in their dances. They are close shaven all over, and shave themselves much more cleanly than we, with nothing but a wooden or stone razor. They believe that souls are immortal, and that those who have deserved well of the gods are lodged in that part of heaven where the sun rises, and the damned in the west. . . .

They have their wars with the nations beyond the mountains, further inland, to which they go quite naked, with no other arms than bows or wooden swords ending in a sharp point, in the manner of the tongues of our boar spears. It is astonishing what firmness they show in their combats, which never end but in slaughter and bloodshed; for as to routs and terror, they know nothing of either.

Each man brings back as his trophy the head of the enemy he has killed, and sets it up at the entrance to his dwelling. After they have treated their prisoners well for a long time with all the hospitality they can think of, each man who has a prisoner calls a great assembly of his acquaintances. He ties a rope to one of the prisoner's arms, by the end of which he holds him, a few steps away, for fear of being hurt, and gives his dearest friend the other arm to hold in the same way; and these two, in the presence of the whole assembly, kill him with their swords. This done, they roast him and eat him in common and send some pieces to their absent friends. This is not, as people think, for nourishment, as of old the Scythians used to do; it is to betoken an extreme revenge. And the proof of this came when they saw the Portuguese, who had joined forces with their adversaries, inflict a different kind of death on them when they took them prisoner, which was to bury them up to the waist, shoot the rest of their body full of arrows, and afterward hang them. They thought that these people from the other world, being men who had sown the knowledge of many vices among their neighbors and were much greater masters than themselves in every sort of wickedness, did not adopt this sort of vengeance without some reason, and that it must be more painful than their own; so they began to give up their old method and to follow this one.

I am not sorry that we notice the barbarous horror of such acts, but I am heartily sorry that, judging their faults rightly, we should be so blind to our own. I think there is more barbarity in eating a man alive than in eating him dead; and in tearing by tortures and the rack a body still full of feeling, in roasting a man bit by bit, in having him bitten and mangled by dogs and swine (as we have not only read but seen within fresh memory, not among ancient enemies, but among neighbors and fellow citizens, and what is worse, on the pretext of piety and religion), than in roasting and eating him after he is dead. . . .

So we may well call these people barbarians, in respect to the rules of reason, but not in respect to ourselves, who surpass them in every kind of barbarity.

Their warfare is wholly noble and generous, and as excusable and beautiful as this human disease can be; its only basis among them is their rivalry in valor. They are not fighting for the conquest of new lands, for they still enjoy that natural abundance that provides them without toil and trouble with all necessary things in such profusion that they have no wish to enlarge their boundaries. They are still in that happy state of desiring only as much as their natural needs demand; anything beyond that is superfluous to them.

They generally call those of the same age, brothers; those who are younger, children;

and the old men are fathers to all the others. These leave to their heirs in common the full possession of their property, without division or any other title at all than just the one that Nature gives to her creatures in bringing them into the world. . . .

Three of these men, ignorant of the price they will pay some day, in loss of repose and happiness, for gaining knowledge of the corruptions of this side of the ocean; ignorant also of the fact that of this intercourse will come their ruin (which I suppose is already well advanced: poor wretches, to let themselves be tricked by the desire for new things, and to have left the serenity of their own sky to come and see ours!)—three of these men were at Rouen, at the time the late King Charles IX[26] was there. The king talked to them for a long time; they were shown our ways, our splendor, the aspect of a fine city. After that, someone asked their opinion, and wanted to know what they had found most amazing. They mentioned three things, of which I have forgotten the third, and am very sorry for it; but I still remember two of them. They said that in the first place they thought it very strange that so many grown men, bearded, strong, and armed, who were around the king (it is likely that they were talking about the Swiss of his guard) should submit to obey a child, and that one of them was not chosen to command instead. Second (they have a way in their language of speaking of men as halves of one another), they had noticed that there were among us men full and gorged with all sorts of good things, and that their other halves were beggars at their doors, emaciated with hunger and poverty; and they thought it strange that these needy halves could endure such an injustice, and did not take the others by the throat, or set fire to their houses.

I had a very long talk with one of them; but I had an interpreter who followed my meaning so badly, and who was so hindered by his stupidity in taking in my ideas, that I could get hardly any satisfaction from the man. When I asked him what profit he gained from his superior position among his people (for he was a captain, and our sailors called him king), he told me that it was to march foremost in war. How many men followed him? He pointed to a piece of ground, to signify as many as such a space could hold; it might have been four or five thousand men. Did all his authority expire with the war? He said that this much remained, that when he visited the villages dependent on him, they made paths for him through the underbrush by which he might pass quite comfortably.

All this is not too bad—but what's the use? They don't wear breeches.

FROM *II.12: Apology for Raymond Sebond*

. . . We recognize easily enough, in most of their works, how much superiority the animals have over us and how feeble is our skill to imitate them. We see, however, in our cruder works, the faculties that we use, and that our soul applies itself with all its power; why do we not think the same thing of them? Why do we attribute to some sort of natural and servile inclination these works which surpass all that we can do by nature and by art? Wherein, without realizing it, we grant them a very great advantage over us, by making Nature, with maternal tenderness, accompany them and guide them as by the hand in all the actions and comforts of their life; while us she abandons to chance and to fortune, and to seek by art the things necessary for our preservation, and denies us at the same time the power to attain, by any education and mental straining, the natural

[26]In 1562. (Frame)

resourcefulness of the animals: so that their brutish stupidity surpasses in all conveniences all that our divine intelligence can do.

Truly, by this reckoning, we should be quite right to call her a very unjust stepmother. But this is not so; our organization is not so deformed and disorderly. Nature has universally embraced all her creatures; and there is none that she has not very amply furnished with all powers necessary for the preservation of its being. For these vulgar complaints that I hear men make (as the license of their opinions now raises them above the clouds, and then sinks them to the antipodes) that we are the only animal abandoned naked on the naked earth, tied, bound, having nothing to arm and cover ourselves with except the spoils of others; whereas all other creatures Nature has clothed with shells, husks, bark, hair, wool, spikes, hide, down, feathers, scales, fleece, and silk, according to the need of their being; has armed them with claws, teeth, or horns for attack and defense; and has herself instructed them in what is fit for them—to swim, to run, to fly, to sing—whereas man can neither walk, nor speak, nor eat, nor do anything but cry, without apprenticeship—

> The infant, like a sailor tossed ashore
> By raging seas, lies naked on the earth,
> Speechless, helpless for life, when at his birth
> Nature from out the womb brings him to light.
> He fills the place with wailing, as is right
> For one who through so many woes must pass.
> Yet flocks, herds, savage beasts of every class
> Grow up without the need for any rattle,
> Or for a gentle nurse's soothing prattle;
> They seek no varied clothes against the sky;
> Lastly they need no arms, no ramparts high
> To guard their own—since earth itself and nature
> Amply bring forth all things for every creature.
>
> —Lucretius [V.222–34]

—those complaints are false, there is a greater equality and a more uniform relationship in the organization of the world. Our skin is provided as adequately as theirs with endurance against the assaults of the weather: witness so many nations who have not yet tried the use of any clothes. Our ancient Gauls wore hardly any clothes; nor do the Irish, our neighbors, under so cold a sky. But we may judge this better by ourselves; for all the parts of the body that we see fit to expose to the wind and air are found fit to endure it: face, feet, hands, legs, shoulders, head, according as custom invites us. For if there is a part of us that is tender and that seems as though it should fear the cold, it should be the stomach, where digestion takes place; our fathers left it uncovered, and our ladies, soft and delicate as they are, sometimes go half bare down to the navel. Nor are the bindings and swaddlings of infants necessary either; and the Lacedaemonian mothers raised their children in complete freedom to move their limbs, without wrapping or binding them. Our weeping is common to most of the other animals; and there are scarcely any who are not observed to complain and wail long after their birth, since it is a demeanor most appropriate to the helplessness that they feel. As for the habit of eating, it is, in us as in them, natural and needing no instruction:

For each one feels his powers and his needs.
—Lucretius [V.1033]

Who doubts that a child, having attained the strength to feed himself, would be able to seek his food? And the earth produces and offers him enough of it for his need, with no other cultivation or artifice; and if not in all weather, neither does she for the beasts: witness the provisions we see the ants and others make for the sterile seasons of the year. These nations that we have just discovered to be so abundantly furnished with food and natural drink, without care or preparation, have now taught us that bread is not our only food, and that without plowing, our mother Nature had provided us in plenty with all we needed; indeed, as seems likely, more amply and richly than she does now that we have interpolated our artifice:

At first and of her own accord the earth
Brought forth sleek fruits and vintages of worth,
Herself gave harvests sweet and pastures fair,
Which now scarce grow, despite our toil and care,
And we exhaust our oxen and our men;
—Lucretius [II.1157–61]

the excess and unruliness of our appetite outstripping all the inventions with which we seek to satisfy it. . . .

I have said all this to maintain this resemblance that exists to human things, and to bring us back and join us to the majority. We are neither above nor below the rest: all that is under heaven, says the sage, incurs the same law and the same fortune,

All things are bound by their own chains of fate.
—Lucretius [V.876]

There is some difference, there are orders and degrees; but it is under the aspect of one and the same nature:

And all things go their own way, nor forget
Distinctions by the law of nature set.
—Lucretius [V.923–24]

Man must be constrained and forced into line inside the barriers of this order. The poor wretch is in no position really to step outside them; he is fettered and bound, he is subjected to the same obligation as the other creatures of his class, and in a very ordinary condition, without any real and essential prerogative or preeminence. That which he accords himself in his mind and in his fancy has neither body nor taste. And if it is true that he alone of all the animals has this freedom of imagination and this unruliness in thought that represents to him what is, what is not, what he wants, the false and the true, it is an advantage that is sold him very dear, and in which he has little cause to glory, for from it springs the principal source of the ills that oppress him: sin, disease, irresolution, confusion, despair.

So I say, to return to my subject, that there is no apparent reason to judge that the beasts do by natural and obligatory instinct the same things that we do by our choice and cleverness. We must infer from like results like faculties, and consequently confess that this same reason, this same method that we have for working, is also that of the animals. Why do we imagine in them this compulsion of nature, we who feel no similar effect? Besides, it is more honorable, and closer to divinity, to be guided and obliged to act lawfully by a natural and inevitable condition, than to act lawfully by accidental and fortuitous liberty; and safer to leave the reins of our conduct to nature than to ourselves. The vanity of our presumption makes us prefer to owe our ability to our powers than to nature's liberality; and we enrich the other animals with natural goods and renounce them in their favor, in order to honor and ennoble ourselves with goods acquired: a very simple notion, it seems to me, for I should prize just as highly graces that were all mine and inborn as those I had gone begging and seeking from education. It is not in our power to acquire a fairer recommendation than to be favored by God and nature. . . .

FROM *III.13: Of Experience*

. . . Nature always gives us happier laws than those we give ourselves. Witness the picture of the Golden age of the poets, and the state in which we see nations live which have no other laws. Here are some who employ, as the only judge in their quarrels, the first traveler passing through their mountains. And these others on market day elect one of themselves who decides all their suits on the spot. What would be the danger in having our wisest men settle ours in this way, according to the circumstances and at sight, without being bound to precedents, past or future? For every foot its own shoe. King Ferdinand, when he sent colonists to the Indies, wisely provided that no students of jurisprudence should accompany them, for fear that lawsuits might breed in this new world, this being by nature a science generating altercation and division; judging, with Plato, that lawyers and doctors are a bad provision for a country. . . .

Philosophical inquiries and meditations serve only as food for our curiosity. The philosophers with much reason refer us to the rules of Nature; but these have no concern with such sublime knowledge. The philosophers falsify them and show us the face of Nature painted in too high a color, and too sophisticated, whence spring so many varied portraits of so uniform a subject. As she has furnished us with feet to walk with, so she has given us wisdom to guide us in life: a wisdom not so ingenious, robust, and pompous as that of their invention, but correspondingly easy and salutary, performing very well what the other talks about, in a man who has the good fortune to know how to occupy himself simply and in an orderly way, that is to say, naturally. The more simply we trust to Nature, the more wisely we trust to her. Oh, what a sweet and soft and healthy pillow is ignorance and incuriosity, to rest a well-made head! . . .

Both kings and philosophers defecate, and ladies too. Public lives are owed to ceremony; mine, obscure and private, enjoys every natural dispensation; soldier and Gascon are also qualities a bit subject to indiscretion. Wherefore I will say this about that action: that we should relegate it to certain prescribed nocturnal hours, and force and subject ourselves to them by habit, as I have done; but not subject ourselves, as I have done as I grew old, to any concern for a particularly comfortable place and seat for this function, and make it a nuisance by slowness and fastidiousness.

And yet in the dirtiest functions is it not somewhat excusable to require more care and cleanliness? *Man is by nature a clean and dainty animal* [Seneca].[27] Of all natural functions that is the one that I can least willingly endure to have interrupted. I have seen many soldiers inconvenienced by the irregularity of their bowels; mine and I never fail

[27] *Letters (Epistulae Morales)* XCII.

the moment of our assignation, which is when I jump out of bed, unless some violent occupation or illness disturbs us. . . .

The constitution of diseases is patterned after the constitution of animals. They have their destiny, limited from their birth, and their days. He who tries to cut them short imperiously by force, in the midst of their course, prolongs and multiplies them, and stimulates them instead of appeasing them. I agree with Crantor,[28] that we must neither obstinately and heedlessly oppose evils nor weakly succumb to them, but give way to them naturally, according to their condition and our own. We should give free passage to diseases; and I find that they do not stay so long with me, who let them go ahead; and some of those that are considered most stubborn and tenacious, I have shaken off by their own decadence, without help and without art, and against the rules of medicine. Let us give Nature a chance; she knows her business better than we do. "But so-and so died of it." So will you, if not of that disease, of some other. And how many have not failed to die of it, with three doctors at their backsides? Example is a hazy mirror, reflecting all things in all ways. If it is a pleasant medicine, take it; it is always that much present gain. I shall never balk at the name or the color, if it is delicious and appetizing. Pleasure is one of the principal kinds of profit. . . .

We must learn to endure what we cannot avoid. Our life is composed, like the harmony of the world, of contrary things, also of different tones, sweet and harsh, sharp and flat, soft and loud. If a musician liked only one kind, what would he have to say? He must know how to use them together and blend them. And so must we do with good and evil, which are consubstantial with our life. Our existence is impossible without this mixture, and one element is no less necessary for it than the other. To try to kick against natural necessity is to imitate the folly of Ctesiphon, who undertook a kicking match with his mule. . . .[29]

Besides, the uncertainty and ignorance of those who presume to explain the workings of Nature and her inner processes, and all the false prognostications of their art, should make us know that she has utterly unknown ways of her own. There is great uncertainty, variety, and obscurity about what she promises us or threatens us with. Except for old age, which is an indubitable sign of the approach of death, in all other ailments I see few signs of the future on which to base our divination. . . .

I hold that this temperateness of my soul has many a time lifted up my body from its falls. My body is often depressed; whereas if my soul is not jolly, it is at least tranquil and at rest. I had a quartan fever for four or five months, which quite disfigured me; my mind still kept going not only peacefully but cheerfully. If the pain is outside of me, the weakness and languor do not distress me much. I know several bodily infirmities that inspire horror if you merely name them, which I should fear far less than a thousand passions and agitations of the spirit that I see prevalent. I have made up my mind to my inability to run any more; it is enough that I crawl. Nor do I complain of the natural decay that has hold of me—

> Who marvels at a goiter in the Alps?
>
> —Juvenal[30]

—any more than I regret that my term of life is not as long and sound as that of an oak. . . .

In truth, I gain one principal consolation in thinking about my death, that it will be

[28]A Greek Academic philosopher of about 300 B.C. quoted in Cicero's *Tusculan Disputations* III.vi.
[29]See Plutarch, "On the Control of Anger" 8, in *Moralia* 457A. [30]*Satires* XIII.162.

normal and natural, and that henceforth in this matter I cannot demand or hope for any but illegitimate favor from destiny. Men enjoy a fond belief that in other days their lives were longer, as their stature was greater. But Solon, who belongs to those old days, nevertheless limits the extreme duration of life to seventy years. Shall I, who in all matters have so worshiped that *golden mean* of the past, and have taken the moderate measure as the most perfect, aspire to an immoderate and prodigious old age? Whatever happens contrary to the course of Nature may be disagreeable, but what happens according to her should always be pleasant. *Everything that happens according to Nature should be considered good* [Cicero].[31] Thus, says Plato, the death that is brought on by wounds or maladies may be called violent, but that which takes us by surprise as old age guides us to it is the easiest of all and in a way delightful. *Young men lose their lives by violence, old men by ripeness* [Cicero].

Death mingles and fuses with our life throughout. Decline anticipates death's hour and intrudes even into the course of our progress. I have portraits of myself at twenty-five and thirty-five; I compare them with one of the present: how irrevocably it is no longer myself! How much farther is my present picture from them than from that of my death! We abuse Nature too much by pestering her so far that she is constrained to leave us and abandon our guidance—our eyes, our teeth, our legs, and the rest—to the mercy of foreign assistance that we have begged, and to resign us to the hands of art, weary of following us. . . .

I, who operate only close to the ground, hate that inhuman wisdom that would make us disdainful enemies of the cultivation of the body. I consider it equal injustice to set our heart against natural pleasures and to set our heart too much on them. Xerxes was a fool, who, wrapped in all human pleasures, went and offered a prize to anyone who would find him others. But hardly less of a fool is the man who cuts off those that nature has found for him. We should neither pursue them nor flee them, we should accept them. I accept them with more gusto and with better grace than most, and more willingly let myself follow a natural inclination. We have no need to exaggerate their inanity; it makes itself felt enough and evident enough. Much thanks to our sickly, kill-joy mind, which disgusts us with them as well as with itself. It treats both itself and all that it takes in, whether future or past, according to its insatiable, erratic, and versatile nature.

> Unless the vessel's pure, all you pour in turns sour.
>
> —Horace[32]

I, who boast of embracing the pleasures of life so assiduously and so particularly, find in them, when I look at them thus minutely, virtually nothing but wind. But what of it? We are all wind. And even the wind, more wisely than we, loves to make a noise and move about, and is content with its own functions, without wishing for stability and solidity, qualities that do not belong to it. . . .

Aristippus defended the body alone, as if we had no soul; Zeno embraced only the soul, as if we had no body. Both were wrong. Pythagoras, they say, followed a philosophy that was all contemplation, Socrates one that was all conduct and action; Plato found the balance between the two. But they say so to make a good story, and the true balance is found in Socrates, and Plato is much more Socratic than Pythagorean, and it becomes him better.

[31]*De Senectute (On Old Age)* XIX. The quotation from Cicero that follows is from the same chapter.
[32]*Epistles* I.ii.54.

When I dance, I dance; when I sleep, I sleep; yes, and when I walk alone in a beautiful orchard, if my thoughts have been dwelling on extraneous incidents for some part of the time, for some other part I bring them back to the walk, to the orchard, to the sweetness of this solitude, and to me. Nature has observed this principle like a mother, that the actions she has enjoined on us for our need should also give us pleasure; and she invites us to them not only through reason, but also through appetite. It is unjust to infringe her laws.

When I see both Caesar and Alexander, in the thick of their great tasks, so fully enjoying natural and therefore necessary and just pleasures, I do not say that that is relaxing their souls, I say that it is toughening them, subordinating these violent occupations and laborious thoughts, by the vigor of their spirits, to the practice of everyday life: wise men, had they believed that this was their ordinary occupation, the other the extraordinary.

We are great fools. "He has spent his life in idleness," we say; "I have done nothing today." What, have you not lived? That is not only the fundamental but the most illustrious of your occupations. "If I had been placed in a position to manage great affairs, I would have shown what I can do." Have you been able to think out and manage your own life? You have done the greatest task of all. To show and exploit her resources Nature has no need of fortune; she shows herself equally on all levels and behind a curtain as well as without one. To compose our character is our duty, not to compose books, and to win, not battles and provinces, but order and tranquility in our conduct. Our great and glorious masterpiece is to live appropriately. All other things, ruling, hoarding, building, are only little appendages and props, at most. . . .

Popular opinion is wrong: it is much easier to go along the sides, where the outer edge serves as a limit and a guide, than by the middle way, wide and open, and to go by art than by nature; but it is also much less noble and less commendable. Greatness of soul is not so much pressing upward and forward as knowing how to set oneself in order and circumscribe oneself. It regards as great whatever is adequate, and shows its elevation by liking moderate things better than eminent ones. There is nothing so beautiful and legitimate as to play the man well and properly, no knowledge so hard to acquire as the knowledge of how to live this life well and naturally; and the most barbarous of our maladies is to despise our being. . . .

As for me, then, I love life and cultivate it just as God has been pleased to grant it to us. I do not go about wishing that it should lack the need to eat and drink, and it would seem to me no less excusable a failing to wish that need to be doubled. *The wise man is the keenest searcher for natural treasures* [Seneca].[33] Nor do I wish that we should sustain ourselves by merely putting into our mouths a little of that drug by which Epimenides took away his appetite and kept himself alive; nor that we should beget children insensibly with our fingers or our heels, but rather, with due respect, that we could also beget them voluptuously with our fingers and heels; nor that the body should be without desire and without titillation. Those are ungrateful and unfair complaints. I accept with all my heart and with gratitude what nature has done for me, and I am pleased with myself and proud of myself that I do. We wrong that great and all-powerful Giver by refusing his gift, nullifying it, and disfiguring it. Himself all good, he has made all things good. *All things that are according to nature are worthy of esteem* [Cicero].[34]

[33] *Letters (Epistulae Morales)* CXIX.
[34] *De Finibus (On Ends)* III.vi.

Of the opinions of philosophy I most gladly embrace those that are most solid, that is to say, most human and most our own; my opinions, in conformity with my conduct, are low and humble. Philosophy is very childish, to my mind, when she gets up on her hind legs and preaches to us that it is a barbarous alliance to marry the divine with the earthly, the reasonable with the unreasonable, the severe with the indulgent, the honorable with the dishonorable; that sensual pleasure is a brutish thing unworthy of being enjoyed by the wise man; that the only pleasure he derives from the enjoyment of a beautiful young wife is the pleasure of his consciousness of doing the right thing, like putting on his boots for a useful ride. May her followers have no more right and sinews and sap in deflowering their wives than her lessons have!

That is not what Socrates says, her tutor and ours. He prizes bodily pleasure as he should, but he prefers that of the mind, as having more power, constancy, ease, variety, and dignity. The latter by no means goes alone, according to him—he is not so fanciful—but only comes first. For him temperance is the moderator, not the adversary, of pleasure.

Nature is a gentle guide, but no more gentle than wise and just. *We must penetrate into the nature of things and clearly see exactly what it demands* [Cicero].[35] I seek her footprints everywhere. We have confused them with artificial tracks, and for that reason the sovereign good of the Academics and the Peripatetics, which is "to live according to nature," becomes hard to limit and express; also that of the Stoics, a neighbor to the other, which is "to consent to nature."

Is it not an error to consider some actions less worthy because they are necessary? No, they will not knock it out of my head that the marriage of pleasure with necessity, with whom, says an ancient, the gods always conspire, is a very suitable one. To what purpose do we dismember by divorce a structure made up of such close and brotherly correspondences? On the contrary, let us bind it together again by mutual services. Let the mind arouse and quicken the heaviness of the body, and the body check and make fast the lightness of the mind. *He who praises the nature of the soul as the sovereign good and condemns the nature of the flesh as evil, truly both carnally desires the soul and carnally shuns the flesh; for his feeling is inspired by human vanity, not by divine truth* [St. Augustine].[36]

There is no part unworthy of our care in this gift that God has given us; we are accountable for it even to a single hair. And it is not a perfunctory charge to man to guide man according to his nature; it is express, simple, and of prime importance, and the creator has given it to us seriously and sternly. Authority alone has power over common intelligences, and has more weight in a foreign language. Let us renew the charge here. *Who would not say that it is the essence of folly to do lazily and rebelliously what has to be done, to impel the body one way and the soul another, to be split between the most conflicting motions?* [Seneca][37]

Come on now, just to see, some day get some man to tell you the absorbing thoughts and fancies that he takes into his head, and for the sake of which he turns his mind from a good meal and laments the time he spends on feeding himself. You will find there is nothing so insipid in all the dishes on your table as this fine entertainment of his mind (most of the time we should do better to go to sleep completely than to stay awake for what we do stay awake for); and you will find that his ideas and aspirations are not worth your stew. Even if they were the transports of Archimedes himself, what of it? I am not

[35]*De Finibus (On Ends)* V.xvi. [36]*City of God* XIV.5. See Chapter 13 above.
[37]*Letters (Epistulae Morales)* LXXIV.

here touching on, or mixing up with that brattish rabble of men that we are, or with the vanity of the desires and musings that distract us, those venerable souls, exalted by ardent piety and religion to constant and conscientious meditation on divine things, who, anticipating, by dint of keen and vehement hope, the enjoyment of eternal food, final goal and ultimate limit of Christian desires, sole constant and incorruptible pleasure, scorn to give their attention to our beggarly, watery, and ambiguous comforts, and readily resign to the body the concern and enjoyment of sensual and temporal fodder. That is a privileged study. Between ourselves, these are two things that I have always observed to be in singular accord: supercelestial thoughts and subterranean conduct.

Aesop, that great man, saw his master pissing as he walked. "What next?" he said. "Shall we have to shit as we run?" Let us manage our time; we shall still have a lot left idle and ill spent. Our mind likes to think it has not enough leisure hours to do its own business unless it dissociates itself from the body for the little time that the body really needs it.

They want to get out of themselves and escape from the man. That is madness: instead of changing into angels, they change into beasts; instead of raising themselves, they lower themselves. These transcendental humors frighten me, like lofty and inaccessible places; and nothing is so hard for me to stomach in the life of Socrates as his ecstasies and possessions by his daemon, nothing is so human in Plato as the qualities for which they say he is called divine. And of our sciences, those seem to me most terrestrial and low which have risen the highest. And I find nothing so humble and so mortal in the life of Alexander as his fancies about his immortalization. Philotas stung him wittily by his answer. He congratulated himself by letter on the oracle of Jupiter Ammon which had lodged him among the gods: "As far as you are concerned, I am very glad of it; but there is reason to pity the men who will have to live with and obey a man who exceeds and is not content with a man's proportions."

Since you obey the gods, you rule the world.

—Horace[38]

The nice inscription with which the Athenians honored the entry of Pompey into their city is in accord with my meaning.

You are as much a god as you will own
That you are nothing but a man alone.

—Amyot's Plutarch[39]

It is an absolute perfection and virtually divine to know how to enjoy our being rightfully. We seek other conditions because we do not understand the use of our own, and go outside of ourselves because we do not know what it is like inside. Yet there is no use our mounting on stilts, for on stilts we must still walk on our own legs. And on the loftiest throne in the world we are still sitting only on our own rump.

The most beautiful lives, to my mind, are those that conform to the common human pattern, with order, but without miracle and without eccentricity. Now old age needs to be treated a little more tenderly. Let us commend it to that god[40] who is the protector of health and wisdom, but gay and sociable wisdom:

[38] *Odes* III.vi.5.
[39] Jacques Amyot (1513–93) was the French translator of Plutarch's *Lives* (1559) and *Moralia* (1572); this quotation is from the "Life of Pompey."
[40] Phoebus Apollo, son of Zeus (Jupiter) and Leto (Latona).

Grant me but health, Latona's son,
And to enjoy the wealth I've won,
And honored age, with mind entire
And not unsolaced by the lyre.
—Horace[41]

The Stay of the Whole World

Of the Laws of Ecclesiastical Polity, by Richard Hooker

Despite its sanguinary beginnings in the turbulent reign of Henry VIII, the Reformed Church of England came to be seen by both defenders and detractors as a compromising (or compromised) middle way between the ceremonialism of the Roman Church and more extreme Calvinists or Puritans, who repudiated such survivals as relics of popery. In his Laws of Ecclesiastical Polity *(Books I-IV were published in 1593, Book V in 1597, and Books VI-VIII long after his death), Richard Hooker (1554–1600) did more than any other writer to formulate the principles by which the Anglican dispensation was justified. The calm and deliberative prose of "the judicious Hooker," directed against the Puritan dissenters, always appeals to reason. And in contrast to the emphasis of most Protestant writers on the incommensurability of a transcendent and unknowable God, for him the function of reason—as Christopher Morris writes in introducing the Everyman's Library edition of 1907, from volume one of which our selections are taken—"was to discover law, particularly the Natural Law, moral and physical, by which God regulates the universe" and on which the polity of God's church must be founded. Insofar as it acts not on "voluntary agents," such as men, who are only partly subject to it, but on natural agents ruled (like the planets) by necessity, "obedience to the law of nature is the stay of the whole world." Though rejecting Plato's theory of forms ("exemplary drafts or patterns" imitated by nature), Hooker affirms the Platonic tradition that things said to be done by nature, as God's creation, "are by divine art performed." His conviction that God, though transcendent, is also immanent in the world enables him, C. S. Lewis writes in* English Literature in the Sixteenth Century *(1954), "to resist any inaccurate claim that is made for revelation against reason, Grace against Nature, the spiritual against the secular. . . . If 'nature hath need of grace,' yet also 'grace hath use of nature' (III.viii.6)." More than any other Renaissance thinker, Hooker thus continues the medieval Christian tradition, descending from Augustine and Boëthius through Thomas Aquinas, of viewing nature as the delegated instrument of God's will.*

I.III.2 Wherefore to come to the law of nature: albeit thereby we sometimes mean that manner of working which God hath set for each created thing to keep; yet forasmuch as those things are termed most properly natural agents, which keep the law of their kind unwittingly, as the heavens and elements of the world, which can do no otherwise than they do; and forasmuch as we give unto intellectual natures the name of voluntary agents, that so we may distinguish them from the other; expedient it will be, that we sever the law of nature observed by the one from that which the other is tied unto. Touching the former, their strict keeping of one tenure, statute, and law, is spoken of by

[41] *Odes* I.xxxi.17–20.

all, but hath in it more than men have as yet attained to know, or perhaps ever shall attain, seeing the travail of wading herein is given of God to the sons of men, that perceiving how much the least thing in the world hath in it more than the wisest are able to reach unto, they may by this means learn humility. Moses, in describing the work of creation, attributeth speech unto God: "God said, Let there be light: let there be a firmament: let the waters under the heaven be gathered together into one place: let the earth bring forth: let there be lights in the firmament of heaven." Was this only the intent of Moses, to signify the infinite greatness of God's power by the easiness of his accomplishing such effects, without travail, pain, or labor? Surely it seemeth that Moses had herein besides this a further purpose, namely, first to teach that God did not work as a necessary but a voluntary agent, intending beforehand and decreeing with himself that which did outwardly proceed from him: secondly, to show that God did then institute a law natural to be observed by creatures, and therefore according to the manner of laws, the institution thereof is described, as being established by solemn injunction. His commanding those things to be which are, and to be in such a sort as they are, to keep that tenure and course which they do, importeth the establishment of nature's law. This world's first creation, and the preservation since of things created, what is it but only so far forth a manifestation by execution, what the eternal law of God is concerning things natural? And as it cometh to pass in a kingdom rightly ordered, that after a law is once published, it presently takes effect far and wide, all states framing themselves thereunto; even so let us think it fareth in the natural course of the world: since the time that God did first proclaim the edicts of his law upon it, heaven and earth have hearkened unto his voice, and their labor hath been to do his will: He "made a law for the rain" [Job 28:26]; He gave his "decree unto the sea, that the waters should not pass his commandment" [Jeremiah 5:22]. Now if nature should intermit her course, and leave altogether though it were but for a while the observation of her own laws; if those principal and mother elements of the world, whereof all things in this lower world are made, should lose the qualities which now they have; if the frame of that heavenly arch erected over our heads should loosen and dissolve itself; if celestial spheres should forget their wonted motions, and by irregular volubility turn themselves any way as it might happen; if the prince of the lights of heaven, which now as a giant doth run his unwearied course, should as it were through a languishing faintness begin to stand and to rest himself; if the moon should wander from her beaten way, the times and seasons of the year blend themselves by disordered and confused mixture, the winds breathe out their last gasp, the clouds yield no rain, the earth be defeated of heavenly influence, the fruits of the earth pine away as children at the withered breasts of their mother no longer able to yield them relief: what would become of man himself, whom these things now do all serve? See we not plainly that obedience of creatures unto the law of nature is the stay of the whole world?

I.III.3 Notwithstanding with nature it cometh sometimes to pass as with art. Let Phidias have rude and obstinate stuff to carve, though his art do that it should, his work will lack that beauty which otherwise in fitter matter it might have had. He that striketh an instrument with skill may cause notwithstanding a very unpleasant sound, if the string whereon he striketh chance to be uncapable of harmony. In the matter whereof things natural consist, that of Theophrastus taketh place, . . . "Much of it is oftentimes such as will by no means yield to receive that impression which were best and most perfect." Which defect in the matter of things natural, they who gave themselves unto the contemplation of nature amongst the heathen observed often: but the true original cause thereof, divine malediction, laid for the sin of man upon these creatures which God had made for the use of man, this being an article of that saving truth which God hath revealed unto his Church, was above the reach of their merely natural capacity and understanding. But howsoever these swervings are now and then incident into the course of nature, nevertheless so constantly the laws of nature are by natural agents observed, that no man denieth but those things which nature worketh are wrought, either always or for the most part, after one and the same manner.

I.III.4 If here it be demanded what that is which keepeth nature in obedience to her own law, we must have recourse to that higher law whereof we have already spoken, and because all other laws do thereon depend, from thence we must borrow so much as shall need for brief resolution in this point. Although we are not of opinion therefore, as some are, that nature in working hath before her certain exemplary drafts or patterns, which subsisting in the bosom of the Highest, and being thence discovered, she fixeth her eye upon them, as travelers by sea upon the pole-star of the world, and that according thereunto she guideth her hand to work by imitation: although we rather embrace the oracle of Hippocrates, that "each thing both in small and in great fulfilleth the task which destiny hath set down"; and concerning the manner of executing and fulfilling the same, "what they do they know not, yet is it in show and appearance as though they did know what they do; and the truth is they do not discern the things which they look on": nevertheless, forasmuch as the works of nature are no less exact, than if she did both behold and study how to express some absolute shape or mirror always present before her; yea, such her dexterity and skill appeareth, that no intellectual creature in the world were able by any capacity to do that which nature doth without capacity and knowledge; it cannot be but nature hath some director of infinite knowledge to guide her in all her ways. Who the guide of nature, but only the God of nature? "In Him we live, move, and are" [Acts 17:28]. Those things which nature is said to do, are by divine art performed, using nature as an instrument; nor is there any such art or knowledge divine in nature herself working, but in the Guide of nature's work.

Whereas therefore things natural which are not in the number of voluntary agents (for of such only we now speak, and of no other), do so necessarily observe their certain laws, that as long as they keep those forms[42] which give them their being, they cannot possibly be apt or inclinable to do otherwise than they do; seeing the kinds of their operations are both constantly and exactly framed according to the several ends for which they serve, they themselves in the meanwhile, though doing that which is fit, yet knowing neither what they do, nor why: it followeth that all which they do in this sort proceedeth originally from some such agent, as knoweth, appointeth, holdeth up, and even actually frameth the same. . . .

The Corrupted Guide

Selections from the Prose Works of John Donne

John Donne (1572?–1632) is best known for his poetry, ranging from satires, elegies, and the erotic poems of his Songs and Sonnets, *most of which probably date from the 1590s, to the later*

[42]Form in other creatures is a thing proportionable unto the soul in living creatures. Sensible it is not, nor otherwise discernible than only by effects. According to the diversity of inward forms, things of the world are distinguished into their kinds. (Hooker)

Divine Poems, *mainly written before he took orders in 1615 and became dean of St. Paul's in 1621. The witty conjunction of opposites* (discordia concors) *and elaborate intellectual "conceits" of his poems helped create a new poetic style later called "metaphysical." Nature in these poems sometimes appears in deliberately forced analogies, like that of "The Flea" in which the intermingled blood of lovers makes the flea a marriage bed. Or Nature may reveal the world's inherent corruption, redeemable only by resurrection, as in "An Anatomy of the World: The First Anniversary," in which the untimely death of a young girl reflects "the frailty and the decay of the whole world." A similarly hyperbolic depiction of the "vicissitudinary transmutation" of an unstable world enters into Donne's prose. The source of our first selection, the somewhat frivolous* Juvenilia: Or Certaine Paradoxes and Problemes, *was published in 1633. Our last two selections are from sermons 80 and 22, respectively, of* LXXX Sermons *(1640). Even here, from the pulpit of St. Paul's, Donne can contrast the "sordid, senseless, nameless dust" of the physical body both with the glory of bodily resurrection in heaven and with the wonder of "ordinary things in nature," such as the daily rising and setting of the sun, that silently reprehend "them who require, or who need, miracles."*

Paradox VIII: That Nature Is Our Worst Guide

Shall she be guide to all creatures, which is herself one? Or if she also have a guide, shall any creature have a better guide than we? The affections of lust and anger, yea even to err is natural, shall we follow these? Can she be a good guide to us, which hath corrupted not us only but herself? was not the first man, by the desire of knowledge, corrupted even in the whitest integrity of Nature? And did not Nature (if Nature did anything) infuse into him this desire of knowledge, and so this corruption in him, into us? If by Nature we shall understand our essence, our definition [our reasonableness], then this being alike common to all (the idiot and the wizard being equally reasonable) why should not all men having equally all one nature, follow one course? Or if we shall understand our inclinations: alas! how unable a guide is that which follows the temperature of our slimy bodies! For we cannot say that we derive our inclinations, our minds, or souls from our parents by any way: to say that it is all from all, is error in reason, for then with the first nothing remains; or is a part from all, is error in experience, for then this part, equally imparted to many children, would like gavelkind lands,[43] in few generations become nothing: or to say it by communication, is error in divinity, for to communicate the ability of communicating whole essence with any but God, is utterly blasphemy. And if thou hit thy father's nature and inclination, he also had his father's, and so climbing up, all comes of one man, and have one nature, all shall embrace one course; but that cannot be, therefore our complexions and whole bodies, we inherit from parents; our inclinations and minds follow that: for our mind is heavy in our body's afflictions, and rejoiceth in our body's pleasure: how then shall this nature govern us, that is governed by the worst part of us? Nature though oft chased away, it will return;[44] 'tis true, but those good motions and inspirations which be our guides must be wooed, courted, and welcomed, or else they abandon us. And that old axiom, *nihil invita*, etc., must not be said "thou *shalt*" but "thou *wilt*" do nothing against Nature; so unwilling he notes us to curb our natural appetites. We call our bastards always our "natural issue," and we define a fool by noth-

[43]Properly "the name of a land-tenure existing chiefly in Kent," the word "gavelkind," the Oxford English Dictionary notes, was from the sixteenth century "often used to denote the custom of dividing a deceased man's property equally among his sons. . . ."
[44]See Horace, *Epistles* I.x.24, quoted in the introduction to Horace in Chapter 11 above.

ing so ordinary, as by the name of "natural." And that poor knowledge whereby we conceive what rain is, what wind, what thunder, we call "metaphysic," "supernatural"; such *small* things, such *no* things do we allow to our pliant nature's apprehension. Lastly, by following her, we lose the pleasant and lawful commodities of this life, for we shall drink water and eat roots, and those not sweet and delicate, as now by man's art and industry they are made: we shall lose all the necessities of societies, laws, arts, and sciences, which are all the workmanship of man: yea, we shall lack the last best refuge of misery, death, because no death is natural: for if ye will not dare to call all death violent (though I see not why sicknesses be not violences), yet causes of all deaths proceed of the defect of that which nature made perfect, and would preserve, and therefore all against nature.

FROM *Sermon Preached at the Funeral of Sir William Cokayne, Knight, Alderman of London. Dec. 12th, 1626*

4 I need not call in new philosophy,[45] that denies a settledness, an acquiescence in the very body of the earth, but makes the earth to move in that place, where we thought the sun had moved; I need not that help, that the earth itself is in motion, to prove this: that nothing upon earth is permanent; the assertion will stand of itself, till some man assign me some instance, something that a man may rely upon, and find permanent. . . . In the elements themselves, of which all subelementary things are composed, there is no acquiescence, but a vicissitudinary transmutation into one another: air condensed becomes water, a more solid body, and air rarefied becomes fire, a body more disputable, and inapparent. It is so in the conditions of men too. . . .

The world is a great volume, and man the index of that book; even in the body of man, you may turn to the whole world; this body is an illustration of all Nature; God's recapitulation of all that he had said before in his *Fiat lux*, and *Fiat firmamentum*,[46] and in all the rest, said or done, in all the six days. Propose this body to thy consideration in the highest exaltation thereof; as it is the temple of the Holy Ghost: nay, not in a metaphor, or comparison of a temple, or any other similitudinary thing but as it was really and truly the very body of God, in the person of Christ, and yet this body must wither, must decay, must languish, must perish. . . . So in this part, where our foundation is, that nothing in temporal things is permanent, as we have illustrated that, by the decay of that which is God's noblest piece in Nature, the body of man, so we shall also conclude that, with this goodness of God, that for all this dissolution, and putrefaction, he affords this body a resurrection. . . .

FROM *Sermon: St. Paul's. Easter Day. March 25th, 1627*

1 There is nothing that God hath established in a constant course of nature, and which therefore is done every day, but would seem a miracle, and exercise our admiration, if it were done but once; nay, the ordinary things in nature would be greater miracles than the extraordinary, which we admire most, if they were done but once. The standing still of the sun, for Joshua's use, was not, in itself, so wonderful a thing as that

[45]Copernican science. In his poem "An Anatomy of the World: The First Anniversary" (1611), Donne had written: "And new philosophy calls all in doubt, / The element of fire is quite put out; / The sun is lost, and th'earth, and no man's wit / Can well direct him where to look for it. . . . / 'Tis all in pieces, all coherence gone; / All just supply and all relation . . ."

[46]"Let there be light" and "Let there be a firmament" (Genesis 1:3, 6).

so vast and immense a body as the sun should run so many miles in a minute; the motion of the sun were a greater wonder than the standing still, if all were to begin again; and only the daily doing takes off the admiration. But then God having, as it were, concluded himself in a course of nature, and written down in the book of creatures, Thus and thus all things shall be carried, though he glorify himself sometimes in doing a miracle, yet there is in every miracle a silent chiding of the world, and a tacit reprehension of them who require, or who need, miracles. . . .

2 . . . When I consider what I was in my parents' loins (a substance unworthy of a word, unworthy of a thought); when I consider what I am now (a volume of diseases bound up together, a dry cinder, if I look for natural, for radical moisture, and yet a sponge, a bottle of overflowing rheums, if I consider accidental: an aged child, a grey-headed infant, and but the ghost of mine own youth); when I consider what I shall be at last, by the hand of death, in my grave (first, but putrefaction, and then, not so much as putrefaction, I shall not be able to send forth so much as an ill air, or any air at all, but shall be all insipid, tasteless, savorless dust; for a while, all worms, and after a while, not so much as worms, sordid, senseless, nameless dust); when I consider the past, and present, and future state of this body, in this world, I am able to conceive, able to express the worst that can befall it in nature, and the worst that can be inflicted upon it by man, or fortune: but the least degree of glory that God hath prepared for that body in heaven, I am not able to express, not able to conceive.

A Harmless Treasure

Of the Historie of Plants, by John Gerard

John Gerard's Elizabethan herbal is representative of an important kind of writing that descended from antiquity through the Middle Ages, but was especially popular during the sixteenth and seventeenth centuries. Ancient writers on plants included Aristotle and Theophrastus (Chapter 8 above), Pliny the Elder, much of whose Natural History *was devoted to plants and their medicinal uses, and Galen (Chapter 12). But the preeminent herbalist was Dioscórides, a Cilician Greek military surgeon of the first century* A.D. *whose herbal, known by its Latin title as* De Materia Medica, *described some 600 plants—illustrated, in its earliest manuscript, by a sixth-century Byzantine artist—and their properties; it remained for a millennium and a half the main source on which other herbals drew. (Though a Latin translation was printed in 1478, followed by printed editions in Greek, Italian, German, Spanish, and French, no English translation was made until that of John Goodyer in 1655, and this remained unpublished until 1933.) Dioscorides' succinct accounts are largely free of the fabulous matter that lards much of Pliny, and his brief introduction sets forth an ideal of careful observation notably absent from the* Physiologus *and its progeny (Chapter 13). "Now it behooves anyone who desires to be a skillful herbalist," he writes, "to be present when the plants first shoot out of the earth, when they are fully grown, and when they begin to fade. For he who is only present at the budding of the herb cannot know it when full-grown, nor can he who hath examined a full-grown herb recognize it when it has only just appeared above ground." Among the most widely read medieval herbals in Latin and the vernaculars, all of which pirated Dioscorides and one another, was a Latin hexameter poem of over two thousand lines,* Macer Floridus De Viribus Herbarum (Macer Floridus on the Powers of Plants), *which was translated into several languages, including Middle English.*

It was not until the sixteenth century, Marie Boas writes in The Scientific Renaissance, *that "herbalists stopped depending* only *on Dioscorides and Theophrastos, zoologists on Aristotle and Pliny; and natural historians began to believe that they could work on their own." Materials from ancient and medieval sources are now increasingly supplemented by firsthand observations or accurate secondhand reports, and by meticulous illustrations drawn not from other manuscripts but from the plants themselves. Major sixteenth-century herbals include, in Germany,* Herbarum Vivae Eicones (Living Images of Plants) *by Otto Brunfels (1530) and* De Historia Stirpium (On the History of Plants) *by Leonhart Fuchs (1542). In England, the anonymous vernacular* Banckes's Herbal *and* Great Herbal *of the early sixteenth century were followed by the* New Herbal *(1551 and 1561) of William Turner (ca. 1510–68), whom Eleanour Sinclair Rohde, in* The Old English Herbals *(1922), calls "the first Englishman who studied plants scientifically."*

But by far the most enduringly popular of English herbals—not for originality, since much of it was plagiarized from an unpublished English translation of the Flemish Rembert Dodoens's Latin herbal Pemptades *(1583), but for its vivid descriptions and engaging style—is* Of the Historie of Plants *(1597) by John Gerard (1545–1607). Gerard's herbal describes an immense variety of plants from the common and familiar, such as meadow grass or wheat, to the exotic and new, including such American imports as potatoes, tobacco, and the Peruvian Marvel of the World. Despite unacknowledged use of other sources—a widespread practice—and despite occasional lapses, such as his solemn account of a tree bearing geese, Gerard is usually punctilious both in observing plants in his own garden and in reporting findings of others. His popularity attests to his success in giving readers "such a harmless treasure of herbs, trees, and plants, as the earth frankly without violence offereth unto our most necessary uses."*

For further reading, besides Rohde, see Gerard's Herbal: The History of Plants, *a one-volume selection, ed. Marcus Woodward (1927; rpt.1994); Agnes Arber,* Herbals: Their Origin and Evolution*(1912; 2nd ed. 1938); and Frank J. Anderson,* An Illustrated History of Herbals *(1977).*

FROM *Epistle Dedicatorie* *to the Right Honorable Sir William Cecil Knight, Baron of Burghley*

Among the manifold creatures of God (right honorable, and my singular good lord) that have all in all ages diversely entertained many excellent wits, and drawn them to the contemplation of the divine wisdom, none have provoked men's studies more, or satisfied their desires so much as plants have done, and that upon just and worthy causes: for if delight may provoke men's labor, what greater delight is there than to behold the earth appareled with plants, as with a robe of embroidered work, set with orient pearls and garnished with great diversity of rare and costly jewels? If this variety and perfection of colors may affect the eye, it is such in herbs and flowers, that no Apelles, no Zeuxis[47] ever could by any art express the like: if odors or if taste may work satisfaction, they are both so sovereign in plants, and so comfortable that no confection of the apothecaries can equal their excellent virtue. But these delights are in the outward senses: the principal delight is in the mind, singularly enriched with the knowledge of these visible things, setting forth to us the invisible wisdom and admirable workmanship of Almighty God.

[47]Famous painters of ancient Greece.

The delight is great, but the use greater, and joined often with necessity. In the first ages of the world they were the ordinary meat[48] of men, and have continued ever since of necessary use both for meats to maintain life, and for medicine to recover health. The hidden virtue of them is such, that (as Pliny noteth) the very brute beasts have found it out: and (which is another use that he observes) from thence the dyers took the beginning of their art.

Furthermore, the necessary use of those fruits of the earth doth plainly appear by the great charge and care of almost all men in planting and maintaining of gardens, not as ornaments only, but as a necessary provision also to their houses. And here beside the fruit, to speak again in a word of delight, gardens, especially such as your honor hath, furnished with many rare simples,[49] do singularly delight, when in them a man doth behold a flourishing show of summer beauties in the midst of winter's force, and a goodly spring of flowers, when abroad a leaf is not to be seen. . . . But my very good lord, that which sometime was the study of great philosophers and mighty princes, is now neglected, except it be of some few, whose spirit and wisdom hath carried them among other parts of wisdom and counsel, to a care and study of special herbs, both for the furnishing of their gardens, and furtherance of their knowledge: among whom I may justly affirm and publish your honor to be one, being myself one of your servants, and a long time witness thereof: for under your lordship I have served, and that way employed my principal study and almost all my time, now by the space of twenty years. To the large and singular furniture of this noble island I have added from foreign places all the variety of herbs and flowers that I might any way obtain, I have labored with the soil to make it fit for plants, and with the plants, that they might delight in the soil, that so they might live and prosper under our climate, as in their native and proper country: what my success hath been, and what my furniture is, I leave to the report of they that have seen your lordship's gardens, and the little plot of mine own especial care and husbandry. . . .

FROM *To the Courteous and Well Willing Readers*

Although my pains have not been spent (courteous reader) in the gracious discovery of golden mines, nor in the tracing after silver veins, whereby my native country might be enriched with such merchandise as it hath most in request and admiration; yet hath my labor (I trust) been otherwise profitably employed, in descrying of such a harmless treasure of herbs, trees, and plants, as the earth frankly without violence offereth unto our most necessary uses. Harmless I call them, because they were such delights as man in the perfectest state of his innocency did erst[50] enjoy: and treasure I may well term them seeing both kings and princes have esteemed them as jewels; sith[51] wise men have made their whole life as a pilgrimage to attain to the knowledge of them: by the which they have gained the hearts of all, and opened the mouths of many, in commendation of those rare virtues which are contained in these terrestrial creatures. . . .

I list[52] not seek the common colors of antiquity, when notwithstanding the world can brag of no more ancient monument than paradise and the garden of Eden; and the fruits of the earth may contend for seniority, seeing their mother was the first creature that conceived, and they themselves the first fruit she brought forth. Talk of perfect happiness or pleasure, and what place was so fit for that as the garden place wherein Adam was set to be the herbarist? whither did the poets hunt for their sincere delights, but into the gardens of Alcinous, of Adonis, and the Orchards of the Hesperides? Where did they

[48]"Food." [49]Medicinal plants. [50]"Formerly." [51]"Since." [52]"Choose," "prefer."

dream that heaven should be, but in the pleasant garden of Elysium? Whither do all men walk for their honest recreation, but thither where the earth hath most beneficially painted her face with flourishing colors? And what season of the year more longed for than the spring, whose gentle breath enticeth forth the kindly sweets, and makes them yield their fragrant smells? Who would therefore look dangerously up at planets, that might safely look down at plants? And if true be the proverb, *Quae supra nos, nihil ad nos* ["what is above us is nothing to us"]; I suppose this new saying cannot be false, *Quae infra nos, ea maxime ad nos* ["what is below us is most to us"]. Easy therefore is this treasure to be gained, and yet precious. The science is nobly supported by wise and kingly favorites; the subject thereof so necessary and delectable, that nothing can be confected, either delicate for the taste, dainty for smell, pleasant for sight, wholesome for body, conservative or restorative for health, but it borroweth the relish of an herb, the savor of a flower, the color of a leaf, the juice of a plant, or the decoction of a root. And such is the treasure that this my treatise is furnished withal. . . .

Chapter 11: Of Ginger

The Description Ginger is most impatient of the coldness of these our northern regions, as myself have found by proof, for that there have been brought unto me at several times sundry plants thereof, fresh, green, and full of juice, as well from the West Indies, as from Barbary and other places; which have sprouted and budded forth green leaves in my garden in the heat of summer, but as soon as it hath been but touched with the first sharp blast of winter, it hath presently perished both blade and root. The true form or picture hath not before this time been set forth by any that hath written; but the world hath been deceived by a counterfeit figure, which the reverend and learned herbarist Matthias Lobel did set forth in his Observations. The form whereof notwithstanding I have here expressed, with the true and undoubted picture also, which I received from Lobel's own hands at the impression hereof. The cause of whose former error, as also the means whereby he got the knowledge of the true Ginger, may appear by his own words sent unto me in Latin, which I have here thus Englished:

How hard and uncertain it is to describe in words the true proportion of plants (having none other guide than skillful, but yet deceitful forms of them, sent from friends or other means) they best do know who have deepliest waded in this sea of simples. About thirty years past or more, an honest and expert apothecary, William Dries, to satisfy my desire, sent me from Antwerp to London the picture of Ginger, which he held to be truly and lively drawn. I myself gave him credit easily, because I was not ignorant, that there had been often Ginger roots brought green, new, and full of juice, from the Indies to Antwerp: and further, that the same had budded and grown in the said Dries' garden. But not many years after I perceived, that the picture which was sent me by my friend was a counterfeit, and before that time had been drawn and set forth by an old Dutch herbarist. Therefore not suffering this error any further to spread abroad (which I discovered not many years past at Flushing in Zeeland, in the garden of William of Nassau, Prince of Orange, of famous memory), I thought it convenient to impart thus much unto Mr. John Gerrard, an expert herbarist, and master of happy success in surgery, to the end he might let posterity know thus much, in the painful and long labored travels which now he hath in hand, to the great good and benefit of his country.

Thus much have I set down, truly translated out of his own words in Latin; though too favorably by him done to the commendation of my mean skill.

THE PLACE Ginger groweth in Spain, Barbary, in the Canary Islands, and the Azores. Our men who sacked Domingo in the Indies, digged it up there in sundry places wild.

THE TIME Ginger flourisheth in the hot time of summer, and loseth his leaves in winter.

THE NAMES Ginger is called in Latin *Zinziber*, and *Gingiber*.

THE VIRTUES Ginger, as Dioscorides reporteth, is right good with meat in sauces, or otherwise in conditures; for it is of an heating and digesting quality, and is profitable for the stomach, and effectually opposeth itself against all darkness of the sight; answering the qualities and effects of Pepper.

Chapter 46: Of the Marvel of the World[53]

THE DESCRIPTION This admirable plant, called the Marvel of Peru, or the Marvel of the World, springs forth of the ground like unto basil in leaves; among which it sendeth out a stalk two cubits and a half high, of the thickness of a finger, full of juice, very firm, and of a yellowish green color, knotted or kneed with joints somewhat bunching forth, of purplish color, as in the female Balsamina: which stalk divideth itself into sundry branches or boughs, and those also knotty like the stalk. His branches are decked with leaves growing by couples at the joints like the leaves of wild Peascods, green, fleshy, and full of joints; which being rubbed do yield the like unpleasant smell as wild Peascods do, and are in taste also very unsavory, yet in the later end they leave a taste and sharp smack of Tobacco. The stalks toward the top are garnished with long hollow single flowers, folded as it were into five parts before they be opened; but being fully blown, do resemble the flowers of tobacco, not ending in sharp corners, but blunt and round as the flowers of Bindweed, and larger than the flowers of Tobacco, glittering oft times with a fine purple or crimson color, many times of an horse-flesh, sometimes yellow, sometimes pale, and sometime resembling an old red or yellow color; sometime whitish, and most commonly two colors occupying half the flower, or intercoursing the whole flower with streaks or orderly streams, now yellow, now purple, divided through the whole, having sometime great, sometime little spots of a purple color, sprinkled and scattered in a most variable order and brave mixture. The ground or field of the whole flower is either pale, red, yellow, or white, containing in the middle of the hollowness a prick or pointal set round about with six small strings or chives. The flowers are very sweet and pleasant, resembling the Narcisse or white Daffodil, and are very suddenly fading; for at night they are flowered wide open, and so continue until eight of the clock the next morning, at which time they begin to close (after the manner of Bindweed) especially if the weather be very hot: but the air being temperate, they remain open the whole day, and are closed only at night, and so perish, one flower lasting but only one day, like the true Ephemerum or Hemerocallis. This marvelous variety doth not without cause bring admiration to all that observe it. For if the flowers be gathered and reserved in several papers, and compared with those flowers that will spring and flourish the next day, you shall easily perceive that one is not like another in color, though you shall compare one hundred which flower one day, and another hundred which you gather the next day, and

[53]The plant *Mirabilis Jalapa*, native of tropical America, with handsome funnel-shaped flowers of various colors which expand towards night. (OED)

so from day to day during the time of their flowering. The cups and husks which contain and embrace the flowers are divided into five pointed sections, which are green, and as it were consisting of skins, wherein is contained one seed and no more, covered with a blackish skin, having a blunt point whereon the flower groweth; but on the end next the cup or husk it is adorned with a little five-cornered crown. The seed is as big as a pepper corn, which of itself fadeth with any light motion. Within this seed is contained a white kernel, which being bruised, resolveth into a very white pulp like starch. The root is thick and like unto a great radish, outwardly black, and within white, sharp in taste, wherewith is mingled a superficial sweetness. It bringeth new flowers from July unto October in infinite number, yes even until the frosts do cause the whole plant to perish: notwithstanding it may be reserved in pots, and set in chambers and cellars that are warm, and so defended from the injury of our cold climate; provided always that there be not any water cast upon the pot, or set forth to take any moisture in the air until March following; at which time it must be taken forth of the pot and replanted in the garden. By this means I have preserved many (though to small purpose) because I have sown seeds that have borne flowers in as ample manner and in as good time as those reserved plants.

Of this wonderful herb there be other sorts, but not so amiable or so full of variety, and for the most part their flowers are all of one color. But I have since by practice found out another way to keep the roots of the year following with very little difficulty, which never faileth. At the first frost I dig up the roots and put up or rather hide the roots in a butter firkin[54] or such like vessel, filled with the sand of a river, the which I suffer still to stand in some corner of an house where it never receiveth moisture until April or the midst of March, if the weather be warm; at which time I take it from the sand and plant it in the garden, where it doth flourish exceeding well and increaseth by roots; which that doth not which was either sown of seed the same year, nor those plants that were preserved after the other manner.

THE PLACE The seed of this strange plant was brought first into Spain, from Peru, whereof it took his name *Mirabilia Peruana*, or *Peruviana:* and since dispersed into all the parts of Europe: the which myself have planted many years, and have in some temperate years received both flowers and ripe seed.

THE TIME It is sown in the midst of April, and bringeth forth his variable flowers in September, and perisheth with the first frost, except it be kept as aforesaid.

THE NAMES It is called in Peru of those Indians there, *Hachal.* Of others after their name *Hachal Indi:* of the high and low Dutch, *Solanum Odoriferum:* of some, *Jasminum Mexicanum:* and of Carolus Clisius, *Admirabilia Peruviana:* in English, rather the Marvel of the World, than of Peru alone.

THE NATURE AND VIRTUES We have not as yet any instructions from the people of India,[55] concerning the nature or virtues of this plant: the which is esteemed as yet rather for his rareness, beauty, and sweetness of his flowers, than for any virtues known; but it is a pleasant plant to deck the gardens of the curious. Howbeit Jacobus Antonius Cortusus of Padua hath by experience found out, that two drams of the root thereof taken inwardly doth very notably purge waterish humors.

[54]A small cask for liquids, fish, butter, etc. (OED)
[55]"The Indies," i.e., the Americas.

Chapter 51: Of Mandrake

THE DESCRIPTION The male Mandrake hath great broad long smooth leaves of a dark green color, flat spread upon the ground: among which come up the flowers of a pale whitish color, standing every one upon a single small and weak foot-stalk of a whitish green color: in their places grow round apples of a yellowish color, smooth, soft, and glittering, of a strong smell: in which are contained flat and smooth seeds in fashion of a little kidney, like those of the Thorn-apple. The root is long, thick, whitish, divided many times into two or three parts resembling the legs of a man, as it hath been reported; whereas in truth it is no otherwise than in the roots of carrots, parsnips, and such like, forked or divided into two or more parts, which Nature taketh no account of. There hath been many ridiculous tales brought up of this plant, whether of old wives, or some runagate[56] surgeons or physic-mongers I know not (a title bad enough for them), but sure some one or more that sought to make themselves famous and skillful above others, were the first broachers of that error I speak of. They add further, that it is never or very seldom to be found growing naturally but under a gallows, where the matter that hath fallen from the dead body hath given it the shape of a man; and the matter of a woman, the substance of a female plant; with many other such doltish dreams. They fable further and affirm, that he who would take up a plant thereof must tie a dog thereunto to pull it up, which will give a great shriek at the digging up; otherwise if a man should do it, he should surely die in short space after. Besides many fable of loving matters, too full of scurrility to set forth in print, which I forbear to speak of. All which dreams and old wives' tales you shall from henceforth cast out of your books and memory; knowing this, that they are all and every part of them false and most untrue: for I myself and my servants also have digged up, planted, and replanted very many, and yet never could either perceive shape of man or woman, but sometimes one straight root, sometimes two, and often six or seven branches coming from the main great root, even as Nature list to bestow upon it, as to other plants. But the idle drones that have little or nothing to do but eat or drink, have bestowed some of their time in carving the roots of Bryony, forming them to the shape of men and women: which falsifying practice hath confirmed the error amongst the simple and unlearned people, who have taken them upon their report to be the true Mandrakes.

The female Mandrake is like unto the male, save that the leaves hereof be of a more swart or dark green color: and the fruit is long like a pear, and the other like an apple.

THE PLACE Mandrake groweth in hot regions, in woods and mountains, as in Mount Garganus in Apulia, and such like places; we have them only planted in gardens, and are not elsewhere to be found in England.

THE TIME They spring up with their leaves in March, and flower in the end of April: the fruit is ripe in August.

THE NAMES Mandrake is called *Circaea,* of Circe the witch, who by art could procure love: for it hath been thought that the root hereof serveth to win love.

THE VIRTUES The wine wherein the root hath been boiled or infused provoketh sleep and assuageth pain.

[56] "Renegade."

The smell of the apples moveth to sleep likewise; but the juice worketh more effectually if you take it in small quantities.

Great and strange effects are supposed to be in Mandrakes, to cause women to be fruitful and bear children, if they shall but carry the same near to their bodies. Some do from hence ground it, for that Rahel desired to have her sister's Mandrakes (as the text is translated),[57] but if we look into the circumstances which there we shall find, we may rather deem it otherwise. Young Ruben brought home amiable and sweet-smelling flowers (for so signifieth the Hebrew word, used *Cantic.* 7.13.[58] in the same sense) rather for their beauty and smell, than for their virtue. Now in the flowers of Mandrake there is no such delectable or amiable smell as was in these amiable flowers which Ruben brought home. Besides, we read not that Rahel conceived hereupon, for Leah, Jacob's wife, had four children before God granted that blessing of fruitfulness unto Rahel. And last of all (which is my chiefest reason), Jacob was angry with Rahel when she said, Give me children or else I die [Genesis 30:1]; and demanded of her, whether he were in the stead of God or no, who had withheld from her the fruit of her body. And we know the Prophet David saith, Children and the fruit of the womb are the inheritance that cometh of the Lord, *Psal.* 127[:3].

He that would know more hereof, may read that chapter of Dr. Turner's book concerning this matter, where he hath written largely and learnedly of this simple.

The Golden Age Gone to the Dogs

Selections from the Works of Miguel de Cervantes Saavedra

For most of his life, Miguel de Cervantes Saavedra (1547–1616) seemed destined to failure, his dreams forever running aground on the intransigent shoals of reality. He was born at Alcalá de Henares, where the university founded in 1508 was for a while a major center of Erasmian humanism, until the Counter Reformation brought tolerance to a sudden halt. His hand was permanently maimed in 1571 at Charles V's great naval victory over the Turks at Lepanto; in 1575 he was captured by Barbary pirates, then imprisoned for five years in Algiers before being ransomed at prohibitive cost to his small family fortune. Those were the glorious years. Thereafter, in his life as government purchasing agent, tax collector, and fundraiser for the Invincible Armada headed for destruction off the coast of England, he was several times imprisoned for shortfall or debt, and not even his petition to migrate to America met with success. Nor did his career as playwright win him great fortune, as the younger and more prolific Lope de Vega (Chapter 17 below) soon surpassed him in popular acclaim.

The book of which he was proudest as he passed age forty was the uncompleted pastoral novel La Galatea *(1585), to which, despite his repeatedly expressed intention, Cervantes never re-*

[57] See Genesis 30:14–17 (Authorized Version): "And Reuben went in the days of wheat harvest, and found mandrakes in the field, and brought them unto his mother Leah. Then Rachel said to Leah, Give me, I pray thee, of thy son's mandrakes. And she said unto her, Is it a small matter that thou hast taken my husband? and wouldest thou take away my son's mandrakes also? And Rachel said, Therefore he shall lie with thee to night for thy son's mandrakes. And Jacob came out of the field in the evening, and Leah went out to meet him, and said, Thou must come in unto me; for surely I have hired thee with my son's mandrakes. And he lay with her that night. And God hearkened unto Leah, and she conceived, and bare Jacob the fifth son."

[58] Song of Solomon 7:13 (Authorized Version): "The mandrakes give a smell, and at our gates are all manner of pleasant fruits, new and old, which I have laid up for thee, O my beloved."

turned. In a restless life lacking, as he writes, "a convenient place, pleasant fields and groves, murmuring springs, and a sweet repose of mind, . . . helps that raise the fancy, and impregnate even the most barren Muses with conceptions that fill the world with admiration and delight," he could hope only to bring forth a "child of disturbance, engendered in some dismal prison, where wretchedness keeps its residence, and every dismal sound its habitation." This child was of course the woebegone knight of his masterpiece, The Ingenious Hidalgo Don Quixote de la Mancha, *the first part of which, published in 1605, when Cervantes was fifty-eight, brought him fame at long last. His* Novelas Ejemplares (Exemplary Novels) *followed in 1613, and the second part of* Don Quixote *in 1615, a year before Cervantes' death.*

Though most of his life was spent in Seville, Madrid, and other cities of Spain, in both Don Quixote *and such* Exemplary Novels *as "La Ilustre Fregona" ("The Illustrious Kitchen Maid"), the "spirit of adventure" and lure of "a roving life on the open road" draw his characters into the wide plains and rugged mountains that offer, for all their barrenness, an alternative to the constricting enclosures of city and court. With this putatively more "natural" world Cervantes associates his perennial dream of imaginative and spiritual freedom. Yet neither the wholly idealized (hence* un*natural) landscape of pastoral romance nor the hardscrabble world of real peasants and shepherds can fulfill that dream; between artifice and indigence, the bounty of nature eludes him.*

In our first selection (revised from Peter Motteux's translation of 1700), Don Quixote imagines a Golden Age of natural abundance, simplicity, and justice reminiscent of Hesiod (Chapter 8 above), Virgil, and Ovid (Chapter 11); but Don Quixote is mad, and the puzzled goatherds who live in contact with real *nature understand not a word of his ravings. In "The Little Gypsy Girl," the Gypsies are portrayed as footloose "lords of the plains" whose landscapes are painted by Nature herself. But this idyll is undercut by its own self-evident artificiality—the little Gypsy girl Preciosa will turn out, like any fairy-tale princess, to be of respectable birth in the end. And, in another* Exemplary Novel, *"The Dogs' Colloquy" ("El Coloquio de los Perros")—translated, like the first, from Cervantes'* Obras completas, *ed. Angel Valbuena Prat (1962)—it is also ruthlessly exposed by the dog Berganza as the stuff of "dreams well written to amuse the idle, not any truth." Life in accord with nature remained an essential object of Cervantes' lifelong yearning, but its realization, in an unnatural world, was no more possible than Don Quixote's restoration of chivalry in an Age of Iron.*

FROM The Ingenious Hidalgo Don Quixote de la Mancha,
translated by Peter Motteux (revised)
I.xi: What Passed between Don Quixote and the Goatherds

All this while the goatherds, who did not understand this jargon of knights-errant, chivalry, and squires, fed heartily, and said nothing, but stared upon their guests, who fairly swallowed whole luncheons as big as their fists with a mighty appetite. The first course being over, they brought in the second, consisting of dried acorns, and half a cheese as hard as a brick; nor was the horn cup idle all the while, but went merrily round up and down so many times, sometimes full, and sometimes empty, like the two buckets of a well, that they made shift at last to drink off one of the two skins of wine which they had there.

And now Don Quixote, having satisfied his appetite, took a handful of acorns, and looking earnestly upon them: "O happy age," cried he, "which the ancients called the age of gold—not because gold, so much adored in this iron age, was then easily purchased, but because those two fatal words, mine and thine, were distinctions unknown to the people of those fortunate times. For all things were in common in that holy age: men, for their sustenance, needed only to lift their hands, and take it from the sturdy oak, whose spreading arms liberally invited them to gather the wholesome savory fruit; while the clear springs and silver rivulets, with luxuriant plenty, offered them their pure refreshing water. In hollow trees, and in the clefts of rocks, the laboring and industrious bees erected their little commonwealths, that men might reap with pleasure and with ease the sweet and fertile harvest of their toils. The tough and strenuous cork-trees did of themselves, and without other art than their native liberality, dismiss and impart their broad light bark, which served to cover those lowly huts, propped up with rough-hewn stakes, that were first built as a shelter against the inclemencies of the air: all then was union, all peace, all love and friendship in the world. As yet no rude plowshare presumed with violence to pry into the pious bowels of our mother earth, for she without compulsion kindly yielded from every part of her fruitful and spacious bosom whatever might at once satisfy, sustain, and indulge her frugal children. Then was the time when innocent beautiful young shepherdesses went tripping over the hills and vales, their lovely hair sometimes plaited, sometimes loose and flowing, clad in no other vestment but what was necessary to cover decently what modesty should always have concealed. The Tyrian dye, and the rich glossy hue of silk, martyred and dissembled into every color, which are now esteemed so fine and magnificent, were unknown to the innocent plainness of that age; yet bedecked with more becoming leaves and flowers, they may be said to outshine the proudest of the vain-dressing ladies of our age, arrayed in the most magnificent garbs and all the most sumptuous adornings which idleness and luxury have taught succeeding pride. Lovers then expressed the passion of their souls in the unaffected language of the heart, with the native plainness and sincerity in which they were conceived, and divested of all that artificial contexture, which enervates what it labors to enforce. Imposture, deceit, and malice had not yet crept in and imposed themselves unbribed upon mankind in the disguise of truth and simplicity. Justice, unbiased either by favor or interest, which now so fatally pervert it, was equally and impartially dispensed; nor was the judge's whim law, for then there were neither judges, nor causes to be judged. The modest maid might walk wherever she pleased alone, free from the attacks of lewd lascivious importuners, and her undoing arose from her own will and desire. But in this degenerate age, no virtue is safe while wanton desires, diffused into the hearts of men, corrupt the strictest watches and the closest retreats, which, though as intricate and unknown as the labyrinth of Crete, are no security for chastity. Thus that primitive innocence being vanished, and oppression daily prevailing, there was a necessity to oppose the torrent of violence; for which reason, the order of Knights Errant was instituted, to defend the honor of virgins, protect widows, relieve orphans, and assist all the distressed in general. Now I myself am one of this order, honest friends; and though all people are obliged by the law of nature to be kind to persons of my order, yet since you, without knowing anything of this obligation, have so generously entertained me, I ought to pay you my utmost acknowledgment, and accordingly return you my hearty thanks for the same."

All this long oration, which might very well have been spared, was owing to the acorns that recalled the Golden Age to our knight's remembrance, and made him thus hold forth to the goatherds, who listened in attentive bewilderment, without answering a word. Sancho too ate his acorns in silence, and frequently visited the second wineskin, which for coolness' sake was hung on a neighboring cork-tree. . . .

FROM *THE EXEMPLARY NOVELS*, TRANSLATED BY ROBERT M. TORRANCE

FROM *The Little Gypsy Girl*

AN OLD GYPSY SPEAKS: "... We are lords of the plains, the fields, the woods, the forests, the fountains, and the rivers. The forests offer us wood without charge, the trees fruit, the vineyards grapes, the gardens vegetables, the fountains water, the rivers fish, the parks game, the rocks shade, their fissures fresh air, and their caves houses. For us the inclemencies of the weather are but breezes, the snows a refreshment, the rain baths, the thunder music, the lightning candles; for us the hard soil is a soft feather-bed. The weather-beaten hide of our bodies is an impenetrable armor protecting us; shackles do not impede our agility, gorges hinder it, nor walls resist it; cords do not restrict our courage, pulleys diminish it, hoods smother, nor the rack subdue it. We make no distinction of "yes" from "no" when it suits us; we always pride ourselves more on being martyrs than confessors; for us, beasts of burden are raised in the fields, and purses are cut in the cities. No eagle or other bird of prey pounces more swiftly on its destined victim than we pounce on opportunities for profit; and finally, we have many talents that promise a favorable end: for in jail we sing, but are silent on the rack; by day we work, and by night we steal—or, to say it better, warn everyone to keep a careful eye on where he puts his belongings. No fear of losing honor troubles us, nor does ambition to increase it keep us awake; we do not form factions, nor rise at dawn to give memorials, attend grandees, or solicit favors. As gilded roofs and sumptuous palaces we esteem these huts and portable camps; as Flemish paintings and landscapes, those that nature provides in the lofty crags and snow-covered rocks, broad meadows, and thick woods that meet our eyes at every step. We are rustic astrologers, for since we almost always sleep under the open sky, at all hours we know the time of day or night; we see how daylight's first blush sweeps the stars from the sky, and how she steps forth with her companion, the dawn, gladdening the air, cooling the water, and bedewing the earth; then, after her, the sun, "gilding the mountaintops," as some poet said, "and ruffling the forests": nor are we afraid of being frozen by his absence when his rays strike us only obliquely, or of being incinerated when they pound perpendicularly upon us. We turn the same face to sun and frost, to dearth and abundance. In conclusion: we are people who live by sharp wits and quick tongues; ignoring the ancient saw, 'Church or Sea or Royal Household,' we have all we want, for we are content with what we have. . . ."

FROM *The Dogs' Colloquy*

THE DOG BERGANZA SPEAKS: . . . But to knot up the thread of my broken story, in the silence and solitude of my siestas I considered, among other things, that what I had heard tell of the lives of shepherds must not be true, not at least what my master's lady used to read in books when I went to her house; for these all told about shepherds and shepherdesses, saying they spent their whole lives singing and playing on flutes and flageolets, rebecs and recorders, and other strange instruments. I would stop to hear her read how the shepherd of Anfriso sang consummately and divinely in praise of the peerless Belisarda, so that in all the forests of Arcadia there was no tree against whose trunk he had not sat to sing from the moment the sun stepped forth in the arms of Aurora till he dropped into those of Thetis, and even after black night had spread her dark black wings over the face of the earth, he did not cease his well-sung and even better-wept laments.[59] Nor did the shepherd Elicio, more lovestruck than bold, remain unspoken, for she said that he neglected both his loves and his flock by entering into the cares of others. She also said that the great shepherd of Filida, a unique portrait painter, had been more confident than lucky. Concerning Sireno's swoons and Diana's repentance, she said she thanked God and the sage Felicia, who with her enchanted water undid that pile of entanglements and cleared up that labyrinth of perplexities. I remembered hearing her read many other such books, not worthy of being brought back to memory. . . .

To continue: all the thoughts I have mentioned, and many more, led me to see how different were the habits and tasks of shepherds I had known, and others of that crew, from those of shepherds I had heard read about in books: for if mine sang, it was not tuneful and well composed songs but "Watch out for the Wolf, Juanica," and suchlike things; not to the sound of recorders, rebecs, or flageolets, but to the clatter of one crook against another, or of tiles placed between the fingers, and not with delicate, sonorous, and euphonious voices, but with rasping voices, which, whether alone or accompanied by others, seemed not to sing, but to holler and grunt. Most of the day they spent picking off lice or patching their sandals; nor were there any among them named Amaryllis, Filida, Galatea, or Diana, nor any Lisardos, Lausos, Jacintos, or Riselos. They were all Antons, Domingos, Pablos, or Llorentes; thus I came to understand, as I think everyone must, that all those books tell of dreams well written to amuse the idle, not any truth: for had it been true, there would have been some trace among my shepherds of that happy life, those pleasant meads, spacious forests, sacred wildwoods, beauteous gardens, clear rivulets, and crystal fountains, those decorous and eloquently declaimed eulogies, and those swoonings, here of a shepherd, there of a shepherdess, as one shepherd's flute resounded yonder, another's pipes here. . . .

. . . I've come to believe that everything we've experienced until now, and are now experiencing, is a dream, and that we are dogs; but let us not therefore cease to enjoy this boon of speech that we possess, and the immense advantage of holding human discourse whenever we can: so do not weary of hearing me describe what happened among the gypsies that hid me in the cave. . . .

What I mostly did among the gypsies at that time was to observe their many rascalities, deceits, and tricks—the thefts that both males and females commit almost from the time they cast off swaddling clothes and know how to walk. Do you see what a multitude of them is scattered throughout all Spain? Well, they all know one other, keep continually in touch, and exchange the stolen goods they transport from one band to another. They give less obedience to their own king than to someone they call Count, who, like all his descendants, is surnamed Maldonado—not because they come from his noble line, but because a knight of this name had a page who fell in love with a beautiful gypsy girl, and she would not return his love unless he became a gypsy and made her his wife. By doing so, the page so pleased the other gypsies that they made him their lord and gave obedience to him; and as a sign of vassalage they present him with a part of all thefts of any significance. To lend color to their idleness they busy themselves in manufactur-

[59]Anfriso and Belisarda are characters from the *Arcadia* of Lope de Vega (1598). In the following sentences, Elicio is from Cervantes' own pastoral novel *La Galatea* (1585); the "shepherd of Filida" is from Luis Gálvez de Montalvo's *El Pastor de Filida* (1582); Sireno, Diana, and Felicia are from Jorge de Montemayor's *Diana Enamorada* (1559).

ing iron objects, and making tools to facilitate their thefts; thus you will see men in the streets hawking the tongs, augers, and hammers they carry, and women their tripods and shovels for cooking. All the women are midwives, and in this they gain an advantage over ours, for they give birth without expense or assistants, and wash the babies in cold water at birth; and from birth to death they toughen themselves and prove they are able to suffer the inclemencies and rigors of the sky; thus you will see that all are high-spirited jumpers, runners, and dancers. They always intermarry, so that their evil ways may not become known to others; women are faithful to their husbands, and few betray them with other men who are not of their lineage. When begging alms, they get more by tricks and mendacities than devotion; and on the pretext that nobody trusts them, do not become servants, but give themselves over to vagrancy; seldom or never, if I remember correctly, have I seen a gypsy girl at the altar taking communion, although I have entered churches many times. Their only thoughts are to imagine how to play tricks and where to steal . . .

17

Great Creating Nature: Renaissance Poetry

A few poets in the rich poetic flowering of the Renaissance make reference to the revolutionary change in outlook set in motion by Copernican astronomy, but most continued to view the world within the framework inherited from the Middle Ages. This widely shared "world picture" assumed not only the Christian religion and Ptolemaic cosmology but also the Aristotelian division between the sublunary and superlunary worlds, and the Platonic/Hermetic vision of man as a microcosm sharing in both and aspiring to rise from the physical to the spiritual plane of existence. In this hierarchical world, most Renaissance poets continued to believe, as C. S. Lewis writes of their medieval predecessors in The Discarded Image, *"that Nature was not everything. She was created. She was not God's highest, much less His only, creature. She had her proper place, below the Moon. She had her appointed duties as God's viceregent in that area. . . . It is precisely this limitation and subordination of Nature which sets her free for her triumphant poetical career": a career as "goddess," dating back to Bernard Silvestris and Alan of Lille in the twelfth and Jean de Meun in the thirteenth century (Chapter 15 above), and passing through Chaucer in the fourteenth to Spenser at the height of the Elizabethan Renaissance.*

But if poets like others implicitly presupposed and often explicitly reaffirmed some such vision of cosmic order, as Shakespeare's Ulysses does in Troilus and Cressida *(I.iii)—*

The heavens themselves, the planets, and this
center
Observe degree, priority, and place,
Insisture, course, proportion, season, form,
Office, and custom, in all line of order—

the view of the world and of man's place within it expressed in their poems was by no means a complacent ratification of previous tradition. Just as Ulysses warned his fellow commanders at Troy that

when degree is shaked,
Which is the ladder of all high designs,
The enterprise is sick,

so the poets of this much shaken time are subject not only to subversion by the new philoso-

phy—which even most scientists were slow to accept—but to the various currents articulated by thinkers of our previous chapter as well. These currents included the immense intellectual stimulus of rediscovered Greek and Roman classics from Homer to Lucretius; the widespread repudiation of Scholastic Aristotelianism in favor of mystical Neoplatonism and the occult Hermetic doctrines and magical practices closely associated with science itself; the impact of unexpected discovery of new worlds and peoples on age-old beliefs about geography and about customs long thought to be God-given and universally valid; and the crisis attendant on the Protestant Reformation and the Catholic reaction. These poets gave voice to a common framework, then, whose tensions threatened to shatter it from within, leaving it, in Donne's words, "all in pieces, all coherence gone."

In Italy, where the Renaissance began, and where any antithesis between Middle Ages and Renaissance is even more arbitrary than elsewhere, the exciting "discovery of the world and of man"—not undiscovered before, but seen now with eyes newly opened to the splendors of nature—is especially prominent in poets of this age; their reawakened sense of wonder at the beauty of the physical world also permeates much poetry of the Spanish Golden Age, the French Renaissance, and the Age of Elizabeth. But tensions are present from the beginning. In Michelangelo's struggle to soar "beyond the visible world" in search of Ideal Forms housed in the heavens above it, or in Tasso's portrayal of a false earthly paradise where nature can only "imitate her imitator art," the disjunctions of the late Renaissance and Baroque are already apparent. The great English poets of the late Elizabethan and early Jacobean years—above all, Spenser, Jonson, and Shakespeare—give unforgettable expression to the conflicting forces of an age that saw nature now, like Spenser, in Platonic or medieval terms as the handiwork or handmaiden of the Creator; now, like Jonson, in neoclassical terms as the pattern with which human constructs must accord; now as the titanic agent or reflection of disorder in the world and man, in King Lear; *or again, in* The Winter's Tale, *as "great creating Nature" from which human art itself (far from being its opposite) derives.*

In the First Youthful Days of Early Spring

Italian Renaissance Poetry

In Italy, where the brilliant galaxy of Dante, Petrarch, and Boccaccio in the "trecento" (fourteenth century) remained an unsurpassable model of vernacular literature in centuries to come, the transition from High Middle Ages to High Renaissance entailed no sharp break with the recent past: the "rebirth," after all, had begun (Vasari and others proclaimed) as far back as Giotto. Thus ***Franco Sacchetti*** *(1332?–1400?), Ragusa-born Florentine author of the* Trecentonovelle (Three Hundred Novellas), *is a medieval at least as much as a Renaissance poet. In his lyrics, including "Mountain Shepherd-Girls" ("O vaghe montanine pasturelle"), even more than in Petrarch's a generation before, George Kay writes in* The Penguin Book of Italian Verse *(1958), which contains the text, "the medieval city with its darkened streets, on which the Madonna of Dante and his friends appeared like a sudden star, is left behind, and the country, which seems so fresh and clean and new, welcomes the poet."*

A century later, ***Matteo Boiardo*** *(1441?–94), count of Scandiano—a prominent courtier at the Este court of Ferrara, governor of Modena and Reggio, humanist translator of Herodotus* *and of Apuleius's* Golden Ass, *and author of the uncompleted verse romance* Orlando Innamorato (Roland in Love)*—traced the course of his love for Antonia Caprara from rapture to jealousy to disenchantment in his* Canzoniere, *or* Song Book. *In the two sonnets here translated, "The Melody of Birds" ("Il canto de li augei di frunda in frunda"; text in L. R. Lind,* Lyric Poetry of the Italian Renaissance *[1954]) and "I've Seen the Sun" ("Già vidi uscir di l'onde una matina"; text in Kay), we are in a High Renaissance world where the poet's lady is associated not (like Dante's Beatrice) with the angelic sphere nor (like Petrarch's Laura) with the conflict between flesh and spirit but with the resplendent beauty of the natural world, whose loveliness only her own surpasses.*

Angelo Poliziano *(1454–94) was born Angelo Ambrogini but took his latinized name of Politianus from his birthplace, Montepulciano. A friend of his fellow poet and ruler of Florence, Lorenzo de' Medici, whose children he tutored, Poliziano was one of the leading humanists and Latin poets of the century and professor of Greek and Latin in Florence after 1480. He was also author of vernacular poems, including the masque* Orfeo, *the unfinished* Stanze per la Giostra (Stanzas for the Joust) *written to celebrate the victory of Lorenzo's brother Giuliano in a tournament, and a number of sprightly lyrics such as the Dance Song here translated ("I' mi trovai, fanciulle, un bel mattino"). "There is in the poetry of Poliziano," Lind writes, "a lusty joy in earthly things and a love of nature combined with a sophisticated polish that reminds one of the later Renaissance."*

Ludovico Ariosto *(1474–1533) was, like Boiardo before and Tasso after him, a courtier at Ferrara. There he wrote, besides some minor poems in Latin and Italian, the long and fabulous verse romance* Orlando Furioso (Roland Mad), *which took up the story where Boiardo had broken off. In the passage given here, the Saracen Ruggiero (who is in love with the Christian Bradamante) journeys astride a winged hippogriff to the island paradise of the sorceress Alcina. After marveling at the luxuriant beauty of nature, he is led into her palace of sensual love and there enjoys both her sexual favors and the delights of country life. Here the artfully "natural" world of the Earthly Paradise, which from Dracontius and Avitus (Chapter 14 above) to Dante (Chapter 15) had often been an object—however unattainable—of Christian longing, is associated with erotic temptation and corruption. (Note how quickly the fearless hares of Canto VI are hunted down in Canto VII.) Yet, as A. Bartlett Giamatti writes in* The Earthly Paradise and the Renaissance Epic *(1966), "Dangerous and corrupting though it certainly was, false and deceptive though its illusions were, Alcina's garden remains as the image of a way of life which man can never wholly reject. He cannot reject it because it is so much a part of himself; it represented something reprehensible but profoundly enjoyable . . . . The garden teaches us that . . . the final illusion is to think life would be at all bearable without illusions." Ariosto's poem has been translated into English by Sir John Harington (1591), William Stewart Rose (1823–31), and Barbara Reynolds (1975–77); my translation, which began as an adaptation of Rose, is based on the Italian text of Lanfranco Caretti (1966).*

In the poems of the great sculptor, painter, and architect **Michelangelo Buonarotti** *(1475–1564), deeply influenced by the Christian Neoplatonism of Ficino and his circle, only earth's loveliness—as in the madrigal written to Tomaso de' Cavalieri and the sonnet "No Mortal Object Did These Eyes Behold" ("Non vider gli occhi miei cosa mortale")—can be the stair by which the soul may seek the ideal form that lies beyond the visible world.*

Finally, **Torquato Tasso** *(1544–95), son of the poet and humanist Bernardo Tasso, was born at Sorrento and studied at Padua, becoming attached to the Este court of Ferrara in 1565. Here in 1575 he finished his epic poem on the First Crusade, the* Gerusalemme Liberata (Jerusalem Delivered), *and in the nearby asylum of Santa Anna was confined for seven years during* *his ensuing madness. He died at the convent of San Onofrio in Rome shortly before the Pope was to have crowned him Petrarch's successor as poet laureate. Some of his finest lyrics evoke the vibrant beauty of the natural world; among these are the short poems "Listen! Waves Murmur" ("Ecco mormorar l'onde"; text in Kay) and "Hushed Are Forests and Streams" ("Tacciono i boschi e i fiumi"; text in Lind). The famous chorus ("O bella età de l'oro") from his pastoral drama of 1573,* Aminta, *in Leigh Hunt's nineteenth-century translation, evokes a happy time when no false honor interfered with the amorous pleasures licensed by nature; its final lines are a translation of lines 4–6 of Catullus 5 ("Let us live, my Lesbia, and love . . ."). His account of the island paradise of the sorceress Armida, by whom the knight Rinaldo is seduced in Canto XVI of* Jerusalem Delivered*—here quoted in the translation by Edward Fairfax (1600)—is clearly modeled on Ariosto's description of Alcina's isle, but Tasso's is a far falser paradise whose "nature" is insidiously counterfeited by art. "This is a hothouse, not a garden," Giamatti writes; "it only imitates, but not duplicates, the world of the 'secoli de l'oro' ['golden ages'], just as finally the garden life, based on false values, provides only the illusion of peace and satisfaction." Here "all distinctions between what is true and false, real and feigned, are collapsed. Beautiful and meretricious no longer can be told apart; and this landscape, of course, is simply a reflection of man's inner condition," the human nature to which outer nature again corresponds.*

Mountain Shepherd-Girls,
by Franco Sacchetti,
translated by Robert M. Torrance

"Mountain shepherd-girls, jauntily debonair,
whence do you come, so sprightly and so fair?

"What is the region you are natives of,
richer in fruits than any other place?
Creatures in my eyes you appear of Love, 5
so radiantly light adorns your face.
Of gold and silver you display no trace,
yet angels you appear, though rags you wear."

"We dwell in mountains, in a forest glade;
our habitation is a simple shed; 10
coming back every evening, we are laid
down by our parents in one little bed.
Nature in flowery fields has always fed
both us and the little flocks for which we care."

"Greatly I grieve that you have shown your grace 15
only to valley and to mountainside;
on all the earth, not even the loftiest place
would fail to honor you with joyful pride.
Oh, tell me, then: can you be satisfied
to live in forests on such meager fare?" 20

"Each of us is far happier to send
our flocks to pasture where bright sunlight falls
than any of you, invited to attend

some feast within the town's confining walls.
Give us, instead of luxury and balls, 25
songs, dances, and flowers in the open air!"

Ballad, were I but what I used to be,
a mountain shepherd-lad would I become,
and, without anyone observing me,
would make their lovely neighborhood my home: 30
calling "Whitey!" and "Martin!" I would roam,
following in their footsteps everywhere.

TWO SONNETS BY MATTEO BOIARDO,
TRANSLATED BY ROBERT M. TORRANCE

The Melody of Birds

The melody of birds from leaf to leaf,
through flowers a fragrantly delicious breeze,
and merrily leaping rivulets that please
gay hearts delightedly discarding grief:
these beauties heaven and nature richly gave 5
her who desires that all the world should love,
till with sweet sounds and scents the air above
is filled, and earth, and every dancing wave.

Where she directs her step, or turns her face,
so radiant a flame of love is seen 10
that winter days are bathed in warmth and light:
at her sweet glance, and by her laughter's grace,
grasses grow green and flowers blossom bright,
the sea is quiet, and the sky serene.

I've Seen the Sun

I've seen the sun, maned round with gold, rise higher
each morning from the waters of the sea,
his face illumined so resplendently
that far and wide he set the waves afire.
And I have seen the dew of early morn 5
open the rose with color so like flame
that, seeing it afar, one might exclaim
fire was upblazing from its bright green thorn.

I've seen the tender newborn grasses rise
in the first youthful days of early spring 10
with blades aspiring heavenward to press:
and seen a lissom lady, gathering
roses in meadows under sunny skies,
vanquish all these in grace and loveliness.

Dance Song, by Angelo Poliziano,
translated by John Heath-Stubbs

I found myself, young girls, while it was May,
In a green garden, at the break of day.

Lilies and violets blossomed all around
On the green turf, and flowers new-sprung and fair—
Yellow, and blue, and red, and white—were found; 5
Then I reached out my hand to pluck them there,
To decorate with them my own brown hair,
And with a wreath confine its disarray.
I found myself, young girls, while it was May,
In a green garden, at the break of day. 10

But when I'd plucked a border-full of blossom,
I saw with various hues the roses bloom,
And so I ran to fill my lap and bosom,
So soft and fragrantly they breathed perfume;
For thence I felt a sweet desire consume 15
My heart, where heavenly pleasure made its way.
I found myself, young girls, while it was May,
In a green garden, at the break of day.

I pondered to myself, "Of all these roses,
How can I tell among them which are fairest, 20
Which of them lately now its bud discloses,
Which are still fresh, and which to fading nearest?"
Then Love said, "Gather those which seem in rarest
And fullest blossom on the thorny spray—"
I found myself, young girls, while it was May, 25
In a green garden, at the break of day.

When first the rose's petals are outspread,
Most lovely and most welcome it appears;
Then weave it in a garland for your head,
In time, before its beauty disappears: 30
Even so, young girls, while its pride still it wears,
Gather the rose that makes your garden gay."
I found myself, young girls, while it was May,
In a green garden, at the break of day.

Alcina's Isle, by Ludovico Ariosto,
translated by Robert M. Torrance
Orlando Furioso

VI.19

Once the bird's flight had reached the farther side
of that vast space, nor once did swerve or bend,
weary of air, and circling far and wide,

down toward an isle he started to descend,
like that where virgin Arethusa's pride,
which caused her lover torments without end,
having long hidden from him, futilely
came by a strange dark path beneath the sea.[1]

20

Of all the isles his wings had overspread,
no lovelier or more pleasant had he seen,
nor would have seen a gentler, had he sped
searching the ends of earth, and all between.
Here the great bird once circled round, then led
Ruggiero downward toward forever green
wide cultivated plains, soft rolling hills,
shadowy banks, lush meadows, sparkling rills.

21

Enchanting groves of sweetly scented bay,
cedar and orange rich in fruit and flower,
palm intermeshed with myrtle brightly gay,
woven in varied shapes to form a bower,
gave cool refreshment from the blazing ray
of summer even at the noonday hour,
while in their branches birds upon the wing
fearlessly flew and ceaselessly did sing.

22

Mid roses darkly red and lilies white
forever freshened by a cooling breeze,
rabbits and hares frisked fearlessly in sight
of stately high-browed stags amid the trees;
no snares or hunters cause them any fright,
grazing and ruminating at their ease.
Fawns and goats leap about with nimble paces
abundantly throughout these rural places.

23

After the hippogriff approached the ground
so near, one leaping down might safely land,
swiftly dismounting there, Ruggiero found
that on the enameled grass he now did stand.
So that his steed might not abruptly bound
skyward, he grips the reins tight in his hand;
to a green myrtle near the surging brine
he ties it then, between a bay and pine.

24

Close to a place where babbled up a fount
girded by fertile palms and cedar trees,

[1]The nymph Arethusa, daughter of Oceanus, when pursued by the river Alphaeus (of Elis, in the Peloponnesus), was transformed into a spring by Diana (Greek Artemis), and flowed underground beneath the sea to escape her lover. His waters, however, mingled with hers, making her efforts vain; they emerged together at the fountain of Ortygia near Syracuse, in Sicily, to which Ariosto here compares Alcina's isle.

he sets his shield and, taking helmet down
from off his brow, his hands from gauntlets frees;
now toward the seashore, now upon the mount
he turns a face cooled by the nurturing breeze
that with exultant murmurings stirs each
tremulous treetop there of fir and beech. . . .

72

There on the threshold of the colonnade
lasciviously sporting damsels run,
and would have seemed still lovelier had they paid
due heed to honors unto ladies done.
In a green gown was each of them arrayed,
and garlands of fresh leaves crowned every one.
These with fair offerings and looks entice
Ruggiero into entering Paradise:

73

for thus one rightly might denote that place,
where I believe Love verily was born.
Here everything is dancing, sport, and grace,
here feasts all hours of every day adorn.
Frosty-haired thought here never shows his face,
or causes any heart to be forlorn:
dearth and disquiet never enter here,
where Plenty, with full horn, stands ever near.

74

April's unclouded graces here abound
with joyful looks and everlasting smile,
where youths and ladies throng: one by the fount
sings in melodiously delightful style;
shaded by trees some dance, or by a mount,
play, or do anything that is not vile;
far from the others, to a faithful friend
one reveals amorous sorrows without end.

75

Through the treetops of pine and fragrant bay,
of lofty beech and of thick shaggy fir,
infant Loves playfully flit round all day,
some reveling in triumphs they incur,
some taking aim at hearts that go astray,
or spreading nets for lovers prone to err:
some temper arrows in a stream, some hone
sharpened points on a swiftly whirling stone. . . .

VII.31

No single pleasure is omitted here
within this amatory precinct's wall.
Twice and thrice daily all of them appear
newly clothed, as occasion may befall.

Frequently banquets, always festive cheer
regale them: wrestling, joust, bath, pageant, ball.
By springs, or in a hillock's shade, each reads
amorous tales of bygone lovers' deeds.

32

O'er many a shadowy vale and pleasant hill
they go a-hunting for the timid hare;
keen-scented hounds with clamorous barking fill
glades, flushing pheasants from their thorny lair;
birdlime and traps for thrushes with great skill
they spread, where juniper perfumes the air:
one with his baited hook, and one with nets,
fish in impenetrable haunts upsets. . . .

TWO POEMS BY MICHELANGELO BUONAROTTI

Madrigal 109: Ravished by All,
translated by George Santayana

Ravished by all that to the eyes is fair,
Yet hungry for the joys that truly bless,
My soul can find no stair
To mount to heaven save earth's loveliness.
For from the stars above 5
Descends a glorious light
That lifts our longing to their highest height
And bears the name of love.
Nor is there aught can move
A gentle heart, or purge or make it wise, 10
But beauty and the starlight of his eyes.

No Mortal Object Did These Eyes Behold,
translated by William Wordsworth

No mortal object did these eyes behold
When first they met the placid light of thine,
And my Soul felt her destiny divine,
And hope of endless peace in me grew bold:
Heaven-born, the Soul a heavenward course must hold; 5
Beyond the visible world she soars to seek
(For what delights the sense is false and weak)
Ideal Form, the universal mold.
The wise man, I affirm, can find no rest
In that which perishes: nor will he lend 10
His heart to aught which doth on time depend.
'Tis sense, unbridled will, and not true love,
That kills the soul: love betters what is best,
Even here below, but more in heaven above.

FROM THE WORKS OF TORQUATO TASSO

Listen! Waves Murmur, translated by Robert M. Torrance

Listen! waves murmur, leaves
quiver with morning's breeze,
shrubs quiver, birds in love
sing, on green boughs above,
a song that beguiles 5
as all the east smiles:
look! already the sea
mirrors dawn splendidly,
skies are growing serene,
frost-pearls color the green 10
fields white, high mountaintops gold.
Aurora, O gorgeously bold!
The breeze heralds you, you the breeze,
setting parched hearts at ease.

Hushed Are Forests and Streams, translated by Robert M. Torrance

Hushed are forests and streams,
hushed the sea without waves;
stilled winds sleep in their caves,
and the moon's pale light
heightens the somber silence of night: 5
let us likewise conceal
the sweet love we feel—
speechless, breathless love's cries,
mute be our kisses, muted my sighs.

The Golden Age, translated by Leigh Hunt

Aminta, *IV.ix*

O lovely age of gold!
Not that the rivers rolled
With milk, or that the woods wept honeydew;
Not that the ready ground
Produced without a wound, 5
Or the mild serpent had no tooth that slew;
Not that a cloudless blue
For ever was in sight,
Or that the heaven, which burns
And now is cold by turns, 10
Looked out in glad and everlasting light;
No, nor that even the insolent ships from far
Brought war to no new lands nor riches worse than war:

But solely that that vain
And breath-invented pain, 15
That idol of mistake, that worshiped cheat,
That Honor—since so called
By vulgar minds appalled—
Played not the tyrant with our nature yet.
It had not come to fret 20
The sweet and happy fold
Of gentle humankind;
Nor did its hard law bind
Souls nursed in freedom; but that law of gold,
That glad and golden law, all free, all fitted, 25
Which Nature's own hand wrote—What pleases is permitted.

Then among streams and flowers
The little wingèd powers
Went singing carols without torch or bow;
The nymphs and shepherds sat 30
Mingling with innocent chat
Sports and low whispers; and with whispers low,
Kisses that would not go.
The maiden, budding o'er,
Kept not her bloom uneyed, 35
Which now a veil must hide,
Nor the crisp apples which her bosom bore;
And oftentimes, in river or in lake,
The lover and his love their merry bath would take.

'Twas thou, thou, Honor, first 40
That didst deny our thirst
Its drink, and on the fount thy covering set;
Thou bad'st kind eyes withdraw
Into constrainèd awe,
And keep the secret for their tears to wet; 45
Thou gather'dst in a net
The tresses from the air,
And mad'st the sports and plays
Turn all to sullen ways,
And putt'st on speech a rein, in steps a care. 50
Thy work it is—thou shade, that wilt not move—
That what was once the gift is now the theft of love.

Our sorrows and our pains,
These are thy noble gains.
But, O, thou Love's and Nature's masterer, 55
Thou conqueror of the crowned,
What dost thou on this ground,
Too small a circle for thy mighty sphere?
Go, and make slumber dear

To the renowned and high; 60
We here, a lowly race,
Can live without thy grace,
After the uses of antiquity.
Go, let us love; since years
No truce allow, and life soon disappears; 65
Go, let us love; the daylight dies, is born;
But unto us the light
Dies once for all; and sleep brings on eternal night.

Armida's Garden, translated by Edward Fairfax
Jerusalem Delivered *XVI*

9

When they had passèd all those troubled ways,
The garden sweet spread forth her green to show,
The moving crystal from the fountain plays,
Fair trees, high plants, strange herbs, and flow'rets new,
Sunshiny hills, dales hid from Phoebus' rays,
Groves, arbors, mossy caves, at once they view;
And that which beauty most, most wonder brought,
Nowhere appeared the art which all this wrought.

10

So with the rude the polished mingled was,
That natural seemed all, and every part
Nature would craft in counterfeiting pass,
And imitate her imitator art.
Mild was the air, the skies were clear as glass,
The trees no whirlwind felt nor tempest's smart,
But ere their fruit drop off the blossom comes;
This springs, that falls, that rip'neth, and this blooms.

11

The leaves upon the selfsame bough did hide,
Beside the young, the old and ripened fig;
Here fruit was green, there ripe with vermeil side,
The apples new and old grew on one twig;
The fruitful vine her arms spread high and wide,
That bended underneath their clusters big;
The grapcs were tender here, hard, young and sour,
There purple, ripe, and nectar sweet forth pour.

12

The joyous birds, hid under greenwood shade
Sung merry notes on every branch and bough;
The wind, that in the leaves and waters played,
With murmur sweet now sang, and whistled now;
Ceasèd the birds, the wind loud answer made,
And while they sung it rumbled soft and low:

Thus, were it hap or cunning, chance or art,
The wind in this strange music bore his part.

13

With parti-colored plumes and purple bill,
A wondrous bird among the rest there flew,
That in plain speech sung lovelays loud and shrill,
Her leden[2] was like human language true;
So much she talked, and with such wit and skill,
That strange it seemèd how much good she knew;
Her feathered fellows all stood hushed to hear,
Dumb was the wind, the waters silent were.—

14

The gently-budding rose (quoth she) behold,
The first scant peeping forth with virgin beams,
Half ope, half shut, her beauties doth up-fold
In their dear leaves, and less seen fairer seems,
And after spreads them forth more broad and bold,
Then languisheth and dies in last extremes:
For seems the same that deckèd bed and bow'r
Of many a lady late and paramour:

15

So in the passing of a day doth pass
The bud and blossom of the life of man,
Nor e'er doth flourish more, but like the grass
Cut down, becometh witherèd, pale, and wan;
O gather then the rose while time thou has,
Short is the day, done when it scant began;
Gather the rose of love while yet thou mayst,
Loving be loved, embracing be embraced.—

16

She ceased; and as approving all she spoke
The choir of birds their heav'nly tunes renew;
The turtles[3] sighed and sighs with kisses broke,
The fowls to shades unseen by pairs withdrew;
It seemed the laurel chaste and stubborn oak,
And all the gentle trees on earth that grew,
It seemed the land, the sea, and heav'n above,
All breathed out fancy sweet and sighed out love. . . .

No White Choir of Naiads

Poetry of the Spanish Golden Age

The Golden Age (siglo de oro) *of Spain, from about the early sixteenth to the late seventeenth century, embraced such disparate poetic strands as the mellifluous Italianate rhythms introduced by the courtier-poet Garcilaso de la Vega, the intense mysticism of St. John of the Cross (San*

[2]"Language." [3]"Turtledoves."

Juan de la Cruz), and the manneristic conjunction of seemingly incompatible elements in Luis de Góngora. Despite the fervent otherworldly faith of its saints—Teresa, Ignatius Loyola, and Francis Xavier as well as John of the Cross—and the lowering shadow of a Counter Reformation intent on extirpating all heresy, interest in nature remained intense. Nature was understood, Otis H. Green observes (in his chapter "Three Aspects of Nature" in volume 2 of Spain and the Western Tradition *[1964]), not only as the creative power of God, called by the Scholastics* natura naturans *("nature naturing"); it was understood also as the created world, or* natura naturata *("nature natured"), and as the divine order by which the universe is ruled.*

*In the writings of Fray—"Friar" or "Brother"—**Luis de León** (1527?–91), classical and Christian dimensions of nature find unusually harmonious expression. A descendant of Jewish converts, Fray Luis joined the Augustinian order in 1544 and in 1561 gained a chair in theology at the University of Salamanca; from 1572 to 1576 he was imprisoned by the Inquisition for questioning the infallibility of the Latin Vulgate Bible and translating the Song of Songs and Book of Job into the vernacular—a practice smacking of Protestantism. He published biblical commentaries in Latin and two Spanish prose works,* La Perfecta Casada (The Perfect Bride) *and* De los Nombres de Cristo (On the Names of Christ), *but revealed his humanistic love of the classics by translating Pindar, Virgil, and Horace. In his poems (mostly written in the five-line* lira *stanza introduced by Garcilaso and used also by St. John of the Cross), first published forty years after his death by Francisco de Quevedo, the Horatian spirit frequently prevails. Thus in "Vida Retirada" ("The Secluded Life")—in the translation of Aubrey F. G. Bell (as "The Life Removed"), published in* Lyrics of Luis de León *(1928) and reprinted with variations in Eleanor Turnbull's* Ten Centuries of Spanish Poetry *(1955)—the poet seeks solitary refuge from the falsity and ambition of the world in the repose of hill, stream, and field. Despite what Pedro Salinas (in Turnbull) calls its "serene exterior, spiritual equilibrium and vision of a calm and tranquil existence," the effect of this and other poems of Fray Luis is far from static. On the contrary, he and his younger contemporary El Greco, Bell suggests, "are the poet and painter of motion. They present no immutable forms and figures, but seize the significant in the fugitive and have caught and rendered permanent the very evanescence of things," the continual change in which the essence of nature lies.*

*In the poetry of **Luis de Góngora** y Argote (1561–1627), who was born in Córdoba, studied at Salamanca, and took minor orders before his worldly temperament led him to the capital of Madrid and its literary feuds, this latent dynamism is a scarcely containable perpetual motion always vibrant but hardly ever serene. Both in the longer and later poems, with their abstruse diction, elaborate wordplay, and obscure mythological allusions, and in simpler poems in popular verse forms such as the* romance *and* letrilla*—from which, however, the convoluted conceits of full-fledged* gongorismo *are seldom wholly absent—for Góngora, as for Fray Luis, "the poetic refuge from the evils and vicissitudes of the Court was Nature," R. O. Jones writes in his edition of the* Poems of Góngora *(1966): "a highly idealized Nature (it goes without saying) deriving from the Renaissance pastoral tradition." Góngora, Jones suggests, contemplates nature in the long poems as "the abode and expression of permanence," in which he finds "an answer to the vicissitudes of life." In earlier poems—such as the* romance *of 1603, "In the Pinewoods of Júcar" ("En los Pinares de Júcar"; text in Jones and in J. M. Cohen's* Penguin Book of Spanish Verse *[1956])—vicissitude and motion, as in the whirl of mountain-girls "dancing / to the sound of streams running / through stones, wind through branches," are inseparable from life in communion with a natural setting whose denizens are "no white choir of Naiads," as in pastoral convention, "but hill-girls of Cuenca, / pride of the sierra."*

*Finally, the poetic works of the prodigiously prolific **Lope** Félix **de Vega** Carpio (1562–1635)—who was born in Madrid, sailed with the Invincible Armada in 1588, became a priest in 1614, and pursued a phenomenally successful career writing poetry of every description, above* *all plays, of which some 500 survive of a purported 1,800—are interspersed with vivid portrayals of peasant life and flashes of lyrical beauty in celebration of nature. Our first selection comes from Act I of* Fuenteovejuna, *a play of about 1613. The peasant girl Laurencia—who will later stir up the villagers of Fuenteovejuna ("Sheep Fountain") to overthrow the vicious Comendador who has subjected their women to rape and their men to flogging—evokes the simple pleasures of village life scorned by those who think, like the Comendador whose advances she spurns, "it's only in the city life is pleasant." Her down-to-earth portrayal of life in accord with nature sharply contrasts with the idealized fantasies of Italian pastoral and romance. The two lyrics that follow—"May Song" ("Maya"), from* El Robo de Dina *XXIII, published in 1638, and "Clover" ("Trébole"), from Act II of* Peribáñez y el Comendador de Ocaña, *a play of about 1610—speak for themselves.*

The Secluded Life, by Fray Luis de León,
translated by Aubrey F. G. Bell

How tranquil is the life
Of him who, shunning the vain world's uproar,
May follow, free from strife,
The hidden path, of yore
Chosen by the few who conned true wisdom's lore! 5

For he, with thoughts aloof,
By proud men's great estate is not oppressed,
Nor marvels at the roof
Of gold, built to attest
The Moor's skill, that on jasper pillars rests. 10

He heeds not though fame raise
His name afar on wings of rumor flung,
He cares not for the praise
Of cunning flatterer's tongue,
Nor for what truth sincere would leave unsung. 15

What boots it my content
That the vain voice of fame should favor me,
If in its service spent
I find myself to be
Vexed by dull care and gnawing misery? 20

O hill, O stream, O field,
O solitary refuge of delight,
Since my bark now must yield
To storm, your solace bright
I seek and flee this sea's tempestuous might. 25

Sleep broken by no fear
Be mine, and a day clear, serene, and free,
Shunning the look severe,
Lofty exceedingly,
Of him whom gold exalts or ancestry. 30

Me may the birds awake
With their sweet, unpremeditated song,
And those dark cares forsake
That e'er to him belong
Who lives not in his independence strong! 35

I so myself would live,
To enjoy the blessings that to heaven I owe,
Alone, contemplative,
And freely love forgo,
And never hope, fear, hate nor envy know. 40

Upon the steep hillside
An orchard I have made with my own hand,
That in the sweet springtide
All in fair flower doth stand
And promise sure of fruit shows through the land. 45

And, as though swift it strove
To see and to increase that loveliness,
From the clear ridge above
A stream pure, weariless,
Hurrying to reach that ground doth onward press. 50

And straightway in repose
Its course it winds there tree and tree between,
And ever as it goes
The earth decks with new green
And with gay wealth of flowers spreads the scene. 55

The air in gentle breeze
A myriad scents for my delight distils,
It moves among the trees
With a soft sound that fills
The mind, and thought of gold or scepter kills. 60

Treasure and gold be theirs
Who to a frail bark would entrust their life:
I envy not the cares
Of those whose fears are rife
When the north wind with the south wind is at strife. 65

In the storm's strain the mast
Groans, and clear day is turned to eyeless night,
While to the skies aghast
Rise wild cries of affright
And they enrich the sea in their despite. 70

But me may still suffice,
Rich only in meek peace, a humble fare;
And the wrought artifice

Be his of gold plate rare
Who dreads not o'er the raging sea to fare. 75

And while in misery
Others are pledged to fierce ambition's throng,
Afire insatiably
For power that stays not long,
May I in pleasant shade recite my song: 80

Yea, lying in the shade,
My brow with ivy and bay immortal crowned,
My ear attentive made
To the soft tuneful sound
Of zither touched by fingers' skill profound. 85

In the Pinewoods of Júcar, by Luis de Góngora,
translated by Robert M. Torrance

In the pinewoods of Júcar
mountain-girls I saw dancing
to the sound of streams running
through stones, wind through branches.
No white choir of Naiads 5
these, dwelling in waters,
nymphs worshiped in forests,
devotees of Diana,
but hill-girls of Cuenca,
pride of the sierra 10
whose foot two streams come kissing,
to kiss their slim ankles.
Holding white hands, together
they weave happy dances
in friendship, half fearful 15
trading partners might change it.
How the hill-girls are dancing!
How well they are dancing!

Their braided locks lavish
light on sun, gold on Arabs, 20
some entangled with flowers,
some with silver cords fastened.
Blue cloth, the sky's color
(if not hope's), they're wearing,
so bright, it eclipses 25
both emerald and sapphire.
Their feet, when their skirt-hems
momentarily flash them,
wear sandals, with mother-
of-pearl and snow sparkling. 30

Girls who, in their movements,
raise upward so chastely
tall columns of crystal
on delicate bases:
How the hill-girls are dancing! 35
How well they are dancing!

Between her white fingers
one, crunching black pebbles—
an ivory musician
the Muses would envy— 40
has silenced birds' chirping
and checked the stream's babble.
No rustling leaf hindered
the song that she sang there:

"Hill-girls of Cuenca 45
to pine-woods advance,
some to pick pine-nuts,
others to dance."

These beautiful hill-girls,
dancing and sharing 50
pine-nuts with each other,
though no pearls they're wearing,
rest now from exchanging
Love's barbs, shot by chance,
some to pick pine-nuts, 55
others to dance.

In between leafy boughs
where blind Love beseeches
Sun to give him new eyes
to see them more clearly, 60
there, on eyes strewn by sunbeams,
see sprightly girls prance,
some to pick pine-nuts,
others to dance.

FROM THE WORKS OF LOPE DE VEGA,
TRANSLATED BY ROBERT M. TORRANCE

A Spunky Village Girl's Life

Fuenteovejuna, *Act I*

Laurencia. Though a spring chick, I'm pretty tough, 215
his reverence will find, to chew!
Lordy, Pascuala, how much higher
I'd value, any blessèd day,
a side of bacon sizzling away
for breakfast on a blazing fire, 220

with a thick slab of hearty bread
from dough I'd kneaded for myself,
out of the jar on mother's shelf
sneaking a glass of wine, dark red;
and how much more, in mid-day's blaze, 225
watching plump veal bob up and down
in cabbage soup and twirl around
frothing melodious hymns of praise;
then, when a long hard day has seen
me laboring with all my might, 230
give me the amorous delight
of marrying ham to aubergine;
sucking grapes toward the end of day,
while supper simmers on the flame,
that straight from my own vineyard came— 235
from which may God keep frost away;
till, after on a tasty ration
of peppery stew at last I've fed,
happily off I go to bed
and "lead me not into temptation" 240
I pray to God with all my heart;
for all these importuning men
who whine of love time and again
practice the same deceiving art:
nothing more do we ever get 245
than grief in giving them delight:
a moment's pleasure in the night;
with dawn, long lingering regret.

May Song

In the month of May
just shortly past dawn
nightingales sing,
fields echo their song.

In the cool of the morning 5
just shortly past dawn
nightingales blanket
the poplars all round.
Fountains laugh merrily
tossing their pearls 10
at flowers near by them
already brimful.
Plants parade gaily
in glorious silks;
for their bright-colored garments 15

cost them not a cent.
Glorious tapestries
gladden the lawn:
nightingales sing,
fields echo their song. 20

May saunters in beauty,
refreshed by cool winds
of beautiful zephyrs
March passed on to him.
Rainshowers of April 25
brought flowers most fair,
so he wreathed with gay garlands
his red-colored hair.
Those already lovers
loved each other anew, 30
and those who were loveless
went seeking love too.
And when they saw May
just shortly past dawn,
nightingales sing, 35
fields echo their song.

Clover

Clover, oh Jesus, what fragrance!
Clover, oh Jesus, how sweet!

Clover for one newly wed
who loves her dear husband well;
or the maid penned up in a cell 5
to safeguard her virgin bed,
yet irresistibly led
astray by lovers' deceit.

Clover, oh Jesus, what fragrance!
Clover, oh Jesus, how sweet! 10

Clover for spinsters' lone lives
enamored of numberless men;
clover for widowed young wives
afire to marry again—
clothed in white garments outside 15
with red undergarments beneath.

Clover, oh Jesus, what fragrance!
Clover, oh Jesus, how sweet!

Admiring Nature's Beautiful Tableau

Poetry of the French Renaissance

The Italian Renaissance had roots in the free communes of the Middle Ages, and flourished at courts of princes such as the Medici in Florence, Este in Ferrara, and Sforza in Milan, and of humanist popes from Nicholas V to Leo X and Clement VII, the patrons of Raphael and Michelangelo. In France, where King Francis I, reigning from 1515 to 1547, continued his predecessors' efforts to dominate Italy—until the Treaty of Cambrai in 1529 acknowledged Spanish and Austrian hegemony—the beginnings of the Renaissance were inseparable from the king and his sister Marguerite de Navarre. Together they patronized artists and writers from Leonardo da Vinci and Benvenuto Cellini to Rabelais and the poet Clément Marot, and established the Collège de France, at the urging of the humanist Guillaume Budé, as a center of learning independent of the Scholastic-dominated University of Paris.

Petrarchan love poetry was a major influence here as elsewhere—Louise Labé in France, like Gaspara Stampa and Vittoria Colonna in Italy, gave powerful voice to a woman's passions—and in Maurice Scève's Délie *(1544) the Petrarchan impulse fuses with a Christian Platonism that affirms (in Poem II) the wondrous beauty of nature formed on the pattern of divine Ideas by God as* natura naturans:

Le Naturant par ses hautes Idées
Rendit de soi la Nature admirable . . .

But it was the younger poets of the "Pléiade," centering around ***Pierre de Ronsard*** *(1524–85), who brought French Renaissance poetry to a culmination. Born at his family château near Vendôme, Ronsard was a page in the household of Francis I from 1536 to 1540, and traveled to England, Scotland, and Flanders. With Joachim du Bellay and Jean-Antoine de Baïf, he formed the core of the Pléiade in the 1540s, published his first poems in the 1550s, and from 1560 to 1574—years when the religious wars in France reached a climax with the St. Bartholomew's Day Massacre of Protestants in 1572—was court poet to Charles IX, the vacillating son of Henri II and Catherine des Medici. His poems range from an abortive epic, the* Franciade, *to odes, hymns, elegies, poems on the troubles of his time, and love sonnets to a series of mistresses. Our first poem, the famous plea of 1553 "To His Mistress" Cassandra (*Odes *I.xvii), "Mignonne, allons voir si la rose," employs the age-old image of the rose to urge his mistress to "pluck youth in its prime" before age fades her beauty forever. The next two poems, also translated from Bernard Weinberg, ed.,* French Poetry of the Renaissance *(1954)—the ode of 1550 "To the Forest of Gastine" (*Odes *II.xv), and passages from an "Elegy" of 1584 (in* Sonnets pour Hélène II*)—proclaim love of the natural world with youthful rapture and with the retrospective contemplation of age.*

Our next selections, from La Sepmaine, ou la Création du Monde *(1578) by* ***Guillaume de Salluste, Sieur du Bartas*** *(1544–90), are very different. Du Bartas, born Guillaume Salustre, was a minor Huguenot noble by virtue of his father's purchase of the château of Bartas in Gascony. He studied law at Toulouse and served—despite revulsion at the wars of religion—the Bourbon King Henri de Navarre, future Henri IV of France. He published his early poems in 1574 in* La Muse Chrestiene (The Christian Muse), *and followed up his epic on the creation of the world,* La Sepmaine (The Week) *of 1578, with an unfinished* Seconde Sepmaine *(1584–1603), narrating later events of sacred history. (For texts, introduction, and commen-*

tary, see his three-volume Works, *ed. Urban Tigner Holmes, Jr., et al. [1935–40].) In 1587, King James VI of Scotland (later James I of England), who had translated du Bartas's early poem* L'Uranie (Urania)*—and whose* Lepanto *du Bartas had translated in turn—invited him to Scotland as the envoy of Henri de Navarre. In England, Spenser was among his admirers, and Sidney undertook an uncompleted translation (now lost) of* La Sepmaine. *The wool merchant Josuah Sylvester (1563–1618) translated* La Sepmaine *and its sequel as* The Divine Weeks and Works of Guillaume de Saluste Sieur Du Bartas *(1605–8); his translations were widely read and often reprinted in Protestant England, and probably influenced Milton's* Paradise Lost. *Sylvester's free translation in a proto-"metaphysical" style, mingling colloquialisms (" 'Sweetheart,' quoth she, and then she kisseth him . . .") with "quaint realistic digressions, and polysyllabic rhymes that seem to be jesting at sublimity," Tucker Brooke writes in* A Literary History of England *(1948), "has no dignity at all. Like his original, he is often tedious and sometimes flat, but as a whole his version still deserves to rank among the most readable of long poems and liveliest of seventeenth-century translations." The following excerpts from volume I of Susan Snyder's edition (1979) are du Bartas's/Sylvester's account—expanding as much on the versions of Hesiod (Chapter 8 above) and Ovid (Chapter 11) as on Genesis (Chapter 3)—of creation on the first day by God ("the Lord High Marshal") of the world out of Chaos, and their description of the nightingale, whose songs the poet has often heard, he tells us, "rapt with delight of their delicious airs."*

THREE POEMS BY PIERRE DE RONSARD, TRANSLATED BY ROBERT M. TORRANCE

To His Mistress

Mignonne, let us see if the rose
that first at break of day disclosed
fresh crimson garments to the sun,
has not relinquished now, with night,
its crimson-pleated robes and bright 5
hues shared, excepting you, by none.

Alas! Mignonne, behold how soon
with fallen robes the ground is strewn,
alas, alas! and beauty shorn.
"Stepmother" Nature is for sure, 10
when such a flower may endure
only till evening follows morn.

Hark, then, mignonne, to what I say
before years confiscate away
the freshness of unfaded hours: 15
hasten to pluck youth in its prime!
Old age shall quickly come, and time
wither your beauty like this flower's.

To the Forest of Gastine

Reclining at ease in the green
of your shadow, I would
sing you, just as the Greeks, Gastine,[4]
sang the Erymanthian wood.[5]

For I cannot malignantly hide (5)
from generations to be
how beneficently you provide
this lovely greenness to me.

Under your sheltering trees
my ravished mind you amuse, (10)
and thus I unfailingly please
my ever-responsive Muse.

Released from the canker of care
I set myself free,
lost in you, only aware (15)
of the book on my knee.

May your groves forever abound
with erotic brigades
of fauns and satyrs who hound
fearful nymphs with their raids! (20)

May the college of Muses now dwell
here, taking your name,
forever immune from the hell
of desecration by flame!

FROM *Elegy*

Now had the seventh year (six having passed)
neared the completion of its course at last
when—keeping love and passion far away—
in perfect liberty I passed each day
and, free of sorrows by which souls are worn, (5)
from evening tranquilly slept on till morn.
Master over myself, I strolled at leisure
where feet directed me, drawn on by pleasure,
having at hand—no better guides than these!—
Aristotle, Plato, wise Euripides, (10)
good silent hosts who never scowl or frown:
just as I pick them up, I put them down,
pleasant companions, helpful, kind, well-bred—
others, by chattering on, would daze my head!

[4]A forest in the Vendômois, near where Ronsard was born.
[5]A forest in the mountains of Erymanthus (Arcadia) where Hercules captured alive a wild boar, as the fourth of his twelve labors. (Weinberg)

Weary of reading, flowers then I view, 15
leaves, branches, stems, of many a brilliant hue—
varied forms intersecting in a maze
painted red, yellow, blue, a hundred ways—
insatiably unable to forgo
admiring Nature's beautiful tableau, 20
saying to flowers newly blossoming:
"This is near God, who fathoms everything,
far from the vulgar herd, and far from palace
hangers-on, artisans of fraud and malice."
Sometimes through savage woods I roamed alone, 25
on riverbanks with rushes overgrown;
sometimes through distant and deserted crags
or thickets, bushy habitat of stags.
Along meandering streams I've often gone
watching wave after wave glide rippling on, 30
one always rising as another sank;
and loved to fish, suspended on a bank,
exulting more, to vex with silent chase
the secret haven of the scaly race
(trembling with rapture when my taut line took 35
a trustfully snagged fish on baited hook),
than a great monarch puffing hard all day,
in hot pursuit, to bring a stag to bay. . . .

Once Vesper's[6] shadowy darkness veils our eyes,
fixed upon heaven, I contemplate the skies,
in which God indicates by luminous features 45
the lots and destinies of all His creatures.
For God, disdaining, as we mortal men
would do, to take up paper, ink, and pen,
bids constellations overhead foretell
good fortunes and catastrophes as well. 50
But mortals, doomed to die, weighed down by earth,
count heavenly characters of little worth. . . .

Now that we've reached, Hélène, the seventh year,
hold me no longer captivated here.
Reason has freed me, and your stringent ways:
Nature's command, at last, old age obeys. 74

FROM *THE DIVINE WEEKS AND WORKS* OF GUILLAUME DE SALLUSTE, SIEUR DU BARTAS, TRANSLATED BY JOSUAH SYLVESTER

FROM *The First Day of the First Week*

That first world (yet) was a most formless form,
A confused heap, a Chaos most difform,
A gulf of gulfs, a body ill compact,

[6]Latin *vesper* (Greek *hésperos*) was the evening star.

An ugly medley, where all difference lacked: 250
Where th'elements lay jumbled all together,
Where hot and cold were jarring each with either:
The blunt with sharp, the dank against the dry,
The hard with soft, the base against the high,
Bitter with sweet; and while this brawl did last, 255
The Earth in Heav'n, the Heav'n in Earth was placed:
Earth, Air, and Fire were with the Water mixed,
Water, Earth, Air, within the Fire were fixed,
Fire, Water, Earth did in the Air abide,
Air, Fire, and Water in the Earth did hide. 260
For yet th'immortal, mighty thunder-darter,
The Lord High Marshal, unto each his quarter
Had not assignèd: the celestial arks
Were not yet spangled with their fiery sparks,
As yet no flowers with odors Earth revived, 265
No scaly shoals yet in the waters dived,
Nor any birds with warbling harmony
Were borne as yet through the transparent sky.
All, all was void of beauty, rule, and light,
All without fashion, soul, and motion, quite. 270
Fire was no fire, the water was no water,
Air was no air, the earth no earthy matter;
Or if one could in such a world spy forth
The fire, the air, the waters, and the earth,
Th'earth was not firm, the fire was not hot, 275
Th'air was not light, the water coolèd not:
Briefly, suppose an earth, poor, naked, vain,
All void of verdure, without hill or plain,
A heav'n unhanged, unturning, untransparent,
Ungarnishèd, ungilt with stars apparent, 280
So may'st thou guess what heav'n and earth was that,
Where, in confusion reignèd such debate.
A heav'n and earth for my base style most fit,
Not as they were, but as they were not yet.
This was not then the world, 'twas but the matter, 285
The nursery whence it should issue after,
Or rather th'embryon that within a week
Was to be born; for that huge lump was like
The shapeless burthen in the mother's womb,
Which yet in time doth into fashion come: 290
Eyes, ears, and nose, mouth, fingers, hands, and feet,
And every member in proportion meet;
Round, large, and long, there of itself it thrives,
And (little-world) into the world arrives.
But that becomes (by Nature's set direction) 295
From foul and dead, to beauty, life, perfection.

But this dull heap of undigested stuff
Had doubtless never come to shape or proof,
Had not th'Almighty with his quick'ning breath
Blown life and spirit into this lump of death. 300

FROM *The Fifth Day of the First Week*

The spink,[7] the linnet, and the goldfinch fill 665
All the fresh air with their sweet warbles shrill.
But all this's nothing to the nightingale,
Breathing so sweetly from a breast so small
So many tunes, whose harmony excels
Our voice, our viols, and all music else. 670
Good Lord! how oft in a green oaken grove,
In the cool shadow have I stood, and strove
To marry mine immortal lays to theirs,
Rapt with delight of their delicious airs!
And (yet) methinks, in a thick thorn I hear 675
A nightingale to warble sweetly clear:
One while[8] she bears the bass, anon the tenor,
Anon the treble, then the counter-tenor,
Then all at once, (as it were) challenging
The rarest voices with herself to sing. 680
Thence thirty steps, amid the leafy sprays,
Another nightingale repeats her lays,
Just note for note, and adds some strain at last
That she had connèd all the winter past:
The first replies, and descants thereupon 685
With divine warbles of division,
Redoubling quavers; and so (turn by turn)
Alternately they sing away the morn,
So that the conquest in this curious strife
Doth often cost the one her voice and life; 690
Then the glad victor all the rest admire,
And after count her mistress of the choir.
At break of day, in a delicious song
She sets the gamut to a hundred young;
And whenas, fit for higher tunes she sees them, 695
Then learnedly she harder lessons gives them,
Which strain by strain they studiously recite,
And follow all their mistress' rules aright.
The Colchian pheasant,[9] and the partridge rare,
The lustful sparrow, and the fruitful stare,[10] 700
The chattering pie, the chastest turtledove,

[7]One or other of the finches; esp. the chaffinch. (OED) [8]"At one moment."
[9]On "Colchian," see note on Bernard Sylvestris's *Cosmographia* I.iii.464 in Chapter 15 above.
[10]A bird of the genus *Sturnus*: = starling. (OED)

The grizzle coist,[11] the thrush that grapes doth love,
The little gnat-sap, worthy princes' boards,
And the green parrot, feigner of our words,
Wait on the Phoenix, and admire her tunes, 705
And gaze themselves in her blue-golden plumes.

As If the World Were Born Anew

Poetry of the English Renaissance

More than any other poet of his time, Chaucer—in Troilus and Criseyde *and* The Canterbury Tales*—assimilated the humane responsiveness to the varied world of nature and man that had made Petrarch and Boccaccio harbingers of a new age; when he died in 1400, England like Italy might well have seemed on the threshold of Renaissance. But fifteenth-century England was mired in the Hundred Years' War with France, followed by the dynastic Wars of the Roses, which ended with the victory of Henry VII over Richard III in 1485. So rapidly had the English language been changing during this time that, unlike Petrarch and Boccaccio (whom Italian writers a hundred fifty years afterward viewed as paradigms of literary style), Chaucer seemed to later English poets such as Sir Philip Sidney to have lived in a "misty time" far removed from their own and to have suffered from faults "fit to be forgiven in so reverent antiquity." The Renaissance that came to England in the stormy reign of Henry VIII (1509–47) was thus both more cut off than in Italy from its native roots and more exclusively the product, in its origins, of a royal court (Henry VIII was a talented poet and musician). Moreover, the turmoil of the Reformation—which cost the leading English humanist, Sir Thomas More, and many others, their heads—further restricted the impact of this courtly phase of the English Renaissance.*

Even so, under the influence of Petrarch and other Italians, a number of poets, notably Sir Thomas Wyatt and ***Henry Howard, earl of Surrey*** *(1517?–47), launched the renewal of English poetry that would make the sixteenth and seventeenth centuries its age of glory. Surrey, who was of royal blood on both his mother's and father's side, enjoyed a humanistic education and residence in the French court in his youth; after a tumultuous career in Henry's service, with repeated imprisonments for violent quarrels, he was beheaded for treason two weeks before Henry's death. His translation into blank verse—a form with a momentous future—of two books of Virgil's* Aeneid *was published, along with other poems, in Richard Tottel's miscellany,* Songes and Sonettes, *of 1559; the sonnet below, "Description of Spring," is adapted from Petrarch's* Canzoniere *CCCX ("Zefiro torna, e 'l bel tempo rimena").*

But it was under the reign (1558–1603) of Henry's younger daughter, Elizabeth—following the troubled interlude of her sister, "Bloody Mary"—that the English Renaissance came to fulfillment, as the courtly humanist tradition merged, in both drama and lyric, with popular and native strands. In the rich poetic flowering of this time, major poets and many who remain anonymous or obscure wrote poems that retain their freshness centuries later, and the beauties of nature are among their foremost themes. These poems were frequently first published in miscellanies of verse, beginning with Tottel's; thus the anonymous ***"In Midst of Woods or Pleasant Grove"*** *appeared in John Mundy's* Songs and Psalms *of 1594, and the four poems that follow*

[11] "Wood-pigeon." (OED)

are from the most important Elizabethan collection, Englands Helicon *of 1600. The first, "As It Fell upon a Day," by* ***Richard Barnfield*** *(1574–1627), is shortened in* Englands Helicon *(where it is attributed to "Ignoto") from the versions published earlier in* Poems in Divers Humors *of 1598 and in* The Passionate Pilgrim *of 1599, where it was attributed to Shakespeare. In the "Palinode" of the Catholic poet* ***Edmund Bolton****, as in many other medieval and Renaissance poems, the melting of snow and withering of flowers are emblems of the transience of human life and of the glories of the world.*

Christopher Marlowe *(1564–93) was the author of* Tamburlaine, The Jew of Malta, Edward II, Doctor Faustus, *and the uncompleted erotic epic* Hero and Leander. *In "The Passionate Shepherd to His Love" (also attributed to Shakespeare in* The Passionate Pilgrim*), he issues an appealing invitation—free of the heavier baggage of pastoral—to the pleasures of love among "valleys, groves, hills, and fields." Of many responses, mainly humorous, to this plea, the best known is that of the courtier poet* ***Sir Walter Ralegh*** *(ca. 1552–1618), whose adventurous life included expeditions up the Orinoco River in Guiana and against the Spanish fleet at Cádiz, and ended with execution on charges of treason against James I. "The Nymph's Reply to the Shepherd"—attributed to "Ignoto" ("Unknown") in* Englands Helicon *but said in Izaak Walton's* Compleat Angler *of 1653 to have been "made by Sir Walter Ralegh in his younger days"—rejects the amorous shepherd's plea in a world where flowers do fade and youth does not last. See also the poem on this theme by John Donne, quoted in Chapter XII of Walton's* Compleat Angler *(Chapter 19 below).*

Thomas Campion *(1567–1620), with the musician Philip Rosseter, published in 1601* A Book of Airs, *for which he wrote all the lyrics and half the music; this was followed by* Two Books of Airs *(ca. 1613) and* The Third and Fourth Books of Airs *(ca. 1617), in which both lyrics and music are his. In "The Peaceful Western Wind," he evokes the revival of love in spring, a world "born anew" from which he alone is excluded by his mistress's rejection.*

Finally, ***Sir John Davies*** *(1569–1626), the author of a philosophical poem of 1599,* Nosce Teipsum (Know Thyself), *was a prominent lawyer who under James I became Speaker of the Irish Parliament and was named Lord Chief Justice of England shortly before his death. Our selections entitled "The Dance of Love" are from the 917-line* Orchestra, or a Poeme of Dauncing *(1596, revised 1622), written in rhyme royal. Suitor Antinoüs, expounding a Pythagorean/Neoplatonic vision of nature, tells "chaste Penelope, Ulysses' queen" that dancing arose from the ordering by Love, "nature's mighty king," of cosmos from chaos; hence dancing, time's twin, "is love's proper exercise." This poem is "as truly 'nature poetry' for the Elizabethans," C. S. Lewis remarks in* English Literature in the Sixteenth Century *(1954), "as Wordsworth was for the Romantics." Love concludes the passage given here by instructing a disordered rabble in the harmony that permeates the dancing universe, through whose movements*

> Kind nature first doth cause all things to love;
> Love makes them dance, and in just order move.

Orchestra *and many other poems are included in* Poetry of the English Renaissance 1509–1660, *ed. J. William Hebel and Hoyt H. Hudson (1929). Among other anthologies are* English Renaissance Poetry, *ed. John Williams (1963);* The Penguin Book of Elizabethan Verse, *ed. Edward Lucie-Smith (1965); and* The Penguin Book of Renaissance Verse 1509–1659, *ed. H. R. Woudhuysen (1993). See also* Englands Helicon, *ed. Hugh Macdonald (1962), and* The Poems of Sir John Davies, *ed. Robert Krueger (1975).*

Description of Spring, Wherein Each Thing Renews Save Only the Lover
by Henry Howard, Earl of Surrey

The soote[12] season that bud and bloom forth brings
With green hath clad the hill and eke the vale,
The nightingale with feathers new she sings,
The turtle to her make[13] hath told her tale.
Summer is come, for every spray now springs, 5
The hart hath hung his old head on the pale,[14]
The buck in brake his winter coat he flings,
The fishes float with new repairèd scale,
The adder all her slough away she slings,
The swift swallow pursueth the flyës smale, 10
The busy bee her honey now she mings,[15]—
Winter is worn, that was the flowers' bale:
And thus I see, among these pleasant things
Each care decays—and yet my sorrow springs.

In Midst of Woods or Pleasant Grove
(Anonymous)

In midst of woods or pleasant grove
Where all sweet birds do sing,
Methought I heard so rare a sound,
Which made the heavens to ring.
The charm was good, the noise full sweet, 5
Each bird did play his part;
And I admired to hear the same;
Joy sprang into my heart.

The blackbird made the sweetest sound,
Whose tunes did far excel, 10
Full pleasantly and most profound
Was all things placèd well.
Thy pretty tunes, mine own sweet bird,
Done with so good a grace,
Extols thy name, prefers the same, 15
Abroad in every place.

Thy music grave, bedeckèd well
With sundry points of skill,
Bewrays thy knowledge excellent,
Engrafted in thy will. 20
My tongue shall speak, my pen shall write,
In praise of thee to tell.
The sweetest bird that ever was,
In friendly sort, farewell.

[12]"Sweet." [13]"The turtledove to her mate." [14]"Paling," "fence."
[15]"Mingles," "mixes" (or possibly "remembers").

As It Fell upon a Day,
by Richard Barnfield

As it fell upon a day,
In the merry month of May,
Sitting in a pleasant shade,
Which a grove of myrtles made,
Beasts did leap, and birds did sing, 5
Trees did grow, and plants did spring.
Every thing did banish moan,
Save the nightingale alone.
She, poor bird, as all forlorn,
Leaned her breast against a thorn, 10
And there sung the dolefull'st ditty,
That to hear it was great pity.
Fie, fie, fie, now would she cry,
Teru, Teru, by and by,
That, to hear her so complain, 15
Scarce I could from tears refrain.
For her griefs so lively shown
Made me think upon mine own.
Ah (thought I) thou mourn'st in vain,
None takes pity on thy pain. 20
Senseless trees, they cannot hear thee,
Ruthless beasts, they will not cheer thee.
King *Pandíon,* he is dead,[16]
All thy friends are lapped in lead.
All thy fellow birds do sing, 25
Careless of thy sorrowing.
Even so, poor bird, like thee,
None alive will pity me.

A Palinode,
by Edmund Bolton

As withereth the primrose by the river,
As fadeth summer's sun from gliding fountains,
As vanisheth the light-blown bubble ever,
As melteth snow upon the mossy mountains:
So melts, so vanisheth, so fades, so withers 5
The rose, the shine, the bubble, and the snow,
Of praise, pomp, glory, joy, which short life gathers,
Fair praise, vain pomp, sweet glory, brittle joy.
The withered primrose by the mourning river,
The faded summer's sun from weeping fountains, 10

[16]Pandíon was the father of Philomela and Procne, who were transformed into the nightingale and the swallow, respectively, to escape the cruelty of Tereus (whose name is suggested by the call "Teru"); see note to *The Vigil of Venus*, line 88, in Chapter 11 above.

The light-blown bubble, vanishèd for ever,
The molten snow upon the naked mountains,
Are emblems that the treasures we uplay[17]
Soon wither, vanish, fade, and melt away.

For as the snow, whose lawn did overspread 15
Th'ambitious hills which giant-like did threat
To pierce the heaven with their aspiring head,
Naked and bare doth leave the craggy seat;
Whenas the bubble, which did empty fly
The dalliance of the undiscernèd wind 20
On whose calm rolling waves it did rely,
Hath shipwreck made where it did dalliance find;
And when the sunshine which dissolved the snow,
Colored the bubble with a pleasant vary,[18]
And made the rathe[19] and timely primrose grow, 25
Swarth[20] clouds withdrawn, which longer time do tarry:
Oh, what is praise, pomp, glory, joy, but so
As shine by fountains, bubbles, flowers, or snow?

The Passionate Shepherd to His Love, by Christopher Marlowe

Come live with me and be my love,
And we will all the pleasures prove
That valleys, groves, hills, and fields,
Woods, or steepy mountain yields.

And we will sit upon the rocks 5
Seeing the shepherds feed their flocks,
By shallow rivers to whose falls
Melodious birds sing madrigals.

And I will make thee beds of roses
And a thousand fragrant posies, 10
A cap of flowers, and a kirtle
Embroidered all with leaves of myrtle;

A gown made of the finest wool
Which from our pretty lambs we pull;
Fair linèd slippers for the cold, 15
With buckles of the purest gold;

A belt of straw and ivy buds,
With coral clasps and amber studs;
And if these pleasures may thee move,
Come live with me, and be my love. 20

[17] "Lay up." [18] "Variation," "iridescence."
[19] "Early" (cf. Milton's *Lycidas* [1638] 142: "the rathe primrose that forsaken dies").
[20] "Swarthy," "dark."

The shepherds' swains shall dance and sing
For thy delight each May morning:
If these delights thy mind may move,
Then live with me and be my love.

The Nymph's Reply to the Shepherd, by Sir Walter Ralegh

If all the world and love were young,
And truth in every shepherd's tongue,
These pretty pleasures might me move
To live with thee and be thy love.

Time drives the flocks from field to field 5
When rivers rage and rocks grow cold,
And Philomel becometh dumb;
The rest complains of cares to come.

The flowers do fade, and wanton fields
To wayward winter reckoning yields; 10
A honey tongue, a heart of gall,
Is fancy's spring, but sorrow's fall.

Thy gowns, thy shoes, thy beds of roses,
Thy cap, thy kirtle, and thy posies
Soon break, soon wither, soon forgotten,— 15
In folly ripe, in reason rotten.

Thy belt of straw and ivy buds,
Thy coral clasps and amber studs,
All these in me no means can move
To come to thee and be thy love. 20

But could youth last and love still breed,
Had joys no date nor age no need,
Then those delights my mind might move
To live with thee and be thy love.

The Peaceful Western Wind, by Thomas Campion

The peaceful western wind
The winter storms hath tamed,
And Nature in each kind
The kind heat hath inflamed:
The forward buds so sweetly breathe 5
Out of their earthy bowers,
That heaven, which views their pomp beneath,
Would fain be decked with flowers.

See how the morning smiles
On her bright eastern hill, 10
And with soft steps beguiles
Them that lie slumbering still.
The music-loving birds are come
From cliffs and rocks unknown,
To see the trees and briars bloom 15
That late were overflown.

What Saturn did destroy,
Love's Queen revives again;
And now her naked boy
Doth in the fields remain, 20
Where he such pleasing change doth view
In every living thing,
As if the world were born anew
To gratify the spring.

If all things life present, 25
Why die my comforts then?
Why suffers my content?
Am I the worst of men?
O, Beauty, be not thou accused
Too justly in this case: 30
Unkindly if true love be used,
'Twill yield thee little grace.

The Dance of Love,
by John Davies

Orchestra, or a Poeme of Dauncing

Dancing, bright lady, then began to be
When the first seeds whereof the world did spring,
The fire, air, earth, and water did agree 115
By Love's persuasion, nature's mighty king,
To leave their first disordered combating,
And in a dance such measure to observe
As all the world their motion should preserve.

Since when they still are carried in a round, 120
And changing come one in another's place;
Yet do they neither mingle nor confound,
But every one doth keep the bounded space
Wherein the dance doth bid it turn or trace.
This wondrous miracle did Love devise, 125
For dancing is love's proper exercise.

Like this he framed the gods' eternal bower,
And of a shapeless and confusèd mass,

By his through-piercing and digesting power,
The turning vault of heaven formèd was, 130
Whose starry wheels he hath so made to pass
  As that their movings do a music frame,
  And they themselves still dance unto the same. . . .

How justly then is dancing termèd new,
Which with the world in point of time began?
Yea, Time itself, whose birth Jove never knew, 150
And which is far more ancient than the sun,
Had not one moment of his age outrun,
  When out leaped Dancing from the heap of things
  And lightly rode upon his nimble wings.

Reason hath both their pictures in her treasure: 155
Where Time the measure of all moving is,
And Dancing is a moving in all measure.
Now, if you do resemble that to this,
And think both one, I think you think amiss;
  But if you judge them twins, together got, 160
  And Time first born, your judgment erreth not.

Thus doth it equal age with Age enjoy,
And yet in lusty youth forever flowers;
Like Love, his sire, whom painters make a boy,
Yet is he eldest of the heav'nly powers; 165
Or like his brother Time, whose wingèd hours,
  Going and coming, will not let him die,
  But still preserve him in his infancy. . . .

When Love had shaped this world, this great fair wight,[21] 190
That all wights else in this wide womb contains,
And had instructed it to dance aright
A thousand measures, with a thousand strains,
Which it should practise with delightful pains
  Until that fatal instant should revolve, 195
  When all to nothing should again resolve;

The comely order and proportion fair
On every side did please his wand'ring eye;
Till, glancing through the thin transparent air,
A rude disordered rout he did espy 200
Of men and women, that most spitefully
  Did one another throng and crowd so sore
  That his kind eye, in pity, wept therefor.

And swifter than the lightning down he came,
Another shapeless chaos to digest; 205
He will begin another world to frame,

[21]"Person," "being."

For Love, till all be well, will never rest.
Then with such words as cannot be expressed
He cuts the troops, that all asunder fling,
And ere they wist[22] he casts them in a ring. 210

Then did he rarefy the element,
And in the center of the ring appear;
The beams that from his forehead shining went
Begot a horror and religious fear
In all the souls that round about him were, 215
Which in their ears attentiveness procures,
While he, with such like sounds, their minds allures:

How doth Confusion's mother, headlong Chance,
Put Reason's noble squadron to the rout?
Or how should you, that have the governance 220
Of Nature's children, heaven and earth throughout,
Prescribe them rules, and live yourselves without?
Why should your fellowship a trouble be,
Since man's chief pleasure is society?

If sense hath not yet taught you, learn of me 225
A comely moderation and discreet,
That your assemblies may well ordered be;
When my uniting power shall make you meet,
With heav'nly tunes it shall be tempered sweet,
And be the model of the world's great frame, 230
And you, earth's children, dancing shall it name.

Behold the world, how it is whirlèd round!
And for it is so whirled, is namèd so;
In whose large volume many rules are found
Of this new art, which it doth fairly show. 235
For your quick eyes in wand'ring to and fro,
From east to west, on no one thing can glance,
But, if you mark it well, it seems to dance.

First you see fixed in this huge mirror blue
Of trembling lights a number numberless; 240
Fixed, they are named, but with a name untrue;
For they are moved and in a dance express
The great long year that doth contain no less
Than threescore hundreds of those years in all,
Which the sun makes with his course natural. 245

What if to you these sparks disordered seem,
As if by chance they had been scattered there?
The gods a solemn measure do it deem

[22]"Know."

And see a just proportion everywhere,
And know the points whence first their movings were, 250
    To which first points, when all return again,
    The axletree of heav'n shall break in twain.

Under that spangled sky five wand'ring flames,[23]
Besides the king of day and queen of night,
Are wheeled around, all in their sundry frames, 255
And all in sundry measures do delight;
Yet altogether keep no measure right;
    For by itself each doth itself advance,
    And by itself each doth a galliard dance. . . .

Only the earth doth stand forever still:
Her rocks remove not, nor her mountains meet,
Although some wits enriched with learning's skill[24]
Say heav'n stands firm and that the earth doth fleet,
And swiftly turneth underneath their feet; 355
    Yet, though the earth is ever steadfast seen,
    On her broad breast hath dancing ever been. . . .

See how those flowers, that have sweet beauty too,
The only jewels that the earth doth wear 380
When the young sun in bravery her doth woo,
As oft as they the whistling wind do hear,
Do wave their tender bodies here and there;
    And though their dance no perfect measure is,
    Yet oftentimes their music makes them kiss. 385

What makes the vine about the elm to dance
With turnings, windings, and embracements round?
What makes the lodestone to the north advance
His subtile point, as if from thence he found
His chief attractive virtue to redound? 390
    Kind nature first doth cause all things to love;
    Love makes them dance, and in just order move. . . .

Eterne in Mutabilitie

The Faerie Queene, by Edmund Spenser

Even in the early Tudor period, questions about nature in relation to man and to God—questions that later occupied English poets such as Davies and Spenser and thinkers such as Dee, Fludd, and Hooker (Chapter 16 above)—were already of intense concern. Thus in Henry Medwall's morality play Nature *(ca. 1495; in* The Plays of Henry Medwall, *ed. Alan H. Nelson [1980]), Nature "syttyth down and sayth":*

[23]The planets Mercury, Venus, Mars, Jupiter, and Saturn. [24]Copernicus and his followers.

Thalmyghty God that made eche creature
As well in heven as other place erthly
By hys wyse ordynaunce hath purveyd[25] me, Nature,
To be as mynyster under hym immedyately
For thencheson[26] that I shold perpetually
His creatures in suche degre mayntayne
As yt hath pleased hys grace for theym to ordeyne. . . .

Who taught the cok hys watche howres to observe
And syng of corage[27] with shryll throte on hye?
Who taught the pellycan her tender hart to carve
For she nolde[28] suffer her byrdys to dye?
Who taught the nyghtyngall to recorde[29] besyly
Her strange entunys[30] in sylence of the nyght?
Certes I, Nature, and none other wyght.

And John Rastell, printer of his own and Medwall's plays and of the writings of Sir Thomas More, in his interlude Four Elements *of around 1520 (in* Three Rastell Plays, *ed. Richard Axton [1979]), makes* Natura Naturata *address philosophical concerns anticipating Spenser's:*

The hye, myghty, most excellent of all,
The fountayn of goodnes, verteu, and connyng,[31]
Which is eterne of power most potencyall,
The perfeccyon and furst cause of every thynge—
I meane that only hye nature naturynge[32]—
Lo, he by his goodnes hath ordeynyd and create
Me here his mynyster, callyd Nature Naturate. . . .

For though the forme and facyon of any thyng
That is a corporall body be distroyed,
Yet every matter remaynyth in his beynge,
Whereof it was furst made and formyd;
For corrupcyon of a body commyxyd
Ys but the resolucyon by tyme and space
Of every element to his owne place.

Their inheritor Edmund Spenser (1552?–99) was born in London and educated at the Merchant Taylors' School and at Pembroke College, Cambridge. In the earl of Leicester's household he met Sir Philip Sidney, author of the sonnet sequence Astrophil and Stella *and the pastoral novel* The Countess of Pembroke's Arcadia, *and the central figure of the courtly Elizabethan Renaissance. In 1580 he left for Ireland as secretary to Lord Grey de Wilton, and there he remained for most of the rest of his life.* A Veue of the Present State of Ireland, *not published until 1633, reveals an English landowner's prejudice—Spenser had acquired an estate in County Cork—against the Irish peasantry as little better than savages. In 1598 he and his family fled Ireland after his castle was burned in an insurrection, and in the following year he died in London and was buried in Westminster Abbey. His major works include a sequence of pastoral elegies,*

[25]"Appointed." [26]"For the reason" or "purpose." [27]"From the heart."
[28]"Would not." On the pelican, see the twelfth-century Latin bestiary in Chapter 13 above.
[29]"Sing." [30]"Songs." [31]"Learning."
[32]*Natura naturans,* Nature as creating power, often identified with God (see introduction to Poetry of the Spanish Golden Age above). Axton notes that "Hawes (*Pastime of Pleasure,* 1509, xliv, 216) plays on the grammar of creation as natural propagation: 'Nature . . . whyche naturynge hath tought Naturately right naturate to make.'"

The Shepheardes Calender *(1579), the love sonnets of the* Amoretti *and the marriage hymn* Epithalamium *(1595), the* Fowre Hymnes *(1596), and the uncompleted allegorical verse romance* The Faerie Queene, *of which three of twelve projected books were published with a letter to Sir Walter Ralegh in 1590, and Books IV–VI in 1596; the "Two Cantos of Mutabilitie," said to be cantos VI and VII (and two stanzas of an eighth) from a fragmentary seventh book,* The Legend of Constancie, *were first published in 1609.*

Already in the playful Muiopotmos *of 1590, the rivalry of nature and art is a theme as the butterfly Clarion (fated for a spider's web) enters a garden where*

lauish Nature in her best attire,
Powres forth sweete odors, and alluring sights;
And Arte with her contending, doth aspire
T'excell the naturall, with made delights.

The conflict becomes crucial in The Faerie Queene *when Guyon, the knight of temperance, enters Acrasia's Bower of Bliss (II.XII) and—ignoring an alluring appeal to*

Gather therefore the Rose, whilest yet is prime,
For soone comes age, that will her pride deflowre—

demolishes the Bower and restores to human shape all (except the swinishly recalcitrant Grill) whom Acrasia's seductions had turned to beasts. The Bower represents, C. S. Lewis writes in The Allegory of Love *(1936), "artifice, sterility, death" as opposed to "nature, fecundity, life" in the Garden of Adonis (III.VI), from which our first selection (in the text of J. C. Smith [1909]) is taken. This "ioyous Paradize" to which Venus leads the foundling Amoretta "to be vpbrought in goodly womanhed," in contrast to the "perfect Maydenhed" in which Diana will rear Amoretta's twin sister Belphoebe, is "so faire a place, as Nature can devize," "the first seminarie / Of all things, that are borne to liue and die."*

Spenser elsewhere affirms a Christian Platonist view of the world according to which, he writes in "An Hymne of Heavenly Beavtie" (from Fowre Hymnes*),*

it plainely may appeare,
That still as euery thing doth vpward tend,
And further is from earth, so still more cleare
And faire it growes, till to his perfect end
Of purest beautie, it at last ascend.

Yet though always repelled, as Lewis remarks, by Nature when opposed not to the artificial and spurious (as in the Bower of Bliss) but to the spiritual or the civil—Nature as "the brutal, the unimproved, the inchoate"—Spenser never falters in his allegiance to Nature in the Aristotelian sense of "unimpeded growth from within to perfection, neither checked by accident nor sophisticated by art." Here is not only "continuall spring" but continual harvest, "both meeting at one time" in a perpetual cycle of life and death, by which the world is repeatedly renewed as forms forever change while substance remains the same. And here sexual love—embodied by Venus's pleasure in Adonis, "the Father of all formes," who, though mortal, is, like everything in nature, "eterne in Mutabilitie"—is exalted, as "franckly each paramour his leman knows." The essential difference between the Bower of Bliss and the Garden of Adonis, Giamatti writes, "is the fact of Time," which "the Bower pretends to do without and . . . the Garden embodies"; Venus's Garden repudiates the denial, in Acrasia's Bower, "of the the fact of death as an element in life; its refusal to face the necessity for decay in the midst of growth." In this place, "at once the garden of earthly delights and the key to eternal life," Leonard Barkan writes in Nature's Work of Art *(1975), "the very mutability of the forms makes them eternal because mutability enables them to return to earth through endless metamorphoses."*

Yet when Mutability herself makes her plea to the androgynous goddess Nature in our second selection, before an assembly on Arlo Hill in Ireland, what she embodies is only part of Nature, which bridges mortal and divine. As Nature declares in her verdict, though all things change, "they are not changed from their first estate," so that Change has no ultimate dominion. There is also, Spenser affirms in the poem's last surviving stanzas, a world eterne in im*mutability:*

When I bethinke me on that speech whyleare,
Of *Mutability*, and well it way:[33]
Me seemes, that though she all vnworthy were
Of the Heav'ns Rule; yet very sooth to say,
In all things else she beares the greatest sway.
Which makes me loath this state of life so tickle,
And loue of things so vaine to cast away;
Whose flowring pride, so fading and so fickle,
Short *Time* shall soon cut down with his consuming sickle.

Then gin[34] I thinke on that which Nature sayd,
Of that same time when no more *Change* shall be,
But stedfast rest of all things firmely stayd
Vpon the pillours of Eternity,
That is contrayr to *Mutabilitie:*
For, all that moueth, doth in *Change* delight:
But thence-forth all shall rest eternally
With Him that is the God of Sabbaoth hight:[35]
O that great Sabbaoth God, graunt me that Sabaoths sight.

Reflecting on Nature's judgment, Lewis finds it "a magnificent instance of Spenser's last-moment withdrawal from dualism. The universe is a battlefield in which . . . the gods, the divine order, stand for Permanence; Change is rebellion and corruption. But behind this endless contention arises the deeper truth—that Change is but the mode in which Permanence expresses itself, that Reality (like Adonis) 'is eterne in mutabilitie,' and that the more Mutability succeeds the more she fails, even here and now," to say nothing of realms in which she can have no place. Yet inasmuch as The Faerie Queene, *Lewis eloquently writes, "is like life itself, not like the products of life, . . . an image of the* natura naturans, *not of the* natura naturata*"—nor of an eternal world beyond—Mutability's "failures" are no less intrinsic to her "success" than death is to life.*

The Garden of Adonis
III.vi

xxix

She brought her to her ioyous Paradize,
Where most she wonnes,[36] when she on earth does dwel.
So faire a place, as Nature can devize:
Whether in *Paphos*, or *Cytheron* hill,
Or it in *Gnidus* be, I wote[37] not well;
But well I wote by tryall, that this same
All other pleasant places doth excell,
And called is by her lost louers name,
The *Gardin* of *Adonis*, farre renowmd by fame.

[33]"Weigh." [34]"Begin."
[35]"Called." Spenser alludes both to Hebrew *Sabaoth* ("Hosts," as in "Lord of Hosts") and *shabbat* ("sabbath"). [36]"Resides." [37]"Know."

xxx

In that same Gardin all the goodly flowres,
Wherewith dame Nature doth her beautifie,
And decks the girlonds of her paramoures,
Are fetcht: there is the first seminarie[38]
Of all things, that are borne to liue and die,
According to their kindes. Long worke it were,
Here to account the endlesse progenie
Of all the weedes, that bud and blossome there;
But so much as doth need, must needs be counted here.

xxxi

It sited was in fruitfull soyle of old,
And girt in with two walles on either side;
The one of yron, the other of bright gold,
That none might thorough breake, nor ouer-stride:
And double gates it had, which opened wide,
By which both in and out men moten[39] pas;
Th'one faire and fresh, the other old and dride:
Old *Genius* the porter of them was,
Old *Genius*, the which a double nature has.

xxxii

He letteth in, he letteth out to wend,
All that to come into the world desire;
A thousand thousand naked babes attend
About him day and night, which doe require,
That he with fleshly weedes would them attire:
Such as him list, such as eternall fate
Ordained hath, he clothes with sinfull mire,
And sendeth forth to liue in mortall state,
Till they againe returne backe by the hinder gate.

xxxiii

After that they againe returned beene,
They in that Gardin planted be againe;
And grow afresh, as they had neuer seene
Fleshly corruption, nor mortall paine.
Some thousand yeares so doen they there remaine;
And then of him are clad with other hew,
Or sent into the chaungefull world againe,
Till thither they returne, where first they grew:
So like a wheele around they runne from old to new.

xxxiv

No needs there Gardiner to set, or sow,
To plant or prune: for of their owne accord
All things, as they created were, doe grow,
And yet remember well the mightie word,
Which first was spoken by th'Almightie lord,
That bad them to increase and multiply:[40]
Ne doe they need with water of the ford,
Or of the clouds to moysten their roots dry;
For in themselues eternall moisture they imply.

[38]"Seedbed." [39]"Might." [40]Genesis 1:28.

xxxv

Infinite shapes of creatures there are bred,
And vncouth formes, which none yet euer knew,
And euery sort is in a sundry bed
Set by it selfe, and ranckt in comely rew:[41]
Some fit for reasonable soules t'indew,
Some made for beasts, some made for birds to weare,
And all the fruitfull spawne of fishes hew
In endlesse rancks along enraunged were,
That seem'd the *Ocean* could not containe them there.

xxxvi

Daily they grow, and daily forth are sent
Into the world, it to replenish more;
Yet is the stocke not lessened, nor spent,
But still remaines in euerlasting store,
As it at first created was of yore.
For in the wide wombe of the world there lyes,
In hatefull darkenesse and in deep horrore,
An huge eternall *Chaos*, which supplyes
The substances of natures fruitfull progenyes.

xxxvii

All things from thence doe their first being fetch,
And borrow matter, whereof they are made,
Which when as forme and feature it does ketch,
Becomes a bodie, and doth then inuade
The state of life, out of the griesly shade.
That substance is eterne, and bideth so,
Ne when the life decayes, and forme does fade,
Doth it consume, and into nothing go,
But chaunged is, and often altred to and fro.

xxxviii

The substance is not chaunged, nor altered,
But th'only forme and outward fashion;
For euery substance is conditioned
To change her hew, and sundry formes to don,
Meet for her temper and complexion:
For formes are variable and decay,
By course of kind, and by occasion;
And that faire flowre of beautie fades away,
As doth the lilly fresh before the sunny ray.

xxxix

Great enimy to it, and to all the rest,
That in the *Gardin* of *Adonis* springs,
Is wicked *Time*, who with his scyth addrest,
Does mow the flowring herbes and goodly things,
And all their glory to the ground downe flings,
Where they doe wither, and are fowly mard:
He flyes about, and with his flaggy wings
Beats downe both leaues and buds without regard,
Ne euer pittie may relent his malice hard.

[41] "Row."

xl

Yet pittie often did the gods relent,
To see so faire things mard, and spoyled quight:
And their great mother *Venus* did lament
The losse of her deare brood, her deare delight:
Her hart was pierst with pittie at the sight,
When walking through the Gardin, them she spyde,
Yet no'te[42] she find redresse for such despight.
For all that liues, is subiect to that law:
All things decay in time, and to their end do draw.

xli

But were it not, that *Time* their troubler is,
All that in this delightfull Gardin growes,
Should happie be, and haue immortall blis:
For here all plentie, and all pleasure flowes,
And sweet loue gentle fits emongst them throwes,
Without fell rancor, or fond gealosie;
Franckly each paramour his leman[43] knowes,
Each bird his mate, ne any does enuie
Their goodly meriment, and gay felicitie.

xlii

There is continuall spring, and haruest there
Continuall, both meeting at one time:
For both the boughes doe laughing blossomes beare,
And with fresh colours decke the wanton Prime,[44]
And eke attonce[45] the heauy trees they clime,
Which seeme to labour vnder their fruits lode:
The whiles the ioyous birdes make their pastime
Emongst the shadie leaues, their sweet abode,
And their true loues without suspition tell abrode.

xliii

Right in the middest of that Paradise,
There stood a stately Mount, on whose round top
A gloomy groue of mirtle trees did rise,
Whose shadie boughes sharpe steele did neuer lop,
Nor wicked beasts their tender buds did crop,
But like a girlond compassed the hight,
And from their fruitfull sides sweet gum did drop,
That all the ground with precious deaw bedight,
Threw forth most dainty odours, and most sweet delight.

xliv

And in the thickest couert of that shade,
There was a pleasant arbour, not by art,
But of the trees owne inclination made,
Which knitting their rancke braunches part to part,
With wanton yuie twyne entrayld athwart,
And Eglantine, and Caprifole emong,
Fashiond aboue within their inmost part,
That neither *Phœbus* beams could through them throng,
Nor *Aeolus* sharp blast could worke them any wrong.

[42]"Could not," "might not." [43]"Lover." [44]"Springtime." [45]"Also at the same time."

xlv

And all about grew euery sort of flowre,
To which sad louers were transformd of yore;
Fresh *Hyacinthus*, *Phœbus* paramoure,
And dearest loue,
Foolish *Narcisse*, that likes the watry shore,
Sad *Amaranthus*, made a flowre but late,
Sad *Amaranthus*, in whose purple gore
Me seemes I see *Amintas* wretched fate,
To whom sweet Poets verse hath giuen endless date.

xlvi

There wont[46] faire *Venus* often to enioy
Her deare *Adonis* ioyous company,
And reape sweet pleasure of the wanton boy;
There yet, some say, in secret he does ly,
Lapped in flowres and pretious spycery,
By her hid from the world, and from the skill
Of *Stygian* Gods,[47] which doe her loue enuy;
But she her selfe, when euer that she will,
Possesseth him, and of his sweetnesse takes her fill.

xlvii

And sooth it seemes they say: for he may not
For euer die, and euer buried bee
In balefull night, where all things are forgot;
All be he[48] subiect to mortalitie,
Yet is eterne in mutabilitie,
And by succession made perpetuall,
Transformed oft, and chaunged diuerslie:
For him the Father of all formes they call;
Therefore needs mote[49] he liue, that liuing giues to all. . . .

Mutability's Plea to Nature
VII.vii

v

Then forth issewed (great goddesse) great dame *Nature*,
With goodly port and gracious Maiesty;
Being far greater and more tall of stature
Then any of the gods or Powers on hie:
Yet certes by her face and physnomy,
Whether she man or woman inly were,
That could not any creature well descry:
For, with a veile that wimpled euery where,
Her head and face was hid, that mote to none appeare.

vi

That some doe say was so by skill devized,
To hide the terror of her uncouth hew,
From mortall eyes that should be sore agrized;[50]

[46]"Is accustomed." [47]Gods of the river Styx in the underworld, i.e., of death.
[48]"Although he be." [49]"Needs must." [50]"Horrified."

For that her face did like a Lion shew,
That eye of wight could not indure to view:
But others tell that it so beautious was,
And round about such beames of splendor threw,
That it the Sunne a thousand times did pass,
Ne could be seene, but like an image in a glass.

vii

That well may seemen true: for, well I weene,
That this same day, when she on *Arlo* sat,
Her garment was so bright and wondrous sheene,[51]
That my fraile wit cannot deuize to what
It to compare, nor finde like stuffe to that,
As those three sacred *Saints*, though else most wise,
Yet on mount *Thabor* quite their wits forgat,
When they their glorious Lord in strange disguise
Transfigur'd sawe; his garments so did daze their eyes.[52]

viii

In a fayre Plaine vpon an equall Hill,
She placed was in a pauilion;
Not such as Craftes-men by their idle skill
Are wont for Princes states to fashion:
But th'earth her self of her owne motion,
Out of her fruitfull bosome made to growe
Most dainty trees; that, shooting vp anon,
Did seeme to bow their bloosming heads full lowe,
For homage vnto her, and like a throne did showe.

ix

So hard it is for any liuing wight,
All her array and vestiments to tell,
That old *Dan Geffrey*[53] (in whose gentle spright
The pure well head of Poesie did dwell)
In his *Foules parley* durst not with it mel,[54]
But it transferd to *Alane*, who he thought
Had in his *Plaint of kindes* describ'd it well:
Which who will read set forth so as it ought,
Go seek he out that *Alane* where he may be sought.

x

And all the earth far vnderneath her feete
Was dight[55] with flowres, that voluntary grew
Out of the ground, and sent forth odours sweet;
Tenne thousand mores[56] of sundry sent and hew,
That might delight the smell, or please the view:
The which the Nymphes, from all the brooks thereby

[51]"Beautiful."
[52]For Mount Tabor, see 1 Samuel 10:3–7. The transfiguration of Christ was witnessed by the apostles Peter, John, and James; see Matthew 17:1–9, Mark 9:2–10, and Luke 9:28–36.
[53]Geoffrey Chaucer ("Dan" is an honorific, like "Sir"), in *The Parliament of Fowls* 316–17 (which Spenser here calls *Foules parley*), drew for his portrayal of the goddess Nature on Alan of Lille's *Complaint of Nature* ("Aleyn, in the Pleynt of Kynde"); see Chapter 15 above. "Gentle spright" = "noble spirit."
[54]"Meddle." [55]"Decked out." [56]"Plants"; "sent and hew" = "scent and hue."

Had gathered, which they at her foot-stoole threw;
That richer seem'd then any tapestry,
That Princes bowres adorne with painted imagery. . . .

xiii

This great Grandmother of all creatures bred
Great *Nature*, euer young yet full of eld,[57]
Sill moouing, yet vnmoued from her sted;[58]
Vnseene of any, yet of all beheld;
Thus sitting in her throne as I have teld,
Before her came dame *Mutabilitie*;
And being lowe before her presence feld,[59]
With meek obaysance and humilitie,
Thus gan her plaintif Plea, with words to amplifie;

xiv

To thee O greatest goddesse, onely great,
An humble suppliant loe, I lowely fly
Seeking for Right, which I of thee entreat;
Who Right to all dost deale indifferently,
Damning all Wrong and tortious[60] Iniurie,
Which any of thy creatures doe to other
(Oppressing them with power, vnequally)
Sith of them all thou art the equall mother,
And knittest each to each, as brother vnto brother.

xv

To thee therefore of this same *Iove* I plaine,[61]
And of his fellow gods that faine[62] to be,
That challenge to[63] themselves the whole worlds raign;
Of which, the greatest part is due to me,
And heauen it selfe by heritage in Fee:[64]
For, heauen and earth I both alike do deeme,
Sith heauen and earth are both alike to thee;
And, gods no more then men thou does esteeme:
For, euen the gods to thee, as men to gods do seeme.

xvi

Then weigh, O soueraigne goddesse, by what right
These gods do claime the worlds whole souerainty;
And that[65] is onely dew vnto thy might
Arrogate to themselues ambitiously:
As for the gods owne principality,
Which *Iove* vsurpes vniustly; that to be
My heritage, *Iove's* self cannot deny,
From my great Grandsire *Titan*, vnto mee,
Deriv'd by dew descent; as is well knowen to thee.

xvii

Yet mauger[66] *Iove*, and all his gods beside,
I doe possesse the worlds most regiment;[67]
As, if ye please it into parts diuide,

[57]"Old age." [58]"Place." [59]"Prostrated." [60]"Wrongful." [61]"Complain."
[62]"Feign," "pretend." [63]"Claim for." [64]"By legal right." [65]"That which."
[66]"In spite of." [67]"Greatest dominion."

And euery parts inholders to conuent,[68]
Shall to your eyes appeare incontinent,[69]
And first, the Earth (great mother of vs all)
That only seems vnmov'd and permanent,
And vnto *Mutability* not thrall;
Yet is she chang'd in part, and eeke in generall.

xviii

For, all that from her springs, and is ybredde,[70]
How-euer fayre it flourish for a time,
Yet see we soone decay; and, being dead,
To turne again vnto their earthly slime:
Yet, out of their decay and mortall crime,
We daily see new creatures to arize;
And of their Winter spring another Prime,
Vnlike in forme, and chang'd by strange disguise:
So turne they still about, and change in restlesse wise.

xix

As for her tenants; that is, man and beasts,
The beasts we daily see massacred dy,
As thralls and vassalls vnto mens beheasts:
And men themselues doe change continually,
From youth to eld, from wealth to pouerty,
From good to bad, from bad to worst of all.
Ne doe their bodies only flit and fly:
But eeke their minds (which they immortall call)
Still change and vary thoughts, as new occasions fall. . . .

lvi

Then since within this wide great *Vniuerse*
Nothing doth firme and permanent appeare,
But all things tost and turned by transuerse:
What then should let,[71] but I aloft should reare
My Trophee, and from all, the triumph beare?
Now iudge then (O thou greatest goddesse trew!)
According as thy selfe doest see and heare,
And vnto me addoom[72] that is my dew;[73]
That is the rule of all, all being rul'd by you.

lvii

So hauing ended, silence long ensewed,
Ne *Nature* to or fro spake for a space,
But with firme eyes affixt, the ground still viewed.
Meane while, all creatures, looking in her face,
Expecting th'end of this so doubtfull case,
Did hang in long suspence what would ensew,
To whether[74] side should fall the soueraigne place:
At length, she looking vp with chearefull view,
The silence brake, and gaue her doome[75] in speeches few.

[68]"Assemble the inhabitants of every part." [69]"Immediately." [70]"Bred."
[71]"Prevent," "hinder." [72]"Adjudge," "award." [73]"What is due to me."
[74]"Which." [75]"Judgment," "verdict."

lviii

I well consider all that ye haue sayd,
And find that all things stedfastnes doe hate
And changed be: yet being rightly wayd[76]
They are not changed from their first estate;
But by their change their being doe dilate:[77]
And turning[78] to themselues at length againe,
Doe worke their owne perfection so by fate:
Then ouer them Change doth not rule and raigne;
But they raigne ouer change, and doe their states maintaine.

lix

Cease therefore daughter further to aspire,
And thee content[79] thus to be rul'd by me:
For thy decay thou seekst by thy desire;
But time shall come that all shall changed bee,
And from thenceforth, none no more change shall see.
So was the *Titaness* put downe and whist,[80]
And *Iove* confirm'd in his imperiall see.[81]
Then was that whole assembly quite dismist,
And *Natur's* selfe did vanish, whither no man wist.[82]

The Art Itself Is Nature

Selections from the Works of William Shakespeare

William Shakespeare (1564–1616) was born into a prosperous burgher's family in Stratford-upon-Avon, married in 1582, and by 1592 had begun his London career with the Lord Chamberlain's (later the King's) Men as actor, playwright, and shareholder in the Globe Theater, where many of his plays were performed; in about 1611–12 he retired to his home in Stratford. His long poems Venus and Adonis *and* The Rape of Lucrece *appeared in 1593–94, and an unauthorized edition of his* Sonnets *in 1609; but not until 1623 were his comedies, tragedies, and histories (many of which had been individually published in quartos) posthumously collected in the First Folio, to which Ben Jonson prefaced a handsome tribute "To the Memory of my Beloved, the Author, Mr. William Shakespeare."*

No poet before the Romantics has been more closely associated with nature (in very different senses) than Shakespeare. Jonson proclaims that

Nature herself was proud of his designs,
And joyed to wear the dressing of his lines!

Milton refers, in "L'Allegro," to Shakespeare "warbling his native woodnotes wild"; Samuel Johnson declares that "Shakespeare is above all writers, at least all modern writers, the poet of nature"; and Samuel Taylor Coleridge asserts that "Nature, the prime genial artist, inexhaustible in diverse powers, is equally inexhaustible in forms; . . . and even such is the appropriate excellence of her chosen poet, of our own Shakespeare." Such exaltations of Shakespeare as a poet of nature by no means depreciate his art; thus Jonson continued,

[76]"Weighed." [77]"Expand," "develop." [78]"Returning." [79]"Content yourself."
[80]"Silenced." [81]"Seat," "throne." [82]"Knew."

Yet must I not give Nature all: Thy Art,
My gentle Shakespeare, must enjoy a part
For though the Poet's matter, Nature be,
His art doth give the fashion.

And Coleridge's essay, "Shakespeare's Judgment Equal to His Genius," was aimed at refuting a view of Shakespeare as "the anomalous, the wild, the irregular genius" ignorant of art and tutored by savage Nature alone.

With this essential caveat, however, praise of Shakespeare as a poet of nature compels assent on several levels. To begin with, no poet's work is richer in concretely sensual images of the sights, sounds, and smells of a natural world with which the human is always intimately connected. He gives us the song of the lark and the flight of the bee, the daisies and violets that "paint the meadows with delight," but also the keen tooth of the winter wind

When icicles hang by the wall,
And Dick the shepherd blows his nail,

the boughs that shake against the autumnal cold, and the oak-cleaving thunderbolts that crack Nature's molds asunder. Such images run through Shakespeare's writings from first to last, but attain a particular purity in the songs; though always appropriate to their dramatic settings, the songs can also stand by themselves as moments of intensely apprehended lyrical beauty.

Beyond these particulars is the larger view of nature incorporated in the Renaissance adaptation of medieval cosmology, which no writer articulated so inclusively as Shakespeare. "Nature rules over three domains, each of which is a reflection of the others, since they are all parts of the same ordered unity," Theodore Spencer writes in Shakespeare and the Nature of Man *(1942): "She rules over the cosmos—the universal world; she rules over the world of created objects on earth; and she rules over the world of human government, of man in society. . . . Thus man, for whom the rest of the world was created, stands in the center of the second of Nature's domains, the world of living beings, his head erect to contemplate the heavens, his soul able to rise from the realm of sense to apprehend the God who made him." To this grand conception—and to the conflict between human dignity and wretchedness that arose from the disjunctions of a world where violation of degree continually threatens the cosmos with chaos—Shakespeare gave eloquent expression throughout his tragedies and histories. Ulysses' speech on order in* Troilus and Cressida *is a* locus classicus.

At least as important as this inherited cosmology is the immemorial conception, implicit in the seasonal imagery of the songs and sonnets but most prominent in the comedies, of the cyclical death and rebirth of nature as an unendingly creative process in which human beings share to the extent that they are in tune with these primordial rhythms. "The mythical or primitive basis of comedy is a movement toward the rebirth and renewal of the powers of nature," Northrop Frye argues in A Natural Perspective *(1965); the "green world" of comedy "is a symbol of natural society . . . associated with things which in the context of the ordinary world seem unnatural, but which in fact are attributes of nature as a miraculous and irresistible reviving power. These associations include dream, magic and chastity or spiritual energy as well as fertility and renewed natural energies." Communion with these powers, in Shakespeare's time, was most closely associated with the festivity that pervades his comedies. "Occasions like May Day and the Winter Revels, with their cult of natural vitality," C. L. Barber writes in* Shakespeare's Festive Comedy *(1959), "were maintained within a civilization whose daily view of life focused on the mortality implicit in vitality. The tolerant disillusion of Anglican or Catholic culture allowed nature to have its day. But the release of that one day was understood to be a temporary license, a 'misrule' which implied rule, so that the acceptance of nature was qualified" by belief that its proper role in human life, though essential, was partial. The movement of Shakespearean comedy is toward reintegration of holiday into the everyday, and thus of nature in culture.*

But Shakespeare goes far beyond expression of commonplaces regarding nature to exploration of the complexities inherent in them. As You Like It *(ca. 1599–1600), for example, entails continual questioning and ironical reversal of the pastoral conventions on which it relies. Its forest of Arden is no blissful Arcadia but a world of adversity pierced by winter winds, by man's ingratitude, by the feigning of friendship and the folly of love. From all of this, however—since in the "natural" forest world all are stripped bare of pretenses—the chastened characters "feelingly" learn what they are. In* King Lear *(ca. 1603–6), the disorder of a world in which, Gloucester laments, "the bond is cracked 'twixt son and father," and "the King falls from the bias of nature," is reflected not only in the microcosm of man but in eclipses of the sun and moon—whose connection with human misconduct only the villainous Edmund (Gloucester's bastard son) denies—and in the raging cataracts and hurricanes to which Lear, fleeing from the ungrateful daughters to whom he has foolishly ceded his kingdom, is exposed, along with his fool and the loyal Kent, on the stormy heath.*

But it is in his late "romances" (ca. 1610–12), The Winter's Tale *and* The Tempest, *that Shakespeare most fully explores nature in relation to human art. In* The Winter's Tale, *the delinquencies of the rogue Autólycus (who enters singing "When Daffodils Begin to Peer"), as E. M. W. Tillyard observes in* Shakespeare's Last Plays *(1938), "keep the earthly paradise sufficiently earthly without disturbing the paradisiac state." In the sheep-shearing scene (IV.iv) occurs a lengthy dialogue between the shepherdess Perdita (who unbeknownst to herself and others is the long-lost daughter of Leontes, king of Sicilia) and Polixenes (disguised king of Bohemia and father of Florizel, who to win the heart of Perdita is playing the role of the shepherd Doricles). Here the Renaissance debate of art versus nature—which we have encountered in Montaigne's "Of Cannibals" (Chapter 16 above) and in Spenser—attains its most subtle expression. In Perdita, Tillyard writes, creative instincts "are implied by her sympathy with nature's lavishness in producing flowers, followed by her own simple and unashamed confession of wholesome sensuality." Her chastity, Frye remarks, "stands not for the purity of nature but for nature as a pulsating power, expressing itself in the miraculous springtime renewal which takes place without the aid of art."*

And because she herself is so artless, she disdains, in distributing flowers appropriate to their age to Polixenes and Camillo (a lord of Sicilia), to include any flower in whose cultivation human art has contributed to "great creating Nature." To this, Polixenes, rejecting any simple dichotomy between nature and art, replies that every art derives from Nature and thus cannot be inferior to her. As Edward William Tayler writes in Nature and Art in Renaissance Literature *(1964), "Polixenes' stand is perhaps the most dignified and carefully argued in the whole history of possible opposition between Nature and Art. Like Aristotle and Plato, Polixenes points out that the 'art itself is nature.'" But this is not to say that Polixenes "wins" the debate. As G. Wilson Knight observes in* The Crown of Life *(1952), he and Perdita "both alike reverence 'great creating nature,' though differing in their conclusions. No logical deduction is to be drawn; or rather, the logic is dramatic, made of opposing statements, which serve to conjure up an awareness of nature as an all-powerful presence and exemplar."*

So, too, in The Tempest, *any simple antithesis between nature and art is continually undercut. Dreams of a golden age such as Don Quixote evoked in his reverie to the goatherds (Chapter 16 above) were a staple of pastoral romances on the pattern of Tasso's* Aminta *or Giambattista Guarini's* Il Pastor Fido *(*The Faithful Shepherd, *1590), in both of which freedom from false conceptions of honor and from restrictions on "natural" impulses of love were defining traits. In*

The Tempest *(II.i), the "honest old councilor" Gonzalo evokes a vision reminiscent both of these and of their classical prototypes in Hesiod, Virgil, and Ovid:*

> All things in common nature should produce
> Without sweat or endeavor. Treason, felony,
> Sword, pike, knife, gun, or need of any engine
> Would I not have; but nature should bring forth,
> Of its own kind, all foison, all abundance,
> To feed my innocent people.

But on the island where Prospero, exiled duke of Milan, has created through magical arts a kingdom to whose laws his usurping brother and his shipwrecked companions are now subject, nature, as embodied in Caliban, is by no means so simply beneficent.

In Shakespeare's variation of pastoral drama, "The main opposition is between the worlds of Prospero's Art, and Caliban's Nature," Frank Kermode writes in his introduction to the Arden Tempest *(6th ed., 1958). "Caliban is the core of the play; like the shepherd in formal pastoral, he is the natural man against whom the cultivated man is measured. But . . . Caliban represents . . . nature without benefit of nurture; nature, opposed to an Art which is man's power over the created world and over himself; nature divorced from grace, or the senses without the mind." In contrast to the inchoate (if ironically tempered) primitivism of Montaigne's "Of Cannibals" (Chapter 16 above), which was one principal "source" of* The Tempest, *Shakespeare's treatment, Kermode observes, "never reaches after a naked opinion of true or false." Caliban "is not, as in pastoral generally, a virtuous shepherd, but a salvage and deformed slave." And yet, in his intransigent defiance of what he sees as Prospero's tyranny (has Prospero not usurped his island as Antonio had usurped Prospero's duchy?), and in his waking dreams of "sounds and sweet airs that give delight, and hurt not," Caliban exerts a profound appeal. He suggests the limits of the "potent art" that Prospero himself will abjure, drowning his book in the elemental waters that will long survive him. Thus the bestial Caliban, in whom nature is never idealized, emerges, in Kermode's words, as "an extraordinarily powerful and comprehensive type of Nature; an inverted pastoral hero, against whom civility and the Art which improves Nature may be measured"—not always to their unequivocal advantage.*

Texts consulted in making the following excerpts include The Riverside Shakespeare, *ed. G. Blakemore Evans (1974), and individual volumes of the Arden and the Signet Classic Shakespeares.*

SIX SONGS

When Daisies Pied

Love's Labor's Lost *V.ii*

Armado. . . . This side is *Hiems*, Winter; this, *Ver*, the Spring; the one maintained by the owl, th'other by the cuckoo. *Ver*, begin.

The Song

Spring. When daisies pied and violets blue
  And lady-smocks all silver-white
And cuckoo-buds of yellow hue
  Do paint the meadows with delight,
The cuckoo then, on every tree, 5
Mocks married men; for thus sings he,

Cuckoo!
Cuckoo, cuckoo! O word of fear
Unpleasing to a married ear![83]

When shepherds pipe on oaten straws, 10
And merry larks are plowmen's clocks,
When turtles tread,[84] and rooks, and daws,
And maidens bleach their summer smocks,
The cuckoo then, on every tree,
Mocks married men; for thus sings he, 15
Cuckoo!
Cuckoo, cuckoo! O word of fear,
Unpleasing to a married ear!

Winter. When icicles hang by the wall,
And Dick the shepherd blows his nail, 20
And Tom bears logs into the hall,
And milk comes frozen home in pail,
When blood is nipped, and ways be foul,
Then nightly sings the staring owl,
Tu-whit, to-who, 25
A merry note,
While greasy Joan doth keel[85] the pot.

When all aloud the wind doth blow,
And coughing drowns the parson's saw,[86]
And birds sit brooding in the snow, 30
And Marion's nose looks red and raw,
When roasted crabs[87] hiss in the bowl,
Then nightly sings the staring owl,
Tu-whit, to-who,
A merry note, 35
While greasy Joan doth keel the pot.

Under the Greenwood Tree
As You Like It *II.v*

Under the greenwood tree
Who loves to lie with me,
And turn his merry note
Unto the sweet bird's throat,
Come hither, come hither, come hither: 5
Here shall he see
No enemy
But winter and rough weather.

[83]Because of its resemblance to "cuckold." [84]"Turtledoves mate."
[85]"Cool," by stirring or skimming. [86]"Wise saying." [87]"Crab apples."

Who doth ambition shun
And loves to live i' the sun, 10
Seeking the food he eats,
And pleased with what he gets,
Come hither, come hither, come hither:
Here shall he see
No enemy 15
But winter and rough weather.

Blow, Blow, Thou Winter Wind

As You Like It *II.vii*

Blow, blow, thou winter wind,
Thou art not so unkind
As man's ingratitude;
Thy tooth is not so keen
Because thou art not seen, 5
Although thy breath be rude.
Heigh-ho! sing heigh-ho, unto the green holly:
Most friendship is feigning, most loving mere folly:
Then heigh-ho the holly!
This life is most jolly. 10

Freeze, freeze, thou bitter sky
That dost not bite so nigh
As benefits forgot:
Though thou the waters warp,
Thy sting is not so sharp 15
As friend remembered not.
Heigh-ho! sing heigh-ho, unto the green holly:
Most friendship is feigning, most loving mere folly:
Then heigh-ho the holly!
This life is most jolly. 20

Hark, Hark! the Lark

Cymbeline *II.iii*

Hark, hark! the lark at heaven's gate sings,
And Phoebus 'gins arise,
His steeds to water at those springs
On chaliced flowers that lies;
And winking Mary-buds begin 5
To ope their golden eyes.
With every thing that pretty is,
My lady sweet, arise:
Arise, arise!

When Daffodils Begin to Peer
The Winter's Tale *IV.iii*

When daffodils begin to peer,
With heigh! the doxy[88] over the dale,
Why, then comes in the sweet o' the year;
For the red blood reigns in the winter's pale.[89]

The white sheet bleaching on the hedge, 5
With heigh! the sweet birds, oh, how they sing!
Doth set my pugging[90] tooth on edge;
For a quart of ale is a dish for a king.

The lark, that tirra-lirra chants,
With heigh! with heigh! the thrush and the jay, 10
Are summer songs for me and my aunts,[91]
While we lie tumbling in the hay.

Where the Bee Sucks
The Tempest *V.i*

Where the bee sucks, there suck I:
In a cowslip's bell I lie;
There I couch when owls do cry.
On the bat's back I do fly
After summer merrily. 5
Merrily, merrily shall I live now
Under the blossom that hangs on the bough.

SIX SONNETS

Sonnet 15

When I consider everything that grows
Holds in perfection but a little moment,
That this huge stage presenteth nought but shows
Whereon the stars in secret influence comment;
When I perceive that men as plants increase, 5
Cheerèd and checked even by the selfsame sky,
Vaunt in their youthful sap, at height decrease,
And wear their brave state out of memory:[92]
Then the conceit[93] of this inconstant stay
Sets you most rich in youth before my sight, 10
Where wasteful Time debateth with Decay
To change your day of youth to sullied night;
And, all in war with Time for love of you,
As he takes from you, I ingraft you new.[94]

[88]"Trollop," "mistress." [89]"Territory" or "pallor." [90]"Thieving." [91]"Sweethearts."
[92]"Their splendid condition until forgotten." [93]"Concept," "idea."
[94]I.e., by the poetry in which I give you new life.

Sonnet 18

Shall I compare thee to a summer's day?
Thou art more lovely and more temperate.
Rough winds do shake the darling buds of May,
And summer's lease hath all too short a date.
Sometime too hot the eye of heaven shines, 5
And often is his gold complexion dimmed;
And every fair from fair sometime declines,
By chance, or nature's changing course, untrimmed.[95]
But thy eternal summer shall not fade,
Nor lose possession of that fair thou owest,[96] 10
Nor shall Death brag thou wand'rest in his shade
When in eternal lines to time thou growest.[97]
So long as men can breathe or eyes can see,
So long lives this, and this gives life to thee.

Sonnet 33

Full many a glorious morning have I seen
Flatter the mountain tops with sovereign eye,
Kissing with golden face the meadows green,
Gilding pale streams with heavenly alchemy;
Anon permit the basest clouds to ride 5
With ugly rack[98] on his celestial face,
And from the forlorn world his visage hide,
Stealing unseen to west with this disgrace:
Even so my sun one early morn did shine
With all-triumphant splendor on my brow; 10
But, out alack! he was but one hour mine,
The region cloud[99] hath masked him from me now.
Yet him for this my love no whit disdaineth;
Suns of the world may stain[100] when heaven's sun staineth.

Sonnet 73

That time of year thou mayst in me behold
When yellow leaves, or none, or few, do hang
Upon those boughs which shake against the cold,
Bare ruined choirs where late the sweet birds sang.
In me thou seest the twilight of such day 5
As after sunset fadeth in the west,
Which by and by black night doth take away,
Death's second self that seals up all in rest.
In me thou seest the glowing of such fire

[95] "Shorn of its beauty." [96] "Ownest." [97] "Art grafted." [98] "Wisp of cloud."
[99] "Cloud in the vicinity." [100] "Be stained."

That on the ashes of his youth doth lie, 10
As the deathbed whereon it must expire,
Consumed with that which it was nourished by.[101]
This thou perceiv'st, which makes thy love more strong,
To love that well which thou must leave ere long.

Sonnet 97

How like a winter hath my absence been
From thee, the pleasure of my fleeting year!
What freezings have I felt, what dark days seen!
What old December's bareness everywhere!
And yet this time removed[102] was summer's time, 5
The teeming autumn, big[103] with rich increase,
Bearing the wanton burden of the prime,[104]
Like widowed wombs after their lords' decease:
Yet this abundant issue[105] seemed to me
But hope of orphans[106] and unfathered fruit; 10
For summer and his pleasures wait on thee,
And, thou away, the very birds are mute.
Or, if they sing, 'tis with so dull a cheer
That leaves look pale, dreading the winter's near.

Sonnet 98

From you have I been absent in the spring,
When proud-pied[107] April, dressed in all his trim,
Hath put a spirit of youth in everything,
That[108] heavy Saturn laughed and leapt with him;
Yet nor the lays of birds, nor the sweet smell 5
Of different flowers in odor and in hue,
Could make me any summer's story tell,
Or from their proud lap pluck them where they grew:
Nor did I wonder at the lily's white,
Nor praise the deep vermilion in the rose; 10
They were but sweet, but figures of delight,
Drawn after you, you pattern of all those.
Yet seemed it winter still, and you away,
As with your shadow I with these did play.

[101]I.e., by life. [102]"Time away." [103]"Pregnant."
[104]I.e., the embryo wantonly engendered in the spring. [105]"Progeny." [106]"Orphaned hope."
[107]"Splendidly dappled." [108]"So that." Saturn was the planet associated with melancholy.

PASSAGES FROM FOUR PLAYS

The Forest of Arden

As You Like It *II.i*

Duke Senior. Now, my co-mates and brothers in exile,
Hath not old custom made this life more sweet
Than that of painted pomp? Are not these woods
More free from peril than the envious court?
Here feel we not the penalty of Adam; 5
The seasons' difference, as the icy fang
And churlish chiding of the winter's wind,
Which, when it bites and blows upon my body
Even till I shrink with cold, I smile and say
"This is no flattery: these are counselors 10
That feelingly persuade me what I am."
Sweet are the uses of adversity,
Which, like the toad, ugly and venomous,
Wears yet a precious jewel in his head;[109]
And this our life, exempt from public haunt, 15
Finds tongues in trees, books in the running brooks,
Sermons in stones, and good in everything.

Nature Disordered

King Lear *I.ii; III.ii*

Gloucester. These late eclipses in the sun and moon portend no good to us. Though the wisdom of Nature can reason[110] it thus and thus, yet Nature finds itself scourged by the sequent effects.[111] Love cools, friendship falls off, brothers divide. In cities, mutinies; in countries, discord; in palaces, treason; and the bond cracked 'twixt son and father. This villain of mine[112] comes under the prediction, there's son against father; the King falls from bias of nature,[113] there's father against child. We have seen the best of our time. Machinations, hollowness, treachery, and all ruinous disorders follow us disquietly to our graves. . . . And the noble and true-hearted Kent banished; his offense, honesty. 'Tis strange. *Exit.*

Edmund. This is the excellent foppery[114] of the world, that when we are sick in fortune, often the surfeits of our own behavior,[115] we make guilty of our disasters the sun, the moon, and stars; as if we were villains on necessity; fools by heavenly compulsion; knaves, thieves, and treachers by spherical predominance;[116] drunkards, liars, and adulterers by an enforced obedience of planetary influence; and all that we are evil in, by a divine thrusting on.[117] An admirable evasion of whoremaster man, to lay his goatish[118] disposition on the charge of a star. My father compounded with[119] my mother under the

[109]The fabulous toadstone. [110]"Explain."
[111]"Actual consequences," as opposed to explanations of natural philosophy ("the wisdom of Nature").
[112]His son Edgar, whom his bastard son Edmund has falsely accused of treachery.
[113]"Natural inclination." [114]"Folly." [115]"Because of our own excesses."
[116]"Traitors because of the ascendancy of a particular star at our birth."
[117]"Compulsion." [118]"Lecherous." [119]"Begot a child by."

Dragon's Tail,[120] and my nativity[121] was under Ursa Major, so that it follows I am rough and lecherous. Fut! I should have been that[122] I am, had the maidenliest star in the firmament twinkled on my bastardizing. . . . (I.ii)

*

Storm still.
Enter Lear and Fool.

Lear. Blow, winds, and crack your cheeks. Rage, blow!
You cataracts and hurricanoes,[123] spout
Till you have drenched our steeples, drowned the cocks.[124]
You sulph'rous and thought-executing fires,
Vaunt-couriers[125] of oak-cleaving thunderbolts, 5
Singe my white head. And thou, all-shaking thunder,
Strike flat the thick rotundity o' th' world,
Crack Nature's molds, all germains spill[126] at once,
That makes ingrateful man.
Fool. O Nuncle, court holy-water[127] in a dry house is better than this rain 10
water out o' door. Good Nuncle, in; ask thy daughters blessing.
Here's a night pities neither wise man nor fools.
Lear. Rumble thy bellyful. Spit, fire. Spout, rain!
Nor rain, wind, thunder, fire are my daughters.[128]
I tax[129] not you, you elements, with unkindness. 15
I never gave you kingdom, called you children.
You owe me no subscription.[130] Then let fall
Your horrible pleasure.[131] Here I stand your slave,
A poor, infirm, weak, and despised old man.
But yet I call you servile ministers, 20
That will with two pernicious daughters join
Your high-engendered battles 'gainst a head
So old and white as this. O, ho! 'tis foul.
Fool. He that has a house to put's head in has a good headpiece.
The codpiece[132] that will house 25
Before the head has any,
The head and he shall louse:
So beggars marry many.[133]
The man that makes his toe
What he his heart should make 30
Shall of a corn cry woe,
And turn his sleep to wake.[134]
But there was never yet fair woman but she made mouths in a glass.[135]

[120]The constellation Draco. [121]"The day of my birth." [122]"What."
[123]"Waterspouts." [124]"Weathercocks." [125]"Heralds." [126]"Destroy all seeds of life."
[127]I.e., even the hypocrisy of the court. [128]Goneril and Regan, who have cast him out.
[129]"Accuse." [130]"Allegiance." [131]"Will," "intention."
[132]Padding over the penis on men's hose, or the penis itself. [133]I.e., many lice.
[134]I.e., the man who, ignoring the fit order of things, elevates what is base above what is noble, will suffer for it as Lear has, in banishing Cordelia and enriching her sisters. (Russell Fraser in Signet Classic *King Lear*)
[135]"In front of a mirror."

Enter Kent.

Lear. No, I will be the pattern of all patience,
I will say nothing. 35

Kent. Who's there?

Fool. Marry, here's grace and a codpiece; that's a wise man and a fool.

Kent. Alas, sir, are you here? Things that love night
Love not such nights as these. The wrathful skies
Gallow[136] the very wanderers of the dark 40
And make them keep their caves. Since I was man
Such sheets of fire, such bursts of horrid thunder,
Such groans of roaring wind and rain, I never
Remember to have heard. Man's nature cannot carry
Th'affliction nor the fear.

Lear. Let the great gods 45
That keep this dreadful pudder[137] o'er our heads
Find out their enemies now. Tremble, thou wretch,
That hast within thee undivulgèd crimes
Unwhipped of justice. Hide thee, thou bloody hand,
Thou perjured, and thou simular[138] of virtue 50
That art incestuous. Caitiff,[139] to pieces shake,
That under covert and convenient seeming[140]
Has practiced on[141] man's life. Close pent-up guilts,
Rive your concealing continents[142] and cry
These dreadful summoners grace.[143] I am a man 55
More sinned against than sinning.

Kent. Alack, bareheaded?
Gracious my lord, hard by here is a hovel;
Some friendship will it lend you 'gainst the tempest.
Repose you there, while I to this hard house
(More harder than the stones whereof 'tis raised, 60
Which even but now, demanding after[144] you,
Denied me to come in) return, and force
Their scanted[145] courtesy.

Lear. My wits begin to turn.
Come on, my boy. How dost, my boy? Art cold?
I am cold myself. Where is this straw, my fellow? 65
The art[146] of our necessities is strange,
That can make vile things precious. Come, your hovel.
Poor fool and knave, I have one part in my heart
That's sorry yet for thee.

[136]"Frighten." [137]"Tumult."
[138]"Simulator." [139]"Wretch."
[140]"Hypocrisy." [141]"Plotted against."
[142]"Containers." [143]"Beg . . . mercy."
[144]"Asking for." [145]"Grudging."
[146]The alchemical art, thought to transmute base metals into gold.

Fool. *[Singing]*
He that has and a little tiny wit, 70
With heigh-ho, the wind and the rain,
Must make content with his fortunes fit,[147]
Though the rain it raineth every day.
Lear. True, my good boy. Come, bring me to this hovel.

Exit [with Kent]. (III.ii)

Fairest Flowers of the Season
The Winter's Tale *IV.iv*

Polixenes. Shepherdess—
A fair one are you—well you fit our ages
With flow'rs of winter.
Perdita. Sir, the year growing ancient,
Not yet on summer's death, nor on the birth 80
Of trembling winter, the fairest flow'rs o' th' season
Are our carnations, and streaked gillyvors,[148]
Which some call Nature's bastards; of that kind
Our rustic garden's barren; and I care not
To get slips of them.
Polixenes. Wherefore, gentle maiden, 85
Do you neglect them?
Perdita. For[149] I have heard it said,
There is an art, which in their piedness shares
With great creating Nature.
Polixenes. Say there be;
Yet Nature is made better by no mean
But Nature makes that mean; so over that art, 90
Which you say adds to Nature, is an art
That Nature makes. You see, sweet maid, we marry
A gentler scion to the wildest stock,
And make conceive a bark of baser kind
By bud of nobler race. This is an art 95
Which does mend Nature, change it rather; but
The art itself is Nature.
Perdita. So it is.
Polixenes. Then make your garden rich in gillyvors,
And do not call them bastards.
Perdita. I'll not put
The dibble[150] in earth, to set one slip of them; 100
No more than were I painted, I would wish
This youth should say 'twere well, and only therefore
Desire to breed by me. Here's flow'rs for you:
Hot lavender, mints, savory, marjoram,

[147]Fortunes befitting his small wit. [148]"Pinks." [149]"Because." [150]A seed-planting tool.

The marigold that goes to bed wi' th' sun, 105
And with him rises, weeping; these are flow'rs
Of middle summer, and I think they are given
To men of middle age. You're very welcome.

Camillo. I should leave grazing, were I of your flock,
And only live by gazing.

Perdita. Out, alas! 110
You'd be so lean that blasts of January
Would blow you through and through.
[To Florizel] Now, my fair'st friend,
I would I had some flow'rs o' th' spring, that might
Become your time of day—
[to the Shepherdesses] and yours, and yours,
That wear upon your virgin branches yet 115
Your maidenheads growing. O Proserpina,
For the flow'rs now, that, frighted, thou let'st fall
From Dis's[151] wagon! Daffodils,
That come before the swallow dares, and take[152]
The winds of March with beauty; violets, dim, 120
But sweeter than the lids of Juno's eyes,
Or Cytherea's[153] breath; pale primroses,
That die unmarried[154] ere they can behold
Bright Phoebus in his strength (a malady
Most incident to maids); bold oxlips, and 125
The crown imperial; lilies of all kinds,
The flower-de-luce being one. O, these I lack
To make you garlands of, and my sweet friend,
To strew them o'er and o'er!

Florizel. What, like a corse?[155]

Perdita. No, like a bank for Love to lie and play on; 130
Not like a corse; or if,[156] not to be buried,
But quick[157] and in mine arms. Come, take your flow'rs;
Methinks I play as I have seen them do
In Whitsun pastorals;[158] sure this robe of mine
Does change my disposition.

Florizel. What you do 135
Still betters what is done. When you speak, sweet,
I'd have you do it ever; when you sing,
I'd have you buy and sell so; so give alms,
Pray so; and for the ord'ring your affairs,
To sing them too. When you do dance, I wish you 140
A wave o' th' sea, that you might ever do

[151]Pluto's. [152]"Captivate." [153]Venus's.
[154]Because it grows in shade, and in spring, Milton has "the rathe primrose that forsaken dies" [*Lycidas* 142]. (Frank Kermode in Signet Classic *Winter's Tale*)
[155]"Corpse." [156]"If so." [157]"Alive."
[158]Whitsun was the season for games related to old spring festivals, and Perdita refers probably to the King and Queen in these games—identified with Robin Hood and Marian. (Kermode)

Nothing but that—move still, still so,
And own no other function. Each your doing,
So singular in each particular,
Crowns what you are doing in the present deeds, 145
That all your acts are queens.

Perdita. O Doricles,[159]
Your praises are too large; but that your youth
And the true blood which peeps fairly through 't,
Do plainly give you out an unstained shepherd,
With wisdom I might fear, my Doricles, 150
You wooed me the false way.[160]

Florizel. I think you have
As little skill[161] to fear, as I have purpose
To put you to 't. But come, our dance, I pray;
Your hand, my Perdita; so turtles[162] pair
That never mean to part.

Perdita. I'll swear for 'em. 155

Polixenes. This is the prettiest low-born lass that ever
Ran on the greensward; nothing she does or seems
But smacks of something greater than herself,
Too noble for this place.

Camillo. He tells her something
That makes her blood look out;[163] good sooth she is
The queen of curds and cream. . . .

This Island's Whose?

The Tempest *I.ii; III.ii; V.i*

Prospero. Thou poisonous slave, got by the devil himself
Upon thy wicked dam,[164] come forth! 320

Enter Caliban.

Caliban. As wicked dew as e'er my mother brushed
With raven's feather from unwholesome fen
Drop on you both![165] A southwest blow on ye
And blister you all o'er!

Prospero. For this, be sure, tonight thou shalt have cramps, 325
Side-stitches that shall pen thy breath up. Urchins[166]
Shall, for that vast of night that they may work,[167]
All exercise on thee; thou shalt be pinched
As thick as honeycomb, each pinch more stinging
Than bees that made 'em.

[159]The shepherd's name by which Perdita knows Florizel, the son of Polixenes, king of Bohemia.
[160]I.e., with flattery, or with dishonorable intentions. [161]"Cause," "reason."
[162]"Turtledoves." [163]"Blush." [164]"Mother." [165]Prospero and his daughter Miranda.
[166]Goblins in the shape of hedgehogs. [167]"That long part of the night when they are active."

Caliban. I must eat my dinner. 330
This island's mine by Sycorax my mother,
Which thou tak'st from me. When thou cam'st first,
Thou strok'st me and made much of me; wouldst give me
Water with berries in't; and teach me how
To name the bigger light, and how the less, 335
That burn by day and night. And then I loved thee
And showed thee all the qualities o' th' isle,
The fresh springs, brine pits, barren place and fertile.
Cursed be I that did so! All the charms
Of Sycorax—toads, beetles, bats, light on you! 340
For I am all the subjects that you have,
Which first was mine own king; and here you sty me
In this hard rock, whiles you do keep from me
The rest o' th' island.

Prospero. Thou most lying slave,
Whom stripes[168] may move, not kindness! I have used thee 345
(Filth as thou art) with humane care, and lodged thee
In mine own cell till thou didst seek to violate
The honor of my child.

Caliban. O ho, O ho! Would't had been done!
Thou didst prevent me; I had peopled else 350
This isle with Calibans.

Prospero. Abhorrèd slave,
Which any print of goodness wilt not take,
Being capable of all ill! I pitied thee,
Took pains to make thee speak, taught thee each hour
One thing or other. When thou didst not, savage, 355
Know thine own meaning, but wouldst gabble like
A thing most brutish, I endowed thy purposes
With words that made them known. But thy vile race,
Though thou didst learn, had that in't which good natures
Could not abide to be with. Therefore wast thou 360
Deservedly confined into this rock, who hadst
Deserved more than a prison.

Caliban. You taught me language, and my profit on't
Is, I know how to curse. The red plague rid you[169]
For learning me your language!

Prospero. Hagseed, hence! 365
Fetch us in fuel. And be quick, thou'rt best,
To answer other business. Shrug'st thou, malice?
If thou neglect'st or dost unwillingly
What I command, I'll rack thee with old cramps,
Fill all thy bones with aches,[170] make thee roar 370
That[171] beasts shall tremble at thy din.

[168]"Lashes." [169]"Destroy you." [170]Pronounced "aitches." [171]"So that."

Caliban. No, pray thee.
[Aside] I must obey. His art is of such pow'r
It would control my dam's god, Setebos,
And make a vassal of him.
Prospero. So, slave; hence!

Exit Caliban. (I.ii)

*

Caliban. Art thou afeard?
Stephano. No, monster, not I.
Caliban. Be not afeard; the isle is full of noises, 135
Sounds and sweet airs that give delight and hurt not.
Sometimes a thousand twangling instruments
Will hum about mine ears; and sometime voices
That, if I then had waked after long sleep,
Will make me sleep again; and then, in dreaming, 140
The clouds methought would open and show riches
Ready to drop upon me, that, when I waked,
I cried to dream again.
Stephano. This will prove a brave kingdom to me, where I shall have my music
for nothing. 145
Caliban. When Prospero is destroyed. (III.ii)

*

Prospero. Ye elves of hills, brooks, standing lakes, and groves,
And ye that on the sands with printless foot
Do chase the ebbing Neptune, and do fly him[172] 35
When he comes back; you demi-puppets that
By moonshine do the green sour ringlets make,[173]
Whereof the ewe not bites; and you whose pastime
Is to make midnight mushrumps,[174] that rejoice
To hear the solemn curfew; by whose aid 40
(Weak masters[175] though ye be) I have bedimmed
The noontide sun, called forth the mutinous winds,
And 'twixt the green sea and the azured vault
Set roaring war; to the dread rattling thunder
Have I given fire and rifted Jove's stout oak 45
With his own bolt; the strong-based promontory
Have I made shake and by the spurs[176] plucked up
The pine and cedar; graves at my command
Have waked their sleepers, oped, and let 'em forth
By my so potent art. But this rough magic 50
I here abjure; and when I have required[177]

[172]"Fly with him."
[173]Make "fairy rings," little circles of rank grass supposed to be made by the dancing of fairies. (Robert Langbaum in Signet Classic *Tempest*)
[174]"Mushrooms." [175]I.e., of supernatural powers. [176]"Roots." [177]"Asked for."

Some heavenly music (which even now I do)
To work mine end upon their senses that
This airy charm is for, I'll break my staff,
Bury it certain fathoms in the earth, 55
And deeper than did ever plummet sound
I'll drown my book. *Solemn music.* (V. i)

As If in Saturn's Reign

Selections from the Works of Ben Jonson

Ben Jonson (1572–1637), the posthumous son of a clergyman, studied at Westminster School under William Camden—"to whom I owe," he later wrote, "all that I am in arts, all that I know"—then practiced his stepfather's trade of bricklaying and served as a soldier in Flanders before entering Philip Henslowe's company as a playwright and actor. During a career in which he was several times imprisoned, both for killing a man in a duel and for imprudent outspokenness, he was recognized as one of the leading dramatists of his age. His plays include comedies such as Every Man in His Humor, Volpone, The Alchemist, The Silent Woman, *and* Bartholomew Fair, *his tragedies* Catiline *and* Sejanus, *and the court masques of which he became the foremost writer. In 1616 he published, in an elegant folio,* The Workes of Benjamin Jonson. *As the leading conversationalist in the gatherings of the Mermaid Tavern, he long reigned over the "Tribe of Ben" whose influence would continue throughout much of the seventeenth century.*

Although the best of his comedies are spirited and even ribald, Jonson increasingly embodied, in both practice and theory, a studied restraint that anticipated the neoclassicism of the coming age. Thus in the critical observations collected in Timber: or Discoveries Made upon Men and Matter *(1641), he commends above all the virtues of perspicuity and simplicity: "Pure and neat language I love, yet plain and customary." Though study, exercise, and imitation are necessary to the poet, "arts and precepts avail nothing, except nature be beneficial and aiding," and "the true artificer will not run away from nature as he were afraid of her, or depart from life and the likeness of truth, but speak to the capacity of his hearers." Jonson's "nature" thus presages the normative or regulative "general nature" of later critics such as Samuel Johnson; it is "the harmony and finish of the scheme of things which is beautiful to Ben Jonson," Ralph S. Walker writes in "Ben Jonson's Lyric Poetry,"*[178] *"and not any incidental development in nature." As Frye remarks in* A Natural Perspective, *"What Shakespeare has that Jonson neither has nor wants is the sense of nature as comprising not merely an order but a power, at once supernatural and connatural, expressed most eloquently in the dance and controlled either by benevolent human magic or by a divine will."*

But Jonson can be lyrically responsive to the splendor of the natural world, as in the song from Cynthia's Revels *(1601) that celebrates both the moon goddess and Queen Elizabeth:*

Queen and huntress, chaste and fair,
Now the sun is laid asleep,
Seated in thy silver chair
State in wonted manner keep;
  Hesperus entreats thy light,
  Goddess excellently bright.

[178]In *Criterion* XIII (1933–34), as reprinted in William R. Keast, ed., *Seventeenth Century English Poetry: Modern Essays in Criticism* (1962).

And in two of his finest poems (both published in 1616), "To Penshurst"—the family home of the Sidney family in Kent—and "To Sir Robert Wroth"—the son-in-law of Sir Philip Sidney's younger brother Robert—he commends the aristocratic life of these country houses not for wealth or grandeur but for closeness to the surrounding world of nature. Blessed is he, Jonson writes in Horatian vein, who can love the country, free from the vices of city and court, and take pleasure not in polished pillars or roof of gold but in soil, air, wood, and water; he in whose hospitable halls "all come in, the farmer and the clown" no less than the king, so that "freedom doth with degree dispense" in a mingling of social classes bound by respect and love; he in whose forests satyrs and dryads throng with Pan, Bacchus, and Silvanus, and Comus, the god of festivity, brings "mirth and cheer / As if in Saturn's reign it were":

Such, and no other, was that age of old
Which boasts t'have had the head of gold.

This is a Golden Age not of the mythical past alone but—as in Jonson's masque The Golden Age Restored *(1615)—of the present, for all those whose lives reflect the closeness to nature that was, and forever is, the* sine qua non *of that Saturnian reign. "The Golden Age is thus naturalized in the hall of an English mansion in a real agricultural setting," Geoffrey Walton writes of the poem to Wroth in* Metaphysical to Augustan *(1955), "and we end with an almost Homeric scene of feasting, in which bounty and humanity have temporarily overthrown the whole social hierarchy."*

Such a saturnalian reversal is the essence of holiday, and the Elizabethan masque was a holiday pageant of music and dance. Among writers of masques at the court of Elizabeth's successor, James I, Jonson "was a special case," Stephen Orgel writes in his introduction to Ben Jonson: Selected Masques *(1970), ". . . because he treated the form seriously as literature," insisting on the priority of the poetic as against the spectacular dimension of which his collaborator Inigo Jones was the master. But "both as Jonson created it and as he received it from his Elizabethan predecessors," the masque "is always about the resolution of discord; antitheses, paradoxes, and the movement from disorder to order are central to its nature." In* Pan's Anniversary, or the Shepherds' Holiday, *performed on King James's birthday, June 19, 1620, the dissonances of the Arcadian shepherds' everyday life are harmoniously resolved in a "true society" by their dancing in celebration of Pan, the archetypal nature god who "makes everywhere the spring to dwell." Exquisitely artificial though the masque may formally be, it is infused with the colors, scents, and sounds of a very concrete natural world.*

Texts and notes consulted for the poems include Poems of Ben Jonson, *ed. George Burke Johnston (1955);* Poetry of the English Renaissance, *ed. Hebel and Hudson (1929); and* Ben Jonson and the Cavalier Poets, *ed. Hugh Maclean (1974). The text and notes of* Pan's Anniversary *are those of the Yale University Press,* Complete Masques *(1969), as reprinted in Orgel's* Selected Masques.

To Penshurst

Thou art not, Penshurst, built to envious show
Of touch[179] or marble, nor canst boast a row
Of polished pillars, or a roof of gold;
Thou hast no lantern[180] whereof tales are told,
Or stairs or courts; but stand'st an ancient pile, 5
And these, grudged at, art reverenced the while.

[179] "Touchstone," a fine-grained dark stone.
[180] A small tower on a roof or dome, with its sides pierced to admit sunlight.

Thou joy'st in better marks, of soil, of air,
 Of wood, of water; therein thou art fair.
Thou hast thy walks for health as well as sport;
 Thy mount, to which the Dryads do resort, 10
Where Pan and Bacchus their high feasts have made
 Beneath the broad beech, and the chestnut shade,
That taller tree, which of a nut was set
 At his great birth,[181] where all the Muses met.
There in the writhèd bark are cut the names 15
 Of many a sylvan,[182] taken with his flames;
And thence the ruddy satyrs oft provoke
 The lighter fauns to reach thy Lady's oak.
Thy copse too, named of Gamage,[183] thou hast there,
 That never fails to serve thee seasoned deer 20
When thou wouldst feast, or exercise thy friends.
 The lower land, that to the river bends,
Thy sheep, thy bullocks, kine, and calves do feed;
 The middle grounds thy mares and horses breed.
Each bank doth yield thee conies;[184] and the tops, 25
 Fertile of wood, Ashore and Sidney's copse,
To crown thy open table, doth provide
 The purpled pheasant with the speckled side;
The painted partridge lies in every field,
 And, for thy mess, is willing to be killed. 30
And if the high-swollen Medway[185] fail thy dish,
 Thou hast thy ponds that pay thee tribute fish,
Fat aged carps that run into thy net,
 And pikes, now weary their own kind to eat,
As loath the second draught or cast to stay,[186] 35
 Officiously[187] at first themselves betray;
Bright eels that emulate them, and leap on land
 Before the fisher, or into his hand.
Then hath thy orchard fruit, thy garden flowers
 Fresh as the air, and new as are the hours. 40
The early cherry, with the later plum,
 Fig, grape, and quince, each in his time doth come;
The blushing apricot and woolly peach
 Hang on thy walls, that every child may reach.
And though thy walls be of the country stone, 45
 They'are reared with no man's ruin, no man's groan;
There's none that dwell about them wish them down,
 But all come in, the farmer and the clown,[188]

[181]On the birth of Philip Sidney [November 30, 1554], an oak tree was planted which survived until 1768. (Hebel and Hudson)
[182]"Forest dweller."
[183]Sir Robert Sidney, Viscount Lisle (the younger brother of Sir Philip), owner of Penshurst at the time Jonson is writing, had married Barbara Gamage. (Hebel and Hudson)
[184]"Rabbits." "Ashore" (or Ashour) in the next line is a grove still in existence in the twentieth century.
[185]The local river. [186]"Await." [187]"Dutifully." [188]"Peasant."

And no one empty-handed,[189] to salute
 Thy lord and lady, though they have no suit.[190] 50
Some bring a capon, some a rural cake,
 Some nuts, some apples; some that think they make
The better cheeses, bring 'em, or else send
 By their ripe daughters whom they would commend
This way to husbands, and whose baskets bear 55
 An emblem of themselves in plum or pear.
But what can this, more than express their love,
 Add to thy free provisions, far above
The need of such, whose liberal board doth flow
 With all that hospitality doth know? 60
Where comes no guest but is allowed to eat
 Without his fear, and of thy lord's own meat;
Where the same beer and bread, and self-same wine
 That is his lordship's shall be also mine.
And I not fain[191] to sit, as some this day 65
 At great men's tables, and yet dine away.[192]
Here no man tells[193] my cups, nor, standing by,
 A waiter doth my gluttony envy,
But gives me what I call[194] and lets me eat;
 He knows below[195] he shall find plenty of meat. 70
Thy tables hoard not up for the next day,
 Nor when I take my lodging need I pray
For fire or lights or livery;[196] all is there
 As if thou then wert mine, or I reigned here;
There's nothing I can wish, for which I stay. 75
 That found King James, when hunting late this way
With his brave son, the prince, they saw thy fires
 Shine bright on every hearth as the desires
Of thy Penates[197] had been set on flame
 To entertain them, or the country came 80
With all their zeal to warm their welcome here.
 What (great I will not say, but sudden) cheer
Didst thou then make 'em! and what praise was heaped
 On thy good lady then! who therein reaped
The just reward of her high huswifery; 85
 To have her linen, plate, and all things nigh
When she was far, and not a room but dressed
 As if it had expected such a guest!

[189]"Without a gift." [190]"Plea," "request." [191]"Obliged," "required."
[192]Jonson told Drummond that "being at the end of my Lord Salisbury's table . . . and he [Jonson] demanded by my Lord why he was not glad: 'My Lord,' said he, 'you promised me I should dine with you, but I do not,' for he had none of his meat—he esteemed only that his meat which was of his own dish. (Hebel and Hudson)
[193]"Counts." [194]"Call for." [195]"Below stairs," in the servants' quarters.
[196]"Provisions." [197]Penátês, Roman household gods.

These, Penshurst, are thy praise, and yet not all.
  Thy lady's noble, fruitful, chaste withal; 90
His children thy great lord may call his own,
  A fortune in this age but rarely known.
They are and have been taught religion; thence
  Their gentler spirits have sucked innocence.
Each morn and even they are taught to pray 95
  With the whole household, and may every day
Read, in their virtuous parents' noble parts,
  The mysteries of manners, arms, and arts.
Now, Penshurst, they that will proportion[198] thee
  With other edifices when they see 100
Those proud, ambitious heaps and nothing else,
  May say, their lords have built, but thy lord dwells.

To Sir Robert Wroth

How blest art thou, canst love the country, Wroth,
  Whether by choice, or fate, or both;
And though so near the city and the court,
  Art ta'en with neither's vice nor sport;
That, at great times, are no ambitious guest 5
  Of sheriff's dinner or mayor's feast,
Nor com'st to view the better cloth of state,
  The richer hangings, or crown-plate,
Nor throng'st, when masquing is, to have a sight
  Of the short bravery[199] of the night, 10
To view the jewels, stuffs, the pains, the wit
  There wasted, some not paid for yet;
But canst at home in thy securer rest
  Live, with un-bought provision blessed,
Free from proud porches, or their gilded roofs, 15
  'Mongst lowing herds and solid hoofs,
Alongst the curlèd woods and painted meads
  Through which a serpent river leads
To some cool, courteous shade, which he calls his,
  And makes sleep softer than it is! 20
Or, if thou list the night in watch to break,
  Abed canst hear the loud stag speak
In spring, oft rousèd for thy master's[200] sport,
  Who, for it, makes thy house his court;
Or with thy friends, the heart of all the year,[201] 25
  Divid'st upon the lesser deer;

[198]"Compare." [199]"Splendor." [200]I.e., the king's. [201]I.e., in summer.

In autumn at the partridge makes a flight,
    And giv'st thy gladder guests the sight;
And in the winter hunt'st the flying hare
    More for thy exercise than fare,[202] 30
While all that follow their glad ears apply
    To the full greatness of the cry;[203]
Or hawking at the river, or the bush,
    Or shooting at the greedy thrush,
Thou dost with some delight the day out-wear, 35
    Although the coldest of the year!
The whilst the several seasons thou hast seen
    Of flow'ry fields, of copses green,
The mowèd meadows with the fleecèd sheep,
    And feasts that either shearers keep, 40
The ripened ears, yet humble in their height,
    And furrows laden with their weight,
The apple-harvest, that doth longer last,
    The hogs returned home fat from mast,[204]
The trees cut out in log; and those boughs made 45
    A fire now, that lent a shade!
Thus Pan and Sylvan[205] having had their rites,
    Comus[206] puts in for new delights,
And fills thy open hall with mirth and cheer
    As if in Saturn's reign it were. 50
Apollo's harp and Hermes' lyre resound,
    Nor are the Muses strangers found;
The rout of rural folk come thronging in
    (Their rudeness[207] then is thought no sin),
Thy noblest spouse affords them welcome grace, 55
    And the great heroes of her race
Sit mixed with loss of state or reverence:
    Freedom doth with degree dispense.[208]
The jolly wassail walks the often round,[209]
    And in their cups their cares are drowned; 60
They think not, then, which side the cause shall leese,[210]
    Nor how to get the lawyer fees.
Such, and no other, was that age of old
    Which boasts t'have had the head of gold;
And such, since thou canst make thine own content, 65
    Strive, Wroth, to live long innocent.

[202]"Food." [203]I.e., the baying of hounds. (Maclean)
[204]Nuts, mashed to serve as food for hogs. (Maclean)
[205]Silvanus, a Roman forest god sometimes equated with the Greek Pan.
[206]In later antiquity, Comus [from Greek *kômos*, "the revels"] was the god of festive mirth and jollification. (Maclean)
[207]"Rusticity." [208]"Does without hierarchy."
[209]I.e., the drinking cups are regularly refilled. (Maclean) [210]"Lose."

Let others watch in guilty arms, and stand
  The fury of a rash command,
Go enter breaches, meet the cannon's rage,
  That they may sleep with scars in age, 70
And show their feathers shot, and colors torn,
  And brag that they were therefore born.
Let this man sweat and wrangle at the bar
  For every price, in every jar,[211]
And change possessions oftener with his breath 75
  Than either money, war, or death;
Let him than hardest sires more disinherit,[212]
  And each where[213] boast it as his merit
To blow up orphans, widow, and their states,[214]
  And think his power doth equal Fate's. 80
Let that go heap a mass of wretched wealth,
  Purchased by rapine, worse than stealth,
And brooding o'er it sit with broadest[215] eyes,
  Not doing good, scarce when he dies.
Let thousands more go flatter vice, and win 85
  By being organs[216] to great sin;
Get place, and honor, and be glad to keep
  The secrets that shall break their sleep,
And, so they ride in purple, eat in plate,[217]
  Though poison, think it a great fate. 90
But thou, my Wroth, if I can truth apply,
  Shalt neither that nor this envy;
Thy peace is made; and when man's state is well,
  'Tis better if he there can dwell.
God wisheth none should wreck on a strange shelf;[218] 95
  To Him, man's dearer than t'himself.
And howsoever we may think things sweet,
  He always gives what He knows meet,[219]
Which who can use is happy: such be thou.
  Thy morning's and thy evening's vow 100
Be thanks to Him, and earnest prayer to find
  A body sound, with sounder mind;[220]
To do thy country service, thyself right,
  That neither want do thee affright
Nor death; but when thy latest sand is spent, 105
  Thou may'st think life a thing but lent.

[211]"Discord," "contention."
[212]Let him disinherit more children than the most severe fathers do. (Maclean)
[213]"Everywhere." [214]"Estates." [215]"Wide-open." [216]"Instruments."
[217]I.e., dine off gold or silver platters. (Maclean) [218]"Shoal." [219]"Fitting," "suitable."
[220]Cf. Juvenal, *Satire* X.356: "Orandum est ut sit mens sana in corpore sano" ("We must pray to have a sound mind in a sound body").

FROM Pan's Anniversary, or the Shepherds' Holiday

The Scene: Arcadia

As it was presented at court before King James. 1620.

The Inventors

Inigo Jones   Ben Jonson

The first presentation is of three nymphs strewing several sorts of flowers followed by an old shepherd with a censer and perfumes.

1st Nymph. Thus, thus begin the yearly rites
Are due to Pan on these bright nights;
His morn now riseth and invites 5
To sports, to dances and delights:
All envious and profane, away;
This is the shepherds' holiday.
2nd Nymph. Strew, strew the glad and smiling ground
With every flower, yet not confound 10
The primrose drop, the spring's own spouse;
Bright day's-eyes and the lips of cows;[221]
The garden star, the queen of May,
The rose to crown the holiday.
3rd Nymph. Drop, drop, you violets, change your hues, 15
Now red, now pale, as lovers use,
And in your death go out as well
As when you lived, into the smell,
That from your odor, all may say
This is the shepherds' holiday. 20
Shepherd. Well done, my pretty ones; rain roses still,
Until the last be dropped. Then hence, and fill
Your fragrant prickles[222] for a second shower;
Bring corn-flag, tulips and Adonis' flower,[223]
Fair ox-eye, goldilocks and columbine, 25
Pinks, goulands, king-cups and sweet sops-in-wine,[224]
Blue harebells, paigles,[225] pansies, calaminth,
Flower-gentle,[226] and the fair-haired hyacinth;
Bring rich carnations, flower-de-luces,[227] lilies,
The checked and purple-ringèd daffodillies, 30
Bright crown-imperial, king's spear,[228] hollyhocks,
Sweet Venus' navel, and soft lady's-smocks;[229]
Bring too some branches forth of Daphne's hair,
And gladdest myrtle for these posts to wear
With spikenard weaved, and marjoram between, 35
And starred with yellow-golds and meadow's queen,[230]
That when the altar, as it ought, is dressed,

[221]Daisies and cowslips. [222]Wicker flower baskets.
[223]"Corn-flag" is gladiolus; "Adonis' flower" is variously the anemone and the rose.
[224]"Goulands" and "king-cups" are types of buttercups; "sops-in-wine" are gillyflowers.
[225]"Blue harebells" are bluebells; "paigles" are cowslips. [226]Amaranth. [227]White iris.
[228]Yellow asphodel. [229]Wall pennywort and cuckooflower. [230]Marigolds and meadowsweet.

More odor come not from the Phoenix' nest;
The breath thereof Panchaia[231] may envy,
The colors China, and the light the sky. . . . 40

Shepherd. And come you prime Arcadians forth, that taught
By Pan the rites of true society,
From his loud music all your manners wrought,
And made your commonwealth a harmony, 140
Commending so to all posterity
Your innocence from that fair fount of light,
As still you sit without the injury
Of any rudeness folly can, or spite;
Dance from the top of the Lycaean mountain[232] 145
Down to this valley, and with nearer eye
Enjoy what long in that illumined fountain
You did far off, but yet with wonder, spy.

HYMN I

1st Arcadian. Of Pan we sing, the best of singers, Pan, 150
That taught us swains how first to tune our lays,
And on the pipe more airs than Phoebus can.
Chorus. Hear, O you groves, and hills resound his name.
2nd Arcadian. Of Pan we sing, the best of leaders, Pan,
That leads the Naiads and the Dryads forth, 155
And to their dances more than Hermes can.
Chorus. Hear, O you groves, and hills resound his worth.
3rd Arcadian. Of Pan we sing, the best of hunters, Pan,
That drives the hart to seek unusèd ways,
And in the chase more than Sylvanus can. 160
Chorus. Hear, O you groves, and hills resound his praise.
4th Arcadian. Of Pan we sing, the best of shepherds, Pan,
That keeps our flocks and us, and both leads forth
To better pastures than great Pales can.[233]
Chorus. Hear, O you groves, and hills resound his worth. 165
And while his powers and praises thus we sing,
The valleys let rebound, and all the rivers ring.
The masquers descend and dance their entry.

HYMN 2

Pan is our all, by him we breathe, we live, 170
We move, we are; 'tis he our lambs doth rear,
Our flocks doth bless, and from the store doth give
The warm and finer fleeces that we wear.
He keeps away all heats and colds,
Drives all diseases from our folds, 175
Makes everywhere the spring to dwell,
The ewes to feed, their udders swell;

[231]A mythical island off the Arabian coast famous for spices and scents.
[232]Lycaeus, Pan's native mountain in Arcadia.
[233]God (or goddess) of the country, and of flocks and shepherds.

But if he frown, the sheep (alas),
The shepherds wither, and the grass.
Strive, strive to please him then by still increasing thus 180
The rites are due to him, who doth all right for us.

The main dance. . . .

Shepherd. End you the rites, and so be eased
Of these, and then great Pan is pleased.

HYMN 4

[Chorus]. Great Pan, the father of our peace and pleasure, 225
Who giv'st us all this leisure,
Hear what thy hallowed troop of herdsmen pray
For this their holiday,
And how their vows to thee they in Lycaeum[234] pay.
So may our ewes receive the mounting rams, 230
And we bring thee the earliest of our lambs;
So may the first of all our fells[235] be thine,
And both the beestning[236] of our goats and kine;
As thou our folds dost still secure,
And keep'st our fountains sweet and pure, 235
Driv'st hence the wolf, the tod, the brock,[237]
Or other vermin from the flock;
That we preserved by thee, and thou observed by us,
May both live safe in shade of thy loved Maenalus.[238]

Shepherd. Now each return unto his charge, 240
And though today you have lived at large,
And well your flocks have fed their fill,
Yet do not trust your hirelings still.[239]
See, yond' they go, and timely do
The office you have put them to, 245
But if you often give this leave,
Your sheep and you they will deceive.

The End.

18

Mapping the Broken Circle: The Scientific Revolution

The scientific revolution and the humanistic renaissance, as noted already, overlapped chronologically in the sixteenth and seventeenth centuries and were closely interdependent. The work of Copernicus in astronomy and Vesalius in anatomy presupposed accurate translation and close study of Ptolemy and Galen. Artists such as Leone Battista Alberti, Leonardo da Vinci (Chapter 16 above), and Albrecht Dürer made important contribu-

[234] Lycaeus, Pan's native mountain. [235] "Fleeces." [236] "First milk."
[237] "Tod" is fox, "brock" is badger. [238] A mountain of Arcadia. [239] "Always."

tions to mathematical perspective and anatomy, and their meticulous observations of the natural world were paralleled by those of horticulturalists such as John Gerard (Chapter 16). And both the Hermetic philosopher Bruno (Chapter 16) and the astronomer Kepler were defenders of Copernicus, their writings deeply imbued with Neoplatonic thought. Natural magic and natural philosophy were by no means clearly distinguished, except as practical and theoretical aspects of the same pursuit, and the rigorous endeavor of humanists such as Lorenzo Valla and Erasmus (Chapter 16) to establish accurate texts even of sacred works (in the teeth of ecclesiastical hostility) was itself a scientific enterprise. Despite its many connections with humanism, however, the revolutionary reexamination of nature that began with Copernicus and culminated in Newton resulted in radical revision of the medieval/Renaissance world picture. Not since the centuries between Thales and Aristotle (Chapter 9), when the notion of physis *as an autonomous power came into being and attained its classic formulation in ancient Greece, had so fundamental a change in the Western conception of nature taken place.*

The most direct consequence of Copernicus's bold heliocentric hypothesis was of course the decentering of the earth, and of man, in a cosmos that no longer circled around them. Beyond this, the vastly larger scale required by the absence of stellar parallax observed from an orbiting earth led thinkers such as Bruno—influenced also by Nicholas Cusanus (Chapter 16) and the rediscovery of Lucretius (Chapter 10)—to speculate on the possible infinity of the universe and of the number of worlds within it. Then in rapid succession, with Tycho Brahe, Kepler, and Galileo, fell a series of dogmas almost equally central to the Aristotelian/Ptolemaic cosmology: the crystalline spheres on which the planets and fixed stars were thought to revolve, spun round by an outermost primum mobile; the perfect circular paths that heavenly bodies were assumed by their nature to follow; the aethereal purity of the superlunary world, including a sun (Copernicus's Lamp, Mind, and Ruler of the universe) now seen to be pocked with black spots; and so forth, leaving little of what, a few generations before, had seemed unassailably certain. What appeared to be unraveling, as the implications (which Copernicus himself, in delaying publication of his system, had foreboded) of the shift from a geocentric to heliocentric cosmos slowly became evident, was no single belief but the tightly interwoven fabric of a world order immutably governed by processes at once natural and divine.

But if the period between the De Revolutionibus *of Copernicus and Newton's* Principia *"is quite rightly regarded as a turning point in the history of the world," as Paolo Rossi affirms in "Hermeticism, Rationality, and the Scientific Revolution,"*[1] *this is "not only because of important discoveries, new theories and novel experiments. It was a time when certain ideas and themes that are inextricably bound up with 'science' came to the fore. These allow us to see the sudden break, the discontinuity that separates the new science from the old and helps us to understand some of the essential and decisive factors of what we call modern thought." Rossi distinguishes fourteen of these factors, including "a new appreciation of technical skills and mechanical arts," "the new importance of scientific instruments," and "the idea of collaboration." Several others, such as "the notion of the world as a machine," "the conception of God as an engineer or watchmaker," "the introduction of the dimension of time," and "the notion of progress" were potentialities of the Cartesian or Newtonian systems that would become increasingly prominent in the post-Newtonian Enlightenment. So, too, would "the theory that man can only know what he does or what he himself constructs."*

But all these were perhaps implicit consequences of Rossi's first theme: "the refutation of the priestly idea of knowledge inherent in hermeticism, in the alchemical literature, and in much of the natural philosophy of the

[1]In M. L. Righini Bonelli and William R. Shea, eds., *Reason, Experiment, and Mysticism in the Scientific Revolution* (1975).

Renaissance"—that is, in most of what it inherited from the Middle Ages or revived, through humanist influence, from the Neoplatonic/Hermetic legacy of antiquity. The scientist was not, like Bruno, an oracle delivering heroically frenzied pronouncements to an elite (even Copernicus considered communicating his "philosophic mysteries" only orally, like Pythagoras, to an esoteric circle). Rather, he was an observer, experimenter, and calculator, demonstrating through empirical evidence and rational deduction truths that any knowledgeable person ought to be able to attain (or refute); in this way, he more resembled a craftsman constructing what he knows than a priest eliciting knowledge from a privileged source beyond him.

A further consequence, as Rossi points out, is the breakdown of the Renaissance opposition between nature and an art that imitates and even—as for Montaigne (Chapter 16), Tasso, Spenser, or Shakespeare's Perdita in The Winter's Tale *(Chapter 17)—perverts or falsifies nature. Bacon's statement, quoted by Rossi from his* History of the Arts, *"that the artificial does not differ from the natural in form or essence, but only in the efficient" (that is, in the agent), parallels Polixenes' defense of art, in Shakespeare's play, as a creation of nature not essentially different from it.*

From Pythagorean, Platonic, and Aristotelian philosophy and Ptolemaic astronomy, the Middle Ages and the Renaissance had derived a cosmology of concentric circles centered on the earth. Now the circle was broken, not because Copernicus (who maintained most of Ptolemy's epicycles, or circles on circles) had shifted the sun to the center, nor because Kepler had shown (very few were attending) that the planets orbited in elliptical paths, but because the dissolution of the outer sphere of the fixed stars into a panorama of infinitely receding worlds left the universe, as Francis Quarles wrote in the sixth of his Emblems *(1635),*

a vast circumference, where none
Can find a center,

and therefore no circle at all. Such a cosmos—or was it, after all, a chaos?—might inspire terror (as in Pascal), or reverence (as in Newton and later in Kant), or fascination (as in Fontenelle), but it could not be compassed or comprehended as a whole. Our modern world of the broken circle, Georg Lukács writes in The Theory of the Novel *(1920; English trans. 1971), "has become infinitely large and each of its corners is richer in gifts and dangers than the world of the Greeks, but such wealth cancels out the positive meaning—the totality—on which their life was based": a vanished totality for which we continually seek some counterpart by "the endless path of an approximation that is never fully accomplished." To apprehend some of the limitless implications of this new, perpetually shifting, and inherently unfinished world picture—to begin to map the broken circle—was the task undertaken in various ways both by leading scientists and by philosophers as diverse as Bacon, Descartes, Pascal, Spinoza, and Leibniz. For all of them, the scientific revolution was a central reality that had forever altered their view of the world.*

We Thus Rather Follow Nature

On the Revolutions of the Celestial Spheres,
by Nicholas Copernicus, translated by Thomas S. Kuhn

Nicholas Koppernigk or Kopernik (latinized as Copernicus) was born at Torun in the Kingdom of Poland and educated at Cracow and at Bologna, Padua, and Ferrara in Italy, studying various subjects including canon law, medicine, and mathematics. In 1512 he settled at Frauenberg in

East Prussia, where he was canon of the cathedral and practiced medicine. His initially tentative revision of the Ptolemaic system was first formulated in a brief sketch circulated among friends, the Commentariolus (Little Commentary) *of 1512, which gained the attention and interest of a larger audience, including a Roman Church then sympathetic to astronomical speculation and calendrical reform. At the urging of a young professor from the Protestant University of Wittenberg, Georg Joachim Rheticus, Copernicus agreed to give his ideas wider circulation (though anonymously) in the* Narratio Prima (First Narration) *of 1540. He finally allowed publication, in the year of his death, 1543, of his life's work,* De Revolutionibus Orbium Coelestium Libri Sex (Six Books on the Revolutions of the Celestial Spheres); *it was prefaced by an unauthorized disclaimer by the Lutheran pastor Andreas Osiander asserting that his system was nothing more than a mathematical hypothesis.*

"Looking through the De Revolutionibus *one is immediately made aware of the fact that, true to his training, Copernicus had studied the* Almagest *very carefully indeed. For the* De Revolutionibus *is the* Almagest*" of Ptolemy (Chapter 12 above), Marie Boas writes in* The Scientific Renaissance, *"book by book and section by section, rewritten to incorporate the new Copernican theory, but otherwise altered as little as might be. Kepler was to remark later that Copernicus interpreted Ptolemy, not nature, and there is some truth in the comment; to Copernicus the way to nature lay in a re-interpretation of Ptolemaic astronomy, wrong in details, but right in conception. . . . Copernicus did not wish to claim novelty, which had no appeal to him; he claimed to be doing no more than revive Pythagorean doctrines . . ." And in doing so, Boas continues, he "was only following humanist precepts: he was trying to replace Aristotelian authority, which to the sixteenth century represented the outmoded intellectual pattern of the Middle Ages, with a system equally derived from Greek authority, which had the added advantage of being consonant with Platonic doctrines, so much more highly esteemed now than those of Plato's pupil Aristotle."*

But this conservative thinker's one consciously bold innovation—displacement of a geocentric by a heliocentric system—which he feared might incur disbelief and scorn, set in motion other, no less fundamental changes. "All unconsciously," Boas remarks, "by denying one essential difference between the heavenly and terrestrial spheres," namely that circular motion pertained to the former and rectilinear to the latter, as Aristotle had thought, "Copernicus began that encroachment on cosmical dualism that was destined to end fatally. . . . Copernicus was the first modern cosmologer to begin to break down the old-established barriers between the Earth and the celestial regions; one by one these barriers were demolished until in the Newtonian universe modern physics allowed a return to the unified and uniform cosmos of the original pre-Socratic conception." In this and other ways, as Thomas S. Kuhn writes in The Copernican Revolution *(1957), from which the following translations are taken, "Copernicus is neither an ancient nor a modern but rather a Renaissance astronomer in whose work the two traditions merge." Perhaps his deepest affinity with Ptolemy, who proposed in uniform circular motion a theory in accord with "the simplicity of heavenly things" as he understood them, is his belief—shared by modern science as well—that we can only "follow Nature" by seeking the simplest explanation consonant with accurate observation of the world.*

FROM THE *Preface: To the Most Holy Lord, Pope Paul III*

I may well presume, most Holy Father, that certain people, as soon as they hear that in this book about the Revolutions of the Spheres of the Universe I ascribe movement to the earthly globe, will cry out that, holding such views, I should at once be hissed off the stage. For I am not so pleased with my own work that I should fail duly to weigh the judgment which others may pass thereon; and though I know that the speculations of a philosopher are far removed from the judgment of the multitude—for his aim is to seek truth in all things as far as God has permitted human reason so to do—yet I hold that opinions which are quite erroneous should be avoided.

Thinking therefore within myself that to ascribe movement to the Earth must indeed seem an absurd performance on my part to those who know that many centuries have consented to the establishment of the contrary judgment, namely that the Earth is placed immovably as the central point in the middle of the Universe, I hesitated long whether, on the one hand, I should give to the light these my Commentaries written to prove the Earth's motion, or whether, on the other hand, it were better to follow the example of the Pythagoreans and others who were wont to impart their philosophic mysteries only to intimates and friends, and then not in writing but by word of mouth, as the letter of Lysis to Hipparchus witnesses.[2] In my judgment they did so not, as some would have it, through jealousy of sharing their doctrines, but as fearing lest these so noble and hardly won discoveries of the learned should be despised by such as either care not to study aught save for gain, or—if by the encouragement and example of others they are stimulated to philosophic liberal pursuits—yet by reason of the dullness of their wits are in the company of philosophers as drones among bees. Reflecting thus, the thought of the scorn which I had to fear on account of the novelty and incongruity of my theory, well-nigh induced me to abandon my project. . . .

That I allow the publication of these my studies may surprise your Holiness the less in that, having been at such travail to attain them, I had already not scrupled to commit to writing my thoughts upon the motion of the Earth.[3] How I came to dare to conceive such motion of the Earth, contrary to the received opinion of the Mathematicians and indeed contrary to the impression of the senses, is what your Holiness will rather expect to hear. So I should like your Holiness to know that I was induced to think of a method of computing the motions of the spheres by nothing less than the knowledge that the Mathematicians are inconsistent in these investigations.

For, first, the mathematicians are so unsure of the movements of the Sun and Moon that they cannot even explain or observe the constant length of the seasonal year. Secondly, in determining the motions of these and of the other five planets, they use neither the same principles and hypotheses nor the same demonstrations of the apparent motions and revolutions. So some use only homocentric circles,[4] while others [employ] eccentrics and epicycles.[5] Yet even by these means they do not completely attain their ends. Those who have relied on homocentrics, though they have proven that some different motions can be compounded therefrom, have not thereby been able fully to establish a system which agrees with the phenomena. Those again who have devised eccentric systems, though they appear to have well-nigh established the seeming motions by calculations agreeable to their assumptions, have yet made many admissions which seem to vio-

[2]This letter, which Copernicus had at one time intended to include in the *De Revolutionibus*, describes the Pythagorean and Neoplatonic injunction against revealing nature's secrets to those who are not initiates of a mystical cult. Reference to it here exemplifies Copernicus' participation in the Renaissance revival of Neoplatonism. (Kuhn)

[3]Some years before the publication of the *De Revolutionibus* Copernicus had circulated among his friends a short manuscript called the *Commentariolus*, describing an earlier version of his sun-centered astronomy. A second advance report of Copernicus' major work, the *Narratio Prima* by Copernicus' student, Rheticus, had appeared in 1540 and again in 1541. (Kuhn)

[4]The Aristotelian system, derived by Aristotle from Eudoxus and Callippus, and revived in Europe shortly before Copernicus' death by the Italian astronomers Fracastoro and Amici. (Kuhn)

[5]On eccentrics and epicycles, see introduction to Ptolemy in Chapter 12 above.

late the first principle of uniformity in motion. Nor have they been able thereby to discern or deduce the principal thing—namely the shape of the Universe and the unchangeable symmetry of its parts. With them it is as though an artist were to gather the hands, feet, head, and other members for his images from diverse models, each part excellently drawn, but not related to a single body, and since they in no way match each other, the result would be monster rather than man. So in the course of their exposition, which the mathematicians call their system, . . . we find that they have either omitted some indispensable detail or introduced something foreign and wholly irrelevant. This would of a surety not have been so had they followed fixed principles; for if their hypotheses were not misleading, all inferences based thereon might be surely verified. Though my present assertions are obscure, they will be made clear in due course.

FROM *Book I, Chapter 10: Of the Order of the Heavenly Bodies*

. . . We therefore assert that the center of the Earth, carrying the Moon's path, passes in a great circuit among the other planets in an annual revolution around the Sun; that near the Sun is the center of the Universe; and that whereas the Sun is at rest, any apparent motion of the Sun can be better explained by motion of the Earth. Yet so great is the Universe that though the distance of the Earth from the Sun is not insignificant compared with the size of any other planetary path, in accordance with the ratios of their sizes, it is insignificant compared with the distances of the Sphere of the Fixed Stars.

I think it easier to believe this than to confuse the issue by assuming a vast number of Spheres, which those who keep Earth at the center must do. We thus rather follow Nature, who producing nothing vain or superfluous often prefers to endow one cause with many effects. Though these views are difficult, contrary to expectation, and certainly unusual, yet in the sequel we shall, God willing, make them abundantly clear at least to mathematicians.

Given the above view—and there is none more reasonable—that the periodic times are proportional to the sizes of the Spheres, then the order of the Spheres, beginning from the most distant, is as follows. Most distant of all is the Sphere of the Fixed Stars, containing all things, and being therefore itself immovable. It represents that to which the motion and position of all the other bodies must be referred. . . . Next is the planet Saturn, revolving in 30 years. Next comes Jupiter, moving in a 12-year circuit; then Mars, who goes round in 2 years. The fourth place is held by the annual revolution [of the Sphere] in which the Earth is contained, together with the Sphere of the Moon as on an epicycle. Venus, whose period is 9 months, is in the fifth place, and sixth is Mercury, who goes round in the space of 80 days.

In the middle of all sits Sun enthroned. In this most beautiful temple could we place this luminary in any better position from which he can illuminate the whole at once? He is rightly called the Lamp, the Mind, the Ruler of the Universe; Hermes Trismegistus names him the Visible God, Sophocles' Electra calls him the All-seeing. So the Sun sits as upon a royal throne ruling his children the planets which circle round him. The Earth has the Moon at her service. As Aristotle says, in his *On [the Generation of] Animals*, the Moon has the closest relationship with the Earth. Meanwhile the Earth conceives by the Sun, and becomes pregnant with an annual rebirth.

So we find underlying this ordination an admirable symmetry in the Universe, and a clear bond of harmony in the motion and magnitude of the Spheres such as can be discovered in no other wise. . . . All these phenomena proceed from the same cause, namely Earth's motion.

That there are no such phenomena for the fixed stars proves their immeasurable distance, because of which the outer sphere's [apparent] annual motion or its [parallactic] image is invisible to the eyes. . . . So great is this divine work of the Great and Noble Creator!

The Miracle of a New Star

On a New Star, Not Previously Seen within the Memory of Any Age Since the Beginning of the World, by Tycho Brahe, translated by John H. Walden

The Danish nobleman Tycho Brahe (1546–1601) took an early interest in astronomy and astrology and studied these subjects and alchemy at Copenhagen, Leipzig, Rostock, and Augsburg. In 1572 the appearance of a "new star" (nova, or supernova) in Cassiopeia, which also attracted the attention of astronomers such as Thomas Digges in England and Michael Maestlin in Germany, seemed a miracle almost without parallel; according to orthodox belief, the aethereal heavens were immutable, so that nothing new was possible there. Tycho's failure to find a parallax convinced him that the new star lay in regions beyond the moon, thus refuting that ancient dogma, just as his observations of the comet of 1577 later led him to reject equally sacrosanct beliefs in crystalline spheres and in circular orbits for all heavenly bodies. After publication of the brief work On a New Star (De Stella Nova) *in 1573—from which our selections are taken, in John H. Walden's translation from a* Source Book in Astronomy *(1929), ed. Harlow Shapley and Helen H. Howarth—Tycho was granted the island of Hveen by King Frederick II of Denmark in 1576. There he constructed a castle, Uraniborg, and an observatory, Stjarneborg, equipped with immense quadrants, astrolabes, and other instruments that permitted him and his helpers to make astronomical observations of unparalleled accuracy. Besides compiling the most complete star catalogue ever made, Tycho meticulously charted the movements of the planets from night to night and year to year, providing the data that would later enable Kepler to calculate the elliptical orbit of Mars. (Kepler had joined Tycho in Prague, where Tycho became imperial mathematician to the Emperor Rudolph II in 1599, two years after the young King Christian of Denmark had cut off his funding.) Tycho never accepted the heliocentric Copernican system but put forth instead a long-influential rival theory (apparently derived, without acknowledgment, from Heracleides of Pontus in the fourth century* B.C.*) in which the other planets orbited the sun while the sun orbited the earth. But he was, as Boas writes, "the greatest observational astronomer since Hipparchos" (see introduction to Ptolemy in Chapter 12 above), and his discarding of the crystalline spheres of Aristotelian and Ptolemaic astronomy—though it left him with no explanation for what caused planetary movements—had immense importance for the new conception of an open universe then coming into being.*

Its First Appearance in 1572. Last year [1572], in the month of November, on the eleventh day of that month, in the evening, after sunset, when, according to my habit, I was contemplating the stars in a clear sky, I noticed that a new and unusual star, surpassing the other stars in brilliancy, was shining almost directly above my head; and since I had, almost from boyhood, known all the stars of the heaven perfectly (there is no great difficulty in attaining that knowledge), it was quite evident to me that there had never before been any star in that place in the sky, even the smallest, to say nothing of a star so conspicuously bright as this. I was so astonished at this sight that I was not ashamed to doubt the trustworthiness of my own eyes. But when I observed that others, too, on having the place pointed out to them, could see that there was really a star there, I had no further doubts. A miracle indeed, either the greatest of all that have occurred in the whole range of nature since the beginning of the world, or one certainly that is to be classed with those attested by the Holy Oracles, the staying of the Sun in its course in answer to the prayers of Joshua, and the darkening of the Sun's face at the time of the Crucifixion.

For all philosophers agree, and facts clearly prove it to be the case, that in the ethereal region of the celestial world no change, in the way either of generation or of corruption, takes place; but that the heavens and the celestial bodies in the heavens are without increase or diminution, and that they undergo no alteration, either in number or in size or in light or in any other respect; that they always remain the same, like unto themselves in all respects, no years wearing them away. Furthermore, the observations of all the founders of the science, made some thousands of years ago, testify that all the stars have always retained the same number, position, order, motion, and size as they are found, by careful observation on the part of those who take delight in heavenly phenomena, to preserve even in our own day. Nor do we read that it was ever before noted by any one of the founders that a new star had appeared in the celestial world, except only by Hipparchus, if we are to believe Pliny. For Hipparchus, according to Pliny (Book II of his *Natural History*), noticed a star different from all others previously seen, one born in his own age. . . .

Its Position with Reference to the Diameter of the World and Its Distance from the Earth, the Center of the Universe. It is a difficult matter, and one that requires a subtle mind, to try to determine the distances of the stars from us, because they are so incredibly far removed from the earth; nor can it be done in any way more conveniently and with greater certainty than by the measure of the parallax [diurnal], if a star have one. For if a star that is near the horizon is seen in a different place than when it is at its highest point and near the vertex, it is necessarily found in some orbit with respect to which the Earth has a sensible size. . . .

In order, therefore, that I might find out in this way whether this star was in the region of the Element[6] or among the celestial orbits, and what its distance was from the Earth itself, I tried to determine whether it had a parallax, and, if so, how great a one; and this I did in the following way. I observed the distance between this star and Schedir of Cassiopeia (for the latter and the new star were both nearly on the meridian), when the star was at its nearest point to the vertex, being only 6 degrees removed from the zenith itself. . . . I made the same observation when the star was farthest from the zenith and at its nearest point to the horizon, and in each case I found that the distance from the above-mentioned fixed star was exactly the same, without the variation of a minute, namely 7 degrees and 55 minutes. Then I went through the same process, making numerous observations with other stars. Whence I conclude that this new star has no diversity of aspect, even when it is near the horizon. For otherwise in its least altitude it would have been farther away from the above-mentioned star in the breast of Cassiopeia than when in its greatest altitude. Therefore, we shall find it necessary to place this star, not in the region of the Element, below the Moon, but far above, in an orbit with respect to which the Earth has no sensible size. For if it were in the highest region of the air, below the

[6]That is, the region of the air, between the earth and the moon.

hollow region of the Lunar sphere, it would, when nearest the horizon, have produced on the circle a sensible variation of altitude from that which it held when near the vertex. . . .

Therefore this new star is neither in the region of the Element, below the Moon, nor among the orbits of the seven wandering stars, but it is in the eighth sphere,[7] among the other fixed stars, which was what we had to prove. Hence it follows that it is not some peculiar kind of comet or some other kind of fiery meteor become visible. For none of these are generated in the heavens themselves, but they are below the Moon, in the upper region of the air, as all philosophers testify; unless one would believe with Albategnius that comets are produced, not in the air, but in the heavens. For he believes that he has observed a comet above the Moon, in the sphere of Venus. That this can be the case, is not yet clear to me. But, please God, sometime, if a comet shows itself in our age, I will investigate the truth of the matter.[8]

Even should we assume that it can happen (which I, in company with other philosophers, can hardly admit), still it does not follow that this star is a kind of comet; first, by reason of its very form, which is the same as the form of the real stars and different from the form of all the comets hitherto seen, and then because, in such a length of time, it advances neither latitudinally nor longitudinally by any motion of its own, as comets have been observed to do. For, although these sometimes seem to remain in one place several days, still, when the observation is made carefully by exact instruments, they are seen not to keep the same position for so very long or so very exactly. I conclude, therefore, that this star is not some kind of comet or a fiery meteor, whether these be generated beneath the Moon or above the Moon, but that it is a star shining in the firmament itself—one that has never previously been seen before our time, in any age since the beginning of the world.

The Formative Playfulness of Nature

Selections from the Works of Johannes Kepler

Born in Württemberg and educated at the University of Tübingen, Johannes Kepler (1571–1630) taught mathematics at Graz, joined Tycho Brahe in Prague in 1600, and succeeded him as imperial mathematician less than a year later. He went to Linz after the Emperor Rudolph's death in 1612, and died in Regensburg. His reputation was based on a series of books from the Mysterium Cosmographicum (Cosmographical Mystery) *of 1596 to the* Harmonices Mundi (Harmony of the World) *of 1619. Influenced by Pythagorean tradition and Plato's* Timaeus *(see introduction to Plato in Chapter 9 above), Kepler, in the* Cosmographical Mystery, *attempted to correlate the intervals between the planetary orbits of the Copernican system—of which Kepler was the first important defender among astronomers—with the five geo-*

[7]The so-called eighth sphere of the heavens was the crystalline sphere in which the fixed stars were set, outside the sphere of the farthest planet, Saturn. See the footnote that follows.

[8]After careful observation of the comet of 1577, and others that followed, Tycho concluded, as he wrote in his *Astronomiae Instauratae Mechanica* of 1598 (translated as *Tycho Brahe's Description of His Instruments and Scientific Work* [1946]), that "all comets observed by me moved in the aetherial regions of the world and never in the air below the moon, as Aristotle and his followers tried without reason to make us believe for so many centuries." Moreover, since the paths of the comets intersected those of the planets, Tycho also rejected the age-old belief that the planets revolved through the heavens on crystalline spheres—a revision, Marie Boas remarks in *The Scientific Renaissance*, "as revolutionary in its own way as the displacement of the earth from the center of the universe."

metrical figures (tetrahedron, cube, octahedron, dodecahedron, and icasohedron) that Euclid had proved were the only possible "perfect solids."

Our first selection, in which Kepler defends Copernican philosophy against the charge of impiety, is taken from William H. Donahue's 1992 translation of the Astronomia Nova (New Astronomy), *or* De Motibus Stellae Martis (On the Motions of the Star Mars), *of 1609. Kepler, after exhaustive calculations from Tycho's observations of the orbit of Mars over many years, established—in what was later called his first law—that planetary paths were not circular, as all previous astronomers had thought, but elliptical. He thus eliminated the cumbersome Ptolemaic system of eccentrics and epicycles (see introduction to Ptolemy in Chapter 12 above) that Copernicus had largely retained, and prepared the way for the grand synthesis of Newton's universal gravitation. Indeed, in postulating a force emanating from the sun (and diminishing with distance from it) that drives the planets round their orbits, Kepler was making a "revolutionary" proposal, as Arthur Koestler writes in* The Sleepwalkers *(1959): "For the first time since antiquity, an attempt was made not only to* describe *heavenly motions in geometrical terms, but to assign them a* physical cause. *We have arrived at the point where astronomy and physics meet again, after a divorce which lasted for two thousand years." Thus despite the Pythagorean/Platonic influences pervading his work, "there is little in Kepler," Marie Boas writes, "of the neo-Platonic number nonsense of the late fifteenth century . . . or of the religio-philosophic pantheism of Giordano Bruno. To Kepler, his newly discovered mathematical harmonies were so many laws which revealed the wonder and order of the world of God; this was a world ruled by mathematical law, which in turn was discoverable by astronomical observation."*

Though his three laws of planetary motions and ratios (buried deep amid other speculations, and thus largely ignored before Newton) were the most significant scientific result of his labors, other aspects of Kepler now seem particularly modern. One is his emphasis on the process as well as the results of discovery: "it is a question not only of leading the reader to an understanding of the subject matter in the easiest way," he writes early in the New Astronomy, *"but also, chiefly, of the arguments, meanderings, or even chance occurrences by which I the author first came upon that understanding." And his view of Nature assumes both regularity and a capricious unpredictability suitable to a pre- (or post-) Newtonian universe: "For the closer we approach to her," he writes (Chapter 58 of the* New Astronomy*), "the more petulant her games become, and the more she again and again sneaks out of the seeker's grasp just when he is about to seize her through some circuitous route." A similar view that a "formative faculty" of Nature is "in the habit . . . of playing with the passing moment" finds expression in a fascinating minor work of 1611,* Strena, seu De Nive Sexangula (A New Year's Gift; or, On the Six-Cornered Snowflake), *from which our second selection is taken in the translation by Colin Hardie (1966).*

FROM *Author's Introduction to* New Astronomy, *translated by William H. Donahue*

I, too, implore my reader, when he departs from the temple and enters astronomical studies, not to forget the divine goodness conferred upon men, to the consideration of which the psalmodist chiefly invites. I hope that, with me, he will praise and celebrate the Creator's wisdom and greatness, which I unfold for him in the more perspicacious explanation of the world's form, the investigation of causes, and the detection of errors of vision. Let him not only extol the Creator's divine beneficence in his concerns for the well-being of all living things, expressed in the firmness and stability of the earth, but also acknowledge His wisdom expressed in its motion, at once so well hidden and so admirable.

But whoever is too stupid to understand astronomical science, or too weak to believe Copernicus without affecting his faith, I would advise him that, having dismissed astronomical studies and having damned whatever philosophical opinions he pleases, he mind his own business and betake himself home to scratch his own dirt patch, abandoning this wandering about the world. He should raise his eyes (his only means of vision) to this visible heaven and with his whole heart burst forth in giving thanks and praising God the Creator. He can be sure that he worships God no less than the astronomer, to whom God has granted the more penetrating vision of the mind's eye, and an ability and desire to celebrate his God above those things he has discovered. . . .

So much for the authority of holy scripture. As for the opinions of the pious on these matters of nature, I have just one thing to say: while in theology it is authority that carries the most weight, in philosophy it is reason. Therefore, Lactantius is pious, who denied that the earth is round, Augustine is pious, who, though admitting the roundness, denied the antipodes, and the Inquisition nowadays is pious, which, though allowing the earth's smallness, denies its motion. To me, however, the truth is more pious still, and (with all due respect for the Doctors of the Church) I prove philosophically not only that the earth is round, not only that it is inhabited all the way around at the antipodes, not only that it is contemptibly small, but also that it is carried along among the stars. . . .

FROM A New Year's Gift; or, On the Six-Cornered Snowflake, *translated by Colin Hardie*

So after examining all the ideas that came into my head I conclude thus: the cause of the six-sided shape of a snowflake is none other than that of the ordered shapes of plants and of numerical constants; and since in them nothing occurs without supreme reason—not, to be sure, such as discursive reason discovers, but such as existed from the first in the Creator's design and is preserved from that origin to this day in the wonderful nature of animal faculties, I do not believe that even in a snowflake this ordered pattern exists at random.

There is then a formative faculty *[facultas formatrix]* in the body of the Earth, and its carrier is vapor as the human soul is the carrier of spirit: so much so that no vapor ever exists without being bound by a formative principle, which others call the craftsman Heat, in the same way as it is by some form of heat that, being turned into what it is said to be, to wit, vapor, it exists and by the same heat is maintained so as to persist in being vapor.

But I will expound what ideas of mine still remain, by meeting two objections. You might argue thus: the ensuing purpose, which is the establishment of a definite natural body, points, in plants, to the pre-existence of a formative principle in some matter; for where the means are adapted to a definite purpose, there order exists, not chance; there is pure mind and pure Reason. But no purpose can be observed in the shaping of a snowflake; the six-cornered shape does not bring it about that the snowflake lasts, or that a definite natural body assumes a precise and durable shape. My reply is: formative reason does not act only for a purpose, but also to adorn. It does not strive to fashion only natural bodies, but is in the habit also of playing with the passing moment, as is shown by many ores from mines. I transpose the meaning of all such from playfulness (in that we say that Nature *plays*) to this serious intention. I believe that the heat, which till then was protecting its matter, is now conquered by the surrounding cold; but just as previously, animated as it is by a formative principle, it had acted and fought in an orderly fashion, so now it displays an order of its own in preparing for retreat and withdrawal, and holds out longer in the selected branches, or outposts, that are distributed in good order over the line of battle than in all the rest of its matter. Thus it takes care "not to fall in an ugly and immodest fashion," as histories relate of Olympias.[9]

Someone else may object: each single plant has a single animating principle of its own, since each instance of a plant exists separately, and there is no cause for wonder that each should be equipped with its own peculiar shape. But to imagine an individual soul for each and any starlet of snow is utterly absurd, and therefore the shapes of snowflakes are by no means to be deduced from the operation of soul in the same way as with plants.

I reply: the likeness is much greater on either side than this objector could believe. Let us grant that each single plant has its own principle; but they are all offspring of one and the same universal principle, inherent in the earth and related to plants as the principle of water is to fish, of the human body to lice, of the bodies of dogs to fleas, and of sheep to some other kind of louse. Not all plants anyhow originated from seed, but most of them arose spontaneously, although they have since propagated themselves by seeding. The faculty of earth is in itself one and the same, but it imparts itself to different bodies and co-operates with them. It engrafts itself on to them, and builds now one design, now another, as the inner disposition of each matter or outer conditions allow. . . .

How true it is that Life without philosophy is Death! If the celebrated adulteress of Aesop's fable[10] had known of this formative faculty in the snowflake, she could have convinced her husband that she had conceived from a snowflake, and would not have been so easily bereaved of her bastard by her husband's cunning. . . .

Nature-Knowledge New and Unheard Of

On the Magnet, Magnetic Bodies Also, and on the Great Magnet the Earth, by William Gilbert, translated by Silvanus P. Thompson

William Gilbert, or Gilberd (1540 or 1544–1603), was born in Colchester, Essex, studied medicine at St. John's College, Cambridge, and went on to become president of the College of Physicians and physician to Queen Elizabeth and King James I. He devoted some seventeen or eighteen years to the study of magnetism, motivated in part by desire to devise aids to navigation, and published his major work, De Magnete . . . (On the Magnet . . .), *in 1600; the following selections are from the Gilbert Club translation of 1900, made largely by Silvanus P. Thompson, as reprinted in 1958 with introduction by Derek J. de S. Price. Gilbert, Marie Boas remarks, "combined rational science and mysticism in a peculiar blend, in which neither interfered with the other,"*

[9]Olympias, wife and then widow of Philip II of Macedonia and mother of Alexander the Great, seized power in the years after her son's death, but was captured and killed by Cassander (son of Alexander's officer and regent, Antipater), who went on to seize the kingship for himself. Hardie notes that she is said by Justin, *Historiae Philippicae Epitome* (after Pompeius Trogus) XIV.vi.12, to have bravely faced the assassins sent to kill her and "to have composed the hairs on her head when dying, and covered her thighs with her garment, so that nothing indecorous could be seen on her body."

[10]This fable, of the adulteress who told her husband on her return from a journey that she had conceived a child from a snowflake, and of her husband's revenge by later taking the child on a journey to the south and returning without the child, saying that the snow-child had melted in the sun, is not found in the collections of fables called Aesop's and derived from the Greek of Babrius or the Latin of Phaedrus, but comes from the *Esopus* of Burchard Waldis, a German who versified "Aesop" and added new fables from other sources. . . . The motif of conception from a snowflake is widespread. (Hardie)

but surely the rational science—the need, as Gilbert stresses in his "Preface to the Candid Reader," for "trustworthy experiments" and "demonstrated arguments," and the assertion that "nothing hath been set down in these books which hath not been explored and many times performed and repeated amongst us"—is more noteworthy. "He was the first," Price writes, "to have the tenacity to work through a whole segment of physics methodically, appealing to experiment and reason throughout."

Gilbert was intensely aware that in contending that "our great mother" the earth was a magnet rotating daily on its axis, he was propounding a "nature-knowledge . . . almost entirely new and unheard of," at variance with the principles of Greek philosophers such as Aristotle. These philosophers, however, Gilbert affirms, would "gladly have accepted" Gilbert's conclusions had they known the new facts revealed by his experiments. In our further selections, Gilbert espouses the common pre-Socratic and Platonic view that "the whole universe is animated," arguing that the lodestone or magnet is "like a living creature." Yet at the same time he anticipates Newtonian gravitation by affirming the universal character of magnetism as attraction at a distance, and repudiates the Aristotelian "monstrosity" of considering the earth alone—in contrast to globes in the pure aethereal realms beyond the moon—to be senseless and lifeless: "paltry, imperfect, dead, inanimate, and decadent." And he categorically rejects the "deep-set error" of believing the heavens are whirled around by the "mad and furious celestial velocity" of the primum mobile, when their apparent motion can be better explained by the diurnal rotation of the earth, "since nature always acts through a few rather than through many."

FROM *Preface to the Candid Reader*

Clearer proofs, in the discovery of secrets, and in the investigation of the hidden causes of things, being afforded by trustworthy experiments and by demonstrated arguments, than by the probable guesses and opinions of the ordinary professors of philosophy: so, therefore, that the noble substance of that great magnet, our common mother (the earth), hitherto quite unknown, and the conspicuous and exalted powers of this our globe, may be the better understood, we have proposed to begin with the common magnetic, stony, and iron material, and with magnetical bodies, and with the nearer parts of the earth which we can reach with our hands and perceive with our senses; then to proceed with demonstrable magnetic experiments; and so penetrate, for the first time, into the innermost parts of the earth. For after we had, in order finally to learn the true substance of the globe, seen and thoroughly examined many of those things which have been obtained from mountain heights or ocean depths, or from the profoundest caverns and from hidden mines, we applied much prolonged labor on investigating the magnetical forces; so wonderful indeed are they, compared with the forces of all other minerals, surpassing even the virtues of all other bodies about us. Nor have we found this our labor idle or unfruitful; since daily during our experimenting, new and unexpected properties came to light; and our philosophy hath grown so much from the things diligently observed, that we have attempted to expound the interior parts of the terrene globe, and its native substance, upon magnetic principles; and to reveal to men the earth (our common mother), and to point it out as if with the finger, by real demonstrations and by experiments manifestly apparent to the senses. And as geometry ascends from sundry very small and very easy principles to the greatest and most difficult, by which the wit of man climbs above the firmament: so our magnetical doctrine and science first sets forth in convenient order the things which are less obscure; from these there come to light others that are more remarkable; and at length in due order there are opened the concealed and most secret things of the globe of the earth, and the causes are made known of those things which, either through the ignorance of the ancients or the neglect of moderns, have remained unrecognized and overlooked. . . .

Whoso desireth to make trial of the same experiments, let him handle the substances, not negligently and carelessly, but prudently, deftly, and in the proper way; nor let him (when a thing doth not succeed) ignorantly denounce our discoveries: for nothing hath been set down in these books which hath not been explored and many times performed and repeated amongst us. Many things in our reasonings and hypotheses will, perchance, at first light, seem rather hard, when they are foreign to the commonly received opinion; yet I doubt not but that hereafter they will yet obtain authority from the demonstrations themselves. Wherefore in magnetical science, they who have made most progress, trust most in and profit most by the hypotheses; nor will anything readily become certain to any one in a magnetical philosophy in which all or at least most points are not ascertained. This nature-knowledge is almost entirely new and unheard-of, save what few matters a very few writers have handed down concerning certain common magnetical powers. Wherefore we but seldom quote ancient Greek authors in our support, because neither by using Greek arguments nor Greek words can the truth be demonstrated or elucidated either more precisely or more significantly. For our doctrine magnetical is at variance with most of their principles and dogmas. Nor have we brought to this work any pretense of eloquence or adornments of words; but this only have we done, that things difficult and unknown might so be handled by us, in such a form of speech, and in such words as are needed to be clearly understood: sometimes therefore we use new and unusual words, not that by means of foolish veils of vocabularies we should cover over the facts with shades and mists (as alchemists are wont to do) but that hidden things which have no name, never having been hitherto perceived, may be plainly and correctly enunciated. After describing our magnetical experiments and our information of the homogenic parts of the earth, we proceed to the general nature of the whole globe; wherein it is permitted us to philosophize freely and with the same liberty which the Egyptians, Greeks, and Latins formerly used in publishing their dogmas: whereof very many errors have been handed down in turn to later authors, and in which smatterers still persist, and wander as though in perpetual darkness. To those early forefathers of philosophy, Aristotle, Theophrastus, Ptolemy, Hippocrates, and Galen, let due honor be ever paid: for by them wisdom hath been diffused to posterity; but our age hath detected and brought to light very many facts which they, were they now alive, would gladly have accepted. Wherefore we also have not hesitated to expound in demonstrable hypotheses those things which we have discovered by long experience. Farewell.

Book V, Chapter XII: Magnetic force is animate, or imitates life; and in many things surpasses human life, while this is bound up in the organic body.

A loadstone is a wonderful thing in very many experiments, and like a living creature. And one of its remarkable virtues is that which the ancients considered to be a living soul in the sky, in the globes and in the stars, in the sun and in the moon. For they suspected that such various motions could not arise without a divine and animate nature, immense bodies turned about in fixed times, and wonderful powers infused into other bodies; whereby the whole universe flourishes in the most beautiful variety, through this primary form of the globes themselves. The ancients, as Thales, Heraclitus, Anaxagoras, Archelaus, Pythagoras, Empedocles, Parmenides, Plato, and all the Platonists, and not only the older Greeks, but the Egyptians and Chaldaeans, seek for some universal life in the universe, and affirm that the whole universe is endowed with life. Aristotle affirms that not the whole universe is animate, but only the sky; but he maintains that its elements are inanimate, whilst the stars themselves are animate. We, however, find this life in globes only and in their homogenic parts; and though it is not the same in all globes (for it is much more eminent in the sun and in certain stars than in others of less nobility) yet in very many the lives of the globes agree in their powers. For each several homogenic part draws to its own globe in a similar manner, and has an inclination to the common direction of the whole in the universe; and the effused forms extend outward in all, and are carried out into an orb, and have bounds of their own; hence the order and regularity of the motions and rotations of all the planets, and their courses, not wandering away, but fixed and determined. Wherefore Aristotle concedes life to the spheres themselves and to the orbs of the heavens (which he feigns), because they are suitable and fitted for a circular motion and actions, and are carried along in fixed and definite courses. It is surely wonderful, why the globe of the earth alone with its emanations is condemned by him and his followers and cast into exile (as senseless and lifeless), and driven out of all the perfection of the excellent universe. It is treated as a small corpuscle in comparison with the whole, and in the numerous concourse of many thousands it is obscure, disregarded, and unhonored. With it also they connect the kindred elements, in a like unhappiness, wretched and neglected. Let this therefore be looked upon as a monstrosity in the Aristotelian universe, in which everything is perfect, vigorous, animated; whilst the earth alone, an unhappy portion, is paltry, imperfect, dead, inanimate, and decadent. But on the other hand Hermes, Zoroaster, Orpheus, recognize a universal life. We, however, consider that the whole universe is animated, and that all the globes, all the stars, and also the noble earth have been governed since the beginning by their own appointed souls and have the motives of self-conservation. Nor are there wanting, either implanted in their homogenic nature or scattered through their homogenic substance, organs suitable for organic activity, although these are not fashioned of flesh and blood as animals, or composed of regular limbs, which are also hardly peceptible in certain plants and vegetables; since regular limbs are not necessary for all life. Nor can any organs be discerned or imagined by us in any of the stars, the sun, or the planets, which are specially operative in the universe; yet they live and imbue with life the small particles in the prominences on the earth. If there be anything of which men can boast, it is in fact life, intelligence; for the other animals are ennobled by life; God also (by whose nod all things are ruled) is a living soul. Who therefore will demand organs for the divine intelligences, which rise superior to every combination of organs and are not restrained by materialized organs? But in the several bodies of the stars the implanted force acts otherwise than in those divine existences which are supernaturally ordained; and in the stars, the sources of things, otherwise than in animals; in animals again otherwise than in plants. Miserable were the condition of the stars, abject the lot of the earth, if that wonderful dignity of life be denied to them, which is conceded to worms, ants, moths, plants, and toadstools; for thus worms, moths, grubs, would be bodies more honored and perfect in nature, for without life no body is excellent, valuable, or distinguished. But since living bodies arise and receive life from the earth and the sun, and grass grows on the earth apart from any seeds thrown down (as when soil is dug up from deep down in the earth, and put on some very high place or on a very high tower, in a sunny spot, not so long after various grasses spring up unbidden) it is not likely that they can produce what is not in them; but they awaken life, and therefore they are living. Therefore the bodies of the globes, as important parts of the universe, in order that they might be independent and that they might continue in that condition, had a need for souls to be united with them, without which there can be neither life, nor primary activity, nor motion, nor coalition, nor controlling power, nor harmony, nor endeavor, nor sympathy; and without which there would be no generation of anything, no alternations of the seasons, no propagation; but all things would be carried this way and that, and the whole universe would fall into wretchedest Chaos, the earth in short would be vacant, dead, and useless. . . . But the magnetic force of the earth and the formate life or living form of the globes, without perception, without error, without injury from ills and diseases, so present with us, has an implanted activity, vigorous through the whole material mass, fixed, constant, directive, executive, governing, consentient; by which the generation and death of all things are carried on upon the surface. For, without that motion, by which the daily revolution is performed, all earthly things around us would ever remain savage and neglected, and more than deserted and absolutely idle. But those motions in the sources of nature are not caused by thinking, by petty syllogisms, and theories, as human actions, which are wavering, imperfect, and undecided; but along with them reason, instruction, knowledge, discrimination have their origin, from which definite and determined actions arise, from the very foundations that have been laid and the very beginnings of the universe; which we, on account of the infirmity of our minds, cannot comprehend. Wherefore Thales, not without cause (as Aristotle relates in his book *De Anima*), held that the loadstone was animate, being a part and a choice offspring of its animate mother the earth.

FROM *Book VI, Chapter III: On the magnetic diurnal revolution of the earth's globe, as a probable assertion against the time-honored opinion of a Primum Mobile*

. . . We must accordingly reject the so deep-set error about this so mad and furious celestial velocity, and the forced retardation of the rest of the heavens. Let theologians discard and wipe out with sponges those old women's tales of so rapid a spinning round of the heavens borrowed from certain inconsiderate philosophers. The Sun is not propelled by the sphere of Mars (if a sphere there be) and by his motion, nor Mars by Jupiter, nor Jupiter by Saturn. The sphere, too, of the fixed stars seems well enough regulated except so far as motions which are in the Earth are ascribed to the heavens, and bring about a certain change of phenomena. The superiors do not exercise a despotism over the inferiors; for the heaven of philosophers, as of theologians, must be gentle, happy, and tranquil, and not at all subject to changes: nor shall the force, fury, swiftness, and hurry of a Primum Mobile have dominion over it. That fury descends through all the celestial spheres, and celestial bodies, invades the elements of our philosophers, sweeps fire along, rolls along the air, or at least draws the chief part of it, conducts the universal aether, and turns about fiery impressions (as if it were a solid and firm body, when in fact it is a most refined essence, neither resisting nor drawing), leads captive the superior. O marvelous constancy of the terrestrial globe, the only one unconquered; and yet one that is holden fast, or stationary, in its place by no bonds, no heaviness, by no contiguity with a grosser or firmer body, by no weights. The substance of the terrestrial globe withstands and sets itself against universal nature. . . . So for these reasons, not only probable but manifest, does the diurnal rotation of the earth seem, since nature always acts through a few rather than through many; and it is more agreeable to reason that the Earth's one small body should make a diurnal rotation, than that the whole universe should be whirled around.

I pass over the reasons of the Earth's remaining motions, for at present the only question is concerning its diurnal movement, according to which it moves round with respect to the Sun, and creates a natural day (which we call a nycthemeron). And indeed Nature may be thought to have granted a motion very suitable to the Earth's shape, which (being spherical) is revolved about the poles assigned it by Nature much more easily and fittingly than that the whole universe, whose limit is unknown and unknowable, should be whirled round; and than there could be imagined an orbit of the Primum Mobile, a thing not accepted by the ancients, which Aristotle even did not devise or accept as in any shape or form existing beyond the sphere of the fixed stars; which finally the sacred scriptures do not recognize any more than they do the revolution of the firmament.

Nature as Interaction of Opposites

On the Nature of Things . . ., by Bernardino Telesio, translated by Arturo B. Fallico and Herman Shapiro

Of the Italian Renaissance thinkers sometimes grouped together as "philosophers of nature," Bernardino Telesio (1509–88)—in contrast to Giordano Bruno (Chapter 16 above) and Tommaso Campanella (1568–1639)—was largely free of mystical Hermetic and Neoplatonic tendencies, and was therefore regarded by Francis Bacon and others as more nearly approaching the spirit of modern science. Born at Cosenza in Calabria, southern Italy, and educated at Milan, Rome, and Padua, Telesio founded the Accademia Cosentina in his native city to further the study of nature, and published the first edition of his major book, De Rerum Natura Iuxta Propria Principia (On the Nature of Things According to Their Own Proper Principles) *in 1565; other, expanded editions followed in 1570 and 1586. Our selections are taken from the passages translated by Arturo B. Fallico and Herman Shapiro in* Renaissance Philosophy, *volume 1:* The Italian Philosophers *(1967). Here he sets forth his guiding, anti-Aristotelian view on the value of sense experience, not reason alone, for the investigation of nature. He considers nature to have both an active, incorporeal aspect that differentiates things by its coming and going, and a passive, corporeal, receptive aspect as matter; the nature of things arises from the interaction of these opposites, which in the form of heat and cold are "the first acting principles of all things." Such a physical theory appears to jettison Aristotle in favor of the pre-Socratics, and Bacon indeed compared Telesio's thought to that of Parmenides.*

Chapter I

That the construction of the world of nature, and the magnitude of bodies therein contained, should be investigated not by reason alone as the ancients believed, but by sense experience.

It appears that those who before us diligently investigated the construction of the world and the nature of things only imagined what they thought they knew: for it is clear that they achieved no certain knowledge. What they said is filled with internal contradiction. These investigators are not even in substantial agreement among themselves. Overly confident of themselves, as if competing with God himself in wisdom, they neither looked upon the things of nature in the light of nature's own laws, nor did they see them in relation to the powers with which things come naturally endowed. Thus, in their daring to search for the principles and causes of this world by the use of the unaided reason, and in their strong desire to succeed in this endeavor, they only invented a world according to their arbitrary will. . . .

We, not so confident of ourselves, endowed with a slower intelligence, and less ambitious lovers and students of all human knowledge, desire more humbly to investigate the nature of this world and every part of it, together with the passions, actions, and operations of the things contained therein. Our knowledge will have reached completion when the things that sense experience now displays are shown to be conformable with things already known. Only then will we have attained perfect knowledge. Indeed, if we really attain to this condition, the magnitude and the form of each thing will be revealed to us; and with these, the property, power, and nature of all things. . . .

Chapter VI

That the nature which appears and disappears is not one and the same, but multiple and endowed with active powers: and that it is incorporeal.

We see that the inducing nature which comes and goes in bodies brings them form, properties, and power; and that these natures are not only different in different entities, but are, as well, opposites, tending to destroy one another. We must not think this nature to be one and the same, lazy, dead, or inactive; but rather as being multiple—such that whatever portion and species of it be endowed with active virtue, this part throws all other parts out of their own proper place. By itself this active part dominates and constitutes things, and works in those already constituted. . . .

This nature, as we said, is incorporeal. Things are made different by its coming and going, in response to which they take on individual natures and distinct powers. Nothing would be truly one or perfectly whole, or endowed with the same virtues, if the nature it had received were not incorporeal. . . . As corporeal, it could never make itself one and the same with the substance into which it enters. If it were not thus, it would be impossible to distinguish between the received nature, the original nature, or the newly mixed nature. . . .

Chapter VII

That the remaining nature is corporeal, all one, without action and operation whatsoever. That it receives and conserves the active natures which operate.

The nature which remains must be taken to be entirely corporeal and one and the same throughout. Although all things manifestly become corrupted and change into other things, or take on other arrangements and other forms, nevertheless, the mass and the body remain. . . .

The said receptive nature, that is, matter, does not appear to be endowed with any active virtue, but seems entirely passive, as if dead. We can thus see that action does not derive from anything except that to which it properly belongs and from which it takes its origin. When another nature supervenes upon the same matter, another action and another operation succeed. There is received in that matter another nature which remains, and in which the action and the operation cooperate, rendering this nature capable of action only in the matter in which active natures operate and conserve themselves.

Being incorporeal, these natures cannot be entirely for themselves, or subsist in themselves. We have this knowledge from observation: that in all things which are made by nature which we seek to understand, no action and no operation depends on any corporeal substance whatsoever. It is the case then, that the remaining nature, that is, matter, is corporeal, all one, and without any action or operation whatever. It only receives and conserves the natures which are active and which operate.

Chapter XIII

That heat and cold are the first acting principles of all things.

. . . Since we are not able to discover any other natures except those which are actively produced by heat and cold, and which are affected and changed by these two principles, we must say that everything is made by them. It is incomprehensible that any other nature could be operating in anything except that nature which is self-constituted. Further, it can be seen that the thing that changes, although it may be endowed with diverse powers, is not altered by any agent except that which is contrary or similar to it. Moreover, no catalyst or active nature which obtains in a subject either abandons its own composition or ceases its own proper action. Those substances which are not similar to it and do not conform with it, it opposes and seeks to change. It thus extends itself into the internal nature of these while always preserving its own equilibrium, and perpetually expands, conserving itself, with all other subjects.

In conclusion, it can truly be restated that the principal agents of all things are heat and cold.

Letters of the Grand Book of Nature

Selections from the Writings of Galileo Galilei

Galileo Galilei (1564–1642) was born in Pisa of a distinguished Florentine family in the same year as Shakespeare. He studied medicine, then mathematics and physics, at the University of Pisa, where he was professor from 1589 to 1592, antagonizing the orthodox Aristotelians of the faculty with his early experiments on motion; from 1592 to 1610 he was professor of mathematics at the University of Padua. In 1609, after hearing of a telescope newly invented in the Netherlands, he manufactured one capable of enlarging objects a thousand times, and turned it on the heavens. His first book, Sidereus Nuncius (The Starry Messenger) *of 1610—his only major work in Latin; see the translations of this and other works by Stillman Drake in* Discoveries and Opinions of Galileo *(1957)—reports his momentous discovery of the mountainous and thus earthlike character of the moon and the existence of four satellites of Jupiter, along with the revelation of innumerable stars unseen by human eye before him, all of which contradicted current astronomical dogma. Later explorations revealed the phases of Venus (confirming the Copernican against the Ptolemaic system) and the rotation of spots on the surface of the sun.*

Although he did not in The Starry Messenger *explicitly declare his advocacy of Copernicus, Galileo promised in a future work on the system of the world to "prove the earth to be a wandering body [i.e., a planet] surpassing the moon in splendor, and not the sink of all dull refuse of the universe; this we shall support by an infinitude of arguments drawn from nature." In 1611, during a visit to Rome, he joined the Accademia dei Lincei ("Academy of the Lynx-eyed," or Lincean Academy), devoted, its later constitution declared, to the "study of nature, especially mathematics." In 1610 he had returned to Florence as philosopher (a title to which he gave great importance) and mathematician to the Grand Duke Cosimo II and as mathematician at the University of Pisa; in 1613 he published the results of his observations of sunspots in a series of letters,* History and Demonstrations Concerning Sunspots and Their Phenomena.

By now he was an open adherent of Copernicanism. And although he seems never to have endorsed or taken notice of Kepler's demonstration of elliptical orbits for the planets (despite correspondence with Kepler), he scoffed at the spheres and orbs of the astronomers as a "farrago," foreign to nature's ways and rejected by "philosophical astronomers" who "seek to investigate the true constitution of the universe—the most important and most admirable problem that there is." "Nature, deaf to our entreaties, will not alter or change the course of her effects," he writes, to fit the theories of those who ignore the evidence of their senses, like Aristotelians who cling to manifestly false conclusions. He himself, in contrast, is willing to change whenever his errors are revealed.

Realizing the growing danger of ecclesiastical opposition to Copernicanism—which the Church had warily tolerated so long as it was taken to be only a device to facilitate computation of planetary positions—in 1615 Galileo composed and circulated in manuscript his Letter to Madame Christina of Lorraine, Grand Duchess of Tuscany, *from which our first selection is taken (translated, like the following, from the text in Galileo Galilei,* Opere, *ed. Ferdinando Flora [1953]). In this letter, not printed until 1636, in Strasbourg, Galileo affirms that discussion of scientific questions ought to begin not from scriptural authority (which cannot always be taken literally) "but from sense-experiences and necessary demonstrations" (which can never contradict scripture rightly interpreted), since Nature "is inexorable and immutable, and never transgresses the terms of the laws imposed on her, like one who cares not at all whether her abstruse reasons and modes of operation are suited or not to the capacities of men." In 1616, however, Pope Paul V, after consultation with Cardinal Roberto Bellarmino, declared the Copernican system heretical and forbade Galileo to advocate it.*

Unable to proceed with his intended book on the system of the world, Galileo nevertheless continued his work, and in 1623 wrote Il Saggiatore (The Assayer). *Stillman Drake, in* Discoveries and Opinions, *calls it "the greatest polemic ever written in physical science"—a "scientific manifesto" in which Galileo speaks not as the experimental scientist of his early books or the theoretical scientist of his later ones, but as the philosopher of science. This book was a reply to an attack by the Jesuit Father Horatio Grassi (writing under the pseudonym of Lothario Sarsi) on Galileo's earlier* Discourse on Comets *(published under the name of Galileo's pupil Mario Guiducci). In it Galileo both reaffirms his lifelong contempt for reliance on a human authority incapable of affecting the facts of nature, which remains inexorably deaf to our wishes, and gives a ringing affirmation of his central belief that this splendid book of the universe "cannot be understood without first learning to understand the language and know the letters in which it is written. It is written in mathematical language, and its characters are triangles, circles, and other geometric figures without which . . . it is only a vain circling about in a dark labyrinth." As Drake observes, Galileo agreed with Telesio that philosophy, in order to become a science, must throw out blind respect for authority, but also saw that observation, reasoning, and mathematics were essential to it: "True philosophy had to be built upon all three, and no combination could supply the absence of any one of them. . . . Finally, he realized that philosophy must learn to be content with pursuing limited objectives, reaching out gradually into the infinity of unknown events and undiscovered laws of nature, without ever achieving complete and exact knowledge of anything at all." Such a view was "calculated to scandalize nearly everyone—not only philosophers, but theologians and rulers too."*

But when, in 1623, Galileo's friend Maffeo Barberini was elected Pope Urban VIII, the outlook seemed greatly improved, and in 1632 appeared his Dialogo sopra i Due Massimi Sistemi del Mondo, Tolemaico, e Copernicano (Dialogue Concerning the Two Chief World Systems, Ptolemaic and Copernican), *from which our final selection is taken in Drake's translation (1953; 2nd ed. 1967). The work was published on condition, however, that it present the rival systems only as mathematical hypotheses without advocating one above the other. When it became evident (as it did to anyone reading the book) that in these exchanges between the Copernican Filippo Salviati, the uncommitted Giovanni Sagredo, and the Aristotelian Simplicio, the arguments overwhelmingly favored the Copernican position, the furious Pope prohibited further sales of the book. By then, six months after publication, it had created a sensation. Galileo was arrested, tried by the Inquisition, and forced to renounce belief in the Copernican system (murmuring under his breath, according to legend, "Eppur si muove"—"And yet it does move"!). For the remainder of his life he was confined to house arrest, first in Rome and Siena, then in seclusion at his home in Arcetri, near Florence, where he went totally blind by 1637.*

In 1638, the year in which he was visited by the young English poet John Milton, he succeeded in publishing in Leyden his last major book, and greatest contribution to theoretical physics, the Discorsi e Dimostrazioni Matematiche, intorno a Due Nuove Scienze (Discourses and Mathematical Demonstrations Concerning Two New Sciences), *in which he set forth the results of his studies in mechanics and the motions of bodies. Galileo's conception of motion—which prepared the way for Newton's laws—was so original, Charles Coulston Gillispie remarks in* The Edge of Objectivity *(1960), "that it may be taken as one of those exceedingly rare events, a true mutation in ideas, a break with the past. It altered man's consciousness of a real world outside himself in nature. This new world is to be grasped rather by measurement than by sympathy. In it an Archimedean science is possible, not just of statics contemplating things at rest, but of dynamics studying to know things in motion." In consequence of this exaltation of a mathematically defined reality, however, man finds himself largely cut off from a natural world to which he relates only as an object of knowledge: a world in which God and nature have other cares than those of the human race alone. "The real world must be the world outside of man; the world of astronomy and the world of resting and moving terrestrial objects," E. A. Burtt writes in* The Metaphysical Foundations of Modern Science *(1924; 2nd ed. 1932); hence "the stage is fully set for the Cartesian dualism . . . Man begins to appear for the first time in the history of thought as an irrelevant spectator and insignificant effect of the great mathematical system which is the substance of reality." The dichotomy of man and nature has become, despite Galileo's passionate interest in every aspect of the world from the stars of the Milky Way to the song of the cicada, at least a distinct possibility.*

FROM Letter to Madame Christina of Lorraine, Grand Duchess of Tuscany,
translated by Robert M. Torrance

. . . The reason they advance for condemning the opinion that the earth moves and the sun is stationary is that many places in Holy Scripture say that the sun moves and the earth stands still, and since Scripture cannot lie or be mistaken, it necessarily follows that the view of anyone who asserts that the sun is immobile and the earth mobile is to be condemned as erroneous.

As to this reasoning, I think in the first place it is very piously said and prudently established that Holy Scripture can never lie—provided that its true meaning be penetrated: but nobody, I think, will deny that it is often abstruse and very different from what the bare significance of its words expresses. From which it follows that anyone who, in expounding it, always tried to stay within the naked literal sense, might by this error make not only contradictions and propositions far from true appear to be in the Bible, but even grave heresies and blasphemies. Thus it would be necessary to give God feet, hands, and eyes, as well as corporeal and human affections such as anger, repentance, hatred, and even sometimes forgetfulness of things past and ignorance of things future. These propositions, uttered by the Holy Ghost, were set down in this manner by the sacred scribes to accommodate the capacities of the vulgar, who are rude and unlearned; but for those worthy to be distinguished from the mob, wise expositors must explicate the true meanings and add the particular reasons for which they were set forth in such words. This doctrine is so well established and accepted by all theologians that it would be superfluous to produce any evidence for it.

Hence I think I may reasonably deduce that whenever Holy Scripture has occasion to advance some natural conclusion, especially those abstruse and difficult to understand, it has made use of this convention, so as not to sow confusion in the minds of the people and make them more contumacious against the dogmas of higher mysteries. For if Scripture (as has been said, and is evident), solely to acccommodate itself to the people's capacity, has not scrupled to obscure important pronouncements, attributing to God himself conditions very remote from, and even contrary to, his essence, who then will assert with confidence that this same Scripture has forgone this purpose when speaking incidentally of earth, water, sun, or other created things, and chosen to confine itself rigorously within the bare and restricted sense of its words? Especially since its pronouncements on these created things do not at all concern the primary foundation of the sacred writings, that is, worship of God and salvation of souls—matters immensely beyond the apprehension of the vulgar.

Such being the case, I think that in disputes about natural problems one should begin not from scriptural authority but from sense-experiences and necessary demonstrations. For Holy Scripture and nature proceed alike from the divine Word, the former as the dictate of the Holy Ghost and the latter as the most observant executrix of God's commands. It is fitting, moreover, for Scripure, in order to accommodate the understanding of everyone, to speak many things apparently differing from absolute truth in the bare meaning of the words. But nature, on the contrary, is inexorable and immutable, and never transgresses the terms of the laws imposed on her, like one who cares not at all whether her abstruse reasons and modes of operation are suited or not to the capacities of men. Thus whatever natural effects sense-experience sets before our eyes, or necessary demonstrations prove to us, should not on any account be called in doubt, much less condemned, by biblical passages whose words might have a different meaning: for not every expression of Scripture is bound by such strict obligations as those that bind every effect of nature, nor is God less excellently revealed in nature's effects than in the sacred sayings of Scripture. Perhaps this is what Tertullian meant by these words: "We affirm that God must be known first from nature, then anew by doctrine: by nature in his works, and by doctrine in his preachings."[11]

From this I do not mean to infer that one should not have the highest respect for Scriptural passages; rather, having reached certainty in natural conclusions, we should

[11]*Adversus Marcionem*, lib. I, cap. 18. (Galileo)

use them as the means most suitable to true explication of Scripture and to investigation of the meanings necessarily contained in it, as being most true and concordant with demonstrated truths. From this I should judge that the authority of Holy Scripture might have been mainly intended to persuade men of those articles and propositions which, surpassing all human discourse, could not be made credible by science or any other means than the voice of the Holy Spirit: all the more so, since in propositions which are not matters of faith, the authority of Holy Scriptures should be preferred to that of all human writings supported not by methodical demonstration but by mere assertion or probability. I think this necessary and fitting to the same extent that divine wisdom surpasses all human judgment and surmise.

But that this same God who has endowed us with senses, discourse, and intellect has wished to devalue their use and give us by other means the knowledge we can attain by them, so that we would be obliged, even in natural conclusions revealed to our eyes or intellect by sense experience or necessary demonstrations, to deny sense and reason—I do not believe we must believe this, especially in sciences of which only the slightest trace, and only in scattered conclusions, is found in Scripture. A case in point is precisely astronomy, of which so little is there that not even names of the planets can be found, except the sun and moon and, once or twice only, Venus, under the name of Lucifer. But if the sacred scribes had meant to instruct people in the dispositions and movements of the heavenly bodies, and thus intended us to learn such knowledge from Holy Scripture, they would not, in my view, have treated it so sparingly: for it is all as nothing compared with the infinite number of admirable conclusions contained and demonstrated in that science. Instead, not only have writers of Scripture not tried to teach us the constitution and movements of the heavens and the stars, and their shapes, sizes, and distances, but though all these things were well known, the most holy and learned Fathers concur that they studiously refrained from speaking of them . . .

I would implore our prudent Fathers to consider carefully the difference between opinionative and demonstrative doctrines; the force of necessary inferences should convince them that professors of demonstrative sciences have no power to change their opinions at will, adopting now one and now another. There is a huge difference between commanding a mathematician or philosopher and disposing a merchant or lawyer: for demonstrated conclusions about natural or celestial phenomena cannot be changed with the same ease as opinions about what is permissible or not in a contract, an invoice, or an exchange. That this difference was well understood by the learned and holy Fathers is made clear by their having devoted much study to refute many arguments or, better, philosophical fallacies, as expressly stated by some of them; in particular, we read in St. Augustine these words: "It should be evident beyond doubt that whatever the wise of this world have been able to demonstrate truly about the nature of things is not contrary to our Scriptures; whatever they teach in their books that is contrary to the Holy Scriptures, however, we may without doubt think wholly false: let us show this in any way we can. And let us thus keep faith in our Lord, in whom are hidden all the treasures of wisdom, so that we shall neither be seduced by the prattle of false philosophy nor terrified by the superstition of counterfeit religion."[12]

From these words I think I may glean this doctrine: that the books of the wise of this world contain some concepts of nature truly demonstrated, and others dogmatically affirmed. It is the duty of wise theologians to show that the former are not contrary to

[12]*Genesis ad literam*, lib. I, cap. 21. (Galileo)

Holy Scriptures; as to the latter, taught but not demonstrated by necessity, if there is anything in them contrary to Scripture, it must be held undoubtedly false, and this should be shown in every possible way. . . .

If, in order to banish from the world this belief [the Copernican theory], it sufficed to shut the mouth of a single man—as those are perhaps persuaded who measure judgments of others by their own, and think it impossible that such a doctrine should persist and find followers—this would be easily done. But the business goes otherwise, since to put in practice such a decision it would be necessary not only to prohibit Copernicus's book and the writings of others who follow the same doctrine, but to ban the whole science of astronomy altogether, forbidding men to look at the heavens, so that they might not see Mars and Venus very near the earth at one moment and at another very distant, with such a difference that Venus seems to be forty times and Mars sixty times larger at one time than another—and so that Venus might not seem now round, now sickle-shaped with slender horns, along with many other sense observations which can in no way be adapted to the Ptolemaic system, but are very solid arguments for the Copernican. But to prohibit Copernicus's position (when day by day, by many new observations and by application of many learned men to reading him, it is being found more true and his doctrine firm), after having permitted it for so many years while it was less followed and less confirmed, would seem in my judgment a contravention of truth, attempting more and more to hide and suppress her as she revealed herself more plainly and clearly. Not to abolish his book entirely, but only to condemn this particular proposition as erroneous, would be, unless I am mistaken, a still greater detriment to souls, giving them occasion to see a proposition proved that it was a sin to believe. And to prohibit the whole science—what else would this be but to deprecate a hundred passages of Scripture which teach that the glory and greatness of almighty God are marvelously perceived in all his deeds and divinely read in the open book of the heavens? Let no one believe that reading the lofty concepts written in those pages leads only to seeing the splendor of the sun and the stars and their rising and setting, which is as far as the eyes of brutes and the vulgar penetrate; for within them are mysteries so deep and concepts so sublime that the vigils, toils, and studies of hundreds upon hundreds of the most acute intellects have still not entirely penetrated them after continual investigations for thousands upon thousands of years. And just as what the eyes of an idiot think they understand when beholding the outward appearance of a human body is a petty thing in comparison with the wonderful artifices which a meticulous anatomist or philosopher discovers in it when he investigates the use of all these muscles, tendons, nerves, and bones, examining the functions of the heart and other principal organs, searching out the seats of the vital faculties, observing the marvelous structures of the sense instruments, and contemplating with ceaseless astonishment and gratification the receptacles of imagination, memory, and discourse: just so, what the visual sense alone represents is nothing in proportion to the exalted marvels that the intellect of the learned perceives in the heavens thanks to long and accurate observation. . . .

FROM The Assayer, *translated by Robert M. Torrance*

6 . . . In Sarsi I think I discern a firm belief that in philosophizing one should lean upon the opinions of some celebrated author, since our mind, if not wedded to someone else's discourse, must remain completely barren in everything. Perhaps he thinks philosophy is a book of fantasy written by a man, like the *Iliad* or *Orlando Furioso*, in which the

least important thing is that what is written be true. No, Signor Sarsi, that is not how it is! Philosophy is written in this splendid book which stands continually open before our eyes (I mean the universe), but it cannot be understood without first learning to understand the language and know the letters in which it is written. It is written in mathematical language, and its characters are triangles, circles, and other geometric figures without which it is humanly impossible to understand it; without these, it is only a vain circling about through a dark labyrinth.

But even supposing, as Sarsi thinks, that our intellect should be enslaved to the intellect of another man, . . . [7] he wants to persuade me that my mind should be quieted and remain satisfied by a little poetic flower which produces no fruit . . . He acts like one unacquainted with either nature or poetry, who does not know that fables and fictions are in a sense necessary to poetry, since without them it could not exist, whereas such falsehoods are so abhorrent to nature that it is no more possible to find one in it than to find shadows in light. . .

21 . . . Through long experience I think I have learned that with respect to intellectual matters the human condition is such that the less one understands and knows of them, the more confidently he wishes to speak about them; whereas the multiplicity of things to be known and understood makes one more hesitant to reach conclusions about anything new.

Once upon a time there was born, in a very solitary place, a man endowed by nature with a sharp mind and extraordinary curiosity; breeding many birds for his own amusement, he took great delight in their singing, and with intense amazement kept observing with what lovely artifice they formed at will, out of the very air they breathed, a diversity of songs, all of them very sweet. One night he happened to hear a delicate sound near his house, and since he could not imagine that it was anything other than some little bird, he set out to catch it. Coming to the road he found a shepherd boy who by blowing into a hollowed stick and moving his fingers over the wood, now closing and now opening some holes that were in it, drew forth varied notes similar to a bird's, though by a very different method. Astonished and impelled by his natural curiosity, he gave the shepherd a calf for the flute; then retiring by himself, and realizing that if the boy had not happened to pass he would never have learned that there were in nature two ways of forming notes and sweet songs, he decided to travel far from home, in expectation of encountering some other adventure.

Next day, passing by a small hut, he chanced to hear similar notes within; to learn whether it was a flute or a blackbird, he went inside, where he found a boy holding a bow in his right hand and sawing some strings stretched over a hollowed piece of wood, while with his left hand he held the instrument and by moving his fingers over it drew from it, without any blowing, a diversity of sweet notes. Now, anyone who partakes in the curiosity of this man's thoughts may judge of his astonishment; for, having seen two new ways of forming notes unexpectedly added to those he had known, he began to think still others might exist in nature.

But what was his amazement when, on entering a certain temple, he looked behind the door to see what had made a sound, and realized that the sound had come from the hinges and fastenings as he opened the door! Another time, prodded by curiosity, he entered an inn thinking he would see someone lightly playing the strings of a violin with a bow, but saw a man rubbing his fingertip around the rim of a glass and producing from it a very sweet sound. But having later observed that wasps, mosquitoes, and flies do not form uninterrupted notes by breathing, as his birds had done to begin with, but made a continuous sound by extremely rapid beating of their wings, as greatly as his astonishment grew did his confidence diminish that he knew how sound was engendered; nor would all his experience of what he had seen have prepared him to believe that crickets, especially since they cannot fly, could scatter their sweetly sonorous shrilling not by breathing but by scraping their wings together.

But when he had come to believe that there could not possibly be more ways of forming notes, after having observed, besides those already mentioned, so many other organs, trumpets, fifes, string instruments of countless kinds, and even that little iron tongue which, when held between the teeth, strangely employs the oral cavity as a sounding box and the breath as a vehicle of sound—when, I say, he thought he had seen everything, he found himself more than ever enveloped in astonished ignorance when he captured in his hand a cicada, and could neither quiet its shrill chirping by closing its mouth or stopping its wings, nor see it move its scales or any other part. Finally, lifting up the casing of its chest, and seeing beneath it some hard thin ligaments, and thinking the noise might come from their scraping, he was reduced to breaking them to make it stop: when all this was in vain, he thrust his needle further into the cicada, and by piercing it through, took away its life along with its voice, so that he could still not ascertain whether the song came from the ligaments. Thus he was reduced to such distrust of his knowledge that, when asked how sounds were created, he would generously answer that he knew a few ways, but was certain that a hundred more could exist which were unknown and unimaginable.

I could explain with many more examples nature's bounteous means of producing effects in ways inconceivable unless sense and experience showed them to us; even these are sometimes insufficient to make up for our incapacity. Therefore I should be excused if I am not able to determine exactly how comets are produced, especially since I have never claimed I could do so, knowing they may be created in some manner beyond our imagining. The difficulty of comprehending how the cicada's song is formed, even while it is singing to us in our hand, more than excuses us for not knowing how comets are formed at such an immense distance. . . .

FROM Dialogue Concerning the Two Chief World Systems, Ptolemaic and Copernican, *translated by Stillman Drake*

From THE FIRST DAY . . . *Salviati.* But to give Simplicio more than satisfaction, and to reclaim him if possible from his error, I declare that we do have in our age new events and observations such that if Aristotle were now alive, I have no doubt he would change his opinion. This is easily inferred from his own manner of philosophizing, for when he writes of considering the heavens inalterable, etc., because no new thing is seen to be generated there or any old one dissolved, he seems implicitly to let us understand that if he had seen any such event he would have reversed his opinion, and properly preferred the sensible experience to natural reason. Unless he had taken the senses into account, he would not have argued immutability from sensible mutations not being seen.

Simplicio. Aristotle first laid the basis of his argument *a priori*, showing the necessity of the inalterability of heaven by means of natural, evident, and clear principles. He afterward supported the same *a posteriori*, by the senses and by the traditions of the ancients.

Salviati. What you refer to is the method he uses in writing his doctrine, but I do not believe it to be that with which he investigated it. Rather, I think it certain that he first obtained it by means of the senses, experiments, and observations, to assure himself as much as possible of his conclusions. Afterward he sought means to make them demonstrable. That is what is done for the most part in the demonstrative sciences; this comes about because when the conclusion is true, one may by making use of analytical methods hit upon some proposition which is already demonstrated, or arrive at some axiomatic principle; but if the conclusion is false, one can go on forever without ever finding any known truth—if indeed one does not encounter some impossibility or manifest absurdity. And you may be sure that Pythagoras, long before he discovered the proof for which he sacrificed a hecatomb, was sure that the square on the side opposite the right angle in a right triangle was equal to the squares on the other two sides. The certainty of a conclusion assists not a little in the discovery of its proof—meaning always in the demonstrative sciences. But however Aristotle may have proceeded, whether the reason *a priori* came before the sense perception *a posteriori* or the other way round, it is enough that Aristotle, as he said many times, preferred sensible experience to any argument. Besides, the strength of the arguments *a priori* has already been examined.

Now, getting back to the subject, I say that things which are being and have been discovered in the heavens in our time are such that they can give entire satisfaction to all philosophers, because just such events as we have been calling generations and corruptions have been seen and are being seen in particular bodies and in the whole expanse of heaven. Excellent astronomers have observed many comets generated and dissipated in places above the lunar orbit, besides the two new stars of 1572 and 1604, which were indisputably beyond all the planets. And on the face of the sun itself, with the aid of the telescope, they have seen produced and dissolved dense and dark matter, appearing much like the clouds upon the earth; and many of these are so vast as to exceed not only the Mediterranean Sea, but all of Africa, with Asia thrown in. Now, if Aristotle had seen these things, what do you think he would have said and done, Simplicio? . . .

Sagredo. It always seems to me extreme rashness on the part of some when they want to make human abilities the measure of what nature can do. On the contrary, there is not a single effect in nature, even the least that exists, such that the most ingenious theorists can arrive at a complete understanding of it. This vain presumption of understanding everything can have no other basis than never understanding anything. For anyone who had experienced just once the perfect understanding of one single thing, and had truly tasted how knowledge is accomplished, would recognize that of the infinity of other truths he understands nothing.

Salviati. Your argument is quite conclusive; in confirmation of it we have the evidence of those who do understand or have understoood some thing; the more such men have known, the more they have recognized and freely confessed their little knowledge. And the wisest of the Greeks, so adjudged by the oracle,[13] said openly that he recognized that he knew nothing. . . .

THE SECOND DAY . . . *Salviati.* It is true that the Copernican system creates disturbances in the Aristotelian universe, but we are dealing with our own real and actual universe.

If a disparity in essence between the earth and the heavenly bodies is inferred by this author[14] from the incorruptibility of the latter and the corruptibility of the former in

[13]Socrates, in Plato's *Apology* 21A.
[14]Scipio Chiaramonti (1565–1652), author of *Anti-Tycho* (1623).

Aristotle's sense, from which disparity he goes on to conclude that motion must exist in the sun and fixed stars, with the earth immovable, then he is wandering about in a paralogism and assuming what is in question. For Aristotle wants to infer the incorruptibility of heavenly bodies from their motion, and it is being debated whether this is theirs or the earth's. Of the folly of this rhetorical deduction, enough has already been said. What is more vapid than to say that the earth and the elements are banished and sequestered from the celestial sphere and confined within the lunar orbit? Is not the lunar orbit one of the celestial spheres, and according to their consensus is it not right in the center of them all? This is indeed a new method of separating the impure and sick from the sound—giving to the infected a place in the heart of the city! I should have thought that the leper house would be removed from there as far as possible.

Copernicus admires the arrangement of the parts of the universe because of God's having placed the great luminary which must give off its mighty splendor to the whole temple right in the center of it, and not off to one side. As to the terrestrial globe being between Venus and Mars, let me say one word about that. You yourself, on behalf of this author, may attempt to remove it, but please let us not entangle these little flowers of rhetoric in the rigors of demonstration. Let us leave them rather to the orators, or better to the poets, who best know how to exalt by their graciousness the most vile and sometimes even pernicious things. Now if there is anything remaining for us to do, let us get on with it. . . .

THE THIRD DAY *Simplicio.* . . . But must we not admit that nothing has been created in vain, or is idle, in the universe? Now when we see this beautiful order among the planets, they being arranged around the earth at distances commensurate with their producing upon it their effects for our benefit, to what end would there then be interposed between the highest of their orbits (namely, Saturn's), and the stellar sphere, a vast space without anything in it, superfluous, and vain? For the use and convenience of whom?

Salviati. It seems to me that we take too much upon ourselves, Simplicio, when we will have it that merely taking care of us is the adequate work of Divine wisdom and power, and the limit beyond which it creates and disposes of nothing. I should not like to have us tie its hand so. We should be quite content in the knowledge that God and Nature are so occupied with the government of human affairs that they could not apply themselves more to us even if they had no other cares to attend to than those of the human race alone. I believe that I can explain what I mean by a very appropriate and most noble example, derived from the action of the light of the sun. For when the sun draws up some vapors here, or warms a plant there, it draws these and warms this as if it had nothing else to do. Even in ripening a bunch of grapes, or perhaps just a single grape, it applies itself so effectively that it could not do more even if the goal of all its affairs were just the ripening of this one grape. Now if this grape receives from the sun everything it can receive, and is not deprived of the least thing by the sun simultaneously producing thousands and thousands of other results, then that grape would be guilty of pride or envy if it believed or demanded that the action of the sun's rays should be employed upon itself alone.

I am certain that Divine Providence omits none of the things which look after the government of human affairs, but I cannot bring myself to believe that there may not be other things in the universe dependent upon the infinity of its wisdom, at least so far as my reason informs me; yet if the facts were otherwise, I should not resist believing in reasoning which I had borrowed from a higher understanding. Meanwhile, when I am told that an immense space interposed between the planetary orbits and the starry sphere would be useless and vain, being idle and devoid of stars, and that any immensity going beyond our comprehension would be superfluous for holding the fixed stars, I say that it is brash for our feebleness to attempt to judge the reason for God's actions, and to call everything in the universe vain and superfluous which does not serve us. . . .

Nature's Servant and Interpreter

Novum Organum (The New Organon; or, True Directions Concerning the Interpretation of Nature), by Francis Bacon, translated by James Spedding

Francis Bacon (1561–1626), son of Sir Nicholas Bacon, Lord Keeper of the Great Seal of Queen Elizabeth, was born in London and studied at Cambridge and Gray's Inn. He entered Parliament in 1584, and under James I rose to Attorney General, Lord Keeper, and Lord Chancellor, becoming Baron Verulam in 1618 and Viscount St. Albans in 1621—when he pleaded guilty of accepting bribes and was barred from holding office. During his remaining years he pursued the scientific studies that had been central for much of his life. Among his many publications were Essays *(1597–1625);* De Sapientia Veterum (On the Wisdom of the Ancients, *1609); and* New Atlantis *(1627), a scientific utopia dedicated to "the knowledge of causes, and . . . the effecting of all things possible." Most important was his unfinished* Instauratio Magna (Great Instauration, *or* Great Renewal), *of which two parts were completed:* The Advancement of Learning, *published in English in 1605 and in an expanded Latin edition in 1623; and the* Novum Organum (New Organon*; "Organon" was the name of Aristotle's logical writings) of 1620, from which our excerpts are taken in the translation by James Spedding from Bacon's* Works *(1857–74), as reprinted in his* Selected Writings, *ed. Hugh G. Dick (1955).*

Bacon was not a major scientist—the only experiment he is remembered for, his stuffing of a chicken with snow to test the properties of refrigeration, led to his death from pneumonia—nor was he in the vanguard of scientific thought of his time. William Harvey "dismissed Bacon as one who 'writes philosophy like a Lord Chancellor,' and Bacon in one of his less fortunate pronouncements," Gillispie remarks in The Edge of Objectivity, *"denied the circulation of the blood," as well as Gilbert's magnetism and Copernicus's heliocentric cosmos. Yet no other thinker played so important a role in propagating the new science as an activity fundamentally distinct in method from the old philosophy that it sought to eliminate: "We must begin anew from the very foundations . . ." Bacon's repudiation of philosophic authority, not only of the Scholastics' Aristotle but also of the humanists' Plato, in the name of experience takes a sometimes stridently reductive form as he endeavors (despite tributes to the "honor" of the Greeks) to eradicate any influence they might have. In his judgment, as expressed in Book I of* The Advancement of Learning, *"the overmuch credit that hath been given unto authors in sciences, in making them dictators, that their words should stand, and not counsels to give advice," has been the principal hindrance to the growth of learning, abetted by "too great a reverence, and a kind of adoration, of the mind and understanding of man; by means whereof men have withdrawn themselves too much from the contemplation of nature and the observations of experience, and have tumbled up and down in their own reason and conceits."*

For Bacon, "if a man will begin with certainties, he shall end in doubts; but if he will be content to begin with doubts, he shall end in certainties." He distinguishes in Novum Organum *four kinds of "Idols," or "empty dogmas," that impede understanding—the Idols of the Tribe, founded in human nature itself; of the Cave, or the individual person; of the Marketplace, the "intercourse and association of men with each other"; and of the Theater, "dogmas of philosophy" and other systems. To establish "progressive stages of certainty," Bacon proposes that the true—though untried—way of discovery is to rise in "a gradual and unbroken ascent" by* a posteriori *reasoning, or induction, from the particulars of sensory perception, verified by careful experiment, to general axioms securely founded upon them.*

Bacon has not unjustly been faulted for his mechanistic epistemology—"In Baconian science the bird-watcher comes into his own," Gillispie writes, "while genius, ever theorizing in far places, is suspect"—and for an instrumental view of science directed toward "progress as it has been understood everywhere in the West since the seventeenth century, progress through technology and the domination of nature." "In an age when nature and the problem of understanding it were being gradually detached from the realm of spirit and from theological presuppositions," Leonard Nathanson observes in The Strategy of Truth *(1967), "it became increasingly difficult to maintain the traditional view that nature, while rationally coherent, was not mechanically uniform and determined." But if Bacon, by categorically distinguishing experimental knowledge of nature from unexaminable religious belief, contributed to a sharpened dichotomy between faith and reason, in other important respects he helped break down the antinomies of philosophy. Thus, despite high position at court, he looked for knowledge—more like Socrates among the artisans of Athens than Plato in the Academy—among those involved in practical pursuits of industry, trade, agriculture, and seafaring. And by viewing the history of arts as a part of natural history, "he departed radically," Paolo Rossi writes in* Francis Bacon: From Magic to Science *(1957; English trans. 1968), "from the traditional opposition of art and nature where the former is only a vain attempt to imitate the latter. . . . For Bacon natural and artificial objects possessed the same kind of form and essence. Art was man added to nature."*

Most importantly, his promotion of scientific inquiry as a business "done as if by machinery" is offset by conviction that "the subtlety of nature is greater many times over than the subtlety of the senses and understanding." Far from simply advocating domination of nature, he knew that "Nature to be commanded must be obeyed"; and far from exalting either the empirical ant or the rational spider one over the other, he favors the middle course of the bee, which transforms whatever it gathers. To portray this passionate interpreter of nature and prophet of science, who would inspire the Enlightenment in Europe for a century and a half after his death, as a mere technocrat of induction would be to forget what Basil Willey emphasizes in The Seventeenth Century Background *(1935): "Bacon was the seer, almost the poet, of the scientific movement in England." Indeed, if matter—as Bacon, who admired Heraclitus (Chapter 9 above), wrote in his essay "Of Vicissitude of Things"—"is in a perpetual flux, and never at a stay," certain knowledge of it is perhaps not attainable even by the most rigorous method.*

FROM THE *Preface*

Those who have taken upon them to lay down the law of nature as a thing already searched out and understood, whether they have spoken in simple assurance or professional affectation, have therein done philosophy and the sciences great injury. For as they have been successful in inducing belief, so they have been effective in quenching and stopping inquiry; and have done more harm by spoiling and putting an end to other men's efforts than good by their own. Those on the other hand who have taken a contrary course, and asserted that absolutely nothing can be known—whether it were from hatred of the ancient sophists, or from uncertainty and fluctuation of mind, or even from a kind of fullness of learning, that they fell upon this opinion—have certainly advanced reasons for it that are not to be despised; but yet they have neither started from true principles nor rested in the just conclusion, zeal and affectation having carried them much too far. The more ancient of the Greeks (whose writings are lost) took up with better judgment a position between these two extremes—between the presumption of pronouncing on everything and the despair of comprehending anything; and though frequently and bitterly complaining of the difficulty of inquiry and the obscurity of things, and like impatient horses champing the bit, they did not the less follow up their object and engage with nature; thinking (it seems) that this very question—viz., whether or no anything can be known—was to be settled not by arguing, but by trying. And yet they too, trusting entirely to the force of their understanding, applied no rule, but made everything turn upon hard thinking and perpetual working and exercise of the mind.

Now my method, though hard to practice, is easy to explain; and it is this. I propose to establish progressive stages of certainty. The evidence of the sense, helped and guarded by a certain process of correction, I retain. But the mental operation which follows the act of sense I for the most part reject; and instead of it, I open and lay out a new and certain path for the mind to proceed in, starting directly from the simple sensuous perception. The necessity of this was felt no doubt by those who attributed so much importance to logic; showing thereby that they were in search of helps for the understanding, and had no confidence in the native and spontaneous process of the mind. But this remedy comes too late to do any good, when the mind is already, through the daily intercourse and conversation of life, occupied with unsound doctrines and beset on all sides by vain imaginations. And therefore that art of logic, coming (as I said) too late to the rescue, and no way able to set matters straight again, has had the effect of fixing errors rather than disclosing truth. There remains but one course for the recovery of a sound and healthy condition—namely, that the entire work of the understanding be commenced afresh, and the mind itself be from the very outset not left to take its own course, but guided at every step; and the business be done as if by machinery. . . .

FROM *Aphorisms Concerning the Interpretation of Nature and the Kingdom of Man*

I Man, being the servant and interpreter of nature, can do and understand so much and so much only as he has observed in fact or in thought of the course of nature: beyond this he neither knows anything nor can do anything.

III Human knowledge and human power meet in one; for where the cause is not known the effect cannot be produced. Nature to be commanded must be obeyed; and that which in contemplation is as the cause is in operation as the rule.

IV Towards the effecting of works, all that man can do is to put together or put asunder natural bodies. The rest is done by nature working within.

V The study of nature with a view to works is engaged in by the mechanic, the mathematician, the physician, the alchemist, and the magician; but by all (as things now are) with slight endeavor and scanty success.

X The subtlety of nature is greater many times over than the subtlety of the senses and understanding; so that all those specious meditations, speculations, and glosses in which men indulge are quite from the purpose, only there is no one by to observe it.

XIII The syllogism is not applied to the first principles of sciences, and is applied in vain to intermediate axioms; being no match for the subtlety of nature. It commands assent therefore to the proposition, but does not take hold of the thing.

XVIII The discoveries which have hitherto been made in the sciences are such as lie close to vulgar notions, scarcely beneath the surface. In order to penetrate into the inner and further recesses of nature, it is necessary that both notions and axioms be derived from things by a more sure and guarded way; and that a method of intellectual operation be introduced altogether better and more certain.

XIX There are and can be only two ways of searching into and discovering truth. The one flies from the senses and particulars to the most general axioms, and from these principles, the truth of which it takes for settled and immovable, proceeds to judgment and to the discovery of middle axioms. And this way is now in fashion. The other derives axioms from the senses and particulars, rising by a gradual and unbroken ascent, so that it arrives at the most general axioms last of all. This is the true way, but as yet untried.

XXII Both ways set out from the senses and particulars, and rest in the highest generalities; but the difference between them is infinite. For the one just glances at experiment and particulars in passing, the other dwells duly and orderly among them. The one, again, begins at once by establishing certain abstract and useless generalities, the other rises by gradual steps to that which is prior and better known in the order of nature.

XXIII There is a great difference between the Idols of the human mind and the Ideas of the divine. That is to say, between certain empty dogmas, and the true signatures and marks set upon the works of creation as they are found in nature.

XXVI The conclusions of human reason as ordinarily applied in matter of nature, I call for the sake of distinction *Anticipations of Nature* (as a thing rash or premature). That reason which is elicited from facts by a just and methodical process, I call *Interpretation of Nature.*

XXXI It is idle to expect any great advancement in science from the superinducing and engrafting of new things upon old. We must begin anew from the very foundations, unless we would revolve forever in a circle with mean and contemptible progress.

XXXVIII The idols and false notions which are now in possession of the human understanding, and have taken deep root therein, not only so beset men's minds that truth can hardly find entrance, but even after entrance obtained, they will again in the very instauration of the sciences meet and trouble us, unless men being forewarned of the danger fortify themselves as far as may be against their assaults.

XXXIX There are four classes of idols which beset men's minds. To these for distinction's sake I have assigned names—calling the first class *Idols of the Tribe*; the second, *Idols of the Cave*; the third, *Idols of the Marketplace*; the fourth, *Idols of the Theater.*

XL The formation of ideas and axioms by true induction is no doubt the proper remedy to be applied for the keeping off and clearing away of idols. To point them out, however, is of great use, for the doctrine of idols is to the interpretation of nature what the doctrine of the refutation of sophisms is to common logic.

XLI The Idols of the Tribe have their foundation in human nature itself, and in the tribe or race of men. For it is a false assertion that the sense of man is the measure of things. On the contrary, all perceptions, as well of the sense as of the mind, are according to the measure of the individual and not according to the measure of the universe. And the human understanding is like a false mirror, which, receiving rays irregularly, distorts and discolors the nature of things by mingling its own nature with it.

XLII The idols of the Cave are the idols of the individual man. For everyone (besides the errors common to human nature in general) has a cave or den of his own, which refracts and discolors the light of nature; owing either to his own proper and peculiar nature; or to his education and conversation with others; or to the reading of books, and the authority of those whom he esteems and admires; or to the differences of impressions, accordingly as they take place in a mind preoccupied and predisposed or in a mind indifferent and settled; or the like. So that the spirit of man (according as it is meted out to different individuals) is in fact a thing variable and full of perturbation, and governed as it were by chance. Whence it was well observed by Heraclitus that men look for sciences in their own lesser worlds, and not in the greater or common world.

XLIII There are also idols formed by the intercourse and association of men with each other, which I call Idols of the Marketplace, on account of the commerce and consort of men there. For it is by discourse that men associate; and words are imposed according to the apprehension of the vulgar. And therefore the ill and unfit choice of words wonderfully obstructs the understanding. Nor do the definitions or explanations wherewith in some things learned men are wont to guard and defend themselves, by any means set the matter right. But words plainly force and overrule the understanding, and throw all into confusion, and lead men away into numberless empty controversies and idle fantasies.

XLIV Lastly, there are idols which have immigrated into men's minds from the various dogmas of philosophies, and also from wrong laws of demonstration. These I call Idols of the Theater; because in my judgment all the received systems are but so many stage-plays, representing worlds of their own creation after an unreal and scenic fashion. Nor is it only of the systems now in vogue, or only of the ancient sects and philosophies, that I speak: for many more plays of the same kind may yet be composed and in like artificial manner set forth; seeing that errors the most widely different have nevertheless causes for the most part alike. Neither again do I mean this only of entire systems, but also of many principles and axioms in science, which by tradition, credulity, and negligence have come to be received. . . .

XLVIII The human understanding is unquiet; it cannot stop or rest, and still presses onward, but in vain. Therefore it is that we cannot conceive of any end or limit to the world; but always as of necessity it occurs to us that there is something beyond. Neither again can it be conceived how eternity has flowed down to the present day: for that distinction which is commonly received of infinity in time past and in time to come can by no means hold; for it would thence follow that one infinity is greater than another, and that infinity is wasting away and tending to become finite. The like subtlety arises touching the infinite divisibility of lines, from the same inability of thought to stop. But this inability interferes more mischievously in the discovery of causes: for although the most general principles in nature ought to be held merely positive, as they are discovered, and cannot with truth be referred to a cause; nevertheless the human understanding being unable to rest still seeks something prior in the order of nature. And then it is that in struggling towards that which is further off it falls back upon that which is more nigh at hand—namely, on final causes; which have relation clearly to the nature of man rather than to the nature of the universe, and from this source have strangely defiled philosophy. But he is no less an unskilled and shallow philosopher who seeks causes of that which is most general, than he who in things subordinate and subaltern omits to do so.

L But by far the greatest hindrance and aberration of the human understanding proceeds from the dullness, incompetency, and deceptions of the senses; in that things which strike the sense outweigh things which do not immediately strike it, though they be more important. Hence it is that speculation commonly ceases where sight ceases, insomuch that of things invisible there is little or no observation. Hence all the working of the spirits enclosed in tangible bodies lies hid and unobserved of men. So also all the more subtle changes of form in the parts of coarser substances (which they commonly call alteration, though it is in truth local motion through exceedingly small spaces) is in like manner unobserved. And yet unless these two things just mentioned be searched out and brought to light, nothing great can be achieved in nature, as far as the production of works is concerned. So again the essential nature of our common air, and of all bodies less dense than air (which are very many), is almost unknown. For the sense by itself is a thing infirm and erring; neither can instruments for enlarging or sharpening the senses do much: but all the truer kind of interpretation of nature is effected by instances and experiments fit and apposite; wherein the sense decides touching the experiment only, and the experiment touching the point in nature and the thing itself.

LI The human understanding is of its own nature prone to abstractions and gives a substance and reality to things which are fleeting. But to resolve nature into abstractions is less to our purpose than to dissect her into parts; as did the school of Democritus, which went further into nature than the rest. Matter rather than forms should be the object of our attention, its configurations and changes of configuration, and simple action, and law of action or motion; for forms are figments of the human mind, unless you will call those laws of actions forms.

LIX But the *Idols of the Market-place* are the most troublesome of all: idols which have crept into the understanding through the alliances of words and names. For men believe that their reason governs words; but it is also true that words react on the understanding; and this it is that has rendered philosophy and the sciences sophistical and inactive. Now words, being commonly framed and applied according to the capacity of the vulgar, follow those lines of division which are most obvious to the vulgar understanding. And whenever an understanding of greater acuteness or a more diligent observation would alter those lines to suit the true divisions of nature, words stand in the way and resist the change. Whence it comes to pass that the high and formal discussions of learned men end oftentimes in disputes about words and names; with which (according to the use and wisdom of the mathematicians) it would be more prudent to begin, and so by means of definitions reduce them to order. Yet even definitions cannot cure this evil in dealing with natural and material things; since the definitions themselves consist of words, and those words beget others: so that it is necessary to recur to individual instances, and those in due series and order; as I shall say presently when I come to the method and scheme for the formation of notions and axioms.

LXII Idols of the Theater, or of Systems, are many, and there can be and perhaps will be yet many more. For were it not that now for many ages men's minds have been busied with religion and theology; and were it not that civil governments, especially monarchies, have been averse to such novelties, even in matters speculative; so that men labor therein to the peril and harming of their fortunes—not only unrewarded, but exposed also to contempt and envy: doubtless there would have arisen many other philosophical sects like to those which in great variety flourished once among the Greeks. For as on the phenomena of the heavens many hypotheses may be constructed, so likewise (and more also) many various dogmas may be set up and established on the phenomena of philosophy. And in the plays of this philosophical theater you may observe the same thing which is found in the theater of the poets, that stories invented for the stage are more compact and elegant, and more as one would wish them to be, than true stories out of history.

In general however there is taken for the material of philosophy either a great deal out of a few things, or a very little out of many things; so that on both sides philosophy is based on too narrow a foundation of experiment and natural history, and decides on the authority of too few cases. For the rational school of philosophers snatches from experience a variety of common instances, neither duly ascertained nor diligently examined and weighed, and leaves all the rest to meditation and agitation of wit.

There is also another class of philosophers, who having bestowed much diligent and careful labor on a few experiments, have thence made bold to educe and construct systems; wresting all other facts in a strange fashion to conformity therewith.

And there is yet a third class, consisting of those who out of faith and veneration mix their philosophy with theology and traditions; among whom the vanity of some has gone so far aside as to seek the origin of science among spirits and genii. So that this parent stock of errors—this false philosophy—is of three kinds; the Sophistical, the Empirical, and the Superstitious.

LXX But the best demonstration by far is experience, if it go not beyond the actual experiment. For if it be transferred to other cases which are deemed similar, unless such transfer be made by a just and orderly process, it is a fallacious thing. But the manner of making experiments which men now use is blind and stupid. And therefore, wandering and straying as they do with no settled course, and taking counsel only from things as they fall out, they fetch a wide circuit and meet with many matters, but make little progress; and sometimes are full of hope, sometimes are distracted; and always find that there is something beyond to be sought. For it generally happens that men make their trials carelessly, and as it were in play; slightly varying experiments already known, and, if the thing does not answer, growing weary and abandoning the attempt. And even if they apply themselves to experiments more seriously and earnestly and laboriously, still they spend their labor in working out some one experiment, as Gilbert with the magnet, and the chemists with gold—a course of proceeding not less unskillful in the design than small in the attempt. For no one successfully investigates the nature of a thing in the thing itself; the inquiry must be enlarged, so as to become more general . . .

XCV Those who have handled sciences have been either men of experiment or men of dogmas. The men of experiment are like the ant; they only collect and use: the reasoners resemble spiders, who make cobwebs out of their own substance. But the bee takes a middle course, it gathers its material from the flowers of the garden and of the field, but transforms and digests it by a power of its own. Not unlike this is the true business of philosophy: for it neither relies solely or chiefly on the powers of the mind, nor does it take the matter which it gathers from natural history and mechanical experiments and lay it up in the memory whole, as it finds it; but lays it up in the understanding altered and digested. Therefore from a closer and purer league between these two faculties, the experimental and the rational (such as has never yet been made) much may be hoped.

XCVI We have as yet no natural philosophy that is pure; all is tainted and corrupted: in Aristotle's school by logic; in Plato's by natural theology; in the second school of Platonists, such as Proclus and others, by mathematics, which ought only to give definiteness to natural philosophy, not to generate or give it birth. From a natural philosophy pure and unmixed, better things are to be expected.

Fruits Never Yet Forbidden

Micrographia: or Some Physiological Descriptions of Minute Bodies Made by Magnifying Glasses, by Robert Hooke

Robert Hooke (1635–1703) attended Westminster School and Oxford, and in 1662 became Curator of Experiments for the Royal Society, founded in 1660. In 1665 he was appointed professor of geometry at Gresham College, and in 1667 City Surveyor. He was known for a wide range of inventions and improvements of scientific instruments, and for contributions to mechanics, physics, physiology, biology, chemistry, optics, and meteorology. As Margaret 'Espinasse writes in Robert Hooke *(1956), he "came very near the early ideal of a scientist whose field is the whole of knowledge." Galileo had been among the claimants to inventing the microscope (in 1610), but Hooke, along with his Dutch contemporary Antony van Leeuwenhoek (1632–1723), pioneered the use of this instrument, by which "we now behold almost as great a variety of creatures as we were able before to reckon up in the whole universe itself." He published his principal results (including study of plant cells), along with superb illustrations, in the* Micrographia *of 1665.*

The preface from which the following selections are taken is a noteworthy statement of essentially Baconian principles of scientific method (to which the Royal Society was dedicated) by a practicing scientist. Like Bacon, Hooke stresses the need, in investigating nature, to rectify the misleading operations of sense, memory, and reason, especially by adding "artificial organs [such as the microscope] to the natural"; also like Bacon, he urges science to "return to the plainness and soundness of observations on material and obvious things" after "wandering far away into invisible notions." He shows both the modesty of the experimental scientist who knows that "infallible deductions" and "certainty of axioms . . . are above my weak abilities" and supreme confidence that by following proper method there is nothing that human wit or industry cannot compass. "As at first, mankind fell by tasting of the forbidden tree of knowledge, so we, their posterity, may be in part restored by the same way, not only by beholding and contemplating, but by tasting too those fruits of natural knowledge, that were never yet forbidden."

FROM THE *Preface*

It is the great prerogative of mankind above other creatures that we are not only able to behold the works of Nature, or barely able to sustain our lives by them, but we have also the power of considering, comparing, altering, assisting, and improving them to various uses. And as this is the peculiar privilege of human nature in general, so is it capable of being so far advanced by the helps of art and experience as to make some men excel others in their observations and deductions almost as much as they do beasts. By the addition of such artificial instruments and methods, there may be, in some manner, a reparation made for the mischiefs and imperfection mankind has drawn upon itself by negligence, and intemperance, and a willful and superstitious deserting the prescripts and rules of Nature, whereby every man, both from a derived corruption, innate and born with him, and from his breeding and converse with men, is very subject to slip into all sorts of errors.

The only way which now remains for us to recover some degree of those former perfections seems to be by rectifying the operations of the sense, the memory, and reason, since upon the evidence, the strength, the integrity, and the right correspondence of all these, all the light by which our actions are to be guided is to be renewed, and all our command over things is to be established. . . .

The first thing to be undertaken in this weighty work is a watchfulness over the failings and an enlargement of the dominion of the senses.

To which end it is requisite, first, that there should be a scrupulous choice and a strict examination of the reality, constancy, and certainty of the particulars that we admit: this is the first rise whereon truth is to begin, and here the most severe and most impartial diligence must be employed; the storing up of all, without any regard to evidence or use, will only tend to darkness and confusion. We must not therefore esteem the riches of our philosophical treasure by the number only, but chiefly by the weight; the most vulgar instances are not to be neglected, but above all, the most instructive are to be entertained; the footsteps of Nature are to be traced, not only in her ordinary course, but when she seems to be put to her shifts, to make many doublings and turnings, and to use some kind of art in endeavoring to avoid our discovery.

The next care to be taken, in respect of the senses, is a supplying of their infirmities with instruments, and, as it were, the adding of artificial organs to the natural; this in one of them has been of late years accomplished, with prodigious benefit to all sorts of useful knowledge, by the invention of optical glasses. By the means of telescopes, there is nothing so far distant but may be represented to our view; and by the help of microscopes, there is nothing so small as to escape our inquiry; hence there is a new visible world discovered to the understanding. By this means the heavens are opened, and a vast number of new stars, and new motions, and new productions appear in them, to which all the ancient astronomers were utterly strangers. By this the Earth itself, which lies so near us, under our feet, shows quite a new thing to us, and in every little particle of its matter we now behold almost as great a variety of creatures as we were able before to reckon up in the whole universe itself.

It seems not improbable, but that by these helps the subtlety of the composition of bodies, the structure of their parts, the various texture of their matter, the instruments and manner of their inward motions, and all the other possible appearances of things, may come to be more fully discovered; all which the ancient peripatetics were content to comprehend in two general and (unless further explained) useless words of "matter" and "form." From whence there may arise many admirable advantages towards the increase of the operative and the mechanic knowledge, to which this age seems so much inclined, because we may perhaps be enabled to discern all the secret workings of Nature, almost in the same manner as we do those that are the productions of art, and are managed by wheels, and engines, and springs, that were devised by human wit. . . .

The truth is, the science of Nature has been already too long made only a work of the brain and the fancy: it is now high time that it should return to the plainness and soundness of observations on material and obvious things. It is said of great empires that the best way to preserve them from decay is to bring them back to the first principles and arts on which they did begin. The same is undoubtedly true in philosophy, that by wandering far away into invisible notions has almost quite destroyed itself, and it can never be recovered or continued but by returning into the same sensible paths in which it did at first proceed.

If therefore the reader expects from me any infallible deductions, or certainty of axioms, I am to say for myself that those stronger works of wit and imagination are above my weak abilities; or if they had not been so, I would not have made use of them in this present subject before me: wherever he finds that I have ventured at any small conjectures at the causes of the things that I have observed, I beseech him to look upon them only as doubtful problems and uncertain guesses, and not as unquestionable conclusions or matters of unconfutable science; I have produced nothing here with intent to bind his understanding to an implicit consent; I am so far from that, that I desire him not absolutely to rely upon these observations of my eyes, if he finds them contradicted by the future ocular experiments of sober and impartial discoverers.

. . . So many are the links, upon which the true philosophy depends, of which, if any one be loose, or weak, the whole chain is in danger of being dissolved; it is to begin with the hands and eyes, and to proceed on through the memory, to be continued by the reason; nor is it to stop there, but to come about to the hands and eyes again, and so, by a continual passage round from one faculty to another, it is to be maintained in life and strength, as much as the body of man is by the circulation of the blood through the several parts of the body, the arms, the fat, the lungs, the heart, and the head.

If once this method were followed with diligence and attention, there is nothing that lies within the power of human wit or (which is far more effectual) of human industry, which we might not compass; we might not only hope for inventions to equalize those of Copernicus, Galileo, Gilbert, Harvey, and of others, whose names are almost lost, that were the inventors of gunpowder, the seaman's compass, printing, etching, graving, microscopes, etc., but multitudes that may far exceed them: for even those discoveries seem to have been the products of some such method, though but imperfect; what may not be therefore expected from it if thoroughly prosecuted? Talking and contention of arguments would soon be turned into labors; all the fine dreams of opinions, and universal metaphysical natures, which the luxury of subtle brains has devised, would quickly vanish, and give place to solid histories, experiments, and works. And as at first, mankind fell by tasting of the forbidden tree of knowledge, so we, their posterity, may be in part restored by the same way, not only by beholding and contemplating, but by tasting too those fruits of natural knowledge, that were never yet forbidden. . . .

So Ambiguous a Word

A Free Enquiry into the Vulgarly Received Notion of Nature,
by Robert Boyle

Robert Boyle (1627–91) exemplifies, Burtt observes in The Metaphysical Foundations of Modern Science, *"all the leading intellectual currents of his day . . . harmonized with considerable success around the foci of his two most dominant enthusiasms, experimental science and religion." Born in Ireland, the fourteenth child of Richard Boyle, first earl of Cork, and educated at Eton, on the Continent, and at Oxford, he was one of the founders of the Royal Society. His earliest writings were on ethics and religion, but it was as a chemist that he established his scientific reputation. He formulated Boyle's law, that the volume of a gas confined at a constant temperature decreases in inverse proportion to the pressure placed on it; distinguished an element from a compound; and, in* The Sceptical Chymist *(1661), rejected both the four elements of the Aristotelian-Scholastic and the three elements of the Paracelsian-alchemical traditions, arguing "that there is not any certain or determinate number of such principles or elements to be met with universally in all mixed bodies."*

What he proposed instead was a "corpuscular philosophy," an early form of modern atomism. In Boyle, though, "as in Democritus [Chapter 9 above], and indeed throughout the seventeenth and eighteenth centuries," Gillispie remarks in The Edge of Objectivity, *"atomism remained rather a precondition than a finding of experimental science." His view of the physical universe (which he opposed, like Descartes, to the noncorporeal realities of God and the soul) was strictly materialistic and mechanical: matter and the accidents of matter, he affirms in* The Origin of Forms and Qualities According to the Corpuscular Philosophy *(1666), are "sufficient to explicate as much of the phenomena of nature as we either do or are like to understand." "And thus in this great* automaton, *the world (as in a watch or clock)," he writes, "the materials it consists of being left to themselves could never at the first convene into so curious an engine: and yet, when the skillful artist has once made and set it a-going, the phenomena it exhibits are to be accounted for by the . . . parts it is made up of."*

At the same time, in contrast to Galileo and Descartes, for whom "the real world was the mathematical and mechanical realm of extension and motion, man being but a puny appendage and irrelevant spectator," Burtt remarks, Boyle reasserts "the factual place of man in the cosmos and his unique dignity as the child of God." The selections that follow are from the long treatise A Free Enquiry into the Vulgarly Received Notion of Nature*—published in January 1686 though composed some twenty years earlier, and abridged in* Selected Philosophical Papers of Robert Boyle, *ed. M. A. Stewart (1979). Boyle critically examines various meanings of "nature"; suggests substitutes to clarify confusions created by these multiple meanings; asks whether nature is a thing or a name, "a real existent being, or a notional entity"; rejects the attribution of agency to such an abstraction; and finally proposes that in the "pregnant automaton" or "compounded machine" of the world created by God, whatever happens according to "the settled order or course of things corporeal . . . may, generally speaking, be said to come to pass according to nature."*

Section II

A considering person may well be tempted to suspect that men have generally had but imperfect and confused notions concerning *nature*, if he but observes that they apply that name to several things, and those too such as have some of them very little dependence on, or connection with, such others. And I remember that in Aristotle's *Metaphysics* I met with a whole chapter expressly written to enumerate the various acceptions of the Greek word *physis*, commonly rendered "nature," of which, if I mistake not, he there reckons up six. In English also we have not fewer, but rather more numerous, significations of that term. For sometimes we use the word nature for that Author of nature whom the schoolmen, harshly enough, call *natura naturans*, as when it is said that nature hath made man partly corporeal and partly immaterial. Sometimes we mean by the nature of things the essence, or that which the schoolmen scruple not to call the quiddity of a thing, namely, the attribute or attributes on whose score it is what it is, whether the thing be corporeal or not, as when we attempt to define the nature of an angel, or of a triangle, or of a fluid body, as such. Sometimes we confound that which a man has by nature with what accrues to him by birth, as when we say that such a man is noble by nature, or such a child naturally forward, or sickly, or frightful. Sometimes we take nature for an internal principle of motion, as when we say that a stone let fall in the air is by nature carried towards the center of the earth, and, on the contrary, that fire or flame does naturally move upwards towards heaven. Sometimes we understand by nature the established course of things, as when we say that nature makes the night succeed the day, nature hath made respiration necessary to the life of men. Sometimes we take nature for an aggregate of powers belonging to a body, especially a living one, as when physicians say that nature is strong or weak or spent, or that in such or such diseases nature left to herself will do the cure. Sometimes we take nature for the universe, or system of the corporeal works of God, as when it is said of a phoenix, or a chimera, that there is no such thing in nature, i.e. in the world. And sometimes too, and that most commonly, we would express by the word nature a semi-deity or other strange kind of being, such as this discourse examines the notion of. . . .

On this occasion I can scarce forbear to tell you that I have often looked upon it as an unhappy thing, and prejudicial both to philosophy and physic, that the word nature hath been so frequently and yet so unskilfully employed, both in books and in discourse, by all sorts of men, learned and illiterate. For the very great ambiguity of this term, and the promiscuous use men are wont to make of it without sufficiently attending to its different significations, makes many of the expressions wherein they employ it (and think they do it well and truly) to be either not intelligible or not proper or not true: which observation, though it be not heeded, may with the help of a little attention be easily verified, especially because the term nature is so often used that you shall scarce meet with any man who, if he have occasion to discourse anything long of either natural or medicinal subjects, would not find himself at a great loss, if he were prohibited the use of the word nature and of those phrases whereof it makes the principal part. And I confess I could heartily wish that philosophers and other learned men (whom the rest in time would follow) would, by common (though perhaps tacit) consent, introduce some more significant and less ambiguous terms and expressions in the room of the too licentiously abused word nature and the forms of speech that depend on it, or would at least decline the use of it as much as conveniently they can; and where they think they must employ it, would add a word or two to declare in what clear and determinate sense they use it. For without somewhat of this kind be done, men will very hardly avoid being led into divers mistakes, both of things and of one another; and such wranglings about words and names will be (if not continually multiplied) still kept on foot as are wont to be managed with much heat, though little use, and no necessity. . . .

But if on this occasion you should be very urgent to know what course I would think expedient, if I were to propose any, for the avoiding the inconvenient use of so ambiguous a word as nature, I should first put you in mind that, having but very lately declared that I thought it very difficult, in physiological discourses especially, to decline the frequent use of that term, you are not to expect from me the satisfaction you may desire in an answer. And then I would add that yet my unwillingness to be altogether silent, when you require me to say somewhat, makes me content to try whether the mischief complained of may not be in some measure either obviated or lessened, by looking back upon the (eight) various significations that were not long since delivered of the word nature, and by endeavoring to express them in other terms or forms of speech.

1 Instead then of the word nature taken in the first sense [for *natura naturans*], we may make use of the term it is put to signify, namely "God," wholly discarding an expression which, besides that it is harsh and needless and in use only among the schoolmen, seems not to me very suitable to the profound reverence we owe the divine majesty, since it seems to make the Creator differ too little by far from a created (not to say an imaginary) being.

2 Instead of nature in the second sense [for "that on whose account a thing is what it is, and is so called"], we may employ the word "essence," which is of great affinity to it, if not of an adequate import. And sometimes also we may make use of the word "quiddity," which, though a somewhat barbarous term, is yet frequently employed and well enough understood in the Schools, and—which is more considerable—is very comprehensive, and yet free enough from ambiguity.

3 What is meant by the word nature taken in the third sense of it [for "what belongs to a living creature at its nativity or accrues to it by its birth"] may be expressed sometimes by saying that a man or other animal is "born" so, and sometimes by saying that a thing has been "generated" such, and sometimes also that it is thus or thus qualified by its "original temper and constitution."

4 Instead of the word nature taken in the fourth acception [for "an internal principle of local motion"], we may say sometimes that this or that body "moves, as it were," or else that it "seems to move, spontaneously (or of its own accord)," upwards, downwards, etc., or that it is put into this or that motion or determined to this or that action by the concourse of such or such (proper) causes.

5 For nature in the fifth signification [for "the established course of things corporeal"], it is easy to substitute what it denotes, "the established order," or "the settled course of things."

6 Instead of nature in the sixth sense of the word [for "an aggregate of the powers belonging to a body, especially a living one"], we may employ the "constitution," "temperament," or the "mechanism," or the "complex of the essential properties or qualities," and sometimes the "condition," the "structure," or the "texture" of that body. And if we speak of the greater portions of the world, we may make use of one or other of these terms, "fabric of the world," "system of the universe," "cosmical mechanism," or the like.

7 Where men are wont to employ the word nature in the seventh sense [for "the universe," or "the system of the corporeal works of God"], it is easy and as short to make use of the word "world" or "universe"; and instead of "the phenomena of nature," to substitute "the phenomena of the universe" or "of the world."

8 And as for the word nature taken in the eighth and last of the forementioned acceptions [for either (as some pagans styled her) "a goddess," or a kind of "semi-deity"], the best way is not to employ it in that sense at all; or at least as seldom as may be, and that for divers reasons which may in due place be met with in several parts of this essay.

But though the foregoing diversity of terms and phrases may be much increased, yet I confess it makes but a part of the remedy I propose against the future mischiefs of the confused acception of the word nature and the phrases grounded on it. For besides the synonymous words and more literal interpretations lately proposed, a dextrous writer may oftentimes be able to give such a form (or, as the modern Frenchmen speak, such a "tour") to his many-ways variable expressions as to avoid the necessity of making use of the word nature, or sometimes so much as of those shorter terms that have been lately substituted in its place. And to all this I must add that, though one or two of the eight forementioned terms or phrases, as "quiddity" and "cosmical mechanism," be barbarous or ungenteel, and some other expressions be less short than the word nature, yet it is more the interest of philosophy to tolerate a harsh term that has been long received in the Schools in a determinate sense, and bear with some paraphrastical expressions, than not to avoid an ambiguity that is liable to such great inconveniences as have been lately, or may be hereafter, represented. . . .

Section IV

. . . It is not unlike that you may expect or wish that on this occasion I should propose some definition or description of nature as my own. But declining (at least at present) to say anything dogmatically about this matter, I know not whether I may not on this occasion confess to you that I have sometimes been so paradoxical, or (if you please) so extravagant, as to entertain as a serious doubt what I formerly intimated, viz. whether nature be a thing or a name—I mean, whether it be a real existent being, or a notional entity, somewhat of kin to those fictitious terms that men have devised that they might compendiously express several things together by one name. . . .

Whilst I was indulging myself in this kind of ravings, it came into my mind that the naturists might demand of me how, without admitting their notion, I could give any tolerable account of those most useful forms of speech which men employ when they say that "nature does this or that," or that "such a thing is done by nature," or "according to nature," or else "happens against nature." And this question I thought the more worth answering, because these phrases are so very frequently used by men of all sorts, as well learned as illiterate, that this custom hath made them be thought not only very convenient, but necessary, insomuch that I look upon it as none of the least things that has procured so general a reception to the vulgar notion of nature, that these ready and commodious forms of speech suppose the truth of it.

It may, therefore, in this place be pertinent to add that such phrases as that nature, or faculty, or suction, "doth this or that" are not the only ones wherein I observe that men ascribe to a notional thing that which indeed is performed by real agents: as when we say that the law "punishes" murder with death, that it "protects" the innocent, "releases" a debtor out of prison when he has satisfied his creditors (and the ministers of justice), on which or the like occasions we may justly say that it is plain that the law, which being in itself a dead letter is but a notional rule, cannot in a physical sense be said to perform these things; but they are really performed by judges, officers, executioners, and other men, acting according to that rule. . . .

Whilst this vein of framing paradoxes yet continued, I ventured to proceed so far as to question whether one may not infer, from what hath been said, that the chief advantage a philosopher receives from what men call nature be not that it affords them on divers occasions a compendious way of expressing themselves: since (thought I), to consider things otherwise than in a popular way, when a man tells me that nature does such a thing, he does not really help me to understand or explicate how it is done. For it seems manifest enough that whatsoever is done in the world, at least wherein the rational soul intervenes not, is really effected by corporeal causes and agents, acting in a world so framed as ours is, according to the laws of motion settled by the omniscient Author of things. . . .

I think it probable (for I would not dogmatize on so weighty, and so difficult, a subject) that the great and wise Author of things did, when he first formed the universal and undistinguished matter into the world, put its parts into various motions, whereby they were necessarily divided into numberless portions of differing bulks, figures, and situations in respect of each other; and that by his infinite wisdom and power he did so guide and overrule the motions of these parts at the beginning of things, as that (whether in a shorter or a longer time, reason cannot well determine) they were finally disposed into that beautiful and orderly frame we call the world, among whose parts some were so curiously contrived as to be fit to become the seeds, or seminal principles, of plants and animals. And I further conceive that he settled such laws or rules of local motion among the parts of the universal matter that, by his ordinary and preserving concourse, the several parts of the universe, thus once completed, should be able to maintain the great construction, or system and economy, of the mundane bodies, and propagate the species of living creatures. So that, according to this hypothesis, I suppose no other efficient of the universe but God himself, whose almighty power, still accompanied with his infinite wisdom, did at first frame the corporeal world according to the divine ideas which he had, as well most freely as most wisely, determined to conform them to. . . . According to the foregoing hypothesis, I consider the frame of the world, already made, as a great and, if I may so speak, pregnant automaton, that, like a woman with twins in her womb, or a ship furnished with pumps, ordnance, etc., is such an engine as comprises or consists of several lesser engines. And this compounded machine, in conjunction with the laws of motion freely established and still maintained by God among its parts, I look upon as a complex principle, whence results the settled order or course of things corporeal. And that which happens according to this course may, generally speaking, be said to come to pass according to nature, or to be done by nature, and that which thwarts this order may be said to be preternatural or contrary to nature. And indeed, though men talk of nature as they please, yet whatever is done among things inanimate, which make incomparably the greatest part of the universe, is really done but by particular bodies, acting on one another by local motion, modified by the other mechanical affections of the agent, of the patient, and of those other bodies that necessarily concur to the effect or the phenomenon produced.

Simple and Always Consonant to Itself

Selections from the Works of Isaac Newton

Sir Isaac Newton (1642–1727) was born on Christmas Day in the farmhouse of Woolsthorpe near Grantham in Lincolnshire. He studied at Trinity College, Cambridge; then during the astonishing years 1665–67, which he spent in virtual solitude at Woolsthorpe while the university was closed because of the danger of plague, he thought through in their earliest form the fundamental concepts on which his fame would be based: the differential and integral calculus, the laws of motion, the theory of universal gravitation, and the composition of light from the colors of the spectrum. He became Lucasian Professor of mathematics at Cambridge from 1669 to 1701, and after inventing a reflecting telescope was made a Fellow of the Royal Society, in whose transactions of 1672 appeared his first published paper, describing his optical experiments. The astronomer Edmund Halley persuaded him to write up the results of his astronomical and physical discoveries leading to the law of gravity, and personally paid for their publication, in 1687, as Philosophiae Naturalis Principia Mathematica (Mathematical Principles of Natural Philosophy), *or* Principia; *a second edition (containing the "General Scholium") followed in 1713 and a third in 1726. In 1699 he became Master of the Mint in London, and from 1703 until his death was president of the Royal Society. His researches on light (followed by sixteen "Queries" of a more speculative nature) were published in English as* Opticks *in 1704; the Latin translation of 1706 added Queries 17–23, and Queries 24–31 appeared in the second English edition of 1717, followed by a fourth edition in 1730. By the time of his death Newton was universally recognized as the greatest scientist of his age, if not of any age, by whom the scientific revolution of the previous two centuries had been brought (it seemed) to completion.*

Our selections from the Principia *(in Florian Cajori's revision [1934] of Andrew Motte's 1729 translation of the third edition of 1726) and* Opticks *(in the fourth edition of 1730, reprinted in 1952) emphasize the general principles underlying Newton's view of nature as "wont to be simple, and always consonant to itself." A major part of his momentous achievement—fulfilling what Copernicus, Tycho, Kepler, and Galileo had begun—was to unite the physics of this earth and the astronomy of the heavens (which Aristotelian science had separated into sublunary and superlunary realms ruled by different laws) into a single science. In this science, as much empirical and experimental as mathematical and theoretical, universal gravitation explained both the fall of the apple and the movement of the planets: this was the ultimate "simplicity" toward which Aristotle and Ptolemy had aimed, and immensely contributed. Thus Newton "was able to unite physics and astronomy in a single science of matter in motion. Finally, by flinging gravity across the void," Gillispie writes, "he reconciled the continuity of space with the discontinuity of matter. This was his resolution of the last of the great Greek philosophical problems which Europe clothed in science, whether the world is a continuum or a concourse of atoms? It is both. In force and motion it is one, in matter the other. And that unites the Platonic-Archimedean tradition with atomism," to a form of which Newton, like Boyle, gave expression in his corpuscular theory of light and his speculations (in Query 31 of the* Opticks) *on the formation of matter by solid hard particles.*

Like Boyle, too, he endeavored to explain the phenomena of nature "by reasoning from mechanical principles," and Newton's theories notably contributed to a mechanistic view of the universe in the century that followed the Principia. *"It was of the greatest consequence for succeeding thought that now the great Newton's authority," Burtt remarks, "was squarely behind that view of the cosmos which saw in man a puny, irrelevant spectator . . . of the vast mathematical system whose regular motions according to mechanical principles constituted the world of nature. . . . The really important world outside was a world hard, cold, colorless, silent, and dead; a world of quantity, a world of mathematically computable motions in mechanical regularity."*

Yet insofar as Newtonianism became identified with this mechanistic view—against which Goethe, Blake, and the Romantics would passionately rebel—Newton himself was not a Newtonian. In his view, as Burtt also recognizes, "space and time were not merely entities implied by the mathematico-experimental method and the phenomena it handles; they had an ultimately religious significance which was for him fully as important; they meant the omnipresence and continued existence from everlasting to everlasting of Almighty God." As Newton writes in the General Scholium to the Principia, *"This most beautiful system . . . could only proceed from the counsel and dominion of an intelligent and powerful Being" before whose mysteries—as before the mystery of gravitation, for which he framed no hypotheses—this mathematical physicist (who devoted enormous time and energy to unending alchemical, biblical, and theosophical studies) stood in awe, well knowing that he had only begun, like a child picking up pebbles on a beach, to penetrate nature's inexhaustible simplicity.*

FROM *MATHEMATICAL PRINCIPLES OF NATURAL PHILOSOPHY*, TRANSLATED BY ANDREW MOTTE (REVISED BY FLORIAN CAJORI)

FROM *Preface to the First Edition*

. . . But I consider philosophy rather than arts and write not concerning manual but natural powers, and consider chiefly those things which relate to gravity, levity, elastic force, the resistance of fluids, and the like forces, whether attractive or repulsive; and therefore I offer this work as the mathematical principles of philosophy, for the whole burden of philosophy seems to consist in this—from the phenomena of motions to investigate the forces of nature, and then from these forces to demonstrate the other phenomena; and to this end the general propositions in the first and second Books are directed. In the third Book, I give an example of this in the explication of the System of the World; for by the propositions mathematically demonstrated in the former Books, in the third I derive from the celestial phenomena the forces of gravity with which bodies tend to the sun and the several planets. Then from these forces, by other propositions which are also mathematical, I deduce the motions of the planets, the comets, the moon, and the sea.

I wish we could derive the rest of the phenomena of Nature by the same kind of reasoning from mechanical principles, for I am induced by many reasons to suspect that they may all depend upon certain forces by which the particles of bodies, by some causes hitherto unknown, are either mutually impelled towards one another, and cohere in regular figures or are repelled and recede from one another. These forces being unknown, philosophers have hitherto attempted the search of Nature in vain; but I hope the principles here laid down will afford some light either to this or some truer method of philosophy. . . .

FROM *Book III: System of the World (in Mathematical Treatment)*

RULES OF REASONING IN PHILOSOPHY

Rule I

We are to admit no more causes of natural things than such as are both true and sufficient to explain their appearances.

To this purpose the philosophers say that Nature does nothing in vain,[15] and more is in vain when less will serve; for Nature is pleased with simplicity, and affects not the pomp of superfluous causes.

Rule II

Therefore to the same natural effects we must, as far as possible, assign the same causes.

As to respiration in a man and in a beast; the descent of stones in Europe and in America; the light of our culinary fire and of the sun; the reflection of light in the earth, and in the planets.

[15]This influential concept (in Latin, *natura nihil agit frustra*) goes back to Aristotle's *On the Heavens (De Caelo)* II.xi (291b): *hê de physis ouden alogôs oude matên poiei.*

Rule III

The qualities of bodies, which admit neither intensification nor remission of degree, and which are found to belong to all bodies within the reach of our experiments, are to be esteemed the universal qualities of all bodies whatsoever.

For since the qualities of bodies are only known to us by experiments, we are to hold for universal all such as universally agree with experiments; and such as are not liable to diminution can never be quite taken away. We are certainly not to relinquish the evidence of experiments for the sake of dreams and vain fictions of our own devising; nor are we to recede from the analogy of Nature, which is wont to be simple, and always consonant to itself. We no other way know the extension of bodies than by our senses, nor do these reach it in all bodies; but because we perceive extension in all that are sensible, therefore we ascribe it universally to all others also. That abundance of bodies are hard, we learn by experience; and because the hardness of the whole arises from the hardness of the parts, we therefore justly infer the hardness of the undivided particles not only of the bodies we feel but of all others. That all bodies are impenetrable, we gather not from reason, but from sensation. The bodies which we handle we find impenetrable, and hence conclude impenetrability to be an universal property of all bodies whatsoever. That all bodies are movable, and endowed with certain powers (which we call the inertia) of persevering in their motion, or in their rest, we only infer from the like properties observed in the bodies which we have seen. The extension, hardness, impenetrability, mobility, and inertia of the whole, result from the extension, hardness, impenetrability, mobility, and inertia of the parts; and hence we conclude the least particles of all bodies to be also all extended, and hard and impenetrable, and movable, and endowed with their proper inertia. And this is the foundation of all philosophy. Moreover, that the divided but contiguous particles of bodies may be separated from one another, is matter of observation; and, in the particles that remain undivided, our minds are able to distinguish yet lesser parts, as is mathemaically demonstrated. But whether the parts so distinguished, and not yet divided, may, by the powers of Nature, be actually divided and separated from one another, we cannot certainly determine. Yet, had we the proof of but one experiment that any undivided particle, in breaking a hard and solid body, suffered a division, we might by virtue of this rule conclude that the undivided as well as the divided particles may be divided and actually separated to infinity.

Lastly, if it universally appears, by experiments and astronomical observations, that all bodies about the earth gravitate towards the earth, and that in proportion to the quantity of matter which they severally contain; that the moon likewise, according to the quantity of its matter, gravitates towards the earth; that, on the other hand, our sea gravitates towards the moon; and all the planets one towards another; and the comets in like manner towards the sun; we must, in consequence of this rule, universally allow that all bodies whatsoever are endowed with a principle of mutual gravitation. For the argument from the appearances concludes with more force for the universal gravitation of all bodies than for their impenetrability; of which, among those in the celestial regions, we have no experiments, nor any manner of observation. Not that I affirm gravity to be essential to bodies: by their *vis insita* [innate force] I mean nothing but their inertia. This is immutable. Their gravity is diminished as they recede from the earth.

Rule IV

In experimental philosophy we are to look upon propositions inferred by general induction from phenomena as accurately or very nearly true, notwithstanding any contrary hypotheses that may be imagined, till such time as other phenomena occur, by which they may either be made more accurate, or liable to exceptions.

This rule we must follow, that the argument of induction may not be evaded by hypotheses.

FROM *General Scholium to Proposition XLII*

. . . This most beautiful system of the sun, planets, and comets, could only proceed from the counsel and dominion of an intelligent and powerful Being. And if the fixed stars are the centers of other like systems, these, being formed by the like wise counsel, must be all subject to the dominion of One; especially since the light of the fixed stars is of the same nature with the light of the sun, and from every system light passes into all the other systems: and lest the systems of the fixed stars should, by their gravity, fall on each other, he hath placed those systems at immense distances from one another.

. . . All that diversity of natural things which we find suited to different times and places could arise from nothing but the ideas and will of a Being necessarily existing. But, by way of allegory, God is said to see, to speak, to laugh, to love, to hate, to desire, to give, to receive, to rejoice, to be angry, to fight, to frame, to work, to build; for all our notions of God are taken from the ways of mankind by a certain similitude, which, though not perfect, has some likeness, however. And thus much concerning God; to discourse of whom from the appearances of things, does certainly belong to Natural Philosophy.

Hitherto we have explained the phenomena of the heavens and of our sea by the power of gravity, but have not yet assigned the cause of this power. This is certain, that it must proceed from a cause that penetrates to the very centers of the sun and planets, without suffering the least diminution of its force; that operates not according to the quantity of the surfaces of the particles upon which it acts (as mechanical causes used to do), but according to the quantity of the solid matter which they contain, and propagates its virtue on all sides to immense distances, decreasing always as the inverse square of the distances. Gravitation towards the sun is made up out of the gravitations towards the several particles of which the body of the sun is composed; and in receding from the sun decreases accurately as the inverse square of the distances as far as the orbit of Saturn,[16] as evidently appears from the quiescence of the aphelion of the planets; nay, and even to the remotest aphelion of the comets, if those aphelions are also quiescent. But hitherto I have not been able to discover the cause of those properties of gravity from phenomena, and I frame no hypotheses;[17] for whatever is not deduced from the phenomena is to be

[16]Saturn was still the outermost planet known in Newton's time; Uranus, the next planet beyond it and the first not visible to the naked eye, was discovered by William Herschel in 1781.

[17]Cajori notes that "Newton does not advance 'hypotheses non fingo' as a general proposition, applying it to all his scientific endeavor; it is used by him in connection with a public statement relating to that special, that difficult and subtle subject, the real nature of gravitation, which was mysterious then and has remained so to our day. Moreover, this 'hypotheses non fingo' is to be taken, not as his private practice, nor his individual habit of thought, but as the position which he took in public print, on the occasion of placing before the scientific world the positive results of his mathematical thinking, which were primarily based on observation and experimentation. Newton's 'hypotheses non fingo' disrupted from its context is a complete misrepresentation of Newton." He further quotes a letter to Gaston Pardies, a Parisian professor of mathematics, in which Newton declares that "the best and safest method of philosophizing seems to be, first diligently to investigate the properties of things and establish them by experiment, and then to seek hypotheses to explain them. For hypotheses ought to be fitted merely to explain the properties of things and not attempt to predetermine them except in so far as they can be an aid to experiments."

called an hypothesis; and hypotheses, whether metaphysical or physical, whether of occult qualities or mechanical, have no place in experimental philosophy. In this philosophy particular propositions are inferred from the phenomena, and afterwards rendered general by induction. Thus it was that the impenetrability, the mobility, and the impulsive force of bodies, and the laws of motion and of gravitation, were discovered. And to us it is enough that gravity does really exist and act according to the laws which we have explained, and abundantly serves to account for all the motions of the celestial bodies, and of our sea. . . .

FROM *Opticks; or, a Treatise of the Reflections, Refractions, Inflections, and Colors of Light* *(4th ed., corrected, London 1730) Book III, Query 31*

. . . And thus nature will be very conformable to herself and very simple, performing all the great motions of the heavenly bodies by the attraction of gravity which intercedes those bodies and almost all the small ones of their particles by some other attractive and repelling powers which intercede the particles. The *vis inertiae* [force of inertia] is a passive principle by which bodies persist in their motion or rest, receive motion in proportion to the force impressing it, and resist as much as they are resisted. By this principle alone there never could have been any motion in the world. Some other principle was necessary for putting bodies into motion; and now they are in motion, some other principle is necessary for conserving the motion. . . .

All these things being considered, it seems probable to me that God in the beginning formed matter in solid, massy, hard, impenetrable, movable particles, of such sizes and figures, and with such other properties and in such proportion to space as most conduced to the end for which he formed them; and that these primitive particles being solids are incomparably harder than any porous bodies compounded of them, even so very hard as never to wear or break in pieces, no ordinary power being able to divide what God himself made one in the first creation. While the particles continue entire, they may compose bodies of one and the same nature and texture in all ages; but should they wear away or break in pieces, the nature of things depending on them would be changed. Water and earth, composed of old worn particles and fragments of particles, would not be of the same nature and texture now, with water and earth composed of entire particles in the beginning. And therefore, that nature may be lasting, the changes of corporeal things are to be placed only in the various separations and new associations and motions of these permanent particles; compound bodies being apt to break, not in the midst of solid particles, but where those particles are laid together and only touch in a few points.

It seems to me further that these particles have not only a *vis inertiae*, accompanied with such passive laws of motion as naturally result from that force, but also that they are moved by certain active principles, such as is that of gravity and that which causes fermentation and the cohesion of bodies. These principles I consider, not as occult qualities supposed to result from the specific form of things, but as general laws of nature by which the things themselves are formed, their truth appearing to us by phenomena, though their causes be not yet discovered. For these are manifest qualities, and their causes only are occult. And the Aristotelians gave the name of "occult qualities," not to manifest qualities, but to such qualities only as they supposed to lie hid in bodies and to be the unknown causes of manifest effects, such as would be the causes of gravity, and of magnetic and electric attractions, and of fermentations, if we should suppose that these forces or actions arose from qualities unknown to us and uncapable of being discovered and made manifest. Such occult qualities put a stop to the improvement of natural philosophy, and therefore of late years have been rejected. To tell us that every species of things is endowed with an occult specific quality by which it acts and produces manifest effects is to tell us nothing: but to derive two or three general principles of motion from phenomena, and afterward to tell us how the properties and actions of all corporeal things follow from those manifest principles, would be a very great step in philosophy, though the causes of those principles were not yet discovered; and therefore I scruple not to propose the principles of motion above mentioned, they being of very general extent, and leave their causes to be found out.

Now by the help of these principles all material things seem to have been composed of the hard and solid particles above mentioned, variously associated in the first Creation by the counsel of an intelligent Agent. For it became him who created them to set them in order. And if he did so, it's unphilosophical to seek for any other origin of the world or to pretend that it might arise out of a chaos by the mere laws of nature, though being once formed it may continue by those laws for many ages. For while comets move in very eccentric orbs in all manner of positions, blind fate could never make all the planets move one and the same way in orbs concentric, some inconsiderable irregularities excepted which may have risen from the mutual actions of comets and planets upon one another, and which will be apt to increase till this system wants a reformation. Such a wonderful uniformity in the planetary system must be allowed the effect of choice. And so must the uniformity in the bodies of animals, they having generally a right and a left side shaped alike, and on either side of their bodies two legs behind and either two arms or two legs or two wings before upon their shoulders, and between their shoulders a neck running down into a backbone and a head upon it, and in the head two ears, two eyes, a nose, a mouth, and a tongue, alike situated. Also the first contrivance of those very artificial parts of animals, the eyes, ears, brain, muscles, heart, lungs, midriff, glands, larynx, hands, wings, swimming bladders, natural spectacles, and other organs of sense and motion, and the instinct of brutes and insects can be the effect of nothing else than the wisdom and skill of a powerful ever-living Agent, who being in all places is more able by his will to move the bodies within his boundless uniform sensorium, and thereby to form and reform the parts of the universe, than we are by our will to move the parts of our own bodies. And yet we are not to consider the world as the body of God, or the several parts thereof as the parts of God. He is an uniform being, void of organs, members, or parts, and they are his creatures subordinate to him, and subservient to his will; and he is no more the soul of them than the soul of man is the soul of the species of things carried through the organs of sense into the place of its sensation, where it perceives them by means of its immediate presence, without the intervention of any third thing. The organs of sense are not for enabling the soul to perceive the species of things in its sensorium, but only for conveying them thither; and God has no need of such organs, he being everywhere present to the things themselves. And since space is divisible *in infinitum* [to infinity] and matter is not necessarily in all places, it may also be allowed that God is able to create particles of matter of several sizes and figures, and in several proportions to space, and perhaps of different densities and forces, and thereby to vary the laws of nature and make worlds of several sorts in several parts of the universe. At least, I see nothing of contradiction in all this.

As in mathematics, so in natural philosophy, the investigation of difficult things by the method of analysis ought ever to precede the method of composition. This analysis consists in making experiments and observations, and in drawing general conclusions from them by induction, and admitting of no objections against the conclusions but such as are taken from experiment, or other certain truths. For hypotheses are not to be regarded in experimental philosophy. And although the arguing from experiments and observations by induction be no demonstration of general conclusions, yet it is the best way of arguing which the nature of things admits of, and may be looked upon as so much the stronger by how much the induction is more general. And if no exception occur from phenomena, the conclusion may be pronounced generally. . . .

The Sun of the Microcosm

On the Movement of the Heart and Blood in Animals,
by William Harvey, translated by R. Willis (revised by Alexander Bouré)

The scientific revolution is associated principally, but by no means only, with the momentous discoveries in astronomy and physics from Copernicus to Newton. Even apart from the inventions that accompanied and fostered it—including printing, pioneered in China, and the telescope and microscope—and from other practical innovations (such as development of navigational instruments), this was an age of great discoveries or important beginnings in other sciences such as chemistry and optics, anatomy and physiology. Indeed, the major work of Renaissance anatomy, the seven-volume De Humani Corporis Fabrica (On the Fabric of the Human Body) *by Andreas Vesalius (1514–64) of Brussels, first appeared in the same year (1543) as Copernicus's* De Revolutionibus. *(Vesalius became professor of anatomy at Padua and physician to the Emperor Charles V and his son Philip II of Spain, and died in Greece while returning from a pilgrimage to Jerusalem.) This ambitious book, with its outstanding illustrations, probably by a pupil of Titian, and its "strongly mechanical concept of bodily function" (in Marie Boas's phrase), was enormously influential—and controversial—for its rejection of hallowed dogmas of Galen (Chapter 12 above), not in large theoretical matters but in important details based on Vesalius's own dissections. Yet "he could not have written his great work," Boas notes, "without Galen; there is a real sense in which Vesalius began with Galen rather than the human body, in the same way in which Copernicus began with Ptolemy rather than with the physical world."*

Galen had believed that blood must pass from the right ventricle of the heart to the left through invisible pores in the muscle wall, or septum, that divides it in half; Vesalius, though dubious, offered no better explanation. Others made important contributions, including descriptions of the pulmonary circulation of the blood—both in a tract by the Spanish physician and theologian Michael Servetus, who escaped from the Catholic Inquisition but was burned at the stake in Calvin's Geneva in 1553, and in a posthumous anatomical treatise of 1559 by Realdus Columbus of Padua, Pisa, and Rome—and the discovery in 1574 of "doorlets" or valves in the veins by Hieronymus Fabricius of Padua. But not until 1618 did the English physician William Harvey (1578–1657), who had studied at Cambridge and with Fabricius at Padua, announce to the Royal College of Physicians his discovery of the circulation of the blood. This discovery united the arterial, venous, and pulmonary systems into one, much as Galileo and later Newton were uniting the movements of heaven and earth. The controversial results of his research (in which only the microscopic capillaries escaped notice until van Leeuwenhoek ascertained them sixty years later) were published in Frankfurt in 1628 as Exercitatio Anatomica de Motu Cordis et Sanguinis in Animalibus; *the following passage is from the translation* On the Movement of the Heart and Blood in Animals *as excerpted in Holmes Boynton's* The Beginnings of Modern Science *(1948). In declaring the heart "the sun of the microcosm," just as the sun is "the heart of the world," Harvey suggests that his view of nature, far from repudiating the ancient and medieval correlation of macrocosm and microcosm, was redefining it in terms derived from, and corroborated by, painstaking empirical observation.*

FROM *Chapter VIII: Of the Quantity of Blood passing through the Heart from the Veins to the Arteries, and of the Circular Motion of the Blood*

Thus far I have spoken of the passage of the blood from the veins into the arteries, and of the manner in which it is transmitted and distributed by the action of the heart; points to which some, moved either by the authority of Galen or Columbus, or the reasonings of others, will give in their adhesion. But what remains to be said upon the quantity and source of the blood, which thus passes, is of a character so novel and unheard of that I not only fear injury to myself from the envy of a few, but I tremble lest I have mankind at large for my enemies, so much doth wont and custom become a second nature. Doctrine once sown strikes deep its root, and respect for antiquity influences all men. Still the die is cast, and my trust is in my love of truth and the candor of cultivated minds.

And sooth to say, when I surveyed my mass of evidence, whether derived from vivisections and my various reflections on them, or from the study of the ventricles of the heart and the vessels that enter into and issue from them, the symmetry and size of these conduits—for nature doing nothing in vain, would never have given them so large a relative size without a purpose—or from observing the arrangement and intimate structure of the valves in particular, and of the other parts of the heart in general, with many things besides, I frequently and seriously bethought me and long revolved in my mind what might be the quantity of blood which was transmitted, in how short a time its passage might be effected, and the like. But not finding it possible that this could be supplied by the juices of the ingested aliment without the veins on the one hand becoming drained, and the arteries on the other getting ruptured through the excessive charge of blood, unless the blood should somehow find its way from the arteries into the veins, and so return to the right side of the heart, I began to think whether there might not be *a motion, as it were, in a circle.*

Now this I afterwards found to be true; and I finally saw that the blood, forced by the action of the left ventricle into the arteries, was distributed to the body at large and its several parts in the same manner as it is sent through the lungs, impelled by the right ventricle into the pulmonary artery, and that it then passed through the veins and along the *vena cava*, and so round to the left ventricle in the manner already indicated.

This motion we may be allowed to call circular, in the same way as Aristotle says that the air and the rain emulate the circular motion of the superior bodies; for the moist earth, warmed by the sun, evaporates; the vapors drawn upwards are condensed, and descending in the form of rain, moisten the earth again. By this arrangement are generations of living things produced; and in like manner are tempests and meteors engendered by the circular motion, and by the approach and recession of the sun.

And similarly does it come to pass in the body, through the motion of the blood, that the various parts are nourished, cherished, quickened by the warmer, more perfect, vaporous, spiritous, and, as I may say, alimentive blood; which, on the other hand, owing to its contact with these parts, becomes cooled, coagulated, and, so to speak, effete. It then returns to its sovereign the heart, as if to its source, or to the inmost home of the body, there to recover its state of excellence or perfection. Here it renews its fluidity, natural heat, and becomes powerful, fervid, a kind of treasury of life, and impregnated with spirits—it might be said with balsam. Thence it is again dispersed. All this depends on the motion and action of the heart.

The heart, consequently, is the beginning of life: the sun of the microcosm, even as the sun in his turn might well be designated the heart of the world; for it is the heart by whose virtue and pulse the blood is moved, perfected, and made nutrient, and is preserved from corruption and coagulation; it is the household divinity which, discharging its function, nourishes, cherishes, quickens the whole body, and is indeed the foundation of life, the source of all action. . . .

Ruins of the Golden Age

The Sacred Theory of the Earth, by Thomas Burnet

If astronomy and physics attained, with Newton, the maturity of a grand synthesis essentially unchallenged until the new revolutions of relativity and quantum mechanics in the twentieth century, and physiology and chemistry began, with Harvey and Boyle, to establish firm principles on which subsequent research could be based, other sciences were still in their infancy at the end of the seventeenth century. Geology would have to await James Hutton's Theory of the Earth *a century later (Chapter 20 below), and the very word "biology" was not coined until 1802. Yet questions concerning the processes that had given rise to the earth as we know it, and to fossils whose presence far from any shore had intrigued Xenophanes two thousand years earlier (Chapter 9 above), were intensely debated in the seventeenth and eighteenth centuries. Thus for Robert Hooke, Paolo Rossi writes in* The Dark Abyss of Time: The History of the Earth and the History of Nations from Hooke to Vico *(1979; English trans. 1984), "the earth and the forms of life on earth have a history. . . . To explain the existence of fossil shells and fossil fish belonging to no known species, Hooke abandoned the idea of immutable and eternal species and formulated the hypothesis of the destruction and the disappearance of living species."*

But for Hooke, as for others, this hypothetical history had to be correlated with sacred history as narrated by Moses. In these discussions no work was more central than Telluris Theoria Sacra (The Sacred Theory of the Earth) *by Thomas Burnet (1635?–1715). The first Latin edition was published in two volumes in 1680, followed in 1684 by its English translation and by a second Latin edition in four volumes, whose English translation (from which our selections are taken in the reprint of 1965) appeared in 1690–91. Nothing could appear farther from the later science of geology than Burnet's theory of an earth that arose from chaos at God's creation as a "smooth, regular, and uniform" world without mountains or seas, a Mundane Egg displaying "the beauty of youth and blooming Nature, fresh and fruitful, and not a wrinkle, scar or fracture in all its body," a world "suited to a Golden Age, and to the first innocency of Nature" in an Eden spread throughout the globe. According to Burnet, this earthly paradise was destroyed when the deluge, springing both from the skies and from waters pent up beneath the earth's crust, laid waste to its beautiful regularity and left behind the disorderly ruins of seas, caverns, and mountains. The evident lack of any design testifies that such a world "was not the work of Nature, according to her first intention, . . . but a secondary work, and the best that could be made of broken materials."*

Yet despite Burnet's dependence on scriptural interpretation and the nearly total absence of empirical evidence for his conclusions, "his belief in the parallel and synchronized working of the scientific laws along with God's dispensations," Basil Willey notes in The Eighteenth Century Background *(1940), "is one of the most significant points in the book," among whose admirers was Isaac Newton. In contrast to the world portrayed by commentators on Genesis in the tradition descending from Saints Basil and Ambrose (Chapter 13 above)—a world whose creation depends solely on God's inscrutable will—Burnet's "was a scientific globe," Marjorie Hope Nicolson writes in* Mountain Gloom and Mountain Glory: The Development of the Aesthetics of the Infinite *(1959), "emerging from chaos by natural principles . . . Although God remained the First Cause, the God of science must have employed secondary causes consistent with the modern laws of nature"; and throughout his book Burnet strives to bring scripture and nature into accord. Stephen Jay Gould, in* Time's Arrow, Time's Cycle *[1987], even finds a parallel between Burnet's conviction that "a ruin can only be a wreck of something once whole, in short a product of history," and Darwin's that "history lies revealed in the quirks and imperfections of modern structures."*

Burnet studied at Cambridge and was deeply influenced by the Cambridge Platonists Ralph Cudworth (Chapter 20 below) and Henry More. In his ascent of the Alps during his grand tour of 1671, Nicolson remarks, he found that "all his idols of proportion, symmetry, and decorum in Nature were suddenly shattered by Nature herself. Like the rude rocks, his ideals lay before him, ruins of a broken world." Hence, in writing his book he "was not so much trying to justify Genesis as attempting to save for himself Plato and Augustine, Ficino and Kepler. 'Assured that all Things were made at first in Beauty and Proportion,' he stubbornly asserted, it was impossible to believe that the world as he had seen it on his travels was the work of the Great Architect." His sacred theory, then, "would justify God and Nature but even more save his own aesthetic and ethical ideals," which were those of the Christian humanist tradition.

Yet his revulsion at the disordered "confusion" of the Alpine landscape was at least equaled by his fascination with its awesome grandeur. The aesthetics to which this "last of the seventeenth-century masters of sonorous prose poetry" most significantly contributed in the century and a half that followed was the new aesthetics of the infinite, or sublime. (This was revived in Boileau's French translation [1674] of the ancient treatise On the Sublime *attributed to Longinus, and later renewed by Addison [Chapter 20 below] and Burke, Kant and Schiller.) "Wherever we look among his passages on wild nature," Nicolson writes, "we find conflict between intellectual condemnation of asymmetry and emotional response to the attraction of the vast. . . . For the first time in England we find a sharp distinction between the emotional effects of the* sublime *and the* beautiful *in external Nature and find, too, awareness of a conflict between old ideas that such qualities as* beauty *exist in objects and a growing realization that they are subjective, residing not in the object but in the 'soul' of the man perceiving the object." Thus Burnet is a writer whose paradoxical vision of nature looks forward to the Romantics at least as much as backward to the Middle Ages.*

FROM *Book I, Chapter XI: Concerning the Mountains of the Earth, their Greatness and Irregular Form, their Situation, Causes, and Origin.*

We have been in the hollows of the earth, and the chambers of the deep, amongst the damps and steams of those lower regions; let us now go air ourselves on the tops of the mountains, where we shall have a more free and large horizon, and quite another face of things will present itself to our observation.

The greatest objects of Nature are, methinks, the most pleasing to behold; and next to the great Concave of the Heavens, and those boundless regions where the stars inhabit, there is nothing that I look upon with more pleasure than the wide sea and the mountains of the earth. There is something august and stately in the air of these things that inspires the mind with great thoughts and passions; we do naturally, upon such occasions, think of God and his greatness: and whatever hath but the shadow and appearance of INFINITE, as all things have that are too big for our comprehension, they fill and overbear the mind with their excess, and cast it into a pleasing kind of stupor and admiration.

And yet these mountains we are speaking of, to confess the truth, are nothing but great ruins; but such as show a certain magnificence in Nature; as from old temples and broken amphitheaters of the Romans we collect the greatness of that people. But the grandeur of a nation is less sensible to those that never see the remains and monuments they have left, and those who never see the mountainous pasts of the earth scarce ever reflect upon the causes of them, or what power in Nature could be sufficient to produce them. The truth is, the generality of people have not sense and curiosity enough to raise a question concerning these things, or concerning the original of them. You may tell them that mountains grow out of the earth like fuzz balls, or that there are monsters under ground that throw up mountains as moles do mole hills; they will scarce raise one objection against your doctrine; or if you would appear more learned, tell them that the earth is a great animal, and these are wens that grow upon its body. This would pass current for philosophy, so much is the world drowned in stupidity and sensual pleasures, and so little inquisitive into the works of God and Nature.

There is nothing doth more awaken our thoughts or excite our minds to inquire into the causes of such things than the actual view of them, as I have had experience myself when it was my fortune to cross the Alps and Apennine Mountains; for the sight of those wild, vast, and indigested heaps of stones and earth did so deeply strike my fancy that I was not easy till I could give myself some tolerable account how that confusion came in Nature. 'Tis true, the height of mountains compared with the diameter of the earth is not considerable, but the extent of them and the ground they stand upon bears a considerable proportion to the surface of the earth; and if from Europe we may take our measures for the rest, I easily believe that the mountains do at least take up the tenth part of the dry land. . . .

'Tis certain that we naturally imagine the surface of the earth much more regular than it is; for unless we be in some mountainous parts, there seldom occur any great inequalities within so much compass of ground as we can, at once, reach with our eye; and to conceive the rest, we multiply the same idea, and extend it to those parts of the earth that we do not see; and so fancy the whole globe more smooth and uniform than it is. But suppose a man was carried asleep out of a plain country, amongst the Alps, and left there upon the top of one of the highest mountains; when he waked and looked about him, he would think himself in an enchanted country, or carried into another world; everything would appear to him so different to what he had ever seen or imagined before. To see on every hand of him a multitude of vast bodies thrown together in confusion, as those mountains are; rocks standing naked round about him; and the hollow valleys gaping under him; and at his feet it may be, an heap of frozen snow in the midst of summer. He would hear the thunder come from below, and see the black clouds hanging beneath him; upon such a prospect, it would not be easy to him to persuade himself that he was still upon the same earth; but if he did, he would be convinced, at least, that there are some regions of it strangely rude, and ruin-like, and very different from what he had ever thought of before. But the inhabitants of these wild places are even with us; for those that live amongst the Alps and the great mountains think that all the rest of the earth is like their country, all broken into mountains and valleys, and precipices; they never see other, and most people think of nothing but what they have seen at one time or another. . . .

As this survey of the multitude and greatness of them may help to rectify our mistakes about the form of the earth, so before we proceed to examine their causes, it will be good to observe farther that these mountains are placed in no order one with another, that can either respect use or beauty; and if you consider them singly, they do not consist of any proportion of parts that is referable to any design, or that hath the least footsteps of art or counsel. There is nothing in Nature more shapeless and ill-figured than an old rock or a mountain, and all that variety that is among them is but the various modes of irregularity; so as you cannot make a better character of them, in short, than to say they are of all forms and figures, except regular. Then if you could go within these mountains (for they are generally hollow), you would find all things there more rude, if possible, than without: and lastly, if you look upon an heap of them together, or a mountainous country, they are the greatest examples of confusion that we know in Nature: no tempest or earthquake puts things into more disorder. 'Tis true, they cannot look so ill now as they did at first; a ruin that is fresh looks much worse than afterwards, when the earth grows discolored and skinned over. But I fancy if we had seen the mountains when they were new-born and raw, when the earth was fresh broken, and the waters of the Deluge newly retired, the fractions and confusions of them would have appeared very ghastly and frightful.

After this general survey of the mountains of the earth and their properties, let us now reflect upon the causes of them. There is a double pleasure in philosophy, first that of admiration, whilst we contemplate things that are great and wonderful, and do not yet understand their causes; for though admiration proceed from ignorance, yet there is a certain charm and sweetness in that passion. Then the second pleasure is greater and more intellectual, which is that of distinct knowledge and comprehension, when we come to have the key that unlocks those secrets, and see the methods wherein those things come to pass that we admired before: the reasons why the world is so or so, and from what causes Nature, or any part of Nature, came into such a state; and this we are now to inquire after as to the mountains of the earth, what their original was, how and when the earth came into this strange frame and structure? In the beginning of our world, when the earth rose from a Chaos, 'twas impossible it should come immediately into this mountainous form, because a mass that is fluid, as a Chaos is, cannot lie in any other figure than what is regular; for the constant laws of Nature do certainly bring all liquors into that form, and a Chaos is not called so from any confusion or brokenness in the form of it, but from a confusion and mixture of all sorts of ingredients in the composition of it. So we have already produced, in the precedent chapters, a double argument that the earth was not originally in this form, both because it rose from a Chaos, which could not of itself, or by any immediate concretion, settle into a form of this nature, as hath been shown in the fourth and fifth chapters; as also because if it had been originally made thus, it could never have undergone a Deluge, as hath been proved in the second and third chapters. . . .

FROM *Book II, Chapter II: The Great Change of the World Since the Flood, from What It Was in the First Ages. The Earth Under Its Present Form Could Not Be Paradisiacal, Nor Any Part of It.*

. . . We may imagine how different a prospect the first world would make from what we see now in the present state of things, if we consider only those generals by which we have described it in the foregoing chapter, and what their influence would be upon mankind and the rest of Nature. For every new state of Nature doth introduce a new civil order, and a new face and economy of human affairs: and I am apt to think that some two planets, that are under the same state or period, do not so much differ from one another as the same planet doth from itself in different periods of its duration. We do not seem to inhabit the same world that our first forefathers did, nor scarce to be in the same race of men. Our life now is so short and vain, as if we came into the world only to see it and leave it; by that time we begin to understand ourselves a little, and to know where we are, and how to act our part, we must leave the stage, and give place to others as mere novices as we were ourselves at our first entrance. And this short life is employed, in a great measure, to preserve ourselves from necessity, or diseases, or injuries of the air, or other inconveniences; to make one man easy, ten must work and do drudgery; the body takes up so much time, we have little leisure for contemplation, or to cultivate the mind. The earth doth not yield us food but with much labor and industry, and what was her freewill offering before, or an easy liberality, can scarce now be extorted from her. Neither are the heavens more favorable, sometimes in one extreme, sometimes in another; the air often impure or infectious, and, for a great part of the year, Nature herself seems to be sick or dead. To this vanity the external creation is made subject as well as mankind, and so must continue till the restitution of all things.

Can we imagine, in those happy times and places we are treating of, that things stood in this same posture? are these the fruits of the Golden Age and of Paradise, or consistent with their happiness? And the remedies of these evils must be so universal, you cannot give them to one place or region of the earth, but all must participate: for these are things that flow from the course of the heavens, or such general causes as extend at once to all Nature. If there was a perpetual spring and perpetual equinox in Paradise, there was at the same time a perpetual equinox all the earth over; unless you place Paradise in the middle of the torrid zone. . . . And as to the fertility of this earth, though in some spots it be eminently more fruitful than in others, and more delicious, yet that of the first earth was a fertility of another kind, being spontaneous, and extending to the production of animals, which cannot be without a favorable concourse of the heavens also. . . .

By this short review of the three general characters of Paradise and the Golden Age, we may conclude how little consistent they are with the present form and order of the earth. Who can pretend to assign any place or region in this terraqueous globe, island, or continent, that is capable of these conditions, or that agrees either with the descriptions given by the ancient heathens of their Paradises, or by the Christian Fathers of Scripture-Paradise? But where, then, will you say, must we look for it, if not upon this earth? This puts us more into despair of finding it than ever; 'tis not above nor below, in the air or in the subterraneous regions: no, doubtless 'twas upon the surface of the earth, but of the primitive earth, whose form and properties as they were different from this, so they were such as made it capable of being truly paradisiacal, both according to the forementioned characters, and all other qualities and privileges reasonably ascribed to Paradise.

Nature in a More Limited Signification

Meditations on the First Philosophy, by René Descartes, translated by Elizabeth S. Haldane and G. R. T. Ross

The scientific revolution gave rise not only to fundamental changes in astronomy, physics, and other sciences but also to major philosophical responses to them; after Bacon's, the most important were those of Descartes, Pascal, Spinoza, and Leibniz. Not that any sharp distinction between science and philosophy can be made: what we call science was then "natural philosophy." Of these four thinkers, Descartes and Leibniz were leading mathematicians (Descartes was also a leading theorist in fields as diverse as physics, optics, and physiology), and Pascal was outstanding in both mathematics and experimental science. But apart from mathematics—Descartes's analytic geometry, Pascal's probability theory, Leibniz's calculus—their lasting contributions were mainly to exploring the philosophical and religious implications of science.

René Descartes (1596–1650) studied at the Jesuit College of La Flèche and the University of Poitiers, then served in several armies before moving in 1628 to the tolerant sanctuary of Holland, where he spent most of the remainder of his life, dying of pneumonia in Sweden shortly after accepting the invitation of Queen Christina to settle there. He had completed a treatise "on the world" in 1633, upholding the Copernican system, but suppressed it on learning of Galileo's recantation; instead, he published in Leyden in 1637 several scientific and mathematical essays along with his Discours de la Méthode (Discourse on Method). *Here he declared his delight in mathematics "because of the certainty of its demonstrations and the evidence of its reasoning"; his determination "to reject as absolutely false everything as to which I could imagine the least ground of doubt"; his attainment of certainty of his own existence through knowledge that "I think, therefore I am"; his adoption of the "general rule, that the things which we conceive very clearly and distinctly"—including the existence of God—"are all true"; and his rejection of "that speculative philosophy which is taught in the Schools" in favor of "a practical philosophy by means of which . . . we can . . . render ourselves the masters and possessors of nature." Other major works include* Meditationes de Prima Philosophia (Meditations on the First Philosophy), *published in Paris in 1641, and* Principia Philosophiae (Principles of Philosophy), *published in Amsterdam in 1644. Our selection is from the second edition of the* Meditations *(Amsterdam, 1642), with bracketed additions from the French edition of 1647, as translated in* Philosophical Works of Descartes *(1911).*

Descartes's philosophy is characterized by sharp duality between the physical and mental worlds and by near-identification of science with mathematics. His physics, in which matter is moved by contiguous "vortices" or whirlpools as if by cogwheels (Descartes rejected the possibility of a vacuum, and of action at a distance), is wholly mechanistic. This fantastic system of intermeshed vortices soon evaporated before the physics of Newton, but the mechanistic vision of nature survived. As Bernard Williams remarks in Descartes: The Project of Pure Enquiry *(1978), "It was the very* idea *of mechanism that gripped Descartes, and there remains something essentially programmatic about his scientific system"; in that system, "there is not a single difference," E. K. Dijksterhuis observes in* The Mechanization of the World Picture *(1950; English trans. 1961), "between a running clockwork and a growing tree." All that nature, in this constricted conception, can teach is to avoid pain and seek pleasure. "It gives pause to think that so subtle a mind should have spun so crude a physics. And on reflection it seems clear (as Descartes, indeed, would wish) that the difficulty," Gillispie writes, "was in his idea of nature. Economy in explanation is, of course, a goal of science. But for Descartes what is simple is nature herself, whereas every neat-handed physicist knows that nature is very complex, and that only the laws of nature are simple. . . . His thought went off into clarity and left the world behind."*

Meditation VI

. . . And first of all there is no doubt that in all things which nature teaches me there is some truth contained; for by nature, considered in general, I now understand no other thing than either God Himself or else the order and disposition which God has established in created things; and by my nature in particular I understand no other thing than the complexus of all the things which God has given me.

But there is nothing which this nature teaches me more expressly [nor more sensibly] than that I have a body which is adversely affected when I feel pain, which has need of food or drink when I experience the feelings of hunger and thirst, and so on; nor can I doubt there being some truth in all this.

Nature also teaches me by these sensations of pain, hunger, thirst, etc., that I am not only lodged in my body as a pilot in a vessel, but that I am very closely united to it, and so to speak so intermingled with it that I seem to compose with it one whole. For if that were not the case, when my body is hurt, I, who am merely a thinking thing, should not feel pain, for I should perceive this wound by the understanding only, just as the sailor perceives by sight when something is damaged in his vessel; and when my body has need of drink or food, I should clearly understand the fact without being warned of it by confused feelings of hunger and thirst. For all these sensations of hunger, thirst, pain, etc. are in truth none other than certain confused modes of thought which are produced by the union and apparent intermingling of mind and body.

Moreover, nature teaches me that many other bodies exist around mine, of which some are to be avoided, and others sought after. And certainly from the fact that I am sensible of different sorts of colors, sounds, scents, tastes, heat, hardness, etc., I very easily conclude that there are in the bodies from which all these diverse sense-perceptions proceed certain variations which answer to them, although possibly these are not really at all similar to them. And also from the fact that amongst these different sense-perceptions some are very agreeable to me and others disagreeable, it is quite certain that my body (or rather myself in my entirety, inasmuch as I am formed of body and soul) may receive different impressions agreeable and disagreeable from the other bodies which surround it.

But there are many other things which nature seems to have taught me, but which at the same time I have never really received from her, but which have been brought about in my mind by a certain habit which I have of forming inconsiderate judgments on things; and thus it may easily happen that these judgments contain some error. . . . But in order that in this there should be nothing which I do not conceive distinctly, I should define exactly what I really understand when I say that I am taught somewhat by nature. For here I take nature in a more limited signification than when I term it the sum of all the things given me by God, since in this sum many things are comprehended which only pertain to mind (and to these I do not refer in speaking of nature) such as the notion which I have of the fact that what has once been done cannot ever be undone and an infinitude of such things which I know by the light of nature [without the help of the body]; and seeing that it comprehends many other matters besides which only pertain to body, and are no longer here contained under the name of nature, such as the quality of weight which it possesses and the like, with which I also do not deal; for in talking of nature I only treat of those things given by God to me as a being composed of mind and body. But the nature here described truly teaches me to flee from things which cause the sensation of pain, and seek after the things which communicate to me the sentiment of pleasure and so forth; but I do not see that beyond this it teaches me that from those diverse sense-perceptions we should ever form any conclusion regarding things outside of us, without having [carefully and maturely] mentally examined them beforehand. For it seems to me that it is mind alone, and not mind and body in conjunction, that is requisite to a knowledge of the truth in regard to such things. . . .

But we not unfrequently deceive ourselves even in those things to which we are directly impelled by nature, as happens with those who when they are sick desire to drink or eat things hurtful to them. . . . And although, considering the use to which the clock has been destined by its maker, I may say that it deflects from the order of nature when it does not indicate the hours correctly; and as, in the same way, considering the machine of the human body as having been formed by God in order to have in itself all the movements usually manifested there, I have reason for thinking that it does not follow the order of nature when, if the throat is dry, drinking does harm to the conservation of health, nevertheless I recognize at the same time that this last mode of explaining nature is very different from the other. For this is but a purely verbal characterization depending entirely on my thought, which compares a sick man and a badly constructed clock with the idea which I have of a healthy man and a well made clock, and it is hence extrinsic to the things to which it is applied; but according to the other interpretation of the term nature I understand something which is truly found in things and which is therefore not without some truth. . . .

Between Two Abysses

Pensées, by Blaise Pascal, translated by W. F. Trotter

Blaise Pascal (1623–62) was the one major philosophical and religious thinker of his age who made important contributions to both mathematics and the practical sciences. Born at Clermont in Auvergne, he lost his mother at age three, and was educated by his father, Etienne, a lawyer who moved to Paris with his son and two daughters in 1631, and then in 1639 to Rouen. He was a child prodigy who, according to his older sister Gilberte, deduced at age twelve that the angles of a triangle equaled two right angles. In 1640 he published a paper on conic sections admired by leading geometers, and in 1642 invented a calculating machine of which a working model was made two years later. His subsequent achievements included formulation of what would become probability theory, and experiments relating to equilibrium of fluids (leading to Pascal's Law, that pressure applied at any point to a confined fluid is transmitted through it in all directions undiminished) and to atmospheric pressure.

But Pascal, who was subject to repeated illness, pondered religious questions as well. In 1646 he and his family (in his "first conversion") fell under the influence of the Jansenists, a pious group within the Catholic church that opposed the moral laxity and politicized casuistry of the dominant Jesuits and stressed an Augustinian sense of man's sinful nature and the need for an "efficacious grace" that (in Pascal's words) "demands hearts pure and disengaged . . . from worldly interests incompatible with the truths of the Gospel." In 1651, after their father's death, Pascal's younger sister Jacqueline entered the Jansenist convent of Port-Royal; when the combined opposition of Jesuits, the Pope, and King Louis XIV (who were rarely united) led to condemnation of Jansenist propositions in general and the leader of Port-Royal, Antoine Arnauld, in particular, Pascal—under the pseudonym of Louis de Montalte—penned in 1656–57 a devastating polemic against the Jesuits, the Lettres écrites à un provincial (Provincial Letters).

Meanwhile he had experienced, on November 23, 1654, a "second conversion" of which he left a memorial, discovered after his death, beginning with the words:

Fire.
"God of Abraham, God of Isaac, God of Jacob" not of the philosophers and scientists.
Certitude. Certitude. Feeling. Joy. Peace.
God of Jesus Christ. . . .
"Thy God shall be my God."
Forgetfulness of the world and of everything, except God. . . .
Joy, joy, joy, tears of joy. . . .

His last years were devoted to a projected "Apology for the Christian Religion," of which fragments—some in his own hand and some dictated, some bound and others unarranged—were published in 1670 as Pensées de M. Pascal sur la religion et sur quelques autres sujets (Thoughts of M. Pascal on Religion and on Some Other Subjects). *The numbering of the fragments in our selections is that of L. Brunschvicg's editions of 1897 and 1904; the translation, that of W. F. Trotter (1904). The overall plan is suggested by Fragment 60:* "First part: *Misery of man without God.* Second part: *Happiness of man with God. Or,* First part: *That nature is corrupt. Proved by nature itself.* Second part: *That there is a Redeemer. Proved by Scripture.*"

"That nature is corrupt" is a sentiment with which the most conventional Christian could concur. But Pascal speaks with an intimate knowledge of experimental physics as well as mathematics—an understanding of natural processes that has held up far better than that of Descartes. (Pascal is reported to have called Cartesianism "the Romance of Nature, something like the story of Don Quixote.") His portrayal of man "ever drifting in uncertainty" between Infinite and Nothing could have been formulated only after the telescope and microscope had revealed unknown worlds of terrifyingly infinite space and eternal silence without and within. And by deploring the rashness of men who rush to examine nature "as though they bore some proportion to her," Pascal is discarding the correspondence of microcosm and macrocosm central to ancient, medieval, and Renaissance thought: to the infinite, nothing is proportionate save God. The scientific revolution had immensely increased comprehension of external nature, with the result that men felt more cut off from it than ever before. Except in moments of fire, certainty is nowhere, God is remote, and—as for Plutarch long before (Chapter 12 above)—"Great Pan is dead."

21 Nature has made all her truths independent of one another. Our art makes one dependent on the other. But this is not natural. Each keeps its own place.

72 Let man then contemplate the whole of nature in her full and grand majesty, and turn his vision from the low objects which surround him. Let him gaze on that brilliant light, set like an eternal lamp to illumine the universe; let the earth appear to him a point in comparison with the vast circle described by the sun; and let him wonder at the fact that this vast circle is itself but a very fine point in comparison with that described by

the stars in their revolution round the firmament. But if our view be arrested there, let our imagination pass beyond; it will sooner exhaust the power of conception than nature that of supplying material for conception. The whole visible world is only an imperceptible atom in the ample bosom of nature. No idea approaches it. We may enlarge our conceptions beyond all imaginable space; we only produce atoms in comparison with the reality of things. It is an infinite sphere, the center of which is everywhere, the circumference nowhere. In short it is the greatest sensible mark of the almighty power of God, that imagination loses itself in that thought.

Returning to himself, let man consider what he is in comparison with all existence; let him regard himself as lost in this remote corner of nature; and from the little cell in which he finds himself lodged, I mean the universe, let him estimate at their true value the earth, kingdoms, cities, and himself. What is man in the Infinite?

But to show him another prodigy equally astonishing, let him examine the most delicate things he knows. Let a mite be given him, with its minute body and parts incomparably more minute, limbs with their joints, veins in the limbs, blood in the veins, humors in the blood, drops in the humors, vapors in the drops. Dividing these last things again, let him exhaust his powers of conception, and let the last object at which he can arrive be now that of our discourse. Perhaps he will think that here is the smallest point in nature. I will let him see therein a new abyss. I will paint for him not only the visible universe, but all that he can conceive of nature's immensity in the womb of this abridged atom. Let him see therein an infinity of universes, each of which has its firmament, its planets, its earth, in the same proportion as in the visible world; in each earth animals, and in the last mites, in which he will find again all that the first had, finding still in these others the same thing without end and without cessation. Let him lose himself in wonders as amazing in their littleness as others in their vastness. For who would not be astounded at the fact that our body, which a little while ago was imperceptible in the universe, itself imperceptible in the bosom of the whole, is now a colossus, a world, or rather a whole, in respect of the nothingness which we cannot reach? He who regards himself in this light will be afraid of himself, and observing himself sustained in the body given him by nature between those two abysses of the Infinite and Nothing, will tremble at the sight of these marvels; and I think that, as his curiosity changes into admiration, he will be more disposed to contemplate them in silence than to examine them with presumption.

For in fact what is man in nature? A Nothing in comparison with the Infinite, an All in comparison with the Nothing, a mean between nothing and everything. Since he is infinitely removed from comprehending the extremes, the end of things and their beginning are hopelessly hidden from him in an impenetrable secret; he is equally incapable of seeing the Nothing from which he was made, and the Infinite in which he is swallowed up.

What will he do then, but perceive the appearance of the middle of things, in an eternal despair of knowing either their beginning or their end. All things proceed from Nothing, and are borne towards the Infinite. Who will follow these marvelous processes? The Author of these wonders understands them. None other can do so.

Through failure to contemplate these Infinites, men have rashly rushed into the examination of nature, as though they bore some proportion to her. It is strange that they have wished to understand the beginnings of things, and thence to arrive at the knowledge of the whole, with a presumption as infinite as their object. For surely this design cannot be formed without presumption or without a capacity infinite like nature.

If we are well informed, we understand that, as nature has graven her image and that of her Author on all things, they almost all partake of her double infinity. Thus we see that all the sciences are infinite in the extent of their researches. For who doubts that geometry, for instance, has an infinite infinity of problems to solve? They are also infinite in the multitude and fineness of their premises; for it is clear that those which are put forward as ultimate are not self-supporting, but are based on others which, again having others for their support, do not permit of finality. But we represent some as ultimate for reason, in the same way as in regard to material objects we call that an indivisible point beyond which our senses can no longer perceive anything, although by its nature it is infinitely divisible. . . .

This is our true state; this is what makes us incapable of certain knowledge and of absolute ignorance. We sail within a vast sphere, ever drifting in uncertainty, driven from end to end. When we think to attach ourselves to any point and to fasten to it, it wavers and leaves us; and if we follow it, it eludes our grasp, slips past us, and vanishes for ever. Nothing stays for us. This is our natural condition, and yet most contrary to our inclination; we burn with desire to find solid ground and an ultimate sure foundation whereon to build a tower reaching to the Infinite. But our whole groundwork cracks, and the earth opens to abysses.

Let us therefore not look for certainty and stability. Our reason is always deceived by fickle shadows; nothing can fix the finite between the two Infinites, which both enclose and fly from it.

If this be well understood, I think that we shall remain at rest, each in the state wherein nature has placed him. As this sphere which has fallen to us as our lot is always distant from either extreme, what matters it that man should have a little more knowledge of the universe? If he has it, he but gets a little higher. Is he not always infinitely removed from the end, and is not the duration of our life equally removed from eternity, even if it lasts ten years longer? . . .

77 I cannot forgive Descartes. In all his philosophy he would have been quite willing to dispense with God. But he had to make Him give a fillip to set the world in motion; beyond this, he has no further need of God.

93 Parents fear lest the natural love of their children may fade away. What kind of nature is that which is subject to decay? Custom is a second nature which destroys the former. But what is nature? For is custom not natural? I am much afraid that nature is itself only a first custom, as custom is a second nature.

119 Nature imitates herself. A seed grown in good ground brings forth fruit. A principle, instilled into a good mind, brings forth fruit. Numbers imitate space, which is of a different nature.

All is made and led by the same master, root, branches, and fruits; principles and consequences.

121 Nature always begins the same things again, the years, the days, the hours; in like manner spaces and numbers follow each other from beginning to end. Thus is made a kind of infinity and eternity. Not that anything in all this is infinite and eternal, but these finite realities are infinitely multiplied. Thus it seems to me to be only the number which multiplies them that is infinite.

129 Our nature consists in motion; complete rest is death.

206 The eternal silence of these infinite spaces frightens me.

229 This is what I see and what troubles me. I look on all sides, and I see only darkness everywhere. Nature presents to me nothing which is not matter of doubt and concern. If I saw nothing there which revealed a Divinity, I would come to a negative conclusion; if I saw everywhere the signs of a Creator, I would remain peacefully in faith. But, seeing too much to deny and too little to be sure, I am in a state to be pitied; wherefore I have a hundred times wished that if a God maintains nature she should testify to him unequivocally, and that, if the signs she gives are deceptive, she should suppress them altogether; that she should say everything or nothing, that I might see which cause I ought to follow. Whereas in my present state, ignorant of what I am or of what I ought to do, I know neither my condition nor my duty. My heart inclines wholly to know where is the true good, in order to follow it; nothing would be too dear to me for eternity.

I envy those whom I see living in the faith with such carelessness, and who make such a bad use of a gift of which it seems to me I would make such a different use.

231 Do you believe it to be impossible that God is infinite, without parts?—Yes. I wish therefore to show you an infinite and indivisible thing. It is a point moving everywhere with an infinite velocity; for it is one in all places, and is all totality in every place.

Let this effect of nature, which previously seemed to you impossible, make you know that there may be others of which you are still ignorant. Do not draw this conclusion from your experiment, that there remains nothing for you to know; but rather that there remains an infinity for you to know.

267 The last proceeding of reason is to recognize that there is an infinity of things which are beyond it. It is but feeble if it does not see so far as to know this. But if natural things are beyond it, what will be said of supernatural?

347 Man is but a reed, the most feeble thing in nature; but he is a thinking reed. The entire universe need not arm itself to crush him. A vapor, a drop of water suffices to kill him. But, if the universe were to crush him, man would still be more noble than that which killed him, because he knows that he dies and the advantage which the universe has over him; the universe knows nothing of this.

All our dignity consists, then, in thought. By it we must elevate ourselves, and not by space and time which we cannot fill. Let us endeavor, then, to think well; this is the principle of morality.

409 *The greatness of man.*—The greatness of man is so evident, that it is even proved by his wretchedness. For what in animals is nature we call in man wretchedness; by which we recognize that, his nature being now like that of animals, he has fallen from a better nature which once was his.

418 It is dangerous to make man see too clearly his equality with the brutes without showing him his greatness. It is also dangerous to make him see his greatness too clearly, apart from his vileness. It is still more dangerous to leave him in ignorance of both. But it is very advantageous to show him both. Man must not think that he is on a level either with the brutes or with the angels, nor must he be ignorant of both sides of his nature; but he must know both.

426 True nature being lost, everything becomes its own nature; as the true good being lost, everything becomes its own true good.

427 Man does not know in what rank to place himself. He has plainly gone astray, and fallen from his true place without being able to find it again. He seeks it anxiously and unsuccessfully everywhere in impenetrable darkness.

579 Nature has some perfections to show that she is the image of God, and some defects to show that she is only His image.

694 *Prophecies.*—Great Pan is dead.

Deus Sive Natura

Selections from the Works of Benedict Spinoza, translated by R. H. M. Elwes

Baruch (in Latin, Benedictus) Spinoza (1632–77) was born in Amsterdam, a descendant of Portuguese Jews who had fled to Holland from the Spanish Inquisition. He received an Orthodox Jewish education which he supplemented by the study of Latin, but because of his highly unorthodox ideas was formally excommunicated from the synagogue in 1656: "The Lord blot out his name under heaven. The Lord set him apart for destruction from all the tribes of Israel . . . There shall no man speak to him" (as Matthew Arnold quotes the decree in Essays in Criticism *[1865]), "no man write to him, no man show him any kindness . . ." He lived frugally as a lens grinder, gaining a reputation for learning that led in 1673 to an offer (which he declined) of a professorship at Heidelberg; he died of consumption in the Hague. During his lifetime he published only the* Principles of Cartesianism Geometrically Demonstrated *under his own name in 1663 and the* Tractatus Theologico-Politicus (Theologico-Political Treatise) *anonymously in 1670. His other works, the* Ethica Ordine Geometrico Demonstrata (Ethics Demonstrated According to Geometrical Order), Tractatus Politicus (Political Treatise), *and* Tractatus de Intellectus Emendatione (Treatise on the Improvement of the Understanding), *were published as* Opera Posthuma (Posthumous Works) *in 1677. Our selections are from* The Chief Works of Benedick de Spinoza *(1883).*

Spinoza's refutation of miracles in the Theologico-Political Treatise *as contrary to the "fixed and immutable order" of nature, whose laws attest "the infinity, the eternity, and the immutability of God," reflects a conception of God equated with nature and reason: "whatsoever is contrary to nature is also contrary to reason, and whatsoever is contrary to reason is absurd, and,* ipso facto, *to be rejected." Ancient Stoicism (Chapter 10 above) maintained a similar view. Spinoza was deeply influenced by Descartes both in his geometrical mode of demonstration in the* Ethics *and in his conviction of a universe ruled by inviolable natural laws. But whereas dualism of mind and body, God and extension, was central to Descartes's philosophy, in which God tended (as in eighteenth-century deism) to become an otiose convenience to which occasional lip service was paid, Spinoza is the strictest of monists. God or Nature* (Deus sive Natura), *whether as extension or thought—the only two of its countless aspects accessible to us—is the sole substance of the universe, and "knowledge and love of God," the* Theologico-Political Treatise *declares, "is the ultimate aim to which all our actions should be directed." "Of anything less than the unique self-creating substance which is the whole of Nature," Stuart Hampshire writes in* Spinoza *(1951), "one cannot say that its existence and attributes can be explained without reference to anything other than itself. Only God or Nature as a whole is self-creating; it follows, therefore, that only God or Nature is absolutely free," and that "everything in the Universe is determined and nothing is contingent."*

Others, including Descartes, had associated God with nature "considered in general," and thinkers from John the Scot (Chapter 13 above) to the Scholastics had identified the divine with the active or creative aspect of nature, natura naturans. *But for Spinoza, God is equally nature in its passive aspect,* natura naturata*: "it is not only correct," Hampshire remarks, "but necessary to attach both of these complementary meanings to the word, neither being complete, or even possible, without the other. This doctrine of the essential identity of the Creator and his Creation, so far from being mystical and anti-scientific in intention, leads logically to the conclusion that every single thing in the Universe necessarily belongs to, or falls within, a single intelligible, causal system." Human freedom, the* Ethics *declares—the concept again parallels that of the Stoics—lies in understanding and acceptance, through "intellectual love of God," of our necessary participation in the divine order.*

A system so uncompromising in its denial of a personal God separate from nature was a scandal and a stumbling block to Jew, Christian, and (more surprisingly) sceptical rationalist alike: the Tractatus *was placed under the ban of both the Protestant States-General of Holland and the Roman Catholic* Index Prohibitorum. *Pierre Bayle in his influential* Dictionnaire historique et critique (Historical and Critical Dictionary), *published in 1697, though admiring this virtuous "atheist" with whom he was nearly obsessed, calls the* Tractatus *"a pernicious and detestable book" and excoriates the doctrine of the* Ethics *as "the most monstrous hypothesis that could be imagined, the most absurd, and the most diametrically opposed to the most evident notions of our mind"; Voltaire, faulting Spinoza's rejection of final causes, seems at times to endorse that judgment. As Paul Hazard observes in* The European Mind 1680–1715 (La Crise de la conscience européenne, *1935; English trans. 1952), "The daring utterances of the* Aufklärung, *of the age of light, pale into insignificance before the aggressive audacities of the* Tractatus theologico-politicus, *the amazing declarations of the* Ethics,*" which influenced even those who reviled them. Only with the Germans of the late eighteenth century—Lessing and Jacobi, Herder and Goethe, and many who followed—did Spinoza come to be seen not only as a seminal thinker ("To be a philosopher," Hegel declares, "one must first be a Spinozist") but also as an inspired poet of nature: in Novalis's phrase,* der Gottvertrunkene Mann *("the God-intoxicated man"). "When we read Spinoza, we have the feeling," Heine writes, "that we are looking at all-powerful Nature in liveliest repose—a forest of thoughts, high as heaven, with green tops ever in motion—while below the immovable trunks are deeply rooted in the eternal earth." The anathema, at long last, was rescinded.*

FROM Theologico-Political Treatise, *Chapter VI: Of Miracles*

. . . Now, as nothing is necessarily true save only by Divine decree, it is plain that the universal laws of nature are decrees of God following from the necessity and perfection of the divine nature. Hence, any event happening in nature which contravened nature's universal laws would necessarily also contravene the divine decree, nature, and understanding; or if anyone asserted that God acts in contravention to the laws of nature, he, *ipso facto*, would be compelled to assert that God acted against His own nature—an evident absurdity. One might easily show from the same premises that the power and efficiency of nature are in themselves the divine power and efficiency, and that the divine power is the very essence of God, but this I pass over for the present.

Nothing, then, comes to pass in nature[18] in contravention to her universal laws, nay, everything agrees with them and follows from them, for whatever comes to pass, comes to pass by the will and eternal decree of God; that is, as we have just pointed out, whatever comes to pass, comes to pass according to the laws and rules which involve eternal necessity and truth; nature, therefore, always observes laws and rules which involve eternal necessity and truth, although they may not all be known to us, and therefore she keeps a fixed and immutable order. Nor is there any sound reason for limiting the power and efficacy of nature, and asserting that her laws are fit for certain purposes, but not for all; for as the efficacy and power of nature are the very efficacy and power of God, and as the laws and rules of nature are the decrees of God, it is in every way to be believed that the power of nature is infinite, and that her laws are broad enough to embrace everything conceived by the divine intellect; the only alternative is to assert that God has created nature so weak, and has ordained for her laws so barren, that He is repeatedly compelled to come afresh to her aid if He wishes that she should be preserved, and that things should happen as He desires: a conclusion, in my opinion, very far removed from reason. Further, as nothing happens in nature which does not follow from her laws, and as her laws embrace everything conceived by the divine intellect, and lastly, as nature preserves a fixed and immutable order; it most clearly follows that miracles are only intelligible as in relation to human opinions, and merely mean events of which the natural cause cannot be explained by a reference to any ordinary occurrence, either by us, or at any rate, by the writer and narrator of the miracle. . . .

It is now time to pass on to the second point, and show that we cannot gain an understanding of God's essence, existence, or providence by means of miracles, but that these truths are much better perceived through the fixed and immutable order of nature.

I thus proceed with a demonstration. As God's existence is not self-evident, it must necessarily be inferred from ideas so firmly and incontrovertibly true that no power can be postulated or conceived sufficient to impugn them. They ought certainly so to appear to us when we infer from them God's existence, if we wish to place our conclusion beyond the reach of doubt; for if we could conceive that such ideas could be impugned by any power whatsoever, we should doubt of their truth, we should doubt of our conclusion, namely of God's existence, and should never be able to be certain of anything. Further, we know that nothing either agrees with or is contrary to these primary ideas; wherefore if we would conceive that anything could be done in nature by any power whatsoever which would be contrary to the laws of nature, it would also be contrary to our primary ideas, and we should have either to reject it as absurd, or else to cast doubt (as just shown) on our primary ideas, and consequently on the existence of God, and on everything howsoever perceived. Therefore miracles, in the sense of events contrary to the laws of nature, so far from demonstrating to us the existence of God, would, on the contrary, lead us to doubt it, where, otherwise, we might have been absolutely certain of it, as knowing that nature follows a fixed and immutable order.

Let us take miracles as meaning that which cannot be explained through natural causes. This may be interpreted in two senses: either as that which has natural causes, but cannot be examined by the human intellect; or as that which has no cause save God and God's will. But as all things which come to pass through natural causes come to pass also solely through the will and power of God, it comes to this, that a miracle, whether it has natural causes or not, is a result which cannot be explained by its cause, that is a

[18]N.B. I do not mean here by "nature" merely matter and its modifications, but infinite other things besides matter. (Spinoza)

phenomenon which surpasses human understanding; but from such a phenomenon, and certainly from a result surpassing our understanding, we can gain no knowledge. For whatsoever we understand clearly and distinctly should be plain to us either in itself or by means of something else clearly and distinctly understood; wherefore from a miracle or a phenomenon which we cannot understand, we can gain no knowledge of God's essence, or existence, or indeed anything about God or nature; whereas when we know that all things are ordained and ratified by God, that the operations of nature follow from the essence of God, and that the laws of nature are eternal decrees and volitions of God, we must perforce conclude that our knowledge of God and of God's will increases in proportion to our knowledge and clear understanding of nature, as we see how she depends on her primal cause, and how she works according to eternal law. Wherefore so far as our understanding goes, those phenomena which we clearly and distinctly understand have much better right to be called works of God, and to be referred to the will of God, than those about which we are entirely ignorant, although they appeal powerfully to the imagination, and compel men's admiration.

It is only phenomena that we clearly and distinctly understand which heighten our knowledge of God, and most clearly indicate His will and decrees. Plainly, they are but triflers who, when they cannot explain a thing, run back to the will of God; this is, truly, a ridiculous way of expressing ignorance. . . . On the other hand, the laws of nature, as we have shown, extend over infinity, and are conceived by us as, after a fashion, eternal, and nature works in accordance with them in a fixed and immutable order; therefore, such laws indicate to us in a certain degree the infinity, the eternity, and the immutability of God.

We may conclude, then, that we cannot gain knowledge of the existence and providence of God by means of miracles, but that we can far better infer them from the fixed and immutable order of nature. By miracle, I here mean an event which surpasses, or is thought to surpass, human comprehension: for in so far as it is supposed to destroy or interrupt the order of nature or her laws, it not only can give us no knowledge of God, but, contrariwise, takes away that which we naturally have, and makes us doubt of God and everything else.

Neither do I recognize any difference between an event against the laws of nature and an event beyond the laws of nature (that is, according to some, an event which does not contravene nature, though she is inadequate to produce or effect it)—for a miracle is wrought in, and not beyond nature, though it may be said in itself to be above nature, and, therefore, must necessarily interrupt the order of nature, which otherwise we conceive of as fixed and unchangeable, according to God's decrees. If, therefore, anything should come to pass in nature which does not follow from her laws, it would also be in contravention to the order which God has established in nature for ever through universal natural laws: it would, therefore, be in contravention to God's nature and laws, and, consequently, belief in it would throw doubt upon everything, and lead to Atheism. . . .

We may, then, be absolutely certain that every event which is truly described in Scripture necessarily happened, like everything else, according to natural laws; and if anything is there set down which can be proved in set terms to contravene the order of nature, or not to be deducible therefrom, we must believe it to have been foisted into the sacred writings by irreligious hands; for whatsoever is contrary to nature is also contrary to reason, and whatsoever is contrary to reason is absurd, and, *ipso facto*, to be rejected. . . .

FROM *Ethics, Part One: Concerning God*

PROPOSITION XV Whatever is, is in God, and without God nothing can be, or be conceived.

Proof.—Besides God, no substance is granted or can be conceived, that is nothing which is in itself and is conceived through itself. But modes can neither be, nor be conceived, without substance; wherefore they can only be in the divine nature, and can only through it be conceived. But substances and modes form the sum total of existence, therefore, without God, nothing can be, or be conceived.

Note.— . . . All things, I repeat, are in God, and all things which come to pass, come to pass solely through the laws of the infinite nature of God, and follow (as I will shortly show) from the necessity of his essence. Wherefore it can in no wise be said, that God is passive in respect to anything other than himself, or that extended substance is unworthy of the divine nature, even if it be supposed divisible, so long as it is granted to be infinite and eternal. But enough of this for the present. . . .

PROP. XXIX Nothing in the universe is contingent, but all things are conditioned to exist and operate in a particular manner by the necessity of the divine nature. . . .

Note.—Before going any further, I wish here to explain what we should understand by nature viewed as active *(natura naturans)*, and nature viewed as passive *(natura naturata)*. I say to explain, or rather call attention to it, for I think that, from what has been said, it is sufficiently clear that by nature viewed as active we should understand that which is in itself, and is conceived through itself, or those attributes of substance which express eternal and infinite essence, in other words God, in so far as he is conceived as a free cause.

By nature viewed as passive I understand all that which follows from the necessity of the nature of God, or of any of the attributes of God, that is, all the modes of the attributes of God in so far as they are considered as things which are in God, and which without God cannot exist or be conceived. . . .

APPENDIX [TO PART I].— . . . It is accepted as certain that God himself directs all things to a definite goal (for it is said that God made all things for man, and man that he might worship him). I will, therefore, consider this opinion, asking first, why it obtains general credence, and why all men are naturally so prone to adopt it? secondly, I will point out its falsity; and, lastly, I will show how it has given rise to prejudices about good and bad, right and wrong, praise and blame, order and confusion, beauty and ugliness, and the like. However, this is not the place to deduce these misconceptions from the nature of the human mind; it will be sufficient here, if I assume as a starting point what ought to be universally admitted, namely, that all men are born ignorant of the causes of things, that all have the desire to seek for what is useful to them, and that they are conscious of such desire. Herefrom it follows first, that men think themselves free, inasmuch as they are conscious of their volitions and desires, and never even dream, in their ignorance, of the causes which have disposed them to wish and desire. Secondly, that men do all things for an end, namely for that which is useful to them, and which they seek. Thus it comes to pass that they only look for a knowledge of the final causes of events, and when these are learned, they are content, as having no cause for further doubt. If they cannot learn such causes from external sources, they are compelled to turn to considering themselves, and reflecting what end would have induced them personally to bring about the given event, and thus they necessarily judge other natures by their own. Further, as they find in themselves and outside themselves many means which assist them not a little in their search for what is useful, for instance, eyes for seeing, teeth for chewing, herbs and animals for yielding food, the sun for giving light, the sea for breeding fish, etc., they come to look on the whole of nature as a means for obtaining such conveniences. Now as they are aware that they found these conveniences and did not make them, they think they have cause for believing that some other being has made them for their use. As they look upon things as means, they cannot believe them to be self-created; but, judging from the means which they are accustomed to prepare for themselves, they are bound to believe in some ruler or rulers of the universe endowed with human freedom, who have arranged and adapted everything for human use. . . . Thus the prejudice developed into superstition, and took deep root in the human mind; and for this reason everyone strove most zealously to understand and explain the final causes of things; but in the endeavor to show that nature does nothing in vain, *i.e.*, nothing which is useless to man, they only seem to have demonstrated that nature, the gods, and men are all mad together. Consider, I pray you, the result: among the many helps of nature they were bound to find some hindrances, such as storms, earthquakes, diseases, etc.: so they declared that such things happen, because the gods are angry at some wrong done them by men, or at some fault committed in their worship. Experience day by day protested and showed by infinite examples that good and evil fortunes fall to the lot of pious and impious alike; still they would not abandon their inveterate prejudice, for it was more easy for them to class such contradictions among other unknown things of whose use they were ignorant, and thus to retain their actual and innate condition of ignorance, than to destroy the whole fabric of their reasoning and start afresh. They therefore laid down as an axiom, that God's judgments far transcend human understanding. Such a doctrine might well have sufficed to conceal the truth from the human race for all eternity, if mathematics had not furnished another standard of verity in considering solely the essence and properties of figures without regard to their final causes. There are other reasons (which I need not mention here) besides mathematics, which might have caused men's minds to be directed to these general prejudices, and have led them to the knowledge of the truth.

I have now sufficiently explained my first point. There is no need to show at length that nature had no particular goal in view, and that final causes are mere human figments. . . .

Hence any one who seeks for the true causes of miracles, and strives to understand natural phenomena as an intelligent being, and not to gaze at them like a fool, is set down and denounced as an impious heretic by those whom the masses adore as the interpreters of nature and the gods. Such persons know that, with the removal of ignorance, the wonder which forms their only available means for proving and preserving their authority would vanish also. . . .

A Perfect Work of God's Making

Selections from the Works of Gottfried Wilhelm von Leibniz

Gottfried Wilhelm, Freiherr (Baron) von Leibniz, or Leibnitz (1646–1716), was born in Leipzig, the son of a professor of moral philosophy, and studied philosophy at Leipzig, mathematics at Jena, and law at Altdorf. As a diplomat for the elector of Mainz, he visited France on a mission to persuade Louis XIV to divert pressure from Germany by attacking Egypt; while in Paris (and briefly in London) between 1672 and 1676, he met many leading thinkers, even visiting Spinoza in Holland. At this time, he developed the infinitesimal calculus, somewhat after but independently of Newton, publishing his results in 1684, three years before his rival. After 1676 he was privy councilor, librarian, and historian to the duke of Brunswick in Hanover, and in 1700 became first president of the scientific academy in Berlin. Though widely known for diplomatic and political as well as mathematical and scientific activities—he tirelessly endeavored to bring about the union of Christian churches and to create a concord of Christian princes—Leibniz published only two philosophical works (both in French) during his lifetime, the Théodicée (Theodicy) *of 1710 and the* Monadologie (Monadology) *of 1714; another major work,* Nouveaux essais sur l'entendement humain (New Essays on Human Understanding), *a response to Locke's denial of innate ideas in his* Essay Concerning Human Understanding *of 1690, was published in 1765, half a century after Leibniz's death. Otherwise, his ideas were disseminated in essays and in letters to his many correspondents throughout Europe.*

Though steeped in Scholastic-Aristotelian philosophy at Leipzig—training which may have stimulated his contributions, much valued in our own century, to symbolic logic—Leibniz generally adopted the rationalistic methods and mechanistic outlook of his century; he strongly rejected, however, the Cartesian-Newtonian view (as he perceived it) that God has no function but "to wind up his watch from time to time." On the contrary, in Leibniz's dynamic and organic conception, "The whole of nature . . . is a perfect work of God's making," in which all things retain "a certain efficacy, form, or force," reflecting God's law, which we call nature, continually active in them at every level. "The organism of animals is a mechanism which supposes a divine preformation. What follows upon it is purely material, and entirely mechanical." In his conception, as Herbert Wildon Carr describes it in Leibniz *(1929), "the natural order is the harmony pre-established by God as the condition of the development of the potencies he has actualized. It involves a hierarchy of forms of activity," in which each constituent "monad"—immaterial and immortal—reflects from a different perspective the universe as a whole and possesses rudimentary perception and appetite, or desire—though only rational creatures have apperception, or self-consciousness. (Leibniz's outlook was clearly influenced by contemporary microscopic discoveries, and his emphasis on generation of living beings from "pre-formed seeds," or animalcules, and on a cycle of life and death not by metempsychosis but by a natural process of metamorphosis demonstrates the prominent biological dimension of his thought. In the same way, his affirmation of the relativity of space and time—contrary to central tenets of the Newtonian synthesis—anticipates a more modern physics.)*

In this harmonious world of correspondences at every level of being, "each mind imitates in its microcosm what God is in the macrocosm," Carr writes, and "all minds enter by reason and by knowledge of eternal truths into fellowship with God, are members of the City of God"; as Leibniz declares in our final selection, nature leads to grace, and grace, making use of nature, perfects it. Therefore (in the phrase travestied by Voltaire in Candide*) this, of all possible worlds, is the best, for it "is not a machine devised for some ulterior utilitarian purpose," Carr observes, but "a world of free individual natures, each nature acting according to the law of its own being. . . . The divine work in creation accordingly was the choice among all possible worlds of that in which the freest and highest expression could be actualized." Nature is a mirror, then, of the perfection of God.*

Of the three selections here included, the first is from a Latin essay of 1698, "De Ipsa Natura sive de Vi Insita Creaturarum" ("On Nature Itself, or on the Inherent Force of Creatures"), as translated by Paul Schrecker and Anne Martin Schrecker in Monadology and Other Philosophical Essays *(1965); the second, "On Newton's Mathematical Principles of Philosophy," is from a series of French "Lettres de Leibniz à Samuel Clarke, 1715–1716," translated by Clarke as "Five Letters to Samuel Clarke" (1717) and revised by Philip P. Wiener in* Leibniz: Selections *(1951); the third is from a French essay of 1714, "Principes de la nature et de la grâce, fondés en raison" ("The Principles of Nature and of Grace, Based on Reason"), translated by Robert Latta in* Leibniz: The Monadology and Other Philosophical Writings *(1898) and revised by Wiener.*

FROM *What Is Nature? Reflections on the Force Inherent in Created Things and on Their Actions, translated by Paul Schrecker and Anne Martin Schrecker*

2 . . . As to the first problem concerning *nature in itself*, . . . I also grant that the admirable events which occur every day and of which we are wont to say, with reason, that the work of nature is the work of an intelligence, are not to be ascribed to some created intelligences, endowed with wisdom and power proportionate to such great achievements. The whole of nature, I would say, is a perfect work of God's making, and this so much so that every natural machine—this is the true but rarely observed difference between nature and art—consists in its turn of an infinity of organs, therefore evincing the infinite wisdom and power of its creator and ruler. . . . I am satisfied with thinking that the machine of the universe is constructed with so much wisdom that all those admirable events are produced by its very operation and that the living beings in particular develop, I would hold, from a certain preformation. . . .

3 So much for what nature is not. Now let us look a bit more deeply into what that nature is which Aristotle defined quite well as the principle of motion and rest. I think, however, that this philosopher has used these words in too wide an acceptation, covering not only local motion and rest, but change in general and *stasin* or persistence. . . .

Robert Boyle, the eminent scientist and expert explorer of nature, has written a little book, *De ipsa natura*, of which the gist, if my memory is correct, amounts to stating that nature should be considered as the very *mechanism* of bodies. This we may accept as a first approach. But if the matter is to be examined with greater precision, it will be necessary to distinguish in mechanism itself the principles from the derivations. To explain clockwork, for instance, it does not suffice to say that it is mechanically driven, unless you distinguish whether its movement is produced by a weight or by a spring. I have contended several times before that the origin of mechanism itself does not flow from the material principle alone nor from mathematical reasons, but from a higher and, if I may say so, metaphysical source. This restriction will serve, I hope, to prevent the mechanical explanations of natural events from being abusively used to the prejudice of piety, as though they implied that matter can subsist by itself and that mechanism does not require any intelligence or spiritual substance. . . .

6 . . . Certainly, if no trace were impressed on created things by the divine order: *Let the earth bring forth*, and *Let the living creatures multiply*, if thereafter the world had moved as though no commandment had occurred, it would follow—since there has to be some connection, immediate or mediated, between cause and effect—either that nothing happens now in conformity to the commandment, or that this commandment operates in the present only and must be constantly renewed in the future. . . . If, however, the law enacted by God has left some vestiges impressed in the things, and if these things are so constructed according to the commandment that they are able to fulfill the will of the legislator, then it must be recognized that these things contain inherent in them a certain efficacy, form, or force. It is this efficacy, form, or force which we are accustomed to call by the name of nature, and from this nature the series of phenomena follows in conformity to the original commandment. . . .

FROM *On Newton's Mathematical Principles of Philosophy (Letters to Samuel Clarke, 1715–16), translated by Samuel Clarke (revised by Philip P. Wiener)*

From Mr. Leibniz's First Paper: Being an Extract of a Letter written in November, 1715

4 Sir Isaac Newton, and his followers, have also a very odd opinion concerning the work of God. According to their doctrine, God Almighty needs to wind up his watch from time to time: otherwise it would cease to move. He had not, it seems, sufficient foresight to make it a perpetual motion. Nay, the machine of God's making is so imperfect, according to these gentlemen, that he is obliged to clean it now and then by an extraordinary concourse, and even to mend it, as a clockmaker mends his work; who must consequently be so much the more unskillful a workman, as he is oftener obliged to mend his work and to set it right. According to my opinion, the same force and vigor remains always in the world, and only passes from one part of matter to another, agreeably to the laws of nature, and the beautiful pre-established order. And I hold that when God works miracles, he does not do it in order to supply the wants of nature, but those of grace. Whoever thinks otherwise, must needs have a very mean notion of the wisdom and power of God.

From Mr. Leibniz's Second Paper

8 I do not say the material world is a machine, or watch, that goes without God's interposition; and I have sufficiently insisted that the creation wants to be continually influenced by its Creator. But I maintain it to be a watch that goes without wanting to be mended by him: otherwise we must say that God wishes to improve upon his own work again. No; God has foreseen everything; he has provided a remedy for everything beforehand; there is in his works a harmony, a beauty, already pre-established.

From Mr. Leibniz's Third Paper

4 As for my own opinion, I have said more than once that I hold space to be something merely relative, as time is; that I hold it to be an order of co-existences, as time is

an order of successions. For space denotes, in terms of possibility, an order of things which exist at the same time, considered as existing together, without inquiring into their particular manner of existing. And when many things are seen together, one perceives that order of things among themselves.

From MR. LEIBNIZ'S FOURTH PAPER

40 The imperfection of our machines, which is the reason why they want to be mended, proceeds from this very thing, that they do not sufficiently depend upon the workman. And therefore the dependence of nature upon God, far from being the cause of such an imperfection, is rather the reason why there is no such imperfection in nature, because it depends so much upon an artist who is too perfect to make a work that wants to be mended. 'Tis true that every particular machine of nature is, in some measure, liable to be disordered; but not the whole universe, which cannot diminish in perfection.

From MR. LEIBNIZ'S FIFTH PAPER

115 As for the motions of the celestial bodies, and even the formation of plants and animals, there is nothing in them that looks like a miracle, except their beginning. The organism of animals is a mechanism which supposes a divine pre-formation. What follows upon it is purely natural, and entirely mechanical.

116 Whatever is performed in the body of man, and of every animal, is no less mechanical than what is performed in a watch. The difference is only such as ought to be between a machine of divine invention, and the workmanship of such a limited artist as man is.

FROM *The Principles of Nature and of Grace, Based on Reason,*
translated by Robert Latta (revised by Philip P. Wiener)

3 All nature is a *plenum*. There are everywhere simple substances, separated in effect from one another by activities of their own which continually change their relations; and each important simple substance, or monad, which forms the center of a composite substance (as, for example, of an animal) and the principle of its unity, is surrounded by a mass composed of an infinity of other monads which constitute the body proper of this central monad; and in accordance with the affections of its body the monad represents, as in a center, the things which are outside of itself. And this body is organic, though it forms a sort of automaton or natural machine, which is a machine not only in its entirety, but also in its smallest perceptible parts. And as, because the world is a *plenum*, everything is connected and each body acts upon every other body, more or less, according to the distance, and by reaction is itself affected thereby, it follows that each monad is a living mirror, or endowed with internal activity, representative according to its point of view of the universe, and as regulated as the universe itself. And the perceptions in the monad spring one from the other by the laws of desires *[appétits]* or of the final causes of good and evil, which consist in observable, regulated or unregulated, perceptions; just as the changes of bodies and external phenomena spring one from another by the laws of efficient causes, that is, of motions. Thus there is a perfect harmony between the perceptions of the monad and the motions of bodies, pre-established at the beginning between the system of efficient causes and that of final causes. And in this consists the accord and physical union of the soul and the body, although neither one can change the laws of the other. . . .

6 Modern researches have taught us, and reason approves of it, that living beings whose organs are known to us, that is to say, plants and animals, do not come from putrefaction or from chaos, as the ancients believed, but from pre-formed seeds, and consequently by the transformation of pre-existing living beings. There are animalcules in the seeds of larger animals, which by means of conception assume a new dress, which they make their own, and by means of which they can nourish themselves and increase their size, in order to pass to a larger theater and to accomplish the propagation of the large animal. It is true that the souls of spermatic human animals are not rational, and do not become so until conception destines these animals to human nature. And as in general animals are not born entirely in conception or generation, neither do they perish entirely in what we call death; for it is reasonable that what does not begin naturally should not end either in the order of nature. Therefore, quitting their mask or their rags, they merely return to a more minute theater, where they can, nevertheless, be just as sensitive and just as well ordered as in the larger. And what we have just said of the large animals takes place also in the generation and death of spermatic animals themselves, that is to say, they are growths of other smaller spermatic animals, in comparison with which they may pass for large; for everything extends *ad infinitum* in nature. Thus not only souls, but also animals, are ingenerable and imperishable: they are only developed, enveloped, reclothed, unclothed, transformed; souls never quit their entire body and do not pass from one body into another which is entirely new to them. There is therefore no metempsychosis, but there is metamorphosis; animals change, take and leave only parts; the same thing which happens little by little and by small invisible particles, but continually, in nutrition; and suddenly, visibly, but rarely, in conception or in death, which cause a gain or loss all at one time. . . .

12 It follows, farther, from the perfection of the supreme author, that not only is the order of the entire universe the most perfect possible, but also that each living mirror representing the universe in accordance with its point of view, that is to say, that each *monad*, each substantial center, must have its perceptions and its desires as well regulated as is compatible with all the rest. Whence it follows, still farther, that *souls*, that is, the most dominating monads, or rather, animals themselves, cannot fail to awaken from the state of stupor in which death or some other accident may put them.

13 For all is regulated in things, once for all, with as much order and harmony as is possible, supreme wisdom and goodness not being able to act except with perfect harmony. The present is big with the future, the future might be read in the past, the distant is expressed in the near. One could become acquainted with the beauty of the universe in each soul, if one could unfold all its folds, which only develop perceptibly in time. But as each distinct perception of the soul includes innumerable confused perceptions, which embrace the whole universe, the soul itself knows the things of which it has perception only so far as it has distinct and clear perceptions of them; and it has perfection in proportion to its distinct perceptions. Each soul knows the infinite, knows all, but confusedly; as in walking on the seashore and hearing the great noise which it makes, I hear the particular sounds of each wave, of which the total sound is composed, but without distinguishing them. Our confused perceptions are the result of the impressions which the whole universe makes upon us. It is the same with each monad. God alone has a distinct knowledge of all, for he is the source of all. It has been well said that he is as center everywhere, but his circumference is nowhere, since everything is immediately present to him without any distance from this center.

14 As regards the rational soul, or *spirit [mind]*, there is something in it more than in the monads, or even in simple souls. It is not only a mirror of the universe of creatures, but also an image of the Divinity. The *spirit [mind]* has not only a perception of the works of God, but it is even capable of producing something which resembles them, although in miniature. For, to say nothing of the marvels of dreams, in which we invent without trouble (but also involuntarily) things which, when awake, we should have to think a long time in order to hit upon, our soul is architectonic also in its voluntary actions, and, discovering the sciences according to which God has regulated things . . . , it imitates, in its department and in its little world, where it is permitted to exercise itself, what God does in the large world.

15 This is why all spirits, whether of men or of genii, entering by virtue of reason and of eternal truths into a sort of society with God, are members of the City of God, that is to say, of the most perfect state, formed and governed by the greatest and best of monarchs; where there is no crime without punishment, no good actions without proportionate recompense; and, finally, as much virtue and happiness as possible; and this is not by a derangement of nature, as if what God prepares for souls disturbed the laws of bodies, but by the very order of natural things, in virtue of the harmony pre-established for all time between the *realms of nature and of grace*, between God as Architect and God as Monarch; so that *nature* itself leads to grace, and *grace*, in making use of nature, perfects it. . . .

The Opera House of Nature

Conversations on the Plurality of Worlds,
by Bernard le Bovier de Fontenelle, translated by H. A. Hargreaves

World-views are slow to change. Despite the momentous transformations in the human understanding of nature that writers of the present chapter had enunciated, elements of the Aristotelian-Scholastic and Platonic-Hermetic versions of ancient, medieval, and Renaissance cosmology remained deeply embedded in the consciousness not only of poets but also of many scientists and philosophers, to say nothing of ordinary people of the age. Our final selections are from H. A. Hargreaves's translation (1990) of Entretiens sur la pluralité des mondes (Conversations on the Plurality of Worlds) *by Bernard le Bovier de Fontenelle (1657–1757), a work, published in 1686—the year before Newton's* Principia*—that did much to popularize the new Copernican-Galilean-Cartesian philosophy among the nonscientific reading public of France and Europe, and to speculate on its imaginative implications. The wit, even the occasional frivolity, of this dialogue between the author and a beautiful and intelligent Marquise whom he instructs in the new conception of nature no doubt facilitated its publication at a time when Copernican views were still considered a threat to the authority of both Church and State armed with formidable powers of censorship.*

Born in Rouen of a distinguished literary family (his mother was the sister of the great dramatist Pierre Corneille), and educated at a Jesuit college, Fontenelle served for over forty years (1699–1741) as secretary of the Académie Royale des Sciences. He published during his long life—he died a month short of his hundredth birthday—a series of lively books examining scien-

tific, religious, and philosophical questions; these include, besides the Entretiens, Dialogues des morts *(*Dialogues of the Dead, *1683),* Histoire des oracles *(*History of Oracles, *1686), and* L'Origine des fables *(*The Origin of Fables, *1724). Reacting against the high-flown sublimity of his uncle's heroic tragedies in a spirit influenced by Descartes and anticipating the Enlightenment, Hazard writes, "Fontenelle's intellect offers an almost ideal example of . . . the power . . . to grasp a thing in all its parts and grasp it quickly, allowing no external influence or inward prompting to mar or distort it."*

This power is evident, in our selections, in the clarity and concision with which he explains the Copernican system (as extended by Descartes, in whose "vortices" Fontenelle—like many others in France—would continue to believe long after Newton's universal gravitation had rendered them superfluous) and discusses the processes of nature in terms accessible to educated laypersons, comparing them, for example, to the backstage machinery of an opera house. "The magnificence is in the design," he succinctly remarks, "the frugality is in the execution." As the dialogue continues on successive evenings, moreover, Fontenelle's exposition becomes more daringly speculative. He envisions an inhabited moon to which, now that "we're beginning to fly a bit," we will go someday, and imagines possible variations among "the infinite multitude of inhabitants of all these planets" in a universe without end. Here the disquiet of Donne and the terror of Pascal have been largely displaced by intense fascination with a cosmos whose explosive enlargement, however troubling at first, liberates the mind from confinement.

Fontenelle's somewhat fanciful conjectures were probably influenced by earlier works—including the English clergyman John Wilkins's Discovery of a World on the Moon *(1638; published in French translation in 1655), Cyrano de Bergerac's imaginary voyages to the moon and sun (1657 and 1662), and Pierre Borel's* Discours nouveau prouvant la pluralité des mondes *(*New Discourse Proving the Plurality of Worlds, *1657)—but he has artfully woven them into his scientific exposition in a way that gives them additional force. For Fontenelle has absorbed and communicated not only the externals but also some of the most fundamental implications of the new philosophy, including its enormous expansion both of space and of time. In this philosophy, change, though unperceived in our momentary lives, is incessant. Nearly a century later, in Diderot's dialogue* d'Alembert's Dream *(Chapter 21 below), Mlle de Lespinasse, attempting to grasp what the raving d'Alembert might mean by "the mayfly's sophism," likens this image of an ephemeral creature's erroneous belief in the immutabiltiy of nature to "Fontenelle's rose, which said no one in the memory of a rose had seen a gardener die." What is gracefully put, Fontenelle and Diderot both knew, need not therefore be less profound.*

FROM *The First Evening*

"Apparently," said I, "the day doesn't inspire sadness and passion like the night, when everything seems to be at rest . . . ."

"I've always felt that," she said. "I love the stars, and I'm almost angry with the Sun for overpowering them."

"I can never forgive it," I cried, "for making me lose sight of all those worlds."

"What do you mean, worlds?" she asked, turning to me.

"Excuse me," I answered. "You've set me onto my weakness, and my imagination is getting the best of me."

"What is this weakness?" she asked, not to be deterred.

"I'm ashamed to admit it," I said, "but I have a peculiar notion that every star could well be a world. I wouldn't swear that it's true, but I think so because it pleases me to think so. The idea sticks in my mind in a most delightful way. As I see it, this pleasure is an integral part of truth itself."

"Well," said the Marquise, "if your idea is so pleasing, share it with me. I'll believe that the stars are anything you say, if I enjoy it."

"Ah, Madame," I answered, "this isn't enjoyment such as you'd find in a Molière comedy; it's enjoyment that involves our reasoning powers. It only delights the mind."

"What?" she cried. "Do you think I'm incapable of enjoying intellectual pleasures? I'll show you otherwise right now. Tell me about your stars!"

"No!" I answered. "It will never be said of me that in an arbor, at ten o'clock in the evening, I talked of philosophy to the most beautiful woman I know. Look elsewhere for the philosophers."

Although I excused myself in this manner several times, I had to give in, but at least, for the preservation of my honor, I made her promise to keep it a secret. Then when I finally had no excuses left and decided to speak, I didn't know where to begin. To someone like the Marquise, who knew nothing of Natural Philosophy, I would have to go a long way to prove that the Earth might be a planet, the other planets Earths, and all the stars solar systems. I told her several times that it would be better to talk about trifles, as all reasonable people would in our place. Finally, however, to give her a general idea of philosophy, here is the proposal into which I threw myself.

"All philosophy," I told her, "is based on two things only: curiosity and poor eyesight; if you had better eyesight you could see perfectly well whether or not these stars are solar systems, and if you were less curious you wouldn't care about knowing, which amounts to the same thing. The trouble is, we want to know more than we can see. Again, if we could really see things as they are, we would really know something, but we see things other than as they are. So true philosophers spend a lifetime not believing what they do see, and theorizing on what they don't see, and it's not, to my way of thinking, a very enviable situation. On this subject I have always thought that nature is very much like an opera house. From where you are at the opera you don't see the stages exactly as they are; they're arranged to give the most pleasing effect from a distance, and the wheels and counter-weights that make everything move are hidden out of sight. You don't worry, either, about how they work. Only some engineer in the pit, perhaps, may be struck by some extraordinary effect and be determined to figure out for himself how it was done. That engineer is like the philosophers. But what makes it harder for the philosophers is that, in the machinery that Nature shows us, the wires are better hidden—so well, in fact, that they've been guessing for a long time at what causes the movements of the universe.

"Imagine all the Sages at an opera—the Pythagorases, Platos, Aristotles, and all those whose names nowadays are dinned into our ears. Suppose that they watched Phaeton lifted by the winds, but they couldn't discover the wires and didn't know how the backstage area was arranged. One of them would say: 'Phaeton has a certain hidden property that makes him lighter.' Another: 'Phaeton is composed of certain numbers that make him rise.' Another: 'Phaeton has a peculiar attraction to the top of the theater, and he is uneasy if he's not up there.' Still another: 'Phaeton wasn't made for flying, but he would rather fly than leave a vacuum in the upper part of the stage.' And there are a hundred other notions which I'm astonished haven't destroyed the reputation of the whole of Antiquity. Finally, Descartes and some other moderns would come along, and they would say: 'Phaeton rises because he's pulled by wires, and because a weight heavier than he is descends.' Nowadays we no longer believe that a body will move if it's not affected by another body and in some fashion pulled by wires; we don't believe that it will rise or fall except when it has a spring or a counter-weight. Whoever sees nature as it truly is simply sees the backstage area of the theater."

"In that case," said the Marquise, "nature has become very mechanical."

"So mechanical," I replied, "that I fear we'll soon grow ashamed of it. They want the world to be merely, on a large scale, what a watch is on a small scale, so that everything goes by regular movements based on the organization of its parts. Admit it! Didn't you have a more grandiose concept of the universe, and didn't you give it more respect than it deserved? Most men esteem it less since they've come to know it."

"Well I hold it in much higher regard," she answered, "now that I know it's like a watch; it's superb that, wonderful as it is, the whole order of nature is based upon such simple things."

"I don't know who has given you such healthy ideas," I said, "but I'm sure few people have them besides you. Most cherish a false notion of mystery wrapped in obscurity. They only admire Nature[19] because they believe she's a kind of magic, and the minute they begin to understand her they lose all respect for her. But Madame," I continued, "you are so much more disposed to hear what I want to say that I need only draw back the curtain and show you the world. . . .

"But . . . you must note, if you please, that we are all naturally like a certain Athenian madman you've heard of, who deluded himself that all the ships entering the harbor at Piraeus belonged to him. Our folly is to believe that all of nature, without exception, is destined for our use, and when one inquires of the philosophers what is the use of the prodigious number of fixed stars, when a fraction would accomplish the same thing, they answer coldly that they serve to please our sight. On this principle one could easily imagine, first of all, that the Earth had to be resting at the center of the universe, while all the heavenly bodies, which were made for her, took the trouble to turn around her and light her. . . ."

"It would seem," the Marquise interrupted, "that your philosophy is a kind of auction, where those who offer to do these things at the least expense triumph over the others."

"It's true," I replied, "and it's only by that means that one can catch the plan on which Nature has made her works. She's extraordinarily frugal. Anything that she can do in a way which will cost a little less, even the least bit less, be sure she'll only do it that way. This frugality, nevertheless, is quite in accord with an astonishing magnificence which shines in all she does. The magnificence is in the design, and the frugality in the execution. There's nothing better than a great design which is executed at little expense. We mortals are often prone to reverse this in our ideas. We look for economy in Nature's design and magnificence in the execution. We credit her with a little design, which she executes with ten times the necessary expense. That's ridiculous."

"I'll be happy," she said, "that the system you are going to tell me of closely imitates Nature, for this good management will aid my imagination, which will then have less trouble understanding what you tell me."

"There are no further unnecessary hindrances," I replied. "Picture a German named Copernicus, who lays violent hands on the different circles and solid spheres which were

[19]I have capitalized nature when Fontenelle personifies it as the goddess, rather than the physical world around us. (Hargreaves)

imagined by Antiquity. He destroys the first and breaks the others in pieces. Seized by a noble astronomical fury, he plucks up the Earth and sends her far from the center of the universe, where she was placed, and puts the Sun at the center, to whom the honor rightly belongs. The planets no longer turn around the Earth and enclose her in the circles they decribe. If they light us, it's more or less by chance as we meet them in their paths. Everything turns around the Sun now, including the Earth, and as punishment for the long rest she was given, Copernicus charges her as much as he can with the same movements she had attributed to the planets and heavens. At last the only thing left of all this celestial train which used to accompany and surround this little Earth is the Moon that turns around her still." . . .

"Surely you don't believe," she cried, "that the vanity of men extends all the way to astronomy. Do you think you've humbled me by telling me the Earth moves around the Sun? I swear to you I don't have any less self-esteem."

"Good Lord, no, Madame!" I said. "I know full well that people are less jealous of their place in the universe than in a drawingroom, and the ranking of two planets will never be as important as that of two ambassadors. However, the same desire which makes a courtier want to have the most honorable place in a ceremony makes a philosopher want to place himself in the center of a world system, if he can. He's sure that everything was made for him, and unconsciously accepts that principle which flatters him, and his heart will bend a matter of pure speculation to self-interest."

"Honestly," said the Marquise, "this is a calumny you've invented against mankind. We should never have accepted Copernicus's system then, because it's so humiliating."

"Well," I answered, "Copernicus himself strongly doubted the success of his opinion. For a long time he didn't want to publish it. Finally he resolved to do it, at the urging of very reputable people, but on the same day that the first proof of his book was brought to him, do you know what he did? He died. He didn't want to rebut all the contradictions he foresaw, and he skillfully withdrew from the affair." . . .

While returning to the château, to exhaust the matter of systems I told her that there was a third, invented by Tycho Brahe, who, absolutely insisting that the Earth be immobile, placed it in the center of the universe, and made the Sun turn around it while all the other planets turned around the Sun, because since the new discoveries there was no longer any means of making the planets turn around the Earth. But the Marquise, who has a lively and prompt discernment, judged that it was too affected to exempt the Earth from turning about the Sun when one could exempt no other large bodies; that it was not so fitting for the Sun to turn about the Earth when all the planets turned about it; that this system couldn't be appropriate for anything but to maintain the immobility of the Earth when one had a great desire to maintain it, and certainly not to persuade one. Finally we resolved to hold to the system of Copernicus, which is more uniform and enticing and free of prejudice. In fact, its simplicity is persuasive and its boldness pleasing.

FROM *The Second Evening*

"Well then," I said to her, "now that the Sun, which is presently motionless, has ceased to be a planet, and the Earth which rolls around him has begun to be one, you won't be surprised to hear that the Moon is a world like the Earth, and that apparently she's inhabited."

"But I've never yet heard anyone say that the Moon was inhabited," she replied, "except as a fantasy and a delusion."

"This may be a fantasy too," I answered. "I don't take sides in these matters . . .

"Even so I could defend myself well enough, if I wished. I've a quite ridiculous thought, which has an air of reasonableness that captivates me; I don't know where it could have come from, audacious as it is. I'll bet that I am going to make you admit, against all reason, that some day there might be communication between the Earth and the Moon. Take your mind back to the state America was in before it was discovered by Christopher Columbus. Its inhabitants lived in extreme ignorance. Far from understanding the sciences, they knew nothing of the simplest, most necessary arts. They went naked, and had no weapons but the bow; they had never conceived that men could be carried by animals; they regarded the sea as a vast place, forbidden to men, which joined the sky, beyond which there was nothing. . . . I don't know, Madam, if you grasp the surprise of these Americans as I do, but never in the world could there have been another to equal it. After that, I would no longer want to swear that there couldn't be communication between the Earth and the Moon some day. Could the Americans have believed anyone who said there could be any between America and a Europe that they didn't even know about? True, it will be necessary to cross the great expanse of air and sky between the Earth and the Moon. But did the great seas seem to the Americans any more likely to be crossed?"

"Really," said the Marquise, staring at me. "You are mad."

"Who's arguing?" I answered.

"But I want to prove it to you," she replied. "I'm not satisfied with your admission. The Americans were so ignorant that they hadn't the slightest suspicion that anyone could make roads across such vast seas. But we, who have more knowledge, would have considered the idea of traveling in the air if it could actually be done."

"We're doing more than just guessing that it's possible," I replied. "We're beginning to fly a bit now; a number of different people have found the secret of strapping on wings that hold them up in the air, and making them move, and crossing over rivers or flying from one belfry to another. Certainly it's not been the flight of an eagle, and several times it's cost these fledglings an arm or a leg; but still these represent only the first planks that were placed in the water, which were the beginning of navigation. From those planks it was a long way to the big ships that could sail around the world. Still, little by little the big ships have come. The art of flying has only just been born; it will be perfected, and some day we'll go to the Moon. Do we presume to have discovered all things, or to have taken them to the point where we can add nothing? For goodness sake, let's admit that there'll still be something left for future centuries to do." . . .

"Have the people on the Moon already come?" she replied, nearly angry.

"The Europeans weren't in America until after six thousand years," I said, breaking into laughter; "it took that much time for them to perfect navigation to the point where they could cross the ocean. Perhaps the people on the Moon already know how to make little trips through the air; right now they're practicing. When they're more experienced and skillful we'll see them, with God knows what surprise."

"You're impossible," she said, "pushing me to the limit with reasoning as shallow as this."

"If you resist me," I replied, "I know what I'll add to strengthen it. Notice how the world grows little by little. . . . Perhaps when the world has finished growing for us, we'll begin to know the Moon. We're not there yet, because all the world isn't discovered yet, and apparently this must be done in order. When we've become really familiar with our home, we'll be permitted to know that of our neighbors, the people on the Moon." . . .

FROM *The Third Evening*

... "My reason is pretty well convinced," said the Marquise, "but my imagination's overwhelmed by the infinite multitude of inhabitants on all these planets, and perplexed by the diversity one must establish among them; for I can see that Nature, since she's an enemy of repetition, will have made them all different. But how can one picture all that?"

"It's not up to the imagination to attempt to picture all that," I answered. "It is not proper for the imagination to go any farther than the eyes can. One may only perceive by a kind of universal vision the diversity which Nature must have placed among all the worlds. All faces in general are made on the same model, but those of two large societies—European, if you like, and African—seem to have been made on two specific models, and one could go on to find the model for each family. What secret must Nature have possessed to vary in so many ways so simple a thing as a face? In the universe we're no more than one little family whose faces resemble one another; on another planet is another family whose faces have another cast. . . .

". . . Our sciences have certain limits which the human understanding has never been able to pass; there's a point at which they suddenly fail us. The rest is for other worlds, where some of what we understand is unknown. This planet enjoys the soft pleasures of love, but it's continually desolated in places by the violence of war. On another planet they enjoy eternal peace; but in the midst of this peace they never know love, and they're bored. To sum up, what Nature does on a small scale in the distribution of happiness or talent among men, she must have done on a grand scale among worlds, and she'll certainly have remembered to put to use this marvelous secret of diversifying things and equalizing them with compensations at the same time. Are you satisfied, Madame?" I added, dropping the serious tone. "Have I spun you enough tall tales?"

"Truly," she replied, "it seems to me I've less difficulty now in grasping the differences of all those worlds. My imagination is working on the plan you've given me. I present myself as best I can with extraordinary characters and costumes for the inhabitants of the planets, and devise completely bizarre shapes for them as well. I couldn't describe them to you, but nevertheless I see something."

"Let me suggest," I answered, "that tonight you give your dreams the task of devising those shapes. We'll see tomorrow if they've served you well, and if they've taught you how the inhabitants of any planet are made."

FROM *The Fifth Evening*

... "But," she replied, "here's a universe so large that I'm lost, I no longer know where I am, I'm nothing. What, is everything to be divided into vortices, thrown together in confusion? Each star will be the center of a vortex, perhaps as large as ours? All this immense space which holds our Sun and our planets will be merely a small piece of the universe? As many spaces as there are fixed stars? This confounds me—troubles me—terrifies me."

"And as for me," I answered, "this puts me at my ease. When the sky was only this blue vault, with the stars nailed to it, the universe seemed small and narrow to me; I felt oppressed by it. Now that they've given infinitely greater breadth and depth to this vault by dividing it into thousands and thousands of vortices, it seems to me that I breathe more freely, that I'm in a larger air, and certainly the universe has a completely different magnificence. Nature has held back nothing to produce it; she's made a profusion of riches altogether worthy of her. Nothing is so beautiful to visualize as this prodigious number of vortices, each with a sun at its center making planets rotate around it. The inhabitants of a planet in one of these infinite vortices see on all sides the lighted centers of the vortices surrounding them, but aren't able to see their planets which, having only a feeble light borrowed from their sun, don't send it beyond their own world."

"You offer me," she said, "a kind of perspective so long that my eyes can't reach the end of it. I see the Earth's inhabitants clearly; next you make me see those of the Moon and the other planets of our vortex clearly enough, it's true, though less clearly than those of the Earth. After them come the inhabitants of the planets of the other vortices who are, I must confess, completely in the dark. Whatever effort I make to see them, I can hardly perceive them at all. And in effect, aren't they nearly annihilated by the phrase you have to use in speaking of them? You're forced to call them 'inhabitants of one of the planets of one of these infinite vortices.' We ourselves, to whom the same phrase applies—admit that you'd scarcely know how to pick us out in the middle of so many worlds. As for me, I'm beginning to see the Earth so frighteningly small that I believe hereafter I'll never be impressed by another thing. Assuredly, if people have such a love of acquisition, if they make up plan after plan, if they go to so much trouble, it's because they don't know about vortices. I can claim that my new enlightenment justifies my laziness, and when anyone reproaches me for my indolence I'll answer: 'Ah, if you knew what the fixed stars are!' " . . .

"Ah, Madame," I replied, "rest assured, it takes time to ruin a world."

"Nevertheless," she said, "isn't time all it takes?"

"I admit it," I said. "All of this immense mass of matter that makes up the universe is in perpetual motion; no part of it is entirely exempt, and the minute there's motion anywhere you can be sure change must come. It comes slowly or quickly, but always in an amount of time proportionate to the effect. The Ancients took pleasure in imagining that the celestial bodies were changeless by nature, because they'd never seen them change. Had they had time to prove it by experience? The Ancients were children compared to us. If roses, which live but a day, wrote histories and left memoirs for one another, the first would have pictured their gardener in a certain fashion, and after more than fifteen thousand rose generations those who had yet to leave the picture to their descendants would have changed nothing. They would say on the subject, 'We've always seen the same gardener; in all the memory of roses we've seen only him, and he's always been exactly as he is. Assuredly he doesn't die like us; he's changeless.' Would the roses' logic be sound? It would actually have more foundation than that of the Ancients concerning celestial bodies, and even though there'd be no change whatever in the skies until today, even though they gave every sign that they were made to last forever without any alteration, I wouldn't believe it yet. I'd wait for a still longer test. Should we make our lifetime, which is a mere instant, the measure of some other? Would that mean that whatever had lasted a hundred thousand times longer than we do must last forever? It's not so easy to be eternal. A thing would have to pass through many generations of man, one after the other, to begin to show some sign of immortality."

"Truly," the Marquise said, "I can see that systems are far from being able to lay claim to it. I wouldn't even give them the honor of being compared to the gardener who lasts so long with respect to the roses; they're more like roses themselves, living and dying one after the other in a garden, for I expect that if the ancient stars disappear, new ones will take their place. Species must replenish themselves."

"Have no fear that they perish," I replied. "Some people will tell us that these are merely suns that are returning to us after having been lost to us for a long time in the depths of heaven. Others will tell us that these are suns that have shaken off the dark crust that began to enclose them. I can easily believe all this, perhaps, but I also believe that the universe could have been made in such a way that it will form new suns from time to time. Why couldn't the proper matter to make a sun, after having been dispersed in many different places, reassemble at length in one certain place, and there lay the foundations of a new system? I've all the more inclination to believe in these new creations, because they correspond better to the high idea I have of the works of Nature. Would she have the secret of making grasses or plants or animals live and die in a continual cycle? I'm convinced, just as you are, that she practices this same secret on systems, and that it costs her no more effort."

"Good heavens," said the Marquise. "I find now that the systems, the heavens, and the celestial bodies are so subject to change I've left them altogether and returned to Earth." . . .

EIGHT

Perilous Balance: Nature, Reason, and Sentiment in the Age of Enlightenment

Every age, in retrospect, is a transition from one to another. Thus, once the Romanticism that burst upon Germany and England in the late eighteenth century had ushered in new attitudes toward nature, the previous two centuries could be seen as a "neoclassical" aftermath to the artistic creativity of the Renaissance, as an "age of reason" consolidating the innovations of the scientific revolution, or, in later phases, as an "age of sentiment" anticipating the Romantic cultivation of feeling. The immense changes in the view of the natural world brought about by the humanist revival of antiquity and the momentous developments in science, though only slowly assimilated into European culture as a whole—Newton's *Principia* was not published until 1687—deeply affected seventeenth- and eighteenth-century thought and poetry; the latter especially, as Marjorie Hope Nicolson shows in *Newton Demands the Muse* (1946), is permeated by images from Newton's *Opticks* of 1704 (whose corpuscular theory of light would be anathema to Blake and Goethe in the age that followed), and the scientific principles of Bacon, Newton, and Locke were the foundation stone on which the French Enlightenment rested.

Yet if the new philosophy gained increasing ascendancy in an age whose culture hero was Newton, older views of nature were major influences as well, not only on Christian poets and writers such as Milton and Herbert, Sir Thomas Browne and the Cambridge Platonists, and on neoclassical critics such as Johnson and Reynolds, but also on French philosophes. While disdaining Christianity, the philosophes continued, as we have seen (Introduction to Part Seven above), the Renaissance "appeal to antiquity." Like their humanist predecessors, they deprecated Aristotle (still fatally associated with Scholasticism) but exalted the practical Cicero and the materialistic Lucretius. Despite the condescension of most philosophes—apart from Rousseau—toward the idealism of Plato, Peter Gay writes in *The Rise of Modern Paganism*, "the Enlightenment was permeated with Platonic ideas." Like that of the Stoics, the predominant view of nature was normative and regulative: nature was virtually identified with reason, which alone could penetrate its secrets. God, for both Newtonian Christian and Voltairian deist, was, like the Zeus of Cleanthes' Stoic hymn (Chapter 10 above), "nature's sovereign, governing all by law." For Enlightenment atheists, too, nature governed by law, but with no need of God.

Both conservative Christians and a thinker as original as Giambattista Vico (Chapter 21 below)—like Blake, Goethe, and the Romantics in future generations—elaborated on Leibniz's strictures (Chapter 18 above) against the mechanized Cartesian/Newtonian universe. Yet even Wordsworth, writing of Cambridge in *The Prelude* (III.58–63), recalls the statue

Of Newton with his prism and silent face,
The marble index of a mind for ever
Voyaging through strange seas of Thought, alone.

And as Newton's pervasive influence suggests, science—far from merely desacralizing a world thought to run like clockwork—"played an all-important part," Basil Willey writes in *The Eighteenth Century Background* (1940), "in producing the divinized 'Nature' of the eighteenth century (and ultimately of the 'romantic' generation)." It could do so because its findings, "to date, could fuse harmoniously with the presuppositions of Christianity, which, though shaken by controversy, still remained as almost unquestioned certainties in men's hearts. For what had science revealed? Everywhere design, order, and law, where hitherto there had been chaos." To this extent, the rational cosmos of the new philosophy merged not only with Christian humanist and Stoic views of nature, but also with those of Aristotle, Plato, and the pre-Socratics that lay behind them.

Thus nature was no less central to the "age of reason" than it would be to the Romantic period that reacted against but, inevitably, grew out of it. "A single word sufficed to put heart into the daring ones who were making ready to begin the task" of building the City of Men in accord with enlightened knowledge and reason, Paul Hazard writes in *European Thought in the Eighteenth Century* (1946; English trans. 1963): "That new, that magic word was Nature. To it they ascribed a virtue more potent than any, since Nature was the source of Knowledge and the touchstone of Reason. . . . Let man but render a willing ear to Nature, and never more would he go astray. All he had to do was to obey her kindly mandate."

Yet this confident credo, reflecting the triumphs of seventeenth century science, entailed a "perilous balance" (as W. B. C. Watkins entitled his book of 1939 on Swift, Johnson, and Sterne) shakily maintained by reason—itself continually threatened by powerful emotions—between faith (whether in religious revelation or in scientific deduction) and refractory experience. Hence "the formula equating Nature with Reason" came into question, Hazard remarks, just when it "seemed to have established Knowledge on an immovable foundation. . . . Nature was too rich in its composition, too complex in its attributes, too potent in its effects to be imprisoned in a formula, and the formula gave way under the strain. Despite all their efforts to elucidate it by analysis, to get possession of it through science, to reduce it to some easily intelligible concept, those same wise and learned men who should have been basking in the warmth of certitude . . . began to behold in Nature the reappearance of that Mystery which they were bent on banishing from their world." The very frustrations and contradictions of the rationalist conception of nature—fully apparent in Diderot and Rousseau—were thus the seeds of the Romantic reaction against its imperial claims.

It was not, indeed, so much the rationalism of the new philosophy that menaced older views of a divinely benevolent nature as it was the empiricism that shadowed it, corrosively undermining the authority of reason itself. Philosophers from Aristotle to Thomas Aquinas to Descartes had after all employed reason to "prove" the existence of God, as Leibniz employed it to refute the mechanistic rationalism of others. But once Locke had declared the mind to be a blank slate *(tabula rasa)*, Bishop Berkeley had affirmed that "to be is to be perceived" *(esse est percipi)*, and Hume's scepticism had demolished the underpinnings even of *in*ductive reason, the floodgates were opened, willy-nilly, to a subjective relativism that made identification of nature's laws with divinely sanctioned reason untenable. Only to the extent that "reason" was equated with will or desire—to which it had been opposed in classical and Christian psychology—could it be seen as "rational" to follow unbridled natural inclination: a path that easily led, by way of Diderot's Tahitians, to the marquis de Sade (Chapter 21) and his inheritors.

Otherwise, reason—the calculating, analytical reason of science by which, in Wordsworth's phrase, "we murder to dissect"—was increasingly seen as antithetical to the organic, ever-changing, spontaneous processes of nature on which, as on human genius, no laws could be imposed without falsifying mysteries knowable only by feeling (*Einfuhlung*, or empathy) and imagination: ambiguous powers. Once nature was defined by subjective responses, it became a mirror reflecting grandeur, sublimity, and benevolence at one moment and emptiness, desolation, and indifference at the next—swift changes of mood to which Rousseau, like his Romantic successors, was often prey. (On this subject, see my study *Ideal and Spleen* [1987].) Yet in the absence of emotional response, reason would lead not to understanding of nature (recently all but identified with it) but to its distortion, if not its destruction. "There was an awful rainbow once in heaven," Keats laments in his *Lamia* of 1820 (II.229–38): but once her woof and texture had been recorded—by those same Newtonian optics that had inspired earlier poets—"in the dull catalogue of common things," cold philosophy could only dissipate her charms, "conquer all mysteries by rule and line," and "empty the haunted air," unweaving the fabric that had made the rainbow, for countless millennia, a thing of wonder and beauty. From the extreme of scientific rationalism, the balance, in the unending endeavor to understand nature, had been tipped—through an alternation that perilously continues in our own time—to an equal and opposite extreme of Romantic unreason, whose corresponding dangers would not be long in becoming apparent.

19

Dialogues with Nature: Seventeenth- and Eighteenth-Century English Poetry and Imaginative Prose

The seventeenth-century scientific revolution and its eighteenth-century aftermath were bound to result, during their confident hegemony, in a relative devaluation of poetic inspiration as a means of understanding. Cartesian thought, Willey writes in The Seventeenth Century Background, *"reinforced the growing disposition to accept the scientific world-picture as the only 'true' one. The criterion of truth which it set up, according to which the only real properties of objects were the mathematical properties, implied a depreciation of all kinds of knowing other than that of the 'philosopher,'" so that "by the beginning of the eighteenth century religion had sunk to deism, while poetry had been reduced to catering for 'delight'—to providing embellishments which might be agreeable to the fancy, but which were recognized by the judgment as having no relation to 'reality.'" For the most part, from the late seventeenth century on, "everybody alike, the English no less than the French, made prose their medium for the communication of ideas," Hazard remarks in* The European Mind (La Crise de la conscience européenne), *so that "poetry died; or at least seemed to die. Strictly logical and matter-of-fact, machine-made, sapless, it lost sight of its true mission." Outside of the drama (where Calderón in Spain and Racine and Molière in France brought their national traditions to a splendid culmination), few major poets for a century or more after the English Restoration and the founding of the Royal Society in 1660 could compare with the great poets either of the Renaissance (or the High Middle Ages) before them or of the rich poetic flowering that would begin in Germany and Britain in the last decades of the eighteenth century.*

In England, the rich poetic legacy of the Elizabethan and early Jacobean Renaissance continued among both Cavaliers and Puri-

tans under Charles I and Cromwell. The classicizing "sons of Ben" who followed Jonson and the religious "metaphysical poets" (as they were later called by Samuel Johnson) of the school of Donne are often remarkably similar in their poetic treatment of nature. The beauties of spring can be celebrated by the libertine Carew, the Catholic Crashaw, and (most sensuously of all) the Puritan Milton, and there is little in Lovelace's poem on the snail or Vaughan's on the bird to pigeonhole the one as a "Cavalier" and the other as a "metaphysical" poem. One touch of nature makes poets of different schools kin, and even in so inwardly religious a poet as Herbert, imagery of the natural world goes far beyond conventional emblems. Deep contrasts of course remain; no one would mistake Herrick's

Gather ye rosebuds while ye may,
Old time is still a-flying;
And this same flower that smiles today,
Tomorrow will be dying,

for Herbert's

Sweet rose, whose hue, angry and brave
Bids the rash gazer wipe his eye:
Thy root is ever in its grave,
And thou must die.

Yet, given such fundamentally opposed outlooks and styles, it is striking how often differences merge in celebration of the glories of nature by Puritan and Anglican, Cavalier and Metaphysical alike.

The fertile scientific thought of seventeenth-century England from Gilbert and Bacon to Newton was paralleled by major contributions to political, religious, and moral philosophy by writers of our next chapter. But three English prose writers of the seventeenth and early eighteenth centuries are included in the present chapter as writers not of philosophical or critical but—loosely speaking—"imaginative" prose. Sir Thomas Browne might be indeed classified as a religious, moral, or scientific thinker except that he willfully defies categorization on every page. By his evocatively sonorous language, continually leaping from subject to subject and weaving paradoxical insights together in a texture of discordantly harmonious images, he far more nearly resembles the metaphysical poets than he does Locke or Hume; his affinities, like Milton's, are more with the Renaissance than with the Enlightenment: with Spenser than with Pope; with Montaigne than with Voltaire. Izaak Walton's lively fisherman's handbook, The Compleat Angler, *long remained a highly popular account of reinvigorating human intercourse with the natural world. And in* Robinson Crusoe, *Defoe created for his practical age the myth of a man who by his own ingenuity turns the wilderness to his uses, making raw nature—as he gradually masters it—an instrument, in the Baconian spirit, of his rudimentary, solitary, but indispensable civilization.*

In poetry itself, the richly varied legacy that had continued into the mid seventeenth century, culminating in Milton and Marvell, was appreciably narrowed after the Restoration by the increasing dominion of neoclassical "regularity" and decorum; of standards of clarity exemplified by the scientific prose of the Royal Society; and of urbane and courtly values that gave far greater prominence to the social and civil dimensions of human existence than to increasingly tenuous connections with the natural world. John Dryden (1631–1700), though he translated Virgil's Georgics *and Horace's "Beatus ille" (Chapter 11 above), and was warmly responsive to the fiery impetuosity of Homer, the godlike plenty of Chaucer, and the copious variety of Shakespeare ("All the images of Nature were still present to him, and he drew them, not laboriously, but luckily . . ."), gives scant attention to external nature in his own plays and satires. Like other neoclassical writers, he viewed the "rules" of classical poetry, for all his generously flexible interpretation of them, as (in Pope's phrase) "Nature methodized."*

Pope himself, Dryden's eighteenth-century successor as "regular" poet and satirist, was not only receptive, like Dryden, to the enlarging influences of Homer (whom he translated) and Shakespeare (whom he edited). Through-

out a poetic career that began with four Pastorals, *he repeatedly addressed widely varied aspects of nature, from evocation of russet plains and wild heaths in* Windsor Forest *to reflection on natural order in* An Essay on Man. *The remainder of a century that would end with publication of Wordsworth and Coleridge's* Lyrical Ballads *of 1798, inaugurating the Romantic movement, would see an extraordinary recrudescence of interest in the natural world—from which the human had never, after all, been wholly cut off; this took form in poems as different as James Thomson's* The Seasons *and Christopher Smart's* A Song to David. *Once again, though driven out with a pitchfork from court and salon, Nature had returned, and dialogue with her continued.*

Among general anthologies consulted for this chapter are Hebel and Hudson's Poetry of the English Renaissance, 1509–1660 *(1929); H. R. Woudhuysen, ed.,* The Penguin Book of Renaissance Verse, 1509–1659 *(1993); Hugh Maclean, ed.,* Ben Jonson and the Cavalier Poets *(1974); Hugh Kenner, ed.,* Seventeenth Century Poetry: The Schools of Donne and Jonson *(1964); Mario A. Di Cesare, ed.,* George Herbert and the Seventeenth-Century Religious Poets *(1978); and Cecil A. Moore, ed.,* English Poetry of the Eighteenth Century *(1935).*

As Fresh and Cheerful as the Light

Lyric Poetry of the Seventeenth Century

In an age whose religious and social divisions culminated in civil war in 1642, followed by the interregnum of Cromwell's Puritan commonwealth between the beheading of Charles I in 1649 and the Restoration of Charles II in 1660, astonishingly many English lyric poets of the first two-thirds of the seventeenth century perpetuate the often rhapsodic celebration of nature—its birds, beasts, and flowers, insects and snails, and the joys of country life, above all in the reawakening of spring—in a spirit inherited from their Renaissance forebears and from the Greek and Roman classics. The following eight poets illustrate both the great variety and the underlying similarities of these inheritors of Donne and Jonson—and of Shakespeare and Spenser, Virgil and Horace—in this rich poetic age.

Sir Henry Wotton *(1568–1639) was educated at Winchester and Oxford, traveled abroad as a diplomat, served as ambassador to Venice, and became provost of Eton in 1624. Only four years Shakespeare's junior, and a friend of Donne, he partly belongs to an earlier age, since much of his poetry probably goes back to the 1590s; but his collected poems, including the description of spring here given, were not published until 1651, as* Reliquiae Wottonianae, *edited by Izaak Walton, who also completed the life of Donne that his friend Wotton had undertaken.*

Henry King *(1592–1669) was the son of John King, Vice-Chancellor of Oxford and later Bishop of London; he attended Westminster School and Christ Church, Oxford, where he took his Doctor of Divinity in 1625, and rose to become Bishop of Chichester in 1642—a position from which he was soon driven out by the Puritan armies until the Restoration almost twenty years later. He too was a close friend of Donne in his younger years, and his poems (of which the most renowned is "The Exequy," an elegy for his wife) were influenced by both Donne and Jonson; most were published in* Poems, Elegies, Paradoxes, and Sonnets *of 1657, but his "Contemplation upon Flowers" is from Harleian Ms. 6917.*

Thomas Carew *(1594/95–1639/40) took his B.A. at Merton College, Oxford, and studied law at the Inner Temple before accompanying the English ambassador Sir Dudley Carleton to*

Venice in 1613 and the Hague in 1616. Dismissed for insulting his patron, he led a profligate life in England that brought about a break with his family; he served Sir Edward Herbert, ambassador to Paris, from 1619 to 1624, and finally became Gentleman of the Privy Chamber and then Sewer to King Charles I. His light verse associates him with the school of Jonson, but he too was influenced by Donne. "The Spring" is from his Poems *of 1640.*

***Richard Lovelace** (1618–56/57), a leading Cavalier poet, was educated at Charterhouse and at Gloucester Hall, Oxford. Imprisoned for defiance of Parliament, he may have fought with the Royalist armies, and was imprisoned a second time; his poem "To Althea, from Prison," in* Lucasta *(1649), contains the famous lines "Stone walls do not a prison make, / Nor iron bars a cage." In "The Snail" (from* Lucasta: Posthume Poems *[1659]), he employs witty conceits of a "metaphysical" kind to portray this "wise emblem of the politic world" as "large Euclid's strict epitome" (because of its varied shapes) and as a "deep riddle of mysterious state" (because it was thought to be self-engendering): an enigmatic epitome, on a small scale, of Nature herself, or of Nature's microcosm, man.*

***Abraham Cowley** (1618–67) was born in London and educated at Westminster School and Trinity College, Cambridge. He left for France during the civil wars in 1646, held various diplomatic positions abroad, and returned as a Royalist spy to England in 1655, where he was released on bail after being arrested and interrogated by Cromwell. He took his M.D. at Oxford in 1657, and after the Restoration returned from France to England, where he spent his last few years in solitude on an estate of the duke of Buckingham. Called a metaphysical poet in Samuel Johnson's "Life of Cowley," he was influenced also by Jonson, and by the scientific thought of the Royal Society. He wrote poems in a wide variety of styles, including a biblical epic, the* Davideis; *"Pindaric" odes; love poems; and many classical imitations and translations, among them the poem of Claudian's included in Chapter 11 above and "The Grasshopper," a delightful adaptation (published in* The Works of Abraham Cowley *[1668]) of the anacreontic poem translated above (in Chapter 11) as "The Godlike Cicada."*

***Richard Crashaw** (1612/13–49) was born in London of a stridently anti-Catholic father and educated at Pembroke College and Peterhouse, Cambridge, becoming curate of Little St. Mary's. He was closely associated with the High Church party of Archbishop Laud and the religious community at Little Gidding. In exile on the Continent during the civil war, he became a Roman Catholic, and died at Loreto in Italy. He is best known for the extravagant baroque style and ecstatic mysticism of poems such as those addressed to St. Mary Magdalene and St. Teresa; but his "Out of Virgil, In the Praise of the Spring"—from* The Delights of the Muses, *published with* Steps to the Temple *in 1646 and revised in 1648—a free translation of* Georgics *II.323–45 (Chapter 11 above), shows his mastery also of the classical style.*

***Henry Vaughan** (1621/22–95) is now considered, along with Herbert (and Crashaw), one of the foremost religious poets of the seventeenth-century metaphysical school; but as Hebel and Hudson remark, "Herbert is drawn to God through the Church, and Vaughan through intimations of divinity from nature. Vaughan is a loyal Anglican, but the institution means little to him compared with his own immediate intuition of the Deity, induced by those phenomena of nature which show the quickening of life—the freshness of the early morning, 'the seed swelling and stirring in the mold, the sap rising, or the branch budding.'" Educated at Jesus College, Oxford, he moved to London to study law, and probably served in the Royalist armies; thereafter he practiced medicine, studied Hermeticism with his twin brother, and wrote the poems collected in* Silex Scintillans *(1650; rev. 1655), including "The Bird."*

Finally, ***Margaret Cavendish, Duchess of Newcastle*** *(1623–73), left for Paris in 1644 as Maid of Honor to Queen Henrietta Maria, wife of Charles I, and in 1645 married William Cavendish, Marquis of Newcastle (thirty years her senior). Returning to England from Paris* *and Antwerp after the Restoration, they lived a "shepherd's life" at Welbeck in Nottinghamshire, where her husband was made a duke. She took an intense interest in chemistry and natural philosophy. In addition to a number of plays, an autobiography, a biography of her husband, collections of essays on various subjects, and a utopian fantasy,* The Description of a New World Called the Blazing World *(1666), she wrote several scientific treatises and many poems on different aspects of nature, including the "Dialogue betwixt Man and Nature" here included from* Poems and Fancies *of 1653.*

On a Bank As I Sat Fishing: A Description of the Spring, by Sir Henry Wotton

And now all Nature seemed in love,
The lusty sap began to move;
Now juice did stir th'embracing vines,
And birds had drawn their valentines;
The jealous trout, that low did lie, 5
Rose at a well-dissembled fly;
There stood my friend, with patient skill
Attending of his trembling quill.
Already were the eaves possessed
With the swift pilgrim's[1] daubèd nest. 10
The groves already did rejoice
In Philomel's triumphing voice.
The showers were short, the weather mild,
The morning fresh, the evening smiled.
Joan takes her neat-rubbed pail, and now 15
She trips to milk the sand-red cow,
Where, for some sturdy football swain,
Joan strikes a sillabub,[2] or twain.
The fields and gardens were beset
With tulip, crocus, violet. 20
And now, though late, the modest rose
Did more than half a blush disclose.
Thus all looked gay, all full of cheer,
To welcome the new-liveried year.

A Contemplation upon Flowers, by Henry King

Brave flowers, that I could gallant it like you
And be as little vain!
You come abroad and make a harmless show,
And to your beds of earth again;
You are not proud, you know your birth, 5
For your embroidered garments are from earth.
You do obey your months and times, but I

[1]Peregrine falcon's.
[2]A drink or dish made of milk (frequently as drawn from the cow) or cream, curdled by the admixture of wine, cider, or other acid, and often sweetened or flavored. (OED)

Would have it ever spring;
My fate would know no winter, never die
Nor think of such a thing; 10
Oh, that I could my bed of earth but view
And smile, and look as cheerfully as you.

Oh, teach me to see death and not to fear,
But rather to take truce;
How often have I seen you at a bier, 15
And there look fresh and spruce;
You fragrant flowers, then teach me that my breath
Like yours may sweeten, and perfume my death.

The Spring, by Thomas Carew

Now that the winter's gone, the earth hath lost
Her snow-white robes, and now no more the frost
Candies the grass, or casts an icy cream
Upon the silver lake or crystal stream;
But the warm sun thaws the benumbèd earth, 5
And makes it tender; gives a sacred birth
To the dead swallow; wakes in hollow tree
The drowsy cuckoo and the humble-bee.
Now do a choir of chirping minstrels bring
In triumph to the world the youthful spring. 10
The valleys, hills, and woods in rich array
Welcome the coming of the longed-for May.
Now all things smile, only my love doth lour;
Nor hath the scalding noonday sun the power
To melt that marble ice, which still doth hold 15
Her heart congealed, and makes her pity cold.
The ox, which lately did for shelter fly
Into the stall, doth now securely lie
In open fields; and love no more is made
By the fireside, but in the cooler shade 20
Amyntas now doth with his Chloris sleep
Under a sycamore, and all things keep
Time with the season; only she doth carry
June in her eyes, in her heart January.

The Snail, by Richard Lovelace

Wise emblem of our politic world,
Sage Snail, within thine own self curled,
Instruct me softly to make haste,
Whilst these my feet go slowly fast.
　Compendious Snail! thou seemst to me 5
Large Euclid's strict epitome;

And in each diagram, dost fling
Thee from the point unto the ring.[3]
A figure now triangular,
An oval now, and now a square; 10
And then a serpentine dost crawl
Now a straight line, now crook'd, now all.
Preventing[4] rival of the day,
Th'art up and openest thy ray,
And ere the morn cradles the moon, 15
Th'art broke into a beauteous noon.
Then when the sun sups in the deep,
Thy silver horns ere Cynthia's[5] peep;
And thou from thine own liquid bed,
New Phoebus, heav'st thy pleasant head. 20
Who shall a name for thee create,
Deep riddle of mysterious state?
Bold Nature that gives common birth
To all products of sea and earth,
Of thee, as earthquakes, is afraid, 25
Nor will thy dire deliv'ry[6] aid.
Thou thine own daughter, then, and sire,
That son and mother art entire,
That big[7] still with thyself dost go,
And liv'st an aged embryo; 30
That like the cubs of India,[8]
Thou from thyself a while dost play:
But frighted with a dog or gun,
In thine own belly thou dost run,
And as thy house was thine own womb, 35
So thine own womb concludes[9] thy tomb.
But now I must (analysed King)
Thy economic[10] virtues sing:
Thou great staid husband[11] still within,
Thou, thee, that's thine, dost discipline; 40
And when thou art to progress bent,
Thou mov'st thyself and tenement,
As warlike Scythians traveled, you
Remove your men and city too;
Then after a sad dearth and rain, 45
Thou scatterest thy silver train;
And when the trees grow nak'd and old,
Thou clothest them with cloth of gold,

[3]"From the center to the circumference"? (Woudhuysen) [4]"Preceding."
[5]I.e., the moon's. Cynthia, or Diana (Greek Artemis, or Phoebe), was often associated with the moon, as her brother Apollo, or Phoebus, was with the sun.
[6]"Giving birth." [7]"Pregnant," "swollen."
[8]The sue, a wild animal, was thought to carry its young in a bag. (Woudhuysen)
[9]"Encloses." [10]"Related to household management" (the literal meaning of Greek *oikonomikos*).
[11]"Household manager."

Which from thy bowels thou dost spin,
And draw from the rich mines within. 50
Now hast thou changed thee[12] saint, and made
Thyself a fane that's cupola'd;[13]
And in thy wreathèd cloister thou
Walkest thine own gray friar too;
Strict, and locked up, th'art hood all o'er 55
And ne'er eliminate[14] thy door.
On salads thou dost feed severe,
And 'stead of beads thou drop'st a tear,
And when to rest each calls the bell,
Thou sleep'st within thy marble cell; 60
Where in dark contemplation placed,
The sweets of Nature thou dost taste;
Who now with time thy days resolve,
And in a jelly thee dissolve,
Like a shot star,[15] which doth repair 65
Upward, and rarify the air.

The Grasshopper, by Abraham Cowley

Happy insect, what can be
In happiness compared to thee?
Fed with nourishment divine,
The dewy morning's gentle wine!
Nature waits upon thee still, 5
And thy verdant cup does fill;
'Tis filled wherever thou dost tread,
Nature self's thy Ganymede.[16]
Thou dost drink and dance and sing,
Happier than the happiest king! 10
All the fields which thou dost see,
All the plants belong to thee,
All that summer hours produce,
Fertile made with early juice.
Man for thee does sow and plow, 15
Farmer he, and landlord thou!
Thou dost innocently joy,
Nor does thy luxury destroy;
The shepherd gladly heareth thee,
More harmonious than he. 20
Thee country hinds[17] with gladness hear,
Prophet of the ripened year!
Thee Phoebus loves, and does inspire;

[12]"Changed yourself into." [13]"A temple with a cupola."
[14]"Come out of," "go beyond the threshold of." (Woudhuysen)
[15]On falling to earth, shooting stars were supposed to turn to jelly. (Woudhuysen)
[16]The cupbearer of Jupiter. [17]"Rustics," "lads."

Phoebus is himself thy sire.
To thee of all things upon earth, 25
Life is no longer than thy mirth.
Happy insect, happy thou,
Dost neither age nor winter know.
But when thou'st drunk, and danced, and sung
Thy fill, the flow'ry leaves among 30
(Voluptuous, and wise withal,
Epicurean animal!),
Sated with thy summer feast,
Thou retir'st to endless rest.

Out of Virgil, In the Praise of the Spring, by Richard Crashaw

All trees, all leafy groves confess the spring
Their gentlest friend: then, then the lands begin
To swell with forward pride, and seed desire
To generation; Heaven's almighty Sire
Melts on the bosom of his love, and pours 5
Himself into her lap in fruitful showers.
And by a soft insinuation, mixed
With earth's large mass, doth cherish and assist
Her weak conceptions; no lone shade, but rings
With chatting birds' delicious murmurings. 10
Then Venus' mild instinct (at set times) yields
The herds to kindly meetings, then the fields
(Quick with warm Zephyr's lively breath) lay forth
Their pregnant bosoms in a fragrant birth.
Each body's plump and juicy, all things full 15
Of supple moisture: no coy twig, but will
Trust his beloved bosom to the sun
(Grown lusty now), no vine so weak and young
That fears the foul-mouthed Auster,[18] or those storms
That the Southwest wind hurries in his arms, 20
But hastes her forward blossoms, and lays out
Freely lays out her leaves: nor do I doubt
But when the world first out of Chaos sprang
So smiled the days, and so the tenor ran
Of their felicity. A spring was there, 25
An everlasting spring, the jolly year
Led round in his great circle; no wind's breath
As then did smell of winter, or of death.
When life's sweet light first shone on beasts, and when
From their hard mother Earth, sprang hardy men, 30

[18]The south(-west) wind.

When beasts took up their lodging in the wood,
Stars in their higher chambers: never could
The tender grove of things endure the sense
Of such a change, but that the heav'ns' indulgence
Kindly supplies sick Nature, and doth mold 35
A sweetly tempered mean, nor hot nor cold.

The Bird, by Henry Vaughan

Hither thou com'st; the busy wind all night
Blew through my lodging, where thy own warm wing
Thy pillow was. Many a sullen storm,
For which course man seems much the fitter born,
Rained on thy bed 5
And harmless head.

And now as fresh and cheerful as the light,
Thy little heart in early hymns doth sing
Unto that Providence whose unseen arm
Curbed them, and clothed thee well and warm. 10
All things that be praise him, and had
Their lesson taught them when first made.

So hills and valleys into singing break,
And though poor stones have neither speech nor tongue,
While active winds and streams both run and speak, 15
Yet stones are deep in admiration.
Thus praise and prayer here beneath the sun
Make lesser mornings, when the great are done.

For each enclosèd spirit is a star,
Enlight'ning his own little sphere, 20
Whose light, though fetched and borrowèd from far,
Both mornings makes and evenings there.

But as these birds of light make a land glad,
Chirping their solemn matins on each tree,
So in the shades of night some dark fowls be, 25
Whose heavy notes make all that hear them sad.

The turtle[19] then in palm trees mourns,
While owls and satyrs howl;
The pleasant land to brimstone turns
And all her streams grow foul. 30

Brightness and mirth, and love and faith, all fly,
Till the day-spring breaks forth again from high.

[19]"Turtledove."

A Dialogue betwixt Man and Nature,
by Margaret Cavendish, Duchess of Newcastle

Man. 'Tis strange
How we do change.
First to live, and then to die,
Is a great misery.

To give us sense, great pains to feel, 5
To make our lives to be death's wheel;
To give us sense, and reason too,
Yet know not what we're made to do.
Whether to atoms turn,[20] or heaven up fly,
Or into new forms change, and never die. 10
Or else to matter prime[21] to fall again,
From thence to take new forms, and so remain.
Nature gives no such knowledge to mankind,
But strong desires to torment the mind:
And senses, which like hounds do run about, 15
Yet never can the perfect truth find out.
O Nature! Nature! cruel to mankind,
Gives knowledge none, but misery to find.
Nature. Why doth mankind complain, and make such moan?
May not I work my will with what's my own? 20
But men among themselves contract, and make
A bargain for my tree; that tree will take:
Most cruelly do chop in pieces small,
And forms it as he please, then builds withal.
Although that tree by me was made to stand, 25
Just as it grows, not to be cut by man.
Man. O Nature, trees are dull, and have no sense,
And therefore feel not pain, nor take offense.
Nature. But beasts have life and sense, and passion strong,
Yet cruel man doth kill, and doth them wrong. 30
To take that life I gave, before the time
I did ordain, the injury is mine.
Man. What ill man doth, Nature did make him do,
For he by Nature is prompt thereunto.
For it was in great Nature's power, and will, 35
To make him as she pleased, either good or ill.
Though beast hath sense, feels pain, yet whilst they live,
They reason want, for to dispute, or grieve.
Beast hath no pain, but what in sense doth lie,
Nor troubled thoughts, to think how they shall die. 40
Reason doth stretch man's mind upon the rack,
With hopes and joys pulled up, with fear pulled back.

[20] "Return."
[21] "Prime matter" *(prima materia)*, the underlying substratum of all matter in Aristotelian physics.

Desire whips him forward, makes him run,
Despair doth wound, and pulls him back again.
For, Nature, thou mad'st made man betwixt extremes, 45
Wants perfect knowledge, yet thereof he dreams.
For had he been like to a stock, or stone,
Or like a beast, to live with sense alone,
Then might he eat, or drink, or lie stone-still,
Ne'er troubled be, either for heaven, or hell. 50
Man knowledge hath, enough for to inquire,
Ambition great enough for to aspire:
And knowledge hath, that yet he knows not all,
And that himself he knoweth least of all:
Which makes him wonder, and thinks there is mixed 55
Two several qualities in Nature fixed.
The one like love, the other like to hate,
By striving both hinders predestinate.
And then sometimes, man thinks, as one they be,
Which makes contrariety so well agree; 60
That though the world were made by love and hate,
Yet all is ruled, and governèd by fate.
These are man's fears; man's hopes run smooth, and high,
Which thinks his mind is some great deity.
For though the body is of low degree, 65
In sense like beasts, their souls like gods shall be.
Nature. Says Nature, why doth man complain, and cry,
If he believes his soul shall never die?

We Have as Short a Spring

Five Poems of Robert Herrick

No "son of Ben" was more devoted than Robert Herrick (1591–1674), among whose poems in his master's honor is a prayer beginning:

When I a verse shall make,
Know I have prayed thee,
For old religion's sake,
Saint Ben, to aid me . . .

Born in London and apprenticed as a goldsmith to his prosperous uncle, Herrick appears to have been at an early age an occasional member of Jonson's circle in London. He was educated at St. John's College and Trinity Hall, Cambridge, taking his B.A. in 1617 and his M.A. in 1620. In 1623 he was ordained a priest, and in 1630 took up residence as vicar of Dean Prior in Devonshire, a living from which he was expelled by the Puritans in 1647 and to which he was reinstated in 1662. He published only a handful of poems before Hesperides *(1648), which included, besides all the poems for which he is remembered, the undistinguished religious poems called* His Noble Numbers.

"*In Herrick Renaissance neo-pagan and belated Elizabethan united to form a pure artist,*"

Douglas Bush writes in English Literature in the Earlier Seventeenth Century *(1945), "... whose poetic creed is to live merrily and trust to good verses, and who in a troubled age is largely content to mirror or create a timeless epicurean Arcadia," echoing reminiscences from Horace and Tibullus to Jonson. His generally lighthearted verses range from epigrams, drinking songs, and love poems to pastorals and epistles, often felicitously phrased ("That liquefaction of her clothes") but within a limited emotional range lacking great heights or depths. Yet few poets have so freshly evoked the country pleasures that Herrick, despite his ambivalence toward rural exile, made his themes in "The Argument of His Book" from* Hesperides*:*

I sing of brooks, of blossoms, birds, and bowers;
Of April, May, of June, and July flowers.
I sing of Maypoles, hock-carts, wassails, wakes,
Of bridegrooms, brides, and of their bridal cakes ...

"He is the poet of strawberries and cream, of fairy lore and rustic customs, of girls delineated like flowers and flowers mythologized into girls," Tucker Brooke remarks in A Literary History of England, *ed. Albert C. Baugh (1948).*

In several of the following poems, including "Corinna's Gone A-Maying"—"one of the most successful poems ever written," Brooke affirms, "in immortalizing a mood and depicting a contemporary scene"—and "The Hock-Cart," Herrick vividly re-creates the festive spirit of popular rural celebrations. Even in the deliberately more artificial "To Phyllis, To Love and Live with Him," with its humorously hyperbolic elaborations of Marlowe's "Passionate Shepherd to His Love" (Chapter 17 above), concretely sensuous images of nature and the tastes and smells of a life close to nature invigorate a poem that is never a mere literary exercise. (Yet "I must confess," he writes in "Art Above Nature: To Julia," "mine eye and heart / Dotes less on nature, than on art.") Throughout his poems, on the surface or just beneath it, runs the theme—also inherited from Horace and other classical poets—of the poignant shortness of life and the need to live it to the full "while time serves, and we are but decaying": the theme immortalized in his best-known poem, "To the Virgins, to Make Much of Time," beginning "Gather ye rosebuds while ye may," and ending:

For having lost but once your prime,
You may for ever tarry.

This is the affinity that he repeatedly finds between human life and all of nature, epitomized in the daffodils: "We have short time to stay as you; / We have as short a spring ..."

Corinna's Gone A-Maying

Get up, get up for shame, the blooming morn
Upon her wings presents the god unshorn.
See how Aurora throws her fair
Fresh-quilted colors through the air;
Get up, sweet slug-a-bed, and see 5
The dew bespangling herb and tree.
Each flower has wept and bowed toward the east
Above an hour since, yet you not dressed;
Nay, not so much as out of bed.
When all the birds have matins said, 10
And sung their thankful hymns, 'tis sin,
Nay, profanation to keep in,
Whenas a thousand virgins on this day
Spring, sooner than the lark, to fetch in May.

Rise and put on your foliage, and be seen 15
To come forth like the springtime, fresh and green,
And sweet as Flora. Take no care
For jewels for your gown or hair;
Fear not, the leaves will strew
Gems in abundance upon you; 20
Besides, the childhood of the day has kept,
Against[22] you come, some orient pearls unwept;
Come and receive them while the light
Hangs on the dew-locks of the night,
And Titan on the eastern hill 25
Retires himself, or else stands still
Till you come forth. Wash, dress, be brief in praying:
Few beads are best when once we go a-maying.

Come, my Corinna, come; and coming, mark
How each field turns[23] a street, each street a park 30
Made green and trimmed with trees; see how
Devotion gives each house a bough
Or branch; each porch, each door, ere this,
An ark, a tabernacle is,
Made up of white-thorn neatly interwove, 35
As if here were those cooler shades of love.
Can such delights be in the street
And open fields, and we not see't?
Come, we'll abroad, and let's obey
The proclamation made for May, 40
And sin no more, as we have done, by staying;
But, my Corinna, come, let's go a-maying.

There's not a budding boy or girl this day
But is got up, and gone to bring in May.
A deal of youth, ere this, is come 45
Back, and with white-thorn laden, home.
Some have despatched their cakes and cream
Before that we have left to dream;
And some have wept, and wooed, and plighted troth,
And chose their priest, ere we can cast off sloth; 50
Many a green-gown has been given,
Many a kiss, both odd and even,
Many a glance too has been sent
From out the eye, love's firmament,
Many a jest told of the keys betraying 55
This night, and locks picked, yet we're not a-maying.

Come, let us go while we are in our prime,
And take the harmless folly of the time.
We shall grow old apace, and die
Before we know our liberty. 60

[22]"Until." [23]"Becomes."

Our life is short, and our days run
As fast away as does the sun;
And as a vapor, or a drop of rain
Once lost, can ne'er be found again,
So when or you or I are made 65
A fable, song, or fleeting shade,
All love, all liking, all delight
Lies drowned with us in endless night.
Then while time serves, and we are but decaying,
Come, my Corinna, come, let's go a-maying. 70

To Phyllis, To Love and Live with Him

Live, live with me, and thou shalt see
The pleasures I'll prepare for thee:
What sweets the country can afford
Shall bless thy bed and bless thy board.
The soft sweet moss shall be thy bed 5
With crawling woodbine overspread,
By which the silver-shedding streams
Shall gently melt thee into dreams.
Thy clothing, next, shall be a gown
Made of the fleece's purest down. 10
The tongues of kids shall be thy meat,
Their milk thy drink; and thou shalt eat
The paste of filberts for thy bread
With cream of cowslips butterèd.
Thy feasting-tables shall be hills 15
With daisies spread, and daffodils,
Where thou shalt sit, and redbreast by,
For meat, shall give thee melody.
I'll give thee chains and carcanets[24]
Of primroses and violets. 20
A bag and bottle thou shalt have,
That richly wrought, and this as brave,[25]
So that as either shall express
The wearer's no mean shepherdess.
At shearing-times, and yearly wakes, 25
When Themilis his pastime makes,
There thou shalt be, and be the wit,
Nay more, the feast, and grace of it.
On holy days when virgins meet
To dance the hays with nimble feet, 30
Thou shalt come forth, and then appear
The queen of roses for that year,
And having danced, 'bove all the best,
Carry the garland from the rest.

[24] "Necklaces" or "headbands." [25] "Fine," "splendid."

In wicker baskets maids shall bring 35
To thee, my dearest shepherdling,
The blushing apple, bashful pear,
And shame-faced plum, all simp'ring there.
Walk in the groves, and thou shalt find
The name of Phyllis in the rind[26] 40
Of every straight and smooth-skin tree,
Where kissing that, I'll twice kiss thee.
To thee a sheep-hook I will send,
Bepranked with ribands, to this end,
This, this alluring hook might be 45
Less for to catch a sheep, than me.
Thou shalt have possets,[27] wassails fine,
Not made of ale, but spicèd wine,
To make thy maids and self free mirth,
All sitting near the glitt'ring hearth. 50
Thou shalt have ribands, roses, rings,
Gloves, garters, stockings, shoes, and strings
Of winning colors, that shall move
Others to lust, but me to love.
These, nay and more, thine own shall be, 55
If thou wilt love and live with me.

His Content in the Country

Here, here I live with what my board
Can with the smallest cost afford.
Though ne'er so mean the viands be,
They well content my Prue and me.
Or pea, or bean, or wort, or beet, 5
Whatever comes, content makes sweet.
Here we rejoice because no rent
We pay for our poor tenement,
Wherein we rest, and never fear
The landlord or the usurer; 10
The quarter-day does ne'er affright
Our peaceful slumbers in the night.
We eat our own and batten more
Because we feed on no man's score;
But pity those whose flanks grow great, 15
Swelled with the lard of other's meat.
We bless our fortunes when we see
Our own beloved privacy,
And like our living, where w'are known
To very few, or else to none. 20

[26] "Bark."
[27] Spiced drinks of hot sweetened milk curdled with wine or ale; "wassails" were drinks of ale or wine spiced with roasted apples and sugar.

The Hock-Cart, or Harvest Home;
To the Right Honorable Mildmay, Earl of Westmoreland

Come, sons of summer, by whose toil
We are the lords of wine and oil;
By whose tough labors and rough hands
We rip up first, then reap our lands.
Crowned with the ears of corn, now come, 5
And to the pipe sing harvest home.
Come forth, my lord, and see the cart[28]
Dressed up with all the country art.
See here a maukin,[29] there a sheet
As spotless pure as it is sweet; 10
The horses, mares, and frisking fillies,
Clad all in linen, white as lilies;
The harvest swains and wenches bound
For joy to see the hock-cart crowned.
About the cart hear how the rout 15
Of rural younglings raise the shout,
Pressing before, some coming after:
Those with a shout, and these with laughter.
Some bless the cart; some kiss the sheaves;
Some prank them up with oaken leaves; 20
Some cross the fill-horse;[30] some with great
Devotion stroke the home-borne wheat;
While other rustics, less attent[31]
To prayers than to merriment,
Run after with their breeches rent. 25
Well on, brave boys, to your lord's hearth,
Glitt'ring with fire, where for your mirth
Ye shall see first the large and chief
Foundation of your feast, fat beef,
With upper stories, mutton, veal, 30
And bacon, which makes full the meal;
With sev'ral dishes standing by,
As here a custard, there a pie,
And here all-tempting frumenty.[32]
And for to make the merry cheer, 35
If smirking wine be wanting here,
There's that which drowns all care, stout beer,
Which freely drink to your lord's health;

[28]The celebration of "harvest home" began with the return from the fields of the hock cart, bringing the last load of the harvest. Herrick's poem, based generally on Tibullus, *Odes*, II.i, bears some slight resemblance to Jonson's "To Penshurst" and "To Sir Robert Wroth" [Chapter 17 above], but scarcely reflects their generous spirit, as lines 1–2 indicate. (Maclean)
[29]I.e., a pole bound with cloth, used as a scarecrow. (Maclean)
[30]I.e., some bless the horse that draws the cart. (Maclean)
[31]"Attentive."
[32]A pudding made of wheat, milk, and spices. (Maclean)

Then to the plow, the commonwealth,
Next to your flails, your fans, your fats;[33] 40
Then to the maids with wheaten hats;
To the rough sickle and crook'd scythe,
Drink, frolic boys, till all be blithe.
Feed and grow fat, and as ye eat
Be mindful that the lab'ring neat,[34] 45
As you, may have their fill of meat.
And know, besides, ye must revoke
The patient ox unto his yoke,
And all go back unto the plow
And harrow, though they're hanged up now. 50
And, you must know, your lord's word's true:
Feed him you must, whose food fills you,
And that this pleasure is like rain,
Not sent ye for to drown your pain
But for to make it spring again. 55

To Daffodils

Fair daffodils, we weep to see
You haste away so soon:
As yet the early-rising sun
Has not attained his noon.
Stay, stay, 5
Until the hasting day
Has run
But to the evensong;
And, having prayed together, we
Will go with you along. 10

We have short time to stay as you;
We have as short a spring;
As quick a growth to meet decay,
As you or anything.
We die, 15
As your hours do, and dry
Away
Like to the summer's rain;
Or as the pearls of morning's dew,
Ne'er to be found again. 20

[33]I.e., your winnowing fans, your vats (for storing grain). (Maclean)
[34]"Cattle." "Meat," in the next line, means "food."

Recovered Greenness

Four Poems of George Herbert

George Herbert (1593–1633) was born in Montgomery, Wales; his mother, Magdalen Herbert, was a friend of John Donne, and his elder brother, Edward, later Lord Herbert of Cherbury, became a noted philosopher and poet. Educated at Westminster School and Trinity College, Cambridge, where he took his B.A. in 1612 and his M.A. in 1616, he too became a friend of Donne, but after attaining the post of Public Orator of Cambridge in 1620 he appears to have abandoned a religious vocation for secular ambition, entering Parliament and cultivating acquaintance with the great, including Bacon. In around 1626 he turned back to religion, taking minor orders, and in 1630 was made rector of Bemerton, near Salisbury, where he wrote many of his poems and led the devout life of a model parson. During his final illness he sent the manuscript of his poems, The Temple, *to Nicholas Ferrar, founder of the religious community at Little Gidding, asking that he either publish or burn this "picture of the many spiritual conflicts that have passed betwixt God and my soul, before I could subject mine to the will of Jesus my Master: in whose service I have now found perfect freedom." Two editions appeared after his death in 1633.*

Herbert's quietly meditative poems are concerned with the love and grace of God as revealed through His law and His Church, and with the soul's struggle to attain the "perfect freedom" of obedience to that law. But for Herbert, as for Christian thinkers from St. Augustine (Chapter 13 above) to Hooker (Chapter 16), as Richard Todd writes in The Opacity of Signs: Acts of Interpretation in George Herbert's "The Temple" *(1986), "the act of reading the Book of Nature is implicit in the act of learning to see God's Providence at work," since "all natural processes and occurrences are answerable in terms of God's having ordained, purposed, and permitted them to take place." The parson, Herbert observes in* A Priest to the Temple, or, The Country Parson *(1652), "scatters in his discourse three sorts of arguments; the first taken from Nature, the second from the Law, the third from Grace. For Nature, he sees not how a house could be either built without a builder, or kept in repair without a housekeeper. . . . He conceives not possibly, how he that would believe a Divinity, if he had been at the Creation of all things, should less believe it, seeing the Preservation of all things, for Preservation is a Creation; and more, it is a continued Creation, and a creation every moment."*

Of the following four poems, the first, "Nature," depicts the refractoriness of human nature to the grace of God, which must tame it if it is ever to achieve salvation. The second, "The Pulley," narrates a fable in which God withholds rest from the gifts given to man at his creation, lest he "rest in Nature, not the God of Nature," from whom alone salvation may come as "repining restlessness" draws him upward, so that weariness may "toss him to my breast." "The Flower" parallels the poet's condition to the sweet return of spring flowers, when the shriveled heart recovers greenness, but also to the ruin they suffer from storms "when the least frown of thine is shown": now, in his later years, "I bud again," by the grace of the Lord of love. In "Virtue," the Christian contrast between the beauties of nature that must all die and the immortality of the "sweet and virtuous soul" attains sublimely simple expression.

Nature

Full of rebellion, I would die,
Or fight, or travel, or deny
That thou hast aught to do with me.
O tame my heart!
It is thy highest art 5
To captivate strongholds to thee.

If thou shalt let this venom lurk
And in suggestions fume and work,
My soul will turn to bubbles straight,
And thence by kind[35] 10
Vanish into a wind,
Making thy workmanship deceit.

O smooth my rugged heart, and there
Engrave thy rev'rend law and fear!
Or make a new one, since the old 15
Is sapless grown,
And a much fitter stone
To hide my dust than thee to hold.

The Pulley

When God at first made man,
Having a glass of blessings standing by;
Let us (said he) pour on him all we can:
Let the world's riches, which dispersèd lie,
Contract into a span. 5

So strength first made a way;
Then beauty flowed, then wisdom, honor, pleasure;
When almost all was out, God made a stay,
Perceiving that alone of all his treasure
Rest in the bottom lay. 10

For if I should (said he)
Bestow this jewel also on my creature,
He would adore my gifts instead of me,
And rest in Nature, not the God of Nature:
So both should losers be. 15

Yet let him keep the rest,
But keep them with repining restlessness:
Let him be rich and weary, that at least,
If goodness lead him not, yet weariness
May toss him to my breast. 20

[35]"By its nature."

The Flower

How fresh, O Lord, how sweet and clean
Are thy returns! Ev'n as the flowers in spring,
To which, besides their own demean,[36]
The late-past frosts tributes of pleasure bring.
Grief melts away 5
Like snow in May,
As if there were no such cold thing.

Who would have thought my shriveled heart
Could have recovered greenness? It was gone
Quite underground, as flowers depart 10
To see their mother-root when they have blown;[37]
Where they together
All the hard weather,
Dead to the world, keep house unknown.

These are thy wonders, Lord of power, 15
Killing and quick'ning,[38] bringing down to hell
And up to heaven in an hour;
Making a chiming of the passing bell.
We say amiss
That this or that is; 20
Thy word is all, if we could spell.

Oh, that I once past changing were,
Fast in thy paradise, where no flower can wither!
Many a spring I shoot up fair,
Off'ring at[39] heaven, growing and groaning thither; 25
Nor doth my flower
Want a spring shower,[40]
My sins and I joining together.

But while I grow in a straight line,
Still upwards bent, as if heav'n were mine own, 30
Thy anger comes, and I decline.
What frost to that? What pole is not the zone
Where all things burn,
When thou dost turn,
And the least frown of thine is shown?[41] 35

And now in age I bud again,
After so many deaths I live and write;
I once more smell the dew and rain,
And relish versing. O my only light,

[36]"Demeanor" or "demesne," "domain." [37]"Blossomed."
[38]"Giving life." [39]"Aiming at." [40]Of tears.
[41]What frost compares to Your anger? Arctic and Antarctic seem like the equatorial zones when you frown! (Di Cesare)

It cannot be
That I am he
On whom thy tempests fell all night. 40

These are thy wonders, Lord of love,
To make us see we are but flowers that glide;[42]
Which when we once can find and prove,[43] 45
Thou hast a garden for us, where to bide.
Who would be more,
Swelling through store,[44]
Forfeit their paradise by their pride.

Virtue

Sweet day, so cool, so calm, so bright,
The bridal of the earth and sky;
The dew shall weep thy fall tonight,
For thou must die.

Sweet rose, whose hue angry and brave 5
Bids the rash gazer wipe his eye;
Thy root is ever in its grave,
And thou must die.

Sweet spring, full of sweet days and roses,
A box where sweets compacted lie; 10
My music shows ye have your closes,
And all must die.

Only a sweet and virtuous soul,
Like seasoned timber, never gives;
But though the whole world turn to coal, 15
Then chiefly lives.

Hesperian Fables True

Selections from the Poems of John Milton

Born in London and educated at St. Paul's School and Christ's College, Cambridge, where he took his B.A. in 1629 and M.A. in 1632, John Milton (1608–74) then retired to his father's estate at Horton in Buckinghamshire for further study and writing. He traveled to Italy in 1638 (where he met the blind Galileo), but returned the following year because of mounting conflict between Parliament and king. In 1641 he published Of Reformation in England, *urging completion of the reformation begun in the previous century, and from 1643 to 1645 he wrote,*

[42] "Slip away unnoticed." (Di Cesare) [43] "Experience." [44] "Abundance" or "overestimation."

among others, The Doctrine and Discipline of Divorce *(occasioned by his unhappy first marriage),* Of Education, *and* Areopagitica, *his famous defense of freedom of the press. His attack on tyranny in* The Tenure of Kings and Magistrates *(1649) was followed by polemical pamphlets defending the execution of the king and the Puritan revolution, which he wrote as Cromwell's Latin secretary; after Cromwell's death, he published in 1660* A Ready and Easy Way to Establish a Free Commonwealth. *Disillusioned by the Restoration, and completely blind since 1652, he withdrew from the public world that had failed his ideals to complete his biblical epic* Paradise Lost *(published in ten books in 1667 and revised in twelve in 1674) and to write the short epic* Paradise Regained, *on the temptation of Christ, and the drama* Samson Agonistes, *published together in 1671.*

Given the moral earnestness of his intention (as he declares in Paradise Lost*) to "assert Eternal Providence, / And justify the ways of God to men," Milton is not obviously a poet of external nature. Yet as the principal seventeenth-century heir of Renaissance Christian humanism, and of the poetic traditions of Ariosto and Tasso, Spenser and Shakespeare (Chapter 17 above), this impassioned revolutionary poet could never leave the physical world—God's handiwork—entirely behind, even when turning increasingly inward in his later years. Three of our four selections, in which Milton's intense response to natural beauty is unmistakable, are indeed early. First published in* Poems of Mr. John Milton, Both English and Latin *(1645), the "Song: On May Morning" and the sonnet "O Nightingale!" were written around 1628 (when Milton was nineteen), and the Latin "Elegia Quinta: In Adventum Veris" ("Fifth Elegy: On the Coming of Spring")*[45] *in 1628–29, at age nineteen or twenty.*

The Latin elegy is perhaps the most astonishing of these. Standing near the end of a long tradition of Latin Renaissance poetry descending from Pontano, Mantuan, Politian, and Sannazaro in the fifteenth and sixteenth centuries, its sensuous evocation of the beauties of spring resurrects—in this late phase of the Renaissance—the youthful vibrancy that had infused its beginnings with wonder at rediscovery of the erotically charged world portrayed by its great pagan model Lucretius (Chapter 10 above). In the Puritan, as in the Epicurean poet, wanton Earth again pants for Phoebus's revivifying embraces. Another Latin poem of this period, "Naturam Non Pati Senium" ("That Nature Does Not Suffer Old Age"), rebuts the contention that the natural world itself has decayed, and continues to decay, as a result of mankind's original sin. Among those who held this view were Luther and, in seventeenth-century England, Donne in the "Anniversaries," Browne in Urn Burial, *and Burnet in* The Sacred Theory of the Earth *(Chapter 18 above).*

In the Christian vision of Milton's early English poems, the beauties of a world permeated by music of the celestial spheres are central manifestations of the divine. The hope—expressed in "On the Morning of Christ's Nativity"—that Nature's "reign had here its last fulfilling" in a renewed age of gold had been deferred by the crucifixion, but the Resurrection promised its future fulfillment. In recognition of this, the pagan oracles fall dumb as

From haunted spring and dale
Edged with poplar pale,
    The parting Genius is with sighing sent;
With flower-inwoven tresses torn
The Nymphs in twilight shade of tangled thickets mourn.

[45]Translated from the text in Merritt Y. Hughes's edition of *Paradise Regained, The Minor Poems, and Samson Agonistes* (1937); other editions consulted include Hughes's of *Paradise Lost* (1935) and *The Student's Milton*, ed. Frank Allen Patterson (1933).

Yet this pagan Genius of Nature lingers on in imagination. In the masque Arcades, *of around 1632, he declares himself (44–49) "the power / Of this fair wood," who nurses its saplings and saves its plants from "noisome winds, and blasting vapors chill."*

For Milton, the sensuous delights of the natural world must be attuned to the Christian moral life, not relinquished. Thus in the masque Comus *of 1634, the Lady resists the blandishments (710–14, 739–44) of Comus, the personification of hedonistic pagan revelry, who asks,*

Wherefore did Nature pour her bounties forth
With such a full and unwithdrawing hand,
Covering the earth with odors, fruits, and flocks,
Thronging the seas with spawn innumerable,
But all to please and sate the curious taste? . . .
Beauty is nature's coin, must not be hoarded,
But must be current, and the good thereof
Consists in mutual and partaken bliss,
Unsavory in th'enjoyment of itself.
If you let slip time, like a neglected rose
It withers on the stalk with languished head.

She defends her virtue not by "puritanical" denial of nature but by protesting (762–67),

Impostor, do not charge most innocent nature,
As if she would her children should be riotous
With her abundance; she, good cateress,
Means her provision only to the good
That live according to her sober laws,
And holy dictate of spare temperance.

She thus affirms (769–72) a "moderate and beseeming share" by which "Nature's full blessings would be well dispensed / In unsuperfluous even proportion."

Joy (or remembered joy) in the beauties of nature—shared in "L'Allegro" by whistling plowman and singing milkmaid amid "Meadows trim with daisies pied, / Shallow brooks, and rivers wide"—though most prominent in his early verse, also pervades Milton's late poetry. It finds a place in countless similes of Paradise Lost *and in the great apostrophe to light at the beginning of Book III. The splendors of nature likewise permeate the archangel Raphael's account of Creation, by which "the vast immeasurable abyss" became a world of teeming shoals of fish and soaring flocks of birds, of grass, flowering herbs, and stately trees that "Rose as in dance . . . and spread / Their branches hung with copious fruit" (VII.324–25). And in Eden's "happy rural seat" (seen, in our last selection, through the envious eyes of Satan, newly arrived from hell), Nature pours forth her bounty*

while universal Pan
Knit with the Graces and the Hours in dance
Led on th'eternal spring . . .

As in the Earthly Paradises of Dracontius and Avitus (Chapter 14 above), Bernard Silvestris, Alan of Lille, and Dante (Chapter 15), this Eden fuses with the classical Golden Age in a world of lost perfection where—if anywhere—the Hesperian fables of ancient poets were true.

Song: On May Morning

Now the bright morning Star, Day's harbinger,
Comes dancing from the East, and leads with her
The flow'ry May, who from her green lap throws
The yellow cowslip, and the pale primrose.
Hail bounteous May that doth inspire 5
Mirth and youth and warm desire!
Woods and groves are of thy dressing,
Hill and dale doth boast thy blessing.
Thus we salute thee with our early song,
And welcome thee, and wish thee long. 10

O Nightingale

O nightingale, that on yon bloomy spray
Warblest at eve, when all the woods are still,
Thou with fresh hope the lover's heart dost fill,
While the jolly hours[46] lead on propitious May,
Thy liquid notes that close the eye of day, 5
First heard before the shallow cuckoo's bill,
Portend success in love; O, if Jove's will
Have linked that amorous power to thy soft lay,
Now timely sing, ere the rude bird of hate
Foretell my hopeless doom in some grove nigh: 10
As thou from year to year hast sung too late
For my relief; yet hadst no reason why,
Whether the Muse, or Love, call thee his mate,
Both them I serve, and of their train am I.

Fifth Elegy: On the Coming of Spring,
translated by Robert M. Torrance

Time, circling round in never-ending ring,
summons new zephyrs back to herald spring.
Casting off cold, rejuvenated Earth,
clad in green, glories in her brief rebirth.
Can it be, sap invigorates my song 5
too, and with spring my Muse again grows strong?—
grows strong and, animated by the joy
of spring, miraculously craves employ!

[46]In classical mythology the Hours (Latin *Horae*, from Greek *Hôrai*, "seasons"), the daughters of Zeus (Jupiter) and Themis, were "the goddesses of the order of nature and of the seasons, who guarded the doors of Olympus, and promoted the fertility of the earth by the various kinds of weather which they gave to mortals. In works of art the Horae are represented as blooming maidens or youths, carrying the different products of the seasons." (E. H. Blakeney, *A Smaller Classical Dictionary* [1910].)

I glimpse Castalia's bifurcated mount,
and dream of visiting Pirene's fount;[47] 10
fired by mysterious stirrings in my breast,
maddened by sacred sounds, I cannot rest!
Phoebus himself arrives, his flowing hair
braided with Daphne's laurel: see him there!
Now my mind penetrates the liquid sky 15
and bodiless through wandering clouds I fly.
Shadowy caves of holy bards I see,
yes, and gods' inmost shrines revealed to me:
everything on Olympus I behold,
as Tartarus' dark secrets, too, unfold. 20
What uproar does my clamorous soul set loose?
What will this frenzied ecstasy produce?
Spring, which inspired it, shall receive my song:
to her who gave it shall this gift belong!
Already, Philomel, you sweetly trill 25
in newborn leaves, while all the woods are still.
I in town, you in forest, let us sing
both welcoming the advent of the spring!
Spring has transformed the world: come, spread her glory;
bid the Muse yearly promulgate her story! 30
Now the sun, fleeing Ethiopian sands,
steers with his golden reins toward northern lands.
Brief is night's journey, brief dark night's delay,
soon with her horrid shadows chased away.
No more the Lycaonian plowman now 35
behind the wagon wearily drives his plow.[48]
Few are the stars that keep their wonted post
around Jove's court, of all the heavenly host;
for murder, fraud, and force recede with night,
and savage giants cause the gods no fright. 40
Some shepherd may—as dawn is shading red
the dewy hilltop where he lies—have said:
"Clearly *last* night no girl beside you lay,
Phoebus, to curb swift horses bringing day!"
Cynthia haunts the woods, intent to try 45
her quiver, watching Lucifer from on high,[49]
glad that her task is shortened (having laid
down her light moonbeams) by her brother's aid.
"Leave the old geezer's bed, Aurora," cries
Phoebus: "cease hoping *he* will ever rise! 50
On green banks Aeolus' son, the hunter, stays.

[47]The Castalian spring of the Muses flowed from the slope of the twin peaks of Mount Parnassus; the spring of Pirênê, near Corinth, was associated with the wingèd horse Pegasus caught by Bellerophon.
[48]The constellation of Boôtês (the Plowman) was thought to drive his plow after the Wagon (or Great Bear), into which King Lycaon of Arcadia's daughter Callisto had been transformed by Zeus.
[49]The huntress Cynthia, or Diana (Greek Artemis), was linked with the moon, as her brother Phoebus Apollo was with the sun; Lucifer (Greek Phôsphoros), the "Lightbringer," was the morning star.

Up! with desire Hymettus soon will blaze."[50]
With blushing face the goddess owns her crime,
urging dawn's steeds more rapidly to climb.
    Reviving Earth throws off abhorrent traces 55
of age, desiring, Phoebus, your embraces:
desires, and wins; for none could be more fair,
with fertile breasts voluptuously laid bare,
as, from soft lips, a spicy fragrance drops
of Paphian[51] rosebuds and Arabian crops. 60
See her high brow crowned with a sacred wood
(so towering pines round Ops[52] on Ida stood),
her moist hair woven with unnumbered flowers
intensifying her seductive powers:
so, weaving flowers in her flowing hair, 65
Proserpine pleased the god who seized her there.[53]
Look, Phoebus, easy loves incite you now;
each breath of spring conveys a honeyed vow;
odorous Zephyr fans his cinnamon wings;
with coaxing blandishments each songbird sings. 70
Not without dowry Earth desires to wed,
nor as a pauper yearns to share your bed;
healing herbs bounteously she heaps on you,
augmenting your renown with titles new.[54]
If costly, glittering gifts accomplish aught 75
with you—for often love with gifts is bought—
she spreads before you all the wealth inside
the seas, and all that massive mountains hide.
How often, when you tired of heaven's steep,
wishing to plunge into the western deep, 80
"Why," she cried, "weary of your daily race,
Phoebus, seek our blue mother's moist embrace?
What's Tethys, or Tartessian waves, to you?[55]
Must filthy salt your holy face imbue?
Cooler within my shadow shall you fare, 85
Phoebus: come, drench with dew your burning hair!
Softer in my fresh grasses you shall nap:
come, and lay down your lusters in my lap!

[50]Aurora (Greek Eôs), the Dawn, obtained from the gods the gift of immortality for her lover Tithonus (Tithônos), brother of Priam of Troy, but neglected to procure eternal youth for him, and he shriveled away in old age. She then fell in love with Aeolus's son Céphalus and stole him away on Mount Hymettus from his wife Procris, whom he loved. See Cephalus's story in Ovid's *Metamorphoses* VII.700–708.
[51]Paphos, on Cyprus, was the site of a temple of Aphrodite (Roman Venus), near the spot where she was said to have been born from the sea, and was the chief center of her worship.
[52]In Latin mythology, Ops, a goddess of fertility and plenty (as her name, from which English "opulence" derives, suggests), was the wife of Saturn.
[53]Literally, in Hughes's translation: "as the Sicanian [Sicilian] goddess [Proserpina] with the flowers plaited in her flowing hair was charming to the Taenarian god [Pluto]." The story of the rape of Proserpine (Greek Persephone) was told in the Homeric Hymn to Demeter (Chapter 8 above).
[54]Phoebus, besides being a god of light and poetry, both inflicted and cured disease.
[55]On Tethys, wife of Oceanus, see note on Virgil's *Georgics* I.31 in Chapter 11 above. Tartessus was an ancient town founded by Phoenicians near the pillars of Hercules (Strait of Gibraltar) in Spain; hence "Tartessian" indicated the western ocean beyond.

Lying on dew-drenched roses, whispering airs
will soothe our mingled bodies of all cares, 90
nor am I terrified by Sémele's lot,
I swear, or Phaëthon's axle smoking hot.[56]
More wisely use that fiery stream you tap:
come, and lay down your lusters in my lap!"
Thus wanton Earth breathes love: one, then another 95
throng rushes forth to emulate their mother.
Now roving Cupid runs through every land
kindling with blazing sun his sputtering brand.
Lethal bow-horns now twang with taut new string,
and steel-tipped shafts are brightly menacing. 100
Invincible Diana he aspires
to wound; chaste Vesta, sitting by her fires.
Venus again repairs her aging form,
rising, renewed each year, from waters warm.
Through marble cities youths "O Hymen!" cry; 105
shores and woods echo "Hymen!" far and nigh.
Elegantly attired from toe to crown
he comes, and purple crocus scents his gown.
Stepping forth, many a maid, with virgin breast
girded with gold, gives lovely spring her best. 110
Each has her prayer, yet each prayer is the same:
may Cytheréa[57] grant each one her flame!
Now on his sevenfold reed the shepherd plays,
while Phyllis harmonizes with sweet lays.
The sailor's night-song sings his stars to rest, 115
calling swift dolphins to the waters' crest.
Jove and Olympian Juno have not ceased
to sport, and servant-gods attend their feast.
Now, as late twilight thickens, satyrs too
through flowery meadows flit, a raucous crew; 120
Sylvánus, too, with cypress branches crowned,
a god half-goat, a goat half-god, is found;[58]
and Dryads,[59] underneath old trees concealed,
range over each high ridge and lonely field.
Where wanton Pan roams orchard, grove, and farm, 125
there Cybele and Ceres risk great harm;[60]
some Oread[61] becomes lewd Faunus' prey
till the nymph leaps with skittish feet away:

[56]On Sémelê, mother of Bacchus—incinerated after making her lover Jupiter (Greek Zeus) reveal his true form to her—see introductory note to Euripides' *Bacchae* in Chapter 8 above. The story of Pháethôn, destroyed by Jupiter's thunderbolt after nearly setting earth and sky afire by reckless driving of the chariot of his father Hêlios (the Sun), is told in Ovid's *Metamorphoses* II.1–366.
[57]Venus (Greek Aphrodite), born (according to one myth) near the island of Cythera.
[58]Silvánus, a god of field and forest, was sometimes conflated with Roman Faunus or Greek Pan; the description here, as editors have noted, more suits a satyr or Pan than it does the Roman Silvanus.
[59]Tree nymphs.
[60]On Cýbelê, the Great Mother goddess, see note on Euripides' *Bacchae* 120 in Chapter 8 above. Ceres (Greek Dêmêtêr) was the goddess of the grain.
[61]Mountain nymph.

she hides, but hopes her lair will be revealed,
and flees, but, fleeing, purposes to yield. 130
Gods prefer woods unhesitantly to heaven,
and to each grove its deity is given.
Long may each grove retain its deity,
never again its forest-home to flee!
Jove, may the golden age to suffering earth 135
bring you back: what are cloud-wrapped missiles worth?
You, at least, Phoebus: drive your rapid team
slowly, and let spring linger like a dream;
retard harsh winter's train of endless nights
and shadows darkening our northern lights! 140

Satan Views the Garden of Eden

Paradise Lost, *Book IV*

Beneath him with new wonder now he views 205
To all delight of human sense exposed
In narrow room Nature's whole wealth, yea more,
A Heaven on Earth; for blissful Paradise
Of God the Garden was, by him in the East
Of Eden planted; Eden stretched her line 210
From Auran[62] eastward to the royal tow'rs
Of great Seleucia,[63] built by Grecian kings,
Or where the sons of Eden long before
Dwelt in Telassar:[64] in this pleasant soil
His far more pleasant Garden God ordained; 215
Out of the fertile ground he caused to grow
All trees of noblest kind for sight, smell, taste;
And all amid them stood the Tree of Life,
High eminent, blooming ambrosial fruit
Of vegetable gold;[65] and next to life 220
Our death, the Tree of Knowledge grew fast by,
Knowledge of good bought dear by knowing ill.
Southward through Eden went a river large,[66]
Nor changed his course, but through the shaggy hill

[62]Auran is probably Aurantis on the Euphrates. Allan H. Gilbert cites Purchas' statement (*Pilgrimage*, p. 19) that Aurantis is "easily declined from Heden [Eden] mentioned after Moses' time in II Kings 19:12 and Isaiah 37:12." (Hughes)

[63]Alexander's general Seleucus founded Seleucia, one seat of the Hellenistic kingdom in Western Asia from about 312 to about 65 B.C., on the Tigris about fifteen miles below modern Baghdad. (Hughes)

[64]Milton must have had in his mind the question put in II Kings 19:12 and repeated in Isaiah 37:12: "Have the gods of the nations delivered them which my fathers have destroyed; . . . the children of Eden which were in Thelasar?" Telassar seems to have been regarded as a city of Eden. (Hughes)

[65]Perhaps there is a vague comparison intended with the golden apples of the Hesperides, for classical tradition put the earthly paradise in those islands. (Hughes)

[66]The marvelous river rests upon the biblical account of a rainless world irrigated by nightly mists and of "a river [which] went out of Eden to water the garden; and from thence it was parted, and became into four heads." (Genesis 2:10) (Hughes)

Passed underneath engulfed, for God had thrown 225
That mountain as his garden mould high raised
Upon the rapid current, which through veins
Of porous earth with kindly[67] thirst updrawn,
Rose a fresh fountain, and with many a rill
Watered the Garden; thence united fell 230
Down the steep glade, and met the nether flood,
Which from his darksome passage now appears,
And now, divided into four main streams,
Runs diverse, wand'ring many a famous realm
And country whereof here needs no account, 235
But rather to tell how, if art could tell,
How from that sapphire fount the crispèd[68] brooks,
Rolling on orient pearl and sands of gold,
With mazy error[69] under pendant shades
Ran nectar, visiting each plant, and fed 240
Flow'rs worthy of Paradise, which not nice art
In beds and curious knots, but Nature boon[70]
Poured forth profuse on hill and dale and plain,
Both where the morning sun first warmly smote
The open field, and where the unpierced shade 245
Imbrowned the noontide bowers: thus was this place,
A happy rural seat of various view:
Groves whose rich trees wept odorous gums and balm,
Others whose fruit burnished with golden rind
Hung amiable, Hesperian fables true, 250
If true, here only, and of delicious taste:
Betwixt them lawns, or level downs, and flocks
Grazing the tender herb, were interposed,
Or palmy hillock, or the flow'ry lap
Of some irriguous[71] valley spread her store, 255
Flow'rs of all hue, and without thorn the rose:[72]
Another side, umbrageous grots and caves
Of cool recess, o'er which the mantling vine
Lays forth her purple grape, and gently creeps
Luxuriant; meanwhile murmuring waters fall 260
Down the slope hills, dispersed, or in a lake,
That to the fringèd bank with myrtle crowned,
Her crystal mirror holds, unite their streams.
The birds their choir apply;[73] airs, vernal airs,
Breathing the smell of field and grove, attune 265
The trembling leaves, while universal Pan
Knit with the Graces and the Hours in dance

[67]"Natural." [68]"Rippling." [69]"Wandering." [70]"Bountiful." [71]"Well-watered."
[72]As Hughes notes, Herrick's epigram "The Rose" repeats the tradition that "Before man's fall the rose was born, / St. Ambrose says, without the thorn."
[73]"Contribute," "make use of." (Hughes)

Led on th'eternal spring. Not that fair field
Of Enna,[74] where Proserpin, gath'ring flow'rs
Herself a fairer flow'r by gloomy Dis 270
Was gathered, which cost Ceres all that pain
To seek her through the world; nor that sweet grove
Of Daphne by Orontes,[75] and th'inspired
Castalian spring might with this Paradise
Of Eden strive . . . 275

Green Thoughts in a Green Shade

Three Poems of Andrew Marvell

Andrew Marvell (1621–78) was educated at Hull Grammar School and Trinity College, Cambridge, where he took his B.A. in 1638–39; it seems he was briefly converted to Catholicism at this time, but returned to the Anglican Church. He took no part in the civil war—"I think the cause was too good to have been fought for," he later declared—but traveled on the Continent after its outbreak, and even contributed commendatory verses to the Royalist Richard Lovelace's Lucasta *in the year of the king's execution, 1649. By 1650, however, in "An Horatian Ode upon Cromwell's Return from Ireland," while admiring the dignity of Charles on "the tragic scaffold," he praised the victorious Cromwell*

Who, from his private gardens, where
He lived reservèd and austere, . . .
Could by industrious valor climb
To ruin the great work of time,
And cast the kingdom old
Into another mold.

Soon afterward, he was appointed tutor to the daughter of the retired Parliamentarian general Lord Thomas Fairfax, about whose home at Nunappleton in Yorkshire he wrote his longest poem, "Upon Appleton House," and whose gardens may have inspired some of his finest verses. Between 1657 and 1659, at Milton's urging, he was appointed assistant Latin secretary, and in 1659 was elected to Parliament from Hull, which he represented for the rest of his life. He is said to have protected Milton from retaliation at the Restoration, and in his later years he argued strongly for religious toleration and attacked the abuses of his age in satirical verse and prose.

Most of the poems for which he is known, first published posthumously as Miscellaneous Poems *in 1681, are tight-knit metrical and argumentative structures that employ the analytic rigor of metaphysical wit without its tortuous paradoxes and sophistries. They elucidate, in a spirit at once playful and deeply serious, an attitude toward the object in question, which is not infrequently a relation between external nature and the human art—and human perception—*

[74]Ovid writes *(Metamorphoses* V.385) that not far from the walls of Henna, or Enna, an ancient town of central Sicily that was a central seat of the worship of Demeter (Roman Ceres), in a grove near the pool of Pergus, where spring was everlasting, Pluto (Dis) snatched away Persephone (Roman Proserpina); see the rest of this passage as quoted in the introduction to Chapter 14 above.

[75]The gardens of Daphne at Antioch in Syria, on the river Orontes, had a temple of Apollo and a spring which was called after the Castalian spring on Mt. Parnassus, near Delphi, in Greece. Milton calls it "inspired" because tradition says that the waters gave oracles by putting marks on the leaves dipped into them by enquirers. (Hughes)

by which it is shaped but cannot be wholly defined. As Douglas Bush writes in English Literature in the Earlier Seventeenth Century, *he unites "a fresh, muscular, agile, and subtle metaphysical wit and the rationality, clarity, economy, and structural sense of a genuine classic, the cultured, negligent grace of a cavalier and something of the religious and ethical seriousness of a Puritan Platonist. To this rare combination of gifts were added, moreover, a feeling for nature at once particular and general, earthly and unearthly . . . His capacity for sensuous self-identification with natural things has a touch of the old symbolic and religious concept of nature as the art of God which appears in so many philosophic writers from Plato to Sir Thomas Browne," along with an awareness of the role of human art, including the poet's own, in this creation. Yet always, Bush adds, "the detached intelligence is there to criticize what it creates. Marvell is aware that he is a man in the world of men, that a golden holiday is not, though it may approach, a mystical vision."*

Scattered couplets of "Upon Appleton House" suggest this range and diversity:

But all things are composèd here
Like Nature, orderly and near . . .
(25–26)

But Nature here hath been so free
As if she said, Leave this to me!
Art would more neatly have defaced
What she had laid so sweetly waste,
In fragrant gardens, shady woods,
Deep meadows, and transparent floods. . . .
(75–80)

Thus I, easy Philosopher,
Among the birds and trees confer:
And little now to make me, wants
Or of the fowls, or of the plants.
Give me but wings as they, and I
Straight floating on the air shall fly:
Or turn me but, and you shall see
I was but an inverted tree. . . .
(561–68)

Out of these scattered Sibyl's leaves
Strange prophecies my fancy weaves:
And in one history consumes,
Like Mexique paintings, all the plumes.
What Rome, Greece, Palestine, ere said
I in this light mosaic read.
Thrice happy he, who not mistook,
Hath read in Nature's mystic book. . . .
(577–84)

Thus, embroidered by oak leaves "between which caterpillars crawl," and by curling ivy, "Under this antic cope I move / Like some great prelate of the grove."

Of the three poems below, the first, "Bermudas," gives praise to God for an Earthly Paradise placed not in the irrecoverable past but (like Prospero's imperfect paradise in Shakespeare's Tempest*) on an island of the New World. The second, "The Mower Against Gardens" (one of a* *series of "mower" poems), is, J. B. Leishman observes in* The Art of Marvell's Poetry *(1966), a "witty elaboration of the antithesis between Nature and Art," which we have met in Montaigne (Chapter 16), Spenser, and Shakespeare (Chapter 17), by which "luxurious man" allures flowers and plants from fields where nature is "most plain and pure," encloses them "within a garden's square," and makes "forbidden mixtures" of the wild and tame. But at the same time, it is "a description, or evocation, of simple natural beauty,"*

> Where willing Nature does to all dispense
> A wild and fragrant innocence.

Finally, "The Garden," Frank Kermode contends in "The Argument of Marvell's 'Garden'," [76] uses the language of the Epicurean locus amoenus *or Christian-Platonic Earthly Paradise "in a formal refutation of the genre." Here "both man and nature are unfallen; it is therefore, for all its richness, not a trap for virtue"—like Alcina's Isle in Ariosto, Armida's Garden in Tasso, or the Bower of Bliss in Spenser (Chapter 17)— "but a paradise of perfect innocence." The poem, with its "inimitable combination of jest and earnest, seriousness and light-heartedness," presents a complex argument maintaining the superiority of the contemplative to the active life, of solitude to society, of the beauty of gardens to the beauty of women, and of the invisible inner world created by the mind to the visible external world perceived by the senses. In the end, nature can have no existence independent of the human mind that creates its own transcendent worlds, "Annihilating all that's made / To a green thought in a green shade."*

Bermudas

Where the remote Bermudas ride
In th'ocean's bosom unespied,
From a small boat, that rowed along,
The list'ning winds received this song.
  What should we do but sing his praise 5
That led us through the wat'ry maze,
Unto an isle so long unknown,
And yet far kinder than our own?
Where he the huge sea-monsters wracks,
That lift the deep upon their backs, 10
He lands us on a grassy stage,
Safe from the storms, and prelate's rage.
He gave us this eternal spring,
Which here enamels everything;
And sends the fowls to us in care, 15
On daily visits through the air.
He hangs in shades the orange bright,
Like golden lamps in a green night;
And does in the pomegranates close
Jewels more rich than Ormus[77] shows. 20
He makes the figs our mouths to meet
And throws the melons at our feet,

[76] From *Essays in Criticism* II (1952), reprinted in William R. Keast, ed., *Seventeenth Century English Poetry: Essays in Criticism* (1962).
[77] Hormuz, on the Persian Gulf, a center for the pearl and jewel trade. (Di Cesare)

But apples[78] plants of such a price,
No tree could ever bear them twice.
With cedars, chosen by his hand, 25
From Lebanon, he stores the land,
And makes the hollow seas that roar
Proclaim the ambergris on shore.
He cast, of which we rather boast,
The Gospel's pearl upon our coast, 30
And in these rocks for us did frame
A temple, where to sound his name.
Oh, let our voice his praise exalt,
Till it arrive at heaven's vault;
Which thence, perhaps, rebounding may 35
Echo beyond the Mexic Bay.[79]
Thus sung they in the English boat
An holy and a cheerful note,
And all the way, to guide their chime,
With falling oars they kept the time. 40

The Mower Against Gardens

Luxurious[80] man, to bring his vice in use,
Did after him the world seduce,
And from the fields the flowers and plants allure,
Where nature was most plain and pure.
He first enclosed within the garden's square 5
A dead and standing pool of air;
And a more luscious earth for them did knead,
Which stupefied them while it fed.
The pink grew then as double as his mind;
The nutriment did change the kind. 10
With strange perfumes he did the roses taint;
And flowers themselves were taught to paint.
The tulip, white, did for complexion seek,
And learned to interline its cheek;
Its onion root they then so high did hold 15
That one was for a meadow sold.
Another world was searched, through oceans new,
To find the marvel of Peru.[81]
And yet these rarities might be allowed
To man, that sov'reign thing and proud, 20
Had he not dealt between the bark and tree,
Forbidden mixtures there to see.
No plant now knew the stock from which it came;
He grafts upon the wild the tame,

[78]"Pineapples." [79]The Gulf of Mexico. [80]"Lecherous," "lustful."
[81]An exotic imported flower. See John Gerard, *Of the Historie of Plants*, in Chapter 16 above.

That the uncertain and adult'rate fruit 25
Might put the palate in dispute.
His green seraglio has its eunuchs too,
Lest any tyrant him outdo;
And in the cherry he does nature vex,
To procreate without a sex.[82] 30
'Tis all enforced; the fountain and the grot,
While the sweet fields do lie forgot,
Where willing nature does to all dispense
A wild and fragrant innocence;
And fauns and fairies do the meadows till 35
More by their presence than their skill.
Their statues, polished by some ancient hand,
May to adorn the gardens stand;
But howsoe'er the figures do excel,
The gods themselves with us do dwell. 40

The Garden

I

How vainly men themselves amaze[83]
To win the palm, the oak, or bays,[84]
And their uncessant labors see
Crowned from some single herb or tree,
Whose short and narrow vergèd[85] shade 5
Does prudently their toils upbraid;
While all flow'rs and all trees do close[86]
To weave the garlands of repose.

II

Fair Quiet, have I found thee here,
And Innocence, thy sister dear! 10
Mistaken long, I sought you then
In busy companies of men.
Your sacred plants, if here below,[87]
Only among the plants will grow.
Society is all but rude, 15
To[88] this delicious solitude.

III

No white nor red was ever seen
So am'rous as this lovely green.
Fond lovers, cruel as their flame,
Cut in these trees their mistress' name. 20
Little, alas, they know, or heed,

[82]I.e., by grafting. [83]"Bewilder," "confuse."
[84]Trees with whose leaves men's heads were crowned in recognition of eminence in war (palm), politics (oak), or poetry (bay, or laurel).
[85]"Edged." [86]"Unite." [87]I.e., if they grow at all here on earth. [88]"Compared to."

How far these beauties hers exceed!
Fair trees! wheres'e'er[89] your barks I wound,
No names shall but your own be found.

IV

When we have run our passions' heat,[90] 25
Love hither makes his best retreat.
The gods, that mortal beauty chase,
Still[91] in a tree did end their race.
Apollo hunted Daphne so,
Only that she might laurel grow. 30
And Pan did after Syrinx speed,
Not as a nymph, but for a reed.[92]

V

What wond'rous life in this I lead!
Ripe apples drop about my head;
The luscious clusters of the vine 35
Upon my mouth do crush their wine;
The nectarine, and curious peach,
Into my hands themselves do reach;
Stumbling on melons, as I pass,
Ensnared with flowers, I fall on grass. 40

VI

Meanwhile the mind, from pleasure less,[93]
Withdraws into its happiness:
The mind, that ocean where each kind
Does straight[94] its own resemblance find;[95]
Yet it creates, transcending these, 45
Far other worlds, and other seas;
Annihilating all that's made
To a green thought in a green shade.

VII

Here at the fountain's sliding foot,
Or at some fruit tree's mossy root, 50
Casting the body's vest[96] aside,
My soul into the boughs does glide:
There like a bird it sits, and sings,
Then whets,[97] and combs its silver wings;
And, till prepared for longer flight, 55
Waves in its plumes the various light.[98]

VIII

Such was that happy garden state,
While man there walked without a mate:

[89]"Wheresoever." [90]"Race," "ardor." [91]"Always."
[92]Syrinx, pursued by Pan, turned into a reed from which he made his pipes, just as Daphne, pursued by Apollo, turned into the laurel tree thenceforth sacred to him.
[93]"Lesser pleasure." [94]"Immediately."
[95]It was commonly thought that the oceans contained counterparts to all earthly creatures. (Di Cesare)
[96]"Vestment." [97]"Preens." [98]Of this world, contrasted to white light of eternity. (Di Cesare)

After a place so pure, and sweet,
What other help could yet be meet![99] 60
But 'twas beyond a mortal's share
To wander solitary there:
Two paradises 'twere in one
To live in Paradise alone.

IX

How well the skillful gard'ner drew 65
Of flow'rs and herbs this dial[100] new;
Where from above the milder sun
Does through a fragrant zodiac run;
And, as it works, th'industrious bee
Computes its time as well as we. 70
How could such sweet and wholesome hours
Be reckoned but with herbs and flowers!

Nature's Curious Mathematics

Selections from the Works of Sir Thomas Browne

Sir Thomas Browne (1605–82) was born in London and educated at Winchester and at Pembroke College, Oxford, receiving his B.A. in 1626 and M.A. in 1629. He thereafter studied medicine at Montpellier, Padua, and Leyden, where he took his M.D. in 1633, and returned to Oxford, from which he received an M.D. by "incorporation" in 1637. At some time before his thirtieth birthday in 1635, whether in Oxford or in Yorkshire, he wrote Religio Medici (A Doctor's Religion), *which after being long circulated in manuscript was first published in two pirated editions in 1642 and in an authorized edition in 1643. His* Pseudodoxia Epidemica, *commonly known as* Vulgar Errors, *was published in 1646, and both* Hydriotaphia, or Urn Burial, *and* The Garden of Cyrus *appeared in 1658.*

"Perhaps no writer is more truly representative of the double-faced age in which he lived," Willey writes in The Seventeenth Century Background, *"an age half-scientific and half magical, half sceptical and half credulous"; in the "great amphibium" of himself he coupled together the natural magic of Renaissance Platonism and Hermetism, with their age-old belief "that this visible world is but a picture of the invisible," and the natural philosophy of Baconian science. Above all, in* Religio Medici, *he expresses the wondrousness of the divinely created natural world that continually reveals "in stenography and short characters something of divinity," manifesting its "curious mathematics" still more in bees, ants, and spiders than in elephants or whales; the ancients, he writes, knew better than the moderns do how to read "these common hieroglyphics" and to "suck divinity from the flowers of nature." In* Pseudodoxia Epidemica *Browne examines and refutes a series of errors concerning a wide variety of natural phenomena (such as the misconception, in the passage here given, that elephants' limbs have no joints). As Leonard Nathanson remarks in* The Strategy of Truth: A Study of Sir Thomas Browne *(1967), "despite his fondness for materials that typically attracted the dabbler and his love of the*

[99]Genesis 2:18: "It is not good that the man should be alone; I will make him an help meet [a helpmate] for him." (Di Cesare)
[100]"Sundial."

curious and puzzling, Browne's basic approach was that of the serious scientist. For like Bacon and Boyle, he sought the underlying and universal principles in nature rather than astonishing prodigies. The prodigies had to be attended to, but their interest always extended beyond themselves. They offered the challenge, perhaps even harbored the secret, to the universal principles which regulate nature."

Such principles Browne found illustrated—following a mode of thinking descending from the Pythagoreans and Plato's Timaeus *(Chapter 9 above) through Kepler (Chapter 18)—in the arrangement of trees in quincunxes, or fivefold patterns, by Cyrus of Persia as described by Xenophon (Chapter 9): a concrete paradigm of the geometrically regular structure of the cosmos.* "The Garden of Cyrus *grew out of his first and last study: botany, biology, God's other manuscript—nature," Frank Livingstone Huntley writes in* Sir Thomas Browne *(1962). "The design of* The Garden, *therefore, is primary, leading up to the very mind of the Infinite Geometrician." For, as* Religio Medici *declares, reaffirming an ancient tradition central to medieval and Renaissance Christianity, "Art is the perfection of nature," and "Nature is the art of God."*

Spelling and punctuation have been modernized from the text in The Prose of Sir Thomas Browne, *ed. Norman J. Endicott (1967).*

FROM Religio Medici, *The First Part*

SECTION 12 ... I have often admired the mystical way of Pythagoras, and the secret magic of numbers: Beware of philosophy is a precept not to be received in too large a sense; for in this mass of Nature there is a set of things which carry in their front, though not in capital letters, yet in stenography and short characters something of divinity, which to wiser reasons serve as luminaries in the abyss of knowledge, and to judicious beliefs as scales and roundles to mount the pinnacles and highest pieces of divinity. The severe schools shall never laugh me out of the philosophy of Hermes [Trismegistus], that this visible world is but a picture of the invisible, wherein as in a portrait, things are not truly, but in equivocal shapes, and as they counterfeit some more real substance in that invisible fabric.

SECTION 15 *Natura nihil agit frustra* [Nature does nothing in vain],[101] is the only indisputable axiom in philosophy; there are no grotesques in nature, nor any thing framed to fill up empty cantons, and unnecessary spaces; in the most imperfect creatures, and such as were not preserved in the ark, but having their seeds and principles in the womb of nature, are everywhere where the power of the sun is; in these is the wisdom of his hand discovered: out of this rank Solomon chose the object of his admiration; indeed what reason may not go to school to the wisdom of bees, ants, and spiders? what wise hand teacheth them to do what reason cannot teach us? ruder heads stand amazed at those prodigious pieces of nature, whales, elephants, dromedaries, and camels; these I confess, are the colossus and majestic pieces of her hand; but in these narrow engines there is more curious mathematics, and the civility of these little citizens more neatly sets forth the wisdom of their maker: who admires not Regiomontanus his fly beyond his eagle,[102] or wonders not more at the operation of two souls in those little bodies than but one in the trunk of a cedar? I could never content my contemplation with those gen-

[101]See note on Rule I of Newton's "Rules of Reasoning in Philosophy" (*Principia*, Book III), in Chapter 18 above, for the source of this Aristotelian maxim.

[102]Regiomontanus (Johann Müller) of Königsberg was supposed to have made an iron fly and a wooden eagle, the fly being naturally more intricate. (Endicott)

eral pieces of wonder, the flux and reflux of the sea, the increase of Nile, the conversion of the needle to the north, and have studied to match and parallel those in the more obvious and neglected pieces of nature, which without further travel I can do in the cosmography of myself; we carry with us the wonders we seek without us: there is all Africa and her prodigies in us; we are that bold and adventurous piece of nature, which he that studies wisely learns in a compendium, what others labor at in a divided piece and endless volume.

SECTION 16 Thus are there two books from whence I collect my divinity; besides that written one of God, another of his servant Nature, that universal and public manuscript, that lies expansed unto the eyes of all: those that never saw him in the one, have discovered him in the other: this was the scripture and theology of the heathens: the natural motion of the sun made them more admire him, than its supernatural station[103] did the children of Israel; the ordinary effects of nature wrought more admiration in them, than in the other all his miracles; surely the heathens knew better how to join and read these mystical letters, than we Christians, who cast a more careless eye on these common hieroglyphics, and disdain to suck divinity from the flowers of nature. Nor do I so forget God, as to adore the name of Nature; which I define not with the Schools,[104] the principle of motion and rest, but that straight and regular line, that settled and constant course the wisdom of God hath ordained the actions of his creatures, according to their several kinds. To make a revolution every day is the nature of the sun, because that necessary course which God hath ordained it, from which it cannot swerve, but by a faculty from that voice which first did give it motion. Now this course of Nature God seldom alters or perverts, but like an excellent artist hath so contrived his work, that with the selfsame instrument, without a new creation, he may effect his obscurest designs. Thus he sweetened the water with a wood, preserved the creatures in the ark, which the blast of his mouth might have as easily created: for God is like a skilful geometrician, who when more easily and with one stroke of his compass, he might describe or divide a right line, had yet rather do this in a circle or longer way, according to the constituted and forelaid principles of his art: yet this rule of his he doth sometimes pervert, to acquaint the world with his prerogative, lest the arrogancy of our reason should question his power, and conclude he could not; and thus I call the effects of nature the works of God, whose hand and instrument she only is; and therefore to ascribe his actions unto her, is to devolve the honor of the principal agent upon the instrument; which if with reason we may do, then let our hammers rise up and boast they have built our houses, and our pens receive the honor of our writings. I hold there is a general beauty in the works of God, and therefore no deformity in any kind or species of creature whatsoever: I cannot tell by what logic we call a toad, a bear, or an elephant, ugly, they being created in those outward shapes and figures which best express the actions of their inward forms; and having passed that general visitation of God, who saw that all that he had made was good, that is, conformable to his will, which abhors deformity, and is the rule of order and beauty; there is no deformity but in a monstrosity, wherein notwithstanding there is a kind of beauty, Nature so ingeniously contriving the irregular parts, as they become sometimes more remarkable than the principal fabric. To speak yet more narrowly, there was never any thing ugly, or misshapen, but the Chaos; wherein notwithstanding, to speak strictly, there was no deformity, because no form, nor was it yet im-

[103]"Standing still," as commanded by Joshua (Joshua 10:12–14).
[104]The Scholastic, or Aristotelian, philosophers.

pregnate by the voice of God: now nature is not at variance with art, nor art with nature; they both being the servants of his providence: art is the perfection of nature: were the world now as it was the sixth day, there were yet a Chaos: Nature hath made one world, and art another. In brief, all things are artificial, for nature is the art of God.

FROM Pseudodoxia Epidemica; or, Enquiries into Very Many Received Tenents, and Commonly Presumed Truths

From The Third Book: Of Divers Popular and Received Tenets Concerning Animals, Which Examined, Prove Either False or Dubious

Chapter 1: Of the Elephant

The first shall be of the elephant; whereof there generally passeth an opinion it hath no joints; and this absurdity is seconded with another, that being unable to lie down, it sleepeth against a tree, which the hunters observing do saw almost asunder; whereon the beast relying, by the fall of the tree falls also down itself, and is able to rise no more.[105] Which conceit is not the daughter of latter times, but an old and grey-headed error, even in the days of Aristotle, as he delivereth in his book *de incessu animalium [The Progression of Animals]*; and stands successively related by several other authors: by Diodorus Siculus, Strabo, Ambrose, Cassiodore, Solinus, and many more. Now herein methinks men much forget themselves, not well considering the absurdity of such assertions.

For first, they affirm it hath no joints, and yet concede it walks and moves about; whereby they conceive there may be a progression or advancement made in motion without inflexion of parts. Now all progression or animal locomotion being (as Aristotle teacheth) performed *tractu et pulsu*; that is, by drawing on, or impelling forward some part which was before in station, or at quiet; where there are no joints or flexures, neither can there be these actions; and this is true, not only in quadrupeds, volatiles [birds], and fishes, which have distinct and prominent organs of motion, legs, wings, and fins, but in such also as perform their progression by the trunk, as serpents, worms, and leeches; whereof, though some want bones, and all extended articulations, yet have they arthritical analogies;[106] and by the motion of fibrous and musculous parts, are able to make progression. Which to conceive in bodies inflexible, and without all protrusion of parts, were to expect a race from Hercules his pillars;[107] or hope to behold the effects of Orpheus his harp, when trees found joints, and danced after his music.

Again, while men conceive they never lie down, and enjoy not the position of rest, ordained unto all pedestrious animals, hereby they imagine (what reason cannot conceive) that an animal of the vastest dimension and longest duration should live in a continual motion, without that alternity and vicissitude of rest whereby all others continue; and yet must thus much come to pass, if we opinion they lie not down and enjoy no decumbence at all. For station is properly no rest, but one kind of motion, relating unto that which physicians (from Galen) do name extensive or tonical; that is, an extension of the muscles and organs of motion maintaining the body at length or in its proper figure; wherein although it seem to be unmoved, it is nevertheless not without all motion; for in this position the muscles are sensibly extended, and labor to support the body; which

[105]See the account of the elephant in the Greek *Physiologus* (Chapter 13 above).
[106]Jointlike parts. (Browne)
[107]The pillars of Hercules, i.e., the cliffs on either side of the Strait of Gibraltar.

permitted unto its proper gravity, would suddenly subside and fall unto the earth, as it happeneth in sleep, diseases, and death. . . .

Moreover men herein do strangely forget the obvious relations of history, affirming they have no joints, whereas they daily read of several actions which are not performable without them. . . .

Lastly, they forget or consult not experience, whereof not many years past, we have had the advantage in England, by an elephant shown in many parts thereof, not only in the posture of standing, but kneeling and lying down. Whereby although the opinion at present be well suppressed, yet from some strings of tradition, and fruitful recurrence of error, it is not improbable it may revive in the next generation again. . . .

FROM The Garden of Cyrus; or, The Quincuncial, Lozenge, or Network Plantations of the Ancients, Artificially, Naturally, Mystically Considered

CHAPTER III . . . Now the number of five is remarkable in every circle, not only as the first spherical number, but the measure of spherical motion. For spherical bodies move by fives, and every globular figure placed upon a plane, in direct volutation, returns to the first point of contaction in the fifth touch, accounting by the axes of the diameters or cardinal points of the four quarters thereof. And before it arriveth unto the same point again, it maketh five circles equal unto itself, in each progress from those quarters absolving an equal circle.

By the same number doth nature divide the circle of the sea-star, and in that order and number disposeth those elegant semicircles, or dental sockets and eggs in the sea hedgehog. And no mean observations hereof there is in the mathematics of the neatest retiary spider, which concluding in forty-four circles, from five semidiameters beginneth that elegant texture.

And after this manner doth lay the foundation of the circular branches of the oak, which being five-cornered in the tender annual sprouts, and manifesting upon incision the signature of a star, is after made circular, and swelled into a round body: which practice of nature is become a point of art, and makes two problems in Euclid. But the bramble which sends forth shoots and prickles from its angles, maintains its pentagonal figure, and the unobserved signature of a handsome porch within it. To omit the five small buttons dividing the circle of the ivy-berry, and the characters in the winter stalk of the walnut, with many other observables, which cannot escape the eyes of signal discerners, such as know where to find Ajax his name in Delphinium, or Aaron's mitre in Henbane. . . .

The sexangular cells in the honey-combs of bees are disposed after this order; much there is not of wonder in the confused houses of pismires, though much in their busy life and actions; more in the edificial palaces of bees and monarchical spirits, who make their combs six-cornered, declining a circle, whereof many stand not close together, and completely fill the area of the place; but rather affecting a six-sided figure, whereby every cell affords a common side unto six more, and also a fit receptacle for the bee itself, which gathering into a cylindrical figure, aptly enters its sexangular house, more nearly approaching a circular figure, than either doth the square or triangle. And the combs themselves so regularly contrived, that their mutual intersections make three lozenges at the bottom of every cell, which severally regarded make three rows of neat rhomboidal figures, connected at the angles, and so continue three several chains throughout the whole comb. . . .

A resemblance hereof there is in the orderly and rarely disposed cells made by flies and insects, which we have often found fastened about small sprigs, and in those cottonary and woolly pillows, which sometime we meet with fastened unto leaves, there is included an elegant network texture, out of which come many small flies. And some resemblance there is of this order in the eggs of some butterflies and moths, as they stick upon leaves, and other substances; which being dropped from behind, nor directed by the eye, doth neatly declare how nature geometrizeth, and observeth order in all things. . . .

Studious observators may discover more analogies in the orderly book of nature, and cannot escape the elegancy of her hand in other correspondencies. The figures of nails and crucifying appurtenances are but precariously made out in the *Granadilla* or flower of Christ's passion: and we despair to behold in these parts that handsome draft of crucifixion in the fruit of the Barbado pine. The seminal spike of Phalaris, or great shaking grass, more nearly answers the tail of a rattlesnake than many resemblances in Porta:[108] and if the man-orchis of Columna be well made out, it excelleth all analogies. In young walnuts cut athwart, it is not hard to apprehend strange characters; and in those of somewhat elder growth, handsome ornamental drafts about a plain cross. In the root of Osmond or water-fern, every eye may discern the form of a half moon, rainbow, or half the character of Pisces. Some find Hebrew, Arabic, Greek, and Latin characters in plants; in a common one among us we seem to read *Aiaia, Viviu, Lilil*. . . .

The Contemplative Man's Recreation

The Compleat Angler, or the Contemplative Man's Recreation: Being a Discourse of Rivers, Fishponds, Fish, and Fishing, by Izaak Walton

Born at Stafford, the son of an alehouse keeper, Izaak Walton (1593–1683) became a prosperous draper in London, where he spent most of his long life. He was a parishioner of John Donne, a Life *of whom (for which Sir Henry Wotton had asked him to gather materials) he hastily wrote after Wotton's death and published in 1640; in expanded form, it was included, along with biographies of Hooker and Herbert, in Walton's* Lives *of 1670. In 1653, during the tumultuous early years of the Puritan Commonwealth—which this peace-loving Anglican despised—he published* The Compleat Angler, *to whose fifth edition of 1676 Charles Cotton (author of* The Compleat Gamester *and* The Planter's Manual, *and future translator of Montaigne) added a second part. Except for his writing, Walton led a largely uneventful life, spending much of his last twenty years in Winchester, where he died and is buried in the cathedral.*

The Compleat Angler, *subtitled* The Contemplative Man's Recreation, *John Buxton remarks in his foreword to the Oxford World's Classics edition of 1982, "has claims to be, as the most often reprinted book in the language (after the Bible and the Book of Common Prayer), the best loved and most popular book" in English over the last three centuries. Reflecting a revival of pastoral poetry in Walton's youth, it not only offers a practical handbook for fishermen but also vividly portrays a countryside "still recognizably English even though one meets, without surprise, a shepherd named Coridon who plays the pipes to an English milkmaid and her cousin*

[108]Giambattista della Porta, 1545–1615, Italian scholar who wrote on plants, mathematics, "natural magic," etc., etc. (Endicott)

Betty, for, in spite of Puritans and politics, Arcadia is not far away." Indeed, this "Arcadia of Angling" (as Walton suggested his book might be called) is interspersed with conversations and poems that make it a continually varied medley of rural or pastoral delights: "the most homespun of idyllic daydreams," in Douglas Bush's words, "the most substantial of poems of escape."

The charm of Walton's temperament also pervades the book, for "though it is known I can be serious at seasonable times," he writes, "yet the whole discourse is, or rather was, a picture of my own disposition, especially in such days and times as I have laid aside business, and gone a fishing with honest Nat. and R. Roe; but they are gone," he adds, "and with them most of my pleasant hours, even as a shadow, that passeth away, and returns not." Thus, in this serene if nostalgic book, as Bush affirms, "a love of angling is an outward and visible sign of an inward and spiritual grace, of a gentle, contemplative benignity of soul which abhors dissension and loves good old ways, whether in the choice of bait or ballads or barley-wine or the worship of the Creator."

In the first of our passages, from the opening chapter, an Angler (Piscator), *contributing to the ancient debate concerning the merits of the contemplative as opposed to the active life, commends "the most honest, ingenuous, quiet, and harmless art of angling" above the rival pursuits of a Hunter* (Venator) *and Falconer* (Auceps); *his eulogy will continue throughout the book as the Hunter becomes the Angler's attentive "scholar" in a leisurely art that brings man and nature into continuous communion. The second selection is a representative chapter, on the perch, illustrating the detailed concreteness of Piscator's instruction, and ending with a poetic recitation that enhances the pleasure of fishing "a little longer under this honeysuckle hedge."*

Notes by Buxton, editor of the World's Classics edition, "are revised and abbreviated," as he indicates, "from those by T. Balston in the Clarendon Press edition of 1915," the basis of the text here modernized.

FROM *Chapter I: Conference betwixt an* Angler [Piscator], *a* Hunter [Venator], *and a* Falconer [Auceps]; *each commending his Recreation.*

Pisc. . . . And for you that have heard many grave, serious men, pity anglers, let me tell you, sir, there be many men that are by others taken to be serious and grave men, which we contemn and pity. Men that are taken to be grave, because Nature hath made them of a sour complexion; money-getting men, men that spend all their time, first in getting, and next, in anxious care to keep it; men that are condemned to be rich, and then always busy or discontented; for these poor-rich-men, we anglers pity them perfectly, and stand in no need to borrow their thoughts to think ourselves so happy. No, no, sir, we enjoy a contentedness above the reach of such dispositions, and as the learned and ingenuous Montaigne says, like himself, freely,[109] "When my cat and I entertain each other with mutual apish tricks (as playing with a garter) who knows but that I make my cat more sport than she makes me? Shall I conclude her to be simple, that has her time to begin or refuse to play as freely as I myself have? Nay, who knows but that it is a defect of my not understanding her language (for doubtless cats talk and reason with one another) that we agree no better: and who knows but that she pities me for being no wiser than to play with her, and laughs and censures my folly, for making sport of her, when we two play together?"

Thus freely speaks Montaigne concerning cats, and I hope I may take as great a liberty to blame any man, and laugh at him too, let him be never so grave, that hath not heard what anglers can say in the justification of their art and recreation; which I may

[109]In Apol[ogy] for Ra[ymond] Sebond. (Walton)

again tell you is so full of pleasure, that we need not borrow their thoughts, to think ourselves happy. . . .

And for that I shall tell you, that in ancient times a debate hath risen (and it remains yet unresolved), whether the happiness of man in this world doth consist more in contemplation or action? . . .

Concerning which two opinions I shall forbear to add a third by declaring my own, and rest myself contented in telling you (my very worthy friend) that both these meet together, and do most properly belong to the most honest, ingenuous, quiet, and harmless art of angling.

And first I shall tell you what some have observed (and I have found it to be a real truth), that the very sitting by the river's side is not only the quietest and fittest place for contemplation, but will invite an Angler to it: and this seems to be maintained by the learned Pet. du Moulin,[110] who (in his discourse of the fulfilling of prophecies) observes that when God intended to reveal any future events or high notions to his prophets, he then carried them either to the deserts, or the seashore, that having so separated them from amidst the press of people and business, and the cares of the world, he might settle their mind in a quiet repose, and there make them fit for revelation.

And this seems also to be intimated by the Children of Israel (Psal. 137), who having in a sad condition banished all mirth and music from their pensive hearts and having hung up their then mute harps upon the willow trees growing by the rivers of Babylon, sat down upon those banks bemoaning the ruins of Sion, and contemplating their own sad condition.

And an ingenuous Spaniard[111] says, that rivers and the inhabitants of the watery element were made for wise men to contemplate, and fools to pass by without consideration. And though I will not rank myself in the number of the first, yet give me leave to free myself from the last, by offering to you a short contemplation, first of rivers, and then of fish; concerning which I doubt not but to give you many observations that will appear very considerable; I am sure they have appeared so to me, and made many an hour pass away more pleasantly, as I have sat quietly on a flowery bank by a calm river, and contemplated what I shall now relate to you. . . .

Chapter XII: Observations of the Perch, *and directions how to fish for him.*

Pisc. The Perch is a very good, and a very bold biting fish; he is one of the fishes of prey, that like the Pike and Trout, carries his teeth in his mouth, which is very large, and he dare venture to kill and devour several other kinds of fish; he has a hooked or hog back, which is armed with sharp and stiff bristles, and all his skin armed or covered over with thick, dry, hard scales, and hath (which few other fish have) two fins on his back; he is so bold that he will invade one of his own kind, which the Pike will not do so willingly, and you may therefore easily believe him to be a bold biter.

The Perch is of great esteem in Italy saith Aldrovandus,[112] and especially the least are there esteemed a dainty dish. And Gesner[113] prefers the Perch and Pike above the Trout,

[110]Peter du Moulin (1568–1658), author of a treatise on *The Accomplishment of Prophecies, Translated out of the French by J. Heath* (Oxford, 1613). (Buxton)

[111]Moses Browne, who edited *The Compleat Angler* (1750) at Dr. Johnson's suggestion, says that the reference is to *The Hundred and Ten Considerations of Signor Valdesso*, of which an English translation by Nicholas Farrar, junior, was published in 1638; but the passage does not appear there. (Buxton)

[112]Ulysses Aldrovandus (1524?–1607), a famous naturalist of Bologna, author of a *Natural History* in thirteen volumes; the present reference is to *De Piscibus [On Fishes]*, v, p. 627, § 1. (Buxton)

[113]Conrad von Gesner (1516–65), *Historia animalium [History of animals]*, iv, p. 824. (Buxton)

or any freshwater fish: he says the Germans have this proverb: More wholesome than a Perch of Rhine; and he says the River Perch is so wholesome that physicians allow him to be eaten by wounded men or by men in fevers, or by women in childbed.

He spawns but once a year, and is, by physicians, held very nutritive, yet by many to be hard of digestion; they abound more in the river Po and in England (says Rondelitius)[114] than other parts, and have in their brain a stone, which is in foreign parts sold by apothecaries, being there noted to be very medicinable against the stone in the reins:[115] these be a part of the commendations which some philosophical brains have bestowed upon the freshwater Perch; yet they commend the Sea Perch, which is known by having but one fin on his back (of which they say we English see but a few) to be a much better fish.

The Perch grows slowly, yet will grow, as I have been credibly informed, to be almost two feet long; for an honest informer told me, such a one was not long since taken by Sir Abraham Williams, a gentleman of worth, and a brother of the angle (that yet lives, and I wish he may): this was a deep-bodied fish, and doubtless durst have devoured a Pike of half his own length; for I have told you, he is a bold fish, such a one as but for extreme hunger, the Pike will not devour; for to affright the Pike and save himself, the Perch will set up his fins, much like as a turkey cock will sometimes set up his tail.

But, my scholar, the Perch is not only valiant to defend himself, but he is (as I said) a bold biting fish, yet he will not bite at all seasons of the year; he is very abstemious in winter, yet will bite then in the midst of the day, if it be warm: and note that all fish bite best about the midst of a warm day in winter, and he hath been observed by some, not usually to bite till the mulberry tree buds; that is to say, till extreme frosts be past the spring; for when the mulberry tree blossoms, many gardeners observe their forward fruit to be past the danger of frosts, and some have made the like observation of the Perch's biting.

But bite the Perch will, and that very boldly; and as one has wittily observed, if there be twenty or forty in a hole, they may be at one standing all catched one after another; they being, as he says, like the wicked of the world, not afraid though their fellows and companions perish in their sight. And you may observe, that they are not like the solitary Pike, but love to accompany one another, and march together in troops.

And the baits for this bold fish are not many; I mean, he will bite as well at some, or at any of these three, as at any, or all others whatsoever: a worm, a minnow, or a little frog (of which you may find many in hay time) and of worms, the dunghill worm called a brandling I take to be best, being well scoured in moss or fennel; or he will bite at a worm that lies under a cow turd with a bluish head. And if you rove for a Perch with a minnow, then it is best to be alive, you sticking your hook through his back-fin; or a minnow with the hook in his upper lip, and letting him swim up and down about mid-water, or a little lower, and you still keeping him to about that depth, by a cork, which ought not to be a very little one; and the like way you are to fish for the Perch with a small frog, your hook being fastened through the skin of his leg, towards the upper part of it: and lastly, I will give you but this advice, that you give the Perch time enough when he bites, for there was scarce any angler that has given him too much. And now I think it best to rest myself, for I have almost spent my spirits with talking so long.

[114]Guillaume Rondelet (1507–66), French naturalist, author of *De Piscibus Marinis [On Sea Fishes]* and *Universae Aquatilium Historiae pars altera [Second Part of the Universal History of Aquatic Animals]*. Of the present passage Buxton writes, "I cannot find that he says so."
[115]"Kidneys."

Venat. Nay, good master, one fish more, for you see it rains still, and you know our angles are like money put to usury: they may thrive, though we sit still and do nothing but talk and enjoy one another. Come, come, the other fish, good master.
Pisc. But, scholar, have you nothing to mix with this discourse, which now grows both tedious and tiresome? shall I have nothing from you that seem to have both a good memory, and a cheerful spirit?
Ven. Yes, master, I will speak you a copy of verses that were made by Doctor Donne, and made to show the world that he could make soft and smooth verses when he thought smoothness worth his labor; and I love them the better, because they allude to rivers, and fish and fishing. They be these:[116]

Come live with me, and be my love,
And we will some new pleasures prove,
Of golden sands, and crystal brooks,
With silken lines, and silver hooks.

There will the river whispering run,
Warmed by thy eyes more than the sun;
And there the enameled fish will stay,
Begging themselves they may betray.

When thou wilt swim in that live bath,
Each fish, which every channel hath,
Most amorously to thee will swim,
Gladder to catch thee, than thou him.

If thou, to be so seen, beest loth
By sun or moon, thou dark'nest both,
And if mine eyes have leave to see,
I need not their light, having thee.

Let others freeze with angling reeds,
And cut their legs with shells and weeds,
Or treacherously poor fish beset,
With strangling snares, or windowy net.

Let coarse bold hands from slimy nest
The bedded fish in banks outwrest,
Let curious traitors sleeve silk flies
To 'witch poor wand'ring fishes' eyes.

For thee, thou need'st no such deceit,
For thou thyself art thine own bait:
That fish that is not catched thereby,
Is wiser far, alas, than I.

Pisc. Well remembered, honest scholar, I thank you for these choice verses, which I have heard formerly, but had quite forgot, till they were recovered by your happy memory. Well, being I have now rested myself a little, I will make you some requital, by telling

[116]This poem appeared (with a few variations) in *Poems* (1633) of John Donne, of whom Walton wrote a *Life*; it parodies Christopher Marlowe's "The Passionate Shepherd to His Love" (Chapter 17 above), which Walton had quoted (as "The Milk Maid's Song") in Chapter IV of *The Compleat Angler*.

you some observations of the Eel, for it rains still, and because (as you say) our angles are as money put to use that thrives when we play, therefore we'll sit still and enjoy ourselves a little longer under this honeysuckle hedge.

Wilderness Squared by Reason

The Life and Strange Surprising Adventures of Robinson Crusoe, of York, Mariner, by Daniel Defoe

Daniel Defoe (1660–1731), with his vigorous prose and practical outlook, embodies the new "bourgeois" mentality of many Englishmen after the overthrow of the Catholic Stuart King James II by the Dutch Protestant William III (with Mary, daughter of Charles I) in the "Glorious Revolution" of 1688–89. Born in London, the son of the tallow chandler James Foe, Defoe (as he later renamed himself) was educated for the Presbyterian ministry at the Newington Dissenting Academy, but turned instead to various commercial ventures, which resulted in bankruptcy in 1692. In 1685 he took part in the abortive rebellion of the Duke of Monmouth, bastard son of Charles I, against James II, and in 1701 defended William against attacks on his foreign birth in the verse satire The True Born Englishman. *In 1703 he was imprisoned and put in the pillory for his satire on Anglican persecution,* The Shortest Way with Dissenters; *in a "Hymn to the Pillory" he declared himself punished "because he was too bold, / And told those truths which should not ha' been told." Released through the influence of Robert Harley (later Prime Minister), he acted as a secret agent in negotiations leading in 1707 to union with Scotland.*

A prolific writer for many periodicals (one of which, the Review, *he seems to have written single-handedly), he was the author of some five hundred works. Among the earliest was his* Essay upon Projects *of 1697, in which, proposing establishment of an academy for women, he denounced the "barbarous" custom of denying to women the advantages of learning. In 1719, at age 59, he published his first novel,* Robinson Crusoe, *based on accounts of Alexander Selkirk, a castaway on the island of Juan Fernandez from 1704 to 1709. Its huge success led to two sequels and another novel,* Captain Singleton, *before the end of 1720, followed by* Moll Flanders, A Journal of the Plague Year, *and* Colonel Jacque *in 1722, and* Roxana *in 1724.* A Tour thro' the Whole Island of Great Britain *appeared in three volumes between 1724 and 1727.*

From the beginning, in An Essay upon Projects, *Defoe proclaims the harsh necessity of making nature serve human ends: "Man is the worst of all God's creatures to shift for himself; no other animal is ever starved to death; Nature without has provided them both food and clothes; and Nature within has placed an instinct that never fails to direct them to proper means for a supply; but man must either work or starve, slave or die." Nowhere is this need more methodically addressed than by Robinson Crusoe, who finds himself stranded on an apparently desert island. (Years later he will discover that he shares it with cannibals from whom he rescues his man Friday, soon teaching him to be "a good Christian, a much better than I.") By "rational judgment of things" and patient "labor, application, and contrivance" he consoles himself for misfortune, makes a comfortable habitation, and even harvests a crop of barley. Self-sufficient individualism is forced upon him, and utilitarian practicality is the lesson experience teaches: "In a word, the nature and experience of things dictated to me upon just reflection that all the good things of this world are no farther good to us than they are for our use."*

Robinson Crusoe, *George Sherburn writes in Baugh's* Literary History of England, *"expressed the eighteenth-century epic theme of the power of the average man to preserve life and to organize an economy in the face of the most unpromising environment," energetically converting nature to culture at every turn. Still more, Bonamy Dobrée remarks in* English Literature in the Early Eighteenth Century *(1959), it is "the myth of Man surviving in an indifferent universe, or of man struggling against circumstance, pitting himself alone against odds. . . . For the first time in a book of this sort the writer was not only telling about the adventure, as happening to himself or somebody else, but seeming to write as he was experiencing the events, and so making the reader experience them too." Through such qualities the novel has appealed not only to pragmatic individualists but also to Rousseau, who in* Emile *makes this "complete textbook of natural education" the first book his young pupil of nature is permitted to read; to Coleridge, who calls it "a happy nightmare" in which "our imagination is kept in full play, excited to the highest; yet all the while we are touching, or touched by, common flesh and blood"; and to countless generations of imaginative children.*

. . . I now began to consider seriously my condition, and the circumstance I was reduced to, and I drew up the state of my affairs in writing, not so much to leave them to any that were to come after me, for I was like to have but few heirs, as to deliver my thoughts from daily poring upon them and afflicting my mind; and as my reason began now to master my despondency, I began to comfort myself as well as I could, and to set the good against the evil, that I might have something to distinguish my case from worse, and I stated it very impartially, like debtor and creditor, the comforts I enjoyed against the miseries I suffered, thus:

EVIL	GOOD
I am cast upon a horrible desolate island, void of all hope of recovery.	But I am alive, and not drowned as all my ship's company was.
I am singled out and separated, as it were, from all the world to be miserable.	But I am singled out too from all the ship's crew to be spared from death; and he that miraculously saved me from death, can deliver me from this condition.
I am divided from mankind, a solitaire, one banished from human society.	But I am not starved and perishing on a barren place, affording no sustenance.
I have not clothes to cover me.	But I am in a hot climate, where if I had clothes I could hardly wear them.
I am without any defense or means to resist any violence of man or beast.	But I am cast on an island, where I see no wild beasts to hurt me, as I saw on the coast of Africa: and what if I had been shipwrecked there?
I have no soul to speak to, or relieve me.	But God wonderfully sent the ship in near enough to the shore, that I have gotten out so many necessary things as will either supply my wants, or enable me to supply myself even as long as I live.

Upon the whole, here was an undoubted testimony that there was scarce any condition in the world so miserable but there was something negative or something positive to be thankful for in it; and let this stand as a direction from the experience of the most miserable of all conditions in this world, that we may always find in it something to comfort ourselves from, and to set in the description of good and evil on the credit side of the account.

Having now brought to mind a little to relish my condition, and given over looking out to sea to see if I could spy a ship, I say, giving over these things, I began to apply myself to accommodate my way of living, and to make things as easy to me as I could.

I have already described my habitation, which was a tent under the side of a rock, surrounded with a strong pale of posts and cables, but I might now rather call it a wall, for I raised a kind of wall up against it of turfs, about two foot thick on the outside, and after some time, I think it was a year and a half, I raised rafters from it leaning to the rock, and thatched or covered it with boughs of trees, and such things as I could get to keep out the rain, which I found at some times of the year very violent.

I have already observed how I brought all my goods into this pale, and into the cave which I had made behind me: but I must observe, too, that at first this was a confused heap of goods, which as they lay in no order, so they took up all my place, I had no room to turn myself; so I set myself to enlarge my cave and works farther into the earth, for it was a loose sandy rock, which yielded easily to the labor I bestowed on it; and so when I found I was pretty safe as to beasts of prey, I worked sideways to the right hand into the rock, and then turning to the right again, worked quite out and made me a door to come out, on the outside of my pale or fortification.

This gave me not only egress and regress, as it were a back way to my tent and to my storehouse, but gave me room to stow my goods.

And now I began to apply myself to make such necessary things as I found I most wanted, as particularly a chair and a table, for without these I was not able to enjoy the few comforts I had in the world; I could not write, or eat, or do several things with so much pleasure without a table.

So I went to work; and here I must needs observe that as reason is the substance and original of the mathematics, so by stating and squaring everything by reason, and by making the most rational judgment of things, every man may be in time master of every mechanic art. I had never handled a tool in my life, and yet in time by labor, application, and contrivance, I found at last that I wanted nothing but I could have made it, especially if I had had tools; however, I made abundance of things even without tools, and some with no more tools than an adze and a hatchet, which perhaps were never made that way before, and that with infinite labor: for example, if I wanted a board, I had no other way but to cut down a tree, set it on an edge before me, and hew it flat on either side with my axe, till I had brought it to be thin as a plank, and then dub it smooth with my adze. It is true, by this method I could make but one board out of a whole tree, but this I had no remedy for but patience, any more than I had for the prodigious deal of time and labor which it took me up to make a plank or board: but my time or labor was little worth, and so it was as well employed one way as another.

However, I made me a table and a chair, as I observed above, in the first place, and this I did out of the short pieces of boards that I brought on my raft from the ship: but when I had wrought out some boards, as above, I made large shelves of the breadth of a foot and a half one over another, all along one side of my cave, to lay all my tools, nails, and iron-work, and in a word, to separate everything at large in their places, that I must come easily at them; I knocked pieces into the wall of the rock to hang my guns and all things that would hang up.

So that had my cave been to be seen, it looked like a general magazine of all necessary things, and I had everything so ready at my hand that it was a great pleasure to me to see all my goods in such order, and especially to find my stock of all necessaries so great. . . .

. . . In the middle of all my labors it happened that rummaging in my things, I found a little bag, which, as I hinted before, had been filled with corn[117] for the feeding of poultry, not for this voyage, but before, as I suppose; when the ship came from Lisbon, what little remainder of corn had been in the bag was all devoured with the rats, and I saw nothing in the bag but husks and dust; and being willing to have the bag for some other use, I think it was to put powder in, when I divided it for fear of the lightning, or some such use, I shook the husks of corn out of it on one side of my fortification under the rock.

It was a little before the great rains, just now mentioned, that I threw this stuff away, taking no notice of anything, and not so much as remembering that I had thrown anything there; when about a month after, or thereabout, I saw some few stalks of something green shooting out of the ground, which I fancied might be some plant I had not seen, but I was surprised and perfectly astonished when, after a little longer time, I saw about ten or twelve ears come out, which were perfect green barley of the same kind as our European, nay, as our English barley.

It is impossible to express the astonishment and confusion of my thoughts on this occasion; I had hitherto acted upon no religious foundation at all, indeed I had very few notions of religion in my head, or had entertained any sense of anything that had befallen me, otherwise than as a chance, or, as we lightly say, what pleases God, without so much as inquiring into the end of providence in these things, or his order in governing events in the world: but after I saw barley grow there, in a climate which I know was not proper for corn, and especially that I knew not how it came there, it startled me strangely, and I began to suggest that God had miraculously caused this grain to grow without any help of seed sown, and that it was directed purely for my sustenance on that wild miserable place.

This touched my heart a little, and brought tears out of my eyes, and I began to bless myself that such a prodigy of nature should happen upon my account; and this was the more strange to me, because I saw near it still all along by the side of the rock some other straggling stalks, which proved to be stalks of rice, and which I knew because I had seen it grow in Africa when I was ashore there.

I not only thought these the pure productions of Providence for my support, but not doubting but that there was more in the place, I went all over that part of the island where I had been before, peering in every corner and under every rock, to see for more of it, but I could not find any; at last it occurred to my thoughts that I had shook a bag of chicken's meat[118] out in that place, and then the wonder began to cease; and I must confess, my religious thankfulness to God's Providence began to abate too upon the discovering that all this was nothing but what was common; though I ought to have been as thankful for so strange and unforeseen Providence, as if it had been miraculous; for it was really the work of Providence as to me, that should order or appoint that 10 or 12 grains of corn should remain unspoiled (when the rats had destroyed all the rest), as if it had been dropped from Heaven; as also, that I should throw it out in that particular place, where it being in the shade of a high rock, it sprang up immediately; whereas if I had thrown it anywhere else, at that time, it had been burnt up and destroyed.

[117]Grain. [118]Chicken feed.

I carefully saved the ears of this corn you may be sure in their season, which was about the end of June; and laying up every corn, I resolved to sow them all again, hoping in time to have some quantity sufficient to supply me with bread: but it was not till the 4th year that I could allow myself the least grain of this corn to eat, and even then but sparingly, as I shall say afterwards in its order; for I lost all that I sowed the first season, by not observing the proper time: for I sowed it just before the dry season, so that it never came up at all, at least, not as it would ha' done: of which in its place. . . .

I had been now in this unhappy island above 10 months; all possibility of deliverance from this condition seemed to be entirely taken from me; and I firmly believed that no human shape had ever set foot upon that place. Having now secured my habitation, as I thought, fully to my mind, I had a great desire to make a more perfect discovery of the island, and to see what other productions I might find, which I yet knew nothing of.

It was the 15th of July that I began to take a more particular survey of the island itself: I went up the creek first, where, as I hinted, I brought my rafts on shore; I found after I came about two miles up that the tide did not flow any higher, and that it was no more than a little brook of running water, and very fresh and good; but this being the dry season, there was hardly any water in some parts of it, at least not enough to run in any stream so as it could be perceived.

On the bank of this brook I found many pleasant savannahs, or meadows, plain, smooth, and covered with grass; and on the rising parts of them next to the higher grounds, where the water, as it might be supposed, never overflowed, I found a great deal of tobacco, green, and growing to a great and very strong stalk; there were divers other plants which I had no notion of, or understanding about, and might perhaps have virtues of their own which I could not find out.

I searched for the cassava root, which the Indians, in all that climate, make their bread of, but I could find none. I saw large plants of aloes, but did not then understand them. I saw several sugar canes, but wild, and for want of cultivation, imperfect. I contented myself with these discoveries for this time, and came back musing with myself what course I might take to know the virtue and goodness of any of the fruits or plants which I should discover; but could bring it to no conclusion; for in short, I had made so little observation while I was in the Brazils that I knew little of the plants in the field, at least very little that might serve me to any purpose now in my distress.

The next day, the 16th, I went up the same way again, and after going something farther than I had gone the day before, I found the brook and the savannahs began to cease, and the country became more woody than before; in this part I found different fruits, and particularly I found melons upon the ground in great abundance, and grapes upon the trees; the vines had spread indeed over the trees, and the clusters of grapes were just now in their prime, very ripe and rich: this was a surprising discovery, and I was exceedingly glad of them; but I was warned by my experience to eat sparingly of them, remembering that when I was ashore in Barbary, the eating of grapes killed several of our English men who were slaves there by throwing them into fluxes[119] and fevers: but I found an excellent use for these grapes, and that was to cure or dry them in the sun, and keep them as dried grapes or raisins are kept, which I thought would be, as indeed they were, as wholesome, as agreeable to eat, when no grapes might be to be had.

I spent all that evening there, and went not back to my habitation, which, by the way, was the first night, as I might say, I had lain from home. In the night I took my first

[119]Dysentery.

contrivance, and got up into a tree, where I slept well, and the next morning proceeded upon my discovery, traveling near four miles, as I might judge by the length of the valley, keeping still due north, with a ridge of hills on the south and north side of me.

At the end of this march, I came to an opening where the country seemed to descend to the west, and a little spring of fresh water which issued out of the side of the hill by me, run the other way, that is due east; and the country appeared so fresh, so green, so flourishing, everything being in a constant verdure, or flourish of spring, that it looked like a planted garden.

I descended a little on the side of that delicious vale, surveying it with a secret kind of pleasure (though mixed with my other afflicting thoughts) to think that this was all my own, that I was king and lord of all this country indefeasibly, and had a right of possession; and if I could convey it, I might have it in inheritance as completely as any lord of a manor in England. I saw here abundance of cocoa trees, orange and lemon, and citron trees; but all wild, and very few bearing any fruit, at least not then: however, the green limes that I gathered were not only pleasant to eat, but very wholesome, and very cool, and refreshing. . . .

When I came home from this journey, I contemplated with great pleasure the fruitfulness of that valley, and the pleasantness of the situation, the security from storms on that side the water, and the wood, and concluded that I had pitched upon a place to fix my abode which was by far the worst part of the country. Upon the whole I began to consider of removing my habitation; and to look out for a place equally safe as where I was now situate, if possible in that pleasant fruitful part of the island.

This thought run long in my head, and I was exceeding fond of it for some time, the pleasantness of the place tempting me; but when I came to a nearer view of it, and to consider that I was now by the seaside, where it was at least possible that something might happen to my advantage, and by the same ill fate that brought me hither, might bring some other unhappy wretches to the same place; and though it was scarce probable that any such thing should ever happen, yet to enclose myself among the hills and woods, in the center of the island, was to anticipate my bondage, and to render such an affair not only improbable but impossible; and that therefore I ought not by any means to remove.

However, I was so enamored of this place that I spent much of my time there, for the whole remaining part of the month of July; and though upon second thoughts I resolved as above, not to remove, yet I built me a little kind of a bower, and surrounded it at a distance with a strong fence, being a double hedge, as high as I could reach, well staked, and filled between with brushwood; and here I lay very secure, sometimes two or three nights together, always going over it with a ladder, as before; so that I fancied now I had my country house, and my seacoast house: and this work took me up to the beginning of August. . . .

A little after this my ink began to fail me, and so I contented myself to use it more sparingly, and to write down only the most remarkable events of my life, without continuing a daily memorandum of other things.

The rainy season and the dry season began now to appear regular to me, and I learned to divide them so as to provide for them accordingly. But I bought all my experience before I had it; and this I am going to relate was one of the most discouraging experiments that I made at all. I have mentioned that I had saved the few ears of barley and rice, which I had so surprisingly found spring up, as I thought, of themselves, and believe there was about thirty stalks of rice, and about twenty of barley; and now I thought it a proper time to sow it after the rains, the sun being in its southern position going from me.

Accordingly I dug up a piece of ground as well as I could with my wooden spade, and dividing it into two parts, I sowed my grain; but as I was sowing, it casually occurred to my thoughts that I would not sow it all at first, because I did not know when was the proper time for it; so I sowed about two thirds of the seed, leaving about a handful of each.

It was a great comfort to me afterwards that I did so, for not one grain of that I sowed this time came to anything; for the dry months following, the earth having had no rain after the seed was sown, it had no moisture to assist its growth, and never came up at all till the wet season had come again, and then it grew as if it had been but newly sown.

Finding my first seed did not grow, which I easily imagined was by the drought, I sought for a moister piece of ground to make another trial in, and I dug up a piece of ground near my new bower, and sowed the rest of my seed in February, a little before the vernal equinox; and this having the rainy months of March and April to water it, sprung up very pleasantly, and yielded a good crop; but having part of the seed left only, and not daring to sow all that I had, I had but a small quantity at last, my whole crop not amounting to above half a peck of each kind.

But by this experiment I was made master of my business, and knew exactly when the proper season was to sow, and that I might expect two seed times and two harvests every year. . . .

Nature Methodized

Selections from the Poems of Alexander Pope

Alexander Pope (1688–1744) is the foremost poet of the English "Augustan" age, which prized social order, civil manners, rational thought, and poetic regularity. Yet—even apart from the incivility that reduces a personal enemy to "a mere white curd of ass's milk," or the ever-looming disorder that bursts forth, at the end of The Dunciad, *when "thy dread Empire,* Chaos! *is restored, . . . And Universal Darkness buries all"—it is remarkable how frequently and variously this urbane wit, trenchant satirist, and master of neoclassic decorum writes of nature. As Wendell Berry observes in "Poetry and Place" (1982),*[120] *"Few poets that I know have been so explicitly appreciative of the human kinship with the natural world as Alexander Pope, and few have been so carefully attentive to the spiritual, moral, and practical implications of that kinship."*

Born in London, where his father was a Roman Catholic merchant, Pope—who suffered from a childhood illness that left him with a permanently twisted spine, and who was barred by his religion from attending a university, or from living within ten miles of London—moved with his family in 1700 to Binfield in Windsor Forest. Our first selection, the Horatian "Ode on Solitude," written when he was twelve years old, gives expression to a desire to live "unseen, unknown" in the innocence of rural seclusion. His first published work, written in 1704, at age sixteen, and published in Tonson's Miscellany *of 1709, was his four* Pastorals *(one for each season), of which the first, "Spring," begins with verses that localize this classical genre in his own familiar surroundings:*

First in these fields I try the sylvan strains,
Nor blush to sport on Windsor's blissful plains . . .

[120]In *Standing by Words* (1983). My thanks to Scott McLean for first pointing out this essay to me.

Such poetry, Pope affirms in his "Discourse on Pastoral Poetry," must portray nature in idealized form: "If we would copy Nature, it may be useful to take this idea along with us, that pastoral is an image of what they call the Golden Age. So that we are not to describe our shepherds as shepherds at this day really are, but as they may be conceived then to have been, when the best of men followed the employment. . . . We must therefore use some illusion to render a pastoral delightful; and this consists in exposing the best side only of a shepherd's life and in concealing its miseries." Theocritus himself (Chapter 11 above) is faulted for depicting his swains as "sometimes abusive and immodest and perhaps too much inclining to rusticity."

Thus the tendency—inherited from ancient usage, especially by the Stoics—to conceive of nature in ideally normative terms (as when Dryden, in "An Essay of Dramatic Poesy," declared that a play, "to be like Nature, is to be set above it") is deeply embedded in Pope's mentality. In An Essay on Criticism (1711), "unerring Nature" is identified with "rules of old," exemplified by Homer and codified by Aristotle, which "are Nature still, but Nature methodized." Nature, Reuben Brower remarks in Alexander Pope: The Poetry of Allusion (1959), is here—as in the Essay on Man—"the personification of the cosmological order, at once the nature of Newton and the natural philosophers and the nature of traditional Platonic cosmology, the scale of created beings in which 'all things' have their assigned rank and powers. . . . The work of art draws its material from nature, and is organized by a similar ordering principle."

By now the young forest prodigy was moving into the orbit of the town, the London literary circles, first of Addison (Chapter 20 below), then of Swift, Gay, and Arbuthnot. In his mock-heroic tour de force, The Rape of the Lock (1712; rev. 1714, 1717), he displayed a total command of "Augustan" wit and urbanity, tinged by rueful sympathy with human foible. The major effort of these years, however, went into his translations of the Iliad (six volumes, 1715–20) and, with assistants, the Odyssey (five volumes, 1725–26); the adaptation of "The Gardens of Alcinous" below was published separately in the Guardian in 1713. The Homeric epics are not, of course, "nature poems." Yet as Douglas Knight contends in "Translation: The Augustan Mode,"[121] one of the "chief means of sustaining coherence and immediacy in his version of Homer is Pope's attention to the life of the natural forces around and beyond man," above all in the similes: "The poetry insists that such a world is one we inhabit, and so the gap between the Homeric figures and ourselves is diminished. They and we are included in one concourse of natural forces."

In 1713 Pope published Windsor Forest, begun in 1704, partly as a descriptive or topographical poem—a genre going back to Ausonius's The Moselle (Chapter 14 above)—on the model of John Denham's then-famous Cooper's Hill (1642), partly as a patriotic celebration of the recently concluded Peace of Utrecht ending the long War of the Spanish Succession between England and France. Yet surely "what is most distinctively Pope's . . . for the modern reader," as Dobrée suggests, "is the delight to be obtained from the vivid descriptions of nature—the chequered shadows, the brilliance of fish, the gorgeous colors of the pheasant with his 'purple crest' and 'painted wings.'" Pope's great model here is Virgil's Georgics (Chapter 11 above), but the tone is strikingly different. "The ancient poet gives a keener sense than the modern," Brower observes, "of closeness to unspoiled nature . . . When Pope longs for 'sequestered scenes,' 'bowery mazes,' and 'surrounding greens,' we see that nature had indeed become a garden in a sense unknown to the Greeks and Romans," a neatly ordered enclosure within which "Virgil's spontaneous natural religion . . . is constantly in danger of being over-shadowed by Nature's mighty plan."

Gardening was a central love of Pope's, especially after he moved with his mother to Twickenham in about 1718, but for Pope the garden, though shaped by art, could never be opposed to nature. "I believe it no wrong observation," he wrote in an essay "On Verdant Sculpture" printed in The Guardian in 1713, "that persons of genius, and those who are most capable of art, are

[121]In Reuben Brower, ed., *On Translation* (1959).

always most fond of nature, as such are chiefly sensible that all art consists in the imitation and study of nature." His motto, as he writes in the fourth of his Moral Essays, *the "Epistle to Burlington" of 1731 (line 50), remained "In all, let Nature never be forgot."*

His increasing disillusionment with the petty squabbles of literary London now found vent in the satires that dominate much of his later poetry, notably the increasingly savage Dunciad *(1728–43). In* An Essay on Man, *however, written in four epistles to Henry St. John, Viscount Bolingbroke, between 1730 and 1732 and published 1733–34, Pope expressed an optimistic, almost Panglossian philosophy—"All are but parts of one stupendous whole, / Whose body Nature is, and God the soul"; "Cease then, nor Order imperfection name"; "All nature is but art, unknown to thee"; "Whatever* IS, *is* RIGHT*"—that is the obverse of his darker satirical vision. It is easy to concur with Samuel Johnson's harsh judgment, in his* Life of Pope, *of this tissue of commonplaces as a poem in which Pope "was in haste to teach what he had not learned": "Having exalted himself into the chair of wisdom, he tells us much that every man knows, and much that he does not know himself. . . . Never were penury of knowledge and vulgarity of sentiment so happily disguised." But as an exemplification of Pope's exiguous conception of wit as "What oft was thought, but ne'er so well expressed," the poem would find admirers from Voltaire to Kant. It weaves together, in the quasi-deistic language of the Newtonian age, Platonic and Christian beliefs in the Great Chain of Being—which "is neither mechanical nor rationalistic," Berry writes, "but at once organic and sacred"—with vivid portrayal of the Lucretian state of nature (Chapter 10 above) as updated by Locke (Chapter 20 below).*

In accord with such views, Pope's intended two-line epitaph for Sir Isaac Newton in Westminster Abbey, composed in 1730, counters the pessimism of the Dunciad*'s Universal Darkness with a no less intensely held faith in the power of enlightened reason to disperse, however momentarily, the shadows that surround and threaten to engulf man:*

Nature, and Nature's laws lay hid in night:
God said, Let Newton be! and all was light.

Apart from such moments of illumination, the Essay on Man *declares (II.10), man hangs irresolutely between God and beast, "born but to die, and reas'ning but to err."*

The text of the following selections is based in part on The Poems of Alexander Pope *(1963), a one-volume edition of the text of the Twickenham Edition, with annotations by John Butt.*

Ode on Solitude[122]

Happy the man, whose wish and care
A few paternal acres bound,
Content to breathe his native air,
In his own ground.

Whose herds with milk, whose fields with bread, 5
Whose flocks supply him with attire,
Whose trees in summer yield him shade,
In winter fire.

Blest! who can unconcern'dly find
Hours, days, and years slide soft away, 10
In health of body, peace of mind,
Quiet by day,

[122]Written ca. 1700, published in *Poems on Several Occasions* (1717). Pope stated that this poem was "written at about twelve years old"; but the earliest extant draft dates from 1709. (Butt)

Sound sleep by night; study and ease
Together mixed; sweet recreation,
And innocence, which most does please, 15
With meditation.

Thus let me live, unseen, unknown;
Thus unlamented let me die;
Steal from the world, and not a stone
Tell where I lie. 20

FROM An Essay on Criticism

First follow Nature, and your judgment frame
By her just standard, which is still the same:
Unerring Nature, still divinely bright, 70
One clear, unchanged, and universal light,
Life, force, and beauty, must to all impart,
At once the source, and end, and test of art. . . .

Those rules of old discovered, not devised,
Are Nature still, but Nature methodized;
Nature, like liberty, is but restrained 90
By the same laws which first herself ordained. . . .

When first young Maro[123] in his boundless mind 130
A work t'outlast immortal Rome designed,
Perhaps he seemed above the critic's law,
And but from Nature's fountain seemed to draw:
But when t'examine every part he came,
Nature and Homer were, he found, the same. 135
Convinced, amazed, he checks the bold design,
And rules as strict his labored work confine
As if the Stagirite[124] o'erlooked each line.
Learn hence for ancient rules a just esteem;
To copy Nature is to copy them. 140

The Gardens of Alcinous[125]

FROM THE SEVENTH BOOK OF HOMER'S ODYSSES

Close to the gates a spacious garden lies,
From storms defended, and inclement skies:
Four acres was th'allotted space of ground,
Fenced with a green enclosure all around.
Tall thriving trees confessed the fruitful mold; 5
The red'ning apple ripens here to gold,
Here the blue fig with luscious juice o'erflows,

[123]Virgil (Publius Vergilius Maro). [124]Aristotle.
[125]See the translation in Chapter 8 above of the passage (*Odyssey* VII.112–35) on which Pope's adaptation is based.

With deeper red the full pomegranate glows,
The branch here bends beneath the weighty pear,
And verdant olives flourish round the year. 10
The balmy spirit of the western gale
Eternal breathes on fruits untaught to fail:
Each dropping pear a following pear supplies,
On apples apples, figs on figs arise:
The same mild season gives the blooms to blow, 15
The buds to harden, and the fruits to grow.
Here ordered vines in equal ranks appear
With all th'united labors of the year,
Some to unload the fertile branches run,
Some dry the black'ning clusters in the sun, 20
Others to tread the liquid harvest join,
The groaning presses foam with floods of wine.
Here are the vines in early flow'r descried,
Here grapes discolored on the sunny side,
And there in autumn's richest purple dyed. 25
Beds of all various herbs, for ever green,
In beauteous order terminate the scene.
Two plenteous fountains the whole prospect crowned;
This through the gardens leads its streams around,
Visits each plant, and waters all the ground: 30
While that in pipes beneath the palace flows,
And thence its current on the town bestows;
To various use their various streams they bring,
The people one, and one supplies the king.

FROM Windsor Forest

To the Right Honorable George [Granville] Lord Lansdown

Thy forests, Windsor! and thy green retreats,
At once the Monarch's and the Muse's seats,
Invite my lays. Be present, Sylvan Maids!
Unlock your springs, and open all your shades.
Granville commands: your aid, O Muses, bring! 5
What Muse for Granville can refuse to sing?
The groves of Eden, vanished now so long,
Live in description, and look green in song:
These, were my breast inspired with equal flame,
Like them in beauty, should be like in fame. 10
Here hills and vales, the woodland and the plain,
Here earth and water seem to strive again;
Not chaos-like together crushed and bruised,
But, as the world, harmoniously confused:
Where order in variety we see, 15
And where, though all things differ, all agree.

Here waving groves a chequered scene display,
And part admit, and part exclude the day;
As some coy nymph her lover's warm address
Nor quite indulges, nor can quite repress. 20
There, interspersed in lawns and op'ning glades,
Thin trees arise that shun each other's shades.
Here in full light the russet plains extend:
There wrapped in clouds the bluish hills ascend.
Ev'n the wild heath displays her purple dyes, 25
And 'midst the desert fruitful fields arise,
That crowned with tufted trees and springing corn,
Like verdant isles the sable waste adorn.
Let India boast her plants, nor envy we
The weeping amber or the balmy tree, 30
While by our oaks the precious loads are borne,
And realms commanded which those trees adorn.
Not proud Olympus yields a nobler sight,
Though gods assembled grace his tow'ring height,
Than what more humble mountains offer here, 35
Where, in their blessings, all those gods appear.
See Pan with flocks, with fruits Pomona[126] crowned,
Here blushing Flora paints th'enameled ground,
Here Ceres' gifts in waving prospect stand,
And nodding tempt the joyful reaper's hand. 40
Rich Industry sits smiling on the plains,
And peace and plenty tell, a Stuart reigns. . . .[127]

See! from the brake the whirring pheasant springs,
And mounts exulting on triumphant wings:
Short is his joy; he feels the fiery wound,
Flutters in blood, and panting beats the ground.
Ah! what avail his glossy, varying dyes, 115
His purple crest, and scarlet-circled eyes,
The vivid green his shining plumes unfold,
His painted wings, and breast that flames with gold? . . .

In genial[128] spring, beneath the quiv'ring shade, 135
Where cooling vapors breathe along the mead,
The patient fisher takes his silent stand,
Intent, his angle trembling in his hand:
With looks unmoved, he hopes the scaly breed,
And eyes the dancing cork, and bending reed. 140
Our plenteous streams a various race supply,
The bright-eyed perch with fins of Tyrian dye,

[126]The Roman goddess of fruit and fruit trees (from Latin *pomum*, "fruit"). Flora (from Latin *flos*, "flower") was the Roman goddess of flowers and of the spring; Ceres, the Roman goddess of the grain.
[127]The poem was written and published during the reign of Queen Anne (1702–14), daughter of James II, and last English monarch of the Stuart line.
[128]"Generative."

The silver eel, in shining volumes rolled,
The yellow carp, in scales bedropped of gold,
Swift trouts, diversified with crimson stains, 145
And pikes, the tyrants of the wat'ry plains. . . .

Here too, 'tis sung, of old Diana strayed, 165
And Cynthus' top[129] forsook for Windsor shade;
Here was she seen o'er airy wastes to rove,
Seek the clear spring, or haunt the pathless grove;
Here armed with silver bows, in early dawn,
Her buskined virgins traced the dewy lawn. . . . 170

Happy the man whom this bright court approves, 235
His sov'reign favors, and his country loves:
Happy next him, who to these shades retires,[130]
Whom nature charms, and whom the Muse inspires;
Whom humbler joys of home-felt quiet please,
Successive study, exercise, and ease. 240
He gathers health from herbs the forest yields,
And of their fragrant physic spoils the fields:
With chemic art exalts[131] the min'ral pow'rs,
And draws the aromatic souls of flow'rs:
Now marks the course of rolling orbs on high; 245
O'er figured worlds now travels with his eye;
Of ancient writ unlocks the learned store,
Consults the dead, and lives past ages o'er:
Or wand'ring thoughtful in the silent wood,
Attends the duties of the wise and good, 250
T'observe a mean, be to himself a friend,
To follow nature, and regard his end;
Or looks on Heav'n with more than mortal eyes,
Bids his free soul expatiate in the skies,
Amid her kindred stars familiar roam, 255
Survey the region, and confess her home!
Such was the life great Scipio once admired,[132]
Thus Atticus, and Trumbull thus retired.[133]
Ye sacred Nine![134] that all my soul possess,
Whose raptures fire me, and whose visions bless, 260

[129]Cynthus was a mountain on the island of Delos, on which Apollo and Artemis (Roman Diana) were said to have been born; from it, the goddess was also known as Cynthia.
[130]Cf. Virgil's *Georgics* II.490 ff. in Chapter 11 above.
[131]In alchemy and early chemistry: to raise (a substance or its qualities) to a higher "degree"; to intensify, render more powerful. (Butt)
[132]After his victory over Hannibal in the second Punic war, Scipio Africanus declined political distinctions offered him. When, many years later, his enemies brought him to trial on charges of misconduct, he reminded the Romans of his past services, set the laws at defiance, and retired to his country seat at Liternum. He never returned to Rome, but passed his life cultivating his estate. (Butt)
[133]Cicero's friend and correspondent, the Epicurean Pomponius Atticus, held aloof from politics and devoted himself to literary and commercial pursuits in Athens (whence his name "Atticus") and on his estate at Buthrotum in Epirus. Pope's mentor Sir William Trumbull had an estate at Easthamstead Park near the young Pope's home at Binfield in Windsor Forest.
[134]The nine Muses.

Bear me, oh bear me to sequestered scenes,
The bow'ry mazes, and surrounding greens:
To Thames's banks which fragrant breezes fill,
Or where ye Muses sport on Cooper's Hill.
(On Cooper's Hill eternal wreaths shall grow, 265
While lasts the mountain, or while Thames shall flow.)
I seem through consecrated walks to rove,
I hear soft music die along the grove:
Led by the sound, I roam from shade to shade,
By godlike poets venerable made: 270
Here his first lays majestic Denham sung;[135]
There the last numbers flowed from Cowley's tongue.[136]
Oh early lost! what tears the river shed,
When the sad pomp along his banks was led!
His drooping swans on ev'ry note expire, 275
And on his willows hung each Muse's lyre. . . .

My humble Muse, in unambitious strains,
Paints the green forests and the flow'ry plains,
Where Peace descending bids her olives spring,
And scatters blessings from her dove-like wing. 430
Ev'n I more sweetly pass my careless days,
Pleased in the silent shade with empty praise;
Enough for me, that to the list'ning swains
First in these fields I sung the sylvan strains.[137]

FROM An Essay on Man

From Epistle I: OF THE NATURE AND STATE OF MAN WITH RESPECT TO THE UNIVERSE

V. Ask for what end the heav'nly bodies shine,
Earth for whose use? Pride answers, " 'Tis for mine:
For me kind Nature wakes her genial pow'r,
Suckles each herb, and spreads out ev'ry flow'r;
Annual for me, the grape, the rose, renew, 135
The juice nectareous, and the balmy dew;
For me, the mine a thousand treasures brings;
For me, health gushes from a thousand springs;
Seas roll to waft me, suns to light me rise;
My footstool earth, my canopy the skies." 140

[135]John Denham (1615–69) was the author of the topographical poem in heroic couplets, *Cooper's Hill*, published in 1642 and much admired for its neoclassical "correctness" by Dryden, Pope, and Johnson. Denham aspired, in famous lines (191–92), to be "Though deep, yet clear, though gentle, yet not dull, / Strong without rage, without o'er-flowing full."
[136]Mr. [Abraham] Cowley died [in 1667, at the age of 49] at Chertsey, on the borders of the Forest, and was from thence conveyed to Westminster. (Pope)
[137]Pope's conclusion was modeled on that of the *Georgics*. As Virgil closed his *Georgics* with the first line of his *Eclogues*, so Pope's final couplet echoes the opening line of *Spring*. (Butt) See introduction to Pope, above.

But errs[138] not Nature from this gracious end,
From burning suns when livid deaths descend,
When earthquakes swallow, or when tempests sweep
Towns to one grave, whole nations to the deep?
"No," 'tis replied, "the first Almighty Cause 145
Acts not by partial, but by gen'ral laws;
Th'exceptions few; some change since all began:
And what created perfect?"—Why then man?
If the great end be human happiness,
Then Nature deviates; and can man do less? 150
As much that end a constant course requires
Of show'rs and sunshine, as of man's desires;
As much eternal springs and cloudless skies,
As men forever temp'rate, calm, and wise.
If plagues or earthquakes break not Heav'n's design, 155
Why then a Borgia, or a Catiline?[139]
Who knows but He whose hand the lightning forms,
Who heaves old ocean, and who wings the storms;
Pours fierce ambition in a Caesar's mind,
Or turns young Ammon[140] loose to scourge mankind? 160
From pride, from pride, our very reas'ning springs;
Account for moral, as for nat'ral things:
Why charge we Heav'n in those, in these acquit?
In both, to reason right is to submit.
Better for us, perhaps, it might appear, 165
Were all there harmony, all virtue here;
That never air or ocean felt the wind;
That never passion discomposed the mind.
But All subsists by elemental strife;
And passions are the element of life. 170
The gen'ral Order, since the whole began,
Is kept in Nature, and is kept in man. . . .
VIII. See, through this air, this ocean, and this earth,
All matter quick, and bursting into birth.
Above, how high progressive life may go! 235
Around, how wide! how deep extend below!
Vast chain of Being! which from God began,
Natures ethereal, human, angel, man,
Beast, bird, fish, insect, what no eye can see,
No glass can reach; from Infinite to thee, 240
From thee to Nothing.—On superior pow'rs
Were we to press, inferior might be ours:
Or in the full creation leave a void,

[138]"Strays."
[139]Cesare Borgia (1476–1507) was famed for his cruelty; Catiline (Lucius Sergius Catilina) in 63 B.C. led a conspiracy, thwarted by Cicero as consul, to overthrow the Roman Republic.
[140]Alexander the Great, who after his conquest of Egypt from the Persians was hailed by the priests of the Egyptian god Ammon (whom the Greeks identified with Zeus) as the god's son.

Where, one step broken, the great scale's destroyed:
From Nature's chain whatever link you strike, 245
Tenth or ten thousandth, breaks the chain alike.
And, if each system in gradation roll
Alike essential to th'amazing whole,
The least confusion but in one, not all
That system only, but the whole must fall. 250
Let earth unbalanced from her orbit fly,
Planets and suns run lawless through the sky;
Let ruling angels from their spheres be hurled,
Being on being wrecked, and world on world;
Heav'n's whole foundations to their center nod, 255
And Nature tremble to the throne of God.
All this dread Order break—for whom? for thee?
Vile worm!—oh madness! pride! impiety!
IX. What if the foot, ordained the dust to tread,
Or hand, to toil, aspired to be the head? 260
What if the head, the eye, or ear repined
To serve mere engines to the ruling Mind?
Just as absurd for any part to claim
To be another, in this gen'ral frame:
Just as absurd, to mourn the tasks or pains 265
The great directing Mind of All ordains.
All are but parts of one stupendous whole,
Whose body Nature is, and God the soul;
That, changed through all, and yet in all the same,
Great in the earth, as in th'ethereal frame, 270
Warms in the sun, refreshes in the breeze,
Glows in the stars, and blossoms in the trees,
Lives through all life, extends through all extent,
Spreads undivided, operates unspent;
Breathes in our soul, informs our mortal part, 275
As full, as perfect, in a hair as heart;
As full, as perfect, in vile man that mourns,
As the rapt seraph that adores and burns:
To him no high, no low, no great, no small;
He fills, he bounds, connects, and equals all. 280
X. Cease then, nor Order imperfection name:
Our proper bliss depends on what we blame.
Know thy own point: this kind, this due degree
Of blindness, weakness, Heav'n bestows on thee.
Submit.—In this, or any other sphere, 285
Secure to be as blest as thou canst bear:
Safe in the hand of one disposing Pow'r,
Or in the natal, or the mortal hour.
All nature is but art, unknown to thee;
All chance, direction, which thou canst not see; 290
All discord, harmony not understood;

All partial evil, universal good;
And, spite of pride, in erring reason's spite,
One truth is clear, Whatever IS, is RIGHT.

From Epistle III: OF THE NATURE AND STATE OF MAN WITH RESPECT TO SOCIETY

IV. Nor think, in Nature's State they blindly trod;
The state of Nature was the reign of God:
Self-love and social at her birth began,
Union the bond of all things, and of man.[141] 150
Pride then was not; nor arts, that pride to aid;
Man walked with beast, joint tenant of the shade;
The same his table, and the same his bed;
No murder clothed him, and no murder fed.
In the same temple, the resounding wood, 155
All vocal beings hymned their equal God:
The shrine with gore unstained, with gold undressed,
Unbribed, unbloody, stood the blameless priest:
Heav'n's attribute was universal care,
And man's prerogative to rule, but spare. 160
Ah! how unlike the man of times to come!
Of half that live the butcher and the tomb;
Who, foe to Nature, hears the gen'ral groan,
Murders their species, and betrays his own.
But just disease to luxury succeeds, 165
And ev'ry death its own avenger breeds;
The Fury-passions from that blood began,
And turn'd on man a fiercer savage, man.
See him from Nature rising slow to Art!
To copy instinct then was reason's part; 170
Thus then to man the voice of Nature spake—
"Go, from the creatures thy instructions take:
Learn from the birds what food the thickets yield;
Learn from the beasts the physic of the field;
Thy arts of building from the bee receive; 175
Learn of the mole to plow, the worm to weave;
Learn of the little nautilus to sail,
Spread the thin oar, and catch the driving gale.[142]
Here too all forms of social union find,
And hence let reason, late, instruct mankind: 180
Here subterranean works and cities see;

[141]I.e., the state of nature was not a state of war, like Hobbes's, and not without society and law, like Lucretius's, but much more like Locke's, which "approximates . . . the Golden Age of the Poets." (Butt)

[142]Oppian, [in his late-second-century-A.D. Greek poem on fishing, the] Halieut[ica], Lib. I, describes this fish in the following manner. They swim on the surface of the sea, on the back of their shells, which exactly resemble the hulk of a ship; they raise two feet like masts, and extend a membrane between which serves as a sail; the other two feet they employ as oars at the side. They are usually seen in the Mediterranean. (Pope)

There towns aërial on the waving tree.
Learn each small people's genius, policies,
The ant's republic, and the realm of bees;
How those in common all their wealth bestow, 185
And anarchy without confusion know;
And these forever, though a monarch reign,
Their sep'rate cells and properties maintain.
Mark what unvaried laws preserve each state,
Laws wise as Nature, and as fixed as Fate. . . ." 190
V. Great Nature spoke; observant men obeyed;
Cities were built, societies were made: 200
Here rose one little state; another near
Grew by like means, and joined, through love or fear.
Did here the trees with ruddier burdens bend,
And there the streams in purer rills descend?
What war could ravish, commerce could bestow, 205
And he returned a friend, who came a foe.
Converse and love mankind might strongly draw,
When love was liberty, and Nature law. . . .

The Raptured Eye

The Seasons ("Spring"), by James Thomson

Nature, as a paramount theme and concern of eighteenth-century sensibility, was of central importance to Pope, the century's foremost poet; yet no one would think of Pope's poetry, as a whole, as "nature poetry." More than anyone else, it was the Scottish poet James Thomson (1700–1748) who in The Seasons *invented that genre by expanding on the descriptive dimensions of such poems as Denham's* Cooper's Hill *and Pope's* Windsor Forest *and praising not the beauties of a particular landscape but those of nature in general, permeated by the "unremitting energy" of a God who "agitates the whole."*

Born in the Scottish lowlands, Thomson attended the University of Edinburgh, intending to pursue the vocation of his father, a Presbyterian minister. After publishing a few poems between 1715 and 1725, he went instead to London as a tutor, and was introduced to the circle of Pope, who remained a friend in later life. In 1726 he published Winter, *followed by* Summer *in 1727,* Spring *in 1728, and* The Seasons *in 1730 (incorporating the earlier three, plus "Autumn" and "A Hymn"). Each seasonal poem continually grew; "Winter," for example, swelled from its original 409 lines to 1069 in the revised* Seasons *of 1740. An ardent Whig but a zealous opponent of Robert Walpole, the powerful Whig minister of George I and George II, Thomson wrote* Britannia *(1729) as an attack on Walpole's policy of peace with France, and* Liberty *(1734–36) as an account of the migration of the Goddess Liberty from her ancient abodes to Great Britain. These chauvinistic poems have long been consigned to oblivion (of* Liberty *Samuel Johnson wrote, in his "Life of Thomson," that "her praises were condemned to harbor spiders, and to gather dust").*

Apart from his patriotic hymn "Rule Britannia" (1740), and perhaps "A Poem Sacred to the Memory of Sir Isaac Newton" (1727) and the pseudo-Spenserian Castle of Indolence *(1748), it is* The Seasons *for which Thomson is remembered, when remembered at all. "The purpose as it seems to have formulated itself gradually in Thomson's mind," Dobrée writes, "was to show the workings of Creative Nature: that, together with the framework of the progression of the year, is what holds the poem together at all, coherence being given by a certain inner rhythm of movement from, in each morsel, the material to the spiritual." Hence description exists not for its own sake, but for the sake of the philosophical and moral arguments to which it gives rise, "since it was only through philosophy that nature could mean anything; nature and philosophy were the same." "His glorification of nature as the embodiment of the divine idea," Cecil A. Moore observes in his edition of* English Poetry of the Eighteenth Century *(1935)—which contains the full 1740 texts of "Spring," "Winter," and "A Hymn"—"is only the poetical adaptation of the Religion of Nature as expounded by various deistic writers, but most engagingly by the third Earl of Shaftesbury [Chapter 20 below] in* The Moralists *(1709). With this reverential feeling for outward and visible nature is associated a belief that human nature is essentially good and that man is instinctively benevolent," despite the corruption both have undergone since the Golden Age, the primordial state of nature when "uncorrupted man" yet "lived in innocence," and "reason and benevolence were law."*

Yet already, in occasional moments of "philosophic melancholy," Thomson also gives voice to a more regretful attitude that will later find expression in Oliver Goldsmith's nostalgic lament, in The Deserted Village *(1770), for "Sweet Auburn, loveliest village of the plain," now deserted by the "rural virtues" of yore:*

Thus fares the land, by luxury betrayed;
In nature's simplest charms at first arrayed;
But verging to decline, its splendors rise,
Its vistas strike, its palaces surprise,
While scourged by famine from the smiling land,
The mournful peasant leads his humble band;
And while he sinks without one arm to save,
The country blooms—a garden, and a grave.

(297–304)

In Thomson's poem, as Raymond Williams remarks in The Country and the City *(1973), "Nature, represented hitherto as a social order, a triumph of law and plenty, is being seen, alternatively, as a substitute order; lonely, prophetic, bearing the love of humankind in just those places where men are not. . . . It will take half a century for this change to work itself fully through, but Thomson is especially interesting because, in* The Seasons, *both versions of Nature, both attitudes to the country and the land, are simultaneously present."*

By deliberately elevating nature to the central theme of serious poetry, Thomson had an immense impact not only on poets inspired to imitate him but also on writers as different in temper as Samuel Johnson. For Johnson—despite reservations about Thomson's "want of method" and "florid and luxuriant" diction ("His fault," he says in Boswell's Life, *"is such a cloud of words sometimes, that the sense can hardly peep through")—Thomson's "descriptions of extended scenes and general effects bring before us the whole magnificence of Nature, whether pleasing or dreadful. . . . The poet leads us through the appearance of things as they are successively varied by the vicissitudes of the year, and imparts to us so much of his own enthusiasm, that our thoughts expand with his imagery, and kindle with his sentiments."*

Judgments of modern readers are likely to be far less favorable, in part because we have greater distaste for the mawkish sentimentality that bids a fisherman

let not on thy hook the tortured worm,
Convulsive, twist in agonizing folds,

("Spring," 388–89)

and for a stilted poetic diction (in which fish are the "finny race") crammed with Miltonic Latinisms and punctuated by rhetorical questions ("But what is this?") and often facile apostrophes to grand abstractions ending in breathless outbursts or imperative verbs, with multiple exclamation points—

Nature! great parent! whose unceasing hand
Rolls round the seasons of the changeful year,
How mighty, how majestic, are thy works!
("Winter," 106–8)

Father of light and life! thou Good Supreme!
O teach me what is good! teach me thyself!
("Winter," 217–18)

Modern dissatisfaction is also due in part, no doubt, to contrast between Thomson and the far subtler nature poetry of Wordsworth and Coleridge that has followed. As Dobrée remarks, disaster befalls Thomson when he assumes "the mantle of the theosophic-rationalist preacher. . . . Thomson, to put it plainly, was no thinker, and he tried to do far too much. There is no driving force in the poem to give it direction such as informs Pope's at least equally complex Essay on Man, *no brilliant mind, no formative energy as we can feel there fusing the incompatible elements. . . . He is so charmingly eager in his discoveries . . . that he has to tell us everything, and the man who tells us everything becomes a bore." But if the poem is indeed "readable* in toto *only by the curious and the scholar," in places—as in the description of "exuberant Nature" unfolding by degrees in springtime as "the raptured eye / Hurries from joy to joy," even now a reader may briefly kindle with the infectious enthusiasm that Thomson's rhapsodic ardor once imparted to countless readers, decades before Rousseau and generations before Wordsworth.*

Come, gentle Spring—ethereal mildness, come;
And from the bosom of yon dropping cloud,
While music wakes around, veiled in a shower
Of shadowing roses, on our plains descend. . . . 4

Be gracious, Heaven! for now laborious man
Has done his part. Ye fostering breezes, blow!
Ye softening dews, ye tender showers, descend! 50
And temper all, thou world-reviving sun,
Into the perfect year! Nor ye who live
In luxury and ease, in pomp and pride,
Think these lost themes unworthy of your ear:
Such themes as these the rural Maro sung[143] 55
To wide-imperial Rome, in the full height
Of elegance and taste, by Greece refined.
In ancient times the sacred plow employed
The kings and awful fathers of mankind;
And some, with whom compared your insect-tribes 60
Are but the beings of a summer's day,
Have held the scale of empire, ruled the storm
Of mighty war; then, with victorious hand,
Disdaining little delicacies, seized

[143]Virgil (Publius Vergilius Maro) in the *Georgics* (see Chapter 11 above).

The plow, and, greatly independent, scorned 65
All the vile stores corruption can bestow.
Ye generous Britons, venerate the plow;
And o'er your hills and long withdrawing vales
Let Autumn spread his treasures to the sun,
Luxuriant and unbounded! As the sea, 70
Far through his azure turbulent domain,
Your empire owns, and from a thousand shores
Wafts all the pomp of life into your ports;
So with superior boon may your rich soil,
Exuberant, Nature's better blessing pour 75
O'er every land, the naked nations clothe,
And be th'exhaustless granary of a world! . . .

From the moist meadow to the withered hill,
Led by the breeze, the vivid verdure runs,
And swells, and deepens, to the cherished eye.
The hawthorn whitens; and the juicy groves 90
Put forth their buds, unfolding by degrees,
Till the whole leafy forest stands displayed,
In full luxuriance, to the sighing gales;
Where the deer rustle through the twining brake,
And the birds sing concealed. At once, arrayed 95
In all the colors of the flushing year
By Nature's swift and secret-working hand,
The garden glows, and fills the liberal air
With lavish fragrance; while the promised fruit
Lies yet a little embryo, unperceived, 100
Within its crimson folds. Now from the town,
Buried in smoke, and sleep, and noisome damps,
Oft let me wander o'er the dewy fields,
Where freshness breathes, and dash the trembling drops
From the bent bush, as through the verdant maze 105
Of sweet-briar hedges I pursue my walk;
Or taste the smell of dairy; or ascend
Some eminence, Augusta,[144] in thy plains,
And see the country, far diffused around,
One boundless blush, one white-empurpled shower 110
Of mingled blossoms; where the raptured eye
Hurries from joy to joy, and, hid beneath
The fair profusion, yellow Autumn spies. . . .

Then spring the lively herbs, profusely wild,
O'er all the deep-green earth, beyond the power
Of botanist to number up their tribes:
Whether he steals along the lonely dale, 225
In silent search; or through the forest, rank

[144]London.

With what the dull incurious weeds account,
Bursts his blind way; or climbs the mountain-rock,
Fired by the nodding verdure of its brow.
With such a liberal hand has Nature flung 230
Their seeds abroad, blown them about in winds,
Innumerous mixed them with the nursing mold,
The moistening current, and prolific rain.
 But who their virtues can declare? Who pierce
With vision pure, into these secret stores 235
Of health, and life, and joy? The food of man,
While yet he lived in innocence, and told
A length of golden years, unfleshed in blood,
A stranger to the savage arts of life,
Death, rapine, carnage, surfeit, and disease— 240
The lord, and not the tyrant, of the world.
 The first fresh dawn then waked the gladdened race
Of uncorrupted man, nor blushed to see
The sluggard sleep beneath its sacred beam;
For their light slumbers gently fumed away, 245
And up they rose as vigorous as the sun,
Or to the culture of the willing glebe,
Or to the cheerful tendance of the flock.
Meantime the song went round; and dance and sport,
Wisdom and friendly talk, successive stole 250
Their hours away. While in the rosy vale
Love breathed his infant sighs, from anguish free,
And full replete with bliss; save the sweet pain,
That, inly thrilling, but exalts it more.
Nor yet injurious act, nor surly deed, 255
Was known among these happy sons of heaven;
For reason and benevolence were law.
Harmonious Nature too looked smiling on.
Clear shone the skies, cooled with eternal gales,
And balmy spirit all. The youthful sun 260
Shot his best rays, and still the gracious clouds
Dropped fatness down; as o'er the swelling mead
The herds and flocks, commixing, played secure.
This when, emergent from the gloomy wood,
The glaring lion saw, his horrid heart 265
Was meekened, and he joined his sullen joy.
For music held the whole in perfect peace:
Soft sighed the flute; the tender voice was heard,
Warbling the varied heart; the woodlands round
Applied their choir; and winds and waters flowed 270
In consonance. Such were those prime of days.
 But now those white unblemished minutes, whence
The fabling poets took their golden age,
Are found no more amid these iron times—

These dregs of life! Now the distempered mind 275
Has lost that concord of harmonious powers,
Which forms the soul of happiness; and all
Is off the poise within: the passions all
Have burst their bounds; and reason half extinct,
Or impotent, or else approving, sees 280
The foul disorder. Senseless and deformed,
Convulsive anger storms at large; or, pale
And silent, settles into fell revenge,
Base envy withers at another's joy,
And hates that excellence it cannot reach. . . . 285
At last, extinct each social feeling, fell 305
And joyless inhumanity pervades
And petrifies the heart. Nature disturbed
Is deemed, vindictive, to have changed her course. . . .

What is this mighty breath, ye curious say,
That, in a powerful language, felt not heard 850
Instructs the fowls of heaven; and through their breast
These arts of love diffuses? What, but God?
Inspiring God! who, boundless spirit all,
And unremitting energy, pervades,
Adjusts, sustains, and agitates the whole. 855
He, ceaseless, works alone, and yet alone
Seems not to work; with such perfection framed
Is this complex, stupendous scheme of things.
But, though concealed, to every purer eye
Th'informing Author in his works appears: 860
Chief, lovely Spring, in thee, and thy soft scenes,
The smiling God is seen; while water, earth,
And air attest his bounty; which exalts
The brute creation to this finer thought,
And, annual, melts their undesigning hearts 865
Profusely thus in tenderness and joy. . . .

Praise Above All

Lyric and Reflective Poetry of the Eighteenth Century

Wordsworth, looking back on the early eighteenth century in his "Essay Supplementary to the Preface of 1815," remarked that "excepting the nocturnal Reverie of Lady Winchelsea, and a passage or two in the 'Windsor Forest' of Pope, the poetry of the period intervening between the publication of 'Paradise Lost' and the 'Seasons' does not contain a single new image of external nature; and scarcely presents a familiar one from which it can be inferred that the eye of the Poet had been steadily fixed upon his object, much less that his feelings had urged him to work upon it in the spirit of genuine imagination." Though other exceptions might certainly be found, the

Restoration and Augustan period, until Thomson, was an age when concrete responsiveness to nature was indeed a rarity among poets.

Anne Finch, Countess of Winchelsea *(1661–1720), born Anne Kingsmill, was a maid of honor to Mary of Modena, Duchess of York and later queen of James II. She belonged to a courtly circle of women poets that included, until their deaths from smallpox in 1685, Katherine Philips and Anne Killigrew, in whose memory Dryden wrote a famous ode. In 1684 she married the soldier and courtier Hineage Finch; after James's expulsion in 1688, Finch's relative, the Earl of Winchelsea (to whose title he succeeded in 1712) provided them with a home. In 1701, Anne Finch published her "Pindaric Poem"* The Spleen, *and in 1713, eighty-one* Miscellany Poems on Several Occasions, Written by a Lady. *These include, besides the "Nocturnal Reverie" admired by Wordsworth, with its evocation of moonlit banks*

> Whence springs the woodbine, and the bramble-rose,
> And where the sleepy cowslip sheltered grows,

a "Petition for an Absolute Retreat" (to be shared with "a partner suited to my mind," and a woman friend), and "To the Nightingale," given below, in which the poet identifies with the bird's wild and unconfined freedom.

After Thomson's Seasons, *many poets of the mid to late eighteenth century—even before Burns and Blake, Wordsworth and Coleridge, in its final decades, inaugurated new poetic styles retrospectively grouped together as Romantic—increasingly turned to the natural world for their themes. Thus William Collins, in his "Ode to Evening" of 1747:*

> Now air is hushed, save where the weak-eyed bat
> With short, shrill shriek, flits by on leathern wing,
> Or where the beetle winds
> His small but sullen horn.

And William Cowper, in The Task *of 1785, proclaimed:*

> God made the country, and man made the town.
> What wonder then that health and virtue, gifts
> That can alone make sweet the bitter draught
> That life holds out to all, should most abound
> And least be threatened in the fields and groves?
> (I.749–53)

In different ways, each of these poets, beginning with Thomson, appears to be striving, with greater or lesser success, to discover or invent poetic styles that will enable them to portray relations between the human and natural worlds more adequately than dominant Restoration and Augustan modes permitted.

In none was this endeavor more self-conscious than in ***Joseph Warton*** *(1722–1800)—whose brother, Thomas Warton, Jr. (1728–90), also wrote poetry and played a key role in reviving interest in Spenser. Their father, Thomas Warton, Sr. (1688–1745), Pope's contemporary and professor of poetry at Oxford from 1728 to 1738, had given an early indication of changes in eighteenth-century taste when he translated a Northern "Runic" poem; in "Retirement: An Ode," he wrote: "Lulled by the lapse of yonder spring, / Of Nature's various charms I sing . . ." But it was Joseph who most influentially challenged what he saw as the artificial correctness of the neoclassical style both in his poems and in his criticism of Pope as a poet consigned by deficiency of imagination to the second rank. Educated at Winchester (whose head master he later became) and at Oriel College, Oxford, from which he received his B.A. in 1744, he published in that year* The Enthusiast; or, the Lover of Nature, *a blank verse poem of 252 lines written four years earlier, when he was eighteen. His description of the Golden Age of "Earth's first infancy" in the passage given below draws heavily on Lucretius and other classical poets, including Virgil (whose*

Eclogues *and* Georgics *he later translated), but also anticipates Romantic sensibility. "In his glorification of the state of nature in contrast with the moral corruptness of civilization," Moore writes, "he foreshadowed Rousseau's central teaching several years before the French sentimentalist had published his initial essay."*

Thomas Gray *(1716–71), the only child of a large family to survive infancy, was sent by his mother, on profits saved from her work as a milliner, to Eton and to Peterhouse, Cambridge, where he remained (except for a change in residence to nearby Pembroke Hall) for the rest of his quiet life. In 1739 he set out on his grand tour to France and Italy with Horace Walpole, his companion from Eton and Cambridge, with whom he quarreled, but was later reconciled. His early "Ode on the Spring," given below, in which direct observation struggles to find expression through "ornately Augustan diction" (in Sherburn's phrase), was written in 1742 and published in 1748. In around 1745–46 he composed his most famous poem, "Elegy Written in a Country Churchyard" (published in 1751), beginning:*

The curfew tolls the knell of parting day,
The lowing herd wind slowly o'er the lea,
The plowman homeward plods his weary way,
And leaves the world to darkness and to me.

Odes by Mr. Gray, *published by Walpole in 1757, gave voice to the new, deliberately obscure and affectedly sublime style of his "Pindaric odes," "The Progress of Poesy" and "The Bard," of which Johnson wrote, in his "Life of Gray": "He has a kind of strutting dignity, and is tall by walking on tiptoe. His art and his struggle are too visible, and there is too little appearance of ease and nature." His last poems, such as "The Fatal Sisters," "The Descent of Odin," and "The Triumphs of Owen," were imitations of Old Norse and Welsh poetry.*

Finally, ***Christopher Smart*** *(1722–71), the son of a steward, attended Pembroke Hall, Cambridge, from which he received his B.A. in 1744 and M.A. in 1747, and of which he was a fellow from 1745 to 1753. His* Poems on Several Occasions *(1752) included a blank verse georgic poem in two books, "The Hop-Garden," and he later made translations of Horace's satires and epistles in both prose and verse. For much of the time between 1756 and 1763 he was confined to hospitals for the insane as a result of his compulsion to pray in public places. ("I do not think he ought to be shut up," Boswell quotes Johnson as saying: "His infirmities were not noxious to society. He insisted on people praying with him; and I'd as lief pray with Kit Smart as any one else. Another charge was, that he did not love clean linen; and I have no passion for it.") After his release in 1763, he published his* Song to David, *a hymn of praise to God in 516 lines, which "in sustained lyric intensity, in bold transitions from the homely to the sublime, in exotic imagery, and in its piercing, mystical piety," Sherburn remarks, ". . . is unique in the century." The stanzas here selected, in their rhapsodic wonder at the glories of God's creation, are reminiscent of the Hebrew psalms (Chapter 3 above), of which Smart made rhymed translations, and of St. Francis's "Praise of Created Things" (Chapter 14). The following year, Smart was confined to debtors' prison, where he died seven years later.*

To the Nightingale, by Anne Finch, Countess of Winchelsea

Exert thy voice, sweet harbinger of spring!
This moment is thy time to sing,
This moment I attend to praise,
And set my numbers to thy lays.
Free as thine shall be my song; 5
As thy music, short, or long.

Poets, wild as thee, were born,
  Pleasing best when unconfined,
  When to please is least designed,
Soothing but their cares to rest; 10
  Cares do still their thoughts molest,
  And still th'unhappy poet's breast,
Like thine, when best he sings, is placed against a thorn.
She begins, let all be still!
  Muse, thy promise now fulfill! 15
Sweet, oh! sweet, still sweeter yet
Can thy words such accents fit,
Canst thou syllables refine,
Melt a sense that shall retain
Still some spirit of the brain, 20
Till with sounds like these it join.
  'Twill not be! then change thy note;
  Let division shake thy throat.
Hark! division now she tries;
Yet as far the Muse outflies. 25
  Cease then, prithee, cease thy tune;
  Trifler, wilt thou sing till June?
Till thy bus'ness all lies waste,
And the time of building's past!
  Thus we poets that have speech, 30
Unlike what thy forests teach,
  If a fluent vein be shown
  That's transcendent to our own,
Criticize, reform, or preach,
Or censure what we cannot reach. 35

FROM The Enthusiast; or, The Lover of Nature, *by Joseph Warton*

  Happy the first of men, ere yet confined
To smoky cities; who in sheltering groves,
Warm caves, and deep-sunk valleys lived and loved,
By cares unwounded; what the sun and showers, 90
And genial earth untillaged, could produce,
They gathered grateful, or the acorn brown,
Or blushing berry; by the liquid lapse
Of murmuring waters called to slake their thirst,
Or with fair nymphs their sun-brown limbs to bathe; 95
With nymphs who fondly clasped their favorite youths,
Unawed by shame, beneath the beechen shade,
Nor wiles, nor artificial coyness knew.
Then doors and walls were not; the melting maid
Nor frown of parents feared, nor husband's threats; 100
Nor had curst gold their tender hearts allured:

Then beauty was not venal. Injured Love,
O! whither, god of raptures, art thou fled? . . .
    In Earth's first infancy (as sung the bard[145]
Who strongly painted what he boldly thought),
Though the fierce north oft smote with iron whip 110
Their shivering limbs, though oft the bristly boar
Or hungry lion woke them with their howls,
And scared them from their moss-grown caves, to rove
Houseless and cold in dark tempestuous nights;
Yet were not myriads in embattled fields 115
Swept off at once, nor had the raging seas
O'erwhelmed the foundering bark and shrieking crew;
In vain the glassy ocean smiled to tempt
The jolly sailor, unsuspecting harm . . .
Yet why should man, mistaken, deem it nobler 135
To dwell in palaces, and high-roofed halls,
Than in God's forests, architect supreme!
Say, is the Persian carpet, than the field's
Or meadow's mantle gay, more richly woven
Or softer to the votaries of ease 140
Than bladed grass, perfumed with dew-dropped flowers?
Oh taste corrupt! that luxury and pomp,
In specious names of polished manners veiled,
Should proudly banish Nature's simple charms!
All beauteous Nature! by thy boundless charms 145
Oppressed, oh where shall I begin thy praise,
Where turn th'ecstatic eye, how ease my breast
That pants with wild astonishment and love!
Dark forest, and the opening lawn, refreshed
With ever-gushing brooks, hill, meadow, dale, 150
The balmy bean-field, the gay-clovered close,
So sweetly interchanged, the lowing ox,
The playful lamb, the distant waterfall
Now faintly heard, now swelling with the breeze,
The sound of pastoral reed from hazel-bower, 155
The choral birds, the neighing steed, that snuffs
His dappled mate, stung with intense desire,
The ripened orchard when the ruddy orbs
Betwixt the green leaves blush, the azure skies,
The cheerful sun that through earth's vitals pours 160
Delight and health and heat; all, all conspire
To raise, to soothe, to harmonize the mind,
To lift on wings of praise, to the great Sire
Of being and of beauty, at whose nod
Creation started from the gloomy vault 165
Of dreary Chaos, while the grisly king
Murmured to feel his boisterous power confined. . . .

[145]Lucretius V.925 ff. (see Chapter 10 above).

Ode on the Spring, by Thomas Gray

Lo! where the rosy-bosomed Hours,
Fair Venus' train appear,
Disclose the long-expecting flow'rs,
And wake the purple year!
The Attic warbler pours her throat, 5
Responsive to the cuckoo's note,
The untaught harmony of spring:
While whisp'ring pleasure as they fly,
Cool Zephyrs through the clear blue sky
Their gathered fragrance fling. 10

Where'er the oak's thick branches stretch
A broader browner shade;
Where'er the rude and moss-grown beech
O'ercanopies the glade,
Beside some water's rushy brink 15
With me the Muse shall sit, and think
(At ease reclined in rustic state)
How vain the ardor of the crowd,
How low, how little are the proud,
How indigent the great! 20

Still is the toiling hand of Care:
The panting herds repose:
Yet hark, how through the peopled air
The busy murmur glows!
The insect youth are on the wing, 25
Eager to taste the honied spring,
And float amid the liquid noon:
Some lightly o'er the current skim,
Some show their gaily-gilded trim
Quick-glancing to the sun. 30

To Contemplation's sober eye
Such is the race of man:
And they that creep, and they that fly,
Shall end where they began.
Alike the busy and the gay 35
But flutter through life's little day,
In fortune's varying colors dressed:
Brushed by the hand of rough Mischance,
Or chilled by age, their airy dance
They leave, in dust to rest. 40

Methinks I hear in accents low
The sportive kind reply:
"Poor moralist! and what art thou?
A solitary fly!
Thy joys no glittering female meets, 45

No hive hast thou of hoarded sweets,
No painted plumage to display:
On hasty wings thy youth is flown;
Thy sun is set, thy spring is gone—
We frolic, while 'tis May." 50

FROM A Song to David, *by Christopher Smart*

The world—the clust'ring spheres he made,
The glorious light, the soothing shade,
  Dale, champaign, grove, and hill;
The multitudinous abyss,
Where secrecy remains in bliss, 125
  And wisdom hides her skill.

Trees, plants, and flow'rs—of virtuous root;
Gem yielding blossom, yielding fruit,
  Choice gums and precious balm;
Bless ye the nosegay in the vale, 130
And with the sweetness of the gale
  Enrich the thankful psalm.

Of fowl—e'en ev'ry beak and wing
Which cheer the winter, hail the spring,
  That live in peace or prey; 135
They that make music, or that mock,
The quail, the brave domestic cock,
  The raven, swan, and jay.

Of fishes—ev'ry size and shape,
Which Nature frames of light escape, 140
  Devouring man to shun:
The shells are in the wealthy deep,
The shoals upon the surface leap,
  And love the glancing sun.

Of beasts—the beaver plods his task; 145
  While the sleek tigers roll and bask,
  Nor yet the shades arouse;
Her cave the mining coney scoops;
Where o'er the mead the mountain stoops,
  The kids exult and browse. 150

Of gems—their virtue and their price,
Which hid in earth from man's device,
  Their darts of luster sheathe;
The jasper of the master's stamp,
The topaz blazing like a lamp 155
  Among the mines beneath. . . .

Tell them I Am, Jehovah said 235
To Moses; while earth heard in dread,
And smitten to the heart,
At once above, beneath, around,
All Nature, without voice or sound,
Replied, O Lord, Thou Art. . . . 240

Praise above all—for praise prevails; 295
Heap up the measure, load the scales,
And good to goodness add:
The gen'rous soul her Savior aids,
But peevish obloquy degrades;
The Lord is great and glad. . . . 300

For adoration seasons change,
And order, truth, and beauty range,
Adjust, attract, and fill:
The grass the polyanthus checks; 310
And polished porphyry reflects,
By the descending rill.

Rich almonds color to the prime
For adoration; tendrils climb,
And fruit-trees pledge their gems; 315
And Ivis[146] with her gorgeous vest
Builds for her eggs her cunning nest,
And bell-flow'rs bow their stems. . . .

The laurels with the winter strive;
The crocus burnishes alive
Upon the snow-clad earth:
For adoration myrtles stay
To keep the garden from dismay, 365
And bless the sight from dearth.

The pheasant shows his pompous neck;
And ermine, jealous of a speck,
With fear eludes offense:
The sable, with his glossy pride, 370
For adoration is descried,
Where frosts the wave condense.

The cheerful holly, pensive yew,
And holy thorn, their trim renew;
The squirrel hoards his nut: 375
All creatures batten o'er their stores,
And careful Nature all her doors
For adoration shuts. . . .

[146]Humming-bird. (Smart)

For adoration, in the skies,
The Lord's philosopher espies
  The dog, the ram, and rose;
The planet's ring, Orion's sword;
Nor is his greatness less adored 395
  In the vile worm that glows. . . .

For adoration, in the dome
Of Christ the sparrows find an home;
  And on his olives perch:
The swallow also dwells with thee,
O man of God's humility, 425
  Within his Savior Church. . . .

Glorious the sun in mid career;
Glorious th'assembled fires appear; 500
  Glorious the comet's train;
Glorious the trumpet and alarm;
Glorious th'almighty stretched-out arm;
  Glorious th'enraptured main:

Glorious the northern lights astream; 505
Glorious the song, when God's the theme;
  Glorious the thunder's roar;
Glorious hosannah from the den;
Glorious the catholic amen;
  Glorious the martyr's gore. 510

Glorious—more glorious, is the crown
Of Him that brought salvation down
  By meekness, called thy Son;
Thou that stupendous truth believed,
And now the matchless deed's achieved, 515
  DETERMINED, DARED, and DONE.

20

States of Nature and Culture: Seventeenth- and Eighteenth-Century British Thought

The "Age of Reason," which was equally an "Age of Nature" (whether the two were identified or opposed), appears in retrospect a time of delicate balance between the revolutionary upheavals of England in the mid seventeenth century and those of France at the end of the eighteenth. In England, where the fiercest sectarian violence lay in the past, religious contention continued; but both Christians and deists (who were sometimes hard to distinguish) took shelter under the cloak of Newton, whose Principia *(Chapter 18 above) appeared just in time to usher in the Glorious Revolution and the age of stability and concord—or at least of compromise and tenacious fudging of differences—that followed. In this time of relative peace, especially under the first two Georges of the new Hanoverian dynasty, when "the humdrum and serviceable succeeded the picturesque but precarious," Basil Willey writes in* The English Moralists *(1964), "in religion, the Lord of Hosts was replaced by The Supreme Being; and churchmen, steeped in deism, were content for a while to let sleeping dogmas lie. . . . The powerful residue of theistic sentiment left over from the Christian centuries could and did fuse naturally with the findings of science, which, as far as nature had then been explored, revealed nothing but what was divinely rational and purposive. 'Things-as-they-are are as they should be'—this is the burden of Augustan optimism; Nature proclaims her divine original, and man is in his proper place in the great chain of Being."*

Yet the most memorable proclamation of this creed, Pope's Essay on Man *(Chapter 19 above), was written by a professing Roman Catholic, and the foundational scripture of deism, Newton's* Principia, *by a fervid Anglican intent (outside of his scientific writings) on plumbing the secrets of biblical prophecy. Such paradoxes suggest both the extreme plasticity, indeed vagueness, of a deism central to the thought of the time yet often difficult to locate, and the latent conflict between beliefs—in Nature, in Reason, in God—whose potential discrepancy was mainly reconciled by being resolutely ignored. Deism, the religion of nature as opposed to revelation—more or less, for no leading thinker was so bold as openly to embrace the scandalous thought of Spinoza (Chapter 18)—dominated much of the* History of English Thought in the Eighteenth Century *chronicled by Leslie Stephen (1876; 3rd ed. 1902). It also exerted, through Voltaire and others, an immense impact on the French and American Enlightenment.*

This English deism—foreshadowed by Lord Herbert of Cherbury (George Herbert's brother) and developed by Samuel Clarke, to whom Leibniz had addressed his reservations on Newton's natural philosophy (Chapter 18), and more forthrightly expounded by John Toland, Anthony Collins, Matthew Tindal, and others now little remembered—sometimes seems no more than Christianity without divine revelation, except in the Book of Nature interpreted by Newton; without miraculous intervention, since God was left with nothing to do after creating his marvelous clockwork world; and therefore without Christ, except as an exemplary moral teacher.

(Only Shaftesbury, of the writers represented below, might be considered a deist, though he did not consider himself one; for brief selections from others, see Peter Gay's Deism: An Anthology [1968].) Deism was thus everywhere and nowhere, and by the latter half of the century, "so far from any philosophy being ready to profit by the victory over the old beliefs," Stephen writes, "the only so-called philosophy was rapidly expiring. The deists might show conclusively that many parts of the biblical narrative were unworthy of the God of nature, as they conceived him, but their conceptions were so faint, and so rapidly decaying, that the discord was of little importance. . . . The traditional religion was absurd; but men must have a traditional religion. . . . In more commonplace minds the same sentiment took a different form. They thought that they could strike out a judicious mean by believing everything, but believing nothing too vigorously."

In autocratic France, Voltaire's intransigently anticlerical deism would have revolutionary consequences; in England, where the boldest thinker of the century, David Hume, applied his radical scepticism not only to revealed religion but also to deism and indeed to science itself, "the religion of nature," Stephen observes, "had expired of inanition." As the century advanced, the tendency "to juxtapose the corruption, injustice and irrationality of human institutions, sometimes of civilization itself," as Willey remarks, "with the divine perfection and simplicity of nature"—thought to be embodied more purely in the instinct of "savage" peoples than in the increasingly suspect reason of European science and art with which it had recently been equated—would come to fruition only in the France of Diderot and Rousseau. In England, Boswell's Johnson, addressing a gentleman who wished to learn from the natives of Tahiti or New Zealand "what pure nature can do for man," retorted that these peoples too had descended from others: "Had they grown out of the ground," but only then, "you might have judged of a state of pure nature."

Yet a putative "state of nature" is deeply rooted in seventeenth- and eighteenth-century English political thought. Thomas Hobbes bleakly portrayed (in Leviathan, published during the turmoil of the English civil war) a primitive state of continual warfare in which the life of man was "solitary, poor, nasty, brutish, and short." Though partly anticipated by Lucretius's vision (Chapter 10) of a childhood of the world when men "could not contemplate the common good / or rule themselves by custom or by law" (V.958–59), Hobbes's account was a sharp reversal of both pagan and Christian idealizations—from Hesiod (Chapter 8) long before to Milton (Chapter 19) shortly afterward—of the Golden Age or Earthly Paradise as a time and place of oneness between the human and the natural worlds. In John Locke's Second Treatise of Government, on the other hand, the state of nature "is so far from resembling the 'ill condition' described by Hobbes," Willey writes in The Seventeenth Century Background, "that it approximates rather to the Eden of the religious tradition, or the golden age of the poets. After Locke, this conception becomes an expression of the current faith that, on the whole, things if left to themselves are more likely to work together for good than if interfered with by meddling man."

In this concept, then, as in the kindred concept of natural law, a profound ambiguity was inherent, for both could act as either a critique or a justification of the status quo. "From the first, . . . Christian Natural Law had two aspects corresponding to the 'original' and the fallen natures of man; there was, on the one hand," Willey notes in The Eighteenth Century Background, "what was natural in Eden, and on the other what was natural in Europe. . . . On the whole, . . . the existing system was sanctioned by God and Nature for the preservation of order, and departures from it would be unnatural. Yet still the idea of the first and purer Natural Law hovered in the background, and man, though severely damaged by the Fall, could feel that he had not *yet lost all his original brightness." The centuries after the Renaissance, Willey continues, "liberated the acquisitive impulses, also in the name of Nature, and severed economic ethics from control by any comprehensive conception of the ultimate purpose of human (not to say Christian) living. . . . The Law of Nature, which in the Middle Ages had been a check on unregenerate impulse, had now been transformed into a sanction for* laissez-faire, *and free competition for the spoils of the world": a justification for the capitalist nation-state and its attendant imperialism. The contradictions and complexities of these equivocal concepts would concern not only English thinkers of the present chapter such as Hobbes, Locke, and Mandeville, and English economists such as Adam Smith, but also continental jurists such as Hugo Grotius and Samuel Pufendorf. Nowhere will the contradictions be more evident than in the brilliantly contorted political thought of Rousseau.*

These intertwined concepts—Natural Religion, State of Nature, Natural Law—are in the background of much late-seventeenth- and eighteenth-century thought even when not directly addressed. And just as each of these has its classical and Christian antecedents, so the great questions debated in antiquity and the Middle Ages concerning the relation of nature to God continued in this age of science triumphant to be of central interest to English thinkers, including Cambridge Platonists such as Ralph Cudworth—and George Berkeley, in whose highly original thought the Christian Platonism of earlier centuries found a bold new expression. Meanwhile, literary critics, men of letters, and artists such as Joseph Addison, Samuel Johnson, and Joshua Reynolds explored different dimensions of the human experience of nature in the context of a classicism broadened by various currents of contemporary thought, including Addison's interests in the psychology of the imagination and in the expansive power of the sublimely grand and infinite.

In some of Johnson's Rambler *essays, and in his* Journey to the Western Islands of Scotland, *we encounter—as we only rarely do in much eighteenth-century literature, despite its ubiquitous concern with different dimensions of nature—concrete descriptions of natural beauties, even of wilderness. To such observations Johnson contrasts the shopworn artifices of conventional pastoral poetry, just as elsewhere in his writing and conversation he deprecates the emotive reveries of Rousseau (Chapter 21 below) and the amorphous effusions of James Macpherson's fake Gaelic bard Ossian, of which Johnson declared, "a man might write such stuff forever, would he* abandon *his mind to it." In descriptive precision, however, few works can compare with Gilbert White's splendidly modest* Natural History of Selborne, *in which a village naturalist, through meticulous observation and precisely cadenced prose, brings the flora and fauna of his native grounds unforgettably to life. Finally, the English scientific tradition again finds expression at the end of the century in James Hutton's* Theory of the Earth; *this work anticipated the geology of Charles Lyell and prepared the way for the evolutionary theory of Charles Darwin in the following century by proposing that the "natural operations of the globe," with "a certain regularity" through long expanses of time, are continually wearing away and building up again the continents of our ever-changing earth.*

Every Man Against Every Man

Leviathan, or The Matter, Forme, and Power of a Commonwealth Ecclesiasticall and Civil, by Thomas Hobbes

Born at Westport, near Malmesbury in Wiltshire, in the year of the Spanish Armada—prematurely, he claimed, because of his mother's fright at the approaching galleons—Thomas Hobbes (1588–1679), in his ninety-one-year life as classicist, mathematician, and philosopher, assimilated many central currents of his "century of genius." After attending Magdalen Hall, Oxford, in 1603, and taking his B.A. in 1608, he became tutor to the son of William Cavendish, Earl of Devonshire, at whose homes he met such leading figures as Lord Herbert of Cherbury and Ben Jonson; for a while he was secretary to Francis Bacon, some of whose essays he translated into Latin. He remained an excellent classical scholar throughout his career, writing works in both Latin and English, publishing a translation of Thucydides in 1629, and late in life writing an autobiography in Latin verse and translating both the Odyssey *(1673) and the* Iliad *(1676). During three trips to the Continent he immersed himself in the geometry of Euclid and in Galileo's physics of motion, meeting Galileo himself, and becoming a convert to the new natural philosophy. In 1640, concerned lest the royalism of his early writings place him in danger in England, where Parliament was ascendant, he went to France, where he joined the scientific circle of Father Marin Mersenne; engaged in controversy with Descartes; published his first major political work,* De Cive (On the Citizen), *in 1642; wrote a treatise on optics (1644); and in 1645 became mathematics tutor of Prince Charles (the future Charles II) in the English court in exile. In 1651, in disfavor with the English court because of the scandalous publication of* Leviathan *in England that year, and with the French because of his outspoken anticlericalism, he returned to England, where he rejoined the Cavendish family circle in 1653, and wrote, among other works,* De Corpore *(On the Body, 1655),* De Homine *(On Man, 1658), and* Behemoth, *a posthumous account (1680) of the English civil wars.*

His political and philosophical works give powerful expression, above all in the vigorous prose of Leviathan, *to a materialistic view of the universe and of man (including the human mind, in sharp contrast to Descartes's dualism of mind and body), in which individual self-interest is the main motive of human action. In consequence, Hobbes asserts in the preface to* De Cive *(which he translated in 1651 as* Philosophical Rudiments Concerning Government and Society*) "that the state of men without civil society, which state we may properly call the state of nature, is nothing else but a mere war of all against all; and in that war all men have equal right unto all things. Next, that all men as soon as they arrive to understanding of this hateful condition, do desire, even nature itself compelling them, to be freed from this misery. But that this cannot be done, except by compact, they all quit that right they have to all things," and thus sacrifice individual freedom to the greater good of security provided by a "supreme power, whether it be one man or an assembly of men." They trade the "natural right" of individual self-interest for voluntary obedience to the "natural law"—defined in* De Cive *(II.i) as "the dictate of right reason, conversant about those things which are either to be done or omitted for the constant preservation of life and members, as much as in us lies"—which is also the moral law and the law of Christ.*

Much reviled for his materialism, supposed atheism, and exaltation of the autocratic nation-state—especially in Leviathan *(the text of which here cited is modernized from C. B. Mac-*

pherson's edition of 1968)—Hobbes exerted an immense influence throughout the seventeenth century and beyond, provoking innumerable refutations of views that placed him, for most contemporaries, in the disreputable company of Machiavelli and Spinoza. His political philosophy, as he conceived it, continued the "knowledge of the Nature of Motion" initiated by Galileo. "The real world, the world of science," as Douglas Bush remarks in English Literature in the Earlier Seventeenth Century, *"was a mathematical mechanism of bodies, vast or minute, moving in space and time in accordance with natural laws. In such a world man himself, a mere aggregate of secondary qualities, was, logically, an insignificant stranger, a superfluous accident." Descartes had attempted to confine the mechanical and mathematical properties to material as opposed to thinking substance: "But it was inevitable that the materialistic cat should swallow the spiritual canary, and in this operation Hobbes was the ruthlessly efficient cause," transforming "the pattern of man's inner and outer world by including* res cogitans *within* res extensa. *Nothing exists except body, and its attributes are extension and motion." Given these axioms concerning nature and human nature, reinforced by empirical observation, the rest, for Hobbes, followed with the force of a geometrical demonstration. For only through submission to the constraints of civil society (which are those of religion rightly conceived) can human beings be saved from the destructive tendencies of their own unregenerate nature. In this vision there is, perhaps, at least as heavy a share of secularized Christianity as there is of science.*

THE FIRST PART: OF MAN

Chapter XIII: Of the NATURAL CONDITION *of Mankind, as concerning their Felicity, and Misery*

Nature hath made men so equal, in the faculties of body and mind, as that though there be found one man sometimes manifestly stronger in body, or of quicker mind than another, yet when all is reckoned together, the difference between man and man is not so considerable as that one man can thereupon claim to himself any benefit to which another may not pretend as well as he. For as to the strength of body, the weakest has strength enough to kill the strongest, either by secret machination or by confederacy with others that are in the same danger with himself.

And as to the faculties of the mind (setting aside the arts grounded upon words, and especially that skill of proceeding upon general and infallible rules, called science, which very few have, and but in few things, as being not a native faculty born with us, nor attained (as prudence) while we look after somewhat else, I find yet a greater equality amongst men than that of strength. For prudence is but experience, which equal time equally bestows on all men in those things they equally apply themselves unto. That which may perhaps make such equality incredible is but a vain conceit of one's own wisdom, which almost all men think they have in a greater degree than the vulgar; that is, than all men but themselves, and a few others, whom by fame or for concurring with themselves they approve. For such is the nature of men, that howsoever they may acknowledge many others to be more witty, or more eloquent, or more learned, yet they will hardly believe there be many so wise as themselves; for they see their own wit at hand, and other men's at a distance. But this proveth rather than men are in that point equal, than unequal. For there is not ordinarily a greater sign of the equal distribution of any thing than that every man is contented with his share.

From this equality of ability ariseth equality of hope in the attaining of our ends. And therefore if any two men desire the same thing, which nevertheless they cannot both enjoy, they become enemies; and in the way to their end (which is principally their own conservation, and sometimes their delectation only) endeavor to destroy or subdue one another. And from hence it comes to pass that where an invader hath no more to fear than another man's single power, if one plant, sow, build, or possess a convenient seat, others may probably be expected to come prepared with forces united to dispossess and deprive him, not only of the fruit of his labor, but also of his life, or liberty. And the invader again is in the like danger of another.

And from this diffidence of one another, there is no way for any man to secure himself, so reasonable as anticipation; that is, by force, or wiles, to master the persons of all men he can, so long, till he see no other power great enough to endanger him: and this is no more than his own conservation requireth, and is generally allowed. Also because there be some, that taking pleasure in contemplating their own power in the acts of conquest, which they pursue farther than their security requires; if others, that otherwise would be glad to be at ease within modest bounds, should not by invasion increase their power, they would not be able long time, by standing only on their defense, to subsist. And by consequence, such augmentation of dominion over men, being necessary to a man's conservation, it ought to be allowed him.

Again, men have no pleasure (but on the contrary a great deal of grief) in keeping company where there is no power able to overawe them all. For every man looketh that his companion should value him, at the same rate he sets upon himself; and upon all signs of contempt, or undervaluing, naturally endeavors, as far as he dares (which amongst them that have no common power to keep them quiet is far enough to make them destroy each other), to extort a greater value from his contemners by dommage,[1] and from others by the example.

So that in the nature of man we find three principal causes of quarrel. First, competition; secondly, diffidence; thirdly, glory.

The first maketh men invade for gain; the second, for safety; and the third, for reputation. The first use violence, to make themselves masters of other men's persons, wives, children, and cattle; the second, to defend them; the third, for trifles, as a word, a smile, a different opinion, and any other sign of undervalue, either direct in their persons or by reflection in their kindred, their friends, their nation, their profession, or their name.

Hereby it is manifest that during the time men live without common power to keep them all in awe, they are in that condition which is called war; and such a war, as is of every man against every man. For WAR consisteth not in battle only, or the act of fighting, but in a tract of time wherein the will to contend by battle is sufficiently known; and therefore the notion of time is to be considered in the nature of war, as it is in the nature of weather. For as the nature of foul weather lieth not in a shower or two of rain, but in an inclination thereto of many days together, so the nature of war consisteth not in actual fighting, but in the known disposition thereto, during all the time there is no assurance to the contrary. All other time is PEACE.

Whatsoever therefore is consequent to a time of war, where every man is enemy to every man, the same is consequent to the time wherein men live without other security than what their own strength, and their own invention, shall furnish them withal. In such condition, there is no place for industry, because the fruit thereof is uncertain: and consequently no culture of the earth; no navigation, nor use of the commodities that

[1] "Damage."

may be imported by sea; no commodious building; no instruments of moving and removing such things as require much force; no knowledge of the face of the earth; no account of time; no arts; no letters; no society; and, which is worst of all, continual fear, and danger of violent death; and the life of man, solitary, poor, nasty, brutish, and short.

It may seem strange to some man, that has not well weighed these things, that nature should thus dissociate, and render men apt to invade, and destroy one another; and he may therefore, not trusting to this inference made from the passions, desire perhaps to have the same confirmed by experience. Let him therefore consider with himself, when taking a journey, he arms himself, and seeks to go well accompanied; when going to sleep, he locks his doors; when even in his house, he locks his chests; and this when he knows there be laws, and public officers armed to revenge all injuries shall be done him: what opinion he has of his fellow subjects, when he rides armed; of his fellow citizens, when he locks his doors; and of his children and servants, when he locks his chests. Does he not there as much accuse mankind by his actions, as I do by my words? But neither of us accuse man's nature in it. The desires, and other passions of man, are in themselves no sin. No more are the actions that proceed from those passions, till they know a law that forbids them: which till laws be made they cannot know; nor can any law be made, till they have agreed upon the person that shall make it.

It may peradventure be thought, there was never such a time, nor condition of war as this; and I believe it was never generally so, over all the world: but there are many places where they live so now. For the savage peoples in many places of America, except the government of small families, the concord whereof dependeth on natural lust, have no government at all, and live at this day in the brutish manner, as I said before. Howsoever, it may be perceived what manner of life there would be, where there were no common power to fear, by the manner of life which men that have formerly lived under a peaceful government use to degenerate into, in a civil war.

But though there had never been any time, wherein particular men were in a condition of war one against another, yet in all times kings and persons of sovereign authority, because of their independency, are in continual jealousies, and in the state and posture of gladiators, having their weapons pointing, and their eyes fixed on one another; that is, their forts, garrisons, and guns upon the frontiers of their kingdoms, and continual spies upon their neighbors, which is a posture of war. But because they uphold thereby the industry of their subjects, there does not follow from it that misery which accompanies the liberty of particular men.

To this war of every man against every man, this also is consequent: that nothing can be unjust. The notions of right and wrong, justice and injustice, have there no place. Where there is no common power, there is no law; where no law, no injustice. Force, and fraud, are in war the two cardinal virtues. Justice, and injustice, are none of the faculties neither of the body, nor mind. If they were, they might be in a man that were alone in the world, as well as his senses, and passions. They are qualities that relate to men in society, not in solitude. It is consequent also to the same condition that there be no propriety, no dominion, no Mine and Thine distinct; but only that to be every man's that he can get, and for so long as he can keep it. And thus much for the ill condition, which man by mere nature is actually placed in: though with a possibility to come out of it, consisting partly in the passions, partly in his reason.

The passions that incline men to peace are fear of death; desire of such things as are necessary to commodious living; and a hope by their industry to obtain them. And reason suggesteth convenient articles of peace, upon which men may be drawn to agreement. These articles are they which otherwise are called the laws of nature: whereof I shall speak more particularly in the two following chapters.

FROM *Chapter XV: Of Other Laws of Nature*

. . . These are the laws of nature, dictating peace, for a means of the conservation of men in multitudes; and which only concern the doctrine of civil society. . . .

And though this may seem too subtle a deduction of the laws of nature to be taken notice of by all men, whereof the most part are too busy in getting food, and the rest too negligent to understand, yet to leave all men unexcusable, they have been contracted into one easy sum, intelligible even to the meanest capacity; and that is, *Do not that to another, which thou wouldest not have done to thyself*; which showeth him that he has no more to do in learning the laws of nature, but, when weighing the actions of other men with his own, they seem too heavy, to put them into the other part of the balance, and his own into their place, that his own passions, and self-love, may add nothing to the weight; and then there is none of these laws of nature that will not appear unto him very reasonable.

The laws of nature oblige *in foro interno* [in the inner forum], that is to say, they bind to a desire they should take place: but *in foro externo* [in the outer forum], that is, to the putting them in act, not always. For he that should be modest, and tractable, and perform all he promises, in such time, and place, where no man else should do so, should but make himself a prey to others, and procure his own certain ruin, contrary to the ground of all laws of nature, which tend to nature's preservation. And again, he that having sufficient security that others shall observe the same laws towards him, observes them not himself, seeketh not peace but war, and consequently the destruction of his nature by violence.

And whatsoever laws bind *in foro interno*, may be broken, not only by a fact contrary to the law but also by a fact according to it, in case a man think it contrary. For though his action in this case be according to the law, yet his purpose was against the law, which where the obligation is *in foro interno*, is a breach.

The laws of nature are immutable and eternal; for injustice, ingratitude, arrogance, pride, iniquity, acception of persons,[2] and the rest, can never be made lawful. For it can never be that war shall preserve life, and peace destroy it.

The same laws, because they oblige only to a desire and endeavor, I mean an unfeigned and constant endeavor, are easy to be observed. For in that they require nothing but endeavor, he that endeavoreth their performance fulfilleth them; and he that fulfilleth the law, is just.

And the science of them is the true and only moral philosophy. For moral philosophy is nothing else but the science of what is good, and evil, in the conversation and society of mankind. Good, and evil, are names that signify our appetites, and aversions; which in different tempers, customs, and doctrines of men, are different: and divers men differ not only in their judgment on the senses of what is pleasant, and unpleasant, to the taste, smell, hearing, touch, and sight, but also of what is conformable, or disagreeable, to reason in the actions of common life. Nay, the same man, in divers times, differs from himself, and one time praiseth, that is, calleth good, what another time he dispraiseth, and calleth evil: from whence arise disputes, controversies, and at last war. And therefore so long a man is in the condition of mere nature (which is a condition of war) as private

[2]The OED defines "acception of persons" (or "of faces") as the receiving of the personal advances of anyone with favor; hence, corrupt acceptance, or favoritism, due to a person's rank, relationship, influence, power to bribe, etc.

appetite is the measure of good, and evil; and consequently all men agree on this, that peace is good; and therefore also the way, or means, of peace, which (as I have showed before) are justice, gratitude, modesty, equity, mercy, and the rest of the laws of nature, are good, that is to say, moral virtues; and their contrary vices, evil. Now the science of virtue and vice is moral philosophy; and therefore the true doctrine of the laws of nature is the true moral philosophy. . . .

Something in Nature Besides Mechanism

"The Digression concerning the Plastick Life of Nature, or an Artificial, Orderly, and Methodical Nature," in *The True Intellectual Systeme of the Universe*, by Ralph Cudworth

Ralph Cudworth (1617–88) was one of the central figures of the group known as the Cambridge Platonists, whose other important members included Benjamin Whichcote (1609–83), Henry More (1614–87), and John Smith (1618–52). Opposition to the mechanical materialism of the new philosophy was at the heart of their endeavor to revive the Platonic and Neoplatonic philosophy of antiquity and the Renaissance and to affirm the reality of spirit, love, and the divinity of man created in the image of God. Henry More was at first sympathetic to the philosophy of Descartes ("a man more truly inspired in the knowledge of Nature, than any that have professed themselves so this sixteen hundred years"), which he initially saw as rebutting materialism, but with time he came to see the Cartesian denial of extension to spirit as "pure mechanism." Asserting that spirit is infinitely extended, he turned increasingly to mystical and occult beliefs.

But Hobbes, still more than Descartes, was the opponent against whom Cudworth took aim in his writings. Born at Aller in Somersetshire, Cudworth attended Emmanuel College, Cambridge (the main center of the new Platonism), taking his M.A. in 1639, his B.D. in 1645, and his D.D. in 1651; in 1645 he became Master of Clare Hall and Regius Professor of Hebrew, and in 1654 Master of Christ's College. In 1647 he delivered a sermon to the House of Commons, in which he affirmed that "all outward forms and models of Reformation, though they be never so good of their kind, yet they are of little worth to us, without this inner Reformation of the heart." The first part of his unfinished philosophical opus, The True Intellectual Systeme of the Universe, *was published in 1678; two other segments appeared long after his death,* A Treatise concerning Eternal and Immutable Morality *in 1731 and* A Treatise of Freewill *in 1838. In the long "Digression concerning the Plastick Life of Nature," from which the following excerpts are taken (in a text modernized from that of C. A. Patrides in* The Cambridge Platonists *[1969]), Cudworth takes up questions discussed for centuries by Christian thinkers such as Thomas Aquinas (Chapter 13 above), for whom "God acts perfectly as first cause; but the operation of nature as second cause is also necessary." He argues that "some energetic, effectual, and operative cause for the production of every effect," which he calls "plastic nature" ("plastic" in the sense of "shaping" or "molding"), must be operative in the universe, since the alternatives—either that everything happens at random or that God intervenes directly in even the most routine matters—are equally inconceivable. Nature is thus a subordinate and intermediary power, but this nature, unlike the "vicar" of Alan of Lille (Chapter 15) or of Chaucer, far from being a goddess, is the "lowest of all lives," corresponding to Aristotle's vegetative soul, and thus inferior not only to the intellectual in man but also to the sensitive in animals.*

FROM *Book I, Chapter III, Section XXXVII*

2 ... Unless there be such a thing admitted as a Plastic Nature, that acts *heneka tou*, for the sake of something, and in order to ends, regularly, artificially, and methodically, it seems that one or other of these two things must be concluded, that either in the efformation and organization of the bodies of animals, as well as the other phenomena, everything comes to pass fortuitously, and happens to be as it is, without the guidance and direction of any mind or understanding; or else, that God himself doth all immediately, and as it were with his own hands, form the body of every gnat and fly, insect and mite, as of other animals in generations, all whose members have so much of contrivance in them, that Galen professed he could never enough admire that artifice which was in the leg of a fly (and yet he would have admired the wisdom of nature more, had he been but acquainted with the use of microscopes). I say, upon supposition of no Plastic Nature, one or other of these two things must be concluded; because it is not conceived by any, that the things of nature are all thus administered, with such exact regularity and constancy everywhere, merely by the wisdom, providence, and efficiency of those inferior spirits, daemons or angels. As also, though it be true that the works of nature are dispensed by a divine law and command, yet this is not to be understood in a vulgar sense, as if they were all effected by the mere force of a verbal law or outward command, because inanimate things are not commandable nor governable by such a law; and therefore besides the divine will and pleasure, there must needs be some other immediate agent and executioner provided for the producing of every effect; since not so much as a stone or other heavy body could at any time fall downward merely by the force of a verbal law, without any other efficient cause; but either God himself must immediately impel it, or else there must be some other subordinate cause in nature for that motion. Wherefore the divine law and command, by which the things of nature are administered, must be conceived to be the real appointment of some energetic, effectual, and operative cause for the production of every effect.

3 Now to assert the former of these two things, that all the effects of nature come to pass by material and mechanical necessity, or the mere fortuitous motion of matter, without any guidance or direction, is a thing no less irrational than it is impious and atheistical. Not only because it is utterly unconceivable and impossible, that such infinite regularity and artificialness, as is everywhere throughout the whole world, should constantly result out of the fortuitous motion of matter, but also because there are many such particular phenomena in nature, as do plainly transcend the powers of mechanism, of which therefore no sufficient mechanical reasons can be devised, as the motion of respiration in animals; as there are also other phenomena that are perfectly cross to the laws of mechanism; as for example, that of the distant poles of the equator and ecliptic, which we shall insist upon afterward. Of both which kinds there have been other instances proposed, by my learned friend Dr. [Henry] More in his *Enchiridion Metaphysicum*, and very ingeniously improved by him to this very purpose, namely to evince that there is something in nature besides mechanism, and consequently substance incorporeal. ...

4 And as for the latter part of the disjunction, that everything in nature should be done immediately by God himself; this, as according to vulgar apprehension, it would render divine providence operose, solicitous, and distractious, and thereby make the belief of it to be entertained with greater difficulty, and give advantage to atheists; so in

the judgment of the writer [of] *De Mundo*,[3] it is not so decorous in respect of God neither, that he should *autourgein hapanta*, set his own hand, as it were, to every work, and immediately do all the meanest and triflingest things himself drudgingly, without making use of any inferior and subordinate instruments. . . .

Moreover, it seems not so agreeable to reason neither, that nature as a distinct thing from the deity should be quite superseded or made to signify nothing, God himself doing all things immediately and miraculously; from whence it would follow also, that they are all done either forcibly and violently, or else artificially only, and none of them by any inward principle of their own.

Lastly: this opinion is further confuted by that slow and gradual process that is in the generations of things, which would seem to be but a vain and idle pomp, or a trifling formality, if the agent were omnipotent; as also by those *hamartêmata* (as Aristotle calls them),[4] those errors and bungles which are committed, when the matter is inept and contumacious; which argue the agent not to be irresistible, and that nature is such a thing as is not altogether uncapable (as well as human art) of being sometimes frustrated and disappointed by the indisposition of matter. Whereas an omnipotent agent, as it could dispatch its work in a moment, so it would always do it infallibly and irresistibly; no ineptitude or stubbornness of matter being ever able to hinder such a one, or make him bungle or fumble in any thing.

5 Wherefore since neither all things are produced fortuitously, or by the unguided mechanism of matter, nor God himself may reasonably be thought to do all things immediately and miraculously, it may well be concluded that there is a Plastic Nature under him, which as an inferior and subordinate instrument, doth drudgingly execute that part of his providence which consists in the regular and orderly motion of matter: yet so as that there is also besides this a higher providence to be acknowledged, which presiding over it, doth often supply the defects of it, and sometimes overrule it; forasmuch as this Plastic Nature cannot act electively nor with discretion. And by this means the wisdom of God will not be shut up nor concluded wholly within his own breast, but will display itself abroad, and print its stamps and signatures everywhere throughout the world; so that God, as Plato (after Orpheus) speaks,[5] will be not only the beginning and end, but also the middle of all things, they being as much to be ascribed to his causality, as if himself had done them all immediately, without the concurrent instrumentality of any subordinate natural cause. Notwithstanding which, in this way it will appear also to human reason that all things are disposed and ordered by the deity, without any solicitous care or distractious providence. . . .[6]

[3]Pseudo-Aristotle, *De mundo [On the World]*, VII. This work, whose authenticity Cudworth rightly doubts, is now dated between 50 B.C. and A.D. 100. (Patrides)

[4]*Physics* II.8 (199a–b); see Chapter 9 above.

[5]"God, as the old tradition declares, holding in His hand the beginning, middle, and end of all that is, travels according to His nature in a straight line towards the accomplishment of His end" (Plato, *Laws*, Book IV, 715E–716A, trans. Jowett). The "old tradition" is probably that of the mythical Orpheus.

[6]The basic flaw in Cudworth's theory of Plastic Nature is that "it explained the unknown by asserting another quite hypothetical unknown" (William B. Hunter, Jr., "The Seventeenth Century Doctrine of Plastic Nature," *Harvard Theological Review*, XLIII [1950], pp. 209 f.). As much was argued by Joseph Glanvill—the platonising friend of Henry More—when he wrote of similar assertions that "The Plastic faculty is a fine word: but what it is, how it works, and whose it is, we cannot learn; no, not by a return into the womb; neither will the Platonic principles unriddle the doubt; for though the soul be supposed to be the body's maker, and the builder of its own house; yet by what kind of knowledge, method, or means, is as unknown" (*The Vanity of Dogmatizing* [1661], pp. 43–44; in Hunter, as before). (Patrides)

7 But because some may pretend, that the Plastic Nature is all one with an occult quality, we shall here show how great a difference there is betwixt these two. For he that asserts an occult quality for the cause of any phenomenon, does indeed assign no cause at all of it, but only declare his own ignorance of the cause; but he that asserts a Plastic Nature, assigns a determinate and proper cause, nay the only intelligible cause, of that which is the greatest of all phenomena in the world, namely the *to eu kai kalôs*,[7] the orderly, regular, and artificial frame of things in the universe, whereof the mechanic philosophers, however pretending to save all phenomena by matter and motion, assign no cause at all. Mind and understanding is the only true cause of orderly regularity, and he that asserts a Plastic Nature, asserts mental causality in the world; but the fortuitous mechanists, who exploding final causes, will not allow mind and understanding to have any influence at all upon the frame of things, can never possibly assign any cause of this grand phenomenon, unless confusion may be said to be the cause of order, and fortune or chance of constant regularity; and therefore themselves must resolve it into an occult quality. Nor indeed does there appear any great reason why such men should assert an infinite mind in the world, since they do not allow it to act anywhere at all, and therefore must needs make it to be in vain. . . .

19 From what hath been hitherto declared concerning the Plastic Nature, it may appear, that though it be a thing that acts for ends artificially, and which may be also called the Divine Art, and the fate of the corporeal world; yet for all that, it is neither God nor Goddess, but a low and imperfect creature. Forasmuch as it is not master of that reason and wisdom according to which it acts, nor does it properly intend those ends which it acts for, nor indeed is it expressly conscious of what it doth, it not knowing but only doing, according to commands and laws impressed upon it. Neither of which things ought to seem strange or incredible, since nature may as well act regularly and artificially, without any knowledge and consciousness of its own, as forms of letters compounded together may print coherent philosophic sense, though they understand nothing at all; and it may also act for the sake of those ends, that are not intended by itself, but some higher being, as well as the saw or hatchet in the hand of the architect or mechanic doth. . . . It is true, that our human actions are not governed by such exact reason, art, and wisdom, nor carried on with such constancy, evenness, and uniformity as the actions of nature are; notwithstanding which, since we act according to a knowledge of our own, and are masters of that wisdom by which our actions are directed, since we do not act fatally only, but electively and intendingly, with consciousness and self-perception, the rational life that is in us ought to be accounted a much higher and more noble perfection than that Plastic Life of Nature. Nay, this Plastic Nature is so far from being the first and highest life that it is indeed the last and lowest of all lives, it being really the same with the vegetative, which is inferior to the sensitive. . . .

[7]"Well and fit" (Cudworth's own translation of Aristotle, *Metaphysics*, I.3 [984b]). (Patrides) The words are translated "goodness and beauty" by W. D. Ross in Chapter 12 above.

A State of Perfect Freedom and Continual Danger

An Essay Concerning the True Original, Extent, and End of Civil Government (The Second Treatise of Government), by John Locke

John Locke (1632–1704), the founder of British empiricist philosophy and of liberal (or Whig) political thought, was along with Bacon and Newton (Chapter 18 above) the most influential of seventeenth-century English thinkers on the European Enlightenment. Born at Wrington in Somersetshire, he was educated at Westminster School and at Christ Church, Oxford, where he took his M.A. in 1658. Beginning in 1660, he became a lecturer in Greek, rhetoric, and philosophy at Oxford (with which he remained affiliated throughout his life), but his interests turned increasingly to medicine and the experimental science of Robert Boyle (Chapter 18) and others associated with the Royal Society, of which Locke was elected a fellow in 1668. In 1666 he attended to the illness of Anthony Ashley Cooper, later first Earl of Shaftesbury, for whom he acted as physician and secretary from 1667 to 1681, residing in France between 1673 and 1679. During the later years of Charles II and the short reign of James II (when Shaftesbury, the anti-Catholic Whig leader satirized in Dryden's Absalom and Achitophel, *was banished), Locke withdrew to Holland (1683–89) and completed his major philosophical works before returning to England after the Glorious Revolution installed William and Mary on the throne. His first important work, the anonymous first* Letter on Toleration, *appeared in 1689, when Locke was fifty-seven (the second and third letters appeared in 1690 and 1692), to be followed in 1690 by his two masterpieces,* An Essay Concerning Humane Understanding *and the anonymous* Two Treatises of Government, *of which a second edition was published in 1694 and a third in 1698. Subsequent publications included his* Thoughts Concerning Education *(1694) and the anonymous* Reasonableness of Christianity *(1695). After about 1695 he was closely associated with John Somers, Lord Chancellor from 1696 and a central member of William III's government; served as a commissioner on the Board of Trade; and played a major role in the recoinage of British money in 1695–96.*

In the Essay on Humane Understanding *Locke decisively rejected the concept of innate ideas held by Leibniz (Chapter 18) and many earlier philosophers, arguing that the mind is a blank slate* (tabula rasa) *on which impressions received through the senses from the external world are imprinted. He also carried to its logical conclusion the distinction, already adumbrated in Galileo and others, between primary qualities (solidity, extension, figure, motion or rest, and number) and secondary qualities such as colors, sounds, and tastes, which "are nothing in the objects themselves," but only in our perception of them. And in the second of his* Two Treatises of Government, *from which the following excerpts are given in a text modernized from Peter Laslett's edition of 1960, Locke, in sharp contrast to Hobbes, conceives of the state of nature as one in which men enjoyed "perfect freedom to order their actions . . . within the bounds of the law of nature" (identified with reason), which prohibited them from harming one other "in life, health, liberty, or possessions"; only the uncertainties of this otherwise blissful condition, and the "fears and continual dangers" inherent in it, led men to contract together to form a body politic for their mutual protection. Yet because this compact does not (like Hobbes's) sacrifice, but essentially preserves, their liberty in modified and modifiable form, the state of nature is not so much abandoned as transformed and carried over into the liberal State of the British constitutional monarchy of the Glorious Revolution—and into its more revolutionary successors in America and France. Both of these would purport, under the influence of Locke, and through Locke of Voltaire, Montesquieu, Rousseau, and Jefferson, to safeguard the inalienable rights of man established by nature and by nature's God.*

FROM *Chapter II: Of the State of Nature*

4 To understand political power right, and derive it from its original, we must consider what state all men are naturally in, and that is, a state of perfect freedom to order their actions, and dispose of their possessions and persons, as they think fit, within the bounds of the law of nature, without asking leave, or depending upon the will, of any other man.

A state also of equality, wherein all the power and jurisdiction is reciprocal, no one having more than another: there being nothing more evident than that creatures of the same species and rank promiscuously born to all the same advantages of nature, and the use of the same faculties, should also be equal one amongst another without subordination or subjection, unless the lord and master of them all should by any manifest declaration of his will set one above another, and confer on him by an evident and clear appointment an undoubted right to dominion and sovereignty. . . .

6 But though this be a state of liberty, yet it is not a state of license; though man in that state have an uncontrollable liberty to dispose of his person or possessions, yet he has not liberty to destroy himself, or so much as any creature in his possession, but where some nobler use than its bare preservation calls for it. The state of nature has a law of nature to govern it, which obliges everyone: and reason, which is that law, teaches all mankind, who will but consult it, that being all equal and independent, no one ought to harm another in his life, health, liberty, or possessions. For men being all the workmanship of one omnipotent and infinitely wise maker—all the servants of one sovereign master, sent into the world by his order and about his business—they are his property, whose workmanship they are, made to last during his, not one another's, pleasure. And being furnished with like faculties, sharing all in one community of nature, there cannot be supposed any such subordination among us, that may authorize us to destroy one another, as if we were made for one another's uses, as the inferior ranks of creatures are for ours. . . .

14 'Tis often asked as a mighty objection, Where are, or ever were, there any men in such a state of nature? To which it may suffice as an answer at present, that since all princes and rulers of independent governments all through the world are in a state of nature, 'tis plain the world never was, nor ever will be, without numbers of men in that state. I have named all governors of independent communities, whether they are, or are not, in league with others: for 'tis not every compact that puts an end to the state of nature between men, but only this one of agreeing together mutually to enter into one community, and make one body politic; other promises and compacts, men may make one with another, and yet still be in the state of nature. The promises and bargains for truck,[8] etc., between the two men in the desert island mentioned by [the Inca] Garcilaso de la Vega, in his *History of Peru*, or between a Swiss and an Indian, in the woods of Amer-

[8]"Barter."

ica, are binding to them, though they are perfectly in a state of nature, in reference to one another. For truth and keeping of faith belongs to men, as men, and not as members of society.

15 To those that say, There were never any men in the state of nature, . . . I moreover affirm that all men are naturally in that state, and remain so, till by their own consents they make themselves members of some politic society; and I doubt not in the sequel of this discourse to make it very clear. . . .

FROM *Chapter III: Of the State of War*

19 And here we have the plain difference between the state of nature and the state of war, which, however some men have confounded [them], are as far distant as a state of peace, good will, mutual assistance, and preservation, and a state of enmity, malice, violence, and mutual destruction are one from another. Men living together according to reason, without a common superior on earth, with authority to judge between them, is properly the state of nature. But force, or a declared design of force upon the person of another, where there is no common superior on earth to appeal to for relief, is the state of war. And 'tis the want of such an appeal gives a man the right of war even against an aggressor, though he be in society and a fellow subject. . . .

FROM *Chapter VIII: Of the Beginning of Political Societies*

95 Men being, as has been said, by nature all free, equal and independent, no one can be put out of this estate, and subjected to the political power of another, without his own consent. The only way whereby anyone divests himself of his natural liberty, and puts on the bonds of civil society, is by agreeing with other men to join and unite into a community for their comfortable, safe, and peaceable living one amongst another, in a secure enjoyment of their properties, and a greater security against any that are not of it. This any number of men may do, because it injures not the freedom of the rest; they are left as they were in the liberty of the state of nature. When any number of men have so consented to make one community or government, they are thereby presently incorporated, and make one body politic, wherein the majority have a right to act and conclude the rest. . . .

FROM *Chapter IX: Of the Ends of Political Society and Government*

123 If man in the state of nature be so free, as has been said: if he be absolute lord of his own person and possessions, equal to the greatest, and subject to nobody, why will he part with his freedom? Why will he give up this empire, and subject himself to the dominion and control of any other power? To which 'tis obvious to answer, that though in the state of nature he hath such a right, yet the enjoyment of it is very uncertain, and constantly exposed to the invasion of others. For all being kings as much as he, every man his equal, and the greater part no strict observers of equity and justice, the enjoyment of the property he has in this state is very unsafe, very unsecure. This makes him willing to quit a condition which, however free, is full of fears and continual dangers: and 'tis not without reason that he seeks out, and is willing to join in society with, others who are already united, or have a mind to unite, for the mutual preservation of their lives, liberties, and estates, which I call by the general name "property."

A Sympathizing of Parts

Characteristicks of Men, Manners, Opinions, Times, Treatise V: *The Moralists, A Philosophical Rhapsody,* by Anthony Ashley Cooper, Third Earl of Shaftesbury

Anthony Ashley Cooper, third Earl of Shaftesbury (1671–1713), was born in London and tutored by John Locke, secretary to his grandfather, the first earl; after attending Winchester, he traveled on the Continent. He was a member of Parliament from 1695 to 1698, then spent a year in Holland. After succeeding to his title in 1700, he sat in the House of Lords, but retired because of ill health to private life, traveling again to Holland in 1703–4, and in 1711 to Naples, where he died. His Inquiry concerning Virtue *was printed (perhaps without authorization) by the deist John Toland in 1699;* A Letter concerning Enthusiasm *in 1708;* Sensus Communis: An Essay on the Freedom of Wit and Humour *in 1709; and, in the same year,* The Moralists, *of which an earlier version,* The Sociable Enthusiast: A Philosophical Adventure, *had been circulated as early as 1705. These four treatises, along with two others, were gathered in three volumes as* Characteristicks of Men, Manners, Opinions, Times, *in 1711. The following selections are modernized from the text of John M. Robertson's edition of 1900, as reprinted, with an introduction by Stanley Grean, in 1964.*

Reacting against the materialism of Hobbes and the scientific philosophy of the Royal Society and of his own tutor, Locke, and influenced by the Cambridge Platonists (his earliest publication was a preface to a collection of Whichcote's sermons in 1698), Shaftesbury affirmed the existence of a "moral sense" in man. He denied that any hypothetically presocial condition of humanity could be considered a "natural" state, and opposed the "enthusiasm" or fanaticism of religious zealotry not by a mechanistic view of the world but by ridicule. More fundamentally, he proposed in its place a different form of enthusiasm—most fully expressed in the "philosophical rhapsody" of The Moralists—*for the glories and beauties of Nature, a "God-governed machine" whose interdependent and mutually sympathetic parts evince a simplicity, harmony, and order that sends Theocles in this dialogue into rapturous transports. "In all Shaftesbury's thinking the ruling idea is that of Nature as a harmonious and beautiful order proclaiming its divine Original," Willey writes in* The English Moralists; *"for Nature in this sense, which includes both the physical and the moral worlds, he felt both religious reverence and aesthetic appreciation," if indeed these can be distinguished. His God, Willey continues, "must be one in whom is no darkness at all, and who is, in fact (as he says), 'the best-natured being in the world,'" and human virtue replicates the harmonious order of the cosmos.*

Shaftesbury's approach to nature, Grean remarks, "was aesthetic and intuitive rather than scientific or mechanistic," for he "derives his model of nature not so much from mechanical structure as from organic form: nature is animate, intelligent, and purposive." He thus both revives the views of classical and Christian thinkers in the Platonic tradition and anticipates the nascent Romanticism of the later eighteenth century. From the deists with whom he otherwise has much in common he differentiates himself by his stress, Grean writes, on the immanence of a God "involved in a constant and living interaction with his creation," and still more importantly by the fervent religiosity of a style in sharp contrast with their "dry and barren reason."

Though Shaftesbury soon fell out of favor in an England philosophically attuned to the commonsensical restraint of Locke and his empiricist successors (whom he nonetheless influenced), his optimistic emphasis on the inherent goodness of man found echoes in the latitudinarian divines of the eighteenth century and in the characters (from Parson Abraham Adams to the goodhearted rapscallion Tom Jones) of Henry Fielding; and his rhapsodic apostrophes to Nature anticipated the exclamatory poetic enthusiasms of James Thomson and Joseph Warton (Chapter 19 above). But it was on the Continent that his influence was greatest. In France, Diderot translated his works, Rousseau praised him, and Montesquieu called him "one of the four great poets of the world" (along with Plato, Montaigne, and Malebranche!). And in Germany, his ideas and style had an immense impact on Herder—who called him "a virtuoso of humanity" and found The Moralists *possibly superior to the dialogues of Plato—and on the young Goethe, Schiller, and Kant. He would thus have for those who reacted against the Enlightenment an importance comparable to that of Locke for those who upheld it—even when, as with Diderot, Rousseau, or Kant, these opposites were combined in one person.*

FROM *Part II, Section IV*

. . . Theocles then proposed we should walk out, the evening being fine, and the free air suiting better (as he thought) with such discourse than a chamber.

Accordingly we took our evening walk in the fields, from whence the laborious hinds were now retiring. We fell naturally into the praises of a country life, and discoursed awhile of husbandry and the nature of the soil. Our friends began to admire some of the plants which grew here to great perfection. And it being my fortune (as having acquired a little insight into the nature of simples) to say something they mightily approved upon this subject, Theocles immediately turning about to me, "O my ingenious friend!" said he, "whose reason in other respects must be allowed so clear and happy, how is it possible that with such insight, and accurate judgment in the particulars of natural beings and operations, you should no better judge of the structure of things in general, and of the order and frame of Nature? Who better than yourself can show the structure of each plant and animal body, declare the office of every part and organ, and tell the uses, ends, and advantages to which they serve? How therefore should you prove so ill a naturalist in this whole, and understand so little the anatomy of the world and Nature, as not to discern the same relation of parts, the same consistency and uniformity in the universe! . . .

"Strange! that there should be in Nature the idea of an order and perfection which Nature herself wants! That beings which arise from Nature should be so perfect as to discover imperfection in her constitution, and be wise enough to correct that wisdom by which they were made! . . .

"Now in this which we call the universe, whatever the perfection may be of any particular systems, or whatever single parts may have proportion, unity, or form within themselves, yet if they are not united all in general, in one system, but are, in respect of one another, as the driven sands, or clouds, or breaking waves, then there being no coherence in the whole, there can be inferred no order, no proportion, and consequently no project or design. But if none of these parts are independent, but all apparently united, then is the whole a system complete, according to one simple, consistent, and uniform design.

"Here then is our main subject insisted on, that neither man nor any other animal, though ever so complete a system of parts as to all within, can be allowed in the same manner complete as to all without, but must be considered as having a further relation abroad to the system of his kind. So even this system of his kind to the animal system, this to the world (our earth), and this again to the bigger world and to the universe.

"All things in this world are united. For as the branch is united with the tree, so is the tree as immediately with the earth, air, and water which feed it. As much as the fertile mould is fitted to the tree, as much as the strong and upright trunk of the oak or elm is fitted to the twining branches of the vine or ivy; so much are the very leaves, the seeds, and fruits of these trees fitted to the various animals: these again to one another and to the elements where they live, and to which they are, as appendices, in a manner fitted and joined, as either by wings for the air, fins for the water, feet for the earth, and by other correspondent inward parts of a more curious frame and texture. Thus in contemplating all on earth, we must of necessity view all in one, as holding to one common stock. Thus too in the system of the bigger world. See there the mutual dependency of things! the relation of one to another; of the sun to this inhabited earth, and of the earth and other planets to the sun! the order, union, and coherence of the whole! and know, my ingenious friend, that by this survey you will be obliged to own the universal system and coherent scheme of things to be established on abundant proof, capable of convincing any fair and just contemplator of the works of Nature. For scarce would any one, till he had well surveyed this universal scene, believe a union thus evidently demonstrable, by such numerous and powerful instances of mutual correspondency and relation, from the minutest ranks and orders of beings to the remotest spheres.

"Now in this mighty union, if there be such relations of parts one to another as are not easily discovered, if on this account the end and use of things does not everywhere appear, there is no wonder, since 'tis no more indeed than what must happen of necessity; nor could supreme wisdom have otherwise ordered it. For in an infinity of things thus relative, a mind which sees not infinitely can see nothing fully; and since each particular has relation to all in general, it can know no perfect or true relation of any thing in a world not perfectly and fully known. . . .

"Now having recognized this uniform consistent fabric, and owned the universal system, we must of consequence acknowledge a universal mind, which no ingenious man can be tempted to disown, except through the imagination of disorder in the universe, its seat. For can it be supposed of any one in the world, that being in some desert far from men, and hearing there a perfect symphony of music, or seeing an exact pile of regular architecture arising gradually from the earth in all its orders and proportions, he should be persuaded that at the bottom there was no design accompanying this, no secret spring of thought, no active mind? Would he, because he saw no hand, deny the handiwork, and suppose that each of these complete and perfect systems were framed, and thus united in just symmetry and conspiring order, either by the accidental blowing of the winds or rolling of the sands?

"What is it then should so disturb our views of nature as to destroy that unity of design and order of a mind, which otherwise would be so apparent? All we can see either of the heavens or earth demonstrates order and perfection, so as to afford the noblest subjects of contemplation to minds, like yours, enriched with sciences and learning. All is delightful, amiable, rejoicing, except with relation to man only, and his circumstances, which seem unequal. Here the calamity and ill arises, and hence the ruin of this goodly frame. All perishes on this account; and the whole order of the universe, elsewhere so firm and entire, and immovable, is here overthrown and lost by this one view, in which we refer all things to ourselves, submitting the interest of the whole to the good and interest of so small a part. . . ."

"What is it then," said the old gentleman, "which we call the state of nature?"

"Not that imperfect rude condition of mankind," said Theocles, "which some imagine; but which, if it ever were in nature, could never have been of the least continuance, or any way tolerable, or sufficient for the support of human race. Such a condition cannot indeed so properly be called a state. For what if, speaking of an infant just coming into the world, and in the moment of the birth, I should fancy to call this a state, would it be proper?"

"Hardly so, I confess."

"Just such a state, therefore, was that which we suppose of man ere yet he entered into society, and became in truth a human creature. 'Twas the rough draft of man, the essay or first effort of Nature, a species in the birth, a kind as yet unformed; not in its natural state, but under violence, and still restless, till it attained its natural perfection."

"And thus," said Theocles (addressing still more particularly the old gentleman), "the case must necessarily stand, even on the supposal that there was ever such a condition or state of men, when as yet they were unassociated, unacquainted, and consequently without any language or form of art. But that it was their natural state to live thus separately can never without absurdity be allowed. For sooner may you divest the creature of any other feeling or affection than that towards society and his likeness. Allowing you, however, the power of divesting him at pleasure, allowing you to reduce even whole parts and members of his present frame, would you transform him thus and call him still a man? Yet better might you do this indeed than you could strip him of his natural affections, separate him from all his kind, and enclosing him like some solitary insect in a shell, declare him still a man. So might you call the human egg or embryo the man. The bug which breeds the butterfly is more properly a fly, though without wings, than this imaginary creature is a man. For though his outward shape were human, his passions, appetites, and organs must be wholly different. His whole inward make must be reversed, to fit him for such a recluse economy and separate subsistence.

"To explain this a little further," continued he, "let us examine this pretended state of nature: how and on what foundation it must stand. For either man must have been from eternity or not. If from eternity, there could be no primitive or original state, no state of nature other than we see at present before our eyes. If not from eternity, he arose either all at once (and consequently he was at the very first as he is now) or by degrees, through several stages and conditions, to that in which he is at length settled, and has continued for so many generations. . . .

"Let us go on, however, and on their hypothesis consider which state we may best call Nature's own. She has by accident, through many changes and chances, raised a creature which, springing at first from rude seeds of matter, proceeded till it became what now it is, and arrived where for many generations it has been at a stay. In this long procession (for I allow it any length whatever) I ask, Where was it that this state of Nature could begin? The creature must have endured many changes; and each change, whilst he was thus growing up, was as natural one as another. So that either there must be reckoned a hundred different states of Nature, or if one, it can be only that in which Nature was perfect, and her growth complete. Here where she rested and attained her end, here must be her state, or nowhere.

"Could she then rest, think you, in that desolate state before society? Could she maintain and propagate the species, such as it now is, without fellowship or community? Show it us in fact anywhere, amongst any of our own kind. For as for creatures which may much resemble us in outward form, if they differ yet in the least part of their constitution, if their inwards are of a different texture, if their skin and pores are otherwise formed or hardened; if they have other excrescences of body, another temper, other natural inseparable habits or affections, they are not truly of our kind. . . .

"To conclude," said he (addressing still to the two companions), "I will venture to add a word in behalf of Philocles: that since the learned have such a fancy for this notion, and love to talk of this imaginary state of Nature, I think 'tis even charity to speak as ill of it as we possibly can. Let it be a state of war, rapine, and injustice. Since 'tis unsocial, let it even be as uncomfortable and as frightful as 'tis possible. To speak well of it is to render it inviting and tempt men to turn hermits. Let it, at least, be looked on as many degrees worse than the worst government in being. The greater dread we have of anarchy, the better countrymen we shall prove, and value more the laws and constitution under which we live, and by which we are protected from the outrageous violences of such an unnatural state. In this I agree heartily with those transformers of human nature who, considering it abstractedly and apart from government or society, represent it under monstrous visages of dragons, leviathans, and I know not what devouring creatures. They would have done well, however, to have expressed themselves more properly in their great maxim. For to say in disparagement of man that he is to man a wolf appears somewhat absurd, when one considers that wolves are to wolves very kind and loving creatures. The sexes strictly join in the care and nurture of the young, and this union is continued still between them. They howl to one another to bring company, whether to hunt, or invade their prey, or assemble on the discovery of a good carcase. Even the swinish kinds want not common affection, and run in herds to the assistance of their distressed fellows. The meaning, therefore, of this famous sentence (if it has any meaning at all) must be, That man is naturally to man as a wolf is to a tamer creature; as, for instance, to a sheep. But this will be as little to the purpose as to tell us that there are different species or characters of men; that all have not this wolfish nature, but that one half at least are naturally innocent and mild. And thus the sentence comes to nothing. For without belying nature and contradicting what is evident from natural history, fact, and the plain course of things, 'tis impossible to assent to this ill-natured proposition when we have even done our best to make tolerable sense of it. . . . But such is mankind! . . ."

FROM *Section V*

. . . The contemplation of the universe, its laws and government, was, I averred, the only means which could establish the sound belief of a Deity. For what though innumerable miracles from every part assailed the sense and gave the trembling soul no respite? What though the sky should suddenly open and all kinds of prodigies appear, voices be heard or characters read? What would this evince more than that there were certain powers could do all this? But what powers, whether one or more, whether superior or subaltern, mortal or immortal, wise or foolish, good or bad; this would still remain a mystery, as would the powers asserted. Their word could not be taken in their own case. They might silence men indeed, but not convince them . . .

"But now," continued I, "since I have been thus long the defendant only, I am resolved to take up offensive arms and be aggressor in my turn . . . I maintain that most of those maxims you build upon are fit only to betray your own cause. For whilst you are laboring to unhinge Nature, whilst you are searching heaven and earth for prodigies, and studying how to miraculize everything, you bring confusion on the world, you break its uniformity and destroy that admirable simplicity of order from whence the one infinite and perfect principle is known. Perpetual strifes, convulsions, violences, breach of laws, variation and unsteadiness of order, show either no control, or several uncontrolled and unsubordinate powers in Nature. We have before our eyes either the chaos and atoms of the atheists, or the magic and demons of the polytheists. Yet is this tumultuous system of the universe asserted with the highest zeal by some who would maintain a Deity. This is that face of things, and these the features by which they represent divinity. Hither the eyes of our more inquisitive and ingenuous youth are turned with care, lest they see anything otherwise than in this perplexed and amazing view. As if atheism were the most natural inference which could be drawn from a regular and orderly state of things! But after all this mangling and disfigurement of Nature, if it happens (as oft it does) that the amazed disciple, coming to himself and searching leisurely into Nature's ways, finds more of order, uniformity, and constancy in things than he suspected, he is, of course, driven into atheism; and this merely by the impressions he received from that preposterous system which taught him to seek for Deity in confusion, and to discover Providence in an irregular disjointed world."

"And when you," replied he, "with your newly-espoused system, have brought all things to be as uniform, plain, regular, and simple as you could wish, I suppose you will send your disciple to seek for Deity in mechanism; that is to say, in some exquisite system of self-governed matter. For what else is it you naturalists make of the world than a mere machine?"

"Nothing else," replied I, "if to the machine you allow a mind. For in this case 'tis not a self-governed but a God-governed machine."

"And what are the tokens," said he, "which should convince us? What signs should this dumb machine give of its being governed?"

"The present," replied I, "are sufficient. It cannot possibly give stronger signs of life and steady thought. Compare our own machines with this great one, and see whether by their order, management, and motions they betoken either so perfect a life or so consummate an intelligence. The one is regular, steady, permanent; the other are irregular, variable, inconstant. In one there are the marks of wisdom and determination; in the other of whimsy and conceit: in the one there appears judgment; in the other, fancy only: in one, will; in the other, caprice: in one, truth, certainty, knowledge; in the other, error, folly, and madness. But to be convinced there is something above which thinks and acts, we want, it seems, the latter of these signs, as supposing there can be no thought or intelligence beside what is like our own. We sicken and grow weary with the orderly and regular course of things. Periods, and stated laws, and revolutions, just and proportionable, work not upon us, nor win our admiration. We must have riddles, prodigies, matter for surprise and horror! By harmony, order, and concord we are made atheists; by irregularity and discord we are convinced of Deity! The world is mere accident if it proceeds in course, but an effect of wisdom if it runs mad!"

Thus I took upon me the part of a sound theist whilst I endeavored to refute my antagonist and show that his principles favored atheism. The zealous gentleman took high offense, and we continued debating warmly till late at night. But Theocles was moderator, and we retired at last to our repose, all calm and friendly. . . .

FROM *Part III, Section I*

". . . Haste then, I conjure you," said I, "good Theocles, and stop not one moment for any ceremony or rite. For well I see, methinks, that without any such preparation some divinity has approached us and already moves in you. We are come to the sacred groves of the Hamadryads, which formerly were said to render oracles. We are on the most beautiful part of the hill, and the sun, now ready to rise, draws off the curtain of night and shows us the open scene of Nature in the plains below. Begin: for now I know you are full of those divine thoughts which meet you ever in this solitude. Give them but voice and accents; you may be still as much alone as you are used, and take no more notice of me than if I were absent."

Just as I had said this, he turned away his eyes from me, musing awhile by himself; and soon afterwards, stretching out his hand, as pointing to the objects round him, he began:—

"Ye fields and woods, my refuge from the toilsome world of business, receive me in your quiet sanctuaries, and favor my retreat and thoughtful solitude. Ye verdant plains, how gladly I salute ye! Hail all ye blissful mansions! known seats! delightful prospects! majestic beauties of this earth, and all ye rural powers and graces! Blessed be ye chaste abodes of happiest mortals, who here in peaceful innocence enjoy a life unenvied, though divine; whilst with its blessed tranquility it affords a happy leisure and retreat for man, who, made for contemplation, and to search his own and other natures, may here best meditate the cause of things, and, placed amidst the various scenes of Nature, may nearer view her works.

"O glorious nature! supremely fair and sovereignly good! all-loving and all-lovely, all-divine! whose looks are so becoming and of such infinite grace; whose study brings such wisdom, and whose contemplation such delight; whose every single work affords an ampler scene, and is a nobler spectacle than all which ever art presented! O mighty Nature! wise substitute of Providence! empowered creatress! Or thou empowering Deity, supreme creator! Thee I invoke and thee alone adore. To thee this solitude, this place, these rural meditations are sacred; whilst thus inspired with harmony of thought, though unconfined by words, and in loose numbers, I sing of Nature's order in created beings, and celebrate the beauties which resolve in thee, the source and principle of all beauty and perfection.

"Thy being is boundless, unsearchable, impenetrable. In thy immensity all thought is lost, fancy gives over its flight, and wearied imagination spends itself in vain, finding no coast nor limit of this ocean, nor, in the widest tract through which it soars, one point yet nearer the circumference than the first center whence it parted. Thus having oft essayed, thus sallied forth into the wide expanse, when I return again within myself, struck with the sense of this so narrow being and of the fulness of that immense one, I dare no more behold the amazing depths nor sound the abyss of Deity.

"Yet since by thee, O sovereign mind, I have been formed such as I am, intelligent and rational, since the peculiar dignity of my nature is to know and contemplate thee, permit that with due freedom I exert those faculties with which thou hast adorned me. Bear with my venturous and bold approach. And since nor vain curiosity, nor fond conceit, nor love of aught save thee alone inspires me with such thoughts as these, be thou my assistant and guide me in this pursuit, whilst I venture thus to tread the labyrinth of wide Nature and endeavor to trace thee in thy works."

Here he stopped short, and starting as out of a dream: "Now, Philocles," said he, "inform me, how have I appeared to you in my fit? Seemed it a sensible kind of madness, like those transports which are permitted to our poets? or was it downright raving?"

"I only wish," said I, "that you had been a little stronger in your transport, to have proceeded as you began, without ever minding me. For I was beginning to see wonders in that Nature you taught me, and was coming to know the hand of your divine Artificer. But if you stop here I shall lose the enjoyment of the pleasing vision. And already I begin to find a thousand difficulties in fancying such a universal genius as you describe."

"Why," said he, "is there any difficulty in fancying the universe to be one entire thing? Can one otherwise think of it, by what is visible, than that all hangs together as of a piece? Grant it; and what follows? Only this, that if it may indeed be said of the world that it is simply one, there should be something belonging to it which makes it one. As how? No otherwise than as you may observe in everything. For to instance in what we see before us: I know you look upon the traces of this vast wood to be different from one another; and this tall oak, the noblest of the company, as it is by itself a different thing from all its fellows of the wood, so with its own wood of numerous spreading branches (which seem so many different trees) 'tis still, I suppose, one and the self-same tree. Now should you, as a mere caviler, and not as a fair sceptic, tell me that if a figure of wax, or any other matter, were cast in the exact shape and colors of this tree, and tempered, if possible, to the same kind of substance, it might therefore possibly be a real tree of the same kind or species, I would have done with you, and reason no longer. But if you questioned me fairly, and desired I should satisfy you what I thought it was which made this oneness or sameness in the tree or any other plant, or by what it differed from the waxen figure, or from any such figure accidentally made, either in the clouds, or on the sand by the sea shore, I should tell you that neither the wax, nor sand, nor cloud thus pieced together by our hand or fancy had any real relation with themselves, or had any nature by which they corresponded any more in that near situation of parts than if scattered ever so far asunder. But this I should affirm, that wherever there was such a sympathizing of parts as we saw here in our real tree, wherever there was such a plain concurrence in one common end, and to the support, nourishment, and propagation of so fair a form, we could not be mistaken in saying there was a peculiar nature belonging to this form, and common to it with others of the same kind. By virtue of this, our tree is a real tree, lives, flourishes, and is still one and the same even when by vegetation and change of substance not one particle in it remains the same."

"At this rate indeed," said I, "you have found a way to make very adorable places of these sylvan habitations. For besides the living genius of each place, the woods too, which by your account are animated, have their Hamadryads, no doubt, and the springs and rivulets their nymphs in store belonging to them, and these too, by what I can apprehend, of immaterial and immortal substances."

"We injure them then," replied Theocles, "to say they belong to these trees, and not rather these trees to them. But as for their immortality, let them look to it themselves. I only know that both theirs and all other natures must for their duration depend alone on that Nature on which the world depends; and that every genius else must be subordinate to that one Good Genius, whom I would willingly persuade you to think belonging to this world, according to our present way of speaking. . . ."

Of Bees and Men

The Fable of the Bees, by Bernard Mandeville

Bernard (de) Mandeville (1670–1733) was born at Dort (Dordrecht), or at nearby Rotterdam, Holland, and studied medicine at Leyden, taking his M.D. in 1691; a year or so later, he came to England, where he remained the rest of his life. His versified Fables after . . . la Fontaine *appeared in 1703, and was expanded in 1704 as* Aesop Dressed, or a Collection of Fables. *In*

1705 he published, in octosyllabic couplets, The Grumbling Hive, or Knaves Turned Honest, *portraying a beehive that prospered as long as the vices of its individual members thrived:*

Thus every part was full of vice,
Yet the whole mass a paradise; . . .
This was the state's craft, that maintained
The whole of which each part complained:
This, as in music harmony
Made jarrings in the main agree.

The hive fell into ruin, however, as soon as a "sermonizing rascal" (who had grown rich "by cheating master, king, and poor") "swore he'd rid / The bawling hive of fraud; and did," until

All arts and crafts neglected lie;
Content, the bane of industry,
Makes 'em admire their homely store,
And neither seek nor covet more.

As a result, poverty soon depopulates the once-prosperous hive. In 1714 Mandeville reissued the poem with prose commentaries as The Fable of the Bees: or Private Vices, Publick Benefits; *the expanded edition of 1723 was declared a public nuisance by the Grand Jury of Middlesex and assailed by eminent writers including Francis Hutcheson, William Law, and George Berkeley. Part II first appeared in 1729, followed by other editions both in Mandeville's lifetime and after. (The standard modern edition, in two volumes, is F. B. Kaye's of 1924; see also the abridgement by Irwin Primer [1962].)*

The design of his Fable, *Mandeville declared in the 1714 edition, "is to show the impossibility of enjoying all the most elegant comforts of life that are to be met with in an industrious, wealthy and powerful nation, and at the same time, be blessed with all the virtue and innocence that can be wished for in a golden age; from thence to expose the unreasonableness and folly of those, that desirous of being an opulent and flourishing people and wonderfully greedy after all the benefits they can receive as such, are yet always murmuring at and exclaiming against those vices and inconveniences that from the beginning of the world to this present day have been inseparable from all kingdoms and states that ever were famed for strength, riches, and politeness, at the same time." The contrast with Shaftesbury's benevolent optimism could hardly have been greater. Mandeville's spokesman Cleomenes declares at the end of Part II that Shaftesbury's ideas "of the goodness and excellency of our nature were as romantic and chimerical as they are beautiful and amiable; that he labored hard to unite two contraries that can never be reconciled together, innocence of manners, and worldly greatness; . . . and, lastly, that by ridiculing many passages of Holy Writ, he seems to have endeavored to sap the foundations of all revealed religion, with design of establishing heathen virtue on the ruins of Christianity." As Kaye observes, "Shaftesbury said, Consider the Whole and the individual will then be cared for; Mandeville said, Study the individual and the Whole will then look after itself. To Shaftesbury, also, the coincidence of public and private good was due to an enlightened benevolence, whereas to Mandeville it was the result of narrow self-seeking—Mandeville believing men completely and inevitably egoistic, Shaftesbury thinking them endowed with altruistic and gregarious feeling . . . Mandeville is on the surface an absolutist, a rationalist, and an ascetic, but is basally a relativist, an anti-rationalist, and a utilitarian; whereas Shaftesbury is superficially a relativist and spokesman for impulse, but is really an absolutist and a rationalist."*

A similar contrast, as Leslie Stephen remarks, characterizes their views of nature. "With Shaftesbury nature is an impersonal deity, of whose character and purpose we can form a conception, inadequate and yet sufficient for our world, by tracing the design manifested in the marvelous order of the visible universe. With Mandeville nature is a power altogether inscrutable to our feeble intelligence. In a certain sense, indeed, we can see that she has formed animals for inhabiting this world; but, in fact, 'every part of her works, ourselves not excepted, are an impenetrable secret to us, that eludes all enquirers.' Nature makes animals to feed upon each other; waste of life, cruelty, voracity, and lust are parts of her mysterious plan; 'all actions in nature, abstractedly considered, are equally indifferent'; and cruelty and malice are words applicable only to our own feelings. Nature, in short, is a dark power, whose action can only be inferred from facts, not from any a priori *theory of design, harmony, and order."*

In the following selections, this view—anticipating Hume's blind nature—finds expression in Cleomenes' admiration of the "order, symmetry, and superlative wisdom to be found in all the works of Nature," including lions, tigers, and wolves ("It is not greater cruelty or more unnatural in a wolf to eat a piece of man than it is in a man to eat part of a lamb or a chicken"), but also in human societies, which require the concurrence of human wisdom and discipline with natural impulse. Like Machiavelli and Hobbes, Mandeville continually directs our attention from the ideal to the real: "One of the greatest reasons why so few people understand themselves," he writes in his introduction, "is that most writers are always teaching men what they should be, and hardly ever trouble their heads with telling them what they really are." For this reason, he scandalized the conventional moralists of his time. But for this reason also, perhaps, Samuel Johnson, despite his fundamentally different view that "the happiness of society depends on virtue," could say to Boswell in 1778: "I read Mandeville forty, or, I believe, fifty years ago. He did not puzzle me; he opened my views into real life very much."

FROM PART II

FROM *The Fourth Dialogue between Horatio and Cleomenes*

Hor. But is not the sociableness of man the work of Nature or rather of the author of Nature, Divine Providence?

Cleo. Without doubt; but so is the innate virtue and peculiar aptitude of everything; that grapes are fit to make wine, and barley and water to make other liquors is the work of Providence; but it is human sagacity that finds out the uses we make of them; all the other capacities of man likewise, as well as his sociableness, are evidently derived from God who made him; everything therefore that our industry can produce or compass is originally owing to the Author of our being. But when we speak of the works of Nature to distinguish them from those of art, we mean such as were brought forth without our concurrence. So Nature in due season produces peas; but in England you cannot have them green in January without art and uncommon industry. What Nature designs, she executes herself; there are creatures of whom it is visible that Nature has designed them for society, as is most obvious in bees, to whom she has given instincts for that purpose, as appears from the effects. We owe our being and everything else to the great Author of the universe; but as societies cannot subsist without his preserving power, so they cannot exist without the concurrence of human wisdom; all of them must have a dependence either on mutual compact or the force of the strong exerting itself upon the patience of the weak. The difference between the works of art and those of Nature is so immense that it is impossible not to know them asunder. Knowing, *a priori*, belongs to God only, and Divine Wisdom acts with an original certainty, of which what we call demonstration is but an imperfect borrowed copy. Amongst the works of Nature, therefore, we see no trials nor essays; they are all complete and such as she would have them at the first production, and, where she has not been interrupted, highly finished beyond the reach of our understanding as well as senses. Wretched man on the contrary is sure of nothing, his own existence not excepted, but from a reasoning *a posteriori*. The consequence of this is that the works of art and human invention are all very lame and defective, and most of them pitifully mean at first; our knowledge is advanced by slow degrees, and some arts and sciences require the experience of many ages before they can be brought to any tolerable perfection. Have we any reason to imagine that the society of bees that sent forth the first swarm made worse wax or honey than any of their posterity have produced since? And again the laws of Nature are fixed and unalterable; in all her orders and regulations there is a stability nowhere to be met with in things of human contrivance and approbation: *Quid placet aut odio est, quod non mutabile credas?*[9] Is it probable that amongst the bees there has ever been any other form of government than what every swarm submits to now? What an infinite variety of speculations, what ridiculous schemes have not been proposed amongst men on the subject of government; what dissensions in opinion and what fatal quarrels has it not been the occasion of! and which is the best form of it, is a question to this day undecided. The projects, good and bad, that have been stated for the benefit and more happy establishment of society are innumerable; but how shortsighted is our sagacity, how fallible human judgment! What has seemed highly advantageous to mankind in one age has often been found to be evidently detrimental by the succeeding; and even among contemporaries, what is revered in one country is the abomination of another. What changes have ever bees made in their furniture or architecture? Have they ever made cells that were not sexangular or added any tools to those which Nature furnished them with at the beginning? What mighty structures have been raised, what prodigious works have been performed by the great nations of the world! Toward all these Nature has only found materials; the quarry yields marble, but it is the sculptor that makes a statue of it. To have the infinite variety of iron tools that have been invented, Nature has given us nothing but the ore, which she has hid in the bowels of the earth.

Hor. But the capacity of the workmen, the inventors of arts and those that improved them, has had a great share in bringing those labors to perfection; and their genius they had from Nature.

Cleo. So far as it depended upon the make of their frame, the accuracy of the machine they had, and no further; but this I have allowed already; and if you remember what I have said on this head, you will find that the part which Nature contributed toward the skill and patience of every single person that had a hand in those works was very inconsiderable. . . .

Hor. Philosophers therefore are very wisely employed when they discourse about the laws of Nature, and pretend to determine what a man in the state of Nature would think and which way he would reason concerning himself and the creation uninstructed.

Cleo. Thinking and reasoning justly, as Mr. Locke has rightly observed, require time and practice. Those that have not used themselves to thinking but just on their present necessities, make poor work of it when they try beyond that. In remote parts and such as are least inhabited, we shall find our species come nearer the state of Nature than it does in and near great cities and considerable towns, even in the most civilized nations. Among the most ignorant of such people you may learn the truth of my assertion; talk to them about anything that requires abstract thinking, and there is not one in fifty that will understand you any more than a horse would; and yet many of them are useful labor-

[9] What pleasure or aversion would you not think subject to change? (Horace, *Epistles* II. i. 101).

ers and cunning enough to tell lies and deceive. Man is a rational creature, but he is not endued with reason when he comes into the world; nor can he afterwards put it on when he pleases, at once, as he may a garment. Speech likewise is a characteristic of our species, but no man is born with it; and a dozen generations proceeding from two savages would not produce any tolerable language; nor have we reason to believe that a man could be taught to speak after five-and-twenty if he had never heard others before that time.

Hor. The necessity of teaching whilst the organs are supple and easily yield to impression, which you have spoke of before, I believe is of great weight, both in speaking and thinking; but could a dog or a monkey ever be taught to speak?

Cleo. I believe not; but I do not think that creatures of another species had ever the pains bestowed upon them that some children have before they can pronounce one word. Another thing to be considered is that though some animals perhaps live longer than we do, there is no species that remains young so long as ours; and besides what we owe to the superior aptitude to learn, which we have from the great accuracy of our frame and inward structure, we are not a little indebted for our docility to the slowness and long gradation of our increases before we are full grown: the organs in other creatures grow stiff before ours are come to half their perfection.

Hor. So that in the compliment we make to our species of its being endued with speech and sociableness, there is no other reality than that by care and industry men may be taught to speak and be made sociable if the discipline begins when they are very young?

Cleo. Not otherwise. A thousand of our species all grown up, that is, above five-and-twenty, could never be made sociable if they had been brought up wild and were all strangers to one another. . . .

FROM *The Fifth Dialogue between Horatio and Cleomenes*

Cleo. . . . Life in all creatures is a compound action, but the share they have in it themselves is only passive. We are forced to breathe before we know it; and our continuance palpably depends upon the guardianship and perpetual tutelage of Nature; whilst every part of her works, ourselves not excepted, is an impenetrable secret to us that eludes all inquiries. Nature furnishes us with all the substance of our food herself, nor does she trust to our wisdom for an appetite to crave it; to chew it, she teaches us by instinct and bribes us to it by pleasure. This seeming to be an action of choice and ourselves being conscious of the performance, we perhaps may be said to have a part in it; but the moment after, Nature resumes her care and, again withdrawn from our knowledge, preserves us in a mysterious manner without any help or concurrence of ours that we are sensible of. Since, then, the management of what we have eat and drank remains entirely under the direction of Nature, what honor or shame ought we to receive from any part of the product, whether it is to serve as a doubtful means toward generation or yields to vegetation a less fallible assistance? It is Nature that prompts us to propagate as well as to eat; and a savage man multiplies his kind by instinct as other animals do without more thought or design of preserving his species than a newborn infant has of keeping itself alive in the action of sucking. . . .

Hor. . . . When I reflect on the condition of man as you have set it before me, naked and defenseless, and the multitude of ravenous animals that thirst after his blood and are superior to him in strength and completely armed by Nature, it is inconceiveable to me how our species should have subsisted.

Cleo. What you observe is well worthy our attention.

Hor. It is astonishing. What filthy, abominable beasts are lions and tigers!

Cleo. I think them to be very fine creatures; there is nothing I admire more than a lion.
Hor. We have strange accounts of his generosity and gratitude, but do you believe them?
Cleo. I do not trouble my head about them. What I admire is his fabric, his structure, and his rage, so justly proportioned to one another. There are order, symmetry, and superlative wisdom to be observed in all the works of Nature, but she has not a machine of which every part more visibly answers the end for which the whole was formed.
Hor. The destruction of other animals.
Cleo. That is true; but how conspicuous is that end, without mystery or uncertainty! That grapes were made for wine and man for society are truths not accomplished in every individual: but there is a real majesty stamped on every single lion, at the sight of which the stoutest animals submit and tremble. When we look upon and examine his massy talons, the size of them and the labored firmness with which they are fixed in and fastened to that prodigious paw, his dreadful teeth, the strength of his jaws, and the width of his mouth equally terrible, the use of them is obvious; but when we consider, moreover, the make of his limbs, the toughness of his flesh and tendons, the solidity of his bones beyond that of other animals, and the whole frame of him, together with his never-ceasing anger, speed, and agility, whilst in the desert he ranges king of beasts! When, I say, we consider all these things, it is stupidity not to see the design of nature and with what amazing skill the beautiful creature is contrived for offensive war and conquest. . . .

Ten troops of wolves with fifty in each would make a terrible havoc in a long winter among a million of our species with their hands tied behind them; but among half that number, one pestilence has been known to slaughter more than so many wolves could have eaten in the same time, notwithstanding the great resistance that was made against it by approved-of medicines and able physicians. It is owing to the principle of pride we are born with, and the high value we all, for the sake of one, have for our species, that men imagine the whole universe to be principally made for their use; and this error makes them commit a thousand extravagancies and have pitiful and most unworthy notions of God and his works. It is not greater cruelty or more unnatural in a wolf to eat a piece of man than it is in a man to eat part of a lamb or a chicken. What or how many purposes wild beasts were made for is not for us to determine; but that they were made we know; and that some of them must have been very calamitous to every infant nation and settlement of men is almost as certain; this you was fully persuaded of, and thought, moreover, that they must have been such an obstacle to the very subsistence of our species as was insurmountable. In answer to this difficulty, which you started, I showed you from the different instincts and peculiar tendencies of animals that in nature a manifest provision was made for our species, by which, notwithstanding the rage and power of the fiercest beasts, we should make a shift, naked and defenseless, to escape their fury so as to be able to maintain ourselves and multiply our kind till by our numbers and arms acquired by our own industry, we could put to flight or destroy all savage beasts without exception, whatever spot of the globe we might have a mind to cultivate and settle on. The necessary blessings we receive from the sun are obvious to a child; and it is demonstrable that without it none of the living creatures that are now upon the earth could subsist. But if it were of no other use, being eight hundred thousand times bigger than the earth at least, one thousandth part of it would do our business as well, if it was but nearer to us in proportion. From this consideration alone I am persuaded that the sun was made to enlighten and cherish other bodies besides this planet of ours. Fire and water were designed for innumerable purposes; and among the uses that are made of them some are immensely different from others. But whilst we receive the benefit of these and are only intent on ourselves, it is highly probable that there are thousands of things, and perhaps our own machines among them, that in the vast system of the universe are now serving some very wise ends, which we shall never know. According to that plan of this globe, I mean the scheme of government in relation to the living creatures that inhabit the earth, the destruction of animals is as necessary as the generation of them. . . .

I never said that wild beasts was designed to thin out our species. I have showed that many things were made to serve a variety of different purposes; that in the scheme of this earth many things must have been considered that man has nothing to do with; and that it is ridiculous to think that the universe was made for our sake. I have said likewise that as all our knowledge comes *a posteriori* it is imprudent to reason otherwise than from facts. That there are wild beasts and that there are savage men is certain; and that where there are but few of the latter, the first must always be very troublesome and often fatal to them is as certain; and when I reflect on the passions all men are born with and their incapacity whilst they are untaught, I can find no cause or motive which is so likely to unite them together and make them espouse the same interest, as that common danger they must always be in from wild beasts in uncultivated countries, whilst they live in small families that all shift for themselves without government or dependence upon one another. This first step to society I believe to be an effect which that same cause, the common danger so often mentioned, will never fail to produce upon our species in such circumstances; what other and how many purposes wild beasts might have been designed for besides, I do not pretend to determine, as I have told you before. . . .

A Grammar for the Understanding of Nature

Selections from the Writings of George Berkeley

Born near Kilkenny, Ireland, of an English immigrant family, George Berkeley (1685–1753) was educated at Kilkenny School and Trinity College, Dublin, where he studied mathematics and the new philosophy of Descartes and Newton, took his B.A. in 1704, and was elected a fellow in 1707. In An Essay towards a New Theory of Vision *(1709), he contended that visual perception is a conventional system of arbitrary signs symbolizing possible tactual experiences, so that what we know is never an external object but only our own perception. In* A Treatise Concerning the Principles of Human Knowledge *(1710), he developed the implications of this theory, maintaining that not only qualities distinguished by Locke as secondary but all qualities of matter are dependent on perception, and thus exist only in the mind—perfectly in God's, imperfectly in the individual human mind. "That neither our thoughts, nor passions, nor ideas formed by the imagination exist without the mind," he writes (I.3), "is what everybody will allow. And it seems no less evident that the various sensations or ideas imprinted on the sense, however blended or combined together (that is, whatever objects they compose), cannot exist otherwise than in a mind perceiving them. . . . For as to what is said of the absolute existence of unthinking things without any relation to their being perceived, that seems perfectly unintelligible. Their* esse *is* percipi *[their being is in being perceived], nor is it possible they should have any existence out of the minds or thinking things which perceive them."*

Though both this book and the more popular account in Three Dialogues between Hylas and Philonous *(1713) were generally ridiculed when noticed at all, Berkeley was admitted to the circle of Swift and Pope during his visit to London in 1713, and gradually became better known. After spending most of the next seven years on the Continent, he returned to Ireland in 1721 and became Dean of Derry in 1724. During these years he conceived his project for founding a college in Bermuda to train Anglican clergy to educate the colonists and natives of America. In the vain hope of receiving funds for his scheme, he lived in Newport, Rhode Island, from 1728 to 1731, convinced—as he wrote in his poem of 1726, "America, or the Muses' Refuge: A Prophecy" (published in his* Miscellany *of 1752)—that "in distant lands now waits a better Time."*

The Force of Art by Nature seems outdone,
And fancied beauties by the true:
In happy Climes, the seat of Innocence,
Where Nature guides and Virtue rules, . . .
There shall be sung another golden Age,

as "westward the course of Empire takes its way." He returned to London in 1731 and in 1732 published anonymously Alciphron: or the Minute Philosopher, *an attack written in Rhode Island on freethinkers and deists such as Shaftesbury. In 1734 he was named Bishop of Cloyne (near Cork, in southern Ireland), and there spent most of the rest of his life, retiring in 1752 to Oxford, where he soon died. His last two publications were* The Querist *(1735–37), on social and political problems in Ireland, and* Siris *(1744), a "chain of philosophical reflections" that begins as a commendation of the curative powers of tar water, a nostrum made in America "by putting a quart of cold water to a quart of tar, and stirring them well together in a vessel," and proceeds to consideration of religious mysteries (including the Trinity) found in writings of ancient Platonic and Pythagorean philosophers.*

Berkeley's immediate philosophical impact (leaving aside those like Samuel Johnson, who thought to refute him by kicking a stone) was mainly on thinkers—above all on David Hume—who saw the sceptical implications latent, against his intentions, in the reduction of matter to perception. But his idealism would later have an important impact also on the English Romantics, notably Coleridge (who named his second son after him), American transcendentalists, and W. B. Yeats; and his thought is perhaps more accessible to a century when matter has become an evanescent tissue of quarks and gluons than it was to a more positivist age. In the selections that follow from the Treatise *(in the revised text of 1734 reprinted in* Principles, Dialogues, and Philosophical Correspondence, *ed. Colin Murray Turbayne [1965]), Berkeley argues against understanding nature as "some blind, unthinking deputy" acting in the place of God, whose guiding hand is evident to the attentive even in the defects of nature. And in the passages from* Siris *(abridged in* Backgrounds of Romanticism *[1967], ed. Leonard M. Trawick; for the full text, see vol. 5 of Berkeley's* Works, *ed. T. E. Jessop [1953]), he proposes "a grammar for the understanding of nature" through the natural connection of signs with things, a view (influenced by Platonic, Pythagorean, and Stoic thought, and more recently by Cudworth) of nature as "the life of the world," in which the divine mind "contains all, and acts all, and is to all created beings the source of unity and identity, harmony and order, existence and stability." Such a vision indeed requires that the phenomenal world not be accepted at its face value.*

FROM A Treatise Concerning the Principles of Human Knowledge

150 But you will say, has nature no share in the production of natural things, and must they all be ascribed to the immediate and sole operation of God? I answer, if by "nature" is meant only the visible *series* of effects or sensations imprinted on our minds, according to certain fixed and general laws, then it is plain that nature, taken in this sense, cannot produce anything at all. But if by "nature" is meant some being distinct

from God, as well as from the laws of nature, and things perceived by sense, I must confess that word is to me an empty sound without any intelligible meaning annexed to it. Nature, in this acceptation, is a vain chimera, introduced by those heathens who had not just notions of the omnipresence and infinite perfection of God. But it is more unaccountable that it should be received among Christians, professing belief in the Holy Scriptures, which constantly ascribe those effects to the immediate hand of God that heathen philosophers are wont to impute to nature. "The Lord he causeth the vapors to ascend; he maketh lightnings with rain; he bringeth forth the wind out of his treasures." Jerem. 10:13. "He turneth the shadow of death into the morning, and maketh the day dark with night." Amos 5:8. "He visiteth the earth, and maketh it soft with showers: He blesseth the springing thereof, and crowneth the year with his goodness; so that the pastures are clothed with flocks, and the valleys are covered over with corn." See Psalm 65. But notwithstanding that this is the constant language of Scripture, yet we have I know not what aversion from believing that God concerns himself so nearly in our affairs. Fain would we suppose him at a great distance off, and substitute some blind, unthinking deputy in his stead, though (if we may believe Saint Paul) "he be not far from every one of us."

151 It will, I doubt not, be objected that the slow and gradual mehods observed in the production of natural things do not seem to have for their cause the immediate hand of an Almighty Agent. Besides, monsters, untimely births, fruits blasted in the blossom, rains failing in desert places, miseries incident to human life, and the like, are so many arguments that the whole frame of nature is not immediately actuated and superintended by a spirit of infinite wisdom and goodness. But the answer to this objection is in a good measure plain . . . ; it being visible that the aforesaid methods of nature are absolutely necessary, in order to working by the most simple and general rules, and after a steady and consistent manner; which argues both the wisdom and goodness of God. Such is the artificial contrivance of this mighty machine of nature that, whilst its motions and various phenomena strike on our senses, the hand which actuates the whole is itself unperceivable to men of flesh and blood. "Verily" (says the prophet) "thou art a God that hidest thyself." Isaiah 45:15. But, though God conceal himself from the eyes of the sensual and lazy, who will not be at the least expense of thought, yet to an unbiased and attentive mind nothing can be more plainly legible than the intimate presence of an all-wise Spirit, who fashions, regulates, and sustains the whole system of being. It is clear, from what we have elsewhere observed, that the operating according to general and stated laws is so necessary for our guidance in the affairs of life, and letting us into the secret of nature, that without it all reach and compass of thought, all human sagacity and design, could serve to no manner of purpose; it were even impossible there should be any such faculties or powers in the mind. . . . Which one consideration abundantly outbalances whatever particular inconveniences may thence arise.

152 We should further consider that the very blemishes and defects of nature are not without their use, in that they make an agreeable sort of variety and augment the beauty of the rest of the creation, as shades in a picture serve to set off the brighter and more enlightened parts. We would likewise do well to examine whether our taxing the waste of seeds and embryos, and accidental destruction of plants and animals, before they come to full maturity, as an imprudence in the Author of Nature, be not the effect of prejudice contracted by our familiarity with impotent and saving mortals. In man indeed a thrifty management of those things which he cannot procure without much

pains and industry may be esteemed wisdom. But we must not imagine that the inexplicably fine machine of an animal or vegetable costs the great Creator any more pains or trouble in its production than a pebble does; nothing being more evident than that an omnipotent spirit can indifferently produce everything by a mere fiat or act of his will. Hence, it is plain that the splendid profusion of natural things should not be interpreted weakness or prodigality in the agent who produces them, but rather be looked on as an argument of the riches of his power. . . .

FROM Siris: A Chain of Philosophical Reflections and Inquiries Concerning the Virtues of Tar Water, and Divers Other Subjects Connected Together and Arising One from Another

252 There is a certain analogy, constancy, and uniformity in the phenomena or appearances of nature, which are a foundation for general rules: and these are a grammar for the understanding of nature, or that series of effects in the visible world, whereby we are enabled to foresee what will come to pass in the natural course of things. Plotinus observes, in his third Ennead, that the art of presaging is in some sort the reading of natural letters denoting order, and that so far forth as analogy obtains in the universe, there may be vaticination.[10] And in reality, he that foretells the motions of the planets, or the effects of medicines, or the result of chemical or mechanical experiments, may be said to do it by natural vaticination. . . .

254 As the natural connection of signs with the things signified is regular and constant, it forms a sort of rational discourse, and is therefore the immediate effect of an intelligent cause. This is agreeable to the philosophy of Plato and other ancients. . . . Therefore, the phenomena of nature, which strike on the senses and are understood by the mind, form not only a magnificent spectacle, but also a most coherent, entertaining, and instructive discourse; and to effect this, they are conducted, adjusted, and ranged by the greatest wisdom. This language or discourse is studied with different attention, and interpreted with different degrees of skill. But so far as men have studied and remarked its rules, and can interpret right, so far they may be said to be knowing in nature. A beast is like a man who hears a strange language, but understands nothing.

255 Nature, saith the learned Doctor Cudworth, is not master of art or wisdom: Nature is ratio mersa et confusa, reason immersed and plunged into matter, and as it were fuddled in it and confounded with it. But the formation of plants and animals, the motions of natural bodies, their various properties, appearances, and vicissitudes, in a word, the whole series of things in this visible world, which we call the course of nature, is so wisely managed and carried on that the most improved human reason cannot thoroughly comprehend even the least particle thereof; so far is it from seeming to be produced by fuddled or confounded reason. . . .

258 Instruments, occasions, and signs occur in, or rather make up, the whole visible course of nature. These, being no agents themselves, are under the direction of one agent concerting all for one end, the supreme good. All those motions, whether in animal bodies or in other parts of the system of nature, which are not effects of particular wills, seem to spring from the same general cause with the vegetation of plants, an aethereal spirit actuated by a mind.

[10]Prophecy.

259 The first poets and theologers of Greece and the east considered the generation of things, as ascribed rather to a divine cause, but the Physici to natural causes subordinate to, and directed still by a divine; except some corporealists and mechanics, who vainly pretended to make a world without a God. The hidden force that unites, adjusts, and causeth all things to hang together, and move in harmony, which Orpheus and Empedocles styled love; this principle of union is no blind principle, but acts with intellect. This divine love and intellect are not themselves obvious to our view, or otherwise discerned than in their effects. Intellect enlightens, Love connects, and the sovereign Good attracts all things. . . .

266 The Pythagoreans and Platonists had a notion of the true system of the world. They allowed of mechanical principles, but actuated by soul or mind: they distinguished the primary qualities in bodies from the secondary, making the former to be physical causes, and they understood physical causes in a right sense: they saw that a mind infinite in power, unextended, invisible, immortal, governed, connected, and contained all things: they saw there was no such thing as real absolute space: that mind, soul, or spirit, truly and really exists: that bodies exist only in a secondary and dependent sense: that the soul is the place of forms: that the sensible qualities are to be regarded as acts only in the cause, and as passions in us: they accurately considered the differences of intellect, rational soul, and sensitive soul, with their distinct acts of intellection, reasoning, and sensation, points wherein the Cartesians and their followers, who consider sensation as a mode of thinking, seem to have failed. They knew there was a subtle aether pervading the whole mass of corporeal beings, and which was itself actually moved and directed by a mind: and that physical causes were only instruments, or rather marks and signs.

267 Those ancient philosophers understood the generation of animals to consist, in the unfolding and distending of the minute imperceptible parts of pre-existing molecules, which passeth for a modern discovery: this they took for the work of nature, but nature animate and intelligent: they understood that all things were alive and in motion: they supposed a concord and discord, union and disunion in particles, some attracting, others repelling each other: and that those attractions and repulsions, so various, regular, and useful, could not be accounted for, but by an intelligence presiding and directing all particular motions, for the conservation and benefit of the whole. . . .

276 Both Stoics and Platonics held the world to be alive, though sometimes it be mentioned as a sentient animal, sometimes as a plant or vegetable. But in this, notwithstanding what hath been surmised by some learned men, there seems to be no atheism. For so long as the world is supposed to be quickened by elementary fire or spirit, which is itself animated by soul, and directed by understanding, it follows that all parts thereof originally depend upon, and may be reduced unto, the same indivisible stem or principle, to wit, a supreme mind; which is the concurrent doctrine of Pythagoreans, Platonics, and Stoics.

277 There is according to those philosophers a life infused throughout all things: the *pyr noeron*, *pyr tekhnikon*, an intellectual and artificial fire,[11] an inward principle, animal spirit, or natural life producing and forming within as art doth without, regulating, moderating, and reconciling the various motions, qualities, and parts of this mundane

[11]See the famous Stoic definition of nature (generally attributed to Zeno) as "an artistically working fire, going on its way to create," quoted in Diogenes Laërtius VII.156 (see introduction to Chapter 10 above), and by Cicero's Lucilius Balbus in *On the Nature of the Gods* II (Chapter 10 above).

system. By virtue of this life the great masses are held together in their orderly courses, as well as the minutest particles governed in their natural motions, according to the several laws of attraction, gravity, electricity, magnetism, and the rest. It is this gives instincts, teaches the spider her web, and the bee her honey. This it is that directs the roots of plants to draw forth juices from the earth, and the leaves and cortical vessels to separate and attract such particles of air, and elementary fire, as suit their respective natures.

278 Nature seems to be not otherwise distinguished from the anima mundi [world soul], than as life is from soul, and, upon the principles of the oldest philosophers, may not improperly or incongruously be styled the life of the world. Some Platonics indeed regard life as the act of nature, in like manner as intellection is of the mind or intellect. As the first intellect acts by understanding, so nature according to them acts or generates by living. But life is the act of the soul, and seems to be very nature itself, which is not the principle, but the result of another, and higher principle, being a life resulting from soul, as cogitation from intellect.

279 If nature be the life of the world, animated by one soul, compacted into one frame, and directed or governed in all parts by one mind, this system cannot be accused of atheism, though perhaps it may of mistake or impropriety. And yet, as one presiding mind gives unity to the infinite aggregate of things by a mutual communion of actions and passions, and an adjustment of parts, causing all to concur in one view to one and the same end, the ultimate and supreme good of the whole, it should seem reasonable to say, with Ocellus Lucanus the Pythagorean, that as life holds together the bodies of animals, the cause whereof is the soul; and as a city is held together by concord, the cause whereof is law; even so the world is held together by harmony, the cause whereof is God. And in this sense, the world or universe may be considered either as one animal or one city. . . .

286 Aristotle hath observed there were indeed some who thought so grossly as to suppose the universe to be one only corporeal and extended nature: but in the first book of his Metaphysics he justly remarks they were guilty of a great mistake; forasmuch as they took into their account the elements of corporeal beings alone; whereas there are incorporeal beings also in the universe; and while they attempted to assign the causes of generation and corruption, and account for the nature of all things, they did at the same time destroy the very cause of motion. . . .

291 Thus much it consists with piety to say that a divine agent doth by his virtue permeate and govern the elementary fire or light, which serves as an animal spirit to enliven and actuate the whole mass, and all the members of this visible world. Nor is this doctrine less philosophical than pious. We see all nature alive or in motion. We see water turned into air, and air rarified and made elastic by the attraction of another medium, more pure indeed, more subtle, and more volatile than air. But still, as this is a moveable extended, and, consequently, a corporeal being, it cannot be itself a principle of motion, but leads us naturally and necessarily to an incorporeal spirit or agent. We are conscious that a spirit can begin, alter, or determine motion, but nothing of this appears in body. Nay the contrary is evident, both to experiment and reflection.

292 Natural phenomena are only natural appearances. They are, therefore, such as we see and perceive them. Their real and objective natures are, therefore, the same: passive without any thing active, fluent and changing without any thing permanent in them.

However, as these make the first impressions, and the mind takes her first flight and spring, as it were, by resting her foot on these objects, they are not only first considered by all men, but most considered by most men. They and the phantoms that result from those appearances, the children of imagination grafted upon sense, such for example as pure space, are thought by many the very first in existence and stability, and to embrace and comprehend all other beings. . . .

295 From the outward form of gross masses which occupy the vulgar, a curious inquirer proceeds to examine the inward structure and minute parts, and from observing the motions in nature, to discover the laws of those motions by the way he frames his hypothesis and suits his language to this natural philosophy. And these fit the occasion and answer the end of a maker of experiments or mechanic, who means only to apply the powers of nature, and reduce the phenomena to rules. But, if proceeding still in his analysis and inquiry, he ascends from the sensible into the intellectual world, and beholds things in a new light and a new order, he will then change his system and perceive that what he took for substances and causes are but fleeting shadows; that the mind contains all, and acts all, and is to all created beings the source of unity and identity, harmony and order, existence and stability. . . .

The Idea of a Blind Nature

Dialogues Concerning Natural Religion, by David Hume

David Hume (1711–76) was born in Edinburgh and grew up on his family estate at Ninewells, near Berwick, Scotland; he spent several years at Edinburgh University without taking a degree, followed by three years in France between 1734 and 1737. While in France, mainly at La Flèche in Anjou, he wrote his first major work, A Treatise of Human Nature, *in three volumes, published in 1739–40. "Never literary attempt was more unfortunate," he later wrote, with some exaggeration: ". . . It fell dead-born from the press, without reaching such distinction as even to excite a murmur among the zealots." He settled at Ninewells between 1740 and 1745, and beginning in 1741 published a series of* Essays Moral and Political, *followed by revision of the first part of his* Treatise *as* An Enquiry Concerning Human Understanding *(originally entitled* Philosophical Essays Concerning Human Understanding*) in 1748, and* An Enquiry Concerning the Principles of Morals *in 1751. In 1745–46 he acted as companion to the Marquis of Annandale, who was declared insane; he then spent two years as aide-de-camp to General St Clair on military missions in Europe during the Seven Years' War. Returning to Edinburgh in 1749, he served as Keeper of the Advocates' Library from 1751 to 1757, and published* The Natural History of Religion *in 1757 and* The History of England *in six volumes between 1754 and 1762. From 1763 to 1766 he lived in Paris as private secretary to the British ambassador and then as chargé d'affaires, and met the leading French philosophes of the age; his befriending of Rousseau, whom he brought back to England, ended badly when Rousseau suspected him, as he did so many others, of plots to malign him. From 1768 until his death he lived in Edinburgh. During his final illness he wrote a short account of his life and made substantial revisions to his* Dialogues Concerning Natural Religion *(first completed around 1751, but withheld on the advice of Adam Smith and others), for whose posthumous publication he left careful instructions; it appeared in 1779.*

Now widely regarded as the greatest British philosopher, Hume, despite the clarity of his style, has been seen very differently by different readers. Just as Hume interpreted Berkeley's demonstration of the perceptual basis of knowledge as leading—against his intention—to scepticism about the possibility of knowledge, so Hume's argument that all reasoning except the purely mathematical is founded in feelings and instincts, or "passions," so that even conclusions concerning causation are inferences undemonstrable by reason, can be seen as extending the philosophical scepticism latent in Locke's empiricism to its all-embracing outcome. On this view, it would seem, as Hazard writes in European Thought in the Eighteenth Century, *that Hume "was a sceptic pure and unadulterated," for whom "Nature is no more . . . There are no laws of Nature, but merely appearances which we interpret erroneously. No more reason, only a chaos of sensations." Yet Hume emphatically asserts that this radical scepticism is left behind when he leaves his study, and has no direct impact on the conduct of life, or even much on the practice of thought, which assumes the existence of a world we cannot know. "We may well ask, What causes induce us to believe in the existence of body?" he writes in the* Treatise *(I.IV.ii), "but 'tis vain to ask, Whether there be body or not? That is a point which we must take for granted in all our reasonings."*

Thus this sceptical philosopher is at the same time the philosopher of the limitations of scepticism: "Most fortunately it happens, that since reason is incapable of dispelling these clouds," he observes (I.IV.vii), "nature herself suffices to that purpose, and cures me of this philosophical melancholy and delirium, either by relaxing this bent of mind, or by some avocation, and lively impression of my senses, which obliterate all these chimeras. I dine, I play a game of backgammon, I converse, and am merry with my friends; and when after three or four hours' amusement, I would return to these speculations, they appear so cold, and strained, and ridiculous, that I cannot find in my heart to enter into them any farther." As Willey remarks in The Eighteenth Century Background, *"He does not in the least wish to impugn the common-sense belief in the existence of an external order—indeed, no one believed in it more firmly than he. He merely denies that this belief is founded upon reason, and that reason alone can account for it." Hume's scepticism, though deflating the false pretenses of inductive as well as deductive reason, far from leading to moral or intellectual relativism, thus validates not only the natural instincts in accord with which we actually live but even the understanding—if accepted as probability, not certainty—to which reasoning grounded on these instincts (it can have no other grounds) leads us. We cannot strictly* know *that the sun will rise each day, often as we have seen it do so, yet the far greater probability of natural laws than of exceptions to them is the basis of Hume's famous refutation of miracles in Section X of* An Enquiry Concerning Human Understanding.

A similar complexity arises with regard to Hume's attitude toward natural religion in the selections from the Dialogues *that follow. The sceptic Philo rebuts both Demea, who holds that the existence of God can be demonstrated by* a priori *reason, and especially Cleanthes, who—like both the deists and Bishop Joseph Butler in his influential* Analogy of Religion, Natural and Revealed, to the Constitution and Course of Nature *(1736)—advances for God's existence an argument from design based on analogies between nature, as God's presumed creation, and human art.*[12] *In Section XI ("Of a Particular Providence and of a Future State") of the* Enquiry, *Hume had endorsed Epicurus's imagined denial of a providential divinity, remarking*

[12]Although Hume took great care to create a genuine dialogue by giving good arguments to each of his interlocutors, and although some have argued that his views are more nearly represented by Cleanthes, the fundamental identification of his outlook with Philo's seems beyond serious dispute. Among those who have affirmed this interpretation are Norman Kemp Smith in his edition of the *Dialogues* (1935; 2nd ed. 1947); Richard Wollheim in his edition of *Hume on Religion* (1963), which contains the full texts of *The Natural History of Religion*, the *Dialogues*, and several shorter essays; Peter Gay in *The Enlightenment: An Interpretation*, vol. 1: *The Rise of Modern Paganism* (1966); and A. J. Ayer in *Hume* (1980).

that "the great source of our mistake on this subject, and of the unbounded license of conjecture, which we indulge, is that we tacitly consider ourselves as in the place of the Supreme Being, and conclude that he will, on every occasion, observe the same conduct which we ourselves, in his situation, would have embraced as reasonable and eligible. But, besides that the ordinary course of nature may convince us that almost everything is regulated by principles and maxims very different from ours; besides this, I say, it must evidently appear contrary to all rules of analogy to reason from the projects and intentions of men to those of a Being so different, and so much superior."

In the Dialogues*—modeled in part, Peter Gay notes, on Cicero's* De Natura Deorum *(Chapter 10 above)—Philo not only expands on this repudiation of anthropomorphic analogy and denies that any inference from experience can legitimately support arguments for God's creation or guidance of the universe; he also starkly portrays a world that presents "nothing but the idea of a blind Nature . . . pouring forth . . . her maimed and abortive children." The refutation is devastating: "To read Hume's* Dialogues *after having read, with sympathetic understanding, the earnest deists and optimistic philosophers of the early century," Carl L. Becker writes in* The Heavenly City of the Eighteenth-Century Philosophers *(1932), "is to experience a slight chill, a feeling of apprehension. It is as if, at the high noon of the Enlightenment, at the hour of the siesta when everything seems so quiet and secure all about, one were suddenly aware of a short, sharp slipping of the foundations, a faint far-off tremor running underneath the solid ground of common sense."*

Yet at the end of the Dialogues *this same Philo declares that no one has "a deeper sense of religion impressed on his mind" than he, "or pays more profound adoration to the Divine Being, as he discovers himself to reason, in the inexplicable contrivance and artifice of Nature." He even affirms that purpose, intention, and design are everywhere apparent, and that Nature (as Aristotle said long before) does nothing in vain. ("The whole frame of Nature bespeaks an intelligent author," Hume likewise affirmed at the beginning of* The Natural History of Religion, *written at about the same time as the* Dialogues; *"and no rational enquirer can, after serious reflection, suspend his belief a moment with regard to the primary principles of genuine Theism and Religion.") What are we to make of this seemingly blatant contradiction? Is Hume, even in death, making prudent obeisance to the pieties of his time, just after demolishing their foundation, or—more probably—is he engaging, as he often does, in ironical assertion of views whose falsity he means the discerning reader to see? Very possibly. Yet possibly, too (Hume leaves each reader to decide), the contradiction is only apparent, for Philo may be saying that reverence arises to the very extent that Nature's contrivance does remain "inexplicable," beyond all demonstration by reason or experience. Such a belief cannot be undermined by rational doubt, since it is based not on rational knowledge but on instinctive feeling, which has no need to prove that its object really exists.*

Hume differentiated himself as sharply from the dogmatic atheism of French philosophes such as Holbach as he did from dogmatic Christianity or dogmatic deism; he was closer, Wollheim remarks, "to an earlier tradition of French thought: to the more thoroughgoing scepticism of Montaigne and Bayle, which was prepared to question all knowledge, even of a negative kind." As Peter Gay writes, "He was willing to live with uncertainty, with no supernatural justifications, no complete explanations, no promise of permanent stability." Yet just as he could turn from the radical scepticism of his study to a game of backgammon or conversation with friends, so perhaps, like Philo, he could turn from demolition of pretended proofs of an anthropomorphically benevolent Nature or God to an unabashedly irrational reverence for the universe in its very inexplicability. "While Newton seemed to draw off the veil from some of the mysteries of nature," he wrote in his History of England *(as quoted by Wollheim), "he showed at the same time the imperfections of the mechanical philosophy; and thereby restored her ultimate secrets to that obscurity in which they ever did and ever shall remain." What Hume here says of the mechanist Newton's restoration of mystery to nature might equally be said, however paradoxically, of the sceptical Hume's.*

FROM *Part II*

... When Cleanthes had assented, Philo, after a short pause, proceeded in the following manner....

But can you think, Cleanthes, that your usual phlegm and philosophy have been preserved in so wide a step as you have taken when you compared to the universe houses, ships, furniture, machines; and from their similarity in some circumstances inferred a similarity in their causes? Thought, design, intelligence, such as we discover in men and other animals, is no more than one of the springs and principles of the universe, as well as heat or cold, attraction or repulsion, and a hundred others which fall under daily observation. It is an active cause, by which some particular parts of nature, we find, produce alterations on other parts. But can a conclusion, with any propriety, be transferred from parts to the whole? Does not the great disproportion bar all comparison and inference? From observing the growth of a hair, can we learn anything concerning the generation of a man? Would the manner of a leaf's blowing, even though perfectly known, afford us any instruction concerning the vegetation of a tree?

But allowing that we were to take the *operations* of one part of nature upon another for the foundation of our judgment concerning the *origin* of the whole (which never can be admitted), yet why select so minute, so weak, so bounded a principle as the reason and design of animals is found to be upon this planet? What peculiar privilege has this little agitation of the brain which we call "thought," that we must thus make it the model of the whole universe? Our partiality in our own favor does indeed present it on all occasions; but sound philosophy ought carefully to guard against so natural an illusion.

So far from admitting, continued Philo, that the operations of a part can afford us any just conclusion concerning the origin of the whole, I will not allow any one part to form a rule for another part, if the latter be very remote from the former. Is there any reasonable ground to conclude that the inhabitants of other planets possess thought, intelligence, reason, or anything similar to these faculties in men? When Nature has so extremely diversified her manner of operation in this small globe, can we imagine that she incessantly copies herself throughout so immense a universe? And if thought, as we may well suppose, be confined merely to this narrow corner, and has even there so limited a sphere of action, with what propriety can we assign it for the original cause of all things? The narrow views of a peasant, who makes his domestic economy the rule for the government of kingdoms, is in comparison a pardonable sophism.

But were we ever so much assured that a thought and reason resembling the human were to be found throughout the whole universe, and were its activity elsewhere vastly greater and more commanding than it appears in this globe, yet I cannot see why the operations of a world constituted, arranged, adjusted, can with any propriety be extended to a world which is in its embryo-state, and is advancing towards that constitution and arrangement. By observation, we know somewhat of the economy, action, and nourishment of a finished animal, but we must transfer with great caution that observation to the growth of a fetus in the womb, and still more, in the formation of an animalcule in the loins of its male parent. Nature, we find, even from our limited experience, possesses an infinite number of springs and principles which incessantly discover themselves on every change of her position and situation. And what new and unknown principles would actuate her in so new and unknown a situation as that of the formation of the universe, we cannot, without the utmost temerity, pretend to determine.

A very small part of this great system, during a very short time, is very imperfectly discovered to us: and do we then pronounce decisively concerning the origin of the whole?

Admirable conclusion! Stone, wood, brick, iron, brass, have not, at this time, in this minute globe of earth, an order or arrangement without human art and contrivance: therefore the universe could not originally attain its order and arrangement without something similar to human art. But is a part of nature a rule for another part very wide of the former? Is it a rule for the whole? Is a very small part a rule for the universe? Is nature in one situation a certain rule for nature in another situation vastly different from the former?

And can you blame me, Cleanthes, if I here imitate the prudent reserve of Simonides, who, according to the noted story, being asked by Hiero, What God was? desired a day to think of it, and then two days more; and after that manner continually prolonged the term, without ever bringing in his definition or description? Could you even blame me, if I had answered at first that I did not know, and was sensible that this subject lay vastly beyond the reach of my faculties? You might cry out sceptic and rallier as much as you pleased: but having found, in so many other subjects much more familiar the imperfections and even contradictions of human reason, I never should expect any success from its feeble conjectures in a subject so sublime and so remote from the sphere of our observation. When two species of objects have always been observed to be conjoined together, I can *infer* by custom the existence of one wherever I *see* the existence of the other: and this I call an argument from experience. But how this argument can have place where the objects, as in the present case, are single, individual, without parallel, or specific resemblance, may be difficult to explain. And will any man tell me with a serious countenance that an orderly universe must arise from some thought and art like the human, because we have experience of it? To ascertain this reasoning, it were requisite that we had experience of the origin of worlds; and it is not sufficient, surely, that we have seen ships and cities arise from human art and contrivance. . . .

FROM *Part X*

. . . Observe too, says Philo, the curious artifices of Nature, in order to embitter the life of every living being. The stronger prey upon the weaker, and keep them in perpetual terror and anxiety. The weaker too, in their turn, often prey upon the stronger, and vex and molest them without relaxation. Consider that innumerable race of insects, which either are bred on the body of each animal, or flying about infix their stings in him. These insects have others still less than themselves, which torment them. And thus on each hand, before and behind, above and below, every animal is surrounded with enemies, which incessantly seek his misery and destruction.

Man alone, said Demea, seems to be, in part, an exception to this rule. For by combination in society, he can easily master lions, tigers, and bears, whose greater strength and agility naturally enable them to prey upon him.

On the contrary, it is here chiefly, cried Philo, that the uniform and equal maxims of Nature are most apparent. Man, it is true, can, by combination, surmount all his *real* enemies, and become master of the whole animal creation: but does he not immediately raise up to himself *imaginary* enemies, the daemons of his fancy, who haunt him with superstitious terrors, and blast every enjoyment of life? His pleasure, as he imagines, becomes, in their eyes, a crime: his food and repose give them umbrage and offense: his very sleep and dreams furnish new materials to anxious fear: and even death, his refuge from every other ill, presents only the dread of endless and innumerable woes. Nor does the wolf molest more the timid flock, than superstition does the anxious breast of wretched mortals.

Besides, consider, Demea: this very society, by which we surmount those wild beasts, our natural enemies—what new enemies does it not raise to us? What woe and misery does it not occasion? Man is the greatest enemy of man. Oppression, injustice, contempt, contumely, violence, sedition, war, calumny, treachery, fraud: by these they mutually torment each other; and they would soon dissolve that society which they had formed, were it not for the dread of still greater ills, which must attend their separation. . . .

And is it possible, Cleanthes, said Philo, that after all these reflections, and infinitely more, which might be suggested, you can still persevere in your Anthropomorphism, and assert the moral attributes of the Deity, his justice, benevolence, mercy, and rectitude, to be of the same nature with these virtues in human creatures? His power we allow infinite: whatever he wills is executed: but neither man nor any other animal are happy: therefore he does not will their happiness. His wisdom is infinite: he is never mistaken in choosing the means to any end: but the course of nature tends not to human or animal felicity: therefore it is not established for that purpose. Through the whole compass of human knowledge, there are no inferences more certain and infallible than these. In what respect, then, do his benevolence and mercy resemble the benevolence and mercy of men?

Epicurus's old questions are yet unanswered.

Is he willing to prevent evil, but not able? then is he impotent. Is he able, but not willing? then is he malevolent. Is he both able and willing? whence then is evil?

You ascribe, Cleanthes (and I believe justly), a purpose and intention to Nature. But what, I beseech you, is the object of that curious artifice and machinery, which she has displayed in all animals? The preservation alone of individuals and propagation of the species. It seems enough for her purpose, if such a rank be barely upheld in the universe, without any care or concern for the happiness of the members that compose it. . . .

FROM *Part XI*

. . . My sentiments, replied Philo, are not worth being made a mystery of; and therefore, without any ceremony, I shall deliver what occurs to me with regard to the present subject. It must, I think, be allowed that if a very limited intelligence, whom we shall suppose utterly unacquainted with the universe, were assured that it were the production of a very good, wise, and powerful being, however finite, he would, from his conjectures, form *beforehand* a different notion of it from what we find it to be by experience; nor would he ever imagine, merely from these attributes of the cause, of which he is informed, that the effect could be so full of vice and misery and disorder as it appears in this life. Supposing now that this person were brought into the world, still assured that it was the workmanship of such a sublime and benevolent Being, he might, perhaps, be surprised at the disappointment, but would never retract his former belief, if founded on any very solid argument, since such a limited intelligence must be sensible of his own blindness and ignorance, and must allow that there may be many solutions of those phenomena which will forever escape his comprehension. But supposing, which is the real case with regard to man, that this creature is not antecedently convinced of a supreme intelligence, benevolent, and powerful, but is left to gather such a belief from the appearances of things: this entirely alters the case, nor will he ever find any reason for such a conclusion. He may be fully convinced of the narrow limits of his understanding, but this will not help him in forming an inference concerning the goodness of superior powers, since he must form that inference from what he knows, not from what he is ignorant of. The more you exaggerate his weakness and ignorance, the more diffident you render him, and give him the greater suspicion that such subjects are beyond the reach of his faculties. You are obliged, therefore, to reason with him merely from the known phenomena, and to drop every arbitrary supposition or conjecture. . . .

In short, I repeat the question: Is the world, considered in general, and as it appears to us in this life, different from what a man or such a limited being would, *beforehand*, expect from a very powerful, wise, and benevolent deity? It must be strange prejudice to assert the contrary. And from thence I conclude that, however consistent the world may be, allowing certain suppositions and conjectures, with the idea of such a deity, it can never afford us an inference concerning his existence. The consistence is not absolutely denied, only the inference. Conjectures, especially where infinity is excluded from the divine attributes, may perhaps be sufficient to prove a consistence, but can never be foundations for any inference.

There seems to be *four* circumstances on which depend all, or the greatest parts of, the ills that molest sensible creatures; and it is not impossible but all these circumstances may be necessary and unavoidable. We know so little beyond common life, or even of common life, that, with regard to the economy of the universe, there is no conjecture, however wild, which may not be just, nor any one, however plausible, which may not be erroneous. All that belongs to human understanding, in this deep ignorance and obscurity, is to be sceptical, or at least cautious, and not to admit of any hypothesis whatever, much less of any which is supported by no appearance of probability. Now this I assert to be the case with regard to all the causes of evil, and the circumstances on which it depends. None of them appear to human reason in the least degree necessary or unavoidable; nor can we suppose them such, without the utmost license of imagination.

The *first* circumstance which introduces evil is that contrivance or economy of the animal creation by which pains as well as pleasures are employed to excite all creatures to action, and make them vigilant in the great work of self-preservation. . . .

But a capacity of pain would not alone produce pain, were it not for the *second* circumstance, viz. the conducting of the world by general laws; and this seems nowise necessary to a very perfect being. . . . A being . . . who knows the secret springs of the universe might easily, by particular volitions, turn all these accidents to the good of mankind, and render the whole world happy, without discovering himself in any operation. . . .

If everything in the universe be conducted by general laws, and if animals be rendered susceptible of pain, it scarcely seems possible but some ill must arise in the various shocks of matter, and the various concurrence and opposition of general laws: but this ill would be very rare, were it not for the *third* circumstance, which I proposed to mention, viz. the great frugality with which all powers and faculties are distributed to every particular being. So well adjusted are the organs and capacities of all animals, and so well fitted to their preservation, that, as far as history or tradition reaches, there appears not to be any single species which has yet been extinguished in the universe. Every animal has the requisite endowments; but these endowments are bestowed with so scrupulous an economy that any considerable diminution must entirely destroy the creature. Wherever one power is increased, there is a proportional abatement in the others. Animals which excel in swiftness are commonly defective in force. Those which possess both are either imperfect in some of their senses, or are oppressed with the most craving wants. The human species, whose chief excellency is reason and sagacity, is of all others the most necessitous and the most deficient in bodily advantages: without clothes, without arms, without food, without lodging, without any convenience of life, except what they owe to their own skill and industry. In short, Nature seems to have formed an exact calculation of the necessities of her creatures; and like a rigid master, has afforded them little more powers or endowments than what are strictly sufficient to supply those necessities. An indulgent parent would have bestowed a large stock in order to guard against accidents and secure the happiness and welfare of the creature in the most unfortunate concurrence of circumstances. Every course of life would not have been so surrounded with precipices that the least departure from the true path, by mistake or necessity, must involve us in misery and ruin. Some reserve, some fund would have been provided to ensure happiness, nor would the powers and the necessities have been adjusted with so rigid an economy. The author of Nature is inconceivably powerful: his force is supposed great, if not altogether inexhaustible; nor is there any reason, as far as we can judge, to make him observe this strict frugality in his dealings with his creatures. It would have been better, were his power extremely limited, to have created fewer animals, and to have endowed these with more faculties for their happiness and preservation. A builder is never esteemed prudent, who undertakes a plan beyond what his stock will enable him to finish. . . .

The *fourth* circumstance, whence arises the misery and ill of the universe, is the inaccurate workmanship of all the springs and principles of the great machine of nature. It must be acknowledged that there are few parts of the universe which seem not to serve some purpose, and whose removal would not produce a visible defect and disorder in the whole. The parts hang all together; nor can one be touched without affecting the rest in a greater or less degree. But at the same time, it must be observed that none of these parts or principles, however useful, are so accurately adjusted as to keep precisely within those bounds in which their utility consists; but they are, all of them, apt, on every occasion, to run into the one extreme or the other. One would imagine that this grand production had not received the last hand of the maker, so little finished is every part, and so coarse are the strokes with which it is executed. . . .

On the concurrence, then, of these *four* circumstances does all, or the greatest part of, natural evil depend. Were all living creatures incapable of pain, or were the world administered by particular volitions, evil never could have found access into the universe; and were animals endowed with a large stock of powers and faculties beyond what strict necessity requires, or were the several springs and principles of the universe so accurately framed as to preserve always the just temperament and medium, there must have been very little ill in comparison of what we feel at present. What then shall we pronounce on this occasion? Shall we say that these circumstances are not necessary, and that they might easily have been altered in the contrivance of the universe? This decision seems too presumptuous for creatures so blind and ignorant. Let us be more modest in our conclusions. Let us allow that, if the goodness of the Deity (I mean a goodness like the human) could be established on any tolerable reasons *a priori*, these phenomena, however untoward, would not be sufficient to subvert that principle, but might easily, in some unknown manner, be reconcilable to it. But let us still assert that as his goodness is not antecedently established, but must be inferred from the phenomena, there can be no grounds for such an inference, while there are so many ills in the universe, and while these ills might so easily have been remedied, as far as human understanding can be allowed to judge on such a subject. I am sceptic enough to allow that the bad appearances, notwithstanding all my reasonings, may be compatible with such attributes as you suppose: but surely they can never prove those attributes. Such a conclusion cannot result from scepticism, but must arise from the phenomena, and from our confidence in the reasonings which we deduce from these phenomena.

Look round this universe. What an immense profusion of beings, animated and organized, sensible and active! You admire this prodigious variety and fecundity. But inspect a little more narrowly these living existences, the only beings worth regarding. How hostile and destructive to each other! How insufficient all of them for their own happiness! How contemptible or odious to the spectator! The whole presents nothing but the idea of a blind Nature, impregnated by a great vivifying principle, and pouring forth from her lap, without discernment or parental care, her maimed and abortive children!

Here the Manichaean system occurs as a proper hypothesis to solve the difficulty: and no doubt, in some respects, it is very specious, and has more probability than the common hypothesis, by giving a plausible account of the strange mixture of good and ill which appears in life. But if we consider, on the other hand, the perfect uniformity and agreement of the parts of the universe, we shall not discover in it any marks of the combat of a malevolent with a benevolent being. There is indeed an opposition of pains and pleasures in the feelings of sensible creatures: but are not all the operations of Nature carried on by an opposition of principles, of hot and cold, moist and dry, light and heavy? The true conclusion is that the original source of all things is entirely indifferent to all these principles, and has no more regard to good above ill than to heat above cold, or to drought above moisture, or to light above heavy.

There may four hypotheses be framed concerning the first causes of the universe: that they are endowed with perfect goodness, that they have perfect malice, that they are opposite and have both goodness and malice, that they have neither goodness nor malice. Mixed phenomena can never prove the two former unmixed principles. And the uniformity and steadiness of general laws seem to oppose the third. The fourth, therefore, seems by far the most probable.

What I have said concerning natural evil will apply to moral, with little or no variation; and we have no more reason to infer that the rectitude of the Supreme Being resembles human rectitude than that his benevolence resembles the human. Nay, it will be thought that we have still greater cause to exclude from him moral sentiments, such as we feel them; since moral evil, in the opinion of many, is much more predominant above moral good than natural evil above natural good.

But even though this should not be allowed, and though the virtue which is in mankind should be acknowledged much superior to the vice, yet so long as there is any vice at all in the universe, it will very much puzzle you anthropomorphites how to account for it. You must assign a cause for it without having recourse to the first cause. But as every effect must have a cause, and that cause another, you must either carry on the progression *in infinitum* [to infinity], or rest on that original principle, who is the ultimate cause of all things. . . .

FROM *Part XII*

I must confess, replied Philo, that I am less cautious on the subject of natural religion than on any other, both because I know that I can never, on that head, corrupt the principles of any man of common sense, and because no one, I am confident, in whose eyes I appear a man of common sense, will ever mistake my intentions. You, in particular,

Cleanthes, with whom I live in unreserved intimacy, you are sensible that, notwithstanding the freedom of my conversation, and my love of singular arguments, no one has a deeper sense of religion impressed on his mind, or pays more profound adoration to the Divine Being, as he discovers himself to reason, in the inexplicable contrivance and artifice of Nature. A purpose, an intention, a design strikes everywhere the most careless, the most stupid thinker; and no man can be so hardened in absurd systems as at all times to reject it. That Nature does nothing in vain is a maxim established in all the schools,[13] merely from the contemplation of the works of Nature, without any religious purpose; and, from a firm conviction of its truth, an anatomist, who had observed a new organ or canal, would never be satisfied till he had also discovered its use and intention. One great foundation of the Copernican system is the maxim that Nature acts by the simplest methods, and chooses the most proper means to any end; and astronomers often, without thinking of it, lay this strong foundation of piety and religion. The same thing is observable in other parts of philosophy: and thus all the sciences almost lead us insensibly to acknowledge a first intelligent Author; and their authority is often so much the greater, as they do not directly profess that intention.

It is with pleasure I hear Galen reason concerning the structure of the human body. The anatomy of a man, says he,[14] discovers above 600 different muscles; and whoever duly considers these, will find that in each of them Nature must have adjusted at least ten different circumstances, in order to attain the end which she proposed: proper figure, just magnitude, right disposition of the several ends, upper and lower position of the whole, the due insertion of the several nerves, veins, and arteries, so that in the muscles alone above 6000 several views and intentions must have been formed and executed. The bones he calculates to be 284; the distinct purposes aimed at in the structure of each, above forty. What a prodigious display of artifice, even in these simple and homogeneous parts! But if we consider the skin, ligaments, vessels, glandules, humors, the several limbs and members of the body, how must our astonishment rise upon us, in proportion to the number and intricacy of the parts so artificially adjusted! The farther we advance in these researches, we discover new scenes of art and wisdom, but descry still, at a distance, farther scenes beyond our reach, in the fine internal structure of the parts, in the economy of the brain, in the fabric of the seminal vessels. All these artifices are repeated in every different species of animal, with wonderful variety, and with exact propriety, suited to the different intentions of Nature in framing each species. And if the infidelity of Galen, even when these natural sciences were still imperfect, could not withstand such striking appearances, to what pitch of pertinacious obstinacy must a philosopher in this age have attained who can now doubt of a Supreme Intelligence?

Could I meet with one of this species (who, I thank God, are very rare) I would ask him: Supposing there were a God, who did not discover himself immediately to our senses, were it possible for him to give stronger proofs of his existence than what appear on the whole face of Nature? What indeed could such a divine Being do, but copy the present economy of things; render many of his artifices so plain that no stupidity could mistake them; afford glimpses of still greater artifices, which demonstrate his prodigious superiority above our narrow apprehensions; and conceal altogether a great many from such imperfect creatures? Now according to all rules of just reasoning, every fact must pass for undisputed when it is supported by all the arguments which its nature admits

[13]Aristotle *On the Heavens* 291b. See note on Rule I of Newton's "Rules of Reasoning in Philosophy" (*Principia*, Book III), in Chapter 18 above.

[14]De Formatione Foetus [On the Formation of the Foetus]. (Hume) See the selections from Galen's *On the Usefulness of the Parts of the Body* in Chapter 12 above.

of, even though these arguments be not, in themselves, very numerous or forcible: how much more, in the present case, where no human imagination can compute their number, and no understanding estimate their cogency!

I shall further add, said Cleanthes, to what you have so well urged, that one great advantage of the principle of theism is that it is the only system of cosmogony which can be rendered intelligible and complete, and yet can throughout preserve a strong analogy to what we every day see and experience in the world. The comparison of the universe to a machine of human contrivance is so obvious and natural, and is justified by so many instances of order and design in Nature, that it must immediately strike all unprejudiced apprehensions, and procure universal approbation. Whoever attempts to weaken this theory cannot pretend to succeed by establishing in its place any other that is precise and determinate: it is sufficient for him, if he start doubts and difficulties, and by remote and abstract views of things, reach that suspense of judgment which is here the utmost boundary of his wishes. But besides that this state of mind is in itself unsatisfactory, it can never be steadily maintained against such striking appearances as continually engage us into the religious hypothesis. A false, absurd system, human nature, from the force of prejudice, is capable of adhering to, with obstinacy and perseverance: but no system at all, in opposition to theory, supported by strong and obvious reason, by natural propensity, and by early education, I think it absolutely impossible to maintain or defend.

So little, replied Philo, do I esteem this suspense of judgment in the present case to be possible, that I am apt to suspect there enters somewhat of a dispute of words into this controversy, more than is usually imagined. That the works of Nature bear a great analogy to the productions of art is evident: and according to all the rules of good reasoning, we ought to infer, if we argue at all concerning them, that their causes have a proportional analogy. But as there are also considerable differences, we have reason to suppose a proportional difference in the causes; and in particular ought to attribute a much higher degree of power and energy to the supreme cause than any we have ever observed in mankind. Here then the existence of a Deity is plainly ascertained by reason; and if we make it a question whether, on account of these analogies, we can properly call him a mind or intelligence, notwithstanding the vast difference which may reasonably be supposed between him and human minds, what is this but a mere verbal controversy? No man can deny the analogies between the effects: to restrain ourselves from inquiring concerning the causes is scarcely possible; from this inquiry, the legitimate conclusion is that the causes have also an analogy: and if we are not contented with calling the first and supreme cause a God or Deity, but desire to vary the expression, what can we call him but Mind or Thought, to which he is justly supposed to bear a considerable resemblance? . . .

The Most Serious and Important Study

An Inquiry into the Human Mind on the Principles of Common Sense, by Thomas Reid

Thomas Reid (1710–96), founder of the Scottish or Common Sense school of philosophy, from 1752 to 1763 was regent master at King's College, Aberdeen, and in 1763 succeeded Adam Smith as Chair of Moral Philosophy at the University of Glasgow, where he taught until 1780. In 1764 he published his Inquiry into the Human Mind on the Principles of Common

Sense. *Deeply influenced by Berkeley and then by Hume's* Treatise of Human Nature, *Reid repudiated the scepticism to which their views seemed to lead. In his philosophy of "common sense" (beliefs common to all rational beings), objects of sense perception are real, not mere mental images as in Locke. The different classes of natural signs are the basis of common sense, of the fine arts or taste, and of philosophy.*

Berkeley and Hume had shown that reason cannot prove the existence of bodies, nor can habit, experience, or education. "At the same time, it is a fact, that such sensations are invariably connected with the conception and belief of external existences. Hence," Reid writes in the Inquiry *(V.III), "by all rules of just reasoning, we must conclude that this connection is the effect of our constitution, and ought to be considered as an original principle of human nature." The process of nature as perceived by the senses "may therefore be conceived as a kind of drama" (VI.XXI), in which "nature is the actor, we are spectators." The language of nature, he affirms in the following passages (from the fourth edition of 1785, ed. Timothy J. Duggan [1970]), never misleads us if rightly interpreted. Children, in their first handling of things, are engaged in this "most serious and important study" of acquiring perceptions from sense experience, which through the inductive principle is the teacher of humanity. But human education continues and reason completes the instruction that nature begins: reason, properly employed, he writes, "will confirm the documents of nature," with which, if no longer identical, it is fully compatible still.*

FROM *Chapter VI, Section XXIV*

. . . Men sometimes lead us into mistakes, when we perfectly understand their language, by speaking lies. But nature never misleads us in this way; her language is always true; and it is only by misinterpreting it that we fall into error. There must be many accidental conjunctions of things, as well as natural connections; and the former are apt to be mistaken for the latter. . . . Philosophers, and men of science, are not exempted from such mistakes; indeed all false reasoning in philosophy is owing to them: it is drawn from experience and analogy, as well as just reasoning, otherwise, it could have no verisimilitude: but the one is an unskilful and rash, the other a just and legitimate, interpretation of natural signs. If a child, or a man of common understanding, were put to interpret a book of science, written in his mother tongue, how many blunders and mistakes would he be apt to fall into? Yet he knows as much of this language as is necessary for his manner of life.

The language of nature is the universal study; and the students are of different classes. Brutes, idiots, and children, employ themselves in this study, and owe to it all their acquired perceptions. Men of common understanding make a greater progress, and learn, by a small degree of reflection, many things of which children are ignorant.

Philosophers fill up the highest form in this school, and are critics in the language of nature. All these different classes have one teacher, Experience, enlightened by the inductive principle. Take away the light of this inductive principle, and Experience is as blind as a mole: she may indeed feel what is present, and what immediately touches her; but she sees nothing that is either before or behind, upon the right hand or upon the left, future or past. . . .

From the time that children begin to use their hands, nature directs them to handle every thing over and over, to look at it while they handle it, and to put it in various positions, and at various distances from the eye. We are apt to excuse this as a childish diversion, because they must be doing something, and have not reason to entertain themselves in a more manly way. But if we think more justly, we shall find that they are engaged in the most serious and important study; and if they had all the reason of a philosopher, they could not be more properly employed. For it is this childish employment that enables them to make the proper use of their eyes. They are thereby every day acquiring habits of perception, which are of greater importance than anything we can teach them. The original perceptions which nature gave them are few, and insufficient for the purposes of life; and therefore she made them capable of acquiring many more perceptions by habit. And, to complete her work, she hath given them an unwearied assiduity in applying to the exercises by which those perceptions are acquired.

This is the education which nature gives to her children. And since we have fallen upon this subject, we may add, that another part of nature's education is that, by the course of things, children must often exert all their muscular force, and employ all their ingenuity, in order to gratify their curiosity, and satisfy their appetites. What they desire is only to be obtained at the expense of labor and patience, and many disappointments. By the exercise of body and mind necessary for satisfying their desires, they acquire agility, strength, and dexterity in their motions, as well as health and vigor to their constitutions; they learn patience and perseverance; they learn to bear pain without dejection, and disappointment without despondence. The education of nature is most perfect in savages, who have no other tutor: and we see that, in the quickness of all their senses, in the agility of their motions, in the hardiness of their constitutions, and in the strength of their minds to bear hunger, thirst, pain, and disappointment, they commonly far exceed the civilized. A most ingenious writer, on this account, seems to prefer the savage life to that of society. But the education of nature could never of itself produce a Rousseau. It is the intention of nature, that human education should be joined to her institution, in order to form the man. And she hath fitted us for human education by the natural principles of imitation and credulity, which discover themselves almost in infancy, as well as by others which are of later growth.

When the education we receive from men does not give scope to the education of nature, it is wrong directed; it tends to hurt our faculties of perception, and to enervate both the body and mind. Nature hath her way of rearing men, as she hath of curing their diseases. The art of medicine is to follow nature, to imitate and to assist her in the cure of diseases; and the art of education is to follow nature, and to assist and to imitate her in her way of rearing men. . . .

The education of nature, without any more human care than is necessary to preserve life, makes a perfect savage. Human education, joined to that of nature, may make a good citizen, a skilful artisan, or a well bred man. But reason and reflection must superadd their tutory, in order to produce a Rousseau, a Bacon, or a Newton. . . .

When reason is properly employed, she will confirm the documents of nature, which are always true and wholesome; she will distinguish, in the documents of human education, the good from the bad, rejecting the last with modesty, and adhering to the first with reverence.

Most men continue all their days to be just what nature and human education made them. Their manners, their opinions, their virtues, and their vices, are all got by habit, imitation, and instruction; and reason has little or no share in forming them.

The Pleasures of the Imagination

The Spectator, by Joseph Addison

British thinkers from Bacon to Reid played a central role in redefining concepts of nature, in relation both to religious beliefs and to scientific discoveries, that profoundly influenced French and German writers of the late eighteenth century and beyond. Others in England, men of letters or "moralists" who made no pretense of being original thinkers, assimilated and popularized these ideas, or reflected upon their implications for the life of their time. None was more characteristic, or more influential, in the early eighteenth century than Joseph Addison (1672–1719). Born at Milston, in Wiltshire, from which he moved as a child when his father was made Dean of Lichfield, he was educated at Charterhouse and at Queen's College, then at Magdalen College, Oxford, where he became a fellow. Having received the patronage of Lord Halifax, between 1699 and 1703 he visited Italy and lived in France, writing in 1701 a verse Letter from Italy, *and in 1704 a poem,* The Campaign, *on Marlborough's victory at Blenheim. He held various government offices thereafter, including Secretary to the Lord Lieutenant of Ireland and Secretary of State, and from 1708 until his death served as a member of Parliament. He was a minor poet and dramatist celebrated for his neoclassical verse tragedy* Cato *(1713), and a leading arbiter of Augustan taste whom Pope, after breaking with his circle, portrayed in his* Epistle to Dr. Arbuthnot *as "Atticus," who could*

> Damn with faint praise, assent with civil leer,
> And without sneering, teach the rest to sneer.

But it was in the periodical essay that Addison found his true form, first in 1709–11 as a contributor to Richard Steele's thrice-weekly The Tatler, by Isaac Bickerstaff, Esq., *then from March 1711 to December 1712, in collaboration with Steele, as principal contributor to the daily* Spectator, *and again in 1714 as sole editor of a second series of* The Spectator. *For these two periodicals he wrote nearly 350 essays, as well as a few for others, such as* The Guardian. *In* Spectator *No. 10, reckoning his "disciples" at some sixty thousand in London and Westminster, he pledged to "endeavor to enliven morality with wit, and to temper wit with morality. . . . It was said of Socrates, that he brought philosophy down from heaven, to inhabit among men; and I shall be ambitious to have it said of me, that I have brought philosophy out of closets and libraries, schools and colleges, to dwell in clubs and assemblies, at tea tables and in coffee houses." He thus took on the role of inculcating Augustan values of common sense, civility, and urbanity, of promoting Whig politics of balance and toleration, and of popularizing the up-to-date science and philosophy of Newton and Locke. "His prose is the model of the middle style," Samuel Johnson writes in his* Life of Addison; *"on grave subjects not formal, on light occasions not groveling; pure without scrupulosity, and exact without apparent elaboration; always equable, and always easy, without glowing words or pointed sentences. . . . His page is always luminous, but never blazes in unexpected splendor."*

Nature is not a frequent theme of this coffee-house moralist, yet "in the opening of the spring, when all Nature begins to recover herself," he can write (in Spectator *393), "the same animal pleasure which makes the birds sing, and the whole brute creation rejoice, rises very sensibly in the heart of man. . . . The cheerfulness of heart which springs up in us from the survey of Nature's works . . . consecrates every field and wood, turns an ordinary walk into a morning or evening sacrifice, and will improve those transient gleams of joy, which naturally brighten up and refresh the soul on such occasions, into an inviolable and perpetual state of bliss and happiness." And the important series of eleven consecutive essays on "The Pleasures of the Imagination"* (Spectator

411–21, June 21 to July 3, 1712), from which the following selections are taken, played a significant part in furthering the change of taste that would later find expression in treatises of Burke, Kant, and Schiller on the sublime, contributing to the Romantic cult of sublimity in nature.

Interest in the sublime as a literary category had been greatly stimulated by Nicolas Boileau-Despréaux's French translation in 1674 of the ancient treatise Peri Hypsous, *or* On the Sublime, *attributed to Longínus (and by Boileau's* Critical Reflections on Longinus *of 1694); and for late neoclassical writers, including Dryden and Pope, the emphasis of "Longinus" on sublimity of thought and language served as a counterbalance to the dominant stress on "correctness" retrospectively ascribed to Aristotle's* Poetics. *An ambivalent awe at the grandeur of nature's wild ruins permeates Thomas Burnet's* Sacred Theory of the Earth *of 1680 (Chapter 18 above). For the most part, sublimity was not especially associated, by either "Longinus" or his neoclassical admirers, with nature; but Chapter 35 of* On the Sublime *(in the translation of T. S. Dorsch in* Classical Literary Criticism *[1965]) proclaims that "the entire universe does not satisfy the contemplation and thought that lie within the scope of human endeavor; our ideas often go beyond the boundaries by which we are circumscribed. . . . This is why, by some sort of natural instinct, we admire, not surely, the small streams, beautifully clear though they may be, and useful too, but the Nile, the Danube, the Rhine, and even more than these the Ocean. The little fire that we have kindled ourselves, clear and steady as its flame may be, does not strike us with as much awe as the heavenly fires, in spite of their often being shrouded in darkness; nor do we think it a greater marvel than the craters of Etna, whose eruptions throw up from their depths rocks and even whole mountains, and at times pour out rivers of that pure Titanian fire."*

Addison, too, writes that "the mind of man naturally hates everything that looks like a restraint upon it," and associates pleasure in "the sight of what is great, uncommon, or beautiful" (the word "sublime" does not appear in these selections) with the "stupendous works of nature," in comparison with which works of art appear deficient. But Addison's principal emphasis is on the imagination or fancy, by whose faculty, he writes in No. 411, "a man in a dungeon is capable of entertaining himself with scenes and landskips more beautiful than any that can be found in the whole compass of Nature." Influenced by the concept of phantasia, *or imagination, in "Longinus" and by Locke's analysis of the subjective nature of perception, Addison affirms in No. 413 that light and colors "are only ideas in the mind, and not qualities that have any existence in matter"; thus the beauty and grandeur of nature do not exist independently of the human mind but only through interaction between them. Nature can be enhanced through association with art, so that the poet "has the modeling of nature in his own hands." Yet art must truly enhance, not distort, nature: the most pleasing gardens are not those of England, artificially neat and elegant, but those of France or Italy, in which art heightens nature's own beauties. (In one of the great reversals of eighteenth-century taste, English gardens of landscape artists such as Charles Bridgeman, Lancelot "Capability" Brown, and Humphry Repton would soon afterwards become paramount models of "natural" gardening throughout Europe.) And nowhere does the sometimes pedestrian Addison approach more nearly to sublimity in his own style than when contemplating—as if in remote reflection of Pascal (Chapter 18 above)—the extent to which the imagination, faced with the "extraordinary degrees of grandeur or minuteness" revealed by the new philosophy, is "swallowed up in the immensity of the void that surrounds it."*

FROM *No. 412: Monday, June 23 [1712]*

I shall first consider those pleasures of the imagination which arise from the actual view and survey of outward objects: and these, I think, all proceed from the sight of what is *great*, *uncommon*, or *beautiful*. There may, indeed, be something so terrible or offensive, that the horror or loathsomeness of an object may overbear the pleasure which results from its *greatness*, *novelty*, or *beauty*; but still there will be such a mixture of delight in the very disgust it gives us, as any of these three qualifications are most conspicuous and prevailing.

By *greatness* I do not only mean the bulk of any single object, but the largeness of a whole view, considered as one entire piece. Such are the prospects of an open champaign country, a vast uncultivated desert, of huge heaps of mountains, high rocks and precipices, or a wide expanse of waters, where we are not struck with the novelty or beauty of the sight, but with that rude kind of magnificence which appears in many of these stupendous works of Nature. Our imagination loves to be filled with an object, or to grasp at anything that is too big for its capacity. We are flung into a pleasing astonishment at such unbounded views, and feel a delightful stillness and amazement in the soul at the apprehension of them. The mind of man naturally hates everything that looks like a restraint upon it, and is apt to fancy itself under a sort of confinement when the sight is pent up in a narrow compass, and shortened on every side by the neighborhood of walls or mountains. On the contrary, a spacious horizon is an image of liberty, where the eye has room to range abroad, to expatiate at large on the immensity of its views, and to lose itself amidst the variety of objects that offer themselves to its observation. Such wide and undetermined prospects are as pleasing to the fancy as the speculations of eternity or infinitude are to the understanding. But if there be a beauty or uncommonness joined with this grandeur, as in a troubled ocean, a heaven adorned with stars and meteors, or a spacious landskip[15] cut out into rivers, woods, rocks, and meadows, the pleasure still grows upon us, as it arises from more than a single principle.

Everything that is *new* or *uncommon* raises a pleasure in the imagination, because it fills the soul with an agreeable surprise, gratifies its curiosity, and gives it an idea of which it was not before possessed. We are indeed so often conversant with one set of objects, and tired out with so many repeated shows of the same things, that whatever is *new* or *uncommon* contributes a little to vary human life, and to divert our minds, for a while, with the strangeness of its appearance: it serves us for a kind of refreshment, and takes off from that satiety we are apt to complain of in our usual and ordinary entertainments. It is this that bestows charms on a monster, and makes even the imperfections of Nature please us. It is this that recommends variety, where the mind is every instant called off to something new, and the attention not suffered to dwell too long, and waste itself on any particular object. It is this, likewise, that improves what is great or beautiful, and makes it afford the mind a double entertainment. Groves, fields, and meadows are at any season of the year pleasant to look upon, but never so much as in the opening of the spring, when they are all new and fresh, with their first gloss upon them, and not yet too much accustomed and familiar to the eye. For this reason there is nothing that more enlivens a prospect than rivers, jetteaus,[16] or falls of water, where the scene is perpetually shifting, and entertaining the sight every moment with something that is new. We are quickly tired with looking upon hills and valleys, where everything continues fixed and settled in the same place and posture, but find our thoughts a little agitated and relieved at the sight of such objects as are ever in motion, and sliding away from beneath the eye of the beholder.

But there is nothing that makes its way more directly to the soul than *beauty*, which immediately diffuses a secret satisfaction and complacency through the imagination,

[15] "Landscape." [16] "Jets of water."

and gives a finishing to anything that is great or uncommon. The very first discovery of it strikes the mind with an inward joy, and spreads a cheerfulness and delight through all its faculties. There is not perhaps any real beauty or deformity more in one piece of matter than another, because we might have been so made that whatever now appears loathsome to us might have shown itself agreeable; but we find by experience that there are several modifications of matter which the mind, without any previous consideration, pronounces at first sight beautiful or deformed. Thus we see that every different species of sensible creatures has its different notions of beauty, and that each of them is most affected with the beauties of its own kind. This is nowhere more remarkable than in birds of the same shape and proportion, where we often see the male determined in his courtship by the single grain or tincture of a feather, and never discovering any charms but in the color of its species. . . .

There is a second kind of beauty that we find in the several products of art and Nature, which does not work in the imagination with that warmth and violence as the beauty that appears in our proper species, but is apt however to raise in us a secret delight, and a kind of fondness for the places or objects in which we discover it. This consists either in the gaiety or variety of colors, in the symmetry and proportion of parts, in the arrangement and disposition of bodies, or in a just mixture and concurrence of all together. Among these several kinds of beauty the eye takes most delight in colors. We nowhere meet with a more glorious or pleasing show in Nature than what appears in the heavens at the rising and setting of the sun, which is wholly made up of those different stains of light that show themselves in clouds of a different situation. For this reason we find the poets, who are always addressing themselves to the imagination, borrowing more of their epithets from colors than from any other topic.

As the fancy delights in everything that is great, strange, or beautiful, and is still more pleased the more it finds of these perfections in the same object, so it is capable of receiving a new satisfaction by the assistance of another sense. Thus any continued sound, as the music of birds, or a fall of water, awakens every moment the mind of the beholder, and makes him more attentive to the several beauties of the place that lie before him. Thus if there arises a fragrancy of smells or perfumes, they heighten the pleasures of the imagination, and make even the colors and verdure of the landskip appear more agreeable; for the ideas of both senses recommend each other, and are pleasanter together than when they enter the mind separately: as the different colors of a picture, when they are well disposed, set off one another, and receive an additional beauty from the advantage of their situation.

No. 414: Wednesday, June 25 [1712]

If we consider the works of *Nature* and *art*, as they are qualified to entertain the imagination, we shall find the last very defective in comparison of the former; for though they may sometimes appear as beautiful or strange, they can have nothing in them of that vastness and immensity which afford so great an entertainment to the mind of the beholder. The one may be as polite and delicate as the other, but can never show herself so august and magnificent in the design. There is something more bold and masterly in the rough careless strokes of Nature than in the nice touches and embellishments of art. The beauties of the most stately garden or palace lie in a narrow compass, the imagination immediately runs them over, and requires something else to gratify her; but in the wide fields of Nature, the sight wanders up and down without confinement, and is fed with an infinite variety of images, without any certain stint or number. For this reason we always find the poet in love with a country life, where Nature appears in the greatest perfection, and furnishes out all those scenes that are most apt to delight the imagination.

> *Scriptorum chorus omnis amat nemus et fugit urbes.*
>
> —Hor.[17]

> *Hic secura quies, et nescia fallere vita,*
> *Dives opum variarum, hic latis otia fundis,*
> *Speluncae, vivique lacus, hic frigida Tempe,*
> *Mugitusque boum, mollesque sub arbore somni.*
>
> —Virg.[18]

But though there are several of these wild scenes that are more delightful than any artificial shows, yet we find the works of Nature still more pleasant, the more they resemble those of art; for in this case our pleasure rises from a double principle: from the agreeableness of the objects to the eye, and from their similitude to other objects; we are pleased as well with comparing their beauties, as with surveying them, and can represent them to our minds either as copies or originals. Hence it is that we take delight in a prospect which is well laid out, and diversified with fields and meadows, woods and rivers; in those accidental landskips of trees, clouds, and cities, that are sometimes found in the veins of marble; in the curious fretwork of rocks and grottoes; and, in a word, in anything that hath such a variety or regularity as may seem the effect of design in what we call the works of chance.

If the products of Nature rise in value according as they more or less resemble those of art, we may be sure that artificial works receive a greater advantage from their resemblance of such as are natural; because here the similitude is not only pleasant, but the pattern more perfect. The prettiest landskip I ever saw was one drawn on the walls of a dark room, which stood opposite on one side to a navigable river, and on the other to a park. The experiment is very common in optics. Here you might discover the waves and fluctuations of the water in strong and proper colors, with the picture of a ship entering at one end, and sailing by degrees through the whole piece. On another there appeared the green shadows of trees, waving to and fro with the wind, and herds of deer among them in miniature, leaping about upon the wall. I must confess, the novelty of such a sight may be one occasion of its pleasantness to the imagination, but certainly the chief reason is its near resemblance to Nature, as it does not only, like other pictures, give the color and figure, but the motion of the things it represents.

We have before observed that there is generally in Nature something more grand and august than what we meet with in the curiosities of art. When, therefore, we see this imitated in any measure, it gives us a nobler and more exalted kind of pleasure than what we receive from the nicer and more accurate productions of art. On this account our English gardens are not so entertaining to the fancy as those in France and Italy, where we see a large extent of ground covered over with an agreeable mixture of garden and forest, which represent everywhere an artificial rudeness much more charming than that neatness and elegancy which we meet with in those of our own country. It might, indeed,

[17]The whole chorus of poets loves the grove and flees the town (Horace, *Epistles* II.ii.77).

[18]Here is carefree repose, and life that knows not how to deceive, rich in varied treasures, and the ease of broad farmsteads, caverns, and living lakes, and cool Tempe, the lowing of cattle and soft sleeps beneath the trees (Virgil, *Georgics* II.467–70; see Chapter 11 above).

be of ill consequence to the public, as well as unprofitable to private persons, to alienate so much ground from pasturage, and the plow, in many parts of a country that is so well peopled, and cultivated to a far greater advantage. But why may not a whole estate be thrown into a kind of garden by frequent plantations, that may turn as much to the profit, as the pleasure of the owner? A marsh overgrown with willows, or a mountain shaded with oaks, are not only more beautiful, but more beneficial, than when they lie bare and unadorned. Fields of corn make a pleasant prospect, and if the walks were a little taken care of that lie between them, if the natural embroidery of the meadows were helped and improved by some small additions of art, and the several rows of hedges set off by trees and flowers that the soil was capable of receiving, a man might make a pretty landskip of his own possessions.

Writers who have given us an account of China tell us the inhabitants of that country laugh at the plantations of our Europeans, which are laid out by the rule and line, because, they say, anyone may place trees in equal rows and uniform figures. They choose rather to show a genius in works of this nature, and therefore always conceal the art by which they direct themselves. They have a word it seems in their language by which they express the particular beauty of a plantation that thus strikes the imagination at first sight, without discovering what it is that has so agreeable an effect. Our British gardeners, on the contrary, instead of humoring Nature, love to deviate from it as much as possible. Our trees rise in cones, globes, and pyramids. We see the marks of the scissors upon every plant and bush. I do not know whether I am singular in my opinion, but, for my own part, I would rather look upon a tree in all its luxuriancy and diffusion of boughs and branches than when it is thus cut and trimmed into a mathematical figure; and cannot but fancy that an orchard in flower looks infinitely more delightful than all the little labyrinths of the most finished parterre. But as our great modelers of gardens have their magazines of plants to dispose of, it is very natural for them to tear up all the beautiful plantations of fruit trees, and contrive a plan that may most turn to their own profit, in taking off their evergreens, and the like movable plants, with which their shops are plentifully stocked.

FROM *No. 418: Monday, June 30 [1712]*

. . . But because the mind of man requires something more perfect in matter than what it finds there, and can never meet with any sight in Nature which sufficiently answers its highest ideas of pleasantness; or, in other words, because the imagination can fancy to itself things more great, strange, or beautiful than the eye ever saw, and is still sensible of some defect in what it has seen: on this account it is the part of a poet to humor the imagination in its own notions by mending and perfecting Nature where he describes a reality, and by adding greater beauties than are put together in Nature, where he describes a fiction.

He is not obliged to attend her in the slow advances which she makes from one season to another, or to observe her conduct in the successive production of plants and flowers. He may draw into his description all the beauties of the spring and autumn, and make the whole year contribute something to render it the more agreeable. His rose-trees, woodbines, and jessamines may flower together, and his beds be covered at the same time with lilies, violets, and amaranths. His soil is not restrained to any particular set of plants, but is proper either for oaks or myrtles, and adapts itself to the products of every climate. Oranges may grow wild in it; myrrh may be met with in every hedge, and if he thinks it proper to have a grove of spices, he can quickly command sun enough to raise it. If all this will not furnish out an agreeable scene, he can make several new species of flowers, with richer scents and higher colors than any that grow in the gardens of Nature. His consorts of birds may be as full and harmonious, and his woods as thick and gloomy as he pleases. He is at no more expense in a long vista than a short one, and can as easily throw his cascades from a precipice of half a mile high as from one of twenty yards. He has his choice of the winds, and can turn the course of his rivers in all the variety of meanders that are most delightful to the reader's imagination. In a word, he has the modeling of Nature in his own hands, and may give her what charms he pleases, provided he does not reform her too much, and run into absurdities, by endeavoring to excel.

FROM *No. 420: Wednesday, July 2 [1712]*

As the writers in poetry and fiction borrow their several materials from outward objects, and join them together at their own pleasure, there are others who are obliged to follow Nature more closely, and to take entire scenes out of her. Such are historians, natural philosophers, travelers, geographers, and, in a word, all who describe visible objects of a real existence. . . .

But among this set of writers, there are none who more gratify and enlarge the imagination than the authors of the new philosophy, whether we consider their theories of the earth or heavens, the discoveries they have made by glasses, or any other of their contemplations on Nature. We are not a little pleased to find every green leaf swarm with millions of animals that at their largest growth are not visible to the naked eye. There is something very engaging to the fancy, as well as to our reason, in the treatises of metals, minerals, plants, and meteors. But when we survey the whole earth at once, and the several planets that lie within its neighborhood, we are filled with a pleasing astonishment, to see so many worlds hanging one above another, and sliding round their axles in such an amazing pomp and solemnity. If, after this, we contemplate those wide fields of ether, that reach in height as far as from Saturn to the fixed stars, and run abroad almost to an infinitude, our imagination finds its capacity filled with so immense a prospect, and puts itself upon the stretch to comprehend it. But if we yet rise higher, and consider the fixed stars as so many vast oceans of flame, that are each of them attended with a different set of planets, and still discover new firmaments and new lights, that are sunk farther in those unfathomable depths of ether, so as not to be seen by the strongest of our telescopes, we are lost in such a labyrinth of suns and worlds, and confounded with the immensity and magnificence of Nature.

Nothing is more pleasant to the fancy, than to enlarge itself, by degrees, in contemplation of the various proportions which its several objects bear to each other, when it compares the body of a man to the bulk of the whole earth, the earth to the circle it describes round the sun, that circle to the sphere of the fixed stars, the sphere of the fixed stars to the circuit of the whole creation, the whole creation itself to the infinite space that is everywhere diffused about it; or when the imagination works downward, and considers the bulk of a human body, in respect of an animal a hundred times less than a mite, the particular limbs of such an animal, the different springs which actuate the limbs, the spirits which set these springs a going, and the proportionable minuteness of these several parts, before they have arrived at their full growth and perfection. But if, after all this, we take the least particle of these animal spirits, and consider its capacity of being wrought into a world, that shall contain within those narrow dimensions a heaven and earth, stars and planets, and every different species of living creatures, in the same analogy and proportion they bear to each other in our own universe; such a speculation, by reason of its nicety, appears ridiculous to those who have not turned their thoughts that way, though at the same time it is founded on no less than the evidence of a demonstration. Nay, we might yet carry it farther, and discover in the smallest particle of this world a new inexhausted fund of matter, capable of being spun out into another universe.

I have dwelt the longer on this subject, because I think it may show us the proper limits, as well as the defectiveness, of our imagination; how it is confined to a very small quantity of space, and immediately stopped in its operations, when it endeavors to take in anything that is very great, or very little. Let a man try to conceive the different bulk of an animal, which is twenty, from another which is a hundred times less than a mite, or to compare, in his thoughts, a length of a thousand diameters of the earth, with that of a million, and he will quickly find that he has no different measures in his mind, adjusted to such extraordinary degrees of grandeur or minuteness. The understanding, indeed, opens an infinite space on every side of us, but the imagination, after a few faint efforts, is immediately at a stand, and finds herself swallowed up in the immensity of the void that surrounds it; our reason can pursue a particle of matter through an infinite variety of divisions, but the fancy soon loses sight of it, and feels in itself a kind of chasm, that wants to be filled with matter of a more sensible bulk. We can neither widen, nor contract the faculty to the dimensions of either extreme: the object is too big for our capacity, when we could comprehend the circumference of a world, and dwindles into nothing, when we endeavor after the idea of an atom. . . .

Just Representations of General Nature

Selections from the Writings of Samuel Johnson

Samuel Johnson (1709–84)—widely known, after receiving an honorary Ll.D. from Trinity College, Dublin, in 1765, as Doctor Johnson—was born at Lichfield in Staffordshire, the son of a bookseller; in infancy he was afflicted with scrofula, which marred his appearance and affected his eyesight. Educated (as Addison had been) at Lichfield Grammar School, he spent an unhappy year or so as a proud but impoverished student at Pembroke College, Oxford, then returned without a degree to Lichfield to read and write verse. In 1731 his Latin translation of Pope's Messiah *was published, and in 1735 he married a forty-five-year-old widow and opened a boarding school that soon failed, having attracted only three scholars. With one of these, David Garrick (who would become the foremost actor of his time), he set out on foot in March 1737 for London, taking with him the manuscript of his tragedy* Irene, *first acted and printed in 1749; here for years he led the marginal existence of a Grub Street writer, as described in his* Life *of his friend Richard Savage, published in 1744. In 1738 he won recognition for his anonymous poem* London, *and in 1749 for* The Vanity of Human Wishes, *both based on satires by Juvenal. In 1746 he undertook his* Dictionary of the English Language, *which he published, after years of single-handed toil, in 1755. When he learned that the Earl of Chesterfield, who had declined his request for assistance seven years earlier, had praised his work, Johnson expressed his indignant contempt in a famous letter, asking: "Is not a patron, my Lord, one who looks with unconcern on a man struggling for life in the water, and, when he has reached ground, encumbers him with help? The notice which you have been pleased to take of my labors, had it been early, had been kind; but it has been delayed till I am indifferent and cannot enjoy it, till I am solitary and cannot impart it, till I am known and do not want it." Single-handedly again, during these years, he wrote 208 twice-weekly numbers of his periodical* The Rambler *(March 1750–March 1752), followed by another 125 or so of* The Adventurer *in 1753–54 and* The Idler *in 1758–60. During eight days in January 1759 he wrote his philosophical romance on the elusiveness of happiness and the dangerous prevalence of imagination,* The History of Rasselas, Prince of Abissinia, *and in 1765 published* The Plays of William Shakespeare *in eight volumes. Between August and November 1773 he toured Scotland with James Boswell, whom he had met ten years before; Johnson's* Journey to the Western Islands of Scotland *appeared in 1775, and Boswell's* Journal of the Tour to the Hebrides *in 1785, to be followed by his* Life of Johnson *in 1791. Johnson's last major work, his* Lives of the Poets, *was published in 1779 and 1781, and revised in 1783; he died in 1784, and was buried in Westminster Abbey.*

Both through his writings and through the force of his personality, immortalized in the conversations recorded by Boswell, Johnson was the dominant literary figure of late-eighteenth-century England. The manners and morals of men in relation to one another in society, and to their Creator in worship, are the constant theme of this deeply humane and religious man, who opposed the preoccupation of many in his age with questions raised by the new philosophy. "The truth is," he affirms in his "Life of Milton," "that the knowledge of external nature, and the sciences which that knowledge requires or includes, are not the great or the frequent business of the human mind." Socrates labored "to turn philosophy from the study of nature to speculations upon life; but the innovators whom I oppose are turning off attention from life to nature. They seem to think that we are placed here to watch the growth of plants, or the motions of stars. Socrates was rather of opinion that what we had to learn was how to do good, and avoid evil."

Yet it is striking how often different dimensions of nature enter into his work. Of the following selections, the first—from the Dictionary *(omitting the citations from earlier authors under each heading)—is a succinct reminder of the range of meanings "nature" had in Johnson's time: definitions 5 ("The regular course of things"), 6 ("The compass of natural existence"), and 8 ("The state or operation of the material world") encapsulate the main philosophical or scientific significations, while definition 1, referring to the mythological figure of Nature, and definitions 7 ("Natural affection or reverence") and 10 ("Sentiments or images adapted to nature") indicate the emotive and imaginative associations the word conveyed.*

Of the essays from The Rambler, *No. 5 evokes the delights and uses that nature's "inexhaustible stock of materials" can bring, especially in springtime, to those who have learned to read her characters, but also touches upon the melancholy of the many who do not know how to take a walk. Nos. 36 and 37 contrast the artificiality of pastoral poets—who "have written with an utter disregard both of life and nature," glorifying the impoverished monotony of rural life by images of the golden age—with the natural order and beauty that appeal to our "inclination to stillness and tranquility." (Johnson's distaste for pastoral poetry, founded on his visceral revulsion from everything false, led him, in his "Life of Milton," to dismiss Milton's deeply felt elegy* Lycidas *as a poem "in which there is no nature, for there is nothing new. Its form is that of a pastoral, easy, vulgar, and therefore disgusting.") In* Rambler *135 he satirically exposes the insincerity of "the annual flight of human rovers" from the city to their country retreats—where, far from "listening to Philomel, loitering in woods, or plucking daisies," they find only the diversity of "doing the same things in a different place," returning to talk "of happiness which they never felt, and beauty which they never regarded." In Chapter XXII of* Rasselas, *the young Abyssinian prince—just after talking with a hermit who has grown weary of examining plants and collecting minerals, and decided to return to society—listens to a philosopher discourse on life led in accord with nature, and realizes the banality of his identification of such a life "with the general disposition and tendency of the present system of things."*

But if Johnson ruthlessly exposed such false understandings, he remained convinced, as he writes in the preface to his edition of Shakespeare, that "nothing can please many and please long, but just representations of general nature"; for though "nature gives no man knowledge," no genuine knowledge is possible without the materials nature profusely provides. Johnson's "general nature" emphasizes qualities common to different persons or things, rather than their individualizing characteristics, just as the philosopher Imlac, in Chapter X of Rasselas, *declares that the poet's business "is to examine, not the individual, but the species; to remark general properties and large appearances: he does not number the streaks of the tulip, or describe the different shades in the verdure of the forest." But though for Johnson the Shakespearean character is "commonly a species," he insists that these characters are never mere types: they are not abstract Romans or kings, but always men. And Shakespeare is sublime: the gardens of "correct" writers are dwarfed by the spreading oaks and towering pines of Shakespeare's unweeded forest.*

It is especially revealing, then, that this London moralist suggested to Boswell (who at first thought it "a very romantic fancy") a journey through the Scottish highlands to the Hebrides, and became one of the few writers of his century to venture—despite his utter disdain for Rousseau's primitivism, for sentimental idealizations of savage life, and for the pseudo-bardic effusions of Macpherson's "Ossian"—into the "rudeness, silence, and solitude" of wild mountainous regions where he could satisfy his need to compare ideas with realities. In this deserted spot near the village of Anoch, where "man is made unwillingly acquainted with his own weakness," Johnson gives tempered expression, as he sits on a bank and conceives the idea of his narration, to something like awe at "the view of an unknown and untraveled wilderness."

FROM A Dictionary of the English Language

NA'TURE. s. [*natura*, Latin]

1. An imaginary being supposed to preside over the material and animal world.
2. The native states or properties of any thing.
3. The constitution of an animated body.
4. Disposition of mind.
5. The regular course of things.
6. The compass of natural existence.
7. Natural affection, or reverence.
8. The state or operation of the material world.
9. Sort; species.
10. Sentiments or images adapted to nature.
11. Physics; the science which teaches the qualities of things.

ESSAYS FROM *THE RAMBLER*

FROM *No. 5: Tuesday, 3 April 1750 [The Pleasures of Spring]*

Et nunc omnis ager, nunc omnis parturit arbos,
Nunc frondent silvae, nunc formosissimus annus.

—Virg. [*Eclogue* III.56–57]

Now ev'ry field, now ev'ry tree is green;
Now genial nature's fairest face is seen.

—Elphinston

. . . There is, indeed, something inexpressibly pleasing in the annual renovation of the world, and the new display of the treasures of nature. The cold and darkness of winter, with the naked deformity of every object on which we turn our eyes, make us rejoice at the succeeding season, as well for what we have escaped, as for what we may enjoy; and every budding flower, which a warm situation brings early to our view, is considered by us as a messenger to notify the approach of more joyous days.

The Spring affords to a mind, so free from the disturbance of cares or passions as to be vacant to calm amusements, almost everything that our present state makes us capable of enjoying. The variegated verdure of the fields and woods, the succession of grateful odors, the voice of pleasure pouring out its notes on every side, with the gladness apparently conceived by every animal, from the growth of his food, and the clemency of the weather, throw over the whole earth an air of gaiety, significantly expressed by the smile of nature.

Yet there are men to whom these scenes are able to give no delight, and who hurry away from all the varieties of rural beauty, to lose their hours, and divert their thoughts by cards, or assemblies, a tavern dinner, or the prattle of the day.

It may be laid down as a position which will seldom deceive, that when a man cannot bear his own company there is something wrong. He must fly from himself, either because he feels a tediousness in life from the equipoise of an empty mind, which, having no tendency to one motion more than another but as it is impelled by some external power, must always have recourse to foreign objects; or he must be afraid of the intrusion of some unpleasing ideas, and, perhaps, is struggling to escape from the remembrance of a loss, the fear of a calamity, or some other thought of greater horror.

Those whom sorrow incapacitates to enjoy the pleasures of contemplation may properly apply to such diversions, provided they are innocent, as lay strong hold on the attention; and those whom fear of any future affliction chains down to misery must endeavor to obviate the danger.

My considerations shall, on this occasion, be turned on such as are burthensome to themselves merely because they want subjects for reflection, and to whom the volume of nature is thrown open, without affording them pleasure or instruction, because they never learned to read the characters.

A French author has advanced the seeming paradox that *very few men know how to take a walk*; and, indeed, it is true that few know how to take a walk with a prospect of any other pleasure than the same company would have afforded them at home.

There are animals that borrow their color from the neighboring body, and, consequently, vary their hue as they happen to change their place. In like manner it ought to be the endeavor of every man to derive his reflections from the objects about him; for it is to no purpose that he alters his position, if his attention continues fixed to the same point. The mind should be kept open to the access of every new idea, and so far disengaged from the predominance of particular thoughts as easily to accommodate itself to occasional entertainment.

A man that has formed this habit of turning every new object to his entertainment finds in the productions of nature an inexhaustible stock of materials upon which he can employ himself, without any temptations to envy or malevolence; faults, perhaps, seldom totally avoided by those whose judgment is much exercised upon the works of art. He has always a certain prospect of discovering new reasons for adoring the sovereign author of the universe, and probable hopes of making some discovery of benefit to others, or of profit to himself. There is no doubt but many vegetables and animals have qualities that might be of great use, to the knowledge of which there is not required much force of penetration, or fatigue of study, but only frequent experiments, and close attention. What is said by the chemists of their darling mercury is, perhaps, true of every body through the whole creation, that, if a thousand lives should be spent upon it, all its properties would not be found out.

Mankind must necessarily be diversified by various tastes, since life affords and requires such multiplicity of employments, and a nation of naturalists is neither to be hoped, or desired; but it is surely not improper to point out a fresh amusement to those who languish in health, and repine in plenty, for want of some source of diversion that may be less easily exhausted, and to inform the multitudes of both sexes, who are burthened with every new day, that there are many shows which they have not seen.

He that enlarges his curiosity after the works of nature demonstrably multiplies the inlets of happiness; and, therefore, the younger part of my readers, to whom I dedicate this vernal speculation, must excuse me for calling upon them to make use at once of the spring of the year, and the spring of life; to acquire, while their minds may be yet impressed with new images, a love of innocent pleasure, and an ardor for useful knowledge; and to remember that a blighted spring makes a barren year, and that the vernal flowers, however beautiful and gay, are only intended by nature as preparatives to autumn fruits.

FROM *No. 36: Saturday, 21 July 1750 [Pastoral Poetry (1)]*

. . . The images of true pastoral have always the power of exciting delight, because the works of nature, from which they are drawn, have always the same order and beauty, and continue to force themselves upon our thoughts, being at once obvious to the most careless regard, and more than adequate to the strongest reason and severest contemplation. Our inclination to stillness and tranquility is seldom much lessened by long knowledge of the busy and tumultuary part of the world. In childhood we turn our thoughts to the country as to the region of pleasure, we recur to it in old age as a port of rest, and perhaps with that secondary and adventitious gladness which every man feels on reviewing those places, or recollecting those occurrences, that contributed to his youthful enjoyments, and bring him back to the prime of life when the world was gay with the bloom of novelty, when mirth wantoned at his side, and hope sparkled before him.

The sense of this universal pleasure has invited "numbers without number" to try their skill in pastoral performances, in which they have generally succeeded after the manner of other imitators, transmitting the same images in the same combination from one to another, till he that reads the title of a poem may guess at the whole series of the composition. Nor will a man, after the perusal of thousands of these performances, find his knowledge enlarged with a single view of nature not produced before, or his imagination amused with any new application of those views to moral purposes.

The range of pastoral is indeed narrow, for though nature itself, philosophically considered, be inexhaustible, yet its general effects on the eye and on the ear are uniform and incapable of much variety of description. Poetry cannot dwell upon the minuter distinctions by which one species differs from another without departing from that simplicity of grandeur which fills the imagination; nor dissect the latent qualities of things without losing its general power of gratifying every mind by recalling its conceptions. However, as each age makes some discoveries and those discoveries are by degrees generally known, as new plants or mode of culture are introduced and by little and little become common, pastoral might receive from time to time small augmentations, and exhibit once in a century a scene somewhat varied.

But pastoral subjects have been often, like others, taken into the hands of those that were not qualified to adorn them, men to whom the face of nature was so little known that they have drawn it only after their own imagination, and changed or distorted her features that their portraits might appear something more than servile copies from their predecessors.

Not only the images of rural life, but the occasions on which they can be properly produced, are few and general. The state of a man confined to the employments and pleasures of the country is so little diversified, and exposed to so few of those accidents which produce perplexities, terrors, and surprises in more complicated transactions, that he can be shown but seldom in such circumstances as attract curiosity. His ambition is without policy, and his love without intrigue. He has no complaint to make of his rival but that he is richer than himself, nor any disasters to lament but a cruel mistress or a bad harvest. . . .

I am afraid it will not be found easy to improve the pastorals of antiquity by any great additions or diversifications. Our descriptions may indeed differ from those of Virgil, as an English from an Italian summer, and, in some respects, as modern from ancient life; but as nature is in both countries nearly the same, and as poetry has to do rather with the passions of men, which are uniform, than their customs, which are changeable, the varieties which time or place can furnish will be inconsiderable: and I shall endeavor to show in the next paper how little the latter ages have contributed to the improvement of the rustic muse.

FROM *No. 37: Tuesday, 24 July 1750 [Pastoral Poetry (2)]*

. . . I cannot indeed easily discover why it is thought necessary to refer descriptions of a rural state to remote times, nor can I perceive that any writer has consistently preserved the Arcadian manners and sentiments. The only reason, that I have read, on which this rule has been founded is that according to the customs of modern life it is improbable that shepherds should be capable of harmonious numbers or delicate sentiments; and therefore the reader must exalt his ideas of the pastoral character by carrying his thoughts back to the age in which the care of herds and flocks was the employment of the wisest and greatest men.

These reasoners seem to have been led into their hypothesis by considering pastoral, not in general as a representation of rural nature and consequently as exhibiting the ideas and sentiments of those (whoever they are) to whom the country affords pleasure or employment, but simply as a dialogue or narrative of men actually tending sheep and busied in the lowest and most laborious offices. From whence they very readily concluded, that either the sentiments must sink to the level of the speakers or the speakers must be raised to the height of the sentiments.

In consequence of these original errors a thousand precepts have been given, which have only contributed to perplex and confound. Some have thought it necessary that the imaginary manners of the golden age should be universally preserved, and have therefore believed that nothing more could be admitted in pastoral than lilies and roses and rocks and streams, among which are heard the gentle whispers of chaste fondness or the soft complaints of amorous impatience. . . .

Other writers, having the mean and despicable condition of a shepherd always before them, conceive it necessary to degrade the language of pastoral by obsolete terms and rustic words, which they very learnedly called Doric, without reflecting that they thus became authors of a mangled dialect which no human being ever could have spoken, that they may as well refine the speech as the sentiments of their personages, and that none of the inconsistencies which they endeavor to avoid is greater than that of joining elegance of thought with coarseness of diction. Spenser begins one of his pastorals with studied barbarity.

Diggon Davie, I bid her good day;
Or, Diggon her is, or I missay.
Dig. Her was her while it was day-light,
But now her is a most wretched wight.

["September"]

What will the reader imagine to be the subject on which speakers like these exercise their eloquence? Will he not be somewhat disappointed when he finds them met together to condemn the corruptions of the church of Rome? Surely, at the same time that a shepherd learns theology, he may gain some acquaintance with his native language. . . .

The facility of treating actions or events in the pastoral style has incited many writers from whom more judgment might have been expected to put the sorrow or the joy which the occasion required into the mouth of Daphne or Thyrsis; and as one absurdity must naturally be expected to make way for another, they have written with an utter disregard both of life and nature, and filled their productions with mythological allusions, with incredible fictions, and with sentiments which neither passion nor reason could have dictated since the change which religion has made in the whole system of the world.

FROM *No. 135: Tuesday, 2 July 1751 [Rustic Seclusion]*

Coelum, non animum mutant.

—Horace [*Epistles* I.xi.27]

Place may be changed; but who can change his mind?

. . . At this time of universal migration, when almost everyone considerable enough to attract regard has retired, or is preparing with all the earnestness of distress to retire, into the country; when nothing is to be heard but the hopes of speedy departure, or the complaints of involuntary delay; I have often been tempted to inquire what happiness is to be gained, or what inconvenience to be avoided, by this stated recession. Of the birds of passage, some follow the summer, and some the winter, because they live upon sustenance which only summer or winter can supply; but of the annual flight of human rovers it is much harder to assign the reason, because they do not appear either to find or seek anything which is not equally afforded by the town and country.

I believe that many of these fugitives may have heard of men whose continual wish was for the quiet of retirement, who watched every opportunity to steal away from observation, to forsake the crowd, and delight themselves with "the society of solitude."[19] There is indeed scarcely any writer who has not celebrated the happiness of rural privacy, and delighted himself and his reader with the melody of birds, the whisper of groves, and the murmur of rivulets: nor any man eminent for extent of capacity, or greatness of exploits, that has not left behind him some memorials of lonely wisdom, and silent dignity.

But almost all absurdity of conduct arises from the imitation of those whom we cannot resemble. Those who thus testified their weariness of tumult and hurry, and hasted

[19]Milton, *Paradise Lost* IX.249 ("For solitude sometimes is best society").

with so much eagerness to the leisure of retreat, were either men overwhelmed with the pressure of difficult employment, harassed with importunities, and distracted with multiplicity, or men wholly engrossed by speculative sciences, who having no other end of life but to learn and teach, found their searches interrupted by the common commerce of civility, and their reasonings disjointed by frequent interruptions. Such men might reasonably fly to that ease and convenience which their condition allowed them to find only in the country. . . .

Such examples of solitude very few of those who are now hastening from the town have any pretensions to plead in their own justification, since they cannot pretend either weariness of labor, or desire of knowledge. They purpose nothing more than to quit one scene of idleness for another, and after having trifled in public, to sleep in secrecy. The utmost that they can hope to gain is the change of ridiculousness to obscurity, and the privilege of having fewer witnesses to a life of folly. He who is not sufficiently important to be disturbed in his pursuits, but spends all his hours according to his own inclination, and has more hours than his mental faculties enable him to fill either with enjoyment or desires, can have nothing to demand of shades and valleys. As bravery is said to be a panoply, insignificancy is always a shelter.

There are however pleasures and advantages in a rural situation, which are not confined to philosophers and heroes. The freshness of the air, the verdure of the woods, the paint of the meadows, and the unexhausted variety which summer scatters upon the earth, may easily give delight to an unlearned spectator. It is not necessary that he who looks with pleasure on the colors of a flower should study the principles of vegetation, or that the Ptolemaic and Copernican system should be compared before the light of the sun can gladden, or its warmth invigorate. Novelty itself is a source of gratification, and Milton justly observes[20] that to him who has been long pent up in cities no rural object can be presented which will not delight or refresh some of his senses.

Yet even these easy pleasures are missed by the greater part of those who waste their summer in the country. Should any man pursue his acquaintances to their retreats, he would find few of them listening to Philomel, loitering in woods, or plucking daisies, catching the healthy gale of the morning, or watching the gentle coruscations of declining day. Some will be discovered at a window by the roadside, rejoicing when a new cloud of dust gathers towards them, as at the approach of a momentary supply of conversation, and a short relief from the tediousness of unideal vacancy. Others are placed in the adjacent villages, where they look only upon houses as in the rest of the year, with no change of objects but what a remove to any new street in London might have given them. The same set of acquaintances still settle together, and the form of life is not otherwise diversified than by doing the same things in a different place. They pay and receive visits in the usual form, they frequent the walks in the morning, they deal cards at night, they attend to the same tattle, and dance with the same partners; nor can they at their return to their former habitation congratulate themselves on any other advantage than that they have passed their time like others of the same rank; and have the same right to talk of the happiness and beauty of the country—of happiness which they never felt, and beauty which they never regarded.

To be able to procure its own entertainments, and to subsist upon its own stock, is not the prerogative of every mind. There are indeed understandings so fertile and comprehensive, that they can always feed reflection with new supplies, and suffer nothing

[20]*Paradise Lost* IX.445–51.

from the preclusion of adventitious amusements; as some cities have within their own walls enclosed ground enough to feed their inhabitants in a siege. But others live only from day to day, and must be constantly enabled, by foreign supplies, to keep out the encroachments of languor and stupidity. Such could not indeed be blamed for hovering within reach of their usual pleasures, more than any other animal for not quitting its native element, were not their faculties contracted by their own fault. But let not those who go into the country merely because they dare not be left alone at home boast their love of nature, or their qualifications for solitude; nor pretend that they receive instantaneous infusions of wisdom from the Dryads, and are able, when they leave smoke and noise behind, to act, or think, or reason for themselves.

FROM The History of Rasselas, Prince of Abissinia

XXII. THE HAPPINESS OF A LIFE LED ACCORDING TO NATURE Rasselas went often to an assembly of learned men, who met at stated times to unbend their minds, and compare their opinions. Their manners were somewhat coarse, but their conversation was instructive, and their disputations acute, though sometimes too violent, and often continued till neither controvertist remembered upon what question they began. Some faults were almost general among them: every one was desirous to dictate to the rest, and every one was pleased to hear the genius or knowledge of another depreciated.

In this assembly Rasselas was relating his interview with the hermit, and the wonder with which he heard him censure a course of life which he had so deliberately chosen, and so laudably followed. The sentiments of the hearers were various. . . .

"This," said a philosopher who had heard him with tokens of great impatience, "is the present condition of a wise man. The time is already come, when none are wretched but by their own fault. Nothing is more idle than to inquire after happiness, which nature has kindly placed within our reach. The way to be happy is to live according to nature, in obedience to that universal and unalterable law with which every heart is originally impressed; which is not written on it by precept, but engraven by destiny, not instilled by education, but infused at our nativity. He that lives according to nature will suffer nothing from the delusions of hope, or importunities of desire: he will receive and reject with equability of temper; and act or suffer as the reason of things shall alternately prescribe. Other men may amuse themselves with subtle definitions, or intricate ratiocination. Let them learn to be wise by easier means: let them observe the hind of the forest, and the linnet of the grove: let them consider the life of animals, whose motions are regulated by instinct; they obey their guide and are happy. Let us therefore, at length, cease to dispute, and learn to live; throw away the incumbrance of precepts, which they who utter them with so much pride and pomp do not understand, and carry with us this simple and intelligible maxim, That deviation from nature is deviation from happiness."

When he had spoken, he looked round him with a placid air, and enjoyed the consciousness of his own beneficence. "Sir," said the prince, with great modesty, "as I, like all the rest of mankind, am desirous of felicity, my closest attention has been fixed upon your discourse: I doubt not the truth of a position which a man so learned has so confidently advanced. Let me only know what it is to live according to nature."

"When I find young men so humble and so docile," said the philosopher, "I can deny them no information which my studies have enabled me to afford. To live according to nature is to act always with due regard to the fitness arising from the relations and qualities of causes and effects; to concur with the great and unchangeable scheme of universal felicity; to co-operate with the general disposition and tendency of the present system of things."

The prince soon found that this was one of the sages whom he should understand less as he heard him longer. He therefore bowed and was silent, and the philosopher, supposing him satisfied, and the rest vanquished, rose up and departed with the air of a man that had co-operated with the present system.

FROM *Preface to* The Plays of William Shakespeare

. . . Nothing can please many, and please long, but just representations of general nature. Particular manners can be known to few, and therefore few only can judge how nearly they are copied. The irregular combinations of fanciful invention may delight awhile, by that novelty of which the common satiety of life sends us all in quest; but the pleasures of sudden wonder are soon exhausted, and the mind can only repose on the stability of truth.

Shakespeare is above all writers, at least above all modern writers, the poet of nature; the poet that holds up to his readers a faithful mirror of manners and of life. His characters are not modified by the customs of particular places, unpractised by the rest of the world; by the peculiarities of studies or professions, which can operate but upon small numbers; or by the accidents of transient fashions or temporary opinions: they are the genuine progeny of common humanity, such as the world will always supply, and observation will always find. His persons act and speak by the influence of those general passions and principles by which all minds are agitated, and the whole system of life is continued in motion. In the writings of other poets a character is too often an individual; in those of Shakespeare it is commonly a species. . . .

Characters thus ample and general were not easily discriminated and preserved, yet perhaps no poet ever kept his personages more distinct from each other. . . . Shakespeare has no heroes; his scenes are occupied only by men, who act and speak as the reader thinks that he should himself have spoken or acted on the same occasion: Even where the agency is supernatural the dialogue is level with life. . . .

His adherence to general nature has exposed him to the censure of critics, who form their judgments upon narrower principles. Dennis and Rymer[21] think his Romans not sufficiently Roman; and Voltaire censures his kings as not completely royal. . . . But Shakespeare always makes nature predominate over accident; and if he preserves the essential character, is not very careful of distinctions superinduced and adventitious. His story requires Romans or kings, but he thinks only on men. . . . These are the petty cavils of petty minds; a poet overlooks the casual distinction of country and condition, as a painter, satisfied with the figure, neglects the drapery. . . .

Shakespeare's plays are not in the rigorous and critical sense either tragedies or comedies, but compositions of a distinct kind, exhibiting the real state of sublunary nature, which partakes of good and evil, joy and sorrow, mingled with endless variety of proportion and innumerable modes of combination, and expressing the course of the world, in which the loss of one is the gain of another; in which, at the same time, the reveler is

[21]John Dennis (1657–1734), author of *An Essay on the Genius and Writings of Shakespeare* (1712), and Thomas Rymer (1643?–1713), author of *The Tragedies of the Last Age Considered and Examined by the Practice of the Ancients and by the Common Sense of All Ages* (1678) and *A Short View of Tragedy* (1693), were leading neoclassical English critics whose narrow principles Johnson often challenged.

hasting to his wine, and the mourner burying his friend; in which the malignity of one is sometimes defeated by the frolic of another; and many mischiefs and many benefits are done and hindered without design. . . . That this is a practice contrary to the rules of criticism will be readily allowed; but there is always an appeal open from criticism to nature. . . .

The work of a correct and regular writer is a garden accurately formed and diligently planted, varied with shades, and scented with flowers; the composition of Shakespeare is a forest, in which oaks extend their branches, and pines tower in the air, interspersed sometimes with weeds and brambles, and sometimes giving shelter to myrtles and to roses; filling the eye with awful pomp, and gratifying the mind with endless diversity. Other poets display cabinets of precious rarities, minutely finished, wrought into shape, and polished unto brightness. Shakespeare opens a mine which contains gold and diamonds in unexhaustible plenty, though clouded by incrustations, debased by impurities, and mingled with a mass of meaner minerals. . . .

But the power of nature is only the power of using to any certain purpose the materials which diligence procures, or opportunity supplies. Nature gives no man knowledge, and when images are collected by study and experience, can only assist in combining or applying them. Shakespeare, however favored by nature, could impart only what he had learned; and as he must increase his ideas, like other mortals, by gradual acquisition, he, like them, grew wiser as he grew older; could display life better, as he knew it more; and instruct with more efficacy, as he was himself more amply instructed. . . .

Nor was his attention confined to the actions of men; he was an exact surveyor of the inanimate world; his descriptions have always some peculiarities, gathered by contemplating things as they really exist. . . . Shakespeare, whether life or nature be his subject, shows plainly that he has seen with his own eyes; he gives the image which he receives, not weakened or distorted by the intervention of any other mind; the ignorant feel his representations to be just, and the learned see that they are complete.

Perhaps it would not be easy to find any author, except Homer, who invented so much as Shakespeare, who so much advanced the studies which he cultivated, or effused so much novelty upon his age or country. . . .

FROM A Journey to the Western Islands of Scotland: *Anoch*

. . . We passed many rivers and rivulets, which commonly ran with a clear shallow stream over a hard pebbly bottom. These channels, which seem so much wider than the water that they convey would naturally require, are formed by the violence of wintry floods, produced by the accumulation of innumerable streams that fall in rainy weather from the hills, and bursting away with resistless impetuosity, make themselves a passage proportionate to their mass.

Such capricious and temporary waters cannot be expected to produce many fish. The rapidity of the wintry deluge sweeps them away, and the scantiness of the summer stream would hardly sustain them above the ground. This is the reason why in fording the northern rivers, no fishes are seen, as in England, wandering in the water.

Of the hills many may be called with Homer's Ida *abundant in springs*, but few can deserve the epithet which he bestows upon Pelion by *waving their leaves*. They exhibit very little variety, being almost wholly covered with dark heath, and even that seems to be checked in its growth. What is not heath is nakedness, a little diversified by now and then a stream rushing down the steep. An eye accustomed to flowery pastures and waving harvests is astonished and repelled by this wide extent of hopeless sterility. The appearance is that of matter incapable of form or usefulness, dismissed by nature from her care and disinherited of her favors, left in its original elemental state, or quickened only with one sullen power of useless vegetation.

It will very readily occur, that this uniformity of barrenness can afford very little amusement to the traveler; that it is easy to sit at home and conceive rocks and heath, and waterfalls; and that these journeys are useless labors, which neither impregnate the imagination, nor enlarge the understanding. It is true that of far the greater part of things, we must content ourselves with such knowledge as description may exhibit, or analogy supply; but it is true likewise that these ideas are always incomplete, and that at least till we have compared them with realities, we do not know them to be just. As we see more, we become possessed of more certainties, and consequently gain more principles of reasoning, and found a wider basis of analogy.

Regions mountainous and wild, thinly inhabited, and little cultivated, make a great part of the earth, and he that has never seen them must live unacquainted with much of the face of nature, and with one of the great scenes of human existence.

As the day advanced towards noon, we entered a narrow valley not very flowery, but sufficiently verdant. Our guides told us, that the horses could not travel all day without rest or meat,[22] and entreated us to stop here, because no grass would be found in any other place. The request was reasonable and the argument cogent. We therefore willingly dismounted and diverted ourselves as the place gave us opportunity.

I sat down on a bank, such as a writer of Romance might have delighted to feign. I had indeed no trees to whisper over my head, but a clear rivulet streamed at my feet. The day was calm, the air soft, and all was rudeness, silence, and solitude. Before me, and on either side, were high hills, which by hindering the eye from ranging, forced the mind to find entertainment for itself. Whether I spent the hour well I know not; for here I first conceived the thought of this narration.

We were in this place at ease and by choice, and had no evils to suffer or to fear; yet the imaginations excited by the view of an unknown and untraveled wilderness are not such as arise in the artificial solitude of parks and gardens, a flattering notion of self-sufficiency, a placid indulgence of voluntary delusions, a secure expansion of the fancy, or a cool concentration of the mental powers. The phantoms which haunt a desert are want, and misery, and danger; the evils of dereliction rush upon the thoughts; man is made unwillingly acquainted with his own weakness, and meditation shows him only how little he can sustain, and how little he can perform. There were no traces of inhabitants, except perhaps a rude pile of clods called a summer hut, in which a herdsman had rested in the favorable seasons. Whoever had been in the place where I then sat, unprovided with provisions and ignorant of the country, might, at least before the roads were made, have wandered among the rocks, till he had perished with hardship, before he could have found either food or shelter. Yet what are these hillocks to the ridges of Taurus, or these spots of wildness to the deserts of America?

It was not long before we were invited to mount, and continued our journey along the side of a lough, kept full by many streams, which with more or less rapidity and noise crossed the road from the hills on the other hand. These currents, in their diminished

[22]"Food," "fodder."

state, after several dry months, afford, to one who has always lived in level countries, an unusual and delightful spectacle; but in the rainy season, such as every winter may be expected to bring, must precipitate an impetuous and tremendous flood. I suppose the way by which we went is at that time impassable.

The General Idea of Nature

Discourses, by Joshua Reynolds

Sir Joshua Reynolds (1723–92) was one of the leading English painters of the eighteenth century, both of portraits and of historical scenes. Born at Plympton-Earl's in Devonshire, the son of a clergyman and grammar-school master, he studied in Italy from 1750 to 1752, and became a member of Samuel Johnson's inner circle, contributing papers on painting to the Idler *and co-founding with Johnson in 1764 The Club (later The Literary Club), which included among its other members David Garrick, Oliver Goldsmith, the Wartons (Chapter 19 above), Edmund Burke, and eventually Boswell. So great was his success as a painter that he became the first president of the Royal Academy, founded in 1768, and in this position delivered between 1769 and 1790 the fifteen* Discourses *in which he formulates his broad neoclassical view of art and nature. (The text of the following selections from Discourse III of 1770 and Discourse VII of 1776 is based on the 1797* Works of Sir Joshua Reynolds, *as reprinted in* Discourses on Art, *ed. Robert R. Wark [1975].)*

Reynolds's classicism firmly rejects any conception of art as simply imitation of external nature. In Discourse III, as Samuel H. Monk writes in The Sublime *(1960), "he develops fully his theory of the sublime in painting, an excellency that goes beyond mere nature and presents the ideal perfection toward which nature strives. This ideality rests not upon technique, nor even upon mere eclecticism, but upon the grandeur of a painter's ideas, his ability to seize our imagination, his 'intellectual dignity.' This is a sure indication of the influence of Longinus." This was an influence evidenced throughout his career, especially in his later years, in his emphasis on Michelangelo's sublimity as a supreme embodiment of art. Yet Reynolds's view that the painter "corrects nature by herself, her imperfect state by her more perfect," harks back to Sir Philip Sidney's* Apology for Poetry *of 1595, long before the rediscovery of "Longinus." "Only the poet, . . . lifted up with the vigor of his own invention, doth grow in effect another nature," Sidney wrote, "in making things either better than Nature bringeth forth, or, quite anew, forms such as never were in Nature. . . . Nature never set forth the earth in so rich tapestry as divers poets have done—neither with pleasant rivers, fruitful trees, sweet-smelling flowers, nor whatsoever else may make the too much loved earth more lovely. Her world is brazen, the poets only deliver a golden."*

Ideal beauty is "the perfect state of nature." And the "general idea of nature," Reynolds proclaims in Discourse VII, is the beginning, middle, and end of everything valuable in taste; this "nature" includes "the nature and internal fabric and organization . . . of the human mind and imagination." In contrast, "deformity is not nature, but an accidental deviation from her accustomed practice. This general idea therefore ought to be called Nature, and nothing else, correctly speaking, has a right to that name."

"So far is Art from being derived from, or having any immediate intercourse with, particular nature as its model," Reynolds writes in Discourse XIII (1786), "that there are many Arts

that set out with a professed deviation from it"; those who adhere too closely to particular nature, as Rembrandt and the Dutch school did in Reynolds's eyes, fail to attain the ideality of general nature. In sum, "The works, whether of poets, painters, moralists, or historians, which are built upon general nature live forever," Reynolds declares in Discourse IV (1771); "while those which depend for their existence on particular customs and habits, a partial view of nature, or the fluctuation of fashion, can only be coeval with that which first raised them from obscurity." This grand conception of general nature, shared by Reynolds with Johnson, epitomizes a neoclassical ideal that would be anathema to writers of a later generation, notably William Blake. In the indignant marginalia to his copy of Reynolds's Discourses *(which he viewed "as the simulations of the hypocrite who smiles particularly where he means to betray"), Blake wrote: "What is General Nature? is there such a thing? what is general knowledge? is there such a thing? Strictly speaking all knowledge"—and of course, all nature—"is particular."*

FROM *Discourse III*

... The first endeavors of a young painter, as I have remarked in a former discourse, must be employed in the attainment of mechanical dexterity, and confined to the mere imitation of the object before him. Those who have advanced beyond the rudiments may, perhaps, find advantage in reflecting on the advice which I have likewise given them, when I recommended the diligent study of the works of our great predecessors; but I at the same time endeavored to guard them against an implicit submission to the authority of any one master however excellent; or by a strict imitation of his manner, precluding themselves from the abundance and variety of Nature. I will now add that Nature herself is not to be too closely copied. There are excellencies in the art of painting beyond what is commonly called the imitation of nature: and these excellencies I wish to point out. The students who, having passed through the initiatory exercises, are more advanced in the art, and who, sure of their hand, have leisure to exert their understanding, must now be told that a mere copier of nature can never produce anything great; can never raise and enlarge the conceptions or warm the heart of the spectator. . . .

. . . Could we teach taste or genius by rules, they would be no longer taste and genius. But though there neither are, nor can be, any precise invariable rules for the exercise, or the acquisition, of these great qualities, yet we may truly say that they always operate in proportion to our attention in observing the works of nature, to our skill in selecting, and to our care in digesting, methodizing, and comparing our observations. There are many beauties in our art that seem, at first, to lie without the reach of precept, and yet may easily be reduced to practical principles. Experience is all in all; but it is not everyone who profits by experience; and most people err, not so much from want of capacity to find their object as from not knowing what object to pursue. This great ideal perfection and beauty are not to be sought in the heavens, but upon the earth. They are about us, and upon every side of us. But the power of discovering what is deformed in nature, or in other words what is particular and uncommon, can be acquired only by experience; and the whole beauty and grandeur of the art consists, in my opinion, in being able to get above all singular forms, local customs, particularities, and details of every kind.

All the objects which are exhibited to our view by nature, upon close examination will be found to have their blemishes and defects. The most beautiful forms have something about them like weakness, minuteness, or imperfection. But it is not every eye that perceives these blemishes. It must be an eye long used to the contemplation and comparison of these forms; and which, by a long habit of observing what any set of objects of the same kind have in common, has acquired the power of discerning what each wants in particular. This long laborious comparison should be the first study of the painter, who aims at the greatest style. By this means, he acquires a just idea of beautiful forms; he corrects nature by herself, her imperfect state by her more perfect. His eye being enabled to distinguish the accidental deficiencies, excrescences, and deformities of things from their general figures, he makes out an abstract idea of their forms more perfect than any one original; and what may seem a paradox, he learns to design naturally by drawing his figures unlike to any one object. This idea of the perfect state of nature, which the artist calls the Ideal Beauty, is the great leading principle by which works of genius are conducted. By this Phidias acquired his fame. He wrought upon a sober principle what has so much excited the enthusiasm of the world; and by this method you, who have courage to tread the same path, may acquire equal reputation.

This is the idea which has acquired, and which seems to have a right to, the epithet of *divine*, as it may be said to preside, like a supreme judge, over all the productions of nature, appearing to be possessed of the will and intention of the Creator, as far as they regard the external form of living beings. When a man once possesses this idea in its perfection, there is no danger but that he will be sufficiently warmed by it himself, and be able to warm and ravish everyone else.

Thus it is from a reiterated experience, and a close comparison of the objects in nature, that an artist becomes possessed of the idea of that central form, if I may so express it, from which every deviation is deformity. But the investigation of this form, I grant, is painful, and I know but of one method of shortening the road; this is, by a careful study of the works of the ancient sculptors, who, being indefatigable in the school of nature, have left models of that perfect form behind them, which an artist would prefer as supremely beautiful, who had spent his whole life in that single contemplation. But if industry carried them thus far, may not you also hope for the same reward from the same labor? We have the same school opened to us, that was opened to them; for nature denies her instructions to none, who desire to become her pupils. . . .

FROM *Discourse VII*

. . . The arts would lie open forever to caprice and casualty if those who are to judge of their excellencies had no settled principles by which they are to regulate their decisions, and the merit or defect of performance were to be determined by unguided fancy. And indeed we may venture to assert that whatever speculative knowledge is necessary to the artist is equally and indispensably necessary to the connoisseur.

The first idea that occurs in the consideration of what is fixed in art, or in taste, is that presiding principle of which I have so frequently spoken in former discourses—the general idea of nature. The beginning, the middle, and the end of everything that is valuable in taste is comprised in the knowledge of what is truly nature; for whatever notions are not conformable to those of nature, or universal opinion, must be considered as more or less capricious.

My notion of nature comprehends not only the forms which nature produces but also the nature and internal fabric and organization, as I may call it, of the human mind and imagination. The terms beauty, or nature, which are general ideas, are but different modes of expressing the same thing, whether we apply these terms to statues, poetry, or picture. Deformity is not nature, but an accidental deviation from her accustomed practice. This general idea therefore ought to be called Nature, and nothing else, correctly speaking, has a right to that name. But we are so far from speaking, in common conversation, with any such accuracy, that, on the contrary, when we criticize Rembrandt and other Dutch painters who introduced into their historical pictures exact representations of individual objects with all their imperfections, we say, Though it is not in a good taste, yet it is nature.

This misapplication of terms must be very often perplexing to the young student. Is not art, he may say, an imitation of nature? Must he not therefore who imitates her with the greatest fidelity be the best artist? By this mode of reasoning Rembrandt has a higher place than Raffaelle. But a very little reflection will serve to show us that these particularities cannot be nature: for how can that be the nature of man in which no two individuals are the same?

It plainly appears, that as a work is conducted under the influence of general ideas, or partial, it is principally to be considered as the effect of a good or a bad taste.

As beauty therefore does not consist in taking what lies immediately before you, so neither, in our pursuit of taste, are those opinions which we first received and adopted the best choice, or the most natural to the mind and imagination. In the infancy of our knowledge we seize with greediness the good that is within our reach; it is by after-consideration, and in consequence of discipline, that we refuse the present for a greater good at a distance. The nobility or elevation of all arts, like the excellency of virtue itself, consists in adopting this enlarged and comprehensive idea; and all criticism built upon the more confined view of what is natural may properly be called *shallow* criticism, rather than false: its defect is that the truth is not sufficiently extensive. . . .

He who thinks nature, in the narrow sense of the word, is alone to be followed, will produce but a scanty entertainment for the imagination: everything is to be done with which it is natural for the mind to be pleased, whether it proceeds from simplicity or variety, uniformity or irregularity; whether the scenes are familiar or exotic; rude and wild, or enriched and cultivated; for it is natural for the mind to be pleased with all these in their turn. In short, whatever pleases has in it what is analogous to the mind, and is therefore, in the highest and best sense of the word, natural. . . .

A More Minute Inquiry

The Natural History and Antiquities of Selborne, by Gilbert White

"Nature" in a wide range of senses was central, as we have seen, to eighteenth-century thought. For the most part, however, these senses tended toward the normative, the ideal, and the general: "The regular course of things," "The compass of natural existence," "The state or operation of the material world," in Johnson's definitions, which reflected the lawful regularities corroborated by the new philosophy. Such a view of nature gave prominence to what can *or* should be *when the painter or poet, as Reynolds wrote, corrects Nature's "imperfect state by her more perfect," through which alone her potential is realized. Given the ubiquity of interest in nature, it is striking how little concrete, detailed, and precise observation of the immediate natural world the literature of this century produced in either poetry or prose.*

In this context, Gilbert White (1720–93), by his "humble attempt to promote a more minute inquiry into natural history," stands sharply apart from most contemporaries, and anticipates writers of the following century as diverse as the Wordsworths (William and Dorothy), John

Clare, Thoreau, Alexander von Humboldt, John Muir, and Charles Darwin. Born in Selborne, Hampshire (a village of some 600 souls), he was educated at Basingstoke Grammar School—whose headmaster, after resigning as Professor of Poetry at Oxford in 1738, was Thomas Warton, father of Thomas, Jr., and Joseph (Chapter 19 above)—and in 1740 went up to Oriel College, Oxford, where in 1743 he received his B.A. and became a fellow, taking his M.A. in 1746. After traveling for years around England, while serving several times as temporary curate of Selborne, he was named curate of nearby Faringdon in 1761, and lived unmarried in Selborne for the remaining thirty-two years of his life.

As early as 1751 he began to keep a diary of the horticultural year, the Garden Kalendar, *which he expanded in 1768 into his* Naturalist's Journal, *a detailed, nearly day-by-day account of his observations in and around his native village. These journals were a principal resource for the series of letters to Thomas Pennant and Daines Barrington published in December 1788 as* The Natural History and Antiquities of Selborne*; selections from the journals were first published as* Gilbert White's Journals, *ed. Walter Johnson, in 1931. "More than any other single book," Richard Mabey writes in his introduction to the Penguin English Library edition (1977) of* The Natural History, *which he calls the fourth most published book in English (presumably after the Bible, the Book of Common Prayer, and Izaak Walton's* Compleat Angler *[Chapter 19]), "it has shaped our everyday view of the relations between man and nature"; for White "was perhaps the first writer to talk of animals—and particularly birds—as if they conceivably inhabited the same universe as human beings." "What is new in Gilbert White, or at least feels new in its sustained intensity," Raymond Williams observes in* The Country and the City *(1973), comparing him to two younger writers from the same region, William Cobbett of the* Rural Rides *(1830) and Jane Austen, "is . . . a single and dedicated observation, as if the only relationships of country living were to its physical facts. . . . While Cobbett and Jane Austen, in their different ways, were absorbed in a human world, Gilbert White was watching the turn of the year and the myriad physical lives inside it: nature in a sense that could now be separated from man"—yet which remained essential, by virtue of its very separation, to man's need for communion with other creatures, both plant and animal, in the universe we share.*

Botany had been a major interest of many writers from Theophrastus (Chapter 9) to Renaissance herbalists such as John Gerard (Chapter 16). Among the most important early eighteenth-century observers of "particular nature" were Jean-Jacques Rousseau (Chapter 21 below) and the great Swedish taxonomist Karl von Linné, or Carolus Linnaeus (1707–78), with whose terminology, as expounded in the monumental Species Plantarum (Species of Plants) *of 1753, White was familiar. Yet in contrast to much of Rousseau's writing—with the notable exception of his eight letters on botany written between 1771 and 1773 and published in 1781—White stayed scrupulously close to observed facts and eschewed sentimental effusions or romantic reveries. And in contrast to Linnaeus, he insisted, in Letter XL to Barrington, that the botanist "should be by no means content with a list of names; he should study plants philosophically . . . Not that system is by any means to be thrown aside; without system the field of nature would be a pathless wilderness: but system should be subservient to, not the main object of, pursuit."*

His journals, besides recording countless isolated facts, contain numerous more general remarks. "Most people know, that have observed at all," he writes on January 4, 1789, for example, "that the swimming of birds is nothing more than a walking in the water, where one foot succeeds the other as on the land; yet no one, as far as I am aware, has remarked that diving fowls, while under water, impel & row themselves forward by a motion of their wings, as well as by the impulse of their feet: but such is really the case, as any person may easily be convinced who will observe ducks when hunted by dogs in a clear pond." The patient accumulation of such observations led to a number of significant contributions to natural history. "A hundred years before Darwin,"

Mabey notes, "he realized the crucial role of worms in the formation of soil," and despite his strange misapprehension that local swallows hibernated rather than migrating, "he was one of the very first naturalists to understand the significance of territory and song in birds. The full list of his discoveries and original insights would cover many pages."

What is still more remarkable in the letters collected in The Natural History of Selborne *(the section on village antiquities is more conventional) is the "quality of his feeling for the life around him," in Williams's phrase, that infuses his observations and inquiries, and makes his book an unfailingly engaging expression of his lively curiosity about the world. Of the selections that follow, "The Naturalist's Summer-Evening Walk" is a poem scarcely equaled in its century for the specificity of its designations of local birds and insects (no "wingèd tribes" here!), and the letters to Barrington display both the dense particularity of his scrutinies (especially of the chimney swallow) and the speculations and conclusions to which these give rise. "Subjects of this kind," as even so small a sampling suggests, "are inexhaustible."*

FROM "THE NATURALIST'S SUMMER-EVENING WALK"

Letter XXIV to Thomas Pennant (Selborne, May 29, 1769)

> . . . *equidem credo, quia sit divinitus illis*
> *Ingenium.*
>
> —Virg. *Georg.*[23]

When day declining sheds a milder gleam,
What time the may-fly[24] haunts the pool or stream;
When the still owl skims round the grassy mead,
What time the timorous hare limps forth to feed;
Then be the time to steal adown the vale, 5
And listen to the vagrant[25] cuckoo's tale;
To hear the clamorous[26] curlew call his mate,
Or the soft quail his tender pain relate;
To see the swallow sweep the dark'ning plain
Belated, to support her infant train; 10
To mark the swift in rapid giddy ring
Dash round the steeple, unsubdued of wing:
Amusive birds!—say where your hid retreat
When the frost rages and the tempests beat;
Whence your return, by such nice instinct led, 15
When spring, soft season, lifts her bloomy head?
Such baffled searches mock man's prying pride,
The GOD OF NATURE is your secret guide!

[23]"Indeed I think they have their wisdom from a divinity . . ." (*Georgics* I.415–16). Virgil's lines (referring to weather-forecasting of rooks) read "haud equidem credo . . ." ("*not* indeed that I think . . .").

[24]The angler's may-fly, the *ephemera vulgata Linn.*, comes forth from its aurelia state, and emerges out of the water about six in the evening, and dies about eleven at night, determining the date of its fly state in about five or six hours. They usually begin to appear about the 4th of June, and continue in succession for bare a fortnight. See Swammerdam, Derham, Scopoli, etc. [This and all following notes, unless bracketed, are by White.]

[25]Vagrant cuckoo; so called because, being tied down by no incubation or attendance about the nutrition of its young, it wanders without control.

[26]*Charadrius aedicnemus.*

While deep'ning shades obscure the face of day
To yonder bench, leaf-sheltered, let us stray, 20
Till blended objects fail the swimming sight,
And all the fading landscape sinks in night;
To hear the drowsy dor come brushing by
With buzzing wing, or the shrill[27] cricket cry;
To see the feeding bat glance through the wood; 25
To catch the distant falling of the flood;
While o'er the cliff th'awakened churn-owl hung
Through the still gloom protracts his chattering song;
While high in air, and pois'd upon his wings,
Unseen, the soft enamored woodlark[28] sings: 30
These, NATURE's works, the curious mind employ,
Inspire a soothing melancholy joy:
As fancy warms, a pleasing kind of pain
Steals o'er the cheek, and thrills the creeping vein!
Each rural sight, each sound, each smell combine; 35
The tinkling sheep-bell, or the breath of kine;
The new-mown hay that scents the swelling breeze,
Or cottage-chimney smoking through the trees.
The chilling night-dews fall: away, retire;
For see, the glow-worm lights her amorous fire![29] 40
Thus, ere night's veil had half obscured the sky,
Th'impatient damsel hung her lamp on high:
True to the signal, by love's meteor led,
Leander hastened to his Hero's bed.[30]

FROM LETTERS TO DAINES BARRINGTON

Letter XI (Selborne, Feb. 8, 1772)

Dear Sir,

When I ride about in the winter, and see such prodigious flocks of various kinds of birds, I cannot help admiring at these congregations, and wishing that it was in my power to account for those appearances almost peculiar to the season. The two great motives which regulate the proceedings of the brute creation are love and hunger; the former incites animals to perpetuate their kind, the latter induces them to preserve individuals; whether either of these should seem to be the ruling passion in the matter of congregating is to be considered. As to love, that is out of the question at a time of the year when that soft passion is not indulged; besides, during the amorous season, such a jealousy prevails between the male birds that they can hardly bear to be together in the same hedge or field. Most of the singing and elation of spirits of that time seem to me to be the effect of rivalry and emulation: and it is to this spirit of jealousy that I chiefly attribute the equal dispersion of birds in the spring over the face of the country.

[27] *Gryllus campestris.*
[28] In hot summer nights woodlarks soar to a prodigious height, and hang singing in the air.
[29] The light of the female glow-worm (as she often crawls up the stalk of a grass to make herself more conspicuous) is a signal to the male, which is a slender dusky *scarabaeus.*
[30] See the story of Hero and Leander.

Now as to the business of food: as these animals are actuated by instinct to hunt for necessary food, they should not, one would suppose, crowd together in pursuit of sustenance at a time when it is most likely to fail: yet such associations do take place in hard weather chiefly, and thicken as the severity increases. As some kind of self-interest and self-defense is no doubt the motive for the proceeding, may it not arise from the helplessness of their state in such rigorous seasons; as men crowd together, when under great calamities, though they know not why? Perhaps approximation may dispel some degree of cold; and a crowd may make each individual appear safer from the ravages of birds of prey and other dangers. . . .

Letter XIV (Selborne, March 26, 1773)

Dear Sir,

The more I reflect on the *storgê* [natural affection of parents for offspring] of animals, the more I am astonished at its effects. Nor is the violence of this affection more wonderful than the shortness of its duration. Thus every hen is in her turn the virago of the yard, in proportion to the helplessness of her brood; and will fly in the face of a dog or a sow in defense of those chickens, which in a few weeks she will drive before her with relentless cruelty.

This affection sublimes the passions, quickens the invention, and sharpens the sagacity of the brute creation. Thus an hen, just become a mother, is no longer that placid bird she used to be, but with feathers standing on end, wings hovering, and clucking note, she runs about like one possessed. Dams will throw themselves in the way of the greatest danger in order to avert it from their progeny. Thus a partridge will tumble along before a sportsman in order to draw away the dogs from her helpless covey. In the time of nidification [nest-building] the most feeble birds will assault the most rapacious. All the *hirundines* [swallows] of a village are up in arms at the sight of an hawk, whom they will persecute till he leaves that district. A very exact observer has often remarked that a pair of ravens nesting in the rock of Gibraltar would suffer no vulture or eagle to rest near their station, but would drive them from the hill with an amazing fury: even the blue thrush at the season of breeding would dart out from the clefts of the rocks to chase away the kestril, or the sparrow-hawk. If you stand near the nest of a bird that has young, she will not be induced to betray them by an inadvertent fondness, but will wait about at a distance with meat in her mouth for an hour together.

Should I farther corroborate what I have advanced above by some anecdotes which I probably may have mentioned before in conversation, yet you will, I trust, pardon the repetition for the sake of illustration.

The fly-catcher of the *Zoology* (the *stoparola* of Ray), builds every year in the vines that grow on the walls of my house. A pair of these little birds had one year inadvertently placed their nest on a naked bough, perhaps in a shady time, not being aware of the inconvenience that followed. But an hot sunny season coming on before the brood was half fledged, the reflection of the wall became insupportable, and must inevitably have destroyed the tender young, had not affection suggested an expedient, and prompted the parent-birds to hover over the nest all the hotter hours, while with wings expanded, and mouths gaping for breath, they screened off the heat from their suffering offspring.

A farther instance I once saw of notable sagacity in a willow-wren, which had built in a bank in my fields. This bird a friend and myself had observed as she sat in her nest; but were particularly careful not to disturb her, though we saw she eyed us with some degree of jealousy. Some days after as we passed that way we were desirous of remarking how this brood went on; but no nest could be found, till I happened to take up a large bundle of long green moss, as it were, carelessly thrown over the nest, in order to dodge the eye of any impertinent intruder.

A still more remarkable mixture of sagacity and instinct occurred to me one day as my people were pulling off the lining of an hotbed, in order to add some fresh dung. From out of the side of this bed leaped an animal with great agility that made a most grotesque figure; nor was it without great difficulty that it could be taken; when it proved to be a large white-bellied field-mouse with three or four young clinging to her teats by their mouths and feet. It was amazing that the desultory and rapid motions of this dam should not oblige her litter to quit their hold, especially when it appeared that they were so young as to be both naked and blind!

To these instances of tender attachment, many more of which might be daily discovered by those that are studious of nature, may be opposed that rage of affection, that monstrous perversion of the *storgê*, which induces some females of the brute creation to devour their young because their owners have handled them too freely, or removed them from place to place! Swine, and sometimes the more gentle race of dogs and cats, are guilty of this horrid and preposterous murder. When I hear now and then of an abandoned mother that destroys her offspring, I am not so much amazed; since reason perverted, and the bad passions let loose, are capable of any enormity: but why the parental feelings of brutes, that usually flow in one most uniform tenor, should sometimes be so extravagantly diverted, I leave to abler philosophers than myself to determine.

Letter XVII (Ringmer, near Lewes, Dec. 9, 1773)

Dear Sir,

I received your last favor just as I was setting out for this place; and am pleased to find that my monography met with your approbation. My remarks are the result of many years' observation; and are, I trust, true on the whole: though I do not pretend to say that they are perfectly void of mistake, or that a more nice observer might not make many additions, since subjects of this kind are inexhaustible.

If you think my letter worthy of the notice of your respectable society, you are at liberty to lay it before them; and they will consider it, I hope, as it was intended, as an humble attempt to promote a more minute inquiry into natural history; into the life and conversation of animals. Perhaps hereafter I may be induced to take the house-swallow under consideration, and from that proceed to the rest of the British *hirundines* [swallows].

Though I have now traveled the Sussex-downs upwards of thirty years, yet I still investigate that chain of majestic mountains with fresh admiration year by year; and think I see new beauties every time I traverse it. This range, which runs from Chichester eastward as far as East-Bourn, is about sixty miles in length, and is called the South Downs, properly speaking, only round Lewes. As you pass along you command a noble view of the wild, or weald, on one hand, and the broad downs and sea on the other. Mr. Ray used to visit a family[31] just at the foot of these hills, and was so ravished with the prospect from Plumpton-plain near Lewes, that he mentions those scapes in his *Wisdom of God in the*

[31]Mr. Courthope, of Danny.

Works of the Creation with the utmost satisfaction, and thinks them equal to anything he had seen in the finest parts of Europe.

For my own part, I think there is somewhat peculiarly sweet and amusing in the shapely figured aspect of chalk-hills in preference to those of stone, which are rugged, broken, abrupt, and shapeless.

Perhaps I may be singular in my opinion, and not so happy as to convey to you the same idea; but I never contemplate these mountains without thinking I perceive somewhat analogous to growth in their gentle swellings and smooth fungus-like protuberances, their fluted sides, and regular hollows and slopes, that carry at once the air of vegetative dilation and expansion. . . . Or was there ever a time when these immense masses of calcareous matter were thrown into fermentation by some adventitious moisture; were raised and leavened into such shapes by some plastic power; and so made to swell and heave their broad backs into the sky so much above the less animated clay of the wild below? . . .

Letter XVIII (Selborne, Jan. 29, 1774)

Dear Sir,

The house-swallow, or chimney-swallow, is undoubtedly the first comer of all the British *hirundines*; and appears in general on or about the thirteenth of April, as I have remarked from many years' observation. Not but now and then a straggler is seen much earlier: and, in particular, when I was a boy I observed a swallow for a whole day together on a sunny warm Shrove Tuesday; which day could not fall out later than the middle of March, and often happened early in February.

It is worth remarking that these birds are seen first about lakes and mill-ponds; and it is also very particular, that if these early visitors happen to find frost and snow, as was the case of the two dreadful springs of 1770 and 1771, they immediately withdraw for a time. A circumstance this much more in favor of hiding than migration; since it is much more probable that a bird should retire to its hybernaculum just at hand, than return for a week or two only to warmer latitudes.

The swallow, though called the chimney-swallow, by no means builds altogether in chimneys, but often within barns and out-houses against the rafters; and so she did in Virgil's time:

. . . Ante
Garrula quam tignis nidos suspendat hirundo.[32]

In Sweden she builds in barns, and is called *ladu swala*, the barn-swallow. Besides, in the warmer parts of Europe there are no chimneys to houses, except they are English-built: in these countries she constructs her nest in porches, and gateways, and galleries, and open halls.

Here and there a bird may affect some odd, peculiar place; as we have known a swallow build down the shaft of an old well, through which chalk had been formerly drawn up for the purpose of manure: but in general with us this *hirundo* breeds in chimneys; and loves to haunt those stacks where there is a constant fire, no doubt for the sake of warmth. Not that it can subsist in the immediate shaft where there is a fire; but prefers one adjoining to that of the kitchen, and disregards the perpetual smoke of that funnel, as I have often observed with some degree of wonder.

[32]["Before the chattering swallow hangs her nest from the rafters" (*Georgics* IV.306–7).]

Five or six more feet down the chimney does this little bird begin to form her nest about the middle of May, which consists, like that of the house-martin, of a crust or shell composed of dirt or mud, mixed with short pieces of straw to render it tough and permanent; with this difference, that whereas the shell of the martin is nearly hemispheric, that of the swallow is open at the top, and like half a deep dish: this nest is lined with fine grasses, and feathers which are often collected as they float in the air.

Wonderful is the address which this adroit bird shows all day long in ascending and descending with security through so narrow a pass. When hovering over the mouth of the funnel, the vibrations of her wings acting on the confined air occasion a rumbling like thunder. It is not improbable that the dam submits to this inconvenient situation so low in the shaft, in order to secure her broods from rapacious birds, and particularly from owls, which frequently fall down chimneys, perhaps in attempting to get at these nestlings.

The swallow lays from four to six white eggs, dotted with red specks; and brings out her first brood about the last week in June, or the first week in July. The progressive method by which the young are introduced into life is very amusing: first, they emerge from the shaft with difficulty enough, and often fall down into the rooms below: for a day or so they are fed on the chimney-top, and then are conducted to the dead leafless bough of some tree, where, sitting in a row, they are attended with great assiduity, and may then be called perchers. In a day or two more they become flyers, but are still unable to take their own food; therefore they play about near the place where the dams are hawking for flies; and when a mouthful is collected, at a certain signal given, the dam and the nestling advance, rising towards each other, and meeting at an angle; the young one all the while uttering such a little quick note of gratitude and complacency, that a person must have paid very little regard to the wonders of nature that has not often remarked this feat.

The dam betakes herself immediately to the business of a second brood as soon as she is disengaged from her first; which at once associates with the first broods of house-martins; and with them congregates, clustering on sunny roofs, towers, and trees. This *hirundo* brings out her second brood towards the middle and end of August.

All the summer long is the swallow a most instructive pattern of unwearied industry and affection; for, from morning to night, while there is a family to be supported, she spends the whole day in skimming close to the ground, and exerting the most sudden turns and quick evolutions. Avenues, and long walks under hedges, and pasture-fields, and mown meadows where cattle graze, are her delight, especially if there are trees interspersed; because in such spots insects most abound. When a fly is taken a smart snap from her bill is heard, resembling the noise at the shutting of a watch-case; but the motion of the mandibles are too quick for the eye.

The swallow, probably the male bird, is the *excubitor* [sentinel] to house-martins, and other little birds, announcing the approach of birds of prey. For as soon as an hawk appears, with a shrill alarming note he calls all the swallows and martins about him; who pursue in a body, and buffet and strike their enemy till they have driven him from the village, darting down from above on his back, and rising in a perpendicular line in perfect security. This bird also will sound the alarm, and strike at cats when they climb on the roofs of houses, or otherwise approach the nests. Each species of *hirundo* drinks as it flies along, sipping the surface of the water; but the swallow alone, in general, washes on the wing, by dropping into a pool for many times together: in very hot weather house-martins and bank-martins dip and wash a little.

The swallow is a delicate songster, and in soft sunny weather sings both perching and flying; on trees in a kind of concert, and on chimney-tops: is also a bold flyer, ranging to distant downs and commons even in windy weather, which the other species seem much to dislike; nay, even frequenting exposed sea-port towns, and making little excursions over the salt water. Horsemen on wide downs are often closely attended by a little party of swallows for miles together, which plays before and behind them, sweeping around, and collecting all the skulking insects that are roused by the trampling of the horses' feet: when the wind blows hard, without this expedient, they are often forced to settle to pick up their lurking prey.

This species feeds much on little *coleoptera*, as well as on gnats and flies: and often settles on dug ground, or paths, for gravels to grind and digest its food. Before they depart, for some weeks, to a bird, they forsake houses and chimneys, and roost in trees; and usually withdraw about the beginning of October; though some few stragglers may appear on at times till the first week in November.

Some few pairs haunt the new and open streets of London next the fields, but do not enter, like the house-martin, the close and crowded parts of the city.

Both male and female are distinguished from their congeners by the length and forkedness of their tails. They are undoubtedly the most nimble of all the species: and when the male pursues the female in amorous chase, they then go beyond their usual speed, and exert a rapidity almost too quick for the eye to follow.

After this circumstantial detail of the life and discerning *storgê* [parental affection] of the swallow, I shall add, for your farther amusement, an anecdote or two not much in favor of her sagacity.

A certain swallow built for two years together on the handles of a pair of garden shears, that were stuck up against the boards of an out-house,[33] and therefore must have her nest spoiled whenever that implement was wanted: and, what is stranger still, another bird of the same species built its nest on the wings and body of an owl that happened by accident to hang dead and dry from the rafter of a barn. This owl, with the nest on its wings, and with eggs in the nest, was brought as a curiosity worthy of the most elegant private museum in Great Britain. The owner, struck with the oddity of the sight, furnished the bringer with a large shell, or conch, desiring him to fix it just where the owl hung: the person did as he was ordered, and the following year a pair, probably the same pair, built their nest in the conch, and laid their eggs.

The owl and the conch make a strange grotesque appearance, and are not the least curious specimens in that wonderful collection of art and nature.[34]

Thus is instinct in animals, taken the least out of its way, an undistinguishing, limited faculty; and blind to every circumstance that does not immediately respect self-preservation, or lead at once to the propagation or support of their species.

Letter XL (Selborne, June 2, 1778)

Dear Sir,

The standing objection to botany has always been that it is a pursuit that amuses the fancy and exercises the memory, without improving the mind or advancing any real knowledge: and where the science is carried no farther than a mere schematic classification, the charge is but too true. But the botanist that is desirous of wiping off this asper-

[33][Toolshed.] [34]Sir Ashton Lever's Museum.

sion should be by no means content with a list of names; he should study plants philosophically, should investigate the laws of vegetation, should examine the powers and virtues of efficacious herbs, should promote their cultivation; and graft the gardener, the planter, and the husbandman on the phytologist. Not that system is by any means to be thrown aside; without system the field of nature would be a pathless wilderness: but system should be subservient to, not the main object of, pursuit.

Vegetation is highly worthy of our attention; and in itself is of the utmost consequence to mankind, and productive of many of the greatest comforts and elegancies of life. To plants we owe timber, bread, beer, honey, wine, oil, linen, cotton, etc., what not only strengthens our hearts, and exhilarates our spirits, but what secures from inclemencies of weather and adorns our persons. Man, in his true state of nature, seems to be subsisted by spontaneous vegetation: in middle climes, where grasses prevail, he mixes some animal food with the produce of the field and garden: and it is towards the polar extremes only that, like his kindred bears and wolves, he gorges himself with flesh alone, and is driven, to what hunger has never been known to compel the very beasts, to prey on his own species.[35]

The productions of vegetation have had a vast influence on the commerce of nations, and have been the great promoters of navigation, as may be seen in the articles of sugar, tea, tobacco, opium, ginseng, betel, paper, etc. As every climate has its peculiar produce, our natural wants bring on a mutual intercourse; so that by means of trade each distant part is supplied with the growth of every latitude. But, without the knowledge of plants and their culture, we must have been content with our hips and haws, without enjoying the delicate fruits of India and the salutiferous drugs of Peru.

Instead of examining the minute distinctions of every various species of each obscure genus, the botanist should endeavor to make himself acquainted with those that are useful. You shall see a man readily ascertain every herb of the field, yet hardly know wheat from barley, or at least one sort of wheat or barley from another.

But of all sorts of vegetation the grasses seem to be most neglected; neither the farmer nor the grazier seem to distinguish the annual from the perennial, the hardy from the tender, nor the succulent and nutritive from the dry and juiceless.

The study of grasses would be of great consequence to a northerly and grazing kingdom. The botanist that could improve the sward of the district where he lived would be an useful member of society; to raise a thick turf on a naked soil would be worth volumes of systematic knowledge; and he would be the best commonwealth's man that could occasion the growth of "two blades of grass where one alone was seen before."

The Natural History of This Earth

Theory of the Earth (1788), by James Hutton

Thomas Burnet's theory of the earth (Chapter 18 above) was explicitly a sacred *theory, striving to reconcile the new emphasis on natural law with a broadly interpreted fidelity to the Mosaic account of the Creation and the Flood. Not until James Hutton (1726–97) a century later read on March 4 and April 7, 1785, the papers published in 1788 in volume 1 of the* Transactions of

[35]See the late *Voyages to the South-seas.*

the Royal Society of Edinburgh as Theory of the Earth*—expanded in 1795 into the two-volume work of the same title—was geology placed on a scientific basis, and the history of the earth made a part of natural history.*

Born in Edinburgh, the son of a merchant who died when he was young, Hutton in 1740 attended the University of Edinburgh, where he studied chemistry and then medicine; after two years in Paris, he took his M.D. in Leyden in 1749. Returning to Edinburgh, he abandoned medicine for agriculture and in 1752 went to Norfolk to learn farming; there, according to the biographical account of 1803 by John Playfair (who in 1802 had published his influential popularization, Illustrations of the Huttonian Theory of the Earth*), Hutton first began to study mineralogy. After touring Flanders and Holland in 1754, he set up a farm in Berwickshire (southeastern Scotland), returning in 1768 to Edinburgh to pursue his scientific interests; in 1774 he extensively toured parts of England and Wales. For about thirty years, Playfair writes, "he had never ceased to study the natural history of the globe, with a view of ascertaining the changes which have taken place on its surface, and of discovering the causes by which they have been produced."*

In his Theory of the Earth*—from whose 1788 version (reprinted in 1970, along with two short works and Playfair's biography) the following selections are taken—Hutton set forth his conclusions that the surface of the land "is made by nature to decay" as its heights are continually leveled and its soil washed away to the depths of the ocean by the "equable and steady" operations of nature over vast expanses of time, which "is to nature endless and as nothing." But in a perpetual cycle of decay and renovation, destruction and production, sediment from the sea bottom is repeatedly consolidated and new land masses formed by volcanoes above the surface of the oceans: a process, Hutton writes in his famous last sentence, with "no vestige of a beginning—no prospect of an end."*

As the science of geology took shape in the early nineteenth century, Hutton was viewed as an advocate of the "Vulcanist" school, which emphasized land formation through volcanic action, as opposed to "Neptunists," such as Abraham Gottlob Werner (1750–1817) in Germany, who proposed that rocks and mountains had emerged by crystallization, sedimentation, and silting of ocean waters. Still more influential, however, on Charles Lyell (whose Principles of Geology*, published in three volumes between 1830 and 1833, was the classic work of nineteenth-century geology), and through Lyell on Charles Darwin, was Hutton's emphasis on the steadiness of these processes. As opposed to the "catastrophism" of Burnet's and most previous accounts of the history of the earth, the "uniformitarianism" of these writers affirmed, in Hutton's words, "a certain regularity, which is not perhaps in nature" (since "it is not given to man to know what things are truly in themselves, but only what those things are in his thought"), "but which is necessary for our clear conception of the system of nature." Newton had bound the movement of planets and the fall of the apple under a single law, but only with Hutton did the equable and steady operations of the Newtonian heavens at last find a counterpart in processes continually re-forming this earth. For Hutton, these processes bore witness, in the very absence of any beginning or end which so shocked the religious sensibilities of the age, "that in nature there is wisdom, system, and consistency" beneath the moon no less than above it.*

FROM *Part I: Prospect of the Subject to be treated of.*

. . . If . . . we employ our skill in research, not in forming vain conjectures; and if *data* are to be found on which science may form just conclusions, we should not long remain in ignorance with respect to the natural history of this earth, a subject on which hitherto opinion only, and not evidence, has decided: for in no subject is there naturally less defect of evidence, although philosophers, led by prejudice, or misguided by false theory, have neglected to employ that light by which they should have seen the system of this world.

But to proceed in pursuing a little farther our general or preparatory ideas. A solid body of land could not have answered the purpose of a habitable world; for a soil is necessary to the growth of plants; and a soil is nothing but the materials collected from the destruction of the solid land. Therefore, the surface of this land, inhabited by man, and covered with plants and animals, is made by nature to decay, in dissolving from that hard and compact state in which it is found below the soil; and this soil is necessarily washed away by the continual circulation of the water, running from the summits of the mountains towards the general receptacle of that fluid.

The heights of our land are thus leveled with the shores; our fertile plains are formed from the ruins of the mountains; and those traveling materials are still pursued by the moving water, and propelled along the inclined surface of the earth. These moveable materials, delivered into the sea, cannot, for a long continuance, rest upon the shore; for, by the agitation of the winds, the tides and currents, every moveable thing is carried farther and farther along the shelving bottom of the sea, towards the unfathomable regions of the ocean.

If the vegetable soil is thus constantly removed from the surface of the land, and if its place is thus to be supplied from the dissolution of the solid earth, as here represented, we may perceive an end to this beautiful machine: an end arising from no error in its constitution as a world, but from that destructibility of its land which is so necessary in the system of the globe, in the economy of life and vegetation.

The immense time necessarily required for this total destruction of the land must not be opposed to that view of future events which is indicated by the surest facts and most approved principles. Time, which measures everything in our idea, and is often deficient to our schemes, is to nature endless and as nothing; it cannot limit that by which alone it had existence; and as the natural course of time, which to us seems infinite, cannot be bounded by any operation that may have an end, the progress of things upon this globe, that is, the course of nature, cannot be limited by time, which must proceed in a continual succession. We are, therefore, to consider as inevitable the destruction of our land, so far as effected by those operations which are necessary in the purpose of the globe, considered as a habitable world; and so far as we have not examined any other part of the economy of nature, in which other operations and a different intention might appear.

We have now considered the globe of this earth as a machine, constructed upon chemical as well as mechanical principles, by which its different parts are all adapted, in form, in quality, and in quantity, to a certain end: an end attained with certainty or success; and an end from which we may perceive wisdom, in contemplating the means employed.

But is this world to be considered thus merely as a machine, to last no longer than its parts retain their present position, their proper forms and qualities? Or may it not be also considered as an organized body? Such as has a constitution in which the necessary decay of the machine is naturally repaired, in the exertion of those productive powers by which it had been formed.

This is the view in which we are now to examine the globe, to see if there be, in the constitution of this world, a reproductive operation by which a ruined constitution may be again repaired, and a duration or stability thus procured to the machine, considered as a world sustaining plants and animals.

If no such reproductive power, or reforming operation, after due inquiry, is to be found in the constitution of this world, we should have reason to conclude that the system of this earth has either been intentionally made imperfect, or has not been the work of infinite power and wisdom.

Here is an important question, therefore, with regard to the constitution of this globe: a question which, perhaps, it is in the power of man's sagacity to resolve, and a question which, if satisfactorily resolved, might add some luster to science and the human intellect.

Animated with this great, this interesting view, let us strictly examine our principles in order to avoid fallacy in our reasoning; and let us endeavor to support our attention in developing a subject that is vast in its extent, as well as intricate in the relation of parts to be stated.

The globe of this earth is evidently made for man. He alone, of all the beings which have life upon this body, enjoys the whole and every part; he alone is capable of knowing the nature of this world, which he thus possesses in virtue of his proper right; and he alone can make the knowledge of this system a source of pleasure and the means of happiness.

Man alone, of all the animated beings which enjoy the benefits of this earth, employs the knowledge which he there receives, in leading him to judge of the intention of things, as well as of the means by which they are brought about; and he alone is thus made to enjoy, in contemplation as well as sensual pleasure, all the good that may be observed in the constitution of this world; he, therefore, should be made the first subject of inquiry.

Now, if we are to take the written history of man for the rule by which we should judge of the time when the species first began, that period would be but little removed from the present state of things. The Mosaic history places this beginning of man at no great distance; and there has not been found, in natural history, any document by which a high antiquity might be attributed to the human race. But this is not the case with regard to the inferior species of animals, particularly those which inhabit the ocean and its shores. We find in natural history monuments which prove that those animals had long existed; and we thus procure a measure for the computation of a period of time extremely remote, though far from being precisely ascertained.

In examining things present, we have data from which to reason with regard to what has been; and, from what has actually been, we have data for concluding with regard to that which is to happen hereafter. Therefore, upon the supposition that the operations of nature are equable and steady, we find, in natural appearances, means for concluding a certain portion of time to have necessarily elapsed in the production of those events of which we see the effects.

It is thus that, in finding the relics of sea-animals of every kind in the solid body of our earth, a natural history of those animals is formed, which includes a certain portion of time; and for the ascertaining this portion of time, we must again have recourse to the regular operations of this world. We shall thus arrive at facts which indicate a period to which no other species of chronology is able to remount.

In what follows, therefore, we are to examine the construction of the present earth in order to understand the natural operations of time past; to acquire principles by which we may conclude with regard to the future course of things, or judge of those operations by which a world, so wisely ordered, goes into decay; and to learn by what means such a decayed world may be renovated, or the waste of habitable land upon the globe repaired.

This, therefore, is the object which we are to have in view during this physical investigation; this is the end to which are to be directed all the steps in our cosmological pursuit. . . .

We are led, in this manner, to conclude that all the strata of the earth, not only those consisting of such calcareous masses, but others superincumbent upon these, have had their origin at the bottom of the sea, by the collection of sand and gravel, of shells, or coralline and crustaceous bodies, and of earths and clays, variously mixed, or separated and accumulated. Here is a general conclusion, well authenticated in the appearances of nature, and highly important in the natural history of the earth.

The general amount of our reasoning is this, that nine tenths, perhaps, or ninety-nine hundredths of this earth, so far as we see, have been formed by natural operations of the globe, in collecting loose materials, and depositing them at the bottom of the sea; consolidating those collections in various degrees, and either elevating those consolidated masses above the level on which they were formed, or lowering the level of that sea. . . .

If this consolidating operation be performed at the bottom of the ocean, or under great depths of the earth of which our continents are composed, we cannot be witnesses to this mineral process, or acquire knowledge of natural causes, by immediately observing the changes which they produce; but though we have not this immediate observation of those changes of bodies, we have, in science, the means of reasoning from distant events; consequently, of discovering, in the general powers of nature, causes for those events of which we see the effects. . . .

It must be evident that nothing but the most general acquaintance with the laws of acting substances, and with those of bodies changing by the powers of nature, can enable us to set about this undertaking with any reasonable prospect of success; and here the science of chemistry must be brought particularly to our aid; for this science, having for its object the changes produced upon the sensible qualities, as they are called, of bodies, by its means we may be enabled to judge of that which is possible according to the laws of nature, and of that which, in like manner, we must consider as impossible.

Whatever conclusions, therefore by means of this science, shall be attained in just reasoning from natural appearances, this must be held as evidence, where more immediate proof cannot be obtained; and, in a physical subject, where things actual are concerned, and not the imaginations of the human mind, this proof will be considered as amounting to a demonstration.

FROM *Part III: Investigation of the Natural Operations Employed in the Production of Land above the Surface of the Sea.*

. . . Has the globe within it such an active power as fits it for the renovation of that part of its constitution which may be subject to decay? Are those powerful operations of fire, or subterraneous heat, which so often have filled us with terror and astonishment, to be considered as having always been? Are they to be concluded as proper to every part upon the globe, and as continual in the system of this earth? If these points in question shall be decided in the affirmative, we can be at no loss in ascertaining the power which has consolidated strata, nor in explaining the present situation of those bodies which had their origin at the bottom of the sea. This, therefore, should be the object of our pursuit; and, in order to have demonstration in a case of physical inquiry, we must again have recourse to the book of nature. . . .

Volcanoes are natural to the globe, as general operations; but we are not to consider nature as having a burning mountain for an end in her intention, or as a principal purpose in the general system of the world. The end of nature in placing an internal fire or power of heat, and a force of irresistible expansion, in the body of this earth, is to consolidate the sediment collected at the bottom of the sea, and to form thereof a mass of permanent land above the level of the ocean, for the purpose of maintaining plants and animals. The power appointed for this purpose is, as on all other occasions where the operation is important, and where there is any danger of a shortcoming, wisely provided in abundance; and there are contrived means for disposing of the redundancy. These, in the present case, are our volcanoes.

A volcano is not made on purpose to frighten superstitious people into fits of piety and devotion, nor to overwhelm devoted cities with destruction; a volcano should be considered as a spiracle to the subterranean furnace, in order to prevent the unnecessary elevation of land and fatal effects of earthquakes; and we may rest assured that they, in general, wisely answer the end of their intention, without being in themselves an end, for which nature had exerted such amazing power and excellent contrivance. . . .

It is not meant to specify every particular in the means employed by nature for the elevation of our land. It is sufficient to have shown that there is, in nature, means employed for the consolidating of strata formed originally of loose and incoherent materials; and that those same means have also been employed in changing the place and situation of those strata. But how describe an operation which man cannot have any opportunity of perceiving? Or how imagine that for which, perhaps, there are not proper data to be found? We only know that the land is raised by a power which has for principle subterraneous heat; but how that land is preserved in its elevated station is a subject in which we have not even the means to form conjecture; at least, we ought to be cautious how we indulge conjecture in a subject where no means occur for trying that which is but a supposition. . . .

FROM *Part IV: System of Decay and Renovation observed in the Earth.*

. . . We are now, in reasoning from principles, come to a point decisive of the question, and which will either confirm the theory, if it be just, or confute our reasoning, if we have erred. Let us, therefore, open the book of Nature, and read in her records if there had been a world bearing plants at the time when this present world was forming at the bottom of the sea. . . .

Having thus ascertained the state of a former earth, in which plants and animals had lived, as well as the gradual production of the present earth, composed from the materials of a former world, it must be evident that here are two operations which are necessarily consecutive. The formation of the present earth necessarily involves the destruction of continents in the ancient world; and, by pursuing in our mind the natural operations of a former earth, we clearly see the origin of that land by the fertility of which we, and all the animated bodies of the sea, are fed. It is in like manner that, contemplating the present operations of the globe, we may perceive the actual existence of those productive causes which are now laying the foundation of land in the unfathomable regions of the sea, and which will, in time, give birth to future continents.

But though, in generalizing the operations of nature, we have arrived at those great events which, at first sight, may fill the mind with wonder and with doubt, we are not to suppose that there is any violent exertion of power such as is required in order to produce a great event in little time; in nature, we find no deficiency in respect of time, nor any limitation with regard to power. But time is not made to flow in vain; nor does there ever appear the exertion of superfluous power, or the manifestation of design, not calculated in wisdom to effect some general end.

The events now under consideration may be examined with a view to see this truth; for it may be inquired, why destroy one continent in order to erect another? The answer is plain: Nature does not destroy a continent from having wearied of a subject which had given pleasure, or changed her purpose, whether for a better or a worse; neither does she erect a continent of land among the clouds, to show her power, or to amaze the vulgar man: Nature has contrived the productions of vegetable bodies, and the sustenance of animal life, to depend upon the gradual but sure destruction of a continent; that is to say, these two operations necessarily go hand in hand. But with such wisdom has nature ordered things in the economy of this world that the destruction of one continent is not brought about without the renovation of the earth in the production of another; and the animal and vegetable bodies, for which the world above the surface of the sea is leveled with its bottom, are among the means employed in those operations, as well as the sustenance of those living beings is the proper end in view.

Thus, in understanding the proper constitution of the present earth, we are led to know the source from whence had come all the materials which nature had employed in the construction of the world which appears; a world contrived in consummate wisdom for the growth and habitation of a great diversity of plants and animals; and a world peculiarly adapted to the purposes of man, who inhabits all his climates, who measures its extent, and determines its productions at his pleasure. . . .

Our object is to know the time which had elapsed since the foundation of the present continent had been laid at the bottom of the ocean, to the present moment in which we speculate on these operations. The space is long; the data for the calculations are, perhaps, deficient: no matter; so far as we know our error, or the deficiency in our operations, we proceed in science, and shall conclude in reason. It is not given to man to know what things are truly in themselves, but only what those things are in his thought. We seek not to know the precise measure of any thing; we only understand the limits of a thing, in knowing what it is not, either on the one side or the other. . . .

To sum up the argument, we are certain that all the coasts of the present continents are wasted by the sea, and constantly wearing away upon the whole; but this operation is so extremely slow that we cannot find a measure of the quantity in order to form an estimate. Therefore, the present continents of the earth, which we consider as in a state of perfection, would, in the natural operations of the globe, require a time indefinite for their destruction.

But, in order to produce the present continents, the destruction of a former vegetable world was necessary; consequently, the production of our present continents must have required a time which is indefinite. In like manner, if the former continents were of the same nature as the present, it must have required another space of time, which also is indefinite, before they came to their perfection as a vegetable world.

We have been representing the system of this earth as proceeding with a certain regularity, which is not perhaps in nature, but which is necessary for our clear conception of the system of nature. The system of nature is certainly in rule, although we may not know every circumstance of its regulation. We are under a necessity, therefore, of making regular suppositions, in order to come at certain conclusions which may be compared with the present state of things. . . .

In thus accomplishing a certain end, we are not to limit nature with the uniformity of an equable progression, although it be necessary in our computations to proceed upon equalities. Thus also, in the use of means, we are not to prescribe to nature those alone which we think suitable for the purpose in our narrower view. It is our business to learn of nature (that is by observation) the ways and means, which in her wisdom are adopted; and we are to imagine these only in order to find means for further information, and to increase our knowledge from the examination of things which actually have been. It is in this manner that intention may be found in nature; but this intention is not to be supposed, or vainly imagined, from what we may conceive to be. . . .

We have now got to the end of our reasoning; we have no data further to conclude immediately from that which actually is: but we have got enough; we have the satisfaction to find that in nature there is wisdom, system, and consistency. For having, in the natural history of this earth, seen a succession of worlds, we may from this conclude that there is a system in nature; in like manner as, from seeing revolutions of the planets, it is concluded that there is a system by which they are intended to continue those revolutions. But if the succession of worlds is established in the system of nature, it is in vain to look for anything higher in the origin of the earth. The result, therefore, of our present inquiry is that we find no vestige of a beginning—no prospect of an end.

21

Myths and Dreams of Reason: The Enlightenment in Italy, France, and America

The militant philosophes of the eighteenth-century French Enlightenment were conscious inheritors of the Renaissance humanist revival of ancient learning (Chapter 16 above), and thus of the humanists' models, prominently including Cicero, Lucretius, and the Stoics (Chapter 10). But by far the greatest influences on their conception of nature were the scientific revolution expounded by Bacon and brought to completion by Newton (Chapter 18), and the British empiricist tradition of Locke (Chapter 20) that descended from them. Neither Spinoza nor Leibniz, nor even their own Descartes (Chapter 18), could rival—for Voltaire and the Encyclopedists who spread the gospel of light in France and abroad—these foundational English thinkers by whom reason was kept within the bounds of experience, and nature never became a mathematical abstraction. Faced with a sporadically repressive government dangerous enough to cause most of Voltaire's major writings to appear anonymously and Diderot's posthumously, the philosophes—above all Voltaire, whose Lettres philosophiques (Philosophical Letters) *drew on his years of English exile—viewed England as embodying a freedom of thought and expression largely absent from the absolutist monarchies and proscriptive principalities of continental Europe. Each associated this freedom, in his different way, with a more nearly natural condition of humanity that the philosophes were striving to bring into being.*

By and large, French Enlightenment thinkers characteristically took positions adumbrated by their English precursors and carried them to more uncompromising, more rigorously logical conclusions. Thus few major English thinkers, as we have seen (introduc-

tion to Chapter 20), adhered to a consistently deistic conception of the world as created by a God who thereafter left its workings wholly to the mechanical processes of nature; instead, they tended—like their great model Newton (or like Locke, Shaftesbury, or Pope)—to combine elements of such a view with more orthodox, if not always fully compatible, Christian notions. But in France, Voltaire combined a somewhat vague reverence for the God revealed through nature—God exists, but if he did not, it would be necessary to invent him!—with an increasingly virulent campaign to "crush the infamous thing" of Christianity. And Rousseau, who set forth his more sentimental rendition of natural religion and natural morality in the "profession of faith of a Savoyard vicar" in Emile, likewise carried deistic beliefs to bold outcomes hardly imaginable in England. (There, the precarious and much-envied freedom was maintained, after decades of religious strife, by fudging issues and not pushing ideologies to an extreme. Hume, to be sure, was a still bolder thinker, whose scepticism demolished the rational basis of natural as well as revealed religion [Chapter 20]; but Hume was a Scot.) And to Voltaire's dismay, many younger philosophes—La Mettrie, Diderot, Helvétius, Holbach—moved more or less belligerently, in the name of enlightened reason, toward overtly materialistic views of the world that discarded God altogether as a hypothesis of which (as the mathematician Laplace was said to have told Napoleon somewhat later)[1] they had no need.

Even within the French Enlightenment, however, and even among its foremost thinkers, there were of course crosscurrents that suggested very different directions than the rationalistic march toward a purely mechanical vision of nature and human nature. Diderot's provocative posthumous dialogues, including D'Alembert's Dream and the Supplement to Bougainville's Voyage, explore the hidden connections between dream and reason, and the uncertainties and contradictions attendant on living life in accord with nature amid a civilization continually in conflict with it. Rousseau, by his quasi-mystical evocation of natural beauty and harsh repudiation of a society whose vaunted arts and sciences place it in opposition to nature's values, anticipated the Romanticism that arose in professed reaction against the Enlightenment's central tenets. Both Bernardin de Saint-Pierre's rhapsodic idyll of an infantile state of nature and the marquis de Sade's exaltation of nature as license to follow brute sexual desire seem distortedly one-sided reflections of Diderot's and Rousseau's complexly ambiguous views.

Outside France, other thinkers of the misguidedly named Age of Reason still more fundamentally assailed the presuppositions on which the Enlightenment unquietly reposed. "The central doctrines of the progressive French thinkers, whatever their disagreements among themselves," Isaiah Berlin writes,[2] "rested on the belief, rooted in the ancient doctrine of natural law, that human nature was fundamentally the same in all times and places." In sharp opposition to this view, the New Science of Giambattista Vico conceived of "gentile" (that is, non-biblical) history as a recurrent cycle of ages characterized not only by different customs and laws, but by different natures as well, since "the nature of things is nothing but their coming into being at certain times and in certain fashions." Revealed through language and myth as expressions of a culture's deepest values, these successive natures are the poetic or creative nature of the theological age; the heroic nature, still thought to be of divine origin, that succeeded it; and our own fully human nature, "intelligent and . . . reasonable, recognizing for laws conscience, reason,

[1]See Stephen Jay Gould, "The Celestial Mechanic and the Earthly Naturalist," in *Dinosaur in a Haystack* (1995).
[2]"The Counter-Enlightenment," in *Dictionary of the History of Ideas* (1968–73) and reprinted in *Against the Current* (1979).

and duty." As Berlin observes, Vico's "revolutionary move is to have denied the doctrine of a timeless natural law the truths of which could have been known in principle to any man, at any time, anywhere. . . . Such historicism was plainly not compatible with the view that there was only one standard of truth or beauty or goodness, which some cultures or individuals approached more closely than others, and which it was the business of thinkers to establish and men of action to realize."

In Germany, J. G. Hamann (and after him Herder, Goethe, and his fellow writers of the Sturm und Drang*), building on Shaftesbury (Chapter 20), Rousseau, and the German mystical tradition, took still further "the conviction that all truth is particular, never general," and maintained that "only love—for a person or an object—can reveal the true nature of anything." For Hamann (who held that "God is a poet, not a mathematician") and his followers, Berlin writes, "Nature is capable of wild fantasy, and it is mere childish presumption to seek to imprison her in the narrow rationalist categories of 'puny' and desiccated philosophers. Nature is a wild dance, and so-called practical men are like sleep-walkers who are secure and successful because they are blind to reality; if they saw reality as it truly is, they might go out of their minds." This repudiation in Germany of the fundamental presuppositions of the Enlightenment, continuing Rousseau's radical scrutiny of them from within, ushers in an inchoate European Romanticism for which the inner processes of nature can be understood not through scientific reason but only through imagination or empathy* (Einfühlung)*—and thus lays the foundation for a new era of nature writing beyond the scope of this volume.*

Finally, in the thirteen colonies of British North America, and the newly United States that succeeded them, currents of European thought as diverse as English Puritanism, scientific inquiry, and Enlightenment rationalism found expression in such varied figures as Jonathan Edwards, Benjamin Franklin, and Thomas Jefferson. The vast and sparsely populated expanses of a largely unexplored continent gave rise to a sense of wonder at the boundless grandeur and continual novelty of untamed nature shared, for all the great differences among them, by men of vastly different tempers—with some exceptions, such as the irredeemably practical Franklin. In descriptions of the New World by Edwards and Jefferson, John Lawson and William Byrd, and above all by William Bartram, the awe-inspiring sublimity of seemingly virgin wilderness—of the "forest primeval" once inhabited by Humbaba or Pan, but since half forgotten—again makes its appearance, and will haunt the imagination of generations to come.

A Nature All Fierce and Cruel

The New Science (third edition, 1744), by Giambattista Vico, translated by Thomas Goddard Bergin and Max Harold Fisch

No important figure of the early eighteenth century more fundamentally questioned the Enlightenment view of the world than Giambattista (or Giovanni Battista) Vico (1668–1744), who repudiated both the rationalist and the empiricist legacies of the previous century, as embodied respectively by Descartes and Locke. For this reason (among others) he was largely ignored, even by the few who read him, until rediscovered by the French historian Jules Michelet in the nineteenth century. Yet Vico, though a deeply religious Catholic, was by no means simply a reactionary. Bacon (Chapter 18 above), Plato (Chapter 9), Tacitus (Chapter 12), and Grotius (whose conception of natural law he later rejected) were the four authors "ever before him in meditation and writing," he wrote in his autobiography of 1725; he considered Bacon "a man of incomparable wisdom both common and esoteric, at one and the same time a universal man in theory and in practice," and fervently admired The Advancement of Learning, The Wisdom of the Ancients, *and the* Novum Organum, *which became one model for Vico's own* New Science. *Among others who shaped his views, even when he opposed their basic assumptions, were Lucretius (Chapter 10) and Hobbes (Chapter 20).*

Born in Naples, the sixth of a bookseller's eight children, he devoted his life to lonely study, above all of the Latin classics, becoming professor of rhetoric at the University of Naples in 1699 and historiographer to the king in 1734. His life's work, Principii di una scienza nuova d'intorno alla natura delle nazioni (Principles of a New Science Concerning the Nature of Nations), *appeared in 1725, with substantially revised editions in 1730 and 1744. In earlier writings, as Isaiah Berlin remarks,*[3] *Vico had argued that mathematical knowledge "is not, as Descartes supposed, discovery of an objective structure, the eternal and most general characteristics of the real world, but rather invention: invention of a symbolic system which men can logically guarantee only because men have made it themselves, irrefutable only because it is a figment of man's own creative intellect." In contrast, "the world—nature—had not been made by men: therefore only God, who had made it, could know it through and through. . . . Men had only an outside view, as it were, of what went on on the stage of nature. Men could know 'from the inside' only what they had made themselves, and nothing else."*

Vico's "boldest contribution" to European thought, Berlin affirms, was to apply this principle to human history itself (since history is made by men) through "the concept of 'philology,' anthropological historicism, the notion that there can be a science of mind which is the history of its development, the realization that ideas evolve, that knowledge is not a static network of eternal, universal, clear truths, either Platonic or Cartesian, but a social process, that this process is traceable through (indeed, is in a sense identical with) the evolution of symbols—words, gestures, pictures, and their altering patterns, functions, structures, and uses," realized above all, among earlier peoples, in poetry and myth. For Vico, as R. G. Collingwood writes in The Idea of History *(1946), "the fabric of human society is created by man out of nothing, and every detail of this fabric is therefore a human* factum, *eminently knowable to the human mind as such. . . . There is a kind of pre-established harmony between the historian's mind and the object which he sets out to study; but this pre-established harmony, unlike that of Leibniz, is not based on a miracle—it is based on the common human nature uniting the historian with the men whose work he is studying." The historian's knowledge is thus fundamentally different from the natural scientist's, since "in addition to the traditional categories of knowledge—*a priori*-deductive,* a posteriori*-empirical, that provided by sense perception and that vouchsafed by revelation—there must now be added," Berlin writes in* Vico and Herder, *"a new variety, the reconstructive imagination. This type of knowledge is yielded by 'entering' into the mental life of other cultures, into a variety of outlooks and ways of life which only the activity of* fantasia*—imagination—makes possible."*

History, understood in this way, is closely connected with nature in a radically different sense from that of Descartes, Boyle, or Newton. Vico "sees the history of mankind as 'an ideal, eternal

[3]This and the following quotations from Berlin are from two essays collected in *Against the Current* (1979): "The Divorce between the Sciences and the Humanities" (1974) and "Vico's Concept of Knowledge" (1969). See also "Vico and the Ideal of the Enlightenment" (1976) in the same volume. For fuller treatment, see Berlin's *Vico and Herder* (1976).

*history traversed in time by every nation in its rise, growth, decline and fall' [*New Science *349]. This is the* idée maîtresse *of his whole thought," Berlin observes. ". . . Indeed this pattern, like a Platonic Idea, is what makes human nature human: it is not a necessity imposed on men's souls or bodies from outside—from above by a deity, or from below by material nature. It is the principle of growth, in terms of which nature herself,* Natura *as* nascimento*—birth and growth—is defined. . . . 'Nature is the* nascimento*—the coming to birth of a thing at certain time and in certain fashions' [*New Science *147]. Nature is change, growth, the interplay of forces that perpetually transform one another; only the pattern of this flow is constant, not its substance, only the most general form of the laws which it obeys, not their content."*

Thus Vico's conception of secular history not as a unilinear progression but as a sequence of ages—an age of barbarism preceding an age of the gods, an age of heroes, and an age of men, followed by relapse into a new "barbarism of reflection"—recurring always in the same order but never in the same way, implies that nature itself will be different for each age. The rationality of the "human" age that succeeds the age of heroes in secular history is also the perpetual condition of sacred history, since "only one history is umana *for its entire span," Paolo Rossi writes in* The Dark Abyss of Time *(1979; English trans. 1984), "and that is Hebraic history," which God's providential guidance exempted from the cycles of the gentiles: "All other histories* became *'human' in different, but in any event more recent ages." In contrast to this rationality, men of the earlier theological and heroic ages, monstrous, doltish, and crude though they were by standards of civilized humanity, saw the world in metaphorical and mythical terms that gave rise to sublime poetry (above all the Homeric epics) wholly beyond the capacities of more abstract and philosophical times. Thus in Vico's view, "change—unavoidable change—rules all man's history," Berlin remarks in "Vico and the Ideal of the Enlightenment"; and "in the course of this process gains in one respect necessarily entail losses in another, losses which cannot be made good if the new values, which are part of the unalterable historical process, are, as indeed they must be, realized, each in its due season." Vico never idealizes those earlier ages, when nature was "all fierce and cruel," but by his capacity to project himself imaginatively into such different ways of perception and thought, he brings them intensely (if confusedly) to life, and anticipates both Romantic and modern understandings of nature as a process continually in the making.*

Selections are from the original (and sometimes more literal) Bergin-Fisch translation (1948) of Vico's third edition of 1744. A revised abridgment was published in 1961, and a revision of the unabridged translation in 1968 and 1984.

FROM *Book I: Establishment of Principles, [Section II:] Elements*

XIV.147 The nature of things is nothing but their coming into being *(nascimento)* at certain times and in certain fashions. Whenever the time and fashion is thus and so, such and not otherwise are the things that come into being.

FROM *Book II: Poetic Wisdom, [Section I: Poetic Metaphysics, Chapter I:] Poetic Metaphysics as the Origin of Poetry, Idolatry, Divination and Sacrifices*

. . . 375 Hence poetic wisdom, the first wisdom of the gentile world, must have begun with a metaphysic not rational and abstract like that of learned men now, but felt and imagined as that of these first men must have been, who, without power or ratiocination, were all robust sense and vigorous imagination, as established in the Axioms. This metaphysic was their poetry, a faculty born with them (for they were furnished by nature with these senses and imaginations); born of their ignorance of causes, for ignorance, the mother of wonder, made everything wonderful to men who were ignorant of everything, as noted in the Axioms. Their poetry was at first divine, because they imagined the causes of the things they felt and wondered at to be gods, as we saw in the passage from Lactantius cited in the Axioms. (This is now confirmed by the American Indians, who call gods all the things that surpass their small understanding. We may add the ancient Germans dwelling about the Arctic Ocean, of whom Tacitus tells us that they spoke of hearing the Sun pass at night from west to east through the sea, and affirmed that they saw the god. These very rude and simple nations help us to a much better understanding of the founders of the gentile world with whom we are now concerned.) At the same time they gave the things they wondered at substantial being after their own ideas, just as children do, whom we see take inanimate things in their hands and play with them and talk to them as though they were living persons, as laid down in an axiom.

376 In such fashion the first men of the gentile nations, children of nascent mankind as we have styled them in the axioms, created things according to their own ideas. But this creation was infinitely different from that of God. For God, in his purest intelligence, knows things, and, by knowing them, creates them; but they, in their robust ignorance, did it by virtue of a wholly corporeal imagination. And because it was quite corporeal, they did it with marvelous sublimity; a sublimity such and so great that it excessively perturbed the very persons who by feigning did the creating, for which they were called "poets," which is Greek for "makers." Now this is the threefold labor of great poetry: (1) to invent sublime fables suited to the popular understanding, (2) to perturb to excess, with a view to the end proposed: (3) to teach the vulgar to act virtuously, as the poets have taught themselves; as will presently be shown. Of this nature of human things there came an eternal property, expressed in a noble phrase of Tacitus: that frightened men vainly "no sooner feign than they believe" (*fingunt simul creduntque* [*Annals* V.10]).

377 Of such natures must have been the first founders of gentile humanity when at last the sky fearfully rolled with thunder and flashed with lightning, as could not but follow from the bursting upon the air for the first time of an impression so violent. As we have postulated, this occurred a hundred years after the flood in Mesopotamia and two hundred after it throughout the rest of the world; for it took that much time to reduce the earth to such a state that, dry of the moisture of the universal flood, it could send up dry exhalations or matter igniting in the air to produce lightning. Thereupon a few giants, who must have been the most robust, and who were dispersed through the forests on the mountain heights where the strongest beasts have their dens, were frightened and astonished by the great effect whose cause they did not know, and raised their eyes and became aware of the sky. And because in such a case, as stated in the Axioms, the nature of the human mind leads it to attribute its own nature to the effect, and because in that state their nature was that of men all robust bodily strength, who expressed their fiery violent passions by shouting and grumbling, they pictured the sky to themselves as a great animated body, which in that aspect they called Jove, the first god of the so-called *gentes maiores* ["greater gentile nations," i.e., Greece and Rome], who by the whistling of his bolts and the noise of his thunder was attempting to tell them something. And thus they began to exercise that natural curiosity which is the daughter of ignorance and the mother of knowledge, and which, opening the mind of man, gives birth to wonder, as we have put it above in the Axioms. This characteristic still persists in the vulgar, who, when they see a comet or sun-dog or some other extraordinary thing in nature, and particularly in the countenance of the sky, at once turn curious and anxiously inquire what

it means, as we have it in an axiom. When they wonder at the prodigious effects of the magnet on iron, even in this age of minds enlightened and made erudite by philosophy, they come out with this: that the magnet has an occult sympathy for the iron; and they make of all nature a vast animate body which feels passions and effects, as we have noted in the Axioms.

378 But the nature of our civilized minds is so detached from the senses, even in the vulgar, by abstractions corresponding to all the abstract terms our languages abound in, and so refined by the art of writing, and as it were spiritualized by the use of numbers, because even the vulgar know how to count and reckon, that it is naturally beyond our power to form the vast image of this mistress called "Sympathetic Nature." Men shape the phrase with their lips but have nothing in their minds; for what they have in mind is falsehood, which is nothing; and their imagination no longer avails to form a vast false image. It is equally beyond our power to enter into the vast imagination of those first men, whose minds were not in the least abstract, refined, or spiritualized, because they were entirely immersed in the senses, buffeted by the passions, buried in the body. That is why we said above that we can scarcely understand, still less imagine, how those first men thought who founded gentile humanity.

379 In this fashion the first theological poets created the first divine fable, the greatest they ever created: that of Jove, king and father of men and gods, in the act of hurling the lightning bolt; an image so popular, disturbing and instructive that its creators themselves believed in it, and feared, revered and worshiped it in frightful religions which we shall shortly describe. And by that trait of the human mind we found noticed by Tacitus in the Axioms, those men attributed to Jove all they saw, imagined or even did themselves; and to all of the universe that came within their scope, to all its parts, they assigned the being of animate substance. This is the civil history of the expression "all things are full of Jove" (*Iovis omnia plena*), by which Plato understood the ether which penetrates and fills the universe. But for the theological poets, as will shortly be seen, Jove was no higher than the mountain peaks. The first men, who spoke by signs, naturally believed that lightning bolts and thunder claps were signs made to them by Jove; whence from *nuo*, to make a sign, came *numen*, the divine will, by an idea more than sublime and worthy to express the divine majesty. They believed that Jove commanded by signs, that such signs were real words, and that nature was the language of Jove. The science of this language the gentiles universally believed to be divination, which by the Greeks was called theology, meaning the science of the language of the gods. Thus Jove acquired the fearful kingdom of the lightning and became the king of men and gods. . . .

FROM *[Section VII, Chapter I:] Poetic Physics*

688 The theological poets considered the physics of the world of nations, and therefore they first defined Chaos as confusion of human seeds in the period of the abominable sharing of women. It was thence that the physicists were later moved to conceive the confusion of the universal seeds of nature, and to express it they took the word already invented by the poets and hence appropriate. [The poetic Chaos] was confused because there was no order of humanity in it, and obscure because it lacked the civil light in virtue of which the heroes were called *incliti*, "illustrious." Further they imagined it as Orcus, a misshapen monster which devoured all things, because men in this infamous community did not have the proper form of men, and were swallowed up by the void because

through the uncertainty of offspring they left nothing of themselves. This [chaos] was later taken by the physicists as the prime matter of natural things, which, formless itself, is greedy for forms and devours all forms. The poets however gave it also the monstrous form of Pan, the wild god who is the divinity of all satyrs inhabiting not the cities but the forests; a character to which they reduced the impious vagabonds wandering through the great forest of the earth and having the appearance of men but the habits of abominable beasts. Afterwards, by forced allegories on which we shall comment later, the philosophers, misled by the name *pan*, "everything," took him as a symbol for the formed universe. Scholars have also held that the poets meant first matter in the fable of Proteus, with whom Ulysses wrestles in Egypt,[4] Proteus in the water and the hero out of it, unable to get a grip on the monster who keeps assuming new forms. But the scholars thus made sublime learning out of what was doltishness and simplicity on the part of the first men, who (just as children, looking in a mirror, will try to seize their own reflections) thought from the various modifications of their own shapes and gestures that there must be a man in the water, forever changing into different shapes.

689 At length the sky broke forth in thunder, and Jove thus gave a beginning to the world of men by arousing in them the impulse which is proper to the liberty of the mind, just as from motion, which is proper to bodies as necessary agents, he began the world of nature. . . .

690 The world of the theological poets was composed of four sacred elements: the air whence Jove's bolts came, the water of the perennial springs whose divinity is Diana, the fire with which Vulcan cleared the forests, and the tilled earth of Cybele or Berecynthia. All four are elements in divine ceremonies: auspices, water, fire and spelt. They are watched over by Vesta, who, as we said before, is the same as Cybele or Berecynthia. She is crowned with the tilled lands protected by hedges and surmounted by the towers of high-placed towns (whence the Latin *extorris*, "exiled," as if *exterris*). This crown encloses all that was signified by the *orbis terrarum*, which is properly the world of men. Thence the physicists were later moved to study the four elements of which the world of nature is composed.

FROM *Book IV: The Course of Nations, [Section I:] Three Kinds of Natures*

916 The first nature, by an illusion of imagination, which is most robust in those weakest in reasoning power, was a poetic or creative nature which we may be allowed to call divine, as it ascribed to physical things the being of substances animated by gods, assigning the gods to them according to its idea of each. This nature was that of the theological poets, who were the earliest wise men in all the gentile nations, when all the gentile nations were founded on the belief which each of them had in certain gods of its own. Furthermore it was a nature all fierce and cruel; but, through that same error of their imagination, men had a terrible fear of the gods whom they themselves had created. From this there remained two eternal properties: one, that religion is the only means powerful enough to restrain the fierceness of peoples; and the other, that religions prosper when those who preside over them are themselves inwardly reverent.

917 The second was the heroic, believed by the heroes themselves to be of divine origin; for, since they believed that the gods made everything, they held themselves to

[4]It was Menelaus, not Odysseus (Ulysses), who wrestled with Proteus in Egypt, as he tells Odysseus's son Telemachus in the fourth book of the *Odyssey*.

be children of Jove, as having been generated under his auspices. Being thus of the human [not a bestial] species, they regarded their heroism as including the natural nobility in virtue of which they were the princes of the human race. And this natural nobility they made their boast over those who had fled from the infamous and bestial communism to save themselves from the strife it entailed, and had taken refuge in their asylums; for, since they had come thither without gods, the heroes regarded them as beasts. We have discussed these two natures above.

918 The third was human nature, intelligent and hence modest, benign and reasonable, recognizing for laws conscience, reason and duty.

At Blind Man's Buff with Nature

Dialogues from the Writings of Voltaire

François-Marie Arouet (1694–1778), who took the name of Voltaire, made himself the leading polemicist of Enlightenment in France and throughout Europe. Born in Paris, the son of a wealthy notary, and educated at the Jesuit College of Louis-le-Grand, he became a well-known poet and wit among freethinking libertines of the regency of Louis XV, and in 1717 spent nearly a year in the Bastille for insults to the regent. In prison he rewrote his first tragedy, Oedipe (Oedipus), *which initiated his fame as France's leading poet, and began his epic on Henri IV, the* Henriade. *Badly beaten in 1726 for witticisms against the young chevalier de Rohan, he was released from the Bastille on condition that he leave for England, where he remained for over two years. His* Letters Concerning the English Nation*—first published in English in 1733, then in French, as* Lettres philosophiques, *in 1734—praised English political liberty and religious toleration in implicit contrast to French repression, and disseminated the scientific and philosophical views of Newton and Locke. Like many books to follow, whose authorship Voltaire routinely denied, it was promptly, though ineffectively, banned in France. Between 1733 and 1749, Voltaire spent much of his time at Cirey in Lorraine with his mistress, the scientifically inclined marquise du Châtelet, who translated Newton while he was writing his influential popular exposition,* Elements of the Philosophy of Newton *(1738). In these years, he became royal historiographer and attained election, after a long campaign, to the French Academy. Between 1749 and 1753 he lived at the court of Frederick the Great of Prussia in Potsdam, then settled, after a period of wandering, at his estate "Les Délices" on the outskirts of Geneva. After the outcry aroused by d'Alembert's article "Genève" in the* Encyclopedia, *he moved in 1758 to Ferney on the French side of the Swiss border, where he lived most of the remainder of his life.*

Among important publications (besides countless poems, tragedies, pamphlets, and letters) were his history Le Siècle de Louis XIV *(*The Century of Louis XIV, *1751); the philosophical tales* Zadig *(1747) and* Candide *(1759), the latter written, after the Lisbon earthquake of 1755, as an assault on what seemed the facile optimism of Leibniz and Pope; and the* Dictionnaire philosophique (Philosophical Dictionary), *first published in 1764. In his late years at Ferney, Voltaire became increasingly committed to the battle against injustice (taking up the cause of the wrongly executed Jean Calas, among others) and against the Church, ending his letters with the words* "écrasez l'infame" *("crush the infamous thing"). In 1778 he returned triumphantly to Paris, where he died.*

Though exploration of nature is not a central concern of Voltaire, who was repelled by Rous-

seau, his dedication to Newtonian physics and to a consistently held deism is a major thread in his writings. Of the selections that follow, the first is from an imagined dialogue of 1756 (anonymously translated in The Writings of Voltaire *[1901]) between the Epicurean Lucretius and the Stoic Posidonius, who sets forth an argument for God's existence from the design of the universe—such as Hume, in the still-unpublished* Dialogues Concerning Natural Religion *(Chapter 20 above), had recently demolished—but concludes (very much in Voltaire's vein) that the Supreme Being must always remain incomprehensible. The second is a dialogue between the Philosopher and Nature from* Questions sur l'Encyclopédie (Questions on the Encyclopedia) *of 1770–72—a work later incorporated into some posthumous editions of Voltaire's* Philosophical Dictionary*—translated from* Oeuvres complètes de Voltaire, *vol. 20 (new edition, 1879). Here Nature herself reveals to the Philosopher that she is "the art of some superlatively potent and ingenious mighty being," and tells him he must be content to play blind man's buff with her forever, since she herself can say nothing about the first principles philosophers vainly seek, but only God knows.*

FROM *Dialogues between Lucretius and Posidonius: First Colloquy*

Lucretius. You allow that nature is eternal, and exists because it does exist. Now if it exists by its own power, why may it not, by the same power, have formed suns, and worlds, and plants, and animals, and men?

Posidonius. All the ancient philosophers have supposed matter to be eternal, but have never proved it to be really so; and even allowing it to be eternal, it would by no means follow that it could form works in which there are so many striking proofs of wisdom and design. Suppose this stone to be eternal if you will, you can never persuade me that it could have composed the "Iliad" of Homer.

Lucretius. No: a stone could never have composed the "Iliad," any more than it could have produced a horse: but matter organized in process of time, and become bones, flesh, and blood, will produce a horse; and organized more finely, will produce the "Iliad."

Posidonius. You suppose all this without any proof; and I ought to admit nothing without proof. I will give you bones, flesh, and blood, ready made, and will leave you and all the Epicureans in the world to make your best of them. Will you only consent to this alternative: to be put in possession of the whole Roman Empire if, with all the ingredients ready prepared, you produce a horse, and to be hanged if you fail in the attempt?

Lucretius. No; that surpasses my power, but not the power of nature. It requires millions of ages for nature, after having passed through all the possible forms, to arrive at last at the only one which can produce living beings.

Posidonius. You might, if you pleased, continue all your lifetime to shake in a cask all the materials of the earth mixed together, you would never be able to form any regular figure; you could produce nothing. If the length of your life is not sufficient to produce even a mushroom, will the length of another man's life be sufficient for that purpose? Why should several ages be able to effect what one age has not effected? . . .

Lucretius. I am at liberty to believe that what is today was yesterday, was a century ago, was a hundred centuries ago, and so on backwards without end. I make use of your argument: no one has ever seen the sun and stars begin their course, nor the first animals formed and endowed with life. We may, therefore, safely believe that all things were from eternity as they are at present.

Posidonius. There is a very great difference. I see an admirable design, and I ought to believe that an intelligent being formed that design.

Lucretius. You ought not to admit a being of whom you have no knowledge.
Posidonius. You might as well tell me that I should not believe that an architect built the capitol because I never saw that architect.
Lucretius. Your comparisons are not just. You have seen houses built, and you have seen architects; and therefore you ought to conclude that it was a man like our present architects that built the capitol. But here the case is very different: the capitol does not exist of itself, but matter does. It must necessarily have had some form; and why will you not allow it to possess, by its own energy, the form in which it now is? Is it not much easier for you to admit that nature modifies itself, than to acknowledge a being that modifies it? In the former case you have only one difficulty to encounter, namely, to comprehend how nature acts. In the latter you have two difficulties to surmount: to comprehend this same nature, and the visible being that acts on it.
Posidonius. It is quite the reverse. I see not only a difficulty, but even an impossibility in comprehending how matter can have infinite designs; but I see no difficulty in admitting an intelligent being, who governs this matter by his infinite wisdom, and by his almighty will.
Lucretius. What? is it because your mind cannot comprehend one thing that you are to suppose another? Is it because you do not understand the secret springs, and admirable contrivances, by which nature disposed itself into planets, suns, and animals, that you have recourse to another being?
Posidonius. No; I have not recourse to a god, because I cannot comprehend nature; but I plainly perceive that nature needs a supreme intelligence; and this reason alone would to me be a sufficient proof of a deity had I no other.
Lucretius. And what if this matter possessed intelligence of itself?
Posidonius. It is plain to me that it does not possess it.
Lucretius. And to me it is plain that it does possess it, since I see bodies like you and me reason. . . .
Posidonius. . . . You are sensible of the weakness of matter, and are forced to admit a supreme intelligent and almighty being, who organized matter and thinking beings. The designs of this superior intelligence shine forth in every part of nature, and you must perceive them as distinctly in a blade of grass as in the course of the stars. Everything is evidently directed to a certain end.
Lucretius. But do you not take for a design what is only a necessary existence? Do you not take for an end what is no more than the use which we make of things that exist? The Argonauts built a ship to sail to Colchis. Will you say that the trees were created in order that the Argonauts might build a ship, and that the sea was made to enable them to undertake their voyage? Men wear stockings: will you say that legs were made by the Supreme Being in order to be covered with stockings? No, doubtless; but the Argonauts, having seen wood, built a ship with it, and having learned that the water could carry a ship, they undertook their voyage. In the same manner, after an infinite number of forms and combinations which matter had assumed, it was found that the humors, and the transparent horn which compose the eye, and which were formerly separated in different parts of the body, were united in the head, and animals began to see. The organs of generation, dispersed before, were likewise collected, and took the form they now have; and then all kinds of procreation were conducted with regularity. The matter of the sun, which had been long diffused and scattered through the universe, was conglobated, and formed the luminary that enlightens our world. Is there anything impossible in all this?

Posidonius. In fact, you cannot surely be serious when you have recourse to such a system: for, in the first place, if you adopt this hypothesis, you must, of course, reject the eternal generations of which you have just now been talking: and, in the second place, you are mistaken with regard to final causes. There are voluntary uses to which we apply the gifts of nature; and there are likewise necessary effects. The Argonauts need not, unless they had pleased, have employed the trees of the forest to build a ship; but these trees were plainly destined to grow on the earth, and to produce fruits and leaves. We need not cover our legs with stockings; but the leg was evidently made to support the body, and to walk, the eyes to see, the ears to hear, and the parts of generation to perpetuate the species. If you consider that a star, placed at the distance of four or five hundred millions of leagues from us, sends forth rays of light, which make precisely the same angle in the eyes of every animal, and that, at that instant, all animals have the sensation of light, you must acknowledge that this is an instance of the most admirable mechanism and design. But is it not unreasonable to admit mechanism without a mechanic, a design without intelligence, and such designs without a Supreme Being?
Lucretius. If I admit the Supreme Being, what form must I give Him? Is He in one place? Is He out of all place? Is He in time or out of time? Does He fill the whole of space, or does He not fill it? Why did He make the world? What was His end in making it? Why form sensible and unhappy beings? Why moral and natural evil? On whatever side I turn my mind, everything appears dark and incomprehensible.
Posidonius. 'Tis a necessary consequence of the existence of this Supreme Being that His nature should be incomprehensible; for, if He exists, there must be an infinite distance between Him and us. We ought to believe that He is, without endeavoring to know what He is, or how He operates. Are you not obliged to admit asymptotes in geometry, without comprehending how it is possible for the same lines to be always approaching, and yet never to meet? Are there not many things as incomprehensible as demonstrable in the properties of the circle? Confess, therefore, that you ought to admit what is incomprehensible, when the existence of that incomprehensible is proved.
Lucretius. What! must I renounce the dogmas of Epicurus?
Posidonius. It is better to renounce Epicurus than to abandon the dictates of reason.

Nature: Dialogue between the Philosopher and Nature,
translated by Robert M. Torrance

Questions on the Encyclopedia

The Philosopher. Who are you, Nature? I live in you; for fifty years I have sought you, and have not yet found you.
Nature. The ancient Egyptians, who lived, it is said, some twelve hundred years, made me the same reproach. They called me Isis; they placed a great veil on my head, and they said that no one could lift it.
The Philosopher. It is for this reason that I address you. I have been able, indeed, to measure some of your globes, to learn their paths, assign the laws of motion; but I have not been able to know who you are.

Are you always active? Are you always passive? Did your elements arrange themselves on their own, as water settles on sand, oil on water, air on oil? Do you have a mind that directs all your operations, as councils are inspired after being assembled, although their members are sometimes dunces? Be so kind as to resolve your enigma for me.

Nature. I am the great everything. I know no more about it. I am not a mathematician; and everything of mine is arranged according to mathematical laws. Guess, if you can, how all this is done.

The Philosopher. Certainly, since your great everything does not know mathematics, and your laws are most profoundly geometrical, there must be an eternal geometer who directs you, a supreme intelligence who presides over your operations.

Nature. You are right. I am water, earth, fire, atmosphere, metal, mineral, stone, vegetable, animal. I clearly sense an intelligence in myself. You have one, you do not see it; I do not see mine either. I sense this invisible power; I cannot know it: why should you, who are only a small part of me, wish to know what I do not know?

The Philosopher. We are curious. I would like to know how one so coarse in your mountains, deserts, and seas appears so ingenious in your animals and vegetables?

Nature. My poor child, do you wish me to tell you the truth? It is this: they have given me a name that does not suit me. They call me Nature, and I am all art.

The Philosopher. That word unsettles all my ideas. What! Nature might only be art?

Nature. Yes, beyond a doubt. Do you not understand that there is an infinite art in those seas, in those mountains that you find so coarse? Do you not understand that all those waters gravitate toward the center of the earth, and only rise by immutable laws; that those mountains that crown the earth are the immense reservoirs of the eternal snows that incessantly produce those fountains, those lakes, those rivers without which my animal kind and my vegetable kind would perish? And as for what are called my animal, vegetable, and mineral kingdoms, you see here only three: learn that I have millions of them. Even if you only consider the creation of an insect or an ear of wheat, of gold or copper, everything will appear to you as a marvel of art.

The Philosopher. It is true. The more I think about it, the more I see that you are only the art of some immensely powerful and ingenious great being, who hides himself while making you appear. All reasoners since Thales, and probably long before him, have played blind man's buff with you; they have said: "I have you!" and they had nothing. We all resemble Ixion: he thought he was embracing Juno, and he was only enjoying a cloud.

Nature. Since I am everything that is, how can a being such as you, so small a part of myself, apprehend me? Be satisfied, atoms who are my children, with seeing some of the atoms around you, with drinking a few drops of my milk, with vegetating a few moments at my breast, and with dying before you have known your mother and your nurse.

The Philosopher. My dear mother, tell me a little why you exist, why there is anything.

Nature. I will answer you as I have answered, for so many centuries, all who have questioned me about first principles: "I know nothing about them."

The Philosopher. Would nothingness not be preferable to this multitude of existences made to be continually dissolved, this throng of animals born and reproduced to devour others and to be devoured, this throng of sentient beings formed for so many painful sensations, and that other throng of intelligent beings who so rarely listen to reason? What good is all that, Nature?

Nature. Oh! go question Him who made me.

The Enlightening Spirit of Method

Preliminary Discourse to the *Encyclopedia*, by Jean le Rond d'Alembert, translated by Richard N. Schwab with the collaboration of Walter E. Rex

The most scientifically gifted of the philosophes associated with the Encyclopedia *was the illegitimate son of the chevalier Destouches and Mme de Tencin (a renegade nun who became a leading Parisian hostess). Abandoned on the steps of the church of St. Jean le Rond, he was christened Jean-Baptiste le Rond by a police officer, and later took the name d'Alembert. His father provided for his early education, and d'Alembert (1717–83) achieved youthful fame for his scientific and mathematical genius; among his writings were important treatises on dynamics (1743) and music (1752), and a posthumously published history of the French Academy, to which he had belonged since 1754 and been secretary since 1772. Along with Diderot, he was co-editor of the* Encyclopedia, *for which he wrote the "Preliminary Discourse" published in its first volume of 1751. After the controversy aroused by his article on Geneva in volume VII in 1757—in which, under Voltaire's influence, he advocated theatrical performances in that city and praised its Calvinist pastors (to their dismay) for abandoning belief in the divinity of Christ—and after the subsequent ban on further publication of the* Encyclopedia, *d'Alembert withdrew from its editorship ("deserted," in Diderot's view) and contributed only occasional articles to later volumes.*

The method he advocates in the "Preliminary Discourse," Richard N. Schwab writes in introducing his translation (1963), "represents an adjustment of the rationalist spirit of Descartes to the empiricism of Locke and Newton—a fusion of traditions which lies at the foundation of the Encyclopedia." *Here, as elsewhere in his writings, Ernst Cassirer observes in* The Philosophy of the Enlightenment *(1932; English trans. 1951), d'Alembert "never deviates from the Newtonian method. He too rejects all questions regarding the absolute nature of things and their metaphysical origin." For the first time "the development of science was approached from the new viewpoint . . . as the methodological self-development of the idea of knowledge itself. He demands that histories of individual subjects be replaced by a philosophical science of principles, and that the history of science be treated according to such principles."*

Nature, d'Alembert affirms, can be known not by "vague and arbitrary hypotheses" but only by "thoughtful study of phenomena," and by methodical reduction of a large number of phenomena to a single principle; only careful observation and experiment can correct excessive "application of algebra to physics" by those who play "intellectual games to which Nature is not obliged to conform." Above all, d'Alembert, epitomizing the Enlightenment, repudiates the rationalistic system-making of the previous century, and of many centuries past—"The spirit of systems is in physics what metaphysics is in geometry"—and insists that all sciences "are confined, as much as possible, to facts and to consequences deduced from them." In this discourse the Encyclopedists could lay claim to a scientific manifesto that replaced the much-admired, but one-sidedly inductive, Novum Organum *of Bacon (Chapter 18 above) nearly a century and a half before.*

. . . In our study of Nature, which we make partly by necessity and partly for amusement, we note that bodies have a large number of properties. However, in most cases they are so closely united in the same subject that, in order to study each of them more thoroughly, we are obliged to consider them separately. Through this operation of our intelligence we soon discover properties which seem to belong to all bodies, such as the faculty of movement or of remaining at rest, and the faculty of communicating movement, which are the sources of the principal changes we observe in Nature. By examining these properties—above all the last one—with the aid of our own senses, we soon discover another property upon which all of these depend: impenetrability, which is to say, that specific force by virtue of which each body excludes all others from the place it occupies, so that when two bodies are put together as closely as possible, they can never occupy a space smaller than the one they filled separately. Impenetrability is the principal property by which we make a distinction between the bodies themselves and the indefinite portions of space in which we conceive them as being placed—at least the evidence of our senses tells us such is the case. Even if they are deceptive on this point, it is an error so metaphysical that our existence and the preservation of our lives have nothing to fear from it, and it continually crops up in our mind almost involuntarily, as part of our ordinary way of thinking. Everything induces us to conceive of space as the place (if not real, at least supposed) occupied by bodies. And indeed, it is by conceiving of sections of that space as being penetrable and immobile that our idea of movement achieves the greatest clarity it can have for us. We are therefore almost naturally impelled to differentiate, at least mentally, between two sorts of extension, one being impenetrable and the other constituting the place occupied by bodies. And thus, although impenetrability belongs of necessity to our conception of the parts of matter, nevertheless, since it is a relative property (that is, we get an idea of it only by examining two bodies together), we soon accustom ourselves to thinking of it as distinct from extension and to considering the latter separately from it.

Through this new consideration we now see bodies only as shaped and extended parts of space, this being the most general and most abstract point of view from which we can envisage them. . . .

Hence we are led to ascertain the properties of extension simply as to shape. This is the object of Geometry, which facilitates its task by considering extension limited first by a single dimension, then by two, and finally by the three dimensions constituting the essence of an intelligible body (that is to say, of a portion of space terminated in every direction by intellectual boundaries).

Thus, by a few successive operations and abstractions of our minds we divest matter of almost all its sensible properties, in order to envisage in a sense only its phantom. . . .

Since our examination of shaped extension presents us with a large number of possible combinations, it is necessary to invent some means of achieving those combinations more easily; and since they consist chiefly in calculating and relating the different parts of which we conceive the geometric bodies to be formed, this investigation soon brings us to Arithmetic or the science of numbers. . . .

Moreover, if we reflect upon these rules we almost inevitably perceive certain principles or general properties of the relationships, by means of which we can, expressing these relationships [numbers] in a universal way, discover the different combinations that can be made of them. The result of these combinations reduced to a general form will in fact simply be arithmetical calculations, indicated and represented by the simplest and shortest expression consistent with their generality. . . . We extend our investigations even to the movement of bodies animated by unknown driving forces or causes, provided the law whereby these causes act is known or supposed to be known.[5]

[5]For example, the movement of the planets, whose ultimate cause is unknown, although the laws of their movement are known, or at least measurable. (Schwab)

Having at last made a complete return to the corporeal world, we soon perceive the use we can make of Geometry and Mechanics for acquiring the most varied and profound knowledge about the properties of bodies. It is approximately in this way that all the so-called physico-mathematical sciences were born. We can put at their head Astronomy, the study of which, next to the study of ourselves, is the most worthy of our application because of the magnificent spectacle which it presents to us. Joining observation to calculation and elucidating the one by the other, this science determines with an admirable precision the distances and the most complicated movements of the heavenly bodies; it points out the very forces by which these movements are produced or altered. Thus it may justly be regarded as the most sublime and the most reliable application of Geometry and Mechanics in combination, and its progress may be considered the most incontestable monument of the success to which the human mind can rise by its efforts.

The use of mathematical knowledge is no less considerable in the examination of the terrestrial bodies that surround us. All the properties we observe in these bodies have relationships among themselves that are more or less accessible to us. The knowledge or the discovery of these relationships is almost always the only object we are permitted to attain, and consequently the only one we ought to propose for ourselves. Thus, it is not at all by vague and arbitrary hypotheses that we can hope to know nature; it is by thoughtful study of phenomena, by the comparisons we make among them, by the art of reducing, as much as that may be possible, a large number of phenomena to a single one that can be regarded as their principle. Indeed, the more one reduces the number of principles of a science, the more one gives them scope, and since the object of a science is necessarily fixed, the principles applied to that object will be so much the more fertile as they are fewer in number. . . .

The only resource that remains to us in an investigation so difficult, although so necessary and even pleasant, is to collect as many facts as we can, to arrange them in the most natural order, and to relate them to a certain number of principal facts of which the others are only the consequences. If we presume sometimes to raise ourselves higher, let it be with that wise circumspection which befits so feeble an understanding as ours.

Such is the plan we must follow in that vast part of physics called General and Experimental Physics.[6] It differs from the physico-mathematical sciences in that it is properly only a systematic collection of experiments and observations. On the other hand, the physico-mathematical sciences, by applying mathematical calculations to experiment, sometimes deduce from a single and unique observation a large number of inferences that remain close to geometrical truths by virtue of their certitude. . . . From a single experiment on the pressure of fluids, we derive all the laws of their equilibrium and their movement. Finally, a single experiment on the acceleration of falling bodies opens up the laws of their descent down inclined planes and the laws of the movements of pendulums.

It must be confessed, however, that geometers sometimes abuse this application of algebra to physics. Lacking appropriate experiments as a basis for their calculations, they permit themselves to use hypotheses which are most convenient, to be sure, but often very far removed from what really exists in Nature. Some have tried to reduce even the art of curing to calculations, and the human body, that most complicated machine, has been treated by our algebraic doctors as if it were the simplest or the easiest one to reduce to its component parts. It is a curious thing to see these authors solve with the stroke of a pen problems of hydraulics and statics capable of occupying the greatest geometers for

[6]Chemistry would fall under this category of General and Experimental Physics. (Schwab)

a whole lifetime. As for us who are wiser or more timid, let us be content to view most of these calculations and vague suppositions as intellectual games to which Nature is not obliged to conform, and let us conclude that the single true method of philosophizing as physical scientists consists either in the application of mathematical analysis to experiments, or in observation alone, enlightened by the spirit of method, aided sometimes by conjectures when they can furnish some insights, but rigidly dissociated from any arbitrary hypotheses. . . .

But while intending to please, philosophy seems not to have forgotten that it is designed principally to instruct. For that reason the taste for systems—more suited to flatter the imagination than to enlighten reason—is today almost entirely banished from works of merit. One of our best philosophers seems to have delivered the death blow to it.[7] The spirit of hypothesis and conjecture formerly was perhaps quite useful and even necessary for the renaissance of philosophy, because at that time judiciousness was less important than acquiring independence of thought. But times have changed, and a writer among us who praised systems would have come too late. The advantages now afforded by that spirit are too small to counterbalance the resulting disadvantages, and if the very small number of discoveries they once occasioned are claimed as proof of the usefulness of systems, one might just as well counsel our geometers to apply themselves to squaring the circle, because the efforts of several mathematicians to do so have given us a few theorems. The spirit of systems is in physics what metaphysics is in geometry. If it may sometimes be required in order to start us on the way, it is almost never capable by itself of leading us to truth. It can glimpse the causes of phenomena when enlightened by the observation of Nature; but it is for calculation to assure, so to speak, the existence of these causes by determining exactly the effects they can produce and by comparing these effects with those revealed to us by experience. Any hypothesis without such a support rarely acquires that degree of certitude which ought always to be sought in the natural sciences, and which is so seldom found in those frivolous conjectures honored by the name of "systems." If all he could have were conjectures of that kind, the principal merit of the physicist would be, properly speaking, to have the spirit of system but never to create one. Thousands of experiments prove how dangerous the use of systems is in the other sciences.

Physics is therefore confined solely to observations and to calculations; medicine to the history of the human body, of its maladies and their remedies; natural history to the detailed description of vegetables, animals, and minerals; chemistry to the composition and experimental decomposition of bodies. In a word, all the sciences are confined, as much as possible, to facts and to consequences deduced from them, and do not concede anything to opinion except when they are forced to. I do not speak of geometry, astronomy, and mechanics, which are destined by their nature always to be perfecting themselves. . . .

[7]M. l'abbé de Condillac, of the Académie Royale des Sciences de Prusse, in his *Traité des Systèmes.* (D'Alembert)

A Term Used in Different Ways

Selections from the *Encyclopedia*, translated by Robert M. Torrance

The central communal enterprise of the French Enlightenment, the Encyclopedia (Encyclopédie, ou Dictionnaire raisonné des arts, des sciences, et des métiers), *was first conceived in 1746 by a group of Parisian publishers as a two-volume French translation of Ephraim Chambers's English* Cyclopaedia, or an Universal Dictionary of the Arts and Sciences; *in the following year, the editorship was assigned to the young Diderot and d'Alembert. Diderot's Prospectus of 1750 projected a work of ten volumes, to be finished by the end of 1754; but the final work, completed in 1772, consisted of seventeen large folio volumes of text plus another eleven of plates. The first volume appeared, with d'Alembert's Preliminary Discourse, in 1751; the publication of the second a year later led to its first suspension of publication by the Council of State in reaction to "maxims that would tend to destroy royal authority, foment a spirit of independence and revolt, and by the use of obscure and equivocal terms lay the foundations for an edifice of errors, for the corruption of morals and religion, and for incredulity."*[8] *Owing to the secret cooperation of the official censor, Chrétien de Lamoignon de Malesherbes, a friend of the philosophes, this suspension was only temporary. But after volume VII, containing d'Alembert's article on Geneva, appeared in 1757 (shortly after Damiens's attempted assassination of Louis XV and public execution by drawing and quartering)—and after publication of Helvétius's materialistic* De l'esprit (On the Mind) *the following year aroused public opinion against the philosophes—a permanent ban on further publication and on sale of existing volumes followed in 1759. In spite of this, Diderot and his colleagues, with the connivance of Malesherbes and his successor, continued to work on in Paris, publishing subsequent volumes with the false imprint of Neuchâtel, Switzerland; the last volume of text, volume XVII, appeared in 1765.*

The following selections illustrate some of the meanings attached to the key word "nature" in the Encyclopedia. *Diderot's article "Nature" is followed by selections from d'Alembert's on the laws of nature; by two short articles on nature in poetry and in sacred criticism by Louis-Jacques Goussier and Louis-Jean-Marie Daubenton; and by selections from a long article on "Beautiful Nature" in the fine arts. The author of the last of these, the chevalier Louis de Jaucourt (1704–79), was one of Diderot's principal collaborators and the author of more articles in the* Encyclopedia *(nearly a fourth of the total) than anyone else, including Diderot himself. Born in Paris of a Huguenot family, he was educated in Geneva, Cambridge, and Leyden, where he studied medicine under Boerhaave; apart from his work on the* Encyclopedia, *he was the author of several medical works in Latin and of a life of Leibniz. Artists, Jaucourt argues, somewhat in the manner of Joshua Reynolds (Chapter 20 above), "have not imitated nature as it is in itself, but such as it can be, and as one can conceive it by the mind"; imitation of nature in its wholeness and perfection will make one more sensible to "the scattered perfections of nature that we see before us." The selections are translated from volume XI of the* Encyclopédie *(1765).*

NATURE, noun fem. (*Philos.*), is a term used in different ways. Aristotle has a whole chapter on the different meanings the Greeks gave to the word *physis*, *nature*; and among

[8]See the brief account in the Introduction to Diderot, d'Alembert, and others, *Encyclopedia: Selections*, ed. Nelly S. Hoyt and Thomas Cassirer (1965).

the Latins its different meanings are so numerous that one author counts up to 14 or 15 of them. Mr. Boyle, in a treatise explicitly on the meanings commonly attributed to the word *nature*, counts eight principal ones.

Nature sometimes signifies the system of the world, the machine of the universe, or the assemblage of all created things. *See* SYSTEM.

It is in this sense that we say "the author of nature," that we call the sun "the eye of the universe," etc., and "the father of nature," because he makes the earth fertile by warming it; in the same way we say of the phoenix or the chimera that there is no such thing in *nature*.

Mr. Boyle proposes that, instead of using the word *nature* in this sense, one avail oneself, to avoid ambiguity or abuse of this term, of the word *world* or *universe*.

Nature is applied, in a less extended sense, to each of the different things created or uncreated, spiritual and corporeal. *See* BEING.

It is in this sense that we say "human nature," understanding in general by this all men who have a spiritual and rational soul. In the same sense theologians say *natura naturans*, and *natura naturata*; they call God *natura naturans*, as having given being and *nature* to all things, to distinguish him from creatures, which they call *natura naturata*, because they have received their *nature* from another's hands.

Nature, in a still more limited sense, is said of the essence of a thing, or of what Scholastic philosophers call the *quiddity*, that is, the attribute that makes a thing to be such or such. *See* ESSENCE.

It is in this sense that the Cartesians say that the *nature* of the soul is to think, and that the *nature* of matter consists in extension. *See* SOUL, MATTER, EXTENSION. Mr. Boyle proposes that one avail oneself of the word *essence* instead of *nature*. *See* ESSENCE.

Nature is more especially used to signify order and the natural course of things, the result of secondary causes, or the laws of movement that God has established. *See* CAUSES and MOVEMENT.

It is in this sense that one says that natural philosophers study *nature*.

Saint Thomas defines *nature* as a sort of divine art communicated to created things to carry them to the end to which they are destined. *Nature* taken in this sense is nothing else but the interlinking of causes and effects, or the order that God has established in all parts of the created world.

It is also in this sense that one says that miracles are above the power of *nature*; that art compels or surpasses *nature* by means of machines, when it produces by this means effects that surpass those that we see in the ordinary course of things. *See* ART, MIRACLE.

Nature is also said of the union of powers or faculties of a body, especially of a living body.

It is in this sense that physicians say that *nature* is strong, weak, or worn out, or that in certain sicknesses *nature* left to itself works their cure.

Nature is further taken in a less extended sense to signify the action of providence, the principle of all things, that is, that spiritual power or being that acts and operates on all bodies to give them certain properties or produce in them certain effects. *See* PROVIDENCE.

Nature taken in this sense, which is that which Mr. Boyle adopts by preference, is nothing other than God himself, acting in accord with certain laws that he has established. *See* GOD.

This appears to accord substantially with the opinion held by several ancients, that *nature* was the god of the universe, the *to pan* [Greek "all"] that presided over all and governed all, though others saw this pretended being as imaginary, understanding nothing else by the word *nature* than the qualities or virtues that God has given to his creatures, and that poets and orators personify.

Father Malebranche claims that everything said in the schools about *nature* is capable of leading us to idolatry, since by this word the ancient pagans understood something that, without being God, acted continually in the universe. Thus the idol *nature* must have been, according to them, an actual principle that was in competition with God, the secondary and immediate cause of all changes that matter undergoes. This [idol] seems to return in the sentiment of those who acknowledged the *anima mundi* [world soul], seeing *nature* as a substitute divinity, a collateral cause, a sort of being between God and creatures.

Aristotle defines nature as *principium et causa motus et eius in quo est primo per se et non per accidens*;[9] a definition so obscure that in spite of all the glosses of his commentators, none of them has been able to succeed in making it intelligible.

This principle, which the Peripatetics called *nature*, acted, according to them, by necessity, and was consequently deprived of knowledge or freedom. *See* Fatality.

The Stoics also conceived of *nature* as a certain spirit or power spread through the universe, which gave each thing its movement, so that everything was compelled by the invariable order of a blind *nature* and by an inevitable necessity.

When one speaks of the action of *nature*, one conveys nothing more than the action of bodies upon one another, in conformity with the laws of movement established by the Creator.

All the sense of this word consists in this, which is only an abridged way to express the action of bodies, and which one would perhaps express better by the word *mechanism* of bodies.

There are, as Mr. Boyle observes, those who understand by the word *nature* only the law that each thing has received from the Creator, and according to which it acts on all occasions; but this sense attached to the word *nature* is inappropriate and figurative.

The same author proposes a definition of the word *nature* more just and exact, in his view, than all the others, and in virtue of which one can easily understand all the axioms and expressions related to this word. To that end he distinguishes between *particular nature* and *general nature*.

He defines *general nature* as the assemblage of bodies that constitute the present state of the world, considered as a principle by virtue of which they act and receive action according to the laws of movement established by the author of all things.

The *particular nature* of a subordinate or individual being is only general *nature* applied to some distinct portion of the universe; it is an assemblage of mechanical properties (like size, shape, order, situation, and local motion) suitable and sufficient to constitute the species and denomination of a particular thing or body, the concurrence of all beings being considered as the principle of movement, of rest, etc. (Diderot)

NATURE, *laws of*, are axioms or general rules of movement and of rest that natural bodies observe in the action they exercise on one another, and in all the changes that befall their natural state.

Although the laws of *nature* are properly the same as those of movement, some dis-

[9]In R. P. Hardie and R. K. Gaye's translation of this definition from *Physics* II.1 (see Chapter 9 above), "Nature is a source or cause of being moved and of being at rest in that to which it belongs primarily, in virtue of itself and not in virtue of a concomitant attribute."

tinctions have nevertheless been made between them. In fact, one finds authors who give the name of "*laws* of movement" to particular laws of movement, and who call "laws of *nature*" the more general and extended laws, which are like the axioms from which the others are deduced.

Of these latter laws, Mr. Newton establishes three. . . . (d'Alembert)

NATURE (*Poetry*) *Nature* in poetry is (1) everything actually existing in the universe; (2) it is everything that has existed before us, etc., that we can know by the history of times, places, and men; (3) it is everything that can exist, but which perhaps has never existed nor ever will exist. We include in History fable and all the poetic inventions to which one accords a supposed existence, which for the arts is worth as much as historical reality. Thus there are three worlds to which poetic existence may go to choose and take what suits it to form its compositions: the real world; the historical world, which includes the fabulous; and the possible: and these three worlds are what one calls *nature*. (Goussier, Daubenton)

NATURE (*Sacred criticism*) The words *nature* and *naturally* are often found employed in Scripture, as well as in the Greek and Latin authors, in opposition to the path of instruction that lets us know certain things. . . . One can say that the Gentiles, who were deprived of revelation, knew by themselves, without this help, the precepts of morality that the natural lights of reason let them discover, and which were the same that the law of Moses taught to the Jews, so that when a pagan acted according to these precepts, he was doing naturally what the law of Moses prescribed. . . . (Goussier, Daubenton)

BEAUTIFUL NATURE (*Fine arts*) *Beautiful nature* is *nature* embellished, perfected by the fine arts for use and amenity. . . .

Wearied by a too uniform enjoyment of the objects that simple *nature* offered them, and finding themselves moreover in a situation suited to receive pleasure, men had recourse to their genius to procure for themselves a new order of ideas and sentiments that might awaken their spirit and revive their taste. But what could this genius make, constricted as it was in its fecundity and in its views, which it could not take farther than *nature*, and having to work, on the other hand, for men whose faculties were bound within the same limits? All its efforts were necessarily reduced to making a selection of the most beautiful parts of *nature* from which to form an exquisite whole that would be more perfect than *nature* itself, without, however, ceasing to be natural. This was the principle on which the plan of the arts was necessarily erected, and which the great artists have followed in every century. Selecting objects and traits, they have presented them with all the perfection of which they are susceptible. They have not imitated *nature* as it is in itself, but such as it can be, and as one can conceive it by the mind. Thus, since the object of imitation of the arts is *beautiful nature*, represented with all its perfections, let us see how this imitation is made. . . .

Either the imitation of *nature* is constricted to a single object, or it brings together in a single work what the artist has observed in several individuals. The former way of imitating produces copies resembling portraits. The latter elevates the spirit of the artist to the beautiful in general, and to ideal notions of beauty. It is this latter path that the Greeks chose, who had the advantage over us of being able to procure these notions both by contemplation of the most beautiful bodies and by frequent opportunities to observe the beauties of *nature*. These beauties, as said elsewhere, revealed themselves to them every day, animated by the truest expression, whereas they offer themselves to us rarely, and still more rarely in the way the artist would desire them to present themselves. . . .

An artist who will let his spirit and his hand be guided by the rule that the Greeks adopted for beauty will find himself on the road that will lead him directly to the imitation of *nature*. The notions of wholeness and perfection, brought together in the *nature* of the ancients, will purify in him and make more sensible to him the scattered perfections of *nature* that we see before us. In discovering the beauties of the latter, he will know how to combine them with perfect beauty; and by means of the sublime forms always present to his mind, he will become a sure rule for himself. . . . *(Le Chevalier de Jaucourt)*

The Mayfly's Sophism in a Madman's World

Selections from the Writings of Denis Diderot, translated by Robert M. Torrance

Denis Diderot (1713–84), the most versatile of the French philosophes, was born at Langres in Champagne, the son of a cutler. Having refused a career in the Church, he studied in Paris, and after secretly marrying, lived a bohemian existence, making friends with d'Alembert, Rousseau (with whom he later broke), and other young thinkers and writers. In 1745 he published the Essai sur le mérite et la vertu, *an adaptation of the* Inquiry Concerning Virtue and Merit *(1699) by the Earl of Shaftesbury (Chapter 20 above), followed by* Pensées philosophiques *(*Philosophical Thoughts, *1746), which was publicly burned by the Paris Parlement because of its deistic tendencies; the erotic novel* Les Bijoux indiscrets *(*The Indiscreet Jewels, *1748); and the* Lettre sur les aveugles à l'usage de ceux qui voient *(*Letter on the Blind for the Use of Those Who See, *1749), in which the dying blind English mathematician Saunderson rejects the deistic argument for a beneficent creator from the design of a universe whose beauty he cannot see. "How many mangled, failed worlds have vanished, are perhaps re-forming themselves and vanishing at any given moment, far off in space, where I cannot touch them and you cannot see them," he exclaims to the clergyman who has advanced the marvels of nature as proofs of God's existence. ". . . You are judging the discontinuous existence of the world as the ephemeral mayfly judges yours. The world is eternal for you just as you are eternal for the being that lives only an instant. Yet the insect is more reasonable than you."*

Mainly in response to this book, Diderot was imprisoned for several months in the château de Vincennes outside Paris, from which he won release by promising to write no more against religion. Meanwhile, he had begun, in 1747 (as recounted in the previous section), his editorship—at first with d'Alembert, then alone—of the Encyclopedia. *His other published works include* De l'interprétation de la nature *(*On the Interpretation of Nature, *1753), from which our first excerpt is taken; two "bourgeois dramas" in prose,* Le fils naturel *(*The Natural Son, *1757) and* Le père de famille *(*The Father of the Family, *1758); and the art criticism of the* Salons *between 1759 and 1781. In 1773–74 he made his only journey abroad, to Holland and to St. Petersburg, where Catherine the Great (to whom he willed his library) proved unreceptive to his philosophical ideals of government.*

By far his most interesting writings to later readers, however—from Goethe and Schiller to Hegel and Marx, and beyond—are those which he wrote for himself or a small circle of friends, and which remained unpublished until after his death. These include the novel La Religieuse *(*The Nun, *1760; pub. 1796), a passionate denunciation of the forced immuring of young* *women;* Le Neveu de Rameau *(*Rameau's Nephew, *from the early 1760s, first published in Goethe's German translation in 1805, and in French in 1823), a dialogue between a philosophe very like Diderot and a bohemian ne'er-do-well very like Diderot's disreputable alter ego; the experimental philosophical novel* Jacques le fataliste et son maître *(*Jacques the Fatalist and His Master, *1773; pub. 1796); and the two philosophical works from which our second and third excerpts are taken, the tripartite dialogue* Le Rêve de d'Alembert *(*D'Alembert's Dream, *1769; pub. 1830) and the* Supplément au voyage de Bougainville *(*Supplement to Bougainville's Voyage, *1772; pub. 1796).*

The general tendency of Diderot's sometimes seemingly self-contradictory thought, as P. N. Furbank writes in Diderot: A Critical Biography *(1992), is away from his early deism toward "materialism, atheism, and determinism." Yet his materialistic conception of nature was far different from those of La Mettrie's* Man a Machine *(1748), Holbach's* System of Nature *(1770), or Helvétius's* On Man *(1772). In sharp contrast to their dogmatic affirmations, Diderot forever remained a questioner if not a sceptic, inquiring and speculating rather than merely pronouncing. The unmethodical methodology of* On the Interpretation of Nature *already discards the mathematical certainties of Cartesian or Newtonian systems in favor of a conjectural natural philosophy for which finality is beyond reach. "His rejection of mathematics was fundamental. He objected to its claim to be the true language of science on all grounds," Gillispie writes, "metaphysical, mechanical, and moral. It is not just that mathematics idealizes. It falsifies, by depriving bodies of the perceptible qualities in which alone they have existence for an empirical, sympathetic science." Nature, in the excerpt given below, remains a continually changing "woman who loves to disguise herself."*

For Diderot, Cassirer writes in The Philosophy of the Enlightenment, *"This infinitely changing universe can only be understood by a flexible manner of thinking, by a kind of thinking which permits itself to be borne and driven from one flight to the next, which does not rest content with what is present and given, but which rather luxuriates in the abundance of possibilities and wants to explore and test them. By virtue of this fundamental characteristic Diderot becomes one of the first to divorce himself from the static philosophy of the eighteenth century and to change this philosophy into a truly dynamic view of the world. . . . Nature knows only diversity and absolute heterogeneity." Nature is not mechanical but organic, not fixed but continually in the process of formation, and it cannot be conceived apart from the human being who perceives it. "If one banishes from the face of the earth the thinking and contemplating entity, man," Diderot writes in the article "Encyclopedia,"*[10] *"then the sublime and moving spectacle of nature will be but a sad and silent scene; the universe will be hushed; darkness and silence will regain their sway. All will be changed into a vast solitude where unobserved phenomena take their course unseen and unheard. It is only the presence of men that makes the existence of other beings significant."*

Nowhere are these ideas so daringly expressed as in D'Alembert's Dream, *where Diderot, in his conversation with d'Alembert, rejects the Cartesian dualism of matter and spirit as "metaphysico-theological mumbo-jumbo" and argues that sentience is "a general property of matter" and that all perceivable forms of nature are indivisible. That night, d'Alembert dreams of a world of perpetual flux in which inanimate matter becomes sentient and "there is nothing distinct in nature," since everything is linked in "one great individual: the whole." "What is obsessing him on his pillow," Furbank writes, "is, precisely, the question of personal identity. . . . We, who think of ourself as an individual, as a unitary member of our species, are in fact a swarm or society; our bodily organs are so many independent animals, held in a 'sympathy, a unity, a general identity' only by the law of 'continuity'" that interconnects nature as a whole.*

[10]From vol. V of the *Encyclopedia* (1755), as translated by Jacques Barzun and Ralph H. Bowen in *Diderot: Rameau's Nephew and Other Works* (1956).

And in the Supplement to Bougainville's Voyage—*elaborating on accounts by Louis-Antoine de Bougainville, and by one of the companions, Philibert Commerson, of his voyage around the world between 1766 and 1769—Diderot's idealized portrayal of a sexually liberated Tahiti threatened with corruption by its European discoverers counters Rousseau's dream of the isolated individual nurtured by nature alone, Carol Blum writes in* Diderot: The Virtue of a Philosopher *(1974), with a visionary "society that would resemble the living organism he described in* Le Rêve de d'Alembert. . . . *The voice of nature and the voice of society were in unison and the word they spoke was: procreate."*

Yet the dialogue is far more than a utopian manifesto of romantic primitivism and sexual promiscuity in contrast with the hypocrisies of European civilization. In the end, as Furbank remarks, "the doctrine that Diderot is underwriting is not primitivism at all, but a more cogent, serious and humane one: that it is madness to create institutions that go against the grain of life and set humans at war with their own selves." Much though we may learn from a seemingly more natural way of life elsewhere, there is no question of going back to nature. "We shall speak out against senseless laws until they are reformed; meantime, submit to them," the dialogue ends. ". . . It is less inconvenient to be mad among madmen than to be wise all alone." The "natural man" of an unnatural eighteenth-century Europe will necessarily be more complex (not to say convoluted) than a romanticized Tahitian, as Diderot knew no less than did Samuel Johnson (introduction to Chapter 20 above): "for anyone who refuses to use his reason," he wrote in the article "Droit naturel" ("Natural Right") in volume V of the Encyclopedia, *"thereby forfeiting his status as a man, ought to be treated as an unnatural being."*

Translations are based on the text of Diderot's Oeuvres philosophiques, *ed. Paul Vernière (1964).*

FROM On the Interpretation of Nature *(XII)*

Nature seems to have taken pleasure in varying the same mechanism in an infinite variety of ways. She abandons any one kind of production only after multiplying its individuals in all possible aspects. When we consider the animal kingdom and perceive that there is no quadruped whose functions and parts, especially internal ones, do not wholly resemble those of another quadruped, might we not readily believe that there was never more than one first animal, prototype of all animals,[11] some of whose organs nature has merely lengthened, shortened, transformed, multiplied, or obliterated? Imagine the fingers of your hand joined together, and the material of the nails so abundant that, by stretching and swelling, it enveloped the whole: instead of a man's hand you have a horse's hoof. When we see the successive metamorphoses of the prototype's outward covering, whatever it may once have been, bring one kingdom by insensible degrees nearer to another, and populate the confines of the two kingdoms (if we may use the term "confines" where there is no real division)—and populate, I say, the confines of the two kingdoms with uncertain, ambiguous beings largely stripped of the forms, qualities, and functions of the one, and garbed in the forms, qualities, and functions of the other, who would not be inclined to believe that there was never more than one first being, prototype of all beings? But whether this philosophical conjecture is accepted as true, with Dr. Baumann

[11]Vernière notes that "prototype" is a central concept in the proto-evolutionary thought of the French Enlightenment, taking different forms, e. g., in Buffon and Maupertuis. Thus Robinet, in *Considérations philosophiques sur la gradation naturelle des formes de l'être* (*Philosophical Considerations on the Natural Gradation of the Forms of Being*, 1768), writes: "A stone, an oak, a horse, an ape, a man, are graduated variations of the prototype that began to be realized by the smallest possible elements."

[Maupertuis], or rejected as false, with M. de Buffon, it must undeniably be embraced as a hypothesis essential to the progress of experimental science and of rational philosophy, and to the discovery and explanation of phenomena dependent on organization. For it is evident that nature could not have preserved so great a resemblance of parts, and assumed so great a variety of forms, without often making perceptible in one organic being what she concealed in another. She is a woman who loves to disguise herself and whose different disguises, disclosing now one part of her, now another, give those who assiduously pursue her some hope that one day they will know her whole person.

FROM *D'ALEMBERT'S DREAM*

Conversation Between D'Alembert and Diderot

Diderot. . . . Suppose that a harpsichord has feeling and memory, and tell me if it will not be able to repeat by itself the tunes that you have performed on its keys. We ourselves are instruments endowed with feeling and memory. Our senses are so many keys that are played by surrounding nature and that often play themselves; this, in my judgment, is all that happens in a harpsichord organized like you or me. There is an impression whose cause is inside or outside the instrument, a sensation is born of this impression, a sensation that lasts: for it is impossible to imagine that it is created and extinguished in one indivisible moment; another impression that succeeds it, likewise caused by something inside and outside the animal; then a second sensation, and voices that designate it by natural or conventional sounds.

D'Alembert. I understand. So, then, if this sentient and animate harpsichord were also endowed with the faculty of feeding and reproducing itself, it would be alive and would engender, by itself or with its female, little living and resonant harpsichords.

Diderot. Undoubtedly. What else, in your opinion, is a finch, a nightingale, a musician, a man? And what other difference do you find between a canary and a canary-organ?[12] Do you see this egg? With this we are overturning all the schools of theology and all the temples of the earth. What is this egg? An insentient mass before the seed is introduced into it; and after the seed is introduced, what is it still? An insentient mass, for that seed itself is nothing but crude and inert fluid. How can this mass be reorganized and pass over to sentience, to life? By heat. And what will produce heat in it? Motion. What will be the successive effects of motion? Instead of answering me, sit down, and let us follow them with our eye from moment to moment. First, there is a vibrating point; a little thread extends and gains color; flesh forms; a beak, wings, eyes, feet appear; a yellowish substance unwinds and produces intestines: it is an animal. This animal moves, tosses, cries; I hear its cries through the shell; down covers it; it sees. The weight of its vibrating head brings its beak repeatedly against the inner wall of its prison; this breaks; it emerges, it walks, it flies, it is angered, it flees, it approaches, it complains, it suffers, it loves, it desires, it enjoys; it feels the same affections as you; all your actions, it too performs. Will you claim, with Descartes, that it is merely an imitative machine? Little children will mock you then, and philosophers will answer you that if this is a machine, you are one too. If you admit that between the animal and you there is no difference but in organization, you will be showing sense and reason, and acting in good faith; but they will charge you with maintaining that from inert matter, arranged in a certain way, impregnated by more inert matter, by heat and by motion, can arise sentience, life, memory, conscious-

[12]As Vernière notes, volume XV of the *Encyclopédie* defines *serinette* as a "little Barbary organ now in use to teach canaries *[serins]* to sing various tunes."

ness, passion, thought. You must then take one of these two courses: either to imagine an element hidden in the inert mass of the egg that was waiting for it to develop before manifesting its presence, or to imagine that this imperceptible element found its way through the shell at a fixed moment of that development. But what is this element? Did it occupy space or not? How did it appear, or vanish, without moving? Where was it? What was it doing there, or elsewhere? Was it created the instant it was needed? Did it exist? Was it awaiting a home? Was it of the same substance as that home, or of a different substance? If of the same substance, it was material; if of a different, we cannot conceive either its inert state before the egg's development or its energy in the developed animal. Listen to yourself, and you will find yourself pitiful. You will realize that to avoid admitting one simple supposition that explains everything—sentience as a general property of matter, or a product of organization—you are renouncing common sense and hurtling into an abyss of mysteries, contradictions, and absurdities.

D'Alembert. A supposition! That's what you like to call it. But what if this were a quality essentially incompatible with matter?

Diderot. And how can you ascertain that sentience is essentially incompatible with matter, you who do not know the essence of anything at all, neither of matter nor of sentience? Do you understand any better the nature of motion, how it exists in a body, and how it is communicated from one body to another?

D'Alembert. Without conceiving the nature of sentience or of matter, I see that sentience is a simple quality, single, indivisible, and incompatible with a divisible subject or agent.

Diderot. Metaphysico-theological mumbo jumbo! What? Can you not see that all the qualities, all the perceivable forms in which nature is robed are essentially indivisible? There can be no greater or lesser impenetrability. There can be half of a round body, but not half of roundness; there can be greater or lesser motion, but these are not more or less motion; there is no more a half, a third, or a quarter of a head, an ear, or a finger, than a half, a third, or a quarter of a thought. If there is no molecule in the universe that is like any other, no point in any molecule that is like any other point, you must agree that even an atom is endowed with a quality, an indivisible form; you must agree that division is incompatible with the essences of forms, since it destroys them. Be a natural philosopher, and admit that an effect is produced when you see it produced, even if you cannot explain the connection between its cause and effect. Think logically, and do not substitute for an existent cause that explains everything another cause that is inconceivable, whose connection with the effect is even less conceivable, and that engenders an infinite multitude of difficulties without resolving a single one. . . .

D'Alembert. Farewell, my friend: good evening, and good night!

Diderot. You're joking; but you will dream of this conversation on your pillow, and if it lacks coherence, so much the worse, for you will be forced to embrace far more ridiculous hypotheses.

D'Alembert. You're mistaken: I shall fall asleep a sceptic, and a sceptic shall arise. . . .

D'Alembert's Dream

Speakers: d'Alembert, Mlle de L'Espinasse, Doctor Bordeu

. . . *Mlle de L'Espinasse.* . . . At about two in the morning, he came back to his drop of water, which he called a mi . . . cro . . .

Bordeu. A microcosm.

Mlle de L'Espinasse. His very word. He was admiring the sagacity of the ancient philosophers. He said, or had his philosopher say, I don't know which: "If, when Epicurus affirmed that the earth contained the seeds of everything, and that the animal kingdom was a product of fermentation, he had proposed to show an image in miniature of what had taken place on a large scale in primordial times, what answer could have been made to him? . . . Yet you have this image under your eyes, and it teaches you nothing. . . . Who knows whether fermentation and its products are exhausted? Who knows what stage in the succession of these animal generations we now are? Who knows whether that deformed biped, only four feet tall, that is still, in the neighborhood of the North Pole, called a man, but would soon lose that name on becoming a little more deformed, is not the image of a vanishing species? Who knows whether it is not the same with all animal species? Who knows whether everything is not tending toward reduction to one enormous, inert and immobile sediment? Who knows how long that inert state would last? Who knows what new race may again arise from so enormous a mass of sentient and living points? Why not one single animal? What was the elephant in the beginning? Perhaps the vast animal we now see, perhaps an atom, for both are equally possible; both require only motion and the various properties of matter. . . . The elephant, that vast organized mass, the unexpected product of fermentation! Why not? The ratio of this huge quadruped to its original matrix is less close than the grub's to the molecule of flour that engenders it, but the grub is only a grub. . . . That is, the smallness that hides its organization from you deprives it of its wonder. . . . But the prodigy is life, sentience itself; and this is a prodigy no longer. . . . Once I have seen inert matter pass to the sentient state, nothing should ever astonish me again. . . . What a comparison, between a small number of elements fermenting in the hollow of my hand, and that immense reservoir of various elements scattered in the bowels of the earth, on its surface, in the depths of the seas, in the vacancy of the air! . . . And yet, since the same causes hold sway, why have their effects ceased? Why do we no longer see the bull piercing the earth with its horns, planting its hooves on the ground, and straining to pull its heavy body free? . . . Let the race of animals still existing pass away; let the immense inert sediment work for several million centuries. Perhaps it requires ten times longer to renew species than is allotted for their duration. Wait, don't be hasty to pass judgment on nature's work. You have two great phenomena: passage from the inert to the sentient state, and spontaneous generations. Let these suffice: draw just conclusions from them, and in an order of things where none is absolutely big or little, lasting or transient, guard against the mayfly's sophism . . ." Doctor, what is the mayfly's sophism?
Bordeu. That of a transient being who believes in the immutability of things.
Mlle de L'Espinasse. Fontenelle's rose, which said no one in the memory of a rose had seen a gardener die?[13]
Bordeu. Precisely; that is graceful and profound . . .

Bordeu. Be quiet.
D'Alembert. I am therefore such as I am, because it was necessary that I be such. Change the whole, and you necessarily change me, yet the whole is ceaselessly changing . . . Man is only a common effect, the monstrosity a rare effect; both equally natural, equally necessary, equally part of the universal and general order. . . . And what is astonishing in that? . . . All beings circulate within one another, and therefore all species . . . everything is in perpetual flux . . . Every animal is more or less man; every mineral is more or less

[13]See Fontenelle's *Conversations on the Plurality of Worlds* (Fifth Evening), in Chapter 18 above.

plant; every plant is more or less animal. There is nothing distinct in nature. . . . And you talk of individuals, poor philosophers! Forget your individuals; answer me: Is there one atom in nature exactly like one other atom? . . . No . . . Do you not agree that everything in nature holds together, and that it is impossible for there to be a break in the chain? What do you mean to say, then, with your "individuals"? There are none, no, there are none . . . There is only one great individual: the whole. In this whole, as in a machine, as in some animal, you will call one part such or such; but when you give the name of individual to that part of the whole, your conception is as false as though you had given the name of individual to a bird's wing, or to a feather on the wing . . . And you talk about essences, poor philosophers! Forget your essences. Consider the general mass, or if your imagination is too narrow to embrace it, consider your first origin and your final end . . . Oh, Archytas![14] You who measured the globe, what are you? A few ashes . . . What is a being? . . . the sum of a certain number of tendencies . . . Can I be anything else but a tendency? . . . No, I advance toward an end . . . And species? . . . Species are only tendencies toward a common end that is their own . . . And life? . . . Life, a series of actions and reactions . . . Alive, I act and react as a mass . . . dead, I act and react as molecules . . . Then I do not die? . . . No, I undoubtedly do not die in that sense, neither I, nor anything that is . . . Birth, life, passing away, is all changing form . . . And what difference does one form or another make? Every form has the happiness and unhappiness proper to it. From the elephant to the louse . . . from the louse to the sentient and living molecule, the origin of everything, there is no point in all nature that does not suffer pain and feel pleasure. . . .

FROM *SUPPLEMENT TO BOUGAINVILLE'S VOYAGE, OR DIALOGUE BETWEEN A AND B*

The Old Man's Farewell

It is an old man who speaks. He was the father of a numerous family. . . . On Bougainville's departure, when the inhabitants were running down in a crowd to the shore, seizing hold of his garments, clasping his companions in their arms, and weeping, this old man advanced with a stern appearance, and said:

"Weep, unhappy Tahitians! weep—but for the arrival, not the departure of these ambitious and wicked men. One day you will know them better. . . . One day you will be their servants, as corrupted, as base, as unhappy as they. But I console myself that I am near the end of my days; I shall not see the calamity I am warning you of. . . ."

Then, addressing Bougainville, he added, "And you, chief of the brigands who obey you, promptly remove your vessel from our shores: we are innocent, we are happy; you can only mar that happiness. We follow the pure instinct of nature; and you have tried to efface its imprint from our souls. Here, everything belongs to everyone; and you have preached to us some kind of distinction between *yours* and *mine*. Our daughters and our wives are common to all; you shared this privilege with us; and you inflamed them with unknown furies. They became madwomen in your arms; you became wild beasts in theirs. They began to hate one another; you slaughtered one another for them; and they came back to us stained with your blood. We were free, and you have now planted in our soil your claim to our future enslavement. You are neither a god nor a demon: what are

[14]Archytas of Taras (Roman Tarentum), a Greek colony in southern Italy, was a famed mathematician and Pythagorean philosopher of the early fourth century B.C.

you, then, to make us slaves? . . . Do you think a Tahitian cannot defend his liberty, and die for it? This Tahitian you are trying to capture like a beast, is your brother. You are both children of nature; what right do you have over him that he does not have over you? . . . Go to your own country to agitate and torment yourselves as much as you wish, but leave us in peace; do not trouble our heads with your artificial needs and chimerical virtues. Look at these men; see how upright, healthy, and robust they are. Look at these women; see how upright, healthy, fresh, and beautiful they are. Take this bow, it is mine; call one, two, three, four of your companions to help you, and try to draw it. I draw it myself. I plow the land, I climb the mountains, I cut through the forest, I run a league on the plain in less than an hour. Your young companions have barely been able to keep up with me, and I am more than ninety years old. Woe to this island! Woe to Tahitians now and to all Tahitians to come, from the day you visited us! We knew only one disease, that to which man, animal, and plant have been condemned, old age; and you have brought us another: you have infected our blood. . . . The idea of crime and the danger of disease first came among us from you. Our delights, once so sweet, are now accompanied by remorse and fear . That man in black, standing near you and listening to me, has talked to our boys; I do not know what he has said to our girls; but our boys hesitate, our girls blush. Plunge if you will into the dark forest with the perverted mistress of your desires, but allow the good and simple Tahitians to reproduce themselves without shame, under the open sky, and in broad daylight. . . .

The Conversation Between the Chaplain and Orou

B. When the Tahitians divided up Bougainville's crew, Orou received the chaplain as his share. The chaplain and the Tahitian were about the same age, thirty-five to thirty-six. Orou then had only his wife and three daughters, Asto, Palli, and Thia. They undressed him, washed his face, hands, and feet, and served him a wholesome and frugal meal. When he was about to go to bed, Orou, who had withdrawn with his family, presented his wife and three daughters, all naked, to him, and said:

"You have dined, you are young and strong; if you sleep alone, you will sleep poorly; a man needs a woman beside him at night. Here is my wife, here are my daughters: choose the one who suits you; but if you wish to oblige me, you will give your preference to my youngest daughter, who has not yet had children."

The mother added: "Alas! I have no reason to complain; poor Thia! it isn't her fault."

The chaplain replied that his religion, his vocation, good morals, and decency did not permit him to accept this offer.

Orou replied: "I do not know what this thing you call religion is, but I can only think ill of it, since it prevents you from enjoying an innocent pleasure to which Nature, the sovereign mistress, invites us all; from giving existence to one of your own kind; from rendering a service that father, mother, and children ask of you; from repaying a host who has given you a good welcome; and from enriching a nation by augmenting it with one more subject. I do not know what this thing you call your vocation is; but your first duty is to be a man, and to be grateful. I do not propose that you take the ways of Orou back to your own country; but Orou, your host and friend, begs you to accommodate yourself to the manners of Tahiti. . . .

At this point, the truthful chaplain admits that Providence had never exposed him to so urgent a temptation. He was young; he was agitated and tormented; he averted his gaze from the lovely suppliants, then turned it toward them; he raised his eyes and hands to heaven. Thia, the youngest, embraced his knees and said, "Stranger, do not afflict my father, do not afflict my mother, do not afflict me! . . . "

The ingenuous chaplain says that she clasped his hands; that she held his eyes with such expressive and touching looks; that she wept; that her father, her mother, and her sisters withdrew; that he remained alone with her; and that, still saying: "But my religion, but my vocation," he found himself on the next day lying beside this young girl, who overwhelmed him with caresses, and invited her father, her mother, and her sisters, when they approached the bed in the morning, to join their gratitude to hers. . . .

Chaplain. Well, then: we believe that this world and what it contains is the workmanship of a workman. . . .
Orou. We've never seen him.
Chaplain. He is never seen. . . . He doesn't grow old; he spoke to our ancestors; he gave them laws; he prescribed the manner in which he wished to be honored; he ordered them to perform certain actions, because they are good, and prohibited them from performing others, because they are bad.
Orou. I understand: and one of the actions he prohibited because they were bad is to go to bed with a woman or a girl? Why did he make two sexes, then?
Chaplain. To be united; but on certain strict conditions, after certain preliminary ceremonies, as a result of which a man belongs to a woman, and belongs only to her, and a woman belongs to a man, and belongs only to him.
Orou. Their whole life long?
Chaplain. Their whole life long.
Orou. So, then, if a woman happened to go to bed with someone other than her husband, or a husband to go to bed with a woman other than his wife . . . but surely that never happens, for, since he is there and that displeases him, he knows how to prevent it.
Chaplain. No; he lets them do it, and they sin against the law of God (for thus we call the big workman) and against the law of their country: they commit a crime.
Orou. It would grieve me if my words offended you; but if you gave me leave, I should tell you my opinion.
Chaplain. Speak.
Orou. I find these singular precepts opposed to nature, contrary to reason, designed both to proliferate crimes and to anger at every moment the old workman who without a head, without hands, and without tools made everything, who is everywhere and is never seen anywhere, who continues today and tomorrow and is never a day older, who commands and is not obeyed, who can prevent and does not prevent. Contrary to nature, because they suppose that a sentient, thinking, free being can be the property of another of the same kind. On what could this right be founded? Don't you see that in your country you have confused something lacking sentience, thought, desire, or will, something cast aside and picked up, kept or exchanged without suffering or complaining, with something that is not acquired, that has freedom, will, and desire, that can give or refuse itself for a moment, give or refuse itself forever, that complains and suffers, and that could not become an article of commerce without its essence being forgotten, and without doing violence to nature? Contrary, too, to the general law of beings. Can you, in fact, conceive of anything more senseless than a precept that proscribes change within us, that enjoins a constancy that we cannot have, and that violates the nature and liberty of male and female by chaining them forever to each other? Or than a fidelity that limits the most capricious of pleasures to one sole individual? Or than a vow of immutability by two beings of flesh and blood in the presence of a sky that is not the same for an instant, in caves threatening collapse, beneath a rock that is falling to powder, beneath a tree that is splitting, on a stone that is tottering? Believe me, you have made the condition of man

worse than that of the animal. I don't know what your big workman is, but I rejoice that he did not speak to our fathers, and I wish he may never speak to our children; for he might happen to repeat the same nonsense, and they might be so nonsensical as to believe him. . . . Do you want to know, in every time and place, what is good and what is evil? Hold fast to the nature of things and of actions, to your relations with your fellow man, to the influence of your conduct on your private advantage and on the general good. You are stark raving mad if you believe that there is anything in the universe, either above or below, that can either add to or subtract from the laws of nature. Her eternal will is that the good should be preferred to the bad, and the general good to the private good. You may ordain the contrary, but you will not be obeyed. You will proliferate wrongdoing and misery by fear, punishment, and guilt; you will deprave people's consciences; you will corrupt their minds; they will no longer know what to do and what to avoid. Troubled in the state of innocence, tranquil in crime, they will have lost sight of the pole star, their true path. . . .

The good chaplain relates that he spent the rest of the day walking over the island and visiting the huts; that in the evening, after supper, the father and mother having begged him to go to bed with the second of their daughters, Palli was presented to him in the same state of undress as Thia; and that he cried several times during the night, "But my religion! but my vocation!"; that on the third night he was agitated by the same remorse with Asto, the eldest; and that he granted the fourth, out of decency, to his host's wife.

Continuation of the Dialogue Between A and B

. . . *A.* But how did it come to pass that an act with such a solemn object, to which nature incites us by the most powerful attraction—that this greatest, sweetest, most innocent of pleasures has become the most fertile source of our depravities and ills?
B. Orou explained it ten times to the chaplain: listen to him again, then, and try to retain it. . . .

How far we are from nature and happiness! The dominion of nature cannot be overthrown: there is no use thwarting and opposing it; it will endure. Write as much as you please on tablets of bronze that—to use the expression of the wise Marcus Aurelius—the voluptuous rubbing of two intestines is a crime; the heart of man will be torn between the threat of your inscription and the violence of its penchants. But this intractable heart will not cease to make its claims, and a hundred times in the course of life your dread letters will disappear from our sight. Carve it in marble: You shall not eat the eagle or the vulture;[15] you shall have carnal knowledge only of your wife; you shall not marry your sister—and do not forget to increase the punishments in proportion to the bizarreness of your prohibitions; savage though you become, you will not succeed in stifling nature within me.
A. How short the law code of nations would be, if we could make it rigorously conform with nature's! How many vices and errors man would be spared!
B. Would you like to learn an abbreviated history of nearly all our misfortunes? Here it is. There was once a natural man; inside this man was introduced an artificial man; in the cavern arose a war continuing through his life. Sometimes the natural man is stronger; sometimes he is laid low by the moral and artificial man. . . . Yet there are extreme circumstances that bring man back to his first simplicity.

[15]See Deuteronomy 14:11–12.

A. Poverty and sickness, two great exorcists.
B. You have named them. . . .
A. . . . But tell me, in short: should we civilize man, or abandon him to his instincts?
B. Do you want a straight answer?
A. Of course.
B. If you propose to be his tyrant, civilize him; poison him as much as you can with a morality opposed to nature; forge all sorts of fetters for him; block his movements with a thousand obstacles; pursue him with terrifying phantoms; perpetuate the war in the cavern, and enchain the natural man always under the moral man's feet. But if you want him happy and free, do not meddle in his affairs—plenty of unforeseen events will lead him to enlightenment or depravity—and always recognize that it is not for you but for themselves that those shrewd lawgivers molded you into such factitious shapes. . . .
A. . . . What shall we do then? Shall we return to nature? Shall we submit to laws?
B. We shall speak out against senseless laws until they are reformed; meantime, submit to them. Anyone who breaks a bad law on his own authority authorizes everyone else to break good ones. It is less inconvenient to be mad among madmen than to be wise all alone. Let us tell ourselves, and never stop shouting, that shame, punishment, and ignominy have been attached to actions innocent in themselves. But let us not commit these actions, for shame, punishment, and ignominy are the greatest evils of all. Let us imitate the good chaplain: a monk in France, a savage in Tahiti. . . .

Nature's Divided Empire

Natural History, General and Particular,
by Georges Louis Leclerc, comte de Buffon,
translated by William Smellie

Georges Louis Leclerc (1707–88), whom Louis XV in 1771 made comte de Buffon, was born at Montbard in Burgundy on the estate of his father, a lawyer in nearby Dijon. Instead of the law, he studied astronomy and geometry. He traveled, between the ages of 20 and 25, in Italy and England, then returned to France, where he divided the rest of his life between Paris and Montbard. His earliest publications were translations of English scientific works, including one by Newton. From 1739 on he was keeper of the botanical garden in Paris, the Jardin du Roi (later the Jardin des Plantes), which provided him, along with his Burgundian estate, with rich material for botanical studies. His monumental Histoire naturelle (Natural History), *compiled in collaboration with several assistants, notably L.-J.-M. Daubenton, was published in forty-four volumes between 1749 and 1804; other important works include his* Discours sur le style (Discourse on Style, *1753) and* Les Epoques de la nature (The Epochs of Nature, *1778).*

Renowned for colossal vanity and belief that "the style is the man," he declared (William Wood remarks in introducing volume I of the twenty-volume English translation of the Natural History *by William Smellie [1812], from the third volume of which the following passages are taken), "I know but five great geniuses, Newton, Bacon, Leibnitz, Montesquieu, and* myself*"; and he repeatedly expressed disdain for his contemporary Linnaeus, the great Swedish taxonomist whose* Systema Naturae (System of Nature) *established the modern classificatory system for plants and animals. Though George Saintsbury, in his* Short History of French Literature

(6th ed., 1901), found Buffon's style so "pompous and inflated" as to "verge on the ridiculous," for Peter Gay, in The Science of Freedom *(1969), "His magnificent, expansive* Histoire naturelle *was an epic effort to write the biography of the world, the work of a man who was at once a skillful mathematician and meticulous stylist. . . . The pages of Buffon's vast and varied production testify to his deep, passionate attachment to nature."*

In Buffon's view, as Cassirer summarizes it in The Philosophy of the Enlightenment, *the mathematical concept of truth "loses its meaning and force as soon as we approach the sphere of the real and try to acclimate ourselves to this sphere," for "only experience can yield that kind of certainty of which the truth of physical objects is capable. . . . We must apply the principle of connection rather than that of analytical differentiation; instead of assigning living creatures to sharply distinguished species, we must study them in relation to their kinship, their transition from one type to another, their evolution and transformations. For these are the things which constitute life as we find it in nature." Buffon drew back from the seemingly evolutionary implications of some early statements and reaffirmed the traditional view—as Arthur O. Lovejoy writes in "Buffon and the Problem of Species"*[16]*—"not only that species are real entities, but also that they are constant and invariable entities"; and his conception of nature as a power animating the universe by God's consent, and of man as dividing "the empire of the world" between himself and nature by "right of conquest," echoes well-worn medieval and Baconian views. Nevertheless, there is in his works, Cassirer affirms, an incipient "transition to a conception of nature which no longer seeks to derive and explain becoming from being, but being from becoming," and which thus looks forward, however hesitantly, to more dynamic conceptions of biology that would culminate in Darwin.*

FROM *Of Nature: First View*

Nature is that system of laws established by the Creator for regulating the existence of bodies, and the succession of beings. Nature is not a body; for this body would comprehend everything. Neither is it a being; for this being would necessarily be God. But Nature may be considered as an immense living power, which animates the universe, and which, in subordination to the first and supreme Being, began to act by his command, and its action is still continued by his concurrence or consent. This power is that portion of the divine power which manifests itself to men. It is at once the cause and the effect, the mode and the substance, the design and the execution. Very different from human art, whose productions are only dead works, Nature is herself a work perpetually alive, an active and never ceasing operator, who knows how to employ every material, and, though always laboring on the same invariable plan, her power, instead of being lessened, is perfectly inexhaustible. Time, space, and matter are her means; the universe her object; motion and life her end.

The phenomena of the universe are the effects of this power. The springs she employs are active forces, which time and space can only measure and limit, but never destroy; forces which balance, mix, and oppose, without being able to annihilate each other. Some penetrate and transport bodies, others heat and animate them. Attraction and impulsion are the two principal instruments by which this power acts upon brute matter. Heat and organic particles are the active principles she employs in the formation and expansion of organized beings.

With such instruments, what can limit the operations of Nature? To render her om-

[16]Reprinted in *Forerunners of Darwin: 1745–1859*, ed. Bentley Glass et al. (1959), from an essay of 1911.

nipotent, she wants only the power of creating and annihilating. But these two extremes of power the Almighty has reserved to himself alone. To create and to annihilate are his peculiar attributes. To change, to destroy, to unfold, to renew, to produce, are the only privileges he has conferred upon another agent. Nature, the minister of his irrevocable commands, the depository of his immutable decrees, never deviates from the laws he has prescribed to her. She alters no part of his original plan; and, in all her operations, she exhibits the zeal of the eternal Lord of the universe. This divine impression, this unalterable prototype of all existence, is the model upon which she operates; a model, all the features of which are expressed in characters so strongly marked that nothing can possibly efface; a model which the number of copies or impressions, though infinite, instead of impairing, only renews.

Everything, therefore, has been created, and nothing is annihilated. Nature vibrates between these two extremes, without ever reaching either the one or the other. Let us endeavor to lay hold of her in some points of this vast space, which she has filled and pervaded from the beginning of ages. . . .

Nature is the external throne of the divine magnificence. Man, who contemplates her, rises gradually to the internal throne of the Almighty. Formed to adore his Creator, he has dominion over every creature. The vassal of heaven, the lord of the earth, he peoples, ennobles, and enriches this lower world. Among living beings he establishes order, subordination, and harmony. To Nature herself he even gives embellishment, cultivation, extension, and polish. He cuts down the thistle and the bramble, and he multiplies the vine and the rose. View those melancholy deserts where man has never resided. Overrun with briars, thorns, and trees which are deformed, broken, corrupted, the seeds that ought to renew and embellish the scene are choked and buried in the midst of rubbish and sterility. Nature, who, in other situations, assumes the splendor of youth, has here the appearance of old age and decrepitude. The earth, surcharged with the spoils of its productions, instead of a beautiful verdure, presents nothing but a disordered mass of gross herbage, and of trees loaded with parasitical plants, as lichens, agarics, and other impure fruits of corruption: all the low grounds are occupied with putrid and stagnating waters; the miry lands, which are neither solid nor fluid, are impassable, and remain equally useless to the inhabitants of the earth and of the waters; and the marshes, which are covered with stinking aquatic plants, serve only to nourish venomous insects, and to harbor impure animals. Between those putrid marshes which occupy the low grounds and the decayed forests which cover the elevated parts of the country, there is a species of lands, or savannas, that have no resemblance to our meadows. There noxious herbs rise and choke the useful kinds. Instead of that fine enameled turf, which appears to be the down of the earth, we see nothing but rude vegetables, hard prickly plants, so interlaced together that they seem to have less hold of the earth than of each other, and which, by successively drying and shooting, form a coarse mat of several feet in thickness. There is no road, no communication, no vestige of intelligence in these savage and desolate regions. Man, reduced to the necessity of following the tract of wild beasts when he wants to kill them, obliged to watch perpetually lest he should fall a victim to their rage, terrified by their occasional roarings, and even struck with the awful silence of those profound solitudes, he shrinks back and says: "Uncultivated nature is hideous and languishing. It is I alone who can render her agreeable and vivacious. Let us drain these marshes; let us animate these waters by converting them into brooks and canals; let us employ this active and devouring element, whose nature was formerly concealed from us; let us set fire to this cumbersome load of vegetables, and to those superannuated forests, which are already half consumed; let us finish the work by destroying with iron what could not be dissipated by fire. Instead of rushes, and water lilies, from which the toad is said to extract his poison, we shall soon see the ranunculus, the truffle, and other mild and salutary herbs; flocks of sprightly cattle will browse upon this land, which was formerly impassable; here they will find abundance of food, a never failing pasture, and they will continue to multiply and to reward us for our labors and the protection we have afforded them. To complete the work, let the ox be subjected to the yoke; let his strength and the weight of his body be employed in plowing the ground, which acquires fresh vigor by culture. Thus will Nature acquire redoubled strength and splendor from the skill and industry of man."

How beautiful is cultivated Nature! How pompous and brilliant, when decorated by the hand of man! He himself is her chief ornament, her noblest production. By multiplying his own species, he increases the most precious of her works. She even seems to multiply in the same proportion with him; for, by his art, he brings to light everything which she concealed in her bosom. What a source of unknown treasures! Flowers, fruits, and grains matured to perfection, and multiplied to infinity; the useful species of animals transported, propagated, and increased without number; the noxious kinds diminished, and banished from the abodes of men; gold, and iron, a more useful metal, extracted from the bowels of the earth; torrents restrained, and rivers directed and confined within their banks; even the ocean itself subdued, investigated, and traversed from the one hemisphere to the other; the earth everywhere accessible, and rendered active and fertile; the valleys and plains converted into smiling meadows, rich pastures, and cultivated fields; the hills loaded with vines and fruits, and their summits crowned with useful trees; the deserts turned into populous cities, whose inhabitants spread from its center to its utmost extremities; open and frequented roads and communications everywhere established, as so many evidences of the union and strength of society. A thousand other monuments of power and of glory sufficiently demonstrate that man is the lord of the earth; that he has entirely changed and renewed its surface; and that, from the remotest periods of time, he alone has divided the empire of the world between him and Nature.

He reigns, however, by the right of conquest only. He enjoys rather than possesses; and preserves his privileges by perpetual vigilance and activity. If these are interrupted, everything languishes, alters, and returns to the absolute dominion of Nature. She resumes her rights, effaces the operations of man, covers with moss and dust his most pompous monuments, which, in the progress of time, she totally destroys, and leaves him only the regret of having lost, by his own fault, what his ancestors had acquired by their industry. . . .

There Exists Nothing Beyond

System of Nature, by Paul Henri Thiry, baron d'Holbach,
translated by Samuel Wilkinson

Paul Heinrich Dietrich, or Paul Henri Thiry, baron d'Holbach (1723–89), was born at Hildesheim in the German Palatinate, but inherited a French barony in his youth and spent most of his life in Paris or at his nearby Château de Grandval; he was a contributor to the Encyclopedia *and a leading patron of the philosophes. He was a prolific propagandist whose many attacks on Christianity and religion in general included* Le Christianisme dévoilé (Christianity Unveiled, *1761). His best-known work was* Système de la Nature (System of Nature), *published in 1770 under the name of Jean-Baptiste de Mirabaud, an academician who had died ten years before; the translation dates from 1820–21. The basic tenets of his militantly atheistic materialism are clear in the passages that follow from Part One, affirming that "man is a being purely physical," and that nothing exists outside of nature, "the great whole that results from the collection of matter."*

Nature, for Holbach, "is a realm of complete determinism; it knows no 'order' and 'disorder,'" Basil Willey writes in The Eighteenth Century Background. *". . . The creator of such a world could only be regarded as a* moral *being by deliberately averting one's eyes from the facts, and arbitrarily attributing to him a number of human qualities. Yet we have only to call the 'actual world'* Nature, *and Holbach's tone changes to one of semi-religious exaltation. What would be odious as divine purpose becomes admirable as natural law." Such doctrinaire determinism is in conflict with Holbach's missionary atheism. His materialism, Cassirer remarks, "is no mere scientific or metaphysical dogma; it is rather an imperative. It not only aims to establish a thesis concerning the nature of things but also to command and to forbid." The* System of Nature *"is not content with theoretical conclusions but rather sets up a norm for human thought and faith."*

As Frederick the Great remarked at the time, "If everything is moved by necessary causes, then all counsel, all instruction, all rewards and punishments are as superfluous as inexplicable; for one might just as well preach to an oak and try to persuade it to turn to an orange tree." Voltaire too pointed out "the contradiction in the fact that Holbach, who dedicated his banner to the fight against dogmatism and intolerance, in turn set up his own thesis as dogma and defended it with fanatical zeal." Even Holbach's "peculiarly harsh and dry" style aimed "to eliminate from the philosophy of nature not only all religious, but all aesthetic elements as well," Cassirer observes, "and to neutralize all the forces of feeling and phantasy." His work, in sharp contrast to Diderot's, therefore repelled a younger generation still more than it did the deists of his own. "Of the System of Nature *Goethe relates that he and his friends could not understand how such a book could be dangerous: 'It seemed to us so grey, so Cimmerian, so deathlike that we could hardly stand its presence, and we shuddered at it as if it were a ghost.'" Against so bleak a vision of nature the Romantics would soon rise in full-throated protest and open rebellion.*

FROM PART I

FROM *Chapter I: Nature and Her Laws*

Man has always deceived himself when he abandoned experience to follow imaginary systems.—He is the work of nature.—He exists in Nature.—He is submitted to the laws of Nature.—He cannot deliver himself from them—cannot step beyond them even in thought. It is in vain his mind would spring forward beyond the visible world: direful and imperious necessity ever compels his return: being formed by Nature, he is circumscribed by her laws; there exists nothing beyond the great whole of which he forms a part, of which he experiences the influence. The beings his fancy pictures as above nature, or distinguished from her, are always chimeras formed after that which he has already seen, but of which it is utterly impossible he should ever form any finished idea, either as to the place they occupy, or their manner of acting—for him there is not, there can be nothing out of that Nature which includes all beings.

Therefore, instead of seeking out of the world he inhabits for beings who can procure him a happiness denied to him by Nature, let him study this Nature, learn her laws, contemplate her energies, observe the immutable rules by which she acts.—Let him apply these discoveries to his own felicity, and submit in silence to her precepts, which nothing can alter.—Let him cheerfully consent to be ignorant of causes hid from him under the most impenetrable veil.—Let him yield to the decrees of an universal power, which can never be brought within his comprehension, nor ever emancipate him from those laws imposed on him by his essence.

The distinction which has been so often made between the *physical* and the *moral* being is evidently an abuse of terms. Man is a being purely physical: the moral man is nothing more than this physical being considered under a certain point of view; that is to say, with relation to some of his modes of action, arising out of his individual organization. But is not this organization itself the work of Nature? . . . All that he does, all that he thinks, all that he is, all that he will be, is nothing more than what Universal Nature has made him. His ideas, his actions, his will, are the necessary effects of those properties infused into him by Nature, and of those circumstances in which she has placed him. In short, art is nothing but Nature acting with the tools she has furnished.

Nature sends man naked and destitute into this world which is to be his abode: he quickly learns to cover his nakedness—to shelter himself from the inclemencies of the weather, first with artlessly constructed huts, and the skins of the beasts of the forest; by degrees he mends their appearance, renders them more convenient; he establishes manufactories to supply his immediate wants; he digs clay, gold, and other fossils from the bowels of the earth; converts them into bricks for his house, into vessels for his use, gradually improves their shape, and augments their beauty. To a being exalted above our terrestrial globe, man would not appear less subjected to the laws of Nature when naked in the forest painfully seeking his sustenance than when living in civilized society surrounded with ease, or enriched with greater experience, plunged in luxury, where he every day invents a thousand new modes of supplying them. All the steps taken by man to regulate his existence ought only to be considered as a long succession of causes and effects, which are nothing more than the development of the first impulse given him by nature. . . .

Nature, therefore, in its most significant meaning, is the great whole that results from the collection of matter, under its various combinations, with that contrariety of motion which the universe presents to our view. Nature, in a less extended sense, or considered in each individual, is the whole that results from its essence; that is to say, the peculiar qualities, the combination, the impulse, and the various modes of action by which it is discriminated from other beings. It is thus that man is, as a whole, or in his nature, the result of a certain combination of matter, endowed with peculiar properties, competent to give, capable of receiving, certain impulses, the arrangement of which is called *organization;* of which the essence is to feel, to think, to act, to move after a manner distinguished from other beings with which he can be compared. . . .

Having described the proper definition that should be applied to the word Nature, I must advise the reader, once for all, that whenever in the course of this work the expression occurs that "Nature produces such and such an effect," there is no intention of personifying that nature which is purely an abstract being; it merely indicates that the effect spoken of necessarily springs from the peculiar properties of those beings which compose the mighty macrocosm. When, therefore, it is said, "Nature demands that man should pursue his own happiness," it is to prevent circumlocution—to avoid tautology;

it is to be understood that it is the property of a being that feels, that thinks, that acts, to labor to its own happiness; in short, that is called *natural* which is conformable to the essence of things, or to the laws which Nature prescribes to the beings she contains, in the different orders they occupy, under the various circumstances through which they are obliged to pass. . . . In short, the *essence* of a being is its particular, its individual nature. . . .

FROM *Chapter III: Of Matter . . .*

. . . If we contemplate a little the paths of Nature—if, for a time, we trace the beings in this Nature, under the different states through which, by reason of their properties, they are compelled to pass—we shall discover that it is to motion, and to motion only, that is to be ascribed all the changes, all the combinations, all the forms, in short, all the various modifications of matter. That it is by motion everything that exists is produced, experiences change, expands, and is destroyed. It is motion that alters the aspect of beings; that adds to, or takes away from their properties; which obliges each of them, by a consequence of its nature, after having occupied a certain rank or order, to quit it, to occupy another, and to contribute to the generation, maintenance, and decomposition of other beings, totally different in their bulk, rank, and essence.

In what experimental philosophers have styled the "three orders of nature," that is to say, the *mineral*, the *vegetable*, and *animal* worlds, they have established, by the aid of motion, a transmigration, an exchange, a continual circulation in the particles of matter. Nature has occasion in one place, for those particles which, for a time, she has placed in another. These particles, after having, by particular combinations, constituted beings endued with peculiar essences, with specific properties, with determinate modes of action, dissolve and separate with more or less facility; and combining in a new manner, they form new beings. The attentive observer sees this law execute itself, in a manner more or less prominent, through all the beings by which he is surrounded. . . .

Animals, plants, and minerals, after a lapse of time, give back to Nature, that is to say, to the general mass of things, to the universal magazine, the elements, or principles, which they have borrowed: the earth retakes that portion of the body of which it formed the basis and the solidity; the air charges itself with those parts that are analogous to it, and with those particles which are light and subtle; water carries off that which is suitable to liquescency; fire, bursting its chains, disengages itself, and rushes into new combinations with other bodies.

The elementary particles of the animal, being thus dissolved, disunited, and dispersed, assume new activity, and form new combinations: thus, they serve to nourish, to preserve, or destroy new beings; among others, plants, which arrived at their maturity, nourish and preserve new animals; these in their turn yielding to the same fate as the first.

Such is the constant, the invariable course of Nature; such is the eternal circle of mutation, which all that exists is obliged to describe. It is thus motion generates, preserves for a time, and successively destroys one part of the universe by the other, whilst the sum of existence remains eternally the same. . . .

FROM *Chapter VI: Moral and Physical Distinctions of Man . . .*

. . . Thus, when it shall be inquired, what is man?

We say, he is a material being, organized after a peculiar manner; conformed to a certain mode of thinking—of feeling; capable of modification in certain modes peculiar to himself—to his organization—to that particular combination of matter which is found assembled in him.

If, again, it be asked, what origin we give to beings of the human species?

We reply that, like all other beings, man is a production of Nature, who resembles them in some respects, and finds himself submitted to the same laws; who differs from them in other respects, and follows particular laws, determined by the diversity of his conformation. . . .

With respect to those who may ask why Nature does not produce new beings, we may inquire of them in turn, upon what foundation they suppose this fact? What it is that authorizes them to believe this sterility in Nature? Know they if, in the various combinations which she is every instant forming, Nature be not occupied in producing new beings, without the cognizance of these observers? Who has informed them that this Nature is not actually assembling, in her immense elaboratory, the elements suitable to bring to light generations entirely new, that will have nothing in common with those of the species at present existing? What absurdity then, or what want of just inference would there be, to imagine that the man, the horse, the fish, the bird, will be no more? Are these animals so indispensably requisite to Nature that without them she cannot continue her eternal course? Does not all change around us? Do we not ourselves change? Is it not evident that the whole universe has not been, in its anterior eternal duration, rigorously the same that it now is? that it is impossible, in its posterior eternal duration, it can be rigidly in the same state that it now is for a single instant? How, then, pretend to divine that to which the infinite succession of destruction, of reproduction, of combination, of dissolution, of metamorphosis, of change, of transposition, may be able eventually to conduct it by their consequence? Suns encrust themselves, and are extinguished; planets perish and disperse themselves in the vast plains of air; other suns are kindled, and illumine their systems; new planets form themselves, either to make revolutions round these suns, or to describe new routes; and man, an infinitely small portion of the globe, which is itself but an imperceptible point in the immensity of space, vainly believes it is for himself this universe is made; foolishly imagines he ought to be the confidant of Nature; confidently flatters himself he is eternal; and calls himself King of the Universe!!! . . .

Let us then conclude that man has no just, no solid reason to believe himself a privileged being in Nature, because he is subject to the same vicissitudes as all her other productions. His pretended prerogatives have their foundation in error, arising from mistaken opinions concerning his existence. Let him but elevate himself by his thoughts above the globe he inhabits, he will look upon his own species with the same eyes he does all other beings in Nature: he will then clearly perceive that in the same manner that each tree produces its fruit, by reason of its energies, in consequence of its species, so each man acts by reason of his particular energy; that he produces fruit, actions, works, equally necessary: he will feel that the illusion which he anticipates in favor of himself arises from his being, at one and the same time, a spectator and a part of the universe. He will acknowledge that the idea of excellence which he attaches to his being has no other foundation than his own peculiar interest; than the predilection he has in favor of himself—that the doctrine he has broached with such seeming confidence bottoms itself on a very suspicious foundation, namely ignorance and self-love.

The Spectacle of Nature and the Inner Voice

Selections from the Writings of Jean-Jacques Rousseau

Jean-Jacques Rousseau (1712–78) was born in Geneva, the son of a watchmaker. His mother died in childbirth, and his father was expelled from Geneva ten years later. Apprenticed to an engraver in 1727, Rousseau fled Geneva the following year and found shelter with Mme Louise Eléonore de Warens; she sent him to Turin, where he became, for a time, a convert to Roman Catholicism. From 1732 to 1740, after several years' wandering as a tutor and music teacher, he spent an idyllic period, as he later portrayed it in his Confessions, *as protégé of Mme de Warens at Les Charmettes near Chambéry. In 1741 he went to Paris, where Diderot invited him to write on music for the* Encyclopedia; *from 1743 to 1744 he was secretary to the French ambassador to Venice; and in 1746 he began his long liaison with the servant girl Thérèse Levasseur, who bore him as many as five children, all of whom he abandoned to orphanages.*

In 1749, on his way to visit Diderot in the château de Vincennes (as Rousseau recounted later), he pulled from his pocket the Mercure de France. *On reading the question posed by the Académie de Dijon as the subject of its prize essay—"Has the progress of the sciences and arts contributed to the corruption or to the improvement of human conduct?"—he was inspired to show "that man is naturally good" and only by the contradictions of the social system becomes evil; the resulting* Discours sur les sciences et les arts (Discourse on the Sciences and Arts) *of 1750 won the prize and made Rousseau famous. His* Discours sur l'origine et les fondements de l'inégalité parmi les hommes (Discourse on the Origin and Foundations of Inequality among Men), *in which he further developed his ideas concerning the social conflicts disrupting man's former oneness with nature, followed in 1754.*

Having quarreled with Diderot and other philosophes, Rousseau spent his most productive years as a writer, from 1757 to 1762, at Mme d'Epinay's estate, the Hermitage, and then with the maréchal de Luxembourg at Montmorency; here he wrote his novel Julie, ou La Nouvelle Héloïse *(1761), his educational romance* Emile *(1762), and his political treatise* Du contrat social *(*The Social Contract, *1762). The latter two were burned by order of the Parlement of Paris and condemned in Geneva, forcing Rousseau to flee arrest; after his house at Motiers near Neuchâtel was stoned by a mob in 1765, he spent several happy months on the small Ile de Saint-Pierre on the Lake of Bienne until he was expelled from the canton of Bern. He then took refuge in England at the invitation of David Hume, but his increasingly paranoid delusions soon led him to accuse his host of treachery. His autobiographical* Confessions *were completed in 1770. In that year he returned to Paris; between 1776 and 1778 he wrote* Les Rêveries du promeneur solitaire *(*Reveries of the Solitary Walker, *1782). He died at Ermenonville, near Paris, in July of 1778.*

No aspect of this tormented genius's seemingly contradictory thought and experience had greater impact than his attitudes toward nature. Often, both by admirers and detractors, Rousseau's ideas were misunderstood as exaltation of the "noble savage" or yearning to go "back to nature" in some prelapsarian state of primal oneness. But the phrase "noble savage" is not Rousseau's. It goes back to John Dryden's heroic tragedy of 1670, The Conquest of Granada *(Part One), in whose opening scene the brave Almanzor declares:*

> I am as free as nature first made man,
> Ere the base laws of servitude began,
> When wild in woods the noble savage ran.

Not Rousseau but Diderot, in his uncorrupted Tahitian, idealized "savage" figures such as the hero of Aphra Behn's Oronooko: or, The History of the Royal Slave *(1698), who maintains his native greatness of soul (partly attributable to European influence!) even in captivity. Rousseau identified, instead, with Defoe's Robinson Crusoe (Chapter 19 above), who strives to make the initially alien surroundings of a desert island his own.*

"Rousseau was neither the only nor the first man in the eighteenth century to coin the motto, 'Back to Nature!'" Cassirer emphasizes in The Question of Jean-Jacques Rousseau *(1932; English trans. 1954). Only sporadically, despite his recurrent nostalgia for an imagined paradise lost, did Rousseau maintain the illusion that return to an earlier oneness with the natural world was possible for civilized man. In his writings as a whole, "it is never entirely clear to what extent his notion of a state of nature is 'ideal' and to what extent it is 'empirical'," as Cassirer remarks in* Rousseau, Kant, Goethe *(English trans. 1945). Yet in his early* Discourse on the Origin of Inequality *(1755), from which our first selections come,*[17] *Rousseau even more explicitly than Hobbes (Chapter 20 above) asserts that the state of nature he portrays is not factual but hypothetical. "Men like me," far from attempting to live in the forest among bears, "will respect the sacred bonds of their respective communities" even while recognizing that "nature makes us pay for the contempt with which we have treated her lessons."*

In the state of nature, in contrast to every society that has succeeded it, "appearance and reality were in perfect equilibrium," Jean Starobinski observes in Jean-Jacques Rousseau: Transparency and Obstruction *(1971; English trans. 1988). As in myths of many tribal peoples (Chapters 1 and 2 above), men shared their existence with the animals. But since it is impossible to return to this condition (if it ever existed), "Rousseau directed his energy," Peter Gay writes in* The Science of Freedom, *"toward discovering not a state of nature without culture but a culture that would realize man's true nature." The natural man is the man whose nature remains uncorrupted by the society to which he belongs: "remember, in the first place," Rousseau affirms in Book IV of* Emile, *"that when I want to train a natural man, I do not want to make him a savage and to send him back to the woods, but that living in the whirl of social life it is enough that he should not let himself be carried away by the passions and prejudices of men; let him see with his eyes and feel with his heart, let him own no sway but that of reason." Thus, as Starobinski affirms, "despite his nostalgia, Rousseau is not a 'primitivist.' While it might have been better if man had never abandoned his primitive condition, the choice has been made."*

Similarly complex attitudes find expression in Rousseau's epistolary novel Julie, or The New Eloïse. *(In the anonymous translation entitled* Eloisa, *published in London in the same year, 1761, as the French original, Rousseau's heroine Julie—associated by his subtitle with the medieval Eloïse loved by Abelard—is renamed "Eloisa.") In our second selection, from a letter from Saint-Preux to Julie evoking the stupendous clouds, vast cascade, and yawning abyss of an Alpine landscape in the Swiss Valais, with its "surprising mixture of wild and cultivated nature," thoughts and memories of the "divine maid" Julie pursue the lover—as thoughts of Laura had pursued Petrarch (Chapter 15 above)—and blend in his mind with the sublimity of nature. Nature thus becomes in part a subjective construct, an outward correlative to his inner mood: "it is in man's heart," Rousseau writes in* Emile, *"that the spectacle of nature lives." As Cassirer observes in* The Question of Jean-Jacques Rousseau, *"Here man no longer simply stands 'over against' nature—nature is not a drama which he enjoys as a mere spectator and observer; he dips into its inner life and vibrates with its own rhythms." (Seen from a more negative perspective, as by Irving Babbitt in* Rousseau and Romanticism *[1919], "The nature over which the Rous-*

[17]In the anonymous translation in the Everyman's Library edition of *The Social Contract and Discourses* (1913).

seauist is bent in such rapt contemplation plays the part of the pool in the legend of Narcissus. It renders back to him his own image. He sees in nature what he himself has put there.")

Elsewhere in this novel, however (in scenes not here included), the ideal is as much cultural as natural, or rather, an inseparable blend of the two. If the Alpine scene reflected the rapture of Saint-Preux's soon-to-be-consummated love for Julie, later scenes—in the locus amoenus *of Julie's Elysian grove (letter 11 of Part IV) and still more in the family life she leads at Clarens on the modest estate of her husband, the wise, fatherly, quasi-divine (though atheistic) M. de Wolmar, to whom she remains faithful until death despite unextinguished love for Saint-Preux—paint a picture of ideal domestic simplicity in accord with the rhythms of the natural world. Such a portrayal is reminiscent as much of classical models from Virgil, Horace, Tibullus, and Claudian (Chapter 11 above) as of Rousseau's own retrospectively happy but isolated existence at Les Charmettes and the Hermitage, as remembered in his* Confessions, *or on the Ile de Saint-Pierre.*

In his idyll of life at Clarens in letter 2 of Part V, "the condition natural to man is to cultivate the earth and live off its fruits," as Virgil's happy farmers had done; in Julie's "simple and modest home," and among the "good and simple" people around her, is a "true magnificence" such as the wealthy can never know. And in the gaiety of the vintage festival glowingly re-created in letter 7 of Part V, the "sweet equality" of peasants and masters "re-establishes the order of nature." "Thanks to Wolmar's wisdom," and to Julie's goodness, Ronald Grimsley observes in Rousseau and the Religious Quest *(1968), "each person finds his place in this ordered world which seems to contain 'all the charms of the golden age.' . . . Through their presence the inhabitants of Clarens are given the innocence, transparence, peace, beauty, and virtue that make them worthy of the 'supreme felicity' of paradise"—a felicity that the "passional ambivalence" (in Starobinski's phrase) of Julie and Saint-Preux's love intensifies by its continual reminder of how precarious this harmony is. Not nature alone is sufficient to restore human oneness, but only nature deliberately shaped by human wisdom and assimilated to human ends. "Nature is restored," Starobinski writes, quoting Kant, "when art and culture attain their highest perfection: 'Consummate art becomes nature anew.'"*

Nowhere are the latent tensions of Rousseau's view of nature more evident than in Emile, *where the elaborate effort to provide the young Emile with a "natural" education will strike most modern readers as almost ludicrously artificial. "Is this attempt not doomed to failure from the outset," Cassirer asks in* The Question of Jean-Jacques Rousseau, *since "the fanatical love of truth, which was to guide this system of education, ends up by degenerating into a curiously complicated system of deceptions, of carefully calculated pedagogical tricks."*

At the heart of this fictionalized treatise stands the Savoyard vicar's deistic profession of faith—from which our third selection is taken, in Arthur H. Beattie's translation, The Creed of a Priest of Savoy *(2nd ed., 1957)—in a governing intelligence, or God, revealed to reason and the inner voice of conscience not by external authority or revelation but by the spectacle of nature's order. Paradoxically, by his emphasis more on the inner voice than on the outward spectacle, Cassirer contends in* Rousseau, Kant, Goethe, *"Rousseau as well as Kant put the ethical aspect so much in the center of religion that both almost lost sight of nature." If Rousseau's "feeling for nature awakens and strengthens in him the feeling for religion, it does not enter immediately into its content. . . . Any kind of mere deification of nature is alien to Rousseau," in contrast, for example, to Shaftesbury (Chapter 20). "The religion that Rousseau is teaching and proclaiming in the* Profession of Faith *does not arise from absorption in the wonders of nature," essential though these are to his conviction of God's existence; for "the real miracle that is central for him is the miracle of human freedom and conscience as the evidence for this freedom." (As*

Starobinski observes of Rousseau's philosophy in general, "He seeks to discover human nature *but will have nothing to do with efforts to discover the reality of physical nature. . . . The only truth accessible to us is in our ideas or sensations or sentiments, that is, in consciousness.")*

Finally, our last selection is the Fifth Walk from Reveries of the Solitary Walker, *translated from John S. Spink's edition of* Les Rêveries du promeneur solitaire *(1948). In this poetic evocation of Rousseau's brief stay on the Ile de Saint-Pierre—an experience also described in Book XII of the* Confessions*—Rousseau's love of nature corresponds, Grimsley remarks in* Jean-Jacques Rousseau *(1961; 2nd ed. 1969), "to a profound need for a kind of primordial unity, security and innocence which overcomes the fear and anguish still associated with his reactions to the hazardous world of human relations. . . . Neither the physical movement nor the 'inner movements' constitute by themselves the state of reverie, but their co-existence, which leads to a 'continuous' movement, is enough to set free the deeper level of the personality for the enjoyment of its own existence." Here the spectacle of nature is one with the inner voice that transmutes it into a memory no change of place or circumstance can destroy. "In immediate contact with himself," as Starobinski writes, Rousseau "is also in immediate contact with all nature," and of this transport "on the wings of imagination" not even his bitterest enemies can deprive him.*

FROM A Discourse on the Origin of Inequality

. . . The philosophers, who have inquired into the foundations of society, have all felt the necessity of going back to a state of nature; but not one of them has got there. . . .

Let us begin then by laying facts aside, as they do not affect the question. The investigations we may enter into in treating this subject must not be considered as historical truths, but only as mere conditional and hypothetical reasonings, rather calculated to explain the nature of things than to ascertain their actual origin: just like the hypotheses which our physicists daily form respecting the formation of the world. . . .

O man, of whatever country you are, and whatever your opinions may be, behold your history, such as I have thought to read it, not in books written by your fellow-creatures, who are liars, but in nature, which never lies. All that comes from her will be true; nor will you meet with anything false, unless I have involuntarily put in something of my own. The times of which I am going to speak are very remote: how much are you changed from what you once were! It is, so to speak, the life of your species which I am going to write, after the qualities which you have received, which your education and habits may have depraved, but cannot have entirely destroyed. There is, I feel, an age at which the individual man would wish to stop: you are about to inquire about the age at which you would have liked your whole species to stand still. Discontented with your present state, for reasons which threaten your unfortunate descendants with still greater discontent, you will perhaps wish it were in your power to go back; and this feeling should be a panegyric on your first ancestors, a criticism of your contemporaries, and a terror to the unfortunates who will come after you. . . .

From THE FIRST PART . . . If we strip this being, thus constituted, of all the supernatural gifts he may have received, and all the artificial faculties he can have acquired only by a long process; if we consider him, in a word, just as he must have come from the hands of nature, we behold in him an animal weaker than some, and less agile than others, but, taking him all round, the most advantageously organized of any. I see him satisfying his hunger at the first oak, and slaking his thirst at the first brook; finding his bed at the foot of the tree which afforded him a repast; and, with that, all his wants supplied. . . .

We should beware . . . of confounding the savage man with the men we have daily before our eyes. Nature treats all the animals left to her care with a predilection that seems to show how jealous she is of that right. The horse, the cat, the bull, and even the ass are generally of greater stature, and always more robust, and have more vigor, strength, and courage, when they run wild in the forests than when bred in the stall. By becoming domesticated, they lose half these advantages; and it seems as if all our care to feed and treat them well serves only to deprave them. It is thus with man also: as he becomes sociable and a slave, he grows weak, timid, and servile; his effeminate way of life totally enervates his strength and courage. To this it may be added that there is still a greater difference between savage and civilized man than between wild and tame beasts: for men and brutes having been treated alike by nature, the several conveniences in which men indulge themselves still more than they do their beasts are so many additional causes of their deeper degeneracy. . . .

Hitherto I have considered merely the physical man; let us now take a view of him on his metaphysical and moral side.

I see nothing in any animal but an ingenious machine, to which nature has given senses to wind itself up, and to guard itself, to a certain degree, against anything that might tend to disorder or destroy it. I perceive exactly the same things in the human machine, with this difference, that in the operations of the brute, nature is the sole agent, whereas man has some share in his own operations, in his character as a free agent. The one chooses and refuses by instinct, the other from an act of free will: hence the brute cannot deviate from the rule prescribed to it, even when it would be advantageous for it to do so; and, on the contrary, man frequently deviates from such rules to his own prejudice. Thus a pigeon would be starved to death by the side of a dish of the choicest meats, and a cat on a heap of fruit or grain, though it is certain that either might find nourishment in the foods which it thus rejects with disdain, did it think of trying them. Hence it is that dissolute men run into excesses which bring on fevers and death, because the mind depraves the senses, and the will continues to speak when nature is silent.

Every animal has ideas, since it has senses; it even combines those ideas in a certain degree; and it is only in degree that man differs, in this respect, from the brute. Some philosophers have even maintained that there is a greater difference between one man and another than between some men and some beasts. It is not, therefore, so much the understanding that constitutes the specific difference between the man and the brute as the human quality of free agency. Nature lays her commands on every animal, and the brute obeys her voice. Man receives the same impulsion, but at the same time knows himself at liberty to acquiesce or resist: and it is particularly in his consciousness of this liberty that the spirituality of his soul is displayed. For physics may explain, in some measure, the mechanism of the senses and the formation of ideas; but in the power of willing or rather of choosing, and in the feeling of this power, nothing is to be found but acts which are purely spiritual and wholly inexplicable by the laws of mechanism.

However, even if the difficulties attending all these questions should still leave room for difference in this respect between men and brutes, there is another very specific quality which distinguishes them, and which will admit of no dispute. This is the faculty of self-improvement, which, by the help of circumstances, gradually develops all the rest of our faculties, and is inherent in the species as in the individual: whereas a brute is, at the end of a few months, all he will ever be during his whole life, and his species, at the end of a thousand years, exactly what it was the first year of that thousand. Why is man alone liable to grow into a dotard? Is it not because he returns, in this, to his primitive state; and that, while the brute, which has acquired nothing and has therefore nothing to lose, still retains the force of instinct, man, who loses, by age or accident, all that his *perfectibility* had enabled him to gain, falls by this means lower than the brutes themselves? It would be melancholy, were we forced to admit that this distinctive and almost unlimited faculty is the source of all human misfortunes; that it is this which, in time, draws man out of his original state, in which he would have spent his days insensibly in peace and innocence; that it is this faculty which, successively producing in different ages his discoveries and his errors, his vices and his virtues, makes him at length a tyrant both over himself and over nature.[18] . . .

Savage man, left by nature solely to the direction of instinct, or rather indemnified for what he may lack by faculties capable at first of supplying its place, and afterwards of raising him much above it, must accordingly begin with purely animal functions: thus seeing and feeling must be his first condition, which would be common to him and all other animals. . . . The only goods he recognizes in the universe are food, a female, and sleep: the only evils he fears are pain and hunger. I say pain, and not death: for no animal can know what it is to die; the knowledge of death and its terrors being one of the first acquisitions made by man in departing from an animal state. . . .

But who does not see, without recurring to the uncertain testimony of history, that everything seems to remove from savage man both the temptation and the means of changing his condition? His imagination paints no pictures; his heart makes no demands on him. His few wants are so readily supplied, and he is so far from having the knowledge which is needful to make him want more, that he can have neither foresight nor curiosity. The face of nature becomes indifferent to him as it grows familiar. He sees in it always the same order, the same successions: he has not understanding enough to wonder at the greatest miracles; nor is it in his mind that we can expect to find that philosophy man needs, if he is to know how to notice for once what he sees every day. His soul, which nothing disturbs, is wholly wrapped up in the feeling of its present existence, without any idea of the future, however near at hand; while his projects, as limited as his views, hardly extend to the close of day. . . .

. . . It is in fact impossible to conceive why, in a state of nature, one man should stand more in need of the assistance of another, than a monkey or a wolf of the assistance of another of its kind: or, granting that he did, what motives could induce that other to assist him; or, even then, by what means they could agree about the conditions. I know it is incessantly repeated that man would in such a state have been the most miserable of creatures; and indeed, if it be true, as I think I have proved, that he must have lived many ages before he could have either desire or an opportunity of emerging from it, this would only be an accusation against nature, and not against the being which she had thus unhappily constituted. But as I understand the word *miserable*, it either has no meaning at all, or else signifies only a painful privation of something, or a state of suffering either in body or soul. I should be glad to have explained to me what kind of misery a free being, whose heart is at ease and whose body is in health, can possibly suffer. I would ask also, whether a social or a natural life is most likely to become insupportable to those who enjoy it. We see around us hardly a creature in civil society who does not lament his existence: we even see many deprive themselves of as much of it as they can, and laws human and divine together can hardly put a stop to the disorder. I ask, if it was ever known that a savage took it into his head, when at liberty, to complain of life or to make

[18]See Appendix (below).

away with himself. Let us therefore judge, with less vanity, on which side the real misery is found. On the other hand, nothing could be more unhappy than savage man, dazzled by science, tormented by his passions, and reasoning about a state different from his own. It appears that Providence most wisely determined that the faculties which he potentially possessed should develop themselves only as occasion offered to exercise them, in order that they might not be superfluous or perplexing to him, by appearing before their time, nor slow and useless when the need for them arose. In instinct alone, he had all he required for living in the state of nature; and with a developed understanding he has only just enough to support life in society. . . .

Above all, let us not conclude, with Hobbes, that because man has no idea of goodness, he must be naturally wicked; that he is vicious because he does not know virtue; that he always refuses to do his fellow-creatures services which he does not think they have a right to demand; or that by virtue of the right he truly claims to everything he needs, he foolishly imagines himself the sole proprietor of the whole universe. Hobbes had seen clearly the defects of all the modern definitions of natural right: but the consequences which he deduces from his own show that he understands it in an equally false sense. In reasoning on the principles he lays down, he ought to have said that the state of nature, being that in which the care for our own preservation is the least prejudicial to that of others, was consequently the best calculated to promote peace, and the most suitable for mankind. He does say the exact opposite, in consequence of having improperly admitted, as a part of savage man's care for self-preservation, the gratification of a multitude of passions which are the work of society, and have made laws necessary. . . .

. . . Mandeville well knew that, in spite of all their morality, men would have never been better than monsters, had not nature bestowed on them a sense of compassion, to aid their reason: but he did not see that from this quality alone flow all those social virtues of which he denied man the possession. . . . Compassion must, in fact, be the stronger, the more the animal beholding any kind of distress identifies himself with the animal that suffers. Now, it is plain that such identification must have been much more perfect in a state of nature than it is in a state of reason. It is reason that engenders self-respect, and reflection that confirms it: it is reason which turns man's mind back upon itself, and divides him from everything that could disturb or afflict him. It is philosophy that isolates him, and bids him say, at sight of the misfortunes of others: "Perish if you will, I am secure." Nothing but such general evils as threaten the whole community can disturb the tranquil sleep of the philosopher, or tear him from his bed. A murder may with impunity be committed under his window; he has only to put his hands to his ears and argue a little with himself to prevent nature, which is shocked within him, from identifying with the unfortunate sufferer. Uncivilized man has not this admirable talent; and for want of reason and wisdom is always foolishly ready to obey the first promptings of humanity. . . .

Men in a state of nature being confined merely to what is physical in love, and fortunate enough to be ignorant of those excellences which whet the appetite while they increase the difficulty of gratifying it, must be subject to fewer and less violent fits of passion, and consequently fall into fewer and less violent disputes. The imagination, which causes such ravages among us, never speaks to the heart of savages, who quietly await the impulses of nature, yield to them involuntarily, with more pleasure than ardor, and, their wants once satisfied, lose the desire. It is therefore incontestable that love, as well as all other passions, must have acquired in society that glowing impetuosity which makes it so often fatal to mankind. . . .

Let us conclude then that man in a state of nature, wandering up and down the forests, without industry, without speech, and without home, an equal stranger to war and to all ties, neither standing in need of his fellow-creatures nor having any desire to hurt them, and perhaps even not distinguishing them one from another; let us conclude that, being self-sufficient and subject to so few passions, he could have no feelings or knowledge but such as befitted his situation; that he felt only his actual necessities, and disregarded everything he did not think himself immediately concerned to notice, and that his understanding made no greater progress than his vanity. If by accident he made any discovery, he was the less able to communicate it to others, as he did not know even his own children. Every art would necessarily perish with its inventor, where there was no kind of education among men, and generations succeeded generations without the least advance; when, all setting out from the same point, centuries must have elapsed in the barbarism of the first ages; when the race was already old, and man remained a child. . . .

From APPENDIX . . . That men are actually wicked, a sad and continual experience of them proves beyond doubt: but, all the same, I think I have shown that man is naturally good. What then can have depraved him to such an extent, except the changes that have happened in his constitution, the advances he has made, and the knowledge he has acquired? We may admire human society as much as we please; it will be none the less true that it necessarily leads men to hate each other in proportion as their interests clash, and to do one another apparent services, while they are really doing every imaginable mischief. What can be thought of a relation in which the interest of every individual dictates rules directly opposite to those the public reason dictates to the community in general—in which every man finds his profit in the misfortunes of his neighbor? . . .

Compare without partiality the state of the citizen with that of the savage, and trace out, if you can, how many inlets the former has opened to pain and death, besides those of his vices, his wants, and his misfortunes. If you reflect on the mental afflictions that prey on us, the violent passions that waste and exhaust us, the excessive labor with which the poor are burdened, the still more dangerous indolence to which the wealthy give themselves up, so that the poor perish of want, and the rich of surfeit; . . . in a word, if you add together all the dangers with which these causes are always threatening us, you will see how dearly nature makes us pay for the contempt with which we have treated her lessons. . . .

What, then, is to be done? Must societies be totally abolished? Must *meum* and *tuum* ["mine" and "thine"] be annihilated, and must we return again to the forests to live among bears? This is a deduction in the manner of my adversaries, which I would as soon anticipate as let them have the shame of drawing. O you, who have never heard the voice of heaven, who think man destined only to live this little life and die in peace; you, who can resign in the midst of populous cities your fatal acquisitions, your restless spirits, your corrupt hearts and endless desires; resume, since it depends entirely on yourselves, your ancient and primitive innocence: retire to the woods, there to lose the sight and remembrance of the crimes of your contemporaries; and be not apprehensive of degrading your species by renouncing its advances in order to renounce its vices. As for men like me, whose passions have destroyed their original simplicity, who can no longer subsist on plants or acorns, or live without laws and magistrates; . . . those, in short, who are persuaded that the Divine Being has called all mankind to be partakers in the happiness and perfection of celestial intelligences, all these will endeavor to merit the eternal prize they are to expect from the practice of those virtues which they make themselves follow in learning to know them. They will respect the sacred bonds of their respective communities; they will love their fellow-citizens, and serve them with all their might; they will scrupulously obey the laws, and all those who make or administer them; they will particularly honor those wise and good princes who find means of preventing, curing, or even palliating all these evils and abuses by which we are constantly threatened; they will animate the zeal of their deserving rulers by showing them, without flattery or fear, the importance of their office and the severity of their duty. But they will not therefore have less contempt for a constitution that cannot support itself without the aid of so many splendid characters, much oftener wished for than found, and from which, notwithstanding all their pains and solicitude, there always arise more real calamities than even apparent advantages.

FROM Julie, or The New Eloïse *(in the contemporary translation entitled* Eloisa*), Part One, Letter XXIII (from Saint-Preux to Julie)*

. . . I set out, dejected with my own sufferings, but consoled with your joy; which held me suspended in a state of languor that is not disagreeable to true sensibility. Under the conduct of a very honest guide, I crawled up the towering hills through many a rugged unfrequented path. Often would I muse, and then, at once, some unexpected object caught my attention. One moment I beheld stupendous clouds hanging ruinous over my head; the next, I was enveloped in a drizzling cloud, which arose from a vast cascade that dashing thundered against the rocks below my feet; on one side, a perpetual torrent opened to my view a yawning abyss, which my eyes could hardly fathom with safety; sometimes I was lost in the obscurity of a hanging wood, and then was agreeably astonished with the sudden opening of a flowery plain. A surprising mixture of wild and cultivated nature points out the hand of man, where one would imagine man had never penetrated. Here you behold a horrid cavern, and there a human habitation; vineyards where one would expect nothing but brambles; delicious fruit among barren rocks, and corn fields in the midst of cliffs and precipices.

But it is not labor only that renders this strange country so wonderfully contrasted; for here nature seems to have a singular pleasure in acting contradictory to herself, so different does she appear in the same place, in different aspects. Towards the east, the flowers of spring; to the south, the fruits of autumn; and northwards the ice of winter. She unites all the seasons in the same instant, every climate in the same place, different soils on the same land, and with a harmony elsewhere unknown, joins the produces of the plains to those of the highest Alps. Add to these the illusions of vision, the tops of the mountains variously illumined, the harmonious mixture of light and shade, and their different effects in the morning and the evening as I traveled; you may then form some idea of the scenes which engaged my attention, and which seemed to change, as I passed, as on an enchanted theater; for the prospect of mountains being almost perpendicular to the horizon strikes the eye at the same instant, and more powerfully, than that of a plane, where the objects are seen obliquely and half concealed behind each other.

To this pleasing variety of scenes I attributed the serenity of my mind during my first day's journey. I wondered to find that inanimate beings should overrule our most violent passions, and despised the impotence of philosophy for having less power over the soul than a succession of lifeless objects. But finding that my tranquility continued during the night, and even increased with the following day, I began to believe it flowed from some other source, which I had not yet discovered. That day I reached the lower mountains, and passing over their rugged tops, at last ascended the highest summit I could possibly attain. Having walked a while in the clouds, I came to a place of greater serenity, whence one may peacefully observe the thunder and the storm gathering below: ah! too flattering picture of human wisdom, of which the original never existed, except in those sublime regions whence the emblem is taken.

Here it was that I plainly discovered, in the purity of the air, the true cause of that returning tranquility of soul to which I had been so long a stranger. This impression is general, though not universally observed. Upon the tops of mountains, the air being subtle and pure, we respire with greater freedom, our bodies are more active, our minds more serene, our pleasures less ardent, and our passions much more moderate. Our meditations acquire a degree of sublimity from the grandeur of the objects around us. It seems as if, being lifted above all human society, we had left every low, terrestrial sentiment behind; and that as we approach the aethereal regions, the soul imbibes something of their eternal purity. One is grave without being melancholy, peaceful but not indolent, pensive yet contented: our desires lose their painful violence, and leave only a gentle emotion in our hearts. Thus the passions which in the lower world are man's greatest torment, in happier climates contribute to his felicity. I doubt much whether any violent agitation, or vapors of the mind, could hold out against such a situation; and I am surprised that a bath of the reviving and wholesome air of the mountains is not frequently prescribed both by physic and morality.

> Quì non palazzi, non teatro o loggia,
> Ma'n lor vece un' abete, un faggio, un pino
> Tra l'erba verde e'l bel monte vicino
> Levan di terra al Ciel nostr'intelletto.[19]

Imagine to yourself all these united impressions: the amazing variety, magnitude and beauty of a thousand stupendous objects; the pleasure of gazing at an entire new scene, strange birds, unknown plants, another nature, and a new world. To these even the subtlety of the air is advantageous, it enlivens their natural colors, renders every object more distinct, and brings it nearer to the eye. In short, there is a kind of supernatural beauty in these mountainous prospects which charms both the senses and the mind into a forgetfulness of one's self and of everything in the world. . . .

But whilst I traversed with delight these regions which are so little known, and so deserving of admiration, where was my Eloisa? Was she banished my memory? Forget my Eloisa! Forget my own soul! Is it possible for me to be one moment of my life alone, who exist only through her? O no! our souls are inseparable, and, by instinct, change their situation together according to the prevailing state of mine. When I am in sorrow, she takes refuge with yours, and seeks consolation in the place where you are; as was the case the day I left you. When I am happy, being incapable of enjoyment alone, they both attend upon me, and our pleasure becomes mutual: thus it was during my whole excursion. I did not take one step without you, nor admire a single prospect without eagerly pointing its beauties to Eloisa. The same tree spread its shadow over us both, and we constantly reclined against the same flowery bank. Sometimes as we sat I gazed with you at the wonderful scene before us, and sometimes, on my knees, I gazed with rapture on an object more worthy the contemplation of human sensibility. If I came to a difficult pass, I saw you skip over it with the activity of the bounding doe. When a torrent hap-

[19] "Here instead of palaces, theater, or loggia, a fir, a beech, a pine, between green grass and the beautiful nearby mountain, raise our intellect from earth to Heaven" (Petrarch, *Canzoniere* X).

pened to cross our path, I presumed to press you in my arms, walked slowly through the water, and was always sorry when I reached the opposite bank. Everything in that peaceful solitude brought you to my imagination; the pleasing awfulness of nature, the invariable serenity of the air, the grateful simplicity of the people, their constant and natural prudence, the unaffected modesty and innocence of the sex, and every object that gave pleasure to the eye or to the heart, seemed inseparably connected with the idea of Eloisa.

O divine maid! I often tenderly exclaimed, that we might spend our days in these unfrequented mountains, unenvied and unknown! Why can I not here collect my whole soul into thee alone, and become, in turn, the universe to Eloisa? Thy charms would then receive the homage they deserve; then would our hearts taste without interruption the delicious fruit of the soft passion with which they are filled: the years of our long elysium would pass away untold, and when the frigid hand of age should have calmed our first transports, the constant habit of thinking and acting from the same principle would beget a lasting friendship no less tender than our love, whose vacant place should be filled by the kindred sentiments which grew and were nourished with it in our youth. Like this happy people, we would practice every duty of humanity, we would unite in acts of benevolence, and at last die with the satisfaction of not having lived in vain. . . .

FROM *The Creed of a Priest of Savoy, translated by Arthur H. Beattie*
Emile, *Book IV*

. . . The first causes of movement are not in matter; it receives movement and communicates it, but it does not produce it. The more I observe the action and reaction of the forces of nature acting on one another, the more I find that, along a chain of effects, one must always go back to some will for the first cause; for to suppose an infinite series of causes is to suppose none at all. In a word, any movement which is not produced by another can be the product only of a spontaneous, willful act; inanimate bodies act only by movement, and there is no real action without will. That is my first principle. I think then that a will moves the universe and animates nature. That is my first dogma, or my first article of faith. . . .

If the movement of matter reveals to me a will, the movement of matter according to certain laws reveals to me an intelligence; that is my second article of faith. Acting, comparing, choosing, these are the operations of an active mind and thinking being: therefore that being exists. "Where do you see him?" you are going to ask me. Not only in the revolving heavens, in the sun which gives us light; not only in myself, but in the grazing sheep, the flying bird, the falling stone, the leaf borne away by the wind.

I recognize the order of the world though I know not the purpose of it, because to recognize that order it suffices for me to compare its parts one with the other, to study the pattern by which they fit together, to note their relationships, to observe their harmony. I do not know why the universe exists; but I do not fail to see how it is modified; I do not fail to perceive the intimate correspondence by which the beings who compose it lend one another a mutual help. I am like a man who should see, for the first time, the movement of a watch, and who would not fail to admire the workmanship, although he did not know the use of the machine and he had not seen the dial. I do not know, he would say, what the purpose of the whole thing is, but I see that each part is formed to work in unison with the others: I admire the workman in the detail of his work, and I am quite sure that all the cogs run thus in harmony only for a common end which it is impossible for me to perceive. . . .

There is not a being in the universe that one cannot regard in some respects as the common center of all others, about which they are all placed, in such a way that they are all mutually both ends and means to one another. The mind is confused and lost in this infinity of relationships, not one of which is confused or lost in the vast number. How many absurd suppositions have been made to deduce all this harmony from the blind mechanism of nature put into motion by chance! Those who deny the unity of intention which is manifest in the relationships of all the parts of this great whole cover in vain their gibberish with abstractions, coordinations, general principles, symbolic terms; whatever they may do, it is impossible for me to conceive a system of beings so constantly regulated without my conceiving an intelligence which governs it. I am not capable of believing that passive and inert matter could produce intelligent beings, that what does not think could produce thinking beings. . . .

. . . Whether matter be eternal or created, whether there be a passive principle or not, still it is certain that the whole is one and reveals a single intelligence—for I see nothing which does not have its place in the same well-ordered system, and which does not contribute to the same end, namely the preservation of the whole in the established order. This being who has will and power, this being active in himself, this being finally, whatever he may be, who moves the universe and governs all things, I call him God. . . .

After discovering those of his attributes by which I conceive his existence, I come back to myself, and I seek what place I occupy in the order of things which he governs, and which I can examine. I find myself incontestably in the first rank by virtue of my belonging to the human race; for by my will and by the instruments which are at my disposal to fulfill it, I am more capable of acting on all the bodies which surround me, whether to receive or to avoid their action as I may wish, than any of them are to act on me in spite of myself by mere physical impulsion; and, by my intelligence, I am the only one who can survey the whole. What being here below, apart from man, can observe all the others, measure, calculate, foresee their movements, their effects, and unite, as it were, the feeling of common existence to that of his own individual existence? What is so ridiculous about believing that everything is made for me, if I am the only one who knows how to consider everything in relation to himself? . . .

You see in what I have set forth only natural religion. It is very strange that any other should be necessary! How can I know the necessity for it? How can I be guilty of wrong in serving God according to the light which he gives to my mind, and according to the sentiments which he inspires in my heart? What purity of morals, what teaching useful to man and honorable to his creator, can I draw from a positive doctrine that I cannot draw without it from the proper use of my faculties? Show me what can be added, for the glory of God, for the good of society, and for my own well-being, to the duties of the natural law, and show me, too, what virtue you will derive from a new mode of worship which is not a consequence of mine. The greatest ideas of Divinity come to us through reason alone. Observe the spectacle of nature; listen to the inner voice. Has God not told everything to our eyes, to our conscience, to our judgment? What more will men tell you? Their revelations only belittle God by ascribing to him human passions. . . .

My son, keep your soul always in the state of desiring that there be a God, and you will never doubt it. Moreover, whatever decision you may take, remember that the real duties of religion are independent of the institutions of men, that a just heart is the real temple of the Divinity, that in every country to love God above all and one's neighbor as oneself is the epitome of the law, that there is no religion which frees one from the duties of morality, that there are no really essential duties other than those, that worship within your own heart is the first of those duties, and that without faith no true virtue exists.

Flee those who, under the pretext of explaining nature, sow in the hearts of men doctrines of despair, and whose apparent skepticism is a hundred times more affirmative and more dogmatic than the confident tone of their adversaries. Under the lofty pretext that they alone are enlightened, true, of good faith, they subject us imperiously to their categorical decisions, and claim to give us as the true principles of things the unintelligible systems which they have built in their imagination. Moreover, upsetting, destroying, trampling under foot everything that men respect, they take away from the afflicted the last consolation for their misfortune, from the powerful and the rich the only brake upon their passions; they snatch away from the depths of the heart remorse for crime, hope of virtue, and yet boast that they are the benefactors of mankind. Never, they say, is the truth harmful to men. I believe it as they; and that is, in my opinion, a great proof that what they teach is not the truth. . . .

FROM Reveries of the Solitary Walker: *Fifth Walk*, *translated by Robert M. Torrance*

Of all the places I have lived in (and I have lived in some charming ones) none has made me so truly happy or left me with such tender regrets as the Island of Saint-Pierre, in the middle of the Lake of Bienne. This little island, which in Neuchâtel is called the "Ile de La Motte," is not well known even in Switzerland. No traveler, to the best of my knowledge, has mentioned it. Yet it is very agreeable, and superbly suited to the happiness of a man who likes to live within limits—for although I may be the only person in the world whose destiny has made this way of living a law, I cannot believe that I am alone in possessing so natural a taste, even if I have not yet encountered it in anyone else.

The shores of the Lake of Bienne are wilder and more romantic than those of Lake Geneva, because the rocks and woods come down nearer the water; but they are no less pleasant. Though there are fewer cultivated fields and vineyards here, fewer towns and houses, there is also more natural greenery, more meadows, secluded places shaded by groves, contrasting scenes and chance encounters. Since there are no wide roads suitable for carriages on these happy shores, the region is little frequented by travelers; but it is perfect for solitary dreamers who love to inebriate themselves at leisure on nature's charms and to meditate in a silence unbroken except by the cry of eagles, the intermittent warbling of birds, and the roar of streams tumbling down from the mountains. This lovely basin, almost round in shape, encloses in its middle two little islands, one inhabited and cultivated, about half a league around, the other smaller, deserted and fallow, which will be worn away in the end by the removal of earth continually taken from it to make up for the ravages inflicted on the larger one by waves and storms. Thus it is that the substance of the weak is always used for the profit of the powerful.

There is only one house on the island, but it is large, agreeable, and comfortable; like the island, it belongs to the Hospital of Bern, and a keeper lives there with his family and servants. He maintains a teeming poultry-yard, an aviary, and ponds stocked with fish. The island, though small, is so varied in soil and aspect that it offers all sorts of sites and is amenable to every sort of cultivation. Here may be found fields, vineyards, woods, orchards, rich pastures shaded by thickets and bordered by every kind of shrub, all of which the shores of the lake keep fresh; a high terrace planted with two rows of trees runs along the length of the island, and in the middle of this terrace a pretty summerhouse has been built where inhabitants of the neighboring shores gather to dance on Sundays in the vintage season.

It is on this island that I took refuge after the stoning at Motiers. I found my stay so charming, and lived a life so suited to my temper, that I resolved to end my days there, and worried only that others might not let me carry out this plan, which was incompatible with that of dragging me off to England—the first hints of which I sensed already. Amid these disquieting presentiments, I might have wished that they had made this refuge my perpetual prison, confining me here for my whole life, and that they had taken away from me all power and hope of leaving, forbidding every kind of communication with the mainland, so that in my ignorance of everything done in the world I might have forgotten its existence and been forgotten by it as well.

They barely let me spend two months on this island, but I might have spent two years, two centuries, and all eternity there without being bored for a moment, though I had, apart from my companion,[20] no society but that of the keeper, his wife, and his servants—who were all certainly very good people—and no more; but this was exactly what I needed. I count these two months the happiest time of my life, so happy that I would have found them enough for my entire existence, without letting desire for any other condition arise in my soul for a single moment.

What then was this happiness, and in what did this rapture consist? I would not expect any man of this century to divine it from the description of the life I led there. Precious *far niente* [doing nothing] was the first and principal rapture that I wished to savor in all its sweetness; and everything I did during my stay was in fact nothing but the delightfully mandatory occupation of a man devoted to idleness.

The hope that others would ask nothing more than to let me alone in this isolated sojourn in which I had ensnared myself, from which it was impossible for me to leave unassisted and unobserved, and where I could have no communication or correspondence with others except by the help of those around me—this, I say, gave me the further hope of ending my days more tranquilly than I had spent them so far; and because I thought I would have time to arrange things at leisure, I began by making no arrangements at all. Transported there suddenly, alone and stripped of everything, I sent in turn for my housekeeper, my books, and my few belongings, which I had the pleasure of not unpacking, leaving my boxes and trunks just as they arrived, and dwelling in the place where I intended to end my days as if in an inn which I was to leave on the morrow. Everything went so well just as it was that to want to order things better would have been to spoil something. One of my greatest delights was above all to leave my books packed up, and to have no writing desk. When forced to take up my pen to answer some wretched letters, I grudgingly borrowed the keeper's writing desk, then hastened to return it in the vain hope of having no further need to borrow it. Instead of these dusty old papers and books, I filled my room with flowers and hay; for I was then feeling my first fervor for botany, for which Doctor d'Ivernois had inspired in me a taste that soon became a passion. Desiring no more laborious work, I needed some pleasant pastime that would give me only such trouble as a loafer is willing to take. I set out to compose a *Flora Petrinsularis [Flora of the Island of Saint-Pierre]* and to describe all the plants on the island, without omitting a single one, in enough detail to occupy me for the rest of my days. They say a German once wrote a book about a lemon-skin; I could have written

[20]His mistress, the servant girl Thérèse Levasseur.

one about every grass in the meadows, every moss in the woods, every lichen carpeting the rocks; in short, I would have left not a single blade of grass, not an atom of vegetation, less than copiously described. In accord with this fine project, every morning after breakfast, which we all took together, I used to go with a magnifying glass in my hand and my *Systema Naturae*[21] under my arm to visit one part of the island, which I had for this purpose divided into small squares, meaning to go through them one by one in each season. Nothing is more extraordinary than the raptures, the ecstasies I would experience at every observation I used to make about the structure and organization of vegetable life and the play of the sexual parts in fructification, the system of which was then wholly new to me. The distinction of genera by their characteristics, of which I had previously had not the least idea, enchanted me as I verified them in common species while waiting for rarer ones to present themselves to me. The forking of the self-heal's two long stamens, the elasticity of the nettle's and pellitory's, the bursting of the balsam's fruit and the box-tree's pod: a thousand little games of fructification which I was observing for the first time filled me brimful with joy, and I went around asking if anyone had seen the horns of the self-heal, just as La Fontaine asked if anyone had read Habakkuk.[22] After two or three hours I would return loaded down with an ample crop, providing amusement at home for the afternoon in case of rain. The rest of the morning I employed in going with the keeper, his wife, and Thérèse, to visit the laborers at the harvest, most often to lend them a hand at their work; often people who came to see me from Bern found me perched in a big tree, filling with fruits a bag round my waist, which I would then lower to the ground on a rope. The exercise I had done in the morning, and the good humor it entailed, made my rest at dinner-time very agreeable, but when this went on too long, and fine weather summoned me, I could wait no longer; and while others were still at table I would slip away and plant myself alone in a boat and row to the middle of the lake when the water was calm; there, stretching out at full length in the boat, eyes turned toward the sky, I let myself float and drift slowly at the water's will, sometimes for several hours on end, plunged in a thousand confused but delightful reveries which, though lacking any determinate or constant object, never ceased in my eyes to be a hundred times preferable to everything that I had found sweetest in what are called the pleasures of life. Often warned by the setting sun that it was time to go home, I found myself so far from the island that I was forced to toil with all my might to arrive before nightfall. At other times, instead of straying into deep water, I chose to skirt the green shores of the island, where the clear waters and cool shadows often invited me to bathe. But one of my most frequent boating trips was from the larger to the smaller island, where I disembarked and spent the afternoon, sometimes in circumscribed strolls among its willows, alders, persicarias, and shrubs of every kind, and sometimes perching myself on the summit of a shady hillock covered with turf, wild thyme, and flowers, even with sainfoin, and with clover that had probably been sown sometime before, very suited for shel-

[21]*System of Nature* (first edition 1735), the first great taxonomical work of the Swedish naturalist Karl von Linné, or Carolus Linnaeus (1707–78). In his *Genera Plantarum [Genera of Plants]* of 1737 and *Species Plantarum [Species of Plants]* of 1753, and in later editions of the *Systema Naturae*, he elaborated his system of classifying plants into genera and species in accord with their sexual characteristics.

[22]It was Baruch (in the Apocrypha), not Habakkuk (in the Old Testament), who aroused La Fontaine's admiration. As reported by Louis Racine (son of the tragic dramatist Jean Racine) in his memoirs, during a lengthy church service his father once handed La Fontaine a copy of the Bible, containing the minor Prophets, to occupy his time, and La Fontaine, struck by the prayer of the Jews in Baruch, could not leave off admiring him: "A beautiful genius this Baruch: who was he?" For days afterward, when he would meet anyone he knew in the street, after the usual greetings, he would raise his voice and say: "Have you read Baruch? He was a beautiful genius." (See note in Spink's edition)

tering rabbits, which could multiply there in peace without fearing anything, and doing harm to nothing. I broached this idea to the keeper, who sent for male and female rabbits from Neuchâtel, and we went with great pomp, his wife, one of his sisters, Thérèse, and I, to install them on the little island, where they were beginning to breed before my departure and where they will no doubt have prospered if they have been able to withstand the rigors of winter. The founding of this little colony was a festive day. The pilot of the Argonauts was not prouder than I, triumphantly leading the company and the rabbits from the large island to the little one; and I proudly noted that the keeper's wife, who was excessively afraid of water and always felt seasick on it, embarked confidently under my guidance and showed no fear during the crossing.

When the lake was too agitated for boating, I would spend my afternoon roving about the island, collecting plants right and left, sitting sometimes in the most pleasing and solitary retreats in order to dream at my ease, and sometimes on the terraces and hillocks to take in with my eyes the superbly ravishing sight of the lake and its shores, crowned on one side by the near-by mountains and stretching out on the other into rich and fertile plains, the view of which extended up to the blue mountains at their far edge.

When evening approached, I liked to come down from the heights of the island, and go sit on the beach in some hidden spot by the edge of the lake; there the noise of the waves and the agitation of the water would steady my senses and chase away all other agitation from my soul, plunging it into a delicious reverie in which night would often take me unawares. The ebb and flow of the water, its continuous yet intermittently swelling noise, as it incessantly struck my ears and eyes, would supplant the inward movements which reverie was quenching within me, and allow me to feel my existence with pleasure, without taking the trouble to think. From time to time there might well up some brief and faint reflection on the instability of the things of this world, whose image the surface of the water offered to me, but these fragile impressions would soon be effaced in the uniformity of the continuous movement that was lulling me, and that never ceased, with no active concurrence of my soul, to hold me fast, to the point that even when summoned by time and the accustomed signal, I could not tear myself away from there without effort.

After supper, when the evening was fine, we would all go out again to take a walk together around the terrace and breathe the cool air from the lake. We would rest in the pavilion, laugh, chat, and sing some old song easily worth as much as any modern medley, and finally we would go to bed satisfied with our day and only wishing the next to be another like it.

Such, leaving aside unforeseen and irking visits, is the way I passed my time on this island during my stay there. At present I wonder if anyone could tell me what was attractive enough about it to arouse such intense, tender, and lasting regrets in my heart that fifteen years later it is impossible for me to think of this cherished place without feeling myself transported every time by pangs of desire.

I have noticed in the vicissitudes of a long life that the periods of the sweetest joys and most intense pleasures are nevertheless not those whose memory attracts and touches me most. These brief moments of madness and passion, intense though they be, are by their very intensity only sparsely scattered points along the line of life. They are too rare and too rapid to constitute a condition, whereas the happiness that my heart regrets is not composed of fugitive moments but is an integral and permanent condition, which has nothing intense in itself, but whose duration increases its charm so that we eventually find in it our supreme felicity.

Everything is in continual flux on this earth. Nothing keeps a constant and stable shape, and our affections, which attach themselves to external things, necessarily change and pass away as these do. Always ahead of us or behind us, they recall a past which exists no longer, or anticipate a future which is often not destined to be; there is nothing solid to which the heart can attach itself. Thus we scarcely have anything but passing pleasure here below; I doubt that lasting happiness is known here. Since there is hardly a moment, amid our most intense raptures, of which the heart can truthfully say: *Would that this moment could last for ever!*, how can one call happiness this fugitive condition that leaves the heart still unquiet and empty, making us long for something past or desire something still to come?

But if there is a condition where the soul finds a solid enough footing to rest and collect its whole being, with no need to recall the past or strive toward the future; where time is nothing to it; where the present lasts forever, leaving no sign of its duration and no trace of its passing, nor any other feeling of deprivation or enjoyment, pleasure or pain, desire or fear, than that of sheer existence, which alone can fill it completely: as long as this condition lasts, one who knows it can be called happy, not with an imperfect, poor, and relative happiness, like that found in life's pleasures, but with a sufficient, perfect, full happiness that leaves no emptiness that the soul feels the need to fill. Such is the state in which I often found myself on the island of Saint-Pierre in my solitary reveries, whether lying in my boat adrift at the will of the water, or seated on the shores of the choppy lake, or elsewhere, on the banks of a lovely river or of a stream babbling over its pebbles.

In what do we find joy in a situation of this kind? Not in anything external to us, not in anything but ourselves and our own existence; as long as this condition lasts we are self-sufficient, like God. The feeling of existence stripped of every other emotion is in itself a precious feeling of contentment and peace, which alone would suffice to make this existence sweet and dear to whoever could cast away all the earthly and sensual impressions that ceaselessly distract us from it and trouble its sweetness here below. But most men, stirred up by continual passion, know little of this condition, and having tasted it but imperfectly for a few moments, they preserve too dim and confused an idea of it to feel its charm. Nor indeed, as things now stand, would it benefit them, in their avid longing for these sweet ecstasies, to take an aversion to the active life which their repeatedly resurrected needs prescribe as their duty. But an unfortunate man who has been cut off from human society, and can do nothing more of use or benefit to himself or others here below, can find in this condition compensations for all human happiness—compensations which neither fortune nor men can take away from him.

It is true that these compensations cannot be felt by every soul, or in every situation. The heart must be at peace and no passion must trouble its calm. The person who experiences these sensations must have a disposition to them, and the objects around him must be in accord with them. There must be neither absolute repose, nor too much agitation, but a moderate and uniform motion, with neither disruptions nor intervals. Without movement, life is mere lethargy. If the movement is unsteady or too violent it awakes us; by recalling us to surrounding objects, it destroys the charm of reverie and tears us from our inner selves, instantly putting us beneath the yoke of fortune and of men, and subjugating us to our feeling of wretchedness. Absolute silence brings on sadness. It presents an image of death. The help of a cheerful imagination is necessary at such times; and it quite naturally assists those whom Heaven has favored with it. Movement which does not arise from outside us then comes from within us. Repose is less complete, it is true—

but it is also more pleasant when sweetly casual ideas barely skim the surface of the soul, so to speak, without agitating its depths. One needs only enough of such ideas to remember one's self while forgetting all one's troubles. This type of reverie can be enjoyed wherever one can be at peace, and I have often thought that in the Bastille, even in a dungeon where no object met my eyes, I could still have agreeably dreamed.

But I must confess that this occurred much more readily and agreeably in a fertile and solitary island, naturally circumscribed and cut off from the rest of the world, where nothing appeared to my eyes but glad images; where there was nothing to recall sorrowful memories; where the society of the small number of inhabitants was pleasant and appealing without being so interesting as to occupy me continually; where I could devote myself all day, in short, without hindrance or care to whatever pastimes I preferred, or to the most luxurious idleness. It was beyond doubt a fine opportunity for a dreamer who, knowing how to sustain himself on agreeable chimeras even in the midst of the most unpleasant objects, could sate himself on them at leisure here, making everything in the real world that struck his senses accord with them. Emerging from a long, sweet reverie, seeing myself surrounded by greenery, by flowers, by birds, and letting my eyes wander far over the picturesque shores that bordered a vast expanse of clear and crystalline water, I merged all these lovely sights with my fictions; and when I was at last aware of being brought gradually back to myself and my surroundings, I could not mark a dividing line between fictions and realities: so much did everything equally conspire to make me cherish the solitary and contemplative life that I led in that beautiful place. Might it not be again reborn? Might I not go and end my days on this beloved island, never leaving it again, or seeing again any inhabitant of the mainland who might bring to mind the calamities of every kind that others have delighted to heap upon me for so many years? They would soon be forgotten for ever; doubtless they would not forget me likewise, but what would that matter to me, so long as they were unable to come here and trouble my repose? Set free from all the earthly passions engendered by the tumult of social life, my soul would often soar above this atmosphere, and would communicate in advance with the celestial intelligences whose number it hopes in a little while to augment. Men will not allow me, I know, to regain so sweet a refuge, in which they had no wish to let me stay. But they will not stop me, at least, from being transported there every day on the wings of imagination, and from tasting for a few hours the same pleasure as if I were living there still. The sweetest thing I could do there would be to dream to my heart's content. In dreaming that I am there, am I not doing the same thing? No, I am doing still more: to the appeal of an abstract and monotonous reverie I join charming images that bring it to life. The objects to which these images corresponded often eluded my senses during my ecstasies, but now, the deeper my reverie is, the more vividly it paints these objects to me. I am often more fully and agreeably in their midst than when I was really there. My misfortune, as imagination cools, is that this takes place with greater and greater effort, and does not last so long. Alas! it is when we are beginning to slough off our mortal skin that we find it most offends us.

Two Children of Nature

Paul and Virginia, by Jacques-Henri Bernardin de Saint-Pierre, translated by Robert M. Torrance

Born at Le Havre in Normandy of a modest bourgeois family, Jacques-Henri Bernardin de Saint-Pierre (1737–1814) traveled to Martinique at age twelve, and returned "disgusted": "A major theme of his life had taken shape," John Donovan writes in introducing his translation of Paul and Virginia *(1982); "all his future voyages would end in disappointment." In Paris, he trained as an engineer, then saw action in the Seven Years' War and served as a soldier of fortune in Holland, Russia, and Germany. He was lieutenant of engineers to Catherine II of Russia, and after amorous entanglement in Poland offered his service to Frederick II of Prussia before returning to France. In 1768 he left for Madagascar as royal engineer with the rank of captain, but quarreled with others, refused to disembark, and sailed on to the Indian Ocean island of Mauritius (then a French colony, the Ile de France), which he later made the site of his novel. Back in France in 1771, he published his* Voyage à l'Ile de France *in 1773, and was an intimate friend of Rousseau until the latter's death in 1778. Thereafter he devoted himself primarily to his* Etudes de la nature (Studies of Nature), *begun as early as 1773 but first published in 1784, "a series of discourses in natural history and philosophy minutely illustrating a single leading idea," Donovan remarks, "that in order to understand Nature aright one must recognize her beneficence, which is nothing other than the manifestation of Divine Providence." His immensely popular and influential novel* Paul et Virginie *was first published in the third edition of* Studies of Nature *in 1788, then separately in 1789. Under the Revolution, Bernardin de Saint-Pierre was named intendant of the Jardin des Plantes and professor of moral philosophy at the Ecole Normale Supérieure.*

The spectacular success of Paul and Virginia, *in which Bernardin de Saint-Pierre took up Rousseau's suggestion to portray "a society made happy by the laws of nature and virtue alone," was owing above all to his artfully naïve portrayal of innocent love between two children reared as "brother" and "sister" by their widowed French mothers on a tropical island far from civilization and its corruptions. As a pastoral idyll of childlike love (which would inspire songs, poems, plays, ballets, operas, engravings, lithographs, paintings, and countless literary echoes throughout the following century), it is somewhat reminiscent of the Greek romance of* Daphnis and Chloe *(Chapter 11 above). But the latter's delightfully playful tone here gives way—especially when civilization and shipwreck, in later parts of the book, claim the life of Virginia—to a tearful sentimentality more immediately influenced by Rousseau's* Nouvelle Héloïse, *whose heroine likewise meets death by water. Most influential of all, however, was his portrayal of the naturally moral (and even naturally enlightened) existence of these two "children of nature" for whom each day was a feast-day and everything that surrounded them a holy temple. Here Rousseau's vision is stripped of its complex ambiguities and reduced to its simplest form, as Paul and Virginia worship in accord with a natural theology consisting entirely of feeling and regulate their lives according to the cycles of nature—like fauns and dryads or, better (since fauns and dryads were never notably innocent), like our first parents in the garden of Eden before the serpent introduced them to the arts and sciences of civilization that constitute, in this world, the disruptive knowledge of good and evil.*

Their only study was to please and help each other; as to the rest, they were as ignorant as Creoles and could neither read nor write. They never bothered themselves with what had happened in faraway times and distant places; their curiosity did not extend beyond this mountain. They thought the world ended where their island ended, and imagined nothing could be pleasant anywhere else. Their affection for each other and for their mothers wholly occupied their souls. Never had useless knowledge made their tears flow; never had lessons of a gloomy morality filled them with torment. They did not know that stealing was prohibited, since their belongings were all held in common; nor intemperance, since they had plenty of simple food; nor lying, since there was no truth they needed to hide. No one had frightened them by telling them that God reserves terrible punishments for ungrateful children; their affection for their mothers was born of their mothers' for them. About religion they had been taught only what makes it loved; if they offered no long prayers in church, yet wherever they were—at home, in the fields, in the woods—they raised toward heaven innocent hands and a heart full of love for their parents.

Thus their early childhood passed like a lovely dawn announcing a lovelier day. . . .

Amiable children! thus you passed the innocence of your earliest days in the practice of kindness. How often your mothers, clasping you in their arms in this place, praised heaven for the consolation you were preparing for their old age, and gave thanks for seeing you come to maturity under such happy auspices! How often, in the shade of these rocks, I have partaken with them of rustic meals that cost no animal its life! Gourds full of milk, fresh eggs, rice-cakes on banana leaves, baskets loaded with sweet potatoes, mangoes, oranges, pomegranates, bananas, sweet-sops, pineapples—these offered at once the healthiest dishes, the gayest colors, and the sweetest juices.

Their conversation was as sweet and innocent as these banquets. Paul often talked of his day's work, and of the morrow's. He was always thinking about something useful for society. In one place the paths were inconvenient; in another there was no place to sit comfortably; these young arbors did not give enough shade; Virginia would be better off over there.

In the rainy season masters and servants spent the day in their hut together, occupied in making grass mats and bamboo baskets. Ranged on the inner walls in perfect order could be seen rakes, axes, and spades; and near these agricultural implements were their fruits: sacks of rice, sheaves of grain, and clusters of bananas. There tastefulness and plenty were always united; there Virginia, instructed by her mother and Marguerite,[23] would prepare sherbets and cordials from the juice of sugar cane, lemons, and citrons. . . .

From time to time Madame de la Tour would read aloud some touching story from the Old or New Testament. They discussed these sacred books but little; for their theology was entirely of feeling, like that of nature, and their morality lay entirely in action, like that of the Gospel. They did not have days designated for pleasure and others for sadness. Each day was a feast-day for them, and everything that surrounded them was a holy temple where they unceasingly admired an infinite, omnipotent, and beneficent Intelligence. This feeling of trust in the Supreme Power filled them with consolation for the past, courage for the present, and hope for the future. Thus these women, compelled by misfortune to return to nature, had developed in themselves and in their children the feelings that nature gives to keep us from falling into misfortune. . . .

[23]Virginia's mother is Madame de la Tour; Marguerite is Paul's mother.

You Europeans, whose minds are filled from childhood with so many prejudices inimical to happiness, cannot conceive that nature can give so much enlightenment and so many pleasures. Your souls, shut up in a narrow sphere of human knowledge, soon reach the limits of their artificial enjoyments: but nature and the heart are inexhaustible. Paul and Virginia had no clocks, no almanacs, no books of chronology, history, or philosophy. The phases of their lives were regulated by nature's. They knew the hours of the day by the shadow of the trees, the seasons by the times when they bring forth flowers or fruits, and the years by the number of their harvests. These sweet images shed the greatest charms on their conversations. "It is dinner-time," Virginia would say to the family, "the shadows of the banana trees are at their feet"; or "Night is coming, the tamarinds are closing their leaves." "When will you come see us?" some friends living nearby would ask. "At sugar-cane time," Virginia would answer. "Your visit will be all the sweeter and more agreeable to us," these girls would reply. When asked about her age and Paul's: "My brother," she would say, "is as old as the big coconut tree by the spring, and I as the small one. The mango trees have given their fruit twelve times and the orange trees their flowers twenty-four since I have been in the world." Like fauns' and dryads', their lives seemed attached to that of the trees. They knew no historical events but those of their mothers' lives, no chronology but that of their orchards, and no philosophy but to do good to everyone and submit to the will of God.

After all, what need did these young people have to be rich and wise in our fashion? Their needs and their ignorance added further to their felicity. Not a day went by when they did not share with each other some help or some enlightenment—yes, enlightenment; and though some errors might be intermingled with it, one pure of heart has no dangerous ones to fear. In such a way these two children of nature grew up. No care had wrinkled their brow, no intemperance had corrupted their blood, no unhappy passion had depraved their heart: love, innocence, and piety manifested each day the beauty of their souls in the ineffable grace of their features, attitudes, and movements. In the morning of life, they had all its freshness; just so our first parents appeared in the garden of Eden when, issuing from the hands of God, they saw each other, drew nearer, and talked for the first time as brother and sister: Virginia, sweet, modest, and trusting like Eve, and Paul, the image of Adam, with the stature of a man and the simplicity of a child. . . .

Thus I pass my days far from men, whom I wished to serve, and who have persecuted me. After traveling through a large part of Europe and some regions of America and Africa, I settled down on this sparsely populated island, seduced by its gentle climate and its solitary places. A cabin that I built in the forest at the foot of a tree, a little field cleared by my own hands, a river that flows past my door, suffice for my needs and pleasures. I add to these the enjoyment of a few good books that teach me to become better. They also make the very world I have abandoned serve my happiness; they present me with pictures of the passions that make its inhabitants so miserable, and by the comparison I make between their lot and mine, they let me enjoy a negative happiness. Like a man saved from shipwreck on a rock, I contemplate from my solitude the storms that rage in the rest of the world, and my calm even redoubles at the distant sound of the tempest. Now that men are no longer in my way, nor I in theirs, I no longer hate them; I pity them. If I encounter some unfortunate person, I try to help him by my advice, just as a man walking along the bank of a rushing stream stretches out his hand toward an unlucky person drowning in it. But innocence alone have I found at all attentive to my voice. Nature vainly calls the rest of men to her; each makes for himself an image of her clothed in his own passions. . . . As for me, I let myself be borne along in peace on the river of time toward the ocean of the shoreless future and, by gazing on the present harmonies of nature, I rise toward its author and hope for a happier destiny in another world. . . .

Natural Causes Sufficing

"Dialogue between a Priest and a Dying Man," by Donatien-Alphonse-François, marquis de Sade, translated by Richard Seaver and Austryn Wainhouse

Donatien-Alphonse-François, comte de Sade (1740–1814), who continued to be known as the marquis de Sade even after he inherited the title of count on his father's death in 1767, was born in Paris and educated at the Jesuit college of Louis-le-Grand. He served as a cavalry officer in the Seven Years' War, then spent large parts of his life, beginning in 1763, in prison, exile, or asylums for various forms of cruelty and debauchery, including charges of flogging, poisoning, and sodomy. During his long imprisonment between 1777 and 1790 in Vincennes, the Bastille of Paris, and the asylum of Charenton (to which he was moved ten days before the storming of the Bastille in 1789), he wrote or began most of the works that he published after being released by the revolutionary government, including Justine, ou les malheurs de la vertu *(*Justine, or the Misfortunes of Virtue, *1791);* Aline et Valcour *(1795);* La philosophie dans le boudoir *(*Philosophy in the Bedroom, *1795);* Histoire de Juliette, ou les prospérités du vice *(*Story of Juliette, or the Prosperities of Vice, *1797); and* La Nouvelle Justine *(*The New Justine, *1797).*

Two of his earliest works remained unpublished in his lifetime. One is the short "Dialogue entre un prêtre et un moribond" ("Dialogue between a Priest and a Dying Man"), written in 1782 but not published until 1926; our selections are taken from the translation by Richard Seaver and Austryn Wainhouse in The Complete Justine, Philosophy in the Bedroom, and Other Writings *(1965). The other is the long, orgiastic novel* Les 120 Journées de Sodome (The 120 Days of Sodom), *which Sade completed in 1785 and thought had been lost in the storming of the Bastille, but whose manuscript survived and was first published in 1904. Though he avoided execution in the Terror—despite his title of nobility and despite his refusal, as a magistrate, to condemn others to death in the name of justice—Sade was imprisoned again in 1801 and committed in 1803 to Charenton, where he wrote theatrical pieces for the other inmates (as brilliantly imagined by Peter Weiss in his play of 1964,* Marat/Sade*), and where he died.*

Sade's "basic philosophy," Maurice Blanchot writes in an essay (from Lautréamont et Sade *[1949]) included in Seaver and Wainhouse's volume, "is one of self-interest, of absolute egoism: Each of us must do exactly as he pleases, each of us is bound by one law alone, that of his own pleasure. . . . Nature wills that we be born alone, there is no real contact or relationship between one person and another. . . . For Sade, the equality of all human beings is the right to equal use of them all, freedom is the power to bend others to his will." But "for some men, inequality is a fact of Nature; certain persons are necessarily slaves and victims, they have no rights, they are nothing, against them any act, any crime is permitted."*

Sade thus takes to the extreme the exaltation of nature in Rousseau and the materialistic reduction of nature and repudiation of God in other philosophes. Diderot, Mario Praz contends in The Romantic Agony *(1933; 2nd ed. 1951), "is one of the greatest exponents of that* Système de la Nature *which, carrying materialism to its logical consequences and proclaiming the supreme right of the individual to happiness and pleasure in opposition to the despotism of morality and religion, paves the way to the justification, in the name of Nature, of sexual perversions." But Diderot's* Supplement to Bougainville's Voyage *was not published until 1796, after Sade's major works had been written; as early as 1783, on the other hand, he wrote to his wife from prison in Vincennes complaining that the authorities had denied him a copy of Rousseau's* Confessions*: "while Rousseau may represent a threat for dull-witted bigots of your species," he defiantly declared, "he is a salutary author for me. Jean-Jacques is to me what* The Imitation of Christ *is for you. Rousseau's ethics and religion are strict and severe to me, I read them when I feel the need to improve myself."*

Rousseau's ambiguous views were undergoing many strange transformations, as the "incorruptible" Robespierre and others turned the worship of Nature into a justification of terror. But nowhere are these distortions more extreme, indeed crude, than in Sade, for whom (as expressed by his spokesmen—and spokeswomen—in Philosophy in the Bedroom*), Nature "is naught else than matter in action," and God is a monstrous criminal "worthy of our hatred and implacable vengeance." Nature becomes a justification for unbounded and completely impersonal sexual promiscuity, whose commands are reducible, in this dialogue, to a single word: "Fuck, in one word, fuck: 'twas for that you were brought into the world; no limits to your pleasure save those of your strength and will; no exceptions as to place, to time, to partner; all the time, everywhere, every man has got to serve your pleasures; continence is an impossible virtue for which Nature, her rights violated, instantly punishes us with a thousand miseries."*

This simplistic message, echoing those of Jean de Meun's Genius and the Roman de Renart's *Bernard the Ass (introduction to* The Romance of the Rose *in Chapter 15 above), becomes in Sade a justification of violence, destruction, and murder, all in the name of Nature. "Destruction being one of the chief laws of Nature,"* Philosophy in the Bedroom *continues, "nothing that destroys can be criminal; how might an action which so well serves Nature ever be outrageous in her?" In a world of total individualism and isolation—"Have we ever felt a single natural impulse advising us to prefer others to ourselves, and is each of us not alone, and for himself in this world?"—our mother Nature, herself the supreme egoist, "never speaks to us save of ourselves; nothing has more of the egoistic than her message, and what we recognize most clearly therein is the immutable and sacred counsel: prefer thyself, love thyself, no matter at whose expense." Every possible cruelty is justified by Nature herself: "what other than Nature's voice suggests to us personal hatreds, revenges, wars, in a word, all those causes of perpetual murder?"*

In consequence, as Blanchot remarks, Sade's infinite hatred of an infinite God "will turn itself with the same intensity and fearlessness against Nature as against the non-existent God it loathes." His nihilistic hatred is ultimately turned not only against God and Nature, and not only against his victims, but against himself and a universe where, as he wrote in the letter to his wife quoted above, "everything is relative." His "philosophy" of nature provides a justification not only for the perversion of human value that bears his name, but also for the many forms of fascism and tyranny that, in one central dimension, are nothing but Sadism writ large.

Priest. Come to this the fatal hour when at last from the eyes of deluded man the scales must fall away, and be shown the cruel picture of his errors and his vices—say, my son, do you not repent the host of sins unto which you were led by weakness and human frailty?
Dying Man. Yes, my friend, I do repent. . . . By Nature created, created with very keen

tastes, with very strong passions; placed on this earth for the sole purpose of yielding to them and satisfying them, and these effects of my creation being naught but necessities directly relating to Nature's fundamental designs or, if you prefer, naught but essential derivatives proceeding from her intentions in my regard, all in accordance with her laws, I repent not having acknowledged her omnipotence as fully as I might have done, I am only sorry for the modest use I made of the faculties (criminal in your view, perfectly ordinary in mine) she gave me to serve her; I did sometimes resist her, I repent it. Misled by your absurd doctrines, with them for arms I mindlessly challenged the desires instilled in me by a much diviner inspiration, and thereof do I repent: I only plucked an occasional flower when I might have gathered an ample harvest of fruit—such are the just grounds for the regrets I have, do me the honor of considering me incapable of harboring any others. . . .

Priest. Who is there can penetrate God's vast and infinite designs regarding man, and who can grasp all that makes up the universal scheme?

Dying Man. Anyone who simplifies matters, my friend, anyone, above all, who refrains from multiplying causes in order to confuse effects all the more. What need have you of a second difficulty when you are unable to resolve the first, and once it is possible that Nature may all alone have done what you attribute to your god, why must you go looking for someone to be her overlord? The cause and explanation of what you do not understand may perhaps be the simplest thing in the world. Perfect your physics and you will understand Nature better, refine your reason, banish your prejudices and you'll have no further need of your god. . . .

Priest. Then you do not believe in God at all?

Dying Man. No. And for one very sound reason: it is perfectly impossible to believe in what one does not understand. . . . My friend, prove to me that matter is inert and I will grant you a creator, prove to me that Nature does not suffice to herself and I'll let you imagine her ruled by a higher force; until then, expect nothing from me, I bow to evidence only, and evidence I perceive only through my senses: my belief goes no farther than they, beyond that point my faith collapses. I believe in the sun because I see it, I conceive it as the focal center of all the inflammable matter in Nature, its periodic movement pleases but does not amaze me. 'Tis a mechanical operation, perhaps as simple as the workings of electricity, but which we are unable to understand. Need I bother more about it? when you have roofed everything over with your god, will I be any the better off? and shall I still not have to make an effort at least as great to understand the artisan as to define his handiwork? By edifying your chimera it is thus no service you have rendered me, you have made me uneasy in my mind but you have not enlightened it, and instead of gratitude I owe you resentment. Your god is a machine you fabricated in your passions' behalf, you manipulated it to their liking; but the day it interfered with mine, I kicked it out of my way, deem it fitting that I did so; and now, at this moment when I sink and my soul stands in need of calm and philosophy, belabor it not with your riddles and your cant, which alarm but will not convince it, which will irritate without improving it; good friends and on the best terms have we ever been, this soul and I, so Nature wished it to be; as it is, so she expressly modeled it, for my soul is the result of the dispositions she formed in me pursuant to her own ends and needs; and as she has an equal need of vices and of virtues, whenever she was pleased to move me to evil, she did so, whenever she wanted a good deed from me, she roused in me the desire to perform one, and even so I did as I was bid. Look nowhere but to her workings for the unique cause of our fickle human behaviors, and in her laws hope to find no other springs than her will and her requirements. . . .

Priest. What are you aiming at?
Dying Man. At proving to you that the world and all therein may be what it is and as you see it to be, without any wise and reasoning cause directing it, and that natural effects must have natural causes: natural causes sufficing, there is no need to invent any such unnatural ones as your god who himself, as I have told you already, would require to be explained and who would at the same time be the explanation of nothing; and that once 'tis plain your god is superfluous, he is perfectly useless: that what is useless would greatly appear to be imaginary only, null and therefore nonexistent; thus, to conclude that your god is a fiction I need no other argument than that which furnishes me the certitude of his inutility. . . .

But I feel my strength ebbing away; preacher, put away your prejudices, unbend, be a man, be human, without fear and without hope forget your gods and your religions too: they are none of them good for anything but to set man at odds with man, and the mere name of these horrors has caused greater loss of life on earth than all other wars and all other plagues combined. Renounce the idea of another world; there is none, but do not renounce the pleasure of being happy and of making for happiness in this. Nature offers you no other way of doubling your existence, of extending it.—My friend, lewd pleasures were ever dearer to me than anything else, I have idolized them all my life and my wish has been to end it in their bosom; my end draws near, six women lovelier than the light of day are waiting in the chamber adjoining, I have reserved them for this moment, partake of the feast with me, following my example embrace them instead of the vain sophistries of superstition, under their caresses strive for a little while to forget your hypocritical beliefs.

Note

The dying man rang, the women entered; and after he had been a little while in their arms the preacher became one whom Nature has corrupted, all because he had not succeeded in explaining what a corrupt nature is.

Oddities and Sublimities of the New World

Four Eighteenth-Century American Writers

Ever since the European discovery of America, Columbus and later explorers (Chapter 16 above) had marveled at the strange forms nature took in this previously unsuspected new world. Despite extensive settlement of the Atlantic coasts of North America by English, French, and Dutch colonists, and by African slaves, in the seventeenth century, vast tracts of wilderness remained even in eastern parts of the future United States, to say nothing of the largely unexplored regions beyond. The following four accounts, three from the southeastern colonies and one from New England, are all by writers who lived most of their lives in America, where three of the four were born and all four died. This is their own world now, no longer one of rivers flowing with gold or of unicorns, three-headed serpents, and self-sacrificing pelicans. Yet it remains, for these men of European descent, a world inspiring awe by its vastness and by its continually surprising revelations of nature's prodigies and sublimities.

Our first selection is from the History of North Carolina, Containing the Exact Description and Natural History of that Country, Together with the Present State Thereof and a Journal of a Thousand Miles Traveled through Several Nations of Indians, Giving a Particular Account of Their Customs, Manners, etc., etc., by ***John Lawson***, *who came to the Carolinas from England in 1700 and explored their unknown regions in the years that followed. He was named surveyor general of North Carolina in 1708, and was captured, tortured, and killed in the Tuscarora Indian uprising of 1711. His book, carefully describing the flora, fauna, and peoples he had come to know, was published in London in 1709 as* A New Voyage to Carolina *and reissued several times; our selections, describing such unfamiliar animals as the buffalo, possum, and alligator, are from the edition of 1714 as reprinted in* Lawson's History of North Carolina, *ed. Frances Latham Harriss (1937).*

A generation later, ***William Byrd II*** *(1674–1744) recounted his survey of 1728, undertaken to settle a boundary dispute between Virginia and North Carolina. Son of a wealthy Indian trader and planter, the younger Byrd was educated in England between the ages of seven and twenty-two. Trained in law at the Middle Temple, he was admitted to the bar in 1695. His scientific interests and friendship with Robert Boyle (Chapter 18 above) gained him election to the Royal Society in 1696; his London literary friends included the leading dramatists William Wycherley and William Congreve. Back in Virginia, he was elected to the House of Burgesses and became a leading figure of the colonial aristocracy on his father's death in 1704. After acting several times as the colony's agent to London, he settled permanently at Westover in 1726; on one of his estates, in 1737, he laid out the future city of Richmond. An avid reader of Greek, Latin, Hebrew, and modern languages, he built up one of the largest libraries in America and kept vivid diaries throughout much of his life, several of which have been recovered and printed.*

His History of the Dividing Line *remained unpublished until 1841. It is a revision of his original diary of the expedition—first published in 1929 as* The Secret History of the Dividing Line—*which contains racy anecdotes and accounts of sexual escapades omitted from the more formal* History. *As Roderick Nash remarks in* Wilderness and the American Mind *(1967; revised ed. 1973), "the original journal . . . did not contain the passages celebrating the wild mountains. Byrd added them as embellishements a decade later when he prepared the manuscript for publication," since "as a well-lettered gentleman Byrd could afford to take delight in wilderness without feeling himself a barbarian or in danger of reverting to one." These passages, enriched though they may have been by memory and fashion, remain, along with Byrd's descriptions of animals found in the wilderness, among the most interesting of the book. For both the* History *and the* Secret History, *see* William Byrd's Histories of the Dividing Line betwixt Virginia and North Carolina, *ed. William K. Boyd (1929; revised ed. 1967), and* The Prose Works of William Byrd of Westover, *ed. Louis B. Wright (1966).*

Very different in background and outlook was ***Jonathan Edwards*** *(1703–58), from whose* Personal Narrative *describing his conversion at age twenty (written perhaps twenty years later) our brief selection is taken. Born in East Windsor, Connecticut, the son of a pastor, Edwards was a precocious youth interested as much in Newton's science and Locke's philosophy as in divinity, to which the works of nature bore witness. In 1716 he attended Yale College, receiving his B.A. in 1720 and his M.A. in 1723. Among his earliest writings, dating back to age eleven, are astonishingly detailed essays "On Insects" (actually on the balloon, or flying, spider) and "Of the Rainbow." Not all New England Puritans were dour Bible-reading, witch-burning fundamentalists. A century before Edwards, Anne Bradstreet had written not only "Of the Vanity of All Wordly Things," but also of the four elements, the four humors, and "The Four Seasons of the Year," including spring, when*

> The Pleiades their influence now give,
> And all that seemed as dead afresh doth live.
> The croaking frogs, whom nipping winter killed,
> Like birds now chirp, and hop about the field,
> The nightingale, the blackbird, and the thrush

Now tune their lays, on sprays of every bush.
The wanton frisking kid, and soft-fleeced lambs
Do jump and play before their feeding dams,
The tender tops of budding grass they crop,
They joy in what they have, but more in hope.

And in 1721, Perry Miller notes in Jonathan Edwards *(1949), Cotton Mather's* Christian Philosopher *affirmed "that the Newtonian world, far from denying, actually proved the existence of God and of design in the cosmos."*

Edwards himself, "with Newton behind him, saw in the phenomena of nature, as employed in Christ's own discourse," Miller writes, "not metaphors to adorn a discourse, but factual embodiments of eternal law. . . . He was prepared to venture in thought where Newton would not tread, into the hiding places of nature, to run down the force that was both the cohesion of atoms and the power of gravity, and to risk the possibility that could he find it or name it, the force might turn out to be simply monstrous." In his career as minister in Northampton, Massachusetts (where his preaching helped bring about the religious revival known as the Great Awakening); as missionary to the Housatonic Indians of Stockbridge; as president for a few months before his death of the College of New Jersey (now Princeton); and as writer of sermons such as "Sinners in the Hands of an Angry God" (1741) and theological tracts such as The Freedom of the Will *(1754), Edwards was known for unbending adhesion to the Calvinist doctrines of predestination and absolute dependence on God's grace. Yet as our excerpt from his* Personal Narrative *shows, Jonathan Edwards—as Miller affirms in* Errand into the Wilderness *(1956)—"was a child of the wilderness as well as of Puritanism." Collections of Edwards's writings include* Representative Selections, *ed. Clarence H. Faust and Thomas H. Johnson (1935);* Basic Writings, *ed. Ola Elizabeth Winslow (1966); and* A Jonathan Edwards Reader, *ed. John E. Smith, Harry S. Stout, and Kenneth B. Minkema (1995).*

Finally, in ***Thomas Jefferson*** *(1743–1826)—principal author of the Declaration of Independence, governor of Virginia during the American Revolution, successor to Benjamin Franklin as minister to France, secretary of state under George Washington, vice president under John Adams, third president of the United States for two terms beginning in 1801, and founder of the University of Virginia—the European Enlightenment attained its fullest expression in America. For Jefferson, who famously affirmed (however imperfectly, as a slave-owning planter, he may have put his principles into practice) that life, liberty, and the pursuit of happiness are inalienable rights bestowed by "the laws of nature and of nature's God," both the concept of nature—as inherited from seventeenth- and eighteenth-century British and French thinkers—and the richly varied particulars of the natural world around him remained central concerns despite almost continuous involvement in politics, diplomacy, and government. His ideal of democracy was an agrarian ideal, founded on the small, independent farmer. "The agrarian doctrines of Jefferson and his contemporaries"—including Hector St. John de Crèvecoeur, author of* Letters from an American Farmer *(1782)—"had been developed out of the rich cluster of ideas and attitudes associated with farming in European cultural tradition," Henry Nash Smith writes in* Virgin Land: The American West as Symbol and Myth *(1950): "the conventional praise of husbandry derived from Hesiod and Virgil by hundreds of poetic imitators, the theoretical teaching of the French Physiocrats that agriculture is the primary source of all wealth, the growing tendency of radical writers like Raynal[24] to make the farmer a republican symbol in-*

[24]The Abbé Guillaume-François-Thomas Raynal, author of the widely read *Philosophical and Political History of the Settlements and Trade of the Europeans in the East and West Indies* (1770), which summed up many ideas of the French Encyclopedists, "viewed these provinces of North America in their true light," Crèvecoeur wrote in dedicating his *Letters from an American Farmer* to him, "as the asylum of freedom, as the cradle of future nations, and the refuge of distressed Europeans," founded on the equality of simple agricultural communities.

stead of depicting him in pastoral terms as a peasant virtuously content with his humble status in a stratified society."

Through his purchase of the vast Louisiana Territory from Napoleonic France, and his sponsorship of its exploration by Meriwether Lewis and William Clark, Jefferson envisaged the possibility, Smith observes, "that settlement beyond the Alleghenies promised an even more perfect realization of the agrarian ideal on a scale so vast that it dwarfed all previous conceptions of possible transformations in human society." Despite an excellent classical and scientific education, and talents for architecture and invention lavishly displayed at his home, Monticello, Jefferson half humorously referred to himself, in a letter of 1785 to Charles Bellini from Paris, as "a savage of the mountains of America." In a letter of the same year to John Jay, he affirmed that "cultivators of the earth are the most valuable citizens. They are the most vigorous, the most independent, the most virtuous." And on preparing to leave the presidency in March 1809, he wrote to Pierre Samuel Dupont de Nemours, "Nature intended me for the tranquil pursuits of science, by rendering them my supreme delight."

His one book-length work, Notes on the State of Virginia *(from which our selections come), was written, Jefferson says in his brief* Autobiography, *in response to queries put to him in 1781 by the marquis de Barbé-Marbois, Secretary of the French Legation in Philadelphia. Revised and enlarged in 1782–83, it was anonymously published in Paris, at Jefferson's expense, in an edition of 200 copies in 1784. Its early sections, on the natural history of Virginia, undertook to describe the mountains and rivers, fauna and flora of his state, and to counter the claims of Buffon that the animals of the New World are smaller and fewer than in the Old. His evocation of the Potomac River's passage through the Blue Ridge as "one of the most stupendous scenes in nature" paints a vivid picture of a landscape no doubt less grand than that of the Alps, yet nearer to the pristine vision of unsettled wilderness to which Jefferson, despite his praise of small farmsteads, was likewise powerfully drawn. Finally, his description of the Natural Bridge as "the most sublime of nature's works" suggests a side of this self-professed Epicurean rationalist that links him, like Burke, Kant, and Schiller, with the Romantic Transcendentalism soon to take shape in England, Germany, and America. For one-volume selections, see* The Life and Selected Writings of Thomas Jefferson, *ed. Adrienne Koch and William Peden (1944), and The Library of America edition of Jefferson's* Writings, *ed. Merrill D. Peterson (1984).*

FROM History of North Carolina, *by John Lawson*

The buffalo is a wild beast of America, which has a bunch on his back as the cattle of St. Lawrence are said to have. He seldom appears amongst the English inhabitants, his chief haunt being in the land of Messiasippi, which is, for the most part, a plain country; yet I have known some killed on the hilly part of Cape-Fair-River, they passing the ledges of vast mountains from the said Messiasippi, before they can come near us. I have eaten of their meat, but do not think it so good as our beef; yet the younger calves are cried up for excellent food, as very likely they may be. It is conjectured that these buffalos, mixed in breed with our tame cattle, would much better the breed for largeness and milk, which seems very probable. Of the wild bull's skin buff is made. The Indians cut the skins into quarters for the ease of their transportation, and make beds to lie on. They spin the hair into garters, girdles, sashes, and the like, it being long and curled, and often of a chestnut or red color. These monsters are found to weigh (as I am informed by a traveler of credit) from 1600 to 2400 weight. . . .

The possum is found nowhere but in America. He is the wonder of all the land animals, being the size of a badger, and near that color. The male's pizzle is placed in retrograde; and in time of coition, they differ from all other animals, turning tail to tail as dog and bitch when tied. The female doubtless breeds her young at her teats; for I have seen them stick fast thereto when they have been no bigger than a small raspberry, and seemingly inanimate. She has a paunch, or false belly, wherein she carries her young, after they are from those teats, till they can shift for themselves. Their food is roots, poultry, or wild fruits. They have no hair on their tails, but a sort of a scale or hard crust, as the beavers have. If a cat has nine lives, this creature surely has nineteen; for if you break every bone in their skin, and mash their skull, leaving them for dead, you may come an hour after, and they will be gone quite away, or perhaps you meet them creeping away. They are a very stupid creature, utterly neglecting their safety. They are most like rats of any thing. I have, for necessity in the wilderness, eaten of them. Their flesh is very white, and well tasted; but their ugly tails put me out of conceit with that fare. They climb trees as the raccoons do. Their fur is not esteemed nor used, save that the Indians spin it into girdles and garters. . . .

The alligator is the same as the crocodile, and differs only in name. They frequent the sides of rivers, in the banks of which they make their dwellings a great way underground; the hole or mouth of their dens lying commonly two feet under water, after which it rises till it be considerably above the surface thereof. Here it is that this amphibious monster dwells all the winter, sleeping away his time till the spring appears, when he comes from his cave, and daily swims up and down the streams. He always breeds in some fresh stream or clear fountain of water, yet seeks his prey in the broad salt waters, that are brackish, not on the seaside, where I never met with any. He never devours men in Carolina, but uses all ways to avoid them, yet he kills swine and dogs, the former as they come to feed in the marshes, the others as they swim over the creeks and waters. They are very mischievous to the weirs made for taking fish, into which they come to prey on the fish that are caught in the weir, from whence they cannot readily extricate themselves, and so break the weir in pieces, being a very strong creature. This animal in these parts sometimes exceeds seventeen foot long. It is impossible to kill them with a gun, unless you chance to hit them about the eyes, which is a much softer place than the rest of their impenetrable armor. They roar and make a hideous noise against bad weather, and before they come out of their dens in the spring. I was pretty much frightened with one of these once, which happened thus: I had built a house about half a mile from an Indian town on the fork of Neus-River, where I dwelt by myself, excepting a young Indian fellow, and a bulldog, that I had along with me. I had not been so long a sojourner in America as to be thoroughly acquainted with this creature. One of them had got his nest directly under my house, which stood on pretty high land and by a creekside in whose banks his entering place was, his den reaching the ground directly on which my house stood. I was sitting alone by the fireside (about nine a clock at night, sometime in March), the Indian fellow being gone to the town to see his relations, so that there was nobody in the house but myself and my dog; when, all of a sudden, this ill-favored neighbor of mine set up such a roaring, that he made the house shake about my ears, and so continued like a bittern (but a hundred times louder if possible) for four or five times. The dog stared as if he was frightened out of his senses; nor indeed could I imagine what it was, having never heard one of them before. Immediately again I had another lesson; and so a third. Being, at that time, amongst none but savages, I began to suspect they were working some piece of conjuration under my house, to get away my goods; not but that at another time I have as little faith in their, or any others', working miracles, by diabolical means, as any person living. At last, my man came in, to whom when I had told the story, he laughed at me, and presently undeceived me by telling me what it was that made that noise. These alligators lay eggs as the ducks do, only they are longer shaped, larger, and a thicker shell than they have. How long they are in hatching I cannot tell; but as the Indians say, it is most part of the summer. They always lay by a spring-side, the young living in and about the same as soon as hatched. Their eggs are laid in nests made in the marshes, and contain twenty or thirty eggs. . . .

FROM The History of the Dividing Line betwixt Virginia and North Carolina, Run in the Year of Our Lord 1728, *by William Byrd II*

. . . [OCTOBER] 25 The air clearing up this morning, we were again agreeably surprised with a full prospect of the mountains. They discovered themselves both to the north and south of us, on either side, not distant above ten miles, according to our best computation.

We could now see those to the north rise in four distinct ledges, one above another, but those to the south formed only a single ledge, and that broken and interrupted in many places; or rather they were only single mountains detached from each other.

One of the southern mountains was so vastly high, it seemed to hide its head in the clouds, and the west end of it terminated in a horrible precipice, that we called the Despairing Lover's Leap. The next to it, towards the east, was lower, except at one end, where it heaved itself up in the form of a vast stack of chimneys.

The course of the northern mountains seemed to tend west-southwest, and those to the southward very near west. We could descry other mountains ahead of us, exactly in the course of the line, though at a much greater distance. In this point of view, the ledges on the right and left both seemed to close, and form a natural amphitheater.

Thus, 'twas our fortune to be wedged in bewixt these two ranges of mountains, insomuch that if our line had run ten miles on either side, it had butted before this day either upon one or the other, both of them now stretching away plainly to the eastward of us.

It had rained a little in the night, which dispersed the smoke and opened this romantic scene to us all at once, though it was again hid from our eyes as we moved forwards by the rough woods we had the misfortune to be engaged with. The bushes were so thick for near four miles together, that they tore the deer-skins to pieces that guarded the bread-bags. Though, as rough as the woods were, the soil was extremely good in all the way, being washed down from the neighboring hills into the plane country. Notwithstanding all these difficulties, the surveyors drove on the line 4 miles and 205 poles.

In the meantime we were so unlucky as to meet with no sort of game the whole day, so that the men were obliged to make a frugal distribution of what little they left in the morning.

We encamped upon a small rill, where the horses came off as temperately as their masters. They were by this time grown so thin, by hard travel and spare feeding, that henceforth, in pure compassion, we chose to perform the greater part of the journey on foot. And as our baggage was by this time grown much lighter, we divided it, after the best manner, that every horse's load might be proportioned to the strength he had left. Though, after all the prudent measures we could take, we perceived the hills began to rise upon us so fast in our front that it would be impossible for us to proceed much further.

We saw very few squirrels in the upper parts, because the wild cats devour them unmercifully. Of these there are four kinds: the fox squirrel, the grey, the flying, and the ground-squirrel. These last resemble a rat in everything but the tail, and the black and russet streaks that run down the length of their little bodies. . . .

26 . . . There was a small mountain half a mile to the northward of us, which we had the curiosity to climb up in the afternoon in order to enlarge our prospect. From thence we were able to discover where the two ledges of mountains closed, as near as we could guess, about 30 miles to the west of us, and lamented that our present circumstances would not permit us to advance the line to that place, which the hand of nature had made so very remarkable.

Not far from our quarters one of the men picked up a pair of elk's horns, not very large, and discovered the track of the elk that had shed them. It was rare to find any token of those animals so far to the south, because they keep commonly to the northward of 37 degrees, as the buffaloes, for the most part, confine themselves to the southward of that latitude.

The elk is full as big as a horse, and of the deer kind. The stags only have horns, and those exceedingly large and spreading. Their color is something lighter than that of the red deer, and their flesh tougher. Their swiftest speed is a large trot, and in that motion they turn their horns back upon their necks, and cock their noses aloft in the air. Nature has taught them this attitude, to save their antlers from being entangled in the thickets, which they always retire to. They are very shy, and have the sense of smelling so exquisite that they wind a man at a great distance. For this reason they are seldom seen but when the air is moist, in which case their smell is not so nice.

They commonly herd together, and the Indians say, if one of the drove happen by some wound to be disabled from making his escape, the rest will forsake their fears to defend their friend, which they will do with great obstinacy, till they are killed upon the spot. Though otherwise, they are so alarmed at the sight of a man that to avoid him they will sometimes throw themselves down very high precipices into the river. . . .

Our men had the fortune to kill a brace of bears, a fat buck, and a wild turkey, all which paid them with interest for yesterday's abstinence. This constant and seasonable supply of all our daily wants made us reflect thankfully on the bounty of Providence.

And that we might not be unmindful of being all along fed by Heaven in this great and solitary wilderness, we agreed to wear in our hats the Maosti, which is, in Indian, the beard of the wild turkey-cock, and on our breasts the figure of that fowl with its wings extended, and holding in its claws a scroll with this motto, "VICE CONTURNICUM," meaning that we had been supported by them in the wilderness "in the room of quails."

27 . . . In the evening we deliberated which way it might be most proper to return. We had at first intended to cross over at the foot of the mountains to the head of James River, that we might be able to describe that natural boundary so far. But, on second thoughts, we found many good reasons against that laudable design, such as the weakness of our horses, the scantiness of our bread, and the near approach of winter. We had cause to believe the way might be full of hills, and the farther we went towards the north, the more danger there would be of snow. Such considerations as these determined us at last to make the best of our way back upon the line, which was the straightest, and consequently the shortest way to the inhabitants. We knew the worst of that course, and were sure of a beaten path all the way, while we were totally ignorant what difficulties and dangers the other course might be attended with. So prudence got the better for once of curiosity, and the itch for new discoveries gave place to self-preservation. . . .

30 . . . At nine a'clock we began our march back toward the rising sun; for though we had finished the line, yet we had not yet near finished our fatigue. We had, after all, 200 good miles at least to our several habitations, and the horses were brought so low that we were obliged to travel on foot great part of the way, and that in our boots, too, to save our legs from being torn to pieces by the bushes and briars. Had we not done this, we must have left all our horses behind, which could now hardly drag their legs after them, and with all the favor we could show the poor animals, we were forced to set seven of them free, not far from the foot of the mountains.

Four men were dispatched early to clear the road, that our lame commissioner's leg might be in less danger of being bruised, and that the baggage horses might travel with less difficulty and more expedition.

As we passed along, by favor of a serene sky we had still, from every eminence, a perfect view of the mountains, as well to the north as to the south. We could not forbear now and then facing about to survey them, as if unwilling to part with a prospect which at the same time, like some rake's, was very wild and very agreeable.

We encouraged the horses to exert the little strength they had, and being light, they made a shift to jog on about eleven miles. We encamped on Crooked Creek, near a thicket of canes. In front of our camp rose a very beautiful hill, that bounded our view at about a mile's distance, and all the intermediate space was covered with green canes. Though, to our sorrow, firewood was scarce, which was now the harder upon us, because a northwester blew very cold from the mountains.

The Indian killed a stately, fat buck, and we picked his bones as clean as a score of turkey-buzzards could have done. . . .

In the evening one of the men knocked down an opossum, which is a harmless little beast, that will seldom go out of your way, and if you take hold of it, it will only grin, and hardly ever bite. The flesh was well tasted and tender, approaching nearest to pig, which it also resembles in bigness. The color of its fur was a goose grey, with a swine's snout, and a tail like a rat, but at least a foot long. By twisting this tail about the arm of a tree, it will hang with all its weight, and swing to anything it wants to take hold of.

It has five claws on the forefeet of equal length, but the hinder feet have only four claws, and a sort of thumb standing off at a proper distance.

Their feet, being thus formed, qualify them for climbing up trees to catch little birds, which they are very fond of.

But the greatest particularity of this creature, and which distinguishes it from most others that we are acquainted with, is the FALSE BELLY of the FEMALE, into which her young retreat in time of danger. She can draw the slit, which is the inlet into this pouch, so close that you must look narrowly to find it, especially if she happens to be a virgin.

Within the false belly may be seen seven or eight teats, on which the young ones grow from their first formation till they are big enough to fall off, like ripe fruit from a tree. This is so odd a method of generation, that I should not have believed it without the testimony of mine own eyes. Besides a knowing and credible person has assured me he has more than once observed the embryo possums growing to the teat before they were completely shaped, and afterwards watched their daily growth till they were big enough for birth. And all this he could the more easily pry into, because the dam was so perfectly gentle and harmless that he could handle her just as he pleased.

I could hardly persuade myself to publish a thing so contrary to the course that nature takes in the production of other animals, unless it were a matter commonly believed in all countries where that creature is produced, and has been often observed by persons of undoubted credit and understanding.

They say that the leather-winged bats produce their young in the same uncommon manner. And that young sharks at sea, and the young vipers ashore, run down the throats of their dams when they are closely pursued.

[31] The frequent crossing of Crooked Creek, and mounting the steep banks of it, gave the finishing stroke to the foundering of our horses: and no less than two of them made a full stop here, and would not advance a foot farther, either by fair means or foul. . . .

Nov. 1 . . . Notwithstanding our being thus delayed, and the unevenness of the ground over which we were obliged to walk (for most of us served now in the infantry), we traveled no less than 6 miles, though as merciful as we were to our poor beasts, another of 'em tired by the way, and was left behind for the wolves and panthers to feast upon.

As we marched along, we had the fortune to kill a brace of bucks, as many bears, and one wild turkey. But this was carrying sport to wantonness, because we butchered more than we were able to transport. We ordered the deer to be quartered and divided amongst the horses for the lighter carriage, and recommended the bears to our daily attendants, the turkey-buzzards.

We always chose to carry venison along with us rather than bear, not only because it was less cumbersome, but likewise because the people could eat it without bread, which was now almost spent. Whereas the other, being richer food, lay too heavy upon the stomach, unless it were lightened by something farinaceous. This is what I thought proper to remark, for the service of all those whose business or diversion shall oblige them to live any time in the woods. . . .

4 . . . One of the young fellows we had sent to bring up the tired horses entertained us in the evening with a remarkable adventure he had met with that day.

He had straggled, it seems, from his company in a mist, and made a cub of a year old betake itself to a tree. While he was new-priming his piece, with intent to fetch it down, the old gentlewoman appeared, and perceiving her heir apparent in distress, advanced open-mouthed to his relief.

The man was so intent upon his game that she had approached very near him before he perceived her. But finding his danger, he faced about upon the enemy, which immediately reared upon her posteriors, and put herself in battle array.

The man, admiring at the bear's assurance, endeavored to fire upon her, but by the dampness of the priming, his gun did not go off. He cocked it a second time, and had the same misfortune. After missing fire twice, he had the folly to punch the beast with the muzzle of his piece; but mother Bruin, being upon her guard, seized the weapon with her paws, and by main strength wrenched it out of the fellow's hands.

The man, being thus fairly disarmed, thought himself no longer a match for the enemy, and therefore retreated as fast as his legs could carry him.

The brute naturally grew bolder upon the flight of her adversary, and pursued him with all her heavy speed. For some time it was doubtful whether fear made one run faster, or fury the other. But after an even course of about 50 yards, the man had the mishap to stumble over a stump, and fell down his full length. He now would have sold his life a pennyworth; but the bear, apprehending there might be some trick in the fall, instantly halted, and looked with much attention on her prostrate foe.

In the meanwhile, the man had with great presence of mind resolved to make the bear believe he was dead by lying breathless on the ground, in hopes that the beast would be too generous to kill him over again. To carry on the farce, he acted the corpse for some time without daring to raise his head to see how near the monster was to him. But in about two minutes, to his unspeakable comfort, he was raised from the dead by the barking of a dog, belonging to one of his companions, who came seasonably to his rescue, and drove the bear from pursuing the man to take care of her cub, which she feared might now fall into a second distress. . . .

6 . . . All the land we traveled over this day, and the day before, that is to say from the river Irvin to Sable Creek, is exceedingly rich, both on the Virginia side of the line, and that of Carolina. Besides whole forests of canes, that adorn the banks of the river and creeks thereabouts, the fertility of the soil throws out such a quantity of winter grass that horses and cattle might keep themselves in heart all the cold season without the help of any fodder. Nor have the low grounds only this advantage, but likewise the higher land, and particularly that which we call the Highland Pond, which is two miles broad, and of a length unknown.

I question not but there are 30,000 acres at least, lying all together, as fertile as the lands were said to be about Babylon, which yielded, if Herodotus tells us right, an increase of no less than 2 or 300 for one. But this hath the advantage of being a higher and consequently a much healthier situation than that. So that a colony of 1000 families might, with the help of moderate industry, pass their time very happily there.

Besides grazing and tillage, which would abundantly compensate their labor, they might plant vineyards upon the hills, in which situation the richest wines are always produced.

They might also propagate white mulberry trees, which thrive exceedingly in this climate, in order to the feeding of silkworms, and making of raw silk.

They might too produce hemp, flax, and cotton, in what quantity they pleased, not only for their own use, but likewise for sale. Then they might raise very plentiful orchards, of both peaches and apples, which contribute as much as any fruit to the luxury of life. There is no soil or climate will yield better rice than this, which is a grain of prodigious increase, and of very wholesome nourishment. In short everything will grow plentifully here to supply either the wants or wantonness of man.

Nor can I so much as wish that the more tender vegetables might grow here, such as orange, lemon, and olive trees, because then we should lose the much greater benefit of the brisk northwest winds, which purge the air, and sweep away all the malignant fevers which hover over countries that are always warm.

The soil would also want the advantages of frost and snow, which by their nitrous particles contribute not a little to its fertility. Besides, the inhabitants would be deprived of the variety and sweet vicissitude of the season, which is much more delightful than one dull and constant succession of warm weather, diversified only by rain and sunshine.

There is also another convenience that happens to this country by cold weather—it destroys a great number of snakes, and other venomous reptiles, and troublesome insects, or at least lays them to sleep for several months, which otherwise would annoy us the whole year round, and multiply beyond all enduring.

Though oranges and lemons are desirable fruits, and useful enough in many cases, yet, when the want of them is supplied by others more useful, we have no cause to complain.

There is no climate that produces everything, since the deluge wrenched the poles of the world out of their place, nor is it fit it should be so, because it is the mutual supply one country receives from another, which creates a mutual traffic and intercourse among men. . . .

FROM Personal Narrative, *by Jonathan Edwards*

. . . Not long after I first began to experience these things, I gave an account to my father of some things that had passed in my mind. I was pretty much affected by the discourse we had together; and when the discourse was ended, I walked abroad alone, in a solitary place in my father's pasture, for contemplation. And as I was walking there, and looking upon the sky and clouds, there came into my mind so sweet a sense of the glorious *majesty* and *grace* of God, as I know not how to express.—I seemed to see them both in a sweet conjunction; majesty and meekness joined together: it was a sweet, and gentle, and holy majesty; and also a majestic meekness; an awful sweetness; a high, and great, and holy gentleness.

After this my sense of divine things gradually increased, and became more and more lively, and had more of that inward sweetness. The appearance of every thing was altered; there seemed to be, as it were, a calm, sweet cast, or appearance, of divine glory in almost every thing. God's excellency, his wisdom, his purity and love, seemed to appear in every thing; in the sun, moon, and stars; in the clouds and blue sky; in the grass, flowers, trees; in the water and all nature; which used greatly to fix my mind. I often used to sit and view the moon for a long time; and in the day, spent much time in viewing the clouds and sky, to behold the sweet glory of God in these things: in the meantime, singing forth, with a low voice, my contemplations of the Creator and Redeemer. And scarce any thing, among all the works of nature, was so sweet to me as thunder and lightning; formerly nothing had been so terrible to me. Before, I used to be uncommonly terrified with thunder, and to be struck with terror when I saw a thunderstorm rising; but now, on the contrary, it rejoiced me. I felt God, if I may so speak, at the first appearance of a thunderstorm; and used to take the opportunity, at such times, to fix myself in order to view the clouds, and see the lightning's play, and hear the majestic and awful voice of God's thunder, which oftentimes was exceedingly entertaining, leading me to sweet contemplations of my great and glorious God. While thus engaged, it always seemed natural for me to sing, or chant forth my meditations; or, to speak my thoughts in soliloquies with a singing voice.

I felt then great satisfaction, as to my good estate; but that did not content me. I had vehement longings of soul after God and Christ, and after more holiness, wherewith my heart seemed to be full, and ready to break; which often brought to my mind the words of the Psalmist, Psal. cxix.28: *My soul breaketh for the longing it hath.* I often felt a mourning and lamenting in my heart, that I had not turned to God sooner, that I might have had more time to grow in grace. My mind was greatly fixed on divine things; almost perpetually in the contemplation of them. I spent most of my time in thinking of divine things, year after year; often walking alone in the woods, and solitary places, for meditation, soliloquy, and prayer, and converse with God; and it was always my manner, at such times, to sing forth my contemplations. I was almost constantly in ejaculatory prayer, wherever I was. Prayer seemed to be natural to me, as the breath by which the inward burnings of my heart had vent. The delights which I now felt in the things of religion were of an exceedingly different kind from those before-mentioned, that I had when a boy; and what then I had no more notion of than one born blind has of pleasant and beautiful colors. They were of a more inward, pure, soul-animating and refreshing nature. Those former delights never reached the heart; and did not arise from any sight of the divine excellency of the things of God; or any taste of the soul-satisfying and life-giving good there is in them. . . .

FROM Notes on Virginia, *by Thomas Jefferson*

From Query IV: A NOTICE OF ITS MOUNTAINS? For the particular geography of our mountains I must refer to Fry and Jefferson's map of Virginia; and to Evans' analysis of his map of America, for a more philosophical view of them than is to be found in any other work. It is worthy notice, that our mountains are not solitary and scattered confusedly over the face of the country; but that they commence at about 150 miles from the sea-coast, are disposed in ridges, one behind another, running nearly parallel with the sea-coast, though rather approaching it as they advance northeastwardly. To the southwest, as the tract of country between the sea-coast and the Mississippi becomes narrower, the mountains converge into a single ridge, which, as it approaches the Gulf of Mexico, subsides into plain country, and gives rise to some of the waters of that gulf, and particularly to a river called the Apalachicola, probably from the Apalachies, an Indian nation formerly residing on it. Hence the mountains giving rise to that river, and seen from its various parts, were called the Apalachian mountains, being in fact the end or termination only of the great ridges passing through the continent. . . . It is in fact the spine of the country between the Atlantic on one side, and the Mississippi and St. Lawrence on the other. The passage of the Patowmac through the Blue Ridge is, perhaps, one of the most stupendous scenes in nature. You stand on a very high point of land. On your right comes up the Shenandoah, having ranged along the foot of the mountain an hundred miles to seek a vent. On your left approaches the Patowmac, in quest of a passage also. In the moment of their junction, they rush together against the mountain, rend it asunder, and pass off to the sea. The first glance of this scene hurries our senses into the opinion that this earth has been created in time, that the mountains were formed first, that the rivers began to flow afterwards, that in this place, particularly, they have been dammed up by the Blue Ridge of mountains, and have formed an ocean which filled the whole valley; that continuing to rise they have at length broken over at this spot, and have torn the mountain down from its summit to its base. The piles of rock on each hand, but particularly on the Shenandoah, the evident marks of their disrupture and avulsion from their beds by the most powerful agents of nature, corroborate the impression. But the distant finishing which nature has given to the picture is of a very different character. It is a true contrast to the foreground. It is as placid and delightful as that is wild and tremendous. For the mountain being cloven asunder, she presents to your eye, through the cleft, a small catch of smooth blue horizon, at an infinite distance in the plain country, inviting you, as it were, from the riot and tumult roaring around, to pass through the breach and participate of the calm below. Here the eye ultimately composes itself; and that way, too, the road happens actually to lead. You cross the Patowmac above the junction, pass along its side through the base of the mountain for three miles, its terrible precipices hanging in fragments over you, and within about twenty miles reach Frederick town, and the fine country round that. This scene is worth a voyage across the Atlantic. Yet here, as in the neighborhood of the Natural Bridge, are people who have passed their lives within half a dozen miles, and have never been to survey these monuments of a war between rivers and mountains, which must have shaken the earth itself to its center. . . .

From Query V: ITS CASCADES AND CAVERNS? . . . The Natural Bridge, the most sublime of nature's works, . . . must not be pretermitted. It is on the ascent of a hill, which seems to have been cloven through its length by some great convulsion. The fissure, just at the bridge, is, by some admeasurements, 270 feet deep, by others only 205. It is about 45 feet wide at the bottom and 90 feet at the top; this of course determines the length of the bridge, and its height from the water. Its breadth in the middle is about 60 feet, but more at the ends, and the thickness of the mass, at the summit of the arch, about 40 feet. A part of this thickness is constituted by a coat of earth, which gives growth to many large trees. The residue, with the hill on both sides, is one solid rock of limestone. The arch approaches the semi-elliptical form; but the larger axis of the ellipsis, which would be the cord of the arch, is many times longer than the transverse. Though the sides of this bridge are provided in some parts with a parapet of fixed rocks, yet few men have resolution to walk to them, and look over into the abyss. You involuntarily fall on your hands and feet, creep to the parapet, and peep over it. Looking down from this height about a minute gave me a violent headache. If the view from the top be painful and intolerable, that from below is delightful in an equal extreme. It is impossible for the emotions arising from the sublime to be felt beyond what they are here; so beautiful an arch, so elevated, so light, and springing as it were up to heaven, the rapture of the spectator is really indescribable! The fissure continuing narrow, deep, and straight for a considerable distance above and below the bridge, opens a short but very pleasing view of the North mountain on one side and Blue Ridge on the other, at the distance each of them of about five miles. This bridge is in the county of Rockbridge, to which it has given name, and affords a public and commodious passage over a valley which cannot be crossed elsewhere for a considerable distance. The stream passing under it is called Cedar creek. It is a water of James river, and sufficient in the driest seasons to turn a grist-mill, though its fountain is not more than two miles above.

An Infinite Variety of Animated Scenes

Travels through North and South Carolina, Georgia, East and West Florida, by William Bartram

William Bartram (1734–1823) was the son of the naturalist John Bartram, who established in Philadelphia the first botanical garden in North America and was elected a member of the Royal Society; he was extolled both by Franklin and Crèvecoeur at home and by Linnaeus abroad. Born in Philadelphia, and reared as a Quaker, William in 1765–66 accompanied his father on a scientific expedition to the St. Johns River region of Florida. Then, from 1773 to 1777, for the purpose of gathering botanical samples, he undertook the extensive travels (sponsored by John Fothergill of England) that led to eventual publication, in 1791, of his Travels through North & South Carolina, Georgia, East & West Florida, the Cherokee Country, the Extensive Territories of the Muscogulges, or Creek Confederacy, and the Country of the Chactaws, Containing an Account of the Soil and Natural Productions of Those Regions, Together with Observations on the Manners of the Indians.

The book was published only a few years after Gilbert White's Natural History of Selborne *(Chapter 20 above), whose neatly ordered world of chalk hills and chimney swallows could hardly appear more different from Bartram's. It remains one of the masterpieces of American nature writing, distinguished both for its punctilious observations and for the splendidly evocative prose that brings its descriptions vividly to life. Bartram blends the older picturesque style with prospects of "the grand sublime" in the new Romantic vein into a verbal symphony that effortlessly harmonizes their varied tonalities. The order of nature manifests "the divine and inimitable workmanship" of God. Yet Bartram describes this order not in pious generalities but with a rationally "inquisitive mind," watching the spider's conquest of the bee (which reveals "premeditation, perseverance, resolution, and consummate artistry") with the intent scrutiny of a trained and unwavering eye, and marveling not only at the vast multitude of ephemeral mayflies but also at their bodily frame and organization, "more delicate, and perhaps as complicated as that of the most perfect human being." In him, Shaftesbury and Mandeville (Chapter 20 above) seem to join, and Gilbert White and Rousseau seem to leave the swallows of Selborne and the self-heal of the Island of Saint-Pierre to venture among the alligators of Florida and the "sublime magnificence" of the American wilderness.*

All these qualities appealed intensely to Romantic writers such as Coleridge and Wordsworth in England and Chateaubriand in France. As John Livingston Lowes demonstrates in The Road to Xanadu *(1927), impressions from reading Bartram "were among the sleeping images in Coleridge's unconscious memory at the time when 'Kubla Khan' emerged from it":*

> And there were gardens bright with sinuous rills,
> Where blossomed many an incense-bearing tree;
> And here were forests ancient as the hills,
> Enfolding sunny spots of greenery.

So, too, in Wordsworth's very different "Ruth," an impressionable English girl is entranced by "a Youth from Georgia's shore" whose tales "in the green shade / Were perilous to hear":

> He told of the magnolia, spread
> High as a cloud, high over head!
> The cypress and her spire;
> —Of flowers that with one scarlet gleam
> Cover a hundred leagues, and seem
> To set the hills on fire. . . .

Chateaubriand's prose romances Atala *and* René *also teem with images culled from this extraordinary writer whom not even rapt contemplation of an inexpressibly magnificent mountain landscape can blind to "the charming objects more within my reach: a new species of rhododendron, foremost in the assembly of mountain beauties; next the flaming azalea, Kalmia latifolia, incarnate Robinia, snowy mantled Philadelphus inodorus, perfumed Calycanthus, etc." Few have observed nature more closely or voiced her beauties more fully. See the editions of Bartram's* Travels *edited by Mark Van Doren (1928), Francis Harper (1958), and Thomas P. Slaughter (Library of America, 1996), and the selections in* John and William Bartram's America, *ed. Helen Gere Cruickshank (1957).*

FROM *Introduction: The Vitality of the Natural World*

The attention of a traveler should be particularly turned, in the first place, to the various works of Nature, to mark the distinctions of the climates he may explore, and to offer such useful observations on the different productions as may occur. Men and manners undoubtedly hold the first rank—whatever may contribute to our existence is also of equal importance, whether it be found in the animal or vegetable kingdoms; neither are the various articles which tend to promote the happiness and convenience of mankind to be disregarded. How far the writer of the following sheets has succeeded in furnishing information on these subjects, the reader will be capable of determining. From the advantages the journalist enjoyed under his father John Bartram, botanist to the king of Great Britain, and fellow of the Royal Society, it is hoped that his labors will present new as well as useful information to the botanist and zoologist.

This world, as a glorious apartment of the boundless palace of the sovereign Creator, is furnished with an infinite variety of animated scenes, inexpressibly beautiful and pleasing, equally free to the inspection and enjoyment of all his creatures.

Perhaps there is not any part of creation, within the reach of our observations, which exhibits a more glorious display of the Almighty hand than the vegetable world. Such a variety of pleasing scenes, ever changing, throughout the seasons, arising from various causes and assigned each to the purpose and use determined. . . .

In every order of nature, we perceive a variety of qualities distributed amongst individuals, designed for different purposes and uses, yet it appears evident that the great Author has impartially distributed his favors to his creatures, so that the attributes of each one seem to be of sufficient importance to manifest the divine and inimitable workmanship. . . .

The animal creation also excites our admiration, and equally manifests the almighty power, wisdom, and beneficence of the Supreme Creator and Sovereign Lord of the universe; some in their vast size and strength, as the mammoth, the elephant, the whale, the lion, and alligator; others in agility; others in their beauty and elegance of color, plumage, and rapidity of flight have the faculty of moving and living in the air; others for their immediate and indispensable use and convenience to man, in furnishing means for our clothing and sustenance, and administering to our help in the toils and labors through life: how wonderful is the mechanism of these finely formed, self-moving beings, how complicated their system, yet what unerring uniformity prevails through every tribe and particular species! the effect we see and contemplate, the cause is invisible, incomprehensible, how can it be otherwise? when we cannot see the end or origin of a nerve or vein, while the divisibility of matter or fluid is infinite. We admire the mechanism of a watch, and the fabric of a piece of brocade, as being the production of art; these merit our admiration, and must excite our esteem for the ingenious artist or modifier, but nature is the work of God omnipotent: and an elephant, even this world, is comparatively but a very minute part of his works. If then the visible, the mechanical part of the animal creation, the mere material part, is so admirably beautiful, harmonious, and incomprehensible, what must be the intellectual system? that inexpressibly more essential principle, which secretly operates within? that which animates the inimitable machines, which gives them motion, empowers them to act, speak, and perform, this must be divine and immortal?

I am sensible that the general opinion of philosophers has distinguished the moral system of the brute creature from that of mankind by an epithet which implies a mere mechanical impulse, which leads and impels them to necessary actions, without any premeditated design or contrivance; this we term instinct, which faculty we suppose to be inferior to reason in man. . . . If we bestow but a very little attention to the economy of the animal creation, we shall find manifest examples of premeditation, perseverance, resolution, and consummate artifice in order to effect their purpose. . . .

As I was gathering specimens of flowers from the shrubs, I was greatly surprised at the sudden appearance of a remarkable large spider on a leaf, of the genus Araneus saliens: at sight of me he boldly faced about, and raised himself up as if ready to spring upon me; his body was about the size of a pigeon's egg, of a buff color, which with his legs were covered with short silky hair; on the top of the abdomen was a round red spot or ocellus encircled with black; after I had recovered from the surprise, and observing the wary hunter had retired under cover, I drew near again, and presently discovered that I had surprised him on predatory attempts against the insect tribes; I was therefore determined to watch his proceedings. I soon noticed that the object of his wishes was a large fat bumble bee (apis bombylicus) that was visiting the flowers and piercing their nectariferous tubes; this cunning intrepid hunter conducted his subtle approaches with the circumspection and perseverance of a Seminole, when hunting a deer, advancing with slow steps obliquely, or under cover of dense foliage, and behind the limbs, and when the bee was engaged in probing a flower he would leap nearer, and then instantly retire out of sight, under a leaf or behind a branch, at the same time keeping a sharp eye upon me; when he had now got within two feet of his prey, and the bee was intent on sipping the delicious nectar from a flower, with his back next the spider, he instantly sprang upon him, and grasped him over the back and shoulder; when for some moments they both disappeared, I expected the bee had carried off his enemy, but to my surprise they both together rebounded back again, suspended at the extremity of a strong elastic thread or web, which the spider had artfully let fall, or fixed on the twig, the instant he leaped from it; the rapidity of the bee's wings, endeavoring to extricate himself, made them both together appear as a moving vapor, until the bee became fatigued by whirling round, first one way and then back again; at length, in about a quarter of an hour, the bee, quite exhausted by his struggles, and the repeated wounds of the butcher, became motionless, and quickly expired in the arms of the devouring spider, who, ascending the rope with his game, retired to feast on it under cover of leaves; and perhaps before night became himself the delicious evening repast of a bird or lizard. . . .

FROM *Part Two, Chapter III: May Flies on the St. Johns River, Florida*

Leaving Picolata, I continued to ascend the river. I observed this day, during my progress up the river, incredible numbers of small flying insects, of the genus termed by naturalists Ephemera, continually emerging from the shallow water, near shore, some of them immediately taking their flight to the land, whilst myriads crept up the grass and herbage, where remaining, for a short time, as they acquired sufficient strength, they took their flight also, following their kindred to the mainland. This resurrection from the deep, if I may so express it, commences early in the morning, and ceases after the sun is up. At evening they are seen in clouds of innumerable millions, swarming and wantoning in the still air, gradually drawing near the river, descend upon its surface, and there quickly end their day, after committing their eggs to the deep; which being for a little while tossed about, enveloped in a viscid scum, are hatched, and the little larvae descend into their secure and dark habitation, in the oozy bed beneath, where they remain, gradually increasing in size, until the returning spring; they then change to a nymph, when the genial heat brings them, as it were, into existence, and they again arise into the world. This fly seems to be delicious food for birds, frogs, and fish. In the morning, when they arise, and in the evening, when they return, the tumult is great indeed, and the surface of the water along shore broken into bubbles, or spurted into the air, by the contending aquatic tribes; and such is the avidity of the fish and frogs, that they spring into the air after this delicious prey.

Early in the evening, after a pleasant day's voyage, I made a convenient and safe harbor in a little lagoon, under an elevated bank, on the west shore of the river, where I shall entreat the reader's patience whilst we behold the closing scene of the short-lived Ephemera, and communicate to each other the reflections which so singular an exhibition might rationally suggest to an inquisitive mind. Our place of observation is happily situated under the protecting shade of majestic live oaks, glorious magnolias, and the fragrant orange, open to the view of the great river and still waters of the lagoon just before us.

At the cool eve's approach, the sweet enchanting melody of the feathered songsters gradually ceases, and they betake themselves to their leafy coverts for security and repose.

Solemnly and slowly move onward to the river's shore the rustling clouds of the Ephemera. How awful the procession! innumerable millions of winged beings, voluntarily verging on to destruction, to the brink of the grave, where they behold bands of their enemies with wide open jaws, ready to receive them. But as if insensible of their danger, gay and tranquil each meets his beloved mate, in the still air, inimitably bedecked in their new nuptial robes. What eye can trace them, in their varied wanton amorous chases, bounding and fluttering on the odoriferous air? with what peace, love, and joy do they end the last moments of their existence?

I think we may assert, without any fear of exaggeration, that there are annually of these beautiful winged beings, which rise into existence, and for a few moments take a transient view of the glory of the Creator's works, a number greater than the whole race of mankind that have ever existed since the creation; and that only from the shores of this river. How many then must have been produced since the creation, when we consider the number of large rivers in America, in comparison with which this river is but a brook or rivulet.

The importance of the existence of these beautiful and delicately formed little creatures, in the creation, whose frame and organization is equally wonderful, more delicate, and perhaps as complicated as that of the most perfect human being, is well worth a few moments' contemplation; I mean particularly when they appear in the fly state. And if we consider the very short period of that stage of existence, which we may reasonably suppose to be the only space of their life that admits of pleasure and enjoyment, what a lesson doth it not afford us of the vanity of our own pursuits.

Their whole existence in this world is but one complete year, and at least three hundred and sixty days of that time they are in the form of an ugly grub, buried in mud, eighteen inches under water, and in this condition scarcely locomotive, as each larva or grub has but its own narrow solitary cell, from which it never travels or moves but in a perpendicular progression, of a few inches, up and down, from the bottom to the surface of the mud, in order to intercept the passing atoms for its food, and get a momentary respiration of fresh air; and even here it must be perpetually on its guard, in order to escape the troops of fish and shrimps watching to catch it, and from whom it has no escape but by instantly retreating back into its cell. One would be apt almost to imagine them created merely for the food of fish and other animals. . . .

FROM *Part Two, Chapter V: The Alligators of Lake Dexter, Florida*

The evening was temperately cool and calm. The crocodiles began to roar and appear in uncommon numbers along the shores and in the river. I fixed my camp in an open plain, near the utmost projection of the promontory, under the shelter of a large live oak, which stood on the highest part of the ground and but a few yards from my boat. From this open, high situation, I had a free prospect of the river, which was a matter of no trivial consideration to me, having good reason to dread the subtle attacks of the alligators, who were crowding about my harbor. Having collected a good quantity of wood for the purpose of keeping up a light and smoke during the night, I began to think of preparing my supper, when, upon examining my stores, I found but a scanty provision; I thereupon determined, as the most expeditious way of supplying my necessities, to take my bob and try for some trout. About one hundred yards above my harbor began a cove or bay of the river, out of which opened a large lagoon. The mouth or entrance from the river to it was narrow, but the waters soon after spread and formed a little lake, extending into the marshes; its entrance and shores within I observed to be verged with floating lawns of the Pistia and Nymphea and other aquatic plants; these I knew were excellent haunts for trout.

The verges and islets of the lagoon were elegantly embellished with flowering plants and shrubs; the laughing coots with wings half spread were tripping over the little coves and hiding themselves in the tufts of grass; young broods of the painted summer teal, skimming the still surface of the waters, and following the watchful parent unconscious of danger, were frequently surprised by the voracious trout, and he in turn, as often, by the subtle, greedy alligator. Behold him rushing forth from the flags and reeds. His enormous body swells. His plaited tail, brandished high, floats upon the lake. The waters like a cataract descend from his opening jaws. Clouds of smoke issue from his dilated nostrils. The earth trembles with his thunder: when immediately, from the opposite coast of the lagoon, emerges from the deep his rival champion. They suddenly dart upon each other. The boiling surface of the lake marks their rapid course, and a terrific conflict commences. They now sink to the bottom folded together in horrid wreaths. The water becomes thick and discolored. Again they rise, their jaws clap together, re-echoing through the deep surrounding forests. Again they sink, when the contest ends at the muddy bottom of the lake, and the vanquished makes a hazardous escape, hiding himself in the muddy turbulent waters and sedge on a distant shore. The proud victor exulting returns to the place of action. The shores and forest resound his dreadful roar, together with the triumphing shouts of the plaited tribes around, witnesses of the horrid combat.

My apprehensions were highly alarmed after being a spectator of so dreadful a battle; it was obvious that every delay would but tend to increase my dangers and difficulties, as the sun was near setting, and the alligators gathered around my harbor from all quarters; from these considerations I concluded to be expeditious in my trip to the lagoon, in order to take some fish. Not thinking it prudent to take my fusee[25] with me, lest I might lose it overboard in case of a battle, which I had every reason to dread before my return, I therefore furnished myself with a club for my defense, went on board, and penetrating the first line of those which surrounded my harbor, they gave way; but being pursued by several very large ones, I kept strictly on the watch, and paddled with all my might towards the entrance of the lagoon, hoping to be sheltered there from the multitude of my assailants; but ere I had halfway reached the place, I was attacked on all sides, several endeavoring to overset the canoe. My situation now became precarious to the last degree: two very large ones attacked me closely, at the same instant, rushing up with their heads and part of their bodies above the water, roaring terribly and belching floods of water over me. They struck their jaws together so close to my ears, as almost to stun

[25]Flintlock musket.

me, and I expected every moment to be dragged out of the boat and instantly devoured; but I applied my weapons so effectually about me, though at random, that I was so successful as to beat them off a little; when, finding that they designed to renew the battle, I made for the shore, as the only means left me for my preservation; for, by keeping close to it, I should have my enemies on one side of me only, whereas I was before surrounded by them, and there was a probability, if pushed to the last extremity, of saving myself by jumping out of the canoe on shore, as it is easy to outwalk them on land, although comparatively as swift as lightning in the water. . . .

It was by this time dusk, and the alligators had nearly ceased their roar, when I was again alarmed by a tumultuous noise that seemed to be in my harbor, and therefore engaged my immediate attention. Returning to my camp I found it undisturbed, and then continued on to the extreme point of the promontory, where I saw a scene, new and surprising, which at first threw my senses into such a tumult that it was some time before I could comprehend what was the matter; however, I soon accounted for the prodigious assemblage of crocodiles at this place, which exceeded everything of the kind I had ever heard of.

How shall I express myself so as to convey an adequate idea of it to the reader, and at the same time avoid raising suspicions of my want of veracity? Should I say that the river (in this place) from shore to shore, and perhaps near half a mile above and below me, appeared to be one solid bank of fish, of various kinds, pushing through this narrow pass of St. Juan's into the little lake, on their return down the river, and that the alligators were in such incredible numbers, and so close together from shore to shore, that it would have been easy to have walked across on their heads, had the animals been harmless. What expressions can sufficiently declare the shocking scene that for some minutes continued, whilst this mighty army of fish were forcing the pass? During this attempt, thousands, I may say hundreds of thousands of them were caught and swallowed by the devouring alligators. I have seen an alligator take up out of the water several great fish at a time, and just squeeze them betwixt his jaws, while the tails of the great trout flapped about his eyes and lips, ere he had swallowed them. The horrid noise of their closing jaws, their plunging amidst the broken banks of fish, and rising with their prey some feet upright above the water, the floods of water and blood rushing out of their mouths, and the clouds of vapor issuing from their wide nostrils, were truly frightful. This scene continued at intervals during the night, as the fish came to the pass. After this sight, shocking and tremendous as it was, I found myself somewhat easier and more reconciled to my situation, being convinced that their extraordinary assemblage here was owing to this annual feast of fish, and that they were so well employed in their own element that I had little occasion to fear their paying me a visit. . . .

FROM *Part Three, Chapter III: Crossing the Mountains*

I waited two or three days at this post [Fort Prince George, South Carolina], expecting the return of an Indian who was out hunting: this man was recommended to me as a suitable person for a protector and guide to the Indian settlements over the hills, but upon information that he would not be in shortly, and there being no other person suitable for the purpose, rather than be detained, and perhaps thereby frustrated in my purposes, [I] determined to set off alone and run all risks.

I crossed the [Keowee] river at a good ford just below the old fort. The river here is just one hundred yards over: after an agreeable progress for about two miles over delightful strawberry plains and gently swelling green hills, began to ascend more steep and rocky ridges. Having gained a very considerable elevation, and looking around, I enjoyed a very comprehensive and delightful view: Keowe, which I had but just lost sight of, appears again, and the serpentine river speeding through the lucid green plain apparently just under my feet. After observing this delightful landscape, I continued on again three or four miles, keeping the trading path which led me over uneven rocky land, crossing rivulets and brooks, and rapidly descending over rocky precipices, when I came into a charming vale, embellished with a delightful glittering river which meandered through it and crossed my road: on my left hand, upon the grassy bases of the rising hills, appears the remains of a town of the ancients, as the tumuli, terraces, posts or pillars, old peach and plum orchards, etc. sufficiently testify. These vales and swelling bases of the surrounding hills afford vast crops of excellent grass and herbage fit for pasturage and hay. . . .

My next flight was up a very high peak, to the top of the Oconee Mountain, where I rested; and turning about, found that I was now in a very elevated situation, from whence I enjoyed a view inexpressibly magnificent and comprehensive. The mountainous wilderness which I had lately traversed, down to the region of Augusta, appearing regularly undulated as the great ocean after a tempest; the undulations gradually depressing, yet perfectly regular, as the squamae of fish or imbrications of tile on a roof: the nearest ground to me of a perfect full green, next more glaucous, and lastly almost blue as the ether with which the most distant curve of the horizon seems to be blended.

My imagination thus wholly engaged in the contemplation of this magnificent landscape, infinitely varied, and without bound, I was almost insensible or regardless of the charming objects more within my reach: a new species of rhododendron, foremost in the assembly of mountain beauties; next the flaming azalea, Kalmia latifolia, incarnate Robinia, snowy mantled Philadelphus inodorus, perfumed Calycanthus, etc.

This species of rhododendron grows six or seven feet high; many nearly erect stems arise together from the root, forming a group or coppice. The leaves are three or four inches in length, of an oblong figure, broadest toward the extremity and terminating with an obtuse point; their upper surface of a deep green and polished, but the nether surface of a rusty iron color, which seems to be effected by innumerable minute reddish vesicles, beneath a fine short downy pubescence; the numerous flexile branches terminate with a loose, spiked raceme, or cluster of large, deep rose-colored flowers, each flower being affixed in the diffused cluster by a long peduncle, which with the whole plant possesses an agreeable perfume.

After being recovered of the fatigue and labor in ascending the mountain, I began again to prosecute my task, proceeding through a shady forest, and soon after gained the most elevated crest of the Oconee Mountain, and then began to descend the other side; the winding rough road carrying me over rocky hills and levels, shaded by incomparable forests, the soil exceedingly rich and of an excellent quality for the production of every vegetable suited to the climate. . . . Now I enter a charming narrow vale, through which flows a rapid large creek. . . . Passed through magnificent high forests, and then came upon the borders of an ample meadow on the left, embroidered by the shade of a high circular amphitheater of hills, the circular ridges rising magnificently one over the other: on the green turfy bases of these ascents appear the ruins of a town of the ancients; the upper end of this spacious green plain is divided by a promontory or spur of the ridges before me, which projects into it; my road led me up into an opening of the ascents through which the glittering brook which watered the meadows ran rapidly down, dashing and roaring over high rocky steps. Continued yet ascending until I gained the top of an elevated rocky ridge, when appeared before me a gap or opening between other yet more lofty ascents, through which continuing as the rough rocky road led me, close by the winding banks of a large rapid brook which, at length turning to the left, pouring down rocky precipices, glided off through dark groves and high forests, conveying streams of fertility and pleasure to the fields below. . . .

This day being remarkably warm and sultry, which, together with the labor and fatigue of ascending the mountains, made me very thirsty and in some degree sunk my spirits. Now past mid-day, I sought a cool shaded retreat, where was water for refreshment and grazing for my horse, my faithful slave and only companion. After proceeding a little farther, descending the other side of the mountain, I perceived at some distance before me, on my right hand, a level plain supporting a grand high forest and groves; the nearer I approach, my steps are the more accelerated from the flattering prospect opening to view; I now enter upon the verge of the dark forest, charming solitude! as I advanced through the animating shades, observed on the farther grassy verge a shady grove, thither I directed my steps; on approaching these shades, between the stately columns of the superb forest trees, presented to view, rushing from rocky precipices under the shade of the pensile hills, the unparalleled cascade of Falling Creek, rolling and leaping off the rocks, which uniting below, spread a broad, glittering sheet of crystal waters, over a vast convex elevation of plain, smooth rocks, and are immediately received by a spacious basin, where trembling in the center through hurry and agitation, they gently subside, encircling the painted still verge, from whence gliding swiftly, they soon form a delightful little river, which continuing to flow more moderately, is restrained for a moment, gently undulating in a little lake, they then pass on rapidly to a high perpendicular steep of rocks, from whence these delightful waters are hurried down with irresistible rapidity. I here seated myself on the moss-clad rocks, under the shade of spreading trees and floriferous fragrant shrubs, in full view of the cascades.

At this rural retirement were assembled a charming circle of mountain vegetable beauties . . . Some of these roving beauties are strolling over the mossy, shelving, humid rocks, or from off the expansive wavy boughs of trees, bending over the floods, salute their delusive shades, playing on the surface, some plunge their perfumed heads and bathe their flexile limbs in the silver stream, whilst others by the mountain breezes are tossed about, their blooming tufts bespangled with pearly and crystalline dewdrops collected from the falling mists, glisten in the rainbow arch. Having collected some valuable specimens at this friendly retreat, I continued my lonesome pilgrimage. My road for a considerable time led me winding and turning about the steep rocky hills; the descent of some of which was very rough and troublesome, by means of fragments of rocks, slippery clay and talc; but after this I entered a spacious forest, the land having gradually acquired a more level surface; a pretty grassy vale appears on my right, through which my wandering path led me, close by the banks of a delightful creek, which sometimes falling over steps of rocks, glides gently with serpentine meanders through the meadows.

After crossing this delightful brook and mead, the land rises again with sublime magnificence, and I am led over hills and vales, groves and high forests, vocal with the melody of the feathered songsters, the snow-white cascades glittering on the sides of the distant hills.

It was now after noon; I approached a charming vale, amidst sublimely high forests, awful shades! darkness gathers around, far distant thunder rolls over the trembling hills;

the black clouds with august majesty and power, moves slowly forwards, shading regions of towering hills, and threatening all the destructions of a thunderstorm; all around is now still as death, not a whisper is heard, but a total inactivity and silence seems to pervade the earth; the birds afraid to utter a chirrup, and in low tremulous voices take leave of each other, seeking covert and safety; every insect is silenced, and nothing heard but the roaring of the approaching hurricane; the mighty cloud now expands its sable wings, extending from North to South, and is driven irresistibly on by the tumultuous winds, spreading his livid wings around the gloomy concave, armed with terrors of thunder and fiery shafts of lightning; now the lofty forests bend low beneath its fury, their limbs and wavy boughs are tossed about and catch hold of each other; the mountains tremble and seem to reel about, and the ancient hills to be shaken to their foundations: the furious storm sweeps along, smoking through the vale and over the resounding hills; the face of the earth is obscured by the deluge descending from the firmament, and I am deafened by the din of thunder; the tempestuous scene damps my spirits, and my horse sinks under me at the tremendous peals, as I hasten on for the plain.

The storm abating, I saw an Indian hunting cabin on the side of a hill, a very agreeable prospect, especially in my present condition; I made up to it and took quiet possession, there being no one to dispute it with me except a few bats and whippoorwills, who had repaired thither for shelter from the violence of the hurricane. . . .

FROM *Part Three, Chapter VII: The Mississippi River*

. . . At evening arrived at Manchac [in East Baton Rouge Parish, Louisiana], where I directed my steps to the banks of the Mississippi, where I stood for a time as it were fascinated by the magnificence of the great sire[26] of rivers.

The depth of the river here, even in this season, at its lowest ebb, is astonishing, not less than forty fathoms, and the width about a mile or somewhat less; but it is not expansion of surface alone that strikes us with ideas of magnificence: the altitude, and theatrical ascents of its pensile banks, the steady course of the mighty flood, the trees, high forests, even every particular object, as well as societies, bear the stamp of superiority and excellence; all unite or combine in exhibiting a prospect of the grand sublime. The banks of the river at Manchac, though frequently overflowed by the vernal inundations, are about fifty feet perpendicular height above the surface of the water (by which the channel at these times must be about two hundred and ninety feet deep) and these precipices being an accumulation of the sediment of muddy waters, annually brought down with the floods, of a light loamy consistence, are continually cracking and parting, present to view deep yawning chasms, in time split off, as the active perpetual current undermines, and the mighty masses of earth tumble headlong into the river, whose impetuous current sweeps away and lodges them elsewhere. There is yet visible some remains of a high artificial bank, in front of the buildings of the town, formerly cast up by the French, to resist the inundations, but found to be ineffectual, and now in part tumbled down the precipice, as the river daily encroaches on the bluff; some of the habitations are in danger, and must be very soon removed or swallowed up in the deep gulf of waters. . . .

[26]Which is the meaning of the word Mississippi. (Bartram)

Acknowledgments

Every reasonable effort has been made to clear the use of materials in this volume with the copyright owners, when these could be located. Copyright material from the following authors, translators, and works has been included by permission:

Claude Colleer Abbott, translator, *Early Mediaeval French Lyrics*, copyright © 1932. Reprinted by permission of the publisher, Constable & Company Limited.

Aelian, *On the Characteristics of Animals*, vols. I, II, and III, translated by A. F. Scholfield, Cambridge, Mass.: Harvard University Press, 1958–1959. Reprinted by permission of the publishers and the Loeb Classical Library.

Alan of Lille (prose selections): Reprinted from Alan of Lille, *The Plaint of Nature*, translated by J. J. Sheridan, pp. 109–127, by permission of the publisher. © 1980 by the Pontifical Institute of Mediaeval Studies, Toronto.

Albert the Great: Reprinted from *Medieval Philosophy: Selected Readings* by Herman Shapiro. Copyright © 1964 by Random House, Inc. Reprinted by permission of Random House, Inc.

Amergin (attributed to), "The Mystery," translated by Douglas Hyde. In *1000 Years of Irish Poetry* (1953), edited by Kathleen Hoagland. Reprinted by permission of Devin-Adair, Publishers.

St. Thomas Aquinas, *On the Power of God*, translated by the English Dominican Friars. Copyright © 1932. Reprinted by permission of Search Press Ltd./Burns & Oates Ltd.

Aristotle: From *The Oxford Translation of Aristotle*, edited by W. D. Ross. From vol. 2 (1930): *Physica* translated by R. P. Hardie and R. K. Gaye and *De Generatione* translated by H. H. Joachim; vol. 4 (1910): *Historia Animalium* translated by D'Arcy Wentworth Thompson; vol. 5 (1912): *De Partibus Animalium* translated by William Ogle; vol. 8 (2nd ed. 1928): *Metaphysica* translated by W. D. Ross. Reprinted by permission of Oxford University Press.

St. Augustine: From *Basic Writings of Saint Augustine* by St. Augustine, edited by Whitney J. Oates. Copyright © 1948 by Random House, Inc. Reprinted by permission of Random House, Inc.

St. Augustine, *On Free Choice of the Will*, translated by Anna S. Benjamin and L. H. Hackstaff, © 1964. Reprinted by permission of Prentice-Hall, Inc., Upper Saddle River, N.J.

Roger Bacon: *The Opus Majus of Roger Bacon*, translated by Robert Belle Burke. Copyright © 1928. Reprinted by permission of the publisher, University of Pennsylvania Press.

Bashō: From *The Narrow Road to the Deep North and Other Travel Sketches* by Matsuo Bashō, translated by Nobuyuki Yuasa (Penguin Classics, 1966), copyright © Nobuyuki Yuasa, 1966. Reprinted by permission of Penguin Books Ltd.

Bernard Silvestris: From *The Cosmographia of Bernardus Silvestris*, translated by Winthrop Wetherbee. Copyright © 1973 by Columbia University Press. Reprinted with permission of the publisher.

Biblical selections from the Revised Standard Version of the Bible, copyright 1946, 1952, 1971, by the Division of Christian Education of the National Council of the Churches of Christ in the USA. Used by permission.

Boethius, *The Theological Tractates* and *The Consolation of Philosophy*, translated by H. F. Stewart, E. K. Rand, and S. J. Tester, Cambridge, Mass.: Harvard University Press, 1918, new edition 1973. Reprinted by permission of the publishers and the Loeb Classical Library.

St. Bonaventura, *The Mind's Road to God*, translated by George Boas, © 1953. Reprinted by permission of Prentice-Hall, Inc., Upper Saddle River, N.J.

Geoffrey Bownas and Anthony Thwaite, translators: From *The Penguin Book of Japanese Verse*, translated by Geoffrey Bownas and Anthony Thwaite (Penguin Books, 1964). Translation copyright © Geoffrey Bownas and Anthony Thwaite, 1964. Reprinted by permission of Penguin Books Ltd.

Life of St. Brendan: From *Lives of the Saints*, translated by J. F. Webb (Penguin Classics, 1965), copyright © J. F. Webb, 1965. Reprinted by permission of Penguin Books Ltd.

John Brough, translator: From *Poems from the Sanskrit*, translated by John Brough (Penguin Classics, 1968), copyright © John Brough, 1968. Reprinted by permission of Penguin Books Ltd.

Giordano Bruno, *Concerning the Cause, Principle, and One*: From *The Infinite in Giordano Bruno*, translated by Sidney Greenberg. Copyright © 1950 by King's Crown Press. Reprinted with permission of the publisher.

Giordano Bruno, *On the Infinite Universe and Worlds*: Excerpts as specified from *Giordano Bruno: His Life and Thought*, by Dorothea Waley Singer. Copyright © 1950 by Harper & Row, Publishers, Inc. Renewed 1978 by Dorothea Waley Singer. Reprinted by permission of HarperCollins Publishers, Inc.

Witter Bynner: From *The Jade Mountain* by Witter Bynner, trans. Copyright 1929 and renewed 1957 by Alfred A. Knopf Inc. Reprinted by permission of the publisher.

Cato, *On Agriculture*, and Varro, *On Agriculture*, translated by William Davis Hooper, revised by Harrison Boyd Ash, Cambridge, Mass.: Harvard University Press, 1935. Reprinted by permission of the publishers and the Loeb Classical Library.

Celsus, *De Medicina*, vol. I, translated by W. G. Spencer, Cambridge, Mass.: Harvard University Press, 1935. Reprinted by permission of the publishers and the Loeb Classical Library.

Kamo no Chomei: From *Anthology of Japanese Literature*, edited by Donald Keene, copyright © 1955 by Grove Press, Inc. Used by permission of Grove/Atlantic, Inc.

Chuang Tzu: From *The Complete Works of Chuang Tzu*, translated by Burton Watson. Copyright © 1968 by Columbia University Press. Reprinted with permission of the publisher.

Cicero, *De Natura Deorum*, translated by H. Rackham, Cambridge, Mass.: Harvard University Press, 1933. Reprinted by permission of the publishers and the Loeb Classical Library.

Columella, *On Agriculture*, vol. I, translated by Harrison Boyd Ash, Cambridge, Mass.: Harvard University Press, 1941. Reprinted by permission of the publishers and the Loeb Classical Library.

Copernicus: From Thomas S. Kuhn, *The Copernican Revolution*, Cambridge, Mass.: Harvard University Press. Copyright © 1957 by the President and Fellows of Harvard College.

Nicholas Cusanus, *Of Learned Ignorance*, translated by Fr. Germain Heron. Reprinted by permission of Routledge.

D'Alembert: From Schwab, *Preliminary Discourse to the Encyclopedia of Diderot*, © 1963. Reprinted by permission of Prentice-Hall, Inc., Upper Saddle River, N.J.

John Dee on Astronomy: Propaedeumata Aphoristica (1558 and 1568), Latin and English, translated/edited by Wayne Shumaker, with an introductory essay by J. L. Heilbron. Copyright © 1978 The Regents of the University of California. Reprinted by permission of the University of California Press.

From *Epicurus: The Extant Remains*, translated by Cyril Bailey (1926). Reprinted by permission of Oxford University Press.

Bernard le B. Fontenelle, *Conversations on the Plurality of Worlds*, translated/edited by H. A. Hargreaves. Copyright © 1990 The Regents of the University of California. Reprinted by permission of the University of California Press.

Galen: Reprinted from *Galen on the Usefulness of the Parts of the Body*, translated and with an introduction and commentary by Margaret Tallmadge May. Copyright © 1968 by Cornell University. Used by permission of the publisher, Cornell University Press.

Galileo Galilei, *Dialogue Concerning the Two Chief World Systems: The Ptolemaic and Copernican*. 2nd revised edition, translated/edited by Stillman Drake. Copyright © 1952, 1962, 1967 Regents of the University of California.

Reprinted by permission of the University of California Press.

Gilgamesh: From *Gilgamesh* by John Gardner and John Maier. Copyright © 1984 by the Estate of John Gardner and John Maier. Reprinted by permission of Alfred A. Knopf Inc. Copyright © 1984 by John Gardner.

Gregory of Nyssa: From *The Later Christian Fathers*, edited and translated by Henry Bettenson (1970). Reprinted by permission of Oxford University Press.

Han-shan: Cold Mountain Poems #1, 2, 3, 6, 7, 8, 9, 12, 17, and 24 from *Riprap and Cold Mountain Poems* by Gary Snyder. Copyright 1990 by Gary Snyder. Reprinted by permission of North Point Press, a division of Farrar, Straus & Giroux, Inc.

David Hawkes, translator: Two songs from *The Songs of the South: An Anthology of Ancient Chinese Poems* by Qu Yuan and other poets, translated by David Hawkes (Penguin Classics, 1985), copyright © David Hawkes, 1985. Reprinted by permission of Penguin Books Ltd.

Hugh of St. Victor: From *The Didascalion of Hugh of St. Victor*, translated by Jerome Taylor. Copyright © 1991 by Columbia University Press. Reprinted with permission of the publisher.

Inanna: Excerpts as specified from *Inanna: Queen of Heaven and Earth*, by Diane Wolkstein and Samuel Noah Kramer. Copyright © 1983 by Diane Wolkstein and Samuel Noah Kramer. Reprinted by permission of HarperCollins Publishers, Inc.

John the Scot: Periphyseon by Uhlfelder, Myra, © 1976. Reprinted by permission of Prentice-Hall, Inc., Upper Saddle River, N.J.

Howard Mumford Jones, translator, "The Philologian and His Cat," from *The Romanesque Lyric*, edited by Philip Schuyler Allen. Copyright © 1928. Reprinted by permission of the publisher, University of North Carolina Press.

Ben Jonson: Selected Masques, edited by Stephen Orgel. Copyright © 1970 by Yale University. Reprinted by permission of the publisher, Yale University Press.

Kenkō: From *Essays in Idleness*, translated by Donald Keene. Copyright © 1967 by Columbia University Press. Reprinted with permission of the publisher.

Kepler: From William H. Donahue, translator, *Johannes Kepler's New Astronomy*. Copyright © 1992. Reprinted with the permission of Cambridge University Press.

Kepler: From *The Six-Cornered Snowflake* by Johannes Kepler, translated by Colin Hardie (1966). Reprinted by permission of Oxford University Press.

Lactantius, *The Divine Institutes*: Reprinted from *Apocalyptic Spirituality* by Bernard McGinn. © 1979 by the Missionary Society of St. Paul the Apostle in the State of New York. Used by permission of Paulist Press.

Brunetto Latini, *The Book of the Treasure (Li Livres dou Tresor)*, translated by Paul Barrette and Spurgeon Baldwin. Copyright © 1993. Reprinted by permission of the publisher, Garland Publishing, Inc.

Leibniz: Selections by Wiener, Philip P., © 1951. Reprinted by permission of Prentice-Hall, Inc., Upper Saddle River, N.J.

Leibniz, *Monadology and Other Philosophical Essays*, translated by Paul Schrecker and Anne Martin Schrecker, © 1965. Reprinted by permission of Prentice-Hall, Inc., Upper Saddle River, N.J.

Leonardo da Vinci: From *The Notebooks of Leonardo da Vinci*, edited and translated by Edward MacCurdy. Copyright © 1938. Reprinted by permission of Random House UK Limited.

Li Ch'ing Chao: From *One Hundred Poems from the Chinese*, by Kenneth Rexroth. Copyright © 1971 by Kenneth Rexroth. Reprinted by permission of New Directions Publishing Corp.

The Lotus of the Wonderful Law, translated by W. E. Soothill (1930). Reprinted by permission of Oxford University Press.

Lyrics of Luis de León, translated by Aubrey F. G. Bell. Copyright © 1928. Reprinted by permission of Search Press Ltd./Burns & Oates Ltd.

The Manyoshu, Nippon Gakujutsu Shinkokai translation. Copyright © 1965 by Columbia University Press. Reprinted with permission of the publisher.

Kuno Meyer, translator. From *Selections from Ancient Irish Poetry*. Reprinted by permission of Constable Publishers.

Montaigne: Reprinted from *The Complete Works of Montaigne: Essays, Travel Journal,*

Letters, translated by Donald M. Frame, with the permission of the publishers, Stanford University Press. Copyright 1943 by Donald M. Frame. © 1948, 1957 by the Board of Trustees of the Leland Stanford Junior University.

Mo Tzu and Hsün Tzu: From *Basic Writings of Mo Tzu, Hsün Tzu, and Han Fei Tzu*, translated by Burton Watson. Copyright © 1964 by Columbia University Press. Reprinted with permission of the publisher.

John G. Neihardt: Reprinted from *Black Elk Speaks*, by John G. Neihardt, by permission of the University of Nebraska Press. Copyright 1932, 1959, 1972, by John G. Neihardt. Copyright © 1961 by the John G. Neihardt Trust.

Isaac Newton: Florian Cajori, ed., *Mathematical Principles of Natural Philosophy and His System of the World*, translated/edited by Andrew Motte. Copyright © 1934 and 1962 Regents of the University of California. Reprinted by permission of the University of California Press.

Raimundo Panikkar, translator, *The Vedic Experience: Mantramañjarī* (University of California Press, 1977). Reprinted by permission of Raimundo Panikkar.

Paracelsus, *Selected Writings*. Copyright © 1958 by Princeton University Press. Reprinted by permission of Princeton University Press.

Donald Philippi, *Norito: A Translation of the Ancient Japanese Ritual Prayers*. Copyright © 1990. Reprinted by permission of Princeton University Press.

Philo, vol. I, translated by F. H. Colson and G. H. Whitaker, and vol. IX, translated by F. H. Colson, Cambridge, Mass.: Harvard University Press, 1929 and 1941. Reprinted by permission of the publishers and the Loeb Classical Library.

Pico della Mirandola: *On the Dignity of Man* by della Mirandola, Pico, © 1985. Reprinted by permission of Prentice-Hall, Inc., Upper Saddle River, N.J.

Pliny, *Natural History*, vols. I and II, translated by H. Rackham, vol. VIII, translated by W. H. S. Jones, and vol. X, translated by D. E. Eichholz, Cambridge, Mass.: Harvard University Press, 1938, 1942, 1963, and 1962. Reprinted by permission of the publishers and the Loeb Classical Library.

Plotinus, vols. III and V, translated by A. H. Armstrong, Cambridge, Mass.: Harvard University Press, 1967 and 1984. Reprinted by permission of the publishers and the Loeb Classical Library.

Plutarch's Moralia, vol. V, translated by F. C. Babbitt, and vol. XII, translated by Harold Cherniss and William C. Helmbold, Cambridge, Mass.: Harvard University Press, 1936 and 1957. Reprinted by permission of the publishers and the Loeb Classical Library.

Angelo Poliziano, "Dance Song," translated by John Heath-Stubbs. Reprinted by permission of the translator.

David Pollack, *Zen Poems of the Five Mountains*. Copyright © 1985 American Academy of Religion. Published by Crossroad Publishing Company and Scholars Press. Reprinted by permission of Scholars Press.

Pseudo-Dionysius the Areopagite, *The Divine Names*: Reprinted from *Pseudo-Dionysius* by Colm Luibheid. © 1987 by Colm Luibheid. Used by permission of Paulist Press.

Ptolemy's Almagest, translated by G. J. Toomer. Copyright © 1984. Used by permission of Gerald Duckworth & Co., Ltd.

Rousseau: From *The Creed of a Priest of Savoy*, by Jean-Jacques Rousseau, translated by Arthur H. Beattie (second, enlarged edition). Copyright © 1956, 1957, by Frederick Ungar Publishing Co. Reprinted by permission of the Continuum Publishing Group.

Ryōkan: From *Dewdrops on a Lotus Leaf: Zen Poems of Ryokan* translated by John Stevens, © 1993. Reprinted by arrangement with Shambhala Publications, Inc., 300 Massachusetts Avenue, Boston, Mass. 02115.

Marquis de Sade: From *The Complete Justine, Philosophy in the Bedroom, and Other Writings* by the Marquis de Sade, translated by Richard Seaver and Austryn Wainhouse. Copyright © 1965 by Richard Seaver and Austryn Wainhouse. Used by permission of Grove/Atlantic, Inc.

Seneca, *Naturales Quaestiones*, vols. I and II, translated by Thomas H. Corcoran, Cambridge, Mass.: Harvard University Press, 1971–1972. Reprinted by permission of the publishers and the Loeb Classical Library.

Sophocles: Excerpt from "Oedipus at Colonus" in *Sophocles: The Oedipus Cycle, An English Version* by Robert Fitzgerald, copyright

1941 by Harcourt Brace & Company and renewed 1969 by Robert Fitzgerald, reprinted by permission of the publisher. CAUTION: All rights, including professional, amateur, motion picture, recitation, lecturing, public reading, radio broadcast, and television are strictly reserved. Inquiries on all rights should be addressed to Harcourt Brace & Company, Permissions Department, Orlando, Florida 32887-6667. Excerpt from *Oedipus at Colonus* by Sophocles translated by Robert Fitzgerald reprinted also by permission of Faber and Faber Ltd.

Sources of Chinese Tradition by William Theodore de Bary. Copyright © 1960 by Columbia University Press. Reprinted with permission of the publisher.

Sources of Indian Tradition by William Theodore de Bary. Copyright © 1958 by Columbia University Press. Reprinted with permission of the publisher.

Surangāma Sūtra: Excerpts from A Buddhist Bible by Dwight Goddard, editor. Copyright 1938, renewed © 1966 by E. P. Dutton. Used by permission of Dutton Signet, a division of Penguin Books USA Inc.

Tacitus: From *The Complete Works of Tacitus* by Tacitus, translated by Alfred J. Church and William J. Brodribb. Copyright © 1942 and renewed 1970 by Random House, Inc. Reprinted by permission of Random House, Inc.

J. R. R. Tolkien, translator, *Sir Gawain and the Green Knight*. Reprinted by permission of HarperCollins Publishers, Ltd.

Colin Turnbull: Reprinted with the permission of Simon & Schuster from *The Forest People: A Study of the Pygmies of the Congo* by Colin M. Turnbull. Copyright © text 1961 by Colin M. Turnbull. Copyright renewed 1989 by Colin M. Turnbull.

From *The Upanishads*, translated by Juan Mascaro (Penguin Classics, 1965), copyright © Juan Mascaro, 1965. Reprinted by permission of Penguin Books Ltd.

Giambattista Vico: Reprinted from *The New Science of Giambattista Vico*, translated from the third edition (1744) by Thomas Goddard Bergin and Max Harold Fisch. Copyright © 1948 by Cornell University. Used by permission of the publisher, Cornell University Press.

Helen Waddell, translator, *Mediaeval Latin Lyrics*. Reprinted by permission of Constable Publishers.

Arthur Waley, translator, *The Book of Songs*; *Chinese Poems*; *Japanese Poetry: 'the Uta'*; and *The Way and Its Power*. Reprinted by permission of HarperCollins Publishers, Ltd.

Wandalbert: From L. P. Wilkinson, *The Georgics of Virgil*. Copyright © 1969. Reprinted with the permission of Cambridge University Press.

Burton Watson, translator: From *The Columbia Book of Chinese Poetry*, translated by Burton Watson. Copyright © 1984 by Columbia University Press. Reprinted with permission of the publisher.

T. H. White, *The Book of Beasts*. Reprinted by permission of Harold Ober Associates Incorporated. Copyright © 1956 by T. H. White.

Edmund Wilson: Excerpts from *Apologies to the Iroquois* by Edmund Wilson. Copyright © 1960 by Edmund Wilson. Reprinted by permission of Farrar, Straus & Giroux, Inc.

Index

Page numbers in **boldface** indicate main selections, and introductory discussions of these and of their authors. Though comprehensive, the index is not complete. Individual items may be selectively indexed or grouped together under a more inclusive heading. Historical figures, place-names, and authors not represented in the anthology are generally indexed only if mentioned at least twice; scholars and editors only if quoted; translators only if main selections are included. Accents added to Greek and Roman names indicate pronunciation.